Advances in Intelligent Systems and Computing

Volume 1272

AF869684

Series Editor

Janusz Kacprzyk, Systems Research Institute, Polish Academy of Sciences, Warsaw, Poland

Advisory Editors

Nikhil R. Pal, Indian Statistical Institute, Kolkata, India

Rafael Bello Perez, Faculty of Mathematics, Physics and Computing, Universidad Central de Las Villas, Santa Clara, Cuba

Emilio S. Corchado, University of Salamanca, Salamanca, Spain

Hani Hagras, School of Computer Science and Electronic Engineering, University of Essex, Colchester, UK

László T. Kóczy, Department of Automation, Széchenyi István University, Gyor, Hungary

Vladik Kreinovich, Department of Computer Science, University of Texas at El Paso, El Paso, TX, USA

Chin-Teng Lin, Department of Electrical Engineering, National Chiao Tung University, Hsinchu, Taiwan

Jie Lu, Faculty of Engineering and Information Technology, University of Technology Sydney, Sydney, NSW, Australia

Patricia Melin, Graduate Program of Computer Science, Tijuana Institute of Technology, Tijuana, Mexico

Nadia Nedjah, Department of Electronics Engineering, University of Rio de Janeiro, Rio de Janeiro, Brazil

Ngoc Thanh Nguyen, Faculty of Computer Science and Management, Wrocław University of Technology, Wrocław, Poland

Jun Wang, Department of Mechanical and Automation Engineering, The Chinese University of Hong Kong, Shatin, Hong Kong

The series "Advances in Intelligent Systems and Computing" contains publications on theory, applications, and design methods of Intelligent Systems and Intelligent Computing. Virtually all disciplines such as engineering, natural sciences, computer and information science, ICT, economics, business, e-commerce, environment, healthcare, life science are covered. The list of topics spans all the areas of modern intelligent systems and computing such as: computational intelligence, soft computing including neural networks, fuzzy systems, evolutionary computing and the fusion of these paradigms, social intelligence, ambient intelligence, computational neuroscience, artificial life, virtual worlds and society, cognitive science and systems, Perception and Vision, DNA and immune based systems, self-organizing and adaptive systems, e-Learning and teaching, human-centered and human-centric computing, recommender systems, intelligent control, robotics and mechatronics including human-machine teaming, knowledge-based paradigms, learning paradigms, machine ethics, intelligent data analysis, knowledge management, intelligent agents, intelligent decision making and support, intelligent network security, trust management, interactive entertainment, Web intelligence and multimedia.

The publications within "Advances in Intelligent Systems and Computing" are primarily proceedings of important conferences, symposia and congresses. They cover significant recent developments in the field, both of a foundational and applicable character. An important characteristic feature of the series is the short publication time and world-wide distribution. This permits a rapid and broad dissemination of research results.

Indexed by SCOPUS, DBLP, EI Compendex, INSPEC, WTI Frankfurt eG, zbMATH, Japanese Science and Technology Agency (JST), SCImago.

All books published in the series are submitted for consideration in Web of Science.

More information about this series at http://www.springer.com/series/11156

A. Pasumpon Pandian · Ram Palanisamy ·
Klimis Ntalianis
Editors

Proceedings of International Conference on Intelligent Computing, Information and Control Systems

ICICCS 2020

Springer

Editors
A. Pasumpon Pandian
Department of Computer Science
and Engineering
KGiSL Institute of Technology
Coimbatore, India

Ram Palanisamy
Department of Business Administration
The Gerald Schwartz School of Business
St. Francis Xavier University
Antigonish, NS, Canada

Klimis Ntalianis
University of Applied Sciences
Aigaleo, Greece

ISSN 2194-5357 ISSN 2194-5365 (electronic)
Advances in Intelligent Systems and Computing
ISBN 978-981-15-8442-8 ISBN 978-981-15-8443-5 (eBook)
https://doi.org/10.1007/978-981-15-8443-5

© The Editor(s) (if applicable) and The Author(s), under exclusive license to Springer Nature Singapore Pte Ltd. 2021
This work is subject to copyright. All rights are solely and exclusively licensed by the Publisher, whether the whole or part of the material is concerned, specifically the rights of translation, reprinting, reuse of illustrations, recitation, broadcasting, reproduction on microfilms or in any other physical way, and transmission or information storage and retrieval, electronic adaptation, computer software, or by similar or dissimilar methodology now known or hereafter developed.
The use of general descriptive names, registered names, trademarks, service marks, etc. in this publication does not imply, even in the absence of a specific statement, that such names are exempt from the relevant protective laws and regulations and therefore free for general use.
The publisher, the authors and the editors are safe to assume that the advice and information in this book are believed to be true and accurate at the date of publication. Neither the publisher nor the authors or the editors give a warranty, expressed or implied, with respect to the material contained herein or for any errors or omissions that may have been made. The publisher remains neutral with regard to jurisdictional claims in published maps and institutional affiliations.

This Springer imprint is published by the registered company Springer Nature Singapore Pte Ltd.
The registered company address is: 152 Beach Road, #21-01/04 Gateway East, Singapore 189721, Singapore

We are honored to dedicate the proceedings of 2nd International Conference on Intelligent Computing, Information and Control Systems to all the participants and editors of ICICCS 2020.

Foreword

This conference proceedings volume contains the written versions of most of the contributions presented during the conference of 2nd ICICCS 2020. The conference provided a setting for discussing recent developments in a wide variety of topics including intelligent control technologies, artificial intelligence [AI], machine learning, intelligent information retrieval, intelligent agents, fuzzy logic control, neuro-fuzzy control, and evolutionary computing. The conference has been a good opportunity for participants coming from various destinations to present and discuss topics in their respective research areas.

ICICCS 2020 conference tends to collect the latest research results and applications on intelligent data communication technologies and Internet of Things. It includes a selection of 82 papers from 264 papers submitted to the conference from universities and industries all over the world. All of accepted papers were subjected to strict peer reviewing by 2–4 expert referees. The papers have been selected for this volume because of quality and the relevance to the conference.

ICICCS 2020 would like to express our sincere appreciation to all authors for their contributions to this book. We would like to extend our thanks to all the referees for their constructive comments on all papers; especially, we would like to thank organizing committee for their hardworking. Finally, we would like to thank Springer publications for producing this volume.

Dr. P. John Paul
Conference Chair
ICICCS 2020
Malla Reddy College of Engineering
Secunderabad, India

Preface

It is with deep satisfaction that I write this preface to the proceedings of the 2nd ICICCS 2020 held in Malla Reddy College of Engineering, Dhulapally, Secunderabad, at June 25–26, 2020.

This conference brings together researchers, academics, and professionals from all over the world and experts in intelligent control technologies, intelligent computing technologies, and intelligent information systems.

This conference particularly encouraged the interaction of research students and developing academics with the more established academic community in an informal setting to present and to discuss new and current work. The papers contributed the most recent scientific knowledge known in the field of intelligent computing models and systems, intelligent information systems, fuzzy sets, intelligent computing technologies control applications on automotive, energy, autonomous systems, big data, and machine learning. Their contributions helped to make the conference as outstanding as it has been. The Local Organizing Committee members and their helpers put much effort into ensuring the success of the day-to-day operation of the meeting.

We hope that this program will further stimulate research in intelligent systems and computing and provide practitioners with better techniques, algorithms, and tools for deployment. We feel honored and privileged to serve the best recent developments in the field of intelligent control and automation, knowledge-based systems, computational and communication constraints, network intelligence and network control, and fuzzy logic control through this exciting program.

We thank all authors and participants for their contributions.

Coimbatore, India

Dr. A. Pasumpon Pandian
Guest Editor
ICICCS 2020

Acknowledgements

ICICCS 2020 would like to acknowledge the excellent work of our conference organizing committee and keynote speakers for their presentation on June 25–26, 2020. The organizers also wish to acknowledge publicly the valuable services provided by the reviewers.

On behalf of the editors, organizers, authors, and readers of this conference, we wish to thank the keynote speakers and the reviewers for their time, hardwork, and dedication to this conference. The organizers wish to acknowledge Dr. A. Pasumpon Pandian, Dr. Klimis Ntalianis, and Dr. Ram Palanisamy for the discussion, suggestion, and finalize the paper of this conference. The organizers also wish to acknowledge for speakers and participants who attend this conference. Many thanks given for all persons who help and support this conference. ICICCS 2020 would like to acknowledge the contribution made to the organization by its many volunteers. Members contribute their time, energy, and knowledge at a local, regional, and international level.

We also thank all the Chair Persons and Conference Committee Members for their support.

Contents

Web Scraping and Naïve Bayes Classification for Political Analysis . . . 1
Noel Varela, Omar Bonerge Pineda Lezama, and Milvio Charris

Proactive Handoff of Secondary User in Cognitive Radio Network Using Machine Learning Techniques . . . 9
Gaurav Wajhal, Vasudev Dehalwar, Ankit Jha, Koki Ogura, and Mohan Lal Kolhe

Inter-device Language Translation Application for Smartphones . . . 23
Ashwini Rao, Abhishek Paradkar, Shruti Gupta, and Sayali Kadam

Real-Time Numerical Gesture Recognition Using MPU9250 Motion Sensor . . . 39
Sathish Raja Bommannan, Chennuru Vineeth, Mylavarapu Uma Hema Sri, Boyanapalli Sri Vidya, and S. Vidhya

Use of Genetic Algorithm Applied to the Optimization of Investments in Financial Actions . . . 57
Noel Varela, Omar Bonerge Pineda Lezama, and Jorge Borda

Development of DWT–SVD based Digital Image Watermarking for Multi-level Decomposition . . . 67
Ehtesham Sana, Sameena Naaz, and Iffat Rehman Ansari

Stability of Model that Makes Automated Comparison Between Market Demands and University Curricula Offer . . . 83
Ylber Januzaj, Artan Luma, Besnik Selimi, and Bujar Raufi

Economic Load Dispatch Using Intelligent Particle Swarm Optimization . . . 93
Nayan Bansal, Rohit Gautam, Rishabh Tiwari, Surendrabikram Thapa, and Alka Singh

Companion: Detection of Social Isolation in Elderly 107
Gayatri Belapurkar, Athul Balakrishnan, Rajpreet Singh Bhengura, and Smita Jangale

Machine Learning Techniques to Determine the Polarity of Messages on Social Networks 117
Jesus Varga, Omar Bonerge Pineda Lezama, and Karen Payares

An Investigation on Hybrid Optimization-Based Proportional Integral Derivative and Model Predictive Controllers for Three Tank Interacting System 125
S. Arun Jayakar, G. M. Tamilselvan, T. V. P. Sundararajan, and T. Rajesh

Building a Land Use and Land Cover (LULC) Classifier Using Decadal Maps 147
D. Bharathi, R. Karthi, and P. Geetha

Role of Artificial Intelligence in Bank's Asset Management 161
Priya Gupta and Parul Bhatia

Detection of Birds Chirping Using Machine Learning 175
B. Sadhana, Neha Shetty, R. K. Ananya, P. Vanditha Shenoy, Chitra Shenoy, Vasath Nayak, and Pragathi Hegde

Methods for Detecting Community Structures in Social Networks 187
Jesus Vargas, Omar Bonerge Pineda Lezama, and Diana Garcia Tamayo

Improving the Performance of an Ultrasonic Sensor Using Soft Computing Techniques for 2D Localization 195
R. Vijay Sunder, S. Venkatachalam, G. Sree Yeshvathi, Kaza Venkat Hruday, and S. Adarsh

Music Information Retrieval and Intelligent Genre Classification 207
Rahul Gupta, Jayesh Yadav, and Cheshtha Kapoor

Hybrid Transform-Based Image Compression Using Adaptive Grid Scanning 225
Venkatateja Jetti and Ram Kumar Karsh

Detection of Proximal Femoral Bone Fracture Using Mask R-CNN . . . 239
Ambarish Moharil and Shreya Singh

Combination of Support Vector Machine (SVM) and Bayesian Model to Identify Criminal Language 255
Amelec Viloria, Omar Bonerge Pineda Lezama, and Juan Hurtado

Higher Education Enrolment Query Chatbot Using Machine Learning 263
B. S. Niranjan and Vinayak Hegde

An Efficient Internet of Things (IoT)-Enabled Skin Lesion Detection Model using Hybrid Feature Extraction with Extreme Machine Learning Model 275
B. Pushpa

Using Convolutional Neural Network to Detect Diabetic Retinopathy in Human Eye 283
Saloni Dhuru and Avinash Shrivas

Use of Ensemblers Learning for Prediction of Heart Disease 297
Meenu Bhatia and Dilip Motwani

Determining the Degree of Relevance of Content on Social Networks Using Machine Learning Techniques and N-Grams 313
Jesus Vargas, Omar Bonerge Pineda Lezama, and Jose Eduardo Jimenez

An XGBoost Ensemble Model for Residential Load Forecasting 321
Karthik Venkat, Tarika Gautam, Mohit Yadav, and Mukhtiar Singh

Breast cancer Analysis and Detection in Histopathological Images using CNN Approach 335
A. L. Prajoth SenthilKumar, Modigari Narendra, L. Jani Anbarasi, and Benson Edwin Raj

Implementation of Stemmer and Lemmatizer for a Low-Resource Language—Kannada 345
G. Trishala and H. R. Mamatha

Election Tweets Prediction Using Enhanced Cart and Random Forest 359
Ambati Jahnavi, B. Dushyanth Reddy, Madhuri Kommineni, H. Anandakumar, and Bhavani Vasantha

Machine Learning Techniques as Mechanisms for Data Protection and Privacy 367
Amelec Viloria, Nelson Alberto, and John Rhenals Turriago

Malware Identification and Classification by Imagining Executable 375
Rupali Komatwar and Manesh Kokare

Augmented Reality in Sports Analysis Using HDM Representation of Players' Data 389
P. Sri HarshaVardhan Goud, Y. Mohana Roopa, R. Sri Ritvik, and Srija Vuyyuru

Shared Access Control Models for Big Data: A Perspective Study and Analysis 397
K. Vijayalakshmi and V. Jayalakshmi

Application Monitoring Using Libc Call Interception 411
Harish Thuwal and Utkarsh Vashishtha

Classification of Clinical Reports for Supporting Cancer Diagnosis 421
Amelec Viloria, Nelson Alberto, and Yisel Pinillos-Patiño

A Review on Swarm Intelligence Algorithms Applied for Data Clustering 429
N. Yashaswini Gowda and B. R. Lakshmikantha

Low Complexity and Efficient Implementation of WiMAX Interleaver in Transmitter 441
S. Anitha and D. J. Chaithanya

Instrument Recognition in Polyphonic Music Using Convolutional Recurrent Neural Networks 449
Bhargav Ram Kilambi, Anantha Rohan Parankusham, and Satya Kiranmai Tadepalli

Prediction of Protein–Protein Interaction as Carcinogenic Using Deep Learning Techniques 461
Rohan Kumar, Rajat Kumar, Pinki Kumari, Vishal Kumar, Sanjay Chakraborty, and Sukhen Das

Network Structure to Estimate Prices of Basic Products: Dairy 477
Noel Varela, Nelson Zelama, and Jorge Otalora

Energy Optimization in WSN Using Evolutionary Bacteria Foraging Optimization Method 485
Shiv Ashish Dhondiyal, Manisha Aeri, Paras Gulati, Deepak Singh Rana, and Sumeshwar Singh

Stock Market Prediction Using Machine Learning 497
Ashfaq Shaikh, Ajay Panuganti, Maaz Husain, and Prateek Singh

Convolutional Neural Network Based Mobile Microorganism Detector 509
Airin Elizabath Shaji, Vinayak Prakashan Choyyan, Sreya Tharol, K. K. Roshith, and P. V. Bindu

Fuzzy-Based Optimization of Evaporation Process in Sugar Industries 521
Sebastian George and D. N. Kyatanavar

Algorithms for Decision Making Through Customer Classification 535
Jesus Vargas, Nelson Alberto, and Oswaldo Arevalo

A Conversational AI Chatbot in Energy Informatics 543
Aparna Suresan, Sneha S. Mohan, M. P. Arya, V. Anjana Gangadharan, and P. V. Bindu

Monitoring Health of Edge Devices in Real Time 555
V. Meghana, B. S. Anisha, and P. Ramakanth Kumar

Statistical Evaluation of Malnutrition Status of Children in Lao Cai Province, Vietnam 567
Mai Van Hung, Nguyen Van Ba, and Dam Thi Kim Thu

Predictive Analysis of Emotion Quotient Among Youth 577
Shrinivas D. Desai, Akula Revathi, S. L. Aishwarya, Aishwarya Mattur, and Aishwarya V. Udasimath

Bots, Internet of Things and Threats to Personal Data in the Technological Era 591
Amelec Viloria, Nelson Alberto, and Carlos Alberto Jiménez Cabarcas

Role of Non-textual Contents and Citations in Plagiarism Detection 601
V. Geetha Lekshmy, R. Athira Krishnan, and S. Aparnna

Leveraging Machine Learning to Augment HR Processes 613
Shweta Jha

A Machine Learning Approach to Analyze Marine Life Sustainability 619
Danish Jain, Shanay Shah, Heeket Mehta, Ayushi Lodaria, and Lakshmi Kurup

A Novel Trespassing Detection System Using Deep Networks 633
Harshit Gupta, Rahul Singh Yadav, Sumith M. Sree Kumar, and M. Judith Leo

Analysis of Malicious DoS Attacks in AES-128 Decryption Module 647
R. Gayatri, Yendamury Gayatri, R. Karthika, and N. Mohankumar

Creating a 3D Model from 2D Images Using Convolution Neural Network 661
K. S. Vani, Rupesh Sapkota, Sparsh Shrestha, and Srujan B

Specular Microscopic Endothelium Image Analysis with Danielsson Morphology 669
Kamireddy Vijay Chandra and Bhaskar Mohan Murari

Sequential Workflow in Production Serverless FaaS Orchestration Platform 681
Urmil Bharti, Deepali Bajaj, Anita Goel, and S. C. Gupta

Practical Implementation and Analysis of TLS Client Certificate Authentication 695
N. Rajathi and Meghna Praveen

Convolutional Neural Networks in the Identification of Benign and Malignant Melanomas 705
Amelec Viloria, Nelson Alberto, and Isaac Kuzmar

Ant Colony Technique for Task Sequencing Problems in Industrial Processes 713
Noel Varela, Nelson Zelama, Ruben Hernandez, and Jeferson Rafael de Avila Villalobos

Classification of Marathi Text Using Hierarchical Attention (HAN)-Based Encoder-Decoder Model 721
Rushali Dhumal Deshmukh and Arvind Kiwelekar

Graph-Based Hybrid Recommendation Model to Alleviate Cold-Start and Sparsity Issue 737
Angira Amit Patel and Jyotindra Dharwa

An Efficient Task Scheduling Using GWO-PSO Algorithm in a Cloud Computing Environment 751
Avinashi Malleswaran Senthil Kumar, Parthiban Krishnamoorthy, Sivakumar Soubraylu, Jeya Krishnan Venugopal, and Kalimuthu Marimuthu

Spike Correlations and Synchrony Affect the Information Encoding of Neurons 763
Manjusha Nair, Richard Laji, and Reshma Mohan

A Novel Design of Dual Band Twin Tag H-Shaped Antenna for the Satellite Applications 775
Aylapogu Pramod Kumar, D. Venkatachari, J. Siddarthavarma, and Donga Madhu

Analysis of Sleep Apnea Considering Biosignals from Peripheral Capillary Oxygen Saturation Level and Electrocardiogram Data 785
Senthilnathan Ramasubbu, R. Vijaya Lakshmi, S. Lakshmi Priya, R. Prakash, A. Balaji Ganesh, A. Lakshmi Sangeetha, and Senthil Kumar Thangavel

Convergence of a Finite-Time Zhang Neural Network for Moore–Penrose Matrix Inversion 797
G. Sowmya and P. Thangavel

Estimation of Differential Code Bias and Local Ionospheric Mapping Using GPS Observations 809
Yogesh Lingwal, Fateyh Bahadur Singh, and B. N. Ramakrishna

Design and Development of a Formation Control Scheme for Multi-robot Environment 825
Harika Pudugosula and M. Rajesh

High Throughput FIR Filter Architecture Using Retiming and Fine-Grain Pipelining . . . 835
K. P. Heena

Line Stability Index-Based Voltage Stability Assessment Using CSO Incorporating Thyristor Controlled Series Capacitor . . . 845
Poonam Upadhyay and S. Ravikumar

Portable Multifunction Tester Design to Check the Continuity of Wires and to Measure the Electrical Parameters . . . 863
A. Kunaraj, J. Joy Mathavan, and K. G. D. R. Jayasekara

Kinematic Performance Analysis of 3-DOF 3RRR Planar Parallel Manipulator . . . 879
Himam Saheb Shaik and G. Satish Babu

Constant Q Cepstral Coefficients and Long Short-Term Memory Model-Based Automatic Speaker Verification System . . . 895
Aakshi Mittal and Mohit Dua

Hardware-in-the-Loop Simulation of Induction Heating System for Melting Applications Using Xilinx System Generator . . . 905
Darshana N. Sankhe, Rajendra R. Sawant, and Y. Srinivasa Rao

A Grid-Connected Distributed System for PV System . . . 919
Md. Fahim Ansari and Anis Afzal

Real-Time Implementation of Multi-model Reference-Based Fuzzy Adaptive PI Controller for a Liquid-Level Process . . . 925
A. Ganesh Ram and S. Meyyappan

Intelligent Learning Control Strategies for Speed Control of DC Motor . . . 937
M. Vijayakarthick, N. Vinoth, S. Sathishbabu, and S. Ramesh

Fabrication of Variable Speed Lamina Cutting Machine . . . 953
Gajavalli Venkata Pavan Gopi Nikhil and Nadendla Srinivasababu

Electrocardiogram QRS Complex Detection Based on Quantization-Level Population Analysis . . . 967
Hamdi M. Mohamed and Akula Rajani

Author Index . . . 989

About the Editors

A. Pasumpon Pandian received his Ph.D. degree in the Faculty of Information and Communication Engineering under Anna University, Chennai, TN, India in 2013. He received his graduation and postgraduation degree in Computer Science and Engineering from PSG College of Technology, Coimbatore, TN, India, in the year 1993 and 2006, respectively. He is currently working as Professor in the Computer Science and Engineering Department of KGiSL Institute of Technology, Coimbatore, TN, India. He has twenty-six years of experience in teaching, research and IT industry. He has published more than 20 research articles in refereed journals. He acted as a Conference Chair in IEEE and Springer conferences and Guest Editor in computers and electrical engineering (Elsevier), soft computing (Springer) and International Journal of Intelligent Enterprise (Inderscience) Journals. His research interest includes image processing and coding, image fusion, soft computing and swarm intelligence.

Prof. Ram Palanisamy is a Professor of Enterprise Systems in the Business Administration Department at the Gerald Schwartz School of Business, St. Francis Xavier University. Dr. Palanisamy teaches courses on Foundations of Business Information Technology, Enterprise Systems using SAP, Systems Analysis and Design, SAP Implementation, Database Management Systems, and Electronic Business (Mobile Commerce). Before joining StFX, he taught courses in Management at the Wayne State University (Detroit, USA), Universiti Telekom (Malaysia) and National Institute of Technology (NITT), Deemed University, India. His research interest includes enterprise systems (ES) implementation; ES acquisition; ES flexibility; ES success; knowledge management systems; healthcare inter-professional collaboration.

Prof. Klimis Ntalianis received his diploma and Ph.D. degrees both from the Electrical and Computer Engineering Department of the National Technical University of Athens (NTUA) in 1998 and 2003, respectively. Between 2004 and 2006, he has written two postdoctoral theses in the areas of multimedia protection and emotion analysis. From 1998 to 2009, he was a Senior Researcher and Projects

Coordinator at the Image, Video and Multimedia Lab of NTUA. During this period, Dr. Ntalianis has participated in the writing, submission and implementation of more than 20 R&D proposals in calls for proposals of the General Secretariat of Research and Technology (GSRT) of Greece (Frameworks: EPEAEK, PABE, EPET, PENED), the Research Promotion Foundation of Cyprus, Information Society S.A. and the European Union (Frameworks: ESPRIT, TMR, IST, Leonardo, FP6 and FP7). In parallel and from 2005 to 2011, he has worked as an Adjunct Lecturer at the University of Peloponnese, the Hellenic Naval Academy, the Hellenic Air Force and the Cyprus University of Technology. Except of his academic activities, Dr. Ntalianis has also worked for the Institute of Communication and Computer Systems (NTUA), Algosystems S.A., Kleidarithmos Publications, Municipality of Egaleo and Informatics and Telematics Institute, Center for Research and Technology Hellas. Additionally, he has served as evaluator of the committee 48/2000 of ASEP for employing staff in the public sector (2000), he was the main writer of horizontal research studies for GSRT's Call 65 (55 MEuro) of the Information Society Programme (2003-2005 and 2008), he has evaluated the competition of The Hellenic Literary and Historical Archive (500 KEuro, 2005), and he has carried out 25 expert consultancies on behalf of Information Society S.A., for 25 proposals of 19 organizations, in the framework of Call 65. He has also worked as an expert evaluator for the Research Promotion Foundation of Cyprus in the framework of the programme "Research for Companies – Product" (February–March 2010) and for the GSRT in the framework of the Action "Collaboration" (March 2010). He was also a regular member of the No. 1 Evaluation Committee of the Ministry of Education for evaluating proposals in the framework of the call «Support of Small and Medium Companies for Research and Development Activities (11 MEuro, 2010–2012). Dr. Ntalianis is an active reviewer of more than 10 International Journals of IEEE, Springer, Elsevier, etc. Additionally, he is an active reviewer and/or participates in the organizing committees of more than 10 International Conferences of IEEE, ACM, etc. From April 2015, he is an Associate Professor at the West Attica University (Department of Marketing, Specialization: "Multimedia over the Internet"). Dr. Ntalianis has participated as Editor in the proceedings of 3 international conferences, he has translated and was responsible for the scientific redaction of two Computer Science books (Kleidarithmos Publications), and he has written more than 150 scientific papers & deliverables and has received more than 650 citations. His main research interests include multimedia processing, social media analysis, crowdsourcing and data mining.

Web Scraping and Naïve Bayes Classification for Political Analysis

Noel Varela, Omar Bonerge Pineda Lezama, and Milvio Charris

Abstract This article reviews different methodologies used to conduct political analysis using various sources of information available on the Internet. In some societies, the use of social networks has a significant impact on the political field with society, and various methodologies have been used to analyze various political aspects and the strategies to be followed. The purpose of this paper is to understand these methodologies in order to provide potential voters with information to make informed decisions. First, the necessary terminology on web scraping is reviewed, and then, some examples of projects for political analysis that have used web scraping are presented. Finally, the conclusions are presented.

Keywords Web scraping · Naïve Bayes · Classification for political analysis

1 Introduction

In the USA, social networks have taken an important role in the political environment: Researchers use them to thoroughly investigate the opposition with specialized teams to find inconsistencies in the opponent. Political parties take advantage of differences and junctures, for example, [1] in the USA, Republicans in Congress opposed the cut in the payroll tax, which would force the USA to cut an average of $40 in each paycheck. One of the arguments used was that $40 was not a lot of money [2].

In less than 12 h, the White House reacted with a strategy that invited citizens to write posts in Twitter, Facebook, and YouTube what $40 meant to them, and then,

N. Varela (✉) · M. Charris
Universidad de la Costa, St. 58 #66, Barranquilla, Atlántico, Colombia
e-mail: nvarela2@cuc.edu.co

M. Charris
e-mail: mcharris7@cuc.edu.co

O. B. P. Lezama
Universidad Tecnológica, San Pedro Sula, Honduras
e-mail: omarp@unit.edu

© The Editor(s) (if applicable) and The Author(s), under exclusive license to Springer Nature Singapore Pte Ltd. 2021
A. P. Pandian et al. (eds.), *Proceedings of International Conference on Intelligent Computing, Information and Control Systems*, Advances in Intelligent Systems and Computing 1272, https://doi.org/10.1007/978-981-15-8443-5_1

Barack Obama took examples of how $40 less a month affects American families by getting the US Congress to reject the payroll tax cut.

This is an example of how some societies take advantage of social networks in the political field. In Mexico, on the other hand, politics is an eternal discussion between voters through the spots they publish in conventional sources (radio, television, even movies), giving a discursive "war" between voters in favor of a party or candidate and those who do not know who to vote for.

Knowing politicians in depth, consulting various sources and analyzing them is a complex task. These are the reasons for developing a tool that, using web scraping and text mining techniques, allows the population to know any politician from various sources (social networks, newspaper websites, Google search), analyze the results of the search, and show indicators of trust based on what is obtained in an analysis of the information in order to give the user a different perspective.

1.1 Politics and the Use of Social Networks

The use of social networks has been an important factor in sharing opinions on various topics that interest citizens, meet candidates running for office, whether it is president, senators or others who work in the political field [3].

In the 2008 elections in the USA, there was a clear example of how the candidates for the presidency of the USA, Mitt Romney and Barack Obama, make use of social networks to make themselves known to people, in addition to giving their points of view on various social issues and thus be able to give their opinion together with society.

The implementation of technology in politics has allowed to engage in a new level of conversation with voters, allowing a campaign where candidates make themselves known to become something much more dynamic, something more than just a dialogue. Thus, Barack Obama, in the 2012 presidential elections, used the web more effectively in order to establish a more insurgent campaign and win the vote of young people, which was working. For example, during this campaign (June 4 to June 17), the Obama campaign made 614 posts through its platform, while the Romney's campaign made only 168, and the Twitter gap was even greater, averaging 29 daily messages from Obama to one from Romney [2–4].

2 Web Scraping

Also known as web harvesting or web data extraction, it is the process of tracking and downloading information from Web sites and extracting understructured or unstructured data into a structured format [5]. To achieve this, human exploration of the World Wide Web is simulated, either by low-level implementation of the hypertext transfer protocol, or by the incorporation of certain Web browsers.

Scraping is done using a program, known as an orchestrator, which organizes and executes the requests to the browser. The elements to be searched must be well defined, and the status of the search to be performed must be indicated (successful search, errors in the search, no results) [6].

The web scraping process is carried out in two stages: The first one is the extraction stage, in which data are queried to a site and saved locally, and then, in the second stage, the analysis of these data is carried out to obtain information [7].

2.1 Information Extraction Techniques

- Web bot, Spider, Crawler, Spiders and Trackers.

These tools inspect Internet Web sites in a methodical and automated way. They are used to track the network. They read the hypertext structure and access all the links referred to on the Web site. They are mostly used to create a copy of all visited web pages so that they can be processed by a search engine, making it possible to index the pages, providing a fast search system [8].

- Vertical aggregation platforms [9].

There are platforms that have the purpose of creating and controlling several robots that are intended for specific vertical markets. Using this, technical preparation is done by establishing the knowledge base for all vertical platforms and then creating it automatically. Platforms are measured by the quality of the information obtained. This ensures that the robustness of the platforms achieves quality information and not just useless data fragments [9].

- Reorganization of semantic annotation.

Web scraping development can be done for web pages that adopt markup and annotation that can be used to locate specific semantic or metadata fragments. Annotations can be embedded in the pages, and this can be seen as structured representation analysis (DOM) [10]. It allows data instructions to be retrieved from any layer of web pages.

2.2 Tools Used in the Extraction

- ScraperWiki. It is a web platform that allows the collaborative creation of scrapers between programmers and journalists to extract and analyze public data contained in the web.
- PHP. It has libraries for performing web scraping such as cURL, which allows the transfer and download of data, files and entire sites through a wide variety of

protocols, and Crawl, which contains several options for specifying the behavior of the extraction as filters Content-Type, handling of cookies, handling of robots, and limiting of options [4].
- Guzzle: It is a framework that includes the necessary tools to create a robust web service client. It includes service descriptions to define the inputs and outputs of an API, iterators to browse paginated Web sites, batch processing to send a large number of requests in the most efficient way. It was created using Symfony2 and uses the PHP cURL library.
- Java Jsoup: It is a library for web scraping that provides a very convenient API for data extraction and manipulation, using the best of DOM, CSS, and similar jQuery methods [6].
- Beautiful Soup: It is a Python library designed for rapid response projects such as screen scraping or web scraping. It offers some simple methods and Python idioms to navigate, search, and modify an analysis tree [11].

2.3 Tools for Analysis

Some examples of tools used for text analysis are:

- MyTrama

A web system that provides its own query language, similar to SQL. It has a visual interface that loads the target web that allows to select the data showing what is needed in screen blocks. The selection process is translated into the construction of a query in the tool's own language, which is called Trama-WQL (Web Query Language) [12].

- Gensim

It is a Python library that provides scalable semantics statistics, analyzes plain text documents for semantic structure, and retrieves semantically similar documents. Gensim's algorithms, such as latent semantic analysis and random projection analysis, discover the semantic structure of documents, by examining co-occurrence patterns within the body of training documents. These algorithms are unsupervised [8].

- Natural Language Toolkit (NLTK)

It is a set of libraries and programs for the processing of the symbolic and statistical natural language (NLP) for the Python language. It provides easy-to-use interfaces to more than 50 lexical bodies and resources, such as WordNet, along with a set of word processing libraries for classification, analysis, and semantic reasoning [13].

3 Related Studies

3.1 Law Projects

Proyectosdeley.pe [4] is a web application that shows, in an orderly and accessible way, the law projects presented in the Peruvian Congress. It is their first attempt to open up state information by creatively using technology to promote transparency. This project tries to present the information about the projects edited by the Congress in a friendly and intuitive interface. In "ProyectosDeLey", they store the data extracted with Beautiful Soup, mainly the titles, authors, publication dates, among other data.

The ProyectosDeLey software is automatically activated every 3 h and starts looking for new projects that have been posted on the Congress Web site. If there are any, it downloads, parse and save them indexed in the local database. When there are no more projects to download or process, it starts generating the HTML files that can be seen in the Web site. It also generates the web pages for each congressman who has authored at least one project.

3.2 Analysis of Political Trends Based on Web Linking Patterns: The Case of Political Groups in the European Parliament

In order to know the political situation in the European Union (EU), in this project [4] various types of data on web links to Web sites of the 96 parties that make up the EU were collected in order to find patterns for their study. Two types of links were used: in-links, which are embedded hyperlinks on one page that point to another page; and co-links, which are embedded links on two or more sites that redirect to the same page [14].

3.3 Extracting Political Positions from Political Texts Using Words as Data

The paper in [2] presents a new way of extracting political positions from political texts that does not see the texts as speeches but as data in the form of words. This approach was compared to previous methods of text analysis and used to replicate published estimates of party political positions in Britain and Ireland, in political, economic, and social dimensions.

The steps to follow for the extraction and analysis of the texts are presented below [15].

- Step 1: The reference texts with known positions are obtained a priori.
- Step 2: Word scores are generated from reference texts (word scoring)
- Step 3: Punctuation of each blank text is obtained using word scores (basic texts)
- Step 4: (optional) Blank text scores are transformed into an original metrics.

For the project, "Word scoring" algorithm techniques were used to successfully replicate the estimated policy publications without the substantial time and labor costs that they require. This algorithm reads text files and calculates a score based on a word sense from an intersection of that word set and chooses the sense with the best scores.

The algorithm takes a reference word and counts how many times it matches in the set of documents or paper and gives it a score according to the match; the lower the match the higher the score it will give and the higher the match the lower the score it will give. If the evaluated word has several meanings in the document, it is considered the closest choice to it and is taken as the best reference [6, 16].

3.4 *Measuring Political Opinions on Blogs*

The project in [11] obtained publications from people who are very involved in politics, as well as from Americans who normally blog about other issues, but for some reason decide to join a political conversation in one or more publications.

All new blog posts were downloaded and analyzed every day. The specific goal is to categorize posts into seven unique categories: extremely negative (−2), negative (−1), neutral (0), positive (1), extremely positive (2), non-opinion (NA), and not a blog (NB). The method proposed in this project is [17]:

First, all publications that are in languages other than English are ignored, since they are considered spam. This project focused on 4303 blog posts about President Bush and 6468 posts about Senator Hillary Clinton.

As a second step, the text of each document was processed, converting everything to lowercase, removing all punctuation, and deriving the words to their primitive origin. For example, "consist," "consisted," "consistency," "consisting" are reduced to its primitive origin word which would be consist, thus reducing the complexity of the information found in the text.

Finally, the pre-processed text was summarized as dichotomous variables, one type for the presence or absence of each word root (or unigram), a second type for each word pair (or bigrama), and a third for each triplet word (or trigram), so that until n-grams are reached, only the presence or absence is measured from the root of the words instead of counting them all (the second appearance of the word "horrific" in a publication does not provide as much information as the first appearance). The usual way to further simplify the variables was to consider only the dichotomous unigrams that come from the root of the indicator variables [12, 18].

4 Conclusions

In this study, a review was made about the state of the art of different methodologies that have been proposed for political analysis in social networks and on the Internet in general. This is the first stage of a research project in which the aim is to obtain information from different sources on the Internet by means of web scraping techniques and to analyze what is obtained by means of text mining techniques [19]. This review will be very useful for defining the necessary indicators, as well as the expected results of the project.

The goal is to provide information to people for promoting informed criticism, with a position based on different sources and on important news regarding Mexican politics.

References

1. Ulbricht, L.: Scraping the demos. Digitalization, web scraping and the democratic project. Democratization **27**(3), 426–442 (2020)
2. Yu, M., Krehbiel, M., Thompson, S., Miljkovic, T.: An exploration of gender gap using advanced data science tools: actuarial research community. Scientometrics, 1–23 (2020)
3. Anglin, K.L.: Gather-narrow-extract: a framework for studying local policy variation using web-scraping and natural language processing. J. Res. Edu. Effectiveness **12**(4), 685–706 (2019)
4. Mahdavi, P.: Scraping public co-occurrences for statistical network analysis of political elites. Polit. Sci. Res. Methods **7**(2), 385–392 (2019)
5. Schrenk, M.: Webbots, spiders, and screen scrapers, a guide to developing internet agent with PHP/CUR, 2nd edn (2012)
6. Mustafaraj, E., Lurie, E., Devine, C.: The case for voter-centered audits of search engines during political elections, January. In: Proceedings of the 2020 Conference on Fairness, Accountability, and Transparency, pp. 559–569 (2020)
7. Rahman, R.U., Wadhwa, D., Bali, A., Tomar, D.S.: The emerging threats of web scrapping to web applications security and their defense mechanism. In: Encyclopedia of Criminal Activities and the Deep Web, pp. 788–809. IGI Global (2020)
8. Jiao, J., Bai, S.: An empirical analysis of Airbnb listings in forty American cities. Cities **99**, 102618 (2020)
9. Aizenberg, E., Hanegraaff, M.: Is politics under increasing corporate sway? A longitudinal study on the drivers of corporate access. West Eur. Polit. **43**(1), 181–202 (2020)
10. De Stefano, D., Fuccella, V., Vitale, M.P., Zaccarin, S.: Using web scraping techniques to derive co-authorship data: insights from a case study. In SIS May 2018. 49th Scientific Meeting of the Italian Statistical Society, pp. 1–6. Pearson (2018)
11. Hopkins, D.J., King, G.: A method of automated nonparametric content analysis for social science. Am. J. Polit. Sci. **54**(1), 229–247 (2010)
12. Maerz, S.F., Schneider, C.Q.: Comparing public communication in democracies and autocracies: automated text analyses of speeches by heads of government. Qual. Quan. 1–29 (2019)
13. Joby, P.P.: Expedient information retrieval system for web pages using the natural language modeling. J. Artif. Intell. **2**(02), 100–110 (2020)
14. Dorle, S., Pise, N.: Political sentiment analysis through social media. In: February 2018 Second International Conference on Computing Methodologies and Communication (ICCMC), pp. 869–873. IEEE (2018)

15. Mitchell, R.: Web scraping with Python: Collecting more data from the modern web. O'Reilly Media, Inc. (2018)
16. Matt, T., Pang, B., Lillian, L.: Get out the vote: determining support or opposition from congressional floor-debate transcripts proceedings of EMNLP, pp 327–335 (2006)
17. Wilkerson, J., Casas, A.: Large-scale computerized text analysis in political science: opportunities and challenges. Annu. Rev. Polit. Sci. **20**, 529–544 (2017)
18. Viloria, A., Varela, N., Lezama, O.B.P., Llinás, N.O., Flores, Y., Palma, H.H., … Marín-González, F.: Classification of digitized documents applying neural networks. In: Lecture Notes in Electrical Engineering, Vol. 637, pp. 213–220. Springer. https://doi.org/10.1007/978-981-15-2612-1_20 (2020)
19. Kamatkar, S.J., Kamble, A., Viloria, A., Hernández-Fernandez, L., García Cali, E.: Database performance tuning and query optimization. In: Lecture Notes in Computer Science (including subseries Lecture Notes in Artificial Intelligence and Lecture Notes in Bioinformatics), Vol. 10943 LNCS, pp. 3–11. Springer. https://doi.org/10.1007/978-3-319-93803-5_1 (2018)

Proactive Handoff of Secondary User in Cognitive Radio Network Using Machine Learning Techniques

Gaurav Wajhal, Vasudev Dehalwar, Ankit Jha, Koki Ogura, and Mohan Lal Kolhe

Abstract Spectrum management always appears as an essential part of modern communication systems. Handoff is initiated when the signal strength of a current user deteriorates below a certain threshold. In cognitive radio network, the perception of handoff is different due to the presence of two categories of users: certified/primary user and uncertified/secondary user. The reason for the spectrum handoff arises when the primary user (PU) returns to one of its band used by the secondary user. The spectrum handoff is of two types: reactive handoff and proactive handoff. There are certain limitations in reactive handoff, such as it suffers from prolonged handoff latency and interference. In the proactive handoff, the operation of handoff is planned and implemented by predicting the emergence of primary user based on the historical data usage. Therefore, proactive handoff boosts the performance of a cognitive radio network. In this work, a spectrum prediction technique is proposed for ensuring the spectrum mobility using machine learning. Machine learning techniques such as decision tree, random forest, stochastic gradient classifier, logistic regression, multilayer perceptron, and support vector machine are researched and implemented. The performance of different techniques is compared, and the accuracy of prediction is measured.

G. Wajhal (✉) · V. Dehalwar · A. Jha
Maulana Azad National Institute of Technology, Bhopal, India
e-mail: gauravwajhal@gmail.com

V. Dehalwar
e-mail: vasudevd@gmail.com

A. Jha
e-mail: jhaankit1200@gmail.com

K. Ogura
Kyushu Sangyo University, Fukuoka, Japan
e-mail: ogura.k@ip.kyusan-u.ac.jp

M. L. Kolhe
University of Agder, Kristiansand, Norway
e-mail: Mohan.L.Kolhe@uia.no

© The Editor(s) (if applicable) and The Author(s), under exclusive license to Springer Nature Singapore Pte Ltd. 2021

A. P. Pandian et al. (eds.), *Proceedings of International Conference on Intelligent Computing, Information and Control Systems*, Advances in Intelligent Systems and Computing 1272, https://doi.org/10.1007/978-981-15-8443-5_2

Keywords Cognitive radio · Spectrum sensing · Spectrum mobility · Spectrum management · Proactive handoff · Machine learning

1 Introduction

More and more applications/services (i.e., smart grid, e-health, security, industrial automation, process control, etc.) are using the cyber-physical system in their working which operates on wireless communication. These applications/services are bandwidth-hungry due to the massive amount of data generated in the network. There is a huge demand for radio spectrum because of big data communication and requirements of high-speed Internet and multimedia services. The scarcity of spectrum is a big bottleneck in the expansion of the radio spectrum [1–3]. The strict timing constraints of different applications such as smart grid and power protection necessitate optimum utilization of the existing spectrum. In wireless regional area network (WRAN), cognitive radio (CR) has come out as smart technology [4, 5]. CR technology allows the use of radio spectrum by unlicensed secondary user (SU) when a licensed primary user (PU) is not present. The transmission is carried out temporarily, which can be claimed at any time by the licensed user. The SU switches between the vacant spectrums when the PU arrives based on the spectrum sensing technique. Finding the vacant spectrum hole is important to continue the transmission. The process of switching between the spectrum hole is called spectrum mobility [6]. Spectrum mobility introduces spectrum handoff, which allows the SU to switch their transmission to unused spectrum holes to maintain continuity in transmission [7].

The spectrum sensing techniques studied are energy detection, cyclostationary features detection, match filter, waveform-based detection, etc. [8, 9]. Energy detection technique is used in this paper for spectrum sensing. Comparing the received energy signal with a pre-determined threshold, the possibility of PU signal can be detected [8].

Spectrum mobility is of two types, namely reactive and proactive [10, 11]. The SU can switch to another vacate channel when the emergence of PU is noticed as a reactive approach. In the proactive handoff, SU predicts the emergence of PU, and handoff is triggered before the true emergence of PU. Once SU detects the arrival of a PU1, it interrupts its data transmission and performs a spectrum sensing in order to search a free channel. SU can find a suitable free channel and switches to the new unused channel to continue transmission as shown in Fig. 1 [12].

As in reactive technique, disclosure of the emergence of PU is available after a certain time, which causes interference between the SU and the PU. Thus, to minimize interference between PU and SU, a proactive handoff approach is needed where SU proactively can predict the emergence of PU and intelligently switches the communication in advance as depicted in Fig. 2. The SU predicts that PU1 may arrive, and thus, it switches its transmission to another vacate channel (i.e., channel 2). It does not cause interference between the SU and PU; however, accurate

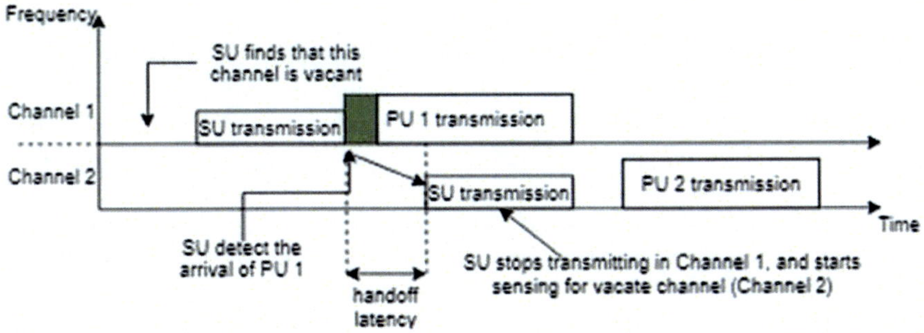

Fig. 1 Reactive handoff in cognitive radio [12]

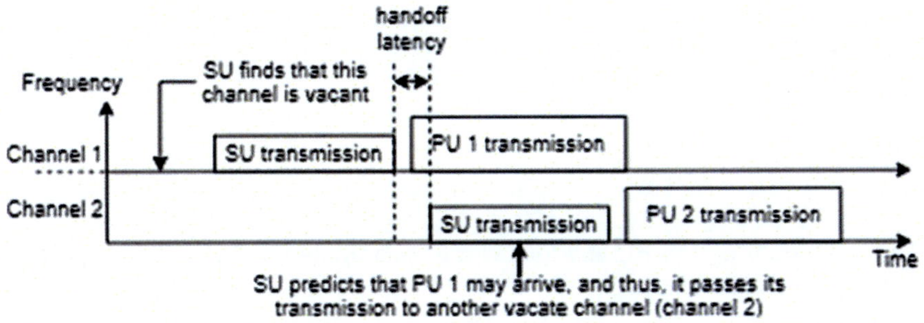

Fig. 2 Proactive handoff in cognitive radio [12]

prediction of emergence of PU is important in this strategy. Otherwise, an inaccurate prediction may lead to degraded performance of spectrum mobility. In this work, machine learning-based prediction technique(s) using historical spectrum usage is proposed.

2 Spectrum Detection Methods in Literature

There has some work [13–21] in the prediction of spectrum utilization by PU. Re-appearance and dis-appearance time point for the PU can be predicted on the channel depending on the historical data. The spectrum forecasting technique using Hidden Markov Model (HMM) with multilayer perceptron (MLP) has given by Tumuluru et al. in [13]. In MLP, past observations are the input data, while output data is future state prediction. HMM-based prediction of channel has been proposed to decrease negative impact to a response time delay. A broad survey based on different reported spectrum prediction approaches is given in [14], which include (i) using Bayesian interference, (ii) autoregressive technique, (iii) moving average based prediction, and (iv) static neighbor graph technique. The paper [15] has concurred that the Bayesian interference technique for spectrum forecasting along with a modified exponential weighted moving average (EWMA) prediction approach (hybrid approach) provides better transmission quality and success rate of transmission for limited wait time of spectrum. It also showed that the Bayesian technique has less computational complexity in comparison with EWMA. The paper [16] has illustrated prediction based on moving average (MA) which generally predicts a trend in a sequence of values. Exponential MA (EMA) can be implemented which exponentially decreasing weighting factors. This paper also described a static neighbor graph (SNG) for future PU locations based on historical topology information of PU mobility. In [17], it has presented the autoregressive model (ARM) for spectrum prediction. Each Yule-Walker equations are used to estimate the status of cognitive radio user to predict next channel. Paper [18, 19] has discussed queuing network for reactive and proactive sensing. Paper [20] describes the advantage of cooperative spectrum prediction in removing local prediction inaccuracy by considering the problems of shadowing, hidden PUs, and multi-path [21].

2.1 Spectrum Usage

The spectrum usage by PU and finding the opportunity for the available spectrum to SU are the main aim of the work. The proactive handoff of SU in the cognitive radio is illustrated in Figs. 3 and 4. In Fig. 3, there are 4 PUs (i.e., PU1, PU2, PU3, PU4) allocated in different channels, and holes are shown as white Gaussian space on which opportunity needs to find for allocating to SU [22]. In Fig. 4, there are 4 PU and 2 SUs (i.e., SU1, SU2), which illustrate allocating the holes (white Gaussian space) to the SU when it is not being used by the PU. Handoff takes place in white Gaussian space, and time taken for handoff is denoted by handoff latency. By allocating the white Gaussian space to SU, they are able to achieve fast and efficient transmission of a signal without any interference between PU and SU.

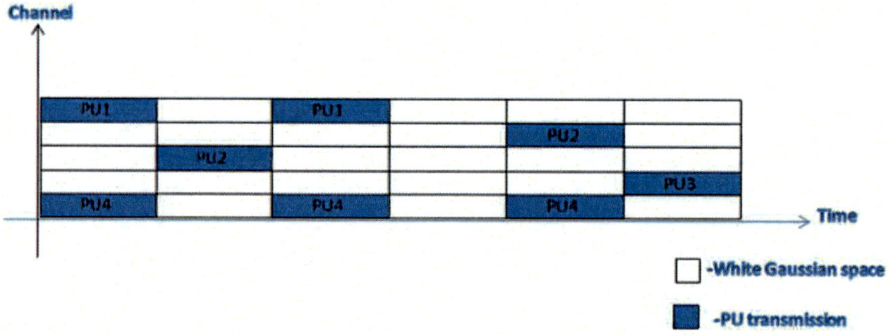

Fig. 3 PUs with white Gaussian space in the channels [10]

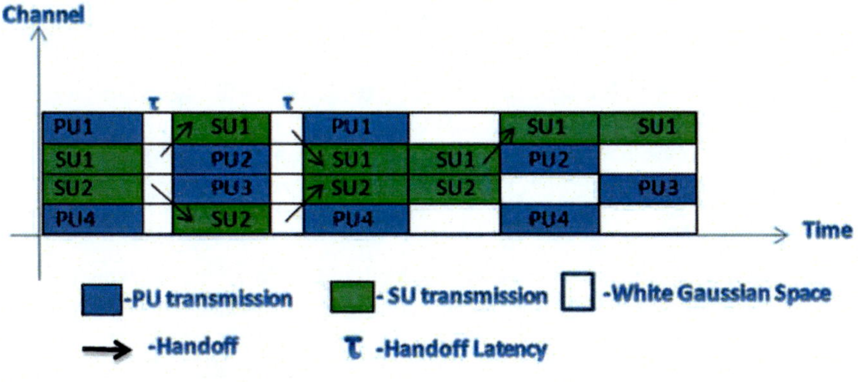

Fig. 4 SU proactively handoff [10]

3 Data Analysis Techniques

In this work, spectrum detection predictions have used machine learning techniques. In the prediction techniques, spectrum usage preparation and then use it as input to machine learning algorithms for predicting availability for SU.

3.1 Spectrum Data Preparation

Energy detection-based spectrum sensing technique is used for preparing a synthetic feature vector for PU [8, 23]. The spectrum dataset has been synthesized considering the spectrum utilization scenarios [8]. In the process, the knowledge vector of the PU as a SNR is used. Local SNR value and receiving signal strength are also calculated and used in the analysis. In this work, the data is prepared for ten cognitive users and ten primary users for full-duplex communication. Let K be noise power sensed over the samples of signal then,

$$K = \sum_{i=n}^{m} |y(n)|^2 \quad (1)$$

The decision of the energy detection is made by comparing energy 'y' with threshold [24]. Information matrix is prepared with energy vector, signal strength, SNR values, and interference. The goal is to predict the emergence of PU, and when the PU emerges, what are the values of different vectors of signal energy, signal strength, SNR, and interference. Binary classification (1 or 0) of the predictor variable is done. When interference has detected the predictors, variable is classified as 1, or else predictor variable is 0.

4 Machine Learning Algorithms

The aim is to classify the predictor variable, so that the ML algorithms, such as stochastic gradient descent (SGD) classifier, decision tree, random forest, logistic regression, and multilayer perceptron (MLP), can be applied. The performance of each machine learning algorithm is compared and analyzed in Sect. 4.

(A) **Decision Tree**

The decision tree is a predictive modeling method for classification, which divide the search space into several subsets through divide and rule. A tree is built for the modeling of the classification process using this method. Once the tree is built, tuple in datasets is applied and obtained the classification results [24]. The decision tree algorithm uses a metric called information gain which depends on the lower in entropy rate; afterward, a dataset is separated on an attribute [24]. Building a decision tree is about searching an attribute that recovers the maximum information gain. 'P_i' is probability of class 'i' in the data, and a positive class and a negative class are considered. Conditional entropy of q is given p $E(q|p)$.

$$E(q) = \sum_{i=0}^{n} -P_i \log_{2i} P_i \tag{2}$$

$$I(q|P) = E(q) + E(q|p) \tag{3}$$

(B) **Random Forest:**

It is a collective learning for classification, regression, and alternative function by designing a group of large decision trees at time of training, and mode of classes of the independent tree is the output class [25]. The features are selected in an unplanned way for each and every decision tree, which is trained and output of random forest is voting result. A random forest is the compilation of the arbitrary regression tree. For ith tree at point x, the predicted value is denoted by $A_n(y, \theta_i)$, where $\theta 1 \ldots \theta_N$ are independent random variables, independent of sample S_0 (Eq. 4) [26].

$$A_{N,n}(y, \theta 1, \ldots, \theta N) = \frac{1}{N} \sum_{i=1}^{N} A_n(y, \theta i) \tag{4}$$

(C) **Stochastic Gradient Descent (SGD) Classifier**

SGD is an elementary, yet very dynamic approach to selective learning of linear classifiers beneath convex loss functions. Multiple binary classifiers are supported by the SGD classifier by multi-class classification by associating in a one versus all (OVA) scheme. A binary classifier is studied that segregates between that and every other $K - 1$ classes for all the K classes [26]. At testing time, the confidence score (i.e., the registered distances from the hyperplane) is calculated for all classifiers and selects the class with the paramount confidence.

(D) **Logistic Regression**

Logistic regression is a binary arrangement design that predicts the variables or the features to binary feedback variable applying function of logistic [26]. It can be defined as Eq. (5), and it is,

$$\sigma(\alpha) = \frac{1}{1 + \exp(-\alpha)} \tag{5}$$

The possibility of special state/class that given input observations or features could be defined as Eq. (6), and it is,

$$P[q_t = \Delta j | O_t |] = \sigma\left(\omega' O_t\right) \tag{6}$$

In this model, ω is the weight that is given to every input feature that is being trained by applying repetitive reweighted least-squares imposed on data of training.

(E) **Multilayer Perceptron (MLP)**

MLP is based on a feed-forward artificial neural network that produces a group of output from a group of input. The multilayer perceptron is portrayed by many layers of input nodes linked as a directed graph among the input and output layer [13]. Backpropagation is used for instructing the network in an MLP. It is a neural network that joins numerous layers in a directed graph, which instruments that the signal direction only goes unidirectional through the nodes. Every node except the input nodes has a nonlinear activation function. The input pattern and its equivalent result pattern are enforced on the network [27]. The MLP gives an output $O(x)$ (Eq. 7), and it is obtained as,

$$O(x) = f\left(\sum_{i=0}^{n-1} (I_i W_i) + b\right) \tag{7}$$

where b is the bias and W_i is the weight companion to input line I, and then the resulted pattern $D(x)$ is compared with the input line i. Resulting an error signal $E(x)$ is given as the error signal is circulated back to the hidden layer (Eq. 8), and it is,

$$E(x) = D(x) - O(x) \tag{8}$$

The threshold and weight related to the network are then enhanced using back-propagation algorithm. Repetition of the above operation is done until the mean square error (MSE) of the network gives a minimum value.

4.1 *Support Vector Machine (SVM)*

Area group of SVM superintended learning approach, which is adopted for classification, regression, and outliner detection. The advantage of the SVM is that it is impressive in high-dimensional space. It is also known for its versatile performance as distinct kernel functions could be specific for the decision function. The SVM is used to implement statistical learning theory. The optimization problem is given by Eq. (9). The normal vector '*w*' is calculated on the hyperplane, and bias '*b*' satisfies optimization problem constraint (Eq. 10)

$$\min \frac{||\omega||^2}{2} + C\sum_{i=1}^{n} \Delta i \tag{9}$$

$$\text{subject to: } y_i(w \cdot x_i + b) > 1 - \Delta i \tag{10}$$

5 Training of Data and Performance Evaluation

5.1 *Train Test Split and Validation*

Before applying machine learning, the dataset has been split into testing data and training dataset. In this work, a synthetic dataset of spectrums has created using MATLAB for the PUs and SUs. The split ratio is 8:2 (i.e., 80% of the data is used for training the model, and the 20% data is used for testing and validation). For validation, the proposed work has used fivefold cross-validation on the data. So, the metrics represented in the tables and the figures have been used for the cross-validated score or accuracy.

5.2 *Performance Evaluation of Classification Models*

This research work has used two evaluation metrics, namely receiver operating characteristics (ROC) curve and confusion matrix and for evaluation. Parameters used for performance analysis are precision, accuracy, recall, and F1 score. Proportion of total number of predictions or percentage of successfully classified instances is called accuracy.

$$\text{Accuracy} = \frac{\text{True Positive} + \text{True negative}}{\text{Positive} + \text{Negative}} \tag{11}$$

Precision quantifies proportion of predicted negatives/positives that are actually positive/negative.

$$\text{Precision} = \frac{\text{True Positive}}{\text{True Positive} + \text{False Positive}} \tag{12}$$

Recall is the proportion of actual negatives/positives, which are predicted positive/negative.

$$\text{Recall} = \frac{\text{True Positive}}{\text{True Positive} + \text{False Negative}} \tag{13}$$

$F1$ score is harmonic average of precision and recall.

$$F1\,\text{score} = \frac{2\text{Precision} * \text{Recall}}{\text{Precision} + \text{Recall}} \tag{14}$$

The decision tree algorithm has been applied to the prepared spectrum dataset. Then training of the model begins and after that validation is performed on the dataset. The confusion matrix is plotted true label against the predicted label. The binary value is been taken where 1 means interference between PU and SU, and 0 means there is no interference. Figure 5 provides the confusion matrix and ROC curve of decision tree.

It can be interpreted from the model that 10,218 times model correctly predicts that PU is not present, and hence, the SUs can transmit using the channel. The model predicted 2338 times the emergence of PU which causes interference. Hence, handoff needs to be performed for communication. Similarly, the model predicts that the PU will not occur 2224 times, but it actually occurs which causes interference between PU and SU. And the last 5220 times predicted that PU could occur, and actually, the PU has occurred. It can be inferred from the confusion matrix that the classifier works pretty well for all the cases, where the primary user has not actually present. The

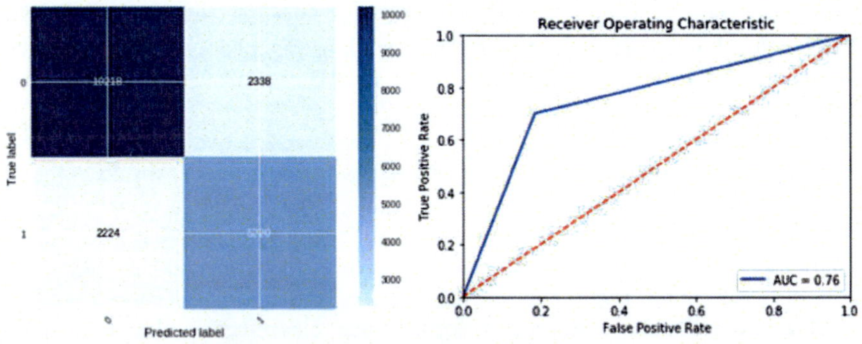

Fig. 5 Confusion matrix and ROC curve for decision tree

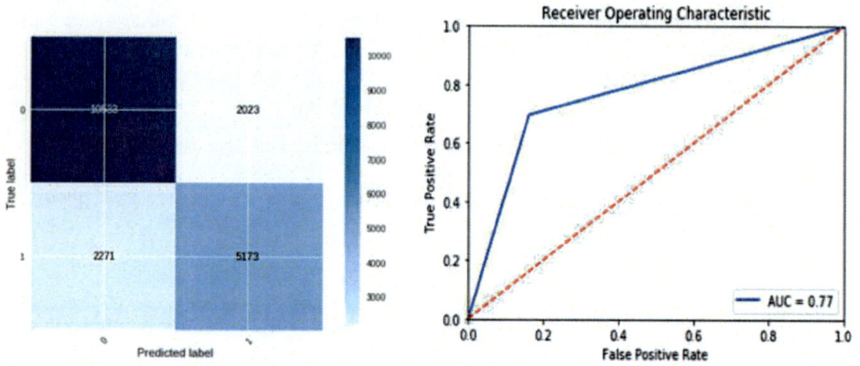

Fig. 6 Confusion matrix and ROC curve for random forest

model classified it correctly as a non-primary user, and hence, the SUs can transmit using the channel. The AUC-ROC curve will give us much better information about the performance. It depicts how much the model is able to differentiate between classes. Higher the AUC, better the model is at predicting between classes. The area under curve (AUC) of decision tree is 0.76. In the decision tree, a false-positive value (predicted interference, but actually may not occur) is more, so random forest can be used.

Figure 6 provides the confusion matrix and ROC of the random forest.

It can be inferred from the confusion matrix that the false-positive value is 2023, which is less than the decision tree. It means that channel is more exploited in this case. The ROC curve of a random forest shows the AUC of is 0.77, which is marginally better than the decision tree.

Random forest reduces the false-positive value, but a false-negative value (predicted that channel is free, but PU can occur) of both curves is more, which causes more interference between the PU and SU. To reduce the number of interference, SGD classifier can be used.

The plot of the confusion matrix and ROC for SDG is given in Fig. 7. It can be deduced from the confusion matrix of SGD that false-negative values are 1488, which is much less than random forest. By applying this model, the interference is further reduced. From the ROC curve of SGD, the AUC is 0.84 which is greater than both random forest and decision tree.

In SGD, false-negative value is reduced, but the false-positive value is increased, so the proposed model has used the logistic regression to get better results.

The plot of the ROC and confusion matrix of logistic regression is shown in Fig. 8. It can be inferred from the confusion matrix of logistic regression that 1387 is false positive and 1671 is false negative which overall gives better results than SGD. The ROC curve of logistic regression provides AUC of 0.83.

MLP and SVM give better results for nonlinear classification. So, the model is trained on the prepared dataset. The confusion matrix of MLP is illustrated in Fig. 9.

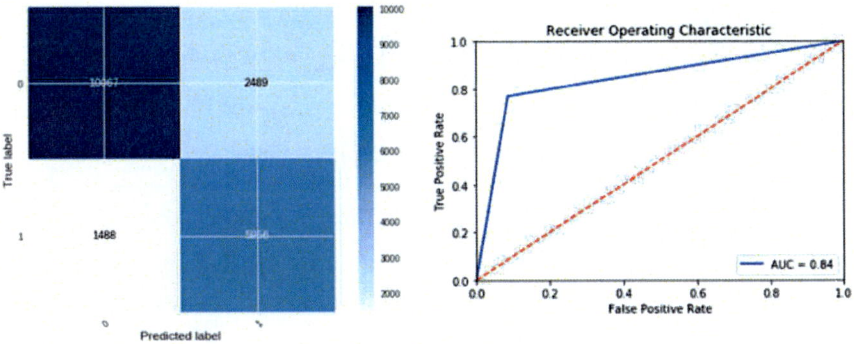

Fig. 7 Confusion matrix and ROC curve for SGD

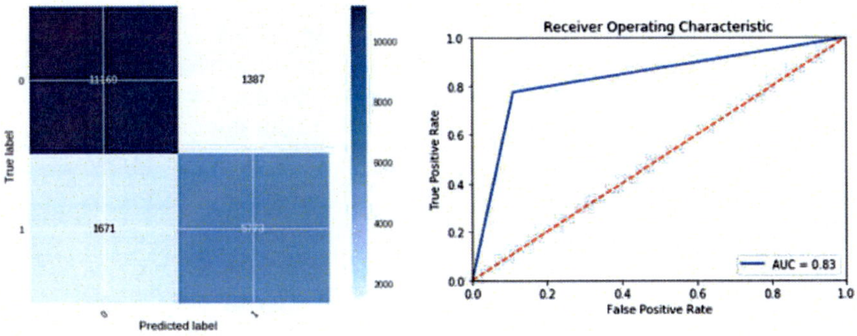

Fig. 8 Confusion matrix and ROC curve for logistic regression

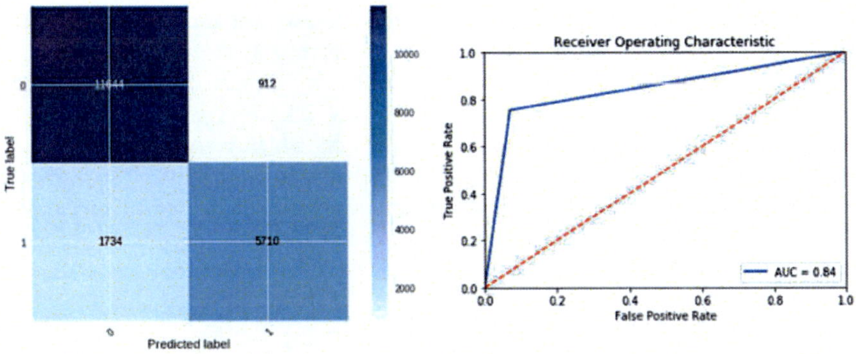

Fig. 9 Confusion matrix and ROC curve for MLP

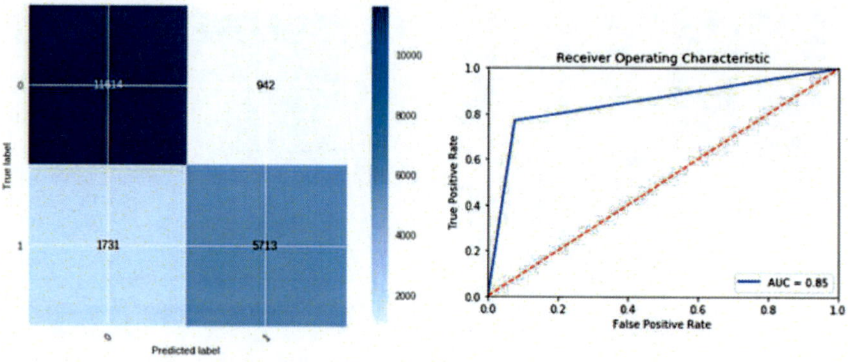

Fig. 10 Confusion matrix and ROC curve for SVM

Table 1 Comparison of machine learning algorithms

Algorithms	Metrics for algorithms			
	Accuracy	Precision	Recall	F1 score
Decision tree	0.7719	0.8138	0.8704	0.8980
Random forest	0.7853	0.8389	0.8226	0.8307
SGD	0.8012	0.8018	0.8712	0.8351
Logistic regression	0.8471	0.8895	0.8699	0.8796
MLP	0.8677	0.9250	0.8704	0.8980
SVM	0.8664	0.9250	0.8703	0.8968

It can be inferred that false positive is 912, which is very less than the other model. The AUC given is 0.84, which is greater than all the models discussed.

The confusion matrix and ROC curve of SVM are illustrated in Fig. 10, which can be inferred that a false-positive value is 942 and a false-negative value is 1731, which gives better accuracy. The AUC given is 0.85 which is the better among all the models used.

Table 1 illustrates comparative study of various algorithms carried out on dataset. It can be concluded that the MLP and SVM can provide better accuracy, precision, recall, and $F1$ score.

6 Conclusion

The demand for high-speed Internet is increasing with the advent of more innovative applications. Cognitive radio in a WRAN can provide better communication for cyber-physical system. For the success of cognitive radio, better spectrum sensing is required. Spectrum efficiency and utilization will increase. Spectrum prediction is one technique that can provide interference-free and low latency communication. In this work, machine learning-based algorithms are presented for predicting spectrum usage by PUs which enables opportunistic use by SU. A prior knowledge of the emergence of PU will minimize the interference in the wireless system. Six machine learning techniques (i.e., decision tree, random forest, stochastic gradient descent classifier, logistic regression, multilayer perceptron, and support vector machine) were studied for the spectrum usage prediction by PUs. The analysis was carried out using a confusion matrix and receiver operating characteristics. The performance of machine learning models was evaluated for accuracy, precision, $F1$ score, and recall. It can be concluded that MLP and SVM can give better results in comparison with other considered machine learning models.

Cognitive radio can be used for effective utilization of unused spectrum (i.e., spectrum hole or white space), but to determine the dynamics of the unused spectrum is challenging. The unused spectrum can be effectively used by SUs. The presented results may help in the planning of spectrum allocation in a better way for the optimal utilization of bandwidth.

References

1. Cabric, D., Mishra, S.M., Brodersen, R.W.: Implementation issues in spectrum sensing for cognitive radios, Vol. 771, pp. 772–776 (2004)
2. Haykin, S.: Cognitive radio: brain-empowered wireless communications. IEEE J Sel Areas Commun **23**(2), 201–220 (2005)
3. Dehalwar, V., Sunita Kolhe, M.K.: Cognitive radio application for smart grid. Int. J. Smart Grid Clean Energy **1**(1) (2012)
4. Dehalwar, V., Kalam, A., Kolhe, M.L., Zayegh, A.: Compliance of IEEE 802.22 WRAN for field area network in smart grid, pp. 1–6 (2016)
5. Dehalwar, V., Kalam, A., Zayegh, A.: Infrastructure for real-time communication in smart grid, pp. 1–4 (2014)
6. Lee, W.Y., Akyildiz, I.F.: Spectrum-aware mobility management in cognitive radio cellular networks. IEEE Trans. Mob. Comput. **11**(4), 529–542 (2012)
7. Mishra, A., Dehalwar, V., Jobanputra, J.H., Kolhe, M.: Spectrum hole detection for cognitive radio through energy detection using random forest. In: Proc. International Conference on Emerging Technology (INCET), IEEE, 05/06/2020 2020 pp. Pages
8. Wyglinski, A.M., Hou, N.: Cognitive radio communications and networks principles and practice (2010)
9. Ridouani, M., Hayar, A., Haqiq, A.: Perform sensing and transmission in parallel in cognitive radio systems: spectrum and energy efficiency. Digit. Signal Proc. **62**, 65–80 (2017)
10. Christian, I., Moh, S., Chung, I., Lee, J.: Spectrum mobility in cognitive radio networks. IEEE Commun. Mag. **50**(6), 114–121 (2012)
11. Ali, A., Hamouda, W.: Advances on spectrum sensing for cognitive radio networks: theory and applications. IEEE Commun. Surv. Tutor. **19**(2), 1277–1304 (2017)
12. Yang, L., Cao, L., Zheng, H.: Proactive channel access in dynamic spectrum networks. Phys. Commun. **1**(2), 103–111 (2008)
13. Tumuluru, V.K., Wang, P., Niyato, D.: A neural network based spectrum prediction scheme for cognitive radio, pp. 1–5 (2010)
14. Xing, X., Jing, T., Cheng, W., Huo, Y., Cheng, X.: Spectrum prediction in cognitive radio networks. IEEE Wirel. Commun. **20**(2), 90–96 (2013)

15. Xing, X., Jing, T., Huo, Y., Li, H., Cheng, X.: Channel quality prediction based on Bayesian inference in cognitive radio networks, pp. 1465–1473 (2013)
16. İ, B., Talay, A.Ç., Altilar, D.T., Khalid, M., Sankar, R.: Impact of mobility prediction on the performance of Cognitive Radio networks, pp. 1–5 (2010)
17. Wen, Z., Luo, T., Xiang, W., Majhi, S., Ma, Y.: Autoregressive spectrum hole prediction model for cognitive radio systems, pp. 154–157 (2008)
18. Wang, C., Wang, L.: Modeling and analysis for proactive-decision spectrum handoff in cognitive radio networks, pp. 1–6 (2009)
19. Zhang, Y.: Spectrum handoff in cognitive radio networks: opportunistic and negotiated situations, pp. 1–6 (2009)
20. Shawel, B.S., Woledegebre, D.H., Pollin, S.: Deep-learning based cooperative spectrum prediction for cognitive networks, pp. 133–137 (2018)
21. Supraja, P., Pitchai, R.: Spectrum prediction in cognitive radio with hybrid optimized neural network. Mobile Netw. Appl. **24**(2), 357–364 (2019)
22. Couturier, S., Krygier, J., Bentstuen, O.I., Le Nir, V.: Challenges for network aspects of cognitive radio (2015)
23. Plata, D.M.M., Reátiga, Á.G.A.: Evaluation of energy detection for spectrum sensing based on the dynamic selection of detection-threshold. Procedia Eng. **35**, 135–143 (2012)
24. Navada, A., Ansari, A.N., Patil, S., Sonkamble, B.A.: Overview of use of decision tree algorithms in machine learning, pp. 37–42 (2011)
25. Wang, X., Liu, Z., Wang, J., Wang, B., Hu, X.: A spectrum sensing method for cognitive network using Kernel principal component analysis and random forest, pp. 5682–5687 (2014)
26. Chen, C.C.M., Schwender, H., Keith, J., Nunkesser, R., Mengersen, K., Macrossan, P.: Methods for identifying SNP interactions: a review on variations of logic regression, random forest and Bayesian logistic regression. IEEE/ACM Trans. Comput. Biol. Bioinf. **8**(6), 1580–1591 (2011)
27. Ettaouil Mohamed, L.M., Ghanou, Y., Abdellah, B.: Architecture optimization model for the multilayer perceptron and clustering (2013)

Inter-device Language Translation Application for Smartphones

Ashwini Rao, Abhishek Paradkar, Shruti Gupta, and Sayali Kadam

Abstract With the recent trends of globalization and diversification in various social environments becoming the norm, along with the advantages of having multiple perspectives, there are some issues that can get introduced in the cliques as well. One such issue is linguistic disparity, which proves to be a bona fide issue while communicating. Due to this hindrance, crucial transfer and exchange of information are affected, and the overall efficacy of the same is altered. The proposed smartphone application is based on the principle of connecting two distinct smartphone devices, having a language that they understand set on their respective devices in an effort to bridge this linguistic communication gap.

Keywords Google cloud · Multipeer connectivity · Translation · Speech to text · Networking · Machine learning · Language barrier

1 Introduction

The key to effective communication is the understanding of common languages. With the trend of increasing diversity across workplaces and other social environments such as educational institutes, people with multiple backgrounds also bring along their own share of linguistic abilities and sensibilities. However, a lack of understanding of each other's languages could prove to be a barrier to effective

A. Rao (✉) · A. Paradkar · S. Gupta · S. Kadam
MPSTME, NMIMS University, Mumbai, India
e-mail: ashwini.rao@nmims.edu

A. Paradkar
e-mail: abhishekparadkar.nmims@gmail.com

S. Gupta
e-mail: shrutigupta.nmims@gmail.com

S. Kadam
e-mail: sayalikadam.nmims@gmail.com

© The Editor(s) (if applicable) and The Author(s), under exclusive license to Springer Nature Singapore Pte Ltd. 2021

A. P. Pandian et al. (eds.), *Proceedings of International Conference on Intelligent Computing, Information and Control Systems*, Advances in Intelligent Systems and Computing 1272, https://doi.org/10.1007/978-981-15-8443-5_3

communication, with an ever-growing emphasis on globalization and diversity. With the increasing importance of globalization, diversity, and collaboration in different social environments, there are also some challenges that people might face along with the benefits that these concepts bring along.

Considering these points, a translation tool is an ideal way to handle language difficulties and attempt to solve them. The most popular tool for translation is in the form of smartphone applications as they are accessible to the masses. But, the limitation of these applications is that most of them work on a single device at a time.

Therefore, for two or more entities wishing to have a conversation has to rely on sharing the smartphone in order to establish communication. This solution may prove to be tiresome after a few exchanges and not prove to be effective in the long run. Therefore, it is still a challenge to be resolved as to how people are enabled from different linguistic backgrounds to effectively and efficiently communicate with each other and understand the information being communicated within the context. The study by Aichhorn and Puck [1] echoes these implications caused due to linguistic differences in the workplace where certain employees were uncomfortable adopting English as the primary language of communication.

In order to resolve such issues, communication with each other needs to be made as natural as possible and gets over any barriers. One way to do it is by connecting the users' devices with one another.

With the proposed solution, a connection between two smartphone devices using inter-device connectivity frameworks is aimed to establish in such a way that the input from one device is passed on to another, in a language that both the entities can understand, which might be different. Using a machine learning-based translation engine based on a cloud platform, the translation can take place instantaneously. This method can make it simpler to communicate effectively as well as help in translating the messages that the entities wish to communicate. Such a solution can help in enabling entities to seamlessly communicate with one another in their respective languages and not let it become a barrier in the conveyance of meaningful information.

2 Literature Review

There have been various studies that have explored the cultural and linguistic barriers that people face in social environments such as workplaces and educational institutes. These barriers prove to be especially significant and detrimental to the work performed in a more professional setting, as highlighted in an explanation provided in a web article by Joseph [2]. Tenzer and Schuster [3] have also contributed toward the language barrier proving to be a hindrance especially in the communication taking place in the work environment among groups or teams when dealing with international business activities and the consequences in terms of team performance. Another study by Tenzer and Pudelko [4] also highlights the impact of power dynamics in multinational teams, as language barriers can often prove to affect the morale of individuals.

Several research publications also highlight the impact of language barriers in other scenarios. The academic thesis by Lou [5] titled "Breaking the vicious cycle of language barriers" describes how language mindsets affect students, especially if they are involved in programs such as foreign exchange and are attending the educational institutions as immigrants. Language barriers can prove to be a barrier to the social and interpersonal processes in intercultural interactions. This can result in affecting the students' performances academically and socially among their peers, as described by Perry et al. [6] in the study titled "International students' perceptions of university life".

Given the trends of diversity and globalization, there are multinational firms that employ people all across the world, and communication between them is essential to the business. In such scenarios, these multinational teams, or with teams who have expatriates who have traveled from another country for work, can face language barriers in communication as documented by Ramlan et al. [7] in their study titled "The impact of language barrier and communication style in organizational culture on expatriate's working performance". Therefore, it could prove to be an issue to effectively communicate not only the key information but also human emotions among the group members and as a result might introduce delays and difficulties in getting their problems resolved correctly and effectively.

In crucial sectors such as health care as well, the disparity between the understanding of languages can prove to be harmful to the operations of the medical institute. As manifested by Van Rosse et al. [8], illustrate the safety risks among patients and doctors that can arise due to language barriers and effective methods of communication between them, titled "Language barriers and patient safety risks in hospital care". This is extremely sensitive as health care is a critical sector and even slight errors in judgment can lead to harmful results.

Additionally, a publication by Evenett and Deltas [9] titled "Language as a trade barrier: Evidence from Georgian public procurement" analyzes the monetary impact of language barriers on trade across countries. In some countries where English might not be the first language or the language of preference might not be much known to international traders, getting a trade done including the various negotiations and finalizations of deals could be a problem without proper communication taking place among the parties.

An interesting study conducted by Giovannoni and Xiong [10] describes the issues in communication in certain teams brought along by language barriers in order to achieve specific goals and how certain other teams who understood the same languages produced better results. de Moissac and Bowen [11] in their study, in Canada, have outlined the difficulties in more specialized professions such as health care over how the inability to understand languages of minorities could impact the quality of health care.

Therefore, the issue of language difficulties becoming a problem in continued communication between entities is one that needs to be addressed sensitively.

3 Methodology

3.1 Limitations of Existing Applications

The solutions existing at the present time are smartphone applications that require the participants in the communication to share one device and use it as the instrument to use a translation application. Though this works in principle, there are some disadvantages and limitations imposed through this solution.

Major limitations of such smartphone applications are that there is only one single smartphone device which is being used for the translation purpose. This brings along several implications to the way users interact with the smartphone application. The smartphone device has to be passed back and forth between the communicating parties. This introduces a constraint that the communicating parties have to maintain a very close distance among themselves. Such a constraint may not solve the issue most effectively in case of a team of members wanting to make use of a translation application occasionally to exchange information, and the process becomes extremely unnatural. Therefore, using such a smartphone application over longer durations is uncomfortable and not effective toward solving the problem to a very high degree.

Another implication brought along with such a solution is that it gives the communicating parties an implicit indication that the person owning the smartphone device and the application is in control of the conversation. This could lead to subliminal intimidation and weaken the morale of the people feeling a lack of control in the conversation and as a result not get their points across effectively.

Given such limitations of existing solutions, the goal of eliminating language barriers or tackling them is left unfulfilled to a major extent.

3.2 Design Goals

The design goals that have set for the proposed smartphone application are with the vision of keeping the users in the center of the system and allowing them to have complete control of the application and be able to use the application intuitively by understanding the functionality in one glance. Among these goals are the following considerations:

- *Usability*—The application interface must be simple to understand and use so that there is no major learning curve when it comes to using the functions of the smartphone application.
- *Visual Appeal*—By making the interface indicative and intuitive, the purpose is served of making it not only appealing and inviting to use the smartphone application, but also sub-serving the goal of making the application more usable.

- *Experience*—Ideally, the application should not ask the user to leave their current context. This means for all of the functionality of the smartphone application, the user must not have to leave the application or have any drastic changes to the way information is presented such that it is entirely a different interaction between the user and the application. Thus, the experience must be kept as consistent as possible throughout the user journey.
- *Closure and Non-Repudiation*—It is crucial that data being transferred has to be displayed to not only the receiver but also the sender. This helps in achieving closure and non-repudiation as both of the parties involved in the communication are aware of their own inputs and the implications of it in such a way that it is not refutable or falsifiable by either party.
- *Control*—Each of the users must be in complete control of their own devices. This means at any given point of time the users can engage with the application in the way they want to perform any functions deemed suitable by them and not rely on any other factors—human or otherwise—to interfere with their usage of the same.

3.3 Limitations of Existing Applications

The activity diagram (Fig. 1) expresses the flow of the application from the communicating users' point of view. From the sender's perspective, the first step is to set up the connection as a host or join an existing connection as a participant.

Due to the way the chosen communication framework operates, it requires a "host" to initiate a network and advertise the device's presence to nearby compatible devices. It also requires the rest of the devices, known in this scenario as "participants" to partake in the communication by choosing to establish communication with the host. By default, when the device radios, viz. Wi-Fi and Bluetooth, are enabled, all the devices will be able to join the host device.

Once the connection is established, the next step is to select the languages that both parties are comfortable with. The list of languages available within the application needs to be pre-programmed in order to be displayed. The languages which are supported by the translation engine are programmed in the application for the users to choose. The input and output will be presented to the user in the chosen language.

Then, the user input is captured either via text or speech which is then received and translated using the translation engine. In the case of text input, the user is expected to type in the message they want to send across in the language they have selected previously. In the case of speech input, when the users speak into the device microphone, their speech sample is first sent to a speech-to-text API, converted to text, and then sent across to the translation engine. This process of conversion takes place in the back end.

As a result, the output is received from the cloud via the translation engine and is presented on the other user's device. In the end, if the users wish to carry on the communication, they may do so without any additional setup as long as the

Fig. 1 Process flow through activity diagram

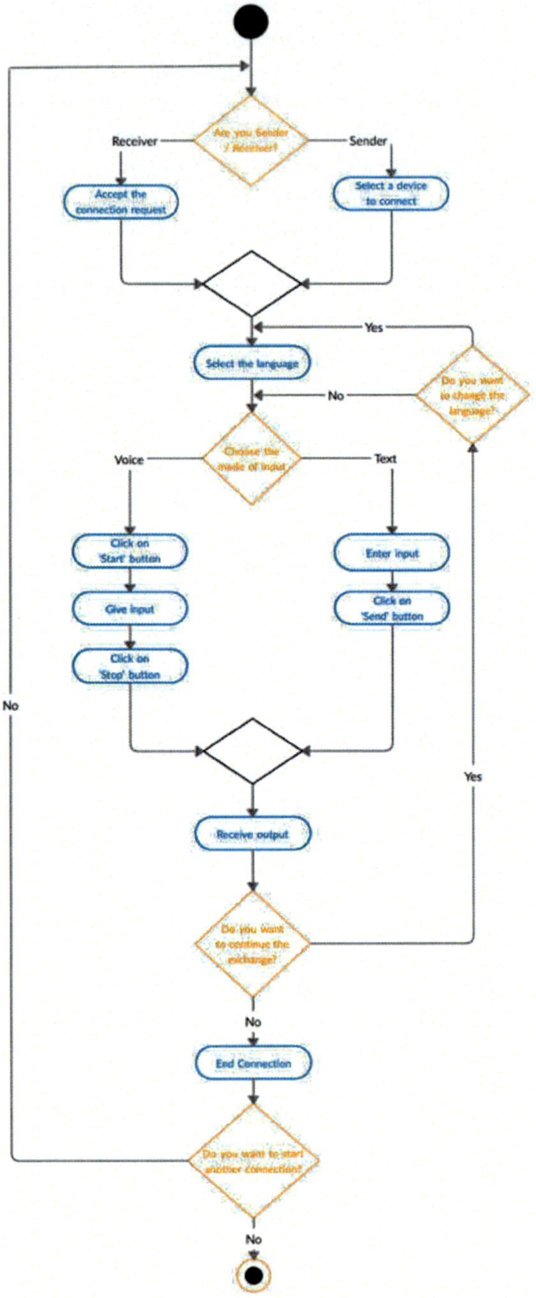

connection is established. Or else, if the users wish to disconnect the session, they may choose the option to do so and the session ends.

3.4 *Software Requirements*

3.4.1 Multipeer Connectivity

The first aspect of the smartphone application follows the principle of connecting two individual, distinct smartphones with one another. This will allow for the individual communicating parties to have complete control over their own smartphone devices. The connectivity frameworks required to make this possible are inter-device connectivity based. This means that the connectivity among the two smartphone devices will be based on the reliance of the device hardware such as Wi-Fi and Bluetooth radios (Fig. 2).

The communication protocol followed is along the lines of the following illustration:

- The first device enables device discovery through the utilized inter-device connectivity framework. This allows that device to advertise its presence to other compatible devices that are within the communicable range.
- The compatible devices shall be able to browse through the list of devices who are advertising their presence at that given point of time. Upon discovering the appropriate device, the second device then sends a connection invitation to the first device that advertised its presence.
- After the first device receives the invitation, the device can choose to respond to it through a simple prompt. If the device accepts the invitation, the session between the two devices begins and they are deemed as connected.
- Once connected, the two devices can send messages in forms of batch data as well as continuous streams.
- Both the devices, once connected, have the same privileges to join additional devices or disconnect from the current session. Thus, both the smartphone devices and by extension the users of those smartphone devices have equal authority to expand or quit the communication taking place at the present time.

In order for the connection to be successful, it requires the communicating parties to accept the connection request. Thus, the connection is based on a request-response mechanism realized by the exchange of a set of temporary keys. This ensures the connection security, and the data transfer is always encrypted based on the set of temporary keys exchanged between the two devices until the session is active. The encryption specification followed is AES, and the encryption itself takes place using *Datagram Transport Layer Security* (DTLS) protocol, which is a variation of *Transport Layer Security* (TLS). Once the session ends, the devices are unpaired from each other. The keys that were exchanged at the beginning of the session are also deemed as invalid and are regenerated during the beginning of subsequent connections. The

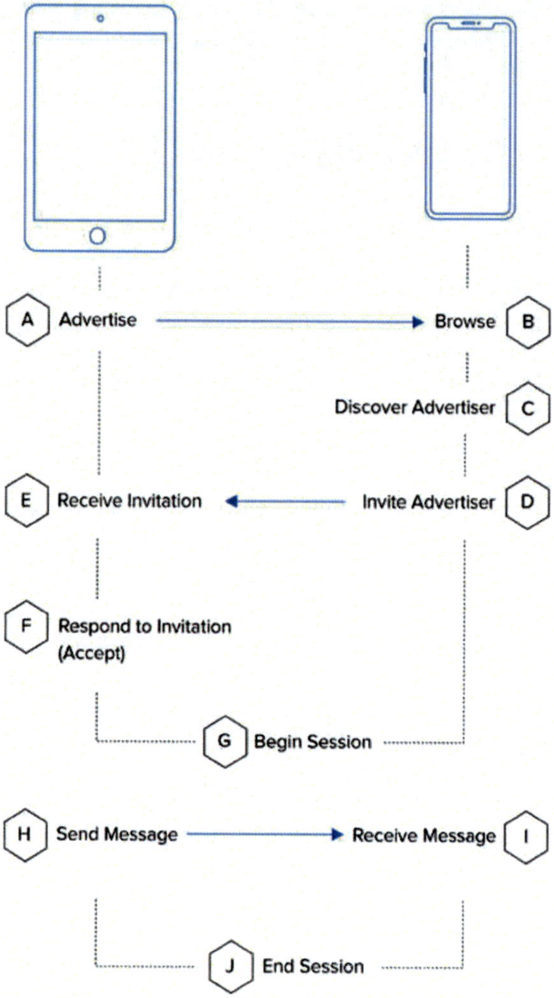

Fig. 2 Block diagram
Source https://www.toptal.com/ios/collusion-ios-multipeerconnectivity [12]

session can be ended manually by the users of the application or when they close the application on their devices.

3.4.2 Speech-to-Text API

The second aspect of the smartphone application is the translation engine. For our smartphone application, a machine learning backed cloud-based translation API fulfills the needs perfectly. The advantage of a machine learning backed algorithm is that it is constantly trained with the help of new data on the server itself with the entities providing the API. This results in constant improvements to the translation engine in the backend and further strengthens the reliability, as opposed to using a translation model built from the ground up. The Google Cloud Speech-to-Text API is based on machine learning algorithms backed by Google Cloud Platform.

It offers machine learning models based on the "Bayesian interpolation method", and the unstructured input is matched against the vocabulary built by Google in the model library dynamically in real time. The Bayesian interpolation algorithm minimizes the irregularities caused by noise in the audio input by normalizing the speech input. The main reason behind choosing Google Cloud Platform over other service providers is that Google Cloud Platform has pre-trained artificial intelligence models that are continually being retrained and upgraded on the backend What this means is that the efficiency and accuracy of the model are being reflected in real time in the application without having to reprogram the API. In a scenario of processing speech to text, an artificial intelligence model serving as the processing engine is essential to keep the latency low and maintain the accuracy level which is expected. Another advantage of using an artificial intelligence-based speech-to-text engine is that there can be numerous ways of predicting inputs that can be tested and converted based on a large amount of data being processed in the pre-trained models used by Google. This allows for higher accuracy and confidence during implementation and there.

The speech sample sent to Google Cloud is AES encrypted and follows the TLS protocol, and it offers "encryption at rest" by default. This means that no additional programming of security protocols such as setting up key exchanges is required to be programmed in the application, and it is provided by Google Cloud itself through cloud encryption key management. Using this function, a highly reliable speech can get to text transcription as the background noise is reduced, and there is a greater emphasis on spoken words and phrases. It also uses correction algorithms to accurately predict correct words and phrases even with background noise. Google Cloud Speech-to-Text API is called in the app, and the audio input is converted to text. The device microphone transfers the audio sample to the API for processing and requires Internet connectivity as the conversion is done in the cloud servers. Thus, device microphone access permission and Internet access permission are required to be granted by the user in order to successfully perform the conversion.

3.4.3 Translation API

The translation API is daisy-chained directly to the inter-device connectivity framework implemented. This means that the input received from the user is directly directed to the translation API through the session pipeline and is not, in any way, shape or form, interfered with or manipulated by the application itself. Thus, the unmodified input from the user is then processed and translated by the translation engine. The output received from the translation API is then sent back across to the inter-device connectivity framework in a similarly daisy-chained fashion.

The difference, however, is that the translation output is displayed to the user on the receiving end of the communication. Thus, the input and output sources are distinct. The Google Cloud Translation API uses machine learning models based on pre-trained neural networks similar to Google Cloud Speech-to-Text API. The difference here is that the model is trained in multiple languages, and the data library is focused on providing the translations of words and phrases based on the text input received. As Google's models rely on Bayesian classification and interpolation techniques, the decision of choosing the correct translated phase takes place in the cloud in real time, again providing with higher accuracy and confidence. Since the task of translation is being computed on text data, the latency of the process is extremely low.

Similar to Google Cloud Speech-to-Text API, the security in Google Cloud Translation API is handled at the backend through cloud encryption key management, AES standard, and TLS encryption protocol. In the application, text output from the speech-to-text API is passed to the translation API, if the users choose speech as the method of input. The translation API is then called, and the data to be translated is passed on to the Google cloud server for processing. The translation is performed dynamically and in real time. Thus, the overhead of processing is on the server and not the device itself. This helps in increasing the speed and efficiency of the application while at the same time reducing the amount of disk space the application requires. Hence, Google Cloud Translation API requires Internet access permission from the user's device.

Therefore, the application works simultaneously on both individual smartphone devices, and the data is passed across both the devices using the inter-device connectivity framework as the medium while the translation engine does the processing of the data and appropriates the languages chosen on both smartphone devices.

4 Results and Discussion

4.1 Application Workflow

This app supports over 200 languages that can be configured using Google Translate. The received message can also be played back in audio format by using the text to speech function of the device if the users prefer. The application requires Internet connectivity in order to use Google Translate and Google Speech-to-Text services. In addition, the application also requires permission from the users to access the device microphone. The flow of the application is as follows:

Screen 1: The initial screen of the application. To start the communication, a connection is needed to be established with other devices by clicking on "connect" (Fig. 3).

Screen 2: The users can select to host or join the communication by clicking on the connect button. One device acts as the host, and the other device can find that device and send a connection request. Once the host accepts the request, the connection is established. The multipeer connectivity framework is end-to-end encrypted. So, the

Fig. 3 Main interface of the application

messages sent across devices cannot be interjected or manipulated. The multipeer connectivity framework allows up to seven devices to join the host, and each user can select their own language on their own devices individually (Fig. 4).

Screen 3a: The users can select the language on their own device which they find suitable. In the selected language, the input can be given in the form of text or speech. The input is then translated by the Google Cloud Translation engine, and translated output is sent to the other device (Fig. 5).

Screen 3b: In the case of speech input, the speech is first converted to text using Google Speech-to-Text engine and then sent to the translation engine. To send a message via speech format, the user has to press the "Start Speaking" button, and when done recording, the user has to press "Stop Speaking", after that they get an option to either send the message or re-record the message. On the receiver end, the user by default will get the message displayed in the language they choose and also an option to hear the message in the chosen language (Fig. 6).

Screen 4: Disconnecting the connection—The connection remains in session if the users simply exit the app for some time, but gets disconnected when either of the users quits the application (Fig. 7 and Table 1).

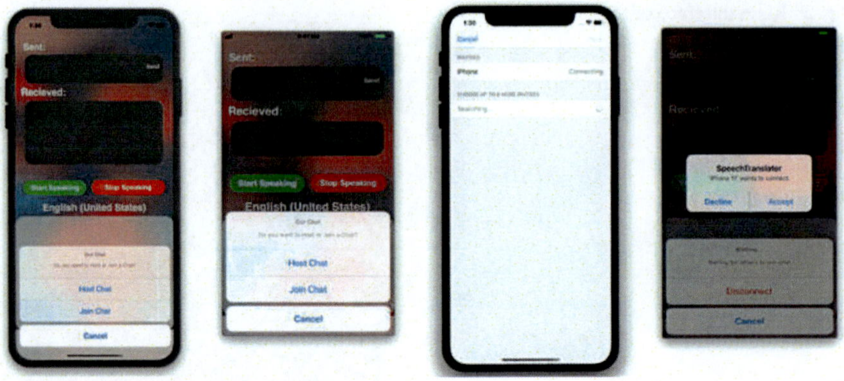

Fig. 4 Connection using multipeer connectivity

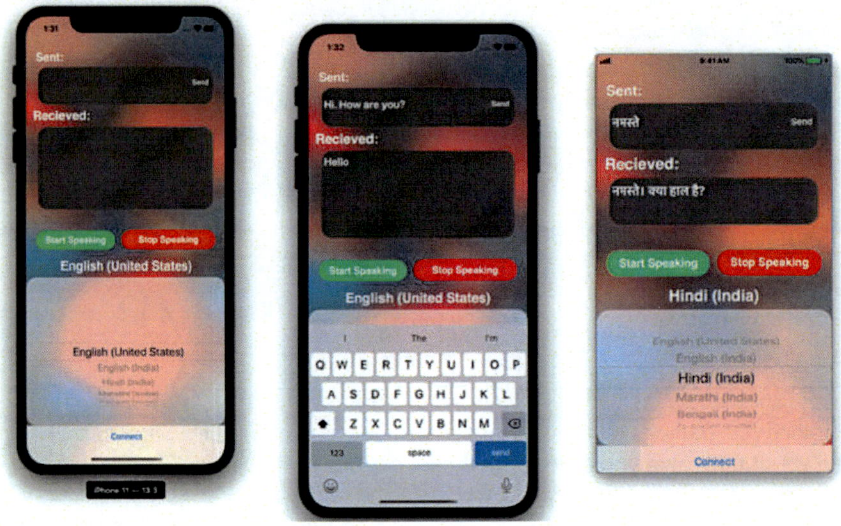

Fig. 5 Language selection and text input

4.2 Limitations of Our Work

Our functional consideration at the time of development of this application has been to not include a database to store messages sent across. This was in light of the data privacy and security restrictions imposed by several regulatory bodies, for example, the GDPR guidelines imposed by the EU. Due to such restrictions, the implementation of a database needs more time to think through and implement. As a result, the limitation presented is that currently, the application does not support features such as storing recent translations and messages, for example, common phrases, or

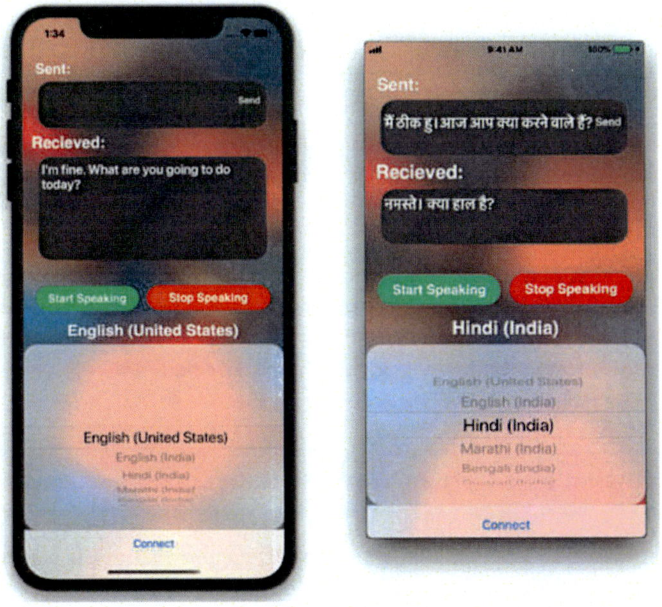

Fig. 6 Language selection and speech input

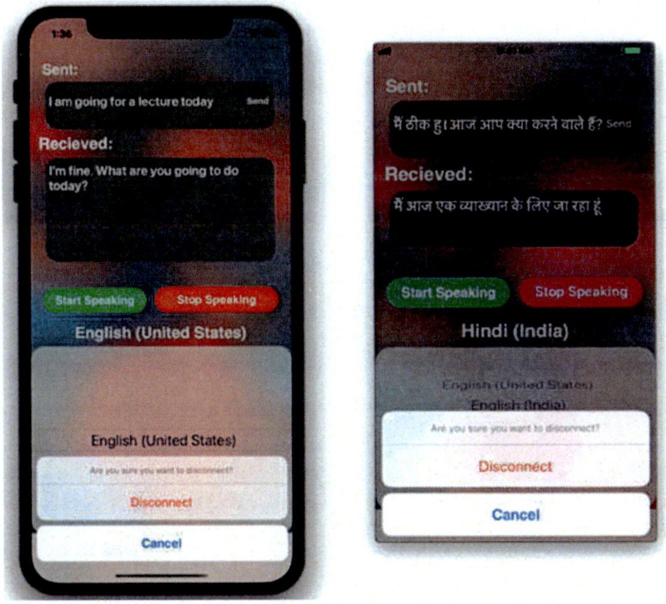

Fig. 7 Disconnecting the devices

Table 1 Accuracy of the API

Name	Request	Error (%)	Latency, median (ms)
Cloud translation API	59	0	30.987
Cloud speech-to-text API	39	7.692	3054.547

Values based on tests conducted as of March 2020 and obtained from Google Cloud Platform Analytics Dashboard

a chat functionality based on user accounts. Another limitation is that currently, the Google Clouds' Speech-to-Text API and Translation API require an active Internet connection to operate. This means that in the case of poor or no network availability, the process of information flow back and forth is affected.

5 Conclusion and Future Work

The language barrier is a crucial issue that can affect information exchange and cause hindrance between the communicating parties. In the current solutions, due to the presence of just one device, there are several limitations such as passing the same device back and forth and therefore both the entities not having complete control over the communication. In the proposed solution, a smartphone application is aimed to be developed that is based on inter-device connectivity frameworks such that each user can use the application on their individual devices which allows them to have complete control of the application. Using cloud-based translation API, a robust translation engine can be incorporated in the application and provide an end-to-end encrypted translated output in the languages of the users' choice. With an application like this, a unique and simple solution is being offered for people facing language-based communication issues by providing an application that has the highest level of encryption, security, efficiency, accuracy, and convenience possible.

In future, the project is aimed to have a robust machine learning-based translation engine that could run on devices without Internet connectivity and use the native speech-to-text implementation to further reduce the turnaround time. A database is planned to build that can store recent translations and common phrases in the languages available in order to make the application more proactive.

References

1. Aichhorn, N., Puck, J.: "I just don't feel comfortable speaking English": Foreign language anxiety as a catalyst for spoken-language barriers in MNCs. Int. Bus. Rev. **26**(4), 749–763 (2017)
2. Joseph, C.: Cultural & Language Barriers in the Workforce. Small Business-Chron.com. Retrieved fromhttp://smallbusiness.chron.com/cultural-language-barriers-workforce-11928.html (n.d.)
3. Tenzer, H., Schuster, T.: Language barriers in different forms of international assignments. In: Expatriate Management, pp. 63–100. Palgrave Macmillan, London (2017)
4. Tenzer, H., Pudelko, M.: The influence of language differences on power dynamics in multinational teams. J. World Bus. **52**(1), 45–61 (2017)
5. Lou, M.-T.: Breaking the vicious cycle of language barriers: growth language-mindsets improve communication experience for migrant university students (2019)
6. Perry, C.J., Lausch, D.W., Weatherford, J., Goeken, R., Almendares, M.: International students' perceptions of university life. Coll. Student J. 279–290 (2017)
7. Ramlan, S., Abashah, A., Samah, I., Rashid, I., Radzi, W.: The impact of language barrier and communication style in organizational culture on expatriate's working performance. Manage. Sci. Lett. **8**(6), 659–666 (2018)
8. Van Rosse, F., et al.: Language barriers and patient safety risks in hospital care. A mixed method study. Int. J. Nurs. Stud. **54**, 45–53 (2016)
9. Evenett, S., Deltas, G.: Language as a trade barrier: evidence from Georgian public procurement. Int. J. Ind. Organ. (2019)
10. Giovannoni, F., Xiong, S.: Communication under language barriers. J. Econ. Theory **180**, 274–303 (2019)
11. De Moissac, D., Bowen, S.: Impact of language barriers on quality of care and patient safety for official language minority francophones in Canada. J. Patient Experience **6**(1). https://doi.org/10.1177/2374373518769008 (2019)
12. Gottlieb, B.: Collusion: nearby device networking with multipeer connectivity in iOS. Retrieved from https://www.toptal.com/ios/collusion-ios-multipeerconnectivity (n.d.)

Real-Time Numerical Gesture Recognition Using MPU9250 Motion Sensor

Sathish Raja Bommannan, Chennuru Vineeth, Mylavarapu Uma Hema Sri, Boyanapalli Sri Vidya, and S. Vidhya

Abstract Human–computer interaction is one of the most exciting areas of research. Hand gesture recognition stands vital for developing better human–computer interaction systems. Most of the existing approaches using camera-based or 3D depth sensors for hand gesture recognition are rather expensive and sensitive to environmental changes. In this paper, we propose a low-cost data glove embedded with MPU9250 motion sensor which overcomes the drawbacks of existing systems. In our work, the primary focus is to develop a numerical gesture recognition system deployable in any real-time application. An extensive comparison of the performance of different machine learning and neural network models is presented. An optimal network model is chosen, and details of deploying the trained model in a real-time unity game application are presented. In our experiment, the highest accuracy achieved is 98.41% with an average real-time inference delay of 2 ms.

Keywords Data glove · ESP8266 NodeMCU · MPU9250 motion sensor · Unity · TensorFlow Lite API · Human–computer interaction · Support vector machines · Decision trees · Simple neural network · Long short-term memory

S. R. Bommannan (✉) · C. Vineeth · M. Uma Hema Sri · B. Sri Vidya · S. Vidhya
Department of Computer Science and Engineering, Amrita School of Engineering, Amrita Vishwa Vidyapeetham, Coimbatore, India
e-mail: cb.en.u4cse16642@cb.students.amrita.edu

C. Vineeth
e-mail: cb.en.u4cse16611@cb.students.amrita.edu

M. Uma Hema Sri
e-mail: cb.en.u4cse16621@cb.students.amrita.edu

B. Sri Vidya
e-mail: cb.en.u4cse16316@cb.students.amrita.edu

S. Vidhya
e-mail: s_vidhya@cb.amrita.edu

© The Editor(s) (if applicable) and The Author(s), under exclusive license to Springer Nature Singapore Pte Ltd. 2021

A. P. Pandian et al. (eds.), *Proceedings of International Conference on Intelligent Computing, Information and Control Systems*, Advances in Intelligent Systems and Computing 1272, https://doi.org/10.1007/978-981-15-8443-5_4

1 Introduction

With the advances in science and technology, extensive research is being conducted on the field of human–computer interaction exploring possible input modalities [1]. Among all other input modalities possible, hands are the easily approachable input modality and can form infinitely different poses [2]. There are several proposed computer-vision-based solutions using cameras [3–5], depth sensors [6, 7], etc. for receiving input from hands. But most of the existing solutions are either expensive or very sensitive to environmental conditions and even have a very limited field of view for recognizing the hand. Also, the existing input hardware like keyboard, mouse, etc. is not natural enough. In order to develop a hand gesture recognition system, which is cost-effective, real-time, natural, insensitive to environmental conditions, and overcome the field of view restrictions, we have chosen the motion-sensor-based approach.

In our motion-sensor-based approach, we have designed a data glove embedded with MPU9250 motion sensor, which consists of an accelerometer and a gyroscope. Hand's acceleration and angular velocity along XYZ axes are collected using an accelerometer and gyroscope, respectively. The hand's acceleration and angular velocity are the super optimal features to better describe the motion of the hand. The motion of the hand can be an input to a system when the motion is performed in a unique way, which is called 'gesture'. Apart from the MPU9250 motion sensor, we have added a tactile switch that stays in the thumb finger in order to find the start frame and end frame of a gesture performed in air. Defining the start frame and end frame of a gesture would drastically reduce false positive and accidental classification triggered by the free movement of the hand in three dimensions.

In our experiment, we have primarily focused classifying the numerical gestures (0 to 9), as shown in Fig. 1. With the designed glove and 20 experimenters, we have collected the dataset for our experiment. Each data sample consists of a temporal accelerometer and angular velocity data from start to end of the gesture. After collecting the dataset, we have trained and evaluated the performance of support vector machines (SVMs), decision trees, simple neural networks, and long short-term memory (LSTM). Also, in order to deploy a trained model in a real-time application, we have chosen TensorFlow Lite [8]. TensorFlow Lite is a deep learning framework used for inference in edge devices. TensorFlow Lite enables us to efficiently

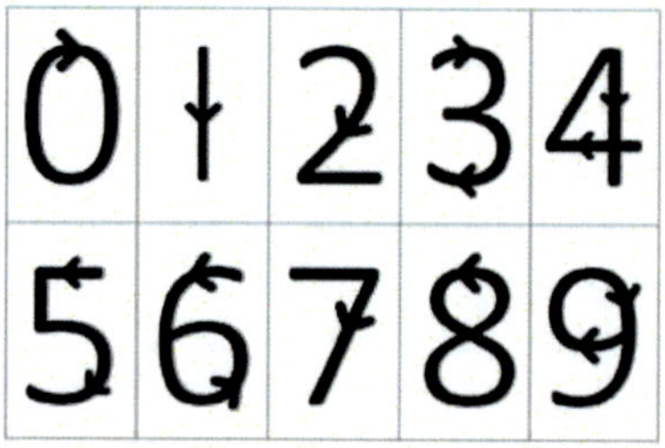

Fig. 1 Training pattern of numerical gestures

run deep learning models on embedded and mobile devices offline. The details of converting a trained model to the TensorFlow Lite model and deploying it in the real-time applications are discussed in this paper.

2 Existing Systems

A lot of research has been published on gesture recognition using data glove. In [9], the experiment used the JY61 6-axis attitude sensor which experimenters wear in their wrists. The JY61 sensor consists of they have collected ten different gesture data from 16 different experimenters. The start and endpoint of the gestures are inferred by the drastic change of accelerometer and gyroscope data received through the serial port. They have normalized the collected data using Z-score normalization and evaluated the performance of recurrent neural network (RNN), LSTM, and gated recurrent units (GRUs). They achieved a recognition rate of 99.75% for LSTM and GRU and 98% for RNN. This work concluded that end-to-end classification of gestures is difficult. In [10], the experiment used an MPU9250 motion sensor. They have attached the sensor to the forearm of the experimenter. The received accelerometer and gyroscope data are reduced to lower dimensions using principal component analysis, and features are extracted using linear discriminant analysis. In their experiment, they got an accuracy of 99.63%.

In [3], they have used the VIVA challenge dataset, where 19 distinct dynamic hand gestures were performed by eight experimenters inside a vehicle. The dataset contains a total of 885 intensity and depth video sequences. Since each dynamic hand gesture sequence had a different duration, using nearest-neighbor interpolation (NNI), they have normalized the sequence lengths of the gestures to be 32 frames by dropping or repeating frames. The proposed 3D convolutional neural network classifier consists of two sub-networks, namely high-resolution network (HRN) and low-resolution network (LRN). The experiment achieved a classification accuracy of 77.5%. In [6], the experiment used a leap motion sensor to detect numerical gestures. This sensor is a USB device consisting of two cameras and three infrared LEDs to detect hand gestures. Several techniques were introduced to detect the start and end of the gesture. They have collected 500 gesture samples where 50 gesture samples for each numeral. They used the geometric template matching method to achieve a classification rate of 70.2%.

In [11], the experiment used the ADXL335 accelerometer sensor. They intended to detect alphabetical gestures excluding ‘J’ and ‘Z’. The accelerometers are attached to the tip of all the five fingers in a hand. They have created a look-up table for all the gestures with the processed data. They have used the Manhattan Distance Algorithm to identify the closeness of a new gesture with gestures in the look-up table and classify the gesture. They have estimated the runtime efficiency of their system. With five accelerometers, the efficiency was estimated to be 95.3%, and the cost of the prototype was estimated at 20 USD. With two accelerometers, the efficiency was estimated to be 87.0%, and the cost of the prototype was estimated at 12.5 USD.

In [12], the experiment uses a pen-type sensing device embedded with triaxial micro-electromechanical systems (MEMS) accelerometer (MMA9551L) and a microcontroller (IAP15W4K58S4). They have considered 24 different gestures which include eight simple and 16 complex gestures. A novel segmentation scheme was introduced to identify the start and endpoints of the gesture. A total of 1600 gestures samples were collected with five experimenters. Features of both the basic and complex gesture samples were extracted and trained with feedforward neural network to classify gestures. For the basic gestures, their experiment achieved an accuracy of 99.88 and 98.88% for user-dependent and user-independent 'basic gestures', respectively. Similarly, for the user-dependent and user-independent 'complex gestures', the experiment achieved an accuracy of 98.88%.

In [13], the experiment used a three-axis accelerometer. With seven experimenters, they have collected 3700 data samples for 18 different gestures. Temporal compression is done on the collected data, and the system uses dynamic time warping and affinity propagation algorithms for training the system. The user-dependent gesture recognition achieved 100% accuracy, and user-independent gesture recognition achieved an accuracy of less than 98% for 18 gestures.

From the previously published research, we infer that the camera-based or depth-sensor-based gesture recognition systems [3, 6] achieve accuracy very less than motion-sensor-based gesture recognition systems. Some proposed systems [9–13] achieved significant recognition rates. But they are not capable of deploying in a real-time application for end-to-end gesture recognition, or they require comparatively costly sensors and devices.

3 Methodology

The proposed system consists of a data glove embedded with ESP8266 NodeMCU microcontroller, MPU9250 motion sensor, and a tactile switch. The architecture of the proposed system is given in Fig. 2. A gesture can be performed by pressing the tactile switch which is attached to the thumb finger. On pressing the button, the NodeMCU microcontroller collects the three-dimensional acceleration and gyroscope data from the MPU9250 sensor. The collected data can be accessed via serial communication with the NodeMCU microcontroller.

The collected data has preprocessed, and features are extracted to train with different machine learning and neural network models. After training, an optimal network model is chosen by evaluating the performance of the trained network models. An optimal trained network model is converted as a TensorFlow Lite model and embedded in a unity windows application for evaluation in real time. With the TensorFlow Lite model, a TensorFlow Lite interpreter instance is created to classify gestures from the data received via serial port.

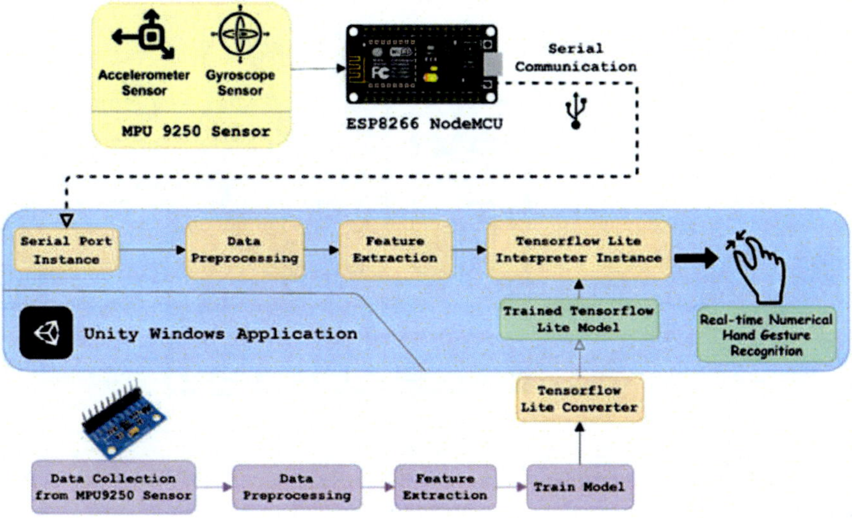

Fig. 2 Architecture diagram of the proposed system

3.1 Hardware in Glove

The data glove consists of ESP8266 NodeMCU microcontroller, MPU9250 sensor, a tactile switch, and a 10 kΩ resistor. Figure 3 shows the designed prototype glove in hand.

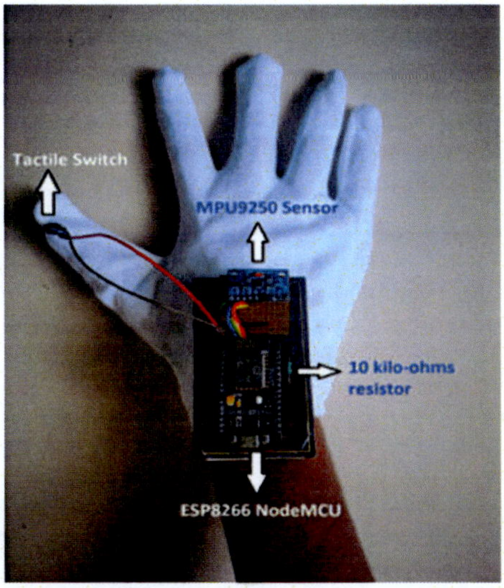

Fig. 3 Designed glove

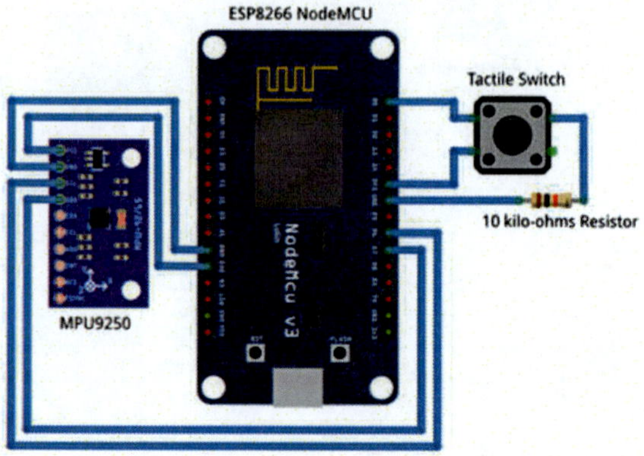

Fig. 4 Connection diagram of the glove hardware

NodeMCU microcontroller consists of CPU—ESP8266 (LX106), memory—128 kilobytes, storage—4 megabytes, and powered by USB [14].

The MPU9250 motion sensor has an inbuilt accelerometer and gyroscope sensor. The accelerometer gives the acceleration of the hand in m/s^2, and gyroscope returns the angular velocity of the hand in rad/s.

When we press the tactile switch, the input pin in the NodeMCU is connected directly to the ground. The input pin reads a low state as the current flows through the resistor to the ground. If the resistor is not there, then the switch would connect the VCC to the ground, which is short. Figure 4 shows the connection between different components in the glove. Once the tactile switch is pressed in the prototype glove, the ESP8266 NodeMCU microcontroller prints the accelerometer and gyroscope data on the serial port.

In our experiment, to avoid possible data loss, we have retrieved the data from the NodeMCU serial port using a micro-USB cable at a speed of 9600 bits per second, and it does not involve any radio waves. It is also possible to retrieve the data via Wi-Fi in ESP8266 NodeMCU.

When mass-producing this data glove, ESP8266 NodeMCU microcontroller can be replaced with surface mounted device (SMD) microcontroller and low energy Wi-Fi or Bluetooth SMD module. The specific absorption rate (SAR) value of the final manufactured data glove will be less than the SAR values of many Smart watches approved by the Federal Communications Commission (FCC).

3.2 *Hardware Components Table*

One of the main focuses of this experiment is to come up with a low-cost hand gesture recognition solution. The list of components used for designing the data glove and its price is given in Fig. 5.

Fig. 5 Cost of components

S.No.	Components	Quantity	Price(INR)
1	ESP8266 NodeMCU	1	400
2	Tactile Switch	1	5
3	10kΩ Resistor	1	2
4	MPU 9250 Sensor	1	400
		Total Cost	807

The total cost of the prototype glove is around 807 Indian rupees which is less than 11 USD. Even a full-fledged final data glove will be cheaper compared to other existing solutions.

3.3 Data Collection

As the tactile switch in the data glove is pressed, it indicates the start of the gesture. Once the user starts performing the gesture, as shown in Fig. 6, the accelerometer and gyroscope data are retrieved via serial communication with the NodeMCU microcontroller. After performing the gesture, the user should release the tactile switch which indicates the end of the gesture.

In order to develop an unbiased gesture recognition system, a total of 20 experimenters (12 males and 8 females) were selected to perform different numerical gestures. Each experimenter performed each gesture ten times at their own pace. All the experimenters have written the numbers in the air as per the trajectory shown in Fig. 1. For each numeral, 200 samples were collected and the final dataset contained 2000 gesture samples. 1400 data samples have been used for training, and 600 data

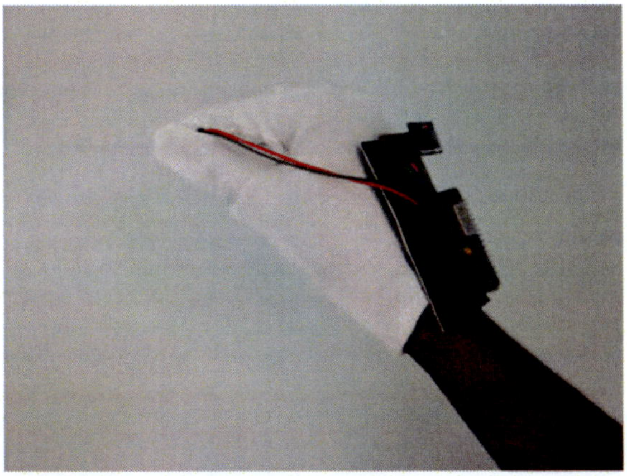

Fig. 6 Pose of hand when performing the gesture

Time Steps	ax	ay	az	gx	gy	gz
1	6.502983	5.257935	-2.576292	-2.813824	0.722741	-0.132109
2	4.223587	5.195683	-1.757434	-2.761359	0.197824	-0.080576
.	.	.	.	.	.	.
.	.	.	.	.	.	.
N	3.969789	4.53006	-0.502808	0.947682	1.028077	-1.019355

Fig. 7 Structure of each temporal gesture data

samples have been used for testing. Figure 7 shows the structure of each temporal gesture data.

3.4 Data Preprocessing and Feature Extraction

Data preprocessing is a data mining technique that is used to transform the raw data into a useful and efficient format. Since we are dealing with temporal motion sensor data, we need to take care of the consistency and format the data.

Usually, any motion can be described using acceleration and rate of change of orientation. Similarly, for hand gesture recognition, acceleration, and rate of change of orientation of the hand are needed. When different gestures are performed by different experimenters, the acceleration and gyroscope values may change greatly. So, we scale the data using standard normalization. Scaling the data increases the performance of the classification algorithms. The formula of standard normalization is represented as:

$$z = \frac{(x - \mu)}{\sigma} \tag{1}$$

Since the time taken for an experimenter to perform a gesture, it changes from gesture to gesture and also experimenter to experimenter. So the window size of data is dynamic. For handling the dynamic window size of data, we transform the data into a constant window size ws_c, which is the average of all window sizes in the dataset. If the window size of a given row in the dataset is greater than ws_c, then we should reduce it to ws_c by downsampling. If the window size of a given row in the dataset is less than ws_c, then we should increase it to ws_c by upsampling using cubic interpolation technique.

3.5 Classification Experiments

This experiment uses two machine learning algorithms, namely support vector machines, decision tree, and two neural network models, namely simple neural network and LSTM.

3.5.1 Support Vector Machine (SVM)

SVM is a supervised machine learning algorithm used for non-linear classifications using kernel trick, implicitly mapping their inputs into high-dimensional feature spaces. The kernels used in our experiment for the classification are linear kernel, radial basis function (RBF) kernel, and polynomial kernel of degree 3,4,5,6, and 7.

The collected dataset can be represented as $(x_1, y_1), (x_2, y_2), \ldots, (x_n, y_n)$, where $x_1, x_2, \ldots, x_n$ are the input vectors containing the interpolated accelerometer and gyroscope data, whereas $y_1, y_2, \ldots, y_n$ are the labels of the corresponding input vectors. We need to find the hyperplane which gives us the boundaries between different classes in our dataset. The equation of the hyperplane is represented as

$$H_0 : w^T x_i + b = 0 \tag{2}$$

where w^T is the weight matrix, x_i is the input vector, and b is the bias.

If the classification cannot be done in the current n-dimensional space, then convert our data into m-dimensional space using a kernel function:

$$K(x, y) = [f(x), f(y)] \tag{3}$$

where K represents kernel function, x and y are n-dimensional inputs, f is a function that maps n-dimension to m-dimensional space, $[x, y]$ denotes the dot product, and

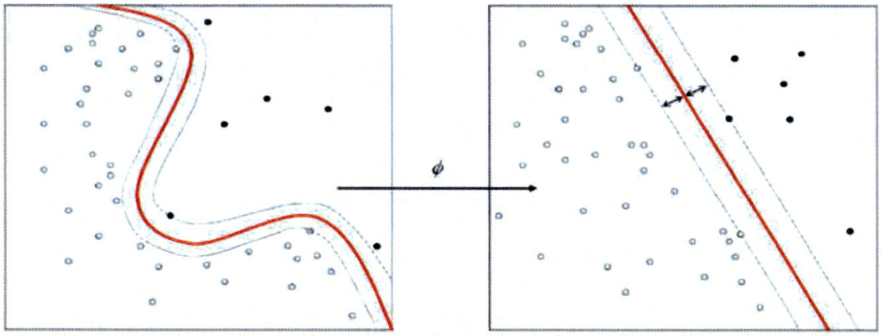

Fig. 8 Result on the application of kernel function

Table 1 Classification accuracies achieved by different SVM kernels with test data

Kernel used	Test accuracy (%)
Linear kernel	85.61
RBF kernel	85.90
Polynomial kernel of degree = 3	89.04
Polynomial kernel of degree = 4	88.35
Polynomial kernel of degree = 5	83.56
Polynomial kernel of degree = 6	80.13
Polynomial kernel of degree = 7	96.40

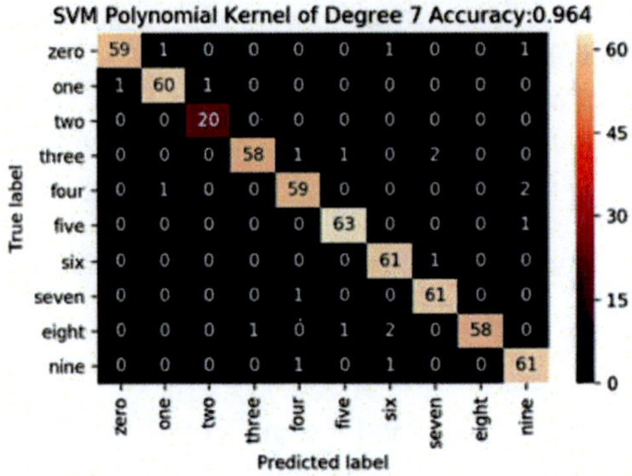

class	precision	recall	f1-score	support
Zero	0.95	0.98	0.97	62
One	0.98	0.95	0.97	62
Two	0.95	0.95	0.95	20
Three	0.94	0.97	0.95	62
Four	0.98	1.00	0.99	62
Five	0.91	0.95	0.93	64
Six	0.95	0.94	0.94	62
Seven	0.98	0.94	0.96	62
Eight	0.98	0.94	0.96	62
Nine	0.95	0.97	0.96	63
avg/total	0.96	0.96	0.96	581

Fig. 9 Confusion matrix and performance metrics of SVM classifier

$m > n$. Different kernels have different kernel functions. Figure 8 depicts an example of what happens to the dataset when kernel transformation is applied.

The classification accuracies achieved by using different kernels are given in Table 1. Among all kernels, polynomial kernel of degree 7 achieved a higher classification accuracy of 96.40%. The confusion matrix and several performance metrics are calculated by classification with test data, and results are shown in Fig. 9.

3.5.2 Decision Trees

Decision trees are a supervised machine learning algorithms. In decision trees, data is split according to a certain parameter at each node, and the outcomes are decided based on which leaf we reach final.

Data is of the form $(X, y) = (X_1, X_2, \ldots, X_n, y)$. Here, X_i is the input data, and y is the target variable to predict. In constructing the decision tree, at each node of the tree, we calculate the information gain, and the variable having the highest information gain is taken for classification at that node. Information gain (IG) measures how the attribute separates the given data according to their target classification. Highest

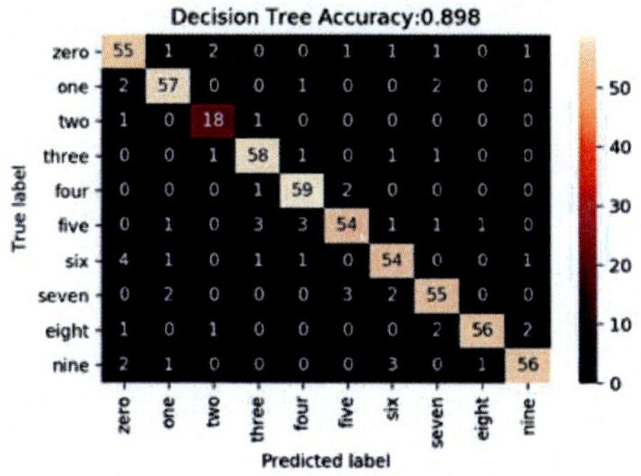

class	precision	recall	f1-score	support
Zero	0.85	0.89	0.87	62
One	0.90	0.92	0.91	62
Two	0.82	0.90	0.86	20
Three	0.91	0.94	0.92	62
Four	0.91	0.95	0.93	62
Five	0.90	0.84	0.87	64
Six	0.87	0.87	0.87	62
Seven	0.89	0.89	0.89	62
Eight	0.97	0.90	0.93	62
Nine	0.93	0.89	0.91	63
avg/total	0.90	0.90	0.90	581

Fig. 10 Confusion matrix and performance metrics of decision trees

information gain has less entropy.

$$IG = \text{Entropy(before)} - \sum_{j=1}^{k} \text{Entropy}(j, \text{after}) \tag{4}$$

Here, entropy gives us how much random our data is. 'before' indicates the dataset before the split, k indicates the number of subsets by splitting, and (j, after) indicates the subset j after the split.

The trained decision tree was able to achieve the highest accuracy of 89.8%. The confusion matrix and performance metrics are given in Fig. 10.

3.5.3 Simple Neural Network

The biological neural network is a network or circuit of biological neurons, whereas an artificial neural network is composed of artificial neurons. The basic unit of an artificial neural network is a neuron. A neuron takes inputs, does some math with them, and produces one output. Figure 11 depicts an n-input neuron. In this experiment, we created a simple neural network with five layers. First layer receives 90 inputs, and the last layer gives ten outputs. The softmax activation function is used in the output layer and ReLu activation function in all other layers. Simple neural network was able to achieve the classification accuracy of 95%. The confusion matrix and performance metrics are given in Fig. 12.

3.5.4 Long Short-Term Memory (LSTM)

The collected sensor data is time series data. For sequence processing and time series problems, LSTM has great success. LSTM is a recurrent neural network architecture.

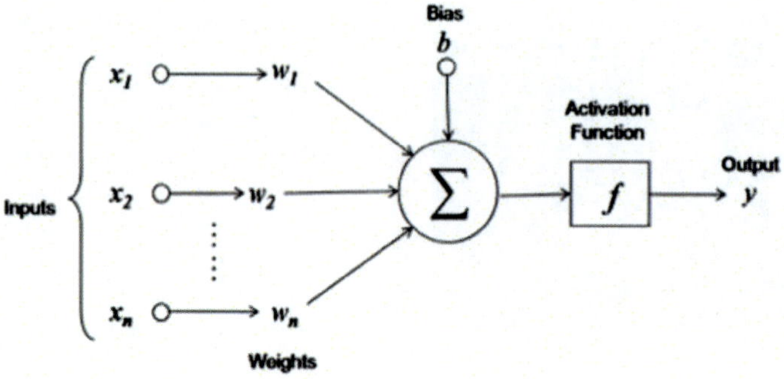

Fig. 11 N-input neuron

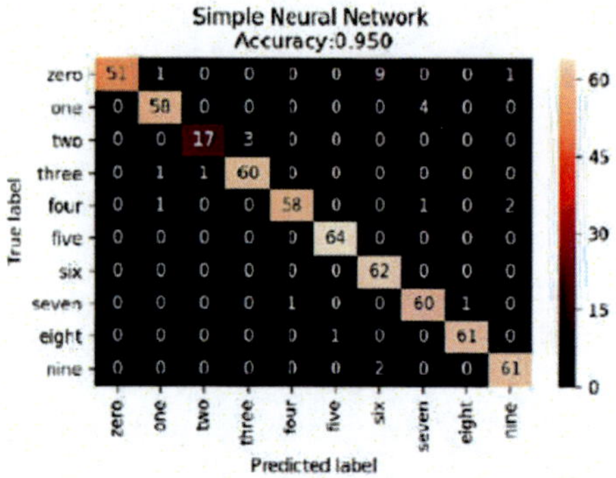

class	precision	recall	f1-score	support
Zero	1.00	0.82	0.90	62
One	0.95	0.94	0.94	62
Two	0.94	0.85	0.89	20
Three	0.95	0.97	0.96	62
Four	0.98	0.94	0.96	62
Five	0.98	1.00	0.99	64
Six	0.85	1.00	0.92	62
Seven	0.92	0.97	0.94	62
Eight	0.98	0.98	0.98	62
Nine	0.95	0.97	0.96	63
avg/total	0.95	0.95	0.95	581

Fig. 12 Confusion matrix and performance metrics of simple neural network

Unlike standard feedforward neural networks, LSTM has feedback connections. The architecture is developed in such a way that they remember data and information for a long period of time.

An LSTM network is composed of LSTM cells. A simple LSTM cell consists of three gates (forget gate, input gate, and output gate), as shown in Fig. 13, to protect and control the cell state. The core concept of LSTMs is the cell state and its various gates. The long-term memory is usually called the cell state, denoted by C_t. The working memory or short-term memory is usually called the hidden state, denoted by h_t. The forget gate f_t discards unnecessary information from the cell state. The input gate i_t decides the values to be updated, and in parallel the tanh layer calculates a vector of new candidate values $\widetilde{C}_t$ that could be added to the state. The values i_t and $\widetilde{C}_t$ are combined to create an update to the state. Then the output layer o_t decides which parts of the cell state we are going to output. The output of the LSTM cell h_t is calculated by multiplying the output of the cell state passed to a tanh layer and the output of the output gate layer o_t.

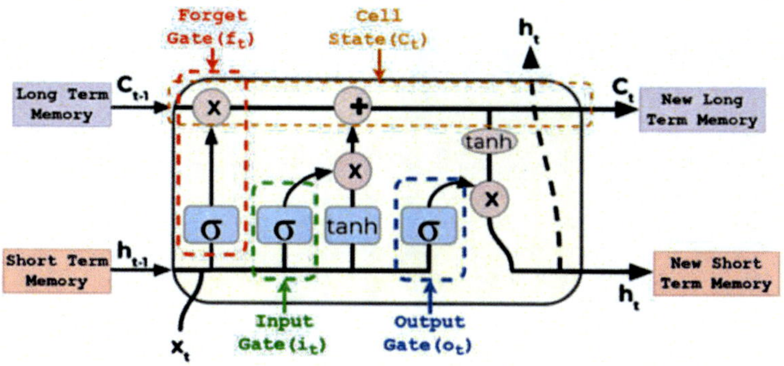

Fig. 13 LSTM cell

In our experiment, after collecting the dataset with 20 different experimenters, it is found that most of the numerical gestures are performed within 20 time steps. Since this work focuses on deploying the data glove in real-time applications, the numerical gestures are not meant to be performed very slowly for more than 20 or 30 time steps. Also, in order to avoid the small computational overhead involved in the preprocessing techniques used for the previous models, the following preprocessing technique is used for the LSTM model. Each gesture data sample is trimmed at the end if it had more than 20 time steps of data. Also, a special mask value of −999 was appended if it had less than 20 time steps of data. The padded mask values are masked by the masking layer as given in Fig. 14, and only the actual time steps of data will be received by the first LSTM layer.

With this preprocessed data, different LSTM architectures were trained and tested by tuning the hyperparameters. As a conclusion, we found that the architecture depicted in Fig. 15 gave the highest accuracy of 98.41%. The confusion matrix and performance metrics of the trained LSTM model are given in Fig. 15. Since the proposed LSTM network model outperformed, all other classification algorithms were tested. Hence, the proposed LSTM network model is chosen for real-time deployment and performance evaluation.

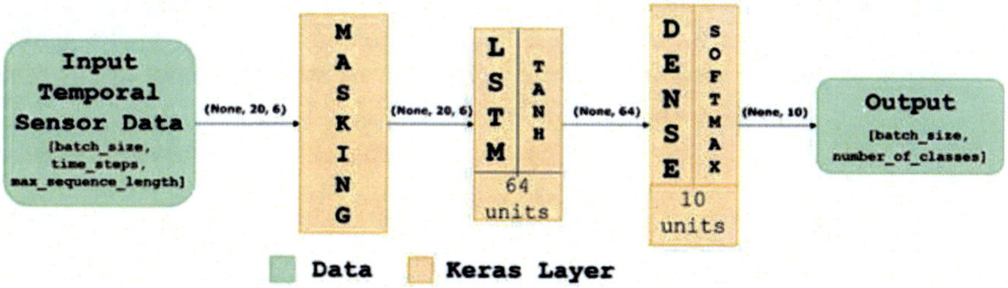

Fig. 14 Optimal architecture of the sequential LSTM model

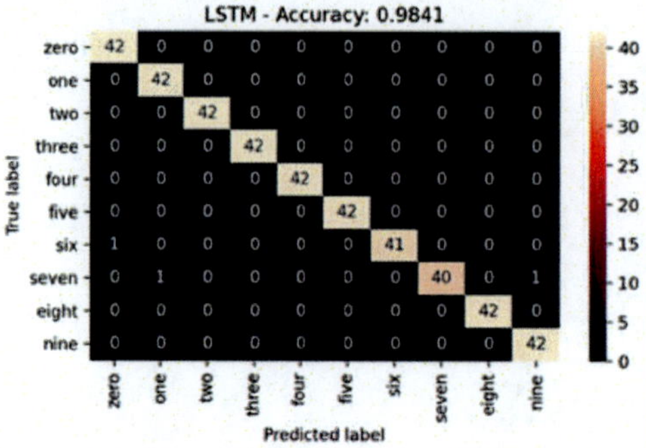

Class	Precision	Recall	F1-Score	Support
zero	0.98	1.0	0.99	42
one	0.98	1.0	0.99	42
two	1.0	1.0	1.0	42
three	1.0	1.0	1.0	42
four	1.0	1.0	1.0	42
five	1.0	1.0	1.0	42
six	1.0	0.98	0.99	42
seven	1.0	0.95	0.98	42
eight	1.0	1.0	1.0	42
nine	0.98	1.0	0.99	42
avg/total	0.99	0.99	0.99	420

Fig. 15 Confusion matrix and performance metrics of LSTM model

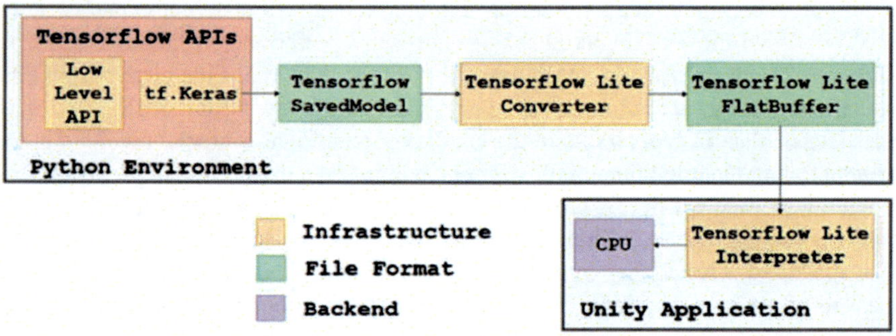

Fig. 16 TensorFlow Lite model creation process

3.6 Conversion to TensorFlow Lite Model

The trained unidirectional sequential LSTM model with an accuracy of 98.41% was converted to TensorFlow Lite model as shown in Fig. 16.

The LSTM model was created and trained using TensorFlow's Keras API. The trained model is saved in the TensorFlow's SavedModel format. Using the TensorFlow Lite converter, the SavedModel is then converted as TensorFlow Lite model, also known as TensorFlow Lite flatbuffer. This TensorFlow Lite model can be used for inference in any end devices offline with any CPU or GPU backend.

3.7 Unity Application

The TensorFlow Lite model is then imported inside the Unity Project. The TensorFlow Lite C library is compiled as a static C library and imported inside the Unity

Fig. 17 Car lane changing game playable with glove

Project. With the TensorFlow Lite C library, a TensorFlow Lite interpreter instance is created with the imported TensorFlow Lite model at runtime. This TensorFlow Lite interpreter instance is used to predict the numerical gestures with the data received from the glove at runtime.

A prototype car lane changer game is designed, as shown in Fig. 17, to test the real-time performance of the trained LSTM model.

The car lane changer game scene consists of a road with ten lanes, signifying ten numbers (0–9). The player car is controlled by the player wearing the data glove. As the game starts, several opponent cars will approach the player car and the player should change lanes by performing appropriate numerical gestures by pressing the button in the data glove to avoid collision with the opponent cars. This is an endless game developed just to evaluate the real-time performance of the glove and the trained LSTM model.

When the user performs a numerical gesture, for real-time gesture classification, the following preprocessing is done on the gesture data received from the glove. If the received gesture data contains more than 20-time-step data, then the last 20-time-step data is taken for inference. If the received gesture data contains less than 20-time-step data, then the remaining time steps are padded with a special mask value of −999.

The preprocessed data is converted to a three-dimensional array of shape [batch_size, time_steps, maximum_sequence_length] and used for real-time inference by the TensorFlow Lite interpreter instance. This interpreter instance would return an output two-dimensional array of shape [batch_size, number_of_classes] which contains the probability values for each class. The class with the highest probability and with probability more than the threshold probability will be inferred as the input for the game. For example, if the inferred class is 5, then the game would receive number 5 as input and car shifts to lane 5 to avoid collision with the opponent's car. If the player collided with the opponent's car, then the game is lost.

On extensive real-time testing of the game, this unity game application is able to predict almost every numerical gesture accurately. When playing the game, the time delay involved in performing real-time inference is tracked. It is found that the average real-time inference delay is not more than 2 ms. This proved that the glove and the trained LSTM network model could be deployed in any real-time application. In our experiment, we used the CPU backend for inference. If a GPU backend is used, then depending on the compute capacity of the GPU, the real-time inference delay can be decreased to a great extent.

4 Conclusion and Future Work

In this paper, an effective real-time numerical gesture recognition pipeline using the MPU9250 motion sensor is proposed. The proposed system uses super optimal features of the hand, namely acceleration and angular velocity of the hand, to predict numerical gestures. The performance of several machine learning and neural network models is analyzed with the features extracted from the data glove. An optimal unidirectional sequence LSTM architecture is proposed for achieving higher classification accuracy. The details of converting a TensorFlow neural network model into TensorFlow Lite flatbuffers for real-time inference on mobile and embedded devices are discussed. Existing gesture recognition systems that use camera or depth sensors need heavy computational resources to detect gestures in real time, which are usually slower than our proposed architecture. A prototype unity game application was developed to test the real-time performance, and our proposed real-time numerical gesture recognition system was able to perform inference within 2 ms in a CPU backend. Through our experiment, we have developed a robust gesture recognition system that is cheaper, real-time, and insensitive to environmental conditions and does not impose any field of view restrictions. Our proposed system with a data glove can be deployed in various real-time games, robot control, augmented reality, and virtual reality applications. This proposed data glove is a more natural input modality and a solid alternative to existing gesture recognition systems. It will easily fit in a pocket, and the user can carry it around without hassle. Still, the form factor of this glove can be reduced to even smaller sizes for mass production at cheaper rates. In the future, the motion sensors in the smartphone can be used for gesture recognition for more convenience.

References

1. Hasan, M., Yu, H.: Innovative developments in HCI and future trends. Int. J. Autom. Comput. **14**. https://doi.org/10.1007/s11633-016-1039-6 (2016)
2. Lin, L., et al.: The effect of hand size and interaction modality on the virtual hand illusion. In: 2019 IEEE Conference on Virtual Reality and 3D User Interfaces (VR), Osaka, Japan, pp. 510–518 (2019)

3. Molchanov, P., Gupta, S., Kim, K., Kautz, J.: Hand gesture recognition with 3D convolutional neural networks. In: 2015 IEEE Conference on Computer Vision and Pattern Recognition Workshops (CVPRW), Boston, MA, pp. 1–7 (2015)
4. Khan, Rafiqul Zaman, Ibraheem, Noor: Hand gesture recognition: a literature review. Int. J. Artif. Intell. Appl. (IJAIA) **3**, 161–174 (2012). https://doi.org/10.5121/ijaia.2012.3412
5. Wang, X., Jiang, J., Wei, Y., Kang, L., Gao, Y.: Research on gesture recognition method based on computer vision. MATEC Web Conf. **232**, 03042 (2018). https://doi.org/10.1051/matecconf/201823203042
6. Sharma, J., Gupta, R., Pathak, V.: Numeral gesture recognition using leap motion sensor, pp. 411–414. https://doi.org/10.1109/cicn.2015.86
7. Sundaram, V., Vasudevan, S., Santhosh, C., Kumar, R.: An augmented reality application with leap and android. Indian J. Sci. Tech. **8**, 678. https://doi.org/10.17485/ijst/2015/v8i7/69907 (2015)
8. https://www.tensorflow.org/lite
9. Du, T., Ren, X., Li, H.: Gesture recognition method based on deep learning. In: 2018 33rd Youth Academic Annual Conference of Chinese Association of Automation (YAC), Nanjing, pp. 782–787 (2018)
10. Liu, F., Wang, Y., Ma, H.: Gesture recognition with wearable 9-axis sensors. In: 2017 IEEE International Conference on Communications (ICC), Paris, pp. 1–6 (2017)
11. Kannan, A., Ramesh, A., Srinivasan, L., Vijayaraghavan, V.: Low-cost static gesture recognition system using MEMS accelerometers. In: 2017 Global Internet of Things Summit (GIoTS), Geneva, pp. 1–6 (2017)
12. Xie, R., Cao, J.: Accelerometer-based hand gesture recognition by neural network and similarity matching. IEEE Sens. J. **16**(11), 4537–4545 (1 June 2016)
13. Ahmad, A., Valaee, S.: Accelerometer-based gesture recognition via dynamic-time warping, affinity propagation, & compressive sensing. In: 2010 IEEE International Conference on Acoustics, Speech and Signal Processing. IEEE 2010, pp. 2270–2273 (2010)
14. https://www.nodemcu.com/

Use of Genetic Algorithm Applied to the Optimization of Investments in Financial Actions

Noel Varela, Omar Bonerge Pineda Lezama, and Jorge Borda

Abstract The optimal form of investment portfolio selection initially proposed by Markowitz (Viloria et al., in Smart innovation, systems and technologies. Springer, Berlin (2020) [1]) in 1952 used a classical statistical analysis approach to find the optimal investment portfolio but did not consider practical limitations related to the cost of resource reallocation and the minimum amount that can be reallocated. If these factors are considered, the complexity of the problem increases (Wang, J Futures Markets: Futures Opt Other Deriv Prod 20(10):911–942, 2000; Li et al., ACM Computing Surveys, vol 46 [2, 3]), and its constraints change considerably, adding to an already complex NP problem, as demonstrated in Oh et al. (Expert Syst Appl 28(2):371–379, 2005 [4]). Multiple heuristic algorithms have demonstrated a good performance when attacking this kind of problems (Kassicieh et al., in Proceedings of the thirtieth hawaii international conference on system sciences, vol 5, pp 484–490, 1997; Fama, J Financ Econ 49:283–306, 1999; Chen and Yang, Neurocomputing 70:4–6, 2007 [5–7]) and do it using genetic algorithms, reinforcement learning and neural networks, respectively, obtaining good results that in turn. In this article, a genetic algorithm is used to extend the Markowitz model and is applied to the stocks that compose the S&P 500, NASDAQ and FTSE index and is compared with the standard model.

Keywords Genetic algorithm applied · Optimization of investments · Financial actions

---

N. Varela (✉) · J. Borda
Universidad de la Costa, St. 58 #66, Barranquilla, Atlántico, Colombia
e-mail: nvarela2@cuc.edu.co

J. Borda
e-mail: jborda1@cuc.edu.co

O. B. P. Lezama
Universidad Tecnológica, San Pedro Sula, Honduras
e-mail: omar@unit.edu

© The Editor(s) (if applicable) and The Author(s), under exclusive license to Springer Nature Singapore Pte Ltd. 2021

A. P. Pandian et al. (eds.), *Proceedings of International Conference on Intelligent Computing, Information and Control Systems*, Advances in Intelligent Systems and Computing 1272, https://doi.org/10.1007/978-981-15-8443-5_5

1 Introduction

A portfolio is a group of financial assets such as stocks, bonds and funds. Portfolios are generally managed by financial professionals, and their purpose is to continually rebuild an investment portfolio according to the associated risk of the assets included and the short- and long-term objectives [8].

In order to determine an adequate resource allocation among the assets included in a portfolio, the individual risk of each asset and the associated equivalent risk of the selected group must be considered. The chosen portfolio must be constructed considering a trade-off between risk and associated return [9]. A portfolio will be better than another if it has a better expectation of return and a lower associated risk [10].

Modern portfolio theory assumes that investors are very risk averse. If two portfolios offer the same expected return on investment, typically the one with the least risk will be selected, however, a reason for selecting a slightly riskier portfolio is that the return expectations are considerably higher [11].

In addition, if the correlation between one asset and another one is high, then the risk of widespread losses is greater, so it is important to select assets with a low level of correlation to maintain the level of controlled volatility [12]. In this paper, a genetic algorithm is used to extend the Markowitz model.

2 Markowitz Model

The Markowitz model can be summarized from the equations:

1. The expected return of each portfolio is defined as [13]:

$$E(R_{\mathrm{p}}) = \sum_i w_i E(R_i) \tag{1}$$

$$\sum_i^n w_i = 1 \tag{2}$$

Where

- R_{p} represents the return of the entire portfolio.
- R_i is the specific return on each asset.
- W_i is the relative importance of each asset relative to the set of resources available.
- n represents the amount of assets available.

2. Interest portfolio volatility is defined as [14]:

$$\sigma_p^2 = \sum_i \sum_j w_i w_j \sigma_{ij} \tag{3}$$

Where

- σij is the covariance of the expected return on assets,
- σ^2 is the volatility and is proportional to the expected risk of the portfolio.

3. Considering that (1) represents the expected return and (3) the associated risk of a specific portfolio, (4) can be considered as the model that should be maximized to obtain the maximum return with the lowest possible risk [15]:

$$(1-\lambda)\sum_i^n w_i R_i - \lambda \sum_i^n \sum_j^n w_i w_j \sigma_{ij} \tag{4}$$

Where: λ represents the desired risk aversion and is limited by $0 < \lambda < 1$.

4. To maximize the return on investment considering transaction costs, it is necessary to take into account the cost of reallocating resources and the amount of resources relocated [16]:

$$\text{maximizar} \quad \left((1-\lambda)\sum_i^n w_i R_i - \lambda \sum_i^n \sum_j^n w_i w_j \sigma_{ij} - \sum_i^n c\Delta w_i \right) \tag{5}$$

Where

- c represents the transaction cost and lambda,
- Δwi represents the reallocated weight of the specific index.

From this modified model, it can be seen that the performance evaluation function of the portfolio should be given by Eq. 5.

3 Genetic Algorithms

Genetic algorithms are inspired by natural selection and are used to optimize and seek solutions to problems through bioinspired operators, mutation, reproduction, selection, etc. [17]. Below are the specific characteristics of the algorithm.

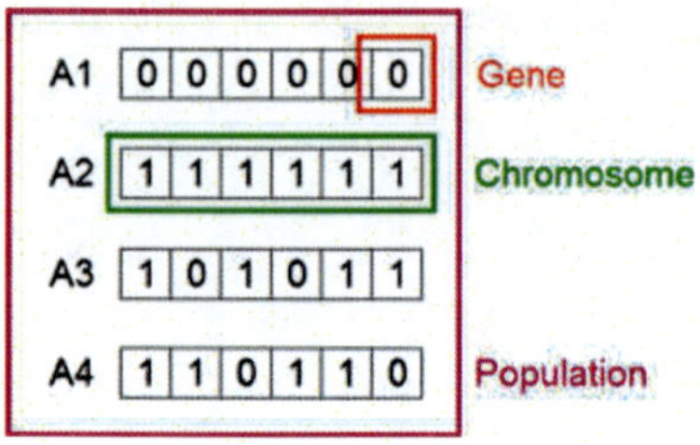

Fig. 1 Representation of the initial population

3.1 *Generation of Initial Population*

By representing the weight of a signal, which is treated as a gene, and by considering all genes as a weight vector, a vector composed of weights constitutes a chromosome that represents a portfolio. As initial population for this algorithm, 100 chromosomes of variable length are presented, depending on the amount of stocks that compose the analyzed index. For S&P 500, 500 genes are used [18], as represented in Fig. 1.

3.2 *Evaluation*

For determining a rating of the performance of each individual in the created population, an appropriate evaluation function proposed by Eq. 5 is used. In addition, as an extra evaluation function, the expected level of return is assigned within a specific moving threshold that ensures the complete generation of the efficiency frontier; it is therefore a multi-objective evaluation function.

3.3 *Selection*

From the entire population, the individuals who have obtained the best score in the evaluation function are selected, and some individuals are randomly assigned to a historical record called the hall of fame. In this case, a selection system based on a "wheel of fortune" is used.

3.4 *Recombination*

Two crossing points are randomly selected (Fig. 2) from each pair of parents, which exchange gene chains with each other and produce the next generation [19]. The probability of each pair of parents having children is commonly proportional to the

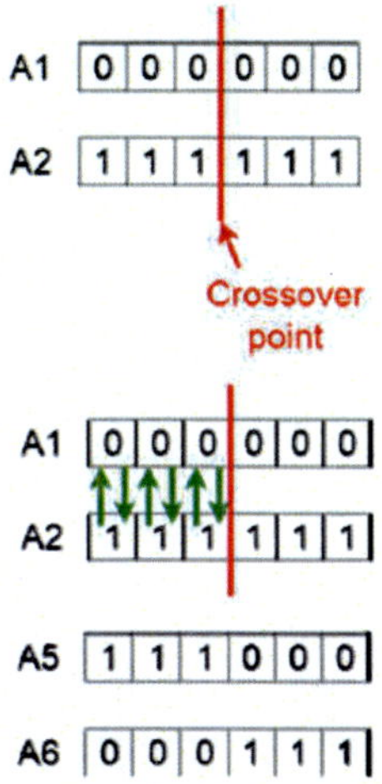

Fig. 2 Simple gene recombination

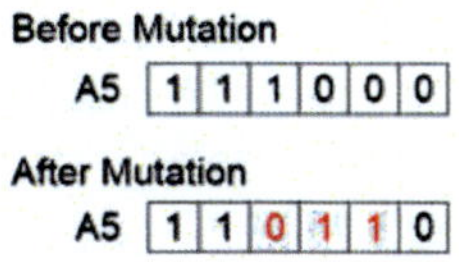

Fig. 3 Simple mutation for binary genes

performance of each parent and replaces the worst evaluated individuals of the next generation.

3.5 *Mutation*

Each member of the new generation has a probability that their genes will be altered with a very low probability, in this case of .3 with a Gaussian function averaged at the same level as the value of the specific gene and $\sigma = 1$ (Fig. 3).

3.6 *Termination*

It is necessary to create a completion criterion, which until it is fulfilled, will imply the repetition of all the previous steps, which will create a generation for each iteration until satisfactory results are obtained. The criterion of the present algorithm is the repetition until the algorithm is not improved in ten consecutive iterations. The general scheme of the genetic algorithm is shown in Fig. 4.

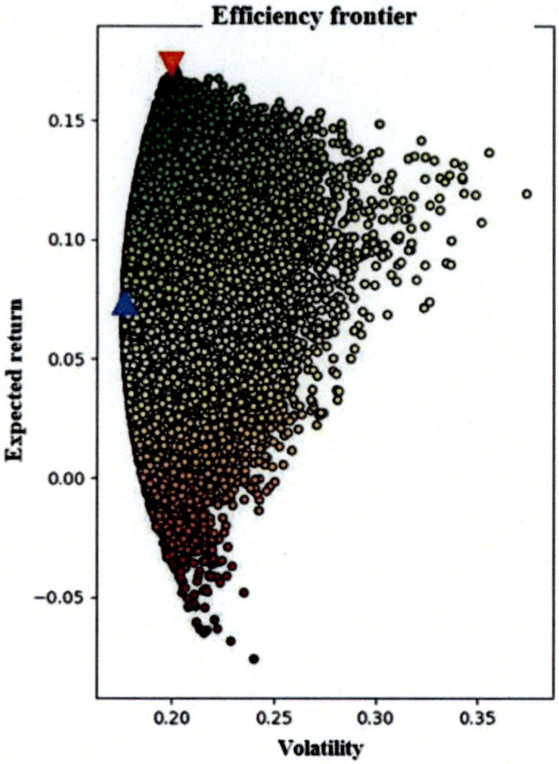

Fig. 4 Efficiency frontier obtained for S&P 500

4 Results

The efficiency frontier draws itself when trying to obtain the combinations of signals where the volatility is minimal at different levels of expected return; this boundary contains the best sets of signals and disables everything that moves away from it. It would not be appropriate to choose combinations with the same level of expected return but with higher risk or the same level of risk but with lower expected return.

Depending on the level of risk aversion, any point (combination obtained) located at the efficiency border between the blue (minimum risk) and red (maximum return) triangles may represent a valid signal selection option within the universe of possible configurations.

Figures 5 and 6 show the results of running the algorithm for ten stock signals included in the S&P 500.

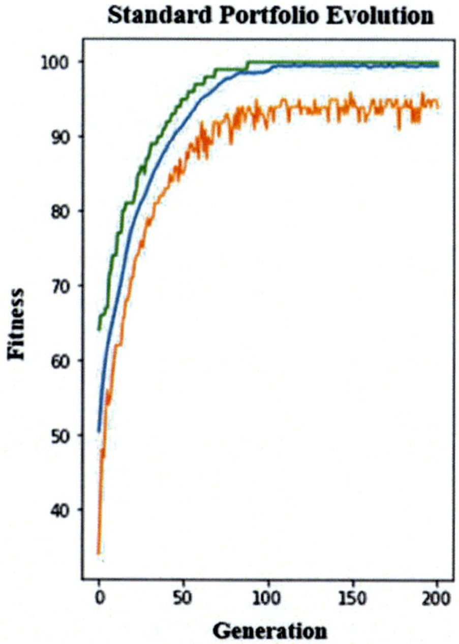

Fig. 5 Standardized portfolio evolution for determined risk

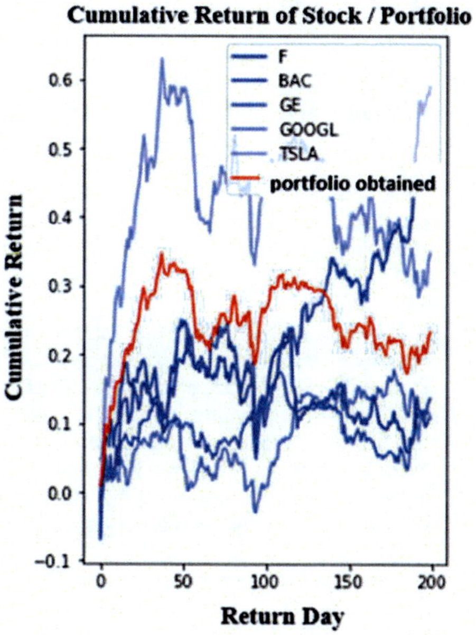

Fig. 6 Return on investment of different stocks and proposed portfolio

5 Conclusions

A method was proposed to explore the possible combinations of a set of financial signals with satisfactory results, and a correct approximation to the efficiency frontier is observed so that the objective is met.

References

1. Viloria, A., Li, J., Sandoval, J. M., Villa, J.V.: Database knowledge discovery in marketing companies. In: Smart Innovation, Systems and Technologies, vol. 164, pp. 65–75. Springer, Berlin (2020). https://doi.org/10.1007/978-981-32-9889-7_6
2. Wang, J.: Trading and hedging in S&P 500 spot and futures markets using genetic programming. J. Futures Markets: Futures Opt. Other Deriv. Prod. **20**(10), 911–942 (2000)
3. Li, B., Hoi, C.H.: Online Portfolio Selection: A survey. ACM Computing Surveys, vol. 46 (2014)
4. Oh, K.J., Kim, T.Y., Min, S.: Using genetic algorithm to support portfolio optimization for index fund management. Expert Syst. Appl. **28**(2), 371–379 (2005)
5. Kassicieh, S.K., Paez, T.L., Vora, G.: Investment decisions using genetic algorithms. In: Proceedings of the Thirtieth Hawaii International Conference on System Sciences, vol. 5, pp. 484–490. IEEE (1997, January)
6. Fama, E.F.: Market efficiency, long-term returns, and behavioral finance. J. Financ. Econ. **49**, 283–306 (1999)
7. Chen, Y., Yang, B.: Abraham, A: Flexible neural trees ensemble for stock index modeling. Neurocomputing **70**, 4–6 (2007)
8. Tsai, T.J., Yang, C.B., Peng, Y.H.: Genetic algorithms for the investment of the mutual fund with global trend indicator. Expert Syst. Appl. **38**(3), 1697–1701 (2011)
9. Deb, K., Agrawal, S., Pratap, A., Meyarivan, T.: A fast and elitist multi-objective genetic algorithm: NSGA-II. IEEE Trans. Evolut. Comput. **6**(2), 182–197 (2002)
10. Shin, K.S., Kim, K.J., Han, I.: Financial data mining using genetic algorithms technique: Application to KOSPI 200. In 한국전문가시스템학회'98 추계학술대회, No. 2, pp. 113–122. Korea Intelligent Information Systems Society (1998)
11. Mahfoud, S., Mani, G.: Financial forecasting using genetic algorithms. Appl. Artif. Intel. **10**(6), 543–566 (1996)
12. Kassicieh, S.K., Paez, T.L., Vora, G.: Data transformation methods for genetic-algorithm-based investment decisions. In: Proceedings of the Thirty-First Hawaii International Conference on System Sciences, vol. 5, pp. 122–127. IEEE (1998, January)
13. Rahimunnisa, K.: Hybrdized genetic-simulated annealing algorithm for performance optimization in wireless adhoc network. J. Soft Comput. Paradigm (JSCP) **1**(01), 1–13 (2019)
14. Parracho, P., Neves, R., Horta, N.: Trading in financial markets using pattern recognition optimized by genetic algorithms. In: Proceedings of the 12th annual conference companion on Genetic and evolutionary computation, pp. 2105–2106 (2010, July)
15. Majhi, R., Panda, G., Sahoo, G., Dash, P.K., & Das, D.P.: Stock market prediction of S&P 500 and DJIA using bacterial foraging optimization technique. In: 2007 IEEE Congress on Evolutionary Computation, pp. 2569–2575. IEEE (2007, September)
16. Lin, X., Yang, Z., Song, Y.: Intelligent stock trading system based on improved technical analysis and Echo State Network. Expert Syst. Appl. **38**(9), 11347–11354 (2011)
17. Simões, C., Neves, R., Horta, N.: Using sentiment from twitter optimized by genetic algorithms to predict the stock market. In 2017 IEEE Congress on Evolutionary Computation (CEC), pp. 1303–1310. IEEE (2017, June)

18. Rout, M., Majhi, B., Mohapatra, U.M., Mahapatra, R.: Stock indices prediction using radial basis function neural network. In: International Conference on Swarm, Evolutionary, and Memetic Computing, pp. 285–293. Springer, Berlin, Heidelberg (2012, December)
19. Rodríguez-Sánchez, J.L., Mercado-Caruso, N., Viloria, A.: Managing human resources resistance to organizational change in the context of innovation. In: Smart Innovation, Systems and Technologies, vol. 167, pp. 330–340. Springer, Berlin (2020) https://doi.org/10.1007/978-981-15-1564-4_31

Development of DWT–SVD based Digital Image Watermarking for Multi-level Decomposition

Ehtesham Sana, Sameena Naaz, and Iffat Rehman Ansari

Abstract In the age of digital data, the safety of digital content has gradually become paramount important. The demand for digital data in various forms such as text, images, video and audio has increased manifold. There are lots of vulnerabilities concerning digital data. It can easily be attacked, forged and manipulated to produce illegal copies. In the watermarking process, information, i.e., watermark image is inserted into a host data and extracted later in case of Copyright violation; owner of the content can demonstrate his ownership by regaining the embedded watermark. A watermark must be capable of being recovered even if a piece of content is modified or changed by several attacks. This paper uses frequency-domain technique, and the multi-level decomposition of the host image is done by using the combination of two-dimensional DWT (2D DWT) and SVD techniques. Here, the host/cover image is decomposed at six different levels separately, and the watermark image is inserted in the region of lowest frequency. After inserting a watermark, the host/cover image is reconstructed using an inverse DWT technique. By exposing the watermarked image to different image processing attacks such as pepper and salt, speckle and rotation, the robustness of various levels has been analyzed. The embedded watermark is extracted afterward and matched with the original one based on PSNR and MSE. The comparative experimental results show that the decomposition of the host image at the third level is found to be more suitable for watermarking among all other levels.

E. Sana · S. Naaz
Department of Computer Science and Engineering, School of Engineering Sciences and Technology, Jamia Hamdard, New Delhi 110062, India
e-mail: 29es17@gmail.com

S. Naaz
e-mail: snaaz@jamiahamdard.ac.in

I. R. Ansari (✉)
Electronics Engineering Section, Faculty of Engineering and Technology, University Women's Polytechnic, A.M.U., 202002 Aligarh, UP, India
e-mail: iffat_rehman2002@yahoo.co.in

© The Editor(s) (if applicable) and The Author(s), under exclusive license to Springer Nature Singapore Pte Ltd. 2021

A. P. Pandian et al. (eds.), *Proceedings of International Conference on Intelligent Computing, Information and Control Systems*, Advances in Intelligent Systems and Computing 1272, https://doi.org/10.1007/978-981-15-8443-5_6

Keywords Singular value decomposition · Digital image watermarking · Discrete wavelet transform · Multi-level decomposition · Peak signal-to-noise ratio · Mean square error

1 Introduction

Security of online contents like audio and images, etc., has become utmost important as these contents transmitted from one source to another in the digital form can be easily replicated without much loss of quality and hence can be efficiently circulated over the Internet and even offline. To counter this problem, digital watermarking is being introduced, and it has significantly overcome the issue.

For the protection of the copyright or integrity of digital products, a digital watermarking method is applied which uses the spare information and unordered characteristics generally found in digital data to add owner's symbol in the digital content.

Hiding proprietary information in digital data like music, video, images or text is very common, and digital watermarking technology plays an important role in information hiding. Copyright infringement issues have been increased manifold due to the ease with which digital data can be sent or received over the Internet.

Over peer-to-peer networks, proprietary information like music, video, images or text can be easily exchanged. This has led to a major headache for the content creators who make these digital contents.

Imperceptibility and robustness are the two fundamental parameters for a digital watermarking technique to be effective to general image manipulations such as scaling, cropping, rotation, filtering, noise, collusion and compression attacks for a watermarked cover image.

Several fields such as cryptography and digital communications are now studying and analyzing the digital watermarking technology and applying in the projects. Different algorithms have been put forward by the analysts from different fields that are different in their research precedence. Generally, these techniques can be categorized as follow:

- Transform domain technique,
- Spatial domain technique.

Frequency-domain technique for watermarking is proved itself to be extremely efficient and effective in attaining the robustness and imperceptibility fundamentals of watermarking (digital) algorithms, whereas transform domain technique for watermarking is proved itself to be not sufficiently efficient as well as ineffective in attaining the robustness and imperceptibility. DCT, DFT and DWT are three of the most broadly used techniques for watermarking of digital content in the frequency domain.

However, due to exceptional multi-resolution and spatial localization characteristics, DWT has been most opted in watermarking (digital image) which are also similar to the human visual system (theoretical model). Further improvements in DWT in terms of performance based on digital data watermarking technique can be achieved by decomposing the image further by using DWT.

Watermarking generally can be understood as hiding secret data such as logo into the host signal itself, thereby protecting the host signal. The watermark is inserted into the interactive media objects, which are commonly referred to as the original image or host image or cover image. Host image should be unchanged after digital watermarking, and afterward, watermark image is extracted to prove the authenticity of the actual person to whom the image belongs.

The digital watermark stays within the image/content in its genuine/original form, and user is not blocked from controlling, analyzing or viewing the object. If compare it to encryption, that is a lot better for transmission. But it does not allow us to analyze the host content in its secured form.

Again, if it is compared to the steganography, in which methods of concealing the message and the message both are kept secret while in digital watermarking, usually the insertion of watermark method is already known, and therefore, there is no compulsion to keep the message secret. The most obvious parameters associated with digital watermarking are

- ***Imperceptibility***: One of the basic conditions for the watermarking of the digital image is imperceptibility. Fidelity or perceptual transparency is the other terms which are frequently used for imperceptibility. Imperceptibility means that there should be a low deformation of the image. Thus, the user cannot distinguish or differentiate between the cover and the watermarked images. The similarities between the host and the watermarked data are characterized by fidelity and transparency. After the watermark insertion into the host data, there should be no distortion in the watermarked content gives the idea that the transparency is higher.
- ***Robustness***: A watermark should be resilient to different attacks (like mean, geometric, Gaussian, rotation, crop, etc.). The inserted watermark content must be extracted back even though the host watermarked image has been attacked. Geometric attacks and removal attacks such as rotation, cropping and translation, are a very good example of image attacks. If a watermark is least affected by various attacks, then it is said to be robust.
- ***Security***: In watermarking methodology, security is one of the most important as well as the most significant prospect. It usually comes handy to explain the ability of a scheme to resist different critical attacks. A technique is only secure when the embedded watermark cannot be extracted or removed by someone who has not authorized person and without having proper knowledge of extraction, embedding procedure and correct proportion of the watermark. An authorized person must only be able to extract the digital image watermark.
- ***Data Payload***: Without a significant change in the quality of an image, the highest quantity of data which can be included in the host image is termed as data payload. It usually refers to the number of bits which can be added to the host image or

the quantity/amount of data which can be hidden within the host image, and this data is the watermark data hidden in the cover/host image.
- ***Computational Complexity***: The time taken to watermark an image and then extract back that watermark; the whole process of extraction process and embedding is termed as computational complexity. Higher computational is demanded more robust security.
- ***Inevitability***: The chance of similarity between the watermarked content and the extracted one is referred to as inevitability. It can be enhancements with the help of certain factors/parameters are reciprocally competitive and obviously, both cannot be performed simultaneously.

2 Literature Review

A literature review was performed on DWT and DWT in combination with SVD techniques to insert watermark in digital images.

Furqan et al. [1] presented a blind digital image watermarking scheme based on SVD–DWT to protect the data against copyright infringement. Here, the host image is divided into four sub-bands, then SVD is practically used to obtain their singular values, and the watermarked image is exposed to various attacks like pixelation, cropping, blurring, rotation, etc. After that, the original watermark is extracted, and its quality is determined based on the values of PSNR and MSE. The results show that if watermarks are inserted in all the frequency bands namely LL, LH, HL, HH, then it will make the watermarked image robust and secure.

Thirumoorthi et al. [2] presented a comparative study of DWT, DCT, FFT, sparse FFT, WHT and KLT. The quality of the watermarked image can be estimated based on MSE, PSNR, compression ratio (CR) and structural content (SC). The result shows that out of all these, DWT fetched the better results in all performance metrics.

Asmara et al. [3] proposed a comparison between 2D DWT, 2D DFT and 2D DCT techniques. Here, analysis using PSNR and MSE shows that DCT inserts on high frequency is better for image watermarking, but on the compressed image (using RIOT application), DWT displays the best results.

Pathak et al. [4] presented a very safe and secure method for medical image watermarking. They have decrypted the watermark before inserting it into the host image. Furthermore, the identical watermark has been inserted in all the four sub-bands. The result shows that the robustness of the image against different attacks is increased manifold, as removing watermark from all the frequency bands is very tough.

Kaur et al. [5] proposed an improved watermarking method based on the concept of SVD and second level DWT. It is found that the method is reliable and robust for the protection of multimedia data. They have used correlation coefficient (CC), MSE and PSNR for qualitative assessment.

Srilakshmi et al. [6] presented a unique scheme for watermarking based on DWT and SVD. The performance of this particular scheme has been evaluated in terms of

PSNR and MSE. This particular scheme is proved to be robust and more secure than the conventional ways of digital image watermarking.

Radouane et al. [7] proposed and analyzed improved method of watermarking for copyright protection. The proposed method is mainly based on DCT, SVD and DWT using optimal block. This scheme also maintains several properties of the watermarked image, such as imperceptibility and capacity.

Choudhary et al. [8] presented a comparative study between the first level and second level DWT. They have used variable visibility factor for embedding watermark in the host/cover image. The qualitative study of the proposed scheme is conducted based on PSNR and NCC, and the second level DWT method is found to be more superior than the first level DWT.

Rajawat et al. [9] worked on RGB components of the image and used second level DWT. The result shows that the security of the image is improved as the value of the PSNR is good, i.e., it is reached up to 55%.

Dubolia et al. [10] presented a comparative study between DWT and DCT. The performance measure for this study is based on PSNR at different threshold values. The result shows that DWT produces a much better quality of the image than DCT.

3 Preliminaries

The watermarking technique aims to hide the private and secret information so that it cannot be easily extracted. The main terminologies used in this work are the host image, DWT and SVD. These are described as follow:

3.1 Host Image

The host image is the one into which the watermark is embedded, and it is the main carrier signal. Thus, various methods/techniques are evolved to guarantee the protection of the signal which will be transmitted over the networks.

3.2 Discrete Wavelet Transform (DWT)

Generally, wavelet transform splits the signal into different components based on different frequency ranges [11].

In this research work, the wavelet transform has been applied by using Haar wavelet function. Wavelets are generally formed or created or generated by dilations of the function which is already fixed, and the translations of that particular function are called mother wavelet. When DWT is applied to an image, it is divided into four

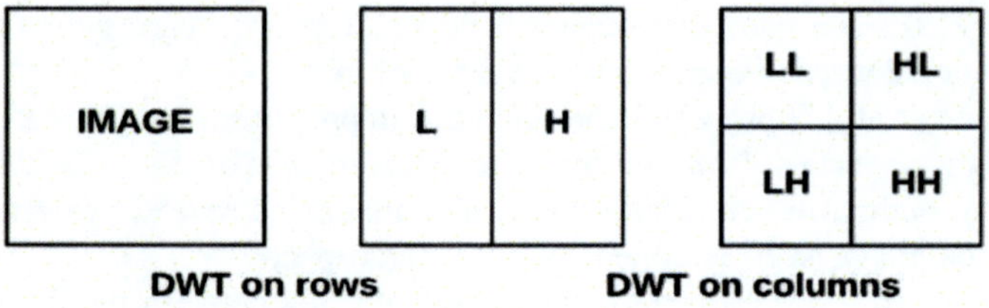

Fig. 1 DWT decomposition of an image

frequency components as shown in Fig. (1), and these are represented as LL, LH, HL and HH [12].

The approximation of the original image is provided by the LL frequency component which is referred to as the lowest frequency component and also represents the lowest resolution level, while the details are provided by the other three frequency components or resolution levels namely LH, HL and HH [13].

3.3 Singular Value Decomposition

Recently, SVD strategy is used in different applications like compression, watermarking, etc. [14], and SVD-based image watermarking scheme discussed in [15] is very robust. On applying SVD on a 2D image (I) of M * N dimensions, the image (I) is fragmented as defined by Eq. (1):

$$I = U * S * V^T \tag{1}$$

where U and V indicate singular orthogonal matrices and S is the diagonal matrix. The singular values (S) provide the algebraic properties of an image which represent the luminance part, while the singular vectors (U and V) provide the geometrical properties or detailed geometry of an image.

4 Proposed System

Here, DWT–SVD-based watermarking scheme is proposed, and the main aim is to analyze the robustness of different levels of decomposition.

To decompose the 2D image, DWT is applied both in the vertical as well as in the horizontal direction. Here, multi-level decomposition is performed that is discussed below:

1. Firstly, the host image is decomposed at a level-1/first level which results in the generation of four frequency components namely LL1, LH1, HL1 and HH1,

respectively, out of which the first one provides the approximated image, whereas the remaining three provide the horizontal edges, vertical edges and diagonal edges of the host/cover image.
2. To perform second level/level-2 decomposition, the lowest frequency component of the first level (LL1) is used as the input which generates four frequency components again namely LL2, LH2, HL2 and HH2, respectively.
3. Similarly, for decomposition at level-3, LL2 is used as the input. After decomposition, four frequency components are again generated, and these are LL3, LH3, HL3 and HH3, respectively.
4. After that, the decomposition is performed at the fourth level, and the resulting frequency components are LH4, LL4, HH4 and HL4.
5. Now, the decomposition is done at a fifth level which fragments LL4 into LL5, HL5, LH5 and HH5.
6. Finally, the decomposition at the sixth level splits the LL5 into HL6, LL6, LH6 and HH6 frequency components.

In this proposed scheme, the sub-band selected for further operations such as SVD and watermark insertion is LL6 frequency sub-band.

Digital image watermarking technique is divided into two major parts namely:

- ***Embedding of the watermark***: method of inserting the watermark in the host/cover image.
- ***Extraction of the watermark***: method of detecting the watermark from the watermarked image.

Figures 2a, b show the flowcharts for both the algorithms.

4.1 Algorithm for Embedding of Watermark

The steps for the embedding of watermark are as follow:

1. Load the watermark image and the host image and read them.
2. Split the host image at the sixth level by using 2D DWT and Haar wavelet to obtain HL6, LL6, LH6 and HH6 frequency components, respectively.
3. Now, apply SVD to the LL6 frequency sub-band of the decomposed cover image as well as on the watermark image that is

$$LL6 = U_{\mathrm{L}} * S_{\mathrm{L}} * V_{\mathrm{L}}^{T}$$
$$W = U_{\mathrm{W}} * S_{\mathrm{W}} * V_{\mathrm{W}}^{T}$$

4. Embed the watermark by adding the matrix S_{L} with the matrix S_{W}.

$$S_{\mathrm{MARK}} = S_{\mathrm{L}} + (Alpha * S_{\mathrm{W}})$$

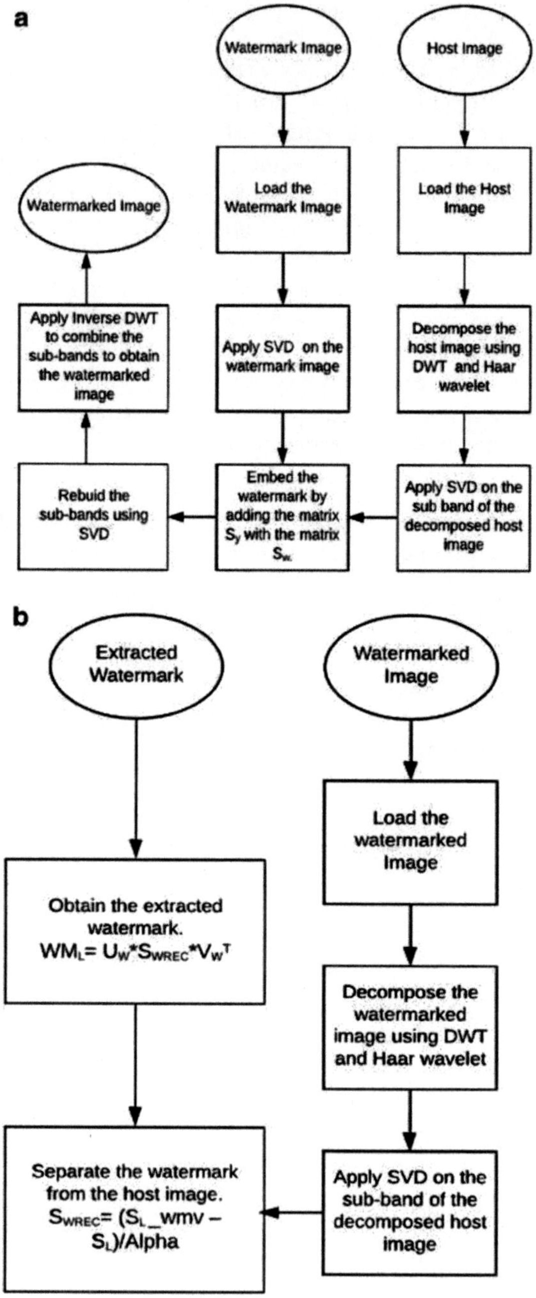

Fig. 2 **a** Process for embedding a watermark. **b** Process for extracting a watermark

5. Rebuild the sub-bands using SVD that is

$$\mathrm{LL6\_1} = U_{\mathrm{L}} * S_{\mathrm{MARK}} * V_{\mathrm{L}}^{T}$$

6. Finally, the IDWT is applied to combine the frequency components so that the watermarked image can be obtained.

4.2 Algorithm for Extraction of Watermark

The steps of the watermarking extraction process are as follow:

1. Load the watermarked image and read it.
2. Decompose the watermarked image at the sixth level by using 2D DWT and Haar wavelet to obtain four frequency components.
3. Then, the SVD is applied to the lowest frequency component which is given by

$$\mathrm{LL6\_wmv} = U_{\mathrm{L}}\mathrm{\_wmv} * S_{\mathrm{L}}\mathrm{\_wmv} * \mathrm{V}_{\mathrm{L}}^{T}\mathrm{\_wmv}$$

4. The watermark is separated from the host/cover image that is

$$S_{\mathrm{WREC}} = (S_{\mathrm{L}}\mathrm{\_wmv} - S_{\mathrm{L}})/\mathrm{Alpha}$$

5. Hence, the watermark is finally extracted as given below:

$$\mathrm{WM_L} = U_{\mathrm{W}} * S_{\mathrm{WREC}} * V_{\mathrm{W}}^{T}$$

5 Performance Evaluation Metrics

Following performance metrics have been used here to find out the strength and vulnerabilities of the watermarked content.

5.1 Peak Signal-to-Noise Ratio

One of the quality measures is PSNR which measures the visual fidelity exists between the host and the watermarked images [16], and it can be expressed in decibels by Eq. (2):

$$\mathrm{PSNR} = 10\log_{10}\left[\frac{255^2}{\mathrm{MSE}}\right] \tag{2}$$

5.2 *Mean Square Error*

The other quality measure MSE measures the mean square error exists between the watermarked and the host images, and it can be defined by Eq. (3) [16]:

$$\mathrm{MSE} = \frac{1}{M \times N} \sum_{i} \sum_{j} [I(i, j) - I_{\mathrm{w}}(i, j)]^2 \tag{3}$$

where, $I\ (i, j)$ is the original host image having $M \times N$ pixels and $I_{\mathrm{W}}\ (i, j)$ is the watermarked image.

6 Results and Discussions

In the proposed work, DWT–SVD method has been developed, and this particular scheme is simulated in MATLAB. The experiment is carried out for the host image "pepper" having dimension 512 × 384, and image "pic" having dimension 181 × 93 is used as the watermark image. The images are shown in Figs. 3a, and b, respectively.

Here, the host/cover image is decomposed by using 2D DWT and Haar wavelet at various levels. The results obtained after decomposition at the first level to the sixth level are shown in Fig. 4a, f.

After decomposition, SVD is applied to the LL6 and also on the watermark image. Then, the watermark is embedded, and finally, the inverse DWT is performed to create the watermarked image.

To extract the watermark image, decomposition of the watermarked image is done at a sixth level, and SVD is applied to the lowest frequency component. Finally, the watermark is separated from the host/cover image, and hence, the watermark images are extracted at first level to the sixth level as shown in Fig. 5a, f.

The comparison is made between host and watermarked images based on PSNR and MSE values before and after the attack at various levels of decomposition as depicted in Table 1.

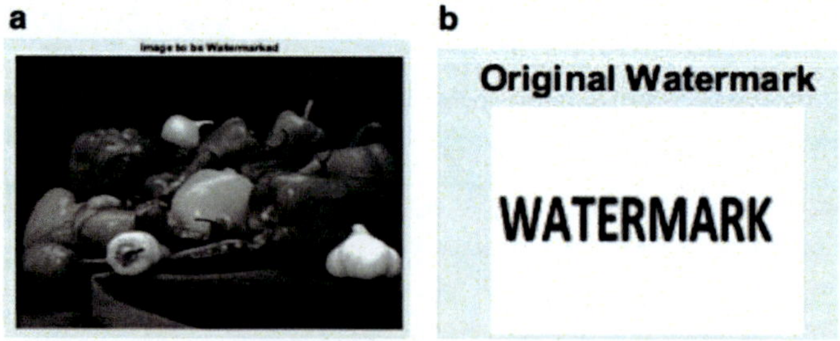

Fig. 3 **a** Host/cover image, **b** watermark image

Fig. 4 DWT decomposition at various levels

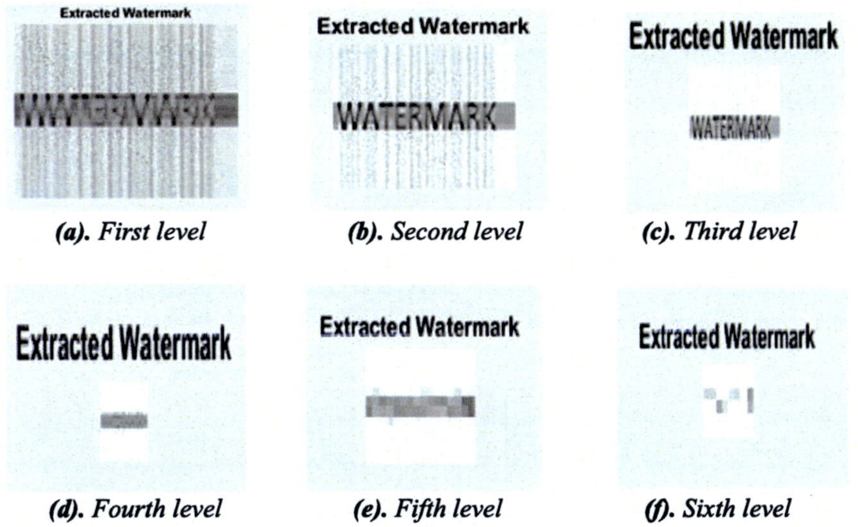

Fig. 5 Watermark images extracted at various levels

The values of PSNR and MSE for the host and watermarked images at various levels of decomposition are shown in Fig. 6a, b.

Table 1 Comparison between host image and watermarked image based on PSNR and MSE

Attacks	First level		Second level		Third level		Fourth level		Fifth level		Sixth level	
	PSNR	MSE	PSNR	MSE	PSNR	MSE	PSNR	MSE	PSNR	MSE	PSNR	MSE
No Attack	8.9246	0.1281	14.9540	0.0320	21.0239	0.0079	27.1322	0.0019	33.1918	0.0005	39.2636	0.0001
Pepper and salt[a]	9.9728	0.1006	14.7202	0.0337	18.6250	0.0137	20.8812	0.0082	21.8005	0.0066	21.9631	0.0064
Rotate[b]	8.5821	0.1386	10.7099	0.0849	11.4538	0.0716	11.6572	0.0683	11.6824	0.0679	11.6740	0.0680
Gaussian[a]	9.5441	0.1111	13.6968	0.0427	16.8874	0.0205	18.6754	0.0136	19.4489	0.0114	19.7201	0.0107
Speckle	10.0508	0.0988	14.4281	0.0361	18.0922	0.0155	20.3358	0.0093	21.3287	0.0074	21.6934	0.0068
Poisson	10.1649	0.0963	15.5227	0.0280	21.2776	0.0075	27.2502	0.0019	33.2732	0.0005	39.2949	0.0001

[a]Density = 0.02, [b]Angle = 22.5

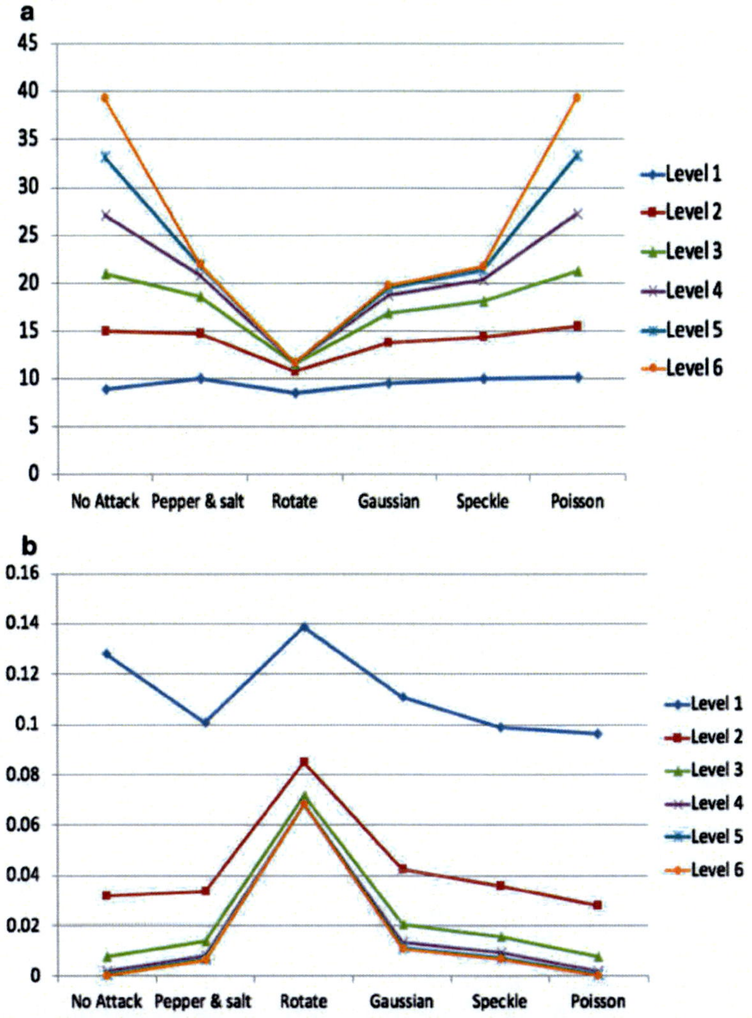

Fig. 6 **a** Graph showing PSNR values for host and watermarked images, **b** graph showing MSE values for host and watermarked images

7 Conclusion

This research paper presents a comparative study of the host image at different levels of decomposition namely first, second, third, fourth, fifth and sixth levels, respectively. Here, DWT–SVD scheme for the multi-level decomposition of the host image has been implemented. The results based on PSNR and MSE show that the MSE is lowest for the sixth level decomposition, while the PSNR is highest, and the sixth level decomposition is also more robust against the majority of attacks, but the extracted watermark is not fairly visible and cannot be recognized. The same visibility issue is with the extracted watermark for the fourth level and fifth level. Whereas third level decomposition is not as robust as the sixth level decomposition, but the extracted watermark can be recognized and can be matched to the original watermark. So, among the various levels of decomposition, third level decomposition is more suitable for watermarking considering all the major factors.

As future work, the watermark can also be inserted in the HH, LH and HL band that would make the watermarked image impervious to attacks.

References

1. Furqan, A., Kumar, M.: Study and analysis of robust DWT-SVD domain based digital image watermarking technique using MATLAB. In: IEEE International Conference on Computational Intelligence and Communication Technology (2015)
2. Thirumoorthi, C., Karthikeyan, T.: A study on discrete wavelet transform compression algorithm for medical images. Biomed. Res. **28**(4), 1574–1580 (2017)
3. Asmara, R.A., Agustina, R.: Hidayatulloh: 'Comparison of Discrete Cosine Transforms (DCT), Discrete Fourier Transforms (DFT), and Discrete Wavelet Transforms (DWT) in Digital Image Watermarking. Int. J. Adv. Comput. Sci. Appl. (IJACSA) 8(2) (2017)
4. Pathak, Y., Dehariya, S.: A more secure transmission of medical images by Two Label DWT and SVD based watermarking technique. In: IEEE International Conference on Advances in Engineering and Technology Research (ICAETR) (2014)
5. Kaur, J., Singh, N., Jain, C.: An improved image watermarking technique implementing 2-DWT and SVD. In: IEEE International Conference on Recent Trends In Electronics Information Communication Technology, India May 20–21 (2016)
6. Srilakshmi, P., Himabindu, Ch.: Image watermarking with path based selection using DWT and SVD. In: IEEE International Conference on Computational Intelligence and Computing Research (2016)
7. Radouane, M., Messoussi, R., Touahni, R., Boujiha, T.: Robust method of digital image watermarking using SVD transform on DWT coefficients with optimal block. In: International Conference on Multimedia Computing and Systems (ICMCS) (2014)
8. Choudhary, R., Parmar, G.: A Robust image watermarking technique using 2-level Discrete Wavelet Transform (DWT). In: IEEE 2nd International Conference on Communication, Control and Intelligent Systems (CCIS) (2016)
9. Rajawat, M., Tomar, D.S.: A secure watermarking and tampering detection technique on RGB Image using 2-Level DWT. In: Fifth International Conference on Communication Systems and Network Technologies (2015)
10. Dubolia, R., Singh, R., Bhadoria, S.S., Gupta, R.: Digital image watermarking by using discrete wavelet transform and discrete cosine transform and comparison based on PSNR. In: International Conference on Communication Systems and Network Technologies (2011)
11. Ansari, I.R., Uddin, S.: Signal denoising using discrete wavelet transform. Int. J. Eng. Technol. Sci. Res. (IJETSR) **03**(11), 23–31 (2016)
12. Naik, N.S., Naveena, N., Manikantan, K.: Robust digital image watermarking using DWT + SVD approach. In: IEEE International Conference on Computational Intelligence and Computing Research (2015)
13. Sawant, N.R., Patil, P.S.: Comparative study of SWT-SVD and DWT-SVD digital image watermarking technique. IJCA, 166(12) (May 2017)
14. Loukhaoukha, K., Chouinard, J.: Hybrid watermarking algorithm based on SVD and lifting wavelet transform for ownership verification. 11th Canadian Workshop on Information Theory, pp. 177–182 (2009)

15. Zhang, H., Wang, C., Zhou, X.: A Robust image watermarking scheme based on SVD in the spatial domain. Future Internet **9**(3), 45 (2017)
16. Ghaderi, K., Akhlghian, F., Moradi, P.: A new digital image watermarking approach based on DWT-SVD and CPPNNEAT. In: 2nd International Conference on Computer and Knowledge Engineering (ICCKE), October 18–19 (2012)

Stability of Model that Makes Automated Comparison Between Market Demands and University Curricula Offer

Ylber Januzaj, Artan Luma, Besnik Selimi, and Bujar Raufi

Abstract The stability of a model that makes automated comparison between labor market demands and university curricula has a great importance. The statistics that will be drawn from our model should be consistent results for the long time. Therefore, during our work, some cases are presented that prove the stability of the model. The first test will be by removing stop words to compare them later with the corpus that contains stop words. Various tests will also be conducted such as increasing the volume of the corps and deleting some contests from the corps. Another test that will be made is cross validation with different results. Finally, our model is compared with another model that compares textual content between different corpuses. All of these tests will prove the stability of our automated model that makes comparison between market demands and university curricula.

Keywords Clusters · Data mining · Job market · University curricula · Web scraping

1 Introduction

In order to test the stability of the model, testing is done to arrive at definitive conclusions about the stability. The tests that will be done on our system are removing stop

---

Y. Januzaj (✉) · A. Luma · B. Selimi · B. Raufi
Faculty of Contemporary Sciences and Technologies, South East European University, Tetovo, Macedonia
e-mail: yj16535@seeu.edu.mk

A. Luma
e-mail: a.luma@seeu.edu.mk

B. Selimi
e-mail: b.selimi@seeu.edu.mk

B. Raufi
e-mail: b.raufi@seeu.edu.mk

© The Editor(s) (if applicable) and The Author(s), under exclusive license to Springer Nature Singapore Pte Ltd. 2021

A. P. Pandian et al. (eds.), *Proceedings of International Conference on Intelligent Computing, Information and Control Systems*, Advances in Intelligent Systems and Computing 1272, https://doi.org/10.1007/978-981-15-8443-5_7

words and comparing with previous results, removing some job offers, increasing the volume of corpus by adding new job offer, cross validation and comparison with other models.

The first start is to remove the stop words from our system and compare with the previous results and present the results that are achieved in order to see if there is a big difference and what is the importance of stop words [1].

The next step is to remove some of the contests from our corpus and compare it with the previous results to confirm the stability of our model [2]. According to our expectations, even after removing or adding competitions to our database, the similarity results should not change much [3].

The next analysis to test the robustness of the system is the addition of some competitions to our database, where the results will change most as new competitions are expected to be added in the future that will increase the volume of our corpus [4].

2 Removing Stop Words

As mentioned above, in order to test the stability of our model, one of the steps is to remove stop words from the body of the competition data. Of course, these words should not be overweight, and based on the experiments, these words will even lower their weight if they are in the corps [5].

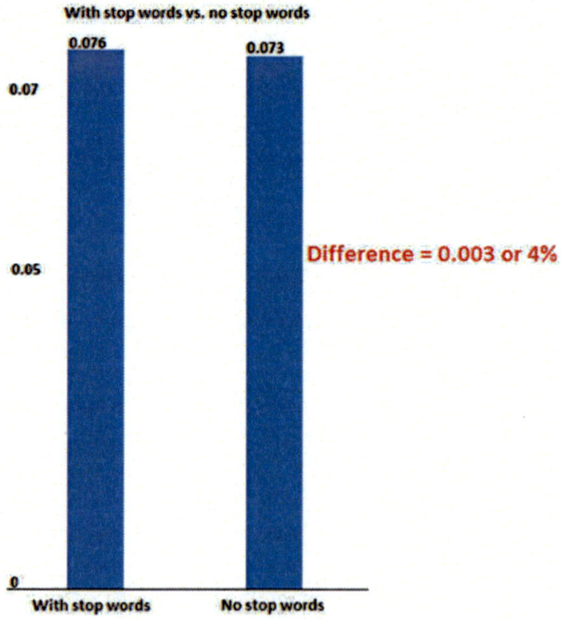

Fig. 1 Difference of corpus with stop words and with no stop words

Figure 1 shows the similarity between labor market requirements and university curricula with stop words and with no stop words. As can be observed when a random comparison is made between labor market requirements and a randomly selected syllabus, the similarity of the textual content acquired is 0.076. By means of an algorithm, the stop words are removed from the corpus containing the data on labor market requirements and again calculated the similarity between the textual content [6]. After removing the stop words from the labor market requirements corpus, we obtained a textural similarity between the labor market requirements and the university curriculum of 0.073. As can be seen in Fig. 1, the difference between stop words and no stop words is 0.003 or 4%. From this, it can be concluded that the importance of stop words is not too great as normalization techniques have been applied in our model which reduce the weight of the most frequently used words [7]. In the following, the results are presented as portion of the contests is removed from our data.

3 Removing Job Vacancies

The content of the corpus with the demands of the labor market is of great importance, but more important is the processing of its content [8]. After processing the labor market requirements corpus, the complete content and part of the contests are compared.

Figure 2 shows the difference of textual similarity between labor market requirements and a randomly selected syllabus. The difference is between the complete content of the corpus with labor market requirements data and the removal of a significant portion of the content contests [9]. As can be seen, the result of the textual similarity between the full content of the labor market requirements and the syllabus is 0.076. After removing a significant portion of the textual content of the labor market corpus, the technique of comparing textual similarity was again applied, and the result obtained was 0.072. As can be seen in Fig. 1, the difference is very small, since we have a difference of 0.004 or if in percentage have a difference of 5%.

Even in this case, the difference is very small, and this gives great importance to our model since despite the changes that the labor market may have for a short time, the results will not change much, and of course, this makes the model very stable. In the following, the case is presented when added some new competitions to the labor market demand corps, which further supports the sustainability of our model [10].

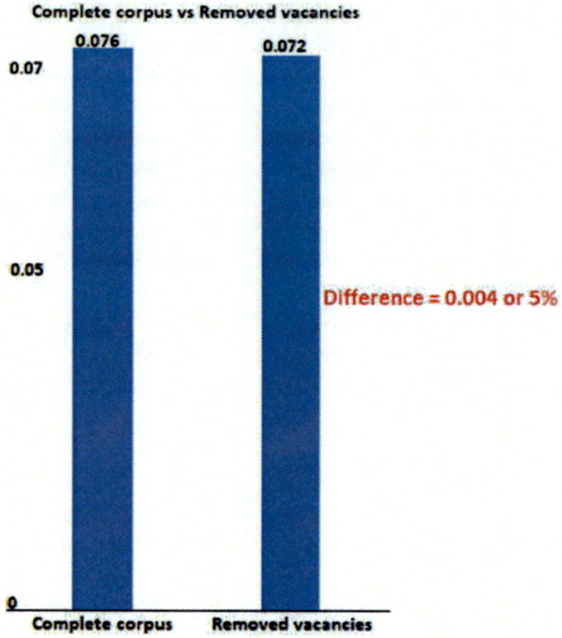

Fig. 2 Difference of complete corpus and removed some vacancies corpus

4 Increasing the Volume of Corpus by Adding New Job Offers

As mentioned in the chapters above, in the coming years, there is expected to be a large increase in labor market demands in the field of technology. What is important about our model is that it is applicable even after a time when there are new competitions published in the field of technology. The next step that will test the viability of our model is to increase the volume of the corpus with data on technology contests.

The next analysis is done again comparing the complete content of the corpus with the labor market requirements with a randomly selected syllabus. After this analysis, the volume has been increased with new contests in our corps, and again, the comparison of textual similarity has been made.

New competitions have also been taken from the Web site which was initially used to launch the published competitions, but also new competitions from other Web sites which are published in the field of technology.

Figure 3 shows the comparison of the similarity of textual content between the corpus created by us, and the corpus after added some new contests. According to analyses, general similarity of textual content between labor market requirements and a randomly selected syllabus is 0.0764. Whereas, after adding some new job offers to our corpus, this result has changed to 0.007 and obtained a textural similarity score between the labor market requirements and the university curriculum of 0.0771. This difference is very small as it is only 1%, and by calculating this small

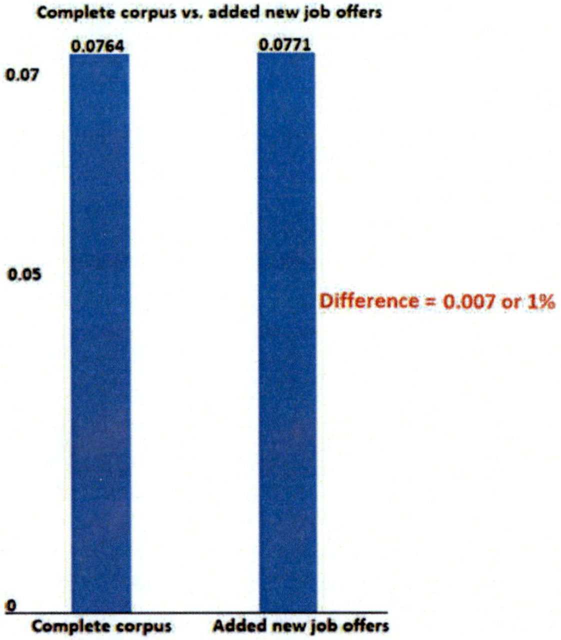

Fig. 3 Difference between complete corpus and added new job offers corpus

difference, concluded that our system is very stable in terms of comparing labor market requirements and curricula offered by the university. Also this difference has a great importance for the fact that almost every day new competitions are published on the website, and those new competitions will not affect us to get results different from those our model is conclude. In the following, the case of cross validation is presented, which is the fifth step of testing our system stability.

4.1 Cross Validation

The next step that confirms the stability of our system is cross validation with the data that have in our body. The way to do this analysis is by dividing our competition corpus into ten different pieces, then make a text mix and remove 10% of each piece from the text. First remove the first 10%, compare, then place the first 10% and remove the second 10% and so on.

Figure 4 shows the cross validation analysis of the job vacancies corps and the placement of new vacancies in our corps. As mentioned above, and as it is known that cross validation analysis works, our body is divided into ten parts. After dividing the competition corpus into ten parts, the text was mixed, and in this way, 10% of the corpus is removed for each part.

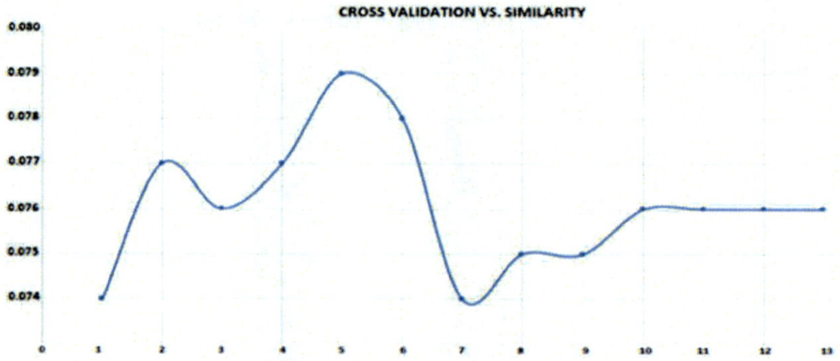

Fig. 4 Cross validation versus similarity

As can be seen in Fig. 4, when the first 10% is removed, a result of textual similarity is obtained between the labor market requirements and university curricula of 0.074. After this step, we return the first 10% to the corpus again, and we remove the second 10% from the corpus, and after the comparison we obtain a similarity score of 0.077. The third part has a score of 0.076, with the fourth part having 0.077. A sharp increase in resemblance is seen since the fifth part of our corps was removed, where gained a similarity of 0.079, and with the removal of the sixth part, a similarity of 0.078 is gained. After the seventh part of corpus is removed,we obtain a sharp decline as the likelihood falls back to 0.074, and with the eighth and ninth part have a stabilization of results as the level of textual similarity increases to 0.075 for the eighth part, and 0.076 for the second part. nine. As can be seen with the completed corpus we have a similarity level of 0.076, and this result begins to stabilize when we add new data in our corps. As can be seen in Fig. 4, with new competition placements, where the graphs show as the eleventh, twelfth, and thirteenth have almost the same results as the completed corps with a small margin of 1%. So, such an analysis supported all the preliminary analysis which was done in order to verify the stability of our system. The next step that underpins the consistency of our model is to compare our model with another model that makes textual content comparisons.

5 Comparison with Other Model

Comparison with an existing model which compares textual similarity between different documents is of great importance in the sustainability of our model. In the following, the differences in results are presented from our model and from another model available online, as well as cross validation analysis of the other model.

Figure 5 shows a comparison of our model with another model which also compares the textual content between different documents. Unlike our model, the

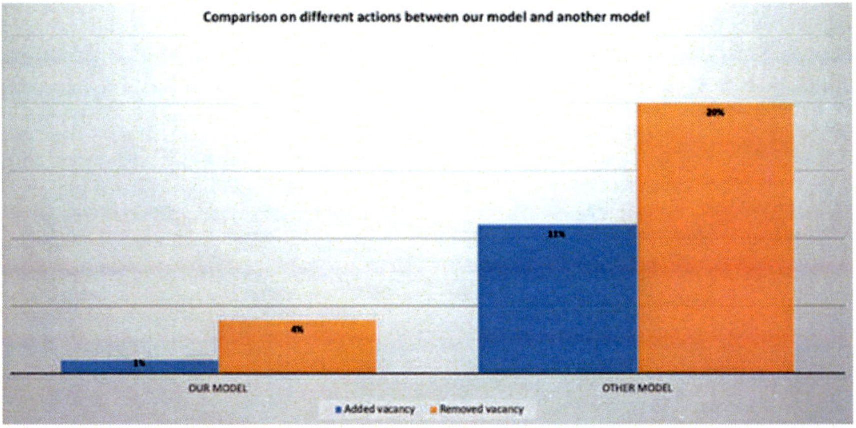

Fig. 5 Comparison on different actions between our model and another model

other model does not use data normalization, which negatively affects the outcome that the model ultimately achieves.

According to the analysis, our model provides accurate and small changes as shown in the graphs above. As can be seen in Fig. 5, our model after the addition of new contests in the competition corps gives us a score of only 1% difference compared to the corpus own. Whereas after leaving the contests from our corps, the model gives us a result which is only 4% difference with the result that the model gives with the original corps.

While the other model that compared to ours, at this point, it yields results that have a big difference with the first results after comparing textual content between labor market requirements and university curricula.

As can be seen in Fig. 5, after adding new contests to our corps, the other model yields a score that is 11% higher than the primary scores the model offers with the original corps. Whereas after removing some contests from our corps, the other model yields a score that is 20% different from the primary scores the model offers with the original corps. Of course, even at this point, our model has an advantage in terms of consistency and accuracy, since text normalization methods have been applied in our model which influence the model to provide more accurate results. In the following, the cross validation analysis of the other model is presented, to see how much the model has stability compared to our model.

5.1 Cross Validation with Other Model

Another analysis that compares our model with another model that compares textual content is cross validation analysis as done in our model. Surely, such an analysis will greatly help us to know how consistent and accurate our model is. The analysis

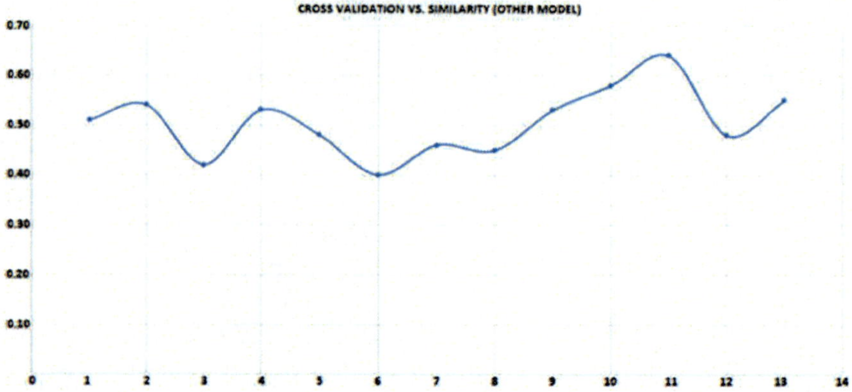

Fig. 6 Cross validation of vacancy corpus (other model)

will be the same as in our model, where the corpus will be divided into ten parts, the textual mix will be removed, and 10% of the corpus will be removed until complete corpus content is reached. Once got to the full content of the corps, new contests are added to see if the system succeeds in stabilizing the results the same as our model.

Figure 6 presents the cross validation analysis of the vacancy corpus of the other model. The same analysis was applied as with our model. Initially, the corpus is split into ten parts and did a text mix to continue with each section later. As can be seen in Fig. 6, the other model has a different level of measurement of textual similarity between different documents. After removing the first part, the similarity level of the other model is 0.50, we return the first part, and remove the second part and the similarity level reached 0.55. We apply the same procedure with other parts of corpus, from first part to the tenth part, where all our corpus is with job vacancies, and the textual similarity between the labor market requirements and the syllabus at this point was reached at 0.58. As can be seen when the placement of new competitions is made, the model has not stabilized, but again has a drastic differences in textual similarity between the two documents.

Certainly comparing our model with such an existing model makes our model much more consistent and accurate, since the existing model when new competitions that are published are added, then the results vary by 10–20%. as shown in different analysis. Therefore, our model provides accurate and consistent data compared to other models that make comparisons of textual content between different documents.

6 Conclusion

The application of methods that increase the accuracy of comparison between different documents has a great importance. During our work, the stability of our model is tested in several forms. First step is testing the accuracy of the system with stop words and without stop words, and the second step is removing some of the competition from the labor market demand corpus. Then, the next test is adding some new competitions to the labor market demand corpus. After these tests, a comparison was made between our model and an online model that compares the textual content. And finally, cross validation in the corpus is done which shows that our system is very stable in the accuracy of the data it provides

Finally, such a methodology of comparing textual content between labor market demands and university curricula will directly contribute to improving the curricula offered by universities.

References

1. Agaoglu, M.: Predicting instructor performance using data mining techniques in higher education. In: IEEE Access (2016)
2. Xie, T., Zheng, Q., Zhang, W., Qu, H.: Modeling and Predicting the Active Video—Viewing Time in a Large—Scale E-Learning System. In: IEEE Access (2017)
3. Njeru, A., Omar, M., Yi, S.: IoTs for capturing and mastering massive data online learning courses. In: IEEE Computer Society, ICIS, Wuhan, China (2017)
4. Heartfield, R., Loukas, G., Gan, D.: You are probably not the weakest link: towards practical prediction of susceptibility to semantic social engineering attacks. In: IEEE Access (2016)
5. Fortuny, E., Martens, D.: Active learning—based pedagogical rule extraction. In: IEEE Transaction on Neural Network and Learning Systems, vol. 26, No. 11 (2015)
6. Mukhopadhyay, A., Bandyopadhyay, S.: A survey of multiobjective evolutionary algorithms for data mining: Part I. In: IEEE Transaction on Evolutionary Computation, vol. 18, No. 1 (2014)
7. Song, Z.H., Kusiak, A.: Optimization of Temporal processes: a model predictive control approach. In IEEE Transaction on Evolutionary Computation, vol. 13, No. 1 (2009)
8. Malgaonkar, S., Soral, S., Sumeet, Sh., Parekhji, T.: Study on big data analytics research domain. In International Conference on Reliability, Infocom Technologies and Optimization ICRITO, Noida, India (2016)
9. Anicic, K., Divjak, B., Arbanas, K.: Preparint ICT graduates for real—world challenges: results of a meta—analysis. In IEEE Transactions on Education, vol. 60, No. 3 (2017)
10. Haskova, A., Merode, D.V.: Professional training in embedded systems and its promotion. In IEEE Transacions on Education (2016)

Economic Load Dispatch Using Intelligent Particle Swarm Optimization

Nayan Bansal, Rohit Gautam, Rishabh Tiwari, Surendrabikram Thapa, and Alka Singh

Abstract This paper introduces a unique and modified method to find the solution of the economic load dispatch (ELD) problem employing intelligent particle swarm optimization. Due to fierce competition in the electric power industry, environmental concerns, and exponentially increasing demand for electric power, it has become necessary to optimize the economic load dispatch problem which includes real-time constraints like valve point effect and operating prohibited zones. Experimental results of the intelligent PSO method and various versions of particle swarm optimization (PSO) are obtained, and comparison is drawn on the basis of their convergence speed and their convergence stability.

Keywords Intelligent particle swarm optimization (IPSO) · Economic load dispatch (ELD) · Variants · Convergence

N. Bansal (✉) · R. Gautam · R. Tiwari · A. Singh
Department of Electrical Engineering, Delhi Technological University, New Delhi, India
e-mail: nayan.7991@gmail.com

R. Gautam
e-mail: rgautam1997@gmail.com

R. Tiwari
e-mail: rishabh240711@gmail.com

A. Singh
e-mail: alkasingh.eed@gmail.com

S. Thapa
Department of Computer Science and Engineering, Delhi Technological University, New Delhi, India
e-mail: surendrabikram_bt2k17@dtu.ac.in

© The Editor(s) (if applicable) and The Author(s), under exclusive license to Springer Nature Singapore Pte Ltd. 2021

A. P. Pandian et al. (eds.), *Proceedings of International Conference on Intelligent Computing, Information and Control Systems*, Advances in Intelligent Systems and Computing 1272, https://doi.org/10.1007/978-981-15-8443-5_8

1 Introduction

The modern power system involves a large number of interconnected electrical networks, and with the constant increase in fuel prices used in thermal power plants, it is necessary to reduce the operating cost of the generating unit. The primary objective of the modern power system is to provide high-quality electrical power to the consumer at the lowest price, keeping into account various constraints of generating units and power systems. This lays the foundation for the ELD problem which emphasizes the minimization of fuel cost and to find the real power generated by each interconnected generating unit. The problem consists of various equality and inequality constraints which makes it a complex problem.

A traditional method such as the Newton method, lambda iteration method, and the gradient method can solve piecewise a linear and a monotonically increasing function [1]. However, the fuel cost curve in the ELD problem becomes highly nonlinear and non-smooth due to factors such as the valve point effect, ramp rate limit inequality constraints, generator efficiency, and prohibited operating zones constraints making ELD a complex non-convex problem whose solution is difficult to be determined using traditional methods. Artificial intelligence, genetic algorithm, and stochastic algorithms that are inspired by nature such as particle swarm optimization can solve these high non-convex problems reaching near to the global optimal solutions [2].

PSO is an evolutionary computational method which is inspired by the group organisms like birds and fishes colonies. PSO is preferred due to the high convergence speed and lesser number of parameters. In this paper, the study of the application of PSO by changing its variants and applying to the ELD problem, comparing each method on the basis of convergence speed and convergence stability has been performed.

2 Particle Swarm Optimization

PSO is one of the modern intelligent evolutionary methods whose computation is inspired by the animal's behaviour that dwells in colonies such as birds and fishes [3]. The principle behind this evolutionary computation technique is the mutual cooperation between the members of the society. Comparing with other computation techniques, PSO has better convergence speed and requires less parameter for its evaluation. Over the years, there has been widespread research on the PSO method, and modified algorithms have been obtained on the basis of adjustment of parameters and improvement in population diversity [4]. The first parameter is used to find equilibrium between local searching and global searching. These include algorithms such as linear weight particle swarm optimization (LWPSO), constriction particle swarm optimization (CPSO), and damped particle swarm optimization (DPSO). The second parameter is used to obtain algorithms to avoid premature convergence. They employ techniques such as natural selection to improve the performance significantly.

In this paper, the focus has been emphasized on the first parameter due to the efficiency of the parameter strategy; less cost of computation, and relatively fewer complexities in it.

In the early modification of PSO, the velocity update equation was reintroduced with an inertia weight coefficient. The inertia weight decreased in a fixed linear manner (LWPSO). This method was useful to increase the speed of convergence and obtaining equilibrium between the global and local search exploitation. However, due to the decrease of the inertia weight in a fixed linear way, there was a compromise with the local exploration. Therefore, further modification in this method and inertia weight was reduced by damping factor instead in a linear manner [5]. This method improved the speed of convergence but compromised the balance between global and local searching of global optimum value without adding any complexities. Further research introduced the constriction factor, eliminating the inertia weight introduced earlier in the research papers in the velocity update equation. Experimentation revealed the value constriction factor to be 0.729 for providing the best optimum solution [6]. This method proved the fact that dynamic updating of the velocity equation improved local searching of optimal solution and speed of convergence without adding complexities to the PSO method. Deeply inspired by the advancement in the modification of the PSO method, this paper has introduced a novel method "intelligent particle swarm optimization" (IPSO). In this method, the reintroduction of the inertia weight eliminated by CPSO has been done. The inertia weight in this method is dependent on the *i*th iteration is an exponential relation. This provides a large relay decay step in the early stage of the algorithm which improves the speed of convergence, and in the later stage, the decay step reduces considerably allowing local exploration which leads to balance in local and global exploration. In this paper, the application of LWPSO, DPSO, CPSO, and IPSO to ELD problem and comparison of the solutions obtained by each algorithm, their convergence speed, and convergence stability has been done.

3 Intelligent Particle Swarm Optimization

Let us consider a population swarm of size n particles with each particle being allowed to move in solution space and have assigned a position vector xi and velocity vector $\mathrm{v_i}$, both of which are P-dimensional vectors which are described as $x_i = (x_{i1}, x_{i2}, \ldots, x_{iP})$ and $v_i = (v_{i1}, v_{i2}, \ldots, v_{iP})$. The position vector x_i represents the possible solution, and $\mathrm{v_i}$ velocity vector affects the convergence speed. The velocity equation affects the convergence speed and the exploration of local and global optimum value which in turn affects the stability of convergence. During the search operation of PSO, each particle obtains a personal best position $P_\mathrm{b} = (P_{\mathrm{b}1}, P_{\mathrm{b}2}, \ldots, P_{\mathrm{bP}})$. The personal best position of the particles is compared with that of the other members of the swarm, and selection of global best position $P_\mathrm{g} = (P_{\mathrm{g}1}, P_{\mathrm{g}2}, \ldots, P_{\mathrm{gP}})$ is done by the algorithm. The personal and global best position are used to update the velocity of the particle which in turn is used in updating the position of the particle. The velocity equation through which it is done can be realized as

$$v_i(n+1) = v_i(n) + c_1 r_1 (P_b(n) - x_i(n)) + c_2 r_2 \big(P_g(n) - x_i(n)\big) \quad (1)$$

The position equation is realized as

$$x_i(n+1) = x_i(n) + v_i(n) \quad (2)$$

where c_1 and c_2 are positive coefficients, r_1 (.) and r_2 (.) are random variable functions [7]. Early research papers introduced inertia weight (LWPSO) in which the inertia constant equation is given by

$$v_i(n+1) = w v_i(n) + c_1 r_1 (P_b(n) - x_i(n)) + c_2 r_2 \big(P_g(n) - x_i(n)\big) \quad (3)$$

and the inertia weight is linearly decreased as follows:

$$w = w_{\max} - \frac{(w_{\max} - w_{\min}) * \mathrm{it}}{\mathrm{MaxIt}} \quad (4)$$

where MaxIt is the total number of iterations and instantaneous iteration is denoted by it. $w_{\max}$ and $w_{\min}$ are constants, and their values are 0.9 and 0.4, respectively. This method improved the speed of convergence and found a balance between global and local searching [8]. Further research work was done on the PSO and was found out that the damping factor in the inertia weight of the updated velocity equation produced better convergence speed [9].

$$w = w * w_{\mathrm{damp}} \quad (5)$$

where w is chosen to be 0.9 and w_{damp} is chosen 0.99.

Further improvement in the algorithm of PSO paved way for velocity update equation in which the inertia weight was eliminated and constriction factor was introduced [10]. The velocity update equation is

$$v_i(n+1) = \chi\big(v_i(n) + c_1 r_1 (P_b(n) - x_i(n)) + c_2 r_2 \big(P_g(n) - x_i(n)\big)\big) \quad (6)$$

$$\text{And, } \chi = \frac{2}{2 - \emptyset - \sqrt{\emptyset^2 - 4\emptyset}} \quad (7)$$

Many experiments were performed to find the value of Φ. The value of Φ was found to be 4.1 which gives $\chi = 0.729$. This was the case when the algorithm performs best in finding the optimal solution. Deeply inspired by this, a new method intelligent PSO (IPSO) is introduced in this paper. In this method, the inertia weight in (3) was modified. The new fixed inertia weight is

$$w = \left(\frac{e}{\text{it}}\right)^{\frac{1}{e}} \tag{8}$$

This algorithm allowed us to have a large step at the beginning of the competition of computation and a small step towards the later stage of computation. This ensured the equilibrium between the global and local searching for the problem. This is achieved without any complexity in the algorithm which is essential for an evolutionary algorithm. With this algorithm, the convergence speed obtained was better than the LWPSO, DPSO, and CPSO, and the convergence stability was founded to the best when the algorithms were applied to the ELD problem. In further section, discussion on the numerical results of these algorithms when applied to ELD has been done.

4 Economic Load Dispatch Problem

The power which is generated in a thermal power plant takes place by the rotation of prime mover in the turbine with the action of steam. The working fluid in a thermal power plant is water. Water is fed to the boiler and super heater which convert it to steam. Steam which carries thermal energy is allowed to expand in the boiler which rotates the rotor shaft of generators. The steam loses energy and is condensed and then pumped to the boiler back to be heated up again. The factors which affect the operating cost include the transmission losses, fuel costs, and efficiency of the generators in action. Usually, labour, maintenance, and operation costs are fixed. A typical cost curve (Fuel) locus of generating unit is depicted below in Fig. 1.

P_{min} is the minimum power that can be drawn from the generating unit below which the operation of the plant is not feasible. P_{max} is the maximum power that can be drawn from the generating unit [11].

The prime objective of ELD is the minimization of thc total fuel cost utilized in the generation of power. This problem can be executed in the following way:

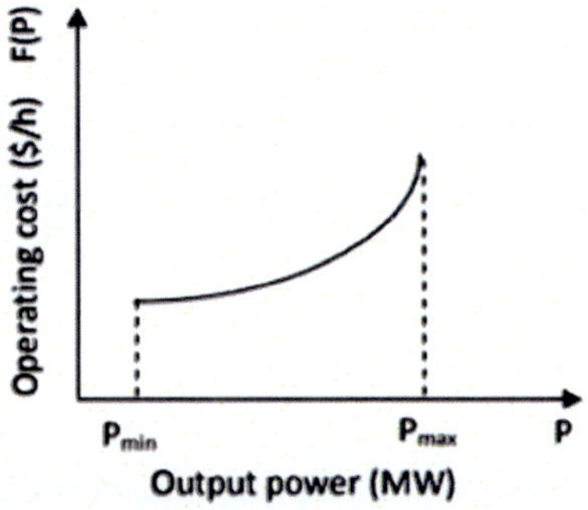

Fig. 1 Cost curve (fuel) locus of a generating unit

$$\text{Minimise}\ X_T = \sum_{i=1}^{n} F_i(P_i) \tag{9}$$

Total cost of generation is denoted by the function X_T, and F_i denotes the cost function of ith generation. The number of generating units is denoted by n, and P_i denotes the real power generated by the generating units. Now, the above function can be approximately expressed as a quadratic equation. The variable in the quadratic function is the real power generated by the generating units [12].

$$F_i(P_i) = \alpha_i + \beta_i P_i + \gamma_i P_i^2 \tag{10}$$

where the ith generating unit has fuel coefficients which are depicted in (10) by α_i, $\beta_{i,}$ and γ_i

4.1 *Equality Constraints*

The total real power generated which is being generated by the generating units in the system under study must be equal to the demand power from the system and the transmission losses which gives rise to the equality constraints.

$$\sum_{i=1}^{n} P_i = P_{\mathrm{D}} + P_{\mathrm{L}} \tag{11}$$

where

P_{D} demand power (MW)
P_{L} transmission losses (MW)

4.2 *Inequality Constraints*

$$P_i^{\min} \leq P_i \leq P_i^{\max} \tag{12}$$

where the minimum real power which has to be generated by ith unit is $P_i^{\min}$ and the maximum which can be generated by ith unit is $P_i^{\max}$.

4.3 *Transmission Losses*

The following equation describes the transmission losses:

$$P_{\mathrm{L}} = P^T B P + B_0^T P + B_{00} \tag{13}$$

where P is the vector of length N which represents power output of each generator, B is loss coefficients square matrix, B_o is another vector whose length is equal to the number of generating units N and B_{oo} is a constant.

4.4 Damp Rate Limit Constants

The power generated in the generating unit is P_i may not be more than the real power generated in the previous interval by a certain amount UR_i, the up ramp rate unit, and may not be less than the amount of real power DR_i, the down rate limit. So, the constraints that arise are as follows:

$$\text{Max}\left(P_i^{\min}, P_i^o - DR_i\right) \leq P_i \leq \text{Min}\left(P_i^{\max}, P_i^o + UR_i\right) \tag{14}$$

4.5 Valve Point Effect

The incremental fuel cost curve of a generating unit in the ELD problem is assumed to be monotonically increasing power is a linear function. Thus, the input–output characteristic is quadratic. However, due to the value point effect, the input–output curve displays discontinuities and linearity of a high order [13]. Thus, in order to take into account, this constraint modification of the original function has been done [14]. The value point effect is demonstrated by modified periodic sinusoidal function mathematically represented as:

$$F_i(P_i) = \alpha_i + \beta_i P_i + \gamma_i P_i^2 + \left|e_i \times \sin\left(f_i \times \left(P_i^{\min} - P_i\right)\right)\right| \tag{15}$$

where the additional fuel cost coefficients that have been included to consider the valve point effect of ith generating unit are e_i and f_i (10).

4.6 Prohibited Operating Zones

The presence of a steam valve in a thermal power plant causes vibration in its shaft bearing which leads to the generation of zones that are prohibited for operation in the fuel cost function [15]. Other reasons may include integrated auxiliary operating equipments like feed pumps and boilers. Prediction of the locus of the fuel cost curve is not possible in these prohibited zones. Preventing the operations of the units in these regions is the best economic solution. The depiction of prohibited zones in a typical cost curve is done in Fig. 2.

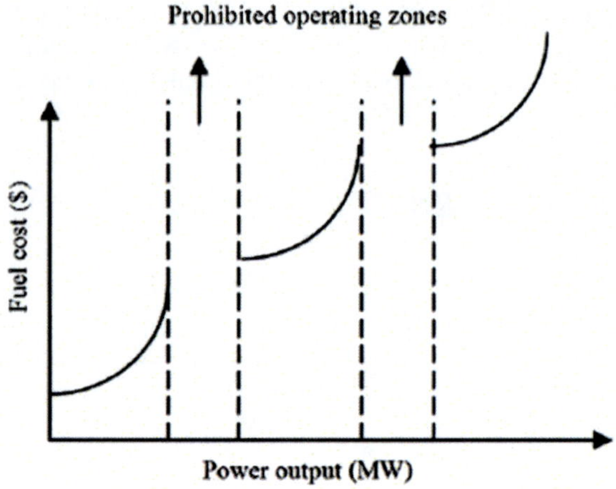

Fig. 2 Prohibited operating zones are shown in cost curve locus

This can be mathematically represented as follows:

$$P_i^{\min} \le P_i \le P_{i,1}^{\text{lower}} \tag{16}$$

$$P_{i,k-1}^{\text{upper}} \le P_i \le P_{i,k}^{\text{lower}}, k = 2, 3, \ldots n_j \tag{17}$$

$$P_{i,n_i}^{\text{upper}} \le P_i \le P_i^{\max} \tag{18}$$

where lower real power limit of prohibited kth zone of ith unit is depicted by $P_{i,k}^{\text{lower}}$. The upper limit of prohibited $k - 1$th zone of ith unit is denoted by $P_{i,k-1}^{\text{upper}}$, and the number of prohibited zones which are present in ith generating unit is equal to n_j [16].

These are the constraints that have been considered in the ELD problem, and solutions have been obtained by various versions of PSO.

5 Numerical Results and Simulation

This paper considers a power system which comprises six generating units. The total demand power is 1200 MW. Each generating unit possesses two prohibited operating zones. The system is used to demonstrate the application of various modified methods of PSO, and results are obtained. The fuel cost coefficients of each generating unit are depicted in Table 1, and the capacity of each unit is depicted below in Table 2. The data corresponding prohibited zones are depicted in Table 3.

The B-coefficients are given below to compute the transmission losses in the given power system:

Table 1 Fuel cost coefficients

Unit	α_i	β_i	γ_i	e_i	f_i
1	240	7	0.0075	250	0.035
2	200	10	0.009	150	0.04
3	220	8.5	0.0095	150	0.038
4	200	11	0.0095	120	0.042
5	220	10.5	0.0085	160	0.037
6	190	12	0.008	120	0.025

Table 2 Generating units characteristics

Unit	$P_i^{\min}$ (MW)	$P_i^{\max}$ (MW)	P_i^o (MW)	UR_i (MW)	DR_i (MW)
1	100	500	450	70	110
2	60	220	160	60	90
3	80	300	220	55	100
4	50	150	140	60	90
5	60	220	170	50	90
6	50	120	120	50	90

Table 3 Generating units prohibited zones

Unit	Prohibited Zone 1		Prohibited Zone 2	
	P_i^{lower} (MW)	P_i^{upper} (MW)	P_i^{lower} (MW)	P_i^{upper} (MW)
1	210	240	350	380
2	90	110	140	160
3	150	170	210	240
4	80	90	110	120
5	90	110	140	150
6	75	85	100	105

$$B = \begin{bmatrix} 0.0017 & 0.0012 & 0.0007 & -0.0001 & -0.0005 & -0.0002 \\ 0.0012 & 0.0014 & 0.0009 & 0.0001 & -0.0006 & -0.0001 \\ 0.0007 & 0.0009 & 0.0031 & 0.0000 & -0.0010 & -0.0006 \\ -0.0001 & 0.0001 & 0.0000 & 0.0024 & -0.0006 & -0.0008 \\ -0.0005 & -0.0006 & -0.0010 & -0.0006 & -0.0129 & -0.0002 \\ -0.0002 & -0.0001 & -0.0006 & -0.0008 & -0.0002 & 0.0150 \end{bmatrix} /50$$

$$B_o = \left[-0.390\ -0.129\ 0.707\ 0.0591\ 0.211\ -0.6635\right]/1000$$

Table 4 Power generated by various units and total costs

Unit	LWPSO (MW)	DPSO (MW)	CPSO (MW)	IPSO (MW)
1	429.75	429.74	429.77	**429.74**
2	220	219.99	219.94	**219.97**
3	218.61	218.71	218.62	**218.66**
4	124.79	124.76	124.76	**124.76**
5	164.91	164.92	164.94	**164.94**
6	73.67	73.66	73.62	**73.65**
Total mean cost (Rs./h)	1512.21	1513.8	1513.4	**1510.3**

$$B_{oo} = [0.055]$$

In this section, the power generated by each generating unit along with the total cost of power generation is calculated using various modified methods of particle swarm optimization [17]. Also, the comparison has been drawn between various modified methods of particle swarm optimization. Evaluation of various optimization methods is done on the basis of two indices: Convergence stability and convergence speed. A better nature-inspired stochastic algorithm is one which has better convergence speed and stability of convergence. The time of computation of the various algorithms is also compared in this paper.

5.1 *Power Generation and Total Cost*

The ELD problem is solved by various modified methods of particle swarm optimization taking all equality and inequality constraints into account. The results are obtained in Table 4. The demand power from the system is 1200 MW. The particle swarm optimization algorithm is performed with a population size of 200, and each algorithm is made to run for 200 iterations.

The total power generated is 1231.72 MW, out of which, 1200 MW is used to meet up demand, and 31.72 MW is wasted in transmission loss. The mean total cost is nearly the same in all the modified versions of PSO as shown in Table 4.

5.2 *Convergence Speed*

An algorithm is said to be convergent if it reaches an optimal region after a certain number of iterations. An algorithm that does not reach the optimal region is said to be divergent. The speed of convergence is determined by the gradient of the convergence

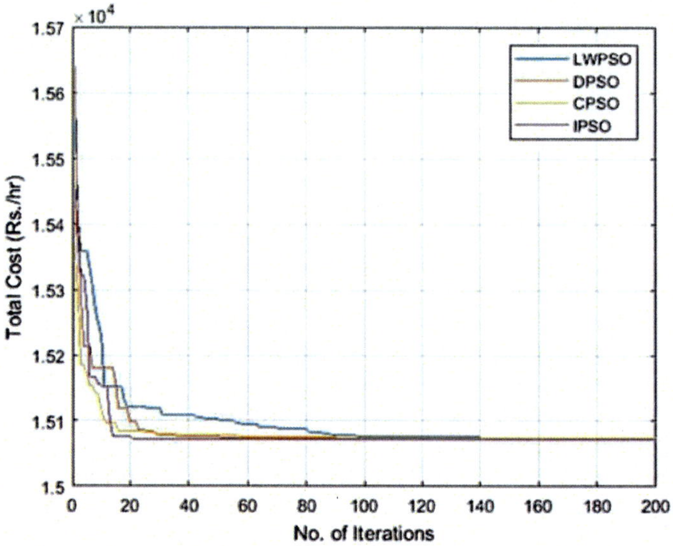

Fig. 3 Convergence curve of different modified PSOs

curve which expresses the speed of convergence [18]. The convergence curves of all four versions of particle swarm optimization are depicted by Fig. 3

In the convergence curve, the vertical axis is represented by the total cost (Rs./hr) in the ELD problem. The horizontal axis represents the number of iterations of the modified algorithms. From the figure, it can be concluded that IPSO outperforms LWPSO, DPSO, and CPSO when each of the algorithms is run for 200 iterations.

5.3 Convergence Stability

Convergence stability refers to the distribution of the global optimum value of the function around the mean value, after it made to run for certain iterations. The concentration of global optimum value is an indicator of convergence stability. The more the concentration the better is convergence stability [18]. The modified algorithms are made to run 40 times, and the global best solution which is the total cost of generation is obtained each time.

To obtain precise digital analysis, mean and standard deviation of the global optima values obtained from the above iterations for various modified PSO methods are calculated. A smaller standard deviation reflects better convergence stability and less divergence. As obtained from the data in Table 5, IPSO has the least standard deviation and mean than LWPSO, CPSO, and DPSO. Hence, the IPSO algorithm has better convergence stability.

Table 5 Digital analysis of various methods of PSO

Criteria	LWPSO	DPSO	CPSO	IPSO
Mean (Rs./h)	1512.21	1513.8	1513.4	**1510.3**
Standard deviation (Rs./h)	64.18	77.63	68.98	**59.1**

6 Conclusion

In this paper, the implementation of PSO and its modified versions have been successfully implemented to the ELD problem. Particle swarm optimization is a nature-inspired stochastic algorithm, and the presence of fewer variants gives it an edge over other nature-inspired evolutionary techniques. This paper successfully implements a new version of particle swarm optimization, and comparison has been made with the existing version of PSO on the basis of convergence speed, convergence stability, and the total mean cost. Digital analysis of convergence stability for various versions has also been performed. The new method (IPSO) has better convergence speed and convergence stability than the existing models (CPSO, DPSO, and LWPSO). However, the total mean cost of the new method is almost the same as the existing models. The iterative weighted term in the velocity equation gives a dynamic step in upgrading of particle velocity. At a later stage, the step becomes small ensuring local exploration, thus establishing equilibrium between global exploration and local exploration. This novel PSO method with better convergence speed and stability can be employed in other applications of power system optimization.

References

1. Chandram, K., Subrahmanyam, N., Sydulu, M.: Equal embedded algorithm for large scale economic load dispatch. In: 2007 IEEE Power Engineering Society General Meeting, Tampa, FL, pp. 1–8 (2007)
2. Chou, K., et al.: Robust feature-based automated multi-view human action recognition system. IEEE Access **6**, 15283–15296 (2018). https://doi.org/10.1109/ACCESS.2018.2809552
3. Koohi, I., Groza, V.Z.: Optimizing particle swarm optimization algorithm. In: 2014 IEEE 27th Canadian Conference on Electrical and Computer Engineering (CCECE), Toronto, ON, 2014, pp. 1–5. https://doi.org/10.1109/ccece.2014.6901057
4. Ding, W., Lin, C., Prasad, M., Cao, Z., Wang, J.: A layered-coevolution-based attribute-boosted reduction using adaptive quantum-behavior PSO and its consistent segmentation for neonates brain tissue. IEEE Trans. Fuzzy Syst. **26**(3), 1177–1191 (2018). https://doi.org/10.1109/TFUZZ.2017.2717381
5. He, M., Liu, M., Jiang, X., Wang, R., Zhou, H.: A damping factor based particle swarm optimization approach. In: 2017 9th International Conference on Modelling, Identification and Control (ICMIC), Kunming, pp. 13–18 (2017)
6. Eberhart, R.C., Shi, Y.: Comparing inertial weights and constriction factor in particle swarm optimization. In: Proceedings of the IEEE Conference on Evolutionary Computation, ICEC. 1., vol. 1., pp. 84–88 (2000) https://doi.org/10.1109/cec.2000.870279

7. Tan, W., Meng, X., Wang, H., Zhao, L.: Notice of retraction: the operating study of circulating water system based on particle swarm optimization. In: 2011 Seventh International Conference on Natural Computation, Shanghai, 2011, pp. 2317–2321
8. Sharma, J., Mahor, A.: Particle swarm optimization approach for economic load dispatch: a review. Int. J. Eng. Res. Appl. **3**, 13–22 (2013)
9. He, M., Liu, M., Jiang, X., Wang, R., Zhou, H.: A damping factor based particle swarm optimization approach. pp. 13–18 (2017). https://doi.org/10.1109/icmic.2017.8321632
10. Eberhart, R.C., Shi, Y.: Comparing inertial weights and Constriction factor in particle swarm optimization. In: Proceedings of the IEEE Conference on Evolutionary Computation, ICEC. 1. pp. 84–88, vol. 1 (2000). https://doi.org/10.1109/cec.2000.870279
11. Alam, M.: State-of-the-art economic load dispatch of power systems using particle swarm optimization (2018)
12. Dihem, A., Salhi, A., Naimi, D., Bensalem, A.: Solving smooth and non-smooth economic dispatch using water cycle algorithm. In: 2017 5th International Conference on Electrical Engineering—Boumerdes (ICEE-B), Boumerdes, pp. 1–6 (2017)
13. Bhullar, P.S., Dhami, J.K.: Particle swarm optimization based economic load dispatch with valve point loading. In: International Journal Of Engineering Research & Technology (IJERT), vol. 04, no. 05 (2015). http://dx.doi.org/10.17577/IJERTV4IS050998
14. Pranava, G., Prasad, P.V.: Constriction coefficient particle swarm optimization for economic load dispatch with valve point loading effects. In: 2013 International Conference on Power, Energy and Control (ICPEC), Sri Rangalatchum Dindigul, pp. 350–354 (2013)
15. Dasgupta, K., Banerjee, S., Chanda, C.K.: Economic load dispatch with prohibited zone and ramp-rate limit constraints—a comparative study. In: 2016 IEEE First International Conference on Control, Measurement and Instrumentation (CMI), Kolkata, pp. 26–30 (2016)
16. Hota, P., Sahu, N.: Non-Convex economic dispatch with prohibited operating zones through gravitational search algorithm. Int. J. Electr. Comput. Eng. (IJECE). 5:1234–1244 (2015). https://doi.org/10.11591/ijece.v5i6
17. Lin, C., Prasad, M., Chang, J.: Designing mamdani type fuzzy rule using a collaborative FCM scheme. In: 2013 International Conference on Fuzzy Theory and Its Applications (iFUZZY), Taipei, pp. 279–282 (2013). https://doi.org/10.1109/ifuzzy.2013.6825450
18. Yan, C., Lu, G., Liu, Y., Deng, X.: A modified PSO algorithm with exponential decay weight. In: 2017 13th International Conference on Natural Computation, Fuzzy Systems and Knowledge Discovery (ICNC-FSKD), Guilin, pp. 239–242 (2017)

Companion: Detection of Social Isolation in Elderly

Gayatri Belapurkar, Athul Balakrishnan, Rajpreet Singh Bhengura, and Smita Jangale

Abstract Elderly isolation is one of the very important issues prevalent in society. Elderly of various communities find it easier to relate to someone of the same age group and communicate with them. There are a number of elderly people who have no one to talk to in their homes, social circle or society. This gradually leads to social isolation which in turn may give rise to depression and/or even suicide. Our system proposes a solution to this problem using data analysis techniques together with the concepts of psychology. The aim is to do so by bringing like-minded people together and forming a group. The collection of points as to why they feel socially isolated is important data in trying to solve the issue. The application goes one step further and suggests cultural gatherings or get-togethers nearby based on their interests, which can act as potential spots for meeting new people and making new friends. This can help decision-makers monitor the mental health of the elderly and help them lead a better life.

Keywords Social isolation · Lubben Social Network Scale · Machine learning · Community detection · K-means clustering · Collaborative filtering · Content-based filtering · Human–computer interaction

G. Belapurkar (✉) · A. Balakrishnan · R. S. Bhengura · S. Jangale
Vivekanand Education Society's Institute of Technology, Mumbai, India
e-mail: 2016.gayatri.belapurkar@ves.ac.in

A. Balakrishnan
e-mail: athul.balakrishnan@ves.ac.in

R. S. Bhengura
e-mail: 2016.rajpreetsingh.bhengura@ves.ac.in

S. Jangale
e-mail: smita.jangale@ves.ac.in

© The Editor(s) (if applicable) and The Author(s), under exclusive license to Springer Nature Singapore Pte Ltd. 2021
A. P. Pandian et al. (eds.), *Proceedings of International Conference on Intelligent Computing, Information and Control Systems*, Advances in Intelligent Systems and Computing 1272, https://doi.org/10.1007/978-981-15-8443-5_9

1 Introduction

As per the Population Census 2011, the count of elderly is nearly 104 million, elderly being of the age 60 or above, in India, 53 million females and 51 million males [21]. There are several physical and mental problems usually faced by these people as they move on to their latter part of life. As people get older, things start to change around them, even in terms of the number of interactions they have with the external world. Their worsening health conditions hold them back from going out and meeting other people as they might be reluctant to cause trouble to people taking care of them. This starts to develop a feeling of loneliness in these older adults. Social relations are associated with good mental health, while their absence is linked to a significant increase in morbidity and mortality [8, 22, 23]. Their families may not be able to give them as much attention as they expect, and this leads to the elderly keeping everything to themselves. Prolonged feeling of loneliness further leads to social isolation which is a state of complete or near-complete lack of contact between an individual and society. This, in turn, may further worsen their already deteriorating health condition. Studies reveal strong evidence that social isolation hurts mortality [24]. The results imply that irrespective of the intensity of isolation faced by the individuals, the risk of mortality persists. Agewell Foundation, a non-profit NGO, based in Delhi, surveyed a sample size of 15,000 elderly individuals [17]. The rates and association are prevalent of social isolation with health issues across different age groups [19]. The results show that social isolation steadily increases with age and is strongly associated with poor health conditions. Increasing age and deteriorating health conditions are both being experienced by the elderly, and hence, they are the most vulnerable to this condition.

Identifying the cause of this problem is important. Several studies have shed light on this context. Similarly, finding solutions to tackle this problem is equally important [4]. This promoted active participation of the elderly in these social activities where several people of their age group are involved showed effective strategies to reduce loneliness and improve social support, mental and physical health.

This paper focuses on tackling this issue by identifying whether a person is likely to experience social isolation using a tool called the Lubben Social Network Scale (LSNS) [14]. Along with this, a social media platform, specifically designed for the elderly, is used to monitor their social connectivity to check how well they respond to meeting new people in the virtual world and to study the effect of expanding social network in reducing social isolation. Various social activities that can be performed by these people according to their capabilities will be suggested by considering several factors. Use of specific machine learning techniques, combined with the concepts of psychology to achieve this, has been made, as discussed in the methodology below.

2 Motivation and Related Work

There have been various studies in the past that show how social isolation has been affecting the elderly population as compared to other age groups. To overcome these, several approaches have been put forward by various researchers some of which are discussed here.

Filipa Landeiro made several investigations on the effect of social isolation on the elderly. One of her works is a systematic review performed to compare the effectiveness of health promotion interventions in alleviating social isolation in older adults [12]. The interventions were grouped to assess the extent of the effect they had in promoting social contacts [15]. These reviews can help policymakers get a better insight on which intervention will be suited for specific groups of people and hence provide customized measures to tackle their issues. But the problem faced by both of these methods is that their execution required physical intervention to conduct the questionnaire which requires time input on the part of the researcher as well as the participants. Moreover, it is not possible in every case due to the widespread reach of the problem of social isolation. Added to that, the elderly sometimes do not even realize that they are socially isolated. Thus, a better way would be to digitize this process to increase its reach and so that the participants can fill them up according to their convenience.

Artificial intelligence has been widely used in healthcare to solve numerous problems. To deal with elderly social isolation, several AI approaches have been suggested. One such approach suggested an ICT-based tool that acts as a bridge between elderly and new technologies to sustain learning ability in older people as this degrades with age [3]. This tool is also suitable to identify variations in the senior ability by evaluating a social isolation parameter. It used the AI approach on learning capacity, interest related to new technologies and proactivity is used to measure social isolation. Elderly are reluctant to use the latest technologies as they require a lot of time and patience to be learned, and these people tend to forget things very frequently. Hence, going through the same process might be very tedious, and they may opt out rather than using them. Research has suggested a design considering this section of the population in mind [26]. The major requirements, as observed in this work, of interpersonal communication applications are ability-aware, flexible interfaces; smart notification; tutorials and guiding transitions. Thus, simplifying the technology can help promote the usage of these by the elderly and help them overcome isolation.

3 Approaches Considered

Multiple approaches were considered and tested before coming up and zeroing in on one approach.

3.1 Social Isolation Detection

The process began by considering the use of machine learning algorithms for the detection of social isolation. This included the collection of data about the user along six different lines [25]. The six categories of data picked were about the user's habits, mental health levels, physical health levels, support from their families, support from their partner (spouse) and social life, which were collected with the help of a chat feature embedded in the mobile application provided to users. These responses were converted into six numerical 'metrics' that would reflect the indulgence of the user in those six criteria. Thereafter, these metrics were used to predict the degree of social isolation of the individual. They were subjected to logistic regression to determine whether a person is isolated or not.

However, this approach was discarded as an elderly would like to click options as opposed to typing out answers to questions. Moreover, text answers would also be vague and would not provide a concrete basis to further process upon. For instance, if a user claims to meet his friends 'often', this can have different meanings for different people. It may mean thrice a week for someone and may mean six times a week for someone else. This will contribute to error in the data collection phase itself. Further, the 'metric' created for conversion of responses is not psychologically tested by a professional and would not represent the degree of isolation in the truest sense.

3.2 Event Suggestions

To solve the problem of suggesting events, selection based on certain attributes associated with an event was done. Each event that was added into the system by various NGOs was given particular 'tags' (which can be considered similar to hashtags in social media) that would specify the terms related to that event. The NGO may also, at times, add a particular threshold value along with one or multiple criteria described as above. These threshold values were added so that any user, whose metric values falling below the specified threshold (if any) along with that criteria, would not be allowed or suggested that particular event. These terms or 'tags' were then used as attributes to select upon. That is, these 'tags' were mapped to the interests or qualities of users and then sampled based on the metric threshold is specified for the event. That way, a user would be able to access an event only if it is beneficial or relatable for him/her. For instance, an NGO has added an intense yoga session that requires a physical level threshold of at least four out of a possible ten. In this case, a user who has a physical level that is below '4' would not be recommended to go for this event as it would not be suitable for him/her.

This method of sampling, however, would not be as efficient or effective as it is based on basic conditions. Thus, arose the need for a recommendation system.

4 Devised Method

The system devised consists of five distinct parts, which are then combined and made available to the user. These steps comprise the data collection step, detection of social isolation, followed by identifying and suggesting prospective friends based on the user's data that is collected and suggesting related and appropriate events nearby and a mobile application to act as an interface that can help the user in overcoming social isolation and yet again feel like a respected member of the society. The basic workflow of the system can be explained as in Fig. 1.

4.1 Data Collection

Data is collected from the user regarding his/her hobbies (user chooses from a list of hobbies), location data, the distance up to which it is possible for the user to commute and income, along with their emails, which is required for account creation in the system. The data is collected using a form in the mobile application provided to users. This information forms our dataset, which is preprocessed and then used for making predictions. Along with this data, all the users in the system are asked to undertake the Lubben Social Network Scale-6 test [9] using the mobile application itself.

Data about the user's habits, mental health levels, physical health levels, support from their families, support from their partner(spouse) and social life was also

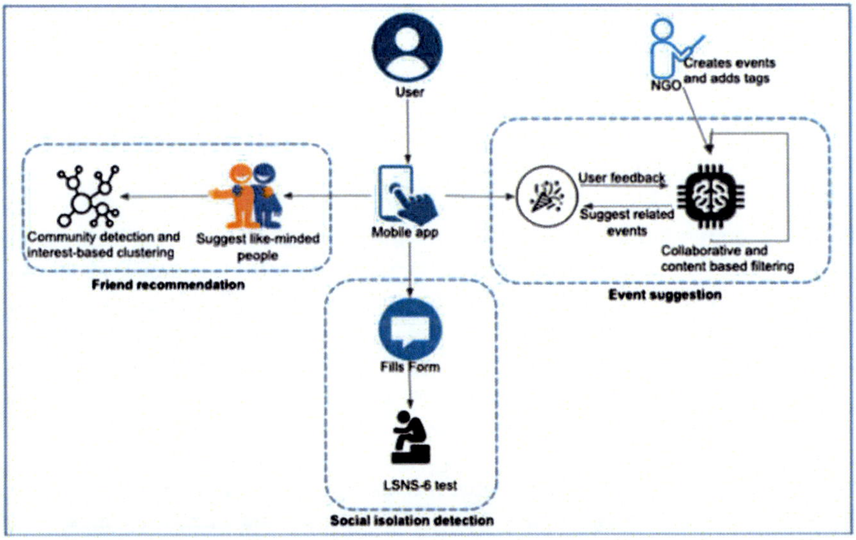

Fig. 1 System workflow

collected [25], which would be the important parameters for clustering to find similar people. This data was collected with the help of a multi-choice form embedded in the mobile application provided to users.

4.2 Social Isolation Detection

To detect the social engagement of a user, the Lubben Social Network Scale (LSNS) [9] has been used, which measures structural social support systems (i.e. social contacts) of a person. It was specifically created for the elderly and has been researched upon by psychologists, and its credibility has been proven. This psychological test, LSNS-6, detects the social isolation of an individual with the help of six questions regarding the person's interaction with relatives and friends. This test has six questions, the LSNS-6 scale, was chosen out of the multiple variations available as it is performed well across a range of settings, and the six questions, though less in number, were found to suffice and in fact proved to be better in the detection of social isolation [18]. It was also appropriate to choose this scale as its brevity may be advantageous in older adult populations that are less accustomed to social surveys. Moreover, this scale was found to be internally more consistent as compared to the other versions of the test [13].

The LSNS-6 test being used scores all individuals out of 30. A higher score implies a higher level of social engagement. Thus, those scoring low on the scale would be subjected to the further process of friend and event recommendation. Individuals with high scores above '24' are considered to be socially active and do not require the help of the system. Users with extremely low scores below '12' [9] experience high levels of social isolation and are immediately suggested to consult a professional. In this way, people with social isolation are detected.

4.3 Recommending 'Friends'

Friend recommendation should be based on the interests of people and their general social network. The user data collected through the form mentioned above was combined with the user's hobby-related information and their location to be the attributes for the K-means clustering model [5]. This is done to introduce people to strangers who share similar interests and ideas and maybe potential friends.

Along with suggesting friends based on the user's data, the user's pre-existing social circle was taken into account, if any. This was done using graph clustering [2]. A graph with users as nodes was created. Once two individuals accept each other as 'friends', an edge is created between the two nodes. Thereafter, the Girvan–Newman algorithm was used for community detection. The formation of communities would then help in recommending friends [20].

Thus, the two methods were combined for the recommendation of friends, to introduce the users to new people who may have similar interests, as well as those friends that people make through a common contact.

4.4 Event Suggestion

To bring the elderly out in the open air, out of their homes and to connect them to new people, suggest nearby events that they can attend. This considers events in a radius that is specified by the user, that is, the geographical limits up to which the user can commute. For this, built a hybrid recommendation engine that used a combination of collaborative filtering and content-based filtering. Collaborative filtering takes advantage of user's past behaviour as well as decisions made by friends of the user, while content-based filtering concentrates on utilizing similarities between events to make relevant recommendations [10, 11, 16].

For content-based filtering, events were clustered based on their similarities. This was followed by the extraction of important features such as event location, event topic or related terms and event popularity. This was then subjected to K-means clustering [5] to find similar events. For collaborative filtering, community detection based on the users' graph was done that is generated above for recommending friends to the users. Thereafter, binary logistic regression [6] and random forest [7] algorithms were used to train the model. The events to be recommended to a user were based on the output of this model.

4.5 Mobile Application

The consideration that the elderly need to be brought out of their isolation zone led to the creation of a social networking platform within the mobile application. Using this platform, they can connect with their known friends or meet new ones in the virtual world. Several existing applications provide the required facilities, but the motive behind creating a new one was to make it friendly enough for the elderly to use and to customize it according to their needs. The mobile application is user-friendly, with guided tutorials and prompts at each step so that it is easy for the elderly to navigate throughout the application. The application has a comfortable interface that would not push away elderly users. Healthcare application apps have been widely using voice-activated commands to make them more user-friendly [1]. The application has voice assistance in it, which further aids the elderly in using the application to the highest possible extent. Every Android phone is equipped with Google assistant which supports voice commands. Google Voice API allows extending this functionality to the app created. For instance, if the user wishes to chat with a particular friend, the user just has to say, 'Ok Google, chat with xxx', and it opens up the chat interface. Studies have shown that as people get older, their eyes get less sensitive towards certain colours, and vision gets clouded preventing them from reading small fonts. Hence, it is necessary to be able to customize the theme and font sizes of the entire app according to user convenience, and this is possible by centralizing the theme properties and making them editable through user's choice. This method has been adopted to cater the app according to individual users' preference and thus making it more user-friendly.

4.6 Addressing Security Concerns

Security is a major concern as the elderly happen to be a set of the population which can be easily exploited. A measure taken to tackle this issue is that all the NGOs entering the system will have to upload certificates which show that they are registered and are recognized by the government. This will ensure that a distrusted entity does not create an event or gather people as that is a potential threat for the elderly.

Another issue may be when the chat option in the application is misused by users for illegal purposes or activities with malicious intent. This is stopped by negative keyword filtering for the chats in the application. Users using these negative keywords are identified and are debarred from using the system after three such occurrences.

5 Conclusion

A mobile phone-based application is devised that can help older people to be accustomed to the latest technologies and be well connected to society. The main focus of this application is to identify, to an extent, whether a person is feeling socially isolated and if yes, suggest them with activities that can help them overcome their issues. A major advantage of this system is that it is specifically designed to minimize the efforts on the user's part. It does not demand any additional hardware apart from a smartphone, which is a common commodity in every household today. Convenience is the major factor considered so that the elderly are motivated in engaging themselves with the application and carrying out the activities suggested. This new process helps save time for both the elderly and the ones who are responsible for taking care of them, be it a caretaker, a doctor or a family member.

6 Future Scope

This application can be further improved to make it comparatively more accessible for the elderly by including multiple regional languages instead of the use of English alone. This will not just increase the reach of the application, but also make users feel more comfortable. Further, a personal assistant can be set up for each user. This personal assistant, or personal chatbot, would act as a friend for the user and chat with the user like it was a human. This would also enable getting a few more data points which may prove to be useful in improving the user's overall experience.

Acknowledgements Would like to extend gratitude to the faculty of the Department of Information Technology for their feedback and support also wish to express our profound thanks to all those who helped us in gathering information for the project.

References

1. Chung, A.E., Griffin, A.C., Selezneva, D., Gotz, D.: Health and fitness apps for hands-free voice-activated assistants: Content analysis. In: JMIR mHealth and uHealth (2017)
2. Developers, N.: Clustering algorithms (October 2019)
3. Di Lecce, V., Giove, A., Quarto, A., Soldo, D., Di Lecce, F.: Social isolation monitoring system via ai approach. In: 2015 IEEE Workshop on Environmental, Energy, and Structural Monitoring Systems (EESMS) Proceedings, pp. 256–261. IEEE (2015)
4. Dickens, A.P., R.S.G.C.C.J.: Interventions Targeting Social Isolation in Older People: A Systematic Review, pp. 204–207. BMC Public Health (August 2011). https://doi.org/10.1186/1471-2458-11-647
5. Guido, A.C.M..S.: Introduction to Machine Learning with Python: A Guide for Data Scientists, pp. 168–181 (2016). 978-1-449-36941-5
6. Géron, A.: Hands-on Machine Learning with Scikit-Learn and Tensorflow, pp. 177–184 (2017). 978-1-491-96229-9
7. Géron, A.: Hands-on Machine Learning with Scikit-Learn and Tensorflow, pp. 232–258 (2017). 978-1-491-96229-9
8. Holt-Lunstad, J., Smith, T.B., Layton, J.B.: Social Relationships and Mortality Risk: A Meta-Analytic Review, vol. 7, pp. 1–1. Public Library of Science (2010). https://doi.org/10.1371/journal.pmed.1000316, https://doi.org/10.1371/journal.pmed.1000316
9. Lubben, J., Blozik, E., G.G.S.I.W.R.V.K.J.C.B., Stuck, A.E.: Lubben Social Network Scale (lsns-6) (2006)
10. Khetwani, J., Sameer Sharma, S.A.: Event-Recommendation-Engine (May 2016)
11. Ji, X., Qiao, Z., Xu, M., Zhang, P., Zhou, C., Guo, L.: Online event recommendation for event-based social networks. In: Proceedings of the 24th International Conference on World Wide Web. p. 45–46. WWW '15 Companion, Association for Computing Machinery, New York, NY, USA (2015). https://doi.org/10.1145/2740908.2742742, https://doi.org/10.1145/2740908.2742742
12. Landeiro, F., Barrows, P., Nuttall Musson, E., Gray, A.M., Leal, J.: Reducing Social Isolation and Loneliness in Older People: A Systematic Review Protocol (2017). https://doi.org/10.1136/bmjopen-2016-013778, https://bmjopen.bmj.com/content/7/5/e013778
13. Lubben, J., Blozik, E., Gillmann, G., Iliffe, S., Kruse, W., Beck, J., Stuck, A.: Performance of an abbreviated version of the lubben social network scale among three european community-dwelling older adult populations. The Gerontologist **46**, 503–13 (2006). https://doi.org/10.1093/geront/46.4.503
14. Lubben, J.E.: Assessing social networks among elderly populations. Family Commun. Health. **11**, 42–52 (1988)
15. López, M.J., Lapena, C., Sánchez, A., Continente, X., Fernández, A.: Community intervention to reduce social isolation in older adults in disadvantaged urban areas: study protocol for a mixed methods multi-approach evaluation. BMC Geriatrics **19**(1) (2019). https://doi.org/10.1186/s12877-019-1055-9

16. Macedo, A.Q., Marinho, L.B., Santos, R.L.: Context-aware event recommendation in event-based social networks. In: Proceedings of the 9th ACM Conference on Recommender Systems, pp. 123–130. RecSys '15, Association for Computing Machinery, New York, NY, USA (2015). https://doi.org/10.1145/2792838.2800187, https://doi.org/10.1145/2792838.2800187
17. Malhotra, N.: India's ageing population is struggling with loneliness but help is available (2018)
18. Myagmarjav, S., Burnette, D., Goeddeke, Jr., F.: Comparison of the 18-item and 6-item lubben social network scales with community- dwelling older adults in mongolia, vol. 14, pp. 1–12. Public Library of Science (2019). https://doi.org/10.1371/journal.pone.0215523, https://doi.org/10.1371/journal.pone.0215523
19. Hämmig, O.: Health Risks Associated with Social Isolation in General and in Young, Middle and Old Age, vol. 14, pp. e0219663. Public Library of Science (2019). https://doi.org/10.1371/journal.pone.0219663
20. Sathiyakumari, K., Vijaya, M.S.: Community detection based on girvan newman al- gorithm and link analysis of social media. In: Subramanian, S., Nadarajan, R., Rao, S., Sheen, S. (eds.) Digital connectivity—social impact, pp. 223–234. Springer, Singapore (2016)
21. Central Statistics Office Ministry of Statistics and Programme Implementation Government of India: Elderly in India (2016)
22. Steptoe, A., Shankar, A., Demakakos, P., Wardle, J.: Social Isolation, Loneliness, and All-cause Mortality in Older Men and Women, vol. 110, pp. 5797–5801. National Academy of Sciences (2013). https://doi.org/10.1073/pnas.1219686110, https://www.pnas.org/content/110/15/5797
23. Tabue Teguo, M., Simo-Tabue, N., Stoykova, R., Meillon, C., Cogne, M., Amiéva, H., Dartigues, J.F.: Feelings of Loneliness and Living Alone as Predictors of Mortality in the Elderly: The Paquid Study, vol. 78 (2016)
24. Tanskanen, J.A.T.: A prospective study of social isolation, loneliness, and mortality in Finland. Am. J. Public Health (AJPH). IEEE (2016)
25. Waite, L., Cagney, K., Dale, W., Hawkley, L., Huang, E., Lauderdale, D., Lau- mann, E.O., McClintock, M., O'Muircheartaigh, C., Schumm, L.P.: National social life, health and aging project (nshap): Wave 3, [united states], 2015–2016 (2019). https://doi.org/10.3886/ICPSR36873.v4
26. Williams, D., Ahamed, S.I., Chu, W.: Designing interpersonal communication soft- ware for the abilities of elderly users. In: 2014 IEEE 38th international computer software and applications conference workshops, pp. 282–287. IEEE (2014)

Machine Learning Techniques to Determine the Polarity of Messages on Social Networks

Jesus Varga, Omar Bonerge Pineda Lezama, and Karen Payares

Abstract With the origin of Web 2.0, the Internet contains large amounts of user-generated information on an unlimited number of topics. Many entities such as corporations or political groups seek to gain knowledge through the opinions expressed by users. Social platforms such as Facebook or Twitter have proven to be successful for these tasks, due to the high volume of real-time messages generated and the large number of users that use them every day (Leis et al., J Med Internet Res 21(6):e14199, 2019, [1]). This paper focuses on the problem of Global Sentiment Analysis. Using the texts that compose the corpus in Spanish built by Sixto et al. (International conference on applications of natural language to information systems. Springer, Cham, 2016 [2]), and selecting the three most used classifiers by the state of the art of Naive Bayes, MSV and J48 through the Weka software.

Keywords Machine learning techniques · Polarity of messages · Social networks

1 Introduction

Currently, the large amount of data stored in electronic media, together with the technological development of computers grouped under the term "data mining," aims to extract information and useful knowledge to apply it in any productive area to support decision-making processes [3]. The area of natural language processing that analyzes and classifies texts into positive, negative, or neutral polarities is called

J. Varga (✉) · K. Payares
Universidad de la Costa, St. 58 #66, Barranquilla, Atlántico, Colombia
e-mail: jvargas41@cuc.edu.co

K. Payares
e-mail: kpayares4@cuc.edu.co

O. B. P. Lezama
Universidad Tecnológica, San Pedro Sula, Honduras
e-mail: omarlezama@unitet.edu

© The Editor(s) (if applicable) and The Author(s), under exclusive license to Springer Nature Singapore Pte Ltd. 2021

A. P. Pandian et al. (eds.), *Proceedings of International Conference on Intelligent Computing, Information and Control Systems*, Advances in Intelligent Systems and Computing 1272, https://doi.org/10.1007/978-981-15-8443-5_10

sentiment analysis, also known as opinion mining, subjectivity analysis, and sentiment orientation [4]. In opinion mining [5], feelings and emotions are expressed in text.

Detecting feelings is considered a difficult task, since the problem involves knowledge of the environment and context where the opinion is executed, which is very broad and complex [6]. It is formally stated that an opinion of a characteristic carries an associated feeling, and the user who issues the opinion is known as the opinion maker. Thus, an opinion is defined as a fivefold opinion (oj, fjk, ooijkl, hi, ti) [7] where:

- oj—is the object of the opinion,
- fjk—is a characteristic of the object on which an opinion is expressed. When no characteristic is detected, it is interpreted as a general opinion, as a characteristic of the object,
- ooijkl—is the polarity of the feeling of opinion about the fjk characteristic of the object oj—positive, negative, or neutral,
- hi—is the issuer of the opinion,
- ti—is the time when opinion is expressed by hi.

The Spanish Society of Natural Language Processing (SEPLN—Sociedad Española del Procesamiento del Lenguaje Natural), a non-profit scientific association that promotes research in all kinds of activities related to the study of natural language processing in Spanish [8], is organizing from 2012 onward a workshop on Sentiment Analysis focused on the Spanish language on texts extracted from the Twitter social network. Twitter is a microblogging platform where users publish messages, opinions, and comments whose contents range from personal feelings to general publications. Publications on Twitter are known as tweets. The main feature of the tweets is that the maximum length of the text is 140 characters [9]. In the last edition of this workshop, two tasks were organized: Global Sentiment Analysis and Aspect Sentiment Analysis. The work presented focuses on the first task of the workshop, which is to determine the global polarity of Twitter messages by categorizing them into six and four categories: (P+, P, NEU, N, N+, NONE) and (P, NEU, N, NONE) [10]. This article summarizes the first experimental approach to determine the polarity of tweets in the Spanish language, using the workshop corpus and Weka's classifiers: J48, Naive Bayes, and support vector machine. The extracted lexical characteristics are included as input data in the mentioned classifiers. The purpose of this study is to analyze the inputs and outputs required by the classifiers, and to identify the adjustments in their parameters that achieved the best results.

2 Task Description and Corpus Analysis

In the studied corpus, there are two different evaluations: one based on six tags of different polarity (P+, P, NEU, N, N+, NONE) and another one based on only four tags (P, NEU, N, NONE) [11]. It consists of a training set of 18,548 tweets labeled

Table 1 Distribution of tweets in the training corpus

Categories	No. of tweets	%
+P	4521	24.37
P	2854	15.39
NEU	1025	5.53
N	3254	17.54
+N	2154	11.61
NONE	4740	25.55

with the polarity corresponding to six tags. The distribution of tweets by their polarity in the training set is shown in Table 1.

There is also a general corpus used as a test set composed of 60,798 tweets. The corpus is coded in XML format, shown in Fig. 2. The evaluation of the developed systems defines the TASS in its Web site, using the accuracy metrics [11], which evaluates the correct polarity assigned to the tweets according to the gold standard. The test corpus is used to assess the accuracy of the learning model, supported by the layout of the labeled test suite. The accuracy of a classification model on a test suite is defined in the following equation [16]:

$$\text{Accuracy} = \frac{\text{Right number of classifications}}{\text{Total number of test cases}} \quad (1)$$

The generated confusion matrix will be used to evaluate accuracy, recall, and measurement—F1 for each individual category. The method used as a first approximation is represented in Fig. 1.

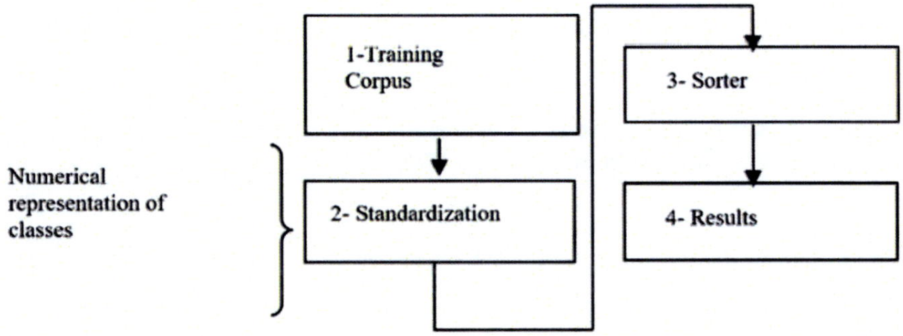

Fig. 1 Proposed method

3 Experimentation

The first solution approach was to use the training corpus by importing the file in XML format into Excel. The information that is not required for sentiment analysis was removed, remaining only the column labeled with the content text, which is the tweet to be analyzed, and the value column, which contains the class labels (+P, P, NEU, N, N+ and NONE). With this format, the classifiers J48 and Naive Bayes were used, and later, the format of the value column was changed so that the classes used a numerical representation as follows: 0 = NONE, 1 = NEU, 2 = N, 3 = N+, 4 = P, and 5 = +P and to be able to experiment with the MSV classifier. The Weka data analysis program version 3.6 was started, and the.csv file containing the labeled training data of the corpus was imported providing a view of the class distribution. The classifiers J48, LibSVM, and Naïve Bayes are selected with the parameter percentage split to 80%, which refers to that, from the entered data, 80% of training will be taken and the remaining 20% as test data. The results of this experiment are shown in Table 2.

The second experiment consisted in using the cross-validation of 10 folders by choosing the same classifiers without pre-processing. The results are shown in Table 3.

For the construction of the second model, the training corpus was divided into six different files, each one containing the texts corresponding to a class. Components were programmed in Python using regular expressions to extract statistical and lexical characteristics, such as the frequency per tweet of hashtags, URLs, and user mentions. From the lexical characteristics, a list of Wikipedia emoticons was used, categorized as smile, laugh, and sadness [17], in order to obtain a frequency per tweet. The state of the art indicates that the use of capital letters represents emphasis when trying to convey an idea, opinion, or feeling, so the frequencies per tweet of words that begin in capital letters were taken into account, as well as words that are written entirely in capital letters. The characteristics file is used to enter it into Weka and build the classification model with the same parameters as the model described above in order to compare the results obtained. It is worth mentioning that the tweets have not received any kind of pre-processing up to this point (Table 4).

Using the same characteristics, but evaluating by means of cross-validation of ten folders, the following results are obtained (Table 5).

Table 2 Results of classification into six categories, without pre-processing, using 80% of division

Classifier	J48	LibSVM	Naive Bayes
Result	.2547	.2514	.2578

Table 3 Results of classification into six categories, without pre-processing, using ten-folder cross-validation

Classifier	J48	LibSVM	Naive Bayes
Result	.2358	.2247	.2358

Table 4 Classification results in six categories, without pre-processing, using 80% division with the lexical characteristics extracted

Classifier	J48	LibSVM	Naive Bayes
Result	.2895	.2884	.2701

Table 5 Classification results in six categories, without pre-processing, using ten-folder cross-validation with the extracted lexical characteristics

Classifier	J48	LibSVM	Naive Bayes
Result	.2954	.2954	.2536

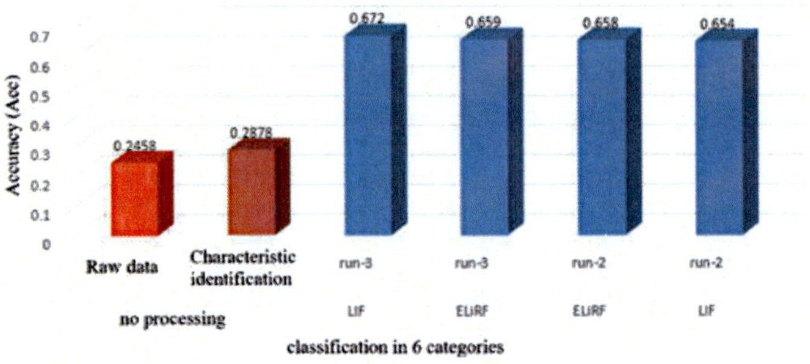

Fig. 2 Comparison of results with those of the state of the art

4 Comparison of Results

Figure 2 shows the results obtained in this study and compares them with those achieved by similar ones [14–16], which obtained the first four places in the workshop and those obtained in the experiments. The experiments are evaluated with the accuracy metrics.

5 Conclusions and Future Research

The state of the art indicates that the special characteristics of the language of Twitter require special treatment to analyze the texts. The particular syntax, mentions to users, URLs, hashtags, emoticons, poor grammar sentences, idioms, and among others lead to a drop in the performance of traditional NLP tools [18]. There is a proposal in the studies [15, 16] for the standardization of the text with these characteristics. This paper summarizes the experimentation carried out with the TASS 2015 corpus data, using the J48, Naive Bayes, and MSV classifiers included in Weka as a first approximation [17, 18]. It was observed that, by extracting feature frequencies, the results have improved. The state of the art indicates that pre-processing is important in this type of texts, so in the first instance, the conclusions indicate that the pre-processing stage can be a key factor in the processing of information to achieve better results. As future research, it is proposed to add the term frequencies and the inverse term frequency (tf-idf) as characteristics represented by unigrams and bigrams experimenting with the same classifiers and the same parameters, comparing the performance of the models based on their results.

References

1. Leis, A., Ronzano, F., Mayer, M.A., Furlong, L.I., Sanz, F.: Detecting signs of depression in tweets in Spanish: behavioral and linguistic analysis. J. Med. Internet Res. **21**(6), e14199 (2019)
2. Sixto, J., Almeida, A., López-de-Ipiña, D.: Improving the sentiment analysis process of Spanish Tweets with BM25. In: International Conference on Applications of Natural Language to Information Systems (pp. 285–291). Springer, Cham (2016)
3. Kauffmann, E., Peral, J., Gil, D., Ferrández, A., Sellers, R., Mora, H.: Managing marketing decision-making with sentiment analysis: an evaluation of the main product features using text data mining. Sustainability **11**(15), 4235 (2019)
4. Zin, T.T.: Sentiment polarity in translation. In 2020 IEEE Conference on Computer Applications (ICCA), pp. 1–6. IEEE (2020)
5. Rodríguez, F.M., Garza, S.E.: TOM: Twitter opinion mining. Comput. Syst. **23**(4) (2019)
6. Solé, M., Giné, F., Valls, M., Bijedic, N.: Real time classification of political tendency of twitter Spanish users based on sentiment analysis. Int. J. Comput. Inf. Eng. **12**(9), 697–706 (2018)
7. Viloria, A., & Pineda Lezama, O.B.: An intelligent approach for the design and development of a personalized system of knowledge representation. In: Procedia Computer Science, vol. 151, pp. 1225–1230. Elsevier B.V. (2019). https://doi.org/10.1016/j.procs.2019.04.176
8. Bibi, M., Nadeem, M.S.A., Khan, I.H., Shim, S.O., Khan, I.R., Naqvi, U., Aziz, W.: Class association and attribute relevancy based imputation algorithm to reduce twitter data for optimal sentiment analysis. IEEE Access **7**, 136535–136544 (2019)
9. Vilares, D., Alonso, M.A., Gómez-Rodríguez, C.: En-es-cs: an english-spanish code-switching twitter corpus for multilingual sentiment analysis. In: Proceedings of the Tenth International Conference on Language Resources and Evaluation (LREC'16), pp. 4149–4153 (2016, May)
10. Alkubaisi, G.A.A.J., Kamaruddin, S.S., Husni, H.: Conceptual framework for stock market classification model using sentiment analysis on twitter based on Hybrid Naïve Bayes Classifiers. Int. J. Eng. Technol. **7**(2.14), 57–61 (2018)
11. Volkova, S., Bell, E.: Identifying effective signals to predict deleted and suspended accounts on twitter across languages. In: Eleventh International AAAI Conference on Web and Social Media (2017, May)
12. Jain, D., Kumar, A., & Garg, G.: Sarcasm detection in mash-up language using soft-attention based bi-directional LSTM and feature-rich CNN. Appl. Soft Comput. 106198 (2020)
13. Ruz, G.A., Henríquez, P.A., Mascareño, A.: Sentiment analysis of Twitter data during critical events through Bayesian networks classifiers. Future Gener. Comput. Syst. **106**, 92–104 (2020)
14. Tavoschi, L., Quattrone, F., D'Andrea, E., Ducange, P., Vabanesi, M., Marcelloni, F., Lopalco, P.L.: Twitter as a sentinel tool to monitor public opinion on vaccination: an opinion mining analysis from September 2016 to August 2017 in Italy. Hum. Vacc. Immunother. 1–8 (2020)
15. Nesi, P., Pantaleo, G., Paoli, I., Zaza, I.: Assessing the reTweet proneness of tweets: predictive models for retweeting. Multimedia Tools Appl. **77**(20), 26371–26396 (2018)
16. Nguyen, H.T., Le Nguyen, M.: An ensemble method with sentiment features and clustering support. Neurocomputing **370**, 155–165 (2019)

17. Thelwall, M.: The Heart and soul of the web? Sentiment strength detection in the social web with SentiStrength. In: Cyberemotions, pp. 119–134. Springer, Cham (2017)
18. Viloria, A., Angulo, M.G., Kamatkar, S.J., de la Hoz-Hernandez, J., Guiliany, J.G., Bilbao, O.R., Hernandez-P, H.: Prediction rules in e-learning systems using genetic programming. In: Smart Innov. Syst. Technol. **164**, 55–63 (2020). https://doi.org/10.1007/978-981-32-9889-7_5

An Investigation on Hybrid Optimization-Based Proportional Integral Derivative and Model Predictive Controllers for Three Tank Interacting System

S. Arun Jayakar, G. M. Tamilselvan, T. V. P. Sundararajan, and T. Rajesh

Abstract This paper deals with the mathematical model for three tank series interacting system, and level control in three tank interacting system is not an easy job, because of the nonlinear behavior due to the interaction of the peer tanks. Here, the accurate transfer function and state space model are obtained by the first principle method. In this process, two cases are considered one with single input in the first tank and the single output in third tank, i.e., single input and single output system (SISO), and other with two inputs in first cum third tank and single output as the third tank, i.e., two input and single output (MISO) system. Utilizing a mathematical model, the best control schemes are designed and implemented for the closed loop feedback control, and the performances are investigated by compared with the dynamic behaviors of the responses. There are three types of control schemes that are selected, and the controllers are personalized based on the mathematical model of the system. PID controller with a hybrid optimization technique and model predictive control (MPC) is designed based on the model. The three tanks interacting system is a higher-order system. In this paper, the exact transfer function model is going to be obtained by the first principle method; the mass balance equation is employed in the process for each the individual tanks in the system, and the state space and

S. Arun Jayakar (✉) · T. Rajesh
Department of Electronics and Instrumentation Engineering, Bannari Amman Institute of Technology Sathyamangalam, Erode, Tamil Nadu, India
e-mail: arunjayakars@bitsathy.ac.in

T. Rajesh
e-mail: rajesme@gmail.com

G. M. Tamilselvan
Department of Information Technology, Sir Krishna College of Technology, Coimbatore, Tamil Nadu, India
e-mail: tamilselvan.gm@skct.ac.in

T. V. P. Sundararajan
Department of Electronics and Communication Engineering, Sri Shakthi Institute of Engineering and Technology, Coimbatore, Tamil Nadu, India
e-mail: sundarajantvp@siet.ac.in

© The Editor(s) (if applicable) and The Author(s), under exclusive license to Springer Nature Singapore Pte Ltd. 2021

A. P. Pandian et al. (eds.), *Proceedings of International Conference on Intelligent Computing, Information and Control Systems*, Advances in Intelligent Systems and Computing 1272, https://doi.org/10.1007/978-981-15-8443-5_11

reduced order modes are obtained and the reduced order model is fine tuned; and the integer and non-integer models' first-order reduced models are obtained through model optimization techniques and the standard PID controller with hybrid optimization technique FminSearch-GA (Fmin-GA), Internal Model Controller and Model Predictive Control (MPC) also implemented, and performances are investigated using time domain specification.

Keywords Nonlinear process · Three tank system · PID controller · Model-based controller · Internal model based controller · Model optimization technique genetic algorithm

1 Introduction

The calculation and control of liquid levels is important for the chemical process industries and the safety of the equipment used there. Whether this process is regulated depends on how these tanks are linked [1]. The tanks are also interconnected so that their level interacts, i.e., the dynamics of another tank are affected by one level of the tank and vice versa. The level and flow control are directly at the core of the chemical industries.

The contact tank level control issue exists in many contexts, such as petrochemical industries, the pharmaceutical industry, and the food and drink industries. Efficient management of these variables is therefore very economically beneficial. Specific computational cases exist depending on the inputs and outputs [2]. Here, two instances are taken into account. The first scenario is the simulation of the three interacting tank system by considering the third development stage of the tank, i.e., vector, and liquid flow into the first tank as data, i.e., controlled element. In this instance, the PID was effectively programmed and implemented using hybrid optimization techniques' internal model control (IMC) fractional order [3]. The second case has two inputs, which is a liquid flow to the first and third tanks, and two outputs, that is, second and third tank volumes, are taken into account.

For nearly all sectors, the processes are nonlinear for nature. Approximate to the order formula is the conical tank and spherical tank structures involved in research [4]. The FOPIC technique allows for level monitoring. The fractional order calculus is an arithmetic discipline with a non-integer order derivative and integral down order calculus [5]. The FOPIC methodology makes victory through useful separation and integration methods and provides additional flexibility in controller design. This paper suggests the use of FOPIC to model nonlinear systems controllers. The MATLAB simulink system partners the conventional Ziegler Nichols (ZNPI) and simulation controller.

In almost all sectors, the processes are nonlinear. Approximate to an order, formula is the conical tank and spherical analytical tank systems [6]. Level control using the FOPIC technique is accomplished. Fractional order calculus is an arithmetical field with an integral and non-integer order derivative. FOPIC methodology offers victory by useful separation and integration methods and provides additional flexibility in controller form [7]. This paper proposes to use FOPIC to model nonlinear systems controllers. MATLAB simulink technology is collaborating with the conventional Ziegler Nichols (ZNPI) controller and simulation.

2 Execution of Three Tank Interacting System

The contact tank level control issue exists in many contexts, such as petrochemical industries, the pharmaceutical industry, and the food and drink industries. Efficient management of these variables is therefore very economically beneficial. Specific computational cases exist depending on the inputs and outputs. Here, two instances are taken into account [8]. The first scenario is the simulation of the three interacting tank system by considering the third development stage of the tank [6], i.e. ,vector, and liquid flow into the first tank as data, i.e. controlled element. In this instance, the PID was effectively programmed and implemented using hybrid optimization techniques' internal model control (IMC) fractional order [9]. The second case has two inputs that are liquid flow to the first and third tanks, and two outputs that are second and third tank volumes which are taken into account. This model was successfully developed and implemented with the predictive control fractional order model. As shown below, the entire modeling process can be divided into three parts.

1. Mathematical simulation of plants focused on the first method of theory.
2. Linearization of the plant to establish the layout of transfer system according to its operating conditions.
3. Request elimination for higher orders (Table 1).

3 Model Estimation of Three Tank Process with Single Input and Single Output (SISO) System

The first level of the tank is regarded as the input of the three tank systems, and the third level of the tank is seen as the output in Fig. 1. For this case, only one input and one output is taken into account (Fig. 2).

Tank 1, Tank 2, and Tank 3 are connected to three identical tanks as shown in figure; head tanks are h_1, h_2, and h_3; the restrictions are R_1, R_2, and R_3; each tank has a cross-sectional area of A_1, A_2, and A_3.

There are generally two types of first principles for initiating the process of mathematical modeling in chemical systems: (i) mass balance equation (ii) energy balance equation [10].

This is the preferable equation of the mass balance system at a liquid level process [11]. The assumption of the mass balancing equation for a single tank with inlet and exit varies between the mass entering a process, and the mass that exits the process must be the same as the amount of mass that the device can handle.

Table 1 Specifications of three tank system

S. No.	Process equipments	Specification
1	Process tank	Total-4 Height-60 cm Diameter-33 cm
2	Reservoir tank	Breadth-60 cm, Length-91 cm, Height-50 cm
3	Pump	Brand: Pearl domestic pump, S. no: MOPBEJ0523, Rpm: 2700, Size: 25 * 25, Power: 370 w/0.5 hp, Insclass: B, Head: 6-24 m, Volt: 230 V, Duty: S1, Current: 2.0 amps, Frequency: 50 Hz, Dis: 2250400LPH, Two pole, IP: 44, Run capacitor: 10 mf, 440 v
4	Level transmitter	Brand name: Switzer, Model: k5750w3sk1r3a01, Range: Length-600 mm S. no: LT1L09, Power: 9232v, Input: capacitance, Output:4-20 mA
5	Air regulator	Brand name: Placka, Model no: FPR, Serial no: 22364, Filter sintered bronze: 15 microns, Maximum input pressure: 18 kg/cm^2, Inlet pressure: 18 kg/cm^2, Output pressure: 2.1 kg/cm^2
6	Rotameter	Brand name: Eureka Equipments Pvt.ltd. S. no: 2009-10/r11075, Model no: Msvs-pg-6c(m)
7	I/P converter	Brand name: ABB, Model: TEIP11 CC No: 09L0477, Input: 4–20 mA, Output: 3–15 psi, Supply Voltage:(11–32 V)
8	Valve positioner	S. no: 09100649, Supply: 2, Signal: 2-1
9	Control valve	Brand name: RK Control Valves, S. no: 09101122 Type: https://doi.org/10.334/a4.35, Body: ¾", Trim: ½", Flange: ans/150rf, Rating: 150, CV: 5, Body Material: a216wcb, Trim material: ss315, Plug char: Linear Travel: 1-1/8", Spring range: -2-1 kg, Max actr. Pressure: 35 psi, On an air failure valve : Close

{Inflow of Mass entering into the system}
− {Outflow of the mass leaving from the system}
= Accumulated Mass in the system.

The mass balance equation by considering Tank 1 in the three tank system is

$$q_{\text{in}} - q_1 = A_1 \frac{\mathrm{d}h_1}{\mathrm{d}t}$$

$$q_1 = \frac{h_1 - h_2}{R_1}$$

$$q_{\text{in}} - \frac{h_1 - h_2}{R_1} = A_1 \frac{\mathrm{d}h_1}{\mathrm{d}t}$$

$$q_{\text{in}} - \frac{h_1}{R_1} + \frac{h_2}{R_1} = A_1 \frac{\mathrm{d}h_1}{\mathrm{d}t}$$

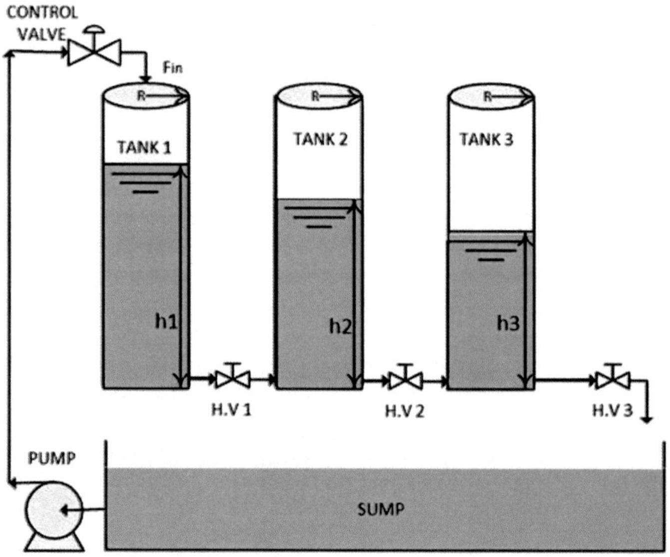

Fig. 1 Construction details of three tanks interacting

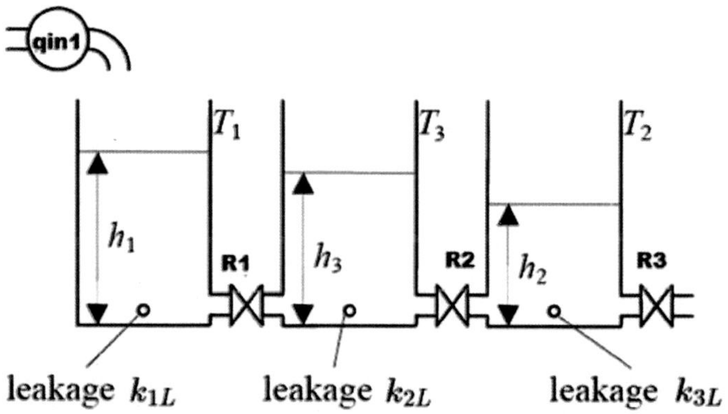

Fig. 2 Three tanks interacting with single input and single output (SISO) system

$$\frac{dh_1}{dt} = -\frac{h_1}{A_1 R_1} + \frac{h_2}{A_1 R_1} + \frac{1}{A_1} q_{in} \tag{1}$$

The mass balance equation by considering Tank 2 in the three tank system is

$$q_1 - q_2 = A_2 \frac{dh_2}{dt}$$

$$q_1 = \frac{h_1 - h_2}{R_1}$$

$$q_2 = \frac{h_2 - h_3}{R_2}$$

$$\frac{h_1 - h_2}{R_1} - \frac{h_2 - h_3}{R_2} = A_2\frac{dh_2}{dt}$$

$$\frac{dh_2}{dt} = \frac{1}{A_2}\left(\frac{1}{R_1} - \frac{1}{R_2}\right)h_1 - \frac{1}{A_2R_1}h_1 + \frac{1}{A_2R_2}h_3 \tag{2}$$

The mass balance equation by considering Tank 3 in the three tank system is

$$q_2 - q_3 = A_3\frac{dh_3}{dt}$$

$$q_2 = \frac{h_2 - h_3}{R_2}$$

$$q_3 = \frac{h_3}{R_3}$$

$$A_3\frac{dh_3}{dt} = \frac{h_2 - h_3}{R_2} - \frac{h_3}{R_3}$$

$$\frac{dh_3}{dt} = \frac{1}{A_3}\left(\frac{1}{R_2} - \frac{1}{R_3}\right)h_1 - \frac{1}{A_3R_2}h_2 \tag{3}$$

$$\begin{bmatrix} \dot{h}_1 \\ \dot{h}_2 \\ \dot{h}_3 \end{bmatrix} = \begin{bmatrix} -\frac{1}{R_1A_1} & \frac{1}{R_1A_1} & 0 \\ \frac{1}{R_1A_2} & \frac{1}{A_2}\left(\frac{1}{R_1} - \frac{1}{R_2}\right) & \frac{1}{A_3R_2} \\ 0 & -\frac{1}{A_3R_2} & \frac{1}{A_3}\left(\frac{1}{R_2} - \frac{1}{R_3}\right) \end{bmatrix}\begin{bmatrix} h_1 \\ h_2 \\ h_3 \end{bmatrix} + \begin{bmatrix} \frac{1}{A_1} \\ 0 \\ 0 \end{bmatrix}U \tag{4}$$

$$Y = \begin{bmatrix} 0 & 0 & 1 \end{bmatrix}\begin{bmatrix} h_1 \\ h_2 \\ h_3 \end{bmatrix} \tag{5}$$

where

h_1, h_2, and h_3 are the level of the Tank 1, Tank 2, and Tank 3, respectively.
$A_1 = A_2 = A_3 = A$ is a cross-sectional tank of the Tank 1, Tank 2 and Tank 3 which is equal to 706.2 cm^2
R_1, Restriction of the valve connecting Tank 1 and Tank 2 = 0.5
R_2, Restriction of the valve connecting Tank 2 and Tank 3 = 0.4
R_3, Restriction of the valve connecting Tank 2 and Tank 3 = 0.35

Hence, the state space model of the system is given below

$$\begin{bmatrix} \dot{h}_1 \\ \dot{h}_2 \\ \dot{h}_3 \end{bmatrix} = \begin{bmatrix} -0.2832 & 0.2832 & 0 \\ -0.2832 & -0.070 & 0.4048 \\ 0 & -0.4048 & -0.049 \end{bmatrix} \begin{bmatrix} h_1 \\ h_2 \\ h_3 \end{bmatrix} + \begin{bmatrix} \frac{1}{706} \\ 0 \\ 0 \end{bmatrix} U$$

$$Y = \begin{bmatrix} 0 & 0 & 1 \end{bmatrix} \begin{bmatrix} h_1 \\ h_2 \\ h_3 \end{bmatrix}$$

$$G(s) = \frac{0.0162\mathrm{e}^{-3t}}{s^3 + 0.4022s^2 + 0.2068s + 0.0494} \tag{6}$$

From the first principle method, the model is obtained, and the developed model is in the third-order system with single input and single output system [12]. The higher order system is approximated as lower order integer order and non-integer order systems using a direct analysis method, and the transfer functions for the integer and fractional order models are tabulated as shown in the table.

4 Model Estimation of Three Tank Process with Multi-Input and Multi-Output (MIMO) System

In the three tank system, two inputs are from qin1 and qin2. qin1 is input in first tank, and the other input qin2 is in the third tank as the second input and the third tank's height as the output. For this case, only one input and one output are considered (Fig. 3).

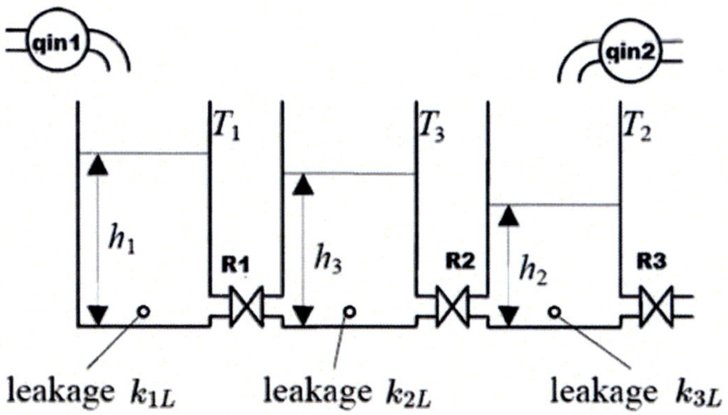

Fig. 3 Three tanks interacting with multi-input and multi-output (MIMO) system

$$\frac{dh_1}{dt} = \frac{h_1}{A_1R_1} + \frac{h_2}{A_1R_1} + \frac{1}{A_1}q_{in} \tag{7}$$

$$\frac{dh_2}{dt} = \frac{1}{A_2}\left(\frac{1}{R_1} - \frac{1}{R_2}\right)h_2 - \frac{1}{A_2R_1}h_1 + \frac{1}{A_2R_2}h_3 \tag{8}$$

$$\frac{dh_3}{dt} = \frac{1}{A_3R_2}h_2 - \frac{1}{A_3}\left(\frac{1}{R_2} - \frac{1}{R_3}\right)h_3 + \frac{1}{A_3}q_{in3} \tag{9}$$

where

h_1, h_2, and h_3 are the level of the Tank 1, Tank 2, and Tank 3, respectively.
$A_1 = A_2 = A_3 = A$ is a cross-sectional tank of the Tank 1, Tank 2, and Tank 3 which is equal to 706.2 cm^2
R_1, Restriction of the valve connecting Tank 1 and Tank 2 = 0.5
R_2, Restriction of the valve connecting Tank 2 and Tank 3 = 0.4
R_3, Restriction of the valve connecting Tank 2 and Tank 3 = 0.35

Hence, the state space model of the system is given below

$$\begin{bmatrix} \dot{h}_1 \\ \dot{h}_2 \\ \dot{h}_3 \end{bmatrix} = \begin{bmatrix} -\frac{1}{A_1R_1} & \frac{1}{A_1R_1} & 0 \\ -\frac{1}{A_2R_1} & \frac{1}{A_2}\left(\frac{1}{R_1} - \frac{1}{R_2}\right) & \frac{1}{A_3R_2} \\ 0 & -\frac{1}{A_3R_2} & -\frac{1}{A_3}\left(\frac{1}{R_2} - \frac{1}{R_3}\right) \end{bmatrix} \begin{bmatrix} h_1 \\ h_2 \\ h_3 \end{bmatrix} + \begin{bmatrix} \frac{1}{A_1} & 0 \\ 0 & 0 \\ 0 & \frac{1}{A_3} \end{bmatrix} U$$

$$\begin{bmatrix} \dot{h}_1 \\ \dot{h}_2 \\ \dot{h}_3 \end{bmatrix} - \begin{bmatrix} -0.2832 & 0.2832 & 0 \\ -0.2832 & -0.070 & 0.4048 \\ 0 & -0.4048 & -0.049 \end{bmatrix} \begin{bmatrix} h_1 \\ h_2 \\ h_3 \end{bmatrix} + \begin{bmatrix} \frac{1}{706.2} & 0 \\ 0 & 0 \\ 0 & \frac{1}{706.2} \end{bmatrix} U$$

$$G_{P2}(s) = \frac{0.0014s + 0.0011}{s^3 + 0.402s^2 + 0.1209s + 0.0435} \tag{10}$$

$$G_{P1}(s) = \frac{0.000399}{s^3 + 0.402s^2 + 0.1209s + 0.0435} \tag{11}$$

5 Model Optimization

The integer order model and non-order models are derived from the first transfer function obtained from the open loop test performed in the real-time method [13]. The models are obtained by using the optimization techniques, by maintaining the open loop transfer function as a reference (Fig. 4).

The optimization techniques used for getting the good models are

- Genetic algorithm

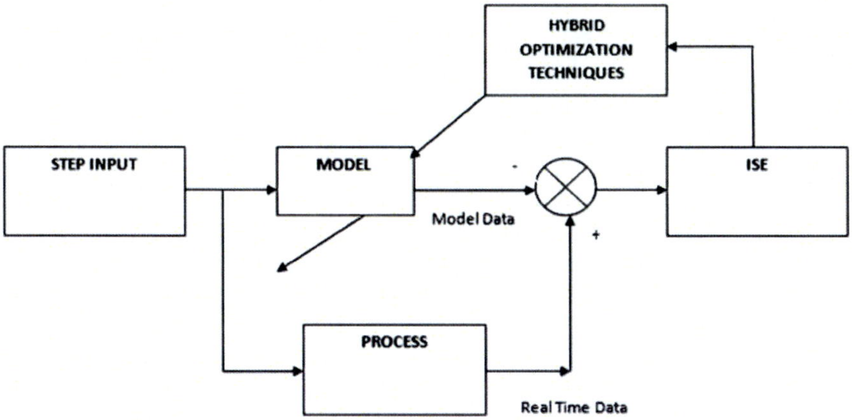

Fig. 4 Model identification using hybrid optimization techniques

Table 2 Transfer functions of three tanks interacting with single input and single output (SISO) system

Nature of the model	Transfer function	ISE
Integer order model	$G_1(s) = \frac{0.034e^{-3.767ts}}{(67.895\,s+1)}$	0.0095
Fractional order model using GA	$G_2(s) = \frac{0.034\,e^{-3.19ts}}{(72.55\,s^{1.015}+1)}$	0.0034

Table 3 Transfer functions of three tanks interacting with multi-input and multi-output (MIMO) system

Nature of the model	Transfer function	ISE
Integer order model	$G_1(s) = \frac{0.17e^{-16.28ts}}{(328.29\,s+1)}$	$G_2(s) = \frac{0.17e^{-12.67ts}}{(353\,s^{1.01}+1)}$
Fractional order model using GA	$G_1(s) = \frac{0.192\,e^{-0.71ts}}{(334.45\,s+1)}$	$G_2(s) = \frac{0.19e^{\ 0.39ts}}{(368\,s^{1.016}+1)}$

- Fminsearch
- Pattern search
- Hybrid optimization techniques (Tables 2 and 3).

6 Controller Optimization

The PID controller gives the output as follows

$$u(t) = k_p e(t) + k_p k_i \int_0^t e(t) + k_p k_d \frac{de(t)}{dt}$$

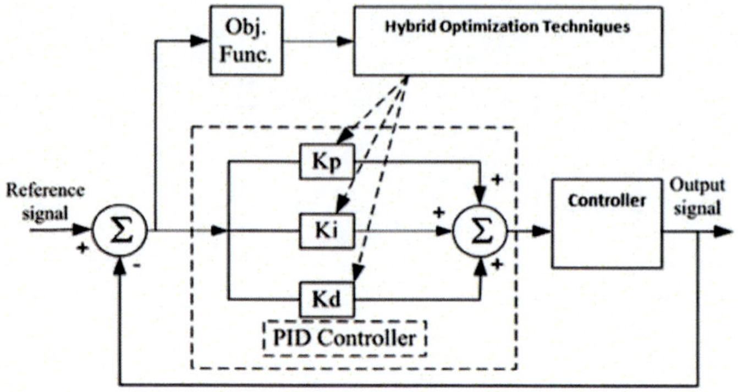

Fig. 5 Block diagram of controller optimization

where

k_p proportional gain,
k_i integral gain
k_d derivative gain.

The expression for non-integer PID controller is given below

$$u(t) = k_p e(t) + k_p k_i \frac{\mathrm{d}^{-\lambda}}{\mathrm{d}t} e(t) + k_p k_d \frac{\mathrm{d}^{\mu} e(t)}{\mathrm{d}t}$$

where

k_p proportional gain,
τ_i integral time,
τ_d derivative time,
λ integral order,
μ derivative order.

Before the working of PID controller, it must be tuned to achieve better performance [14]. Distinctive sorts of tuning techniques like auto tuning and Zeigler Nichols tuning method are created to tune the PID controllers and require much consideration from the administrator to choose best values of proportional, integral, and derivative gains [15] (Fig. 5).

6.1 Fmin-Genetic Algorithm (Fmin-GA)

The initial reference for tuning the PID controller is collected using MATLAB. As the initial value of Fmin search, unregulated optimization offers the optimum tuning value on the basis of the value that is selected by the lower and upper range for GA.

Only three parameters (k_p, τ_i, τ_d) for the integer PID are considered [16] and five parameters (k_p, τ_i, τ_d, λ, μ) for the non-integer PID (fractional PID). The lower and the upper limits are set; these limits are called limits [17]. The limitations can help to minimize the population size, so that the search space and the number of iterations are reduced. The limitations eradicate the population of the PID parameter when the process variable reaches the set point and the simulated response.

This GA algorithm can solve both constrained and unconstructed optimization problems in various systems such as linear, nonlinear, time-invariant, differential, and non-differential systems [17].

The genetic algorithm uses three key types of rules to establish the next generation of the present population at any given stage: the selection rules pick individuals called parents who contribute to the next generation of people.

- Selection rules pick people, called parents who contribute to the next-generation population.
- Crossover rules combine two parents to create the next generation of children.
- Random modifications to individual parents in order to form children adhere to mutation laws.

GA is an intrinsic heuristic quest set of rules whose fundamental lies in the primary concept of the herbal evolution of living beings that favors the reality of more powerful individuals, which enables them to battle against all odds and to extinct them more and more in a competitive world. While certain events occur where disadvantaged individuals aspire and may excel, these instances are controlled to meet standards [18].

The flow diagram shows GA's method to optimize PID controls, the initial population is first chosen on a random basis based on the limit, and fitness method assessment is carried out on the basis of the integral square errors (ISE) and the integral absolute error (IAE) feature [19] (Fig. 6).

In the controller optimization problem, the following syntax is employed to reduce the integral time error.

$$x = ga(\text{fitnessfcn}, \text{nvars}, A, b, \text{Aeq}, \text{beq}, \text{LB}, \text{UB})$$

Defines a set of lower and upper bounds on the design variables, x, so that a solution is found in the range $\text{LB} \leq x \leq \text{UB}$. (Set Aeq = [] & Beq = []) if no linear equalities exist.

6.2 Model Predictive Control (MPC)

This model predictive control (MPC), the most widely used in the oil and chemical industries, is a groundbreaking process. This is used to control processes and to eliminate weaknesses and issues from multivariable devices. It can simultaneously change the different parameters inside the processing station [20]. The advance is

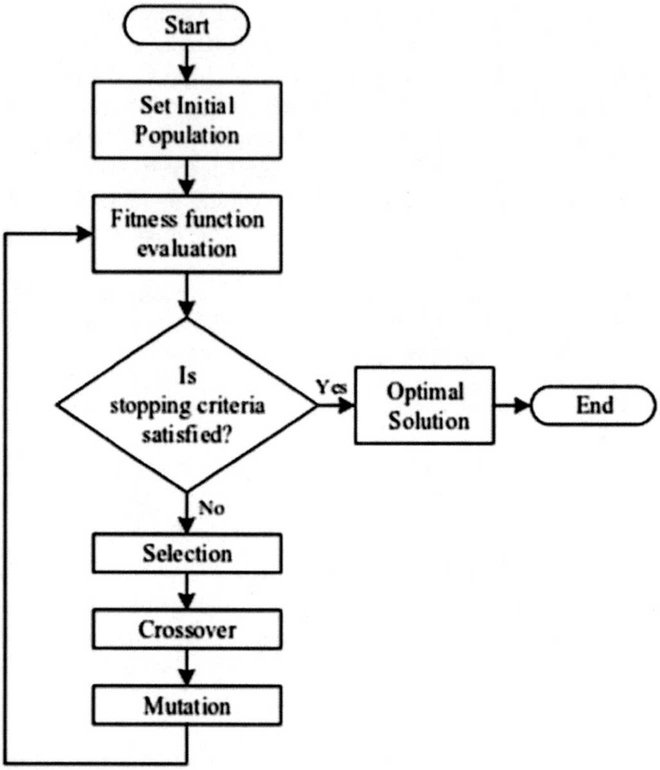

Fig. 6 Flowchart for CGA based simulation

that MPC had confidential demands at present. MPC is used to monitor multiple inputs and multiple outputs to meet the limitations. The model and current effects can be calculated as well as potential output values (Fig. 7).

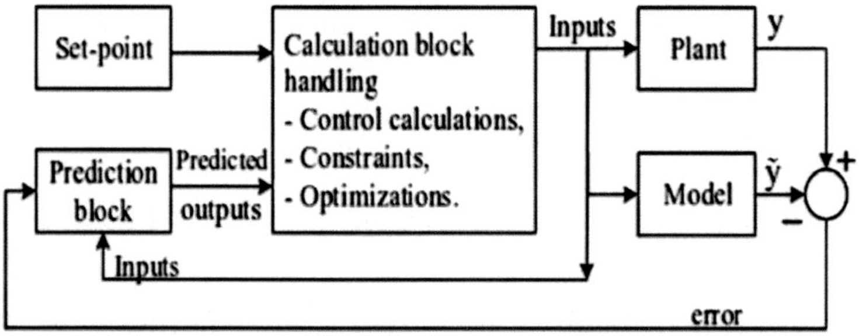

Fig. 7 MPC structure

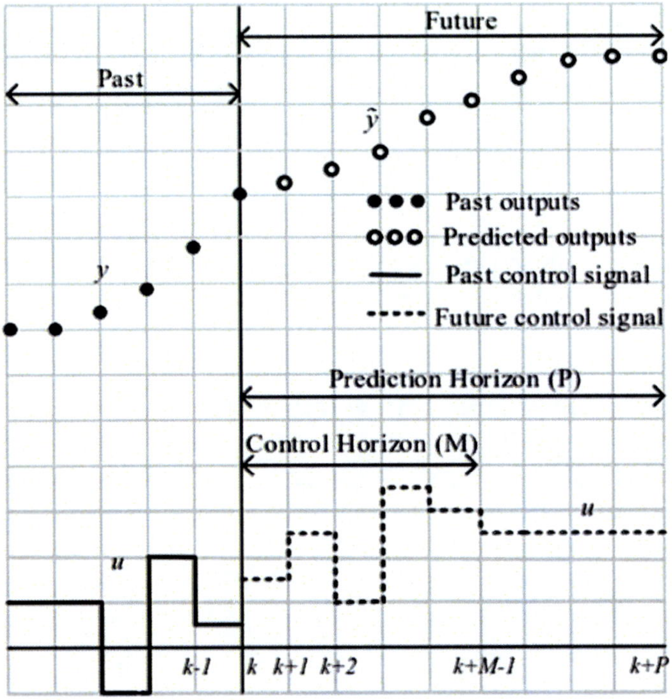

Fig. 8 Horizon-based approach for MPC

Within this MPC, you can compare model output and process output, generate the error, and enter the prediction chain. The mistake exists. The future output value is calculated, depending on the error and input value of the crop [21]. The calculation block is calculated using the fixed point and expected value; on the basis of the information supplied at each sample time on the calculation block, it generates the predefined objective functions (Fig. 8).

Optimization provides the best value in a certain success metric and is referred to as an objective function. It may be a square minimum standard or a square objective function [22]. It is simply the sum of normal error squares (variations between the set point and the model's predicted outputs) and the control movements (changes in control action step by step).

The general equation is

$$\phi = \sum_{i=1}^{P} \left| r_{k+i} - \hat{y}_{k+i} \right| + w \sum_{i=0}^{M-1} \left| \Delta u_{k+i} \right|$$

Optimization of the control signal and the modeling equations is feasible [23]. The least quadratic formulations are primarily employed in the predictive control

Table 4 MPC setting for three tank interacting system

S. No.	Parameters	Values
1	Sampling time (Ts)	0.25
2	Prediction horizon (P)	6000
3	Control horizon (M)	2
4	Objective function	Quadratic

model. It offers theoretical explanations for unconstrained problems and eliminates large errors as opposed to minor ones.

7 MPC for Three Tank Interacting System

Model predictive control (MPC) is designed for each case based on the integer and non-integer order model of the three tank interacting system as defined in Table 1. Frequent simulations were carried out using MATLAB SIMULINK in order to test the control competence of the model MPC controllers [24] (Table 4).

8 Result and Discussion

The mathematical model for the three tank interacting system is obtained as transfer function and state space model which is developed by means of first principle method, and also, the integer and non-integer order models are obtained by fine-tuning the offered process parameters by means of model optimization, and the transfer functions are tabulated in Tables 1 and 2.

The classic standard PID are designed based on the developed model. The controller parameters of integer and fractional order PID such as proportional gain (k_p), integral constant (τ_i), derivative constant (τ_d), and proportional gain (k_p), integral constant (τ_i), derivative constant (τ_d), integral derivative order (λ), and derivative order (μ), respectively, are optimized based on the hybrid optimization techniques such as Fmin-GA for the integer and non-integer order transfer function models, and the closed loop responses and controller dynamics are plotted and shown in Figs. 4.2 and 4.3 for integer order model and in Figs. 4.4 and 4.5 for non-integer model (Figs. 9, 10, 11, 12, 13, 14, 15, 16, 17, 18, 19; Tables 5, 6, 7).

9 Conclusion

The level of the three tank interacting system is a very important parameter, and it is also difficult to be control; hence, the out flow is depending on the level. And

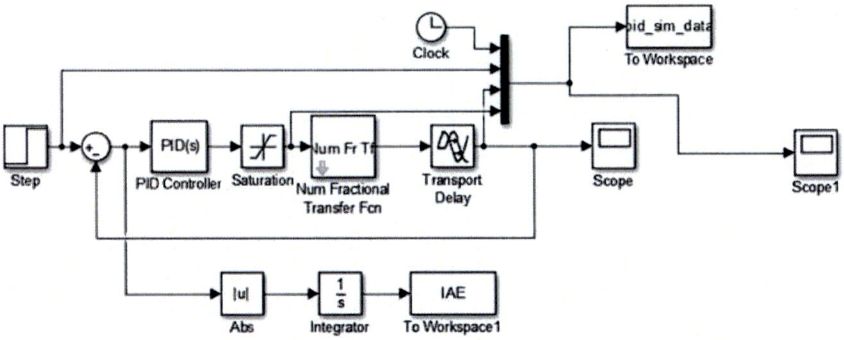

Fig. 9 Closed loop control system diagram

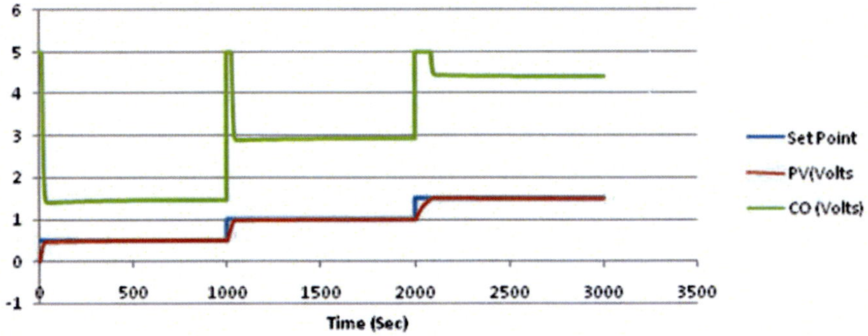

Fig. 10 Response of three tank process Fmin-GA method for IO-PID controller for set point tracking

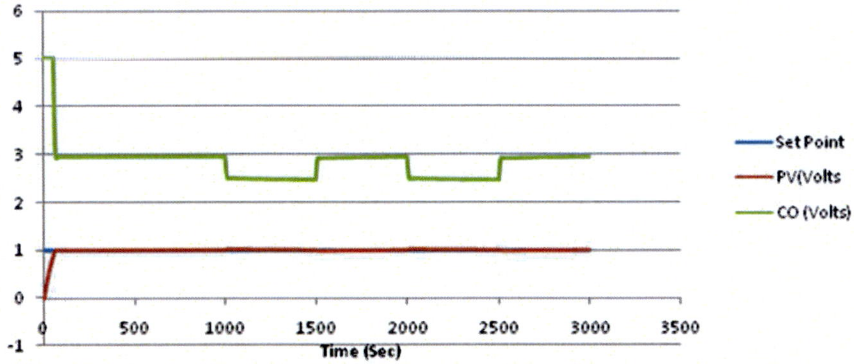

Fig. 11 Response of three tank process Fmin-GA method for IO-PID controller for load tracking

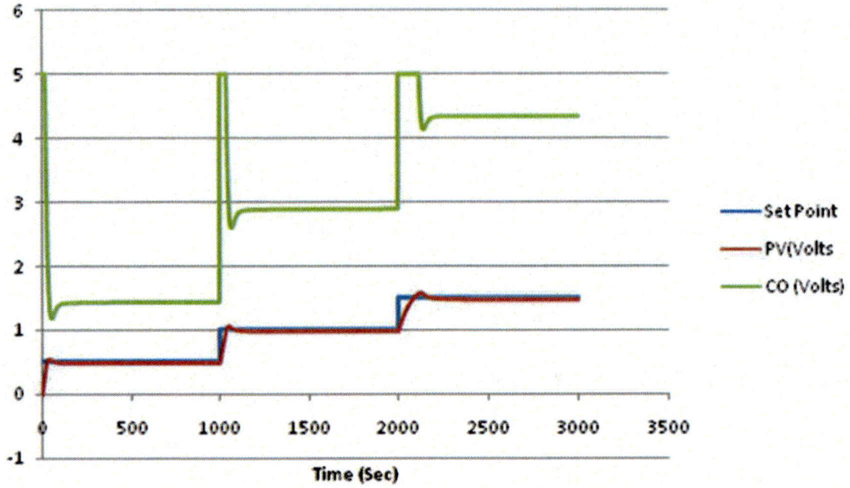

Fig. 12 Response of three tank process Fmin-GA method for FO-PID controller for set point tracking

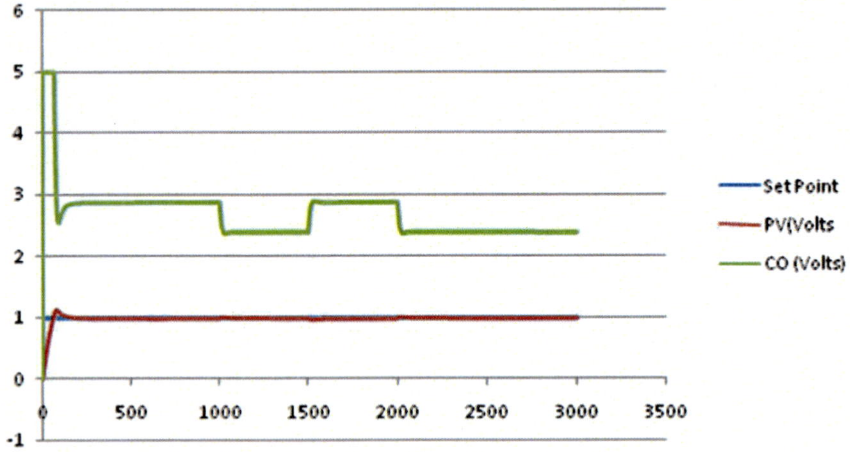

Fig. 13 Response of three tank process Fmin-GA method for FO-PID controller for load tracking

it is also expenditure process because for pump a certain liquid in industries need more energy. By implementing effective control action, the energy consumption could be optimized in the three tank interacting system. The classic integer order and non-integer order PID controller for integer order and non-integer order models are optimized based on the hybrid optimization techniques Fmin–genetic algorithm (Fmin-GA), and model predictive controllers (MPC) are implemented, respectively.

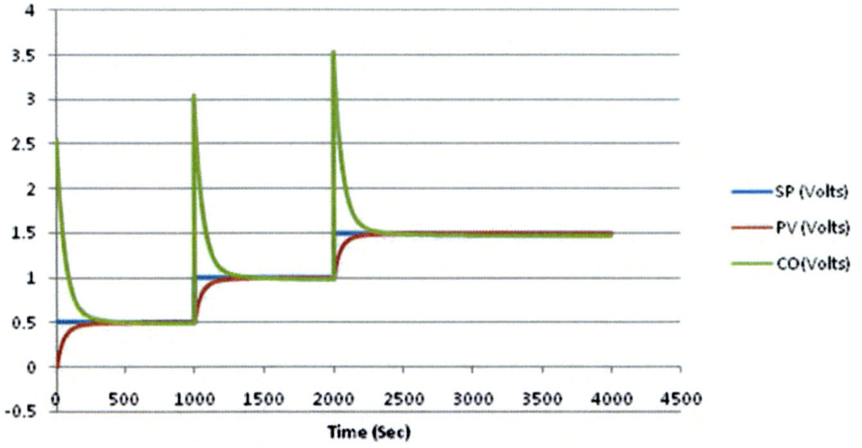

Fig. 14 Response of three tank (MIMO) process Fmin-GA method for FO-PID controller for set point change

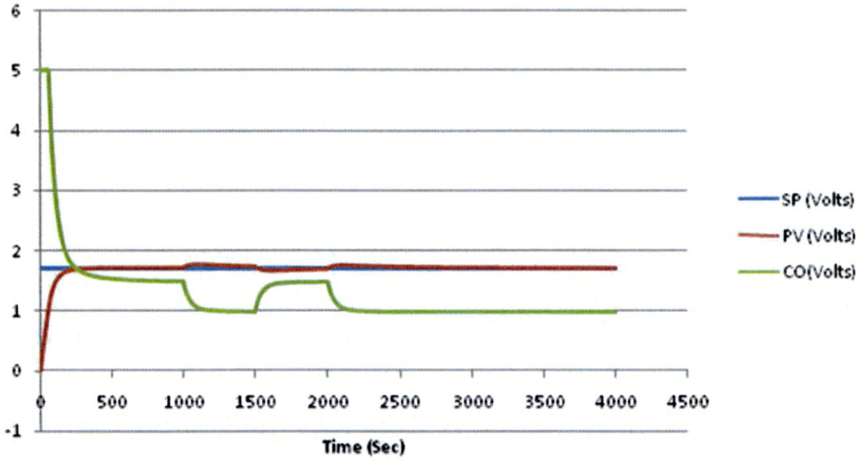

Fig. 15 Response of three tank (MIMO) process Fmin-GA method for FO-PID controller for load change

10 Future Scope

The level control in three tank system is a challenging task because of high nonlinearity due to the interaction between the tanks. In this work, one pair of hybrid optimization technique is used to optimize integer order proportional integral derivative (IO-PID), fractional order proportional integral derivative (PID), and model predictive controller (MPC). In future, it may extent for other combinations of hybrid

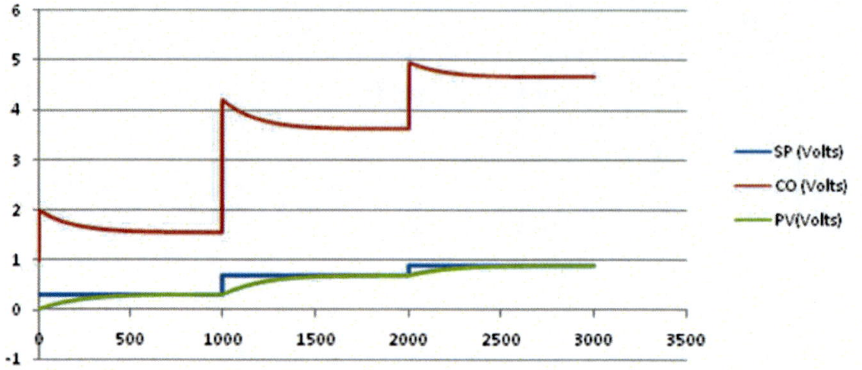

Fig. 16 Response of closed loop MPC for three tank interacting process (MIMO) for set point tracking

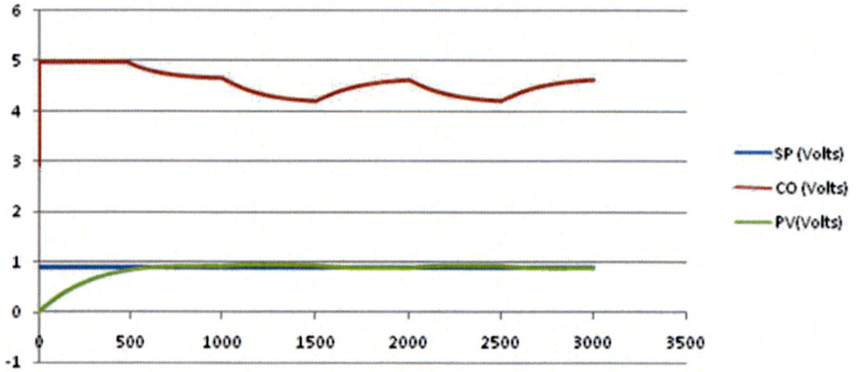

Fig. 17 Response of closed loop MPC for three tank interacting process (MIMO) for load tracking

optimization techniques for various nonlinear processes such as interacting conical tank systems, spherical tank systems, and continues stirred tank reactor (CSTR) process

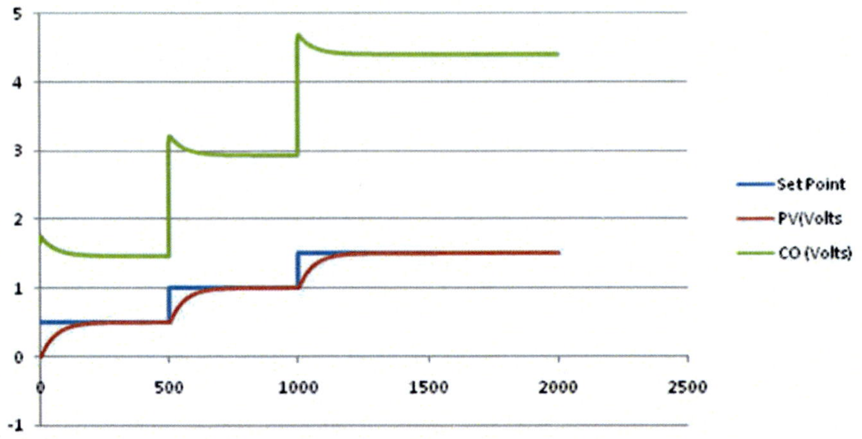

Fig. 18 Response of closed loop MPC for three tank interacting process (SISO) for set point tracking

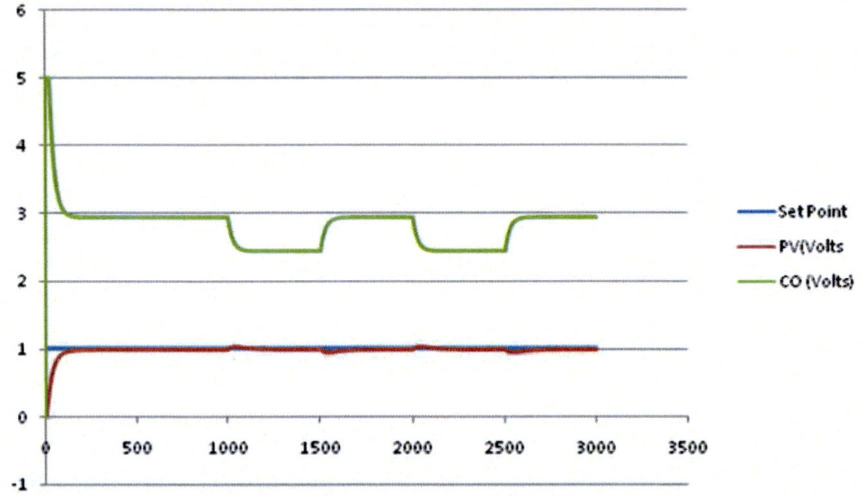

Fig. 19 Response of closed loop FO-MPC for three tank interacting process (SISO) for set point tracking

Table 5 Time domain specifications for three tank interacting system (SISO)

Optimization techniques	Maximum peak overshoot Mp%	Rise time 's'	Settling time 's'	Offset
Integer order transfer function				
F_{min} – Genetic algorithm (F_{min} – GA)	0%	15	15	0.000016
Non-integer order transfer function				
F_{min} – Genetic algorithm (F_{min} – GA)	0%	12	8	0.000012

Table 6 Time domain specifications for three tank interacting system (MIMO)

Optimization techniques	Maximum peak overshoot Mp%	Rise time 'Sec'	Settling time 'Sec'	Offset
Integer order transfer function				
F_{min} – Genetic algorithm (F_{min} – GA)	0%	15	15	0.000016
Non-integer order transfer function				
F_{min} – Genetic algorithm (F_{min} – GA)	0%	12	8	0.000012

Table 7 Time domain specifications of MPC-three tank interacting process (SISO)

Process model transfer function	Maximum peak overshoot Mp%	Rise time 'Sec'	Settling time 'Sec'	Offset
Integer order	0	67	224	0.0065
Non-integer order	0	42	288	0.00023

References

1. Muniraj, L., Masilamani, A.: Temperature control water bath system using PID controller. Int. J. Appl. Eng. Res. (2015)
2. Yadav, E., Indiran, T.: Servo mechanism technique based anti-reset windup PI controller for pressure process station. Ind. J. Sci. Technol. **9**(11) (2016). https://doi.org/10.17485/ijst/2016/v9i11/89298
3. Bingi, K., Ibrahim, R., Karsiti, M.N., Hassan, S.M., Harindran, V.R.: Real-time control of pressure plant using 2DOF fractional-order PID controller. Arab. J. Sci. Eng. Received: 22 January 2018. Accepted: 17 May 2018. King Fahd University of Petroleum & Minerals (2018)
4. Heliot, F.: Low-complexity energy-efficient joint resource allocation for two-hop MIMO-AF systems. IEEE Trans. Wirel. Commun. **13**(6), 3088–3099 (2014)
5. Tan, R.W.C., Xu, J., Wang, Z., Jin J., Man, Y.: Pressure control for a hydraulic cylinder based on a self-tuning PID controller optimized by a hybrid optimization algorithm. Algorithms **10**, 19 (2017)
6. Correia, L.M., Zeller, D., Blume, O., Ferling, D., Jading, Y., Godor, I., Auer, G., Perre, L.V.D.: Challenges and enabling technologies for energy-aware mobile radio networks. IEEE Commun. Mag. **48**(11), 66–72 (2010)

7. Saxena, S., Hote, Y.V.: A simulation study on optimal IMC based PI/PID controller for mean arterial blood pressure. Biomed. Eng. Lett. The Korean Society of Medical & Biological Engineering and Springer 7 December (2012)
8. Srivastava, N., Tanti, D.K., Ahmad, Md.A.: Matlab simulation of temperature control of heat exchanger using different controllers. Science Publishing Group (2014)
9. Ferrari, A., Mittica, A., Pizzo, P., Jin, Z.: PID controller modeling and optimization in CR systems with standard and reduced accumulators. Int. J. Autom. Technol. **19**(5), 771–781 (2018)
10. Astrom, K. J., Hagglund, T.: Revisiting the Ziegler Nichols step response method for PID control. J. Process Control **14**, 635–650 (2004)
11. Pan, I., Das, S., Gupta, A.: Tuning of an optimal fuzzy PID controller with stochastic algorithms for networked control systems with random time delay. ISA Trans. **50**(1), 28–36 (2011)
12. Shah, P., Agashe, S.: Experimental analysis of fractional PID controller parameters on time domain specifications. Progr. Fract. Differ. Appl. Int. J. **2**, 141–154 (2017)
13. Doyle III, F.J., Pearson, R.K., Ogunnaike, B.A.: Identification and Control Using Volterra Models. Springer, Berlin (2001). ISBN 978- 1852331498
14. Chen, D., Seborg, D.E.: Relative gain array analysis for uncertain process models. AIChE J. **48** (2002)
15. Pilatasig, M., Chacon, G., Silva, F.: Airflow station controlled by PID and fuzzy controllers using a low cost card for didactic uses in controllers' evaluation. In: The 9th International Multi-conference on Complexity, Informatics and Cybernetics IMCIC (2018)
16. Das, S., Mullick, S.S., Suganthan, P.N.: Recent advances in deferential evolution {an updated survey. Swarm Evolut. Comput. **27**, 1–30 (2016)
17. Seborg, D.E., Edgar, T.E., Mellichamp, D.A.: Process Dynamics and Control, 2nd edn. Wiley, Hoboken (2004)
18. Prakash, J., Srinivasan, K.: Design of nonlinear PID controller and nonlinear model predictive controller for a continuous stirred tank reactor. ISA Trans. **48**, 273–282 (2009)
19. Kurilla, J., Hubinsky, P.: Model predictive control of room temperature with disturbance compensation. J. Electr. Eng. **68** (2017)
20. Cuadros, M.A.d.S.L., Munaro, C.J., Munareto, S.: Improved stiction compensation in pneumatic control valves. Comput. Chem. Eng. **38**, 106114 (2012)
21. Gaing, Z.L.: A particle swarm optimization approach for optimum design of PID controller in AVR system. Energy Convers. IEEE Transact. **19**(2), 384–391 (2014)
22. Tan, W., Liu, K., Tam, P.K.S.: PID tuning based on loop-shaping H-Infinity control. IEE Proc. Control Theor. Appl. **145**(6), 485–490 (1998b)
23. Begum, Y., Venkata Marutheswar, G.V., Ayyappa Swamy, K.: Tuning of PID Controller for superheated steam temperature system using modified Zeigler-Nichols tuning algorithm. Int. J. Eng. Adv. Technol. (IJEAT) **2**(5) (2013)
24. Ogunnaike, B.A., Ray, W.H.: Process Dynamics, Modelling, and Control (Topics in Chemical Engineering). Oxford University Press, (1994)
25. Russo, L.P., Bequette, B.W.: Impact of process design on the multiplicity behaviour of a jacketed exothermic CSTR. AIChE J. **41**, 135–147 (1995)

Building a Land Use and Land Cover (LULC) Classifier Using Decadal Maps

D. Bharathi, R. Karthi, and P. Geetha

Abstract Global satellite program for remote sensing and earth observation has yielded huge volume of images with rich information. Geoportals such as Bhuvan and USGS host large number of satellite images and tools for analysis. The objective of this paper is to investigate the potential of using freely available satellite images to build a Land Use Land Cover classifier model (LULC) using machine learning approaches. The historically available LULC map is proposed to be used for identifying ground truth labels during classifier construction. Multispectral Landsat-8 images are used as input for classification by decision tree (DT), random forest (RF) and support vector machines (SVM). The data was mapped to six classes according to the International Geosphere-Biosphere Program (IGBP) classification scheme. The performance metrics used for evaluating the classifiers are accuracy, kappa coefficient, user, and producer accuracies. The study infers that random forest is able to classify LULC data with higher accuracy. The study provides a mean of building LULC maps from available data for multiple terrains. These classifiers can be used as an automated tool for generation of LULC maps.

Keywords Landsat · Classifiers · Land use land cover (LULC) · Decadal maps · Multiclass · Remote sensing

---

D. Bharathi · R. Karthi (✉)
Department of Computer Science and Engineering, Amrita School of Engineering, Amrita Vishwa Vidyapeetham, Coimbatore, India
e-mail: r_karthi@cb.amrita.edu

D. Bharathi
e-mail: d_bharathi@cb.amrita.edu

P. Geetha
Center for Computational Engineering and Networking (CEN), Amrita School of Engineering, Amrita Vishwa Vidyapeetham, Coimbatore, India
e-mail: p_geetha@cb.amrita.edu

© The Editor(s) (if applicable) and The Author(s), under exclusive license to Springer Nature Singapore Pte Ltd. 2021
A. P. Pandian et al. (eds.), *Proceedings of International Conference on Intelligent Computing, Information and Control Systems*, Advances in Intelligent Systems and Computing 1272, https://doi.org/10.1007/978-981-15-8443-5_12

1 Introduction

Remote sensing is a technique of deriving information about the earth surface using electromagnetic radiation based on the reflected and emission properties of spectrum from the earth surface. The information embedded in remote sensing images is used in many applications like agriculture, forestry, urban monitoring, robot navigation, etc. The advantage of using satellite images is that it provides repetitive images of the same area which can be used for analysis. Different satellites such as Landsat, Sentinel, MODIS, and ASTER are used for capturing data at different spatial and temporal resolution. The huge volume of remote-sensed images is available for researchers to focus on better understanding of images and extract patterns for various applications. Petabytes of data have become freely available from U.S. Geological Survey (USGS), NOAA, European Space Agency, and Bhuvan platform [1]

Land cover describes the physical surface cover of the land, and land use describes the purpose of use. Examples of land cover classes include water, snow, grassland, deciduous forest, and bare soil. Land use examples include: wildlife habitat, agricultural land, urban, recreation area, etc. The major challenge in building a LULC classifier is in identifying similar classes as the spectral variations are less for these classes as the number of classes increases. LULC mapping algorithms extract and use pixel, sub-pixel, and object-based information as features for classification. Pixel-level classification is assigning class labels to each pixel in an image. Sub-pixel method assumes that multiple land cover types are mapped in one pixel and employ fuzzy, spectral mixture analysis techniques for classification. Object-based image analysis (OBIA) uses objects as basic units for analysis where geographical objects need to be segmented from the image [2]. Landsat satellites continuously monitor the earth and are providing images of the earth surface for four decades. Land cover classification was developed using Landsat multispectral scanner (MSS), thematic mappers (TM), enhanced thematic mappers (ETM+), and observation land images (OLI) images. For Landsat image classification using pixel-based approaches, maximum likelihood classifier is commonly used by many researchers and GIS products for various applications. Building a pixel-based classification model is investigated in this work for a multiclass problem.

The paper focuses on building a LULC classifier using machine learning techniques and finds a suitable model for the multiclass problem. The efficiency of the classifier model is evaluated using standard measures and by visualization. The paper is organized as follows: Section 2 discusses the related work, and Sect. 3 discusses the problem of LULC and approaches to build a classifier model using machine learning methods. Section 4 presents the results and discusses the effects on classifier by varying parameters of classifier. Section 5 summarizes the insights from the analysis and conclusion from the study.

2 Literature Survey

The first Landsat mission initiated by NASA was Landsat 1 launched in 1972. It has four bands that occupied visible spectrum and near-infrared wavelength. The Landsat 4 and 5 with thematic mapper had seven bands with two shortwave infrared channels. The thematic mapper + in Landsat 7 onward had additional 15-m spatial resolution panchromatic channel. The operational land images (OLI) in Landsat 8 have an enhanced blue and cirrus channel [3]. Landsat classification methods are broadly classified into pixel-based; subpixel-based, and object-based approaches. The object-based image analysis has been reported superior in most studies. A comparative analysis of each method with the advantage and disadvantages was discussed in the study [2]. Machine learning techniques such as SVM, decision tree, random forest, boosted DT, ANN, and KNN for classification on two bench mark datasets are compared using choice of algorithm, number of training data, parameter selection, computational cost, and feature reduction [4]. Landscape types were identified automatically using ANN, SVM, and logistic regression and validated using georeferenced map were class labels are assigned to each pixel by a semi-automated software. Logistic regression with nonlinear kernel has accuracy 92.82%, while SVM and ANN have 92.72 and 92.20% accuracy, respectively [5]. ANN with sigmoid activation function and nonlinear SVM and logistic regression with both linear and nonlinear kernels were used to classify data from six-dimensional spaces out of which SVM outperformed the other classifiers in terms of accuracy [6]. The spectral indices such as normalized difference water index (NDWI), enhanced vegetation index (EVI), index-based build up (IBI), enhanced build up and bareness index (EBBI) were used for automatic land covers classification from Landsat 8 images using QGIS and K-means, random forest, and maximum likelihood [6]. In addition to the spectral values, the texture indices from the gray level co-occurrence matrix were identified to improve the performance of the classifiers such as KNN, SVM, and hybrid classifiers for smaller part of Bolivia in particular homogeneity, and entropy indices were used to improve the accuracy of the classification from Landsat 7 images [7].

Tirupati, a region in India, was classified into four classes using NDVI that uses visible and near-infrared band and classifies the data into four main classes, namely water bodies, dense vegetation, vegetation, and bare soil. The images of Tirupati taken at three different dates are given as input to the classifier. The study indicates that there is a change in NDVI value of vegetation and its dynamics, even though there is not much change for dense vegetation [8]. Changes in LULC pattern of Sikkim were analyzed using Landsat 5 and Sentinel 2A data. Maximum likelihood classifier was used to prepare the LULC map. The major classes identified are water, built-up area, dense forest, open forest, and barren land. Three Landsat 5 and one sentinel 2A images were used for change detection [9]. The vegetation types and land use of Rajasthan were mapped using IRS P6 LISS III satellite images. The entire region was classified into 26 classes by visual interpretation of vegetation types. The

IRS LISS IV multispectral data was used for land cover mapping where the two classifiers maximum likelihood and SVM were compared for analyzed [10].

Deep learning methods are recently used in remote sensing applications. Convolutional neural networks have better performance than other conventional machine learning implementations. CNN classifiers have outperformed SVM for existing benchmark datasets, with accuracies surpassing 99% [11]. Feature extraction methods and image resolution play a major role in land cover monitoring, and change detection was identified by experimentation with various soft computing techniques [12, 13]. Literature review reveals that in several works, the number of classes considered is 3–4 classes which are easily identifiable such as water, grass land, barren land, and built-up area. Many of the works focus on change detection among these four classes for a specific region of Interest. Work related to LULC focus on techniques for validation using visualization and field surveying techniques which is cumbersome. This work focuses on building an automated LULC classifier for multiple land cover classes using machine learning techniques.

3 Methodology

3.1 Study Area

New Delhi is the capital of India and has been experiencing one of the fastest urban expansions in the world. Vast areas of croplands and grasslands are being turned into urban landscape with buildings, streets, parking lots, attracting an unprecedented amount of new residents. The latitude and longitude of Delhi are 3177013.60 N, 705388.86 E, and the image bounded by an extent of 7852515° N, 7591485° S, 587415° E, 327885° W is selected for the study.

A true-color image combines measurements of red, green, and blue lights which can be visualized by the humans. A false-color image is used for identifying features that are not present in the visual spectrum. Vegetation is more prominently visible in false-color composite image as they reflect the near-infrared more than red, green, or blue wavelength of the spectrum. Figure 1 shows the downloaded image in its true color composite and false-color composite form. The number of pixels in the image is 8651 * 8701. The various land cover types in the satellite image include water bodies, cropland, bare land, grassland, waste land, and urban cover which provided us with a representative study area to test various land covers.

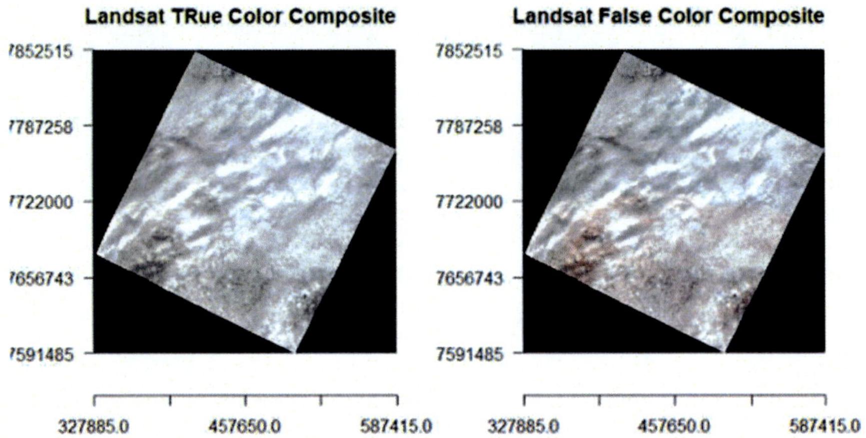

Fig. 1 Downloaded image from USGS in true and false-color composite

3.2 Data Set

Landsat image

Clear (<10% cloud cover) Landsat 8 OLI spectral image (path/row: 147/40) obtained from the United States Geological Survey (USGS) Web site (https://glovis.usgs.gov/) is used in this study. Table 1 summarizes the characteristics of the of Landsat 8 sensors.

Reference data

Reference data was generated from LULC decadal India map. This data set provides LULC classification products at 100-m resolution for India for 3 years (1985, 1995,

Table 1 Landsat 8 operational land imager (OLI) sensor

Spectral band	Wavelength (μm)	Resolution (m)
Coastal/aerosol	0.433–0.453	30
Blue	0.450–0.515	30
Green	0.525–0.600	30
Red	0.630–0.680	30
Near-infrared	0.845–0.885	30
Short infrared	1.560–1.660	30
Short infrared	2.100–2.300	30
Panchromatic	0.500–0.680	15
Cirrus	1.360–1.390	30
Long wavelength infrared	10.3–11.3	100
Long wavelength infrared	11.5–12.5	100

and 2005). The LULC maps were generated using classification techniques, visual interpretation, and validation using field study with ground truth data to produce a 2005 national LULC map. LULC 2005 map of India from Oak Ridge National Laboratory Distributed Active Archive Center has different classes as per the IGBP Classifications scheme. The 19 classes specified are mainly based on forest types, barren and waste lands, built-up areas, crop and plantation areas, water and aquatic bodies, snow and ice areas.

3.3 Algorithm for LULC Classification

The proposed algorithm for LULC classification is given below in Fig. 2. Landsat 8 image is downloaded from USGS Earth Explorer for the selected study area. From the downloaded image, a subset of the image is obtained by clipping the image. The subset is obtained for all the bands of the image where each band represents the spectral value of that band. Each pixel in the subset has eleven spectral values from all the different bands of the image and seven bands are used to construct the feature vector. The LULC decadal Indian image is used as a reference image for obtaining

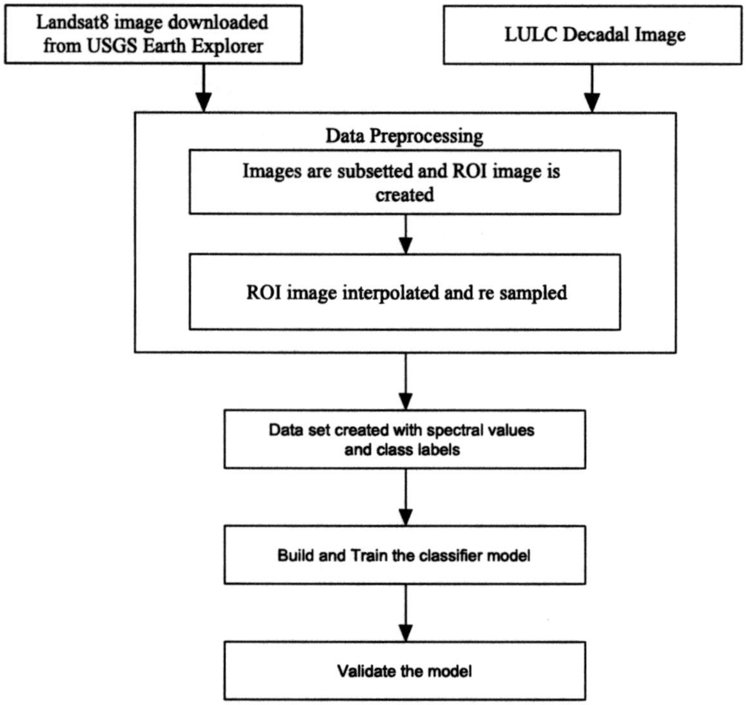

Fig. 2 Flowchart of the algorithm proposed for LULC classification

the class labels of each pixel. The LULC image is also clipped with the same extent to obtain the class value of each pixel in the image. The images are preprocessed using interpolation techniques so that they have the same pixel resolution.

There are 19 classes identified from the LULC decadal Indian image out of which six classes are selected for this study. The classes which are similar are merged into a single class to reduce the number of classes. The six classes identified for the study were cropland, built-up land, shrub land, fallow land, water, and grassland. As the satellite image and LULC image are georeferenced, each pixel from satellite image is matched with its class label from LULC image. The feature vector is appended with the corresponding class labels, and the training data set is generated for building the classification models. The training data is used to build the classifier model, and the model is validated using testing data set. The performance of the classifiers is validated using standard metrics specified for remote sensing.

The algorithmic steps involved for building a classifier are outlined below:

Step 1: Data preprocessing: Preprocess of LANDSAT 8 image

The downloaded Landsat 8 image is preprocessed using QGIS. The downloaded Landsat 8 image has eleven bands from which seven bands: Band 1 to band 7 are used for land classification. The image is clipped using an extent from which all the seven bands are clipped. The selected seven bands have pixels with the resolution of 30 m. The subset ROI images created are re-sampled using bilinear interpolation and georeferenced to remove noise in the image. Spatial interpolation is the process of using points with known values to estimate values at other unknown points. It is used as preprocessing step in LULC application. Using the four nearest neighboring cells, bilinear interpolation assigns the output cell value by taking the weighted average of nearest neighbors.

Speckle noise is commonly found in radar images and satellite images. The spatial filters like arithmetic mean, geometric mean, harmonic mean, median, max, and min can be used to filter the noise and make image enhancement at high quality [14]. Remote-sensed images are already prepossessed for noise using geometric and radiometric correction by the sensors. There may be noise during clipping images and during overlaying, so the boundary pixels are eliminated to remove the noise. In our model, pixels with class value 127 do not correspond to any class, so the pixels with spectral values 127 are eliminated. Figure 3 shows the clipped ROI image in the true- and false-color composite form.

Step 2: Data preprocessing: Preprocess of LULC image

The LULC decadal image is also clipped to the same extent defined in step 1 to obtain class information for every pixel of the subset ROI clipped image. The LULC has pixel resolution of 100 m which is different from Landsat image resolution. To map pixels from LULC to Landsat, the image is resampled to the resolution 30 m using nearest neighbor technique. Nearest neighbor resampling can produce adjacent pixels with identical values. After the process of interpolation, the processed LULC image is georeferenced with Landsat image and stored.

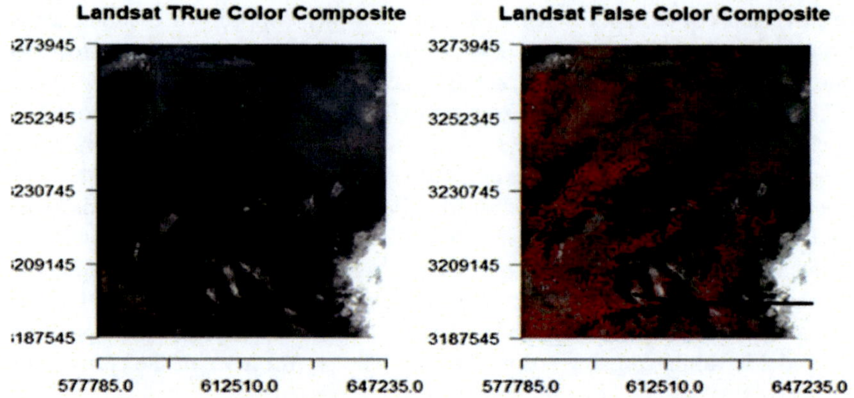

Fig. 3 ROI image in true and false-color composite form

Step 3: Defining the classes

The study area has classes which were mapped to 19 classes according to IGBP classification scheme. The downloaded image has 7405398 pixels from 10 classes. Few pixels were assigned to class 127, which is considered as noise. Table 2 shows the proportion of pixel to each of the classes.

For the study, similar classes are merged and are mapped to six classes as listed in Table 3. The noise data with class label 127 is assigned to the maximum probable class crop land to remove noise.

The LULC decadal image is clipped and resampled, and the ROI image specifying the class labels after merging is shown in Fig. 4. The data set generated is unbalanced as there is no uniform distribution of pixels to each of the classes. Table 4 infers that a vast majority of the pixels are in crop land class and a minority of pixels belongs

Table 2 Classes with pixel count for the subset image

Class No.	Class name	Pixels
1	Deciduous forest	7636
2	Crop land	7,042,349
3	Built-up area	156,153
5	Shrub land	17,699
6	Barren land	1905
7	Fallow land	79,379
8	Waste land	10,454
9	Water bodies	34,752
10	Plantations	2748
14	Grassland	52,244
127	None	79

Table 3 List of classes after merging similar classes

Class No.	Class name	Merged classes	No. of pixels
2	Crop land	{127}	7042428
3	Built-up land		156153
5	Shrub land	{1,10}	28083
7	Fallow land	{6,8}	91738
9	Water bodies		34752
14	Grassland		52244

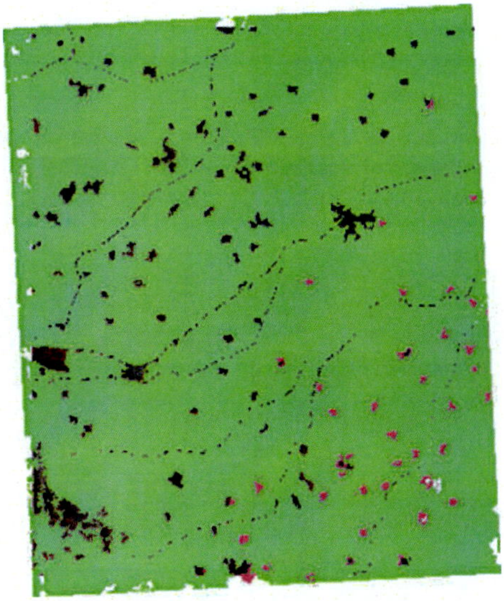

Fig. 4 LULC subset image with class labels

Table 4 Performance metrics for classifiers

Measure	Formula
Accuracy	$\frac{TP+TN}{P+N}$
User accuracy	$\frac{TP}{TP+FP}$
Producer accuracy	$\frac{TP}{TP+FN}$
Kappa coefficient	$\frac{P_0-P_e}{1-P_e}$

to other classes. Balancing of classes is to be done before providing data as input to the classifier model. A balance in the number of instance for all class is achieved by over or undersampling strategy. In this study, undersampling is done for balancing where pixels from the majority classes are removed. The number of pixels selected from each class after balancing is 24152.

Step 4: Creating training samples for classifier

Each pixel of the preprocessed Landsat 8 image has seven spectral values from seven bands of the ROI, which forms the feature vector for the classifier. The class labels are assigned to each corresponding pixel from the LULC decadal subset. The feature vector with class labels forms the dataset and is used for training and testing the classifier. The training data set is created by sampling pixels with the corresponding class labels. The dimension of the feature vector is seven, and the classification problem is formulated as a six-class problem.

Step 5: Building classifier models

Classifier model is built using by random forest (RF), support vector machines (SVM), and decision tree (DT) algorithm. The classifier is tuned by testing with various hyperparameters to find their optimal value. Each of the classifier is stored and validated using test data for its performance. A brief review conventional classifiers models is given below:

Support vector machines (SVM): SVM finds a hyperplane in high-dimensional space that separates training data into classes so that class members are separated with maximum margin. The data points on the margin are called support vectors.

Random forest (RF): RF classifier is composed of a set of classification trees trained using boot strapped sampling. Trees are constructed by partitioning the samples into nodes until the stopping criterion was met. The results of random forest classifier are generated based on majority voting calculated from all trees.

Decision tree (DT): DT is a tree-based framework which splits the input data samples into independent regions based on selected attributes at each node of tree. It iteratively builds the tree using criteria measure for each attribute and chooses the attribute with optimal value.

Step 6: Validate the model

The build models are stored and tested with test samples generated in step 4. The performance of the classifiers is evaluated using the metrics: overall accuracy, kappa coefficient, user and producer accuracies. These measures are used by researches to measure how good a classifier is in predicting the class label, for untrained tuple.

4 Results and Discussions

4.1 Experiment Setup

The algorithms in this study were implemented using RStudio 3.4.3 and QGIS 10.3.1 and ran over a PC with Intel i-7 2.5 GHz processor and 8 GB RAM. R offers many packages for building the classifiers and for visualization of multispectral images. QGIS is used for preprocessing the Landsat image and to create the training data set for the models and for validation of the results. Holdout method was used for sampling and partitioning the data into two independent sets, for training and testing the classifiers. The training data set was used to build the classifier model, and test data was used for evaluation of the accuracy of the classifier. In this study, ML model was built on 80% of the data, and the residual 20% was used to test the performance of the classifier.

4.2 Performance Analysis of Classifiers

Three different classifier models were trained to classify the ROI into six different classes. Different performance metrics such as overall accuracy, user accuracy, producer accuracy, and kappa coefficient were used to determine how good each classifier was in predicting the class label, for untrained tuple. Producer accuracy referred as the error of omission is the accuracy calculated from the perspective of a map producer. It is the measure of how the real features on the ground are correctly classified by the algorithm. It is a measure of number of data points classified by algorithm to a class divided by the total number of data points in that class. User accuracy referred as the error of inclusion is the accuracy from the perspective of user, where the accuracy of map generated by classifier is measured. It is a measure of number of correct data points of that class to that classified by algorithm to that class.

Table 4 shows the performance measures considered in this study. Performance metrics of classifiers are calculated from the confusion matrix and user accuracy, producer accuracy, and kappa coefficients measures specific to remote sensing are used for comparison. From the experimental analysis, it was found that optimization of classifier parameters was required to improve the performance of the algorithms. For each of the classifier models, a series of parameter values is tested to obtain the optimal parameters which is determined by the highest overall classification accuracy.

Table 5 shows the hyperparameters that were set for different algorithms in the study. To compare the performance of classifiers, all the models are trained and validated with the same data set.

Table 6 shows the overall accuracy and kappa coefficient obtained by various classifiers on the test data. The table infers that random forest algorithm has higher accuracy among the trained models with 68.08%. Kappa coefficient values lie in the range between 0 and 1, where 1 represents the perfect agreement between the actual

Table 5 Algorithmic hyperparameters

Algorithm	Hyperparameter
RF	Ntree – 1000
SVM	Kernel = radial, gamma = 1.0
DT	Tune length – 1000

Table 6 Accuracy and kappa coefficient of RF, SVM, and DT classifiers

	RF	SVM	DT
Accuracy	68.08	64.50	59.44
Kappa coefficient	0.617	0.574	0.513

Table 7 User and producer accuracies of RF, SVM, and DT classifiers

	Grassland	Crop land	Built-up land	Shrub land	Fallow land	Water bodies
	User accuracy					
SVM	68.9	42.3	60	65.8	82.1	67.8
RF	73.5	45.9	64	71.8	83.5	69.9
DT	61.4	40.8	55.4	64.6	76.4	58.1
	Producer accuracy					
SVM	64.7	55.9	61.1	66.6	77.1	59.3
RF	67.7	58.5	64.1	73.4	79.5	63.2
DT	61.5	44.9	55.5	63.2	73.9	55.8

and predicted results. Random forest has higher kappa value of 0.617 compared to other classifiers.

Table 7 shows the user and producer accuracies of the three classifiers considered in this study. The user accuracy table infers that fallow land has the highest accuracy of 83.5% among all classes and RF classifier produces the best results for all classes. From the producer accuracy table, it is inferred that RF algorithm has best producer accuracy for fallow land class with an accuracy of 79.5% among all classes. From the results, it is shown that fallow land class is identified by all the classifiers with higher accuracy compared to other classes for both the metrics. The user accuracy for crop land is 42.3% from which inferred that only 42.3% of data mapped to crop land is identified to belong to this class. Crop land class has the least producer accuracy which shows that DT has not identified data belonging to crop land class correctly.

5 Conclusion and Future Scope

Convention machine learning algorithms are popularly used for LULC analysis on Landsat data. Research on building a classifier to predict and understand how the classifiers perform for multiclass problem, with more than four classes, is limited. Experiments on building these classifiers with optimized parameter are a challenge as many remote sensing software work with default parameters. In this paper, random forest, SVM, and decision tree classifiers are built and tested for a multiclass problem with six classes. The classifiers are tested with the benchmark reference data from

LULC decadal images. Accuracy is estimated for the different classification algorithms, and from the results, it is concluded that random forest is able to identify the LULC classes from Landsat data. The work can be enhanced by applying deep learning and probabilistic models for LULC classification.

References

1. Gorelick, N., et al.: Google earth engine: planetary scale geospatial analysis for everyone. Remote Sens. Environ. 18–27 (2017)
2. Phiri, D., Morgenroth, J.: Developments in landsat land cover classification methods: a review. Remote Sens. **9**, 967 (2019)
3. Wulder M.A., et al.: Current status of Landsat program, science, and applications. In: Remote Sensing of Environment (2019), pp. 127–147.
4. Maxwell, A.E., Warner, T.A., Fang, F.: Implementation of machine-learning classification in remote sensing: an applied review. Int. J. Remote Sens. 2784–2817 (2018)
5. Draganstevic, et al.: Automatic identification of land cove types using multi spectral satellite images. Energy Build. **115**, 131–137(2016)
6. Gasparovica, M., et al.: Automatic cost-effective method for land cover classification (ALCC). Comput. Environ. Urban Syst. 1–10 (2019)
7. Paneque-Gálvez, J., et al.: Enhanced land use/cover classification of heterogeneous tropical landscapes using support vector machines and textural homogeneity. Int. J. Appl. Earth Observ. Geoinf. 372–383 (2013)
8. Jeevalakshmi, D., Narayana Reddy, S., Manikiam, B.: Land cover classification based on NDVI using LANDSAT8 time series: a case study Tirupati region. In: International Conference on Communication and Signal Processing, pp. 1332–1335 (2016)
9. Mishra, P.K., Rai, A., Rai, S.C.: Land use and land cover change detection using geospatial techniques in the Sikkim Himalaya, India. Egypt. J. Remote Sens. Space Sci. (accepted and in press)
10. Sudhakar Reddy, C., et al.: Mapping the vegetation types of Rajasthan, India using remote sensing data. J. Environ. Res. Manage. **2**(1), 001–009 (2011)
11. Khatami, R., Mountrakis, G., Stehman, S.V.: A meta-analysis of remote sensing research on supervised pixel-based land-cover image classification processes: General guidelines for practitioners and future research. Remote Sens. Environ. 89–100 (2016)
12. Thyagharajan, K.K., Vignesh, T.: Soft computing techniques for land use and land cover monitoring with multispectral remote sensing images: a review. Arch Comput. Methods Eng. 275–301 (2019)
13. Rithin Paul Reddy, K., Sai Srija, S., Karthi, R., Geetha, P.: Evaluation of water body extraction from satellite images using open-source tools. Adv. Intel. Syst. Comput. 129–140 (2020)
14. Loganathan, A., Bharathi, D.: Sparsification of graph Laplacian for image indexing using multi-dimensional Spectral Hashing. Int. J. Imaging Robot. 43–56 (2015)

Role of Artificial Intelligence in Bank's Asset Management

Priya Gupta and Parul Bhatia

Abstract Banking operations have been supported by latest technology-based systems along with the traditional processes which prevailed in earlier banking era. With the large number of smart phone users, there has been a paradigm shift in increasing demand for digital products like mobile banking, Internet banking, E-wallets, etc. A common trend in banking technology is using an application programming interface (API) that enables a third-party application to use an interface through which the customer can access a variety of services. Besides APIs, technologies like blockchain and artificial intelligence (AI) have great impact in changing the face of banking industry. However, this exposes the banks to a variety of cybersecurity threats which may cause service disruptions to the customers. Therefore, this study discusses the pros and cons of these AI techniques along with their scope of application for banks' asset management.

Keywords Artificial intelligence · Bank efficiency · Non-performing assets · Asset management · Creditworthiness

1 Introduction

The non-performing assets (NPAs) in banks pose a serious threat to profitability of banks. AI-based models and indicators can be of great help to help improve the existing banking processes. The development of forecast models using both financial and non-financial indicators in an AI-based setup may be useful for the banking industry, and its scope may be tested in various banking applications like

P. Gupta (✉)
Finance Department, Lal Bahadur Shastri Institute of Management, New Delhi, India
e-mail: priyagupta.1704@gmail.com

P. Bhatia
Finance Department, Apeejay School of Management, New Delhi, India
e-mail: parul.bhatia84@gmail.com

© The Editor(s) (if applicable) and The Author(s), under exclusive license to Springer Nature Singapore Pte Ltd. 2021

A. P. Pandian et al. (eds.), *Proceedings of International Conference on Intelligent Computing, Information and Control Systems*, Advances in Intelligent Systems and Computing 1272, https://doi.org/10.1007/978-981-15-8443-5_13

loan approval, handling NPAs, marketing of various banking products and other banking operations.

The evaluation of customer's creditworthiness for their loan approvals may tend to tamper the relationship marketing concept which underlies the basic theory of bankers' association with people. However, the rising competition from FinTech companies and biasfree approach after usage of artificial intelligence (AI) in various fields of management poses a thought for exploring AI in banking operations [1]. Advancing loans and accepting deposits have been the major task for a bank which justifies its existence as per the regulatory act. The progress and track record of its assets performance which are majorly loans classified into various categories may be tapped with AI in future times. The ongoing rise of non-performing assets (NPAs) requires an automated system and technology-driven mechanism to replace the physical follow-up entirely from the banking system.

The advent of information technology (IT) and online transactions may not be figured in an introductory phase for the banking sector. The 2000 IT boom had promoted all sectors to adopt and adapt with the Internet of things (IOT) and speedily increase the business volume. The public and private sector banks both have equipped themselves with Internet and mobile applications for their basic and advanced services offered to the masses. The traditional branch banking has now been replaced by quick application-based service on a click of a mouse.

An AI model using customer loyalty factors for the banking services has been developed with various factors (customer satisfaction, service quality, perceived value and trust) for Islamic banks [2]. The loyalty of the customers toward a particular bank may also force him to avail loan from the same bank. There may be a possibility of bias and subjective approach, while such a loan is sanctioned. The initial loan request by the customer and its further approval by the banking authorities has been a critical stage for monitoring of bank's asset performance and its quality in future. In case of a term loan where the progress has to be furnished on yearly basis to regulatory authorities and monthly monitoring stays as a prerequisite for a healthy asset of the bank, introduction of AI-based measures may lead to higher level of efficiency and transparency.

The recognition of patterns, classifying them and modeling them in a time series data for forecasting the output related to loan performance of banks can be better performed by artificial neural networks. The Jordanian cases of banks have shown that AI may be a successful methodology to make decisions related to multiple loan applications received by the banks and may be cost effective and efficient way of approving or disapproving them [3].

The espousal of robotic process automation (RBA) in retail banking processes has been premeditated with various dynamics that may lay emphasis on its usefulness and attainment. The findings portrayed that security, privacy, reliability along with usage benefits of this system may have a significant impact on its implementation within banking operations [4].

The intelligent protection system (IPS) tested for Italian branch banking has signaled using Internet of things (IoT) for banking physical security. The study suggested a broader scope for testing the concept for other parts of the world as the findings may not be generalized for all types of banking operations [5].

The examination of technological (AI-based) and non-technological factors affecting mobile banking and AI-enabled mobile banking usage revealed that the comparative advantage has been the most important force that led to mobile banking adoption. However, the AI-based mobile banking operations have not been found to be significantly influenced by relative advantage. The complexity in such services has been much higher in contrast with general banking operational services over mobiles. Trust, risk and accuracy while using AI-based applications may be additional considerations for customers [6].

The forecast of banking related measures and operations may be better analyzed with advanced machine learning than the traditional systems. There may be an advanced planning relating unknown with unknown factors than only known with unknown forces. Machine learning software may be capable of explaining the unknown behavior, security or fraud issues associated with multiple banking operations [7].

2 Usage of Artificial Intelligence in Finance Domain

AI has provided the world a whole gamut of services to cater to the customized needs of the customers with the help of which not just they but the enterprises utilizing it can take more wiser and efficient decisions to save, invest their money. The predictive modeling used by various AI platforms has solved the problem of uncertainty up to a larger extent and has helped to foresee the volatility in the capital and money markets. The platforms suggest not just a viable option for maximizing returns but also envisage the safer ways to access them. Some of the most essential contributions made by the AI platforms are defined as under:

A. **Credit decisions**: AI help the banks and credit lenders to assess the credibility of the borrowers. It also helps the banks to assess their underwriting capabilities with such borrowers. For this, there are lot of AI techniques available that collect innumerable unstructured and structured data, transform it and then learn from each interaction with the customer and offer contextual underwriting intelligence.
B. **Analytical decisions**: There are AI software like "Kensho" used by top financial institutions like J.P. Morgan, Bank of America, Morgan Stanley and Standard and Poor Global which use a combination of cloud computing and natural language processing (NLP) to provide answers to difficult questions of the traders and investors. Fund managers of the hedge funds are also implementing predictive modeling techniques with the help of machine learning to identify market changes earlier than is possible with traditional investment models.
C. **Virtual financial assistance**: There are some virtual financial assistants available like "Abe" that integrates all the data of the customer through various mediums

like Amazon Alexa, Facebook, SMS, Google home, etc., and then integrates this information to provide a convenient banking experience. Along with this, this app also helps customers to track their expenses and plan for their investments.

As per a Survey conducted by PwC on the US investors in 2016, the following results were presented:

Image Source: PwC 2016 Global Data and Analytics Survey: Big Decisions [8].

This survey also clearly pointed out that there is going to be great surge in the demand of machine learning algorithms by the FinTech companies for providing customized services to their clients.

3 Artificial Intelligence for Bank's Efficiency

Artificial intelligence means artificially developed intellectual ability. It started in the decade 1950. Through this, a computer system or robotic system is developed, which is run on same arguments based on which the human brain works. According to John McCarthy, the father of artificial intelligence, it is the science and engineering of intelligent machines, especially intelligent computer programs, i.e., the intelligence displayed by machines. Artificial intelligence is rapidly evolving around the world. This technology is getting better day by day, thereby increasing its use in various sectors around the world. The banking sector is also not insulated from it and is using it in various sectors.

The AI-based indices (DEA like mathematical programming techniques) have been used in various studies to determine credit rating and prediction of failure of parties related to branch banking. The limitation of these studies has been usage of financial indicators, and thus, non-financial indicators and qualitative factors may be an attractive area of research in this filed [9]. The classification of regression trees and neural networks outperforms the traditional credit scoring systems for analyzing the new prospective customers' risk profiles. The major hurdle for any banker may be to pass through the initial hiccups of right kind of customers whose default probability would be minimum so that the asset management may not become a tedious and disastrous task [10].

The text mining and latent Dirichlet allocation technique have shown rising trend of credit management in banks through intelligent business system. This has been further extended to prediction of bankruptcy and fraud detection by customers. The loan assets continue to be the major focus for all bankers where innovative techniques may play a vital role and provide remarkable solutions [11].

There may be inestimable vulnerabilities associated with cloud computing networks related to a service-based industry. The study identified 22 such vulnerabilities and categorized them into eight broad categories. The comparative approach of these factors has shown that mobile-based services may be the most immature systems as regards to security concerns. This area may have maximum percentage of concerns, while customers adopt them in their day to day transactions [12].

The software as a service (SaaS) has emerged as one of the key elements in cloud computing-based customer relationship management (CRM). The advancement of technology has further raised many issues related to business size, service deployment, efficiency, ease and availability of technique for CRM [13].

The multi-agent systems and cloud computing can have a positive impact on the construction of business processes. However, it has become challenging to use these cloud-based services from the Internet due to infinite problems. The study extended a solution surrounding management aspects and competence. It suggested categorizing of similar services and coding them according to service category and quality of services [14].

The cloud-based cyber physical system may improvise the quality of service when adopted with big data analytics and such other key-enabling techniques. However, the security concerns remain the main cause of concern which hinder the way for its easy and convenient implementation. There may be opportunities as well as challenges with this framework which may generate security as well as quality issues [15].

It may be true that cloud computing poses a serious threat to security issues as the data usage over Internet has got various concerns. This technique has characteristics which reflect an obvious threat to individual data used for forecasting and other operative measures. The study proposed a hypervisor layer named hypervisor detector that may use a hybrid approach of fuzzy C-means algorithm with neural networks based on AI. The proposed measure may detect anomalies with high accuracy, and the false alarm rate may be very low [16].

The advent of cloud-based services (CBS) in the rapidly increasing information technology milieu requires an intermediation for launch of these services effectively. A cloud service broker entity in the name of virtual cloud bank has been proposed through the study, and it has been strongly felt that such channel may accelerate the growth of CBS. The cloud service broker shall remain a keen interest area for study by industry and academia [17].

The risk analysis with an AI technology-based forecasting model may be minimal and can provide intelligent system-based output to bankers for planning banking operations and asset management. The weighted risk to assets may be computed using system-driven factors, and their further modeling may be decided based on

Table 1 Stepwise loan approval and monitoring with AI

Initial request by customer	Banker's AI mechanism replaced with traditional setup
Document verification	Assigned with Banker's experience in loan portfolio
Creditworthiness check	AI-based indicator to replace physical checks
Initial amount dispatch	Automatic alignment with customer–banker statement
EMI due	System generated message and communication
Tracking payments	Physical monitoring to be 100% replaced with AI
Default action	Track customer beforehand and banker to peep in

Note Authors' own compilation

input–output ratio obtained from the model. The decision making may be improved using CBR-based systems than statistical techniques like logistic regression [18].

The general algorithm model proposed in the study has shown that the loan screening time can reduce from 12 to 50%, and the profitability for the banks from its asset management may rise from 3.9 to 8.1% as compared to the traditional statistical scanning techniques. The increase in the efficiency of assets using this model can lead to better loan accounts handling and profit margins [19] (Table 1).

The similar process can be depicted with the help of the following image:

Machine learning provides Hassle free loan processing

Image Source: Let's Talk Payments (https://gomedici.com/investor-profiles/lets-talk-payments).

The discrimination aware data mining (DADM) tool may be used by bankers to detect or prevent the frauds from loan accounts which can help them in making decisions related to loan applications. There has been a strong evidence for using these techniques over the traditional ones which further depends on their intended usage [20]. The Weka application suggested that among the three algorithms, J48 has been found to be the best fit for assessing the risk related to bank loans. Data mining may now be a motivating and vital research area in this field of financial risk exposure related to loans [21]. The decision support system from artificial intelligence neural networks may act as a useful model predictor for analyzing bad credit risks in the banking domain. The assessment of financial risk measures can be automated using the system of vector quantization networks based on customers' demographic, biographic and behavioral characteristics [22].

The employer–employee point of view may vary across different parts of the globe. The use of AI neural networks has different complexities gazing to a service industry. Nigerian service-based industry had shown negative output in terms of efficiency of employees when conclusions were drawn based on resources theory. SEM modeling has shown a different analysis than usual for service industry along with neural networks [23].

The innovation with AI-based operations has got various legal and security issues. The inculcation of complex algorithms in countries like Canada, UK, US, South Korea and China may depend on investment policies, legislation rules and ongoing regulatory developments. The initiatives in this field have not been enough to promote the entire shift toward AI-based operations. There may be a requirement to optimize risk management operations and foundational principles for laying AI-based processes by human professionals [24].

The response to new business needs in the banking sector has been well recognized in the AI-based customer management system. The Internet-based services and cloud-based networking have attracted banker's attention with little concern of users' perspective. The study has focused on customers slant with a construct on trust, security, privacy and associated cost. A primary survey of 162 bank customers in Malaysia using partial least square disclosed that the construct had a strong association with behavioral intention to embrace the AI-based services [25].

Further, the new paradigm of cloud computing which has been blockchain may not be able to replace the cloud-based applications in an organization perspective. The number of transactions that can take place with Visa-based systems may be nearly 10 times of blockchain capacity to roll over requests. Though the framework for such solutions has been coming from corporations like Microsoft, IBM, JP Morgan to name a few, their adaptability may still be a question [26].

The organizations these days work in a complex and dynamic environment with no longer managing and generating business only. It has been more about retaining customers with quality service during each transaction. The electronic banking industry with a case of Eghtesad Novin Bank has replicated the usage of cloud computing for customer relationship management. The conceptual model depicted security, computing power, lack of dependency on location/device and decreased dependence on hardware as possible factors influencing the acceptance of cloud-based techniques for managing customers [27].

The trust and privacy remain the first order concern while using Internet-based business solutions in managing customers and providing them quality service. The entry of big new entrants in this field may be short lived if appropriate solution to transparency and safety does not get formed. Thus, the success of cloud-based network and AI-enabled techniques may be dependent upon the building of trust of customers for these technologies [28].

4 Current Status of Artificial Intelligence in Indian Banking System

The State Bank of India, which is India's largest bank, organized the "Code for Bank" National Hackathon in 2017 and "Predict for Bank" in 2019 to encourage developers of technology such as artificial intelligence and blockchain in the banking sector. State Bank of India has facilitated its customers with artificial intelligence-driven chat assistants that address customer enquiry problems. It helps in everyday activities just like the representative of the bank. Major private banks like HDFC Bank and ICICI Bank have already introduced chatbots to serve customers. Some banks are also using robots to serve customers. Canara Bank has installed "Mitra" and "Candy" robots in some of its offices in recent years. Some payment companies are using artificial intelligence to provide consumers with a personalized payment experience.

Axis Bank recently launched the artificial intelligence and natural language processing-enabled App "AXIS AHA" to respond to FAQs for financial and non-financial transaction features and to contact customers for marketing loan products. It is currently available on Facebook and Axis Bank's Web site which will soon be expanded to mobile banking channels as well. The app aims to provide 24 * 7 assistance to customers in an intuitive manner. To reduce turn around time (TAT), the bank has also implemented artificial intelligence in more than 125 areas. As per the report of Accenture Banking Technology Vision 2018, 83% of Indian bankers believe that artificial intelligence will work with human beings in the banking system in the next two years. 93% of bankers have said that they use the data heavily to make important and automated decisions. They also asserted that banks have been able to meet the expectations of their customers using artificial intelligence.

Chatbots:

Chatbot is an automated chat program that is either run automatically or follows a predefined path. Many banks are using artificial intelligence for chatbots. This app is very helpful for the banking industry as nowadays customers lack time due to hectic routines, and it is not possible for them to physically attend the bank. At such times, chatbots along with other digital banking services seem to be very helpful as they are available 24 * 7. Chatbots are a smart way to provide efficient customer service. It helps customers get their transaction details and additional services. Banks can understand the needs of each customer in a better and quick way through chatbots.

5 Benefits of Artificial Intelligence in Banking

5.1 Customer Satisfaction

Customer satisfaction is the first and most important goal of any bank. Banks strive to provide personalized and efficient services to each customer. Artificial intelligence helps to make fast decision-making, maintain effective customer relationships and ultimately increase profitability. Artificial intelligence not only ensures that customers are happy with the bank's facilities and services but also helps banks maintain a streamlined back office.

5.2 Detection of Frauds

Banking frauds are one of the greatest concerns to customers and the banking industry in today's time. Especially when it is a financial fraud, the affected customer must incur financial losses. However, in case of fraud relating to accounts, banks are taking necessary steps at the earliest and necessary actions in an organized manner. This quick process of the bank has increased the confidence of customers in the banking system, but their unveiling takes time and thus reflects that are also shortcomings somewhere. The use of artificial intelligence in banking has ensured immediate action. Fraud is detected by tracking irregular behavioral patterns by scanning huge data of transactions available in banks in which artificial intelligence has an important role to play. ATM-related frauds can be detected and prevented using images of cameras and advanced artificial intelligence technology in real time.

5.3 Risk Management

Chatbots can provide suitable products to the customers by taking old data as a base for risk analysis and removing human errors from the hand-crafted models. This analysis is done based on the data available in the back office, and artificial intelligence has a special contribution to this process. The risk assessment process is very complex and important while lending. This requires both accuracy and privacy. Artificial intelligence makes this process simple by analyzing relevant data from a potential borrower. It can analyze the activities of the account to identify the latest transactions, market trends and potential risks in lending.

5.4 Personalized Financial Guidance

Artificial intelligence helps customers make easy, accurate and quick financial decisions. It also provides updated stock market information as well as suggestions on stocks and bonds in which customers can invest. Artificial intelligence can also be used in the field of digital wallet and interactive voice response system.

6 Possible Threats of Artificial Intelligence

Although artificial intelligence has demonstrated its remarkable usefulness in the banking sector, will the banking industry be saved from the potential threats of artificial intelligence? It is one of the most important questions arising out of its usage.

6.1 Limited Exposure to the Staff

A limited number of staff who can effectively use artificial intelligence and a lack of usable data can limit the usefulness of artificial intelligence. The artificial intelligence system performs its work by analyzing huge volume of data and is adapted through continuous drafting of interactive data and user feedback. But the big question is how this scenario will ensure how to train the algorithm of artificial intelligence to be impartial and not prejudiced.

6.2 *Technology-Generated Unemployment*

Technology-generated unemployment is the unemployment that arises from the coming of a new technology. The coming of machines or systems containing artificial intelligence is likely to be an employment replacement. This will bring about significant changes in the workforce and markets. Some kinds of work roles and employment will be out of circulation. This can increase unemployment in many areas as human space will take machines equipped with artificial intelligence. Recently, a research by Oxford University says that around 1.5 lakh people will be unemployed in the next two decades due to artificial intelligence in the USA.

6.3 *Security*

Security is an important aspect in terms of artificial intelligence. If any caution will not be taken in the use of artificial intelligence, it can have disastrous consequences, especially when the threat of cyber security is increasing day by day. Increasing confidence in the AI tools and techniques with increased spending on cyber security solutions proves that the banks are also relying a lot on them. However, this also indicates an opportunity for the hackers to use disruptive ways to turn the tables and deploy AI-powered attacks. Hackers may alter machine learning algorithms and may change the inputs from the way the system recognizes them. Once the input information is changed, it becomes extremely difficult to correct such problems and return the network to its original state.

7 Conclusion

Of course, like any other technology, artificial intelligence has both advantages and disadvantages, but artificial intelligence is gaining popularity in the banking sector day by day, and banks are using this technology to provide sophisticated and quick service to customers. In one of its reports, McKinsey summarized the areas which are emerging as new trends in banking and showcasing the new trends of the FinTech industries:

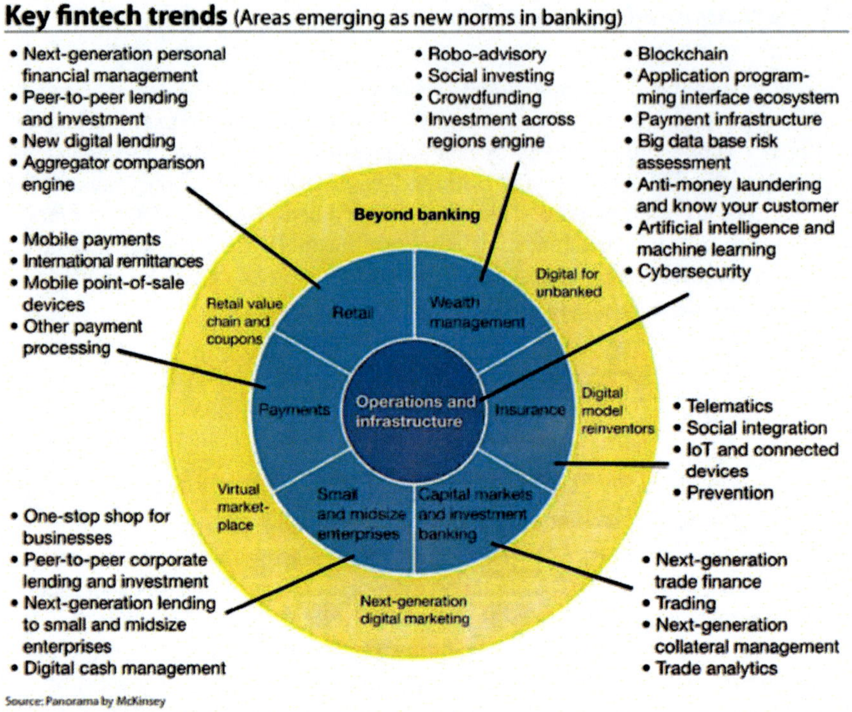

Image Source: Panorma, McKinsey & Company (2016) [29].

Its future in banking is very bright. After demonetization in the banking sector, digital services have come up, banking has evolved, and more and more banks are using it to cut their operating costs. Though it is still in its nascent phase, artificial intellectual revolution can be felt in the banking sector. The development of artificial intelligence in the banking sector will pave the way for low-cost high-quality services.

References

1. Jaksic, M., Marinc, M.: Relationship banking and information technology: the role of artificial intelligence and FinTech. Risk Manag. **21**(1), 1–18 (2019)
2. Kishada, Z.M., Wahab, N.A., Mustapha, A.: Customer loyalty assessment in Malaysian Islamic banking using artificial intelligence. J. Theor. Appl. Inf. Technol. **87**(1), 80–91 (2016)
3. Eletter, S.F., Yaseen, S.G.: Applying neural networks for loan decisions in the Jordanian commercial banking system. Int. J. Comput. Sci. Netw. Secur. **10**(1), 209–214 (2010)
4. Milkau, U., Bott, J.: Active management of operational risk in the regimes of the "Unknown": what can machine learning or heuristics deliver? Risks **6**, 2 (2018)
5. Elizabeth, M.P., Peltier, J.W., Barger, V.A.: Mobile banking and AI-enabled mobile banking: an international journal. J. Res. Inter. Market. **12**(3), 328–346 (2018)
6. Ammirato, S., Sofo, F., Felicetti, A.M., Raso, C.: A methodology to support the adoption of IoT innovation and its application to the Italian bank branch security context. Eur. J. Innov. Manag. **22**(1), 146 (2019)

7. Kumar, K.N., Balaramachandran, P.R.: Robotic process automation—a study of the impact on customer experience in retail banking industry. J. Internet Bank. Comm. **23**(3), 1–27 (2018)
8. Rar, T.: Scopes of machine learning and artificial intelligence in banking & financial services | ML & AI—The Future of Fintechs (2017). [Online https://www.stoodnt.com/blog/scopes-of-machine-learning-and-artificial-intelligence-in-banking-financial-services-ml-ai-the-future-of-fintechs/]
9. Duygun-Fethi, M., Jackson, G.: Assessing bank performance with operational research and artificial intelligence techniques: a survey, Working Paper Series (2009.02). http://citeseerx.ist.psu.edu/viewdoc/download;jsessionid=9DB0586DD7C1955356A60E75A477FF23?doi=10.1.1.473.4002&rep=rep1&type=pdf
10. Ince, H., Aktan, B.: A comparison of data mining techniques for credit scoring in banking: a managerial perspective. J. Bus. Econ. Manag. **10**(3), 233–240 (2009)
11. Moro, S., Cortez, P., Rita, P.: Business intelligence in banking: a literature analysis from 2002 to 2013 using text mining and latent Dirichlet allocation. Expert Syst. Appl. **42**(3), 1314–1324 (2015)
12. Kouatli, I.: A comparative study of the evolution of vulnerabilities in IT system and its relation to the new concept of cloud computing. J. Manag. Hist. **20**(4), 409–433 (2014)
13. Souri, A., Asghari, P., Rezaei, R.: Software as a service based CRM providers in the cloud computing: challenges and technical issues. J. Serv. Sci. Res. **9**(2), 219–237 (2017)
14. Lacheheub, M.N., Maamri, R.: Pr. Towards a construction of an intelligent business process based on cloud services and driven by degree of similarity and QoS. Inf. Syst. Front. **18**(6), 1085–1102 (2016)
15. Mourtzis, D., Vlachou, E.: Cloud-based cyber-physical systems and quality of services. TQM J. **28**(5), 704–733 (2016)
16. Pandeeswari, N., Kumar, G.: Anomaly detection system in cloud environment using fuzzy clustering based ANN. Mobile Netw. Appl. **21**(3), 494–505 (2016)
17. Park, J., An, Y., Kang, T., Yeom, K.: Virtual cloud bank: Consumer-centric service recommendation process and architectural perspective for cloud service brokers. Comput. Arch. Inf. Numer. Comput. **98**(11), 1153–1184 (2016)
18. Charlo, M.J.: The most relevant variables to support risk analysts for loan decisions: an empirical study. Reg. Sect. Econ. Stud. **10**(1), 61–70 (2010)
19. Metawa, N., Hassan, M.K., Elhoseny, M.: Genetic algorithm based model for optimizing bank lending decisions. Expert Syst. Appl. **80**, 75–82 (2017)
20. Berendt, B., Preibusch, S.: Better decision support through exploratory discrimination-aware data mining: foundations and empirical evidence. Artif. Intell. Law. **22**(2), 175–209 (2014)
21. Hamid, A.J., Ahmed, T.M.: Developing prediction model of loan risk in banks using data mining. Mach. Learn. Appl. Int. J. (MLAIJ). **3**(1), 1–9 (2016)
22. Moonasar, V.: Credit risk analysis using artificial intelligence: evidence from a leading South African Banking Institution. Doctoral dissertation, University of South Africa (2007). https://s3.amazonaws.com/academia.edu.documents/3456049/Credit_risk_analysis_using_artificial_intelligence_evidence_from_a_leading_South_African_banking_institution
23. Abubakar, A.M.: Using hybrid SEM—artificial intelligence. Personnel Review **49**(1), 67–86 (2019)
24. Caron, M.S.: The transformative effect of AI on the banking industry. Bank. Finance Law Rev. **34**(2), 169–214 (2019)
25. Asadi, S., Nilashi, M., Abd Razak, C.H., Yadegaridehkordi, E.: Customers perspectives on adoption of cloud computing in banking sector. Inf. Technol. Manage. **18**(4), 305–330 (2017)
26. Cearnău, D.: Block-cloud: The new paradigm of cloud computing. Acad. Econ. Stud. Econ. Informat. **19**(1), 14–22 (2019)
27. Ghane, F., Gilaninia, S., Homayounfar, M.: The effect of cloud computing on effectiveness of customer relation management in electronic banking industry: a case study of Eghtesad Novin Bank. Kuwait Chapter Arab. J. Bus. Manag. Rev. **5**(8), 50–61 (2016)

28. Sarma, A., Girao, J.: Supporting trust and privacy with an identity-enabled architecture. Fut. Internet. **4**(4), 1016–1025 (2012)
29. Dietz, M., HV, V., Lee, G.: Bracing for seven critical changes as fintech matures. Panorma. Mc Kinsey & Co., (2016) [Online: https://www.mckinsey.com/industries/financial-services/our-insights/bracing-for-seven-critical-changes-as-fintech-matures]

Detection of Birds Chirping Using Machine Learning

B. Sadhana, Neha Shetty, R. K. Ananya, P. Vanditha Shenoy, Chitra Shenoy, Vasath Nayak, and Pragathi Hegde

Abstract Birds are said to be the natural indicators regarding the condition of the environment and the changes that happen in the surrounding. Thus, ecologists strive a lot in monitoring these birds to understand their trending population, and doing this manually for each and every bird becomes a slow never-ending process. The main motive of this computerized system is to track down these bird kinds with the help of their sound. This automated system is of great benefit to the bird watchers in the study of wildlife, civic science, and data signals library management and to find out their evolution within time. From further research, it has been known that bird sound detection has become a usual necessary first step before classification of input data and also manages large datasets by straining down the data of interest. This has now become one of the potential papers in acoustic observation of sound-emitting birds. This gives us quick real-time information depending upon the dataset fed into the system without any delay in managing the dataset which would otherwise happen if the classification of the birds was done manually and gives a general estimation regarding the biodiversity of the surrounding.

B. Sadhana (✉) · N. Shetty · R. K. Ananya · P. V. Shenoy · C. Shenoy · V. Nayak · P. Hegde
Department of ISE, Canara Engineering College, Mangalore, India
e-mail: sadhana26rai@gmail.com

N. Shetty
e-mail: nehashetty731@gmail.com

R. K. Ananya
e-mail: ananyahrk@gmail.com

P. V. Shenoy
e-mail: vanditha1997@gmail.com

C. Shenoy
e-mail: shenoymangalore153@gmail.com

V. Nayak
e-mail: vasanthnayak01@gmail.com

P. Hegde
e-mail: pragathi.nitte@gmail.com

© The Editor(s) (if applicable) and The Author(s), under exclusive license to Springer Nature Singapore Pte Ltd. 2021
A. P. Pandian et al. (eds.), *Proceedings of International Conference on Intelligent Computing, Information and Control Systems*, Advances in Intelligent Systems and Computing 1272, https://doi.org/10.1007/978-981-15-8443-5_14

Keywords Feature extraction · Classification · Mel · SVM

1 Introduction

The environment is a place that supports our lives. There are a variety of species in this environment, and birds are kind of animal species with few common parameters among all species [1]. Monitoring of such species needs to be done in an effective way. Every bird has a unique way of chirping, but it is not an easy task to identify birds from its voice even when they have different voice [2]. Therefore, main aim of this project is to identify birds from its voice. Identifying or monitoring the birds from its sound will be an efficient way to nurture the species, for bird population monitoring, to know more about its living.

Few attempts were done to monitor birds by its voice manually, but it included lot of efforts and maintenance. Some methods exhibit average accuracy in small surveys and unfeasible results in large surveys [3]. There is an extreme growth in the field of machine learning in recent decades. In this project, some techniques of machine learning are made use of to identify birds from its voice [4].

The main aim of the project is to automatically identify birds from its voice using machine learning techniques [5]. In this system, a web application is also used to display details of the identified bird so that the user need not waste his time on searching the details of the bird in search engine.

1.1 Purpose

The purpose of the system is to invent a structure that would accept Aves species or identify each by their chirping sound. Feature extraction can be adapted by audio processing and machine learning techniques [6]. The characteristics of Aves such as size, species, and types of call will be extracted. The proposed technique extracts the audio of a bird from the stored database and then stores it in the form of a matrix [7]. This feature matrix further helps us in recognition of the bird species with the help of support vector machines (SVMs).

So, the main purpose of developing this system is that

- It would help in identifying the birds from their sound.
- It would help in getting few main features of that bird.

1.2 Problem Statement

The main aim of this system is to help in identifying the birds based on their sound or songs using the concept of machine learning.

Identifying bird manually based on their sound seemed to be a difficult task and time-consuming [8]. More human resources were required to do this task, and it was not efficient. Therefore, this proposed system helps user to identify birds with less time and is efficient.

1.3 Scope

Machine learning is one of the vastly growing technologies and will soon see itself housing. The recent reduction in animal species has urged the necessity to protect ecosystem biodiversity [9]. One useful and effective way to ensure that there will not be any harm to the biodiversity is by learning more about the ecosystem which could possible lead to less extinction [10]. Identifying the birds has been one of the most common problems faced by the ornithologists which leaves them rather clueless regarding the taxonomy of the species and also making it not possible to predict the rate at which those species are getting extinct as days pass [11].

Recognition of voice other than birds will be difficult as this system is being trained for purpose of serving clients with clarity issues of bird species [12]. There are possibilities in degradation of the performance if massive datasets will be used. If the inputs collide with each other, then this system fails to provide the actual output.

1.4 Objective

The objective of this project is

- To reduce the need of volunteers in biological project.
- To help bird watchers in identifying birds.
- To take count of the warbler in a specific area.
- To approximate long-term population drifts.

2 Literature Survey

In the paper [1], they have suggested a superintend, protuberant presentation-based audio hashing structure for passerine species categorization. The suggested system categorizes the raptors vocal sound with the help of convolutional code space, and it does not need any explicit categorization or reestablishment fallacy computations. The execution of these frameworks is differentiated with existing raptor bird categorization frameworks on the vocal recordings of 50 different passerine breeds.

The paper [2] has proposed an approach for identification of passerine from audio field recordings. First, the audible segment is broken down into secluded parts. Each part is constituted by a series of the frequency and balanced magnitude values.

The carnal advancement in the characteristics is modelled with the help of hidden Markov models. A narrative technique of an unverified modelling in a passerine's sound section is presented.

Suggested system in the paper [4] presents a magnified approach for fowl species grouping from their documented audio signals. Noticing that appearance of syllables in audio spectrograms has observable discriminating potentiality amid distinct raptor breeds, which acquire those quality attributes of songbird race categories. They determine a photographic record of spectrum from documented voice sound. They offer an amplified syllable extraction methodology to recognize the syllables in the spectrogram.

The proposed system [6] is used to detect bird species from its sound by segmenting and classification. Since the previously available systems used only segmentation for detecting bird species, it led to many errors. In order to overcome the errors present in the previous system, initially segmentation of the bird species is done and later classification method is applied using the convolutional neural network (CNN) with encoder-decoder architecture. In order to obtain feature hierarchies from input spectrogram, each encoder of the encoder network undergoes convolution. Decoder of the decoder network is used to map the downsampled features to input size features.

In this proposed system [7], hierarchical classification is used to identify bird species. Classification construction cycle of machine learning model is used for the task of bird species recognition. Flat classification approach, local-model hierarchical classification approach, and global-model hierarchical classification approach are the types of hierarchical classification used in this proposed system. Hierarchical precision, hierarchical recall, and hierarchical F-measure are obtained from the evaluation measures for bird species recognition.

In this proposed system [8], main concept used is audio feature extraction and support vector machine to automatically detect bird from its voice. Feature matrix was built from extracted MFCC features. Obtained matrix was then trained and tested in 7:3 ratio. In this system [9], hierarchical subset learning method is used for bird species recognition. Major concept involved in this technique is division of main parent task into subtask if they were primitive actions. Local expert classifier learning is used to learn from each subset. Hybrid decision system is used to take decisions in this proposed system.

In this proposed system [10], wavelet transform is used for achieving the goals. Main steps involved are initially bird sound is sent for pre-processing; later, feature extraction is done followed by recognition of bird. Key concept in wavelet transform is that even when there is a frequency change, time resolution remains unchanged. Morlet wavelet of cosine type and Morlet wavelet of sine type are drawn from wavelet functions. For the identification process, three-layer neural networks have been used.

In this paper [11], they explored the conducting of raptor breed recollection with the help of neural networks with non-identical pre-processing techniques and distinct sets of qualities. Context neural network architecture was outlined to implant the vigorous nature of bird songs into inputs. They implemented a clamour reduction algorithm and successfully put in it to intensify passerine species detection. The conducting of the context neural network architecture was relatively assessed with linear/mel-frequency cepstral coefficients, and assuring exploratory outcome was attained.

The paper [12] is associated with the evolution of signal processing approaches for automatic recollection of raptor breed. Three contrast framework depictions are compared. The first portrayal is based on sinusoidal modelling which has been before time found functional for highly vocal bird sounds. Mel-cepstrum frameworks are used since they have been established very useful in the speech recognition.

3 System Requirements Specification

Humans can grasp any kind of sounds, and they are capable of identifying individual sounds. But sometimes, it is a major issue to identify the types of sources. Our naturally existing birds chirping sound differentiation is a major challenge. Using software-based products, this problem can be solved.

Now, many number of researches are going on in this field of environmental system which will be more beneficial to study wild field and respective development with time. Considering such systems an automatic system for bird category identification using sound as parameter is developed. Wild bird protection is a key factor in the field of environmental perspective.

3.1 Product Functions

The main function of this project is

- Users will be able to provide the feedback regarding the system after determining the bird species. Users are provided with the facility of knowing more information about the identified bird species.
- To provide information regarding the birds by using the voice signals.

3.2 Proposed System Design and Implementation Constraints

See Fig. 1.

System has following design and implementation constraints system should allow the user to get information about various birds' species. User needs to register themself to use our system. Administrator of the system will add, delete, and update bird dataset. Administrator of the system is responsible for managing the system. Software development life cycle agile method is used, which combines the advantages of waterfall approach and an iterative model.

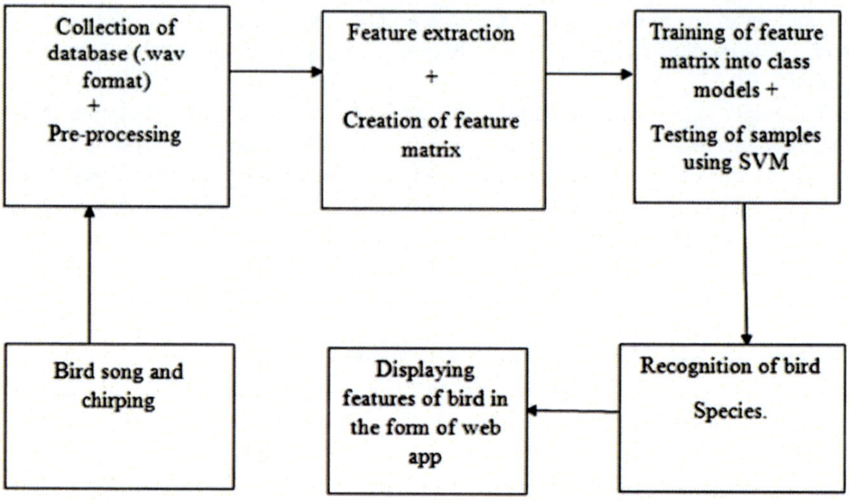

Fig. 1 Prposed system design

Pre-Processing

See Figs. 2, 3, and 4.

The inputs from the web application will have unwanted noise in it and needed only the bird sounds from the mixed environmental noise or the least is to reduce the noise contained in it. The pre-processing is necessary to get better results from

Fig. 2 Steps in pre-processing

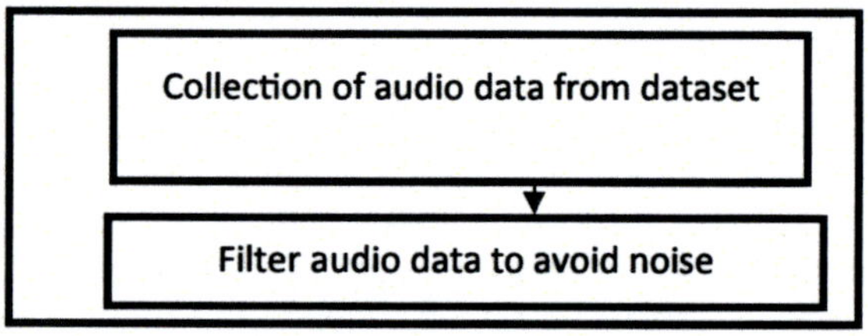

Fig. 3 Steps in feature extraction

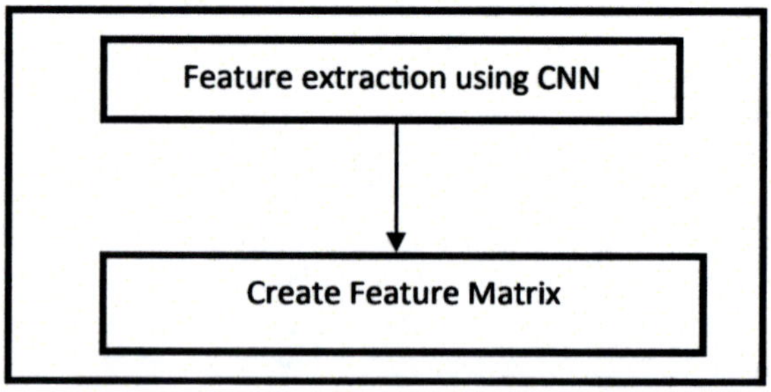

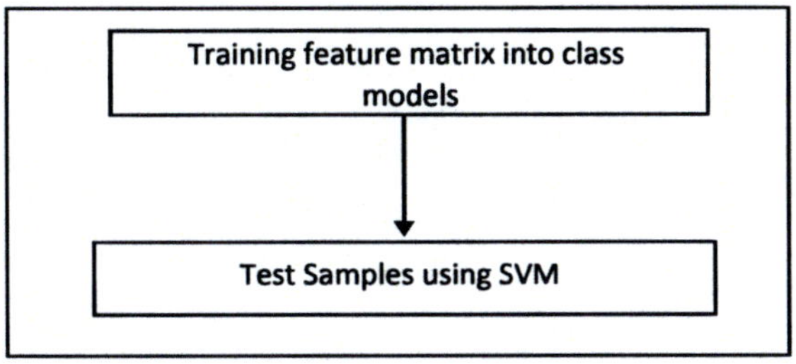

Fig. 4 Steps in classification

the system with better accuracy. This can be achieved with the help of Praat tool. This software is used for speech signal analysis. The filtering of noise from the input signal is done using this software. The output of the processed input will be in the form of waveform.

In our system, CNN is made used for extracting the features from the audio signals taken from a website. CNN has a step-by-step process of doing this. The first step wherein Xeno-Canto helps in getting the audio signals which can be converted into sound waves for further process of feature extraction using CNN. Our system gets the signal from this website in mp3 format. In the second step, these sound waves further can be divided into training data and testing data. The next step is the prior processing of these sound waves by dividing them into pieces of similar length so that it becomes easy during processing the data samples. Further, these samples are sent for filtering wherein unwanted samples get removed. The fourth step is creating spectrograms where the output that got in converted into an image. During the next step of classification, all these converted images are gathered and create a graph. Finally, the output is got by evaluating these different pictures acquired from the spectrograms.

4 Results and Discussions

The proposed software helps in identifying the bird species by using their sound. **This system is a web application which is trained with 1761 data samples.** When the user submits the audio signal, it will get compared with these data samples, and the user will be able to access the features of the bird from the given system. **Frame the bird call signal with a frame size of 20 ms and frame rate of 10 ms. Since the system gives output to most of the audio signals sent by the user, 95% of accuracy is got from the system.**

Figure 5 shows the waveform of the input signal before filtering process using Praat tool (Fig. 6).

Figure 7 shows the login page for existing user. At first, the user is requested to login to the system using valid username and password if the user had already registered. If not, the user has to register himself by providing certain details into the

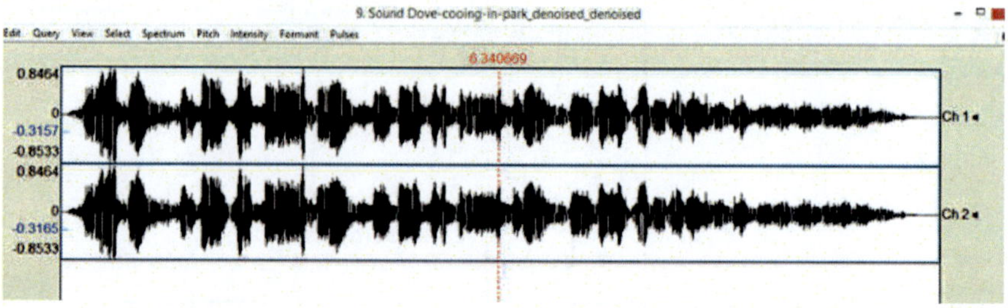

Fig. 5 Original waveform

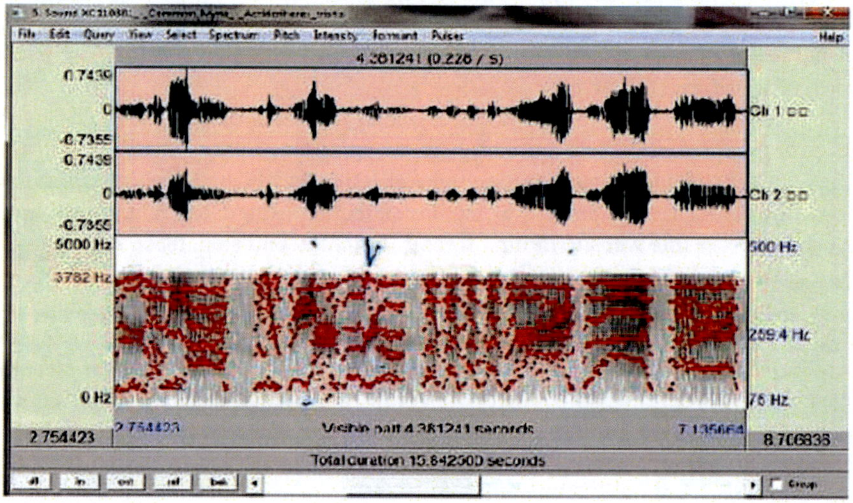

Fig. 6 The processed waveform after the filtering process is as shown in this figure

Fig. 7 Login page of web application

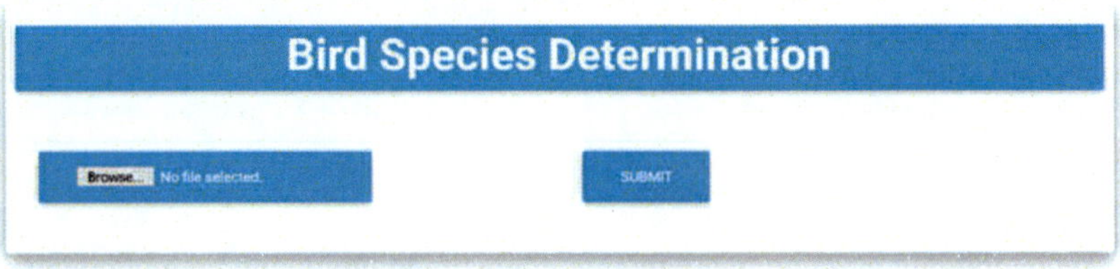

Fig. 8 Home page of web application before selecting input voice

system. The user details are stored into the database so that he gets authorized by the system every time he logins by using valid credentials.

Figure 8 represents home page of web application before selecting the input voice from local server.

Figure 9 shows selection of bird voice as input from local system. Local server will contain many voice files in FLAC format. User has to select any one from the collection for further processing.

Figure 10 shows home page of web application after inputting the voice. Once the user has logged into the system, the user can input the file which contains the recorded sound of the bird. The user can browse the input file from multiple files available in the local server. As soon as the user submits the input file containing the bird sound, the software sends it to the database and starts to compare it with the dataset available.

In Fig. 11, based on training dataset, the machine will identify the bird and display its name and features on the screen helping the user identifying the bird.

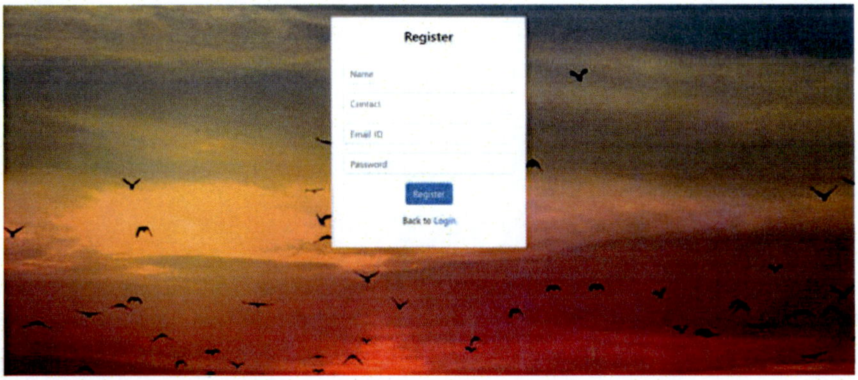

Fig. 9 Collection of bird audio files

Fig. 10 Home page after uploading input voice

Fig. 11 Final result with bird name

5 Conclusion

A desired way of segregating or sorting the bird species is put forth by using the actual voices or calls coming from the syrinx of the bird. Machine learning techniques were executed, and the numerous recordings of bird voices from different species of birds around were considered to make a better impact on the performance factor of the project, which help the user to identify the bird features through the web application.

References

1. Thakur, A., Sharma, P., Abrol, V., Rajan, P.: Conv-codes: audio hashing for bird species classification. In: Proceedings of the ICASSP 2019 –2019 IEEE International Conference on Acoustics, Speech and Signal Processing (ICASSP), Brighton, United Kingdom, United Kingdom, 12–17 May 2019
2. Jancovic, P., Kokuer, M.: Bird species recognition using unsupervised modeling of individual vocalization elements. IEEE/ACM Trans Audio Speech Lang Process (TASLP) **27**(5), 932–947 (2019)
3. Thakur, A., Abrol, V., Sharma, P., Rajan, P.: Compressed convex spectral embedding for bird species classification. In: Proceedings of the ICASSP 2018 IEEE International Conference on Acoustics, Speech and Signal Processing (ICASSP), April 2018
4. Shamim Towhid, Md., Mijanur Rahman Md.: Spectrogram segmentation for bird species classification based on temporal continuity. In: Proceedings of the 2017 20th International Conference of Computer and Information Technology (ICCIT), Dhaka, Bangladesh, 22–24 Dec 2017
5. Qiao, B., Zuofeng, Z., Zhou, Z., Yang, H., Cao, J.: Bird species recognition based on SVM classifier and decision tree. In: Proceedings of. 2017 First International Conference on Electronics Instrumentation & Information Systems (EIIS), Harbin, China, 3–5 June 2017
6. Narasimhan, R., Fern, X.Z., Raich, R.: Simultaneous segmentation and classification of bird song using CNN. In: Proceedings of the 2017 IEEE International Conference on Acoustics, Speech and Signal Processing (ICASSP), New Orleans, LA, USA, 5–9 Mar 2017
7. Silla, C.N., Kaestner, C.: Hierarchical classification of bird species using their audio recorded songs. Int. J. Speech Techn. **19**(4), 791–804 (2016)
8. Rai, P., Golchha, V., Srivastava, A., Vyas, G., Mishra, S.: An automatic classification of bird species using audio feature extraction and support vector machines. In: Proceedings of the 2016 International Conference on Inventive Computation Technologies (ICICT), Coimbatore, India, 26–27 Aug 2016
9. Ge, Z.Y., McCool, C., Sanderson, C., Bewley, A., Chen, Z., Corke, P.: Fine-grained bird species recognition via hierarchical subset learning. In Proceedings of the 2015 IEEE International Conference on Image Processing (ICIP), Quebec City, QC, Canada, 27–30 Sept 2015
10. Sun, R., Yihenew, M., Zhao, H.A.: Wavelet transform digital sound processing to identify wild bird species. In: Proceedings of the 2013 International Conference on Wavelet Analysis and Pattern Recognition, Tianjin, China, 14–17 Jul 2013
11. Zhang, J., Roe, P., Pham, B., Ee, D., Cai, J.: Sensor network for the monitoring of ecosystem: bird species recognition. In: Proceedings of the 2007 3rd International conference on intelligent networks, sensor networks and information, Melbourne, Qld., Australia, 25 April 2008
12. Somervuo, P., Härmä, A., Fagerlund, S.: Parametric representations of bird sounds for automatic species recognition. IEEE Trans. Audio Speech Lang. Process. **14**(6), 2252–2263 (2006)

Methods for Detecting Community Structures in Social Networks

Jesus Vargas, Omar Bonerge Pineda Lezama, and Diana Garcia Tamayo

Abstract The use of social networking platforms in political contexts around the world has not only become a prominent means of spreading social movements, but also become an important modality of socialization and information dissemination. In the case of Colombia, the integration of Information and Communication Technologies (ICTs) and access to digital social networks allow young people and citizens in general to have an alternative means of communication and to be able to interact, organize and make their causes visible. This paper proposes the use of the "Hierarchical Maps" method for the simplification of nodes from a set of interactions in a social network.

Keywords Interactions simplification · Detection of communities · Social network

1 Introduction

Twitter is one of the main social networks, whose main feature is that the messages sent have a maximum length of 140 characters, called tweets, which can be stored and categorized into topics from the use of tags preceded by the symbol # (Hashtag) [1], allowing the user to follow, search and find conversations related to a common topic. The generation of internal communities in the social network is a natural phenomenon that occurs when there are many actors who generate information related to a specific topic [2]. The identification of communities within a social network is an important issue since, in addition to simplifying the large amount of information contained in

J. Vargas (✉) · D. G. Tamayo
Universidad de la Costa, St. 58 #66, Barranquilla, Atlántico, Colombia
e-mail: jvargas41@cuc.edu.co

D. G. Tamayo
e-mail: dgarcia34@cuc.edu.co

O. B. P. Lezama
Universidad Tecnológica, San Pedro Sula, Honduras
e-mail: omar@unitet.edu

© The Editor(s) (if applicable) and The Author(s), under exclusive license to Springer Nature Singapore Pte Ltd. 2021

A. P. Pandian et al. (eds.), *Proceedings of International Conference on Intelligent Computing, Information and Control Systems*, Advances in Intelligent Systems and Computing 1272, https://doi.org/10.1007/978-981-15-8443-5_15

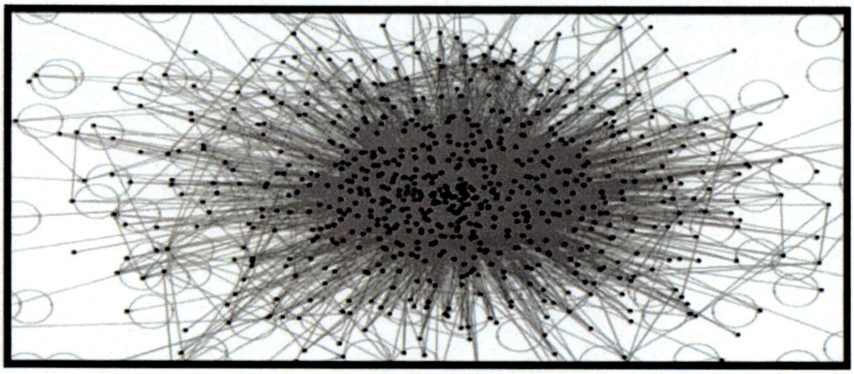

Fig. 1 Network of interactions #DolarCol on January 16, 2020 [4]

the network, it allows for the detection of important groups or sectors that interact with each other [3]. The study analyzes the devaluation of the Colombian peso in Colombia on January 16, 2020. Among the multiple elements of analysis that were generated in social networks under the #DolarCol Hashtag, the analysis focused on the main communities that were generated within the social network, thus showing, in a concise and simplified way, the large amount of information generated on the network. Next, Fig. 1 shows the network generated by the interventions around the mentioned event and using the #DolarCol Hashtag.

2 Objective and Method

The objective of the study is to analyze and simplify the interactions that emerge from a trending topic in a social network. For this purpose, the detection of communities is an important element to simplify the interactions. The method used consists of extracting information generated through Twitter on the day of the mentioned event, using the #DolarCol Hashtag, and the first extraction was made by using the NodeXL tool [5], which helped to display the result shown in Fig. 1. The algorithm based on the hierarchical maps [6], which will be explained later, was used to make the groupings and simplifications on the initial network corresponding to the extracted information.

Finally, an identification of relevant users in each community is proposed, showing the simplification of interactions [7]. The proposed method is schematized in Fig. 2. The corpus is composed of 12,666 Twitter messages in Spanish written by journalists, politicians and general public, issued from Colombia on January 16, 2020 from 18:00 and 18:05 h. For each message, the information contained in the corpus corresponds to: user identifier in the social network, name of the user who wrote the tweet, name of the user it mentions, date and time, the tweet with textual description of the message, hashtags referred to in the tweet, number of followers (Followers) of each

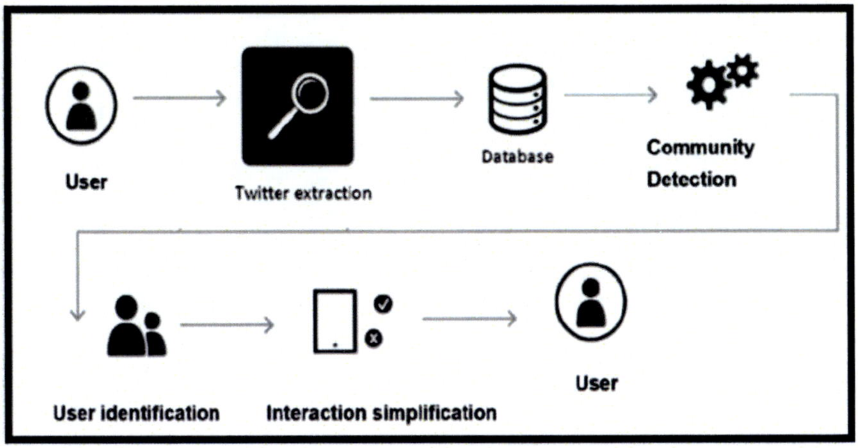

Fig. 2 Proposed method for simplifying interactions

user, number of tweets issued by each user, description in the user profile, location from which the tweet was issued and the date in which the user joined Twitter. Of all this information contained in the corpus, the main interest is on the mentions made, translated later as the interactions made by users, in addition to the number of followers and frequency of publication of tweets [8, 9].

3 Hierarchical Maps

The most general definition of community is that there is a group of nodes in the network that are densely interconnected. In the meantime, from the point of view of information propagation, a community is a group of nodes where information is more likely to be retained rather than spread. Considering that the fundamental information propagation model is the "random walk" [10], then the structure of a community can be identified by a local search in the structure. Recent studies [11, 12] have shown that modularity [9], which is a quality function, is used to find communities where there are densely connected groups of nodes. In this paper, "hierarchical maps" are used to describe the dynamics of links and targeted nodes [13, 14].

Consequently, it is important to understand the full flow of information on the Web. This grouping lies in the duality between the search for the community structure in the networks and the problem of coding: finding an efficient code, an M partition of n nodes is then looked for within the m modules to minimize the distance of the random walk. Using the partition module M, the average distance for information to pass from one node to another is given by Eq. 1 [15], which is formed by two terms: The first one is the entropy of movement between the modules, and the second one is the entropy of movement in the modules (Fig. 3).

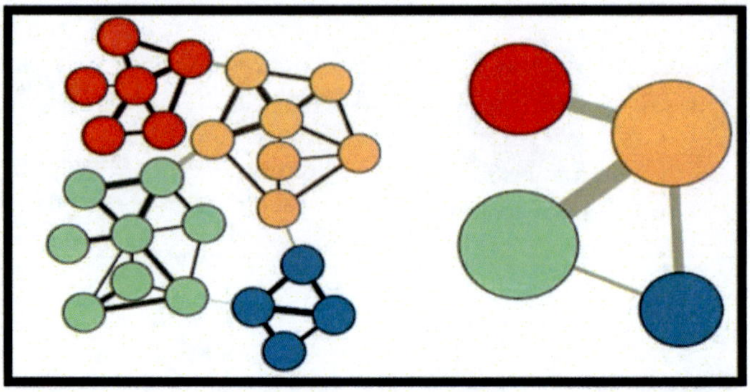

Fig. 3 Schematization of node aggregations in a network [16]

$$L(\mathsf{M}) = q_{\curvearrowright} H(\mathcal{Q}) + \sum_{i=1}^{m} p^{i}_{\circlearrowright} H(\mathcal{P}^{i}). \tag{1}$$

The results of this method are shown using the network generated in the #DolarCol movement.

4 Simplification of the Social Network

The detection of communities is of great interest for analyzing the interactions that are generated in the network. It allows predicting the intensity of interaction, the stage in which it is and the most important actors. This section shows the application of the hierarchical mapping algorithm described in the previous section to simplify the initial social network, which causes a considerable decrease in the number of visible nodes. Of the 850 nodes (tweeters), 30 large communities were grouped and categorized. These were selected because the percentage of interaction was greater than 1.0, which reflects an important set of interactions in the social network. All interactions less than 1.0 were discarded. Of these 30 identified communities, the most representative actor was selected to lead the community based on number of followers and number of tweets performed. In this way, 30 communities are shown in Table 1 which presents the grouping module, the most influential user (Fig. 4), the amount of information shared between them, the ones corresponding to the percentage of interactions in the network and the number of nodes that are grouped or that are agglutinated within this large community.

In this scenario, it is impossible to identify the users with which the representative actors interact and identify the importance in the relationships. Figure 5 shows a simplification using the hierarchical maps where only the communities with which they interact are shown. This visualization is clearer and more concise and allows for better analysis of the existing interactions within the network.

Table 1 Main communities identified with the #DolarCol Hashtag

Module	Representative actor	Percentage of interaction	Nodes in the community
1	@ekj	14.0	78
2	@cnnbog	17.2	59
3	@cnnee	11.9	67
4	@aristeguionline	7.2	79
5	@adnpolitico	7.3	45
6	@bogotapn	1.4	2
7	@bogtoline	1.9	3
8	@lDolarCol	5.5	21
9	@jernacoro	2.6	5
10	@andrespe	5.6	44
11	@lopezdoriga	4.2	23
12	@viscaya	2.2	3
13	@nytimes	2.8	88
14	@erp	1.9	3
15	@pedroandaluz	2.4	4
16	@washingtonpost	1.9	78
17	@el_pais	1.1	70
18	@anatorres4	3.4	6
19	@bqlla	3.1	15
20	@clarotox	2.2	13
21	@hyg425	1.0	2
22	@claipas	1.4	10
23	@cali	1.3	9
24	@el_universal	1.3	25
25	@lasiva	2.4	14
26	@jorgenews	2.8	44
27	@reform	1.3	80
28	@the_sheet	2.9	2
29	@joseamo	3.0	2

Applying this simplification, it can be immediately noted that, in this interaction, there was a very important presence of the media. Moreover, it can be determined who were the most active users with the most weight (defined by the percentage of interaction found) in the network. In this way, it is possible to identify the users who were influencers in this interaction.

Another important feature to highlight is the possibility of analyzing sub-networks of information; that is, they can analyze what is happening within the community, the important interactions, the relevant actors in the sub-network and the forces

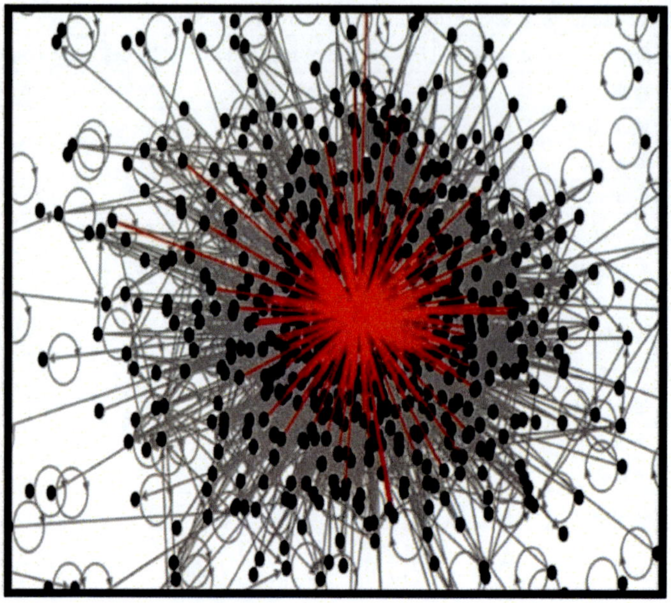

Fig. 4 User @cnbog links in the original network

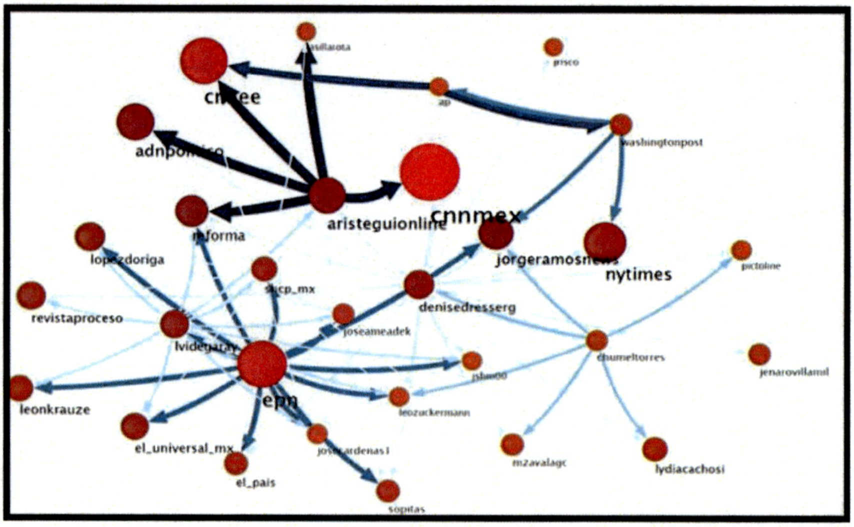

Fig. 5 Simplification of the #DolarCol network in 30 modules

interacting within the community. This analysis is of great help to determine the importance of each community and, above all, the power of influence not just of a user, but of the community [17].

5 Conclusions

The development of methods for detecting community structures in social networks is very important element in order to carry out a proper analysis. Besides, it can reveal underlying relationships between the elements of a network that can hardly be seen in the original network. In this paper, a simplification of a social network is generated from a set of interactions on a major event (trending topic), #DolarCol in this case, which took great importance at the national level on January 16, 2020. The detection of communities within the social network allowed to identify a set of important characteristics: (i) the number of communities that are generated internally, (ii) the importance of each community determined by the number of interactions and (iii) the identification of important or influential actors within the community.

References

1. Fani, H., Jiang, E., Bagheri, E., Al-Obeidat, F., Du, W., Kargar, M.: User community detection via embedding of social network structure and temporal content. Inf. Process. Manage. **57**(2), 102056 (2020)
2. Devi, J.C., Poovammal, E.: An analysis of overlapping community detection algorithms in social networks. Proc. Comput. Sci. **89**, 349–358 (2016)
3. Huang, M., Zou, G., Zhang, B., Liu, Y., Gu, Y., Jiang, K.: Overlapping community detection in heterogeneous social networks via the user model. Inf. Sci. **432**, 164–184 (2018)
4. Campos Leão, J., Amaral Brandão, M., Vaz de Melo, P. O., Laender, A.H.: Improving community detection by mining social interactions. arXiv preprint (2018) arXiv:1810.02002
5. Chouchani, N., Abed, M.: Online social network analysis: detection of communities of interest. J. Intell. Inf. Syst. **54**(1), 5–21 (2020)
6. Kim, Y., Son, S.W., Jeong, H.: Finding communities in directed networks. Phys. Rev. E **81**, 016103 (2010)
7. Viloria, A., et al.: Integration of data mining techniques to postgreSQL database manager system. Proc. Comput. Sci. **155**, 575–580 (2019)
8. Yao, K., Papadias, D., Bakiras, S. (2019, August). Density-based community detection in geo-social networks. In: Proceedings of the 16th International Symposium on Spatial and Temporal Databases, pp. 110–119
9. Moscato, V., Picariello, A., Sperlí, G.: Community detection based on game theory. Eng. Appl. Artif. Intell. **85**, 773–782 (2019)
10. Li, X., Zhou, S., Liu, J., Lian, G., Chen, G., Lin, C.W.: Communities detection in social network based on local edge centrality. Physica A **531**, 121552 (2019)
11. Sathesh, A.: Enhanced soft computing approaches for intrusion detection schemes in social media networks. J. Soft Comput. Parad. (JSCP) **1**(2019), 69–79 (2019)
12. Mohammadmosaferi, K.K., Naderi, H.: Evolution of communities in dynamic social networks: an efficient map-based approach. Expert Syst. Appl. **147**, 113221 (2020)

13. Chaudhary, L., Singh, B.: Community detection using maximizing modularity and similarity measures in social networks. In: Smart Systems and IoT: Innovations in Computing, pp. 197–206. Springer, Singapore (2020)
14. Adraoui, M., Retbi, A., Idrissi, M.K., Bennani, S.: A new algorithm to detect and evaluate learning communities in social networks: Facebook groups. Int. J. Emerg. Technol. Learn. (IJET) **14**(23), 165–179 (2019)
15. Viloria, A., Lezamab, O.B.P.: Improvements for determining the number of clusters in k-means for innovation databases in SMEs. Proc. Comput. Sci. **151**, 1201–1206 (2019)
16. Fatima, S., Badugu, S.: A study on overlapping community detection for multimedia social network. In: Advances in Decision Sciences, Image Processing, Security and Computer Vision, pp. 572–578. Springer, Cham (2020)
17. Viloria, A. et al.: Classification of digitized documents applying neural networks. In: Bindhu, V., Chen, J., Tavares, J. (eds) International Conference on Communication, Computing and Electronics Systems. Lecture Notes in Electrical Engineering, vol 637. Springer, Singapore (2020)

Improving the Performance of an Ultrasonic Sensor Using Soft Computing Techniques for 2D Localization

R. Vijay Sunder, S. Venkatachalam, G. Sree Yeshvathi, Kaza Venkat Hruday, and S. Adarsh

Abstract Ultrasonic sensors (HCSR04) are used for distance measurements whose range extends up to a distance of 4 m. These kinds of sensors are found being used in car reverse parking mechanisms and also helping blind people in navigation. But, there always exists a trade-off between distance and accuracy. This paper presents a method to improve the accuracy of an ultrasonic sensor using soft computing techniques for 2-D indoor localization. Advanced neuro fuzzy inference system (ANFIS) was used to minimize the error in distance measurement. Finally, the simulated FIS model was implemented in hardware for real-time evaluation. The error analysis showed that the distance estimation accuracy was improved by 77%.

Keywords Soft computing · ANFIS · 2-D localization

1 Introduction

Most important parameter for any sensor is its accuracy and precision. For any real-time application such as mobile robot navigation or reverse parking mechanisms, distance measurement must be accurate to avoid any treacherous situation. So, this paper explores techniques to improve the distance measurement accuracy using ANFIS. ANFIS is an intelligent soft computing technique which uses neural networks and fuzzy logic to provide better performance.

Fuzzy logic is used in situations where the input variable can take any value between 0 and 1. Neural networks work significantly on a unit called as neurons.

R. Vijay Sunder (✉) · S. Venkatachalam · G. Sree Yeshvathi · K. Venkat Hruday · S. Adarsh
Department of Electronics and Communication Engineering, Amrita School of Engineering, Amrita Vishwa Vidyapeetham, Coimbatore, India
e-mail: rvsunder@gmail.com

S. Venkatachalam
e-mail: venkatsrinivasan1998@gmail.com

S. Adarsh
e-mail: s_adarsh@cb.amrita.edu

© The Editor(s) (if applicable) and The Author(s), under exclusive license to Springer Nature Singapore Pte Ltd. 2021

A. P. Pandian et al. (eds.), *Proceedings of International Conference on Intelligent Computing, Information and Control Systems*, Advances in Intelligent Systems and Computing 1272, https://doi.org/10.1007/978-981-15-8443-5_16

ANFIS is a combination of the above two systems which can recognize the pattern and make decisions for optimizing the system. The main objective of developing an intelligent system is to make it think and manage complicated situations the same way as a human does. Basically, there are two kinds of fuzzy systems, commonly known as Mamdani and Sugeno models. ANFIS manages the inaccurate measurements by adjusting the weights associated with the input and output membership functions. This kind of system is widely used in many major fields such as improving the return loss of an antenna, hydrological modeling and a lot more.

2 Literature Survey

The commonly used ultrasonic sensors for distance measurement are the NXT sensors and Parallax Ping sensors. So, deciding on which sensor suits the best for our system is of at most importance. Performance evaluation of the sensors suggested that parallax ping was more accurate for indoor localization [1]. Also, analyzing the behavior of the sensor is very much important before using it for real-time applications. Analysis could be carried out both computationally and statistically since each senor responds uniquely at different conditions [2].These kinds of sensors are widely used for millimeter range measurements due to the fact that it has high precision. But a major disadvantage is that at this range, there are high possibilities for the accuracy to be affected by a number of environmental parameters. So, determining the factors which have a major influence in affecting the accuracy of any sensor and formulating an equation to estimate the distance of the obstacle which takes all these into account is a necessity [3]. The most common technique or the principle on which an ultrasonic sensor works is the time of flight (TOF). Several algorithms have been designed for developing a pulse-echo function which keeps track of the time duration from which the distance can be calculated [4]. Another globally accepted technique for estimating the distance is by evaluating the phase difference between the transmitted signal and received signal. This incorporates a sampling algorithm which completely relies on the bandwidth allocated to the system [5]. Thus, by utilizing a small bandwidth, a well-crafted system can be developed which has a better response time than the traditional method. One of the key factors which must be dealt with while estimating the distance is the knowledge of speed of sound and the necessity in converting temporal to spatial information. For this purpose, a signal processing technique is used which can compensate the variations in the speed of sound with the help of an extended Kalman filter [6]. Even though by compensating factors, a better accuracy system could be developed, and it is not cost effective. Also, the memory occupancy is on the higher side. Thus, with the use of software whose weight is growing exponentially in today's world, many software-based algorithms have also been developed to improve the accuracy. One of the disadvantages of this technique is that it can be used only for low-cost ranging sensors which operate in a millimeter range [7]. Generally, the sensors are used in real time within an environment filled with lot of clutters and background disturbances. This kind of unwanted signals leads to signal degradation. Thus, the effect of noise from clutters can be reduced by an adopting interpolation algorithm along with wavelet transform technique. This concept can be integrated and can be used to improve the accuracy of any time-based applications [8].

Recent trends in machine learning have taken embedded systems way beyond expectations. A low-cost high accuracy system can be designed using soft computing techniques like ANFIS, SVR and NN. But each technique performs uniquely at different conditions. The results provided by the ANFIS system are different for indoor and outdoor velodromes. Due to signal interference with clutters, the indoor localization gave better results compared to the outdoor environment [9]. But, when the obstacle is dynamic or when there are a cluster of static obstacles, determining the exact location of the target could be challenging. For this purpose, real-time collision-free path can be achieved by controlling the parameters associated with navigation. This can be done by an ANFIS controller, which provides with more accurate lanes based on the obstacle distance information given as an input to the system [10]. Developing an apt ANFIS system depends on the type of membership chosen and on the number of epochs to which the system is trained. Thus, performance evaluation of each function with various epochs needs to be done [11].

3 ANFIS Structure

ANFIS is a hybrid model which uses the learning abilities of ANN and decision handling of fuzzy logic in order to achieve a desired output. The prediction involves membership functions, if-then rules and fuzzy operators. This algorithm uses two methods namely least square method and back propagation for training. Before the start of the training process, the number of epochs or iterations must be determined. Each iteration consists of two passes—a forward pass, where identification of the parameters associated with the input is carried out and a backward pass, where the corresponding weights get updated. This process is repeated until the desired output is obtained.

The algorithm works by altering the tunable specifications which are the membership functions and the input parameters. The first and the fourth layers are the only ones which can be trained, while the other layers are fixed. The primary procedure which the ANFIS model follows is dividing the input–output data into rules. This can be done by employing any one of the methods like grid-partitioning, subtractive clustering and fuzzy c-mean clustering. In order to generate an FIS model, the data set is divided into grids. For developing an efficient FIS model, the grids must have a finer depth. This can be achieved by choosing adequate cross validation. Once the above steps have been successfully done, it leads to the generation of a Sugeno-type FIS structure.

Figure 1 depicts the ANFIS architecture with two inputs and a single out, with each having two membership functions.

Layer1: All are square nodes with a node function attached to it.

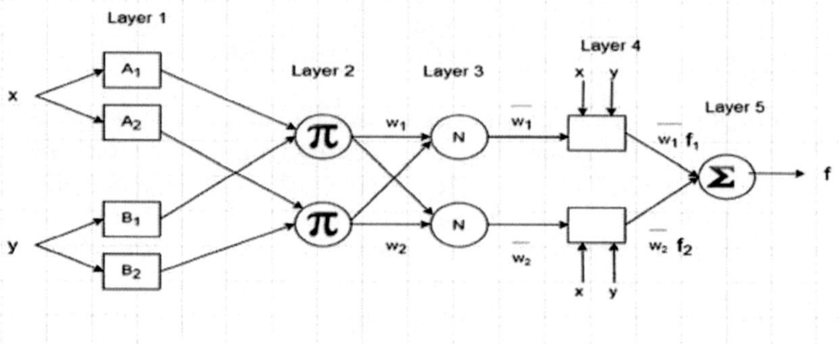

Fig. 1 ANFIS architecture

Layer2: Every node is a circular node which is named as (π) indicating that it acts as a multiplier. The output is a firing strength of a rule.

Layer3: This layer is particularly intended for normalization. All nodes are again circular, and the output calculates the ratio between the current firing strength to the sum of all firing strength.

Layer4: It is similar to layer 1 where each node is a square node with a unique node function.

Layer5: The output of this layer is just the summation of all incoming signals labeled as ($\sum$).

4 Membership Functions

Membership functions are those functions which determine the degree to which the values are mapped to an input in a space referred to as a universe of discourse. Each membership function has a unique curve where it is assumed that the mapping is done in such a way that the input data set satisfies the curve.

4.1 Trapezoid

Trapezoidal membership function has four parameters, two define its feet, and the other two define its shoulders giving it a truncated shape. Due to the fact that the graph of the function is evenly poised on both axes, it is widely being used in medical industry. Advanced diagnosis systems have been developed using the ANFIS structure integrated with stimulus algorithms and are being tested with the trapezoidal function stated in (1) wherein the parameters($x1$, $x2$, $x3$, $x4$) correspond to the x coordinate off our corners of the trapezoidal function, which proved to have a high convergence rate [12].

$$\text{trapezoid}(X, x1, x2, x3, x4) = \max\left(\min\left(\frac{X - x1}{x2 - x1}, 1, \frac{x4 - X}{x4 - x3}\right), 0\right) \quad (1)$$

4.2 GBell

A GBell membership function has a symmetrical structure with three parameters wherein one represents the width, the second is any positive integer, and the last one sets the center of the curve. Since the GBell curve has a gradual increase, it can compensate the effect of very minute parameters like temperature change. Thus, temperature control systems could be developed using the GBell function stated in (2) where the parameters ($x1$, $x3$) are the two ends of the function and $x2$ is just an arbitrary positive integer, which can handle complex data dependency [14].

$$\text{gbell}(X, x1, x2, x3) = \frac{1}{1 + \text{mod}\left(X - \frac{x3}{x1}\right)(2 * x2)} \quad (2)$$

4.3 Triangular

The triangular membership function is the elementary among all other membership functions wherein two of the three parameters represents the shoulders, while the last one is the tip of the curve. The graph of the triangular function has a pointed spike at one of its edges. Thus, systems with high precision and accuracy could be achieved with very less reaction time. Precision in detection of magnetic coating on metals could be achieved by an ANFIS model with triangular function stated in (3) in which the parameters($x1$, $x2$, $x3$) correspond to the x coordinate of three corners of the underlying triangular function, due to the fact that the function can adapt to uneven deviations [15].

$$\text{triangle}(X, x1, x2, x3) = \max\left(\min\left(\frac{X - x1}{x2 - x1}, 1, \frac{x3 - X}{x3 - x2}\right), 0\right) \quad (3)$$

4.4 Gaussian

The Gaussian membership function is widely used because of its smoothness and adapting nature to any kind of data set. It has two parameters in which one stands for the center of the curve and the other defines the width. Since this function can be used in places where there needs to be a high focus on something particular, image detection based on the area of interest could be carried out by developing an ANFIS model using the Gaussian function stated in (4) where μ and σ correspond to the mean and standard deviation of the function, which is capable of pixel degeneration [13].

$$\text{Gaussian}(X, \mu, \sigma) = \mathrm{e}^{\frac{-(X-\mu)^2}{2\sigma^2}} \tag{4}$$

5 Algorithm

The algorithm of the proposed system is given in Fig. 2.

The detailed algorithm is provided below;

STEP1: Appropriate sensors for 2D-localization were chosen. The sensor chosen was HCSR04 whose beam-width was 30° and operates in the range of 10–400 cm.

STEP2: A square-shaped model of dimension 240 cm × 240 cm was constructed wherein the sensors were placed in the midpoint of each side of the square whose coordinates were (0, 120), (240, 120), (120, 240) and (120, 0). The pictorial representation of the experimental setup consisting of sensors is depicted in Fig. 3.

STEP3: Sensors were placed at the right position in order to have maximum area of localization.

STEP4: The area under the test was designed to a dimension of 60 cm × 60 cm which was further divided into 100 identical squares having dimension 6 cm × 6 cm.

STEP5: The obstacle was placed in each of these coordinates, and the distance from the sensors was measured using Arduino wherein the real-time data captured by the Arduino are processed at a sampling rate of 10 ms, and the corresponding coordinates were calculated.

STEP6: Repeat step five for six iterations using different set of sensors, and a total of 726 discrete data points were collected.

STEP7: The data set was divided in the ratio 60:20:20 for training, testing and validation, respectively.

STEP8: The data set that was allocated for training was given as an input to the ANFIS model.

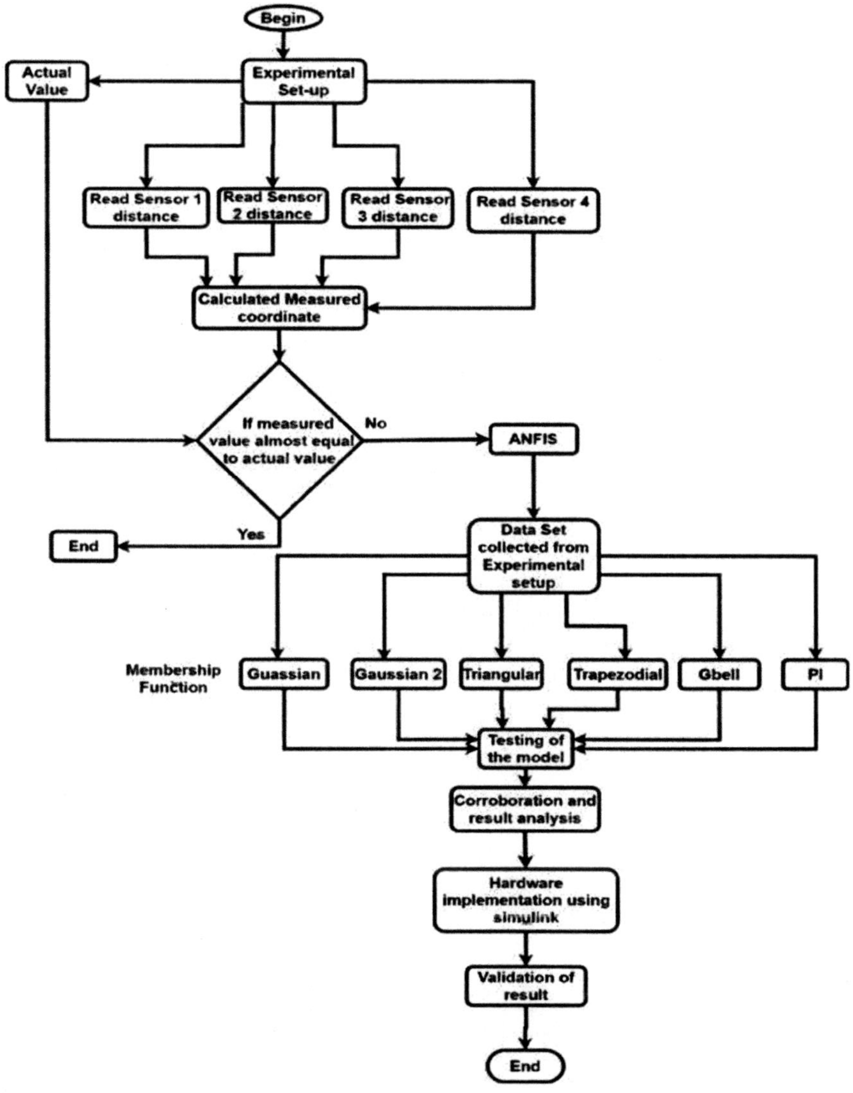

Fig. 2 Flowchart for error optimization for 2-D localization in ANFIS

STEP9: The training process was carried out with six different membership functions (Gaussian, Gaussian2, triangular, trapezoidal, GBell and pi) where the number of epochs was varied as 100 and 500.

STEP10: The data set that was allocated for testing was given as an input to these trained models.

STEP11: The error analysis was done before and after training.

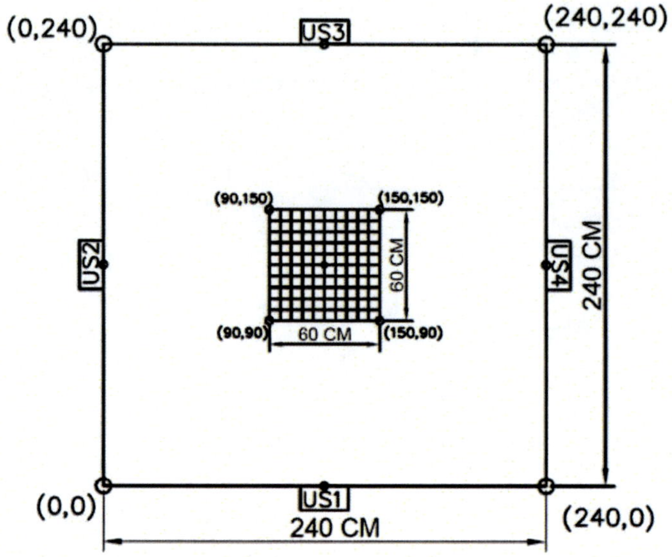

Fig. 3 Experimental setup

STEP12: Hardware implementation was done by converting the .fis file which was created during simulation to .c file and was burned onto Arduino.

STEP13: Finally, validation on the grounds of analysis made earlier was done.

6 Results and Discussion

This section contains the simulation results which were obtained using MATLAB software and is finally implemented in hardware for real-time validation. The data collected from the sensors were given as the input to the system. Determining the appropriate membership function plays a vital role for ensuring better results. Thus, the system is trained with six different membership functions namely Gaussian, Gaussian2, triangular, GBell, pi and trapezoidal, and the number of membership function is fixed at three. Furthermore, the system was trained for 100 and 500 epochs. The RMSE, MSE and MAPE were calculated for all the trained systems in a systematic approach using the expression (5), (6) and (7), respectively. Finally, validation was done on the grounds of error analysis done earlier. The results clearly indicated better performance over independent input variables which had a nonlinear relationship.

$$\text{RMSE} = \frac{\sqrt{\sum (\text{predicted}(i) - \text{actual}(i))^2}}{n} \tag{5}$$

Table 1 Error analysis for different membership function for *X* coordinate

Membership function	RMSE	MAPE	MSE
Gaussian	0.7007	0.4732	0.4909
Gaussian2	1.7695	1.1497	3.1312
Triangular	0.716	0.488	0.5126
GBell	0.7295	0.4894	0.5322
Trapezoidal	2.2423	1.5277	5.0279
Pi	2.8302	1.9462	8.0102

Table 2 Error analysis for different membership function for *Y* coordinate

Membership function	RMSE	MAPE	MSE
Gaussian	0.88249	0.573983	0.778778
Gaussian2	1.922267	1.361307	3.695112
Triangular	0.781398	0.509965	0.610583
GBell	1.037359	0.696878	1.076114
Trapezoidal	2.312404	1.632465	5.347212
Pi	2.979672	2.154552	8.878444

$$\text{MSE} = \frac{1}{n}\sum (\text{predicted}(i) - \text{actual}(i))^2 \tag{6}$$

$$\text{MAPE} = \sum \frac{\text{actual}(i) - \text{predicted}(i)}{\text{actual}(i)} * 100 \tag{7}$$

In accordance with the results obtained in the previous section, Tables 1 and 2 compare the system wise response for different membership functions for *X* and *Y* coordinates, respectively.

Figure 4 depicts a graph plotted between the different membership functions versus the mean absolute error for simulation and hardware implementation in both *X* and *Y* coordinates.

7 Conclusion

This paper explains the effect of various membership functions on the improvement of the accuracy of sensors. A total of six membership functions were taken into consideration. The RMSE for each system was calculated and recorded. The results were compared and analyzed, and it showed that triangular member function with number of epochs as 500 offered the least error. On the other hand, GBell and Gaussian almost had a similar response on the grounds of RMSE. The distance estimation accuracy was improved by 77% where the error was reduced from 1.144 to 0.259 in the case of *X*-axis coordinate and 0.99–0.233 in *Y* coordinate. Finally,

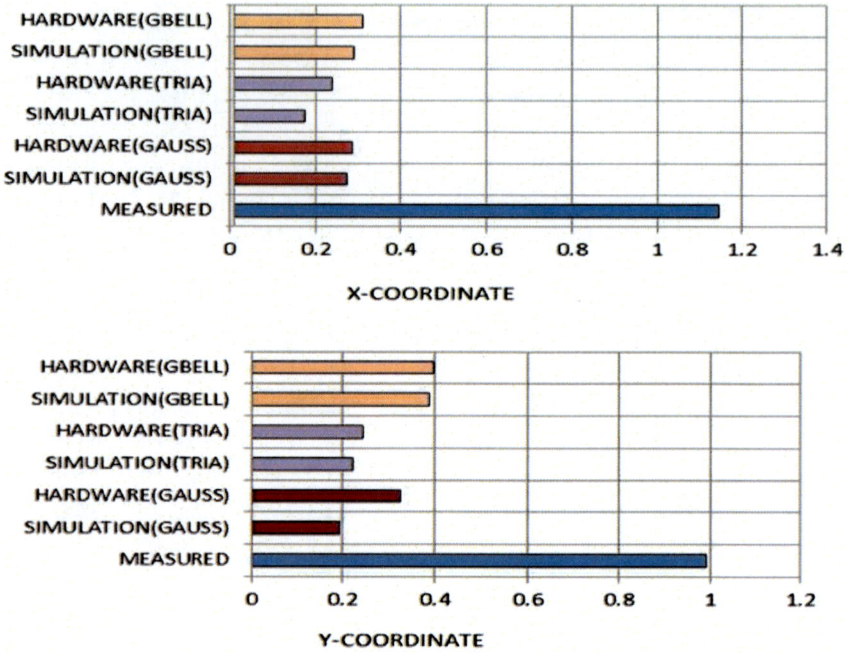

Fig. 4 Comparison of the results from simulation and hardware implementation

this ANFIS system can be used even for outdoor velodromes where the obstacle is dynamic and is intervened by clutters.

The proposed ANFIS model can be used in an environment for dynamic obstacles. Also, the above system can be trained with other soft computing techniques like neural networks and support vector regression.

References

1. Găşpăresc, G., Gontean, A.: Performance evaluation of ultrasonic sensors accuracy in distance measurement. In: 2014 11th International Symposium on Electronics and Telecommunications (ISETC), Timisoara, pp. 1–4 (2014)
2. Sankar, J., Adarsh, S., Ramachandran, K.I.: Performance evaluation of ultrasonic and infrared waves on human body and metal surfaces for mobile robot navigation. Mater. Today Proc. **5**, 16516–16525 (2018)
3. Panda, K.G., Agrawal, D., Nshimiyimana, A., Hossain, A.: Effects of environment on accuracy of ultrasonic sensor operates in millimetre range. Persp. Sci. **8**, 574–576 (2016). ISSN 2213-0209
4. Marioli, D., Narduzzi, C., Offelli, C., Petri, D., Sardini, E., Taroni, A.: Digital time-of-flight measurement for ultrasonic sensors. IEEE Trans. Instrum. Meas. **41**(1), 93–97 (1992)

5. Gueuning, F.E., Varlan, M., Eugne, C.E., Dupuis, P.: Accurate distance measurement by an autonomous ultrasonic system combining time-of-flight and phase-shift methods. IEEE Trans. Instrum. Meas. **46**(6), 1236–1240 (1997)
6. Sabatini, A.M.: A digital signal-processing technique for compensating ultrasonic sensors. IEEE Trans. Instrum. Meas. **44**(4), 869–874 (1995)
7. Khoenkaw, P., Pramokchon, P.: A software based method for improving accuracy of ultrasonic range finder module. In: 2017 International Conference on Digital Arts, Media and Technology (ICDAMT), Chiang Maipp. 10–13 (2017)
8. Andria, G., Attivissimo, F., Lanzolla, A.: Digital measuring techniques for high accuracy ultrasonic sensor application. In: IMTC/98 Conference Proceedings. IEEE Instrumentation and Measurement Technology Conference. Where Instrumentation is Going (Cat. No.98CH36222), St. Paul, MN, USA, vol. 2, pp. 1056–1061 (1998)
9. Gharghan, S.K., Nordin, R., Jawad, A.M., Jawad, H.M., Ismail, M.: Adaptive neural fuzzy inference system for accurate localization of wireless sensor network in outdoor and indoor cycling applications. IEEE Access **6**, 38475–38489 (2018)
10. Geedhu, K.V., Ramachandran, K.I., Adarsh, S.: Canfis based robotic navigation. In: 2017 International Conference on Advances in Computing, Communications and Informatics (ICACCI), Udupi, pp. 1660–1664 (2017)
11. Adil, O., Ali, A., Sumait, B.: Comparison between the effects of different types of membership functions on fuzzy logic controller performance. Int. J. Emerg. Eng. Res. Technol. **3**, 76–83 (2015)
12. Hsieh, S.-T., Lin, C.L.: Work-in-progress: an intelligent diagnosis influenza system based on adaptive neuro-fuzzy inference system. In: 2015 1st International Conference on Industrial Networks and Intelligent Systems (INISCom), Tokyo, pp. 177–180 (2015)
13. Pongpanitanont, P., Sittiprapaporn, W., Prasertsakul, T., Charoensuk, W.: ROI detection of reading process using ANFIS method. In: The 4th 2011 Biomedical Engineering International Conference, Chiang Mai, pp. 258–261 (2012)
14. Oluwo, A., Khan, M.R., Salami, M.J.E.: Intelligent temperature control of a tropical post-harvest storage system. In: 2015 10th Asian Control Conference (ASCC), Kota Kinabalu, pp. 1–6 (2015)
15. Basari, A.S.H., Jaya, A.S.M., Muhamad, M.R., Rahman, M.N.A., Hashim, S.Z.M., Haron, H.: Application of ANFIS in predicting TiAlN coatings flank wear. In: 2011 Third International Conference on Computational Intelligence, Modelling & Simulation, Langkawi, pp. 57–62 (2011)

Music Information Retrieval and Intelligent Genre Classification

Rahul Gupta, Jayesh Yadav, and Cheshtha Kapoor

Abstract Music is a sound that arises sensations in human mind and body. Music, since the beginning of time, has been present in every culture and life form, in audio and symbolic form, and in physical or digital mode of communication. While availability and scope for advancements increase exponentially, so does the need to search, compare, and organise music. Music industry has been striving towards finding the best possible approach to categorise music whether classification on the basis of emotions, instrumentation, genres, or any other music information will be most efficient as well as useful to listeners and music service providers. With the aim to support the best music experience, the current study statistically shows, with the help of prior research in music information retrieval and implementation of several powerful machine learning-based technologies, that genre classification, that too, ensemble-based, can be as accurate as 73.17%. The study analyses the performances of all models towards genre classification and concludes by proving max-voting ensemble-based models to be more accurate than each component classifier and advanced ensemble models and also optimal for real-world music genre classification as compared to prior experiments on GTZAN database, which is the novel contribution of the study.

Keywords Music · Genre · Machine learning · Features · Pre-processing · Categorisation · Classification · Accuracy · Ensemble

---

R. Gupta (✉) · J. Yadav (✉) · C. Kapoor
Department of Computer Science and Engineering, Delhi Technological University, New Delhi, India
e-mail: rahulgupta100689@gmail.com

J. Yadav
e-mail: jayeshyadav2004@gmail.com

C. Kapoor
e-mail: cheshtha.kapoor10@gmail.com

© The Editor(s) (if applicable) and The Author(s), under exclusive license to Springer Nature Singapore Pte Ltd. 2021

A. P. Pandian et al. (eds.), *Proceedings of International Conference on Intelligent Computing, Information and Control Systems*, Advances in Intelligent Systems and Computing 1272, https://doi.org/10.1007/978-981-15-8443-5_17

1 Introduction

In 2012, the worldwide music industry revenue was valued at $45.5 billion [1]. Since then, a significant growth has been observed in online and offline streaming, live music, as well as downloading revenues. In 2019, the worldwide music industry revenue was amounted $56.7 billion and is anticipated to rise as high as $65.17 billion by 2023 [1]. The revenue of offline or physical music industry in India was estimated at around $9 million in 2017 [1], significantly less than $13 million revenue generated in the previous year [1]. In contrast, in 2016, the revenue of global streaming went up by 45.2%, driving 3.2% of global growth [1].

Due to the increasing interest of consumers as well as music industrialists in the growth of digital music streaming market, the goal is to simplify music listening, searching, storing, organising, and recommending experience of users.

The study understands the needs and advancements of music information retrieval (MIR) via research on prior art and establishes why genre classification is the best method for retrieval of music information. With the deterministic aim of developing intelligent genre classifiers using advanced machine learning, the study describes the dataset and genres considered (Country, Classical, Blues, Disco, Hip-Hop, Pop, Jazz, Rock, Metal, and Reggae), as well as the methodology adopted for modelling different kinds of classifiers, such as k-nearest neighbour (KNN), support vector machines (SVMs), random forest, Naïve Bayes, logistic regression, and artificial neural networks (ANNs). Here, the dataset is made up of 1000 audio music files, each of 30-s duration and 22.050 kHz 16-bit mono audio.WAV format. Various statistical features such as timbral features (spectral centroid, spectral bandwidth, zero-crossing rate, and spectral rolloff), mel-frequency cepstral coefficients (MFCCs), and chroma features for each audio signal are extracted from their spectrograms and compiled in a.CSV file to undergo pre-processing (handling categorical and erroneous data, dimensionality reduction, and feature scaling) and act as input for modelling the classifiers.

Each of the classifiers mentioned above is trained and tested using the same sets of data, and their independent performances are reported using confusion matrix estimates (like precision, recall, and $F1$-score) and k-cross-validation. The study also develops an ensemble model comprising of the best capabilities of each of the above-mentioned classification models. The ensemble model votes from the predictions of component classifiers (using max-voting mechanism) and develops the final result of genre classification with a better accuracy than any of the individual models (evaluated using k-cross-validation), as hypothesised.

The study also compares performances of various basic and advanced ensemble models (random forest, max-voting, AdaBoost, and XGBoost) and assesses the results of the experiments when compared to prior art in the field. Lastly, the study discusses the future scope and capabilities of ensemble-based genre classification as well as MIR, beneficial to the forthcoming era of enriched musical experience.

2 Literature Review

To understand MIR needs of people, study [2] considered 502-real-world queries on music art via Google ask-an-expert out of which 81.3% queries involved metadata (title of work, performers, date of composition, instruments, etc.), whereas 32.7% were genre-based and 28.9% included lyric excerpts. An extension of this study [3] reflects the paradigm shift from text-based queries to 86% queries resolved using audio techniques, and 47% resolved using music information retrieved, such as melody and mood. The study also discussed various tasks of MIR, such as artist, beat, composer, and cover song identification, similarity and mood detection, genre classification, and query by humming. In study [4], statistical analysis on over 400 K playlists, managed over 20 years, was carried out to categorise music on the basis of popularity and diversity indexes, designed by the researchers. The study discussed MIR application in playlist creation and continuation from user and algorithmic end.

Study [5] took timbral texture, rhythmic and pitch features of audio signals from GTZAN dataset to perform 61% accurate statistical pattern recognition by using Gaussian mixture model (GMM) (to accurately estimate the probability density function) and KNN (with multiple random starting points which acts as initialiser for GMM). Study [6] utilised SVM, principal component analysis (PCA), and Marsyas software for integrating audio processing with machine learning to detect similarities between five prevailing genres of Sri Lankan music, namely Ragadari, Classical, Western, Baila, and Calypso. MFCCs, zero-crossing rate (ZCR), chroma features, and rhythm histograms, to represent music/silence in the audio, were extracted, and prediction accuracy of the model was 74.5%. The study established Classical music roots in Indian-origin Ragadari music via large similarities.

Study [7] adopted fast Fourier transform (FFT) to generate features of spectral centroid, spectral rolloff, spectral flux, and ZCR along with MFCC and Daubechies wavelength coefficient histograms (where audio signal was split into frames, based on frequency, to analyse each and generate histograms by averaging out the probability distribution. For 1000 audios distributed on 10 genres, the study used five-KNN, SVM, and Gaussian mixture algorithms plus linear discriminant analysis to result in 72% model accuracy. Study [8] utilised 400 60-s audios distributed on four genres to extract ZCR, MFCC, spectral flux, and linear predictive cepstral coefficients features and implemented 97.8% accurate deep neural network on dataset distributed into 60% training data, 20% four-cross-validation data, and 20% test data. Study [9] utilised 84.7% accurate dictionary learning algorithm on GTZAN database, where signals were split into 5-s frames and a dictionary of code words was obtained to train using SVM algorithm.

3 Methodology

The methodological approach for the current study firstly includes a statistical analysis from prior art in the field to discuss the importance of genre classification. This sets a founding stone for our study to develop and assess established as well as novel and advanced techniques to classify genres. The study defines GTZAN dataset of 1000 audio files, each of 30-s duration and 22.050 kHz 16-bit mono audio, distributed uniformly across the most commonly listened to genres considered in the study, namely Country, Classical, Blues, Disco, Hip-Hop, Pop, Jazz, Rock, Metal, and Reggae. Next, various features were extracted from the generated spectrograms of each audio file using *Librosa* library of Python (spectral centroid, spectral bandwidth, zero-crossing rate, spectral rolloff, MFCCs, and chroma features), and respective feature vectors were stored simultaneously into an empty.CSV file.

The obtained statistical dataset was then pre-processed to remove any possible variation in formats, ranges, or containment of erroneous or missing features. In the study, categorical data (genre labels) were transformed into numerical data using label encoding, and all dimensions of the dataset were scaled using standardisation. Finally, the dataset was split into training and testing data at 70% mark for modelling the classifier. Each classifier model (KNN, SVM, random forest, Naïve Bayes, logistic regression, and ANN) was then developed using *scikit-learn* tools of Python and trained using the same dataset to learn patterns.

To draw the best of the abilities of each classifier into one model, which not only performs better in terms of prediction accuracy, but is also more efficient and reliable, an ensemble of the above-mentioned classifiers was developed, which produces cumulative result using max-voting mechanism. Apart from this, advanced ensemble techniques (AdaBoost and XGBoost) were also implemented.

Lastly, performances of individual classifiers and basic and advanced ensemble models were evaluated using confusion matrix estimates and k-cross-validation. The findings were then compared with prior art findings to obtain novel results.

4 Genre Classification in MIR

According to [3], MIR includes tasks of identification of artists, beats, tempo, moods, instruments, genres, etc. Some of these tasks can be accomplished via text-based queries, while others using audio-based techniques (86% of queries studied), such as similarity detection and genre classification. Moreover, findings of study [2] on people's needs of MIR showed that 81.3% of the queries involved metadata of music (like performers, title of work, instrumentation, language, date of composition, etc.), whereas the next best technique for MIR was querying by genres (32.7%).

While querying music with the help of bibliographical metadata can be convenient for users, development end for this requires huge inventories. Moreover, comparison of musical tracks, playlist creation, and online or offline organisation with metadata may not be useful for listeners or service providers. Hence, the study infers that classification of music based on which genre (or style/form) a music audio belongs to can be most beneficial to categorise, compare, as well as store and manage music by personal choice.

5 Dataset

Genre of music is a category that can be assigned to a music track or piece of it. Since the invent of music, these categories have been created by human experts of music to represent unique traditions, conventions, sensations, and style. As music gets accessed on physiological effects, such as sensation and stimulation (thrilling or calming), emotional effects, such as moods and influence (happy or depressing), and depth of music (inspiring or casual partying), genres can be described on various combinations of these factors and hence give rise to a wide variety of genre sets.

Majority of prior studies based on music information retrieval worked upon a genre set of 3–4 genres. However, limited genres may not be able to exhaustively define all forms of music. The current study employs ten distinguishable genres, namely Country, Classical, Blues, Disco, Hip-Hop, Pop, Jazz, Rock, Metal, and Reggae. For each of the ten genres considered, 100 distinct music audio files were collected (a total of 1000 music audio files for all genres) from Marsyas website. These audio files were initially collected by G. Tzanetakis and P. Cook during 2000–2001 from various sources, such as CDs, microphone recordings, and radio for their research study [5]. The dataset is also available for download and is known popularly as the GTZAN dataset [10]. Each of these audio files was of 30-s duration and 22.050 kHz frequency, 16-bit mono audio kind, and available for creditable use in.WAV format.

It is assumed that the genre labels assigned to each audio file were by human music experts. Additionally, only the most suitable genre has been assigned to each music file in the current study, eliminating the consideration for multilabel classifications. The research agrees that human's music perception is subjective; however, our aim is to ensure that the labelling input as well as the results tends to be as close as the ground truth.

6 Feature Extraction

Since classifier systems work best upon numerical data and direct modelling of audio signals is tricky, spectrograms of each audio file in the dataset were iteratively generated using *Librosa* library of Python and subsequently stored. Later, spectrograms for each individual audio file were invoked, and all necessary corresponding features were extracted to obtain a feature vector corresponding to each audio signal. In prior experiments, for similar extraction and classification techniques as well as for similarly large dataset, 3–6 features have been employed. Including MFCC features, size of feature vectors for such experiments ranges from 20 to 25. On parallel lines, various features extracted are described.

1. **Timbral features**. Standard features of audio music that differentiates tracks on the basis of instrumentation and compositions are known as timbral features. These represent musical roughness, flatness, density, resolution, brightness, loudness, and tension.
    a. **Spectral centroid**. It is the measure of the location of the centre of mass for a sound. It is computed as the weighted mean of all frequencies comprising the sound signal.
    b. **Spectral bandwidth**. It is the measure of the resolution of an instrument or a signal, that is the ability of a musical instrument to separate sound into different wavelength regions and distinguish them. It is the interval of wavelength in which spectral energy is greater than or equal to half of the spectral quantity's maximum value.
    c. **Zero-crossing rate**. It is the measure of the rate of change of signs (positive to negative or vice versa) along a signal, that is, zero-crossings of the audio signal. For highly percussive music (composing of instruments that are struck to produce sounds, such as drums and gongs), ZCR has high values.
    d. **Spectral rolloff**. It represents the configuration of the music signal as the measure of the frequency below which a pre-determined percentage (by default, 85%) of the total spectral energy or amplitude concentration lies.
2. **Mel-frequency cepstral coefficients**. MFCCs are a set of features (cardinality 20) that describe the overall shape of the signal's spectral envelop (one brief window or point in time that has the ability to represent the features of the vocal tract of humans, like, teeth, tongue, windpipe, etc.). The estimates for each frame or window are aggregated.
3. **Chroma features**. Representing the tone, melodic and harmonic characteristics of an audio, this feature is a 12-elemental vector that projects all the energy occurring in the music into one of the 12 defined classes or bins of pitches. These representations are then aggregated using standard deviation.

These feature vectors were simultaneously appended into an empty.CSV file with the help of *CSV* library. Various columns in the data file represented the features extracted from spectrograms, while each row was named as the particular audio filename and represented the feature vector for that audio file <filename, chroma_stft, spectral_centroid, spectral_bandwidth, rolloff, zero_crossing_rate, mfcc1, mfcc2, …, mfcc19, mfcc20, and label>. The data file for feature vectors for 1000 audio files shall now acts as the dataset input to developed classifiers.

7 Pre-processing

As the data being collected for analysis can be from various sources, the data may be varying in format, range, or contain corrupted or missing values. To convert input raw data into cleaner and more manageable format, pre-processing of data is done before feeding into the model. This is because machine learning algorithms are incapable of handling garbage values to be worked upon statistically. This, therefore, becomes the first step towards creating a machine learning model.

Pre-processing is a multi-step process as described below.

1. **Importing necessary libraries**. To handle data (*Pandas*) and data file (*CSV*), work upon multidimensional data (*Numpy*), module to process audio signals (*Librosa*), and libraries necessary for modelling classifiers (*sklearn, keras*).
2. **Importing dataset**. "*read_csv*" function of *CSV* library used to import copy of the dataset, to perform required analysis without altering original dataset.
3. **Handling missing and erroneous data**. Imported dataset was to be cleaned for missing or erroneous data by either entirely removing the tuple or replacing the noisy data by average of the feature values for other data points. Since the data file of features was prepared in the study itself, it was taken care that no missing or erroneous data exists.
4. **Handling redundant data**. To eliminate redundancy caused by possible identical audios, "*drop_duplicate*" function of *Pandas* library was employed with default parameter values to in-place remove duplicate occurrences.
5. **Handling categorical data**. Machine learning models, expect for text analysis, prefer to work upon numeric data for statistical analysis. The categorical data (genre labels) were therefore transformed alphabetically into numeric values using label encoding technique and *LabelEncoder* class function "*fit_transform*" from *sklearn.preprocessing* library.
6. **Splitting the dataset**. Using "*train_test_split*" function of *sklearn.model_selection* library, the dataset thus pre-processed was split into two subsets—training dataset (to train the model to learn generalisations from existing classification patterns) and testing dataset (to test model performance before presenting into real-world problems) at a 70% mark.
7. **Feature scaling**. In computations involving statistical distances of closeness between data points, large ranges of numeric data may cause extreme values or noise-inducing outliers to dominate the prediction results. Hence, in the current study, using standardisation technique, each dimension *(X)* was scaled to have mean (μ) as 0 and standard deviation (σ) as 1. For this, *StandardScaler* class functions "*fit_transform*" and "*transform*" of *sklearn.preprocessing* library were used for training and testing datasets, respectively.

$$X_{\text{updated}} = \frac{X_{\text{original}} - \mu}{\sigma} \tag{1}$$

8. **Dimensionality reduction**. The dataset currently contains large number of dimensions out of which some features may not contribute new insight towards predictions, while others may be correlated such that more than one features can be represented as a single feature. To improve computational efficiency as well as prediction clarity and capability, while retaining most of the core information, linear discriminant analysis (LDA) for dimensionality reduction was adopted. Using *LinearDiscriminantAnalysis* class function "*fit_transform*" of *sklearn.discriminant_analysis*, 26-dimensional dataset was reduced to 13-dimensional dataset (*n_components* parameter = 13). LDA also reduces remnant noise in the data by eliminating the axes containing noise.

8 Modelling Classifiers

Once the dataset had been pre-processed, next comes modelling of classifiers. Six machine learning classifier models were developed using *scikit-learn* tools and subsequently trained using the same 70% of original dataset of audio features. These were then tested on the remaining 30% dataset, and individual performances were evaluated using confusion matrix and *k*-cross-validation techniques.

Various classification algorithms implemented in the study are as follows.

1. **K-nearest neighbour algorithm**. This adopts majority voting mechanism to make classification for a data point based on its nearest *k* data points' labels.
2. **Gaussian support vector machine**. This adopts modelling using non-linear hyperplane by mapping data points to higher dimensions and then reducing them to original Cartesian *(kernel trick)*.
3. **Random forest algorithm**. An ensemble model that classifies by majority voting of predictions of defined number of decision trees passed as hyperparameters.
4. **Naive Bayes algorithm**. This adopts probabilistic approach while deciding prediction by using Bayes theorem of conditional probability.
5. **Logistic regression**. In multiclass classifiers, it is the continuous adaptation of multiple *One-vs-Rest* binary logistic regression classifiers on the same data point until a particular class is predicted as its label.
6. **Artificial neural networks**. Large interconnections are present between input feature vectors and output class labels; amidst several hidden layers [computing upon input features and assigned weights using *sigmoid* activation function, Formula (2)], it uses backpropagation technique to obtain minimal cost function (*root mean square error* of actual and predicted results).

$$S(x) = \frac{1}{1 + e^{-x}} \tag{2}$$

where x is the input or feature vector value.

The general procedure followed to model each classifier individually is as below.

Table 1 Table for libraries, classes, and hyperparameters used for various classifiers

Algorithm	Python library	Library class	Hyperparameters passed
KNN	sklearn.neighbours	KNeighboursClassifier	*n_neighbours = 5, metric = 'minkowski', p = 1* (Manhattan measure)
SVM (Gaussian kernel)	sklearn.svm	SVC	*kernel = 'rbf', random_state = 4*
Random forest	sklearn.ensemble	RandomForestClassifier	*n_estimators = 10, criterion = 'entropy', random_state = 4*
Naïve Bayes	sklearn.naive_bayes	GaussianNB	
Logistic regression	sklearn.linear_model	LogisticRegression	*random_state = 0*
ANN	keras, keras.wrapper. scikit_learn	Sequential, Dense, KerasClassifier	*output_dim = {256, 128, 64, 10}, activation = 'relu', input_dim = 13 create_model (batch_size = 128, epoch = 25)*
Max voting ensemble	sklearn.ensemble	VotingClassifier	*voting = 'hard'*
AdaBoost ensemble	sklearn.enemble	AdaBoostClassifier	*base_estimator = classifier_i, n_estimators = 20, algorithm = "SAMME"*
XGBoost ensemble	xgboost	XGBClassifier	

1. Necessary libraries and classes of Python are imported for each classifier, and respective object classifiers are created using dataset-based set of hyperparameters (as described in Table 1).
2. Each classifier was fitted over the training data using "*fit*" function to learn generalisation of classification patterns.
3. The trained model was tested on test data to predict genre labels on the basis of feature vectors of test data using "*predict*" function.
4. The predicted results for each classifier were stored and compared with actual genre labels of test data points using them as parameters for *confusion_matrix* class of *sklearn.metrics* library, and respective classification report containing performance metrics (precision, recall, and $F1$-score) is generated (Fig. 1).
5. Using "*cross_val_score*" class of library *sklearn.model_selection*, objects for each model were trained on testing data using ten-cross-validation technique to obtain respective mean and maximum possible accuracies (Fig. 2).

Confusion matrix is a *nXn* matrix (*n* refers to the number of genre classes in the study, 10), where each entry $X_{i,j}$ refers to the count of how many cases occurred where the audio file actually belongs to *i*th class/genre, but was predicted to be of *j*th class/genre.

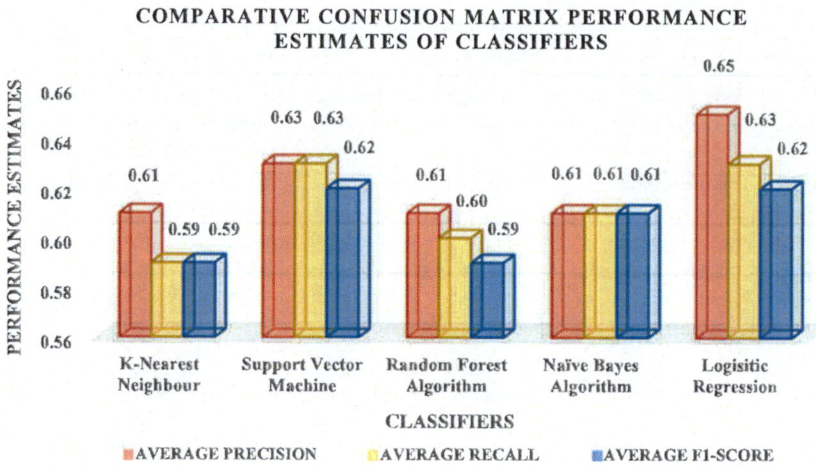

Fig. 1 Comparative confusion matrix performance estimates of classifiers

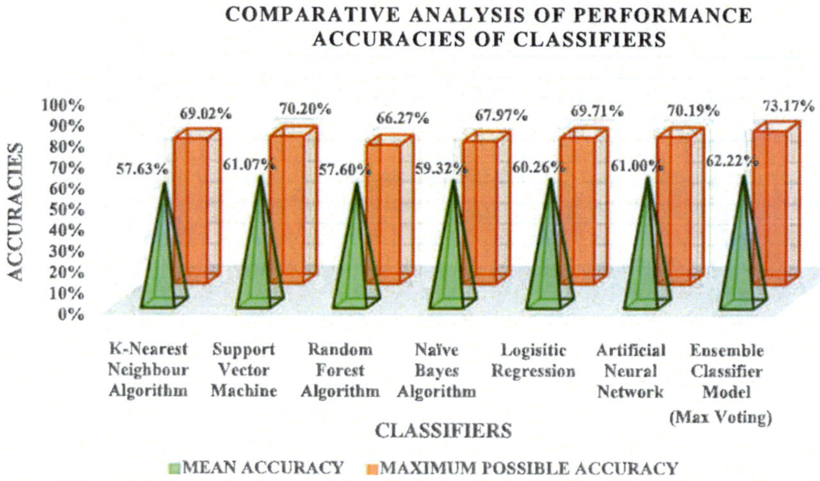

Fig. 2 Comparative analysis of performance accuracies of classifiers

Performance measures extracted from confusion matrices are listed below:

1. **Precision of predicting a particular class C or positive prediction rate.**

$$\text{Precision} = \frac{\text{Correct predictions for class C}}{\text{Sum of correct predictions and false positives}} \tag{3}$$

2. **Recall or sensitivity for a particular class C or true positive rate.**

$$\text{Recall} = \frac{\text{Correct predictions for class C}}{\text{Sum of all instances where actual class is C}} \tag{4}$$

3. **F1-score of a particular class C**. It weighs *Recall* and *Precision* to contribute equally as a performance measure. Best value for $F1$-score is '1', whereas the worst value is '0'.

$$F1\text{-Score} = \frac{2 * \text{Precision} * \text{Recall}}{\text{Precision} + \text{Recall}} \tag{5}$$

A closer to '1' value of each of the above estimates proves that the model is good. To compute their respective overall estimates, weighted average (of their values for each class) is calculated. Here, weights refer to *Support*, that is the number of instances in test dataset for each class label.

In k-cross-validation method, the initial testing dataset is divided into k equal parts. One of these parts is held back for testing, whereas the other $k - 1$ parts are used to train the model. The model is then tested with one of the k parts that was held back. This step is repeated k number of times to provide each segment of the testing dataset an opportunity to train as well as test the data. This introduces randomness and also increases the possibility for better training and prediction accuracy as a model's performance is highly biased on how the initial dataset has been split into training and testing datasets.

The "*cross_val_score*" function used above provides a list of k accuracies, one for each iteration performed. To compute the overall accuracy, the mean of these accuracies is taken. Also, adding the computed standard deviation to the mean, maximum possible accuracy of the learning model is derived.

9 Ensemble Classifier

An ensemble classifier model operates on a finite set of component classifiers by combining their individual predictions (in weighted or unweighted manner) to give a classification model exhibiting more reliable predictions. Six component algorithms, as stated earlier, have been modelled individually, and using a defined voting mechanism, the predictions and respective accuracies of each model are taken into account to predict the final ensemble-based genre class.

According to [11], the necessary and sufficient condition for an ensemble classifier to be more accurate in prediction than any of its component classifiers is that each of the component classifiers must be individually accurate (better than random guessing 50%) and diverse (for an unknown data point, different errors are made by different algorithms). Now, the voting mechanism will be able to filter out the best possible classification result while being more than 50% sure. Ensemble learning, not only statistically for underfit data, but also computationally for overfit data, is found to be more efficient. It is able to initiate the algorithm from multiple starting points to introduce distributed randomness, increasing the representational domain of ensemble classifier.

For development of the ensemble classifier model, following steps were followed.

1. Label dictionary to store component classifier objects was defined to be used as list of estimators.
2. An object of imported *VotingClassifier* class from *sklearn.ensemble* library was created passing label dictionary as estimators and "*hard*" as the voting mechanism (referring to max-voting process conducted using absolute, not probabilistic, majority vote amongst predictions of individual classifiers).
3. The classifier was fitted over the training data using "*fit*" function to learn generalisation of classification patterns.
4. The trained model was tested on test data to predict genre labels on the basis of feature vectors of test data using "*predict*" function.
5. The predicted results for ensemble classifier were stored and compared with actual genre labels of test data points using ten-cross-validation technique to obtain resultant mean and maximum possible accuracy for the classification model (Fig. 2).
6. For XGBoost boosting ensemble model, classifier object of imported *XGBClassifier* class from *xgboost* library was created with default parameters and trained on original training dataset (without voting on component classifiers) with the help of inherent decision tree modelling (Table 1).
7. As for AdaBoost boosting ensemble model, each component model was individually boosted for accuracy using object of imported *AdaBoostClassifier* class from *sklearn.ensemble* library with parameters *base_estimator* (component classifiers iteratively), *n_estimators* (number of decision trees for inherent modelling), and other default parameters. The boosted models were then ensembled using steps 1–5 (Table 1).

10 Performance Analysis and Findings

The performance reports of each of the nine classifiers (six component classifiers and three ensemble models) with respect to genre classification prediction of the GTZAN test dataset has been presented graphically in this section.

Figure 1 graphically depicts the comparative confusion matrix performance estimates of five of the component classifiers, namely k-nearest neighbour, support vector machine, random forest algorithm, Naïve Bayes algorithm, and logistic regression. Please note that the remaining classifiers (artificial neural network and basic and advanced ensemble models) have been evaluated for prediction accuracy using *k*-cross-validation technique along with the formerly mentioned algorithms as shown in Fig. 2.

From Fig. 1, the study observes that logistic regression classification model turns out to be the most precise amongst the set which attributes to its continuous iterative binary classification approach until a single genre label is predicted for a test data point (*One-vs-Rest* technique for multiclass classifications). By doing so, the algorithm tends to result in a larger amount of correct predictions than false positives. Also, the average recall and average $F1$-score are the highest and coincide for support vector machine and logistic regression algorithms describing that these can be used interchangeably for large datasets, to provide the correct prediction for a data point from all data points available for the related class. The study infers that for GTZAN dataset, SVM models turn out to be second best approach, followed by Naïve Bayes model. In the latter, equal estimates of average precision and recall depict that on an average, the algorithm results in equal amount of incorrect predictions, that is in the favour of a class 'C' (false positives) or against it (false negatives). Amongst the five classifiers noted above, random forest model and k-nearest neighbour algorithm perform the least beneficially for the GTZAN dataset in terms of confusion matrix performance estimates.

Next, Fig. 2 graphically depicts the comparative estimates of performance accuracies of all six component classifiers, along with max-voting basic ensemble model. Initially, the study implemented and assessed a basic ensemble classifier model that worked upon the above-mentioned component classifiers using simple majority voting mechanism, max-voting. Later, advanced ensemble models were also implemented, and their performance accuracies compared (as in Fig. 3).

Please note that these classifier models have been independently evaluated for prediction accuracy using tenfold cross-validation technique on GTZAN test dataset.

From Fig. 2, the study observes that the developed basic ensemble model (max-voting) outperforms all component classifier algorithms that were implemented on

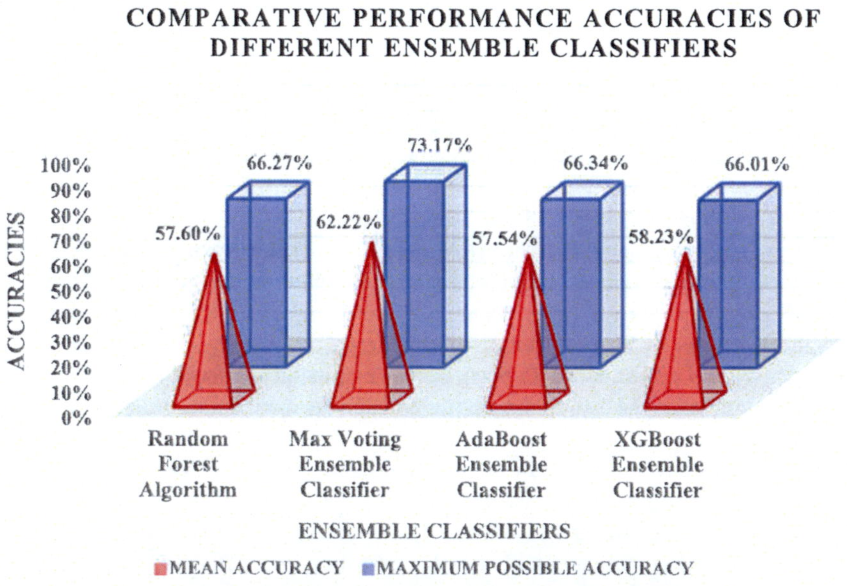

Fig. 3 Comparative performance accuracies of different ensemble classifiers

the same training and testing datasets as well as the same modelling environment, in terms of both mean and maximum possible accuracies. This duly fulfils the anticipation of the researchers of the study that ensemble-based machine learning models can be successfully as well as efficiently be implemented for modern-day contemporary classification of music genres to create more accurate and reliable genre classification systems, for a better and more useful retrieval of music information. Even though component classifier algorithms performed relatively good, yet the slight improvement in performance accuracies of ensemble model over the others illustrates that ensemble-based machine learning techniques, when amalgamated with the appropriate values of hyperparameters and estimators (component classifiers), can induce larger improvement in genre class prediction performance. Considering that max-voting ensemble technique takes into account equally weighted individual votes of each component classifier, instead of opting for a probabilistic approach, proves to perform better in the current case of large number of classification labels.

As observed from the comparative chart (Fig. 2), artificial neural network and Gaussian SVM immediately follow max-voting ensemble classifier in performance accuracy and perform rather good in comparison with other four classification algorithms implemented. This is because SVM and ANN, when performed independently on the same dataset, display and prove the powerful characteristic nature of these algorithms—kernel trick (for SVM) and backpropagation (for ANN). On the other hand, as observed previously in Fig. 1, KNN and random forest classifier models again portray the lowest mean accuracies amongst the seven classifiers set considered in Fig. 2. The study can now substantially infer that KNN and random forest algorithms performed least accurately than any other classifier implemented in the study. As for the subordinate prediction performance of Naïve Bayes classifier algorithm, it is attributed to the non-linear nature of GTZAN dataset and large number of classification labels considered in the study. Since the data points are known to be distributed, for any data point, the posterior probability computed in Naïve Bayes algorithm implementation will be low due to the large number of deciding labels. This is confirmed with the better prediction performance accuracy of SVM model that employs higher dimensional mapping using kernel trick.

Moreover, since accuracy and other metrics of a model's performance are highly biased on how the initial dataset has been shuffled and split into training and testing datasets, a more random technique ought to be applied for training and testing as well. The *k*-fold cross-validation technique used to measure the final accuracy of each individual as well as their ensembled voting classifier gives a better insight of the accuracy and performance as compared to confusion matrix report considering that each segment of the testing dataset has been given equal opportunity to train and test the model.

Figure 3 graphically depicts the comparative estimates of performance accuracies of different kinds of ensemble classifier models developed, trained, and tested in the same modelling environment and dataset. These include basic ensemble model max-voting, advanced boosting ensemble models AdaBoost and XGBoost classifiers, and at last random forest classification algorithm implemented as a component classifier earlier. Please note that all the mentioned classifier models have been independently evaluated for prediction accuracy using tenfold cross-validation technique on the GTZAN test dataset.

Figure 3 also shows that the earlier presented max-voting basic ensemble model outperforms each of the other ensemble classifiers in the set (in terms of both mean and maximum possible accuracies) for large multiclassification GTZAN dataset. This invokes the idea that although XGBoost and AdaBoost are powerful accuracy boosting algorithms, however, for a complex, distributed, and large multiclassification dataset, these do not improve the accuracy of the ensemble mode, but only reduce it further. Max-voting mechanism, which is a comparatively simpler voting technique, nullifies the complexity induced by the current underfitting dataset and proves to be more powerful in terms of performance and accuracy when compared with other advanced ensemble techniques.

XGBoost advanced ensemble model, which is a powerful inherent decision tree-based framework, performs least accurately in the current study. This can be attributed to its inability to build on the component classifiers as estimators, unlike AdaBoost and max-voting models, although working upon the same training and testing dataset. On the other hand, AdaBoost, which is another advanced boosting algorithm, was used to boost or improve the performance of each individual estimator (classifier) and in turn used to develop the ensemble classifier using majority voting mechanism. However, attributing to the complexity of multiclass dataset, AdaBoost was unable to boost the performance accuracy of individual models as well as the ensemble classifier.

Lastly, random forest model, which was one of the six component classifiers utilised for max-voting ensemble model, is in itself an ensemble of several decision trees inherently created by the algorithm to perform classifications. Here, the algorithm predicts the outcome class of a test data point on the basis of equally weighted votes received by individual decision trees. However, the performance for random forest model was reported least accurate in the set and noted as the lowest performing classifier amongst all nine classifiers considered in the study.

In recent as well as prominent prior art in the field of MIR, appreciable results have been achieved. This includes advanced mechanisms for feature extraction, different from those adopted in the current study, to implement genre classification. For instance, supervised online dictionary learning algorithm was employed by study [9] to create a dictionary of code words extracted from 5-s frames of audio signals. These keywords were then aggregated and used as features to classify music into various genres using SVM training. The study obtained successful accuracy of 84.7% for GTZAN dataset. However, the study of online dictionary learning limits when working upon noisy training dataset and tends to exhibit significant sensitivity towards the same. With the lack of feasibility to store and transfer large datasets, dictionary learning classification algorithms cause difficult interpretation and computational complexity [12]. Apart from being a highly intricate algorithm, lack of necessary auxiliary tools, sufficient related literature, base models, and methodology knowledge raises the question that although being more accurate than the current study, online dictionary learning cannot be efficiently and realistically applied to everyday real-world classification problems, quite unlike tried and tested ensemble-based machine learning models.

Another approach adopted for genre classification of GTZAN dataset was the use of Daubechies wavelet coefficient histogram for feature extraction [7] exhibiting 78.5% accuracy with SVM training. Similar issues arise as described previously for online dictionary learning, along with the presence of vanishing moments, limiting the ability of polynomial representation done by wavelets. Also, repeated periodicity must be considered to handle edge problems, which is not generally the case of real-world music.

For models that considered substantially smaller datasets (approximately 100 audio files distributed over 3–5 distinct genres), resulting in faintly better accuracies than the current study, such as 74.5% accurate [6], the presented ensemble model proves to exhibit optimum results considering underfitted data with 1000 audio files distributed over ten genres in a multiclassification problem.

Nevertheless, the current study and methodology have performed significantly optimally, in terms of both accuracy and computational convenience, when compared with previous approaches. This includes study [5] that used Marsyas software for integrating machine learning and audio processing on large GTZAN dataset, but with lower accuracy 61%. Also, the current work on ensemble-based machine learning classification model is more optimum than 97.8% exceptionally accurate deep learning neural network study [8] in terms of computational complexity and efficiency, the size of dataset considered (100 audio files for only four distinct genres), as well as real-world applicability of continuous stream of music data and user demands.

11 Conclusion

As music service providers, especially streaming services, like Spotify and YouTube, invest hundreds of millions of dollars into research and development of their music services, expectations of listeners, as well as the scope for intelligent advancements in music information retrieval systems keep on increasing. To validate the advancing prospects of music information retrieval and its usage for categorising music for computerised searches, organisation, automatic playlist creation, and recommendation systems and to enhance musical experience, the study performed detailed genre classification on GTZAN dataset of 1000 audio files distributed across ten broad genres (Country, Classical, Blues, Disco, Hip-Hop, Pop, Jazz, Rock, Metal, and Reggae). Music genre classification of pre-processed dataset (of feature vectors of a 1000 audio signals) was performed parallelly, yet independently, on six different machine learning algorithms, namely k-nearest neighbour, support vector machine, random forest algorithm, Naïve Bayes algorithm, logistic regression, and artificial neural network. Next, to increase the reliability of accurate classification predictions, basic as well as advanced boosting ensemble models (random forest, max-voting, AdaBoost, and XGBoost) of component classifier models were developed, and performance reports of all implemented classifiers were compared and analysed.

With the help of above-conducted experiments, the study was successfully able to validate the hypothesis that an ensemble-based classification model proves to be more accurate than any component classifier. Here, the maximum possible accuracy achieved for our best performing max-voting ensemble classifier was 73.17%. Moreover, it was observed that max-voting basic ensemble classifier outperformed advanced boosting models in terms of both mean and maximum possible accuracies for the large GTZAN dataset of music signals. In addition, the study compared the current work with recent as well as prominent works in the field of MIR and smart genre classification on parameters of methodologies for data collection, feature extraction, pre-processing, modelling and evaluation, and performances. It is observed that the presented ensemble-based machine learning classifier performs optimally as compared to prior art models; in terms of computational complexity and efficiency, the size of dataset considered, necessary auxiliary tools, as well as real-world applicability of continuous stream of music data and user demands. The study concludes that the methodologies adopted by the study and findings accomplished thereupon are novel to the field of MIR.

12 Future Scope

Indeed, ensemble-based techniques for machine learning purposes, especially genre classification, are relatively intricate yet accurate, and such systems can be made more intelligent and reliable using additional contemporary mechanisms for selection of hyperparameters, such as grid search. Moreover, convolutional neural networks can be employed as contributing components in ensemble classifiers for extracting music information directly from the spectrogram images of audio signals using computer vision tools. This mechanism also inherently considers various combinations of music features for intelligent categorisation.

By selecting more diverse genres to eliminate parent–child relationship in origin of genres, and also features being considered for learning classification patterns, performance of a classifier model may improve. In addition, use of equivalent sized dataset consisting of limited number of genre labels can remove any possibility of underfitting of data that causes erroneous prediction accuracy.

Music information retrieval (MIR) finds extensive application in the forthcoming era of playlist creation, automated recommendation systems, similarity and emotion detection, as well as audio-based query redressal systems, such as query by singing or humming. Such audio-based systems may demand extremely little knowledge of desired music from the listener as well as provide an enriched musical experience with numerous available intelligent facilities. In this direction, leading edge technologies of big data analytics, such as direct computations on audio files, are hypothesised by the current study to provide intriguing improved results in the application field of MIR, contributing to the capable future of MIR.

Acknowledgements The authors are grateful to George Tzanetakis for availability of his dataset for our experimental work which was encouraged by Mr. Rahul Gupta, and the authors thank them for their perpetual guidance and vision.

References

1. Global Music Industry Revenue 2023|Statista. https://www.statista.com/statistics/259979/global-music-industry-revenue/. Accessed 14 Feb 2020
2. Cunningham, S.J., Bainbridge, D., Downiw, J.S.: How people describe their image information needs: a grounded theory analysis of visual arts queries. In: Proceedings of ACM IEEE International Conference Digit Library JCDL, pp. 47–54 (2004)
3. Downie, J.S.: The music information retrieval evaluation exchange (2005–2007): a window into music information retrieval research. Acoust. Sci. Tech. **29**, 247–255 (2008). https://doi.org/10.1250/ast.29.247
4. Porcaro, L., Gomez, E.: 20 years of playlists : a statistical analysis on popularity and diversity. In: Proceedings of 20th International Symposium on Music Information Retrieval, ISMIR 2019 4–11 (2019)
5. Tzanetakis, G., Cook, P.: Musical genre classification of audio signals using geometric methods. Eur. Sign. Process. Conf. **10**, 497–501 (2010)
6. Peiris, R., Jayaratne, L.: Supervised learning approach for classification of Sri Lankan Music (2016)
7. Li, T., Ogihara, M.: Toward intelligent music information retrieval. IEEE Trans Multimed **8**, 564–574 (2006). https://doi.org/10.1109/TMM.2006.870730
8. Jawaherlalnehru, G., Jothilakshmi, S.: Music genre classification using deep neural networks. **4**, 935–940 (2018)
9. Yeh, C.C.M., Yang, Y.H.: Supervised dictionary learning for music genre classification. In: Proceedings of 2nd ACM International Conference on Multimedia Retrieval, ICMR 2012 (2012). https://doi.org/10.1145/2324796.2324859
10. Data Sets|MARSYAS music analysis, retrieval, and synthesis for audio signals. http://marsyas.info/downloads/datasets.html. Accessed 14 Feb 2020
11. Dietterich, T.G.: Ensemble methods in machine learning. Lect. Notes Comput. Sci. (including Subser Lect Notes Artif Intell Lect Notes Bioinformatics) **1857** LNCS: 1–15 (2000). https://doi.org/10.1007/3-540-45014-9_1
12. Gangeh, M.J., Farahat, A.K., Ghodsi, A., Kamel, M.S.: Supervised dictionary learning and sparse representation-a review (2015)

Hybrid Transform-Based Image Compression Using Adaptive Grid Scanning

Venkatateja Jetti and Ram Kumar Karsh

Abstract A lot of attention has been paid by the researchers for lossy color image compression. Due to the high uses of multimedia, a high image compression ratio while keeping good image quality is still a research issue. In this paper, an image compression is proposed using adaptive scanning, in which discrete wavelet transform (DWT) is performed to source image, discrete cosine transform (DCT) is performed to obtained DWT elements. Next, nonzero quantized discrete cosine transform (DCT) coefficients are represented in less number of bits. First, the source image is divided into 16×16 grids and then DWT transformed, later the obtained elements are divided into 8×8 grids and then DCT is performed, quantized, differential encoded, and an indexed vector is formed from the quantized matrix using adaptive scanning. Furthermore, the indexed vector is divided into two indexed vectors, namely $nonzero$ vector and *zerocount* vector. The nonzero coefficients are stored in $nonzero$ vector and the number of zeros preceding a nonzero coefficient is stored in *zerocount* vector. A new block is introduced to store the number of bits required to store the highest value of nonzero coefficient in $nonzero$ vector so that all the nonzero elements can be represented in that many bits rather than a fixed number of bits. The experimental results show that the proposed method has achieved more compression ratio with slight reduction in image quality as compared to existing recent methods.

Keywords Lossy color image compression · DWT · DCT · Adaptive grid scanning · Peak signal-to-noise ratio · Compression ratio

1 Introduction

In multimedia application, there is a requirement for huge amount of data. Using compression, actual amount of data can be represented into smaller quantity which

V. Jetti (✉) · R. K. Karsh
National Institute of Technology Silchar, 788010 Assam, India
e-mail: Venkatateja12345@gmail.com

© The Editor(s) (if applicable) and The Author(s), under exclusive license to Springer Nature Singapore Pte Ltd. 2021

A. P. Pandian et al. (eds.), *Proceedings of International Conference on Intelligent Computing, Information and Control Systems*, Advances in Intelligent Systems and Computing 1272, https://doi.org/10.1007/978-981-15-8443-5_18

yields less storage and efficient transmission. Actual data can be compressed by removing redundant information.

Image compression approaches may be typically categorized into two types such as lossless and lossy. In lossless, the actual image has to be reconstructed without loss in information. This approach is mainly used in medical diagnosis. In some applications, loss of some information can be tolerated. This type of compression is called lossy, which is typically applied in television applications. The lossy image compression is widely studied and is typically classified [1] as direct and indirect methods. In first one, image samples are directly controlled as discussed in vector quantization (VQ) [2] and block truncation coding (BTC) [3]-based methods. In later one, generally the mathematical transformations are used to map energy of image into lesser number of coefficients. Some of the transform-based methods are discrete wavelet transform (DWT) [4–6], discrete cosine transform (DCT) [7–10], and principal component analysis (PCA) [11].

DCT-based approaches are computationally efficient; however, a small loss in signal-to-noise ratio is observed compared with DWT [7, 12]. It has been observed from the experimental results [4] that the important factors in the image encoding are entropy coder and quantizer, irrespective of difference between DCT and wavelet transform. This is the main factor, the existing video and image coding standards favor DCT above DWT [12, 13]. For many image and video processing applications, DCT is the basic building block. So DCT finds its application in many image and video processing standards like JPEG [14–16].

In this paper, lossy image compression based on hybrid transform, i.e., both DWT-DCT, with novel encoder is proposed. First, the source image is divided into 16 $\times$ 16 blocks. Next, DWT is performed, and the elements are divided into 8 $\times$ 8 blocks. Later, DCT, quantization, and thresholding are performed successively. Finally, using the proposed encoder adaptive block scanning is performed to obtain the compressed image.

The remaining of the manuscript is arranged as follows. Section 2 reveals the steps of lossy image compression method. In Sect. 3, the proposed encoder is presented. Section 4 deals with experimental results and their observations. Finally, the conclusion and future works are drawn in Sect. 5.

2 Steps of Lossy Image Compression Methods

The compression and decompression phases are shown in Fig. 1, and the steps are as follows:

Compression phase:

- Reading the image.
- Color conversion of image from RGB to YCbCr.
- Segregating pixels into small 16 $\times$ 16 matrices.

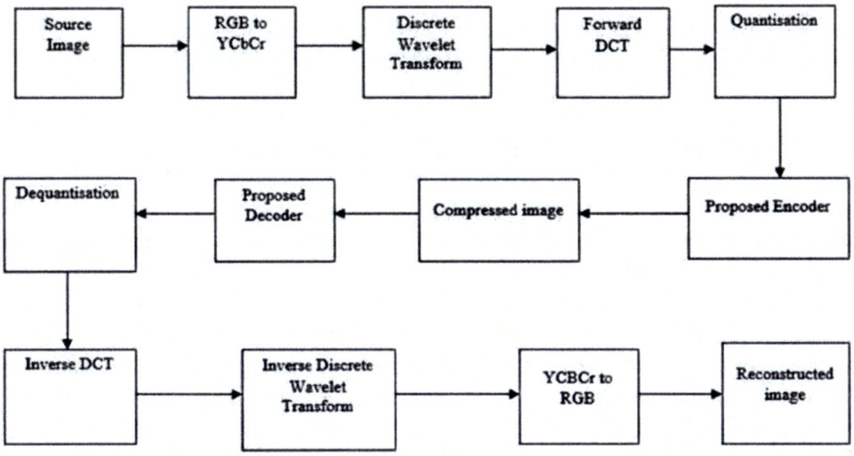

Fig. 1 Block diagram showing the steps of compression and decompression

- Discrete wavelet transform of segregated 16 × 16 matrices.
- Segregating pixels into small 8 × 8 matrices.
- Discrete cosine transform of segregated 8 × 8 matrices.
- Quantization of obtained DCT matrices.
- Adaptive block scanning using the following scans:

  1. Zigzag scan
  2. Horizontal scan
  3. Vertical scan
  4. Hilbert scan

- Determining the $zerocount_i$, $nonzero$ vectors.
- Using the vectors generates a bitstream (i.e., compressed image).

Decompression phase:

- Reconstructing the quantized matrix from the bitstream.
- Dequantizing the reconstructed quantized matrices.
- Inverse discrete cosine transform of the obtained dequantized matrices.
- Inverse discrete wavelet transform of the constructed DCT matrices.
- Color conversion from YCbCr to RGB.
- Reconstruct the source image.

3 Proposed Encoder

The proposed approach used for encoding is simple and efficient. In this method, after first level DWT decomposition of the source image, perform DCT to the LL subband and discard remaining subbands. Later quantization is done yields quantized DCT coefficients. The quantization procedure is discussed in Sect. 3.1 Next the whole image is divided into 8 × 8 grids. Further, for every grid an indexed vector V_i ($i = 1, 2, \ldots, 4$) is generated from the differential encoded 8 × 8 grid following four different types of scan orders. The four different scan orders are zigzag, horizontal, vertical, Hilbert. The vector formed using zigzag, horizontal, vertical, and Hilbert scan orders are V_1, V_2, $V_{3,}$ and $V_{4,}$ respectively. All indexed V_i ($i = 1, 2, \ldots, 4$) subsequently divided into two indexed vectors namely *nonzero* vector and *zerocount* vector. Both the *nonzero* vector and *zerocount* are also indexed vectors.

The *nonzero* vector contains all the nonzero elements present in the quantized DCT coefficients. The *zerocount* vector stores the number of zeros preceding a nonzero coefficient. The *nonzero* vector and *zerocount* vector are generated for all the four scans, and better scan order is determined. The better scan order is obtained as follows: The maximum value of *zerocount* vector is determined for all four scans. Next, the maximum values of all four scan orders are compared and the minimum of the maximum values is found out. The scan order which has minimum of the maximum value is the best scan order because every element in the *zerocount* vector can be represented in binary in least number of bits compared to other scans. Also, a variable AS of 2 bits is used to store which scan order is used while encoding so that it can be successfully decoded in the decompression phase. The AS value is stored as 00,01,10,11 for zigzag scan, horizontal scan, vertical scan, Hilbert scan, respectively. Further for the best scan order, the maximum value of nonzero element in *nonzero* vector is found out and the minimum number of bits required to store that value is stored in third block in Table 1. As depicted in Table 1, for every 8 × 8 grid all the six blocks are constructed orderly. Every grid got six blocks while encoded. The generation of bitstream is discussed as follows

- The first block is termed as nonzero count, and it stores the number of nonzero elements in the grid.
- The second block is termed AS block, and it stores the information of which scan is stored. 00 for zigzag scan, 01 for horizontal scan, 10 for vertical scan, and 11 for Hilbert scan.

Table 1 Construction of bitstream

Number of nonzero elements in each 8 × 8 grid	The selected scan (AS) represented in binary	Value of 'a' represented in binary	Nonzero elements each represented in 'a' bits	Value of 'b' represented in binary	Zero count vectors elements each represented in 'b' bits
P bits	2 bits	3 bits	$m * a$ bits	3 bits	$m * b$ bits

'a': least number of bits required to represent highest pixel value in *nonzero* vector
'b': least number of bits required to represent highest value in *zerocount* vector
'm': number of nonzero elements in each 8 × 8 grid

- The third block stores the least number of bits required to store the highest pixel value in each grid. If the highest pixel value is represented in k bits, then every pixel can be presented in k bits. Let the value be *a*.
- The fourth block stores all the elements of best scan's *nonzero* vector in *a* bits.
- The fifth block stores the least number of bits required to store the highest value in $zerocount_i$ vector for selected scan. If the highest value is represented in k bits, then every pixel can be represented in *k* bits. Let the value be *b*.
- The sixth block stores all the elements of best scan's *zerocount* vector in *b* bits.

3.1 Quantization

Quantization is done to the obtained DCT matrix. It is the process of converting real DCT coefficients into discrete. Quantization is done by dividing each element of the matrix by respective element of quantization matrix. The quantization matrix for luminance (Q_l) and chrominance (Q_c) is defined as follows:

$$Q_c = \begin{bmatrix} 16 & 11 & 10 & 16 & 24 & 40 & 51 & 61 \\ 12 & 12 & 14 & 19 & 26 & 58 & 60 & 55 \\ 14 & 13 & 16 & 24 & 40 & 57 & 69 & 56 \\ 14 & 17 & 22 & 29 & 51 & 87 & 80 & 62 \\ 18 & 22 & 37 & 56 & 68 & 109 & 103 & 77 \\ 24 & 35 & 55 & 64 & 81 & 104 & 113 & 92 \\ 49 & 64 & 78 & 87 & 103 & 121 & 120 & 101 \\ 72 & 92 & 95 & 98 & 112 & 100 & 103 & 99 \end{bmatrix}$$

$$Q_l = \begin{bmatrix} 17 & 18 & 24 & 47 & 99 & 99 & 99 & 99 \\ 18 & 21 & 26 & 66 & 99 & 99 & 99 & 99 \\ 24 & 26 & 56 & 99 & 99 & 99 & 99 & 99 \\ 47 & 66 & 99 & 99 & 99 & 99 & 99 & 99 \\ 99 & 99 & 99 & 99 & 99 & 99 & 99 & 99 \\ 99 & 99 & 99 & 99 & 99 & 99 & 99 & 99 \\ 99 & 99 & 99 & 99 & 99 & 99 & 99 & 99 \\ 99 & 99 & 99 & 99 & 99 & 99 & 99 & 99 \end{bmatrix}$$

3.2 Illustrative Example for Compression

In order to clarify the idea, for example, the 16×16 intensities matrix *B* depicted from arbitrary image of size 512×512.

$$B = \begin{bmatrix} 80&80&80&80&80&80&79&79&81&81&78&78&81&81&80&80 \\ 80&80&80&80&80&80&79&79&81&81&78&78&81&81&80&80 \\ 81&81&80&80&80&80&80&80&81&81&78&78&81&81&80&80 \\ 81&81&80&80&80&80&80&80&81&81&78&78&81&81&80&80 \\ 80&80&81&81&80&80&79&79&81&80&79&78&81&80&80&80 \\ 80&80&81&81&80&80&79&79&80&81&79&78&81&80&80&80 \\ 81&81&81&81&79&79&79&79&81&80&79&78&81&81&80&81 \\ 81&81&81&81&79&79&79&79&80&81&79&78&80&80&81&80 \\ 80&80&80&81&80&79&81&78&81&81&78&79&81&81&81&80 \\ 80&80&80&81&80&78&81&78&80&80&79&78&80&80&80&81 \\ 81&81&81&81&80&81&81&80&80&81&80&81&79&80&80&79 \\ 81&81&81&81&81&80&79&77&79&77&79&80&80&79&79&80 \\ 80&80&79&77&81&81&79&79&80&81&79&81&80&81&79&80 \\ 80&81&79&77&81&80&79&78&79&80&80&79&81&80&81&79 \\ 79&79&79&78&78&79&79&78&79&79&80&78&81&80&80&79 \\ 79&81&78&79&79&78&78&79&80&81&80&77&79&81&79&79 \end{bmatrix}$$

After the DCT transform of B, the obtained matrix is B_{dwt}

$$B_{\mathrm{dwt}} = \begin{bmatrix} 160&160&160&159&161&156&161&159 \\ 161&160&159&159&161&155&161&159 \\ 160&161&160&158&161&156&162&159 \\ 161&161&159&159&161&156&161&159 \\ 160&160&160&159&161&156&162&159 \\ 162&162&157&154&159&159&157&159 \\ 159&159&162&157&159&161&158&155 \\ 157&157&155&155&157&158&155&157 \end{bmatrix}$$

After the DCT transform of B_{dwt}, the obtained matrix is $\mathrm{B}_{\mathrm{dct}}$

$$\mathrm{B}_{dct} = \begin{bmatrix} 1271.5&3.6&2.8&1.3&0&-1.5&-4.1&5.8 \\ 5.8&-0.1&0.5&-3.1&2.4&2.4&-4.9&3.2 \\ -4.4&0.5&-1.5&0.5&-0.3&0.3&2.5&-1.5 \\ 2.1&1.0&0.8&0.7&-0.5&-0.7&0.8&0.2 \\ -1.2&-1.5&-0.5&-1.2&1.7&0.9&-1.3&-0.2 \\ 1.8&1.0&-1.7&0.2&-2.2&0.3&1.3&0.2 \\ -1.6&-0.5&2.7&0.9&0.4&-2.5&-0.1&-0.9 \\ 1.5&-0.7&-2.5&-1.5&-1.7&2.0&0&-1.5 \end{bmatrix}$$

After quantization with a quantizer matrix, the obtained matrix is:

$$Q = \begin{bmatrix} 63 & -1 & -1 & 0 & 0 & 0 & 0 & 0 \\ 0 & -1 & -1 & 0 & 0 & 0 & 0 & 0 \\ -1 & 0 & 0 & 0 & 0 & 0 & 0 & 0 \\ 0 & 0 & 0 & 0 & 0 & 0 & 0 & 0 \\ 0 & 0 & 0 & 0 & 0 & 0 & 0 & 0 \\ 0 & 0 & 0 & 0 & 0 & 0 & 0 & 0 \\ 0 & 0 & 0 & 0 & 0 & 0 & 0 & 0 \\ 0 & 0 & 0 & 0 & 0 & 0 & 0 & 0 \end{bmatrix}$$

Four scans are generated by the adaptive block scanning. The best scan is the scan that gives the lowest value of maximums of the index vectors $zerocount_i$ (the last zeros sequence is not considered). The *zerocount* vector contains length of the zero-run sequence preceding a nonzero DCT quantized coefficient. The different generated scans are:

1. **Zigzag scan**:

   63 −1 0 −1 −1 −1 0 1 1 0 0 0 0 0 0 0 0 0 0 0 0 0 0 0 0 0 0 0 0 0 0 0 0 0 0 0 0 0 0 0 0 0 0 0 0 0 0 0 0 0 0 0 0 0 0 0 0 0 0 0 0 0 0 0

   nonzero: 63 −1 −1 −1 −1 1 1
   zerocount: 0 0 1 0 0 1 0
   AS: 00

2. **Horizontal scan**:

   63 −1 −1 0 0 0 0 0 0 −1 1 0 0 0 0 0 −1 1 0 0 0 0 0 0 0 0 0 0 0 0 0 0 0 0 0 0 0 0 0 0 0 0 0 0 0 0 0 0 0 0 0 0 0 0 0 0 0 0 0 0 0 0 0 0

   nonzero: 63 −1 −1 −1 −1 1 1
   zerocount: 0 0 0 6 0 5 0
   AS: 01

3. **Vertical scan**:

   63 0 −1 0 0 0 0 0 −1 −1 1 0 0 0 0 0 −1 1 0 0 0 0 0 0 0 0 0 0 0 0 0 0 0 0 0 0 0 0 0 0 0 0 0 0 0 0 0 0 0 0 0 0 0 0 0 0 0 0 0 0 0 0 0 0

   nonzero: 63 −1 −1 −1 −1 1 1
   zerocount: 0 1 5 0 0 5 0
   AS: 10

4. **Hilbert scan**:

   63 −1 −1 0 −1 0 0 1 0 0 0 0 0 1 −1 0 0 0 0 0 0 0 0 0 0 0 0 0 0 0 0 0 0 0 0 0 0 0 0 0 0 0 0 0 0 0 0 0 0 0 0 0 0 0 0 0 0 0 0 0 0 0 0 0

   nonzero: 63 −1 −1 −1 −1 1 1
   zerocount: 0 0 0 1 2 5 0
   AS: 11

Here the best scan is the zigzag scan as the maximum of $zerocount_i$ vector is 1 and is minimum of the all 4 scan vectors. So the encoded bitstream for this 8×8 grid is

000111 00 110 111111101111110111111011111101111110000011000001 001 0010010.

In this way, bitstream is calculated for all the grids in the image matrix.

3.3 Encoding and Decoding Algorithm

Encoding algorithm:

Algorithm 1: Encoding algorithm (the algorithm used to generate the image into a bitstream) Input: F (source image gone through quantization, followed by the discrete wavelet transformation and discrete cosine transformation, size $M \times N$

Output: Compressed image in the form of bitstream.

1. The image is divided into 16×16 grids, and every 16×16 grid is discrete wavelet transformed; it forms a 8×8 grid.
2. After the discrete wavelet transformed image is obtained; it is quantized, followed by discrete cosine transform, the image is divided into 8×8 grid and each grid is read.
3. Perform the scan operations (zigzag, horizontal, vertical, and Hilbert) and determine the indexed vector V_i for all the four scan orders.
4. From the obtained vectors V_i determine, the *nonzero* vector and *zerocount* vector for all the four scans.
5. The *nonzero* vector stores the non zero elements in respective scan order. The *zerocount* vector for each scan order stores number of zeros before a nonzero coefficient for respective scan order.
6. $a = 0$;
7. f = 0;
8. ***for*** $b = 1$ **to** length(V_i) **do**
9. **If** V (b) ! = 0 **then**
10. *nonzero* (a) = V (b);
11. $zerocount_i$ (a) = f;
12. f = 0;
13. $a = a + 1$;
14. Else
15. $f = f+1$;
16. ***endif***
17. *Endfor*
18. Finding AS the index of the best scan

    AS = '00' if $zerocount_1$ is the min(max($zerocount_1$), ..., max($zerocount_4$))

AS = '01' if $zerocount_2$ is the min(max($zerocount_1$), ..., max($zerocount_4$))
AS = '10' if $zerocount_3$ is the min(max($zerocount_1$), ..., max($zerocount_4$))
AS = '11' if $zerocount_4$ is the min(max($zerocount_1$), ..., max($zerocount_4$))

19: Find the value of '*a*' and '*b*' where '*a*': least number of bits required to represent highest pixel value in *nonzero* vector
'*b*': least number of bits required to represent highest value in *zerocount* vector and form the bitstream as shown in Table 1.

Decoding algorithm: To determine the DCT vector

Algorithm 2: Decoding algorithm (the algorithm used to reconstruct the compressed image from obtained bitstream) Input: Bitstream of a compressed image

Output: The DCT coefficients are reconstructed and are obtained as a matrix.

1. Get *AS* bit, *nonzero, zerocount* vectors, value of *a, b* from the bitstream.
2. Determine the scan order used from AS bits
3. Initialize a new block to store decoded quantized (DQ) values and initiate each value to zero
4. DQ = 0;
5. ***for*** $b = 1$ **to** length(non*zero)* **do**
6. $a = a + zerocount(\text{b})$;
7. DQ(a) = *nonzero*(b);
8. **End *for***
9. Construct a matrix from the reconstructed DQ vector doing the inverse of respective scan order, the differential decoding is performed.
10. Thus from the reconstructed quantized matrix, the DCT matrix is obtained.
11. Inverse DCT is performed to the obtained DCT matrix, and thus, the discrete wavelet transformed matrix is obtained.
12. Inverse DWT is performed to the obtained DWT matrix, the reconstructed image is obtained.

4 Experimental Results and Discussion

To show the efficacy of the proposed compression method with the state-of-the-art methods, experiments have been conducted. In this experiment, six mostly used standard images are used. These images are peppers, Lena, Airplane, Zelda, Couple and Baboon. The resolution of all the test images is 512×512 each. The six test images are compressed with the proposed method, and results are compared with the previous works. The bits per pixel (Bpp) and peak signal-to-noise ratio (PSNR) of the reconstructed images are taken for the comparison.

Let the source image be $f(x, y)$ and the reconstructed image is $f'(x, y)$.

Error between two images:

$$e(x, y) = f'(x, y) - f(x, y) \tag{1}$$

Total error between two images is:

$$\sum_{x=0}^{M-1}\sum_{y=0}^{N-1}[f'(x, y) - f(x, y)] \tag{2}$$

RMS error is given by:

$$e_{\text{rms}} = \left[1/MN\sum_{x=0}^{M-1}\sum_{y=0}^{N-1}[f'(x, y) - f(x, y)]^2\right]^{\frac{1}{2}} \tag{3}$$

Peak signal-to-noise ratio (PSNR) is given by:

$$\text{PSNR} = 20\log_{10}\frac{255}{\sqrt{\text{MSE}}} \tag{4}$$

where MSE is mean squared error e_{rms}.

Also to determine the compression ratio, a term known as bits per pixel (Bpp) is used. Bpp is the number of bits required to store one pixel.

$$\text{Bpp} = \frac{\text{Total number of bits in Bitstream}}{\text{number of pixels in image}} \tag{5}$$

So, lower the Bpp less number of bits are required to store the image, higher the compression ratio. Compression rate can be measured with the help of Bpp. In the experiment, the Bpp is calculated using (5).

The compression is performed on the six test images, and the results are compared with the [23] method and JPEG method [24] as shown in Table 2. For the test image

Table 2 Performance comparison between the proposed method [23] method and JPEG [24] algorithms

Image	Proposed method		[23]		JPEG [24]	
	PSNR	Bpp	PSNR	Bpp	PSNR	Bpp
Peppers	26.73	0.43	31.60	0.90	35.63	3.50
Lena	28.84	0.38	31.55	0.73	34.39	3.76
Airplane	27.13	0.39	32.17	0.64	35.63	3.50
Zelda	32.73	0.30	32.24	0.72	37.94	3.28
Couple	27.57	0.39	33.65	0.58	40.31	3.60
Baboon	21.68	0.57	29.79	0.80	31.93	5.25
Average	27.44	0.41	31.83	0.72	35.97	3.81

peppers, the PSNR was better in the JPEG [24] method which is 35.63, whereas in the proposed method, it was 26.73, but the Bpp was better in the proposed method which is 0.43, whereas the Bpp in the JPEG [24] method was 3.50 which is too high when compared with proposed method's Bpp. When the average results of six test images are compared, the PSNR was still better in JPEG [24] method with value of 35.97, whereas the PSNR of proposed method is 27.44, but the Bpp was still better in proposed method with a value of 0.41, whereas the Bpp in the JPEG [24] method is 3.81 which is still higher than the proposed method's Bpp.

The results of proposed method are compared with state-of-the-art methods [23, 24]. In all the cases, even though the PSNR is slightly lesser in the proposed method, the Bpp is far better and it compensates the little loss in image quality. All the experiments were performed with the MATLAB (R2018 a) on a Intel® core™ I5-5200U CPU @2.20 GHz with a RAM of 8 GB.

5 Conclusion and Future Work

In this paper, an image compression techniques are proposed using adaptive scanning, in which the nonzero quantized DCT coefficients are represented in less number of bits. In the proposed method, the source image is divided into 16×16 grids and then DWT transformed, later the obtained elements are divided into 8×8 grids and then DCT is performed, quantized, differential encoded, and an indexed vector is formed from the quantized matrix using adaptive scanning. Furthermore, the indexed vector is divided into two indexed vectors, namely *nonzero* vector and *zerocount* vector. The nonzero coefficients are stored in *nonzero* vector, and the number of zeros preceding a nonzero coefficient is stored in *zerocount* vector. A new block is introduced to store the number of bits required to store the highest value of nonzero coefficient in *nonzero* vector so that all the nonzero elements can be represented in that many bits rather than a fixed number of bits. The experimental results show that the proposed method gives high compression ratio and better than the state-of-the-art methods with slight reduction in image quality.

In the future work, some other transformation approaches may be explored for further improvement in compression ratio.

References

1. Khalid, S.: Introduction to Data Compression, 4th edn. Elsevier, San Francisco (2006)
2. Feng, Y.S., Nasrabadi, N.M.: Dynamic address-vector quantization of RGB colour images. IEE Proc. I Commun. Speech Vision **138**(4), 225–231 (1991)
3. Kurita, T., Otsu, N.: A method of block truncation coding for color image compression. IEEE Trans. Commun. **41**(9), 1270–1274 (1993)
4. Xiong, Z., Ramchandran, K., Orchard, M.T., Zhang, Y.Q.: A comparative study of DCT and wavelet-based image coding. IEEE Trans. Circuits Syst. Video Tech. **9**(5), 692–695 (1999)
5. Pearlman, W., Islam, A., Nagaraj, N., Said, A.: Efficient, low complexity image coding with a set partitioning embedded block coder. IEEE Trans. Circ. Syst. Video Tech. **14**(11), 1219–1235 (2004)
6. Shoitan, R., Nossair, Z., Isamil, I., Tobal, A.: Hybrid wavelet measurement matrices for improving compressive imaging. Signal Image Video Process. **11**(1), 65–72 (2017)
7. Wallace, G.K.: The JPEG still picture compression standard. IEEE Trans. Consum. Electr. **38**(1), 18–34 (1992)
8. Dagher, I.: Highly-compacted DCT coefficients. Signal Image Video Process. **4**(3), 303–307 (2010)
9. Ponomarenko, N., Lukin, V., Egiazarian, K., Astola, J.: DCT based high quality image compression. In: Scandinavian Conference on Image Analysis, pp. 1177–1185. Springer, Berlin (2005)
10. Ponomarenko, N., Lukin, V., Egiazarian, K., Astola, J.: ADCTC: advanced DCT-based image coder. In: Proceedings of LNLA, Switzerland (2008)
11. Clausen, C., Wechsler, H.: Color image compression using PCA and back propagation learning. Patt. Recogn. **33**(9), 1555–1560 (2000)
12. Ohm, J.R., Sullivan, G.J., Schwarz, H., Tan, T.K., Wiegand, T.: Comparison of the coding efficiency of video coding standards including high efficiency video coding (HEVC). IEEE Trans. Circ. Syst. Video Tech. **22**(12), 1669–1684 (2012)
13. Sullivan, G.J., Ohm, J.R., Han, W.J., Wiegand, T.: Overview of the high efficiency video coding (HEVC) standard. IEEE Trans. Circ. Syst. Video Tech. **22**(12), 1649–1668 (2012)
14. Bauermann, I., Steinbach, E.: Further L ossless C ompression of JPEG Images. Picture Coding Symposium. In: Proceedings of PCS San Francisco, December 15–17 (2004)
15. Ponomarenko, N., Egiazarian, K., Lukin, V., Astola, J.: Additional lossless compression of JPEG images. In: Proceedings of 4th Symposium on Image and Signal Processing and Analysis, Zagreb, Croatia, September, pp 117–120 (2005)
16. Silveira, T.L.T.D., Oliveira, R.S., Bayer, F.M., Cintra, R.J., Madanayake, A.: Multiplierless 16-point DCT approximation for low-complexity image a nd video coding. Signal Image Video Process. **11**(2), 227–233 (2017)
17. Messaoudi, A., Benchabane, F., Srairi, K.: DCT-based color image compression algorithm using adaptive block scanning
18. Rahul, K., Tiwari, A.K.: Saliency enabled compression in JPEG framework. IET Image Process. **12**(7), 1142–1149 (2018). https://doi.org/10.1049/iet-ipr.2017.0554
19. Dhara, B.C., Chanda, B.: Color image compression based on block truncation coding using pattern fitting principle. Patt. Recogn. **40**(9), 2408–2417 (2007)

20. Douak, F., Benzid, R., Benoudjit, N.: Color image compression algorithm based on the DCT transform combined to an adaptive block scanning. AEU Int. J. Electron. Commun. **65**(1), 16–26 (2011)
21. Boucetta, A., Melkemi, K.E.: DWT based-approach for color image compression using genetic algorithm. In: Elmoataz, A., Mammass, D., Lezoray, O., Nouboud, F., Aboutajdine, D. (eds.) Image and Signal Processing. ICISP 2012. Lecture Notes in Computer Science, vol. 7340. Springer, Berlin, Heidelberg (2012)
22. Messaoudi, A., Srairi, K.: Colour image compression algorithm based on the DCT transform using difference lookup table. Electr. Lett. **52**(20), 1685–1686 (2016)
23. Nigam, A.K., Khare P., Srivastava, V.K.: Image compression using hybrid approach and adaptive scanning for color images. In: Dutta D., Kar H., Kumar C., Bhadauria V. (eds) Advances in VLSI, Communication, and Signal Processing. Lecture Notes in Electrical Engineering, vol 587. Springer, Singapore (2020)
24. Rabbani, M., Joshi, R.: Eastman Kodak Company, Rochester, NY 14650, USA. An overview of the JPEG2000 still image compression standard. Signal Process. Image Commun. **17**, 3–48 (2002)

Detection of Proximal Femoral Bone Fracture Using Mask R-CNN

Ambarish Moharil and Shreya Singh

Abstract Detection of femoral fractures from X-ray images requires multiple scans which can compromise on the procedure being cost effective as well as delay the need for immediate surgery. Bone mass or bone mineral content is currently assessed most commonly via dual-energy X-ray absorptiometry (DEXA). Procedures and cross-sectional techniques like magnetic resonance imaging (MRI) and quantitative computed tomography (qCT) have been proved to provide useful information along with the DEXA scans like the metrics of the bone structure, density, and quality of the bone. However, such procedures are extremely costly and unaffordable for large masses. To make the process cost effective, faster, and efficient, it is proposed to extract a lot of information from the X-ray scans itself as they are cheap and economically affordable. The large population which suffers from osteoporosis is the older generation with an average age greater than 60 years, undergoing and bearing the cost of frequent MRI, qCT, or DEXA scans which is economically difficult for a greater chunk of this population. Hence, it is very necessary to compute and detect these fractures in early stages and from economically viable options like X-rays. Hence, this paper intends to analyze X-ray scans using CNNs and Mask R-CNNs and extract regions of fractures mainly intracapsular and extracapsular which are the major regions of occurrences of proximal femoral fractures.

Keywords Mask R-CNN · Convolution · Deep learning · Fast R-CNN

A. Moharil (✉) · S. Singh
Department of Instrumentation and Control Engineering, Vishwakarma Institute of Technology, Pune, India
e-mail: ambarish.moharil17@vit.edu

S. Singh
e-mail: shreya.singh17@vit.edu

© The Editor(s) (if applicable) and The Author(s), under exclusive license to Springer Nature Singapore Pte Ltd. 2021

A. P. Pandian et al. (eds.), *Proceedings of International Conference on Intelligent Computing, Information and Control Systems*, Advances in Intelligent Systems and Computing 1272, https://doi.org/10.1007/978-981-15-8443-5_19

1 Introduction

Osteoporosis is a public health problem and is growing across the world at a rapid rate. Prolonged osteoporosis makes the hip bone extremely vulnerable to fractures, and detecting its onset at a stage is extremely crucial. Currently, there are technologies such as the dual-energy X-ray absorptiometry (DEXA) and quantitative computed tomography which aid in calculating the bone mineral density which may help to detect early onset of osteoporosis [1]. In a DEXA scan, when soft tissue absorption is subtracted out, the bone mineral density can be determined from the absorption of each beam by bone. Meanwhile, CT scans have resolution 100 times more than an X-ray and can measure volumetric density, and the results are given in g/cm [2]. While these procedures are though effective, they are yet extremely costly. With the growing need for orthopedic units, comes the demand for automation. In the United Kingdom alone there are 65,000 hip fractures that are detected every year. The fractures in X-ray images is detected including the region of fracture; this method aims to prevent further delay in administering treatment to the patient.

2 Proximal Femoral Fractures

Proximal femoral fractures form the subset of fractures that occur in the hip region. They tend to occur in older patients, and in those who have osteoporosis. In this condition, a bone becomes weak and brittle with age. This increases the risk of fracture in patients of osteoporosis. In this group of patients, fracture is usually the result of low-impact trauma although, the femoral fracture can also take place in younger patients as a result of high impact injuries such as those caused during an accident.

Femoral fractures are ideally classified into two kinds:

1. Intracapsular fractures: These fractures damage the intracapsular blood vessels. These vessels are the source of blood supply to the femoral head and occur at the neck and the head of the femoral bone. They require immediate and appropriate treatment to prevent further escalation of the condition.
2. Extracapsular fractures: Fractures outside the capsule (the soft tissue consisting of the lubricant that nourishes the hip joint), their treatment is administered differently as the degree of the vessel damage is not the same as in the case of an intracapsular femoral fracture.

Types of extracapsular fractures:

(2.1) Intertrochanteric
(2.2) Subtrochanteric (Fig. 1).

In a study carried out in the Rohtak district of North India, the number of cases of hip fractures above the age of 50 years was 129 per 100,000 people for both

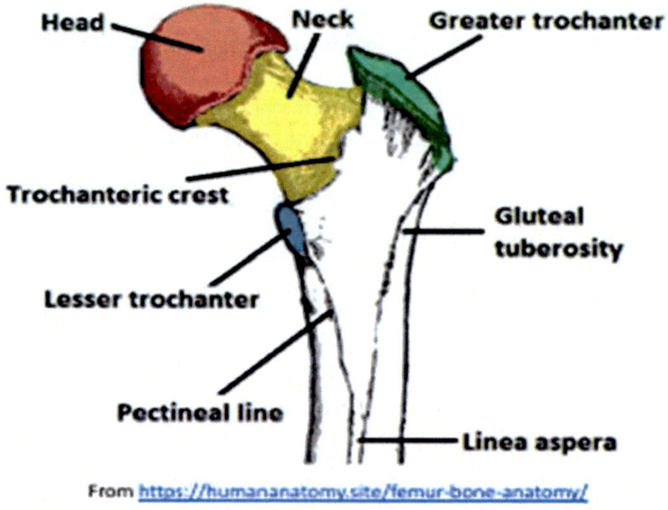

Fig. 1 Anatomy of a femur bone (*Source* humananatomy.site/femur-bone-anatomy/)

men and women. However, it was found that above the age of 55 years the numbers significantly increased for women. This is a global concern for both doctors and agencies such as the World Health Organization [2, 3]. As patient cases continue to grow exponentially, the need of orthopedic units and testing facilities is also in high demand. For a long time now, X-rays have been used to analyze fractures, these X-rays are scanned by the doctors, and appropriate treatment is then administered and intend to completely automate the process of reading an X-ray of the patient and predicting the region of fracture of the proximal femur utilizing the latest convolutional neural network which is the Mask R Convolutional Neural Network. Using this technique, the region of fracture is masked by allowing the neural network to train on the Regional Proposal Network (RPN). Several techniques have been used in the past for fracture detection such as the Convolutional Neural Network and various image processing techniques which include intensity gradient calculation, segmentation, and back propagation to determine the presence of a fracture [4–6]. Our neural network is a unique Faster R-CNN as it detects the fracture and masks the probable region of occurrence. Image processing techniques are very useful and significant in the medical field such as in X-ray radiographs.

3 Uniqueness and Flexibility of Approach

Our method uses the Mask R-CNN, which extends the Faster R-CNN by putting predictive segmentation masks on the region of interest (RoI). Our region of interest is built in correspondence to the bounding boxes which enclose the probable region of fracture [7]. A Fully Convolutional Network (FCN) is applied to each region of interest which forms our mask as the predictive mask that is developed in a pixel by pixel manner. A large part of utilization of Mask R-CNNs have been for instance segmentation and to identify various independent objects in an image.

In our dataset, which contained several images of proximal femoral patient fractures, mapping was done using the annotations exported in coco format which contained the attributes such as the width, height, type of fracture, and the bounding box coordinates. The Mask R-CNN framework is constructed over the Faster R-CNN. Mask R-CNN returns an object mask for every target along with the bounding boxes and class labels as compared to Faster R-CNN for a given input image.

4 Mask R-CNN

Convolutional Neural Networks have proved to be a boon in the field of medical imaging. They have helped researchers in classifying medical images into various categories. Mask R-CNNs are a recent development in the field of image processing and machine learning, and they are an upgrade to the widely known Faster R-CNN algorithm. Mask R-CNN was developed in 2017 and can be applied to perform numerous image processing tasks like image segmentation, object localization, and object instance segmentation of natural images [2, 5]. Using Convolutional Neural Networks, it is quite possible to detect any specific natural image as a whole and classify it into its specific category. But when we have a detection problem like this, Convolutional Neural Networks do not help much as an individual algorithm. Scenarios to detect regions while classifying them into different classes cannot be done independently using Convolution Neural Networks. Bone fracture X-ray images can be efficiently classified into their respective class using CNNs but When we intend to extract the regions of fracture, or detect the regions of maximum impact something is needed that is faster and applies various segmentation techniques along with the standard CNN architecture [5, 8]. Mask R-CNNs are a perfect solution for such problems, where they help in instance segmentation and classification of independent objects from an entire image. The independent objects are detected from a specific image at pixel level using Mask R-CNNs.

Convolutional Neural Networks are robust feature extractors. CNNs are the most used and are a state-of-the-art algorithm used for the purpose of classification. When it comes to image classification, CNNs have proven their merit flawlessly but have a few shortcomings. CNNs are restricted toward classification of an image as a whole or a single entity and fail to classify different images present in a certain image. As proposed in this paper, CNNs limit the usability in classifying the fractured image as a whole but fail to locate the region of fracture.

The robust convolutional layers with the help of other feature extracting layers present in a CNN can perform instance segmentation of objects present in an image. Mask R-CNNs exploit the feature extractors present in a CNN by performing intra-class instance segmentation of various objects and entities.

Mask R-CNN can be defined as a general framework for detecting localized objects, instance segmentation from an entire image or to perform semantic segmentation on a particular natural image [9, 10]. The Region Proposal Networks or the RPNs form the core of Mask R-CNNs. Mask R-CNNs are an extension to the Faster R-CNN algorithm which returns a masked output along with the bounding boxes for the detected region. Mask R-CNNs depend on something called as "proposals" which are generated by the Regional Proposal Network (RPN) [8]. The RPN acts as a feature extractor and generates these proposals, detecting whether the desired object is present in that particular region or not. In order to extract the features from an image, the ResNet 101 architecture is used. ResNet forms the backbone of the Mask R-CNN architecture and is an excellent feature extractor. Above the ResNet 101 architecture, a new layer of the Regional Proposal Networks (RPNs) is built, so the features get extracted in the initial layer which is the ResNet layer. These extracted features act as an input to the RPN. Using these feature maps, RPN predicts whether the target object (fracture in this scenario) is present in the proposed region or not. [2, 8] RPN extracts these desired regions which are different in shape. In order to convert all these extracted regions into a universal shape, a pooling layer is applied at the end of the RPN. And just like the classical CNN, the outputs from the pooling layer are given as input to the fully connected layer to learn on these features and then predict the bounding boxes and the class labels. This architecture is quite similar to that of the Faster R-CNNs (Figs. 2, 3).

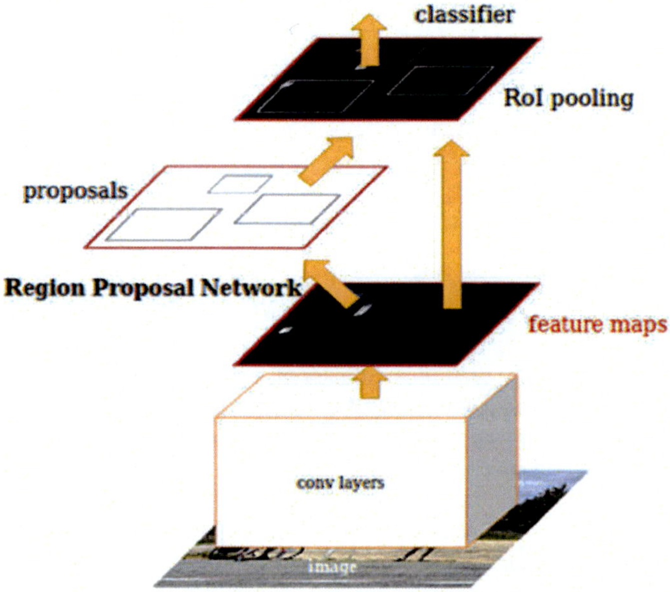

Fig. 2 Architecture of the region proposal network (*Source* "Faster R-CNN: Towards Real-Time Object Detection with Region Proposal Networks" by Shaoqing Ren, Kaiming He, Ross Girshick, and Jian Sun)

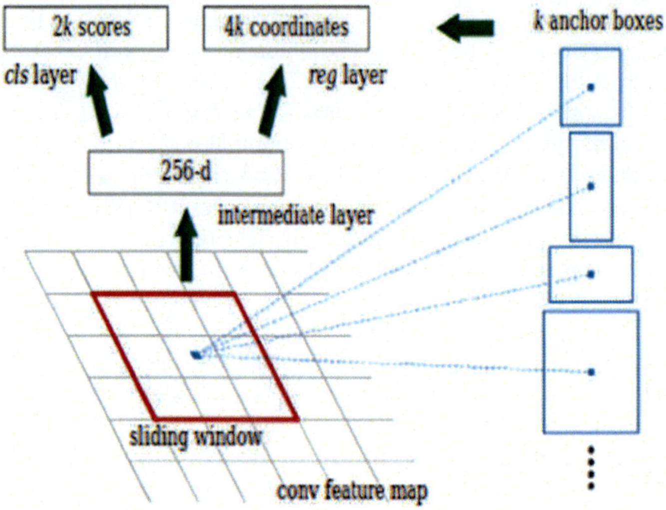

Fig. 3 Feature pyramidal network layer (*Source* "Faster R-CNN: Towards Real-Time Object Detection with Region Proposal Networks" by Shaoqing Ren, Kaiming He, Ross Girshick, and Jian Sun)

The part that makes it different is that, along with this, a segmentation mask is generated which is a new and a unique approach. In order to generate a segmentation mask, the Mask R-CNNs compute a region of interest (ROI) on the input image [8, 11]. This reduces the computation time of the algorithm and makes it faster and efficient. For all the regions predicted by the RPN, the intersection over union (IoU) is computed with all the truth boxes or the ground truth boxes. It can be computed from the formula stated below (Fig. 4).

$$\text{IoU} = \text{Area of Intersection}/\text{Area of Union}$$

It makes Mask R-CNN efficient in such a way that, only those IoUs are selected which are greater than 0.5 and drop any value which is less than 0.5. This process is repeated for all the regions that are present, and regions having an IoU value greater or equal to 0.5 are selected and masked. Once these ROIs are computed on an image, a segmentation mask is added to the existing architecture and extract the region of interest. Hence, Mask R-CNNs provide an edge in automated segmentation of medical images. When viewing a problem of detecting fractures in the proximal femur, the intracapsular fractures in the neck region of the femur are extremely minute and thin. Our architecture successfully detected that these small regions reducing the computation and detection cost while increasing the efficiency, speed, and reliability.

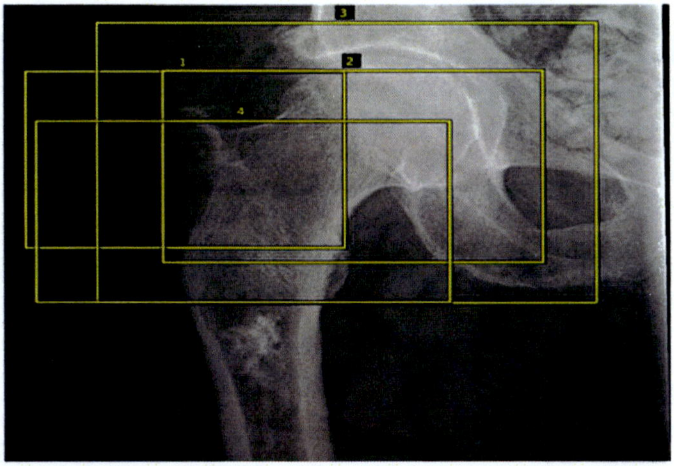

Fig. 4 Boxes 3,1, & 4 have an IoU greater than 50, and hence, the probability of the target being in that region is higher

5 Architecture

The Regional Proposal Network (RPN) forms the foundation of the architecture. Our architecture consists of ResNet 101 a deep residual network as the bottom layer to extract maximum number of features from the input image [8]. This ResNet 101 network is used as a pathway for the feature pyramid network (FPN) to form a pyramid of layers over the existing input image [10]. This pyramidal structure of layers helps in extracting features from the input image efficiently and at multiple scales. Before feeding the images to this ResNet 101 architecture, the images are annotated accordingly to predict the location of the fractures in the proximal femoral bone region. These annotations act as a set of predefined anchors. These anchors then bind with the feature map according to the corresponding levels and locations. This pyramid of features is then passed through the convolution network to predict the classes or generate the classes of the anchors (foreground and background) and also to predict the refinements of the bounding boxes. The predefined anchors are further filtered by the Regional Proposal Network [8, 11] The RPN selects the anchors which possess the highest probability of consisting of the region of interest or the region to be masked and predicted. A ROI align algorithm is applied before running the bounding box prediction by the ROI regressor to align the input and the feature maps properly using interpolation. Once the feature maps are aligned with the input, the segmentation branch takes all the positive ROIs selected by the classifier and then generates the mask accordingly (Fig. 5).

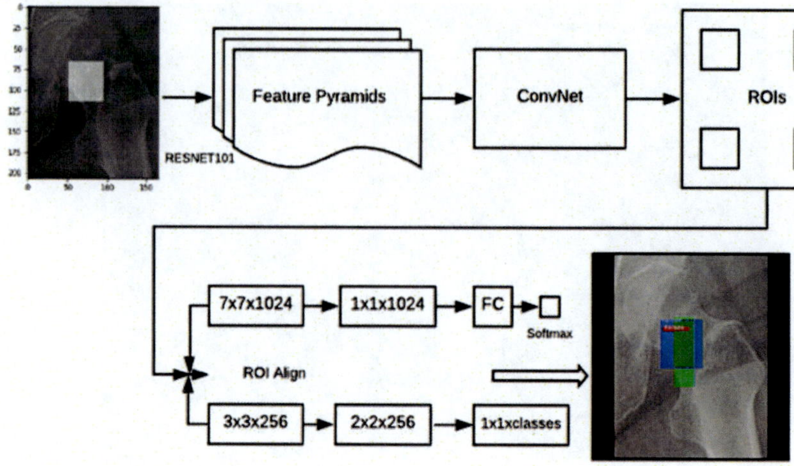

Fig. 5 Architecture consisting of step by step layers

6 Dataset

A dataset of images was acquired from sources like www.radiopedia.org and through real-time consultations with orthopedics and physicians. Our dataset comprises of X-ray images of fractures of the proximal femoral region which forms a subset of fractures that occur in the hip region. All the X-ray images present in the dataset were converted to the standard JPEG format. The average resolution of the images present in the dataset was 2345 pixels $\times$ 2310 pixels. Efforts were made to keep the dataset extensive and vast covering different patient cases and different types of proximal fractures like intracapsular and extracapsular fractures. The network is trained in a ratio of 75:25 meaning, 75% of the dataset was used in training the network, and 25% of the dataset was used for testing purpose. All the images comprised in the dataset had to be annotated. The annotation of these images was performed under the careful observation of experienced and practicing orthopedics and medical consultants. These images comprise of the X-rays of the acetabulum of the pelvis, neck of the femoral bone joint, acetabular sclerosis, shallow acetabulum, delayed ossification of the femoral head, and the hip joint (Figs. 6 and 7).

7 Results

7.1 Comparison with Other Algorithms

The purpose of the paper is to highlight an efficient way of fracture detection, particularly of the proximal femoral bone in X-ray images. Previous algorithms

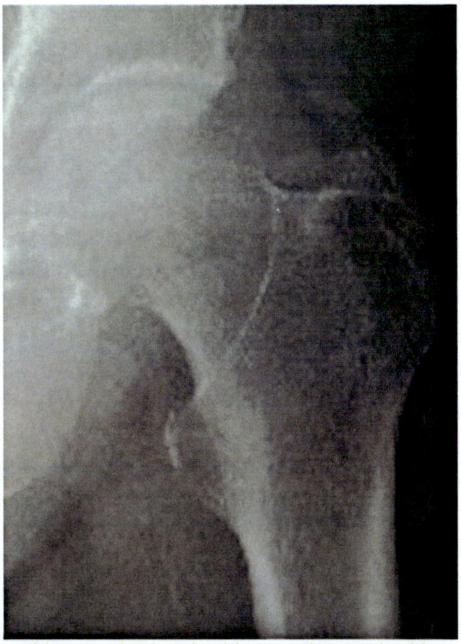

Fig. 6 Above image is from the dataset and depicts the fractured neck region of the femur

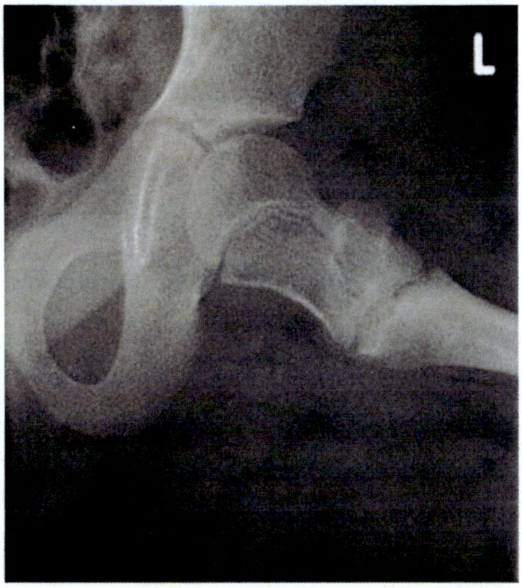

Fig. 7 Above image is from the dataset and depicts the fractured region of the hip joint

applied in cases of fracture detection, such as Convolutional Neural Networks, Haar Cascade, You Only Look Once: an algorithm for real-time object detection have greatly improved methods in assessing fracture classification, detection, and manual segmentation of images. However, when compared to Mask R-CNN, an extension of the faster R-CNN, the former proves to be powerful and has a branch for classification and bounding box regression that utilizes the powerful ResNet 101 architecture to extract important features from the region of interest (RoI). In a Mask R-CNN, segmentation is a little more difficult than simply drawing a box around the object as in Haar Cascading, but results with Mask R-CNN are usually very accurate, and the integration with a neural network adds to the uniqueness of the chosen algorithm.

While an alternative algorithm could have been considered for the purpose of fracture detection such as YOLO, the Mask R-CNN was found to be suitable for the purposes of object detection and segmentation, as YOLO is primarily an object detector. In comparison with an Region-Based Fully Convolutional Network (R-FCN), the number of proposals generated can impact Faster R-CNN (FRCNN) significantly without decreasing the accuracy to a great extent. For example, with the use of the inception ResNet, Faster R-CNN can improve the speed three times more when using 50 proposals instead of 300. The drop in accuracy was only found to be 4%. The R-FCN has much less work per ROI; therefore, the improvement in speed is not found to be extremely significant. To predict multiple objects or instances of objects in an image, Mask R-CNN makes thousands of predictions. The anchor boxes are generated and only those with intersection over union (IoU) > 0.5 are used to output bounding boxes, resulting in higher accuracy shown by the Mask R-CNN framework.

7.2 *Observed Results*

A cumulative set of results were observed after the implementation of the algorithm. Using these results, it was possible to test the usability and scalability of the algorithm on X-rays of the proximal femur. Our dataset is trained on 18 epochs consisting of 35 cycles each. The training loss, validation loss, mask_r_cnn class loss, and the mask_r_cnn_bbox_loss was recorded. The following images show the nature of the loss curves (Fig. 8).

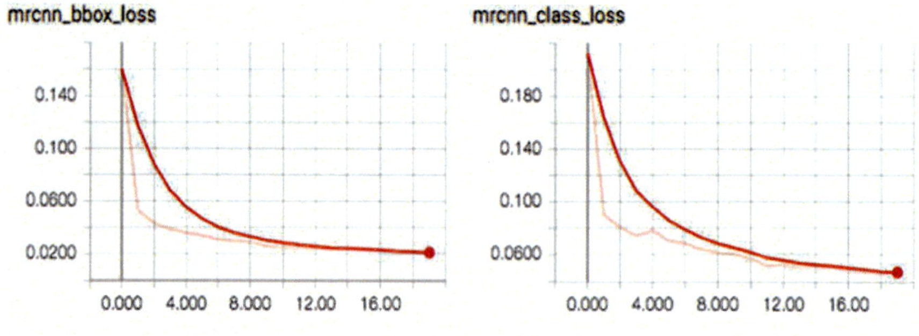

Fig. 8 mrcnn bounding box and mrcnn class loss curves

Table 1 Confusion matrix of the results observed

Confusion matrix	(Actual data) Positives	(Actual data) Negatives
(Predicted) Positives	149	29
(Predicted) Negatives	18	128

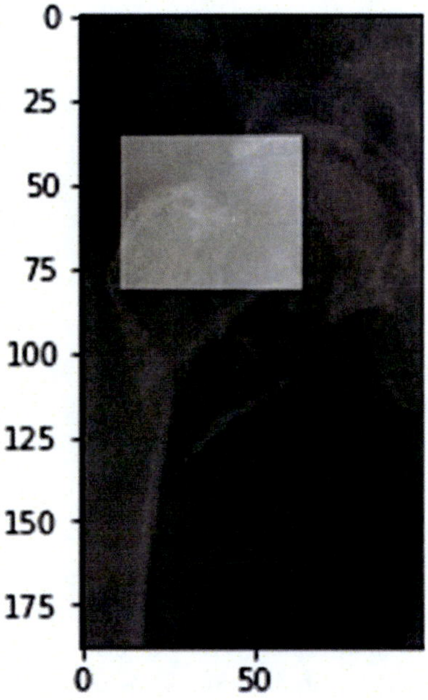

Fig. 9 Fracture region detected by the CNN and FPN layer in the neck region of the femur

The output images were recorded with a considerable accuracy. With 324 high-resolution X-ray images trained on NVIDIA 940MX 4 Gigahertz GPU, an accuracy of 85.4% is obtained (Table 1).

The output of the algorithm was obtained as follows (Figs. 9, 10, 11, and 12).

8 Conclusion

From the results obtained, it can be concluded that Mask R-CNNs form a very robust architecture in detecting fractures in the proximal femoral bone region. These fractures are detected with an accuracy of 85.4% with 149 true positive images, 28 true negative images, 18 false positive, and 29 false negative images. These results show the scalability and effectiveness of the architecture. Minute cracks or bone fractures are detected in the X-rays using this architecture which extends the possibility of

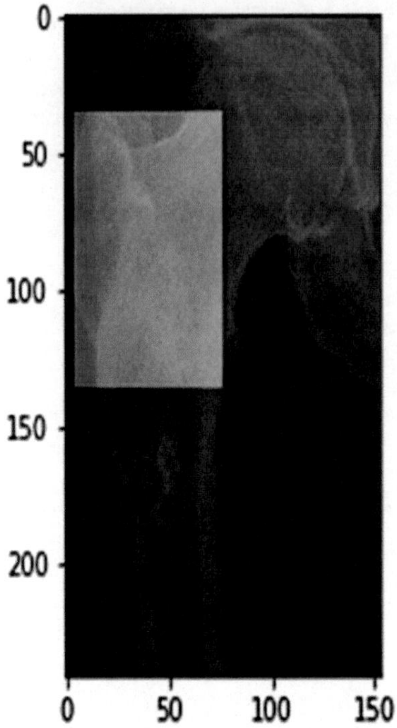

Fig. 10 Fracture region detected by the CNN and FPN layer in the neck region of the femur

its use. This paper also concludes that Mask R-CNNs are better and versatile than Faster R-CNNs. With the fractures masked from the X-ray images, the door is open to the possibility of performing quantitative analysis on the segmented masks of the fractured region. This algorithm and approach definitely prove to be a big step toward eradicating or eliminating the major problem of osteoporosis. From this set of experimentation, with a bigger dataset and more efficient training resources, a major global health problem can be solved.

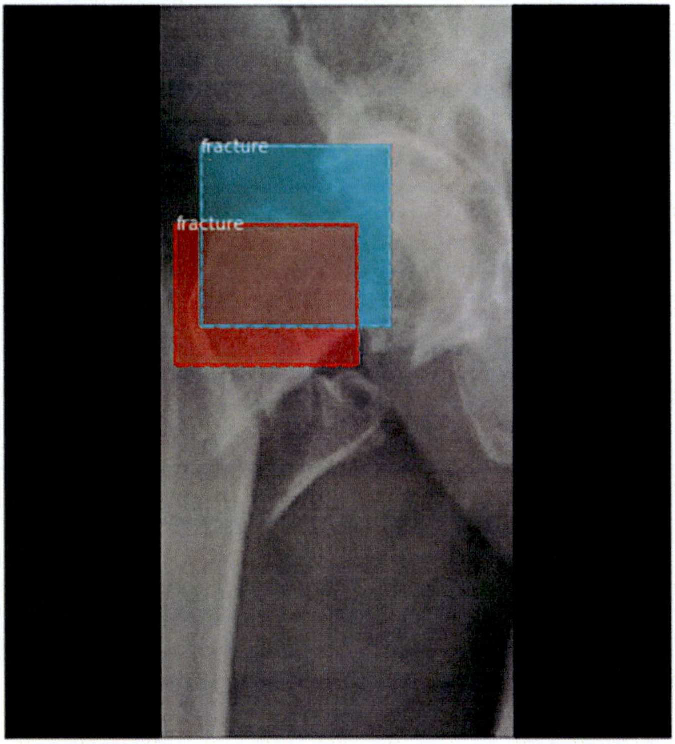

Fig. 11 Detected region of fracture segmented and masked by the Mask R-CNN

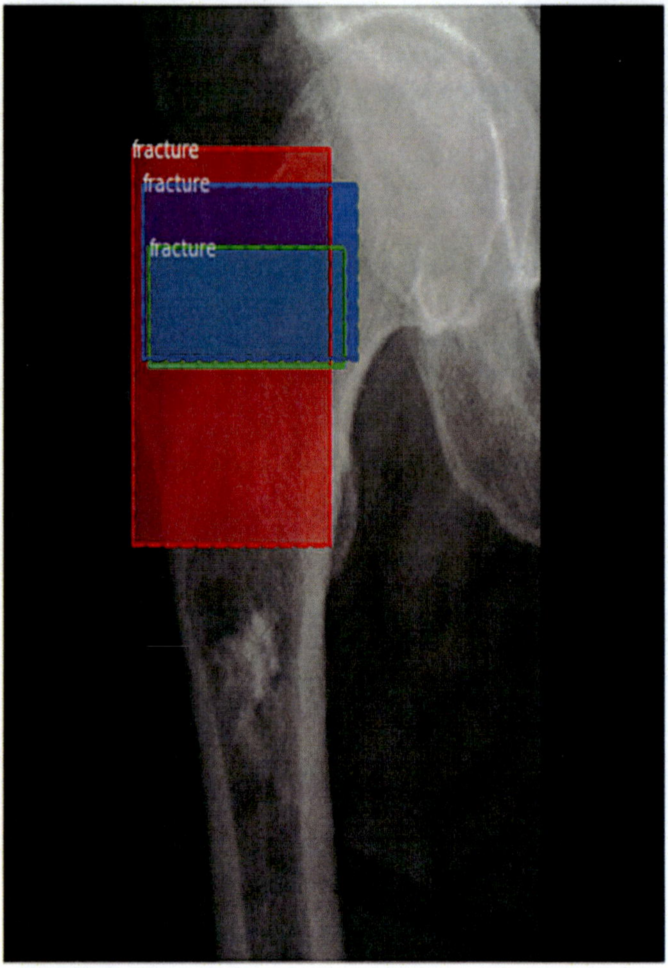

Fig. 12 Detected region of fracture segmented and masked by the Mask R-CNN

References

1. Deniz, C.M., Xiang, S., Hallyburton, R.S., et al.: Segmentation of the proximal femur from MR images using deep convolutional neural networks. Sci Rep **8**, 16485 (2018)
2. He, K., Gkioxari, G., Dollár, P., Girshick, R.: Mask R-CNN, 2017 IEEE International Conference on Computer Vision (ICCV), Venice, pp. 2980–2988 (2017)
3. Dimililer, K.: IBFDS: intelligent bone fracture detection system. Procedia Comput. Sci. **120**, 260–267 (2017)
4. Zeng, G. et al.: 3D U-net with multi-level deep supervision: fully automatic segmentation of proximal femur in 3D MR images. In: Wang, Q., Shi, Y., Suk, H.-I. Suzuki, K. (eds.) Machine Learning in Medical Imaging, pp. 274–282. Springer International Publishing (2017)
5. Girshick, R.: Fast R-CNN. InICCV (2015)

6. Ren, S., He, K., Girshick, R., Sun, J.: Faster R-CNN: To-wards Real-Time Object Detection with Region Proposal Net-Works. InNIPS (2015)
7. Shrivastava, A., Gupta, A., Girshick, R.: Training Region-Based Object Detectors with Online Hard Example Mining. 2016 IEEE Conference on Computer Vision and Pattern Recognition (CVPR) (2016)
8. Ren, S., He, K., Girshick, R., Sun, J.: Faster R-CNN: Towards Real-Time Object Detection with Region Proposal Networks
9. Girshick, R.: Fast R-CNN. In ICCV (2015)
10. Krizhevsky, A., Sutskever, I., Hinton, G.: ImageNet classification with deep convolutional neural networks. In NIPS (2012)
11. Milletari, F., Navab, N., Ahmadi, S.-A.: V-Net: Fully Convolutional Neural Networks for Volumetric Medical Image Segmentation (2016). *arXiv* 1–11, http://arxiv.org/abs/1606.04797
12. Ren, S., He, K., Girshick, R., Sun, J.: Faster R-CNN: To-wards real-time object detection with region proposal net-works. In NIPS (2015)

Combination of Support Vector Machine (SVM) and Bayesian Model to Identify Criminal Language

Amelec Viloria, Omar Bonerge Pineda Lezama, and Juan Hurtado

Abstract In the last few years, different investigations have been developed related to the use of Natural Language Processing (NLP) and especially of information extraction (IE) in criminal matters (Abbass et al., in 2020 IEEE 14th international conference on semantic computing (ICSC). IEEE, pp 363–368, 2020 [1]). This paper discusses the creation and characterization of a specialized corpus in criminology. The corpus is made up of plain text news divided into five classes of crimes: killing, attack, abduct, sexual abuse, and blackmail. The classifiers used are the classic support vector machine (SVM) and a Bayesian model. The importance of automatic classification of criminological texts lies in the fact that it can be used for criminalistic analysis (Viloria et al., in Intelligent computing, information and control systems. Springer, Cham, 2020 [2]) and for the detection of criminal entities (Peersman, in Detecting deceptive behaviour in the wild: text mining for online child protection in the presence of noisy and adversarial social media communications. Lancaster University, 2018 [3]). News classification can even help to find patterns in crime reports or other criminalistic aspects (Chan et al., Theor Criminol 20:21–39, 2016 [4]).

Keywords Classification of a criminological corpus · Spanish · Support vector machine (SVM) · Bayesian model

---

A. Viloria (✉) · J. Hurtado
Universidad de la Costa, St. 58 #66, Barranquilla, Atlántico, Colombia
e-mail: aviloria7@cuc.edu.co

J. Hurtado
e-mail: jhurtado4@cuc.edu.co

O. B. P. Lezama
Universidad Tecnológica, San Pedro Sula, Honduras
e-mail: omar@unitet.edu

© The Editor(s) (if applicable) and The Author(s), under exclusive license to Springer Nature Singapore Pte Ltd. 2021

A. P. Pandian et al. (eds.), *Proceedings of International Conference on Intelligent Computing, Information and Control Systems*, Advances in Intelligent Systems and Computing 1272, https://doi.org/10.1007/978-981-15-8443-5_20

1 Introduction

Based on the research presented by Schmidt and Wiegand [5], it can be concluded that most of the works looking for criminal patterns use specific elements such as firearms or knives. However, there are other investigations that use the NLP, more specifically the IE, to carry out this task. IE techniques are varied and are often performed based on the frequency of terms in the texts. In [6], the TF-IDF model is used for vectorizing the input texts, in the same way.

The same technique is used in the study of [7], but the TF-IDF model is used in this case for feature extraction. Subsequently, the extracted characteristics are used to classify the texts used by occupying the cosine similarity measure.

Analyzing the results of the mentioned studies, it was decided to implement these techniques for the extraction of characteristics, with the difference of proposing a model designed for short texts. This is to allow maintaining the same quality in the extracted characteristics, but increasing the processing speed. It was noted that the work of [8, 9] could be adapted. In the first one, an adaptation of the TF-IDF model is made for short texts, and in the second one, rare bigrams are taken as EN candidates. With these techniques, it will be possible to search for relevant terms in the news and then consider them as characteristics that could describe each class that is being studied in this research.

2 Textual and Statistical Descriptive Corpus

The following elements describe the process followed for the conformation of the corpus presented in this study. The statistical information about it is also detailed.

2.1 Construction of the Corpus

For the conformation of the corpus, it was necessary to download news from various local journalistic Web sites in the Colombian State. Newspapers of local and national circulation were chosen, since it is in these media that criminal activity in the area is mostly reported.

A total of 1000 news items were downloaded and stored in plain text in utf8 format. The recovered documents meet the condition of reporting at least one of the following crimes:

- killing,
- attack,
- abduct,
- rape.

Table 1 Keywords of the corpus

Class	Keywords
Attack	Robbery, theft, burglary, threatening, stripping, intercepting, removing, to plunder, dismantle, surprise in possession, hide
Killing	Finding no life, lynching, killing, running over, finding dead, dead body, shoot, shot, riddled, attacked, shooting, lifeless person, dying, to shoot
Abduct	Abduct, forced entry, struggle, freedom, lifting person, rescued victim, deprivation, release victim, forced to go, taken, rescue
Rape	Rape, sexual attack, intimidation, statutory rape
Blackmail	Extorting, beating, coercing, intimidating, threatening

It should be noted that at the time of labeling the news, the note-takers noticed an additional class present in a significant number of newspaper articles: blackmail. Therefore, this class was added as one of the possible crimes of the corpus.

2.2 Class Keywords

In addition to the manual annotation made by the four note-takers, they were asked to make a list of key words that would allow them to classify the news. In this way, not only was a manual annotation of the corpus obtained, but also a set of terms, single or multi-word, which are used recurrently in the corpus. Table 1 shows the terms, in their canonical form, found by the note-takers for each class.

As can be seen, the vast majority of key words found in each class correspond to verbs (e.g., strip, struggle, shoot) and nouns (deprivation, rape, corpse). However, there are also some formations such as verb-substantive (raise-person, liberate-victim), verb-adjective (find-dead), and adjective (intimate, shot).

3 Methodology

The method proposed in this paper is divided into two parts: extraction of characteristics (Sect. 3.1) and news classification by content (Sect. 3.2).

3.1 Extraction of Characteristics

This process was based on the conclusions and results obtained by Brennan [10]. This study argues that it is the nominal syntagma that best describe the information in a text. In this particular case, the identification of the type of crime reported in the note is of interest. The present study considers that the acts detailed in the text can be analyzed by identifying the verbs, which coincides with the conclusions of the note-takers, who observed that nouns and verbs provide the most information for the detection of the type of crime. Therefore, verbal syntagma will also be studied [11].

The process followed for the extraction of the characteristics is described below:

First, each of the texts is annotated with labels indicating the grammatical category of the words (POS or part of speech), using the FreeLing tool [12].

Then, from the POS tags of the words in each text, the following syntax patterns are extracted:

- Verbal syntagm (VP–verb phrases),
- Nouns,
- Verbs.

Once these syntactic patterns are extracted, the degree of importance of each word that appears in the syntagma is calculated, which is done based on the research of [13]. To do this, the number of words in the syntagm (VP) is multiplied by the frequency of each word in the syntagm (UF or unigram frequency), and this operation is formalized in Eq. 1:

$$\mathrm{UF(VP)} = \sum_{i=0}^{|\mathrm{VP}|} \text{Unigram Frequency } (w_i) \tag{1}$$

Then, according to the model proposed by Kamatkar et al. [14], the result obtained in Eq. 1 is multiplied by the frequency of the syntagm in the paper, (VPF(VP)). The result is divided by the number of words in the syntagm (|VP|):

$$\mathrm{Score(VP)} = \frac{\mathrm{UF(VP)} \cdot \mathrm{VPF(VP)}}{|\mathrm{VP}|}. \tag{2}$$

Once the groups and their elements have been determined, their scores are calculated. The calculation of the above is only an arithmetic average. More specifically, this is done by adding up the score of each element, belonging to the group, over the number of elements belonging to the same group. Equation 3 presents the formula used to calculate the score of each group [15]:

$$\mathrm{Score(Grupo)} = \frac{\sum_{i=0}^{|\mathrm{Grupo}|} \mathrm{Score}(\mathrm{VP}_i)}{|\mathrm{Grupo}|}. \tag{3}$$

Finally, a threshold was established to delimit the most important words or syntagma of the processed text. These words are the ones that will make up the pockets of words or characteristics that will give rise to each class of crimes. The threshold was established with respect to the average values of each class, thus eliminating those group scores that are too high. Such is the case of stop words or words that do not give any relevant description about the document. Similarly, they represent those classes with very low scores, such as names of people, places, and dates (the named entities).

3.2 Corpus Baseline Classification

In order to establish a basic measure of performance, it was decided to use two classifiers on the annotated corpus. The corpus was divided into a learning corpus of (LC) and a proof corpus (PC). The distribution of news in each sub-corpus was obtained randomly with a uniform distribution. This was done to ensure the same distribution as in the corpus. The task was then to determine which class each news item belongs to.

The LC learning corpus consists of a subset of news from 70% of the total corpus, and the test corpus from the remaining 30%. These news subsets were subjected to the feature extraction process described in Sect. 3.1. As training data, the features extracted from the manually annotated news set were used (see Sect. 2.1). These characteristics were used to analyze the class to which the news items of the PC corpus correspond.

For the classification of the PC, the Weka8 platform was used, which allows working with different classification algorithms. In this study, tests were carried out with a Bayesian model (Naïve Bayes) and with a support vector machine (SVM).

4 Results and Evaluation

The results presented below consider three experiences. In the first one, the full story was analyzed: Tables 2 and 3. In the second one, only the title of the news and the first paragraph were used: Tables 4 and 5. Finally, the last test was done considering only the title of the news: Tables 6 and 7. In all cases, the classical measure F-Score was used for the evaluation, of nest by Eq. 4 [16]:

$$\mathrm{F} - \$Score\$ = \frac{2 \times (\text{Precisió n} \times \text{Recall})}{\text{Precisió n} + \text{Recall}}. \quad (4)$$

Table 2 Results of the classification (Full news—SVM)

Classes	Attack	Killing	Abduct	Rape	Blackmail	Mean
Precision	0.7094	0.7359	0.6597	0.5746	0.7778	0.6915
Recall	0.9072	0.6423	0.7105	0.4147	0.9328	0.7201
F-Score	0.7871	0.6657	0.6706	0.4542	0.8376	0.683

Table 3 Results of the classification (Complete news—Naïve Bayes)

Classes	Attack	Killing	Abduct	Rape	Blackmail	Mean
Precision	0.7496	0.7891	0.6001	0.5483	0.6981	0.6773
Recall	0.89636	0.46406	0.81086	0.47606	0.97756	0.72496
F-Score	0.79458	0.5524	0.66944	0.48297	0.79485	0.6588

Table 4 Results of the classification (Title of the news and first paragraph—SVM)

Classes	Attack	Killing	Abduct	Rape	Blackmail	Mean
Precision	0.9025	0.6178	0.50583	0.50196	0.8494	0.6755
Recall	0.71456	0.73936	0.49606	0.42486	0.71826	0.61856
F-Score	0.76977	0.65118	0.4759	0.4335	0.75147	0.61622

Table 5 Results of the classification (Title of the news and first paragraph—Naïve Bayes)

Classes	Attack	Killing	Abduct	Rape	Blackmail	Mean
Precision	0.9168	0.65952	0.50544	0.5136	0.6577	0.6505
Recall	0.75096	0.60176	0.74426	0.50176	1.14616	0.74896
F-Score	0.79827	0.60343	0.58262	0.48267	0.81947	0.65728

Table 6 Results of the classification (Title of the news—SVM)

Classes	Attack	Killing	Abduct	Rape	Blackmail	Mean
Precision	0.8793	0.39893	0.56552	0.7022	0.7822	0.6656
Recall	0.54436	0.86146	0.59326	0.18136	0.61056	0.55816
F-Score	0.71124	0.5896	0.61947	0.32832	0.72487	0.5946

In the mean column of Tables 2, 3, 4, 5, 6 and 7, the average of Precision, Recall, and F-Score of each experiment is indicated.

It can be seen that there is not a great difference between the results given by the SVM and the Bayesian models. This may be due to the fact that the SVM were not optimized on the corpus. Despite this, the F-Score average using SVM is the highest, with F-Score = 0.70254. By changing the SVM parameters, even better results could be obtained. As expressed by the scorers, who in several moments

Table 7 Results of the classification (News title—Naïve Bayes)

Classes	Attack	Killing	Abduct	Rape	Blackmail	Mean
Precision	0.9247	0.4158	0.6113	0.7115	0.6739	0.6674
Recall	0.70795	0.74205	0.72975	0.24505	0.78975	0.64295
F-Score	0.828	0.5621	0.6925	0.3916	0.7543	0.6457

found more than one crime in a single news item, it is considered that one way to improve the results (in Precision, Recall, and F-Score) could be to analyze the news with a multi-class classifier. Such a classifier would allow the same news to be included in two or more classes. It could even improve the F-Score in news related to killing, abduct, and sexual abuse which have the lowest percentages, due to the great coverage that exists between them.

5 Conclusions

In this paper, the Annotated Corpus of Crimes in Colombia has been introduced. This corpus is characterized, and some baseline measures for the automatic classification of five categories of crimes were presented. The corpus can be used in crime classification tasks using NLP tools. These analysis tools could be useful for different government bodies (police, youth institutes, etc.) as well as for decentralized (human rights commissions) or non-governmental organizations.

Some examples of possible tools are maps indicating high-impact crimes, crime news search engines, documentation tools, and the generation of crime summaries [17, 18], among others.

References

1. Abbass, Z., Ali, Z., Ali, M., Akbar, B., Saleem, A.: A framework to predict social crime through twitter tweets by using machine learning. In: 2020 IEEE 14th International Conference on Semantic Computing (ICSC), pp. 363–368. IEEE (2020)
2. Viloria, A. et al.: Big data marketing during the period 2012–2019: a bibliometric review. In: Pandian, A., Ntalianis, K., Palanisamy, R. (eds.) Intelligent Computing, Information and Control Systems. ICICCS 2019. Advances in Intelligent Systems and Computing, vol. 1039. Springer, Cham (2020)
3. Peersman, C.: Detecting deceptive behaviour in the wild: text mining for online child protection in the presence of noisy and adversarial social media communications. Doctoral dissertation, Lancaster University (2018)
4. Chan, J., Bennett Moses, L.: Is big data challenging criminology? Theor. Criminol. **20**(1), 21–39 (2016)
5. Schmidt, A., Wiegand, M.: A survey on hate speech detection using natural language processing. In: Proceedings of the Fifth International Workshop on Natural Language Processing for Social Media, pp. 1–10 (2017)
6. Milivojevic, S., Radulski, E.M.: The 'future Internet' and crime: towards a criminology of the Internet of Things. Current Issues in Criminal Justice, pp. 1–15 (2020)
7. Kaity, M., Balakrishnan, V.: An automatic non-English sentiment lexicon builder using unannotated corpus. J. Supercomput. **75**(4), 2243–2268 (2019)
8. Bollen, J., Mao, H., Pepe, A.: Modeling public mood and emotion: Twitter sentiment and socio-economic phenomena. ICWSM **11**, 450–453 (2011)
9. Abbass, Z., Ali, Z., Ali, M., Akbar, B., & Saleem, A.: A framework to predict social crime through Twitter Tweets by using machine learning. In: 2020 IEEE 14th International Conference on Semantic Computing (ICSC), pp. 363–368. IEEE (2020)

10. Brennan, T.: An alternative scientific paradigm for criminological risk assessment: closed or open systems, or both? In: Handbook on Risk and Need Assessment, pp. 180–206. Routledge (2016)
11. Raj, J.S., Vijitha Ananthi, J.: Recurrent neural networks and nonlinear prediction in support vector machines. J. Soft Comput. Paradigm (JSCP) **1**(01), 33–40 (2019)
12. Tollenaar, N., Van der Heijden, P.G.M.: Which method predicts recidivism best? a comparison of statistical, machine learning and data mining predictive models. J. R. Stat. Soc. Ser. A (Stat. Soc.) **176**(2), 565–584 (2013)
13. Brennan, T., Oliver, W.L.: Emergence of machine learning techniques in criminology: implications of complexity in our data and in research questions. Criminol. Pub. Poly **12**, 551 (2013)
14. Kamatkar, S. J., Kamble, A., Viloria, A., Hernández-Fernandez, L., Cali, E.G.: Database performance tuning and query optimization. In: International Conference on Data Mining and Big Data. Springer, Cham, pp. 3–11 (2018)
15. Bhosale, D., Ade, R.: Feature selection-based classification using naive bayes, j48 and support vector machine. Int. J. Comput. Appl. **99**(16), 14–18 (2014)
16. Zaid, A., Alqatawna, J.F., Huneiti, A.: A proposed model for malicious spam detection in email systems of educational institutes. In: 2016 Cybersecurity and Cyberforensics Conference (CCC), pp. 60–64. IEEE (2016, August)
17. Dzhumaliev, M.: Detection of damage and failure events of road infrastructure using social media. In: Web Services–ICWS 2018: 25th International Conference, Held as Part of the Services Conference Federation, SCF 2018, Seattle, WA, USA, June 25–30, 2018, Proceedings, vol. 10966, p. 134. Springer, Berlin (2018, June)
18. Viloria, A., Lezamab, O.B.P.: Improvements for determining the number of clusters in k-means for innovation databases in SMEs. Procedia Comput. Sci. **151**, 1201–1206 (2019)

Higher Education Enrolment Query Chatbot Using Machine Learning

B. S. Niranjan and Vinayak Hegde

Abstract The automation has taken over many of the works that were done by humans because of the accuracy of the machine which is not found in humans. One such technology is a chatbot, also recognized as a communications agent. Chatbots are widely used in industry, banking, consultancy, and many more sectors. Education is one of the areas a chatbot is being used in. By the use of techniques like Natural Language Processing (NLP), Natural Language Understanding (NLU), Pattern Matching Techniques, and Retrieval of Information (IR), a chatbot can be built faster, easier, and more effectively. Incorporating a bot in an educational institution can help make rendering information faster than normal, and processes such as registration, application for ID, etc., can be automated, so people who are anxious about talking to others can use the bot to access the information they need, and also prevent confusion.

Keywords Chatbot · Vector space model · Question–answer system · Information retrieval

1 Introduction

Chatbots are a piece of software that mimics a human being's interaction, i.e., they are talking to a user as if another person is talking to them. ELAIZA was the first-ever chatbot to be created [1] when its history and classification are talked about and now, there are many bots like Alexa, Google Assistant, and Cortana. Chatbots are of two types, one is a static bot that gives a response based on the answers that the user has stored in it and the other is a dynamic bot that leans based on the interaction it is

B. S. Niranjan (✉) · V. Hegde
Department of Computer Science, Amrita School of Arts and Sciences, Amrita Vishwa Vidyapeetham, Mysuru Campus, Bhogadi, Karnataka, India
e-mail: niranjan.bs.1997@gmail.com

V. Hegde
e-mail: h_vinayak@asas.mysore.amrita.edu

© The Editor(s) (if applicable) and The Author(s), under exclusive license to Springer Nature Singapore Pte Ltd. 2021

A. P. Pandian et al. (eds.), *Proceedings of International Conference on Intelligent Computing, Information and Control Systems*, Advances in Intelligent Systems and Computing 1272, https://doi.org/10.1007/978-981-15-8443-5_21

experiencing. A chatbot is an Artificial Intelligence (AI) program that can mimic the translation of a person into a natural language, i.e., a language that is understandable to the user. From a technological point of view, however, the platform reflects only the natural evolution of a questionnaire response using Natural Language Processing (NLP) [2]. Formulating natural language responses to questions is one of the most common examples of natural language used in various businesses.

1.1 Chatbot

A chatbot is a piece of software that replicates or mimics the conversation of two human beings, i.e., it replies to the user of the bot as a normal person would. It can be powered by an AI engine or it can just be a question and answer chatbot or it can be based on machine learning concepts [1].

Chatbot is found in all the sectors of work, some are industries, banking, and now slowing in education. The best example of chatbots can be query answering machines which can be commonly found in many Web sites of various companies or organizations which will help the users to get information regarding their company or organization.

1.2 Challenges Faced by a Chatbot

The technology of chatbot is a very vast and complicated topic and there are yet many improvements possible in this technology. Mimicking the conversation between two human beings is not an easy task and a need lot of complex calculations to be performed to predict what might be the possible response that was to be given to a user who is interacting with the chatbot to get some information.

But when take a look at question answering chatbot [3], the problem arises which is returning a response that has to be related to the question which is been given by the user. There are some challenges faced and they are

- The query which is given by the user, i.e., the question which is issued by the user needs to be properly preprocessed so that the system may understand the query.
- A deadlock scenario where there is no response generated or there is more than one response found then what should be done [3].

2 Motivation

In a quest for additional support, insights, and opinions, it is not that inept, but that the society today with the notion of co-working or support has made all possible possibilities of self-sufficiency. Only with this can an individual successfully meander through our professional or personal ecosystem's complex dynamics. But assistance will not always be accessible to a person who needs support, and also that technology has taken over the world and all major activities such as automotive manufacturing, electronic device construction, and much more are automated work that has to be done quickly. So why should not it automate activities such as registration, gathering information, and other small tasks?.Why should all the operations be carried out manually? As step into an age of automation, there should be a method for accomplishing all such things more efficiently and more precisely.

3 Literature Survey

F. O. Adibe, E. C. Nwokorie, and J. N. Odii start to speak about chatbot and it were based on pattern matching. The evolution was from simple pattern matching to object-oriented analysis and design methodology (OOADM) and also discusses turning test which is an evaluation for AI chatters. Further, they speak about the early system which is ELIZA, PARRY, and ALICE which were the first of its kind. They also describe pattern matching which was used by the systems. They also mentioned search engines like Special search Engine and Matrix Search Engine for Keyword searching [1].

Endang Wahyu Pamungkas starts by introducing what a chatbot/chatter is and explain about another version of a chatbot which is an Emotionally aware chatbot (EAC). He then speaks about the evaluation of the EAC which he chose the turning test which is the default test for any chatbot. Then, he starts to explain the history of EAC and also explains how an EAC is built. Next, he speaks about techniques like classification and Natural Language Processing (NLP) which play a very important role in building a chatbot [4].

Sameera A. Abdul-Kader and Dr. John Woods speak about the trend in Human–Computer Speech (HCS) and how NLTK with Python will help in developing an HCS. Then, they speak about the strategies used in their developed and also the type of input and output can be in the form of text or a speech. Next, they start explaining the techniques and methodologies used which are Parsing, Pattern matching, Artificial Intelligent Markup Language (AIML), SQL and Relational DB, Markov chain, and Language Tricks [5].

Amit Singhal speaks about the vague history and the evolution of information storing and retrieval and gradually starts the discussion about the various models which are used in IR; some of them are Boolean, vector space, and probabilistic models which are also the most famous models which are being still used. Further, he speaks about in-depth details of each of these models and how it can be implemented. He next moves on to describing the data structures(DS) which are used in these models one of which is the inverted list and also elaborates about the various sub-processes which are involved in the process of IR, i.e., Stop-word Removal, Stemming, Noun group Identification [6].

Vinayak Hegde speaks about a dashboard to enhance the process of admission in higher education at private universities. He gives processes for identifying the individual's intellect, logical and problem-solving skills. This paper shows an overall academic dashboard that is used to maintain all the activities related to admission such as walk-in, follow-up, prospectors sold, etc., and with the help of SQL and google API reports are generated for visual aid [7].

Sushma Rao H.S., Suresh A., Hegde V. speak about an automated way of dealing with the data which is being generated in any institute to overcome the traditional way of storing information which is files. They also speak of attaining pattern from the huge data by using ML. They compare the total admission rate and frequency of students across the state to aid in decision making for the organization [8].

4 Proposed System

The basic flow of the system goes like this. There are three main phases: first is the preprocessing, the next is the search engine, and the last phase is the representation. All the phases will be explained in the next section (Fig. 1).

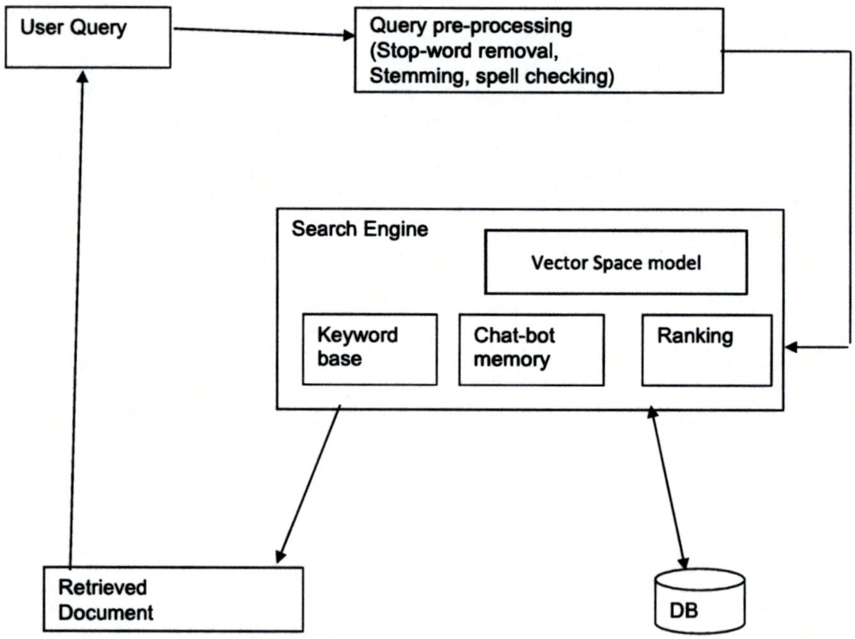

Fig. 1 Proposed system architecture diagram

When a user enters a query, the first basic step which has to be done is the preprocessing, i.e., the input needs to be cleaned of all the unwanted, junk, and redundant information that the system will not require is taken out or removed.

After preprocessing, the result is given to the search engine so that it can match the result to a corresponding response in the database/knowledge base. After a match is found, the response is represented to the user in both forms of text and speech. The bot uses the vector space model to compare the cosine similarity [9, 10] between the documents in the corpus and the query given by the user.

5 Implementation

5.1 Phase One (Preprocessing)

In the first phase as said above, the user enters a query and preprocessing happens so that the system can understand the query and give the user response. There are several steps which take place here, and they are

- Tokenization: The first step in preprocessing is the tokenization of the given query or the document [11]. Tokenization refers to the process of breaking down the string into a list of token, i.e., list of a word that is present in the string. One can imagine that the token as words when there is a string and token as strings/sentences when there is a paragraph.

  word_tokens = word_tokenize(example_sent)

- Stop-word Removal: This is the process of removing stopwords, i.e., removal of the unnecessary word that will not help in any at the time of search of a response [12]. These words are commonly used in a sentence such as (as, a, the, in) which are left because it can take up a lot of space and also take up precise processing time/power. The removal of these words is not restricted to a certain number of words but the list can be customized according to the user's needs.

  stop_words = set(stopwords.words('english'))

  words = ["a", "about", "above", "above", "across", "after", "afterwards", "again", "against", "all", "almost", "alone", "along", "already"]

- Noun Group: In certain cases, identification of noun groups can help in speeding up the search in the system because when a noun is identified ,it eliminates all the adjectives, adverbs, and verbs present in the query giving an even more refined query to the system to process.
- Stemming: In this section, the root words present in the query, i.e., a common grammatical root of certain words is found so that the query can be reduced even more and the process becomes even faster. One example of stemming is consider the words fish, fishing stick, fisherman, fish bait are present in a query then the

root word would be fish or fishing because all the words in the query refer to the root word.

```
ps = PorterStemmer()
words = ["program", "programs", "programer", "programing", "programers"]
for w in words:
    print(w, " : ", ps.stem(w))
```

- Indexing: When all the above operations are applied to the query, the final step is to index the remaining word of the query or a document. Index terms are noting but noun groups, root words, etc., which can be used to refer a context in the knowledge base.

5.2 *Phase Two (Search Engine)*

In this phase, the actual searching takes place where the results from phase one will use to get a response to the user's request. There are mainly two parts in this; there are the search engine and the knowledge/database.

5.2.1 Search Engine

- Vector Space Model: Vector space model(VSM) is an algebraic model where it mainly contains two steps in it; first, the document is represented in a vector [13] of terms and the second, the vector is transformed into a numerical format so it can be used to measure the relevance between the document and the query [14]. Consider there are two document d1 and d2 and the query q ,then the relevant document to that query id calculated by

$$d_r() = \text{MAX}(\text{Sim}(d1, q), \text{Sim}(d2, q)) \tag{1}$$

$$\text{Sim}\left(d_j, q\right) = \frac{q.d1}{|q| * |d1|} = \frac{\sum_{i=1}^{t} Wiq.Wij}{\sqrt{\sum_{j=1}^{t} (Wiq)^2 * \sum_{i=1}^{t} (Wij)^2}} \tag{2}$$

A team document matrix is used to maintain document vectors in a 2 × 2 matrix where the row represents documents(d_i) and the columns the vector terms and the coinciding cells represent the frequency of the term in that particular document. After the matrix is done, term weights are calculated to find out the term which will define a document uniquely.

Term frequency–inverse document frequency (tf–idf) [15, 16], which gives higher weights to terms that occur more in one document but rarely occurs in all other documents, lower weights to terms that generally occur within and among all documents are used because not always do the higher weighted terms relate to a relevant document and sometimes, lower weighted can relate to the relevant document. And it is calculated by a logarithmic function shown below.

$$\text{Tf-idf} = \text{tf} \times \log(N/\mathrm{d}f) \quad (3)$$

where

tf term frequency
N Number of documents

- Ranking: As said is the challenge of a bot one challenge is a deadlock scenario where there is no response or more than more response retrieved. If there is no response found, then the knowledge has to be updated or has to be looked up. But, in the other case, a ranking model/method is used to decide which response can be given to the user. Since this work uses a smaller size of knowledge base, a simple hit count ranking can be used to rank the retrieved document and at the time of a deadlock, the ranking can be compared and the document with the highest-ranking can be retrieved.
- Keyword Base: This is a database that stores all of the found keywords or that will be found in the future because it is not feasible to run pre-processing steps every time in the knowledge base on the document, and it does not make sense because it gives the same result every time so it is better to store the keywords of the documents in a DB with an Id so that it can be used as many times as wanted.

5.2.2 Knowledge Base

This is a database that contains all the documents, i.e., the knowledge the search engine will use to retrieve information for the user's query. This database is present outside the main search engine

5.3 Phase Three (Representation)

This is phase mainly deals with the representation of the retrieved response to the user's query. Since it is a chatbot, just a textual representation is not enough so another way of representation technique or way is through the ways of speech, i.e., converting the text into a speech of a certain language and presenting it to the user so that even if the user cannot read or cannot see, they can get the information they needed.

6 Results

As mentioned above, the user's query is accepted by two ways which are by text and the other by speech which is then converted into text by a speech recognizer

- In this chabot, there are two types of queries; they are general greeting queries and responses like (hello, Welcome, thanks, bye, hi, nice talking to you). The below conversation with the chatbot shows the results of the greeting query and its responses.
- The other types of queries are the actual query which will be used to get the information from the knowledge base so that the user asking for the information can be provided with the information he needs (Figs. 2 and 3).

When there is no response found from the knowledge base, the user is given back a message saying "I am sorry! I don't understand you" so that the user can refine his search keys. Also, another option is when a user does a spelling mistake, the user is prompted back to verify and change the mistake. An example of the chatbot is shown below (Fig. 4).

As in the above results that in a query, the word duplicate was misspelled so it prompted the user back asking what did the user mean. And the other two queries responded with a sorry message because the course MBA and journalism are not available in the college.

And also all these responses which are being generated by the chatbot are presented to the user both textually and through audio.

Fig. 2 Results of greeting query and responses

```
>hello

BOT: Hi

>hi

BOT: Hello

>hey how you doing

BOT: I am doing fine Thank you

>It was nice talking to you

BOT: It was nice talking to you as well

>thank you

ROBO: You are welcome..
```

Fig. 3 Results of query acting on the knowledge base

>what time the college will start

BOT: the college starts at 8:50 am everyday from Monday to Saturday.

>how many departments are there in the college

BOT: there are 4 departments in the college.

>what are the departments present in the college

BOT: the departments present in the college are visual media, computer science, department of sciences and commerce.

> courses offered by visual media

ROBO: the courses offered by the department visual media department are b.sc(vm), integrated m.sc.

Fig. 4 Results of exception cases in the chatbot

>What course are offered by journalism department

BOT: I am sorry! I don't understand you

>How to apply for dupliate id card

BOT: Did you mean : duplicate

>What is the fee structure for MBA

BOT: I am sorry! I don't understand you

As said in the above examples that there are mainly three cases in the chatbot and each case records different response times due to its different type of information retrieval.

- In the first case that is greeting query, it uses a simple string matching algorithm used in the bot ELAIZA to return a response to the user, i.e., all the greeting responses are stored in an array of responses and the time taken to retrieve a response is O(n) where n is the size of the array.
- In the second case, the user query is acting on the corpus and it also needs time for preprocessing, and to calculate the similarity between the documents and the query as recorded it took 3–4 s to fetch a document relevant to the user's query.
- In the third case, i.e., the exception case, the time to respond to the user is the same as in the before case because after the query is compared with all the documents in the corpus, then it gives a message saying "I am sorry! I don't understand you."

7 Discussion

As saw in this paper that a chatbot is so useful in its way apart from the uses mentioned above, there is more to discuss regarding this topic. Firstly, the advantages of the chatbot in any institution, there will be many sources of information about what a person can go to, but the question is "Are those sources accurate?". Instead of having more number of sources, centralizing the source can help remove redundant/false information. Also since there will be holidays in an institution not always a person will be able to get the information, but in the case of a chatbot, it will be available to the user's use all the time so that there will be no delay in getting information. When speak about the time of retrieving information, it may be a time-consuming task if the gathering of information is done manually but when a chatbot is used to get the information, it is quick and easy task to accomplish.

8 Future Work

Since chatbot supports speech input and output, there should be a support of multiple languages by which a variety of users can access the system. For now, the support the English language is given and a plan to add support for other languages in the next iterations of the system. Also with the existing static knowledge base, a Web crawler link the Web site of the institution so that if there is no information found in the knowledge base, the Chatbot can try to find the information on the Web site.

9 Conclusion

As proposed in the above experiment, since the number of documents or the knowledge base is comparatively smaller than any other database, the time taken to retrieve the relevant document to the user's query will be less and also because of the storage of the vector terms of every document in the knowledge in a separate database, the pre-processing time of each document can be reduced every time a query is passed. As lived in the age of automation, it is efficient that all the work should be fast, easy, and effectively done without any errors and any confusion. Introducing a chatbot in an educational institute will help a lot of people in various tasks and help in eliminating redundant information.

References

1. Adibe, F.O., Nwokorie, E.C., Odii, J.N.: Chatbot Technology and Human Deception. ISSN 1119-961 X, 286
2. Gupta, P., Gupta, V.: A survey of text question answering techniques. Int. J. Comput. Appl. **53**(4) (2012)
3. Jovita, L., Hartawan, A., Suhartono, D.: Using vector space model in question answering system. Procedia Comput. Sci. **59**, 01 (2015)
4. Pamungkas, E.W.: Emotionally-Aware Chatbots: A Survey (2019). arXiv preprint arXiv:1906.09774
5. Abdul-Kader, S.A., Woods, J.C.: Survey on chatbot design techniques in speech conversation systems. Int. J. Adv. Comput. Sci. Appl. **6**(7) (2015)
6. Singhal, A.: Modern information retrieval: a brief overview. IEEE Data Eng. Bull. **24**(4), 35–43 (2001)
7. Hegde, V.: An academic framework for designing dashboard and enhancing the quality of higher education admission process through Java Enterprise Edition. In: 2017 IEEE International Conference on Computational Intelligence and Computing Research (ICCIC), Coimbatore, 2017, pp. 1–5. https://doi.org/10.1109/iccic.2017.8524221
8. Sushma Rao H.S., Suresh A., Hegde, V.: Academic dashboard—descriptive analytical approach to analyze student admission using education data mining. In: Mishra, D., Nayak, M., Joshi, A. (eds.) Information and Communication Technology for Sustainable Development. Lecture Notes in Networks and Systems, vol. 10. Springer, Singapore (2018)
9. Jain, A., Jain, A., Chauhan, N., Singh, V., Thakur, N.: Information retrieval using cosine and jaccard similarity measures in vector space model. Int. J. Comput. Appl. **164**(6), 28–30 (2017)
10. Yan, Z., Duan, N., Bao, J., Chen, P., Zhou, M., Li, Z., Zhou, J.: Docchat: an information retrieval approach for chatbot engines using unstructured documents. In: Proceedings of the 54th Annual Meeting of the Association for Computational Linguistics (Volume 1: Long Papers), pp. 516–525 (2016, August)
11. Thannaing, M., Hlaing, A.: Improving Information Retrieval Based on Query Classification Algorithm
12. Kaur, J., Buttar, P.K.: A systematic review on stopword removal algorithms. Int. J. Future Revol. Comput. Sci. Commun. Eng. (IJFRSCE) **4**(4), 207–210
13. Saini, B., Singh, V., Kumar, S.: Information retrieval models and searching methodologies: Survey. Inf. Retrieval **1**(2), 20 (2014)
14. Wong, S.M., Ziarko, W., Wong, P.C.: Generalized vector spaces model in information retrieval. In: Proceedings of the 8th Annual International ACM SIGIR Conference on Research and Development in Information Retrieval (pp. 18–25). ACM (1985, June)

15. Gholap, D.A., Gumaste, S.V.: Information retrieval using keyword search technique. Int. J. Innov. Res. Comput. Commun. Eng. **3**(5)
16. Wahyudi, E., Sfenrianto, S., Hakim, M.J., Subandi, R., Sulaeman, O.R., Setiyawan, R.: Information retrieval system for searching JSON files with vector space model method. In: 2019 International Conference of Artificial Intelligence and Information Technology (ICAIIT) (pp. 260-265). IEEE (2019, March)

An Efficient Internet of Things (IoT)-Enabled Skin Lesion Detection Model using Hybrid Feature Extraction with Extreme Machine Learning Model

B. Pushpa

Abstract At present times, Internet of Things (IoT)-based healthcare diagnosis models become more popular and applicable in diverse scenarios. Skin lesion segmentation acts as an important part in the early identification of skin cancer using automated diagnosis models. The automated detection and classification of skin lesion is a critical task because of the constraints such as artefacts, unsure boundaries, and different shapes of the lesion images. This study introduces a new IoT based automated skin lesion detection model. The proposed model involves a series of processes, namely data acquisition using IoT devices, bilateral filtering-based preprocessing, K-means clustering-based segmentation, hybrid feature extraction, and extreme learning machine (ELM)-based classification. The HF-ELM model determines the identification of lesions exist in the dermoscopic images. The HF-ELM model undergo simulation using skin image dataset, and the simulation results indicated the effective performance of the presented model over the other methods.

Keywords IoT · Segmentation · Feature extraction · Skin lesion · Segmentation

1 Introduction

The advanced study in the application of Internet of Things (IoT) has been utilized in the healthcare sector to analyze the health status of the person. This kind of examination utilizes the data acquired from diverse sensors, lime cameras, smart watches, and so on. The developers concentrated on the application of image processing approaches that guides poor and old age people. The employment of machine learning (ML) models needs a finite database where the classification method is trained. The collection and isolation of data in modern homes are examined in [1]. In order to

B. Pushpa (✉)
Department of Computer and Information Science, Annamalai University, Chidambaram, Tamil Nadu, India
e-mail: pushpasidhu@gmail.com

© The Editor(s) (if applicable) and The Author(s), under exclusive license to Springer Nature Singapore Pte Ltd. 2021

A. P. Pandian et al. (eds.), *Proceedings of International Conference on Intelligent Computing, Information and Control Systems*, Advances in Intelligent Systems and Computing 1272, https://doi.org/10.1007/978-981-15-8443-5_22

verify the newly developed model, the researchers have gathered information around 25 days, once reaching better accuracy.

An applicable device installed is named as smart phone which combines different sensors, determined data, and provide the results in a remarkable form. Every sensor and whole operating system (OS) require few processing energy for the derived results and a database to save the data. The solution is a said to be particular server or the application of cloud computing (CC).

Prediction of skin diseases is computed by image processing is one of the significant modules in sensors and sensing models and for computational intelligence as well as image processing. In addition, developers have focused on the signs of skin diseases and markers for chemical investigation. Several skin lesion diagnosis models have presented in the literature. Esteva et al. [2] proposed a Deep Convolutional Neural Network (DCNN) for the categorization of the skin lesion. The presented method has been trained from images, by employing the pixels and disease labels on input. In [3], researchers applied a method that integrates deep learning (DL) to the skin lesion ensemble model. Schwarz et al. [4] employed an opto-acoustic dermoscopy approach according to the performance of excitation energy and inducing depth metrics. A model for predicting skin lesion and melanoma using DL was established in [5].

In [6], diverse machine learning (ML) techniques used in skin diseases examination, whereas in [7], an extensive comparison has been carried out in mobile application for skin observation as well as melanoma prediction. Then, various apps have been compared and defined the efficiency in image processing model in smart phone camera as well as sensing models which provides legal factors of ethical, quality, and visual deployment of applications for medical purpose. On the other hand, the study on image analysis models has been carried out in sensing mechanism.

This paper develops an efficient IoT based automated skin lesion detection model using hybrid feature extraction and extreme learning machine (ELM) model, called HF-ELM model. The HF-ELM model includes a set of IoT devices to capture the dermoscopic image of the patient using cameras and sensing devices. Then, the captured image undergoes several processes namely acquisition, preprocessing, K-means clustering-based segmentation, hybrid feature extraction, and ELM based classification. The proposed HF-ELM model determines the identification of lesions exist in the dermoscopic images. The proposed model is tested using a skin image dataset, and the results are validated under several measures.

2 Proposed Work

The working process involved in the HF-ELM model is shown in Fig. 1. As shown in figure, the skin image of the person is captured by the use of IoT devices namely cameras and sensors. The images gathered by the IoT devices undergoes further processing to detect and classify the skin lesion. The captured input image undergo preprocessing technique for discarding the noise exist in the dermoscopic image.

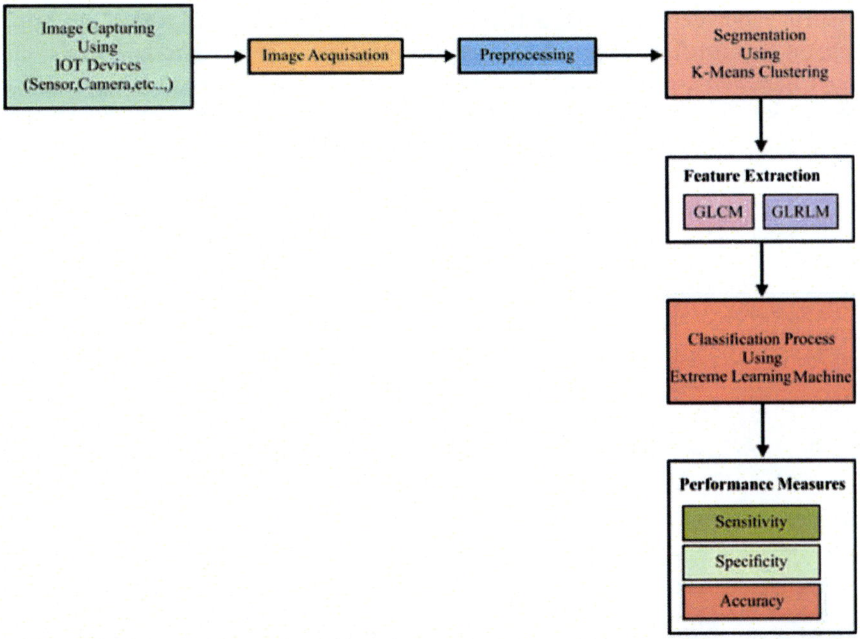

Fig. 1 Working process of HF-ELM model

Then, the preprocessed image will be segmented by K-means clustering technique to identify the diseased region in the image followed by gray-level co-occurrence matrix (GLCM) and gray-level run length matrix (GLRLM) features which will be filtered from the segmented image. Finally, ELM based classification process is carried out to identify the different classes involved in the dermoscopic images.

2.1 Preprocessing

Once the dermoscopic images were captured by the IoT devices, BF model is applied for preprocessing operation which removes the noise present in the dermoscopic images. Actually, the existence of noise leads to the ineffective image classification. In line with this, it is used for noise discrimination at primary image that can be classified. The probability of BF is based on specific weight of linking pixels to eliminate noise. The BF is defined as a simple model, which have distance-driven filter portion (p, p'), and a gray-value is based on the range filter part $r\left((fa), (fa')\right)$:

$$\tilde{f}(a) = \frac{1}{N(u)} \int_{-\infty}^{\infty} f(a')d(a, a')r(f(a), f(a))da' \quad (1)$$

where a and a' define the position of intermittent and neighboring pixels, and $N(a)$ denotes a normalized factor. Based on the local mean of neighboring pixels, the range filter part executes a value-based method to eliminate the noise reagarding the boundaries. The field and range filter areas, and Gaussian function are applied which depends upon Euclidean pixel distance as shown below,

$$d(a, a') \propto \exp\left(-\frac{(a - a')^2}{2\sigma_d^2}\right) \tag{2}$$

$$r(f(u), f(u')) \propto \exp\left(-\frac{(f(u) - f(u'))^2}{2\sigma_f^2}\right) \tag{3}$$

where σ_d is the width parameter of filter kernel and σ_f implies the noise and standard deviation (SD) of reproduced value.

2.2 *Segmentation*

Once the input dermoscopic images were preprocessing, K-means clustering technique is utilized for identifying the skin lesions exist in the applied dermoscopic image. Clustering is defined as the model which classifies a group of data to a particular set of clusters [8]. It is meant to be the familiar techniques such as k-means clustering. In this approach, it partitions a set of data to k number of disjoint clusters. Initially, it is applied for calculating the k centroid while the second class assumed all points to a cluster that have closer centroid from corresponding data point. Also, diverse models were applied in representing the distance of closer centroid and prominently employed model is Euclidean distance. After the clustering is processed, it re-validates the novel centroid of every cluster, and according to the calculated centroid, Euclidean distance is also determined among every cluster center and data point which assign the points in a cluster with lower Euclidean distance. The clusters are divided by corresponding member objects and centroid. K-means is referred to be an iterative method where the the total distance between object to cluster centroid is limited. Suppose an image has the definition of $x \times y$, and the image should be clustered into k numbers. Then, $p(x, y)$ is input pixels that has to be clustered and ck implies the cluster center. The process involved in it is listed below:

- Upload the cluster count k and center.
- For all pixels, estimate the Euclidean distance d among the center and pixels of the image with the help of function provided in Eq. (4):

$$d = \|p(x, y) - c_k\| \tag{4}$$

- Allocate each pixel to the closest center according to distance d.
- Once the pixels are assigned, re-determine the novel position of a center with the help of given function.

$$c_k = \frac{1}{k} \sum_{y \in c_k} \sum_{x \in c_k} p(x, y) \tag{5}$$

- Repeat the steps till it meets the error rate.
- Reform the clustering pixels into the image.

2.3 *Feature Extraction*

In this study, GLCM and GLRLM features are extracted from the segmented images. GLCM is an applicable model to examine the texture which has spatial relationship between the pixels in GLCM named as gray-level spatial dependency matrix. The GLCM shows the ascertain pixel with explicit features in a predetermined spatial association which is carried out in the image, and compute GLCM. Once the procedure is computed, the factual values are isolated from the matrix. The GLCM features used are autocorrelation, contrast, correlation, dissimilarity, cluster prominence, cluster shade, entropy, energy, homogeneity, and maximum probability. GLRLM approach is a technique to eliminate maximum order statistical texture data [9]. Most of the gray dimensions G image is often limited by earlier re-quantized to collect the system as denoted in Eq. (6):

$$K(\theta) = \left(\frac{r(u, v)}{\theta}\right), 0 \le u \le N_r, 0 \le v \le K\max \tag{6}$$

where N_r is the maximum gray-level values, $K\max$ implies maximum length, and (u, v) denotes the sizes of matrix values. The GLRLM features involved are gray-level non-uniformity (GLN), run length non-uniformity (RLN), run percentage (RP), short-run emphasis (SRE), and long-run emphasis (LRE).

2.4 *Classification*

The extracted features are fed into the ELM to classify the existence of different classes of skin lesions exist in the dermoscopic images. ELM is defined as feed forward neural network (FFNN) with a three-layer architecture such as input, hidden, and output layers [10]. Assume $n, r\, and\, c$ be the input, hidden and output layer nodes. For N diverse samples $(x_i, l_i)\, 1 \le i \le N$ in which $x_i = [x_{i1}, x_{i2}, \ldots, x_{in}]^T \in R^n, l_i = [l_{i1}, l_{i2}, \ldots, l_{ic}]^T \in R^c$, the arithmetic function of ELM is depicted in Eq. (7):

$$o_k = \sum_{i=1}^{r} \beta_i g(w_i \cdot x_k + b_i), k = 1, 2, \ldots, N \quad (7)$$

where $0_k = [0_{k1}, 0_{k2}, \ldots, 0_{kc}]^T$ denotes the network output number, $w_i = [w_{i1}, w_{i2}, \ldots, w_{in}]$ implies weight vector which links the *ith* hidden node and input nodes, $\beta_i = [\beta_{i1}, \beta_{i2}, \ldots, \beta_{ic}]^T$ defines the weight vector that links *ith* hidden and output nodes, $g(x)$ means activation function, and named as Sigmoid function, and b_i signifies threshold of the *ith* hidden node. At the initial point of training, w_i and b_i have been produced arbitrarily and remains the same. Then, β is a special parameter that undergoes training. The numerical expression is illustrated in Eqs. (8) and (9):

$$H = \begin{bmatrix} g(w_1 \cdot x_1 + b_1) & \cdots & g(w_r \cdot x_1 + b_r) \\ \vdots & & \ldots \vdots \\ g(w_1 \cdot x_N + b_1) & \cdots & g(w_r \cdot x_N + b_r) \end{bmatrix}_{N \times r} \quad (8)$$

$$\beta = H^{\dagger} L \quad (9)$$

where H denotes the hidden layer output matrix of the NN, $\beta = [\beta_1, \beta_2, \ldots, \beta_r]^T$ implies the output weight vector, and $L = [l_1, l_2, \ldots, l_N]^T$ represents the selected output vector. While β is resolved, then ELM network training process has been completed.

3 Performance Validation

The performance of the HF-ELM model has been determined using a benchmark skin image database from https://isic-archive.com/. The applied database includes seven classes namely Anigioma, Nevus, lentigo NOS, Solar Lentigo, Melanoma, Seborrheic Keratosis, and Basal cell carcinoma. Figure 2 visualizes the results offered by the HF-ELM model on the recognition of skin lesion in the applied test images.

Figure 3 demonstrates the comparison of the outcome offered by the HF-ELM and existing models interms of sensitivity, specificity, and accuracy. The table value indicated that the ResNet-50 model has demonstrated ineffective outcome compared

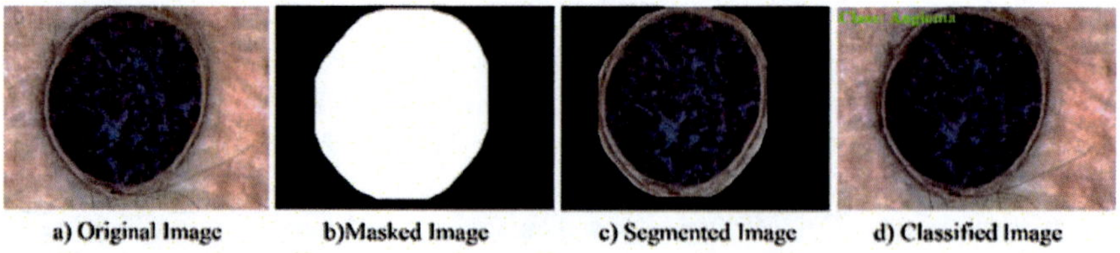

Fig. 2 Visualization of the HF-ELM model

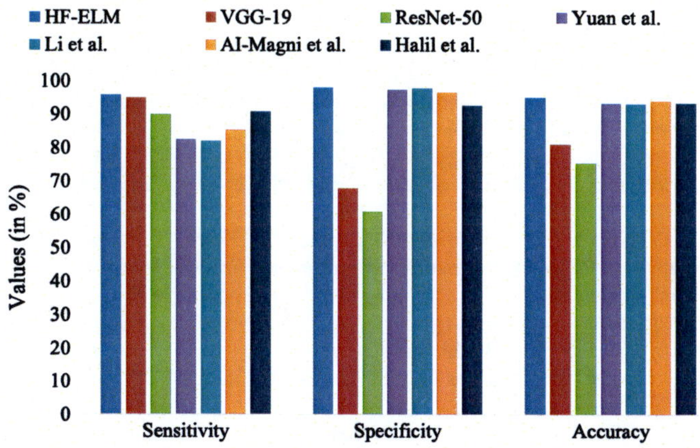

Fig. 3 Classification results analysis of different models

to other methods. Simultaneously, the VGG-19 model has exhibited slightly higher classifier results over the earlier model. Then, the models devised by Yuan et al., Li et al., and Halil et al. have provided moderate and closer classifier results. At the same time, the model by AI-Magni et al has shown better results over the other methods except the HF-ELM model. Finally, the HF-ELM model has outperformed the earlier methods with the sensitivity of 95.87%, specificity of 98.08%, and accuracy of 95.14%.

4 Conclusion

This paper has developed an efficient IoT based automated skin lesion detection using HF-ELM model. Once the IoT devices captures the dermoscopic image, it is preprocessed by BF technique to remove the noise exist in the dermoscopic image. Then, the preprocessed image will be segmented, and then, features will be extracted from the segmented image. Finally, ELM based classification process is carried out to identify the different classes involved in the dermoscopic images. An extensive validation of the HF-ELM model takes place using benchmark skin image dataset. The obtained simulation outcome pointed out that the HF-ELM model has outperformed the previous model with the sensitivity of 95.87%, specificity of 98.08%, and accuracy of 95.14%. In future, the outcome of the HF-ELM model can be improved by the use of deep learning approaches.

References

1. Verma, P., Sood, S.K.: Fog assisted-IoT enabled patient health monitoring in smart homes. IEEE Internet of Things J. **5**, 1789–1796 (2018)
2. Esteva, A., Kuprel, B., Novoa, R.A., Ko, J., Swetter, S.M., Blau, H.M., Thrun, S.: Dermatologist-level classification of skin cancer with deep neural networks. Nature **542**, 115 (2017)
3. Codella, N.C., Nguyen, Q.B., Pankanti, S., Gutman, D., Helba, B., Halpern, A., Smith, J.R.: Deep learning ensembles for melanoma recognition in dermoscopy images. IBM J. Res. Dev. **61**, 1–5 (2017)
4. Schwarz, M., Soliman, D., Omar, M., Buehler, A., Ovsepian, S.V., Aguirre, J., Ntziachristos, V.: Optoacoustic dermoscopy of the human skin: tuning excitation energy for optimal detection bandwidth with fast and deep imaging in vivo. IEEE Trans. Med. Imaging **36**, 1287–1296 (2017)
5. Li, Y., Shen, L.: Skin lesion analysis towards melanoma detection using deep learning network. Sensors **18**, 556 (2018)
6. Pathan, S., Prabhu, K.G., Siddalingaswamy, P.: Techniques and algorithms for computer aided diagnosis of pigmented skin lesions—a review. Biomed. Signal Process. Control **39**, 237–262 (2018)
7. Chao, E., Meenan, C.K., Ferris, L.K.: Smartphone-based applications for skin monitoring and melanoma detection. Dermatol. Clin. **35**, 551–557 (2017)
8. Dhanachandra, N., Manglem, K., Chanu, Y.J.: Image segmentation using K-means clustering algorithm and subtractive clustering algorithm. Procedia Comput. Sci. **54**, 764–771 (2015)
9. Raj, R.J., Shobana, S.J., Pustokhina, I.V., Pustokhin, D.A., Gupta, D., Shankar, K.: Optimal feature selection-based medical ımage classification using deep learning model in ınternet of medical things. IEEE Access **17**, 58006–58017 (2020)
10. Han, M., Liu, B.: A remote sensing image classification method based on extreme learning machine ensemble. In: International Symposium on Neural Networks, pp. 447–454. Springer, Berlin (2013)

Using Convolutional Neural Network to Detect Diabetic Retinopathy in Human Eye

Saloni Dhuru and Avinash Shrivas

Abstract A medical condition in diabetic patients is recognized as diabetic retinopathy; it is a primarily involved disease in human eye. This disease needs to be detected by ophthalmologist which can cause incorrect results and also in a tedious process. Hence, a need for automated detection of these diseases without human intervention is necessary such that drawbacks from misconception from the humans are prevented. The data set used for the same is being downloaded from Kaggle website [11]. The systems detect this disease by scanning the retinal images of human to check whether the patient is having diabetic retinopathy or no. Also, a fewer pre-processing of the data set's images needs to be done in order to get the images to a standardized form. Later, the convolutional neural network (CNN) is being used because of the fact that neural networks acquire the approach of biological brain which understands and learns the various patterns and nerves in human eye to conclude whether the patient has diabetic retinopathy or not. This system uses 1600 and 400 images as training and testing data, respectively. The accuracy of this system comes out to be 80%.

Keywords Diabetic retinopathy · Greater pixels data set images · Deep learning · Convolutional neural network (CNN)

1 Introduction

As there is an upliftment in the ever-increasing human lifestyle, there is increase in different diseases too. It has been seen that most of the population is being affected due to changing lifestyle leading to obesity, hypertension, diabetes, etc. According to recent estimates, 438 million people (7.8%) of the adult population is expected to

S. Dhuru (✉) · A. Shrivas
Computer Engineering Department, Vidyalankar Institute of Technology, Mumbai, India
e-mail: salonidhuru26@gmail.com

A. Shrivas
e-mail: avinash.shrivas@vit.edu.in

© The Editor(s) (if applicable) and The Author(s), under exclusive license to Springer Nature Singapore Pte Ltd. 2021

A. P. Pandian et al. (eds.), *Proceedings of International Conference on Intelligent Computing, Information and Control Systems*, Advances in Intelligent Systems and Computing 1272, https://doi.org/10.1007/978-981-15-8443-5_23

have diabetes by the year 2030. One such disease caused due to diabetes is called diabetic retinopathy. Person having diabetes for 20 years or more are said to have 85% chances of being affected by diabetic retinopathy. It causes spillage of veins and expanding, limiting blood from fleeting through and also constantly surge of irregular new blood vessels in the retina. Pigmentation or dull strand happens to make eyesight, fluctuating eyesight, hindered shading eyesight, obscured vision, dim or void regions in view, and eyesight misfortune that are few of typical side effects of the disease [1]. Regular clue and signals of the same are small-scale aneurysms, veins spillage, expanding of retina, irregular new blood vessels and harmed nerve tendon. Diabetic retinopathy is efficient of evaluation with techniques, for example, focal laser therapy, scatter laser therapy and vitrectomy. Medical procedure frequently deteriorates or disallows the advancement of diabetic retinopathy, yet it is anything but a total fix. It is fundamental factor, forthcoming retinal harm, and eyesight misfortune has additionally conceivable [2]. Key side effect of DR is exudates, which might be showed in the picture of the eyes and is a gesture of generating DR within the patient or the patient may by that time got it.

Hence, a relevant screening of this disorder is a compelling. Diagnosis strategies are fluorescein angiography and optical cognizance tomography that consist of outside liquid to cover on to the patient's eye afterwards the image is gathered. In any case, this framework can naturally and promptly foresee diabetic retinopathy with no outside operator, and that is a progressively advantageous technique both for specialists and patients [3].

To address the above-stated problems, there is a need to develop a model that aims on introducing diabetic retinopathy diagnosis that naturally determines features which are crucial in diagnosing the stage of the disease without certain or manual feature extraction. For this, CNN is being used because of the fact that neural networks acquire the approach of biological brain. These processes are being pursued by image pre-processing. Pre-processing of image is commenced by rotations, resizing the images to carry on the further analysis. To distinguish the images according to the listed features, the image classification steps work the best, thus getting the desired results. The following work should involve minimization of noise for obtaining an enhanced image and accurate data.

2 Literature Review

[1] **Shailesh Kuman, Basant Kumar** have exhibited two attributes, i.e. count and area of microaneurysm. Techniques exhibit initially as extraction of green channel, equalization of histogram and method of morphological are exhibited. It views morphological execution, contrast narrow flexible histogram equalization (CLAHE), PCA, averaging filter phase to expose microaneurysm. Sensitivity is 96% and specificity is 92% to DR identifying system [4].

[2] **Mohamed Chetoui, Moulay A. Akhloufi, Mustapha Kardouchi** have proposed favour of distinct texture attributes mode for diabetic retinopathy,

principally local ternary pattern (LTP) and local energy-based shape histogram (LESH). It does exhibit SVM for the distribution of refined histogram which is studied for binning scheme for representing attributes. Their primary outcome provides that LESH is the performing better with technology with accuracy of 0.904 approves SVM and SVMRBF. Studying ROC curve that exhibits LESH including SVM-RBF gives finest area under curve performance gives 0.931 [5].

[3] **Tajbia Karim, Md. Salehin Riad** proposed a new technique for fundus image, and it will be identified the diabetic retinotherapy from rear end image [6].

[4] **Jayant Yadav, Manish Sharma** proposed aims to resolve this problem by using computer vision not only to diagnose this disorder, but also by automating this process using a neural network to provide results to several patients within a short timeframe [7].

[5] **Preethi Patil, Savita Sheelavant** proposed a technique detection for diabetic retinopathy (DR) and it is most efficient part in automated screen system [8].

[6] **Asti Herliana, Toni Arifin, Sari Susanti and Agung Baitul Hikmah**, proposed that research was conducted by implementing particle swarm optimization (PSO) mode to prefer the finest diabetic retinopathy attribute establishing on diabetic retinopathy data. Next selected aspect is again derived using any classification mode considering neural network (NN). Outcome of study exhibit increases in outcome by applying NN-based PSO of 76.11%. The outcome also exhibits building up in outcome by emphasizing 4.35% of selection phase as to old outcome of 71.76% [9].

[7] **Enrique V. Carrera, Andres Gonzalez, Ricardo Carrera** proposed a computerized analysis established on the digitally altering process of retinal figure considering to encourage people for identifying diabetic retinopathy in before it propels. Their ambition is classifying naturally grade of non-proliferative diabetic retinopathy. Primarily original image before processing phase, they isolate blood vessel, microaneurysms and hard exudates in an ordered way to select attributes which can be worn by SVM to consider the retinopathy grade of every single retinal figure. The technique was tried on almost 400 figures of retinal which are labelled according scale of grades non-proliferative diabetic retinopathy. Final outcome obtained is 95% sensitivity and 94% predictive gap [10].

3 Problem Statement

Diabetes arises when our body is not being able to produce acceptable insulin; therefore, it leads towards high glucose level which causes damage to various body organs like heart, brain and also leads in slower healing of any wounds. Also, diabetes can affect the eye and seen in the blood vessels of retina, which may cause blindness and this process is what known as diabetic retinopathy. Diabetic retinopathy is a condition which can be seen in human eyes who are suffering from diabetes over a long period of time. Often the eye gets affected after 20 years of prolonged diabetes.

Diabetic retinopathy can be classified as NPDR (known as non-proliferative diabetic retinopathy) and PDR (known as proliferative diabetic retinopathy). NPDR (known as non-proliferative diabetic retinopathy) can be again further be broken down into mild, moderate and severe diabetic retinopathy. The traditional methods focused on the frequent visits to the screening laboratories for getting the eye screened for images where retina was scanned and then these retinal images where scanned by the ophthalmologist in order for the proper detection of it. But it can many times result in inconsistent results and also make a fewer more frequent trips to the ophthalmologist.

3.1 Objectives

In this system, the user can give input to the system if the form of scanned retinal images and the system tells us whether the patient has diabetic retinopathy or not. This helps the patients to know what the condition of their vision is and also gets to know the exact remedy to be taken.

- Provides unique approach to concealed patterns in the data.
- Helps avoid human biasness.
- To implement neural network that classifies the disease as per the input of the user.
- Reduce the cost of medical tests.

4 Proposed System

The neural network system uses the training data to train models to observe patterns and uses the test data which evaluates the predictive nature of the trained model. Splitting the data into two ways—one for training and other for testing sets—is an important part of assessing neural network models. When classifying a data set in two parts, 80% of the data is used for training purpose and 20% data is used for testing. The known output can be used as training set and the model begins to learn on this data set in order to be induced to other data afterwards. The model is tested by predictions against the testing set after the model has been processed by using the training set. It is easy to determine whether the model that will be guessed are correct or not because the testing set data already contains familiar values for the attribute that needs to be predicted.

In today's world, scanned images of the retina of the eye are playing an important role to carefully diagnose the problems associated with it. Most of the study systems lack the data set pre-processing steps which have a potency to give faulty outputs. For studying purpose, pre-processing of images is done through fewer rotations of images and also resizing them which enlarges the features of the image. In addition, pre-processing done is of resizing the image to a pixel of 200×200 which makes the images to a standardized size which makes it easier for the system to process the images with same efficiency.

4.1 Convolution in Images

As a single input image in Fig. 1, some filter or mask or window is applied to them. Applying these value x in the red square corresponds for all these values are pixels of the image in that position. Applying the filter that represented by "w", these nine positions obtained an output value by combining the input image weighted by all values in the sides of the filter. For example, the above image would have the following calculations:

$w = (-1, -1)$ $w = (-1, 0)$ $w = (-1, 1)$

$w = (0, -1)$ $w = (0, 0)$ $w = (0, 1)$

$w = (1, -1)$ $w = (1, 0)$ $w = (1, 1)$

$X = 224 * w = (-1, -1)$

$224 * w = (-1, 0)$

$223 * w = (-1, 1)$

$206 * w = (0, -1)$

$206 * w = (-1, 0)$

$206 * w = (-1, 1)$

$189 * w = (0, -1)$

$189 * w = (-1, 0)$

$189 * w = (-1, 1)$

As shown in Fig. 2, the first layer is composed of one or more convolutions and after these one more step of pooling can be applied thereafter, designed the fully connected layer and at the end can apply sum of all layers called SoftMax. CNNs are the most representative supervised deep learning model. Deep learning is also showing best results in path recognition.

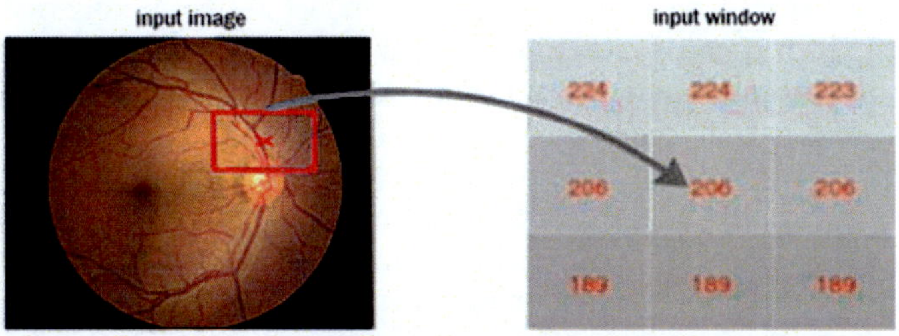

Fig. 1 Example of an image

Fig. 2 CNN steps

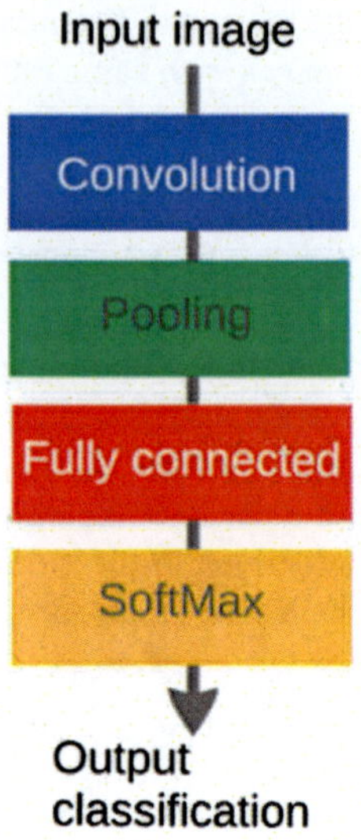

4.2 Proposed System Architecture

As shown in Fig. 3, the input to the system is in form of scanned retinal images. These are high-resolution images which are downloaded from Kaggle website [9]. Next these images are pre-processed using rotations and resizing to get them to a standardized form. Later, CNN is applied to this pre-processed image. Next the system tells whether the image is having retinopathy or the eye is no retinopathy which means that the eye is healthy.

The neural network system uses the training data to train models to observe patterns and uses the test data which evaluates the predictive nature of the trained model. Splitting the data into two ways—one for training and other for testing sets—is an important part of assessing neural network models. When classifying a data set in two parts, 80% of the data is used for training purpose and 20% data is used for testing. The known output can be used as training set and the model begins to learn on this data set in order to be induced to other data afterwards. The model is tested by predictions against the testing set after the model has been processed by using the training set. It is easy to determine whether the model that will be guessed are correct or not because the testing set data already contains familiar values for the attribute that needs to be predicted.

Most of the study systems lack the data set pre-processing steps which have a potency to give faulty outputs. For studying purpose, pre-processing of images is done through fewer rotations of images and also resizing them which enlarges the features of the image. In addition, pre-processing done is of resizing the image to a pixel of 25×25 which makes the images to a standardized size which makes it easier for the system to process the images with same efficiency.

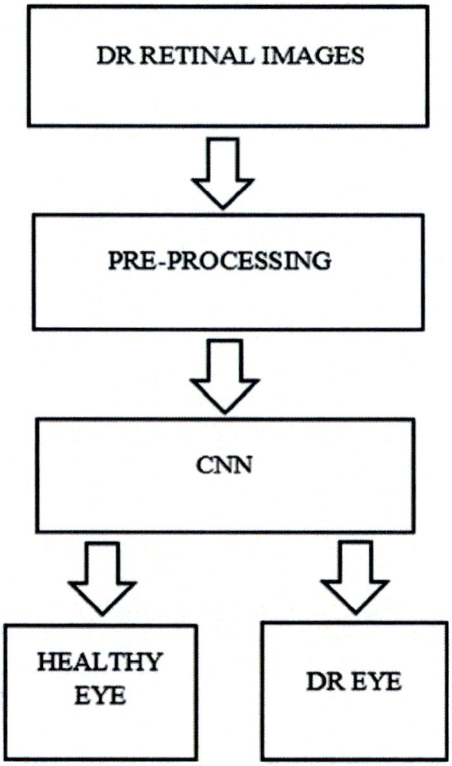

Fig. 3 Flow of data

a. The Data Set

The data set used is the diabetic retinopathy fundus images [11] of retina having pixels over varied resolutions. The data set is downloaded from Kaggle website, which is openly available in it, having around 32,000 images for training set and 18,000 images for testing set. The inconsistency in the website's data lies in the image resolution that can vary for distinct figures and the figures encloses noise therefore proper filtering is necessary to get proper information from the data set. For this study, all the images have been chosen at 6:2 training and testing ratio. The data set has labels assigned to them as per the severity of the retinopathy. Figure 4

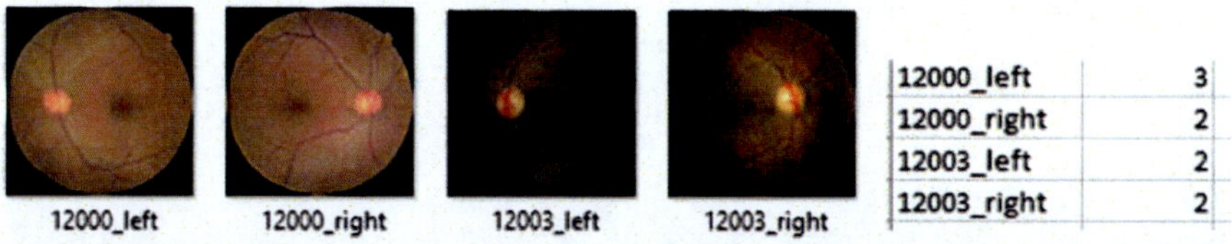

Fig. 4 Samples of data set

shows the images which are present in the data set. As it can be seen in the samples of data set, all images are taken by distinct people, using distinct devices, and of distinct intensity. Pertaining to the pre-processing section, this data is outrageously disrupted and wishes for numerous pre-processing steps in order to train the model and to get all figures to a processable form.

b. **Data Pre-processing**

The images in the data set are already labelled with the levels of severity and hence pre-processing involves fewer rotations and resizing of the images.

- Rotate and resize all images: All images were scaled to 25 × 25. Despite taking longer to train, the detail present in images of this size is much greater than at 128 by 128. Moreover, 403 images were deselected from the training set. Scikit-image prompts multiple warnings during resizing, due to these images having no colour space. Because of these problems, any images that were completely black were removed from the training data.
- Rotate and mirror images: All images were rotated and mirrored including retinopathy images, no retinopathy images and also middle staged images were mirrored, and rotated at 90, 120, 180 and 270°.

In Fig. 5, the first images show two pairs of eyes, along with the black borders. Notice at in the resizing and rotations how the majority of noise is removed.

For rectification of class imbalance, few rotations and mirroring are done with a few more images having retinopathy. In total, there are 106,385 images being processed by the neural network.

CNN-1st convolution: An input image whose size is 25 × 25, composed of three channels and all these three channels are perceived as filters being applied to all the three channels. In this configuration, I have decided to fit of the size 5 × 5. By applying the first filter to all the three channels, the max function or rectified linear unit as activation function and then applying this activation function to the second filter. We will obtain the second output without applying any activation function. These two images that are formed are called tensors, two tensors of size 25 × 25.

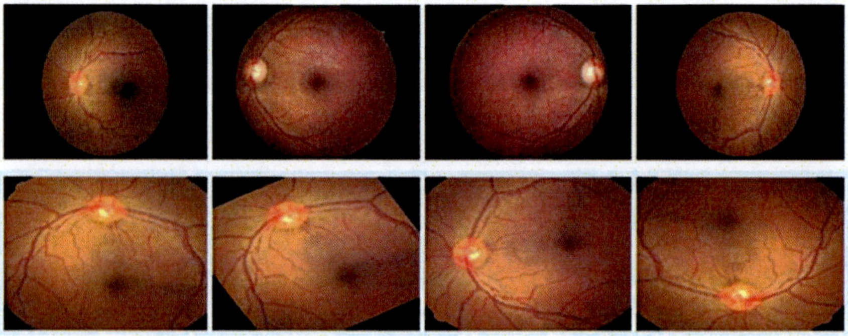

Fig. 5 Samples of data pre-processing

CNN-2nd Convolution

In the next layer, the same steps are applied; the input will be exactly same as the previous output so the two things of the previous output will be input to this layer, but now decided three filters of size 3 × 3 followed by the same max function as activation function. If there are three filters, then it will have three tensors of the output, and then in the next step, a new layer is defined [3].

CNN Pooling

The next layer is called pooling operation that integrates some values and the output will be single value. For example, here there are four pixels and the function that applied here is to get the four pixels and obtain the maximum of these four pixels. So, in our example, there are three tensors from the convolution and a filter of 4 × 4 is applied so we get four pixels four times and they are chosen as the max or maximum value of the pixel to be the output. So, the size of the image is reduced to have three input tensors and the output will also have three tensors but the size of these tensors will be different as pooling is applied [3].

CNN-Fully Connected

The next step is fully connected layer and it also gets the tensors from the previous convolution besides 6 × 6 and got the three images with size 6 × 6 and created a single built composite of all the pixels and now gets the values of the pixels so this is what called fully connected. These input values are got and connected to a strip stride of well-known neural networks—weighted sum of inputs [10]. In our example, the output of fornerons is decided.

CNN—SoftMax

So, these fornerons will be input of next layer that called SoftMax. This is an example of an equation that outputs a value relating to a similarity of the input to a single class. The input value are composed by the values in x and have all the elements in omega which are the weights of the neurons. So, the SoftMax algorithm will play with equation that show a fraction and denominator will make a computation considering these input values and in the case of class of elements considered it will be product of these two values plus the bias that have been used here and divided by the sum of all these values. In this case since we have input as retinal images we expect the output will be produced as whether the patient is having retinopathy or no as it solely depends upon the input image so the output classified by the system should be the highest value of the SoftMax as per the image presented. The correct classification is produced by all these processes that is whether the patient is having diabetic retinopathy or not.

4.3 System Implementation

Figure 6 shows the UI of the system which takes the retinal images as input. There are several other options which help us to get a more depth about the system. You can select the image from the list which needs to be checked for diabetic retinopathy.

In Fig. 7, the preview option is used wherein the selected images are being viewed to see if the selected image is being selected properly or no. Also, it makes sure that whether the system gives correct results with the prescription of the ophthalmologist.

Figure 8 is giving the output depending upon the image presented. It gives the output based on the CNN algorithm applied and this output is supposed to be almost accurate and to its best form.

Figure 9 is giving the output depending upon the image presented. It gives the output based on the CNN algorithm applied, and this output is supposed to be almost accurate and to its best form.

Figure 10 shows the graph of percentage of patient having healthy eye to the percentage of retinopathy eye. This analysis is also done using the different parameters present in the retinal image which gets to a conclusion of how much percentage the eye is affected by retinopathy and how much of the eye is healthy.

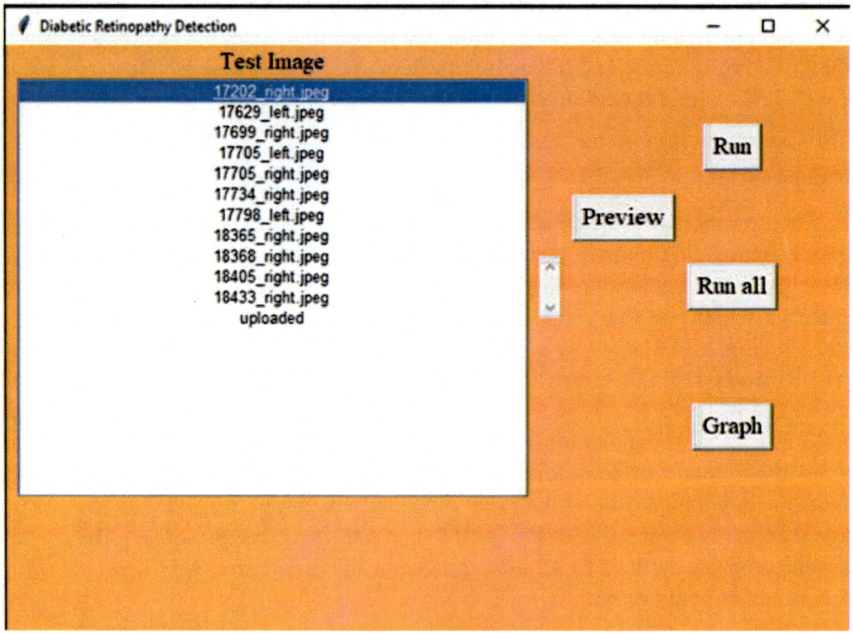

Fig. 6 System UI

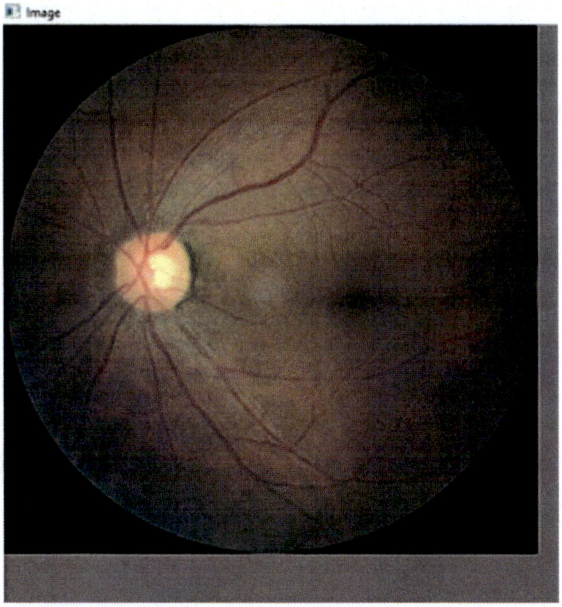

Fig. 7 Preview option of the system

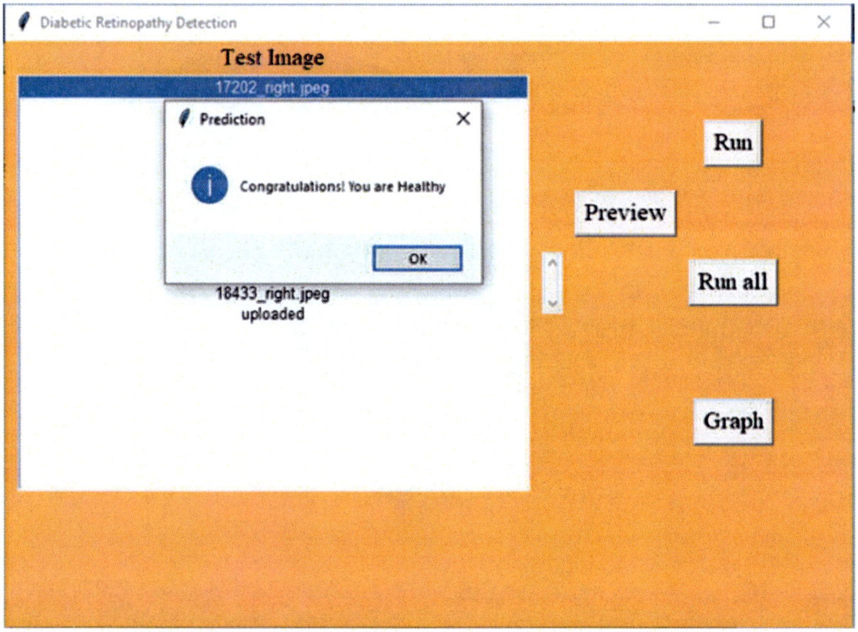

Fig. 8 System output for no retinopathy

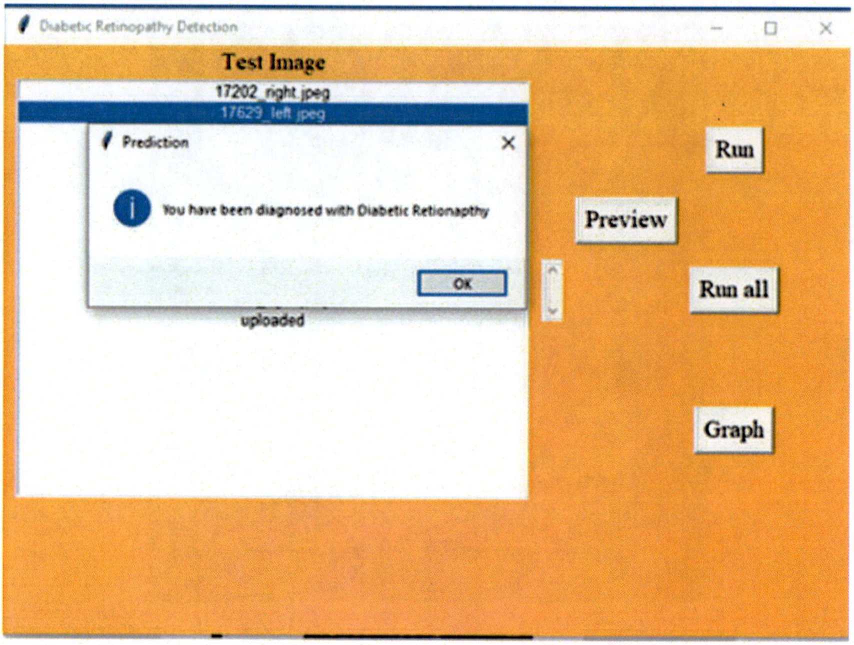

Fig. 9 System output for retinopathy

Fig. 10 Patient graph

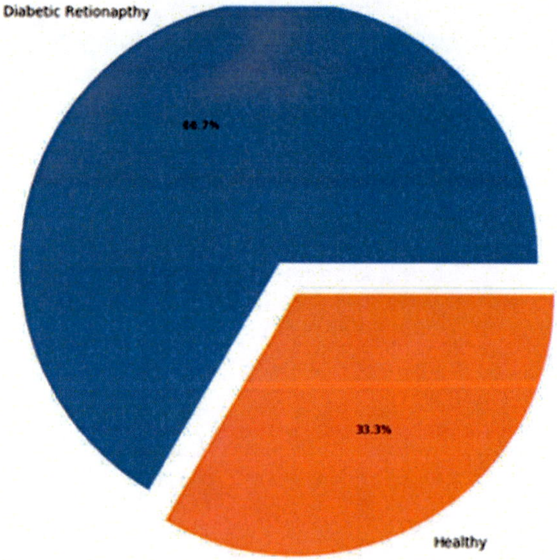

5 Results

The consequence of an image pre-processing step is to get enhanced image with proper features as shown in Fig. 5. Hence, to get the enhanced image, a step-by-step working of few rotations and resizing needs to be done which is fed as an input to the CNN model. The convolutional neural network model that has been presented in this paper is developed using TensorFlow deep learning framework. Pre-processing was done on the images (Fig. 5) with rotations and resizing and used for making the learning and decisions. The accuracy of the system comes out to be 80%. The sensitivity of the system comes out to be 95%.

5.1 Future Scope

This project does not differentiate the levels of diabetic retinopathy. For future work, the model can not only differentiate the image of being diabetic retinopathy or not but the system will also predict the levels of diabetic retinopathy which is: no retinopathy, mild, moderate, severe and proliferate retinopathy with efficiency using GPU systems. A single system with higher accuracy is proven to be good recognition of the disease.

6 Conclusion

The proposed paper is on artificial neural network approach to diabetic retinopathy with the use of typical convolutional neural network (CNN) architecture. Alterations in diabetic retinopathy data set images are fundamental to get fitting lineaments. Also, resizing these images makes the system to efficiently and uniformly process the images. Statistical values predict the severity level, but when the data is noisy or inconsistent leading to the chances of having poor data set, it will let down the accuracy and would give incorrect results. Accuracy will be obtained with the best of the knowledge and the aim is to have a high rate of accuracy reached by the automation of diabetic retinopathy detection system.

References

1. https://www.mayoclinic.org/disease-conditions/diabeticretinopathy/symptoms-causes/sys/20371661
2. https://www.advancedeyecareny.com/retinopathy/

3. Chakrabarty, N.: A deep learning method for the detection of diabetic retinopathy. In: 2018 5th IEEE Uttar Pradesh Section International Conference on Electrical, Electronics and Computer Engineering (UPCON), p. 978
4. Kuman, S., Kumar, B.: Diabetic retinopathy detection by extracting area and number of microaneurysm from colour fundus image. In: 5th International Conference on Signal Processing and Integrated Networks (SPIN) (2018)
5. Chetoui, M., Akhloufi, M.A., Kardouchi, M.: Diabetic retinopathy detection using machine learning and texture features. In: 2018 IEEE Canadian Conference on Electrical & Computer Engineering, (CCECE) IEEE
6. Karim, T., Riad, Md.S.: Symptom analysis of diabetic retinopathy by micro-aneurysm detection using NPRTOOL. In: 2019 International Conference on Robotics, Electrical and Signal Processing Techniques (ICREST)
7. Yadav, J., Sharma, M.: Diabetic retinopathy detection using feedforward neural network. In: 2017 Tenth International Conference on Contemporary Computing (IC3) (2017)
8. Patil, P., Sheelavant, S.: Detection and classification of microaneurysms and haemorrhages from fundus images for efficient grading of diabetic retinopathy. The International Conference on Electrical, Electronics, Communication, Computer, and Optimization Techniques (ICEECCOT) (2018)
9. Herliana, A., Arifin, T., Susanti, S., Hikmah, A.B.: Feature selection of diabetic retinopathy disease using particle swarm optimization and neural network. The 6th International Conference on Cyber and IT Service Management (CITSM 2018) (2018)
10. Carrera, E.V., Gonzalez, A., Carrera, R.: Automated detection of diabetic retinopathy using SVM. In: 2017 IEEE XXIV International Conference on Electronics, Electrical Engineering and Computing (INTERCON) (2017)
11. High Resolution Fundus Retinal Image Database: https://www.kaggle.com/c/diabetic-retinopathy-detection/data

Use of Ensemblers Learning for Prediction of Heart Disease

Meenu Bhatia and Dilip Motwani

Abstract Among all fateful disease, heart diseases are considered as the utmost prevalent. Medical specialist oversight various audit of disease related to heart and segregate reports of heart patients, their symptoms and illness. Progressively reported about patients with common diseases who have typical symptoms. The current scenario people in globe want people to choose luxurious life rather than peaceful life a so they work harder trying to equalize with the machine to secure high capital and to comfort their livelihoods which in result gives no rest and high pressure regards distinct health issues, because of this there intake of healthy and balanced diet modifies and also their complete to-do and not a to-do list; in this type of living style, there are distinct issues like getting depressed and suffering from blood pressure, sugar level issue at an initial age and not enough rest for themselves and eating habits as an outcome contributing with minor negligence leads to an extensive threat of disease like heart disease. Term 'heart disease' includes the divergent diseases that affect the heart. Heart predictor system will use the data mining intelligence to give a user adapting outcome to new and hidden design in the data. The technology used to implement is serviceable for skilful healthcare to get the finest quality of utility and to shorten the extent of conflicting medicine consequence. Heart disease can be handled productively stabilizing the livelihood, remedy, medication and final stage of surgeries.[1] With the right analysis, the manifestation of heart disease can be shortened, and the operating of the heart gets improved. The proposed work focuses on ensembles techniques which combine the weak learners to implement a hybrid model and provides better results than individual classifier.

Keywords Data mining · Heart disease · Healthcare

M. Bhatia (✉) · D. Motwani
Computer Engineering Department, Vidyalankar Institute of Technology, Mumbai, India
e-mail: meenubhatia810@gmail.com

D. Motwani
e-mail: dilip.motwani@vit.edu.in

© The Editor(s) (if applicable) and The Author(s), under exclusive license to Springer Nature Singapore Pte Ltd. 2021
A. P. Pandian et al. (eds.), *Proceedings of International Conference on Intelligent Computing, Information and Control Systems*, Advances in Intelligent Systems and Computing 1272, https://doi.org/10.1007/978-981-15-8443-5_24

1 Introduction

Heart maladies are more well known in men than in ladies. As per insights saw from WHO, it has been seen that 24% passing is incited by heart sicknesses. 33% among the entirety of the worldwide passing is because of heart sicknesses. Half of the passing in the United States, and other created nations are owed to heart sicknesses. Around 17 million individuals bite the dust because of cardiovascular infection (CVD) consistently around the world, and this is profoundly basic in Asia. The various sorts of heart ailments are, for example, coronary illness, angina pectoris, congestive cardiovascular breakdown, cardiomyopathy, innate coronary illness, arrhythmias and myocarditis. It is trying to physically discover the chances of obtaining coronary illness hazard dependent on a few elements. Along these lines, AI strategy is valuable to conjecture the yield from present information. Henceforth, this research relates one such AI system called ensembles for foreseeing coronary illness chance from the hazard factors. This One improves accuracy by combining weak learners. It is illogical for a typical man to spend much on exorbitant tests like the ECG, and in this manner, there should be a framework set up which is convenient and simultaneously dependable, in foreseeing the odds of coronary illness. Thus, it will be useful to build up an application which considers factors like age, alcohol, smoke, cholesterol, lifestyle, gender, stress, etc.[2] The algorithms demonstrate to be accurate and precise than using individual algorithms and hence utilized in the recommended framework. The main aim is to suggest a system/application which will consist of two modules: Doctor Login and Patient Login. The doctor can record the case details along with the case history of the patient. Whereas patient can view its entire medical history. The proposed model will work on ensembles learning that is Bagging, Boosting and Stacking algorithms. Bagging, Boosting and Stacking works on a hybrid algorithm which tests and trains the dataset.

Thus, machine learning algorithms are suitable to determine the output from current information. Hence, this research affixes one such machine learning algorithm called ensembles learning to predict heart disease risk from the given factors. It also endeavours to enhance the accuracy of forecasting heart disease risk.

2 Literature Review

"A Comparison of Several Ensemble Methods for Text Categorization" Supported by Motorola Labs, China Research Centre, Yan-Shi Dong, Shanghai Jiao Tong University, Ke-Song Han Motorola Labs, China Research Centre [3]

Two sorts of classifiers, naïve Bayes classifiers and SVM classifiers, are utilized as base classifiers for gathering. The credulous Bayes classifiers anticipate the joint probabilities of highlights to foresee the class probabilities following the Bayesian equation. The NB classifiers have been effectively applied in report characterization. There are two distinctive generative models in like manner use of the NB classifiers.

One model spoke to by a vector of paired characteristics showing which words happen and which not happen in the record. This is called as two-fold NB classifiers. The subsequent model indicates that an archive is spoken to by the arrangement of word events from the document. This is called as multinomial NB classifiers.

"Analysis and Prediction of Cardio Vascular Disease using Machine Learning Classifiers" N. Komal Kumar, G. Sarika Sindhu, D. Krishna Prashanthi, A. Shaeen Sulthana, 2020, ICACCS [4]

This paper explains progressing works in foreseeing consistent and incredible diseases having AI classifiers. It researches the AI classifiers for clinical even similar to their authenticity and exactness, choice tree, arbitrary woods, bolster vector machine, neural system and strategic relapse calculations were additionally used to analyse. The work inspected the cardiovascular breakdown rate with the help of heartbeat travel time variable examination, separation appropriation lattice; convolution neural system model and bolster vector AI models were used. Bolster vector machine beats every one of the classifiers in the trial of cardiovascular breakdown acknowledgement. An approach to manage grouping cardiovascular ailments using coordinated AI classifiers were given in, the programmed enlarged cardiomyopathy (DCM) and Atrial Septal Defect (ASD) illness disclosure have been structured the isolated features are masterminded using controlled assist vector with machining calculation.

"Efficient Heart Disease Prediction System using Optimization Technique" Chaitanya Suvarna, Abhishek Sali, Sakina Salmani, 2017, IEEE [5]

Concealed examples and connections can be separated from huge information sources utilizing information mining. Information mining combines factual investigation, AI and database innovation. Information mining has been applied in a few zones of clinical administrations such has disclosure of connections among analysis information and put away clinical information. The current clinical conclusion is a composite procedure which requires exact patient information, numerous long stretches of clinical experience and decent information on clinical writing.

"Prediction of Heart Disease Based on Decision Trees" [6]

The prediction of coronary illness is most entangled undertaking in the field of clinical sciences which is not seen with a naked eye and can come instantly anywhere, anytime. So there emerges a need to build up a choice emotionally supportive network for identifying coronary illness. A coronary illness expectation model utilizing information mining system, called choice tree calculation which helps clinical specialists in identifying the infection dependent on the patient's clinical information. Right now propose a proficient choice tree calculation strategy for coronary illness expectation. To accomplish right and practical treatment, PC-based frameworks can be created to make great choice. Information mining is a ground-breaking innovation for the extraction of concealed prescient and significant data from huge databases; the fundamental target of this task is to build up a model which can decide and separate obscure information (patterns and relations) related with coronary disease from a past coronary ailment database record. It can fathom confounded questions for distinguishing coronary illness, and in this manner help clinical specialists to settle on brilliant clinical choices.

"A Review on Prediction And Diagnosis of Heart Failure" G B Dept. of ECE Sathyabama University Chennai, India, Ebenezar Jebarani M.R. Dept. of ECE Sathyabama University Chennai, India [7]

Quick and rapid development is found in human services benefits over recent years. Coronary illness causes a great many passing around the world. Numerous remote correspondence advancements have been created for coronary illness expectation. Information mining calculations are valuable in the location and finding of coronary illness. Right now is completed on a few single and cross-breed information mining calculations to distinguish the calculation that best suits the coronary illness expectation with elevated level of precision.

3 Problem Statement

Heart illness can be groomed adequately with a combination of a way of life changes, medication and at times with medical procedure. With the proper treatment, the side effects of heart illness can be decreased, and the working of the heart can be improved. The point of consolidating various classifiers is to get improved execution as contrasted than individual classifier. The overall objective of our work will be to predict accurately with few tests and attributes with the presence of heart disease. Attributes considered form the primary basis for tests and give accurate results. Many more input attributes can be taken but our goal is to predict with few attributes and faster efficiency the risk of having heart disease. Chances are now and again made dependent on masters' nature and experience rather than on the data rich data concealed with the enlightening assortment and databases. This preparation helps unfortunate tendencies, botches and superfluous clinical costs which impact the idea of organization provided for patients [8]. The executives of cardiovascular breakdown can be mind boggling and are frequently one of a kind to every patient; nonetheless, there are general rules that ought to be followed. Avoidance of intense intensifications can slow the movement of cardiovascular breakdown just as expands the security and generally speaking prosperity of the patient. At the point when a patient who has intense congestive cardiovascular breakdown is readmitted, the expense and weight to the patient are incremented.

3.1 Objectives

The fundamental objective of this system is to develop a heart prediction system. The system and the framework can find and concentrate concealed information related to ailments from a verifiable heart informational collection. Heart disease prediction system intends to use information mining systems on clinical informational index to aid the forecast of the heart infections.

3.2 *Specific Objectives*

- Provides a new approach to concealed patterns in the data.
- Helps avoid human biases.
- To implement ensembles learning that classifies the disease as per the input of the user.
- Reduce the cost of medical tests (Fig. 1).

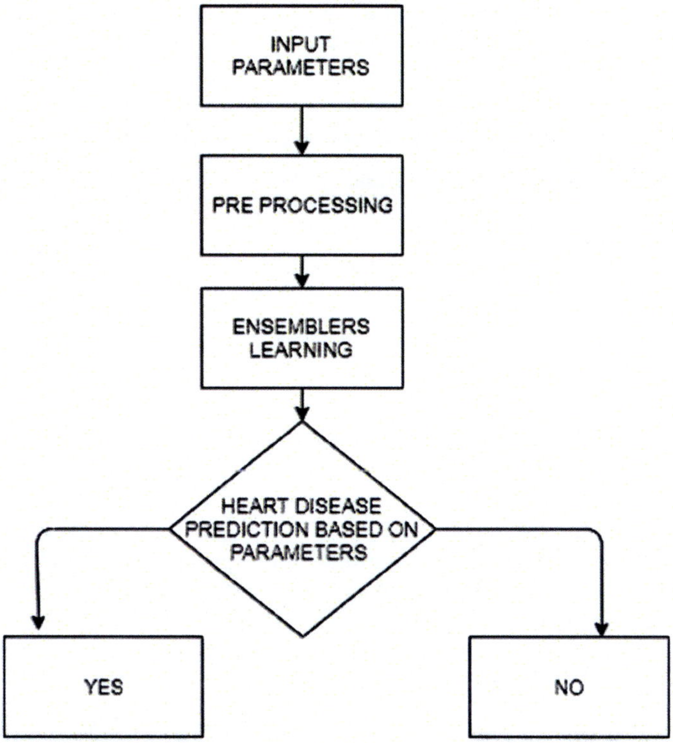

Fig. 1 Flowchart of the proposed system

4 Proposed System

A. **Description of the dataset**:

The analysis is done on the sets of cardiovascular heart records taken form Kaggle. The dataset comprises of eight traits. There are six downright traits and two numeric characteristics. The portrayal of the dataset appears in Table — Patients Id, age have been chosen in this dataset. Male patients are signified by sexual orientation esteem one and female patients are meant by sex esteem two. Height and weight are considered in cm and kgs, respectively. The next attribute ap_hi, ap_lo is the comprehension value of venous pressure. Cholesterol can be either normal or above normal. 1 denotes normal and 2 denotes abnormal as shown in Fig. 2.

(https://www.kaggle.com/sulianova/cardiovascular-disease-dataset)

The ensemble is a strategy that can be utilized to upgrade the precision of a classifier. It is an amazing meta grouping system that consolidates powerless students to upgrade the exhibition of the feeble student by joining crossover students for better outcomes. In this framework, the ensemble method is utilized for coronary illness expectation. The point of interfacing numerous classifiers is to grow better execution when contrasted with an individual classifier. As illustrated in Fig. 3

B. **The system has two main modules**:

- Patient Login and Doctor Login.
- Patient login includes only where the patient can view all the case history if the patient is registered.

Doctor login includes generating case paper of patient, patient's case history, medications, updates and patient details (Fig. 4).

Limitations of Proposed Work:

- The reports are necessary as cholesterol is not the only factor to predict heart disease. Factors such as hereditary can also be considered as it is not predictable through hereditary.

Attributes	Description	Values
Id	No	
Age	In days	
Gender	Men/ Women	1-Women, 2-Men
Height	cm	
Weight	kg	
ap_hi	Systolic blood pressure	
ap_lo	Diastolic blood pressure	
Cholesterol	Normal/AboveNormal	1-Normal ,2-Abnormal

Fig. 2 Dataset attributes

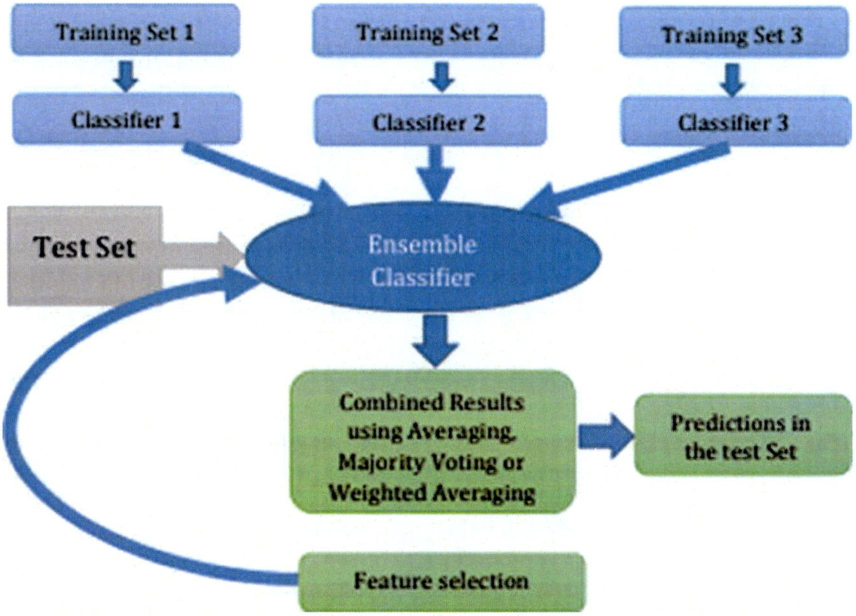

Fig. 3 Ensemble process

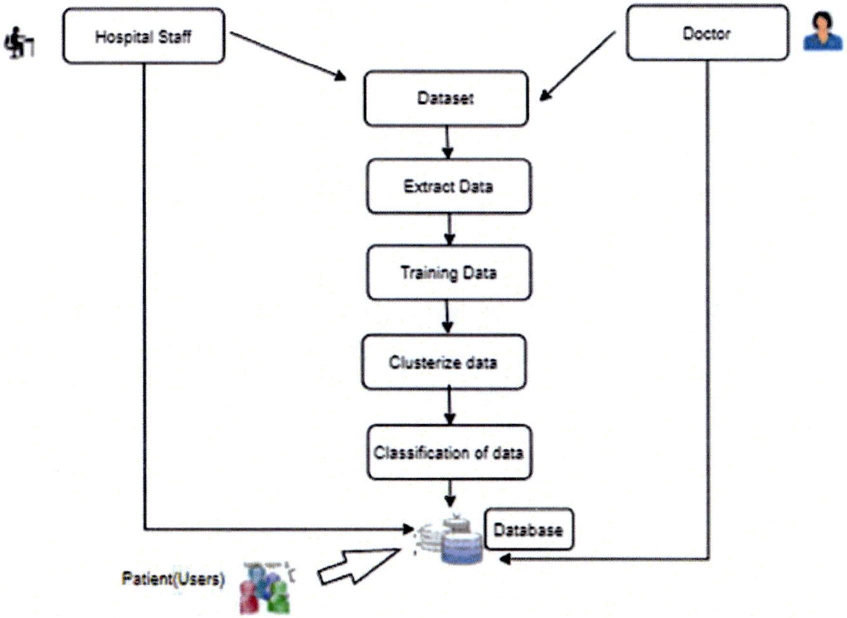

Fig. 4 System architecture

- The lifestyle is also a major drawback in the proposed work as it is not included in this system whereas it is a major factor nowadays. Depending on your lifestyle, the chance to predict heart disease becomes easier

C. **Methodology**:

Bagging (Bootstrap Aggregating)

Multiple models of similar learning algorithm prepared with different subsets of dataset randomly picked from the skilled dataset. Also known as bootstrap aggregation. The newly developed training set will have the same number of patterns as the original training set with a few deletions and repetitions. The new preparing set is known as Bootstrap recreate as shown in Fig. 5. Bootstrap tests are looked from the information, and the classifier is prepared with each example. The democratic from every classifier is consolidated, and the order result is chosen two techniques majority casting a ballot or aggregate averaging (mean or middle). Research shows that sacking can be utilized to upgrade the presentation of a powerless classifier ideally. Stowing diminishes the difference of expectation, since it creates various arrangements of information from irregular examples of the first dataset, with substitution. The algorithm can be trained on models or decision tree models can be used. It works with either row sampling with replacement or feature selection. Random forest algorithm is used in bagging shown in Fig. 6. By training a given weak learning algorithm and training sets for several rounds (the training sets obtained through the use of bootstrap techniques), then all of the base classifiers are combined to learn [9].

Bagging Algorithm

(1) Set the number of individual classifiers M
(2) Set the training set $t = \{(x_1, d_1), \ldots (x_n, d_n)\}$
(3) For $i = 1$ to M:

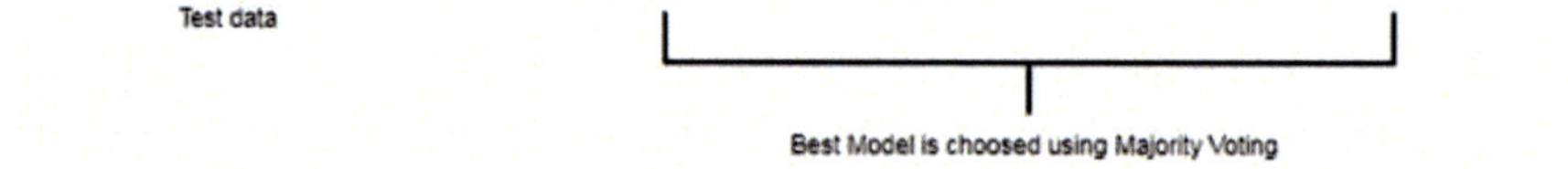

Fig. 5 Bagging process

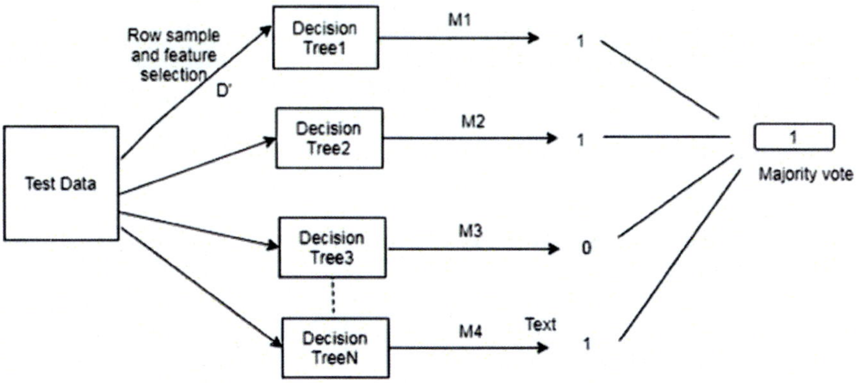

Fig. 6 Random forest process

(a) Obtain new training sets t bag by bootstrap technique; the number of training set samples is N;
(b) Train the classifiers by obtaining training sets [10].

(4) Output results are predicted by:

$$f_{\text{bag}}(x) = \frac{1}{M}\sum_{i}^{M} f_i(x)$$

Random forest uses a test data in which there are d records and m columns ($m < n$) the test data uses row sample and feature selection where data (D') is trained on decision tree, where $D > d$ The output of the decision tree, respectively, decides the majority by voting classifier or regressor. Classification is done as: For i=1 to k do Predict/Classify a testing data X using model Mi. Endfor Return class that is predicted most often [11].

Decision tree has two parameters:

(a) Low bias—It gets trained for all training datasets so that it results to low training error.
(b) High variance—New test data are prone to give large error (Fig. 7).

Boosting

In boosting, initially the dataset is classified into several subsets. The base learner model is trained with one of the subset to get the outcome the classification. If the elements which were incorrectly classified by the previous base learner, then are passed to the next base learner model. This process continues depends on the base learners created. Then, the ensemble process enhances their exhibition by consolidating the feeble models together utilizing a cost work (Illustrated in Fig. 8) [13].

Let $D = \{d_1, d_2, d_3, \ldots\ d_n\}$ *be the given dataset*
$E = \{\}$, *the set of ensemble classifiers*
$C = \{c_1, c_2, c_3, \ldots c_n\}$, *the set of classifiers*
X = *the training set*, $X \in D$
Y = *the test set*, $Y \in D$
$L = n(D)$
Let $init = 1$
$S(init)$ = *A random subset of* X; $S(init) \subset X$
$M(0) = \{\ \}$
fori $= 1$ *to* L *do*
if $i > 1$
$s(i)$ = *Set of incorrectly classified instances of* $M(i-1) + S(i)$
$M(i)$ = *Model trained using* $C(i)$ *on* $S(i)$
$E = E \cup C(i)$
end if
next i
for $i = 1$ *to* L
$R(i)$ = Y *classified by* $E(i)$
next i
Result $= max(R(i)$: $i = 1, 2, \ldots, n)$

Fig. 7 Boosting algorithm [12]

Boosting technique works on:

(a) AdaBoost
(b) Gradient Boosting
(c) XgBoost.

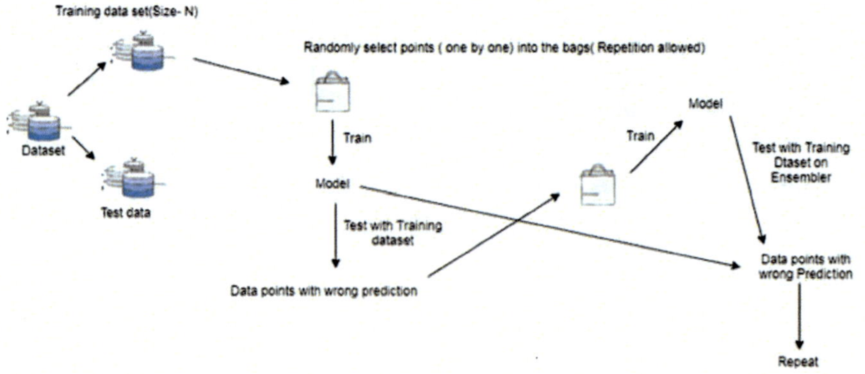

Fig. 8 Boosting process

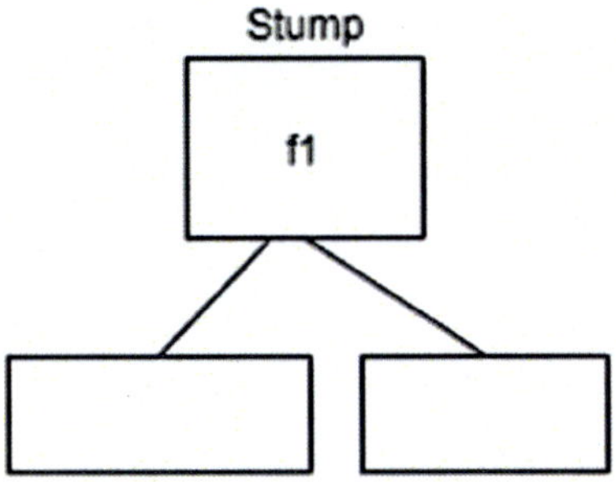

Fig. 9 Decision tree (stumps) with features

(1) AdaBoost

In this, dataset has several features with some output and sample weight. To initiate the sample weight, it is given as:
$W = 1/n$ (where n is the number of records in the dataset). Here, the base learners are also decision tress but not as in random forest algorithm. Here, decision tress are only of 1 depth. This decision trees are basically called as stumps as shown in Fig. 9.

To select a decision tree base learner model:

It has two properties: Entropy or Gini's coefficient [14]. The lesser the entropy of that stump that will be selected as the sequential base learning model decision tree. If the record is incorrectly classified, then the total error is calculated, i.e.
Total Error = Sum of all sample weights
To classify the stump, i.e. to predict the performance of the stump, which is:
Performance of the stump = 1/2 {log e (1 − Total Error/Total Error)}
Now, only the wrong prediction set is then passed to the next decision tree. To update the new sample weight, the formula can be given as:

$$\text{For Incorrect Classification: New Sample Weight} = \text{weight} * \hat{e}(\text{Performance weight})$$
$$\text{For Correct Classification New Sample Weight} = \text{weight} * \hat{e} - (\text{Performance weight})$$

The new sample weight is the updated weight and then obtains normalize values by: Updated weight/Total sum of all the values in the updated weight. Now, this normalize weight will be divided into buckets; then, the value which fits in that bucket is the wrong row and then given to the other iteration [15].

Stacking

Stacking is a gathering strategy where various arrangement models are joined through a meta classifier. Various layers are put in a steady progression, where every one of the models pass their anticipated results to the model in the layer above, and the model in the highest layer settles on choices dependent on the models beneath. The base layer models get input highlights from the first dataset. The top layer model takes the yield from the base layer and makes the forecast. In stacking, the first information is given as contribution to a few individual models [16]. At that point, the meta classifier is utilized to evaluate the information together with the yield of each model, and the loads are assessed. The best execution model is chosen, and the others are disposed of. Stacking consolidates various base classifiers prepared by utilizing distinctive learning calculations L on a solitary dataset S, by methods for a meta classifier (Figs. 10, 11).

D. **Snapshots of the Proposed System**:

The snapshots of the proposed system describe:

1. The patient login which has patient details like name, email id, gender, mobile number, status

Let $D = \{d_1, d_2, d_3, \ldots\ d_n\}$ *be the given dataset*
$E = \{E_1, E_2, E_3, \ldots E_n\}$, *the set of ensemble classifiers*
$C = \{c_1, c_2, c_3, \ldots c_n\}$, *the set of classifiers*
X = *the training set*, $X \in D$
Y = *the test set*, $Y \in D$
K = *meta level classifier*
$L = n(D)$
for $i = 1$ *to* L *do*
$M(i)$ = *Model trained using* $E(i)$ *on* X
Next i
$M = M \cup K$
Result = Y *classified by* M

Fig. 10 Stacking algorithm [12]

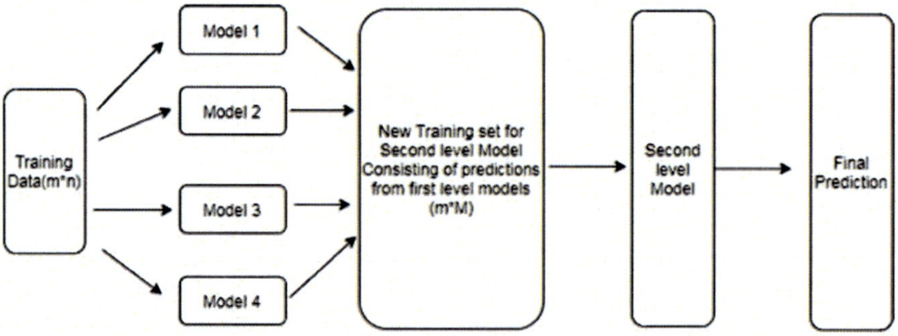

Fig. 11 Stacking process [17]

2. The doctor login which has date, diagnostic of patient, Remarks and prescription.
3. Final output which predicts heart disease based on input parameters.

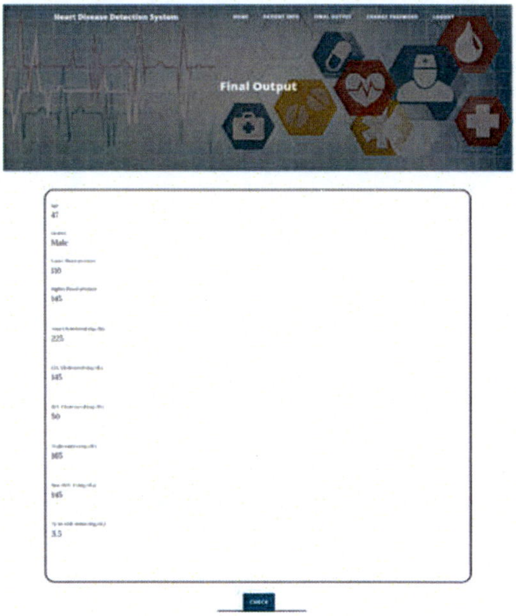

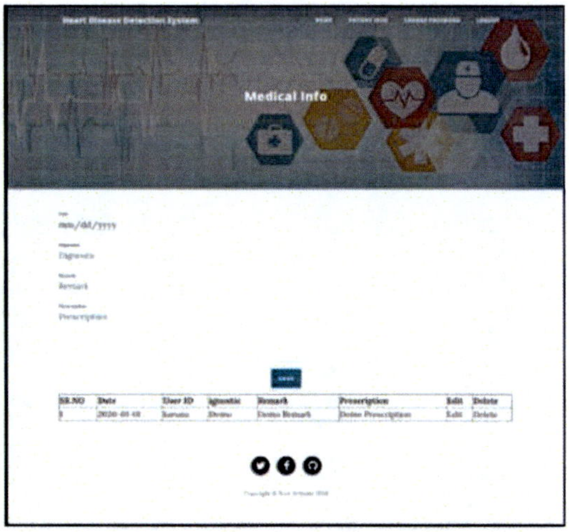

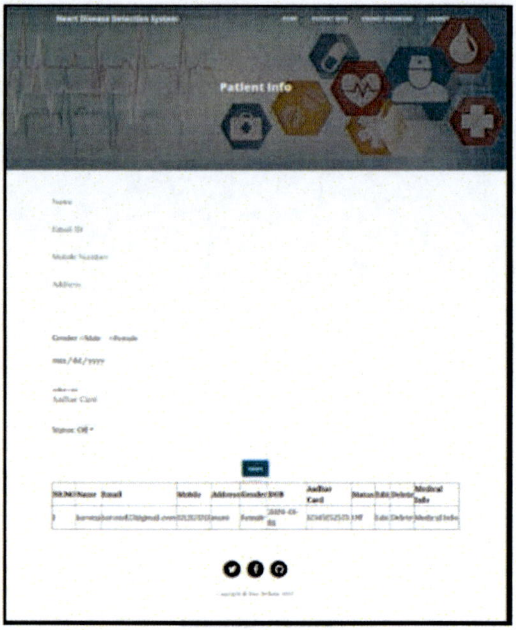

E. **Result Analysis and Comparison**:

Accuracy of various ensemble techniques is illustrated in Fig. 12. As observed, majority voting classifiers give the highest accuracy. Bagging, boosting, stacking also give better results of accuracy.

Various classifiers are compared with accuracy, sensitivity, specificity. The proposed ensembles technique gives better accuracy as compared to other classifiers. The table in Fig. 13 shows percentage wise details for each classifier, and the proposed technique gives more accuracy. The same is represented in graph-wise structure in Fig. 14.

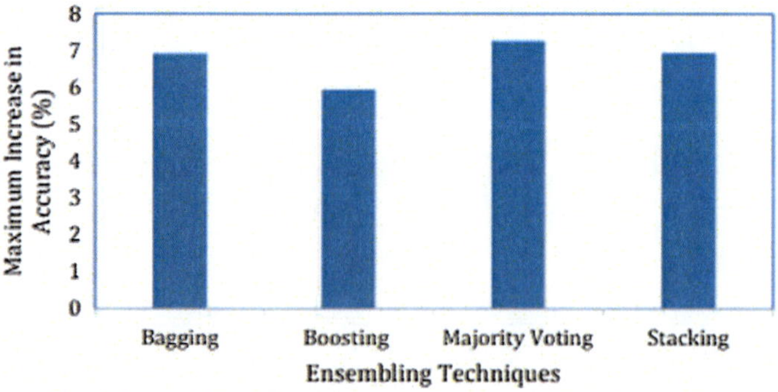

Fig. 12 Graph of accuracy against ensemblers techniques

Classifiers	Accuracy	Sensitivity	Specificity
Naïve Bayes	78.79%	68.42%	92.86%
Decision Tree	72.73%	63.16%	85.71%
SVM	75.76%	73.68%	78.57%
Proposed	81.82%	73.68%	92.86%

Fig. 13 Comparison of accuracy, sensitivity, specificity for various classifiers

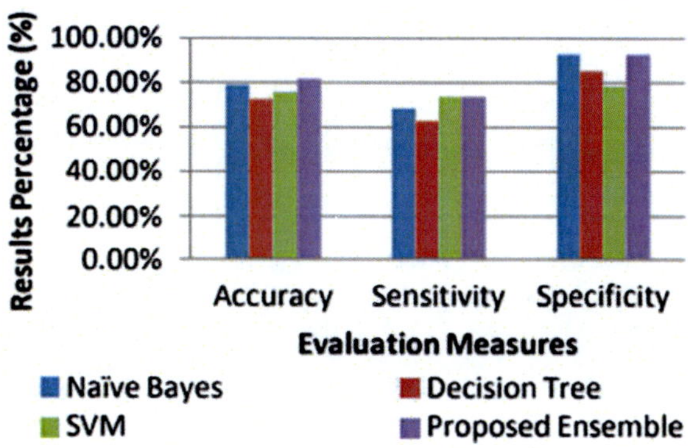

Fig. 14 Graph of accuracy, sensitivity, specificity for various classifiers

5 Conclusion

This framework deciphers the exactness of expectation of coronary illness utilizing an ensemble classifiers. The cardiovascular heart dataset from Kaggle was utilized for preparing and testing purposes. The Ensemble calculations are packing, boosting, stacking, and lion's share casting a ballot were utilized for testing and preparing. When stowing calculation is applied, the exactness was improved by a limit of 6.92%. While boosting calculation is utilized, the precision was improved by a limit of 5.94%. At the point when the feeble classifiers are ensembled with larger part casting a ballot, the exactness was improved by a limit of 7.26%, and stacking improved the precision by a limit of 6.93%. The correlation of results demonstrated that lion's share casting a ballot creates the most noteworthy improvement inexactness. The exhibition was additionally upgraded utilizing highlight choice systems. The element determination systems assisted with improving the precision of the troupe calculations.

Acknowledgements I want to stretch out my genuine gratitude to all who helped me for the undertaking work. I want to earnestly express gratitude towards Dr. Dilip Motwani for their input and steady direction for giving critical data with respect to the commission likewise, for their help in completing this task work. I want to offer my express gratitude towards individuals and folks from Vidyalankar Institute of Technology for their thoughtful co-activity and support.

References

1. Subha, R., Anandakumar, K., Bharathi, A.: Study on Cardiovascular Disease Classification Using Machine Learning Approaches
2. Pouriyeh, S., Vahid, S., Sanninoy, G., De Pietroy, G., Arabnia, H., Gutierrez, J.: A Comprehensive Investigation and Comparison of Machine Learning Techniques in the Domain of Heart Disease. Department of Computer Science, University of Georgia, Athens, USA, ISCC (2017)
3. Dong, Y.-S.: A Comparison of Several Ensemble Methods for Text Categorization. Supported by Motorola Labs, China Research Center, Shanghai Jiao Tong University Ke-Song Han Motorola Labs, China Research Center
4. Princy, T., Thomas, J.: Human Heart Disease Prediction System using Data Mining Techniques. Theresa Princy. R Research Scholar Department of Information Technology Christ University Faculty of Engineering, Bangalore, India, 560060, J. Thomas, Department of Computer Science and Engineering Christ University faculty of engineering, Bangalore, India, 560060, 2016
5. Komal Kumar, N., Sarika Sindhu, G., Krishna Prashanthi, D., Shaeen Sulthana, A.: Analysis and Prediction of Cardio Vascular Disease using Machine Learning Classifiers. ICACCS (2020)
6. Pattekari, S.A., Parveen, A.: Prediction System for heart disease using Naïve Bayes. Department of Computer Sci & Engg Khaja Nawaz College of Engineering
7. Gnaneswar, B., Ebenezar Jebarani, M.R.: A Review on Prediction and Diagnosis of Heart Failure. Gnaneswar B. Dept. of ECE Sathyabama University Chennai, India, Ebenezar Jebarani M.R. Dept. of ECE Sathyabama University Chennai, India
8. Das, R., Turkoglu, I., Sengur, A.: Effective diagnosis of heart disease through neural networks ensembles. Exp. Syst. Appl. Elsevier **36**(4), 7675–7680 (2009)
9. Zhang, J., Zhang, H.: Paper Bagging Ensemble Based on Fuzzy c-means. Department of Information Science and Engineering, Shandong Normal University Jinan, China CISP
10. Wan, S., Yang, H.: Comparison Among Methods of Ensemble Learning. School of Information and Safety Engineering Zhongnan University of Economics and Law Wuhan, China IEEE (2013)
11. Tu, M.C., Shin, D., Shin, D.K.: Effective Diagnosis of Heart Disease Through Bagging Approach. Department of Computer Science, Sejong University, Korea
12. Wan, S., Yang, H.: Comparison Among Methods of Ensemble Learning, IEEE
13. Opitz, D.W., Maclin, R.E.: An Empirical Evaluation of Bagging and Boosting for Artificial Neural Networks. Department of Computer Science, University of Montana, IEEE
14. Shouman, M., Turner, T., Stocker, R.: Integrating decision tree and k-means clustering with different initial centroid selection methods in the diagnosis of heart disease patients. In: Proceedings of International Conference on Data Mining, Australia Defence Force Academy Northcott Drive, Canberra, 2012, pp. 1–7
15. Lakhmishree, J., Paramesha, K.: Prediction of Heart Disease Based on Decision Trees. IJRASET (2017)
16. Santos, S.P., Costa, J.A.F.: Comparison Between Hybrid and Non Hybrid Classifiers in Diagnosis of Induction Motor Faults. IEEE
17. https://www.packtpub.com/product/ensemble-machine-learning-techniques-video/9781788392716

Determining the Degree of Relevance of Content on Social Networks Using Machine Learning Techniques and N-Grams

Jesus Vargas, Omar Bonerge Pineda Lezama, and Jose Eduardo Jimenez

Abstract Today, the use of social networks has revolutionized the way users exchange ideas, opinions, and information. Thanks to this paradigm shift in the way users interact, large companies and public characters have begun to pay particular attention to the opinion generated about their products and/or services, acts and/or events within social networks. This activity is known as online reputation analysis (ORA); an activity carried out mainly by expert users in image analysis who are at the same time able to make strategic decisions that help improve the reputation of a company or public character. The relevance of this activity in recent years has motivated the scientific community to propose automatic methods that support the work of an ORA. In this study, an automatic method is proposed to determine when a tweet is important within a predefined category of messages. The proposed method is based on the use of n-grams to establish the importance of the contents generated in Twitter. The experiments carried out show that the occurrence of certain terms allows an automatic classification model to effectively determine (F-measure = 0.7) when a tweet is or is not important for an ORA.

Keywords Degree of relevance · Social networks · Machine learning techniques · N-grams

J. Vargas (✉) · J. E. Jimenez
Universidad de la Costa, St. 58 #66, Barranquilla, Atlántico, Colombia
e-mail: jvargas41@cuc.edu.co

J. E. Jimenez
e-mail: jjimenez25@cuc.edu.co

O. B. P. Lezama
Universidad Tecnológica, San Pedro Sula, Honduras
e-mail: omar@unitet.edu

© The Editor(s) (if applicable) and The Author(s), under exclusive license to Springer Nature Singapore Pte Ltd. 2021
A. P. Pandian et al. (eds.), *Proceedings of International Conference on Intelligent Computing, Information and Control Systems*, Advances in Intelligent Systems and Computing 1272, https://doi.org/10.1007/978-981-15-8443-5_25

1 Introduction

Today, the extensive use of social networks has generated a new way of exchanging ideas, opinions, and information in real time and from anywhere on the planet, even from outside. These exchanges of ideas and information take place not only among common users, but also involve users representing large companies, artists, public characters, political figures, to name a few [1].

Through the use of Twitter, for example, a consumer can almost instantly contact a product company to report a bug or ask for a component they need. Likewise [2], the company can immediately know how a certain group of users is responding to the presence of one of its products in the market; or if a public character is thought of, he or she can know how his or her followers are reacting to a certain campaign, etc. [3].

Although the advantages that social networks offer to companies are great, the amount of information produced on them is gigantic [4]. Consequently, a reputation analyst faces the great problem of reviewing and analyzing all the information produced in social networks regarding his product/service or public character, which results in an extremely overwhelming activity that is carried out with little efficiency. It is important to mention that only on Twitter there are about 500 million messages (tweets) per day [4]. Manual handling of such an amount of information is practically impossible. In this context, proposing automatic systems to perform an online reputation analysis (ORA) is necessary to process and understand the large amounts of content generated on Twitter associated with a particular company or entity [5].

This online reputation analysis is based on the constant monitoring of the messages produced in social networks. Online reputation monitoring is done in stages. The first stage of the process is to filter only the messages that belong to the entity (company or person) [6]; subsequently, the analyst must decide the polarity of those messages, i.e., the positive or negative implications that those messages could have on the company or person. Since a company does not only handle a single product, as well as public figures in different fields, the next step in the ORA process is to group the messages according to themes related to a certain product or service [7]. Finally, once the issues addressed in the messages are identified, the analyst must determine which of those messages are really important for the company or which ones should be given more attention due to the implications that could be generated within the company, or in the case of a person, implications in the public image of that character [8].

Given the complexity involved in performing all the tasks involved in ORA, this paper focuses only on the automatic detection of the importance of each message; that is, once the tweets are grouped by subject, the interest is in automatically detecting the highest priority tweets that an analyst should consider for the analysis. This last process of online reputation monitoring is also called priority detection [9, 10].

The method proposed in this paper is based on a traditional text classification approach. The main hypothesis is that the importance of a tweet can be determined from the words contained in it and that they somehow reflect the form and use of certain terms that are dependent on the line of business or category of the company and/or character being analyzed. The results obtained in four different categories show that a traditional text classification approach provides sufficient elements to determine when a tweet is important within a certain category.

2 Proposed Method

As mentioned in previous sections, this paper addresses the problem of detecting important tweets by applying the text classification paradigm. Under this paradigm, a first step is the indexing (i.e., representation) of the training documents (Tr), an activity that denotes the mapping of a dj document in a compact form of its content [11].

2.1 *The Weighing Schemes*

The wkj weight can be calculated using different approaches, the simplest of which is the Boolean approach of assigning a value of 1 to the term if it appears in the document and 0 if not. This weighing scheme intuitively captures the presence and/or the absence of terms within a document [12].

Another weighing scheme is known as relative frequency, in which the weight of a tk term in the dj document is determined in direct proportion to the number of times the term appears in the document, and inversely proportional to the number of documents in which the tk term appears in the total training set. In particular, the weight is given by [13]:

$$w_{kj} = \mathrm{TF}(t_k) \times \mathrm{IDF}(t_k) \quad (1)$$

where TF $(t_k) = f_{kj}$ is the frequency of the term t_k in the d_j document. And the inverse frequency or IDF(t_k) is one way of measuring the rarity of the t_k term. The following formula [14] is applied to calculate the value of IDF:

$$\mathrm{IDF}(t_k) = \log \frac{|Tr|}{\{d_j \in D : t_k \in d_j\}} \quad (2)$$

This relative frequency will be called Tf-Idf throughout this document.

2.2 *Processing*

A tweet is a text of no more than 140 characters that may contain components inherent to that social network such as: words out of some dictionary, informal text, hashtags, mentions to other users, URLs, abbreviations of words, repetitions of letters, repetitions of punctuation marks. Most of these components are not useful in a traditional text classification approach [15] or in the definition of message priority. Therefore, each of the messages has been pre-processed, applying the following steps [6]:

- Tweets are converted to lowercase in order to standardize the vocabulary.
- Any sequence of blank spaces becomes a single space. Any mention of users found is removed.
- Any link (URL) found is removed.
- Punctuation is removed. Consequently, any emoticons are also removed.
- Empty and/or functional words are removed.

3 Experimental Evaluation

The measures used in the evaluation of the system, the dataset used and the results obtained in the experiments carried out are described below.

3.1 Evaluation Measures

To evaluate the proposed method, the traditional measures for evaluating text classification systems, such as precision, recall, and F-measurement, were used. Recall (R), the proportion of texts correctly classified in a ci class with respect to the number of texts actually belonging to that class. Thus, precision can be seen as a measure of the system's correctness, while the reminder gives a measure of coverage or completeness. Measure F is normally used to describe classification behavior, which is defined as

$$F = \frac{(1 + \beta^2)P * R}{\beta^2(P + R)} \tag{3}$$

where β represents the harmonic mean between precision and memory. The function of β is to control the relative importance between precision and recall measurements. It is common to assign a value of 1 indicating equal importance to both measures.

3.2 Dataset

The four categories are: automotive, which includes entities for which reputation is determined on the basis of their products; banking, for which transparency and ethics of their activities are factors to be considered in assessing their reputation;

Table 1 Number of MI and UN tweets existing in each of the different categories in the subset of data considered

	Automotive	Banking	Musical	Universities	Total
MY	973	561	1518	162	3214
UN	1242	997	1448	65	3752
Total	2224	1567	2975	236	7002

universities, where their reputation hangs on the variety of intangible products; and music, where their entities base their reputation on both the quality of their products and personal qualities equally.

Table 1 shows information about the subset of data used in the experiments. It should be mentioned that the original data collection mentions that the tweets are classified in three categories, namely important (I), medium important (MI), and not important (UN).

3.3 *Classifiers*

Since the proposal of identifying the importance of the content generated in Twitter does not depend on any particular learning algorithm, practically any classifier can be used to address the problem. For the experiments, two different learning algorithms were selected, which are representative algorithms within the great variety of learning algorithms currently available in the computational learning field [4], specifically the following: naïve Bayes (NB) and support vector machine (SVM).

The Weka [5] implementation of each of these algorithms with the default parameters was used in the experiments. It is important to mention that the ten-fold cross-validation strategy was used for all experiments.

4 Experimental Setup and Results

To achieve the objective, two sets of experiments, which are described below, were proposed:

> **Experiment 1**: Evaluate the impact of using simple words as a form of representation, i.e., word charts, on the classification of important tweets. The main hypothesis of this experiment suggests that there are some characteristic terms of the domain of interest (Automotive, Banking, Music, and Universities) that allow to efficiently identify when a tweet is important.

Experiment 2: Evaluate the impact of word sequences of length two, i.e., word bigrams, on the ranking of important tweets. The main hypothesis of this experiment suggests that, for certain domains, there are word sequences, frequently used by users, which would allow an ORA for identifying important tweets.

Finally, two different weighing schemes were used for both experiments: Boolean and Tf-Idf. The objective of using both weighting schemes was to assess the impact of the presence of certain terms against the frequency of their appearance.

The results of the proposed experiments can be seen in Tables 2 and 3. Note that in both tables the contribution of the shape of the Boolean weighing scheme and Tf-Idf was evaluated with both a Bayesian classifier (NB) and the support vector machines (SVM). The results of the experiments are shown in terms of their precision (P), recall(R), and F-measurement (see Sect. 3.1).

The results obtained in the automotive domain make believe that the type of language used among the users is more varied in this specific domain when referring their opinions; the bigrams with greater gain of information for two domains, in particular, is the music and automotive domain. It is clear that in the case of the automotive domain the opinions do not directly mention the entity of which they are expressed, some of its terms may well represent the banking or university domain, which may be the cause of the poor performance in this domain (Fig. 1).

Table 2 Classification results using word unigrams

Categories of Tweets	Boolean						TF-IDF					
	NB			SVM			NB			SVM		
	P	R	F	P	R	F	P	R	F	P	R	F
Automotive	0.62	0.63	0.62	0.54	0.55	0.53	0.64	0.59	0.59	0.55	0.49	0.53
Banking	0.69	0.70	0.72	0.82	0.82	0.82	0.30	0.73	0.70	0.80	0.78	0.80
Musical	0.70	0.69	0.71	0.72	0.72	0.72	0.66	0.65	0.65	0.68	0.67	0.67
Universities	0.74	0.74	0.76	0.74	0.75	0.74	0.74	0.73	0.73	0.71	0.72	0.71
Average	0.71	0.71	0.69	0.69	0.68	0.69	0.60	0.70	0.69	0.69	0.69	0.68

Table 3 Classification results using word bigrams

Categories of Tweets	Boolean						TF-IDF					
	NB			SVM			NB			SVM		
	P	R	F	P	R	F	P	R	F	P	R	F
Automotive	0.53	0.56	0.53	0.52	0.53	0.49	0.64	0.54	0.50	0.53	0.55	0.53
Banking	0.75	0.74	0.75	0.81	0.82	0.80	0.73	0.69	0.72	0.79	0.79	0.78
Musical	0.70	0.69	0.70	0.72	0.72	0.69	0.65	0.66	0.65	0.70	0.70	0.69
Universities	0.73	0.73	0.74	0.71	0.70	0.68	0.75	0.66	0.66	0.70	0.70	0.66
Average	0.68	0.69	0.69	0.69	0.68	0.69	0.70	0.64	0.64	0.69	0.67	0.68

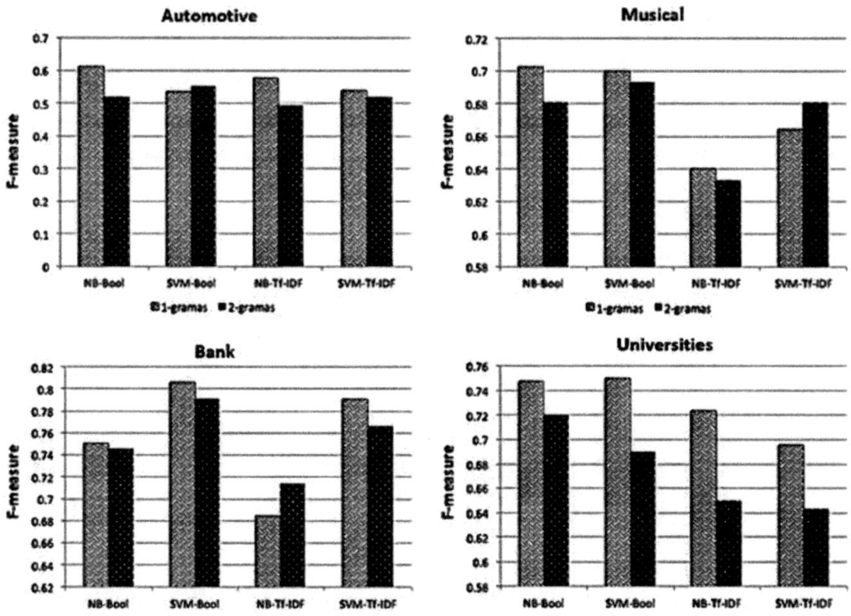

Fig. 1 Comparison of results obtained in terms of measure *F*

5 Conclusions and Future Research

A text classification method for detecting the priority of tweets was presented, concluding that: i) it was shown that the presence of the terms is sufficient for a classifier to learn to distinguish important tweets from those that are not; ii) it was found that the form of representation proposed is the one that allows the problem to be solved with some effectiveness, and that the results do not depend on the learning algorithm; iii) finally, it was found that the use of word bigrams is not sufficiently descriptive to overcome the performance obtained by a representation based on a chart.

References

1. Al Hamoud, A., Alwehaibi, A., Roy, K., & Bikdash, M.: Classifying political tweets using Naïve Bayes and support vector machines. In: International Conference on Industrial, Engineering and Other Applications of Applied Intelligent Systems, pp. 736–744. Springer, Cham (2018)
2. Lim, S., Tucker, C.S.: Mining Twitter data for causal links between tweets and real-world outcomes. Exp. Syst. Appl. X **3**, 100007 (2019)
3. Chen, H., Mckeever, S., Delany, S.J.: Presenting a labelled dataset for real-time detection of abusive user posts. In: Proceedings of the International Conference on Web Intelligence, pp. 884–890 (2017, August)

4. Mishra, N., Singh, A.: Use of twitter data for waste minimisation in beef supply chain. Ann. Oper. Res. **270**(1–2), 337–359 (2018)
5. Viloria, A., Lezama, O.B.P.: Improvements for determining the number of clusters in k-means for innovation databases in SMEs. In: Procedia Computer Science, vol. 151, pp. 1201–1206. Elsevier B.V. (2019). https://doi.org/10.1016/j.procs.2019.04.172
6. Kiprono, K.W., Abade, E.O.: Comparative Twitter sentiment analysis based on linear and probabilistic models. Int. J. Data Sci. Technol. **2**(4), 41–45 (2016)
7. Priyoko, B., Yaqin, A.: Implementation of Naive Bayes algorithm for spam comments classification on Instagram. In: 2019 International Conference on Information and Communications Technology (ICOIACT), pp. 508–513. IEEE (2019, July)
8. Sathesh, A.: Enhanced soft computing approaches for intrusion detection schemes in social media networks. J. Soft Comput. Paradigm (JSCP) **1**(2019), 69–79 (2019)
9. López-Chau, A., Valle-Cruz, D., Sandoval-Almazán, R.: Sentiment analysis of twitter data through machine learning techniques. In: Software Engineering in the Era of Cloud Computing, pp. 185–209. Springer, Cham (2020)
10. Nauze, F., Kissig, C., Zarafin, M., Villada-Moiron, M. B., Genet, R.: U.S. Patent No. 9,678,946. U.S. Patent and Trademark Office, Washington, DC (2017)
11. Viloria, A., Pineda Lezama, O.B.: An intelligent approach for the design and development of a personalized system of knowledge representation. In Procedia Computer Science, vol. 151, pp. 1225–1230. Elsevier B.V. (2019). https://doi.org/10.1016/j.procs.2019.04.176
12. Savyan, P.V., Bhanu, S.M.S.: UbCadet: detection of compromised accounts in twitter based on user behavioural profiling. Multimed. Tools Appl. 1–37 (2020)
13. Sanchez, H., Kumar, S.: Twitter bullying detection. ser. NSDI **12**(2011), 15 (2011)
14. Altawaier, M.M., Tiun, S.: Comparison of machine learning approaches on arabic twitter sentiment analysis. Int. J. Adv. Sci. Eng. Inf. Technol. **6**(6), 1067–1073 (2016)
15. Mohammad, A.S., Jaradat, Z., Mahmoud, A.A., Jararweh, Y.: Paraphrase identification and semantic text similarity analysis in Arabic news tweets using lexical, syntactic, and semantic features. Inf. Process. Manage. **53**(3), 640–652 (2017)

An XGBoost Ensemble Model for Residential Load Forecasting

Karthik Venkat, Tarika Gautam, Mohit Yadav, and Mukhtiar Singh

Abstract An accurate residential load forecast benefits a consumer domain energy management system extensively, as it provides minimum threshold energy to the consumers instead of completely shedding supply by the electricity distribution companies. In this paper, a model is proposed using the extreme gradient boosting ensemble algorithm for forecasting of residential loads. The publicly available UCI dataset is utilized, which is based on the real-life electric power consumption of a residence in France. Lag-based features capturing power consumption for different periods are added to the dataset. Correlation analysis of the features is done to filter the redundancy of features. The dataset was resampled for different time resolutions and used it for forecasting the power consumption of the residence for a day and a week ahead. The results from experimentation strongly indicate that the proposed model outperforms the existing machine learning models as to the accuracy of forecast and computational time.

Keywords Consumer energy management system · Correlation analysis · Ensemble learning · Extreme gradient boosting · Residential load forecasting

K. Venkat (✉) · T. Gautam · M. Yadav · M. Singh
Department of Electrical Engineering, Delhi Technological University, New Delhi, India
e-mail: vkarthik_bt2k16@dtu.ac.in

T. Gautam
e-mail: tarika.gautam8@gmail.com

M. Yadav
e-mail: mohitrao085@gmail.com

M. Singh
e-mail: mukhtiarsingh@dce.ac.in

© The Editor(s) (if applicable) and The Author(s), under exclusive license to Springer Nature Singapore Pte Ltd. 2021

A. P. Pandian et al. (eds.), *Proceedings of International Conference on Intelligent Computing, Information and Control Systems*, Advances in Intelligent Systems and Computing 1272, https://doi.org/10.1007/978-981-15-8443-5_26

1 Introduction

Residential load forecasting has undergone extensive research, as it plays a major role in power systems management, scheduling, and dispatch. With the ever-growing demand for electricity in the residential space due to increased usage of gadgets or the uptake of electric vehicles [1], rises electrical energy costs. This has sent alarm ringing across the sector and piqued the researchers' interest to find out ways to minimize the cost of electric energy on the demand side. Accurate forecasting plays a vital role in the planning and operation of power systems [2] and is also a basic link for smart grid (SG) construction. Knowledge of consumer electrical power usage is central to the smart grid (SG) [3]. In recent times, the rollout of Advanced Metering Infrastructure (AMI), a building block of SG, has been responsible for collecting data of consumer loads [4]. Smart meters, a sub-system of AMI, have been widely adopted in many countries and utilized for forecasting the consumer electric power demand in the residential space. Smart meters provide sub-metering for separate rooms to appliances in a household, enabling for bifurcation between the controllable as well as uncontrollable energy consumption in a residence. The data accumulated by the smart meter paves the way for residential load forecasting. An accurate load forecast at the demand side (consumers) is essential for energy scheduling. Residential load forecasting also requires computation time to be minimum [5], and the ability to run accurate forecast methods that do not call for high computational requirement enables consumers to engage in electric energy markets, allowing the demand curve to be modulated according to particular consumer profile.

Short-term load forecasting (STLF) includes residential load forecasting and has been subject to numerous methods that researchers have proposed over the years. Various approaches used for STLF include linear regression introduced by Song et al. (2005) [6], support vector regression [7], autoregressive models [8], and fuzzy logic [9]. Lachut et al. (2014) [10] have investigated conventional models such as k-nearest neighbor, support vector machines, ARMA, naive Bayes for prediction of power consumption. Studies have also been conducted for cluster-based models [11] to utilize the probabilistic nature of residential load aggregation for forecasting. One main approach to STLF has long been the usage of artificial neural networks (ANN). The ANN approach to STLF has seen criticism in the aspect of "overfitting" of the model and an increase in the layers of ANN by the increment in the input variables. The advent of deep learning [12] methods has circumvented this issue by using stacks of multi-layers of neural networks, with the ability to transform the input data into a lower dimension. This portrays a composite representation of the input, which is used for learning by stacks of neural networks. Deep learning network architectures such as convolutional neural network (CNN) [13] and long short-term memory (LSTM) [14] have proven to be effective models for different applications including time-series-based electric load forecasting. The idea of increasing the generalization capability of a model framework by an ensemble of techniques that reinforce one another in a combination has gone mainstream in recent times. Nowadays, the smart home energy management systems framework is being proposed [15] using intelligent systems for STLF in the household.

Publicly available individual household electric power consumption dataset released by Hebrail and Berard [16] is used on the UCI Machine Learning Repository, for residential load forecasting. This dataset is made up of real-life measurements of electric power consumption for a French residence. This dataset has been subjected to the prediction of electric power consumption using deep learning solutions. Kim and Cho [17] have used a hybrid model [18] combining CNN-LSTM for forecasting power consumption. This was later improved by Le et al. [19] by extending the CNN-LSTM framework by introducing additional convolution and LSTM layers to increase the forecasting performance. While increasing the overall accuracy of electric power consumption prediction, it also introduces computational complexity due to training and model tuning. This dataset was also used by Wu et al. [5] in a scheme of transfer learning where the multiple kernel learning-based transfer regression model was proposed using a gradient boosting-based framework. This approach has shown to reduce the computational cost successfully but has a significant forecast performance degradation.

In this paper, the issue is addressed with the aim of accurate residential load forecasting with minimum computational cost by proposing a novel approach using an ensemble of extreme gradient boosting tree ensemble learning method (XGBoost) introduced by Chen and Guestrin [20] with correlation analysis for removing features that are strongly correlated while calculating feature score after iteratively training the XGBoost algorithm over the training set, this framework is named RLF-XGB.

2 Materials and Methods

2.1 Dataset

The UCI dataset of household electric consumption was released by Hebrail et al. [16]. Dataset attributes include the date, time, global active power (in kilowatt), global reactive power (in kilowatt), voltage (in volt), global intensity (in ampere), sub-metering 1 (in active energy watt-hour), sub-metering 2 (in active energy watt-hour), sub-metering 3 (in active energy watt-hour). The measurements for electric power consumption are taken for four years. The dataset consists of sub-metering readings for power consumed by the connected loads in the household. The dataset is split by the 3rd of January, 2010 which is the first Sunday for the year. The dataset consisting of minutely reading of electric power consumption is resampled for daily and weekly time resolution. The dataset is split into three years for a train set and one year for the test set. The two resampled datasets to train are utilized and validate our proposed model for forecasting electric power consumption for a day and a week ahead. These steps are also taken to maintain a similar measure taken by Le et al. [19], as our model in the experimental result section is compared.

2.2 XGBoost Regression Learning Model

XGBoost regression model uses the lag-based features having different lags to learn the temporal relation in the energy consumption dataset to forecast the residential power consumption for a short or long term. XGBoost regression is an optimized gradient boosting ensemble technique that produces a model of prediction by aggregating the predictions of weak prediction models, generally decision trees. With the boosting technique, weak predictors are applied sequentially to the set with each one trying to improve the efficiency of the whole ensemble.

When implementing the XGBoost regression model, the dataset with n training samples consisting of a lag version of time-series features as input x_i and expected power consumption output y_i in terms of kilowatts, a tree ensemble model $\phi(x_i)$ is defined as the sum of K regression trees $f_k(x_i)$:

$$\hat{y}_i = \phi(x_i) = \sum_{k=1}^{K} f_k(x_i) \tag{1}$$

To evaluate the output of a given model, a loss function $l\left(\hat{y}_i, y_i\right)$ is selected to calculate the error between the target value and expected value, and add a regularization term $\Omega(f_k)$ to penalize excessively complex trees, with leaf weight ρ:

$$L(\phi) = \sum_{i}^{n} l(\hat{y}_i, y_i) + \sum_{k}^{K} \Omega(f_k) \tag{2}$$

where $\Omega(f_k) = \gamma T + \frac{1}{2} \cdot \lambda\left(||\rho||^2\right)$.

The loss function applied to new prediction and compared to the result to determine if our predictions are improving are not. The XGBoost algorithm minimizes $L(\phi)$ by adding increasing f_k iteratively. Suppose the ensemble now includes K trees. A new tree f_{K+1} is added which minimizes:

$$\sum_{i}^{n} l\left(\hat{y}_i, y_i + f_{k+1}(x_i)\right) + \Omega(f_k) \tag{3}$$

The term γT is used for the pruning of trees built by the model, where γ is the penalty factor for pruning, and T denotes the number of terminal nodes or the leaves of the tree. XGBoost can prune even when γ is set to zero. Pruning takes place after the full tree is built and plays no role in deriving the optimal predicted value or similarity score for the residual which is the difference between the expected and target values of power consumption. The tree that strengthens the current model most as defined by L is greedily added. Then a new tree is trained using the function defined in (2); this is done by using Taylor series approximation by taking the first and second gradient of loss function $l(\hat{y}_i, y_i)$. The score of leaves (ρ) in XGBoost regression tree is given by:

$$\rho = \frac{\sum_{i=1}^{n} (y_i - \hat{y}_i)}{n + \lambda} \tag{4}$$

where the numerator term is the summation of residuals and the denominator denotes the sum of the number of residuals and regularization term λ for regression in (4).

$f_0(x)$ for $k = 0$ in (1) gives the predictions from the first stage of our model. The residual for each instance is given by $(y_i - f_0(x))$. To build a XGBoost regression tree, for the forecasting used the residuals from $f_0(x)$. The regression tree built will attempt to reduce the residuals from the previous step. The output of the regression tree will not be the prediction of electric energy consumed; instead, it will help to predict the successive function $f_1(x)$ which will bring down the residuals. The additive model of regression tree calculates the residual mean $(y_i - f_0(x))$ at each tree leaf.

The boosted function $f_1(x)$ is obtained by the summation of $f_0(x)$ with the regression tree. This means the regression tree learns from $f_0(x)$ residuals and suppresses them in $f_1(x)$. This is repeated to compute an ensemble of regression trees for n more iterations. The residuals from the previous function $f_{k-1}(x)$ will be used by each of these additive learners. Each learner is trained on the residuals. At every stage, all the additive learners in boosting are modeled on the residual errors. The boosting learners make use of patterns that are found in the residual errors. At the stage where optimum accuracy is reached by boosting, the residuals tend to be distributed randomly without a pattern.

2.3 Evaluation Metrics

Different criteria for determining the efficiency of the proposed model are used. Mean squared error, root mean squared error, mean absolute error, mean absolute percentage error are the most widely used time-series forecasting metrics for calculating the accuracy of the forecasting models.

Mean square error (MSE) calculates the error square average, that is, the average of the square difference between the predicted values and the actual values. The MSE is calculated using (5), where y_i is the actual electric energy consumption vector and $\hat{y}_i$ is the predicted vector of values of electric energy consumption by the forecasting model.

$$\text{MSE} = \frac{1}{n} \sum_{i=1}^{n} (y_i - \hat{y}_i)^2 \tag{5}$$

The root of mean squared error (RMSE) has the advantage of being in the same units as the response variable. It is used as a quantitative measure for comparing the forecasts for the same time series on different models. The smaller the error, the better the model can forecast according to the RMSE criterion. The RMSE measure is outlined as follows.

$$\text{RMSE} = \sqrt{\frac{1}{n}\sum_{i=1}^{n}(y_i - \hat{y}_i)^2} \tag{6}$$

Also, the MAE measure is determined using (7). The MAE measure is less prone to large deviations than the squared loss.

$$MAE = \frac{1}{n}\sum_{i=1}^{n}\frac{|y_i - \hat{y_i}|}{|y_i|} \tag{7}$$

The last evaluation measure is MAPE, which also measures the accuracy and used as a loss function in forecasting models. It expresses the accuracy of the forecasting model in terms of percentage. Equation (8) is used to calculate the MAPE measure.

$$\text{MAPE} = \frac{1}{n}\sum_{i=1}^{n}\frac{|y_i - \hat{y}_i|}{|y_i|} * 100 \tag{8}$$

2.4 Proposed Model

Figure 1 shows the architecture used to forecast residential power consumption using the RLF-XGB model. The proposed model uses the lag method an analog of sliding window method for power consumption prediction. The date–time index of the dataset is used to add 8 lag features ($\boldsymbol{l}_i$) with different periods $\boldsymbol{i}$, on the basis that the calculation of features is fast, the number of features does not increase for the 2 datasets, and the features can well reflect the power consumption for the specific time resolution of the dataset. The 8 lag features are shown in Table 1. The proposed model takes the Pearson correlation coefficient $\zeta_{l'l}$ between every two features say l' and l that is calculated using (9).

$$\zeta_{l'l} = \frac{\sum_{i=1}^{n}\left(l'_i - \overline{l'}\right)\left(l_i - \overline{l_i}\right)}{\sqrt{\sum_{i=1}^{n}\left(l'_i - \overline{l'}\right)^2 \sum_{i=1}^{n}\left(l_i - \overline{l}\right)^2}} \tag{9}$$

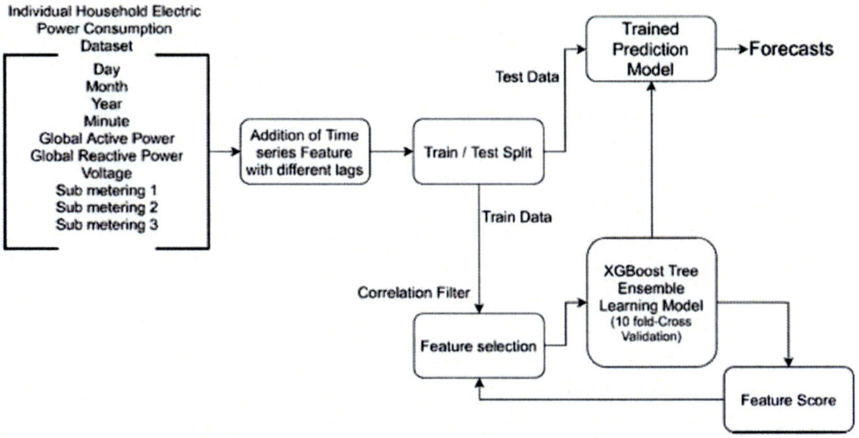

Fig. 1 RLF-XGB model architecture

Table 1 Lag features for the datasets

Features	Description
L_1	Hourly electric power consumption
L_2	Electric power consumption for a day of a week
L_3	Consumption of power for one quarter of a year
L_4	Electric power consumption for a month
L_5	Electric power consumption for the year
L_6	Electric power consumption for one day of the year
L_7	Daily electric power consumption
L_8	Electric power consumption for one week of the year

The value of $\zeta_{l'l}$ varies from -1 to 1. The correlation for every pair of features is evaluated and features with high correlations magnitudes above a threshold are removed by filtering; this takes care of the redundancy of features. The correlation heatmap of features of the resampled dataset is outlined in Fig. 2.

The XGBoost algorithm is trained with the initial model hyperparameters on the train set. Feature importance is calculated by the XGBoost algorithm. Feature importance of the subset of features used by the proposed model is evaluated against the feature score (F-score), which is the summation of the number of times each feature is split on. F-score reveals the discriminatory power of each feature independently of the others. The F-score for features utilized in the model for daily and weekly datasets is shown in Fig. 3. Features with high F-score are selected for the successive forecasts. The initial subset of features is selected after correlation filtering and further feature elimination is done based on F-score. Random search cross-validation was used to optimize the model hyperparameters. The XGBoost model is run for 200 iterations with tenfold split amounting to 2000 fits and ranked the hyperparameters

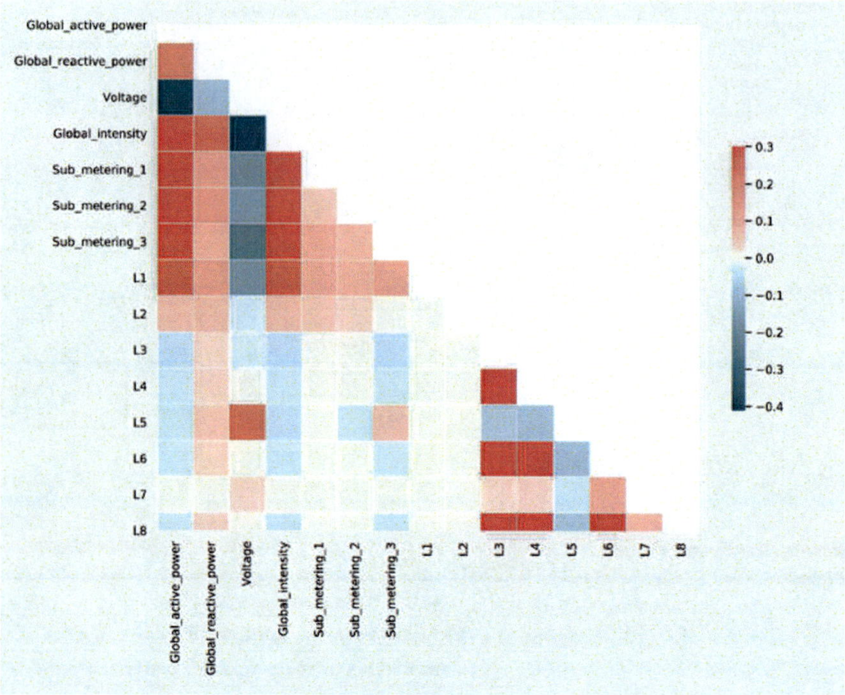

Fig. 2 Correlation between the features of the dataset

according to mean squared scoring and generated a parameter list with the best score as shown in Table 2. The utilization of the correlation filter with a feature score for the feature selection process reduces the computational time for successive runs of the proposed model.

The optimized parameters values from Table 2 were used in the RLF-XGB model to forecast on the test set. The performance of the model was evaluated using measures defined in (5)–(8). The steps followed by the RLF-XGB model are defined in Fig. 4. The model stops training when there is no improvement after validating the testing set split (t) of the train set (T) for 100 iterations.

3 Experimental Results and Discussion

The proposed model, named RLF-XGB, was used to forecast the electrical power consumption of the re-sampled datasets. Our proposed model was used to forecast the global active power consumption for a day and a week ahead, for the individual household. Our model is run on a Quad-core Intel i7 CPU laptop with 16 GB of memory. The experimental results for the daily and weekly datasets are shown in

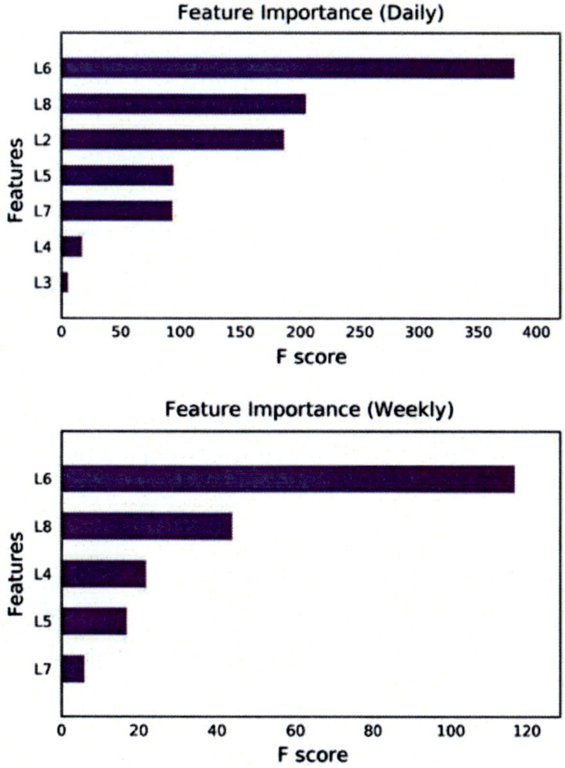

Fig. 3 The plot of feature score for daily and weekly datasets

Table 2 Optimal values for tunable model parameters

Model Parameter	Description	Value
n_estimators	Number of trees in XGBoost model	422
Learning rate	Learning rate/eta	0.0361
Gamma	Minimum loss reduction to create new tree split	0.199
Alpha	L1 regularization of leaf weights	0.5
Lambda	L2 regularization of leaf weights	4.5
max_depth	Maximum depth per tree	2
Subsample	Percent of samples used per tree	0.638
colsample_bytree	Percent features used per tree	0.953

Table 3. The results of our proposed method are compared with the existing models using the same dataset. The models compared to are the multi-step linear regression, long short-term memory, and CNN Bi-LSTM models proposed by Le et al. [19].

Initially, the electric power consumption is forecasted for a day ahead. This is done using the daily resampled dataset. Our proposed model shows a relative improvement

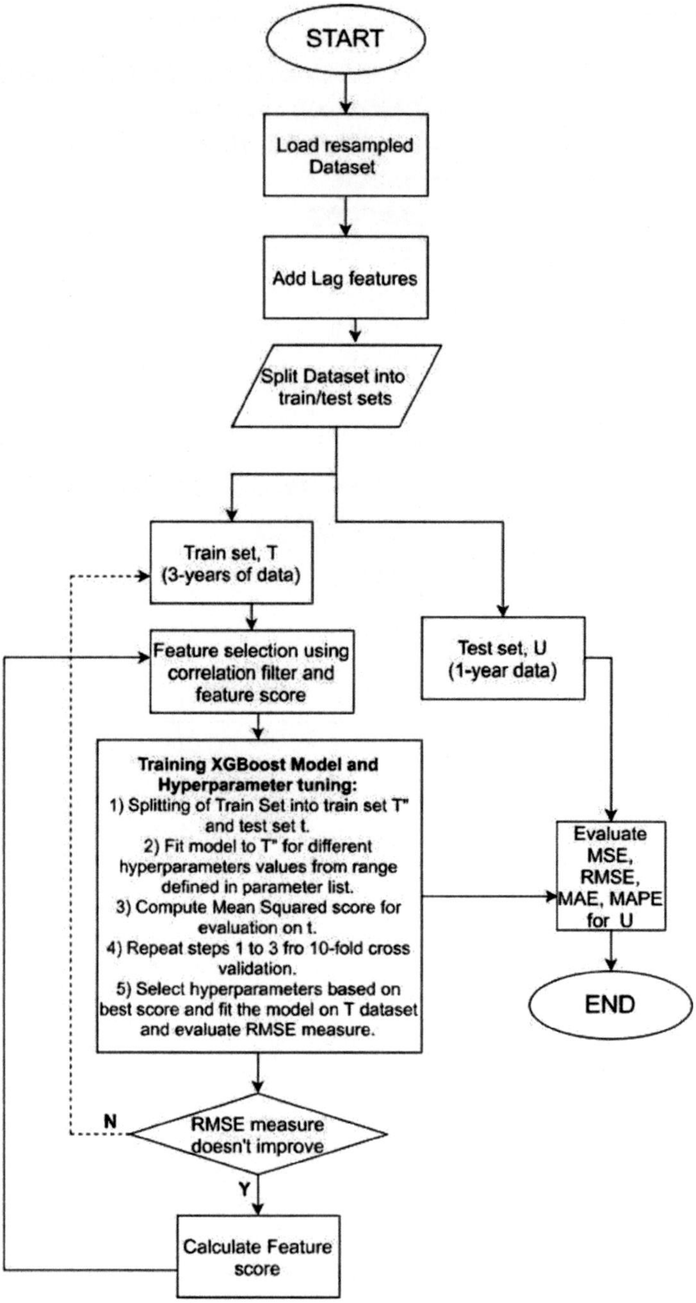

Fig. 4 RLF-XGB load forecasting process

Table 3 Forecasting performance for different time resolutions

Method	Dataset	MSE	RMSE	MAE	MAPE	Training Time (sec)	Prediction Time (sec)
Multi-step linear regression	Daily	0.253	0.503	0.392	52.69	27.83	1.32
	Weekly	0.148	0.385	0.32	41.33	11.23	1.48
LSTM	Daily	0.241	0.491	0.413	38.72	106.06	2.97
	Weekly	0.105	0.324	0.244	35.78	24.42	3.66
CNN Bi-LSTM	Daily	0.065	0.255	0.191	19.15	61.36	0.71
	Weekly	0.049	0.22	0.177	21.28	20.7	0.4
XGBoost	Daily	0.126	0.355	0.297	30.86	34.5	1.02
	Weekly	0.038	0.195	0.134	16.03	17.89	0.89
RLF-XGB	Daily	**0.06**	**0.245**	**0.184**	**19.12**	**16.24**	**0.56**
	Weekly	**0.025**	**0.158**	**0.116**	**13.88**	**9.07**	**0.2**

of 4% for root mean square error (RMSE) and mean absolute error (MAE) over the baseline score set by the CNN Bi-LSTM model. The computations required to reach this result are vastly less as our model takes 74% and 21% less time for training and making forecasting. Figure 5 shows a day-ahead forecast of electric power consumption on the test dataset which starts from January of 2010.

In a week-ahead forecast using the weekly dataset, the forecast performance of our proposed method has a significant improvement of 28% and 34% for the RMSE and MAE, respectively, over the CNN Bi-LSTM model while making a forecast taking half the time required by the latter. When compared to the baseline score of multi-step linear regression, our approach improves by 44 and 60% for RMSE and MAE and cuts down on training and forecasting time by 30 and 72%, respectively. Furthermore, in comparison to the LSTM approach, our model has an improvement of 50% for both RMSE and MAE. The RLF-XGB model takes 74 and 88% less time

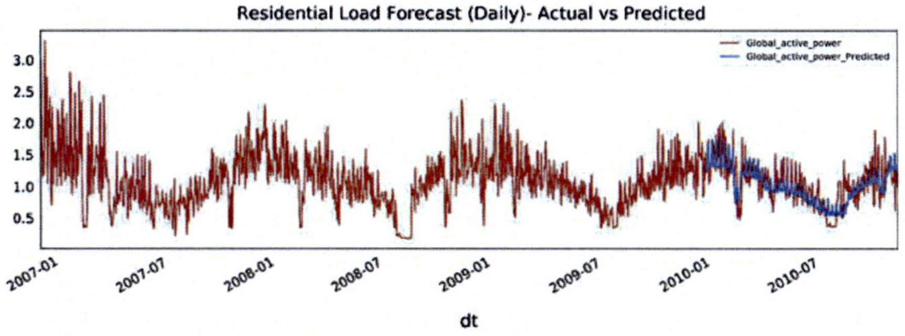

Fig. 5 Proposed model forecast for day-ahead power consumption

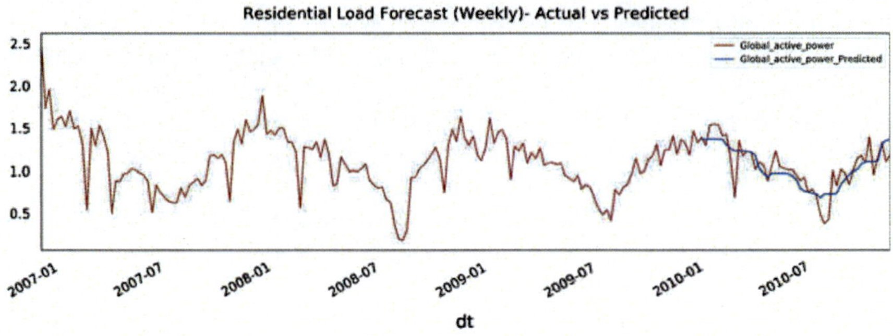

Fig. 6 Proposed model forecast for week-ahead power consumption

for training and making a forecast in comparison to the LSTM approach. Figure 6 shows a week-ahead forecast of electric power consumption for the residence.

As a sanity check, our approach is also compared with the XGBoost algorithm without measures taken for feature selection as in the proposed model. The results show the proposed model has improved by an average of 25% for both RMSE and MAE. Our proposed model takes 50 and 60% less time to train and make a forecast on average over both the daily and weekly datasets.

4 Conclusion and Future Work

In this paper, the model is proposed using an extreme gradient boosting algorithm (XGBoost) for residential load forecasting and utilized publicly available UCI dataset of electric power consumption measurements of an individual household, resampled it to make daily and weekly electric power consumption datasets. A lag-based features in the datasets are introduced to cover the electric power consumption for different time resolutions and capture the time variations according to total power consumption. The initial features subset for the proposed model was produced using a correlation filter to remove the redundancy of features. The selected subset of features was used for training the XGBoost algorithm with the training set. The features with a high F-score generated by the XGBoost algorithm are selected for the trained model. Optimized parameters for the model were found using the cross-validation method. The trained model with optimized parameters was then used to make power consumption forecast for a day ahead and week ahead using daily and weekly datasets, respectively. The experimental result indicates that our approach improves by 25% over the XGBoost algorithm for both RMSE and MAE. Furthermore, our proposed method takes less than half the time for training and prediction than the XGBoost algorithm. The experimental results show that our approach outperforms existing models for forecasting electric power consumption in an individual household while keeping the computation time to a minimum.

For the future, our study can be extended by using a different ensemble of deep learning-based methods with our model to analyze the performance improvement on other electric power consumption datasets. Looking forward to applying our model in the consumer energy management system and utilize the real-time energy consumption from the smart meter for allowing an accurate forecast of the electric power consumption in the residential space. This would allow households to control their energy expenditure through the modification of energy demand patterns in the house.

References

1. Marcincin, O., Medvec, Z., Moldrik, P.: The impact of electric vehicles on distribution network. In: 2017 18th International Scientific Conference on Electric Power Engineering (EPE), Kouty nad Desnou, 2017, pp. 1–5, doi: https://doi.org/10.1109/EPE.2017.7967344
2. Senjyu, T., Takara, H., Uezato, K., Funabashi, T., Member, S. : One-Hour-Ahead Load Forecasting Using Neural Network (2002)
3. Yu, C.N., Mirowski, P., Ho, T.K.: A sparse coding approach to household electricity demand forecasting in smart grids. IEEE Trans. Smart Grid **8**(2), 738–748 (2017). https://doi.org/10.1109/TSG.2015.2513900
4. Rashed Mohassel, R., Fung, A., Mohammadi, F., Raahemifar, K.: A survey on advanced metering infrastructure. Int. J. Electr. Power Energy Syst. **63**, 473–484 (2014). doi: https://doi.org/10.1016/j.ijepes.2014.06.025
5. Wu, D., Wang, B., Precup, D., Boulet, B.: Multiple Kernel learning-based transfer regression for electric load forecasting. IEEE Trans. Smart Grid **11**(2), 1183–1192 (2020). https://doi.org/10.1109/TSG.2019.2933413
6. Bin Song, K., Baek, Y.S., Hong, D.H., Jang, G.: Short-term load forecasting for the holidays using fuzzy linear regression method. IEEE Trans. Power Syst. **20**(1), 96–101 (2005). doi: https://doi.org/10.1109/TPWRS.2004.835632
7. Jiang, H., Zhang, Y., Muljadi, E., Zhang, J.J., Gao, D.W.: A short-term and high-resolution distribution system load forecasting approach using support vector regression with hybrid parameters optimization. IEEE Trans. Smart Grid **9**(4), 3331–3350 (2018). https://doi.org/10.1109/TSG.2016.2628061
8. Lopez, J.C., Rider, M.J., Wu, Q.: Parsimonious short-term load forecasting for optimal operation planning of electrical distribution systems. IEEE Trans. Power Syst. **34**(2), 1427–1437 (2019). https://doi.org/10.1109/TPWRS.2018.2872388
9. Rejc, M., Pantoš, M.: Short-term transmission-loss forecast for the Slovenian transmission power system based on a fuzzy-logic decision approach. IEEE Trans. Power Syst. **26**(3), 1511–1521 (2011). https://doi.org/10.1109/TPWRS.2010.2096829
10. Lachut, D., Banerjee, N., Rollins, S.: Predictability of energy use in homes. In: International Green Computing Conference, Dallas, TX, 2014, pp. 1–10. doi: https://doi.org/10.1109/IGCC.2014.7039146
11. Zhang, Y., Chen, W., Xu, R., Black, J.: A cluster-based method for calculating baselines for residential loads. IEEE Trans. Smart Grid **7**(5), 2368–2377 (2016). https://doi.org/10.1109/TSG.2015.2463755
12. Goodfellow, I., Bengio, Y., Courville, A.: Deep Learning. MIT Press (2016)
13. Dong, X., Qian, L., Huang, L.: A CNN based bagging learning approach to short-term load forecasting in smart grid. In: 2017 IEEE SmartWorld, Ubiquitous Intelligence & Computing, Advanced & Trusted Computed, Scalable Computing & Communications,

Cloud & Big Data Computing, Internet of People and Smart City Innovation (SmartWorld/SCALCOM/UIC/ATC/CBDCom/IOP/SCI), San Francisco, CA, 2017, pp. 1–6, doi: 10.1109/UIC-ATC.2017.8397649
14. Kim, N., Kim, M., Choi, J.K.: LSTM based short-term electricity consumption forecast with daily load profile sequences. In: 2018 IEEE 7th Global Conference on Consumer Electronics (GCCE), Nara, pp. 136–137 (2018). doi: https://doi.org/10.1109/GCCE.2018.8574484.
15. Ozturk, Y., Senthilkumar, D., Kumar, S., Lee, G.: An intelligent home energy management system to improve demand response. IEEE Trans. Smart Grid **4**(2), 694–701 (2013). https://doi.org/10.1109/TSG.2012.2235088
16. Hebrail, G., Berard, A.: Individual household electric power consumption data set. UCI Machine learning Repository.University of California, School of Information and Computer Science, Irvine, CA (2012). https://archive.ics.uci.edu/ml
17. Kim, T.Y., Cho, S.B.: Predicting residential energy consumption using CNN-LSTM neural networks. Energy **182**, 72–81 (2019). https://doi.org/10.1016/j.energy.2019.05.230
18. Tan, M., Yuan, S., Li, S., Su, Y., Li, H., He, F.: Ultra-short-term industrial power demand forecasting using LSTM based hybrid ensemble learning. In: IEEE Transactions on Power Systems, pp. 1–1 (Dec. 2019). doi: https://doi.org/10.1109/tpwrs.2019.2963109
19. Le, T., Vo, M.T., Vo, B., Hwang, E., Rho, S., Baik, S.W.: Improving electric energy consumption prediction using CNN and Bi-LSTM. Appl. Sci. (Switzerland) **9**(20) (2019). doi: https://doi.org/10.3390/app9204237
20. Chen, T., Guestrin, C.: XGBoost: a scalable tree boosting system. In: Proceedings of the ACM SIGKDD International Conference on Knowledge Discovery and Data Mining, vol. 13–17, pp. 785–794 (Aug. 2016)

Breast cancer Analysis and Detection in Histopathological Images using CNN Approach

A. L. Prajoth SenthilKumar, Modigari Narendra, L. Jani Anbarasi, and Benson Edwin Raj

Abstract The detection of cancer from the histology images with hematoxylin and eosin stained is significant, requires more analysis, and often leads to disagreement among pathologists. Computerized diagnostic systems help pathologists to enhance diagnostic accuracy and efficiency. Recent advances in this field have led them to successfully process histology images using convolution neural networks (CNNs). The classification from histology images into malignant and benign depends upon the cell density, variability, and tissue structure. The state and art of review of various research works using histology images are detailed in this work. The VGG16 model is also used to classify the breast cancer histology images into benign and malignant which achieved 88% accuracy on the testing data. The findings are likewise compared to other approaches.

Keywords Breast cancer detection · Histology images · Image classification · CNN · Deep learning techniques

A. L. Prajoth SenthilKumar (✉) · L. Jani Anbarasi
School of Computer Science and Engineering, Vellore Institute of Technology, Chennai 600127, India
e-mail: prajoth0698@gmail.com

L. Jani Anbarasi
e-mail: janianbarasi.l@vit.ac.in

M. Narendra
Department of Computer Science and Engineering, VFSTR deemed to be University, Guntur, India
e-mail: narendramodigari@gmail.com

B. E. Raj
Higher Colleges of Technology, Fujairah, United Arab Emirates
e-mail: braj@hct.ac.ae

© The Editor(s) (if applicable) and The Author(s), under exclusive license to Springer Nature Singapore Pte Ltd. 2021

A. P. Pandian et al. (eds.), *Proceedings of International Conference on Intelligent Computing, Information and Control Systems*, Advances in Intelligent Systems and Computing 1272, https://doi.org/10.1007/978-981-15-8443-5_27

1 Introduction

Cancer is the world's second largest cause of death for women. Approximately 9.6 million deaths in 2019 were due to cancer and about 627,000 people died due to breast cancer. The pathologist is the one who performs the diagnosis of whether breast cancer or not. Without the pathologist, a diagnosis of breast cancer cannot be made. The pathologist determines the histological type of breast cancer tumor and the size of the tumor. The pathologist would examine the cells under the microscope to find any signs of malignant activity. The pathologist will perform the immune-histochemical tests to check the hormone-level status of the breast. Typically, a biopsy preceded by electron microscopic examination is essential in diagnosing breast cancer. The microscopic analysis is to analyze the cellular-level components of the breast tissue. Breast cancer is classified into two types: benign and malignant. Generally, benign lesions are normal tissue of the breast parenchyma. They are the region which is not cancerous. The benign tumors develop gradually and tend to be located fairly harmless. On the other hand, the malignant lesions or carcinoma tissue type are cancerous. In malignant regions, the abnormal cells divide uncontrollably and destroy other body tissues too. The malignant tumor is made of cancer cells that can invade the nearby tissues too.

Imaging techniques such as mammograms (X-rays), magnetic resonance imaging, ultrasound, and thermography can be used to detect and treat breast cancer. Imaging techniques for cancer detection have been identified as the best solution. But using biopsy techniques is the only way to diagnose the cancer cells. Among the biopsy techniques, the data collected from the 'BreaKHis' is by using surgical open biopsy (SOB) method. In general, the collected samples or cells are attached to a glass microscopic slide for the detection process. Generally, before the visual examination of cancer by the doctor/scientist, the tissue samples of the pathological images obtained during the biopsy are typically treated with 'hematoxylin and eosin' (H&E). The pathologist analyzed the microscopic images of the tissue samples in different magnification factors. But the result of the manual process sometimes leads to misclassifications. The success in digital imaging methods was applied at some microscopic magnification for analyzing the pathological images [4, 5]. These techniques applied to the pathology images showed a better performance in segmentation, classification, and detection. For these medical imaging techniques needs a large volume of annotated data but it is not accessible in large amount in this domain of applications. This paper discusses the deep learning approach to the detection of breast cancer using the 'CNN' model and is validated for the 'BreaKHis' dataset.

2 Literature Review

Aswathy et al. [5] paper implemented an image processing technique on histopathological images for breast cancer detection. The proposed system added color to the images by H&E staining at the cytoplasmic stroma and nucleus regions of the biopsy regions of the breast and immunohistochemistry staining of the images to identify the presence and absence of the particular proteins. ROI was computed using marker-controlled watershed algorithm, active contour model, and intuitive segmentation using vectorial data.

Bayramoglu et al. [3] in 2016 proposed a model that analyzed BreaKHis dataset using two different CNN architectures; i.e., CNN's single task is to predict malignancy, and multiple task CNN is performed to forecast malignancy and magnification levels. Multi-task CNN learns the magnification factor and the malignancy and classifies as either benign or malignant along with the magnification factor. Single-task CNN achieved an average recognition rate of about 83.25%, the multi-task CNN for the classification of the benign or malignant task achieved average recognition rate of about 82.13%, and the magnification of images resulted in an average recognition rate of about 80.10%.

For the breast cancer histopathological image classification, Wei et al. [1] presented a 'BiCNN' model based on deep convolution neural networks and performed two class classifications. With the help of the BreaKHis dataset, this paper describes the classification of the histopathology images. The paper describes the usage of advanced augmented techniques to detect edge features of the cancerization region in the image. For better training accuracy, transfer learning and fine-tuning methods have been opted. The training of the images used BiCNN model using Caffe framework. The BiCNN model considers breast cancer class and subclass categories as prior knowledge. BiCNN has a string ability to feature images. The main role of the BiCNN is that the convolution will happen as the feature detector with the input image. The proposed method used the Google Net model as the base model in comparison with the BiCNN model. To achieve better output with the highest probability, the model uses Softmax classifier for multi-class classification of the targets. For the classification of the images, the BiCNN model achieved an accuracy of '97%'.

In 2018, Adeshina et al. [2] proposed a model using deep convolution neural networks (DCNNs) combined with the ensemble learning method (AdaBoost) to achieve the automated classification of images. Steve et al. classified the pathology images into multi-class classification, i.e., into 8 different classes. To increase the accuracy of the model, by minimizing the error, the optimization algorithms are used where the weights are updated during the backward pass of the network. This DCNN model achieved an accuracy of 91.5%.

In 2019, Alom et al. [4] suggested a model for classification of breast cancer by using the deep convolutional neural network (DCNN) model which is inception recurrent residual convolutional neural network [IRRCNN]. The IRRCNN model shows better performance when compared to inception and residual networks. IRRCNN includes three layers of recurrent convolution layers, inception block, and residual block. Different kernel sizes have been applied to perform different recurrent convolution operations in the inception unit. The input and output dimension does not change due to the residual block in IRRU layer. '96.84%' is achieved for binary classification and '97.65%' for multi-class classification on the BreaKHis dataset. It also achieved '97.51%' for binary classification and '97.11%' for multi-class classification on the breast cancer classification challenge dataset by patch-wise classification. '99.05%' is achieved for binary classification and '98.59%' for multi-class classification on the breast cancer classification challenge dataset.

Jonathan de Matos et al., in 2019 [6], implemented a double transfer learning for breast cancer classification in the histopathological images using Inception v3 pre-trained on ImageNet dataset. Training using SVM classifier was performed to filter the patches from the histopathology images that removed the irrelevant features from the images. BreaKHis dataset was first trained using the SVM on the CRC dataset to classify the tissues with patches and tissues without patches and then classified as benign and malignant. Benhammou et al. [7] in 2019 proposed a state-of-the-art CNN model for breast cancer histopathological images into benign and malignant. The proposed scheme is analyzed using 'Inception V3' on the BreaKHis dataset and provided a comparison results with the existing models.

Idowu et al. in 2015 [8] predicted the breast cancer risk using the predictive data mining classification techniques such as Naïve Bayes and J48 decision tree with the help of the 'Weka software' and achieved 82.6% using Naïve Bayes model and 94.2% using J48 decision tree. Chiao et al., in 2019 [9], classified the breast tumors on sonogram data using mask R-CNN and got an accuracy of about 85%. Table 1 details the comparative analysis of the proposed algorithms along with its accuracy and tools used.

Table 1 Comparative analysis of the existing algorithms

Ref. No.	Proposed technique	Tools used	Accuracy
[1]	BiCNN deep learning model	Caffe framework	97%
[2]	Deep learning combined with ensemble learning method (AdaBoost)	TensorFlow	91.5%
[5]	Applied image processing techniques: (1) Marker-controlled watershed algorithm (2) Using active contour model (3) Intuitive segmentation using vectorial data	By using charged couple device camera with microscope	Improved classification accuracy
[7]	Inception v3 deep learning model on BreaKHis dataset	TensorFlow and Nvidia GTX 1080Ti GPU with 11 Gb of VRAM	Accuracy (1) 40× → 93.0% (2) 100× → 88.9% (3) 200× → 89.4% (4) 400× → 86.9%
[8]	Data mining classification technique Naïve Bayes classifier and J48 Decision Tree	Weka software	(1) Naïve Bayes classifier → 82.6% (2) J48 decision tree → 94.2%

3 Proposed System

The proposed work included transfer learning-based approach to the histology dataset for the improvement of the classification process. The data is fine-tuned and tested using VGG16 a pertained model in ImageNet classification data with 1000 categories of images. The breast cancer histology image referred as BreaKHis dataset [11] includes 9109 microscopic images of breast tumor tissues collected from 82 patients used for experimental analysis. The data is collected from these 82 patients at different magnification factors like 40×, 100×, 200×, and 400×; there are around 5429 malignant samples and 2480 benign images as of today. The images are of size 700 × 460, and color RGB images of 8-bit depth in red, blue, green channel. The image formats of these images are PNG (Fig. 1).

3.1 Preprocessing

Dataset was preprocessed before the training procedure. Images are normalized to have a better qualitative analysis. Image normalization not only eliminates noise but also modifies the intensity of the image measured on a different scale to a common normalized scale using an average or any other modes [14]. The normalized images are shown in Fig. 2.

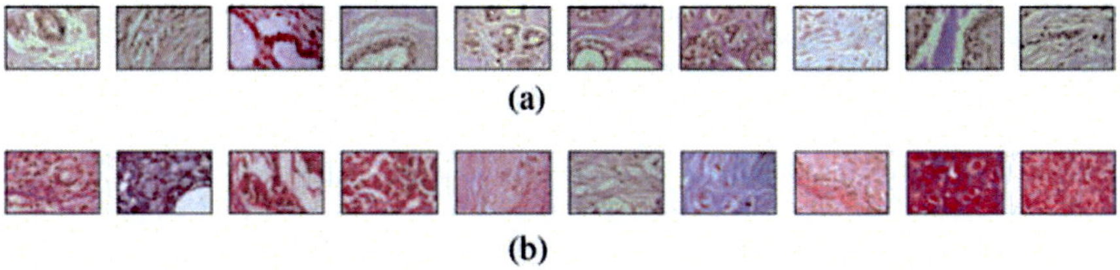

Fig. 1 Histology image. **a** Benign and **b** malignant

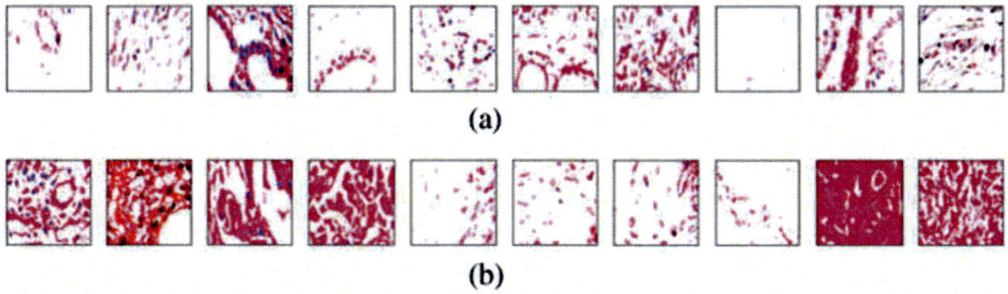

Fig. 2 Normalized images. **a** Benign and **b** malignant

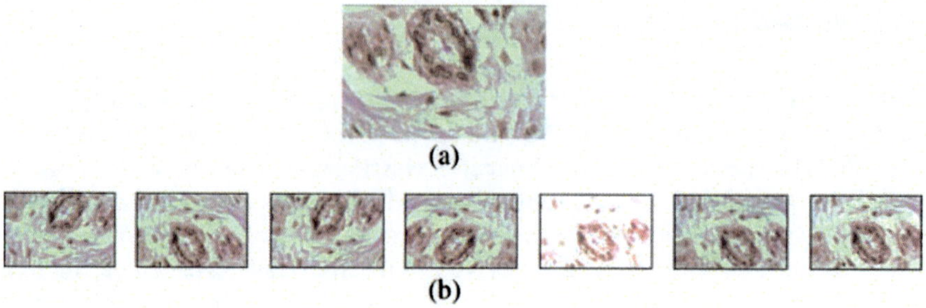

Fig. 3 Sample augmented images

3.2 Data Augmentation

The dataset that is used [10] for our proposed analysis included the data from the 400× magnification scale. These images are very less for the deep learning algorithms and transfer learning process. So, the augmentation technique is used to increase the dataset so that the effect of overfitting can be reduced. Rotation range included 15, width shifting is performed at a rate of 0.05, and height shifting is performed at a rate of 0.05, shearing at 0.05 along with horizontal and vertical flipping is performed. The sample augmented data is shown in Fig. 3.

3.3 Model Training

The transfer learning on VGG16 architecture was performed using the augmented and actual dataset. The training was performed with a size of 128 × 128. Then, layers of the architectures were modified according to the transfer learning context described in the transfer learning section. The overall flow of the proposed work is given in Fig. 4.

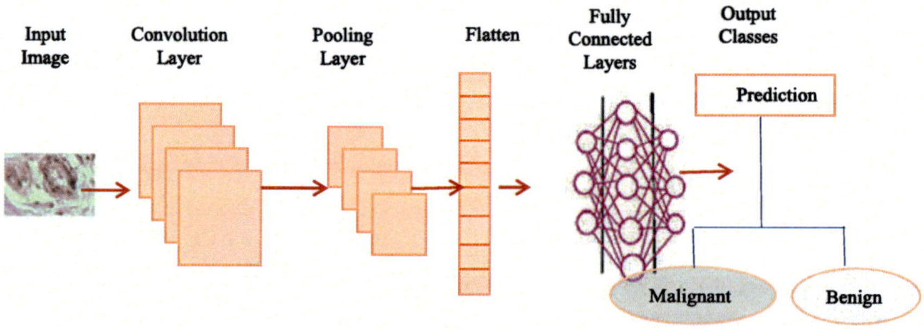

Fig. 4 Process flow of the breast cancer detection process

4 Performance Evaluation

The performance evaluation was studied based on the accuracy and sensitivity of the obtained results [15]. Accuracy depends on the number of samples, i.e., true negative and true positive, which are correctly categorized and calculated as in Eqs. 1, 2, and 3.

$$\text{accuracy} = \frac{\text{Number of TP} + \text{Number of TN}}{\text{Total number of images 'S'}} \times 100 \tag{1}$$

where 'S' is the number of data samples present in the dataset used for processing.

Sensitivity refers to the probability of positive diagnosis among the patients those are confirmed with disease and is computed as

$$\text{Sensitivity} = \frac{\text{Number of TP}}{\text{Number of TP} + \text{Number of TN}} \tag{2}$$

The values range between 0 and 1 where 0 means worst and 1 refers to the best case.

Precision refers to the probabilistic measure that determines whether the positive case belongs to the positive class. Recall refers to the probabilistic measure to identify whether an actual positive case is determined correctly or not. The F1 score is computed as the geometric mean between precision and recall.

$$\text{F1 Score} = 2 * \frac{\text{Precision} * \text{Recall}}{\text{Pecision} + \text{Recall}} \tag{3}$$

The model is trained using the pertained VGG16 model, and the parameters of the model are given in Fig. 5.

The model was trained for 30 epochs where steps per epoch are also considered as 50 steps. Dropout [12, 13] is specified to avoid the model from learning the same features. The fully convolutional layer included the activation function as sigmoid. The RMSprop is used as the optimizer and accuracy as included as the metrics, and the loss function used is binary_crossentropy. The training accuracy and loss achieved for the proposed model are given in Fig. 6.

This deep learning algorithm model 'VGG16' was found to be useful in detecting the cancer regions in the digital histopathology images and able to classify the images a benign or malignant and able to generate an accuracy of about 89%.

5 Conclusion

The proposed work presented a review of various methods proposed for the detection of breast cancer using histology images that have been analyzed. This paper also

Model: "sequential_1"

Layer (type)	Output Shape	Param #
vgg16 (Model)	(None, 7, 7, 512)	14714688
flatten_1 (Flatten)	(None, 25088)	0
dropout_1 (Dropout)	(None, 25088)	0
dense_1 (Dense)	(None, 1)	25089
Total params:	14,739,777	
Trainable params:	25,089	
Non-trainable params:	14,714,688	

Fig. 4 Model parameters of the proposed system

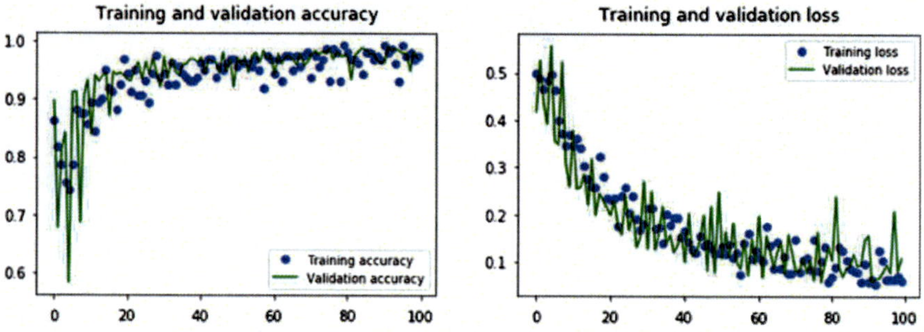

Fig. 5 **a** Training and validation accuracy, and **b** training and validation loss

presented the use of transfer learning for the classification of breast cancer using histopathological images. The images are preprocessed and augmented for better results. The transfer learning model applied in the 400× magnification factor has achieved an accuracy of 89% and is further working to enhance the accuracy by modeling other architectures and the preprocessing schemes.

References

1. Weil, B., Han, Z., He, X., Yin, Y.: Deep learning model based breast cancer histopathological image classification. In: 2017 the 2nd IEEE International Conference on Cloud Computing and Big Data Analysis
2. Adeshina, S.A., Adedigba, A.P., Adeniyi, A.A., Aibinu, A.M.: Breast cancer histopathology image classification with deep convolutional neural networks. In: 14th International Conference On Electronics Computer and Computation ICECCO 2018
3. Bayramoglu, N., Kannala, J., Heikkila, J.: Deep learning for magnification independent breast cancer histopathology image classification. In: 2016 23rd International Conference on Pattern Recognition (ICPR) Cancún Center, Cancún, México, December 4–8, 2016
4. Alom, Md.Z., Yakopcic, C., Taha, T.M., Asari, V.K.: Breast Cancer Classification from Histopathological Images with Inception Recurrent Residual Convolutional Neural Network. arXiv preprint arXiv:1803.01164
5. Aswathy, M.A., Jagannath, M.: Detection of breast cancer on digital histopathology images: present status and future possibilities. Informatics in Medicine Unlocked. http://dx.doi.org/10.1016/j.imu.2016.11.001
6. de Matos, J., Britto Jr., A.de.S., Oliveira, L.E.S., Koerich, A.L.: Double transfer learning for breast cancer histopathologic image classification. In: Conference: 2019 International Joint Conference on Neural Networks (IJCNN)
7. Benhammou, Y., Tabik, S., Achchab, B., Herrera, F.: A first study exploring the performance of the state-of-the art CNN model in the problem of breast cancer. In: ACM LOPAL Conference, Rabat, Morocco, May 2018 (LOPAL'18)
8. Idowu, P.A., Williams, K.O., Balogun, J.A., Oluwaranti, A.I.: Breast cancer risk prediction using data mining classification techniques. Transactions on Networks and Communications, vol 3 No 2, April (2015)
9. Chiao, J.-Y., Chen, K.-Y., Ken, M.D., Liao, Y.-K., Hsieh, P.-H., Zhang, G., Huang, T.-C.: Detection and classification the breast tumors using mask R-CNN on sonograms. ncbi.nlm
10. Shukla, U., Mishra, A., Jasmine, S.G., Vaidehi, V., Ganesan, S.: A deep neural network framework for road side analysis and lane detection. Procedia Comput. Sci. **1**(165), 252–8 (2019)
11.) https://web.inf.ufpr.br/vri/databases/breast-cancer-histopathological-database-breakhis/
12. Zahangir Alom, B.Md., Yakopcic, C., Nasrin, S., Taha, T.M., Asari, V.K.: Breast cancer classification from histopathological images with inception recurrent residual convolutional neural network. J. Digit. Imag. (2019)
13. Ahmad, H.M., Ghuffar, S., Khurshid, K.: Classification of breast cancer histology images using transfer learning. In: 2019 16th International Bhurban Conference on Applied Sciences and Technology (IBCAST) (2019)
14. Sharon, J.J., Anbarasi, L.J., Raj, B.E.: DPSO-FCM based segmentation and classification of DCM and HCM heart diseases. In: 2018 Fifth HCT Information Technology Trends (ITT), Dubai, United Arab Emirates, 2018, pp. 41–46
15. Sharon, J.J., Anbarasi, L.J.: Diagnosis of DCM and HCM heart diseases using neural network function. Int. J. Appl. Eng. Res. **13**(10), 8664–8668 (2018)

Implementation of Stemmer and Lemmatizer for a Low-Resource Language—Kannada

G. Trishala and H. R. Mamatha

Abstract Stemming and lemmatization are two basic modules used for text normalization in Natural language processing (NLP) which qualifies text, words, and documents for further processing. Stemming is the process of eliminating the affixes from the inflectional word to generate root word. The extracted stem or root word may not be a valid. Lemmatization is also a process of removing the affixes from the word but returning the word in dictionary form which is known as lemma. This lemma will always be meaningful word. Hence, while developing the Lemmatizer semantic knowledge is considered. In this paper, Unsupervised Stemmer and Rule-Based Lemmatizer have been proposed for Kannada. Experimentation is done by building a dataset of 17,825 root words with the help of Kannada dictionary.

Keywords Natural language processing (NLP) · Lemmatization · Stemming · Inflectional words · Linguistic rules

1 Introduction

India is a multilingual country, with different cultures. So, to communicate with the people across the country, it is essential for one to have knowledge of different languages. Kannada is one of the four major languages of the Dravidian family. There are 29 regional languages spoken in India. Among 40 most spoken languages in the world, Kannada is one language with 70 million speakers. As the number of Internet users is increasing exponentially, the users who are creating online materials for different languages are also increasing and for applications built on natural language processing like sentimental analysis, chatbot, text summarization, spell checking, keyword search, information extraction, advertisement matching, etc.;

G. Trishala (✉) · H. R. Mamatha
Department of Computer Science, PES University, Bangalore, India
e-mail: trishalag.96@gmail.com

H. R. Mamatha
e-mail: mamathahr@pes.edu

© The Editor(s) (if applicable) and The Author(s), under exclusive license to Springer Nature Singapore Pte Ltd. 2021

A. P. Pandian et al. (eds.), *Proceedings of International Conference on Intelligent Computing, Information and Control Systems*, Advances in Intelligent Systems and Computing 1272, https://doi.org/10.1007/978-981-15-8443-5_28

there is requirement of some form of preprocessing of the data for the future use. This fact creates many challenges in developing applications in different languages for markets which rely on the understanding of text to function such as call center, social listening, search, virtual agents, and market research. Hence, it turns out to be critical to break the language boundary among Indian languages and make simpler correspondence among the individuals from various states and various countries. Online materials for Kannada are increasing day to day, and amount of work done to build preprocessing tools like Stemmer and Lemmatizer for future use for Kannada compared to other regional languages is very less. So, building Stemmer and Lemmatizer for Kannada language would be of great help for preprocessing of the data and for development of NLP-based Kannada language applications. This is the motivation to build Stemmer and Lemmatizer for regional languages especially Kannada which is low-resource and structurally complex language.

The main aim of NLP is to develop software which is mainly used for analyzing, understanding, and generating languages that are used by humans. Machine translation system (MTS) is one of the areas of NLP, where input in source language is analyzed for its correctness in grammar generally referred as source analysis phase which uses Stemmer and Lemmatizer as an initial module. This module separates the given input word into affix and root word. Affix is the combination of prefixes and suffixes.

In stemming process, inflected words or derived words are reduced to their word stem, base, or root form in linguistic morphology. The stem does not need to be identical with the word's morphological origin. For example, if words "likely," "like," "likes," "liked" are given as input to Stemmer it is stemmed as "like." But, the two major drawbacks of stemming are over-stemming and under-stemming. Over-stemming occurs when two input words with a different context and meaning are stemmed to the same root word. For example, consider words "car," "cares" which are completely different in their meaning but are stemmed as "car" because of the morphological similarity in words. Under-stemming occurs when two words belong to same group but are stemmed to different roots. For example, "study," "studies" and "go," "went," consider the words "go" and "went" are belonging to same group, but Stemmer returns "go" and "went" as two different stem words due to deficiency in the similarity between the words.

In this paper, both designs of Unsupervised Stemmer and Rule-Based Lemmatizer for Kannada inflectional words are proposed. Developed Stemmer can be used in applications such as retrieval of information, search engines, domain analysis, document clustering, text categorization, and applications of text mining. Developed Lemmatizer can be used in applications like chatbots, tagging systems, entity relation modeling, morphological analyzer, etc.

2 Literature Survey

Stemming and lemmatization are two approaches to handle inflections in search queries. The literature shows that amount of work done on several Stemmer and Lemmatizer for Indian languages.

Mishra et al. [1] proposed a Stemmer for Hindi language called "MAULIK" using hybrid method which is the combination of brute-force and suffix removal approaches. Shambhavi et al. [2] proposed Kannada morphological analyzer and generator using trie which divides language into three categories, namely: (i) declinable words, (ii) conjugable words or verbs, and (iii) uninflected words based on which stemming of the word was carried out. Here, the author claims that rule-based with paradigm approach is used. Bhat [3] proposes a Statistical Stemmer for Kannada, different techniques like truncation of affixes, formation of clusters, and an unsupervised morpheme segmentation algorithm on sample text. Distance measure is carried out between the set of strings, and threshold distance is measured to achieve better performance. Padma and Prathibha [4] proposed morphological stemmer, analyzer, and generator for nouns in Kannada language using stripping of affix methods. The performance is tested against a set of nouns randomly picked from a Kannada dictionary named "Kannada Rathna Kosha." Padma and Prathibha [5] proposed a new technique for Kannada language, and it corrects lemma from the inflectional words. It provides accuracy of 85% from the four different Kannada datasets. Kasturi et al. [6] proposed Stemmer mainly for information retrieval systems which is independent of language. Using dynamic programming addresses unsupervised stemming is hybridized with partial lemmatization for four languages, namely English, French, Tamil, and Hindi which is morphologically different. Standard NLP techniques like Levenshtein distance and longest common subsequence are used for stemming process. 98.39% is the accuracy generated in terms of producing correct stems. Bharti et al. [7] proposed a Rule-Based Stemmer for Sindhi Devanagari scripts. Inflectional Stemmer Module uses affix stripping approach. Paul et al. [8] proposed a Rule-Based Lemmatizer for Hindi language using paradigm approach. Hindi is also highly inflected language like Kannada, so word structure should be studied in deep. The accuracy obtained was 89%. Deepamala et al. [9] proposed Kannada Stemmer and also briefed its effects on classification of Kannada document. Stemming is carried out on Kannada words using unsupervised suffix array approach, and effect of Stemmer on classification of Kannada documents is done with the help of Naive Bayes approach and comparison of maximum entropy methods. The performance was enhanced from 58 to 68% by creating of 18,804 stem words dictionary and suffix list. Thangarasu et al. [10] proposed Stemmer for Tamil language is the combination of unsupervised k-means clustering and rule-based approaches. Suffix list was created manually, and stemmed Tamil words are clustered with the help of clustering approach. The accuracy is 99% which is the highest accuracy recorded for regional-language Stemmer. This is one of the papers that motivated to develop Stemmer using k-means clustering approach.

An immense amount of work has been carried out, and different stemming procedures have been designed and implemented in various languages to enhance the efficiency of information retrieval (IR) especially in European languages than Indian languages.

3 Kannada Overview

Kannada language has 49 characters in its alphabet set or Aksharamala, and each of the letters in alphabet set is called Akshara. This alphabet set is further divided into 15 vowels and 34 consonants. Kannada script does not have uppercase and lowercase alphabet concepts unlike English alphabet set. Each of the consonants is combined with diacritic as shown in Fig. 1 to form a new letter termed Kagunita.

Conjunct consonant or subscript is combination of consonants which are also used within the consonant system referred as vathu in Kannada which is applied only on consonants and not on vowels shown in Fig. 2.

Building Stemmer and Lemmatizer for regional languages like Hindi, Kannada, Telugu, Tamil, etc., is difficult when compared to European languages because of the character complexity. In English language, having 26 alphabets and any kind of modification is not performed on these letters. Kannada alphabet set includes 49 characters, but diacritic is used on consonants to form Kagunita which implies in total 610 letters. So, complexity involved in forming words out of these 610 letters along with conjunct consonants and preprocessing of the words formed to collect the affixes from the words is a difficult task.

ಅ	ಆ	ಇ	ಈ	ಉ	ಊ	ಋ	ಎ	ಏ	ಐ	ಒ	ಓ	ಔ	ಅಂ	ಅಃ
	◌ಾ	◌ಿ	◌ೀ	◌ು	◌ೂ	◌ೃ	◌ೆ	◌ೇ	◌ೈ	◌ೊ	◌ೋ	◌ೌ	◌ಂ	◌ಃ
ಕ	ಕಾ	ಕಿ	ಕೀ	ಕು	ಕೂ	ಕೃ	ಕೆ	ಕೇ	ಕೈ	ಕೊ	ಕೋ	ಕೌ	ಕಂ	ಕಃ

Fig. 1 Formation of Kagunita

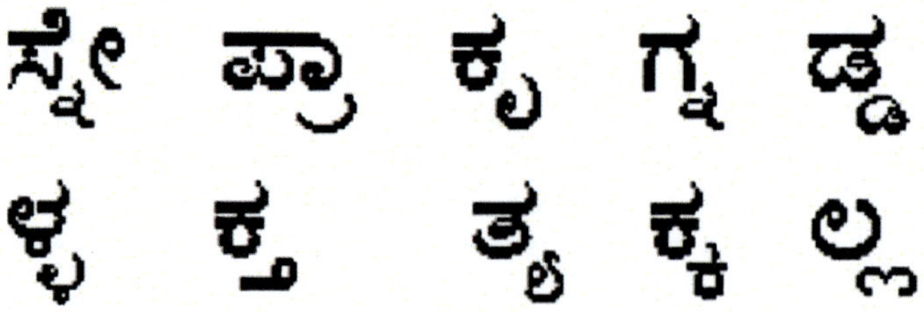

Fig. 2 Conjunct consonant

Fig. 3 Root_dic for Lemmatizer

ಅಂಕ
ಅಂಕಣ
ಅಂಕಾರ
ಅಂಕಿ
ಅಂಕಿತ
ಅಂಕೀಯ
ಅಂಕುಶ

4 Dataset Creation

Dataset is generated manually for development of Lemmatizer. "Kn_IN.dic" dictionary referred in [11] consists of 60,985 Kannada words. All the words listed in "Kn_IN.dic" dictionary are valid words, and no filtering is required. Hence, "Kn_IN.dic" dictionary is chosen over EMILLE corpus. The dictionary consists of combination of different words in root form or in deflected form. For example, ಅಂಕ,ಅಂಕಗಳನ್ನು,ಅಂಕಗಳಲ್ಲಿ,ಅಂಕಗಳಿಕೆ,ಅಂಕಗಳು,ಅಂಕಅಂಶಗಳನ್ನು, ಅಂಕಗಣಿತೀಯವಾಗಿ,ಅಂಕಣ,ಅಂಕಣಕಾರ.. By considering only root words from the Kn_IN.dic dictionary, dataset is created manually and stored in.xlsx file format as represented in Fig. 3. This dataset is a root dictionary which consists of 17,825 root words.

5 Implementation of Unsupervised Stemmer

Unsupervised Stemmer is built using unsupervised k-means clustering approach for clustering of input words based on the similarity in sequence of letters in input word. These are the steps followed to generate stem word for input Kannada document.

5.1 Data Wrangling

Data wrangling is used to clean the input Kannada document for further processing of data which contains around 3500 Kannada words, digits, and duplicates. It is carried in three steps:

1. Tokenization of input file: Tokenization is the process of splitting a sentence into words via unique space character. Input is a Kannada document in text file format. Each sentence in the input file is tokenized by removing punctuations

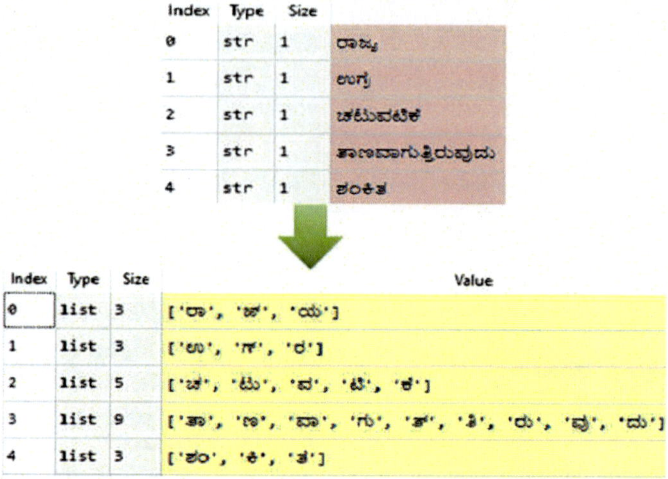

Fig. 4 Input after data wrangling

and storing each word of sentence as tokens in list. After tokenization, there are around 3500 words.

2. Removal of digits and duplicates from the input: In this step, duplicate words and digits present in list of tokens are removed which are not necessary to be stemmed by Stemmer.
3. Converting Kagunita format: In Kannada, Kagunita is a combination of consonants and diacritic as shown in Fig. 1. Diacritic is considered as different letters because each diacritic has its own Unicode. This might take a large amount of time for further processing and also forms a hurdle in forming the cluster. Hence, each Kagunita in the word is converted into a letter as shown in Fig. 4.

5.2 *Clusters*

Clustering of the input words is carried out with the help of unsupervised k-means clustering algorithm. There are around 610 letters in Kannada Varnamala. Before clustering, each letter in the Kannada Varnamala is given a unique number starting from digit 1. This numbering of letter is done because k-means clustering algorithm will operate only on numbers (Fig. 5).

Figure 6 represents the clusters formed out of words in tokens using k-means clustering algorithm.

k-means clustering algorithm for Kannada document

Input: List of Kannada words is given as input

Output: k number of clusters where k value is specified by Elbow method

Step 1: Specify the optimal k value generated from Elbow method.

Step 2: Determine the centroid from a given input.

Step 3: Find the distance of numbers assigned to Kannada words using Euclidean distance metrics.

Step 4: Grouping takes place based on minimum distance between the Kannada words in cluster.

Step 5: Repeat the process from Step 2, until centroid value is equal.

Fig. 5 Clustering algorithm for Stemmer

list	2	['ವಸ್ತ್ರೋಪಕರಣ', 'ವಸ್ತ್ರೋಪಕರಣವನ್ನು']
list	2	['ತಲೆಗೂದಲನ್ನು', 'ತಲೆ']
list	3	['ಅಡಿಯಲ್ಲಿ', 'ಅಡಿಗೆ', 'ಅಡಿಗೆಯ']
list	1	['ರಾಧಾ']
list	3	['ರಾಜಾಧಿಕಾರಿ', 'ರಾಜಾಧಿಕಾರಿಗಳಿಗೆ', 'ರಾಜಾ']

Fig. 6 Formation of clusters

5.3 String Comparison

String comparison is carried out on each cluster formed, and longest substring match will be returned as stem word. First column is cluster formed, and second column is the stem word generated for out of that cluster as shown in Fig. 7.

5.4 Elbow Method

Instead of assigning the k-value, i.e., the number of clusters to be formed manually with trial and error method, elbow method is used. The elbow method is a heuristic method used in data-driven models to truncate the number of parameters. There are

['ವಸ್ತ್ರಾಪಹರಣ', 'ವಸ್ತ್ರಾಪಹರಣವನ್ನು']	ವಸ್ತ್ರಾಪಹರಣ
['ತಲೆಗೂದಲನ್ನು', 'ತಲೆ']	ತಲೆ
['ಅಡಿಯಲ್ಲಿ', 'ಅಡಿಗೆ', 'ಅಡಿಗೆಯ']	ಅಡಿ
['ರಾಧಾ']	ರಾಧಾ
['ಠಾಣಾಧಿಕಾರಿ', 'ಠಾಣಾಧಿಕಾರಿಗಳಿಗೆ', 'ಠಾಣಾ']	ಠಾಣಾ

Fig. 7 Stem word generation

around 2518 unique input words. The range to find the optimal k-value starts with 1100 based on behavior observed on different numbers of unique input words from Kannada document. Graphs below are plotted to obtain optimal k-value where x-axis indicates different k-values and y-axis indicates distortion.

In Fig. 8 a, graph is plotted with the range 1100–1500 and can observe the curve is pretty straight after point marked. Have a close look at those points by varying the range in as show in Fig. 9.

From Fig. 10, it is observed that elbow point is near to the point 1220. So, trying with values from values between 1210 and 1217. Here, good accuracy for k-value 1215 is achieved; hence, optimal k-value is considered to be the same.

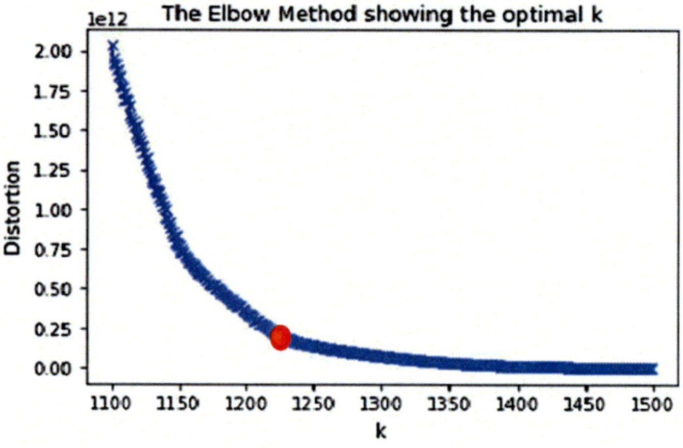

Fig. 8 Elbow method (range: 1100–1500)

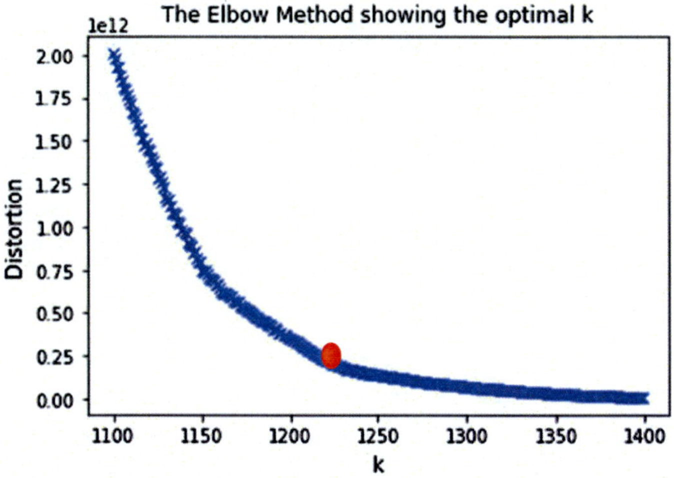

Fig. 9 Elbow method (range: 1100–1400)

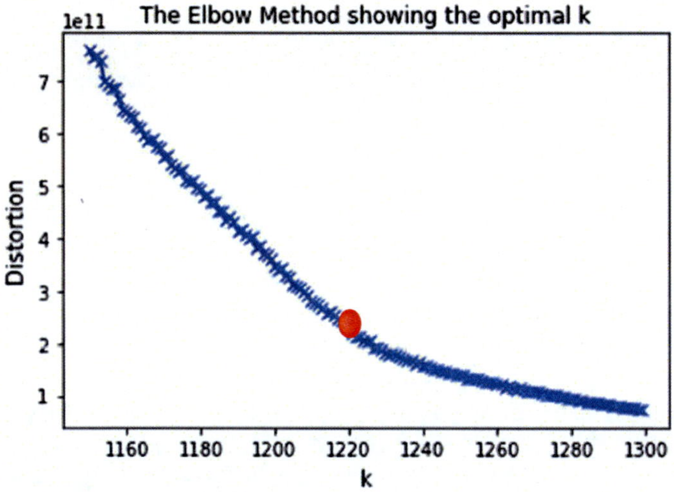

Fig. 10 Elbow method (range: 1160–1300)

6 Implementation of Rule-Based Lemmatizer

A Rule-Based Lemmatizer designed is capable of generating lemma by calculating the similarity between the words in the Root_dic represented in Fig. 1 and input file in Fig. 4.

['ಮ', 'ಹಾ', 'ಭಾ', 'ರ', 'ತ']
['ಜ', 'ಯ']
['ಎಂ', 'ಬ']
['ಗ್', 'ರಂ', 'ಥ', 'ದಿ','ಂ', 'ದ']
['ನಿ', 'ಷ್', 'ಪ', 'ನ್', 'ನ', 'ಪಾ', 'ಗಿ', 'ದ್', 'ದೆ','ಂ', 'ದು']
['ಕೆ', 'ಲ', 'ಪ', 'ರ']

Fig. 11 Final input after data wrangling

0	1	similarity
['ಅಂ', 'ಕ']	ಅಂಕ	0.142857
['ಅಂ', 'ಕ', 'ಣ']	ಅಂಕಣ	0.125
['ಅಂ', 'ಕಾ', 'ರ']	ಅಂಕಾರ	0.125
['ಅಂ', 'ಕಿ']	ಅಂಕಿ	0.142857
['ಅಂ', 'ಕಿ', 'ತ']	ಅಂಕಿತ	0.125
['ಅಂ', 'ಕೀ', 'ಯ']	ಅಂಕೀಯ	0.285714
['ಅಂ', 'ಕು', 'ಶ']	ಅಂಕುಶ	0.125
['ಅಂ', 'ಕೆ']	ಅಂಕೆ	0.142857

Fig. 12 Similarity generation

6.1 Data Wrangling

Data wrangling is carried out in the same way as in the Stemmer to convert raw data to required format for further processing. Input is a text file which is tokenized, duplicates and digits are removed, and tokens are stored in list after converting to Kagunita format as shown in Fig. 11. Root_dic contains 17,825 lemma words generated manually out of Kn_IN.dic dictionary, and each word is converted to its Kagunita format and stored in data frame named lemma as shown in first column of Fig. 12.

6.2 Generating Similarity

Each word from the final input referred in Fig. 11 is compared with words in the Root_dic letter by letter, and similarity is generated as shown in Fig. 12. Here, the second column of Root_dic is same as the first column but not converted to Kagunita form just to keep in readable form.

Similarity column in data frame will be cleared to generate new similarity for each input word in the list after calculating the similarity for an input word, and formulae of similarity calculation are as follows:

Similarity calculation =
(length of each word in Root_ dic AND with length of each word in Final input)/
(length of each word in Root_ dic OR with length of each word in Final input)

6.3 Algorithm

Input: Set of input words after data wrangling.

Output: Lemma for input words.

Begin

Step 1: For each input word in list, match first letter of the word with the first letter of Root_dic.

Step 2: If match occurs calculate the similarity.

Step 3: Find the highest similarity.

Step 4: Generate the word with highest similarity as lemma for the given input word.

Step 5: Repeat above steps for all the input words.

End.

The proposed Lemmatizer first checks for the match between the first character in the input word and first character of Root_dic; if there is a match, similarity will be calculated. In this way, the time taken is reduced to calculate similarity for all 17,825 words. The word with the highest similarity is the lemma word for the input word and is shown in Fig. 13.

7 Experimental Result Analysis and Discussion

7.1 Results of Unsupervised Stemmer

One of the Tamil Stemmer mentioned in [10] with high accuracy of 99% motivated us to build a Stemmer using unsupervised clustering approach, i.e., by clustering

word	lemma
['ಮ', 'ಹಾ', 'ಭಾ', 'ರ', 'ತ']	['ಮ', 'ಹಾ', 'ಭಾ', 'ರ', 'ತ']
['ಜ', 'ಯ']	['ಜ', 'ಯ']
['ಎಂ', 'ಬ']	['ಎಂ', 'ಬ']
['ಗ್', 'ರಂ', 'ಥ', 'ದಿ','ಂ', 'ದ']	['ಗ್', 'ರಂ', 'ಥ']
['ನಿ', 'ಷ್', 'ಪ', 'ನ್', 'ನ', 'ವಾ', 'ಗಿ', 'ದ್', 'ದೆ','ಂ', 'ದು']	['ನಿ', 'ಷ್', 'ಪ', 'ನ್', 'ನ']
['ಕೆ', 'ಲ', 'ವ', 'ರ']	['ಕೆ', 'ವ', 'ಲ']
['ಪ್', 'ರ', 'ತೀ', 'ತಿ']	['ಪ್', 'ರ', 'ತೀ', 'ತಿ']
['ಇ', 'ದ', 'ರ', 'ಲ್', 'ಲಿ']	['ಇ', 'ರ', 'ದ']
['ಉ', 'ಲ್', 'ಲೇ', 'ಖಿ', 'ಸಿ', 'ರು', 'ವ']	['ಉ', 'ಲ್', 'ಲೇ', 'ಖಿ', 'ತ']
['ಘ', 'ಟ', 'ನೆ', 'ಗ', 'ಳ']	['ಘ', 'ಟ', 'ನೆ']
['ನಿ', 'ಜ', 'ವಾ', 'ದ']	['ನಿ', 'ಜ']

Fig. 13 Lemma generation

the input words with the help of unsupervised k-means algorithm. Steps involved in accuracy calculation:

1. Input file with 3977 words was considered which was reduced to 2518 words after removal of duplicates, digits, etc.
2. Clusters were formed for these 2518 words with the help of k-means clustering algorithm, and the optimal k-value was 1215 from elbow method. The optimized k-value is as shown in Fig. 9.
3. Each of these 1215 clusters generated a stem word.
4. These 1215 stem words generated are matched with Kn_IN.dic dictionary for the calculation of accuracy:

$$\text{Accuracy} = \text{Number of Matched Words}/\text{Total Number of Input Words}$$

The number of matched words was 671 words out of 1215 input words. The accuracy generated by Unsupervised Stemmer when compared with meaningful Kannada words is 55.5%.

7.2 *Results of Rule-Based Lemmatizer*

To build Lemmatizer, a dataset of 17,825 root words is used which is manually generated and is referred as Root_dic once converted to required format. Lemma word is generated with the help of similarity calculation.

Steps involved in accuracy calculation:

1. Similarity between the word in Root_dic and input word is calculated.
2. Word with highest similarity is generated as lemma.
3. Once lemma is generated for all input words. Length of lemma will be always less than or equal to the length of actual input word. Hence, the difference operation is carried on list of lemma words generated and list of input words.
4. Only if the length of input words is zero after difference, then the number of correct lemma words will be incremented.
5. Accuracy is calculated using the formulae:

$$\text{Accuracy} = \text{number of Correct Lemma Words for Input Words/number of Input Words}$$

Accuracy of Lemmatizer is **73.48%**.

8 Conclusion and Future Work

8.1 Conclusion

Developing Stemmer and Lemmatizer for Kannada language is relatively demanding and prone to errors due to its structural complexity, increased character set with 610 letters, and lack of resources. An effort is made in this way by understanding formation of Kannada character set and the word structure. The accuracy obtained from the proposed methods is condensed mainly in case of Stemmer which is developed with unsupervised approach. This lag in accuracy is because operating directly on Kannada character set instead of Kannada words in English transliterated form.

The drawbacks of base papers have successfully overcome such as over-stemming which was the major drawback of [3], multiple comparisons with multiple datasets in case of Lemmatizer [5, 8], and Kannada words in English transliterated form [5].

8.2 Future Work

The results are found to be reasonable for the methods, rules, and the algorithms used. Developed Unsupervised Stemmer has very less over-stemming, since optimal k-value is used for clustering words. The accuracy is the only restriction faced in the proposed solutions which can be considered as future work. The accuracy of the Stemmer can be increased with list of prefixes and suffixes along with a set of rules. Developed Lemmatizer accuracy and speed can be enhanced by framing and adding few more rules.

References

1. Mishra, U., Prakash, C.: MAULIK: an effective stemmer for Hindi language. Int. J. Comput. Sci. Eng. **4**, 711717 (2012)
2. Shambhavi, B.R., Ramakanth Kumar, P., Srividya, K., Jyothi, B.J., Kundargi, S., Varsha Shastri G.: Kannada morphological analyser and generator using trie. IJCSNS Int. J. Comput. Sci. Network Security **11**(1), (2011)
3. Bhat, S.: Statistical stemmer for Kannada. In: International joint conference on natural language processing, Nagoya, Japan, pp. 25–33 (2013)
4. Padma, M.C., Prathibha, R.J.: Development of morphological stemmer, analyzer and generator for Kannada nouns. Lecture Notes in Electrical Engineering, (LNEE) Series, Springer India, vol. 248, pp. 713–723 (2014). ISSN 18761100
5. Prathibha, R.J., Padma, M.C.: Development of morphological analyzer for Kannada Verbs. IEEE Xplore Digital Library, pp. 22–27 (2013). ISBN-9781-84919-842-4
6. Kasthuri, M., Britto Ramesh Kumar, S., Khaddaj, S.: PLIS: Proposed Language Independent Stemmer for Information Retrieval Systems Using Dynamic Programming. Kingston University, UK (2017)
7. Nathani, B., Purohit, G.N., Joshi, N.: A Rule Based Light Weight Inflectional Stemmer for Sindhi Devanagari Using Affix Stripping Approach. Department of Computer Science Banasthali Vidyapith Banasthali India (2018)
8. Paul, S., Tandon, M., Joshi, N., Mathur, I.: Design of a rule based Hindi lemmatizer. © Computer Science and Information Technology, pp. 67–74 (2013)
9. Deepamala, N., Ramakanth, P.: Kannada stemmer and ıts effect on Kannada documents classification. Computational Intelligence in Data Mining, vol. 3, Smart Innovation, Systems and Technologies 33, © Springer India (2015). doi: https://doi.org/10.1007/978-81-322-2202-6_7
10. Thangarasu, M., Manavalan: Design and development of stemmer for Tamil language: cluster analysis. Int. J. Adv. Res. Comput. Sci. Softw. Eng. **3**, 812–818 (2013)
11. https://raw.githubusercontent.com/santhoshtr/silpa/master/src/silpa/modules/spellchecker/dicts/kn_IN.dic
12. Majumder, P., Mitra, M., Pauri, S.K., Kole, G., Mitra, P., Datta, K.: YASS: yet another suffix stripper. ACM Trans. Inf. Syst. **25**(4), 18–38 (2007)
13. Kasthuri, M., Ramesh Kumar, B.: An improved rule based iterative affix stripping stemmer for Tamil language using K-mean clustering. Int. J. Comput. Appl. **94**, 36–41 (2014)
14. Thangarasu, M., Manavalan: Stemmers for Tamil language: performance analysis. Int. J. Comput. Sci. Eng. Techonol. **4**, 902–908 (2013)

Election Tweets Prediction Using Enhanced Cart and Random Forest

Ambati Jahnavi, B. Dushyanth Reddy, Madhuri Kommineni, H. Anandakumar, and Bhavani Vasantha

Abstract Political system of a country has always been complicated in nature, and this complexity can be due to various factors such as number of parties, policies and, most notably, mixed public opinion. The advent of social media has given people around the world the ability to converse and discusses with a very wide audience; the sheer amount of attention gained from a tweet or a post is unimaginable. Recent advances in the area of profound learning have contributed to their use of many different verticals. Techniques such as long-term memory (LSTM) allow a sentiment analysis of the posts to be carried out. This can be used to determine the masses' overall feelings towards a political party or person. Several experiments have shown how to forecast public sentiment loosely by examining consumer behaviour in blogging sites and online social networks, such as in national elections. Machine learning has a rapid growth in recent years and has been applied from self-driving cars to e health sectors in every technology. A model of machine learning is proposed to predict the chances of winning the upcoming election based on consumer or supporter views on the web of social media. The supporter or user share their opinion or suggestions for the group or opposite group of their choice in social media. The text posts are needed to be collected about election and political campaigns; the models of machine learning are developed to predict the outcome.

Keywords Sentiment analysis · Decision tree · Random forest and logistic regression

---

A. Jahnavi (✉) · B. Dushyanth Reddy · M. Kommineni · B. Vasantha
Department of Computer Science and Engineering, Koneru Lakshmaiah Education Foundation, Vaddeswaram, AP, India
e-mail: jahnavi1499@gmail.com

M. Kommineni
e-mail: madhuri.cbit@gmail.com

H. Anandakumar
Department of Computer Science and Engineering, Sri Eshwar College of Engineering, Coimbatore, Tamil Nadu, India

© The Editor(s) (if applicable) and The Author(s), under exclusive license to Springer Nature Singapore Pte Ltd. 2021
A. P. Pandian et al. (eds.), *Proceedings of International Conference on Intelligent Computing, Information and Control Systems*, Advances in Intelligent Systems and Computing 1272, https://doi.org/10.1007/978-981-15-8443-5_29

1 Introduction

The online platform has become an enormous course for individuals to communicate their preferences, and there is a wealth of transparent data on assessment of electronic life. Using evaluation assessment, the ultimate intent of finishes can be found, for example, by eviscerating the content of the tendency, positive, negative or truthful. Assessment appraisal has been noteworthy for relationship to hear their client's insights on their things imagining eventual outcomes of races and getting ends from film ponders. The data snatched from feeling evaluation is helpful for affiliations picking future choices. Rather of associating individual terms, the relation between the set of words is considered. While selecting the general assumption, each word's ending is settled and united using a cap. Pack of words also ignores word demand, which prompts phrases with invalidation in them to be erroneously described. In the past scarcely any years, there has been a massive improvement in the use of little-scale blogging stages, for instance, Twitter. Nudged by that advancement, associations and media affiliations are continuously searching for ways to deal with burrow Twitter for information about what people ponder their things and organizations. Associations, for instance, Twitter, tweetfeel, and social mention are just an uncommon sorts of individuals who advance Tweet presumption examination as one of their organizations.

Although a significant proportion of work has been performed on how emotions are expressed in forms such as academic studies and news reports, significantly less study has been done on how findings are imparted given the easy-going language and message-length requirements of small scale blogging. Features, for instance, customized linguistic component marks and resources, for instance, idea vocabularies have exhibited accommodating for supposition examination in various spaces and anyway will they also show significant for evaluation assessment in Twitter? In this paper, this request is analysed.

2 Literature Survey

Notwithstanding the character goals on tweets, working out the concept of Twitter messages is basically close to the sentence level assumption evaluation; the welcoming and express language used in tweets, as well as the general idea of the local micro blogging allows Twitter's thinking evaluation an all-around undertaking. It's an open solicitation how well the highlights and procedures utilized on continuously well-shaped information will move to the micro-blogging space.

It involves measures such as data collection, pre-processing of documents, sensitivity identification and classification of emotions, training and model testing. This research subject has grown over the last decade with the output of models hitting approximately 85–90% [1].

In this paper, Vector Machine Support, Random Forest Support and Random Forest Support Vector Machine Algorithms (RFSVM) are compared, which are very suitable for the generation of rules on classification techniques. From the implementation results, it is said that the algorithm random forest support vector machine appears better than the other algorithms Amazon offers for product reviews [2].

Firstly, in this paper, they presented the method of sentiment analysis to identify highly unstructured data on Twitter. Second, they discussed in detail various techniques for carrying out an examination of sentiments on Twitter information [3].

They suggested a novel approach in this paper: Hybrid Topic Based Sentiment Analysis (HTBSA) for the task of predicting election by using tweets. They first extracted the most common latent topics using BTM over rich corpus, discussed them on Twitter, and then used pre-existing lexicons to find polarity of sentiment and score of each subject [4].

Using two separate versions of SentiWordNet and evaluating regression and classification models across tasks and datasets, offering a new state of the art method for sentiment analysis while computing the prior polarity of terms. Conclude their investigation finding interesting differences in the measured prior polarity scores when considering word part of speech and annotator gender [5].

They proposed a novel hybrid classification algorithm in this paper that explains the conventional method of predictive sentiment analysis. They also integrated qualitative analyses along with data mining techniques to make sentiment analysis method descriptive [6].

They chose to use two automated classification learning methods in this paper: Support Vector Machines (SVM) and Random Forest and incorporate a novel hybrid approach to classify Amazon's product reviews [7].

The proposed research aims to build a hybrid sentiment classification model that explores the basic features of the tweet and uses domain-independent and domain-related lexicons to provide a domain-oriented approach, thus analysing and extracting consumer sentiment towards popular smartphone brands in recent years [8]. Observed how the opinions are generated, so that overall decisions can be made from the sentiment analysis [9]. Observed the classification of opinions into text by using opinion mining and appraisal extraction [10]. They proposed that how tweets are belonging to different telecom companies using sentimental analysis [11]. It is observed that they have proposed a system by defining and classifying opinions or sentiments which are represented as electronic context [12].

3 Methodology

The following figure shows the steps which the proposed model follows (Fig. 1).

Decision Tree

As the implementation of machine learning algorithms to solve problems at the industry level grew, the need for more complex and iterative algorithms became

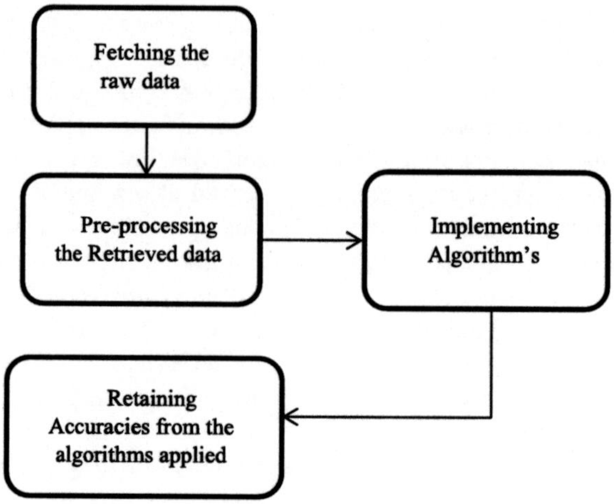

Fig. 1 Flow chart of steps and techniques utilized

a requirement. The decision tree algorithm is one such algorithm used to solve problems in both regression and classification.

Decision tree is considered one of the most useful algorithms in machine learning because it can be used to solve a number of problems. Here are a few reasons why decision tree should be used:

1. It is considered the most comprehensible machine learning algorithm and can easily be interpreted.
2. This can be used for problems with classification and regression.
3. It deals better with nonlinear data as opposed to most machine learning algorithms.
4. Building a decision tree is a very quick process since it uses only one function per node to divide the data.

Recursive partitioning is an important instrument in data mining. It lets us explore the structure of a collection of data while making decision rules simple to imagine for predicting a categorical (classification tree) or continuous (regression tree) outcome. This section explains the modelling of the CART, conditional inference trees (Fig. 2).

Random Forest

The random forest algorithm works by aggregating the predictions from different depths of multiple decision trees. Decision tree in the forest will be trained on a dataset subset called the bootstrapped dataset.

The portion of samples left out when constructing each decision tree in the forest is referred to as the Out-Of-Bag (OOB) dataset. As seen later, the model can automatically determine its own output by running each of the samples through the forest in the OOB dataset.

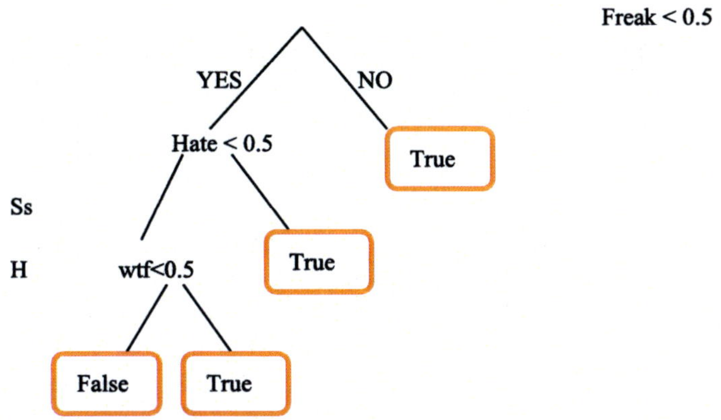

Fig. 2 A sample tree that appears after the implementation of cart algorithm

Remember how the impurity generated by each feature is measured using the Gini index or entropy when deciding on the criteria with which to split a decision tree. Nonetheless, randomly choose a predefined number of features as candidates in random forest. The above would result in a greater difference between the trees which would otherwise have the same characteristics.

If the random forest is used for classification, and a new sample is provided; the final prediction is made by taking most of the predictions produced in the forest by each individual decision tree. In the event, it is used for regression and a new sample is provided; the final prediction is made by taking the average of the predictions produced in the forest by each individual decision tree.

Logistic Regression

Key backslide is a genuine model which uses a determined ability to display a parallel subordinate variable in its main structure, but there are many dynamically complex developments. Determined backslide is surveying the parameters of a critical model in backslide analysis.

Numerically, a double-determined model has a dependent variable with two possible characteristics, for example, pass/bomb, which is represented by a marker variable called “0” and “1”. In the vital model, the log risks (the odds logarithm) for the value “1” tested is an immediate mix of one independent variable (“markers”) in any event the free factors can be either a double factor (two classes, coded by a pointer variable) or a constant variable (any authentic value). The relative probability of the value called “1” will shift from 0 (irrefutably the value “0”) to 1; from now on the denoting; the limit that changes to probability over log opportunities is the defined limit, hence the name.

The unit of estimation for the scale of the log chances is known as a logit, from a given unit, hence the elective names. Essentially proportionate to models with a different sigmoid limit as opposed to the defined limit, the probit model for example can be used similarly; the usual explanation for the main model is that increasing

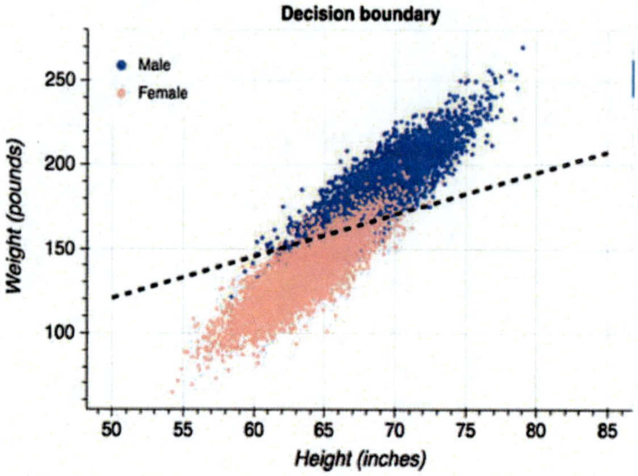

Fig. 3 Classification of the retrieved data using logistic regression, represented in a form of plot

one of the self-sufficient factors multiplies the odds of the result at a predictable rate, with each free factor having its own parameter; this summarizes the odds magnitude for a double bad variable.

The twofold determined backslide model has extensions to different degrees of the dependent variable: straight out yields with various characteristics are shown by multinomial vital backslide, and if the various classes are mentioned, by ordinal vital backslide, for example the comparing chances ordinal key model.

The model itself fundamentally models probability of yield similar to information, and doesn't perform verifiable gathering (it's definitely not a classifier), anyway it might be used to make a classifier, for instance, by picking a cutoff regard and orchestrating commitments with probability more noticeable than the cutoff as one class, underneath the cutoff as the other; this is a commonplace technique to make a two-fold classifier. The coefficients are overall not handled by a shut structure verbalization, not in the least-like straight least squares (Fig. 3).

4 Performance and Result Analysis

Considering that election outcomes are very difficult to predict using other methods, including public opinion polls, and with social media such as Facebook and Twitter increasingly prevalent, the authors chose to use Twitter's sentiment analysis to forecast Indian general election results (Table 1).

From the above table, the overall accuracies of the tweets using cart algorithm is 91.13%; random forest algorithm is 93.24%; and logistic regression is 90.56%. So, according to the results obtained random forest algorithm works better on election tweets data considered.

Table 1 Accuracies of algorithm's

Algorithm	Accuracy of tweets	
	Neg	Pos
Cart	0.8789	0.9437
Random forest	0.9155	0.9493
Logistic regression	0.8845	0.9268

5 Conclusion

To order to expand the dataset size, there may be several other prospective fields in which to perform this analysis, including data from other major social networking sites, such as Twitter and have more space to work with the training dataset, such as considering the model dataset which already specifies a certain number of algorithm features. Our research downside is that the emoticons are not recognized that are also significant when defining polarity of a tweet. Since the data was labelled manually, the volume was not high enough to have more precise information, so got more tweets and then marks them. In the future too, the size of net words will be increased.

References

1. Gupta, B., Negi, M., Vishwakarma, K., Rawat, G., Badhani, P.: Study of Twitter Sentiment Analysis using Machine Learning Algorithms
2. Umadevi, V.: Sentiment analysis using weka. IJETT Int. J. Eng. Trends Technol. **8**(4), 181–183 (2014)
3. Techniques for sentiment analysis of Twitter data: A comprehensive survey. 2016 International Conference on Computing, Communication and Automation (ICCCA)
4. Bansala, B., Srivastavaa, S.: On predicting elections with hybrid topic based sentiment analysis of tweets. Department of Applied Sciences, The NorthCap University, Gurugram, India
5. Guerini, M., Gatti, L., Turchi, M.: Sentiment Analysis: How to Derive Prior Polarities from SentiWordNet
6. Chen, X., Vorvoreanu, M., Madhavan, K.: Mining social media data to understand students' learning experiences. IEEE Trans. **7**(3), 246–259 (2014)
7. Al Amrani, Y., Lazaar, M., El Kadiri, K.E.: Random Forest and Support Vector Machine based Hybrid Approach to Sentiment Analysis. Author links open overlay panel
8. Venugopalan, M., Gupta, D.: Exploring Sentiment Analysis on Twitter Data, IEEE 2015
9. Malik, A., Habiba, S., Agarwala, P.: International Conference on Computational Intelligence and Data Science (ICCIDS 2018) A Novel Approach to Web-Based Review Analysis Using Opinion Mining. Department of Computer Science and Engineering, Jamia Hamdard, New Delhi 110062, India
10. Agarwal, A., Xie, B., Vovsha, I., Rambow, O., Passonneau, R.: Sentiment Analysis of Twitter Data
11. Alwagait, E.: Popularity analysis for Saudi Telecom companies based on Twitter data. Res. J. Appl. Sci. Eng. Technol. (2013)
12. Liu, B: Sentiment Analysis and Opinion Mining. Morgan & Claypool Publishers (2012)

Machine Learning Techniques as Mechanisms for Data Protection and Privacy

Amelec Viloria, Nelson Alberto, and John Rhenals Turriago

Abstract The exponential use of the Internet in the world has changed the way that data are collected, exchanged, and manipulated. Processing speeds, information volumes, and content relevance have changed rapidly to unimaginable levels. This is associated with the interest in creating personalized services based on the information available. Different organizations have data about people, representing a critical element, and it is estimated that up to 80% of all data stored in organizations can be classified as big data. The data stored by organizations usually comes from multiple sources (devices, people, organizations, among others), and these sources become the producers of the data. These data can be published or shared with other organizations or individuals, so the recipients of these publications become the consumers of the information. This research exposes problems in measuring the usefulness of data and relates them to the preservation of privacy in publishing. Machine learning models are developed to determine the risk of predicting sensitive attributes and as a means of verifying the utility of the data.

Keywords Machine learning techniques · Mechanisms for data protection · Privacy

1 Introduction

A data publication usually contains information from multiple sources, and many organizations use data mining to extract relevant knowledge [1]. From an individual

A. Viloria (✉) · J. R. Turriago
Universidad de la Costa, St. 58 #66, Barranquilla, Atlántico, Colombia
e-mail: aviloria7@cuc.edu.co

J. R. Turriago
e-mail: jrhenals2@cuc.edu.co

N. Alberto
Universidad Tecnológica, San Pedro Sula, Honduras
e-mail: nelson@unitet.edu

© The Editor(s) (if applicable) and The Author(s), under exclusive license to Springer Nature Singapore Pte Ltd. 2021

A. P. Pandian et al. (eds.), *Proceedings of International Conference on Intelligent Computing, Information and Control Systems*, Advances in Intelligent Systems and Computing 1272, https://doi.org/10.1007/978-981-15-8443-5_30

perspective, this scenario makes raise questions about how much knowledge can be gathered about a person's life, for example, about their current situation or location [2]. Once the data are published, the consequences for the owner or producer of the data can be very varied, exposing the individual to discrimination or public scorn, among others [3]. Laws and standards barely keep up with the potential for invasion of privacy. As a consequence of these regulations, the Preservation of Privacy in Data to be Published (PPDP) has become an area of interest for researchers and practitioners. The PPDP assumes that those who attempt to discover confidential information about individuals [4, 5] may be among the recipients of the data. Therefore, the goal of PPDP techniques is to modify data by making them less specific so that the privacy of the data owners is protected while maintaining the usefulness of the data being processed. Preserving data privacy requires consideration of the balance between respecting the privacy wishes or preferences of multiple data owners, whether or not the inference is authorized [6], and the possibility of reidentifying each owner or linking sensitive information about the published dataset [7]. Regulations such as the general data protection regulation define that tasks must be limited, for example, by requesting informed consent of all possible use scenarios for the data collected. It is important to note that this is extremely difficult, for example, in open data initiatives; it is almost impossible to identify all recipients and their possible uses of the data [8].

This paper proposes the study of [9], where a multidimensional variant known as the Mondrian algorithm [10, 11] is applied in a practical way and a methodology for evaluating the usefulness of information, and the resulting preservation of information privacy is analyzed

2 Related Studies

The selection of an appropriate algorithm to protect privacy when disseminating data is a general concern for the preservation of the privacy of data to be published or PPDP. As a result, comparing multiple privacy preservation algorithms from the perspective of data utility and effectiveness in preserving privacy represents important research work. Some authors argue that the effectiveness of privacy preservation is best assessed by the utility it provides to the target applications and that metrics may behave differently with different privacy preservation algorithms [12]. In addition, an evaluation framework is presented in [13] to estimate and compare different types of privacy preservation algorithms specifically in data mining. Although these studies are closely related to PPDP, the data mining tasks considered are closely related to the proposed solutions.

Privacy preservation metrics help improve the privacy of data owners [14]. Some studies [15, 16] have conducted or analyzed such experiments, although in general practice, some specific metrics are used [17].

This approach is not addressed when analyzing privacy metrics and models. This underpins the need for a privacy preservation assessment methodology such as the one presented in this paper.

3 Experimental Evaluation

The experimental evaluation is intended to provide a methodology for assessing utility and preserving privacy. The steps of an experimental protocol to evaluate a privacy preservation algorithm with respect to a dataset are described, and a case study is developed to compare the results of several machine learning models created from publications built with different levels of privacy and different configurations of quasi-identifiers.

3.1 Methodology

One way to evaluate the quality of privacy preservation is to create a predictive model using the anonymized data and then evaluates the model for the accuracy of the prediction of both the sensitive attribute and the attribute of interest to the study. To achieve this purpose, the following methodology for assessing utility and privacy preservation is presented [2, 6, 18]:

1. Pre-processing of the data.
2. Categorization of data into identifying, quasi-identifying, sensitive, and non-sensitive attributes.
3. Definition of privacy preservation levels and identification of quasi-identifiers to be used.
4. Anonymization of data by means of the privacy preservation algorithm (Mondrian basic).
5. Dataset construction comprising anonymized data, non-sensitive data, interest attribute, and sensitive attribute.
6. Pre-processing and recoding of the generated dataset to be used in the machine learning algorithm.
7. Categorization of attributes as predictive attributes and target attributes
8. Segmentation of the set into training and test data.
9. Generation of the prediction model by means of the selected machine learning algorithm.
10. Evaluation of the prediction accuracy for the generated model.

This proposed methodology for assessing data usefulness and preserving data privacy is used in a case study to demonstrate its usability. The case study developed maintains the interest attribute and the sensitive attribute in the data to be published during the privacy preservation phases and during the creation of the machine learning model.

The basic Mondrian algorithm implemented in the Python language is used as a privacy preservation algorithm, as well as implementations of the machine learning algorithms provided by the "scikitlearn" software package [19], specifically, decision trees, random forests, and logistic regression. The test dataset is the AdultDatabase [20], which contains 34589 rows with 20 attributes, after pre-processing 32541 rows were left. The sensitive attribute (not quasi-identifier) was defined as: "race", the attribute of interest (not quasi-identifier) as: "income" and the non-sensitive attribute were defined as: "education" (to ensure a constant degree of usefulness, which is equivalent to the "education-num" attribute). The other attributes are used as possible quasi-identifiers. In the experiments, the size of the set of quasi-identifying attributes was varied in a range of 1, 6, and 14 [21]. When the number of quasi-identifier attributes is modified, the attribute that is no longer quasi-identifier automatically becomes non-sensitive in order to keep the same number of entries in the machine learning algorithms. Generalization hierarchies for categorical attributes are available in GitHub open-source developments. Also, the privacy level k was defined as the range of values: 2, 5, 10, 25, 50, and 100.

3.2 *Analysis of Results*

The case study on the proposed methodology includes assessing the usefulness of the data and the preservation of the privacy of the data to be published. For this purpose, prediction models were created using the selected machine learning algorithms.

There are two types of models, those that predict the variable of interest to check the usefulness of the data and those that predict the sensitive variable to check the preservation of privacy. As can be seen in Fig. 1, decision trees present a loss in accuracy in the utility of the data (interest variable prediction) as the level of privacy k increases with twelve quasi-identifiers. This behavior is normal considering that the data are generalized from one of the equivalence classes. What is not consistent is that the loss of accuracy does not continue to increase. In general, the accuracy of the prediction of the decision tree does not suffer variations as the level of privacy is changed.

The prediction accuracy behavior of models created with the random forest algorithm is presented in Fig. 2. It is observed that the variation in accuracy for measuring the usefulness of the data is similar to that of models created with the decision tree algorithm. The models created with the logistic regression algorithm share similarity with the previously evaluated models. Figure 3 shows that all models created have an accuracy above 75%, which is consistent with the results of the previously evaluated models and allows to theorize that the data is still useful. At the same time, it can be stated that the preservation of privacy is very low, since neither the number of quasi-identifiers used nor the level of privacy preservation seems to influence the result of the prediction for the sensitive data.

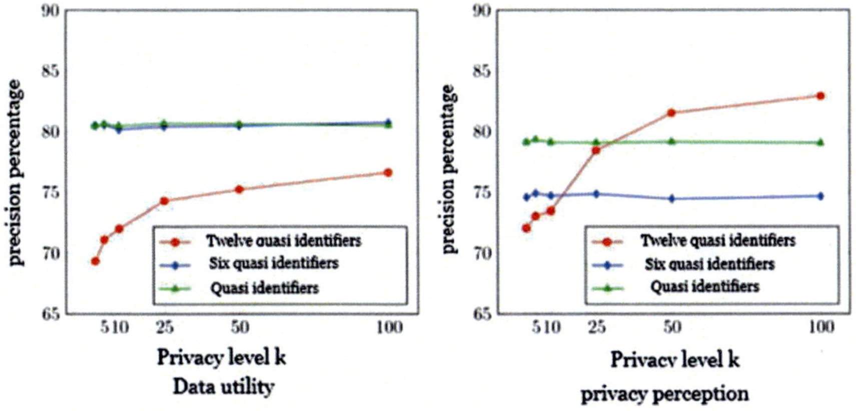

Fig. 1 Evaluation of models created with decision trees

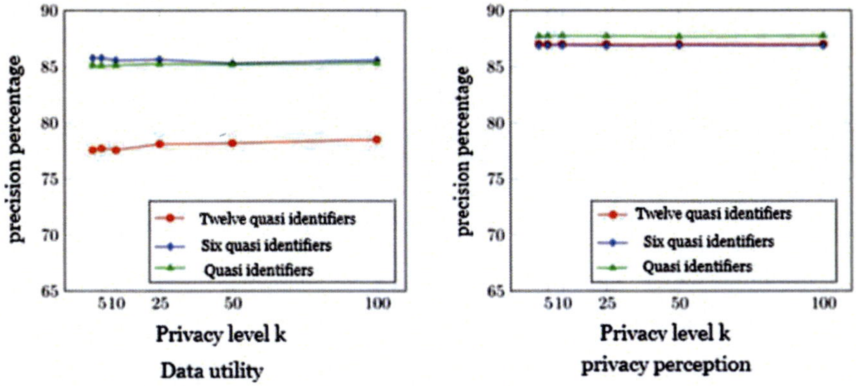

Fig. 2 Evaluation of models created with random forests

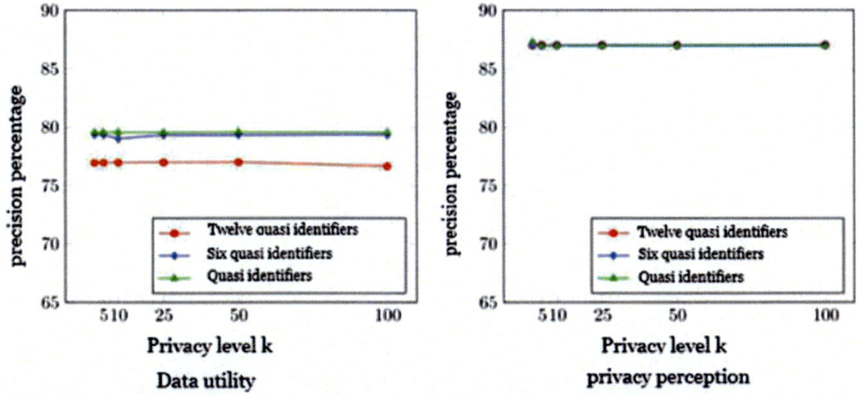

Fig. 3 Evaluation of models created with logistic regression

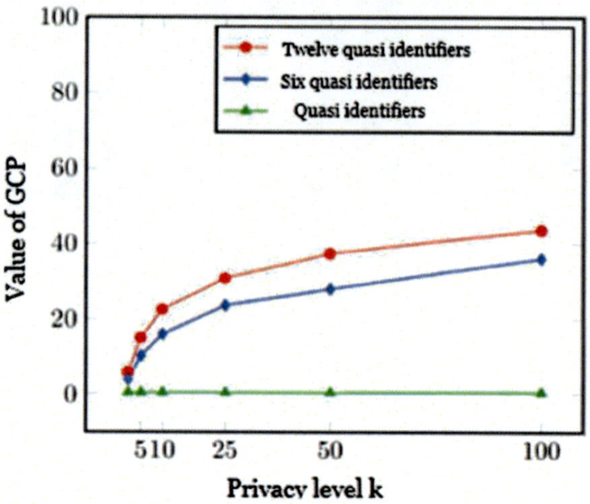

Fig. 4 Relationship between the privacy level k, the number of quasi-identifiers, and the value obtained by the GCP metrics

The GCP metrics was evaluated for comparing the results obtained by the machine learning models. The results of the GCP metrics in Fig. 4 indicate that the loss of information can reach 40%, and this result is not consistent with the results of the models created with the machine learning algorithms.

The GCP metrics reports an increase in the loss of information when modifying the k privacy level, a situation that is not reflected in the predictive accuracy of the machine learning algorithms. It is likely that the reduction of attribute domains influenced the preservation of predictive accuracy.

4 Conclusions

The usefulness of the data is a fundamental issue for motivating the use of privacy preservation techniques, but at the same time, it may be the key to respecting the wishes or rights of the data owners. Privacy preservation algorithms for data to be published are widely studied in the quest to protect privacy, but it is difficult to indicate which data are more at risk of re-identification. This paper provides indications of the existence of difficulties in establishing good parameters in privacy algorithms such as Mondrian basic. The experimental study generates evidence about the low effectiveness of GCP metrics with respect to the utility of anonymized data in machine learning models that make predictions. Furthermore, the low effectiveness of GCP is present both in the prediction of the attribute of interest (measuring utility) and in the sensitive attribute (measuring privacy preservation).

References

1. Gupta, R., Anand, P., Chaudhury, S., Lall, B., Singh, S.: Compressive Sensing-Based Privacy for Fall Detection (2020). arXiv preprint arXiv:2001.03463
2. Bhave, D.P., Teo, L.H., Dalal, R.S.: Privacy at work: a review and a research agenda for a contested terrain. J. Manag. **46**(1), 127–164 (2020)
3. Shateri, M., Messina, F., Piantanida, P., Labeau, F.: Privacy-Cost Management in Smart Meters Using Deep Reinforcement Learning (2020). arXiv preprint arXiv:2003.04946
4. Bernabe, J.B., David, M., Moreno, R.T., Cordero, J.P., Bahloul, S., Skarmeta, A.: Aries: evaluation of a reliable and privacy-preserving European identity management framework. Futur. Gener. Comput. Syst. **102**, 409–425 (2020)
5. Silva, J., Solano, D., Fernandez, C., Romero, L., Villa, J.V.: Privacy preserving, protection of personal data, and big data: a review of the Colombia case. Procedia Comput. Sci. **151**, 1213–1218 (2019). https://doi.org/10.1016/j.procs.2019.04.174
6. Alraja, M.N., Farooque, M.M.J., Khashab, B.: The effect of security, privacy, familiarity, and trust on users' attitudes toward the use of the IoT-based healthcare: the mediation role of risk perception. IEEE Access **7**, 111341–111354 (2019)
7. Beaulieu-Jones, B.K., Wu, Z.S., Williams, C., Lee, R., Bhavnani, S.P., Byrd, J.B., Greene, C.S.: Privacy-preserving generative deep neural networks support clinical data sharing. Circ. Cardiovasc. Qual. Outcomes **12**(7), e005122 (2019)
8. Wang, Y., Herrando, C.: Does privacy assurance on social commerce sites matter to millennials? Int. J. Inf. Manage. **44**, 164–177 (2019)
9. Sweeney, L.: Achieving k-anonymity privacy protection using generalization and suppression. Int. J. Uncertain. Fuzziness Knowl. Based Syst. **10**(05), 571–588 (2002). https://doi.org/10.1142/S021848850200165X
10. LeFevre, K., DeWitt, D.J., Ramakrishnan, R.: Mondrian multidimensional K-anonymity. In: Proceedings—International Conference on Data Engineering, vol. 2006, p. 25. IEEE (2006a)
11. Le Fevre, K., De Witt, D.J., Ramakrishnan, R.: Workload-aware anonymization. In: Proceedings of the 12th ACM SIGKDD International Conference on Knowledge Discovery and Data Mining—KDD '06, p. 277. ACM Press, New York, USA (2006b). https://doi.org/10.1145/1150402.1150435
12. Bhatia, J., Breaux, T.D.: Empirical measurement of perceived privacy risk. ACM Trans. Comput.-Human Interact. (TOCHI) **25**(6), 1–47 (2018)
13. Duraipandian, M.: Performance evaluation of routing algorithm for MANET based on the machine learning techniques. J. Trends Comput. Sci. Smart Technol. (TCSST) **1**(01), 25–38 (2019)
14. Thuraisingham, B., Kantarcioglu, M., Bertino, E., Clifton, C.: Towards a Framework for Developing Cyber Privacy Metrics: A Vision Paper. In: 2017 IEEE International Congress on Big Data (BigData Congress), pp. 256–265. IEEE (2017, June)
15. Korshunov, P., Melle, A., Dugelay, J.L., Ebrahimi, T.: Framework for objective evaluation of privacy filters. In: Applications of Digital Image Processing XXXVI, vol. 8856, p. 88560T. International Society for Optics and Photonics (2013)
16. Silva, J., Cubillos, J., Villa, J. V., Romero, L., Solano, D., Fernández, C.: Preservation of confidential information privacy and association rule hiding for data mining: a bibliometric review. Procedia Comput. Sci. **151**, 1219–1224 (2019). https://doi.org/10.1016/j.procs.2019.04.175
17. Al-Asmari, H.A., Saleh, M.S.: A conceptual framework for measuring personal privacy risks in Facebook online social network. In: 2019 International Conference on Computer and Information Sciences (ICCIS). pp. 1–6. IEEE (2019, April)
18. Li, X., Yang, Y., Chen, Y., Niu, X.: A privacy measurement framework for multiple online social networks against social identity linkage. Appl. Sci. **8**(10), 1790 (2018)
19. Karapiperis, D., Gkoulalas-Divanis, A., Verykios, V.S.: FEDERAL: a framework for distance-aware privacy-preserving record linkage. IEEE Trans. Knowl. Data Eng. **30**(2), 292–304 (2017)

20. El Makkaoui, K., Ezzati, A., Beni-Hssane, A., Motamed, C.: Cloud security and privacy model for providing secure cloud services. In: 2016 2nd International Conference on Cloud Computing Technologies and Applications (CloudTech), pp. 81–86. IEEE (2016, May)
21. Lis-Gutiérrez, J.P., Zerda Sarmiento, Á., Viloria, A.: Intellectual property in Colombian Museums: an application of machine learning. In: Lecture Notes in Computer Science (including subseries Lecture Notes in Artificial Intelligence and Lecture Notes in Bioinformatics), vol. 11936, LNCS, pp. 289–301. Springer (2019). https://doi.org/10.1007/978-3-030-36204-1_24

Malware Identification and Classification by Imagining Executable

Rupali Komatwar and Manesh Kokare

Abstract Malware classification is a significant criterion while designing any operating system. Due to the recent outbreak of malware like Wannacry and Petya, world security is at stake. This malware usually affects Windows-based operating systems due to the open security model used by the OS. This paper proposed a technique to identify malware based on their EXEcutable file signature. Presented work of this paper based on pattern analysis of malware by converting executables to bitmaps and then perform complex feature analysis, to evaluate the type of malware present in the Exe file.

Keywords Wannacry · Malware · Bitmap · Feature analysis · Patterns

1 Introduction

Malware termed malicious software, which destroys, corrupts, or disrupts the normal functioning of other software. Malware classification and analysis have been an area of research for decades now, but the evolution of malware has both slowed down and inspired the innovation in malware research. Corporate companies spend more than 20% of their revenues on malware analysis and find out ways to keep the malware out of their systems. But, the recent outbreaks of Wannacry and Petya have demonstrated that much of this research has proven unsuccessful. So it needs much improvement in terms of accuracy and speed of malware classification, especially in Windows-based OSes.

Malware analysis is usually carried out by using static and dynamic approaches. The static approach majorly deals with the bit patterns of the malware, and then based

R. Komatwar (✉)
Government Polytechnic, Mumbai, Maharashtra, India
e-mail: rupalikomatwar@gmail.com

M. Kokare
S.G.G.S. Institute of Engineering & Technology, Nanded, Maharashtra, India
e-mail: mbkokare@sggs.ac.in

© The Editor(s) (if applicable) and The Author(s), under exclusive license to Springer Nature Singapore Pte Ltd. 2021

A. P. Pandian et al. (eds.), *Proceedings of International Conference on Intelligent Computing, Information and Control Systems*, Advances in Intelligent Systems and Computing 1272, https://doi.org/10.1007/978-981-15-8443-5_31

on some linear classification technique, maps these bit patterns to malware, which in turn creates a classification engine for detecting trained and non-blind testing malware signatures.

Dynamic malware analysis has performed in various ways but is not limited to, behavior analysis, pattern analysis, execution analysis, and among others [1]. From the existing literature given in [1], observed that dynamic malware analysis is superior in terms of malware analysis and recognition than static analysis.

Our approach of detecting malware comes under the pattern analysis version of the dynamic malware analysis domain. Our approach focuses on converting the malware executables into images and then performing pattern analysis on these images to classify the malware into one of many categories. Executable file to image conversion is done by reading the bytes of the Exe file, and then arranging it in a 2D row, column format. This method gives a visualization of the byte level content variations in the executable files. After that, a complex set of feature extraction techniques applied to the image, which assists in finding out the color and shape variation of the malware structure. These variations are then given to a linear KNN classifier to classify the malware into one of k number of categories.

2 Literature Review

Prior, much research is carried on the classification and clustering of malware. The classification split toward three types: static analysis, dynamic analysis, and hybrid analysis. Schultz et al. [2] used static components of program executable metadata to start with the static analysis. The system utilized three distinct elements—DLL implied and functions alluded, strings, and byte clusters. A proposed method utilized machine learning computation, such as Naïve Bayes and tried strategy upon an informational index comprised of 4266 documents (3265 malware and 1001 clean records), where the accomplished classification correctness about 97.11%.

In [3], the author exhibited a characterization technique on function length and recurrence of function length. Several bytes of code present within the methods used to calculate the length of a function. To extricating above-mentioned qualities of unloaded code, IDA [4] employed. The method used k-overlay cross approval to computing closeness among two examples and accomplished 88% average precision.

The system presented in [5] is a unique approach to deriving mutable length instruction series, which can recognize worms from clean applications by applying data mining techniques. The investigation was promoted on the application control movement learning carried within the instruction series. Based on comprehensive statistics collected through those instruction series, the system expressed the difficulty of a binary classification issue and developed tree-based classifiers, like bagging, decision tree, and random forest. This strategy conferred a 95.6% exposure scale toward new worms whose information was not practiced during the system development process.

Kolmogorov complexity and program executable file components are used for building a non-cryptographic hash for the clustering of malware [6]. Wicherski G. utilized program executable's essential attributes like Section Characteristics, Subsystem, Heap Commit Size with Virtual Address, Image Characteristics, Stack Commit Size, Raw Size, utilized for estimate Kolmogorov Complexity of every area in PE document. Bzip2 pressure proportion utilized being an estimate of Kolmogorov complexity [7]. The technique carried about three informational indexes carrying 184,538, 90,105, and 322 examples gathered of mwcollect Alliance Database, Arbor Networks, and Windows XP establishment separately. The system has 0 false positives toward malware documents and seven false positives toward records gathered from the Windows XP establishment.

The second category of malware classification introduced is dynamic analysis to overwhelm the weaknesses of static analysis. Biley et al. [8] propose newly classification method that represents malware behavior in terms of operation status variations (e.g., files written, processes designed) preferably than in series or models of system calls. To direct the altogether quantity of malware and heterogeneity of its performance, the author presents a system for automatically classifying those outlines of malware in different categories that show the same sets of functions and describe whereby behavior-based clustering gives a further immediate and powerful style of analyzing and classifying Internet malware. For classifying malware, a single pairwise linkage various leveled calculation is utilized with normalized compression distance (NCD) being a separation metric. One constraint of the work above is that a few malware does not indicate absolute behavior/power in virtual machine.

A clustering strategy utilizes behavior profiles, presented by Bayer et al. [9]. Malware is administered in an observed condition to gather behavior outlines. Specific outlines speak to malware conduct regarding operating systems (OS) articles and OS services. During clustering, the system utilized local-sensitive hashing (LSH). All could bunch 75,000 examples in under 3 h.

Firdausi et al. [10] introduced a record of the idea of malware classification in light of malware behavior. Malware was administered under a sandbox situation termed Anubis [11]. Meager vector forms made to every Anubis record. The technique utilized 5 distinctive classification calculations, such as support vector machine (SVM), J48 Decision Tree, naive Bayes, k-Nearest Neighbors (KNN). The system accomplished 93.6% accuracy by J48.

In [12], the technique introduced a malware arrangement in light of behavior charts utilizing maximal regular subgraph identification. A behavior chart produced by observing framework calls alongside parameter esteems and the performance of an unusual specimen in an observed condition. Likeness estimation about two dynamic framework call reliance (DSCD) charts figured utilizing maximal normal subgraph (MCS). All assessed the offered technique upon 300 unique specimens and could achieve 95.7% test accuracy.

Finally, the hybrid analysis is also applied in multiple applications. Cesare et al. [13] introduced a Malwise framework that utilizes both dynamic and static examination methods. At first, entropy investigation was performed to decide whether a given record pressed or not if stuffed at that point of concealed code, which uncovered utilizing dynamic examination. A static investigation was conducted at unloaded code toward creating marks to measure stream charts in every strategy. CRC64 was utilized to create a hash of string marks. The technique assessed the framework about 15,000 genuine malware, and 86% of the specimens could classify under 1.3 s.

Santos et al. [14] exhibited OPEM, a hybrid of obscure malware indicator that utilization recurrence of opcodes and execution hint of executable. The malware primarily dismantled to generate opcodes, and its frequency figured utilizing phase frequency. Malware is executed by the sandbox for extracting the mentioned resources. The above technique assessed toward 1000 vindictive and 1000 clean documents were utilizing decision tree, support vector machine, K-closest neighbor, Bayesian system.

Edward [15] introduced a network architecture to actively treat a raw byte sequence of above 2 Million tracks and train such a model successfully. This system learns a broader breadth of information varieties correlated to other domain-knowledge open methods to malware apprehension. The presented method further highlights a breakdown situation for batch-normalization, which initially executed this system unable to learn.

Venkatraman [16] exhibited the novel design, and design's principal goal is to cope with packed and polymorphic alterations and metamorphic obfuscations of malware for delivering competent solutions to direct the zero-day malware exposure difficulty. The method utilizes the information of standard behavior designs of the application programming interface (API). It examines patterns of obfuscated code more on using different data mining methods of extracted features and statistical n-gram opcode analysis, including innovative methods to distinguish whether the binary content is malicious or benign. The information of API function calls features simultaneously among multiple distance measures of vector designs are applied in a visual way to detect obfuscated malware families.

Donahue et al. [17] put two visualization systems executed toward a portable executable file. One method provides a navigational portable executable file structure, through improving the ability of a hex editor. The second method uses the Markov Byte Plot to identify packed parts of portable executable files while attempting to reverse engineer it.

Kumar et al. [18] introduced a machine learning-based resolution to classify a unit as malware or benign by great accuracy and fewer estimated expenses. A combined feature set has combined as a compound of portable executable header domains raw value and obtained values. The experimental results show that for the integrated feature set 89.23% accuracy is achieved on the novel test database.

In 2011, Nataraj et al. [19] offered a novel and a distinct strategy to visualize and analyze the malware using the image processing concept. Malware is executable visualized as binary images, and by observing the significant visual similarities in image textures, the system classifies the malware in different families. This technique overcomes the number of limitations of static and dynamic analysis.

In [20], a hybrid customized convolutional neural network (CCNN) is proposed with a K-nearest neighbor (KNN) classification scheme. CNN is executed with a leaky ReLu activation function, and then KNN classifier is used to classify the different malware. An investigation result proves that the CNNN-KNN scheme classifies malware varieties into their particular families with 99.35% accuracy. Thus, the offered hybrid scheme outlines the most comprehensive malware classification systems.

3 Proposed Approach

The tremendous increase in the amount of malware is due to mutation. A mutation is a technique of creating new variants of malware from existing ones. Hackers perform a small change in existing malware to create new variants. Only images can capture the small changes from the malware binary yet preserve the global structure. Observations of different malware families from [21] conclude that malware belongs to one family with the same visual features, and malware belongs to two different families with various visual features. Motivated by this observation, this paper is introducing a novel approach for analyzing and classifying malware using image processing.

The proposed malware detection approach divided into the following research components,

- Conversion of executables into bitmaps
- Complex feature extraction from bitmaps
- Observing difference in bitmap features between standard executables and malware executables
- Application of linear classification for categorizing malware.

The first step of the malware identification and classification technique is converting the malware executable into the bitmap as shown in Fig. 1.

Conversion of executables into bitmaps done using the following procedure,

1. Select a max executable file size, say 192 kB, make sure the file size is divisible into symmetric or non-symmetric image size of M × N × 3, where M is the number of rows in the image, N is the number of columns in the image, and 3

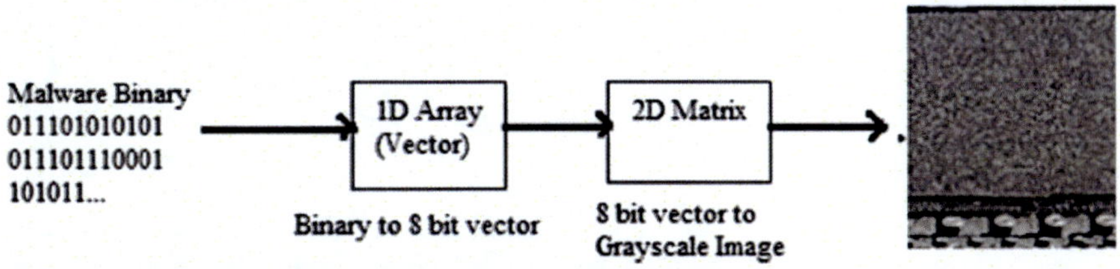

Fig. 1 Visualization of malware images

is the number of color components (namely RGB) in the image. In our case of 192 kB, the image size is 256 × 256 × 3, which denotes the asymmetric image.
2. Create an empty array of M × N × 3 sizes, and initialize it with all zeros (initializing with zeros is important, as the feature extraction technique skips zeros while extracting features).
3. Read all the bytes (8-bit format) from the executable file in binary mode and store them sequentially in the image. Start by filling the columns of the first row, and go to the kth row, where k will vary based on the size of the input malware image.

Visualization as grayscale images with a detailed taxonomy of different primitive binary fragments can found in [22]. Malware images belong to different malware families are obtained from [21].

Here are some examples of exe to bitmap conversion: first one is a non-malware file (Fig. 2 Notepad executable), and second is a malware file (Fig. 3 Ammyy Admin executable).

The side-by-side comparison of these two executables demonstrates that the notepad executable (Fig. 2) has symmetric image patterns, while the ammyy executable (Fig. 3) has non-symmetrical patterns and does not repeat any pattern in the entire executable bitmap. Upon exe to bitmap conversion, a complex set of features is evaluated from the bitmap. These features include a color map and an extended edge map of the bitmaps.

Color Map Detection: The color map denotes the variation of color levels in the image, which is evaluated by finding out a quantization table of 128 bit, finding out the number of pixels of each quantization level, and then plotting them against the

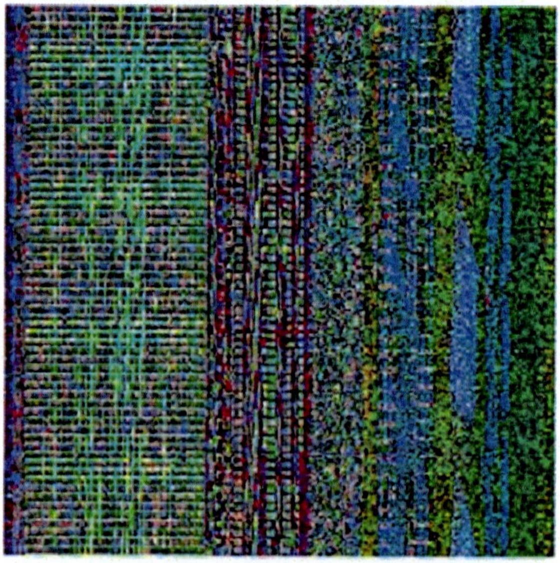

Fig. 2 Notepad executable

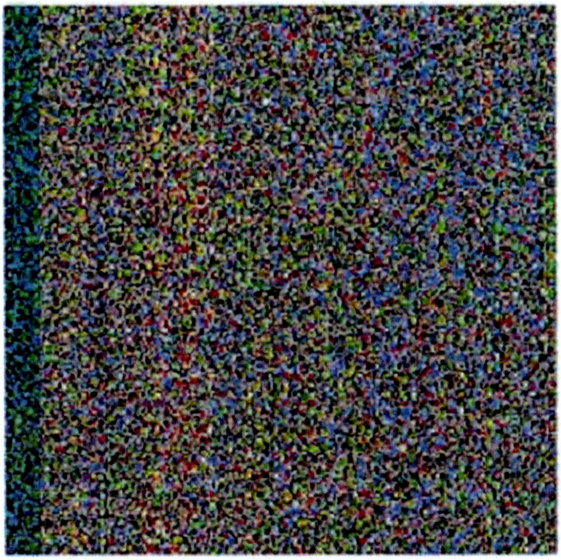

Fig. 3 Ammyy Admin executable

quantization levels themselves. It is similar to a histogram but denotes quantized color levels instead of grayscale levels. The color composition histogram is built under the HMMD (hue-min-max-difference) color space to ensure interoperability. The primary spectral element color is described by the hue (H) in its classic appearance, being into red, green, or yellow. The minimum and maximum between the values of R, G, B are represented by min and max element, respectively. The difference between max and min is specified by different elements. The conversion from a bitmap (the values of every spectral element scale of 0–1) toward the different HMMD color spaces is shown in Algorithm 1.

Algorithm 1: Conversion of bitmaps to HMMD color space

```
Max = max (R, G, B);
Min = min (R, G, B);
Diff = Max – Min
Sum = (max + min) /2
if (Max == Min) Hue is undefined (achromatic color)
otherwise;
if (Max == R && G > B) Hue = 60 * (G –B) / (Max – Min)
else if( Max == R && G < B) Hue = 360 * 60 * (G – B) / (Max - Min)
else if (G = Max) Hue = 60 * (2.0 +(B – R) / (Max - Min ))
            else Hue =60 * (4.0 + (R – G) / (Max –Min))
```

If the color map represents the color variation, then a feature vector is also needed which can denote the change of shape of the bitmap. This change of shape is represented using the edge map of the bitmap. In the case of the edge map, first evaluate the probability of occurrence of the edge at each pixel, and then plot it against the pixel intensity, thereby obtaining a map which denotes the probability of edge occurrence of edges in the bitmap.

Edge Map Detection: The spatial distribution of edges captured by the edge histogram descriptor. The edges distribution is a vital texture feature that is helpful during an image matching, yet the underlying texture is heterogeneous. The steps for creating an edge map are as followed:

Step 1: The bitmap partitioned in 16 sub-bitmaps effects in 64 bins.
Step 2: Initialize the edge detection mask as follows:

$$\begin{bmatrix} 1 & 1 & 1 \\ 0 & 0 & 0 \\ -1 & -1 & -1 \end{bmatrix} \quad \begin{bmatrix} 1 & 0 & -1 \\ 1 & 0 & -1 \\ 1 & 0 & -1 \end{bmatrix} \quad \begin{bmatrix} 0 & 1 & 1 \\ -1 & 0 & 1 \\ -1 & -1 & 0 \end{bmatrix} \quad \begin{bmatrix} 1 & 1 & 0 \\ 1 & 0 & -1 \\ 0 & -1 & -1 \end{bmatrix}$$

(a) Horizontal Mask, (b) Vertical Mask, (c) Diagonal Mask, (d) Anti-Diagonal Mask.
Step 3: Complete the filtering by utilizing the above masks to build the four edge bitmaps.
Step 4: Estimate the edge histograms by dividing each 16 sub-bitmaps into bitmap blocks. The size of these bitmap blocks scale with the bitmap size and is assumed to be a power of 2. The bitmap blocks quantity per sub-bitmap remained consistent, independent of the initial bitmap dimensions, by scaling their size correctly.
Step 5: A single edge mask is next employed to every macro-block, employing the macro-block being a pixel bitmap. The average of the intensity values of corresponding pixels is utilized to compute the pixel intensities for the bitmap block distributions.
Step 6: These bitmap blocks with edge strengths exceeding a specific maximum threshold are applied while estimating the histogram. The four edge strengths are calculated toward a bitmap block, with every four masks of Step 2. The specific preset threshold is exceeded by the maximum of specific edge strengths, then the corresponding bitmap block is deemed an edge block.
Step 7: Group the bitmap blocks (and the corresponding bins) to accomplish the extended histogram. The global and semi-global histograms are nothing but these extended bins.
Step 8: Accomplish the global histogram through consolidating all the 16 bitmap blocks. And then accomplish the semi-global histogram by pooling the image blocks/bins by rows (four rows), columns(four columns), and in groups of 2×2 (four groups). It results in four bins for the global histogram and 13×4 for the semi-global histograms from the 64 local histogram bins. The total number of bins is thus 120.

Figures 4 and 5 visualize the color maps and edge maps for malware executable(ammyy.exe) and safe executable (notepad.exe), respectively.

Surprisingly, the edge map denotes a better visual difference than the colormap for the RGB bitmap. It is because if the edges of the malware bitmap are evaluated, it

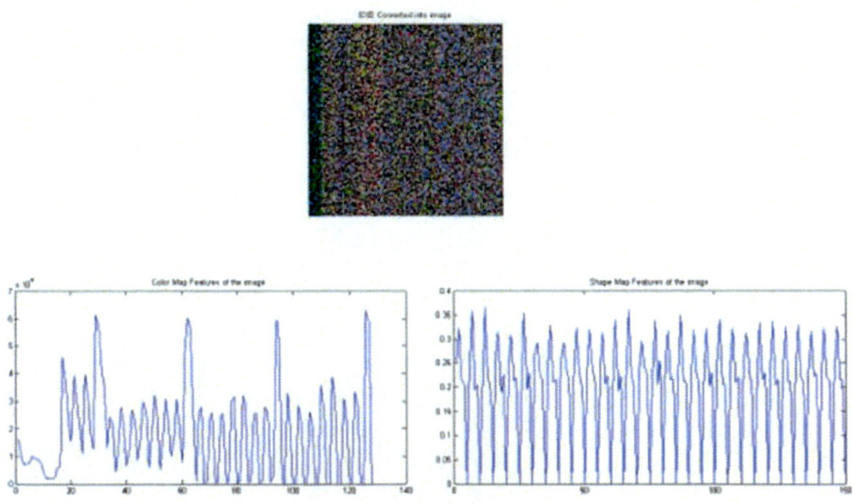

Fig. 4 Color map and edge map of executable malware

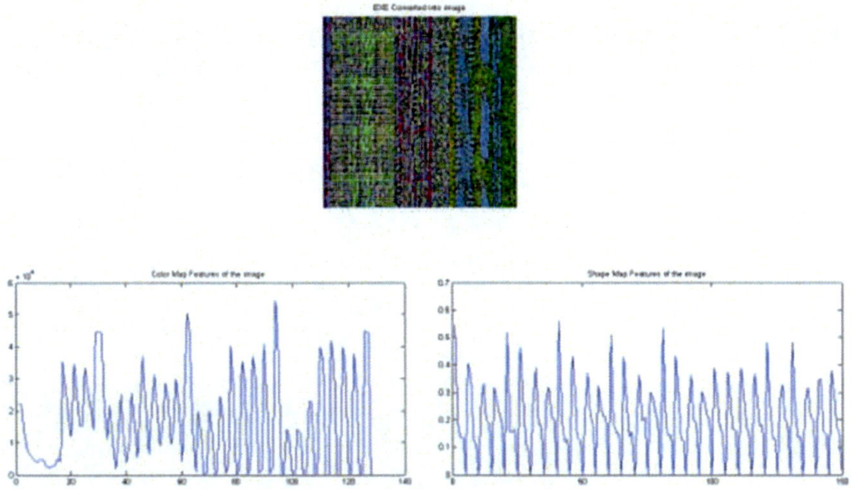

Fig. 5 Color map and edge map of safe executable

shows a much larger variety of edges than a standard bitmap; the results are shown in Figs. 6 and 7 for malware and normal executables, respectively. The color maps also show some variation, but that is not visually appealing, because both color bitmaps have similar color components.

These color and edge variations from 11,500 executables are evaluated and given to a linear k-nearest neighbor classifier for categorizing the malware into one of the k categories. The performance of any classifiers initially depends on the dataset and

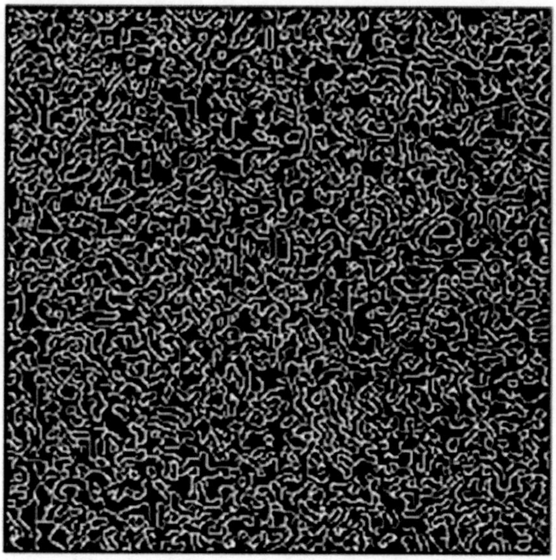

Fig. 6 Edge variation in malware executable bitmap

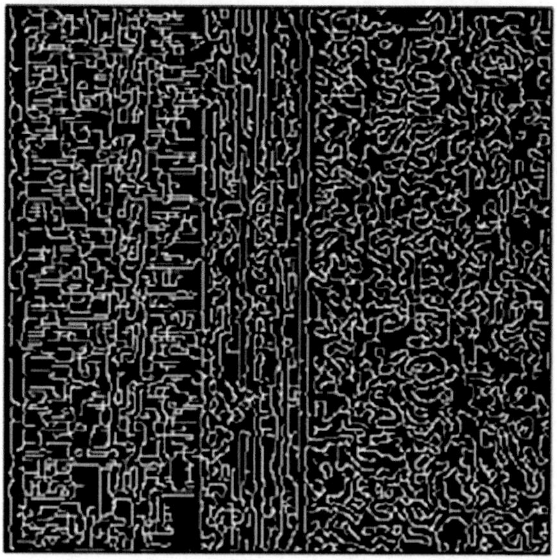

Fig. 7 Edge variation in standard executable bitmap

dimensional space. In our case, the dataset has low-dimensional space, and the KNN classifier performs better with low-dimensional space. One more reason to use a KNN classifier, it is automatically nonlinear; it can detect linear or nonlinear distributed data; it tends to perform very well with many data points. The fixed data is divided it into training and testing, then KNN may perform better. The k-nearest neighbor

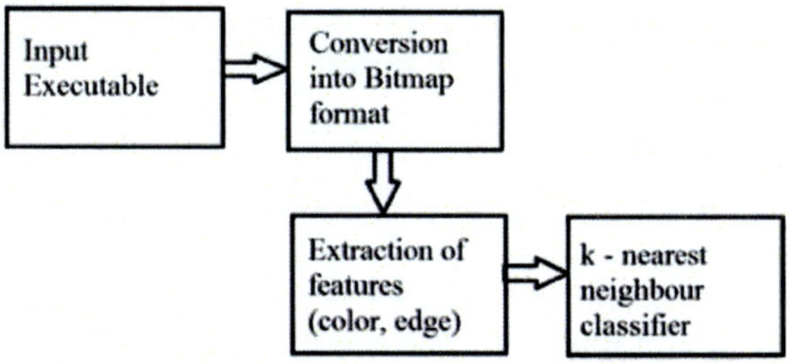

Fig. 8 Overall system block diagram

classifier gives an accuracy between 70% (for blind malware testing), and 100% (for non-blind malware testing). The classification accuracy can be further improved by training the system with a higher number of images and reducing the number of almost similar pattern executables. The following block diagram demonstrates the overall process of visualization and identification of malware when executable is given as an input to the system (Fig. 8).

4 Results

The proposed methodology is tested for 11,500 test executables, in intervals of 5 and 10 executables per set, and obtained the following results for 80% non-blind and 20% blind samples (blind samples are the executables not present in the database, while non-blind are the executables trained in the classifier). The result presented in Table 1.

The classifier accuracy grows linearly and saturates around 94.8%, which can be considered as the classifier training threshold. This accuracy is relatively high and can use for small practical applications like OS firewall related tasks.

Table 1 Experimental evaluation

Sample trained	Sample tested	Correct classification	Accuracy (%)
1000	1000	900	90
2000	2000	1800	90
3000	3000	2800	93.3
5000	5000	4700	94
7000	7000	6600	94.2
11,500	11,500	10,900	94.7

5 Conclusion and Future Scope

The novel methodology is proposed to visualize and identify malware based on their EXEcutable file signature. The system is based on texture analysis of malware by converting executables to bitmaps and then computes complex feature analysis, to evaluate the type of malware present in the Exe file. The accuracy of the proposed malware classifier is good enough for small practical applications, but cannot be used in mission-critical applications like ransomware attack detection at a global scale. This is because the k-nearest neighbor classifier usually saturates for a high number of training feature sets. To overcome this drawback, a novel classifier is made to work based on PSO and SVM, which can be tested in a wide variety of applications and thereby be used at a much larger real-time global scale for pre-cognitive detection and if possible removal of malware from the system under operation.

References

1. Egele, M., Scholte, T., Kirda, E., Kruegel, C.: A survey on automated dynamic malware analysis techniques and tools. ACM Comput. Surv. **40**(2), 1–42 (2012)
2. Schultz, M.G., Eskin, E., Zadok, F., Stolfo, S.J.: Data mining methods for detection of new malicious executables. In: Proceedings of 2001 IEEE Symposium on Security and Privacy, pp. 1–12 (2001)
3. Tian, R., Batten, L.M., Versteeg, S.: Function length as a tool for malware classification. In: 3rd International Conference on Malicious and Unwanted Software, pp. 69–76. IEEE (2008)
4. Hex-rays: The IDA Pro disassembler and debugger, Hexrays, [Online]. Available: http://www.hex-rays.com/idapro/. Accessed 12 Feb 2014
5. Siddiqui, M., Wang, M.C., Lee, J.: Detecting internet worms using data mining techniques. J. Syst. Cybern. Inform. **6**, 48–53 (2008)
6. Wicherski, G.: pehash: a novel approach to fast malware clustering. In: Proceedings of the 2nd USENIX conference on Large-Scale Exploits and Emergent Threats (LEET) (2009)
7. NIST: Kolmogorov Complexity. Available: http://xlinux.nist.gov/dads//HTML/kolmogorov.html. Accessed 12 Jan 2014
8. Bailey, M., Oberheide, J., Andersen, J., Mao, Z. M., Jahanian, F., Nazario, J.: Automated classification and analysis of internet malware. In: Recent Advances in Intrusion Detection, pp. 178–197. Springer, Berlin (2007)
9. Bayer, U., Comparetti, P. M., Hlauschek, C., Kruegel, C., Kirda, E.: Scalable, behavior-based malware clustering. In: Network and Distributed System Security Symposium (NDSS) (2009)
10. Firdausi, I., Lim, C., Erwin, A., Nugroho, A.S.: Analysis of machine learning techniques used in behavior-based malware detection. In: 2nd International Conference on Advances in Computing, Control and Telecommunication Technologies (ACT) (2010)
11. ANUBIS.: Analysis of unknown binaries. Available: http://anubis.iseclab.org/. Accessed 20 June 2014
12. Park, Y., Reeves, D., Mulukutla, V., Sundaravel, B.: Fast malware classification by automated behavioral graph matching. In: Proceedings of the Sixth Annual Workshop on Cyber Security and Information Intelligence Research (2010)
13. Cesare, S., Xiang, Y., Zhou, W.: Malwise—an effective and efficient classification system for packed and polymorphic malware. IEEE Trans. Comput. **62**(6), 1193–1206 (2013)
14. Santos, I., Devesa, J., Brezo, F., Nieves, J., Bringas, P.G.: Opem: A static-dynamic approach for machine-learning-based malware detection. In: International Joint Conference CISIS'12-ICEUTE' 12-SOCO' 12 Special Sessions, pp. 271–280. Springer, Berlin (2013)

15. Edward, R., Barker, J., Sylvester, J., Brandon, R.: Malware detection by eating a whole EXE. In: AAAI Conference on Artificial Intelligence, pp. 268–276 (2018)
16. Venkatraman, S., Alazab, M.: Use of data visualisation for zero-day malware detection. Security and Communication Networks, pp. 1–13. Hindawi (2018)
17. Donahue, J., Paturi, A., Mukkamala, S.: Visualization Techniques for Efficient Malware Detection. RiskSense Technical White Paper Series, pp. P 1–P 9 (2018)
18. Kumar, A., Kuppusamy, K.S., Aghila, G.: A learning model to detect maliciousness of portable executable using integrated feature set. J. King Saud Univ. Comput. Inform. Sci. **31**(2), 252–265 (2019)
19. Nataraj, L., Karthikeyan, S., Jacob, G., Manjunath, B. S.: Malware images: visualization and automatic classification. In: Proceedings of the 8th International Symposium on Visualization for Cyber Security, VizSec'1, USA, pp. 4:1–4:7. ACM (2011)
20. Komatwar, R., Kokare, M.: Customized convolutional neural networks with K-nearest neighbor classification system for malware categorization. J. Appl. Secur. Res. 2–22 (2020)
21. Malware Images.: (2013) *Retrieved from* http://vision.ece.ucsb.edu/~lakshman/malware_images/album/
22. Conti, G., Bratus, S., Shubina, A., Lichtenberg, A., Ragsdale, R., Perez-Alemany, R., Sangster, B., Supan, M.: A Visual Study of Binary Fragment Types. Black Hat, New York (2010)

Augmented Reality in Sports Analysis Using HDM Representation of Players' Data

P. Sri HarshaVardhan Goud, Y. Mohana Roopa, R. Sri Ritvik, and Srija Vuyyuru

Abstract Augmented reality acts as a digital add-on to the user's world. The augmented reality is a mixture of computer-generated or mediated perceptions and the real world. In this paper, the augmented reality advancement will help in the development of sports and performances of players are shown. How this technology will help in the broadcasting of sports and games and also data representation for real-time performance analysis. Hierarchically distributed data matrix (HDM) representation is used.

Keywords Augmented reality · Sports and games · Player performance analysis · Data processing

1 Introduction

The development of technology which helps you see more than what others see, hear more than what others hear and even touch that others cannot [1]. This helped in the development of augmented reality and mixed reality in the year 1994 by Paul Milgrim [2]. Since then, there has been a rapid progress and development in different applications like education, health sector, sports and gaming, and broadcasting. Augmented reality and mixed reality are helping in the development of sports and gaming. At present, augmented reality is acting like an integrated part in every sector. With the development of head-mounted displays (HMDs) [3] and goggles and specialized helmets, it is improving the user view and experience toward the game (Fig. 1). Augmented reality is also helpful in making decisions in a game which may be critical and may change the result of the game (Fig. 2). In many of the sports, the ultimate goal of the players and teams is to improve the skill set and performance at

P. Sri HarshaVardhan Goud (✉) · Y. Mohana Roopa · R. Sri Ritvik · S. Vuyyuru
Institute of Aeronautical Engineering, Hyderabad, Telangana, India
e-mail: Sriharsha27299@gmail.com

Y. Mohana Roopa
e-mail: ymohanaroopa@gmail.com

© The Editor(s) (if applicable) and The Author(s), under exclusive license to Springer Nature Singapore Pte Ltd. 2021

A. P. Pandian et al. (eds.), *Proceedings of International Conference on Intelligent Computing, Information and Control Systems*, Advances in Intelligent Systems and Computing 1272, https://doi.org/10.1007/978-981-15-8443-5_32

Fig. 1 Augmented reality for analysis of lawn tennis

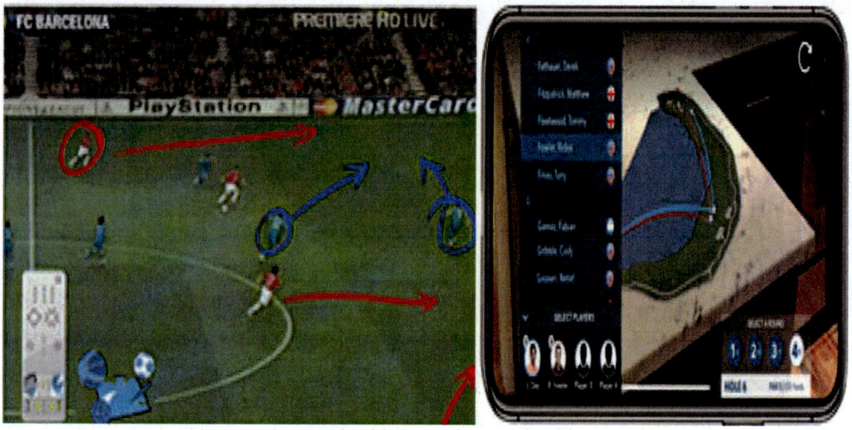

Fig. 2 Figure explaining the positions of the players

global level. The teams are modern means to improve the player performances [4]. Augmented reality is one such technology that can be used for a better understanding of the game by the players and the training staff and improvement in the performance of players by augmenting the computer-generated objects to the real world [2] and creating the experience of a real game. Augmented reality also has its part in the broadcasting sector which has evolved over time and can see the development of the sector with these modern technologies.

2 Augmented Reality in Sports

So far, the augmented reality and mixed reality have been used in sports to enhance the user's experience of the game. Now, the augmented reality can also be used in training sessions of the players and for the individuals who are seeking sports as their profession by starting at early stages. The augmented reality will be helpful for these people to get the real-time experience of the game by using head-mounted displays, special helmets, and also goggles. This creates the user the experience and also to improve his performance, especially in spatiotemporal games [5]. The augmented reality which can be created by using certain devices like head-mounted displays, special helmets, and goggles can be termed as wearable technology. With the advent of wearable technology [6], it has become easier to analyze the performances and get better results [4]. The devices cannot only help to get the real-time experience of the game, but they can be developed in such a way that they can get the insights of the player performances which can help in the analysis of their game. These devices can create the data sets with these insights and help overtime to understand the strengths and weaknesses of the players which help in making different game strategies and tactics for a better performance [4]. The data can be trained to obtain patterns of the game by machine learning algorithms. Thus, the overall performances of the players and teams are an amalgamate of all the modern technologies. The modern technology is also helping to understand the patterns of the game and strategies of the opponents in order to know their strengths and weaknesses. Augmented reality (AR) is one such technology. In the figure below, the analysis of a tennis player is seen where the strike is analyzed using AR. By taking a set of these observations, the strategy of the opponent is understood. In the same way, can have a number of observations that are sufficient to get the analysis of players depending on the sports.

2.1 For Training in Team Sports

Augmented reality, if applied to any sector like sports, can be a great add-on. The technology can be used for training in sports. Augmented reality gives the complete picture of how a training session can be fulfilled with its advent. The game of hockey is considered for example. Since real-time accurate registration between both spaces and players is crucial for the collaboration, a video-rate registration algorithm is implemented with magnetic headtrackers and video cameras attached to optical see-through head-mounted displays (HMDs) [3, 7]. It also helps to understand the game-play of the team and to analyze individual player's performance which helps to improve themselves. This is possible by implementing computer-generated objects at the positions of the co-players and opponents for a player standing in different positions and makes the player get into that particular situation and play accordingly. All this is possible when a player is equipped with special goggles and helmets. The

Fig. 3 Player's statistics

same augmented reality can be used in other spatiotemporal games [5] like football (Fig. 3) and American football. Based on the type of game, the training of the players varies and the positions of the players change and also the accuracy rate and calculation of the metrics change [8].

2.2 *For Performance Analysis*

Augmented reality, as it is already explained as the mixture of computer-generated objects and the real-world objects [1]. In sports like cricket, football, and many others, this augmented reality can be used to analyze the performances of the players and keep a record of them, which can be displayed in a way understandable by everyone. For example, the statistics of a player can be viewed by augmenting the computer-generated image of that player in the real world (Fig. 4). These statistical data help in the calculation of metrics in every game and also make the viewers understand it easily. While performing the analysis of a player in any sport, in order to improve his performance, it is necessary for him to understand and rectify the mistakes that he might have committed in his previous games. So, in order to do that, a player needs to have the recordings of his game which are recorded with wearable computers [3, 9] and have some special augmented reality devices. This helps him to have a view of his performance with all required 3-D graphics.

This will be helpful to calculate the different parameters of his performance and understand the mistakes which will be helpful for his future games. The parameters can be implemented in the same way using the same principles of augmented reality, mixed reality [10], augmented virtuality [4], virtual reality [11] (Fig. 5).

For the performance analysis, augmented reality which utilizes the NUBOMEDIA [12] pass for the multimedia applications development and the augmented reality sports service identifies the athlete and sensor based on an augmented reality marker

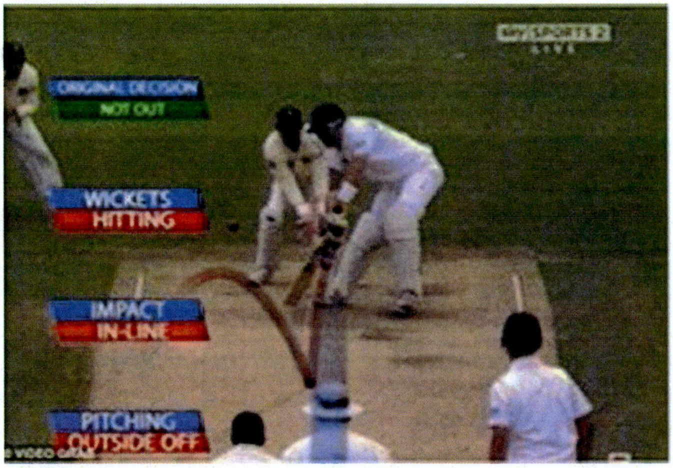

Fig. 4 Augumented reality in cricket

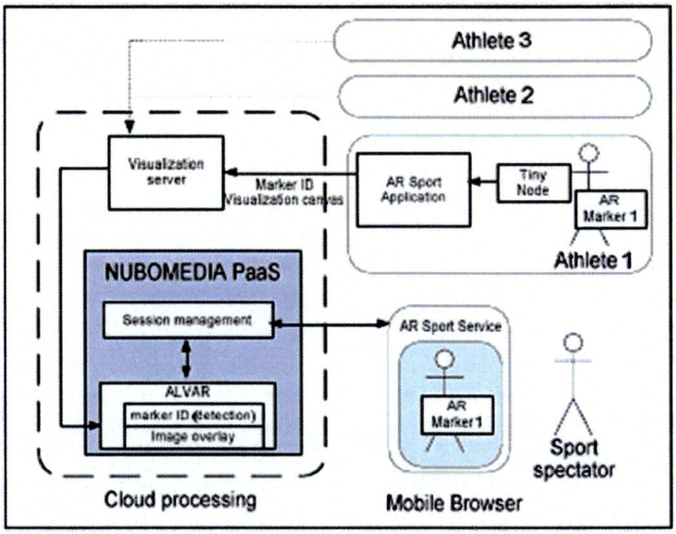

Fig. 5 Sports analysis architecture of nubomedia

connects the visual tag with visualization [13] canvas that which created by the Internet of things sports data analysis software on an Android device. [14] The sports event spectator can see the analysis results of the athlete in an augmented reality video stream in real time. The cloud-based augmented reality services are created for the sports events for creating the totally new type of engaged experiences for the enthusiastic fans. The sensors which worn by the athletes will provide new opportunities, and this provides the visualization applications for the spectators, and sensors create a sports fans, to become prosumers in sports events. The linking of information allows tags the video streams with the data of local sensor and also provides the additional information which includes the athlete name, nationality, ranking, etc., and enhances the user interface based on wider testing and integration of advanced performance analysis on the AR sports application.

2.3 For Sports Broadcasting

Augmented reality is used in ice hockey to make the puck visible as it is not visible to the viewers watching on television [15]. In the same way, it can also be used in the American football to identify the first down line are inserted into the broadcast in real time [1]. To display the statistics of the players and understand the condition of the game, the augmented reality can be used. In general, this is mainly used to make the broadcaster to get the attention of the viewers. Thus, augmented reality can be used for sports analysis, training, and broadcasting.

3 Players Data Representation

The processed data of the players will be stored in the databases. The real-time processing and analysis of the data are the only way to analyze player performance, and these data sets should be composable and reusable for subsequent development [16]. By using the big data frameworks like MapReduce framework and Apache Spark, real-time analysis cannot be performed due to the abstraction offered by these frameworks. Thus, there is a need of introducing unconventional methods in this regard. hierarchically distributed data matrix (HDM) is one such representation of data which will be used for data processing. The data is represented in the form of matrix which can be processed natively in real time. By using HDM, the data can be processed efficiently.

4 Conclusion

The advancement in augmented reality can be used for the improvement of many fields not only like sports, but also in the entertainment industry. The data representation in the form HDM helps for the real-time analysis of the data. By using HDM representation, the benefit is that it can be composed natively and the data is secure and encrypted and cannot be accessed to everyone. At the same time, HDM's ability to provide access only to those who are permitted also helps for the successful data analysis of sports' data. Thus, augmented reality in sports industry with the real-time analysis using HDM, helped to get better results in time of speed, performance, and security compared to the other big data frameworks like Spark and MapReduce.

Acknowledgements I would like to thank my professor Dr Y. Mohana Roopa who has helped me to understand the importance of learning new technologies and always encouraged me to choose a topic of my interest and relate it to my field of study.

References

1. van Krevelen, D.W.F., Poelman, R.: A survey of augmented reality technologies, applications and limitations. In: Systems Engineering Section, Delft University of Technology, Delft, The Netherlands
2. Milgram, P., Kishino, F.: A taxonomy of mixed reality visual displays,. IEICE Trans. Inform. Syst. **E77-D**(12), 1321–1329 (1994)
3. Spitzer, M.B., et al.: Eyeglass-based systems for wearable computing. In: Proceedings of 1st International Symposium on Wearable Computers, (ISWC 97), pp. 48–51. IEEE CS Press, Los Alamitos (1997)
4. Sri HarshaVardhan Goud P, Mohana Roopa Y (2019) Player performance analysis in team sports-with fusion of machine learning and wearable technology, pp. 616–619. In: 3rd International Conference on Computing Methodologies and Communication (ICCMC 2019). ISBN: 978-1-5386-7807-7
5. Gudmundsson, J., Horton, M.: Spatio-temporal analysis of team sports. J. ACM Comput. Surv. **50**(2), Article no. 22. ACM, New York
6. Raina, A., Lakshmi, G.T.: CoMBaT: wearable technology based training system for novice badminton players. In: 2017 IEEE 17th International Conference on Advanced Learning Technologies (ICALT 2017)
7. Ohshima, T. et al.: AR2 hockey: a case study of collaborative augmented reality. In: Proceedings of IEEE Virtual Reality Annual International Symposium (VRAIS 98), pp. 268–275. IEEE CS Press, Los Alamitos (1998)
8. Srinivasa, R.R., Veluchamy, U.P., Bose, J.: Augmented Reality adaptive web content. In: 2016 13th IEEE Annual Consumer Communications & Networking Conference (CCNC), Las Vegas, NV, pp 107–110 (2016). https://doi.org/10.1109/ccnc.2016.7444740
9. Jebara, S.T., et al. (1997) Stochasticks: augmenting the billiards experience with probabilistic vision and wearable computers. In: Proceeding of the 1st International Symposium on Wearable Computers (ISWC 97), pp. 138–145. IEEE CS Press, Los Alamitos (1997)
10. Regenbrecht, H., Ott, C., Wagner, M.: An augmented virtuality approach to 3-D video conferencing. In: Proceedings of the Second IEEE and ACM International Symposium on Mixed and Augmented Reality (ISMAR'03), ISBN 0-7695-2006-5/03
11. Brooks Jr., F.P.: What's real about virtual reality? IEEE Comput. Graph. Appl. **16**(6), 16–27
12. Makela, S-M., Plaviainen, M.: Use of augmented realityin sports performance visualization: media tools for prosumers. In: AMBIENT-2017: Teh Seventh International Conference on Ambient Computing, Applications, Services and Technologies
13. Stein, M., Janetzko, H.: Bring it to the pitch: combining video and movement data to enhance team sport analysis. In: IEEE Trans. Visual. Comput. Graph. (2018)
14. www.thinkmind.org
15. Cavallaro, R.: The FoxTrax hockey puck tracking system. IEEE Comput. Graph. Appl. **17**(2), 6–12 (1997)
16. Wu, D., Sakr, S., Zhu, L., Lu, Q.: HDM: a composable framework for big data processing. IEEE Trans. Big Data. https://doi.org/10.1109/tbdata.2017.2690906

Shared Access Control Models for Big Data: A Perspective Study and Analysis

K. Vijayalakshmi and V. Jayalakshmi

Abstract With the rapid development of information technology and the Internet, the data owners can avail of the cloud storage to store their vast amounts of data. Data and all resources are outsourced and distributed to make available to everyone who needs them. Outsourcing of big data is more beneficial. It reduces the tension of data owners from implementing infrastructure and maintaining software to manage the massive amount of their data. The cloud service providers take all these responsibilities. But outsourcing big data causes many security challenges that the outsourced big data should be protected from the malicious intruders. The cloud service providers implement various access control models to protect the data from anonymous users. This paper presents multiple existing access control models with theoretical evidence. The primary goal of writing this paper is to study and analyze various access control models, particularly analyze attribute-based access control and role-based access control models. This paper examines the reliability, scalability, and efficiency of some existing access control models theoretically. This work may be useful for researchers and practitioners who are in the field of big data security.

Keywords Access control models · Attribute-based access control model · Cloud storage · Big data security · Role-based access control model

K. Vijayalakshmi (✉)
Arignar Anna Government Arts College, Vels Institute of Science, Technology and Advanced Studies, Cheyyar, Chennai, India
e-mail: vijiyuvavelan@gmail.com

V. Jayalakshmi
School of Computing Sciences, Vels Institute of Science, Technology and Advanced Studies, VISTAS, Chennai, India
e-mail: jayasekar1996@yahoo.co.in

© The Editor(s) (if applicable) and The Author(s), under exclusive license to Springer Nature Singapore Pte Ltd. 2021

A. P. Pandian et al. (eds.), *Proceedings of International Conference on Intelligent Computing, Information and Control Systems*, Advances in Intelligent Systems and Computing 1272, https://doi.org/10.1007/978-981-15-8443-5_33

1 Introduction

Even in the well-developed information technology era, one of the essential goals to be achieved is providing correct data to the right people [1]. Data owners are simply outsourced their big data in cloud storage, but securing these data from unauthorized users or malicious intruders is a big challenge. The shared resources can be categorized as data and other resources. The data resources are operating systems, software, applications, and confidential information stored as distributed files, databases, and data warehouses. Additional resources are network and hardware. Cloud service providers establish various access control models to accomplish the following tasks. Those tasks are: (i) maintaining confidentiality, integrity, and privacy of resources (data), (ii) preventing these resources against the illegal access by the unauthorized resources, and (iii) providing correct data to the right users (preventing denial of service) [2]. Access control is a software module or function used to protect the outsourced resources or to prevent the illegal access of shared resources by unauthorized users [3]. The access control model is framed with a set of security policies. That security policy determines the category of authorized and unauthorized users based on the rules in which it contains. Establishing and applying an efficient access control model prevent many illegal accesses of data. Some examples of unauthorized access are: a terrorist is reading a military or defense department record (protect the confidentiality of data), a hacker is using the duplicate account in a bank (preserves the integrity of data), a hospital receptionist is reading a patients' personal medical report (preventing unauthorized access of data), and an attack on the server may deny the access of data to the regular authorized clients (prevent the denial of service) [2, 3].

In the late 1960s, many access control models have been implemented to prevent the use of data by unauthorized users. Some models have got great success in securing the data in the distributed storage environment while some were a failure. Discretionary Access Control (DAC) and Mandatory Access Control (MAC) models were proposed in the 1970s. Both are well good at performing all security tasks, and these models satisfy the security requirements. But these can overcome the security issues when the rate of data generation is very slow, the number of users is limited, and the network bandwidth rate is low [4]. As information technology and the development and use of social networks are grown rapidly, the generation of data, the number of users and resources are also increased exponentially. In this situation, MAC and DAC are not efficient in satisfying security requirements. The access control model Bell-Lapadula is mainly used in government and military systems. The next evolution of the access control model is the Role-Based Access Control Model (RBAC). RBAC establish an interface called role between authorized users and access right. RBAC can prevent the above-mentioned illegal access to resources when the size of data and the number of users are large. As information technology is growing unimaginably, and new computing technologies (cloud computing, mobile computing, IoT, edge computing, fog computing) are evolved, this emerging field complicates the mechanisms of the access control model. In this situation, the next access control model, Attribute-Based Access Control (ABAC) model, has been proposed to meet the complex security requirements. The primary goal of this paper is to review and analyze the access control models. Our paper is organized as follows: Sect. 2 says shortly about the research done on the access control models, and Sect. 3 presents the fundamental concepts of the four access control models DAC, MAC, RBAC, and ABAC. The analysis of access control models is presented based on theoretical evidence in Sect. 4 and concluded in Sect. 5.

2 Related Work

In previous research, Ebrahim Sahafizadeh and Saeed Parsa presented a survey on various access control models in 2010. They described the following access control models: MAC, Bell-Lapadula (B-L) Model, Multi-level secure (MLS) database management systems, RBAC, and View-Based Access Control (VBAC). MAC works on labels categorized upon the level of sensitivity of data. MAC policy verifies the sensitivity label of the object being requested by the user. If the verification is succeeded, MAC allows the user to work with the object (shared resource). Otherwise, the request of the user is denied. Ebrahim Sahafizadeh said multi-level secure (MLS) database management systems are useful to protect the data in multi-level database systems. This model is developed based on the idea of MAC. Based on research [5], they told RBAC is viewed as a variation of the access matrix. The RBAC policy assigns rights (privileges) to the roles, and the subjects are assigned with the roles. The subject can access the object after playing the role which has rights required for the access. RBAC model can protect shared resources in large systems with less complexity and cost. They finally conclude that RBAC is flexible, efficient, and fine-grained access control models than others.

In 1992, David F. Ferraiolo and D. Richard Kuhn made a research on the RBAC model [6]. They conclude DAC is the appropriate and correct one for the single-level government, civil, and military systems. In contrast, MAC is suitable for the security needs of multi-level military systems, but both these two models are used very rarely in other applications. They argued that RBAC is the flexible, appropriate, centric, and fine-grained model for all applications includes government, civil, commercial, and military organizations. In 1996, Ravi S. Sandhu, Edward J. Cope, and their team defined the RBAC model as it assigns the users to the roles and permissions to the roles. In RBAC, the subject gets permission to access an object through the roles. RBAC manages two mapping cardinalities: user-role and role-permission [7]. Ninglekhu and Krishnan [8] introduced a new access control model, Attribute-Based Administration of Role-Based Access Control (AARBAC). It differs from a traditional access control model by using attributes for the entities (user, role, and permission) to manage the access control model flexibly and efficiently. They developed two models Attribute-based User-Role Assignment (AURA) model and Attribute-based Role-Permission Assignment (APRA) model to manage user-role mapping constraints and role-permission mapping constraints. AARBAC is capable of providing the features of many previous access control models ARBAC97 [9],

ARBAC99 [10], ARBAC02 [11], and Uni-ARBAC [12]. In 2011, V. Suhendra [13] surveyed various access control models and also analyzed deployment issues of them.

Every access control model has three essential components, namely authentication, authorization, and accountability. The authentication-component finds the legitimate users, and the authorization component makes a decision (allow or deny) on the user request by using a security policy. The accountability-component stores (logging) the information about all operations done by the user in a log file (log entry). The authorization component covers the major task of the access control model. This research made the deployment of the access control model in the manner that the security policies are defined first; second potential users are considered; third, the security issue necessary to prevent is analyzed; and finally, the priority of rules is considered in emergencies. Finally, he concludes that based on this deployment, anyone can choose and implement the suitable access control model to protect the system from the security breaches. In 2018, Qi and Di [4] made an analysis of access control models on RBAC and ABAC. They said RBAC is flexible and efficient for large-scale applications and businesses. RBAC meets security needs efficiently, even if the number of resources and users grows rapidly. But today, the different computing technologies cloud computing, fog computing, mobile computing, and IoT are evolved. These computing technologies increase the complexity of access control models.

In this situation, the new access control model ABAC is emerged to protect the shared resources. The traditional model RBAC is good for large-scale applications, while ABAC is suitable for today's new computing technologies. They proposed a hybrid approach called the Role and Attribute-Based Access Control model (RABAC) with the combination of RBAC and ABAC. RABAC uses RBAC to manage the static mappings between users and permissions and users ABAC to manage the dynamic mappings between users and permissions. This research proposed this new model RABAC after carefully monitoring the drawbacks of the traditional approach RBAC based on efficiency, flexibility, accuracy, and security needs. They made an analysis of the new model RABAC with theoretical evidence. Table 1 shows the actual work done at each paper from [4–13], including this paper.

3 Access Control Models

Access control is a software component that controls the access of the object (data or other resources) by allowing or denying the user access of the object based on setting constraints. Simply access control is a mechanism used for protecting the shared resources from unknown users. Access control restricts the single user to access multiple resources and also allows only authorized user to access the shared resources [13]. The organization, application, or resource-owner use access control to prohibit the unauthorized access of their resources. The access control model sets rules and constraints to specify which user can access the resource and what type of operations the user can perform on the resources. Rules or constraints are expressed

Table 1 Research contribution toward access control models of [4–13] including this paper

References	Proposed work
[4]	Developed a new hybrid model RABAC by combining the feature of RBAC and ABAC. RBAC is used for static, and ABAC is used for dynamic mappings
[5]	Implemented RBAC approach using prime numbers and made a minimal change in the source code
[6]	Analyzed MAC, DAC, and RBAC and concluded that DAC is suitable for single-level government and military organizations. In contrast, MAC is for multi-level, but RBAC is a centric and flexible approach for all large-scale applications, including commercial, business, government, and military
[7]	Described the RBAC model in detail
[8]	Proposed a new model AARBAC, which has two sub-models, AURA and APRA. AURA model uses attributes to manage user-role mapping constraints, and APRA uses attributes to manage role-permission mapping constraints
[9]	Defined a new role-based model ARBAC97 with three components URA97, PRA97, and RRA97 to manage user-role, permission-role, and role-role mappings, respectively
[10]	Introduced a new model called ARBAC99 (enhanced version of ARBAC97) to implement the feature of decentralized administration of RBAC
[11]	Proposed a model ARBAC02 that resolves the deficiencies of ARBAC97. ARBAC02 is a bottom-up approach, while ARBAC97 is a top-down approach
[12]	Presented a unified model called Uni-ARBAC with many new features of previous administrative RBAC models. This model administrates the permission-role assignments as a unit, not an individual assignment
[13]	Presented a survey and analysis of four access control models. This paper concluded that ABAC is suitable for today's new computing platforms

as the security policies to regulate access over the resources. The term subject refers to process or entity running on behalf of the organization or the user, and the access rights of those entities are managed based on security policies. The term object refers to the shared resources, which are requested by the subject for access [14].

Figure 1 describes the simple mechanism of the access control model. Whenever the subject made a request for access to the object, the intermediate mechanism made a decision (allow or deny) based on access control policy, which regulates the access of the subject. The access control can be classified as discretionary and non-discretionary. The discretionary access control gives full rights to the owner of

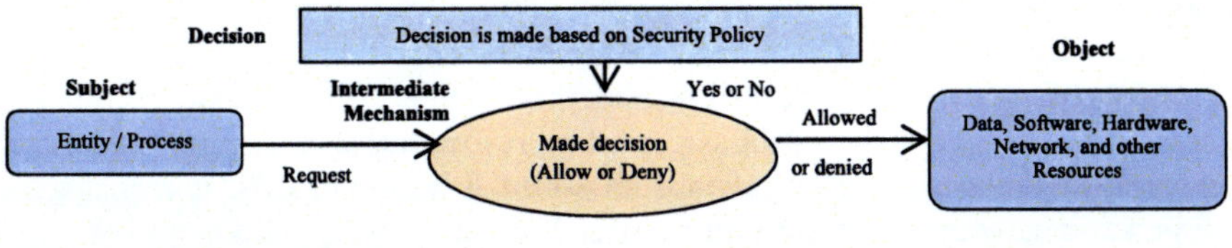

Fig. 1 Access control mechanism

the object that they decide to grant or deny the user access of the object. Thus, the decision on the subject's request is made based on the owner's discretion. The non-discretionary access made the decision (allow or deny) on the request based on the policy rules. A rule specifies the nature of the request, and access like the user's role, environmental conditions, the value of attributes of the subject, and object. DAC is an example of a discretionary access control model, and MAC, RBAC, and ABAC are non-discretionary access control models.

3.1 Discretionary Access Control Model (DAC)

It is a common and simple access control model where the owner of the object controls all access of his object. They decide which user can access the object and what level of access the user has. DAC gives full control to the user who created the object or the owner of the object. The owner of the object can grant any access privilege (Example: read-only, write, print, download, etc.) to any user with his own discretion, but the user cannot control the access of the object, which is owned by another user. The main aim of the sue of DAC is to prohibit the access of the object from unauthorized subjects. DAC maintains Access Control List (ACL) for each object to protect from illegal access. ACL is the list of records, and each record has the users or groups and access privileges of the users/groups on the object [3]. For example, $ACL_o = \{rs_1, rs_2, \ldots, rs_n\}$is the ACL list contains records $rs_1, rs_2, \ldots, rs_n$ for the object o. The record rs_i specifies all the access privileges granted to the subject s_i. Let $rs_i = \{ap_1, ap_2, \ldots, ap_n\}$, $ap_1, ap_2, \ldots, ap_n$ are the access privileges given to the subject. Hence, ACL can be expressed as $ACL_o = \{rs_1\{ap_1, ap_2, \ldots, ap_n\}, rs_2\{ap_1, ap_2, \ldots, ap_n\}, \ldots, rs_n\{ap_1, ap_2, \ldots, ap_n\}\}$. Example of ACL list of the object file:Patient-Blood_Report (Object) is $ACL_{Patient\text{-}Blood\_Report}$ = {Chief-microbioligist{full-control}, Chief-doctor{read,write,print,share}, Duty-Doctor{read, write, print}, Nurse{read}}. This list shows the access rights of subjects who are authorized.

DAC may use option GRANT to implement the delegation feature. If the owner of the object sets the right GRANT to the recipient user, then this user itself can give this right to the other subject, and this may be a chain activity. The task of managing the delegation feature in the distributed system is a big challenge. In SQL databases, the grantor cannot control all the subjects who receive this right. For managing the delegation feature properly, there should be a function to revoke the rights back. After a revocation process, the security state should be the same as before the delegation feature applied. The removing operation should be a transitive process between the subject who propagates these rights to the other subject and the grantee. This operation is known as cascading revocation [15]. Figure 2 describes the cascading revocation. Delegation is an essential feature of DAC to manage the cooperation between the owner of the object and users.

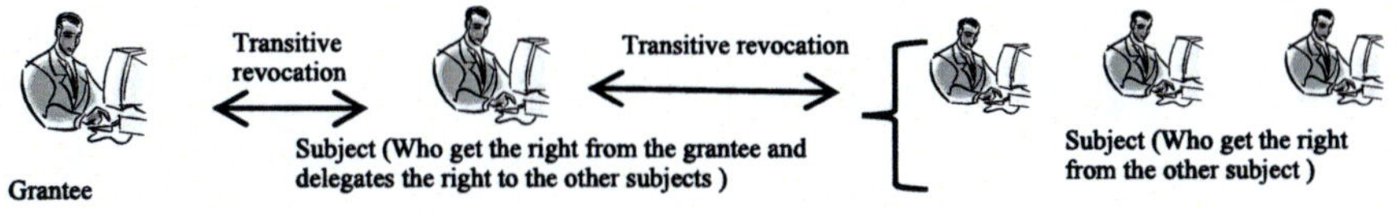

Fig. 2 Cascading revocation in DAC

Efficiency and Applications. One of the features of DAC is the flexibility that DAC allows the owner of the object to fix the security mechanism individually. DAC simplifies the usability where policies are easily updatable. As DAC is a distributed administrative one, it reduces the complexity of the administrator. This access control is efficient if the number of users and the size of the data are low. DAC is the easier access control to manage than all other access controls. Most networking operating systems (UNIX, WINDOWS, and Linux) use this access control to meet the security requirements for shared resources [16].

Limitations. Today, new computing technologies increase the security requirements and complexity of access control. This model fails in protecting shared resources if communication is very high. DAC allows the feature delegation. So, if a subject made any mistake, then confidentiality, integrity, or availability of data are lost. There must be an owner in DAC, but this is not possible in today's new computing technologies. DAC provides less security than other access controls. This access control is stateless that it does not know what happens after permission has been given.

3.2 Mandatory Access Control (MAC)

Unlike DAC, MAC is a central authority control. The system administrator has full control over the access of the object. The operating system manages the access of the object based on the configuration specified by the administrator. The user does not have any rights to change the access configuration like DAC. MAC assigns security labels to every object and subject. Every security label has two parts of information, namely classification and category. The classification-part specifies the nature of the object (private, top-secret, confidential, public), and the category-part specifies the nature of the place, user, project, application, or operation to which the resource is available. MAC allows the subject to access the requested object if there is a match in the security labels of the subject and object. Thus, the subject's classification-part should be matched with the object's classification-part, and the subject's category-part must be matched with the object's category-part.

Figure 3 describes the security mechanism of MAC. Subject-A is allowed to access Object-1 due to the match in the security labels of both. MAC denied the request of subject-B because the category-part of subject-B is not matched with the category-part of object-1. Thus, access of subject to object is determined by the attributes of subjects and objects. The attributes of subjects and objects are defined by the

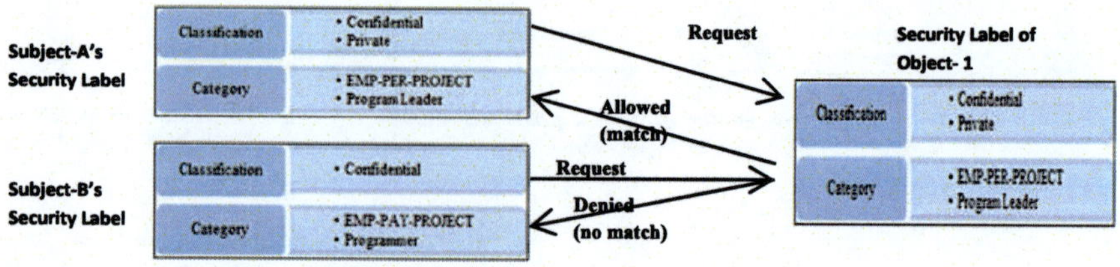

Fig. 3 MAC mechanism with security labels

system administrator and cannot be changed by the subjects or objects. Businesses, government organizations, and users require increased security on sensitive data, but DAC fails to meet these security requirements, and MAC restricts malicious intruders from accessing sensitive data [17].

Efficiency and Applications. MAC can handle trojan horses to prevent unauthorized access. DAC does not adequately address the security requirements of integrity and privacy of the data, but MAC is an access control mechanism to fulfill these security requirements. No unauthorized subject can do the modification in the protection state, and only the administrator or a trusted authority can define security labels of subjects and objects. MAC enforces the mechanism called reference monitor which manage all request and allow or deny the request based on the security policy. The reference monitoring mechanism uses a database to store all security policies, labels, and transition states. MAC gives more security than DAC. Most operating systems like Security-Enhanced Linux (SE-Linux) use MAC to overcome the weakness of DAC.

Limitations. MAC requires high implementation effort and needs system management for updating the security labels of existing subjects and objects and new subjects and objects. It is challenging to maintain and manage all security policies, labels, and transition states if the number of subjects and objects is high. Thus, MAC complicates the task of the administrator to accomplish all these security states in large applications or business organizations [3].

3.3 Role Based Access Control (RBAC)

RBAC is a non-discretionary access control model. Thus, access rights of all the subject to the object are determined by the central authority. The decision on a request of the subject is made based on the subject's job or duty (Role) in the organization. RBAC assigns permissions to the roles and roles to the subjects. This access control model manages two categories of mappings, namely permission-role association and role-user association. The necessary permissions to access a requested object are determined based on the subject's role [7]. The assigned permissions (access privileges) for a particular object vary based on the subject's role. The first mapping

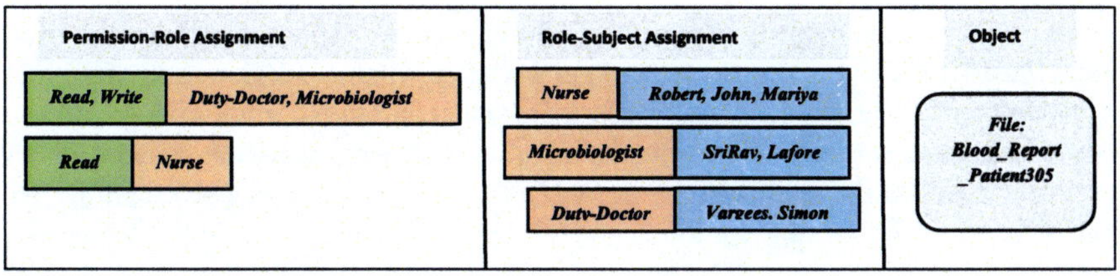

Fig. 4 RBAC mapping associations

permission-role assigns all permissions desired for a concerned job or function to the corresponding role, and the second mapping assigns a role to the subject. The subject can not have any additional access right over what he had based on his role. For example, the permission 'Read' is assigned to the role 'Nurse', and the permissions 'Read and write' are assigned to the role 'Duty-Doctor' for the object 'File_Blood-Report_Patient305'. The subject which has the role 'Nurse' can only have access to reading a file 'Blood-Report_Patient305', and the Duty-Doctor can read and write the same file. Hence, the permissions are assigned based on the role of the subject. Figure 4 shows the RBAC with permission-role and role-user relationships.

RBAC introduced two important features of least privilege and separation of jobs or duties. Least privilege is referred to as granting the access rights desired for a concerned role only, and the subject is given no additional access rights (not desired for his role). Separation of jobs is the process of assigning or distributing tasks or operations mutually [1]. The extended model introduced two new features hierarchy and constraints. One role (Senior role) can inherit the permissions of multiple roles (Junior roles). If the role 'chief-doctor' is allowed to inherit the permissions of the role 'duty-doctor' and the senior role 'chief-doctor' is assigned to any user, then he can have all access rights of duty-doctor. The security configurations of a role are accepted based on constraints [3]. In RBAC, each subject can be associated with more than one role based on central authority decisions, and multiple transactions or permissions can be assigned to a single role [6]. A security policy is expressed as follows.

$R_s\{r_1, r_2, \ldots, r_n\}$, R_s is the set of roles $r_1, r_2, \ldots, r_n$ assigned to the subject (s) and any one rule is active at a time.
$\mathrm{Pr_i} = \{p_1, p_2, \ldots, p_n\}$, $\mathrm{Pr_i}$ is the set of permissions or privileges $p_1, p_2, \ldots, p_n$ assigned to the role r_i.

Efficiency and Applications. This access control meets the security requirements of the non-military organizations while the government and military use DAC and MAC to protect confidential information. RBAC offers a central administration security system, and it is flexible that security requirements can be easily implemented as security policies. The feature of least privilege meets the security requirement of integrity. RBAC allows the subject to perform only the transactions granted to him and nothing he can do. Hence, the unauthorized transactions are not performed, and it preserves the integrity of the data. Another feature separation of duties avoids fraud intruders. The fraud intruder will get a chance when a collaboration of various jobs is allowed. The implementation of the feature separation of duties never allows a single individual subject to perform all transactions. The discretionary access control never ensures that the end-users are capable of performing allowed transactions, but RBAC is a central administration security system to make the end-users perform their allowed transactions. The aggregation process (grouping roles) simplifies the complexity of security administration. Many government and commercial organizations use RBAC at all levels of computing, such as database management systems, operating systems, networking, and Web-services.

Limitations. Even many organizations use RBAC, there are no complete and meaningful definitions of roles specifications and implementation in RBAC. Sometimes, the roles are expressed with the set of permissions that are not part of the role. Thus, RBAC has the facility of monitoring and analyzing the subjects and their allowed permissions, but the expressive power of security policies is poor. ABAC is such a model that has the capacity to express security policies well. But there is no facility for analyzing the subject's permissions. A rich organization needs the access control model with the good expressive power of policies and facility of analysis of the subject's permissions [3].

3.4 *Attribute Based Access Control (ABAC)*

As RBAC implements only permanent mappings for permissions-roles and roles-user, it fails to meet the security requirements of the environment where dynamic mappings and role-independent policies are necessary. In this situation, ABAC was introduced. ABAC uses attributes of subjects, objects, and environmental conditions(Time, date, economic value) to establish the rules and policies. ABAC has the capacity to achieve dynamic associations between permissions and users [4]. Figure 5 shows the simple ABAC model. ABAC made the decision (allow or deny) on the subject's request for accessing the object based on the valid attributes specified in the rules of security policies [18, 19]. Generally, ABAC rules are formed with the attributes of three categories (Subject, Object, and Environment conditions). Attributes are referred to as characteristics or properties of the subject, object. Rules are nothing but a set of conditions formed with a set of pairs of information (name and value of the attribute). A rule of ABAC policy can be expressed as follows

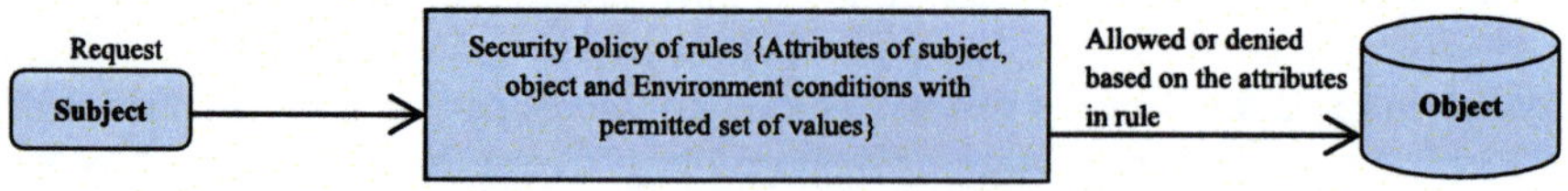

Fig. 5 ABAC model

R_1 = {allowread|Designation = {Chief-Doctor, Microbiologist, Duty-doctor, Nurse, Nutritionist}, Department = {Pedeatric}, FileName = {Blood_Reprots}, Time = {08.00–18.00}}
R_2 = {allowwrite|Designation = {Chief-Doctor, Microbiologist}, Department = {Pedeatric}, FileName = {Blood_Reprots}, Time = {08.00–18.00}}.

The above rule R_1 uses the attributes Designation and Department for the subject, Filename for Object, and Time for Environmental conditions. The attribute 'Designations' has the permitted set of values {Chief-Doctor, Microbiologist, Duty-doctor, Nurse, Nutritionist}, the attribute 'Department' has the permitted set of values {Pedeatric}, and the attribute 'Time' has the permitted set of values {08.00–18.00}. The rule R_2 also specifies the decision to be made based on given attributes and values. Based on the above rules, if the subject 'Nurse' is trying to write the object (File: Blood_Reprots), then his request will be rejected because he will be allowed for the operation 'reading a file' only. Hence, the decision is made in ABAC based on the attributes and values used in the security policies. In our ongoing research on access control models, a new parameter priority-level is introduced in ABAC rules in addition to the attributes of the above categories [14]. The inclusion of the parameter priority-level in ABAC rules avoids the anomaly of conflict-demand. The anomaly conflict-demand means that many subjects request access for the limited object.

Efficiency and Applications. RBAC is achieved great success in meeting the security requirements, even if the number of users and objects is high. But today, new computing technologies (Cloud, fog, and edge computing) increase the level of security needs and the complexity of security mechanisms. In this situation, ABAC provides great flexibility, fine-grained, coarse-grained access, and dynamic decision efficiency in protecting shared objects in large-scale applications. Today, most of the organizations which implement the new computing technologies such as cloud or IoT use ABAC to address the security issues.

Limitations. RBAC has a set of rules to ensure the authorized access of the objects, permission-role, and role-user assignments. But there are no such rules in ABAC. Due to the use of a larger number of security policies and rules in large-scale applications, it complicates the security mechanism and analysis [4]. Thus, the major drawback of ABAC is its complexity in managing a huge number of security policies. Hence, new many kinds of research are accomplished to introduce a new access control model by combining the efficient and necessary features of the previous access control models such as MAC, RBAC, and ABAC.

4 Analysis of Access Control Models

Although there are many access control models, four primary models are only chosen. In the above section, the basic concepts are described, mechanism, efficiency, applications of each model. This section is used to evaluate the above four models DAC,

Table 2 Summary of analysis of access control models

Access control model	Granularity	Flexibility	Efficiency	Security level
DAC	Good at small-scale applications	Good	Poor	Low
MAC	Good at small-scale applications	Good	Poor	Better than DAC
RBAC	Good at large-scale applications	Good	Good	Good
ABAC	Good at today's computing technologies and big data	Good	Good	Good

MAC, RBAC, and ABAC, up to the level of our knowledge gained by surveying many research papers. The properties flexibility, granularity, efficiency, and security level are used to achieve a better analysis of the four models. Table 2 gives a summary of the analysis of access control models.

Granularity. The granularity of access control is referred to as the accuracy level of the model. It depends on the number of permissions, users, mappings, or attributes used to specify the ACL or rules. DAC uses ACL to specify the relationship between user and object where MAC uses security labels. The granularity of DAC and MAC is good at only small-scale applications [6]. RBAC uses rules with two mappings (permission-role and role-user). The granularity of RBAC is based on the number of mappings used in the security rules and good at large-scale applications but poor in new computing technologies. The granularity of ABAC is measured based on the number of attributes used [4].

Flexibility. In this context, flexibility is referred to as how the security requirements are expressed, implemented, and controlled easily and efficiently. In DAC, the security concepts are controlled by the owner of the object with the use of ACL. The owner of the object can protect his object in his own way [3]. MAC uses security labels for protection of the object, but it is implemented and controlled by the central administrator [17]. Both models complicate the implementation if the number of associations between the subject and object is high. RBAC and ABAC provide flexibility in implementing security policies but complicates at large-scale and new computing technologies [4].

Efficiency. The efficiency of the access control model means that how quickly and correctly made the decision on the user request. DAC and MAC made the decision efficiently if the number of users and resources is small. RBAC made the decision (allow or deny) correctly based on the mappings (permission-role, role-user). The decision efficiency of ABAC depends on the number of attributes used in the policies. RBAC and ABAC made better decisions, but this task will be complicated in the cloud and new computing platforms due to the use of a large number of mappings or rules.

Security level. The security level of an access control model is determined by how the objects are protected efficiently from unauthorized access. DAC protects the object efficiently, but the improper use of the delegation feature affects the integrity and confidentiality of the data. MAC overcome the above issue, but it fails if the number of users and objects is high. The least-privilege feature of RBAC preserves the integrity of the data, and the feature separation of duties prevents the fraud intruder. RBAC and ABAC meet the security requirements of today large computing platforms, but expressing, implementing, and managing a large number of policies are very difficult.

5 Conclusion

The four primary and widespread access control models are analyzed and described. Each has its features and drawbacks. Either RBAC or ABAC can be used to meet the security requirements of large-scale computing organizations. ABAC is a suitable one to address the security issues of today's new computing technologies, which handles a large number of resources and big data. The major drawbacks of ABAC policies are anomalies, such as redundancy, conflict-decision, and conflict-demand. The efficiency of the ABAC model can be increased by accurately detecting and removing all the anomalies in the policies. The removal of all anomalies in ABAC policies reduces the complexity in implementing and managing security policy and improves the performance of the access control mechanism. In our previous work, an approach to detect anomalies in ABAC policies using the clustering technique is proposed, and one new parameter priority-level is introduced to avoid the anomaly conflict-demand. In our future research, all possible anomalies is found and analyzed in ABAC policies and propose an approach to remove those anomalies.

References

1. Sahafizadeh, E.: Survey on Access Control Models, pp. 1–3 (2010)
2. Vijayalakshmi, K., Jayalakshmi, V.: Big data security challenges and strategies in cloud computing: a survey. In: International Conference on Soft Computing and Optimising Techniques, Aug 2019
3. Tarigan, P.B.: Encyclopedia of Cryptograpghy and Security, 2nd ed. Springer, Berlin (2013) (J. Chem. Inform. Model. **53**(9), 1689–1699). https://doi.org/10.1017/cbo9781107415324.004
4. Qi, H., Di, X., Li, J.: Formal definition and analysis of access control model based on role and attribute. J. Inform. Secur. Appl. **43**, 53–60 (2018). https://doi.org/10.1016/j.jisa.2018.09.001
5. Sahafizadeh, E., Sartoly, S., Chamkoori, A.: Role-based access control implementation using prime numbers. In: 2009 International Conference on Computer and Electric Engineering (ICCEE 2009), vol. 1, pp. 234–237 (2009). https://doi.org/10.1109/iccee.2009.154
6. Ferraiolo, D., Kuhn, D.R.: Role-based access controls, Mar 2009
7. Sandhu, R.S., et al.: Role based access control models. Inform. Secur. Tech. Rep. **6**(2), 21–29 (1996). https://doi.org/10.1016/S1363-4127(01)00204-7
8. Ninglekhu, J., Krishnan, R.: Attribute based administration of role based access control : a detailed description. In: CoRR 2017 vol. abs/1706.03171 (2017)

9. Sandhu, R., Bhamidipati, V., Munawer, Q.: The ARBAC97 model for role-based administration of roles. ACM Trans. Inf. Syst. Secur. **2**(1), 105–135 (1999). https://doi.org/10.1145/300830.300839
10. Sandhu, R., Munawer, Q.: The ARBAC99 model for administration of roles. In: Proceedings of Annual Computer Security Application Conference (ACSAC), pp. 229–238 (1999). https://doi.org/10.1109/csac.1999.816032
11. Oh, S., Sandhu, R., Zhang, X.: An effective role administration model using organization structure. ACM Trans. Inf. Syst. Secur. **9**(2), 113–137 (2006). https://doi.org/10.1145/1151414.1151415
12. Bishop, M., Nascimento, A.C.A.: Information security example policy. In: Uni-ARBAC: A Unified Administrative Model for Role-Based Access Control, vol. 1, pp. 218–230. Springer, Berlin (2016). https://doi.org/10.1007/978-3-319-45871-7
13. Suhendra, V.: A survey on access control deployment. Commun. Comput. Inf. Sci. **259**, 11–20 (2011). https://doi.org/10.1007/978-3-642-27189-2_2
14. Vijayalakshmi, K., Jayalakshmi, V.: A priority-based approach for detection of anomalies in ABAC policies using clustering technique. In: 2020 Fourth International Conference on Computing Methodologies and Communication (ICCMC), pp. 897–903. IEEE (2020)
15. Griffiths, P.P., Wade, B.W.: An authorization mechanism for a relational database system. ACM Trans. Database Syst. **1**(3), 242–255 (1976). https://doi.org/10.1145/320473.320482
16. Faircloth, J., Faircloth, J.: Information Security (2014)
17. Amin, M., Nauman, M., Ali, T.: A comprehensive analysis of MAC enhancements for leveraging distributed MAC. Lect. Notes Eng. Comput. Sci. **2168**(1), 293–300 (2008)
18. Crampton, J., Morisset, C.: Monotonicity and completeness in attribute-based access control. In: LNCS 8743, pp. 33–34. Springer International Publishing (2014)
19. Tran Thi, Q.N., Si, T.T., Dang, T.K.: Int. Publ. A **10018**, 305–315 (2016). https://doi.org/10.1007/978-3-319-48057-2

Application Monitoring Using Libc Call Interception

Harish Thuwal and Utkarsh Vashishtha

Abstract Generally available application monitoring solutions comprise basic aggregated metrics. This paper collects such metrics in the form of raw data. We have worked out a very primitive piece written in `Golang`, which manifests results from `HTTP/HTTPS` calls from/to the system being monitored. This process can be easily extended and applied to various architectures that use the `libc` library to make low-level system calls to interact with sockets or, in a more basic sense, files. The experimentation and the proof of concept were carried out on various languages, viz. `C, Java, Python`, and `Golang` itself using Docker and some classic open-source compilers.

Keywords Process-level monitoring · `ld_preload` · gccgo · Golang monitoring · Language-agnostic monitoring

1 Introduction

The central problem started with trying to develop a `Golang` monitoring system which worked agnostic of the underlying architecture and was a stable solution but ultimately ended with the **development of a language-agnostic, platform-agnostic solution to collect data at the process level**, allowing us to essentially collect data from sockets at the application layer.

The study falls under application monitoring but instead of highly sophisticated techniques available today, it provides for a crude but novel technique to be taken and transformed into production-ready software. This mitigates challenges that one faces while trying to study their application which gets mostly fixed at compile time,

H. Thuwal (✉)
Department of Computer Science and Engineering, IIT, Delhi, India
e-mail: hthuwal@gmail.com

U. Vashishtha
Department of Computer Science and Information Systems, BITS, Pilani, India
e-mail: f2015130@alumni.bits-pilani.ac.in

© The Editor(s) (if applicable) and The Author(s), under exclusive license to Springer Nature Singapore Pte Ltd. 2021

A. P. Pandian et al. (eds.), *Proceedings of International Conference on Intelligent Computing, Information and Control Systems*, Advances in Intelligent Systems and Computing 1272, https://doi.org/10.1007/978-981-15-8443-5_34

instead of being flexible enough to be mutated at run time. To cite an example, the Go language with a compiler that uses the `gc toolchain` resolves its `bind()` (the socket bind operation) method at compile time, the memory location of this particular method gets ingrained in the binary(.o) file and the corresponding symbol in the generated symbol table gets resolved.

The work performed as part of this paper not only shows that we can perform monitoring of applications programmed in compiled languages but also opens up various possible use cases. The ability to intercept low-level calls allows us to noninvasively monitor, verify, restrict, and even modify a program's behavior.

When it comes to application monitoring, we show that any application can be monitored irrespective of the language used to build it, whereas traditional monitoring systems usually require different solutions for different programming languages. Any I/O operation, be it a network call or a file system access, occurs via low-level calls which when intercepted gives us access to the data being read and written to the network or the file system. This allows us to monitor the I/O performance of an application and also reinforce the security of the system by discarding suspicious or malicious data.

A simple illustration of restricting a program's behavior is limiting its ability to allocate/deallocate memory to a specific number of bytes by intercepting the `malloc/free` system calls. This will prevent any program from using more than a definitive amount of memory. Another example is preventing sensitive and special files from being accessed irrespective of the user access level. This can be achieved by tapping in the `fopen` system call and discarding it if the path of the file being accessed matches a sensitive location.

The following sections of the paper discuss the core idea of low-level call interception and its application to perform basic monitoring of a Golang server, but it can easily be extended to languages like Python and Java.

2 Literature Review

Sahasrabudhe et. al in [6] demonstrate application monitoring methodologies. These involve the deployment of an application performance monitoring tool, generally referred to as the agent, alongside the application being monitored in the production environment. It collects the adequate metrics and thresholds and sends these over to a data store which creates and evaluates various important KPIs. Solutions based on the aforementioned methodology usually do not have a single universal tool but rather have different solutions for every programming language. The work in the paper as will be described ahead provides a way to overcome this constraint for many cases.

Saito [2] in his work proposed an execution record/replay tool for debugging Linux programs. The tool can record the invocation of system calls and CPU instructions and later replay them in a deterministic manner. This approach is limited to instances where the link step of the application can be modified, as the symbols of interest need to be specified at link time. At its core, the tool consists of a compiled shared object `libjockey.so` which is used to intercept every system call in `libc`. This is done by preloading the shared object before the target application. During recording, the shared object intercepts and logs the values generated by the system calls which are then returned using the replay. To ensure that the target sees the same set of environment variables and command-line parameters during both record and replay, the tool also creates checkpoints.

Shi et al. [4] show that the same principle of intercepting/hooking the system calls can be used to create a Lightweight File System Management Framework. The main idea is to hook and modify file requests sent by applications from management service running in user space. One thing that they do differently from the record/replay tool is that they use `LD_PRELOAD` in two stages, once to hook the file system management framework to a shell program such as `bash` to get access to all processes being created and then to preload a dynamic library based on the type of process. Former interception targets all `exec()` functions which are the entry point for all user process creation, while the latter targets process-specific calls (specified by a Rule lib) depending on the use case.

Lee et al. [5] in their paper show that system call and standard library interception attacks are possible on the ARM-based Linux system. They preload a `libhook.so` object and demonstrate that they can modify the behavior of `getuid()` function. Though the goal here is to demonstrate interception attacks such as rootkit.

Brendel et al. [8] demonstrate a simple and robust way to record performance data on library function calls in *C/C++*. To demonstrate this, they eventually need to create wrappers around the functions they aim to benchmark. They use `-wrap` option of the GNU to do this statically. But the static approach is limited to instances where the link step of the application can be modified, as the symbols of interest need to be specified at link time.

All the aforementioned techniques, from various sources, end up utilizing `LD_PRELOAD` to either inject a shared object into the main application or for wrapping functions dynamically.

3 Tools Used

The design and working code was developed for a small-scale application using specific Docker images running on a couple of machines and consequently excludes the performance aspects of this solution for very high-intensity processes. Some type of degradation in performance corresponding to the operations involved in call interception and generation of any custom metrics, if any, is bound to happen, the intensity of which may vary depending upon the use case and how efficient the solution pans out taken forward from its current crude form. We primarily used the following tools for our study:

3.1 LD_PRELOAD

Before the execution of any program, the Linux dynamic loader locates and loads all the necessary shared libraries to perform symbol resolution and prepare the program for execution.

For this study, we made extensive use of the `LD_PRELOAD` [10] environment variable which provides one with an option to communicate with the loader to load custom logic in the form of shared objects before any other library. This is generally referred to as preloading a library that allows the preloaded library methods to be used before others of the same name in any other library being loaded later on. This provides us with the capability to intercept commonly used methods and replace them allowing for a subtle modification without re-compilation of the user code. The following listings demonstrate this through an example of a `C` program.

```
#include <stdio.h>
#include <stdlib.h>
#include <time.h>

int main(){
  srand(time(NULL));
  int i = 10;
  while(i--) printf("%d\n",rand()%100);
  return 0;
}
```

Listing 1.1 random_num.c [9]

```
int rand(){
    return 42; //the most random number in the
   universe
}
```

Listing 1.2 unrandom.c Overriding rand [9]

The main application `random_num.c` is programmed to print 10 random numbers on each execution. But when the compiled shared object of `unrandom.c` is preloaded before its execution using

```
LD_PRELOAD=unrandom.so ./random_nums
```

the injected implementation of the `rand` is used and ergo the output contains the number 42 printed 10 times. We leverage this technique to inject our monitoring code into the main application.

3.2 Gccgo

The creators/maintainers of Golang support two different types of compilers: `gc` and `gccgo` [3] which use two separate `toolchains`, namely the `gc toolchain` and the `gcc toolchain`. Go is a compiled language and needs to be assembled to a binary format before it is executed. This binary contains assembly code from either of those `toolchains`, depending upon the compiler you choose.

`gccgo` is a `Go` front end connected to the `GCC` back end. Compared to `gc`, `gccgo` compiled code contains optimizations implemented in `GCC` over the years, and while it may be slower to compile, it supports more powerful optimizations and consequently in many cases may run faster [12].

Due to its long and successful history, GCC, and hence the `gccgo` compiled code, supports many more processors and architectures than `gc`. This may further be utilized to create a better architecture-agnostic framework supporting even `Solaris`.

The primary consequence to be drawn out from this study is how to build a multi-purpose monitoring/ data collection system and put it into a binary that can be easily used. As aforementioned as part of this study, we would be using Docker images to create a Go server and multiple users in multiple programming languages, basic Linux tools to figure out various `libc` calls being made during server–user interaction, an ingestion service, using a back-end DB for storing any data/metrics that would be created.

4 Experimentation and Implementation

After developing the call interception mechanism using `LD_PRELOAD` 3.1, we used the `strace`[11] and `ptrace`[1] system calls to figure out which `libc` calls to intercept to get the required information, essentially the socket level data, which amounts to reading from a simple file descriptor. We realized during this exercise that, for writing the interception class, we would need to override all the basic calls related to file interactions used while working with a socket. For instance, the call used to accept an incoming connection, which is an `accept()` `libc` call, is overloaded into many types such as the `accept4()` method. Different programming languages use different forms of the `accept()` method. We would also need to maintain access to the original `libc` handles so that we can perform the actual task after we are done gathering data. This was done by using `dl GoLang` package [7], a runtime dynamic library loader.

This multi-call interception class was then compiled into a shared object(.so) file that would be further injected using the aforementioned `LD_PRELOAD` technique.

```
//export accept4
func accept4(s C.int, a unsafe.Pointer, st C.size_t,
    flags C.int) *C.int {
     t := (*syscall.RawSockaddr)(a)
     lib, _ := dl.Open("libc", 0)
     defer lib.Close()

     var old_accept func(s C.int, a *syscall.
    RawSockaddr, st C.size_t, fl C.int) *C.int
     lib.Sym("accept", &old_accept)
     oa := old_accept(s, t, st, flags)
     return oa
}
```

Listing 1.3 Intercepting accept4 libc call in Golang

We also realized, using the `nm` command, that the symbol table created when compiling a Go program using the go toolchain generates already resolved symbols for the basic system calls that we initially intended to override for the interception. Next we stumbled upon the gccgo compiler that uses the gccgo toolchain explained in Sect. 3.2 which renders these symbols, for instance the `bind()`, `listen()`, etc., unresolved. We needed to now ensure that the target application uses gccgo for compilation.

Using `LD_PRELOAD`, `GccGo`, and `libc` interception, we were able to successfully override `libc` system calls comprehensive enough to encompass most of the mainstream programming languages. Our final goal was to intercept socket interaction calls that were related to any of the HTTP verbs. We wanted to capture the data being put into/read by a socket, specifically a file descriptor (FD), since at a particular time a process would be allocated a socket, and thereby an FD, creating somewhat of a session identifier that could be useful to study the pattern of data being sent from a particular IP.

```
//export read
func read(fd C.int, buf unsafe.Pointer, count C.size_t
    ) *C.size_t {
     // Similar to accept, read the data using the libC
     read
     // into strdata

     // Perform monitoring/reporting tasks only for
    network traffic
     if socket != nil && strings.Contains(strdata, "
    HTTP") {
          // It is a response of earlier outgoing
    request
          if !strings.Contains(strdata, "Host:") {
               socketstr := getPortFromSocket(socket)
               if guid, ok := outgoingcalls[socketstr];
    ok {
```

```
                    // do whatever you want here
                }
            } else { // New Incoming request that contains
    singularity header
                if strings.Contains(strdata, "singularity"
    ) {
                    // do whatever you want to do.
                }
            }
        }
        return oa
}
```

Listing 1.4 Intercepting read libc call in Golang

We published this (IP, FD, socket data) tuple into a back-end database (Fig. 1) which could be queried according to the type of data/IP to get a kind of cumulative information for any further analysis.

5 Results and Future Work

Figure 1 depicts the overall architecture of the final setup that we implemented.

Element descriptions are as follows:

- **Source Application**: The target application written in `Golang` that will ultimately be monitored. The application can both receive and send HTTP traffic.
- **Interception Code**: The shared object code that contains the `libc` call interception, filtering, and monitoring logic.

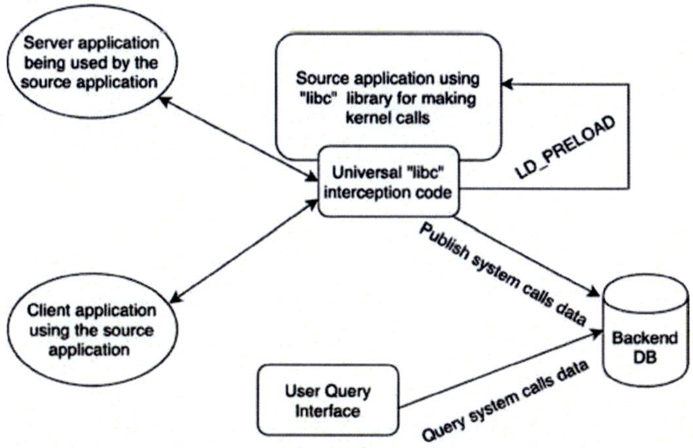

Fig. 1 Overview

- **Server Application**: The application that the source application uses for its services.
- **Client Application**: The application that uses source application for its services.
- **Back-end DB**: The database to populate the collected data from the source application.
- **User Query Interface**: The framework to query back-end DB.

The source application preloaded with our interception code receives the `HTTP` request from the client application and sends an `HTTP` request to a Web service. The incoming connections from client applications are monitored by intercepting the `read libc` call, while the `write libc` call interception is used to monitor the outgoing requests to the Web server. The monitoring data and statistics collected by the interception code are periodically published to the back-end database which can be later queried to analyze the source application's behavior.

Note that for demonstration purposes, to simulate rather simple real-time systems, we used Docker images to create the aforementioned source/client/server applications across multiple machines and captured metrics/socket data on a separate one.


```
"_source" : {
    "data1" : {
        "Port": 37474,
        "Addr": [172,17,0,2],
        "Content": "HTTP/1.1 200 OK\nContent-Encoding:
     gzip\nAccept-Ranges: bytes\nCache-Control: max-age
    =604800\nEtag: \"3147526947+gzip\"\nExpires: Thu,
    30 Jan 2020 19:56:53 GMT\nServer: ECS (phd/FD69)
    .....more data"
     },
    "data2" : {
        "Port": 37500,
        "Addr": [172,17,0,2],
        "Content": "HTTP/1.1 200 OK\nContent-Encoding:
     gzip\nAccept-Ranges: bytes\nCache-Control: max-age
    =604800\nEtag: \"3147526947+gzip\"\nExpires: Thu,
    30 Jan 2020 19:56:54 GMT\nServer: ECS (phd/FD69)
    .....more data"
     },
    "data3" : {
        "Port": 37488,
        "Addr": [172,17,0,2],
        "Content": "HTTP/1.1 200 OK\nContent-Encoding:
     gzip\nAccept-Ranges: bytes\nCache-Control: max-age
    =604800\nEtag: \"3147526947+gzip\"\nExpires: Thu,
    30 Jan 2020 19:56:58 GMT\nServer: ECS (phd/FD69)
    .....more data"
     }
}
```


Listing 1.5 Sample monitoring data

Listing 1.5 depicts sample data collected from the source application, and this is the raw data obtained from a socket's file descriptor. The application as of now only captures the raw data but can be further developed to create a full-fledged socket monitoring application.

Our implementation mainly focuses on the HTTP traffic but since we can intercept calls associated with opening a file, closing a file, and thereby have access to every byte being written to or read from the associated file descriptor one can easily extend the same idea to perform file system monitoring.

Apart from the `libc` call interception approach, we also pondered over the idea of ELF file manipulation which if successful might be more performant. `Golang` applications generally use the net/http library to send and receive HTTP requests. Similar to `libc` call interception, this approach also involves preloading a shared object. The shared object will contain a modified implementation of the net/http library's `httpServeAndListen` method which is essentially used by all `goLang` written code to serve http calls. The task will then be to modify the target ELF in such a way that the `httpServeAndListen` calls now go to the memory address of our modified implementation. This path can be explored as part of further research.

6 Conclusions

Through this work, we demonstrated that by intercepting `libc` calls we can non-invasively monitor an application irrespective of the programming language used, given it makes use of those for basic open/read/write operations. The potential implications open up future work to easily create in-house monitoring systems that could span across multiple tiers from files to cross-service communications.

The exponential pace at which every sector of the world is becoming digital has made application monitoring the need of the hour. But since modern applications span across different platforms and programming languages, the task of monitoring all of them comprehensively becomes difficult. The ability to perform language-agnostic monitoring simplifies this task by requiring just one system that can monitor all.

Acknowledgements This work was done as part of a hackathon organized by AppDynamics, Cisco. We would like to thank and acknowledge AppDynamics for providing us with the opportunity to take part in the hackathon which enabled us to research, design, and implement the proposed work.

References

1. Sahasrabudhe, M., Panwar, M., Chaudhari, S.: Application performance monitoring and prediction. In: 2013 IEEE International Conference on Signal Processing, Computing and Control (ISPCC) (2013), pp. 1–6
2. Saito, Y.J.: A user-space library for record-replay debugging. In: Proceedings of the Sixth International Symposium on Automated Analysis-Driven Debugging (2005), pp. 69–76

3. Shi, Z., Feng, D., Zhao, H., Zeng, L. USP: A lightweight file system management framework. In: 2010 IEEE Fifth International Conference on Networking, Architecture, and Storage (2010), pp. 250–256
4. Lee, H., Kim, C.H., Yi, J.H.: Experimenting with system and Libc callinterception attacks on ARM-based Linux kernel. In: Proceedings of the 2011 ACM Symposium on Applied Computing (2011), pp. 631–632
5. Brendel, R., et al.: Programming and Performance Visualization Tools, pp. 21–37. Springer, Berlin (2017)
6. ld.so, ld-linux.so - dynamic linker/loader. http://man7.org/linux/man-pages/man8/ld.so.8.html
7. Dynamic linker tricks, in 5.7.1. https://rafalcieslak.wordpress.com/2013/04/02/
8. Lance Taylor, I.: The Go frontend for GCC in Proceedings of the GCC Developers' Summit pp. 115–128 (2010)
9. The Go Blog, Gccgo in GCC 4.7.1. https://blog.golang.org/gccgo-in-gcc-471
10. strace - trace system calls and signals. http://man7.org/linux/man-pages/man1/strace.1.html
11. Padala, P.: Playing with ptrace. Part I (2002)
12. S.L., R. C. Runtime dynamic library loader https://github.com/rainycape/dl (2015)

Classification of Clinical Reports for Supporting Cancer Diagnosis

Amelec Viloria, Nelson Alberto, and Yisel Pinillos-Patiño

Abstract Currently, in the clinical field, large amounts of clinical texts or reports expressed in natural language (unstructured data) are generated in the form of pre-operative notes, discharge notes, radiological reports, examination and findings reports, admission notes, among others. The automatic processing of this information is complex and costly since it does not have a semantic and computer-processable structure that can make its retrieval, categorization, and automatic analysis possible. This paper presents a supervised classification approach to clinical reports using the support vector machine (SVM) algorithm. Linguistic information from the texts is used to support the diagnosis of four types of cancer: stomach, lung, breast, and skin.

Keywords Classification of clinical reports · Cancer diagnosis · Support vector machine (SVM) algorithm

1 Introduction

The accelerated growth of large amounts of unstructured clinical data generated today in the field of medicine is due to the widespread adoption of the "Electronic Health Records (EHR)," and the appropriate use of this information has the potential to deliver assisted clinical care [1].

Oncology is a clinical specialty that generates large amounts of medical notes or reports in unstructured text. This information, in many cases, is not analyzed in its

A. Viloria (✉)
Universidad de la Costa, St. 58 #66, Barranquilla, Atlántico, Colombia
e-mail: aviloria7@cuc.edu.co

N. Alberto
Universidad Tecnológica, San Pedro Sula, Honduras
e-mail: nelson@unitet.edu.hn

Y. Pinillos-Patiño
Universidad Bolivar, Barranquilla, Atlántico, Colombia
e-mail: ypinillos@unisimonbolivar.edu.co

© The Editor(s) (if applicable) and The Author(s), under exclusive license to Springer Nature Singapore Pte Ltd. 2021

A. P. Pandian et al. (eds.), *Proceedings of International Conference on Intelligent Computing, Information and Control Systems*, Advances in Intelligent Systems and Computing 1272, https://doi.org/10.1007/978-981-15-8443-5_35

entirety, nor is it processed to support decision making for cancer early diagnosis, timely treatment, or constant monitoring of diagnosed cancer patients [2].

In this approach, the patient becomes a key factor and focal point of any tool or information processing system. In the future, clinical decision makers can rely on these clinical text analysis tools to improve diagnoses and avoid errors in this critical decision. The idea is to improve the quality of life in patients through appropriate and timely clinical decision making, whether inpatient or outpatient [3]. For both kinds of patients, the error-free and timely diagnosis and automated monitoring of their health status are of paramount importance. The treatment of this unstructured information involves a great challenge for the Natural Language Processing (NLP) due to the great diversity of structures and language phenomena that present in these reports or clinical notes. However, supervised learning supported by NLP techniques can make possible the classification of clinical notes or reports to support the diagnosis of cancer. An admission note is generated when a patient already has a clinical record open and describes his or her current condition [3].

The early diagnosis and treatment of cancer, as well as its constant and timely monitoring, significantly improves the quality of life of inpatients and outpatients. Computational tools and approaches have been proposed to automatically analyze the texts of clinical notes and reports. For this reason, several studies are presented below to extract information, categorize clinical texts, and classify them [4].

The extraction of information from clinical texts is a task that serves as a basis for various classification, grouping or processing tasks on clinical data. In this area, there are approaches for extracting symptoms for various diseases such as depression, hypertension, diabetes [5–8]; drugs or medications for the treatment of a disease or drug iterations [9–11]; names of diseases and relationships between them to find inheritable family histories in clinical notes [12, 13], and patient monitoring [14, 15].

This paper presents a supervised classification approach to clinical reports, specifically, oncology specialty admission notes. The entire process involves processing the text of the admission note through preprocessing, extracting linguistic characteristics (grammatical word labels), and finally, weighting the characteristics that will be the basis of the supervised learning algorithm to determine the type of cancer described in the admission note.

2 Preprocessing of Admission Notes

Admission notes created by physicians have a current condition section that describes the evaluation made on the patient under study. It describes the patient's signs and symptoms, appearance, and findings. This current condition is described as unstructured text and may contain irrelevant information. Therefore, a number of tasks are performed as preprocessing in order to improve the quality of these texts.

The texts in the admission note describing the patient's current condition are segmented, i.e., divided into words (tokens), while special characters (# $ % & *), periods, commas, and signs (¿, ¡) are removed. The resulting sentences are tagged with the corresponding grammatical categories (noun, verbs, adjectives, pronouns, and determinants) using the TreeTagger tool [3]. Additionally, a lemmatization is carried out for reducing the words to their roots, eliminating suffixes, inflections, and conjugations of the words.

Additionally, a normalization of the sentences is carried out by means of conversion to lower case and eliminating the stopwords which are words that do not provide meaning and therefore are not functional for the classification of the type of cancer. Stopwords include articles (un, la, los), prepositions (a, con, de, para), and non-functional verbs (ser, estar).

3 Representation and Weighting of Characteristics

A set of characteristics was used for the representation of admission notes. Two types of characteristics are in this paper: linguistic characteristics and grammatical word unigrams.

The linguistic characteristics correspond to the number of verbs, nouns, conjunctions, determinants, adjectives, adverbs, personal pronouns, and prepositions found in an admission note. In addition, the following simple characteristics are added:

(a) Average length of sentences in terms of words.
(b) Presence of denials (no/no, negado/denied, ni/neither, nunca/never). This is a binary characteristic.
(c) Presence of verbs that indicate a symptom (presenta/present, tiene/has, acude con/comes with).

The unigrams correspond to the list of words without repetition that an admission note contains. The grammatical category is preserved because it has been experimented with verbs, nouns, adjectives, and their combination.

The weighting of the unigrams is carried out using the bag of words model as a vector, $V_j = (v_{1j}, v_{2j}, v_{3j})$ which consists of a lexicon or dictionary of the words in the admission notes. The weighting consists of determining the values of each word, so that the component represents the importance produced by characteristic i, in admission note j with the words in the whole set of notes. The importance of a word (unigram) is determined by the TF-IDF formula (word frequency in the admission note in relation to the word frequency in the whole note set). To do this, it is necessary to obtain, through Eq. 1, the value of term frequency (TF), which consists of the number of times that a word (t) appears in an admission note (S) [13, 15].

$$\mathrm{TF}(t_i, S_j) = f(t_i, S_j) \tag{1}$$

Then, it obtained the inverse frequency that indicates if the term is common in the collection of clinical admission notes and that is obtained by Eq. 2.

$$\mathrm{IDF}(t_i, S_j) = \log \frac{|S|}{1 + |s \in S: t_i \in s} \quad (2)$$

This information is then used to calculate the final value of TF-IDF using Eq. 3.

$$W_{ij} = \mathrm{TF}(t_i,\quad S_j) \times \mathrm{IDF}(t_i, S_j) \quad (3)$$

Finally, a normalization phase is carried out from the matrix obtained by applying Eq. 4.

$$W_{\mathrm{norm}} = \frac{W_{ij}}{\sqrt{\Sigma_{i=0}^{n} |W_{ij}|^2}} \quad (4)$$

where n represents the total number of admission notes and j expresses each note.

4 Classification of Admission Notes

The identification of the type of cancer described in the admission note, specifically in the section on the patient's current condition, corresponds to a typical text classification task. The idea is to determine the category of cancer expressed in the note. Four types of cancer are considered in the admission notes: stomach, lung, breast, and skin cancer.

The classification of texts corresponding to the current condition is based on the weighted and standardized word vector of each of the admission notes. These vectors are the input for the support vector machines supervised classification algorithm, which aims to predict the category of the note based on a set of previously labeled training notes.

The task of classifying admission notes is carried out by means of the support vector machine (VSM) algorithm [2], which has been widely used in the classification of texts with simple labels. This classifier constructs a set of hyperplanes in n-dimensional space with the texts of the training notes, and these hyperplanes are used to predict the class of the new admission notes.

The idea is to evaluate the classification task, combining the various characteristics (linguistic and n-grams) with the SMV algorithm and the word weighting (TF-IDF) to find the best configuration in terms of accuracy and coverage. The implementation of the classification algorithm was carried out using the WEKA tool [4].

5 Results

In this paper, the experimentation consists in the task of classifying clinical reports (admission notes) for early cancer detection in order to support decision making with respect to diagnosis. This experimentation is based on word n-grams and linguistic characteristics of the notes. The evaluation of the admission notes classification approach is carried out with the database called MIMIC-II [5]. The MIMIC-II database contains, among other data, clinical notes with unstructured text in English from approximately 40,000 stays into intensive care units (ICU) of nearly 33,000 patients at Beth Israel Deaconess Medical Center (BIDMC) in Boston, Massachusetts, from 2010 to 2018.

For each hospitalization, the study included all the ICU notes that describe the doctor's admission note, thus conforming a set of admission notes. These notes correspond to patients, who are identified through the field HADM_ID of the hospitalizations during which a diagnosis of interest was assigned or made. This diagnosis is of utmost importance to this approach since it provides a single label for each admission note, which is solved with the support vector machine (SVM) sorter with simple labels.

Only those belonging to four categories (stomach, lung, breast, or skin cancer) are extracted from the total set of admission notes, because they have a higher number of notes. A total of 1078 admission notes are extracted and distributed as given in Table 1, which sets out the number of notes for each diagnostic category. These notes will then be divided into two sets: training and testing.

The characteristic vectors are extracted from this set of 1371 admission notes. These vectors are then divided into a classifier training set and a test set to validate the efficiency of the classification task. The training set corresponds to 70% of the scores for a total of 960, while the rest corresponds to the test set, which is made up of 411 admission notes.

The experiment consists of using the classification algorithm called support vector machines with the TF-IDF weighting. All the experiments were carried out with the following parameters: Complexity (number of hyperplanes to be built):—C 1; Range (type of kernel to be used):—K PolyKernel; Size of the cache memory to be used:—C 250007; Tolerance:—L 0.001.

For the classifier evaluation stage, a set of mutually exclusive evidence of the training set is provided, consisting of 324 clinical reports (admission notes). Two practical metrics are used for the analysis of admission score classification: percentage of correctly classified instances (C) for each category and the percentage of incorrectly classified instances (I). All experiments were performed on the uniquely labeled set in four types of cancer diagnoses: stomach, lung, breast, and skin.

Table 1 Distribution of admission notes by diagnosis

Diagnosis	Number of admission notes
Stomach cancer	330
Lung cancer	451
Breast cancer	289
Skin cancer	301
Total	1371

Table 2 Summary of classification of medical notes using linguistic characteristics

Types of diagnostics	% Correct	% Incorrect
Stomach cancer	46.4	55.4
Lung cancer	60.2	41.5
Breast cancer	40.2	63.0
Skin cancer	48.6	54.2
Average	48.0	53.2

Table 3 Summary of classification of medical notes with grammatical characteristics

Grammatical feature	Noun		Verb		Adjective	
	% C	% I	% C	% I	% C	% I
Stomach cancer	70.2	32.4	57.2	44.2	65.1	34.2
Lung cancer	75.3	24.1	63.1	39.1	72.3	29.4
Breast cancer	72.1	27.0	53.5	48.1	62.8	39.1
Skin cancer	71.6	30.2	51.5	50.2	64.3	37.1
Average	73.2	29.2	54.5	45.2	66.2	35.0

An exhaustive and comparative experimentation between linguistic characteristics and word unigrams by grammatical category is performed. The grammatical categories to be evaluated are nouns, verbs, and adjectives. Table 2 shows the results of the classification using linguistic characteristics for each category, which were described in Sect. 2.

Table 3 shows the results of the classification using the lexicon of unigrams of verbs, nouns, and adjectives as characteristics for the classification of clinical notes based on the diagnosis of type of cancer.

6 Discussion

The results presented in Tables 2 and 3 show, in summary, that the classification of admission notes using grammatical characteristics is better than using simple linguistic characteristics such as the length of sentences, the presence of negations, and the presence of verbs indicating symptoms. In addition, Table 3 shows that from the grammatical categories, the best results are obtained using nouns, with which an average percentage of 73.2% of instances correctly classified for the four categories is achieved. Therefore, it is verified that this behavior is due to the fact that nouns describe disease names, drug names, symptom names, and any entity named within the admission notes. This means that by detecting any entity named as a noun, the algorithm better divides the four categories of cancer diagnoses.

7 Conclusions

In this paper, an approach to the classification of cynical notes was presented using the automatic learning algorithm called support vector machines (SVM) to predict the category or type of cancer described in the current condition of each of the admission notes.

The processing of admission notes includes various tasks. First, a preprocessing of the clinical notes that involves lemmatization, elimination of empty words, and grammatical labeling of parts of speech. In addition, admission notes are characterized as a set of linguistic and grammatical characteristics, in addition to using the TF-IDF measure for term weighting or characteristics. Finally, a classification of the admission scores by means of the SVM algorithm is performed on the set of admission notes.

The main contributions of this study are: (a) the approach of classification of admission notes by means of the support vector machines algorithm; (b) the characterization of notes by means of the length of sentences, the presence of negation words, and the presence of verbs that indicate symptoms, in addition to presenting, as another characterization alternative, lexicons grouped by grammatical categories of verbs, nouns, and adjectives; (c) the discovery of the best alternative for classifying clinical notes to support the diagnosis of cancer, showing that nouns better characterize the chosen cancer groups, due to their property for expressing disease names, drug names, and symptom names. The approach obtained can be very useful when generating a tool for the support of cancer diagnoses, and it is possible to extend it to diverse diseases such as diabetes and hypertension, among others.

References

1. Sukhai, M.A., Craddock, K.J., Thomas, M., Hansen, A.R., Zhang, T., Siu, L., Kamel-Reid, S.: A classification system for clinical relevance of somatic variants identified in molecular profiling of cancer. Genetics Med. **18**(2), 128–136 (2016)
2. Wu, W., Li, B., Mercan, E., Mehta, S., Bartlett, J., Weaver, D.L., Shapiro, L.G.: MLCD: A Unified Software Package for Cancer Diagnosis. JCO Clin. Cancer Inform. **4**, 290–298 (2020)
3. Chandrasekaran, S.T., Hua, R., Banerjee, I., Sanyal, A.: A fully-integrated analog machine learning classifier for breast cancer classification. Electronics **9**(3), 515 (2020)
4. Sun, Y., Reynolds, H., Wraith, D., Williams, S., Finnegan, M.E., Mitchell, C., Haworth, A.: Predicting prostate tumour location from multiparametric MRI using Gaussian kernel support vector machines: a preliminary study. Australas. Phys. Eng. Sci. Med. **40**(1), 39–49 (2017)
5. Viloria, A., et al.: Determination of dimensionality of the psychosocial risk assessment of internal, individual, double presence and external factors in work environments. In: Tan, Y., Shi, Y., Tang, Q. (eds.) Data Mining and Big Data. DMBD 2018. Lecture Notes in Computer Science, vol. 10943. Springer, Cham (2018)
6. Kocbek, S., Cavedon, L., Martinez, D., Bain, C., Mac Manus, C., Haffari, G., Verspoor, K.: Text mining electronic hospital records to automatically classify admissions against disease: measuring the impact of linking data sources. J. Biomed. Inform. **64**, 158–167 (2016)
7. Vijayakumar, T.: Classification of brain cancer type using machine learning. J. Artif. Intell. **1**(02), 105–113 (2019)

8. Gharibdousti, M.S., Haider, S.M., Ouedraogo, D., Susan, L.U.: Breast cancer diagnosis using feature extraction techniques with supervised and unsupervised classification algorithms. Appl. Med. Inform. **41**(1), 40–52 (2019)
9. Moccia, S., De Momi, E., Guarnaschelli, M., Savazzi, M., Laborai, A., Guastini, L., Mattos, L.S.: Confident texture-based laryngeal tissue classification for early stage diagnosis support. J. Med. Imag. **4**(3), 034502 (2017)
10. Schelb, P., Kohl, S., Radtke, J.P., Wiesenfarth, M., Kickingereder, P., Bickelhaupt, S., Maier-Hein, K.H.: Classification of cancer at prostate MRI: deep learning versus clinical PI-RADS assessment. Radiology **293**(3), 607–617 (2019)
11. Nakhleh, M.K., Amal, H., Jeries, R., Broza, Y.Y., Aboud, M., Gharra, A., Glass-Marmor, L.: Diagnosis and classification of 17 diseases from 1404 subjects via pattern analysis of exhaled molecules. ACS Nano **11**(1), 112–125 (2017)
12. Banizs, A.B., Silverman, J.F.: The utility of combined mutation analysis and microRNA classification in reclassifying cancer risk of cytologically indeterminate thyroid nodules. Diagn. Cytopathol. **47**(4), 268–274 (2019)
13. Aubreville, M., Knipfer, C., Oetter, N., Jaremenko, C., Rodner, E., Denzler, J., Maier, A.: Automatic classification of cancerous tissue in laserendomicroscopy images of the oral cavity using deep learning. Sci. Rep. **7**(1), 1–10 (2017)
14. Doig, K.D., Fellowes, A., Bell, A.H., Seleznev, A., Ma, D., Ellul, J., Lara, L.: PathOS: a decision support system for reporting high throughput sequencing of cancers in clinical diagnostic laboratories. Genome medicine **9**(1), 38 (2017)
15. Esteva, A., Kuprel, B., Novoa, R.A., Ko, J., Swetter, S.M., Blau, H.M., Thrun, S.: Dermatologist-level classification of skin cancer with deep neural networks. Nature **542**(7639), 115–118 (2017)

A Review on Swarm Intelligence Algorithms Applied for Data Clustering

N. Yashaswini Gowda and B. R. Lakshmikantha

Abstract Nature has inspired researchers in many ways, and swarm intelligence (SI) algorithms are also results of nature's inspiration. The coordination, food search techniques, and fighting for survival techniques from birds, animals, and also insects have given researchers many areas to think upon. These algorithms are results of ants, bats, fireflies, fishes, cuckoos, and many more. Swarm means being together, so these algorithms are a result of species which live together in a large number. Clustering means separating, today data is available in abundance, but segregating the data accurately is necessary before working on it. So, different SI algorithms which are used for data clustering are discussed. SI algorithms give better clustering of data than the traditional clustering algorithms. This paper gives the reader a timely analysis of different SI algorithms applied in data clustering.

Keywords Swarm intelligence · Data clustering · SI algorithms · Optimization · PSO · ACO · Swarms · Bat algorithm

1 Introduction

1.1 Swarm Intelligence

Study of swarms deals with collective behavior of swarming organisms. Organisms which live in groups enable to solve problems that are impossible to resolve individually. Hence, swarm intelligence is a solution which one can use to solve

N. Yashaswini Gowda (✉)
Department of Electronics & Communication Engineering, Visvesvaraya Technological University, Belagavi, Karnataka, India
e-mail: yashaswini.gn@gmail.com

B. R. Lakshmikantha
Dayananda Sagar Academy of Technological & Management, Visvesvaraya Technological University, Belagavi, Karnataka, India

© The Editor(s) (if applicable) and The Author(s), under exclusive license to Springer Nature Singapore Pte Ltd. 2021

A. P. Pandian et al. (eds.), *Proceedings of International Conference on Intelligent Computing, Information and Control Systems*, Advances in Intelligent Systems and Computing 1272, https://doi.org/10.1007/978-981-15-8443-5_36

cognitive problems. Swarm intelligence is capable of handling complex problems with minimal interaction with the immediate neighbors and hence produces a global emergent behavior. Swarms do not have a leader neither they have a global plan, any member in the group can act as a leader. Swarm intelligence is an area in artificial intelligence which has significant applications in various areas of engineering and technology such as multi-robot systems, mobile sensor networks, unmanned aerial vehicles, and flocking.

Swarm intelligence (SI) is derived from artificial intelligence [1, 2] which is the result of motivation acquired by the biological habits of birds and is an originative quick-witted breakthrough. SI approach is resultants from the unified behavior of swarms of bees, schools of fishes, and insect colonies which are applied in their food search process, getting around with each other and connecting in their colonies. These flocks mainly birds and fishes show structural behavior, so even when they change their direction they appear as a single entity [1, 2]. The swarm models are self-organizing, decentralized, communicative, and cooperative among themselves within the team [3, 4].

The main principles of swarms as shown in Fig. 1 are [5, 6]

- Homogeneity: All birds in the flock follow the same behavior model, and there may be temporary leaders for the flock to move but it does not have a fixed leader.
- Locality: The movement of each bird depends only by its nearest neighbor, and the sense of vision is the most important factor to keep up the flock organized.
- Collision avoidance: Avoid collision with the nearest flock members
- Velocity matching: All members keep up the same speed to maintain the same flocking velocity.
- Flock centering: Members try to stay close with flock mates.

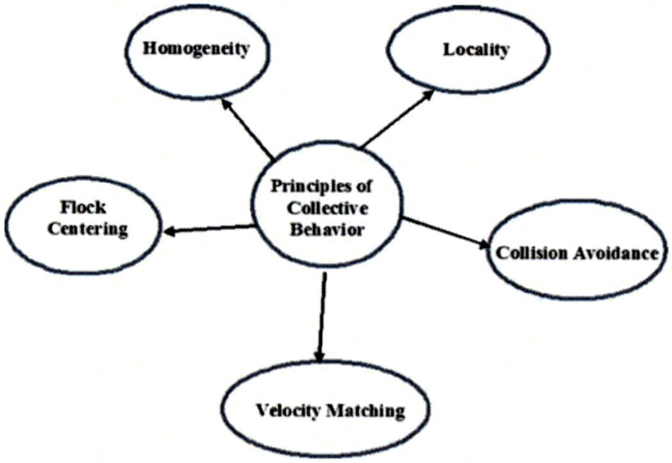

Fig. 1 Principles of swarms

Different swarm intelligence algorithms that are used include genetic algorithms, ant colony optimization, particle swarm optimization, differential evolution, bee colony optimization, glow worm swarm optimization, cuckoo search algorithm, firefly optimization algorithm, bat optimization, gray wolf optimization, lion Optimization, monkey optimization, and many more algorithms are present. There are algorithms which are resultants of merging to SI algorithms. In this paper, a survey on algorithms like particle swarm optimization, ant colony optimization, and bat algorithm is presented.

1.2 Data Clustering

The world comprises a huge set of unstructured data which has to be analyzed before using it. Hence, segregating the data with a common parameter is essential to analyze the data. Data clustering is grouping of unlabeled data into groups called clusters [3, 7], a cluster will have similar objects within it, and the data within each cluster will have a similar objects in it, but different clusters will have data that are similar within them but different from the other clusters [3, 8]. Clustering involves processes like feature selection, similarity measure, grouping of similar data, and output assessment. Clustering can be performed in three different ways supervised, semi-supervised, and unsupervised [4, 9]. In supervised learning, there is a teacher who gives inputs and outputs, and the rule is to map inputs and outputs whereas in unsupervised learning there is no teacher and hence no predefined structure.

Clustering algorithms have wide applications that include machine learning, pattern recognition, image analysis, bioinformatics, data analytics, data mining, image segmentation, and mathematical programming [1, 4, 7, 10].

There are mainly four kinds of clustering methodologies

1. Distance-based methods which includes
   - Partitioning algorithms: K-means, K-medians, K-medoids
   - Hierarchical algorithms: Agglomerative (bottom-up) vs divisive methods (top-down)
2. Density-based and grid-based methods
   - Density-based
   - Grid-based
3. Probabilistic and generative models
   - Assume a particular form of generative model
   - Parameters of the model are estimated with the expectation maximization (EM) algorithm
   - Later the generative probability of the underlying data points are estimated.

4. High-dimensional clustering
   - Subspace clustering: bottom-up, top-down, correlation-based methods vs δ-cluster methods
   - Dimensionality reduction [cluster columns; or cluster columns and rows together (co-clustering)]
   - Probabilistic latent semantic indexing (PLSI) then LDA, topic modeling of text data; a cluster (topic) is associated with a set of words (dimensions) and a set of documents (rows) simultaneously
   - Non-negative matrix factorization (NMF).

2 Swarm Algorithm and Techniques

2.1 K-Means

K-means clustering algorithm was developed by Hartigan (1975), and it is a simple also a prime clustering method that is widely practiced [8, 11]. K is the number of clusters that are formed from the data, which has to be specified by the user. So, based on the number specified, the data will be divided into K clusters. Each cluster will have a single center point [4, 7]. It uses Euclidean distance to measure similarity between the data, and this distance should be minimum to obtain good clustering results [12, 13]. K-means is very simple and easy to understand algorithm, which is used by most of the beginners to test data clustering but it has certain drawbacks as well. First, the user should specify the number of clusters, so if the number specified is incorrect, then clustering will also be inefficient.

2.2 Particle Swarm Optimization

Swarm intelligence technique was used to develop a mathematical principle by Kennedy and Eberhart in 1995. This model was mainly developed by the inspiration of social behavior of birds and fishes, and it was called the particle swarm optimization (PSO). Swarms are a large number of individual particles which interact locally within themselves. This model works on the principle of self-organization that describes the dynamics of complex systems. PSO works on an extremely simplified structure of social behavior to take care of the optimization problems, in a collective and intuitive structure. Each particle in the group follows simple rules to communicate and cooperate with each other. The particles are supposed to search and explore the space in a direction guided by inertia which is the previous velocity of the particle, cognitive force also known as p best or personal best which is an outcome of the particles own experience and social force known as g best or global best which is the outcome of the swarms best experience. The p best is the best solution the particle has achieved until now and g best is the best solution the complete swarm has achieved [4, 9]. Table 1 discusses some work that is carried out using PSO algorithms.

Table. 1 PSO algorithms

Author details	Algorithm/techniques	Purpose	Datasets/benchmark functions	Results
van der Merwe and Engelbrecht [10]	Hybrid (PSO with K-means)	To improve the quality of quantization error, inter-cluster, and intra-cluster distances	Iris Wine Breast cancer Automobiles Artificial 1 Artificial 2	Improved convergence toward quantization error, inter-cluster distance is increased, and intra-cluster distance is reduced
Gajawada and Toshniwal [7]	PCPSO	To find subspace clusters	Synthetic datasets	Efficient subspace clustering
Dhote et al. [12]	BEE + PSO	Robustness in intra-cluster distances	Iris Cancer Thyroid Wine	The intra-cluster distance is reduced. Error rate is reduced when compared to other algorithms like K-means
Tiwari and Gajbhiye [8]	KPSO (PSO + K-means)	Efficient clustering	Iris Glass Wine Sonar pima WDBC	Satisfactory efficiency and robustness
Gong et al. [14]	PSO, Bat, Cuckoo, Firefly	Data size, dimensionality, number of clusters	SEMG Arrhythmia Mice protein Expression Heart disease Arcene Dorothea	PSO & BAT are faster and their accuracy is better Cuckoo is slowest
Malik et al. [13]	K-mean PSO-K-mean PCA-K-mean	To improve clustering	Iris Liver Pima Indian diabetes Breast cancer wine	PCA-K-mean gives better clustering in high-dimensional data

(continued)

Table. 1 (continued)

Author details	Algorithm/techniques	Purpose	Datasets/benchmark functions	Results
Das et al. [2]	PS-BCO-K K-PS-BCO	Efficient data clustering	Iris Wine Breast cancer CMC Hill valley	Results in better clustering compared with other methods
Alswaitti et al. [15]	DPSO algorithm PSO + KDE (kernel density estimation)	To enhance the accuracy in classification and compactness of clusters	Iris Wine Breast cancer Glass Vowels Seeds New thyroid Haber man Dermatology Heart Landsat	Better accuracy, cluster compactness, repeatability, and reduced execution time for high-dimensional datasets
IOS Press & Authors 2017 [16]	IFAPSO PSO + Firefly algorithm	Effective data cluster	Iris Wine Glass Diabetes Sonar	Better results in cluster compactness and improved accuracy
Skaik [11]	PSO + IWC	Improve the clustering process	Iris Cancer Wine Liver disorder SET I SET II	Faster convergence is obtained
Mukhopadhyay [17]	hybrid PSO-fuzzy-based algorithm	To work on better results in intra-cluster, inter-cluster distance, and quantization error	2014 × 7 stock market dataset	The intra-cluster distance and quantization error values are minimized in each generation by the proposed algorithm whereas the inter-cluster distance is maximized

2.3 Ant Colony Optimization

Ant colony optimization was found by Dorigo and Gambardella in 1997, which simulates ants behavior, its cooperation, and mechanism followed during food search. The ant colony algorithm is used to find optimized path based on the conduct of ants food search. Initially, ants move randomly, when an ant finds food it travels back to its colony leaving 'markers' known as pheromone. The path with pheromone indicates that there is food in that path hence more ants are likely to follow that path for food. As more ants travel to find the food, they choose the shortest path hence leave pheromones on that path. So, the path with a strong pheromone trail is likely to be the shortest path. Based on this principle, ant colony algorithm is developed, which is being used to solve many real-time problems. It is also observed that ants leave different flavors of pheromone based on the food type which they are carrying hence the other ants can decide upon taking up that path. The ant colony system (ACS) has found to be versatile and robust in finding solutions to a wide range of optimization problems. The ants in the ACS look for a cheaper path on the graph, then it is compared with ants that have found a comparative costlier path and then decided upon which path is cheapest. It is found that the behavior of ants in the ACS is inspired by the ones with the real ants. Table 2 discusses some work that is carried out using ant colony algorithms.

2.4 Bat Algorithm

Bat algorithm was developed by Xin-She Yang in 2010, which is inspired by the echolocation behavior of the bats. Echolocation means knowing the location of objects by reflection of sounds, which is used by animals like bats and dolphins to find their preys. Bats find it difficult to find their food in nighttimes; hence, they use this technique called as echolocation in which they generate a sound and based on the echo's generated back they get to know the location of the target. The sound they generate is based on different hunting strategies.

In bat algorithm, the nature of the bat is understood and designed as follows [25]:

(1) All bats use echolocation to find the distance of target, and it gathers information about the food and the surrounding obstacles in a way that only they can understand.
(2) The bat algorithm uses λ for waves, A_0 for loudness, and $f_{\min}$ for frequency, to search at a location X_i with speed V_i. The bats can effectively adjust the wavelength of the pulse depending on the distance from the prey and adjust the frequency $r \in (0, 1)$ of the transmitted pulse when it is nearing to the prey.
(3) It works on the loudness change from the maximum A_0 (positive) to the minimum Amin (constant).

Upon these three assumptions, the bat algorithm randomly generates a set of solutions and then searches the optimal solution in the process of searching the optimal solution to strengthen the local search.

Table 3 discusses some work that is carried out using bat algorithms.

Table. 2 ACO algorithms

Author Details	Algorithm/techniques	Purpose	Datasets/benchmark functions	Results
Gao [17]	Improved ant colony optimization	To improve the computational efficiency and accuracy	Iris data Animal data Soyabean data	Improved computational efficiency and accuracy when compared with ACO algorithm
Menendez [18]	METACOC algorithm METACOC-K algorithm	Efficient clustering	Multiple synthetic & Real-time data	Both algorithm performs better than existing one
Pacheco [19]	Anthill algorithm	Automatic data clustering	Wine Iris Breast cancer Wisconsin Pima Indians diabetes	Algorithm obtains significant partitioning over traditional ant algorithm
Yang [20]	ACOCC	Efficient clustering techniques	Iris Wines Thyroid UKD	ACOCC can result in better solution which is faster and stable
Wijayanto [6]	FGWCACO	Better clustering quality	Synthetic dataset	This algorithm outperforms FGWC algorithm and provides a better analytical clustering
DKourav [21]	CCACRO	Better clustering and classification techniques	Text data	Obtained better results in correlation to other systems
Kharche [22]	ACPSO	Optimal clustering process	Iris	In most of the evaluation metrics, performance of this algorithm is good when compared with the existing algorithm

(continued)

Table. 2 (continued)

Author Details	Algorithm/techniques	Purpose	Datasets/benchmark functions	Results
Subhadra [23]	ACO + Medoids using hybrid distance	Effective and efficient clustering	Documentary data	Better clusters with precision, recall rate in less time is achieved
Nagarajan 2017 [24]	ACO	Semantics-based clustering foe documents	Documentary data	Efficient clustering is obtained with less time

Table. 3 Bat algorithm

Author details	Algorithm/techniques	Purpose	Datasets/benchmark functions	Results
Sharma [26]	DFBPKBA	To improve cluster quality and speed	Iris Glass Wine Magic Pokerhand	Good speed up is achieved for large datasets. Efficient cluster is achieved on most of the datasets
Ashish [25]	PBA	To handle massive data clustering	Iris Glass Wine Magic Poker hand	Makes efficient clusters
Jensi [27]	Modified bat algorithm	Efficient clustering	Iris Thyroid Wine Cancer CMC Glass Crude oil Liver disorder	Proposed algorithm gives efficient clustering results
Aboubi [28]	BAT-CLARA	Efficient clustering	Concrete Wisconsin breast cancer database	Better clustering results are obtained
Senthilnath [29]	Bat algorithm	Crop-type classification	Image segmentation multispectral crop (satellite Images)	Data classification and time complexity are efficiently achieved

(continued)

Table. 3 (continued)

Author details	Algorithm/techniques	Purpose	Datasets/benchmark functions	Results
Vellaichamy [30]	Hybrid collaborative movie recommender system (FCM + BA)	Better scalability and improved clustering	Movie lens datasets	The proposed algorithm has reduced MAE, improved precision, and recall, hence gives better accuracy
Gupta [31]	Location optimization using FCM + BA	Identifying nearest Kiosk in rural areas	Real-time location data in India	Reduced traveling time by the users
Nguye [32]	Compact bat algorithm	Saving variable memory and unequal clustering	10 test functions Rosenbrock Quadric Ackley Rastrigin Griewangk Spherical Shubert Quartic noisy Schwefel Langermann	The proposed algorithm makes better use of limited memory and works better on unequal clusters

3 Conclusion

The different swarm intelligence algorithms are studied and their applications in data clustering and have understood how data clustering can be performed efficiently using SI algorithms. First, the different types of SI algorithms and their principle of working are discussed, later discussed data clustering and the important methodologies in them. Different research papers are discussed in particle swarm optimization algorithm, ant colony optimization, and bat algorithm used for clustering. These papers give a better insight about the different algorithms their applications and advantages over traditional methods.[33], [34]

References

1. Ghorpade-aher, J., Metre, V.A.: Scope of research on particle swarm optimization based data clustering. **6**, 1–6 (2014)
2. Alswaitti, M., Albughdadi, M., Isa, N.A.M.: Density-based particle swarm optimization algorithm for data clustering. Expert Syst. Appl. **91**, 170–186 (2018)
3. Bharne, P.K., Gulhane, V.S., Yewale, S.K.: Data clustering algorithms based on swarm intelligence. In: ICECT 2011—2011 3rd International Conference on Electronics Computer

Technology **4**, 407–411 (2011)
4. Yan, Z., Ge, H., Pan, C., Mei, L.: The study on face detection strategy. 391–396 (2014). https://doi.org/10.1007/978-3-642-55038-6
5. Grosan, C., Abraham, A., Chis, M.: Swarm intelligence in data mining. Stud. Comput. Intell. **34**, 1–20 (2006)
6. Wijayanto, A.W., Mariyah, S., Purwarianti, A.: Enhancing clustering quality of fuzzy geographically weighted clustering using Ant Colony optimization. In: Proceedings of 2017 International Conference on Data and Software Engineering (ICoDSE 2017), pp. 1–6 (2018)
7. Gajawada, S., Toshniwal, D.: Projected clustering using particle swarm optimization. Procedia Technol. **4**, 360–364 (2012)
8. Tiwari, S., Gajbhiye, S.: Algorithm of swarm ıntelligence using data clustering. **4**, 549–552 (2013)
9. Alam, S., Dobbie, G., Koh, Y.S., Riddle, P., SUr Rehman : Research on particle swarm optimization sbased clustering: A systematic review of literature and techniques. Swarm Evol. Comput. **17**(1), 13 (2014)
10. Van Der Merwe, D.W., Engelbrecht, A.P.: Data clustering using particle swarm optimization. In: 2003 Congress on Evolutionary Computation (CEC 2003), vol. 1, pp. 215–220 (2003)
11. Bhattacharjee, D., Bhola, P., Dan, P.K.: Computational intelligence. Commun. Bus. Anal. **776**, 72–83 (2017)
12. Dhote, C.A., Thakare, A.D., Chaudhari, S.M.: Data clustering using particle swarm optimization and bee algorithm. In: 2013 4th International Conference on Computer and Communications and Networking Technologies (ICCCNT 2013) (2013). https://doi.org/10.1109/ICCCNT.2013.6726828.
13. Das, P., Das, D.K. Dey, S.: PSO, BCO and K-means based hybridized optimization algorithms for data clustering. In: Proceedings of 2017 International Conference on Information Technology (ICIT 2017), pp. 252–257 (2018). https://doi.org/10.1109/ICIT.2017.58.
14. Malik, H., Laghari, N.U.Z., Sangrasi, D.M., Dayo, Z.A.: Comparative analysis of hybrid clustering algorithm on different dataset. In: Proceedings of 2018 IEEE 8th International Conference on Electronic Information and Emergency Communication (ICEIEC 2018), pp. 25–30 (2018). https://doi.org/10.1109/ICEIEC.2018.8473568.
15. Danesh, M., Shirgahi, H.: A novel hybrid knowledge of firefly and pso swarm intelligence algorithms for efficient data clustering. J. Intell. Fuzzy Syst. **33**, 3529–3538 (2017)
16. Ahmadyfard, A., Modares, H.: Combining PSO and k-means to enhance data clustering. In: 2008 International Symposium on Telecommunications (IST 2008), vol. 4, pp. 688–691 (2008)
17. Gao, W.: Improved ant colony clustering algorithm and its performance study. Comput. Intell. Neurosci. **2016** (2016)
18. Menéndez, H.D., Otero, F.E.B., Camacho, D.: Medoid-based clustering using ant colony optimization. Swarm Intell. **10**, 123–145 (2016)
19. Pacheco, T.M., Gonçalves, L.B., Ströele, V., Soares, S.S.R.F.: An ant colony optimization for automatic data clustering problem. In: 2018 IEEE Congress on Evolutionary Computation (CEC 2018) (2018). https://doi.org/10.1109/CEC.2018.8477806
20. Yang, L., et al.: An ımproved chaotic ACO clustering algorithm. In: Proceedings of 20th International Conference on High Performance Computing and Communications. 16th International Conference on Smart City. 4th International Conference on Data Science and System (HPCC/SmartCity/DSS 2018), pp. 1642–1649 (2019). https://doi.org/10.1109/HPCC/SmartCity/DSS.2018.00267
21. Kourav, D., Khilrani, A., Nigam, R.: Class clustering with ant colony rank optimization for data categorization. In: 2015 International Conference on Applied and Theoretical Computing and Communication Technology (iCATccT), pp. 201–206 (2015)
22. Kharche, D., Thakare, A.: ACPSO: Hybridization of ant colony and particle swarm algorithm for optimization in data clustering using multiple objective functions. Glob. Conf. Commun. Technol. GCCT **2015**, 854–859 (2015). https://doi.org/10.1109/GCCT.2015.7342783
23. Subhadra, K., Shashi, M., Das, A.: Extended ACO based document clustering with hybrid distance metric. In: Proceedings of 2015 IEEE International Conference on Electrical,

Computer and Communication Technology (ICECCT 2015) (2015). https://doi.org/10.1109/ICECCT.2015.7226090

24. Nagarajan, E., Saritha, K., Madhugayathri, G.: Document clustering using ant colony algorithm. In: Proceedings of 2017 International Conference on Big Data Analytics and Computational Intelligence (ICBDACI 2017), vol. 80, pp. 459–463 (2017)
25. Ashish, T., Kapil, S., Manju, B.: Parallel Bat Algorithm-based clustering using mapreduce. 73–82 (2018). https://doi.org/10.1007/978-981-10-4600-1_7
26. Tripathi, A.K., Sharma, K., Bala, M.: Dynamic frequency based parallel k-bat algorithm for massive data clustering (DFBPKBA). Int. J. Syst. Assur. Eng. Manag. **9**, 866–874 (2018)
27. Jensi, R., & Jiji, W.: MBA-LF: a new data clustering method using modified Bat Algorithm and levy flight. ICTACT J. Soft Comput. **06**, 1093–1101 (2015)
28. Aboubi, Y., Drias, H., Kamel, N.: BAT-CLARA: BAT-inspired algorithm for clustering LARge applications. IFAC-PapersOnLine **49**, 243–248 (2016)
29. Senthilnath, J., Kulkarni, S., Benediktsson, J.A., Yang, X.S.: A novel approach for multispectral satellite image classification based on the Bat Algorithm. IEEE Geosci. Remote Sens. Lett. **13**, 599–603 (2016)
30. Vellaichamy, V., Kalimuthu, V.: Hybrid collaborative movie recommender system using clustering and bat optimization. Int. J. Intell. Eng. Syst. **10**, 38–47 (2017)
31. Gupta, R., Muttoo, S.K., Pal, S.K.: BAT algorithm for improving fuzzy C-means clustering for location allocation of rural kiosks in developing countries Under E-Governance **40**, 77–86 (2016)
32. Nguyen, T. T., Pan, J. S. & Dao, T. K. A compact bat algorithm for unequal clustering in wireless sensor networks. Appl. Sci. **9** (2019)
33. Zhu, L.F., Wang, J.S.: Data clustering method based on bat algorithm and parameters optimization. Eng. Lett. **27**, 241–250 (2019)
34. Gong, X., et al.: Comparative research of swam intelligence clustering algorithms for analyzing medical data. IEEE Access **7**, 137560–137569 (2019)

Low Complexity and Efficient Implementation of WiMAX Interleaver in Transmitter

S. Anitha and D. J. Chaithanya

Abstract In this paper, an efficient and low complexity method has been proposed for implementation of an address generation circuit of the two-dimensional deinterleaver which is used in the Worldwide Interoperability for Microwave Access (WiMAX) transmitter and receiver. An algorithm with simple method along with its mathematical operations for address generation has been proposed in this paper of the wireless multiple access channel deinterleaver which supports all possible code rates with different modulation techniques like QPSK and 16-QAM has been implemented and also the necessity of floor function has been eliminated. The resources shared for a quadrature phase shift keying and 16 quadrature amplitude modulation algorithm with different code rates make this method to be efficiently high when compared with conventional lookup table (LUT).

Keywords WiMAX · QAM · QPSK · Lookup table

1 Introduction

An enormous use of Internet increased from the last two decades. Broadband wireless access emerged as the access solution and competitor to new technologies which are existed in today's world, and it is gaining more popularity as an replaceable technology solution to DSL or an modem for a cable in Internet access. Broadband wireless access has requirements like fast design turnaround time, high processing speed, and flexibility. These necessity requirements make the designers to select field programmable gate array (FPGA) as an reconfigurable hardware platform. The digital subscriber line can cover about three miles and a range of 18,000 ft which

S. Anitha (✉)
Department of BME, ACS College of Engineering, Bengaluru, India
e-mail: dranithasammilan@gmail.com

D. J. Chaithanya
ACS College of Engineering, Bengaluru, India
e-mail: Chaithanyadj9@gmail.com

© The Editor(s) (if applicable) and The Author(s), under exclusive license to Springer Nature Singapore Pte Ltd. 2021

A. P. Pandian et al. (eds.), *Proceedings of International Conference on Intelligent Computing, Information and Control Systems*, Advances in Intelligent Systems and Computing 1272, https://doi.org/10.1007/978-981-15-8443-5_37

means that urban, rural, and suburban areas not possible to serve up to maximum extend. Due to its coverage limitations, the wireless fidelity broadband connection may solve this problem a small enough but not much possible to serve everywhere to use. But wireless multiple access, the metropolitan area wireless standard, can solve these problems as a replicable to wireless fidelity. Wireless multiple access is a new technology which provides an broadband and IP connectivity. In wireless communication, it selects both line of sight (LOS) wireless communication and a non-line of sight (NLOS). To add an additional bits at the transmitter end to its physical layer, an orthogonal frequency-division multiple (OFDM) access is used by wireless multiple access network. WiMAX uses the concept of cyclic prefix, and it is a new technology which has gained a large popularity due to broadband wireless access systems and rapid growing interest among people.

The authors Martina and Masera [1] have explained about the basic concept of architecture method of today's wireless multiple access decoders in VLSI and its physical layer related to adaptive modulation new techniques. Gyan Prakash and Sadhana Pal [2] discussed the main important features, application of the improved mobile wireless multiple access and they also explain about the wireless multiple access variant of the system for delay tolerant that developed new mobility applications which support secure seamless handovers everywhere. Sanyal and Upadhyaya [3] discussed the basic interleaving block in encoder method. In a minimum bandwidth, the author also explained about how to achieve a maximum data transfer, maintaining an acceptable QoS. In reducing or minimizing the burst error, the channel interleaver in the wireless multiple access transmitter and receiver plays a very important role in recent technologies. A new technique is proposed for low complexity, very high speed, and high efficient of an address generator for the deinterleaver used in the wireless multiple access method. In WiMAX technology, transmitter and receiver have been newly developed in recent technology which terminates the necessity of large amount of floor function.

The various blocks of wireless multiple access transceiver are shown in Fig. 1. Before being encoded, data inputs which are received from a source are being randomized by two methods of error correction coding techniques namely convolutional coding and Reed–Solomon [2]. Randomization method is used to approximate a sequence which is being randomized, in which it rearranges the data components in a serial bit order. After completion of randomizer, the obtained output result is passed on to the RS encoder part. In order to add abundance to the data sequence, Reed–Solomon code is used. While transmission of the data or signal, this abundance addition helps in block errors to correct. Modulation and construction of orthogonal frequency-division multiplexing symbols are performed by two subsequent blocks namely modulation and IFFT as shown in Fig. 1. Modulation technique involves mapping of discrete or digital information on to continuous analog form such that it can be transmitted over the channel in the encoder part. To convert input stream from frequency domain to time domain, inverse fast Fourier transform is used. Basically, to work in frequency domain, convert it from Fourier domain or time domain to frequency domain in fast Fourier transform (FFT). In the decoder [3], the various blocks are arranged in reverse order to enable the original data sequence to be restore.

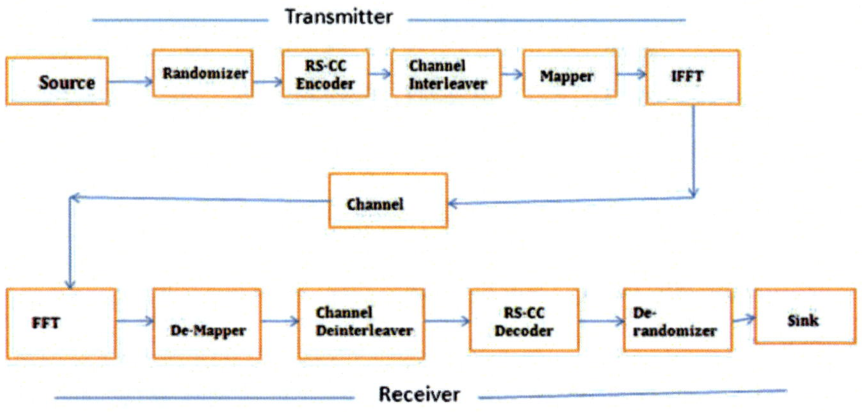

Fig. 1 Block diagram of the wireless multiple access transceiver

2 Interleaver/Deinterleaver Structure in WiMAX

Block of 2-D interleaver in encoder and deinterleaver structure in decoder is used as channel interleaver and deinterleaver in wireless multiple access network, as mentioned in Fig. 2. An interleaving which is considered as a block represents memory block as one which is being written, the other one is read. When selection results which is equal to one, write enabled signal of M-1 is active which is written as WE. During this period, the data input is written in M-1 as it receives the write addresses. Similarly channel interleaver data stream is read from M-2 as it is considered with the read addresses. After the memory blocks are written or read up to the desired location as specified by interleaver depth, the status of signal which is considered as select input is changed to swap the read or write operation.

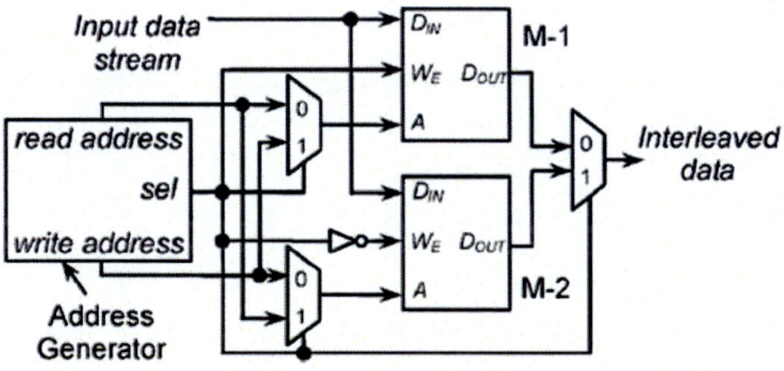

Fig. 2 Block diagram of interleaver/deinterleaver structure [3]

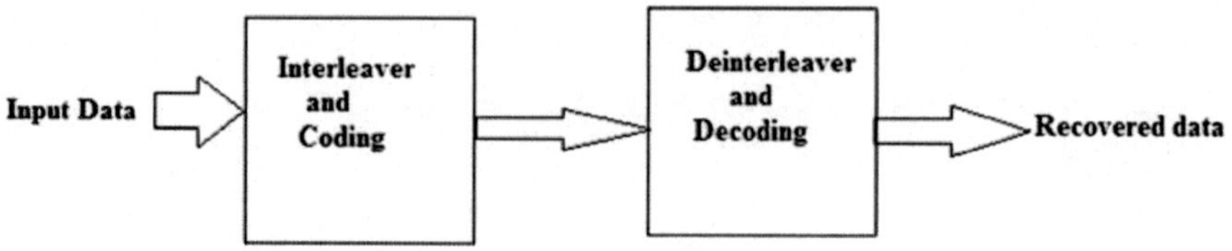

Fig. 3 Interleaving block diagram

3 Interleaving in WiMAX System

The channel interleaver role in encoder design to distribute the sequence of bits in a bitstream to reduce or to minimize the burst errors in interleaver introduced in transmission. An interleaver usually used with some type of error correcting code to generate different sequence of same width but in different order. The channel interleaver in wireless multiple access takes a sequence of information bits with a fixed amount width. Interleaving in WiMAX transmitter is implemented by a two-dimensional array buffer which is a storage element such that the data enters the storage element in rows which specify the number of steps in channel interleaver, then it is read out in columns (Fig. 3).

4 Implementation of Deinterleaver

Recently in communication systems, there possibly occurs consecutive errors depending on data transmission paths. In recent communication systems, there possibly occur errors in data depending on type of environments of transmission paths and likely to result invalidate the effect of the error correcting code randomly. The data is sent from a transmitter side and then it rearranges the random data sequence to the correct order in decoder which is known as deinterleaver (Tables 1 and 2)

Table 1 Implementation results of WiMAX deinterleaver

Logic utilization	Used	Available	Utilization (%)
Number of slices	52	3584	1
Number of slice flipflops	39	7168	0
Number of four input LUTs	101	7168	1
Number of bonded IOBS	16	141	11
Number of GCLKs	1	8	12

Table 2 Deinterleaver sample of first four rows and five columns of addresses for three code rates and modulation types

N_{cbps}, code rate and modulation type	Deinterleaver addresses				
N_{cbps}= 96 bits, ½ code rate, QPSK	0	16	32	48	64
	1	17	33	49	65
	2	18	34	50	66
	3	19	35	51	67
N_{cbps}= 192 bits, ½ code rate, 16-QAM	0	16	32	48	64
	17	1	49	33	81
	2	18	34	50	66
	19	3	51	35	83

4.1 Simulation Results

The algorithm is developed in this paper for different modulation schemes and code rates. Finally, the design is being synthesized using Xilinx ISE and the obtained results are validated against design specifications. Simulation results are obtained from two permissible modulation and code rates. Then, obtained simulation output is then verified through the MATLAB program in order to ensure the correct working of the circuit. Using the Xilinx ISE, the WiMAX deinterleaver architecture has been designed in HDL. The results have been simulated to satisfy or to ensure the correct working of the main concept in the algorithm tested and then verified through the MATLAB results. Finally, the test obtained in QPSK, 16-QAM, and 64-QAM has provided satisfactory results in WiMAX deinterleaver (Figs. 4, 5, and 6).

5 Conclusion

A new algorithm with its simple method along with its mathematical background has been implemented in final output, which terminates the necessity of floor function in encoder design. For different modulation schemes, different values of code or information rates in MATLAB have been implemented, and simultaneously entire system design has been coded in Verilog and then it is simulated in Xilinx ISE. Finally, in this paper, different modulation schemes 16-QAM and QPSK have been developed, implemented, and finally it is verified or tested through a MATLAB in order to check the correct working of the newly concept design. The final implemented method has enabled us to provide more satisfactory results. Using Xilinx ISE, the design is synthesized and the obtained results are verified through specifications of the design. Simulation results for 16-QAM and QPSK have been obtained for different modulation techniques and permissible code rates. Each and every blocks are implemented in Verilog and then synthesized. Finally, simulation was done in

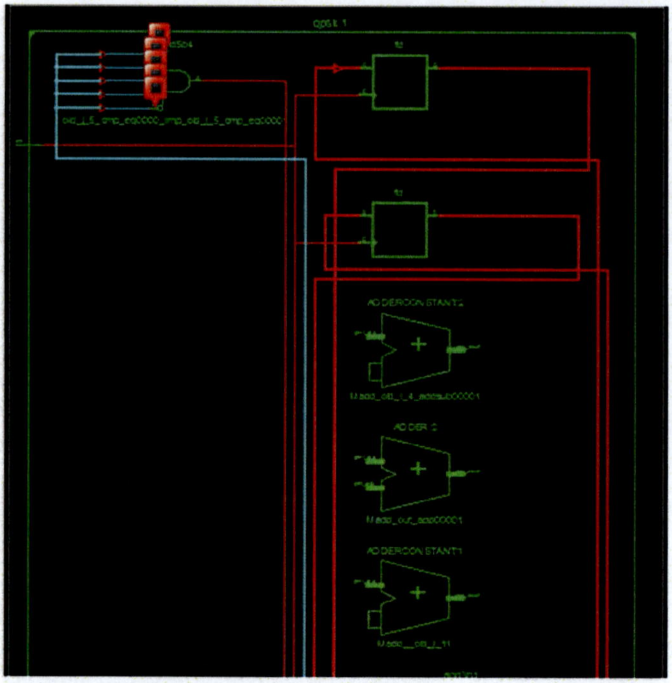

Fig. 4 Internal block diagram of QPSK modulation scheme

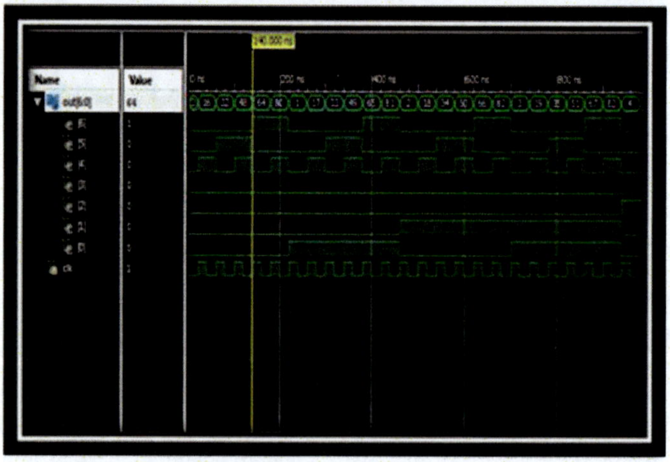

Fig. 5 Simulation diagram of QPSK modulation scheme

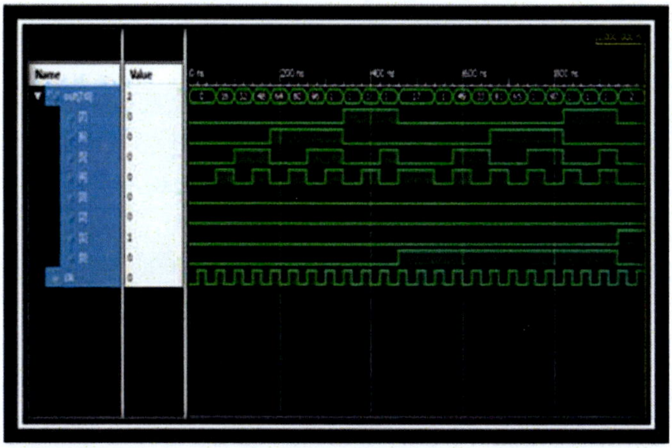

Fig. 6 Simulation diagram of 16-QAM modulation scheme

Xilinx ISE tool. In future, ASIC implementation can be done for both WiMAX interleaver and deinterleaver and also FPGA implementation can be done for 64-QAM algorithm.

References

1. Martina, M., Masera, G.: VLSI architectures for WiMAX channel decoders. Int. J. Commun. Issue **2** (2003)
2. Prakash, G., Pal, S.: WiMAX technology and its applications. Int. J. Eng. Res. Appl. (IJERA) **1**(2) (2006)
3. Sanyal, S.K., Upadhyaya, B.K.: Efficient FPGA implementation of address generator for WiMAX deinterleaver. IEEE Trans. Circ. Syst. (2013)

Instrument Recognition in Polyphonic Music Using Convolutional Recurrent Neural Networks

Bhargav Ram Kilambi, Anantha Rohan Parankusham, and Satya Kiranmai Tadepalli

Abstract Sounds or music usually occurs in an unstructured environment where their frequency varies from time to time. These temporal variations are one of the major problems in the music information retrieval. Additionally, polyphonic music or polyphony is simultaneous combination of two or more tones or melodic line, where each line is an independent melody of an instrument. As a result, identifying various instruments from recordings of polyphonic music is difficult and inaccurate using conventional methods. In this paper, a framework is presented for predominant instrument recognition in real-world polyphonic music. The framework consists of both convolutional neural networks (CNNs) and recurrent neural networks (RNNs). CNNs are used to pull out important features that are unchanging to local spectral and temporal variations. Similarly, RNNs are used as they quickly learn the long-term dependencies in the audio signals. The results obtained by the convolutional recurrent neural networks (CRNNs) showed an improved performance when compared to network built using only convolutional neural networks.

Keywords Convolutional neural networks · Recurrent neural networks · Convolutional recurrent neural network · Long short-term memory

1 Introduction

Human beings can differentiate the instruments in a music that they listen to. Whereas it is still an arduous task for machines such as computers to do the same. This is mainly

B. R. Kilambi (✉) · A. R. Parankusham · S. K. Tadepalli
Chaitanya Bharathi Institute of Technology, Hyderabad, India
e-mail: bhargavramkilambi@gmail.com

A. R. Parankusham
e-mail: anantharohan299@gmail.com

S. K. Tadepalli
e-mail: tadepallikiranmai84@gmail.com

© The Editor(s) (if applicable) and The Author(s), under exclusive license to Springer Nature Singapore Pte Ltd. 2021

A. P. Pandian et al. (eds.), *Proceedings of International Conference on Intelligent Computing, Information and Control Systems*, Advances in Intelligent Systems and Computing 1272, https://doi.org/10.1007/978-981-15-8443-5_38

because music in general is polyphonic, which makes gathering of information from it a very difficult task. Furthermore, sound from an instrument varies in quality, timber, and playing style, which makes it even more challenging for identification.

Knowing the instruments that are being played in music has always been the most desirable task in music information retrieval (MIR) industry. Perhaps, the information related to the instruments is useful to those who seek, and it can be achieved through audio tags. The necessity of music search has increased with the increase of number of music files. Unlike image or text search, it is difficult to search for music as the input given by the user is in the form of text. To overcome this problem, songs are assigned with tags which contain the names of the instruments that are played in it. In addition, the retrieved instrument information can be used in other applications. Instrument recognition can be obtained through various forms [1, 2]. In this paper, neural network-based approach is made used for providing name tags to the instruments for a given polyphonic music. A model is developed by integrating convolutional neural networks and recurrent neural network to produce a better result compared to the existing system.

2 Literature Survey

2.1 Deep Convolutional Neural Networks for Predominant Instrument Recognition in Polyphonic Music

For instrument recognition in deep convolutional neural networks, a model of ConvNet architecture is trained with smaller stride and repetitive window size [1]. The dataset, IRMAS, consisting 6705 audio files, was used for training the model after down-sampling the audio files to Nyquist frequency and then converting the linear frequency scale from STFT to mel-scale. After training the model with optimal hyper-parameters, they achieved 0.619 for micro- and 0.513 for macro-F1-measure.

The vital factors for improvement in performance were analysis window size, max-pooling, and model ensemble whereas using different activation functions did not have a significant impact.

The testing configuration comprised of a series of ConvNets making predictions on windows of sizes ranging from 0.5 to 3.0 s and the collective output is normalized for multi-class labels using sigmoid outputs. Instrument wise result analysis proved that their approach was successful in terms of identifying single predominant instrument. The limitation in this approach was conventional ConvNets were used, although they were dealing temporal data. The results and the F1-scores could have been better if RNN was included in the architecture.

2.2 CRNN for Polyphonic SED

Cakır, Emre, et al. have achieved polyphonic sound event detection by training a model of CRNN architecture [3]. The main motive for incorporating RNNs along with ConvNet was to temporally locate the sound event in the audio snippet. This was achieved in two stages, sound representation, i.e., locating the significant sound event in the input and sound classification, i.e., predicting the source of the sound in the audio signal. The audio was preprocessed into frame level features such as Mel band and Mel Frequency Cepstral Coefficients (MFCC) [4].

The datasets used were TUD-SED Synthetic 2016, TUT-SED 2009, and TUT-SED 2016 all of which comprise of real-life or synthetic sound events such as dog barking, alarms, sirens, etc. For experimenting purposes, the datasets after down-sampling to Nyquist frequency were trained against several network architectures such as FNN, CNN, RNN and CRNN. The training configurations and hyper-parameters were adjusted accordingly. F1-measure and equal error rate (EER) were used as the evaluation metrics for this approach.

An astounding portrayal of comparison between different network architectures with properly tuned configurations over different datasets was presented by the authors. The CRNN proved to have achieved best of the results in all the outcomings. It has achieved an F1-measure of 0.691 ± 0.04 for TUT-SED 2009 dataset.

2.3 Polyphonic Sound Event Detection by Using Capsule Neural Networks

The main motive of Vesperini, Fabio et al. for using combination of capsule neural networks with ConvNet instead of conventional methodologies like DCNN, RNN, etc., was to overcome the loss of information on features during max-pooling and normalizations by using local units and routing procedures that produce a vector output, which are commonly termed as capsules [5]. In contrast with conventional neurons in the neural network, the total input s_j of a capsule j is calculated as follows:

$$s_j = \sum_i \alpha_{ij} W_{ij} u_i = i\alpha_{ij} u_{j|i} \tag{1}$$

The datasets in use for this approach are TUT-Sound Events 2016 and 2017 for polyphonic conditions and TUT Rare Sound Events 2017 for monophonic SED. The audio signals were converted into two acoustic spectral representations, namely *STFT* and *LogMel* coefficients. The performance of their approach proved to be in the range of 0.58–0.72 across different scenarios among the datasets in polyphonic SED case study. The merit of this approach was that the bottleneck of losing vital information during different series of steps involving data convolution was recovered by using capsules.

3 System Overview

3.1 Architecture

The proposed system, as shown in Fig. 1, uses a CRNN [6] which consists a pool of convolutional layers followed by a recurrent neural network and a fully connected dense network. The CNN [7–9] extracts vital features from the audio input which in turn are used by the RNN [10, 11] to identify multiple instruments present in the audio input. The fully connected neural network at the end ensures to allocate class labels in the order of their probabilities.

Convolution Layers. The model has four sets of convolutions that consists of convolutional layers, max-pooling, activation functions, and batch normalization. The convoluted layers are reshaped from 3D to 2D before feeding it to the recurrent network. After reshaping [12] from the convolution layers, it is passed on to LSTM layers which in turn return class labels through a fully connected neural network (Dense Layers).

Recurrent Layers. These CNN algorithms are used for feature identification and these features are forwarded to the next layers, i.e., to the recurrent neural networks for identifying the multiple instruments present in the input polyphonic audio file. LSTMs can be used with RNNs to produce better results. The LSTM model is trained to identify the class labels based on the predominant features of the audio, such as

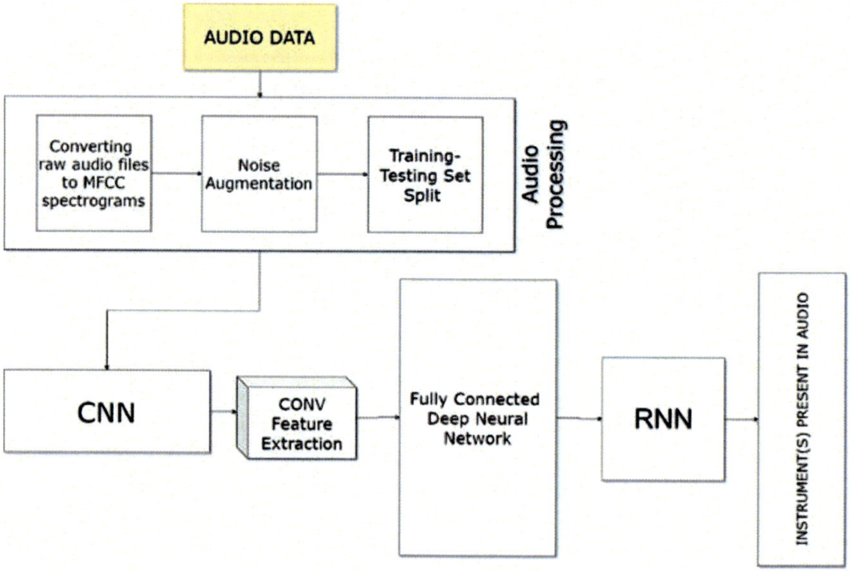

Fig. 1 Proposed CRNN architecture

STFT. Each output by the memory cell gives out a vector, which is used to predict the instruments based on the highest probability and "forwarded" again as input to the memory cell.

3.2 *Dataset*

The dataset used for our experiment setup is IRMAS instrument dataset which comprises of 6705 audio files with excerpts of 3 s from more than 2000 distinct recordings. The dataset consists of 11 instruments—Cello, Clarinet, Flute, Acoustic guitar, Electric Guitar, Organ, Piano, Saxophone, Trumpet, Violin, and Voice.

3.3 *Audio Processing*

The audio processing step involved modifying the dataset by augmenting linear noise into a smaller portion of the dataset without disturbing the actual features of the audio data. This step is to ensure that the training steps involve the model to be trained against data which is almost like the real-time scenario.

3.4 *Building the Dataset*

The dataset is processed through a series of steps before feeding it as the input to the neural network. They are as follows:

One Hot Encoding. One Hot Encoding or Label Encoding is to assign numeric values to the class labels. After assigning numeric values, they are converted to an array of zeroes of ones for each label with the index of the label being one. Meaning the output shape from the neural network, i.e., output label representation is an array of 11, each index representing an instrument. Manual label encoding is opted because it provides us flexibility to modify the dataset like shuffling, test-train split, augmenting noise, etc.

Noise Augmentation and Training–Testing split. Building the dataset from .npy files that were exported in the previous step. It involves splitting the dataset into training and testing sets by 4:1 ratio. Furthermore, a small portion (35%) of only the training set is augmented with linear noise in such a way that it adds noise to the pitch but frequency and the STFT remain unchanged, as shown in Fig. 2. Processing

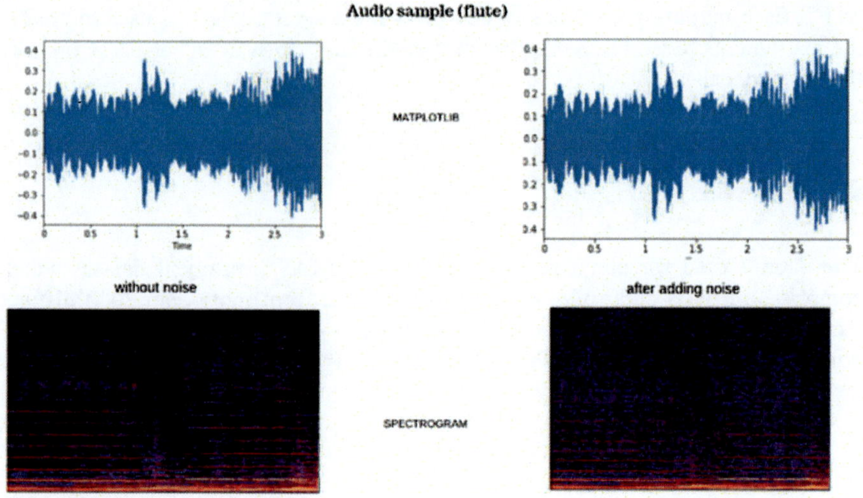

Fig. 2 MFCC before and after adding noise

the audio files with librosa [13] and saving the processed py objects as .npy files for training and testing.

Training Configuration

Batch Size and Epochs. Batch size is defined as the number of training examples the model uses for each iteration. Similarly, epochs refer to the number of times the entire training dataset has been passed to the algorithm as a whole. Through trial and error, a batch size of 128 and 100 as number of epochs was ideal for our approach and implementation.

Hyper-parameters. The different hyper-parameters [14] used in a neural network implementation include learning rate, optimizers, dropout rate, etc. The optimizer parameters were set to (initial $\text{lr} = 0.0001$, decay $= 1 \times 10^{-6}$), dropout rate variable between 0.25 and 0.5.

Activation Function. Activation Function is simply an equation to determine if the neuron is allowed to pass further through the hidden layers. The usage of the different activation functions like ReLU, Sigmod, LReLU, Tanh, etc., depends on the problem statement. In our case study, Rectified Linear Unit (ReLU) as found to be optimal for hidden layers. Softmax activation function was used for the output layer since it is suitable for normalizing probabilities. The mathematical equations for ReLU and Softmax functions, respectively, are as follows (Table 1).

$$f(x) = \max(x, 0) \tag{2}$$

$$\sigma(z)_j = \frac{e^{zj}}{\sum_{k=1}^{K} e^{z_k}} \tag{3}$$

Table 1 Proposed model structure

Layer (type)	Output shape	Param #
conv2D 1	128 × 259 × 64	1664
max-pooling2D 1	42 × 86 × 64	0
batch-normalization 1	42 × 86 × 64	256
dropout 1	42 × 86 × 64	0
conv2D 2	48 × 86 × 64	102,464
max-pooling2D 2	14 × 28 × 64	0
batch-normalization 2	14 × 28 × 64	256
dropout 2	14 × 28 × 64	0
conv2D 3	14 × 28 × 128	204,928
max-pooling2D 3	4 × 9 × 128	0
batch-normalization 3	4 × 9 × 128	512
dropout 3	4 × 9 × 128	0
conv2D 4	4 × 9 × 128	409,728
max-pooling2D 4	1 × 3 × 128	0
batch-normalization 4	1 × 3 × 128	512
dropout 4	1 × 3 × 128	0
permute 1	3 × 1 × 128	0
reshape 1	3 × 128	0
lstm 1	3 × 32	20,608
lstm 2	3 × 32	8320
dropout 5	3 × 32	0
time-distributed 1	3 × 128	4224
dropout 6	3 × 128	0
time-distributed 2	3 × 64	8256
dropout 7	3 × 64	0
flatten1	192	0
dense 3	11	2123
Total params: 763,851		
Trainable params: 763,083 Non-trainable params: 768		

4 Results and Discussion

In this section, the results are presented for the IRMAS dataset and the experiment setup mentioned in Sect. 3. The evaluation is done based on Precision, Recall, Support, and F1-score for each individual classes as well as for the entire dataset as a whole.

Confusion Matrix. From the confusion matrix in Fig. 3, the performance of the model was highest for violin, organ and piano is inferred while the multi-label performance was moderate for the same. Electric guitar (gel) had the most formidable performance in terms of both individual and multi-label predictions.

$$\text{Precision} = \frac{\text{True positive}}{\text{True Positive} + \text{False Positive}} \tag{4}$$

$$\text{Recall} = \frac{\text{True positive}}{\text{True Positive} + \text{False Negative}} \tag{5}$$

$$\text{F-Measure} = \frac{2 * \text{Precision} * \text{Recall}}{\text{Precision} + \text{Recall}} \tag{6}$$

Support of a rule is a measure of how frequently the items involved in it occur together. Using probability notation: support (A implies B) $= P(A, B)$.

Table 2 infers that F1-scores of classes with highest support as well as lowest support are lying in the same range, which suggests that our model was successful in making predictions even for those classes with limited information in dataset. With our proposed system, a macro-F1-measure of 0.743, micro-average of 0.75, weighted average of 0.752, and average accuracy of 0.75 is achieved.

Receiver Operating Characteristics Curve. ROC curve is a probability curve and area under ROC curve, or simply AUC represents the degree of separability between classes. The commonly used terms while calculating AUC are as follows:

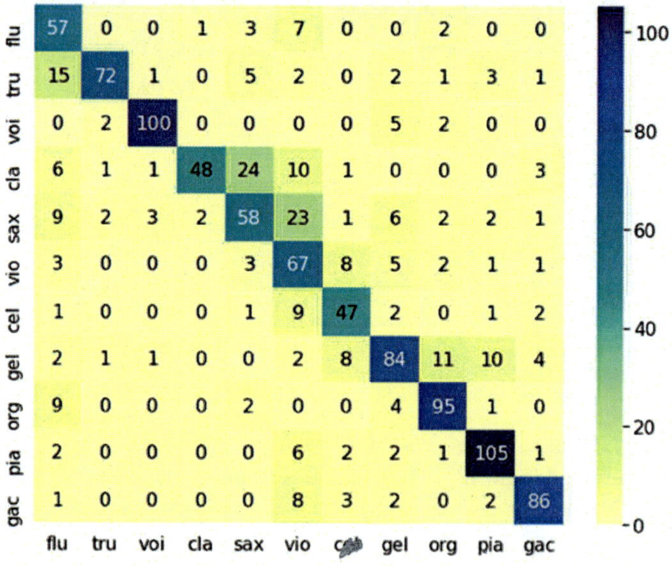

Fig. 3 Confusion matrix

Table 2 Evaluation results after testing phase

Class	Evaluation metrics			
	Precision	Recall	F1-score	Support
Flute	0.54	0.81	0.65	70
Trumpet	0.92	0.71	0.80	102
Voice	0.94	0.92	0.93	109
Clarinet	0.94	0.51	0.66	94
Saxophone	0.60	0.53	0.57	109
Violin	0.50	0.74	0.60	90
Cello	0.67	0.75	0.71	63
Electric guitar	0.75	0.68	0.71	123
Organ	0.82	0.86	0.84	111
Piano	0.84	0.88	0.86	119
Acoustic guitar	0.87	0.84	0.86	102
Accuracy			0.75	1092
Macro-Avg.	0.76	0.75	0.74	1092
Weighted Avg.	0.78	0.75	0.75	1092

$$\text{Sensitivity} = \frac{\text{True positive}}{\text{True Positive+False Negative}} \tag{7}$$

$$\text{Specificity} = \frac{\text{True Negative}}{\text{True Negative+False Positive}} \tag{8}$$

$$\text{False Positive Rate} = \frac{\text{False positive}}{\text{True Negative} + \text{False Positive}} \tag{9}$$

In general, AUC and ROC curves are plotted for models which are binary classifiers in nature. ROC curves can be used to analyze how much the network is able to distinguish between zeros as zeros and ones as ones. When it comes to plotting ROC curves for multi-label or multi-class classifiers, such as our network, One-versus-All strategy is incorporated in which ROC curve for each class is plotted against all the other classes.

Figure 4 depicts the ROC curve plotted for the all the individual classes and micro- and macro-averages in one-versus-all fashion. The common understanding is that higher the AUC, the higher the network can distinguish between classes. In our implementation, AUC for micro-average and macro-average was 0.96 and AUC for individual classes against the rest was lying in the range of 0.95–0.97.

Figure 5 is same as the above in Fig. 4, zoomed in the top left corner. It can be inferred that the ROC curves for micro- and macro-averages lie between the ROC curves of rest of the classes.

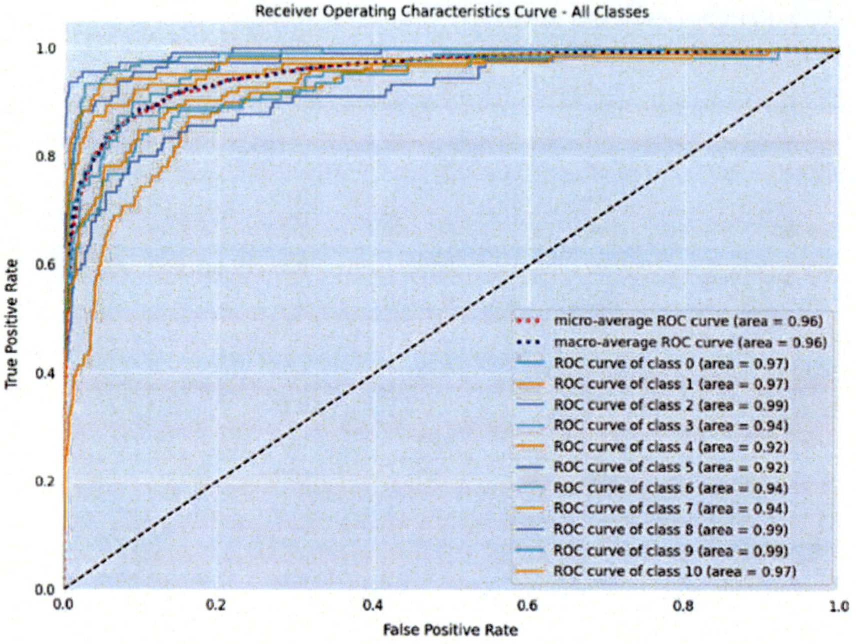

Fig. 4 ROC and AUC—one versus all strategy

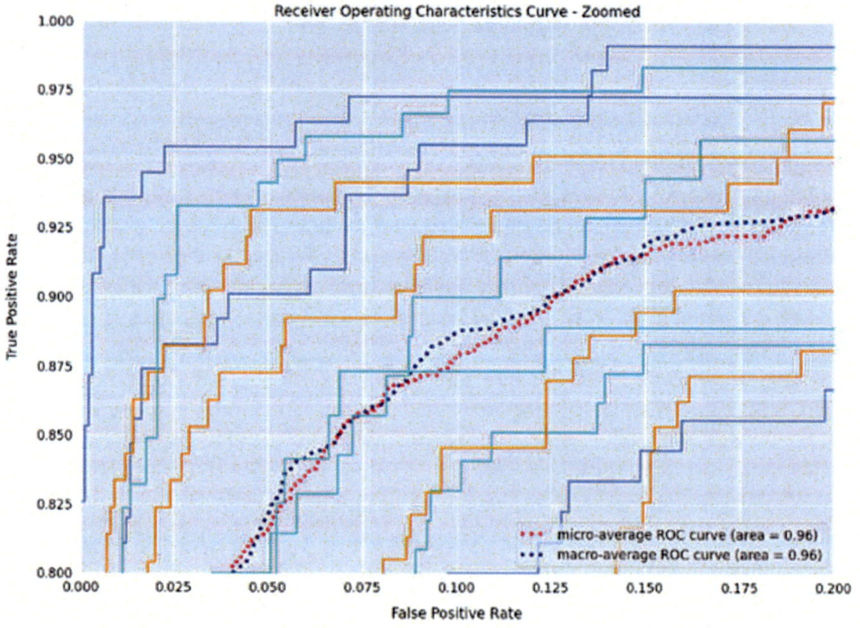

Fig. 5 ROC and AUC—zoomed

5 Current Limitations and Future Scope

Our model is limited to identifying only ten instruments and voice that are present in IRMAS dataset. With the same network architecture, the model can be trained further for it to identify more instruments.

The training configuration in our approach was to train the model over the entire dataset. Our dataset being limited in size, opted out to just augment linear noise into the audio snippets. A further efficient method would be to train the model with a larger dataset and then train the output layer with smaller datasets (transfer learning) to achieve better results.

Our current approach being a prototype/experimental setup in a python notebook lacks the element of being modular in terms of accessibility. The trained model can be exported and used in a medium in such a way that it could be used for real-time predictions and the micro-surveys that can be done by collecting real-time data can be used to further train the network with everlasting diverse inputs.

6 Conclusion

This research work explains the process of CRNN model and it is implemented to analyze the predominant instrument in the real-world music. From the music clip, fixed length single labeled data can identify the predominant instruments with a variable length. Experimental results observed that the convergence of CNN and RNN provides a good results based on the input data fetched from the appropriate feature. The proposed neural network architecture provides an effective performance in predominant instrument identification from the IRMAS dataset. Mel-spectrogram was used as an input to the network, and any source separation in the preprocessing is not used unlike in existing networks.

Acknowledgements It is our privilege to acknowledge with deep sense of gratitude and devotion for keen personal interest and invaluable guidance rendered by our Head of the Department, Dr. Suresh Pabboju, Professor, Department of Information Technology, Chaitanya Bharathi Institute of Technology.

References

1. Han, Y., Kim, J., Lee, K.: Deep convolutional neural networks for predominant instrument recognition in polyphonic music. IEEE/ACM Trans. Audio Speech Lang. Process. **25**(1), 208–221 (2016)
2. Li, P., Qian, J., Wang, T.: Automatic instrument recognition in polyphonic music using convolutional neural networks. arXiv preprint arXiv:1511.05520 (2015)

3. Cakır, E., Parascandolo, G., Heittola, T., Huttunen, H., Virtanen, T.: Convolutional recurrent neural networks for polyphonic sound event detection. IEEE/ACM Trans. Audio Speech Lang. Process. **25**(6), 1291–1303 (2017)
4. Hossan, M.A., Memon, S., Gregory, M.A.: A novel approach for MFCC feature extraction. In: 2010 4th International Conference on Signal Processing and Communication Systems. IEEE (2010)
5. Vesperini, F., Gabrielli, L., Principi, E., Squartini, S.: Polyphonic sound event detection by using capsule neural networks. IEEE J. Select. Topics Signal Process. **13**(2), 310–322 (2019)
6. Zuo, Z., Shuai, B., Wang, G., Liu, X., Wang, X., Wang, B., Chen, Y.: Convolutional recurrent neural networks: Learning spatial dependencies for image representation. In: Proceedings of the IEEE Conference on Computer Vision and Pattern Recognition Workshops (2015)
7. LeCun, Y., Bengio, Y., Hinton, G.: Deep learning. Nature **521**(7553), 436–444 (2015)
8. O'Shea, Keiron, and Ryan Nash. "An introduction to convolutional neural networks." arXiv preprint arXiv:1511.08458 (2015)
9. Albawi, S., Abed Mohammed, T., Al-Zawi, S.: Understanding of a convolutional neural network. In: 2017 International Conference on Engineering and Technology (ICET). IEEE (2017)
10. Aaron van den, O., Kalchbrenner, N., Kavukcuoglu, K.: Pixel recurrent neural networks. arXiv preprint arXiv:1601.06759 (2016)
11. Lichao, M., Ghamisi, P., Zhu, X.X.: Deep recurrent neural networks for hyperspectral image classification. IEEE Trans. Geosci. Remote Sensi. **55**(7), 3639–3655 (2017)
12. Ghosh, S., Das, N., Nasipuri, M.: Reshaping inputs for convolutional neural network: Some common and uncommon methods. Pattern Recogn. **93**, 79–94 (2019)
13. Brian, M., Raffel, C., Liang, D., Ellis, D.P.W., McVicar, M., Battenberg, E., Nieto, O.: librosa: audio and music signal analysis in python. In: Proceedings of the 14th Python in Science Conference, vol 8 (2015)
14. Bergstra, J.S., Bardenet, R., Bengio, Y., Kégl, B.: Algorithms for hyper-parameter optimization. Adv. Neural Inform. Process. Syst. (2011)

Prediction of Protein–Protein Interaction as Carcinogenic Using Deep Learning Techniques

Rohan Kumar, Rajat Kumar, Pinki Kumari, Vishal Kumar, Sanjay Chakraborty, and Sukhen Das

Abstract Protein–protein interactions be talking one of the fundamental utilitarian pertinence among qualities and integrative genomic information with available proteomic information that may oblige us with a thorough rough approximation of unimportant human issues. Amidst the strategies directed by numerous protein variations of human ailments, a large portion of them has unusual systems. Here, the profound neural system tends to utilize protein–protein interaction (DeepPPI) which selects the profound neural system to expert the portrayal of protein data refined from the protein–protein interaction (PPI) network. An out-of-the-container method of looking over the protein–protein interaction has been presented by the utilization of the PPI systems that will approve one to comprehend cell pathways and making practical medicines for the treatment of human ailments. In any case, our proposed dataset has likewise the amino acids constituent rate that permits one to anticipate the genuine site where the anticipated association can happen. The exploratory results exhibit that DeepPPI achieves pervasive execution in the test set with an exactness of 83.50%, an accuracy of 82.49%, and review of 83.89%. Along these lines, compared with the past profound neural system, our model achieved better or identically incredible execution. The PPI network is being broken down using Cytoscape.

Keywords Protein–protein interaction · Amino acids · Carcinogenicity prediction · PPI classification · Deep learning · Supervised learning

1 Introduction

Protein–protein interaction systems are exceptionally noteworthy for examining biomedicinal capacities and investigating buildings that cause maladies [1]. Using

R. Kumar (✉) · R. Kumar · P. Kumari · V. Kumar · S. Chakraborty · S. Das
Information Technology Department, Techno-Main Saltlake, Kolkata, India
e-mail: rohan.kraug20@gmail.com

© The Editor(s) (if applicable) and The Author(s), under exclusive license to Springer Nature Singapore Pte Ltd. 2021

A. P. Pandian et al. (eds.), *Proceedings of International Conference on Intelligent Computing, Information and Control Systems*, Advances in Intelligent Systems and Computing 1272, https://doi.org/10.1007/978-981-15-8443-5_39

the corresponding pattern, trailing PPI networks leads to the investigation of disease-related protein–protein interfaces and the topological assets of the malignant growth network. Given a PPI network, tentatively comprehending protein–protein communication complex structures [2] is costly and troublesome and calculation forecasts of protein–protein cooperations are turning out to be increasingly vital. High throughput techniques bear an eloquent amount of data for fabricating primary databases. Moreover, due to the required biases, the outcomes created by these strategies are inclined to trial mistakes. Besides, contrasted and a lot of protein arrangement information, the utilitarian units that have been recognized are generally constrained. Various relevant machine learning techniques [3] such as decision trees (DT), Naive Bayes (NB) [4], support vector machines (SVM) [5], and nearest neighbor (NN) [6] have been ending up being proficient in information mining assignments, but not that significant. As of late, deep learning-based methodologies extricating nonlinear and high-dimensional features from the protein successions are another inclination [7]. In this way, the value of a protein is encoded in its amino corrosive course of action and that is productive in deciding protein–protein connection. In theory, a deep learning structure is elected to use particular concealed layers to harvest protein gathering comparability. This framework facilitates both social and physicochemical information from proteins and viably fuses them using an element combination system. These linked features will be given to the affiliation classifier and protein family identifier for getting ready and testing. To the best support of our understanding, this is one of the most satisfying precise deep learning frameworks for separating protein–protein connection systems. Specifically, the grant of this work is listed below.

- No cancer human PPI database has been constructed till now, to the best of our knowledge, which provides cancer (breast cancer [8, 9], brain tumor, leukemia) human protein interactions centrally. Motivated by this, here a small database is introduced.
- A deep learning system is introduced that coordinates supervised learning to figure out how to anticipate protein–protein interactions and distinguish protein families.
- In the supervised learning model, coordinate the features which are score or confidence of protein–protein interaction, the corresponding contribution of each breed of an amino acid component in protein–protein interaction to combine structural [10] highlights, and the physicochemical succession characteristics for every protein.
- In our experimental analysis, the number of hidden layers is kept on changing and certain significant tools legitimate in the supervised model and accomplish higher PPI prediction and protein family distinguishing accuracy.

The rest of the paper is mannered as follows. In Sect. 2, a brief literature overview of various learning strategies to extract significant features of the PPI network is given. In Sect. 3, the proposed dataset and its preprocessing strategies are elaborately discussed and also describe some theoretical background along with the analysis of our proposed algorithm. Furthermore, Sect. 4 presents the structure of parameter settings, dataset portrayal, and test results investigation. Section 5 listed the overall benefits of our proposed work. Finally, the paper is concluded.

2 Literature Review

The long short-term memory networks—–usually just called "LSTMs" —–are a special kind of recurrent neural networks (RNN), capable of learning long-term dependencies. The VCP data obtained from the biogrid dataset are converted into computational qualities, protein marks and ProtVec plans successfully. At last, the LSTM model [11] was deployed over the dataset for prediction, then the upshot of both the marks and ProtVec-based features was juxtaposed for the best reveal. The act of the LSTM was resolved at log-score, ROC curve values, and classification accuracy. But "***LSTM has a few bad marks like it is shallow prompting a decent presentation just in 2 to 3 layers and consequence, the consistent performance drops***." Compared with LSTM, the multimodal approach was a lot distinct. Here, the framework grasps both the relational and physicochemical data integrating them into feature fusion techniques [12]. The joining is given to the association classifier and purposed for training and testing. So, it coordinates both unsupervised learnings and supervised learning on how to anticipate protein–protein cooperation. "***The methodology was bulky as it included both supervised and unsupervised methods alongside CBOW prompting plenty of authentic calculations***." By referring to a biclustering-based association rule mining approach [13], it is showing the prediction power of the association rule mining algorithm over the traditional classifier model without choosing a negative dataset. The time complexity of the biclustering-based association rule mining is also analyzed and compared to traditional association rule mining. "***This model only involved the positive data means the proteins that were already cancerous and not the non-cancerous ones, leading to a lesser accuracy***." Making it more challenging, this motivated us to use a randomized mixture of cancerous and non-cancerous to predict the new anticipated cancerous proteins with satisfying accuracy, fewer calculations unlike the multimodal approach and efficient, unlike the LSTM model.

3 Proposed Work

3.1 Proposed Dataset Generation

In this paper, the biological database of search tool is used for the retrieval of interacting genes/proteins (STRING) datasets (https://string-db.org/). This dataset includes the precisely found proteins accounted for assorted cancer diseases (breast cancer, brain tumor, leukemia) on Homo sapiens [14, 15]. For the diverse sequence lengths obtained from the Swiss-Prot database for a particular protein, the feature extraction method is discussed in the next section where it diminishes them to 20 such features for each protein in a PPI. The overall pictorial view is given in Fig. 1 for our proposed dataset generation. The STRING database contains more viable data than other optional databases such as biogrid and DIP. The information for carcinogenic

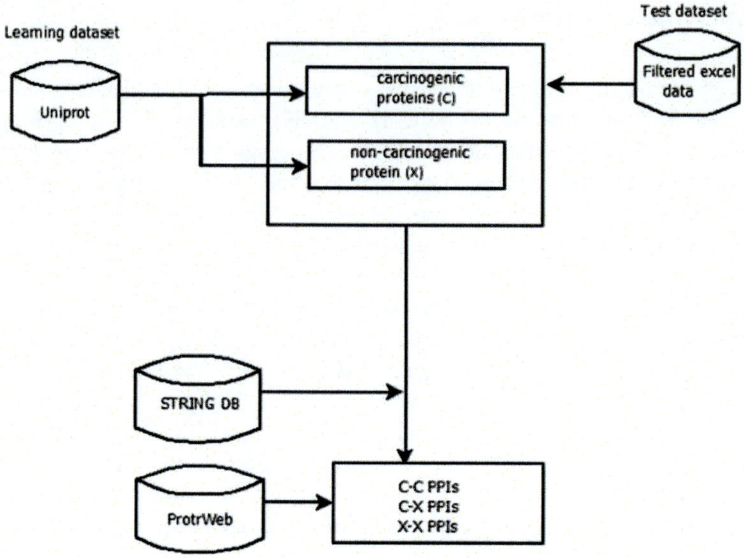

Fig. 1 Dataset generation

proteins is extricated from Swiss-Prot "http://www.bmrb.wisc.edu/data_library/Diseases/" and "https://www.uniprot.org." After peeling out the duplex protein interactions and the self-synergy, our dataset contains a total of 466 protein interactions which includes 298 positive interactions and 168 negatives.

The annotation type plays a crucial part in forecasting. A cell counters to catalysts via distinct expressways. These signals are broadcasted through different interactions of proteins in the cell. So, it gives information which protein activates/reacts/expresses/catalyzes or post-translates which protein [14]. The responsible proteins which interact with human proteins are denoted positive interactions and vice versa. After the completion of the information development process, the proposed dataset is blended and rearranged according to these positive and negative examples and it is further divided into a ratio of 75% of the training set and 25% of the testing set.

3.2 Feature Extraction

There is a fixed number of data sources our profound learning model can oblige for preparing and testing. These sources of info are gotten from variable length protein successions utilizing physicochemical and basic grouping inferred highlights which can be good for portraying and deciding proteins of a different assistant, valuable, and associating properties.

(1) Amino acid architecture: Amino acids are the structure squares of protein development. The amino acid architecture is the fraction of each amino acid type {*A*, *N*, ..., *Y*, *V*} within a protein. The amino corrosive structure gives 20 highlights and the divisions of each of the 20 regular amino acids are determined as:

$$fr(t) = \frac{N_t}{N}, \quad t \in \{A, R, N, \ldots.., Y, V\} \tag{1}$$

Here, in condition (1), N_t is the number of amino acid kind t in a protein arrangement with length N. $\{A, N, \ldots, Y, V\}$ symbolizes twenty types of amino acids.

(2) Score or weight: The combined score or weight is identified by blending the probabilities from the diverse proof channels and improved for the likelihood of haphazardly watching an interaction. "***They are basically, the 'edge weights' in each network adjoined by various proteins.***" They are in the range from 0 to 1.

Our dataset comprises 46 columns, where the first two are the interacting proteins, followed by score of the interaction and then the domain sequence code and 20 amino acids constituent percent for first protein and then the same for the second protein mentioned there.

3.3 Deep Neural Network

An artificial neural framework, gained by neural frameworks in the psyche, includes layers of neurons. It comprises of shrouded layers that are compared to the profundity of the neural system and width to the most outrageous number of neurons in one of its layers. At the point when the counterfeit neural system is prepared with numerous layers (at least two shrouded layers), it advances into a profound neural system. The arrangement of the neural system gets data through the info layer, which is changed in a non-straight way through a movement of various concealed layers. They are finally handled into the yield layer. The neurons in this arrangement of concealed layers or yield layers are completely connected with neurons in the past layer. The neurons consist of a mathematical function that computes the weighted sum of inputs $x_i w_i$ and the weights w_i.

$$y = f\left(\sum_i^n x_i w_i + b\right) \tag{2}$$

Then, the result of the weighted sum is used as a nonlinear activation function f to calculate its output y. The most notorious activation function is rectified linear unit (ReLU). Others are sigmoid, tanh. "***To test out the impact of the different neural network architectures, our neural network is designated to two parts: (a) Each***

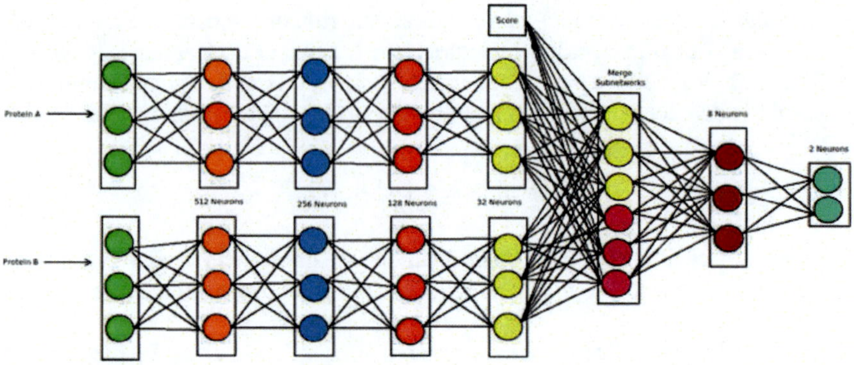

Fig. 2 Deep neural network architecture

protein undergoes as different input, hence two separate networks. (b) Directly connecting the two proteins along with 'score' into a single neural network. The hidden layers for both subnetworks are described in 512, 256, 128, 32 fashion and applied ReLU as our activation function. ReLU is less computationally costly than tanh and sigmoid because it includes more straightforward scientific activities. **So, A(x) = max(0,x). It gives x if x is positive else 0."**

A basic deep learning network architecture with multiple hidden layers is shown in Fig. 2.

Numerous substitutes to such deep neural systems have been created for explicit applications, which contrast in the direction of neurons. Convolution neural network is a competent algorithm extensively used in pattern recognition [6] and image processing. Recurrent neural network is primarily used in language modeling or sequential data [16], restricted Boltzmann machines [17, 18], and autoencoders [19] for unsupervised learning. The decision of network and various parameters ought to be kept in a piece of information-driven and fair-minded route by dissecting the model execution on the training and approval dataset.

3.4 PPI Identification of Neural Network

For the PPI ID task, two separate inputs profound neural frameworks. One is for protein A, and the other is for protein B. The structures for both of the neural frameworks are undefined. From the last layer, the high-dimensional features of the two proteins are connected and combined with the third data layer which suits the scores of PPI. The completely associated layer is yielded by the backpropagation calculation during the investigation procedure, and the double cross-entropy is utilized as the misfortune work.

$$L(\hat{y}, y) = \frac{-1}{N} \sum_{i=1}^{N} \left[y_i \log \hat{y}_i + (1 - y_i) \log(1 - \hat{y}_i) \right] \tag{3}$$

"Here, ŷ represents predicted output, y represents actual output and N is the total number of predictions, that is how computed value of double cross-entropy." Subsequently, the last order results are the likelihood that the two given proteins A and B interact with each other.

Proposed Pseudocode

1. *Start;*
2. *Obtain Dataset;*
3. *Normalize and scale the Dataset;*
4. *Split the dataset in the ratio of 80:20 as a training set and testing set;*
5. *Initialize the input_size, hidden_layers, hidden_neurons, batch_size and epochs;*
6. ***mergeNN****= Define the deep neural network with definite parameters and return "model";*
7. *Initialize and randomize weight as "w" and bias as "b" for "model";*
8. *Set the loss function = binary_crossentropy, learning_rate(LR) for the "model";*
9. *For each epoch do*

   a. List the "losses" as an entity;
   b. Select the batch;
   c. For each batch do
      i. *X, Y = retrieve from the given batch;*
      ii. *protein A, protein B, score < - X*
      iii. *Y_pred < - Prediction from "model" for a given batch;*
      iv. *L = loss(Y, Y_pred);*
      v. *Append L to "losses";*
      vi. *Set w.grad and b.grad to zero;*
      vii. *grad < - compute gradient of the loss with respect to "model" framework;*
      viii. *w = w + w.grad * LR;*
      ix. *b = b + b.grad * LR;*
      x. *end For;*

      Obtain the scores and update the mean of scores as accuracy;
      Evaluate the deep neural network for test dataset;
      Obtain the accuracy, precision, recall, F1-score for test dataset;
      End;

Module: mergeNN

1. *left_model, right_model = define two separate models for protein A and B*
2. *Describe hidden layer for both subnetworks as 512-256-128-32*
3. *model_weight = input combined score between protein A and B*

4. *merge model_weight to both subnetworks*
5. *model = concatenate left_model and right_model*
6. *separate high dimensional highlights of two proteins*
7. *return mode*

4 Result Analysis

4.1 Evaluation Metrics

The model is set up with different hyperparameters and the one which brings the best is picked. The readiness set is used to learn models with different hyperparameters and a short time later appointed authority is subject to different evaluation estimations. The hypothesis execution of the model is analyzed and a short time later diverged from different AI portrayal techniques for the testing set. In our examinations, the informational collection is proportioned to 75% preparing set and 25% testing set.

The following metrics are accounted for valuation: prediction, accuracy, precision, recall, and F_1-score. These metrics account for evaluation and deviation of accuracy in the PPI dataset. These are defined as the following:

$$\text{Accuracy} = \frac{\text{TP} + \text{TN}}{\text{TP} + \text{TN} + \text{FP} + \text{FN}} \tag{4}$$

$$\text{Recall} = \frac{\text{TP}}{\text{TP} + \text{FN}} \tag{5}$$

$$\text{Precision} = \frac{\text{TP}}{\text{TP} + \text{FP}} \tag{6}$$

$$F_1 = \frac{2\text{TP}}{2\text{TP} + \text{FP} + \text{FN}} \tag{7}$$

- **True positive (TP)**: Correctly classified as the class of interest
- **True negative (TN)**: Correctly classified as not the class of interest
- **False positive (FP)**: Incorrectly classified as the class of interest
- **False negative (FN)**: Incorrectly classified as not the class of interest.

Precision is the checking of true positive in the predicted positives dataset while recall is computed by the number of true positives in the realistic positives dataset. In the statistical analysis, the F_1-score (also F-score) is the proportion of test exactness that can be deciphered as the consonant mean of the precision and recall, where F_1 arrives at its best an incentive even from a pessimistic standpoint at 0.

4.2 *Experimental Analysis*

4.2.1 Comparison with Traditional Methods

The training datasets are qualified with 5-CV (cross-validation) and the aced model is evaluated with the test set. Our existing deep learning framework is contrasted with several traditional algorithms, including logistic regression, random forest, Gaussian Naive Bayes, decision trees, AdaBoost, SVM, nearest neighbors, and linear discriminant analysis (LDA). Figure 3 shows a brief contrast results with our contemplated model in terms of accuracy, precision, recall, and F-score on the proposed input human-cancer PPI dataset.

In Fig. 3, the highest prediction accuracy obtained is 83.5% for our contemplated deep learning model. Prediction accuracy values of other methods are 80.34, 79.49, 80.34, 68.38, 77.78, 64.10, 77.78, 79.49, and 81.20%. Post an eyesight observation, such pioneer accuracy convinces that the thoroughness of our predictors is notable.

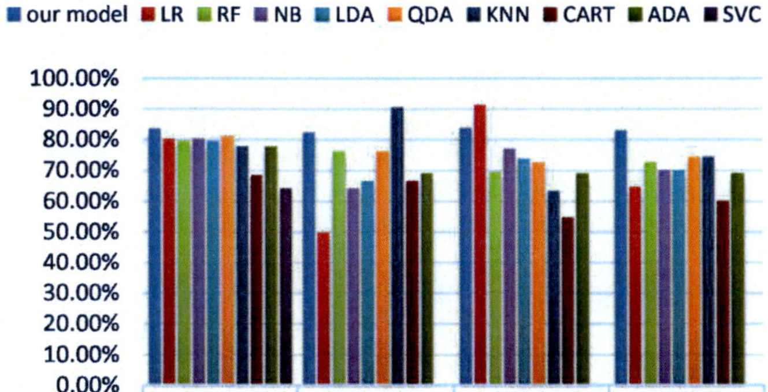

	accuracy	precision	recall	f1-score
our model	83.50%	82.49%	83.89%	83.18%
LR	80.34%	50.00%	91.30%	64.62%
RF	79.49%	76.19%	69.57%	72.73%
NB	80.34%	64.29%	77.14%	70.13%
LDA	79.49%	66.67%	73.68%	70.00%
QDA	81.20%	76.19%	72.73%	74.42%
KNN	77.78%	90.48%	63.33%	74.51%
CART	68.38%	66.67%	54.90%	60.22%
ADA	77.78%	69.05%	69.05%	69.05%

Fig. 3 Performance comparison of different metrics

Table 1 Contrasted parameters of our model on other datasets

	Accuracy (%)	Precision (%)	Recall (%)	F_1 (%)	Specificity (%)
Test 0	72.36	74.4	80.1	77.1	61.4
Test 1	60.8	65.32	69.83	67.5	48.1
Test 2	74.8	78.9	77.5	78.26	78.2
Test 3	88.9	89.1	92.2	90.6	84.3

4.2.2 Comparison with Other Public Datasets

This zone for the most part contains the unpredictability of various datasets available transparently. Most unpretentiously, an open disease dataset is taken and our computation is chipped away at the controlled dataset realizing the going with characteristics that showed up in the table underneath. On a relationship with another dataset in Table 1, an astounding exactness of 88.9% is cultivated that legitimizes that our computation is adequately changed following envision protein connections.

4.2.3 Hyperparameter Analysis

This subsection mainly accords with the hyperparameters which have used so far. Table 2 describes all the central parameters suitable for our model. The learning rate that taken is mentioned but the recommended is 0.01. The batch sizes used are 512,256,128, and 32 for the best accuracy possible [15].

The hyperparameters adaptive learning methods that used are mainly the stochastic gradient descent for the least possible loss and the activation function used is mainly ReLU, Sigmoid, and Softmax at their most suitable positions.

Table 3 briefs about the various proteins corresponding to the respective cancers, the research is manipulated on, these contain both the carcinogenic and non-carcinogenic proteins, as the study is mainly about predicting whether the PPI interaction is carcinogenic or not.

Table 2 Central parameters of our model

Name	Range	Recommendation
Learning rate	1, 0.1, 0.01, 0.001, 0.0001	0.01
Batch size	512, 256, 128, 32	128
Learning rate method	SGD, RMSprop, Adam, Adagrad	SGD
Activation function	ReLU, tanh, Sigmoid, Softmax	ReLU

Table 3 Cancer and their responsible proteins

Various types of cancers	Responsible proteins
Breast cancer	NCOA6, BRCA1, CXCR1, DMTF1, DEFB106B, LMO4, ABCG2, BRCA2, BCAR1, BCAR3, AGR3, SNCG, ZNHIT6, ANKRD17
Leukemia	LIFR, PBXl, PML, TET1, HLF, KMT2A, MLFl, PICALM, LIF, FLVCRl, KMT2E, RUNX1, MME, MCL1, TRIM13, TCL1A
Brain tumor	MLH1, EXT1, CSF3, ETV1, PTK6, SDHB, VHL, RHEBL1, HIC1\xa0, CHEK1, KHDRBS3, REST, TP53, NR4A3, DMRT1, CDKN2A, CKAP5, TPD52, TRIM13, ATM

4.3 Statistical Analysis

The entire PPI information has been given a role as a network in which hubs speak to the proteins, and edges speak to the collaborations between the proteins. Then, the network is analyzed using Cytoscape [20] network analyzer. They behave almost like the other neural networks [21]. Before auditing the networks, the parameters should be acknowledged and are explained underneath. Upon comparing the degrees of all 430 proteins, quirky proteins present in the database got 13 hub proteins. They are TRIM13, TP53, ATM, BCAR1, CDKN2A, RUNX1, CHEK1, BRCA1, BRCA2, KHDRBS3, PTK6, LIF, and TET1. The center point proteins are significant in malignant growth to investigate as they are profoundly communicated in ailing cells.

- *Network density*: It explains how populated the system is, with edges, barring self-circles, and copy edges. In the graph with the set of edges N and the set of vertices X.
  For an undirected graph,

$$\text{Density}_1 = \frac{2|N|}{|X|(|X|-1)} \tag{8}$$

  and for a directed graph,

$$\text{Density}_2 = \frac{|N|}{|X|(|X|-1)} \tag{9}$$

- *Clustering coefficient*: In a system, the clustering coefficient of a hub is characterized as the proportion of the number of edges among the neighbors of the hub to the greatest number of edges that might exist among the neighbors. It is measured as

$$C_i = \frac{\text{Number of triangles connected to node } i}{\text{Number of triples centered around node } i} \tag{10}$$

  The same for the thorough network can be manipulated as

Table 4 PPI network statistics

Parameters	Results (s)
Network density	0.005
Clustering coefficient	0.037
Avg. no. of neighbors	2.103

$$\mathrm{C} = \frac{1}{n}\sum_{i=1}^{n} C_i, \tag{11}$$

where $n =$ total number of vertices in the graph and $I \in \{1, 2, \ldots, n\}$. The statistical analysis of the carcinogenic PPI network [1] shows in Table 4 where the scale-less feature of the degree of nodes as the majority of proteins participate in a few interactions.

Figure 4 shows a sample of PPI network fetched using Cytoscape [20] where the network density is 0.005 and the clustering coefficient is 0.037. The network is sparse as density is close to 0. In any case, it likewise encourages the way that the proteins responsible for a comparative kind of ailment, similar to cancer, will in general interface with one another.

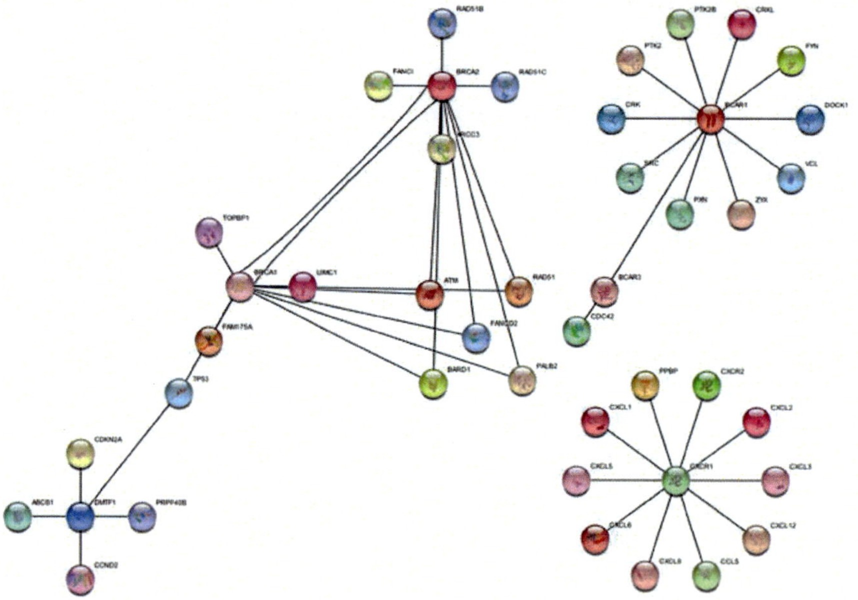

Fig. 4 Major hub proteins of breast cancer database. These five major hubs BRCA1, BRCA2, DMTF1, BCAR1, and CXCR1 are interacting with possible human proteins

5 Benefits of Proposed Work

Our contemplated method is contrasted with various other frameworks. Correlations show that our proposed model accomplishes aces in class execution with almost no intricacy. The most significant social impact of our research are:

- Many of the researches that have been done involve a huge amount of computational data that is not embedded in our thesis, so that makes it less cumbersome.
- Our test results uncover great forecast execution and show extraordinary consistency on a huge scale of human PPI datasets.
- Our test results uncover great forecast execution and show extraordinary consistency on a huge scale of human PPI datasets. It is imperative that the quirky conduct of a neural network can give extra ordered data to researchers.
- Thereafter, a database Web site is deployed using a free hosting service that contains a dataset of major cancer proteins of various fatal diseases like brain tumor, breast cancer, and leukemia.
- The Web site allows any analyst to get different features like score or confidence of the interaction and amino acid constituent of carcinogenic proteins in a single place so, it will be easier for them to reanalyze them with other proteins removing the overhead of wandering at conflicting sites.
- The Web site also has an interaction network of different cancers that authorized them to comprehend cell pathways and creating compelling treatments for the therapies of human ailment and content heads.

6 Conclusion

In this paper, a profound neural system-based PPI extraction model is discussed. Our model lines progressively shrouded layers in a profound neural system driving the model to take off the precision unequivocally diminishing the multifaceted nature. Our model essentially broadens the utilization of fundamental highlights like the amino corrosive rate constituents and the scores in such a computational method to frustrate the computational blunders. These highlights helped us get a communication system of various cancer-causing and non-cancer-causing PPIs which encourages us to foresee, guide, and imagine the interactomes and give a layout of presumably the most huge interactomes resources. Any examination on arranging arrangement intends to improve the exhibition assessment metric and time multifaceted nature. Datasets expect a critical activity in the show of aligners. Distinguishing the protein species in these very much contemplated species assists with revealing the edifices present in inadequately examined species. Making a system for estimating the diverse aligners and the assessment metric alongside the informational collection to be utilized will assist the scientist with categorizing the best aligner to utilize. This work fills in as a significant survey of the field and counselors for help more masters adequately use these mind-boggling approaches in their review. It will permit analysts to foresee a protein whether that is cancer-causing or non-cancer-causing with a precision of 83.50%. In the future, proposing to bring association rule mining (ARM) calculation on this dataset that will help one with anticipating any future protein that may interface with these proteins and can be viewed as cancer-causing.

Supporting Links
Contemplated Web site link for supporting PPI Dataset:

1. "https://datappi.000webhostapp.com/."
2. Cancer database
   - https://drive.google.com/drive/folders/1-jnesm2JZVUNvWot11ymQW7avRqDgwpr?usp=sharing.

References

1. Ekbal, E., Saha, S., Bhattacharyya, P.: A deep learning architecture for protein–protein interaction article identification. In: 2016 23rd International Conference on Pattern Recognition (ICPR), pp. 3128–3133, Dec 2016
2. Anooja Ali, V.R.: Alignment of protein interaction networks and disease prediction: a survey. Int. J. Adv. Trends Comput. Sci. Eng. 1300–1307 (2019). https://doi.org/10.30534/ijatcse/2019/42842019
3. Du, X., Sun, S., Hu, C., Yao, Y., Yan, Y., Zhang, Y.: DeepPPI: boosting prediction of protein–protein interactions with deep neural networks. J. Chem. Inf. Model. **57**(6), 1499–1510 (2017)
4. Dey, L., Mukhopadhyay, A.: Biclustering-based association rule mining approach for predicting cancer-associated protein interactions. IET Syst. Biol. **13**(5), 234–242 (2019)
5. Huang, C.H., Peng, H.S., Ng, K.L.: Improving protein complex classification accuracy using amino acid composition profile. Comput. Biol. Med. **43**(9), 1196–1204 (2013)
6. Zhang, D., Kabuka, M.R.: Multimodal deep representation learning for protein interaction identification and protein family classification. BMC Bioinform. **20**(S16), 595–602 (2019)
7. Zhang, H., Guan, R., Zhou, F., Liang, Y., Zhan, Z.-H., Huang, L., Feng, X.: Deep residual convolutional neural network for protein–protein interaction extraction. IEEE Access **7**, 89354–89365 (2019)
8. Taylor, I.W., Linding, R., Warde-Farley, D., Liu, Y., Pesquita, C., Faria, D., Bull, S., Pawson, T., Morris, Q., Wrana, J.L.: Dynamic modularity in protein interaction networks predicts breast cancer outcome. Nat. Biotechnol. **27**(2), 199–204 (2009)
9. Kar, G., Gursoy, A., Keskin, O.: Human cancer protein–protein interaction network: a structural perspective. PLoS Comput. Biol. **5**(12), e1000601 (2009)
10. Sujatha, M.M., Srinivas, K., Kumar, R.K.: Construction of breast cancer-based protein–protein interaction network using multiple sources of datasets. Soft Comput. Med. Bioinform. 11–16 (2019)
11. Kourou, K., Exarchos, T.P., Exarchos, K.P., Karamouzis, M.V., Fotiadis, D.I.: Machine learning applications in cancer prognosis and prediction. Comput. Struct. Biotechnol. J. **13**, 8–17 (2015)
12. Chiang, J.-H., Lee, T.-L.M.: In Silico prediction of human protein interactions using fuzzy–SVM mixture models and its application to cancer research. IEEE Trans. Fuzzy Syst. **16**(4), 1087–1095 (2008)
13. Alakus, T.B., Turkoglu, I.: Prediction of protein–protein interactions with LSTM deep learning model. In: 2019 3rd International Symposium on Multidisciplinary Studies and Innovative Technologies (ISMSIT), pp. 1–5 (2019)

14. Shannon, P.: Cytoscape: a software environment for integrated models of biomolecular interaction networks. Genome Res. **13**(11), 2498–2504 (2003)
15. Ji, Y., Yu, S., Zhang, Y.: A novel Naive Bayes model: packaged hidden Naive Bayes. In: 2011 6th IEEE Joint International Information Technology and Artificial Intelligence Conference, vol. 2, pp. 484–487 (2011)
16. Wang, B., Zeng, Y., Yang, Y.: Generalized nearest neighbor rule for pattern classification. In: 2008 7th World Congress on Intelligent Control and Automation, pp. 8465–8470 (2008)
17. Bock, J.R., Gough, D.A.: Predicting protein–protein interactions from primary structure. Bioinformatics **17**(5), 455–460 (2001)
18. Lipton, Z.C., Berkowitz, J., Elkan, C.: A critical review of recurrent neural networks for sequence learning (2015). arXiv preprint arXiv:1506.00019
19. Hinton, G.E.: A practical guide to training restricted Boltzmann machines. In: Lecture Notes in Computer Science, pp. 599–619 (2012)
20. Salakhutdinov, R., Larochelle, H.: Efficient learning of deep Boltzmann machines In: Proceedings of the Thirteenth International Conference on Artificial Intelligence and Statistics, pp. 693–700 (2010)
21. Vijayakumar, T.: Comparative study of capsule neural network in various applications. J. Artif. Intell. **1**(01), 19–27 (2019)

Network Structure to Estimate Prices of Basic Products: Dairy

Noel Varela, Nelson Zelama, and Jorge Otalora

Abstract Price prediction is one of the key issues in sector analysis. However, both the level of production and the prices of the agricultural sectors are highly variable characteristics with strong dependencies on eventualities, since they are subject to climatic and political shocks, making complex the modeling of their behavior and, consequently, the task of predicting or forecasting their future evolution. The present study aims to adjust a neural network model to make predictions applied to the Colombian dairy sector and to compare the results with the predictions provided by a multivariate time series model.

Keywords Neural networks · Price prediction · Dairy sector

1 Introduction

The Colombian dairy complex is not immune to such complexity and problems of variability throughout its structure. It is a sector that is integrated by levels, namely primary production on the one hand; industrialization on the other; internal and external wholesale marketing; and finally, final consumption [1]. The flow of products and inputs throughout the chain is coordinated by the price system, which, in addition to being key to the distribution of sectoral income, is one of the critical elements for the incentives in the chain and for determining its competitiveness [2].

However, there are constant conflicts between the agents that result mainly from the lack of predictability and transparency in the formation of prices in the sector. A

N. Varela (✉) · J. Otalora
Universidad de la Costa, St. 58 #66, Barranquilla, Atlántico, Colombia
e-mail: nvarela@cuc.edu.co

J. Otalora
e-mail: jotalora@cuc.edu.co

N. Zelama
Universidad Tecnológica, San Pedro Sula, Honduras
e-mail: nelson@unit.edu.hn

© The Editor(s) (if applicable) and The Author(s), under exclusive license to Springer Nature Singapore Pte Ltd. 2021

A. P. Pandian et al. (eds.), *Proceedings of International Conference on Intelligent Computing, Information and Control Systems*, Advances in Intelligent Systems and Computing 1272, https://doi.org/10.1007/978-981-15-8443-5_40

possible solution could be the political decision to establish clear rules of the game in the sector and to make the price formation process transparent, which would allow reducing the variability and unpredictability of the behavior of the key variables in decision making, namely prices [3]. Having estimates on the future evolution of prices in the sector would help in making both production and investment decisions, since these would be made on more probable scenarios. Therefore, having reliable prediction tools for these prices in Colombia would be extremely useful for all agents in the chain. Even policy decisions would be based on real foundations and not on mere assumptions [4].

In theoretical terms, there is no single prediction tool; for multivariate economic problems, error correction vector models (ECVM) are the most used in time series since, generally, the variables are not stationary and this method takes into account this issue so as not to generate spurious results [5–7]. However, there is a growing use of artificial neural networks (ANN) in time series modeling with predictive objectives, given their flexibility to recognize both linear and nonlinear patterns [8–11].

The study aims at using this last tool in the Colombian dairy market, and then, at analyzing its performance with respect to results of other commonly used methodologies, under the hypothesis that it could be competent and even better at making predictions.

2 Materials and Methods

2.1 Methodology

Predicting time series data using ANN involves designing the processing of patterns that extend over time. A particular value of a variable in time t may depend not only on the past values of that variable, but also on the t-value of other related variables (and their respective past values). The type of ANNs contemplated by this architecture are those invented by [2] called multilayer perception, since they contain several layers with nonlinear structure [12].

A network of this type has three parts, an input layer, the hidden part (which can be one or more layers), and the output layer. In each layer, there are nodes, which are interconnected with the following layers based on weights, which are estimated through an iterative process in order to reduce the cost function. Therefore, in order to estimate an ANN, it is first necessary to define its different components, the relationships between them, as well as the estimation procedures and parameters, namely (a) the architecture and interconnections; (b) the trigger or transfer function(s); (c) the cost function, which evaluates the network output(s) with their original values; (d) the training algorithm, which iteratively changes the weights to minimize the cost function [13].

A general formulation of a pro-fed (or feedforward) ANN with four layers (two hidden) offers the possibility of including a second hidden layer with a total of e nodes [14]:

$$y_u = f_o\left(\sum_{d=1}^{e} w_d f_{h_2}\left(\sum_{j=1}^{s} w_j f_{h_1}\left(\sum_{i=1}^{n} w_{ij} x_i\right)\right)\right) \quad (1)$$

With regard to the design, an ad-hoc is used, where the input variables arise from the economic theory and from the knowledge and availability of information of the economic sector to recognize the factors that can be related along the Colombian dairy production chain. The variable of predictive interest is used as an output neuron. The number of hidden layers and their corresponding nodes is selected through a trial-and-error procedure, leaving the model with the greatest predictive capacity, safeguarding the form requirements of the ANNs [15].

The learning procedure is supervised and a backpropagation algorithm is used to estimate the weights that interconnect the network nodes.

The results will be analyzed and compared with those obtained by means of error correction vector models (ECVM) [16]. With each methodology, the last available year is predicted, and from the differences between the predictions and the observed real values, three error measures are calculated, namely the root mean square error (RMSE), the mean absolute deviation (MAD), and the maximum percentage error (MPE) [17].

2.2 *Variables and Sources*

The variables used are: (a) the monthly prices paid to the milk producer (P_{prod}), P measured in \$/lt. (of predictive interest); (b) monthly retail milk mix prices ($P^{\text{mix}}_{\text{min}}$), measured in \$/kg; (c) the monthly wholesale dairy mix prices ($P^{\text{mix}}_{\text{may}}$), measured in \$/kg; (d) the mixed milk export prices ($P^{\text{mix}}_{\text{exp}}$), measured in u\$s/kg.; (e) the quantities (lts.) produced monthly of raw milk (Q_{lts}) an indicator of the high seasonality of producer prices (High.prices). Mixed prices are a price index based on a basket of relevant goods at each level of the chain, weighted by their importance at that level; and the variable related to seasonality is included through a dummy indicating the seasons with high price levels or above the average level, when prices are constant.

All the variables have a monthly periodicity that goes from 2015 to 2019; and the same (except the seasonality) are transformed into logarithm, such as lp.prod, lp.min, lp.may, lp.exp, and lq.liters, respectively. This transformation, usually used in economics, allows the analysis of estimates in terms of elasticities; and in the context of NNA, it allows the correction of large-scale differences between variables.

3 Results

3.1 General Description

The price to the primary producer is the variable to predict. Figure 1 shows its evolution.

As for the wholesale and retail prices of the weighted mix, the evolution has also been positive and with similar behavior to that of raw milk (Fig. 2). The growth rate of retail prices has been more pronounced than that of wholesale prices, especially from 2015 onwards.

The quantity of liters of raw milk produced is another of the variables considered relevant, since the relative scarcity in the short term can generate friction in the prices of dairy goods.

Figure 3 shows the evolution of the quantity of milk produced in Colombia, in the period of analysis. A markedly seasonal behavior can be observed, but with a

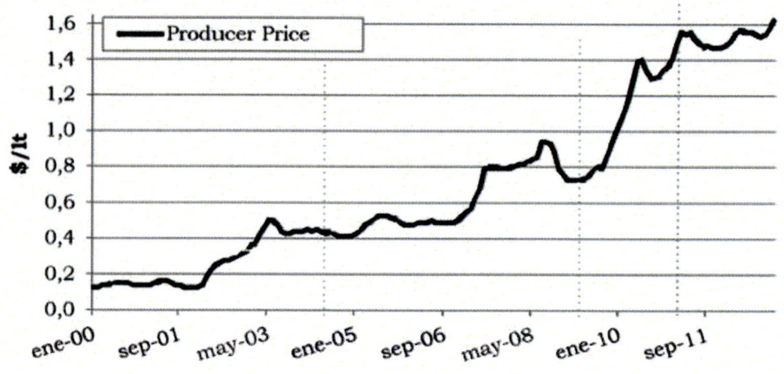

Fig. 1 Price paid to producer. Period 2015–2019

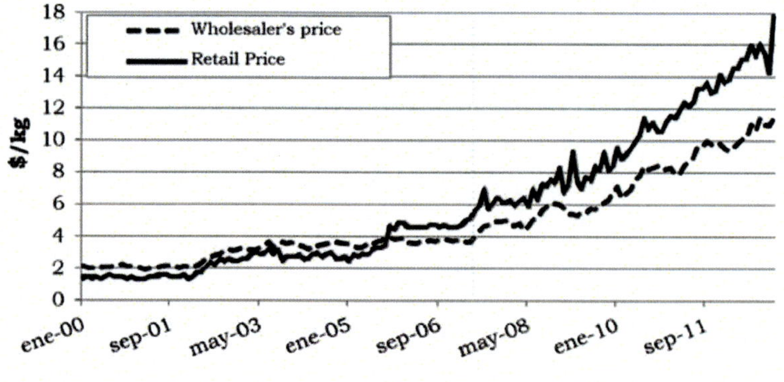

Fig. 2 Wholesale and retail price (index). Period 2015–2019

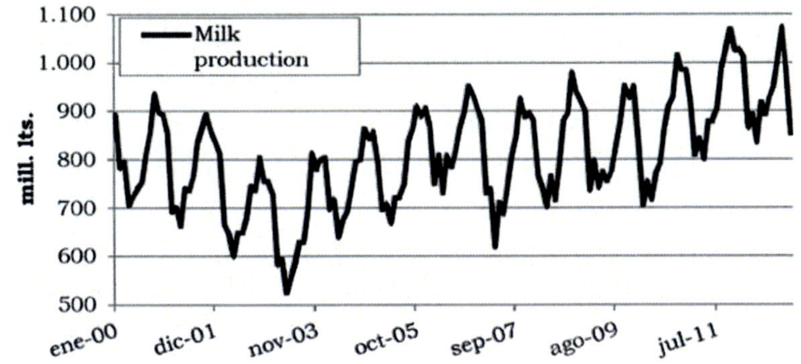

Fig. 3 Quantity (mill. lts.) of raw milk produced in the period 2015–2019

clear positive trend from the last three years. On the other hand, export prices are determined outside the country, so they are usually considered an exogenous variable in the determination of local prices.

3.2 Price Estimates Paid to the Producer

With the variables set out in the methodology, an ECVM has been estimated, the estimated parameters of which are given in the Annex. Given that it is a multiequation system, the number of parameters estimated is high, since not only the long-term relationships (or co-integration parameters) are estimated, but also the short-term ones.

As for the ANN model, Fig. 4 shows the selected architecture with the estimated coefficients for the connections. An ANN with three layers has been estimated, with four input nodes, two in the hidden one and one only in the output.

The network estimation algorithm has converged fairly quickly (less than 100 iterations) to training errors relatively close to zero, both in the training and test sets.

The comparative predictions of both models are shown in Fig. 5. On average, the VECM predictions have higher errors than those corresponding to the ANN (RMSE: 0.0387 vs. 0.0186; MAD: 0.034 vs. 0.0153, and MPE (%): 7.64 vs. 3.5, respectively). The figure shows the evolution of the predictions of both models and the real values of the price paid to the Colombian raw milk producer.

The greater similarity of the ANN predictions to the real values can also be highlighted, as well as their evolution and behavior. However, the evolution of the two predictions, in general, agrees with the real one, but with a greater dispersion than those of the ECVM.

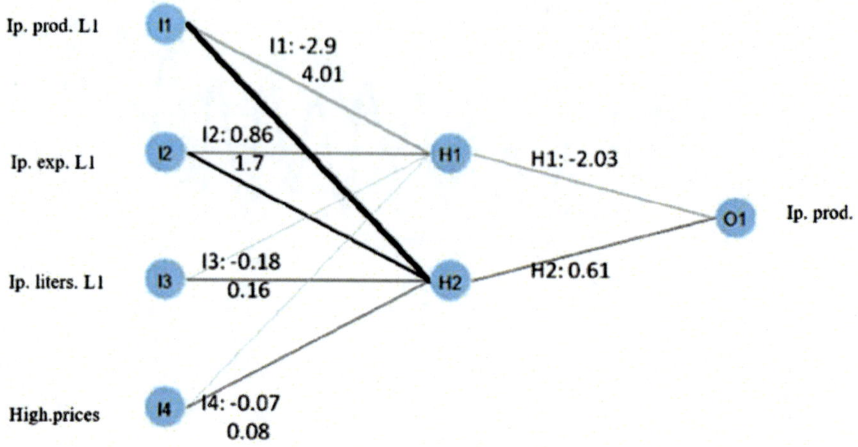

Fig. 4 Network structure for predicting the lp.prod

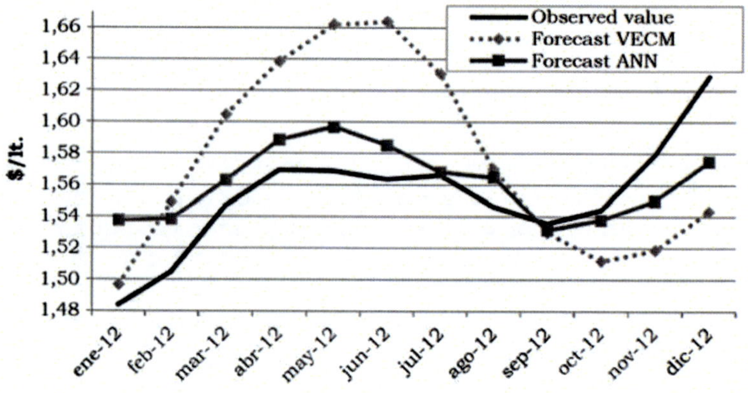

Fig. 5 Observed values and realized predictions for dairy producer price with static models. Year 2020

4 Conclusions

Making forecasts on price behavior is of utmost importance in the economic area in general, and in the dairy sector in particular. Having reliable predictions that allow making decisions with less uncertainty is precisely highly valued by the agents that interact in the sector, given that it would allow a more serious planning on their activities. However, it is not always easy to have a methodology that allows to reach such an objective. The degree of reliability of predictions often depends on the stability of the sector, and in particular, the agricultural and livestock sector is often tied to unforeseen events, both climatic and political, which make it difficult to project. The Colombian dairy sector is made up of different levels, all of which are coordinated by the price system, which is the object of constant conflict between the different levels of the sector. In particular, primary producers are the ones who are often dissatisfied with the prices paid. They turn out to be the link with the least bargaining power and with problems in obtaining information on those prices received. For these reasons, the predictive interest of this thesis is focused on prices to the primary dairy producer.

To this end, a relatively new methodology in the field of economic forecasting was analyzed in order to evaluate its performance. Different variables were used to represent the sector and are relevant when modeling primary prices. The majority corresponds to the prices of each level, although both quantities produced and indications of the seasonality of the level with predictive interest were also incorporated. The index variables try to summarize the price behavior at a level where there are multiproducts. With a monthly database from 2015 to 2019, each methodology, namely VECM and ANN, was modeled and predictions were made for 2020 and performance measures were calculated for both methodologies. It could be observed that although both predictions follow a similar evolution to the real data, the prediction errors of the ANN model were lower than those of the ECVM, concluding that in predictive terms, the use of the ANN allows for fewer errors than the ECVM with a more parsimonious and simpler model. Therefore, in order to predict the prices paid to primary producers of raw milk, such a model is the most appropriate. In this particular case, an NNA model is more appropriate for making predictions, but these results do not imply that it is appropriate for all cases. The present is an attempt to begin to evaluate this methodology, and it is of fundamental importance to emphasize that it must continue to be improved in order to study more complex models that involve a more mechanized search process and architecture, and that are subject to more severe evaluation processes.

References

1. da Rosa Righi, R., Goldschmidt, G., Kunst, R., Deon, C., da Costa, C.A.: Towards combining data prediction and internet of things to manage milk production on dairy cows. Comput. Electron. Agric. **169**, 105156 (2020)
2. Khamaysa Hajaya, M., Samarasinghe, S., Kulasiri, G.D., Lopez Benavides, M.: Detection of dairy cattle Mastitis: modelling of milking features using deep neural networks (2019)
3. Yang, Q.: Prediction of global value chain based on cognitive neural network-take Chinese automobile industry as an example. Transl. Neurosci. **10**(1), 81–86 (2019)
4. Ragni, L., Iaccheri, E., Cevoli, C., Berardinelli, A.: Spectral-sensitive pulsed photometry to predict the fat content of commercialized milk. J. Food Eng. **171**, 95–101 (2016)
5. Alamin, Y., Castilla, M.D.M., Alvarez, J.D., Jimenez, M.J., Perez, M., Ruano, A.: Prediction of wall thermal transfer properties using Artificial Neural Networks (2019)
6. Nguyen, Q.T., Fouchereau, R., Frenod, E., Gerard, C., Sincholle, V.: Comparison of forecast models of production of dairy cows combining animal and diet parameters. Comput. Electron. Agric. **170**, 105258 (2020)
7. Bashar, A.: Survey on evolving deep learning neural network architectures. J. Artif. Intell. **1**(02), 73–82 (2019)

8. Viloria, A., Lezamab, O.B.P.: Improvements for determining the number of clusters in k-means for innovation databases in SMEs. Procedia Comput. Sci. **151**, 1201–1206 (2019)
9. Silva, N., Siqueira, I., Okida, S., Stevan, S.L., Siqueira, H.: Neural networks for predicting prices of sugarcane derivatives. Sugar Tech. **21**(3), 514–523 (2019)
10. Shahriary, G., Mir, Y.: Application of artificial neural network model in predicting price of milk in Iran. Mod. Appl. Sci. **10**(4), 173–178 (2016)
11. Silva, J., Mojica Herazo, J.C., Rojas Millán, R.H., Pineda Lezama, O.B., Morgado Gamero, W.B., Varela Izquierdo, N.: Early warning method for the commodity prices based on artificial neural networks: SMEs case (2019)
12. Abdollahi-Arpanahi, R., Gianola, D., Peñagaricano, F.: Deep learning versus parametric and ensemble methods for genomic prediction of complex phenotypes. Genet. Select. Evol. **52**(1), 1–15 (2020)
13. Gonzalez-Fernandez, I., Iglesias-Otero, M.A., Esteki, M., Moldes, O.A., Mejuto, J.C., Simal-Gandara, J.: A critical review on the use of artificial neural networks in olive oil production, characterization and authentication. Crit. Rev. Food Sci. Nutr. **59**(12), 1913–1926 (2019)
14. Zhang, J., Meng, Y., Wu, J., Qin, J., Yao, T., Yu, S.: Monitoring sugar crystallization with deep neural networks. J. Food Eng. **280**, 109965 (2020)
15. Silva, J., Varela, N., Caraballo, H.M., Guiliany, J.G., Vásquez, L.C., Beltrán, J.N., Castro, N.L.: An early warning method for basic commodities price based on artificial neural networks. In: International Symposium on Neural Networks, pp. 359–369. Springer, Cham (2019)
16. Marsot, M., Mei, J., Shan, X., Ye, L., Feng, P., Yan, X., Li, C., Zhao, Y.: An adaptive pig face recognition approach using convolutional neural networks. Comput. Electron. Agric. **173**, 105386 (2020)
17. Amelec, V.: Carmen, Vasquez: Relationship between variables of performance social and financial of microfinance institutions. Adv. Sci. Lett. **21**(6), 1931–1934 (2015)

Energy Optimization in WSN Using Evolutionary Bacteria Foraging Optimization Method

Shiv Ashish Dhondiyal, Manisha Aeri, Paras Gulati, Deepak Singh Rana, and Sumeshwar Singh

Abstract Wireless sensor network is still one of the widest research areas for the researchers, which is made of self-sufficient sensor nodes in a group called cluster to screen physical or ecological conditions of the environment. Energy utilization and routing are the central issues in remote sensor systems. Therefore, numerous conventions have been proposed to minimize the energy utilization of these sensor nodes. The fundamental objective of this exploration is to enhance the directing procedure. This paper presents evolutionary bacteria foraging optimization method, named as EBFO, which uses a fitness function to measure the combined energy of the all sensor nodes in the cluster to make sure overall energy of cluster lesser energy than combined energy. It provides better field placement and cluster head selection. The results show the improvement in the energy efficiency and routing process in WSN as compared to standard BFO.

Keywords WSN · BFO · Cluster head · Leach protocol · EBFO

1 Introduction

A wireless sensor network is a communication infrastructure between nodes. WSN connection is made directly or indirectly with the help of other nodes. Each node itself is an autonomous body. The wireless network is composed of many autonomous nodes that communicate with each other without any infrastructure. Since the topology of the network is dynamic, any node can easily join or leave at any time. A wireless sensor network is flexible that means it can change its topology according to environmental condition. Wireless networking is used to connect the various devices wirelessly. Wireless communication networks usually work and monitored using

S. A. Dhondiyal (✉)
Graphic Era Deemed To Be University, Dehradun, Uttarakhand, India
e-mail: shivashish1234@gmail.com

M. Aeri · P. Gulati · D. S. Rana · S. Singh
Graphic Era Hill University, Dehradun, Uttarakhand, India

© The Editor(s) (if applicable) and The Author(s), under exclusive license to Springer Nature Singapore Pte Ltd. 2021

A. P. Pandian et al. (eds.), *Proceedings of International Conference on Intelligent Computing, Information and Control Systems*, Advances in Intelligent Systems and Computing 1272, https://doi.org/10.1007/978-981-15-8443-5_41

radio frequency communication. The integrity of information or data is one of the important aspects of a wireless network, also the battery power consumption of sensor devices is important.

1.1 Properties of Wireless Network

- In WSN, the node itself is an autonomous system, that is, it works as both host and router.
- if the distance is very large between source and destination, then the multihop system is used.
- WSN has a dynamic topology.
- WSN can easily be deployed.
- WSN is distributed in nature.
- In WSN, each node acts as both host and router.
- Nodal connectivity is intermittent.
- Sensor nodes are equipped with small memory, small-power unit, and lightweighted features.
- High user density and large level of user mobility.

1.2 Wireless Sensor Node Architecture

There are four basic units in the sensor node:

- Detecting unit
- Evaluation unit
- Communication unit
- Battery unit

Detecting Unit: Detecting unit is used to detect the environment condition. The sensor is a basic element in detecting unit which is used to translate physical climate condition to electrical signs. Sensors can be named either basic or propelled contraptions. There exists a blended pack of sensors that measure natural parameters, for instance, temperature, light compel, sound, attractive fields, picture, etc. The simple signs produced by the sensors in the perspective of the watched wonder are changed over to computerized motions by the ADC and a short time later reinforced into the handling.

Evaluation unit: The evaluation unit is used to process the information get from the detecting unit. The evaluation unit consists of a processor chip which processes the data, and it is one of the important units of WSN which schedules the process execution also.

Communication unit: The communication unit is to form the connection between the source and the destination nodes. Once the data is collected from different nodes to the head node, then it is transferred to the destination node.

Battery unit: The battery supplies vitality to the entire sensor center point. It expects a fundamental part of choosing a sensor center point lifetime. The measure of force drawn from a battery should be correctly watched. Sensor center points are generally little, light, and shabby, and the degree of the battery is obliged.

2 Related Work

Kulkarni et al. explained WSNs, which present PSO and examine its suitability for WSN applications. It additionally exhibits a brief review of particle swarm optimization (PSO) which is customized to address these issues [1]. Bhattacharya et al. exhibit such a configuration, which minimizes cost and energy consumption, in this way upgrading the lifetime of proficiency in addition to and system lifetime individually. The simulation result shows that major improvement in Qos parameters [2]. Fu et al. in this paper proposed an advanced leach algorithm based on new technology. Algorithm balanced out the energy consumption and improved the overall lifetime of the network. Simulation results show that the proposed algorithm is better than leach protocol [3]. Bai et al. explain distance among the nodes which is calculated, and based on that distance, availability of the link is predicted. experimental results clearly show that the current approach is far better than the existing approach [4]. Dhondiyal proposed sleeping mode mod-leach protocol which modification of model each protocol. The simulation result shows that the proposed protocol is far better than the original one [5]. Panigrahi proposes the simulation result which shows that the proposed protocol is far better than the original one Anar [6]. Hady propose a low-energy adaptive clustering hierarchy centralized sleeping protocol (LEACH-CS) for wireless sensor networks and compare the results of the proposed protocol with LEACH protocol [7].

3 System Design

Wireless sensor networks (WSNs) comprise of little hubs with detecting, calculation, and remote correspondence capacities. Many directing, control administration, and information spread conventions have been particularly intended for WSNs where vitality utilization and steering are two fundamental plan issues. Vitality utilization is trying since all sensor hubs work on batteries, so vitality is restricted. Steering in WSNs is additionally testing because of the intrinsic qualities that recognize these systems from different remote systems like versatile impromptu systems or cell systems. In the first place, because of the moderately extensive number of sensor hubs, it is impractical to construct a worldwide tending to conspire for the arrangement of countless hubs as the overhead of ID upkeep is high. In this manner, conventional IP-based conventions may not be connected to WSNs. In our proposed work vitality utilization and steering, the issue will be comprehended utilizing extemporized BFO calculation by giving better field arrangement of sensor hubs. Efficient field deployment strategies should achieve the following:

- Good connectivity
- Maximal network coverage
- Reduced construction and communication cost
- Good event detection probability
- Reduced energy consumption
- Accurate positioning
- Less position estimation error
- Low complexity.

Much of the time, the sensors shaping these systems are sent arbitrarily and left unattended to and are required to play out their central goal appropriately and productively. Because of this irregular organization, the WSN has generally differing degrees of hub thickness along its zone. Sensor systems are likewise vitality compelled since the individual sensors, which the system is framed with, are to a great degree vitality obliged too. The specialized gadgets on these sensors are little and have constrained power and range. Both the likely contrast of hub thickness among a few areas of the system and the vitality limitation of the sensor hubs cause hubs gradually kick the bucket making the system less thick. Additionally, it is very regular to send WSNs in cruel condition, which makes numerous sensors inoperable or broken. Therefore, these systems should be blamed tolerant so that the requirement for upkeep is limited. Normally the system topology is consistently and powerfully changing, and it is not a wanted answer to renew it by mixing new sensors rather the drained ones. A genuine and fitting answer for this issue is to actualize directing conventions that perform proficiently and using the less measure of vitality as feasible for the correspondence among hubs.

3.1 Methodology Used

Remote sensor networks (WSNs) comprise of little hubs with detecting, calculation, and remote correspondence capacities. Many directing, control administration, and information spread conventions have been particularly intended for WSNs where vitality utilization and steering are two fundamental plan issues. Vitality utilization is trying since all sensor hubs are work on batteries, so vitality is restricted. Steering in WSNs is additionally testing because of the intrinsic qualities that recognize these systems from different remote systems like versatile impromptu systems or cell systems. In the first place, because of the moderately extensive number of sensor hubs, it is impractical to construct a worldwide tending to conspire for the arrangement of countless hubs as the overhead of ID upkeep is high. In this manner, conventional IP-based conventions may not be connected to WSNs. In our proposed work vitality utilization and steering, issue will be comprehended utilizing extemporized BFO calculation by giving better field arrangement of sensor hubs. Efficient field deployment strategies should achieve the following:

- Good connectivity.
- Maximal network coverage.
- Reduced construction and communication cost.
- Good event detection probability.
- Reduced energy consumption.

3.2 Evolutionary BFO Algorithm

Notations used

P: Probability that a sensing node becomes cluster head.
e1: Remaining energy of sensing element node.
e2: MList_Level1ost energy a sensing element node will have.
CH: Cluster head.
BS: BList_Level1ase station CM.
F(x): Threshold energy.

Initialization phase

Initialize no. of sensor nodes (N).
Initialize length of sensor field (L).
Initialize width of sensor node (W).
Initialize the no. of rounds (Ri).
Plot X and Y location of all sensor nodes (X, Y).
Initialization phase sets up the field for the distribution of sensor nodes. Sensor nodes are randomly distributed over the field, with the same energy initially.

Execution phase

Start
while Ri is not equal to 0 do find source node.
if source node not failed then.
soc = 1 and the process is executed.
Else
soc = 0 and the process is aborted.
End
if the destination node is not failed then.
des = 1 and the process are executed.
Else
des = 0 and the process is aborted.

End
Calculate minimum threshold F(x)

$$f(x) = e1/e2 \quad (1)$$

$$g(x). = P/Q \quad (2)$$

$$F(x) = f(x) * g(x) \quad (3)$$

End
if cluster head fails then
implement BFO and get optimized sensor nodes.
Else
continue with the current cluster head.
End
Each cluster head sends request to their nearby sensor nodes to join their cluster.
Sensor node joins their respective clusters based on their distance from CH.
After the formation of clusters, each member node sends the data to their CH.
CH collects the data from all members and sends to the sink node.
Find the final route through which the data will send.
Continue the above steps whenever the data is required to be sent.
Stop.

3.3 Simulation and Result

The initial parameters taken for the simulation are shown in Table 1.

Figure 1 shows node deployment 1000 * 1000 length and width sensor field. Green dot describes the source and its coordinated nodes through which data will travel. Yellow node describes the sink node. Red dot describes the sensor nodes which do not participate in the next round. Blue node describes the sensor nodes which are not currently involved in the current round. In every round, position of sink node and

Table 1 Parameter initialization for E-BFO protocol

Parameters	Values
Network field	150 * 150
Number of sensors	200
starting node energy	0.7 j
Packet size	3000 bits
Aggregation energy	56 nj per bit
Number of rounds	4000

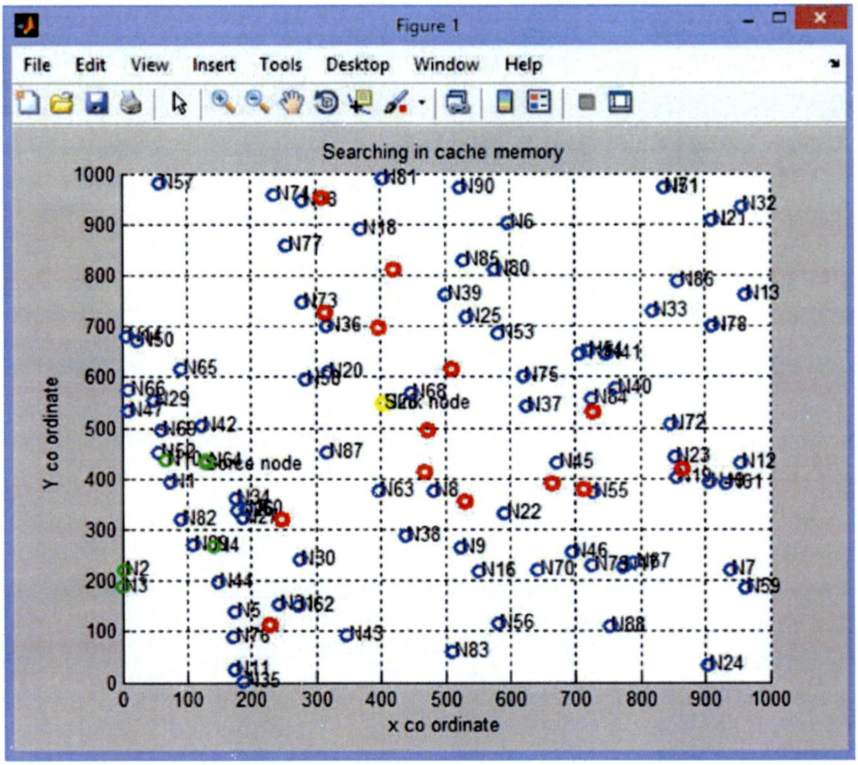

Fig. 1 Distribution of sensor nodes in the field

source node will be changed according to the requirement of data need by the base station.

Figure 2 shows graph plotted between end to end delay and throughput against a number of rounds of data transfer on which end delay occurs and throughput gain. End to End Delay signifies the total amount of time taken by a packet from source to destination. Throughput is the number of packets sent over the network in a given time or it can also be defined as the average rate of successful messages delivered over a communication channel. According to the above diagram end to end delay for E-BFO is less and throughput is high as a comparison to standard BFO and LEACH protocol.

Figure 3 shows graph plotted in between a round of data transfer and energy consumed by the nodes after utilizing E-BFO, standard BFO algorithm, and LEACH protocol. The error rate is defined as the no. of errors found when packet delivery takes place from source to destination. Energy consumption is defined as the total energy required for transmission between nodes in the network. It is measured in J. It has been seen that energy consumption and error rate for E-BFO are less by LEACH and standard BFO.

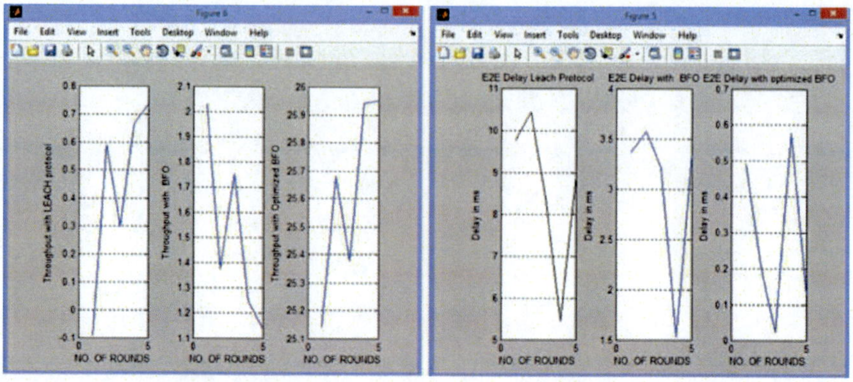

(a) Throughput (b) End to End Delay

Fig. 2 Comparison of throughput and end to end delay

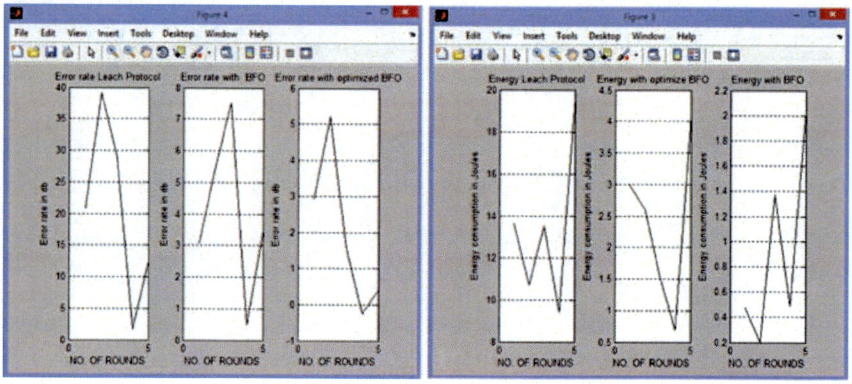

(a) Error rate. (b) Energy optimization.

Fig. 3 Comparison of error rate and energy consumption

Table 3 shows that E-BFO performs better than standard BFO as shown in Table 2 in terms of all the parameters like energy consumption, end-to-end delay, throughput, and error rate. Results clearly show that E-BFO is far better than BFO. It is very clear that as the number of round increases the difference between optimization level also increases, and this shows that optimized minimize every parameter to some extent. Energy consumption in E-BFO is 5% of original energy in overall rounds while in BFO take 14%. EBFO consumes 9% less energy than BFO.

Figure 4 represents the performance EBFO for optimization of QoS parameters with BFO. Here, graph has been drawn for different rounds cycles like a green bar is for 7, red is for 5, and green is for 2 rounds, respectively. Error rate and the energy consumption are pretty high in BFO as compared to optimized BFO and throughput

Table 2 Performance of standard BFO

Rounds	Energy consumption	End delay	Throughput	Error rate
2	12	3.5	1.5	10
5	16	6.5	5	14
7	25	9	8	16.5
9	30	11.5	7	18
11	39	14	9	23.5

Table 3 Performance of evolutionary BFO

Rounds	Energy consumption	End delay	Throughput	Error rate
2	10	2.5	2	8
5	14	4	8	7
7	18	6.5	9	11.5
9	20	9	11	17
11	22	11.5	14	18

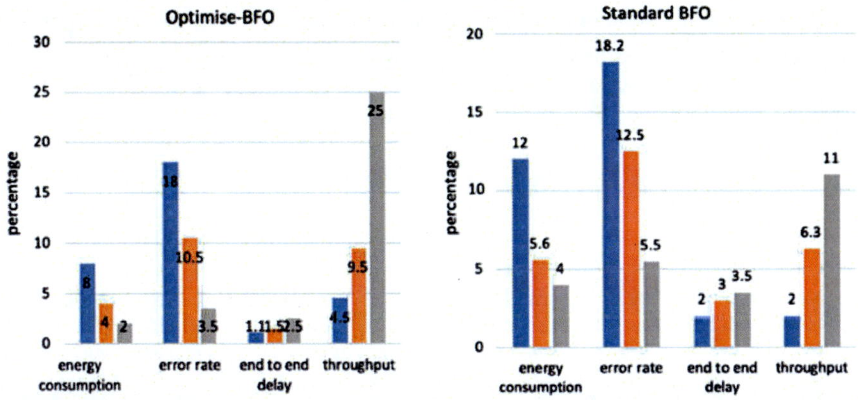

Fig. 4 Comparison between BFO and E-BFO

are less. Also, end-to-end delay is high in standard BFO than optimized BFO. Figure 4 represents that EBFO has better performance than standard BFO in terms of all QoS parameters.

4 Conclusion and Future Work

In this paper, an advanced enhanced BFO algorithm has been implemented to minimize energy consumption as compared to standard BFO. Main concern in WSN is to minimize energy to maximum extent. Since size of sensor node is small, all of its component is also small like battery, processor, and memory unit. So, due to limited resources, it is very difficult to increase the life time of the WSN network. Proposed schemes show that very effective improvement in network energy consumption in EBFO as compared to standard BFO. Simulation results prove that proposed scheme is far better than existing scheme (Fig. 4).

5 Conclusion and Future Scope

In this paper, an advanced enhanced BFO algorithm has been implemented to minimize energy consumption as compared to standard BFO. The main concern in WSN is to minimize energy to the maximum extent. Since the size of sensor, node is small so all of its components is also small like battery, processor, and memory unit. So due to limited resources, it is very difficult to increase the lifetime of the WSN network. Proposed schemes show that very effective improvement in network energy consumption in EBFO as compared to standard BFO. Simulation results prove that proposed scheme is far better than existing scheme.

References

1. Kulkarni, R.V., Venayagamoorthy, G.K.: Particle swarm optimization in wireless-sensor networks: a brief survey. IEEE Trans. Syst. Man Cybern. **41**(2), 262–267 (2011)
2. Bhattacharya, D., Krishna Moorthy, R.: Power optimization in wireless sensor networks. IJCSI **8**, 415–419 (2011)
3. Fu, C., Jiang, Z., Wei, W., Wei, A.: An energy balanced algorithm of LEACH protocol in WSN. IJCSI **10**, 354–359 (2013)
4. Bai, Y., Huang, J., Han, Q., Qian, D.: Link availability based mobility-aware max-min multi-hop clustering (M4C) for mobile ad hoc networks. IEICE Trans. (2009)
5. Dhondiyal, S.A., Rana, D.S.: Sleeping mode MODLEACH protocol for WSN. IJARCCE. **7**, 112–116 (2018). https://doi.org/10.17148/IJARCCE.2018.7823
6. Panigrahi, B.K., Ravi Kumar Pandi, V.: Bacterial foraging optimisation: Nelder–Mead hybrid algorithm for economic load dispatch. IET Gener. Transm. Distrib. **2**(4), 556–565 (2008)
7. Ahady, A., Abd El-Kader, S.M., Eissa, H.S.: Intelligent sleeping mechanism for wireless sensor network", Egypt. Inform. J. (2013)
8. Mahmood, D., Javaid, N., Mehmood, S., Qureshi, S., Memon, A.M., Zaman, T.: MODLEACH: a variant of LEACH for WSNs. In: Twenty sixth IEEE Canadian Conference on Electrical and pc Engineering (CCECE 2013), Regina, Canadian province, Canada (2013)
9. Singh, D., Nayak, S.K.: Enhanced modified Leach (EMODLEACH) protocol for WSN. In: 2015 International Symposium on Advanced Computing and Communication (ISACC) (2015)

10. Cook, D.J., Das, S.K.: Smart environments: technologies, protocols, and applications. Wiley, New York (2004)
11. Bhattacharya, D., Krishnamoorthy, R.: Control optimisation in wireless sensor networks. IJCSI **8**, 415–419 (2011)
12. Goldberg, D.E.: Genetic algorithms in search, optimization, and machine learning, pp. 170–174. Addision-Wesley, Boston (1989)
13. Bouhafs, F., Merabti, M., Mokhtar, H.: A hub recuperation plot for information scattering in remote sensor systems. In: IEEE International Conference on Correspondences 2007 (ICC'07), p 3876 (2007)
14. Gogu, A., Nace, D., Dilo, A.: Optimisation Problems in Wireless Sensor Networks, pp. 302–309. IEEE (2011)
15. Pouwelse, J., Langendoen, K., Tastes, H.: Dynamic voltage scaling on a low-control microchip. In: Proceedings of the seventh yearly global gathering on mobile registering and systems administration, pp. 251–259. ACM Press, New York (2001)

Stock Market Prediction Using Machine Learning

Ashfaq Shaikh, Ajay Panuganti, Maaz Husain, and Prateek Singh

Abstract Stock market prediction is the act of trying to determine the future value of a company's stock. The successful prediction of a stock's future price could yield significant profit. The main objective of this project is to predict the stock prices of any particular company using the foremost machine learning techniques. The machine learning model uses historical prices and human sentiments as two different inputs, and the output is distinguished as a graph showing the future prediction and a label (positive neutral and negative), respectively. The machine learning techniques used for prediction are the recurrent neural network (RNN), long short-term memory (LSTM) model and sentimental analysis. The machine learning model is then trained with several data points, and the results are evaluated. As for sentimental analysis, the public's opinion from a social media platform is scraped and then a label is generated.

Keywords Stock market prediction · RNN · LSTM · Sentimental analysis

1 Introduction

There is a notion that the prices of the stock market cannot be predicted, it is somewhat true to a degree, but the main purpose of prediction in the stock market is getting a result in a binary form, i.e., whether it is going up or down. Burton Malkiel in his

A. Shaikh (✉) · A. Panuganti · M. Husain · P. Singh
Information Technology, M.H. Saboo Siddik College of Engineering, Mumbai, India
e-mail: ashfaq.shaikh@mhssce.ac.in

A. Panuganti
e-mail: ajaypan1998@gmail.com

M. Husain
e-mail: maazhussain2905@gmail.com

P. Singh
e-mail: sin.prateek03@gmail.com

© The Editor(s) (if applicable) and The Author(s), under exclusive license
to Springer Nature Singapore Pte Ltd. 2021

A. P. Pandian et al. (eds.), *Proceedings of International Conference on Intelligent Computing, Information and Control Systems*, Advances in Intelligent Systems and Computing 1272, https://doi.org/10.1007/978-981-15-8443-5_42

book Random Walk Down the Wall Street said that even a monkey's guess would be as good as a stockbroker because of the nature of the stock market [1]. However, nowadays there are lot of tools available, which can be used to get an idea of the direction of the motion of the stock price. Even knowing a direction of motion of the stock price will be helpful because at the end of the day investors are aiming to make money in the stock market and it is enabled by the forecast of the stock price. If the investor gets to know that the stock price is going to go down, then they can get their stocks sold which can make the loss as less as possible, and if there is a forecast that the price is going to go up then the investor can wait and sell it at a higher price.

Machine learning has enabled lot of tools that can use the historical prices of the as well as the sentiment of the people. There is a study which used simple moving average model to predict using the gold, silver, forex price, previous-day price to predict the stock price of NIFTY 50 for a day after [2] but they do not account for people's sentiment which is proven in this study that accuracy of the model increases when included with people's sentiment [3].

Sentiment of the people is important; when it comes to the stock price, it can really soar or tank the price of an organization. For example, when a company is involved in a major scandal then it can result in a bad sentiment among the public and that is going to result in the downfall of the stock price [4]. It has been proven in a study that the sentiment of the people is strongly correlated to the stock price of the company [3], which concludes that the sentiment of people cannot be ignored when it comes to stock prices.

Prediction for stock price can be best done using LSTM model [5]. In this [6] research, they used LSTM and ARIMA for prediction of bitcoin price and concluded that LSTM outperforms ARIMA which is a very popular time series forecasting tool. LSTM is part of the recurrent neural network, which is an essential part of the prediction process when it comes to the stock prices. It has the ability to learn from the previous result and improve on it, and ignoring the redundant values, which are not affecting the result on itself, this results in higher accuracy.

2 Literature Review

See Table 1.

3 Proposed System

3.1 LSTM Model

An LSTM network is a type of RNN architecture, which has feedforward, and feedback connections which makes the model proficient to remember the order in a time

Table 1 Literature review

Paper name	Authors	Description
Predicting stock and stock price index movement using Trend Deterministic Data Preparation and machine learning techniques (2015) [7]	Jigar Patel, Sahil Shah, Priyank Thakkar, K Kotecha	The study compares four prediction models, artificial neural network (ANN), support vector machine (SVM), random forest and naive Bayes with two approaches for input to these models
Time series forecasting using neural networks (2014) [8]	Bogdan Oancea, Ştefan Cristian Ciucu	This paper compared the performances of different feedforward and recurrent neural networks and training algorithms for predicting the exchange rate EUR/RON and USD/RON
News impact on stock price return via sentiment analysis (2014) [9]	Xiaodong Li, Haoran Xie, Li Chen, Jianping Wang, Xiaotie Deng	Sentiments were used for prediction; textual news articles are then quantitatively measured and projected onto the sentiment space. Accuracy increased by 6.07% by using sentiment of the people
Stock market prediction of s&p 500 via a combination of improved bco approach and bp neural network (2009) [10]	Yudong Zhang, Lenan Wu	Back-propagation (BP) artificial neural network to develop an efficient forecasting model for prediction of various stock indices
Stock market prediction using machine-learning techniques [11]	Syed Hasan Adil, Kamran Raza and Syed Saad Azhar Ali	The prediction model uses different attributes as an input and predicts the market as positive and negative. They used this for the Karachi stock exchange

series prediction problem. It can also process multiple data points, which perfectly fits for our task of handling multiple attributes of our dataset. Figure 1 shows an overview of the LSTM model.

The LSTM network consists of the following terms that are listed below:

Cell—Naturally, the cell is liable for monitoring the conditions and dependencies between the data points.

Input gate—The input gate controls the degree to which the new value streams into the cell.

Forget gate—The forget gate controls the degree to which an input stays inside the cell.

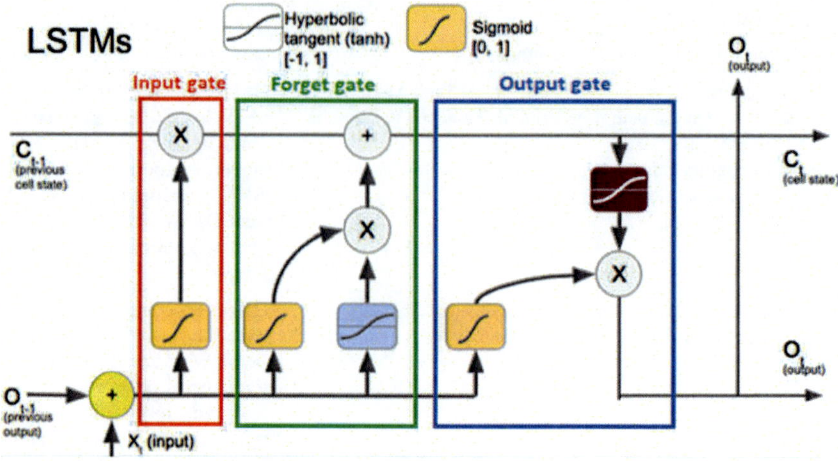

Fig. 1 LSTM model. Reprinted from Nandankumar et al. [12, p. 2]

Output gate—The output gate controls the degree to which the input in the cell is used to evaluate the output.

The LSTM network is mostly befitting for deep learning problems like classification and making predictions depending on time series data.

3.2 *Why LSTM?*

- LSTM addresses long time lags by using constant error back-propagation units, which are present within its cells.
- LSTM networks can provide the best results for sequential data.
- Conventional RNNs experience memory fading and exploding gradient problems. Hence, new set of gates like input, forget and output gates are introduced to let the model learn long-term dependencies.
- According to the study, it has obtained better results compared to the other neural networks.

3.3 *Stock Prediction Algorithm*

The following steps show how the algorithm is utilized for prediction.

(1) Start
(2) INPUT: Import the historical stock price data.
(3) Split the dataset into train and test data:

   (a) The data is split in a ratio of 9:1.
   (b) That means 90% of our data is used for training.
   (c) The remaining 10% of our data is used for testing.

(4) Normalize the train and test set in the range (0, 1).
(5) Model Construction:

   (a) A library called TensorFlow is used to set the parameters for construction of our LSTM model.
   (b) The number of LSTM layers is decided that will be stacked on top of one another.
   (c) The number of neurons in each layer is decided followed by a dense (output) layer having one neuron.
   (d) An activation function called ReLu is used.
   (e) A dropout value is set at each layer.

(6) Compile the model with a chosen optimizer:

   (a) Adam optimizer is used for adaptive learning.

(7) Train the constructed model on data:

   (a) Set the number of epochs, i.e., the training cycles.
   (b) Set the batch size used for training the model.

(8) Make prediction on test data using trained model.
(9) OUTPUT: DE normalizes the results into actual value.
(10) Plot the prediction results along with actual results.
(11) Evaluate accuracy by calculating the RMSE value between predicted values and actual values.
(12) End.

3.4 Terminologies

The terminologies that are used in our proposed system are listed below with its explanation.

(1) Units—A unit is a neuron. Neural network is established by forming a connection between multiple neurons. These neurons take inputs, and then it is multiplied by its weight and passed through an activation function to other neurons as an input.
(2) Activation function—An activation function is responsible for the output of a neuron. It decides whether a neuron should be activated or not. It bounds the output value by doing a nonlinear transformation before it is sent to the next layer of neurons.

   (a) ReLu—It is a type of activation function which stands for rectified linear unit.

It is formulated as shown in Eq. 1.

$$f(x) = \max(0, x) \tag{1}$$

The ReLu function will not activate the neuron if the input is negative by converting it to zero. It will only pass positive values.

(3) Dropout—The term dropout refers to dropping out units. It ignores a set of neurons during the training phase. This is done so that a reduced network is left.
(4) Adam—It is an optimization algorithm used for adaptive learning while training the model.
(5) Epochs—It is the no. of times the algorithm addresses the dataset during the training phase.
(6) Batch size—It is the size of the data, set for one iteration.
(7) RMSE value—It stands for root-mean-square error. It results in a numeric value which indicates the variance of how near the observed value is from the model's predicted value. It varies depending on the size of your data. It goes by the formula shown in Eq. 2.

$$\text{RMSE} = \sqrt{\sum_{i=1}^{n} \frac{(\hat{y}_i - y_i)^2}{n}} \tag{2}$$

where $\hat{y}$ = predicted value, y = actual value, n = number of observations.

Secondly, as the stock market is greatly affected by the news and incidents that involve the company, the opening prices can be predicted by accessing the public's opinion about the company through a social media platform. Top headlines/posts are scrapped and performed sentimental analysis on these posts. It has been proven that sentimental analysis can improve the accuracy of the system [9]. Among the headlines/posts, the phrases have positive meaning or imply positive behavior segregated from the neutral and negative ones. Later, the frequency of all the positive, neutral and negative phrases is calculated and a percentage value of each is generated. Relatively, a graph is plotted comparing the positive, neutral and negative sides about a company.

4 System Architecture

The stock prediction system shown in Fig. 2 has following important steps

- Data collection and preprocessing.
- Construction of model.
- Evaluation of model.

A brief explanation of each is given below.

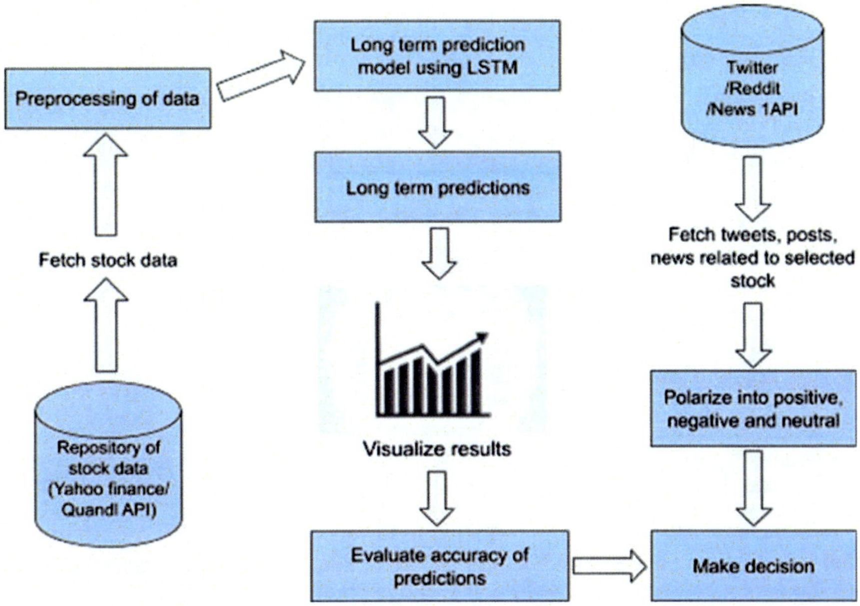

Fig. 2 System architecture

4.1 Data Collection and Preprocessing

The stock prediction model shown above is divided into two parts and collects both the historical data of company stock price and public sentiments regarding the company or the market. The historical data is collected from Yahoo finance, which also provides an API for live data collection. The historical data contains years of company stock price data and contains features such as daily opening, closing, adjusted closing, highest and lowest prices along with the volume of shares traded every day.

The public sentiment data is collected from social media Web sites or news posts such as Twitter and Reddit. Use of Reddit API (praw) is being done to collect this data from various tweets and subreddits that involve the company or the market in general. Various news APIs can also be integrated to collect more data.

The historical data collected from Yahoo finance is preprocessed before feeding into the model. The preprocessing involves cleaning of data, filling missing values, normalization, removing unnecessary features and reshaping the data into model input shape. The public sentiment data collected from Twitter and Reddit is also preprocessed for removing all types of irrelevant information like emojis, special characters and extra blank spaces. Sentiment analysis is the process of identifying or extracting the public opinion hidden in a text.

4.2 LSTM Model Construction

The input data is divided into train and test datasets. A ratio of 9:1 is used; i.e., 90% of data will be used to train the model, and 10% will be used to test the accuracy of the model. The LSTM network is constructed with an input layer that takes five features (open, close, high, low and volume). The input layer is followed by 'h' hidden layers having 'n' neurons per layer, which is followed by a single output layer having one neuron. The parameters of the model are tuned for optimal values such as the number of hidden layers 'h', number of neurons 'n' in each hidden layer, batch size and time steps (Fig. 3).

The public sentiment data available in the form of text from Twitter and Reddit is classified into three labels (Fig. 4):

1. Positive
2. Negative
3. Neutral.

Natural language processing (NLP) is used for this purpose. By classifying the text into positive, negative and neutral, we can understand what the majority of the

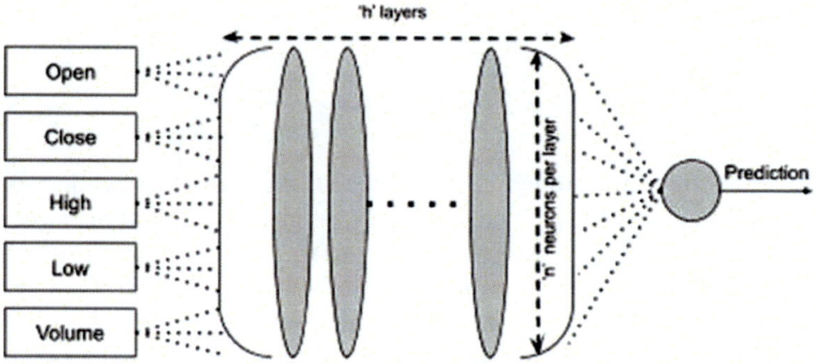

Fig. 3 LSTM model construction. Reprinted from Nandankumar et al. [12, p. 4]

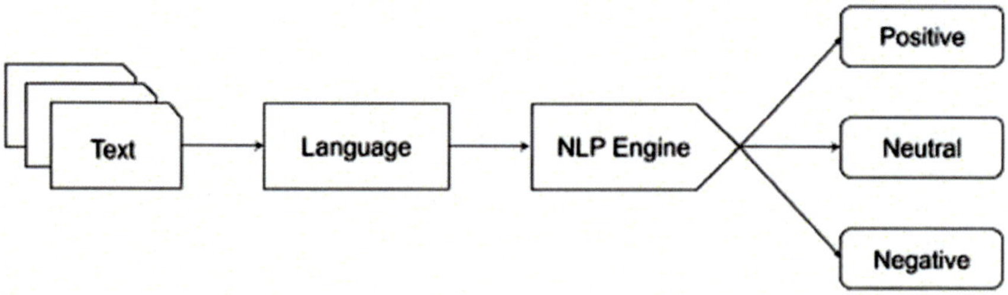

Fig. 4 Natural language processing of text

public thinks about the company, its stock or market in general. This understanding of public sentiment is used to forecast the market on the next day.

4.3 Evaluation of Model

Once the model is fitted with the training dataset, the model can be used to predict the next day's opening price of any arbitrary stock. Subsequently, the sentiment analysis can also run on public opinion regarding the company or market and have an idea of whether the market will go up or down on the next opening day.

The accuracy of the LSTM model is estimated by calculating the root-mean-square error (RMSE) score. The model can be tuned to get lower RMSE scores, i.e., increased accuracy, by changing the parameters such as time steps, batch size, number of hidden layers and neurons in each layer. The size of data used to train the model also plays a huge role in getting better accuracy.

The model is reconstructed repeatedly to get the lowest RMSE score and has then taken the average of all RMSE scores as it can slightly vary every time the model is trained. The RMSE is inferred by testing the predictions against the real stock price movement.

5 Results

In Figs. 5, 6 and 7, the blue line and the red line show the predicted and actual value of the stock prices, respectively.

Figure 5 shows the comparison of actual and predicted stock price of the company Tata Steel. Further, a RMSE score is calculated for the same, which is equal to 16.33.

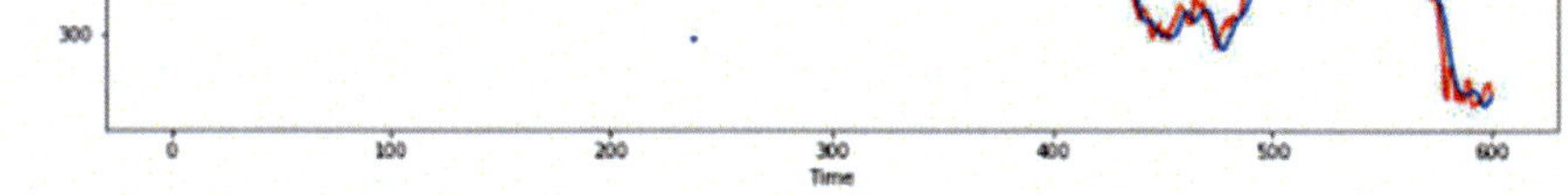

Fig. 5 Predicted graph of Tata Steel

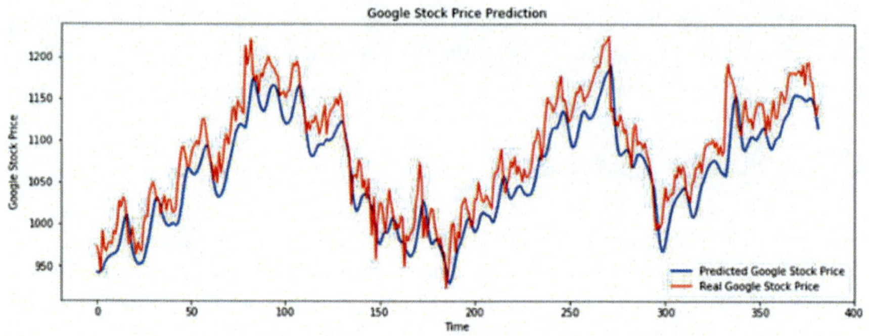

Fig. 6 Predicted graph of Google

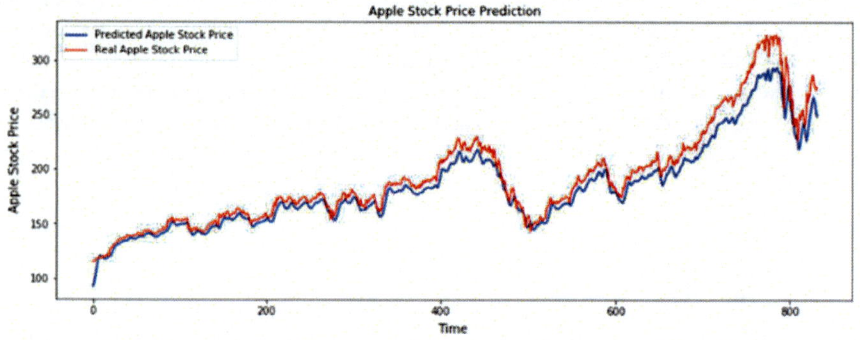

Fig. 7 Predicted graph of Apple

A total of 25 years of data has been taken as an input. The parameters for model have been repetitively changed as a trial and error method for the best prediction. It was found out that with 256 batch sizes, 50 epochs and 3 hidden layers the model has given optimum results. After acquiring the most suitable parameters, the model has been trained with stock prices of different companies. Subsequently, the well-trained model is set for prediction for other companies as well to evaluate the model for accurate prediction.

Here are some of the predictions for other companies as well.

Figure 6 shows the predicted graph for Google. The RMSE value is 30.01. Ten years of data has been taken as input for the trained model but with the batch size of 126, 50 epochs and 0.3 dropout.

Figure 7 shows the predicted graph for Apple. The RMSE value is 11.18. Similarly, 10 years of historical data was taken with the batch size of 126, 50 epochs and 0.3 dropout.

Thus, we can conclude from our study that the prediction accuracy is directly proportional to the size and accuracy of the data. The size of the data substantially increases the accuracy of the model.

Fig. 8 Sentimental analysis of Apple

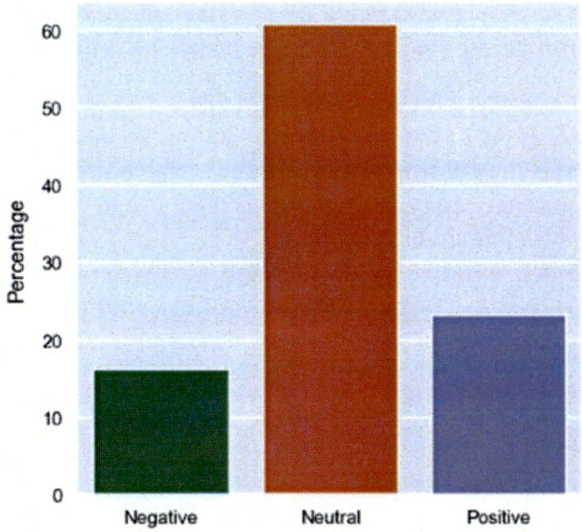

Figure 8 shows the public sentiments that our system has calculated for the company Apple. Recent Reddit headlines/posts are scrapped, and its polarity is calculated. The graph has three colored bars representing negative, neutral and positive sentiments. One can figure out if the stock prices for a particular company will fall or rise by analyzing what people have to say about that company. Social media platform like Reddit is a very reliable source to grab relevant information.

Therefore, by looking at the results the addition of sentimental analysis to the system makes the prediction more reliable.

6 Conclusion

A model has been created, which predicts the direction in which the stock price is going by taking into account the historical prices as well as the sentiment of the public toward that stock. This model is affected by the negative as well as the positive sentiment of the public, which can be concluded that people's opinion matters when it comes to the stock market. There can be an instance in which the stock suddenly crashes for an organization when a negative news is gathered about a company. The negative news can be something that the people do not like about the company, and it can be the comments of the CEO of the company, or plan to close the facility which results in the job losses for the public, not delivering on the promises made, etc. Positive news can be the plan to develop a new office and manufacturing plant; also, it can be hiring of employees which results in the job creation for the public which ultimately results in the positive behavior of the public. Historical behavior of the company plays an important role too. If the company is not been able to recover from a financial meltdown in the past, then there is a large possibility that the company will not be able to recover this time either. Also, the effects of market on the stock of the company determine how much of an impact the market has on that particular organization. Good companies bounce back from tough situations and have a good historical record which can be seen by the historical stock prices of the company. Therefore, in conclusion it can be stated that the behavior of the stock price of the company can be predicted using the historical prices as well as the sentiment of the public toward that company.

Reference

1. Malkiel, B.: A Random Walk Down Wall Street. W.W. Norton & Co., New York (1973)
2. Kale, A., Khanvilkar, O., Jivani, H., Kumkar, P., Madan, I., Sarode, T.: Forecasting Indian stock market using artificial neural networks: In: 2018 Fourth International Conference on Computing Communication Control and Automation (ICCUBEA), Pune, India, pp. 1–5 (2018)
3. Deng, S., Mitsubuchi, T., Shioda, K., Shimada, T., Sakurai, A.: Combining technical analysis with sentiment analysis for stock price prediction. In: 2011 IEEE Ninth International Conference on Dependable, Autonomic and Secure Computing, Sydney, NSW, pp. 800–807 (2011)
4. Bhardwaj, S.: Which stocks fell the most due to corporate scandals in past one year? The Economic Times. M.economictimes.com (2020). [Online]. Available https://m.economictimes.com/wealth/invest/which-stocks-fell-the-most-due-to-corporate-scandals-in-past-one-year/amp_articleshow/72082456.cms. Accessed 27 April 2020
5. Zhao, Z., Rao, R., Tu, S., Shi, J.: Time-weighted LSTM model with redefined labeling for stock trend prediction. In: 2017 IEEE 29th International Conference on Tools with Artificial Intelligence (ICTAI), Boston, MA, pp. 1210–1217 (2017)
6. McNally, S., Roche, J., Caton, S.: Predicting the price of bitcoin using machine learning. In: 2018 26th Euromicro International Conference on Parallel, Distributed and Network-based Processing (PDP), Cambridge, pp. 339–343 (2018)
7. Patel, J., Shah, S., Thakkar, P., Kotecha, K.: Predicting stock and stock price index movement using trend deterministic data preparation and machine learning techniques. In: Expert Systems with Applications, 05 Aug 2014. [Online]. Available https://www.sciencedirect.com/science/article/abs/pii/S0957417414004473. Accessed 29 April 2020
8. Oancea, B., Ciucu, Ş.C.: Time series forecasting using neural networks. arXiv.org, 07 Jan 2014 [Online]. Available https://arxiv.org/abs/1401.1333. Accessed 29 April 2020
9. Li, X, Xie, H., Chen, L., Wang, J., Deng, X.: News impact on stock price return via sentiment analysis. In: Knowledge-Based Systems, 26 April 2014. [Online]. Available https://www.sciencedirect.com/science/article/abs/pii/S0950705114001440. Accessed 29 April 2020
10. Zhang, Y., Wu, L.: Stock market prediction of S&P 500 via combination of improved BCO approach and BP neural network. In: Expert Systems with Applications, 28 Nov 2008.[Online].Available https://www.sciencedirect.com/science/article/abs/pii/S095741740800852X. Accessed 29 April 2020
11. Usmani, M., Adil, S.H., Raza, K., Ali, S.S.A.: Stock market prediction using machine learning techniques. In: 2016 3rd International Conference on Computer and Information Sciences (ICCOINS), Kuala Lumpur, pp. 322–327 (2016)
12. Nandankumar, R., Uttamraj, K,R., Vishal, R., Lokeswari, Y.V.: Stock price prediction using long short term memory. IRJET (2018)

Convolutional Neural Network Based Mobile Microorganism Detector

Airin Elizabath Shaji, Vinayak Prakashan Choyyan, Sreya Tharol, K. K. Roshith, and P. V. Bindu

Abstract Disease detection and prediction are of paramount interest in the area of health-care. The better the diagnosis and prediction, the better ease the treatment would provide and the doctors will be able to cover a larger number of patients in a limited time. Advances in technology have made an impact in the field of medical care from the early detection to assisting in several complicated treatments. However, numerous studies have revealed the lack of health workers in the greater part of the world where patients lack the basic diagnosis and treatment. In addition, cases have been reported where human errors made by health workers have been creating havoc in the early diagnosis of many infectious diseases. A considerable population has been kept unrecognized of infection because of the lack of appropriate access to laboratories. Another problem that persists is the inaccessibility to laboratories by the rural people. This paper explores the possibilities of detecting microorganisms in blood smear by using image processing and Convolutional Neural Networks in Artificial Intelligence. A reliable and easy-to-use portable system called Mobile Microorganism Detector is developed to bring out a better diagnosis of diseases.

A. E. Shaji (✉) · V. P. Choyyan · S. Tharol · K. K. Roshith · P. V. Bindu
Department of Computer Science & Engineering,
Government College of Engineering, Kannur, Kerala, India
e-mail: airinelizabathshaji@gmail.com

V. P. Choyyan
e-mail: pcvinayak1234@gmail.com

S. Tharol
e-mail: tharolsreya@gmail.com

K. K. Roshith
e-mail: roshithk23@gmail.com

P. V. Bindu
e-mail: bindupv@gcek.ac.in

© The Editor(s) (if applicable) and The Author(s), under exclusive license to Springer Nature Singapore Pte Ltd. 2021

A. P. Pandian et al. (eds.), *Proceedings of International Conference on Intelligent Computing, Information and Control Systems*, Advances in Intelligent Systems and Computing 1272, https://doi.org/10.1007/978-981-15-8443-5_43

1 Introduction

In recent years, computerized microscopy image analysis [2] has become inevitable in medical diagnosis. Machine learning and artificial intelligence techniques has boosted up several areas of medical researches and clinical practices. Deep learning is an important machine learning tool in computer vision which can be employed in biomedical image analysis and interpretation. Till recent years, microscopic image data analysis have involved human interventions to a large extent. The manual procedures are found to be inconsistent and are often complicated processes that require a huge amount of correlative data. It is therefore essential to decrease the manual interventions to handle the massive volume of data and scale back the cost of the process, besides building a detection system of high accuracy.

Efficient researches on automatic pathogen classification have been going on, particularly in the field of machine learning. The researches so far conducted are not accessible for the rural areas. The people in rural areas face difficulty in analyzing microscopic data, be it for testing blood or for other medical or educational purposes. In order to alleviate this problem, in this work, we have built a machine learning-based cheap, portable, and convenient device that when attached to the ocular of the microscope will accomplish fundamental to advance report generation on the image it captures.

A portable system called Mobile Microorganism Detector has been developed to facilitate the early detection of morphologically identifiable and differentiable microorganisms present in the blood. The infected blood cells may transform, replicate, or form certain structures that could be easily identified, which are recorded by a camera affixed at the ocular. Supervised learning methodology on the pre-processed images serves for the better identification of the pathogens and thus presents to be a reliable and portable device with increased performance from its predecessors. This work also plans to incorporate machine learning to improve prediction accuracy which primarily works on pure computational power rather than the chemicals needed for analysis. All these make an adept project to be taken up for improving the diagnosis in rural areas.

2 Related Work

Automated micro-organism detection in medical diagnosis is an area where significant studies have to be conducted for efficient health care practices. The papers [5, 6, 12] reflect some of the related works in the field under consideration.

Altun and Taghiyev [5] shine a light on the latest ways of examining biological objects. Their work explains a general-purpose automated system of processing and possibilities of using confocal microscope for analysis of biological objects. The confocal approach simplified the process of obtaining images of living micro-objects by making use of multiple simultaneous fluorescent markers. The automated image processing system (AIPS) records, stores, displays, and processes the stored images, where image recording comprises of two-steps. The AIPS consist of photosensors and a converter that makes up the digital form. Confocal microscopy is another topic under discussion. It has its basic unit in the optical section. Scanning over the desired area of the prepared sample, several images are extracted from Z-series. The deconvolution method reconstructs the three-dimensional structure of the micro-object using the optical sections from Z-series.

John et al. [6] present an accurate method for locating cell nuclei. This is an efficient method for segmentation. A feature similarity index measure (FSIM) is used for detecting the position of nuclei in the image. A Fuzzy C-Means clustering algorithm has been applied for separating background and foreground regions of nuclei further. This can be further used for segmentation tasks. The matched regions belong to nuclei regions, using the DCN-FSIM method, where the gradient magnitude and phase congruency are used as measures for identifying the texture features.

Rajaraman et al. [12] proposed an automated malaria detection system using deep learning techniques that persists as an effective diagnostic method. The performance evaluation, correlation with customized models, and cross-validation among the predictive models at the patient level of the pre-trained DL models are concerned in the paper. These are done to reduce every sort of error that could show up at any phase whether it is the selection of optimal layers for pre-trained DL models or calculating the performance mentioned.

3 Problem Statement

Given a patient's blood smear, the aim of this work is to develop a fast and accurate system called Mobile Microorganism Detector (MMD) for detecting the presence of pathogens in the blood.

4 Solution Methodology

Mobile Microorganism Detector system has a set of hardware for capturing the image and set of software, viz, Convolutional Neural Network (CNN), for predicting the presence of pathogens. In his work, we have implemented the system for detecting malarial parasite [3], of which *P. falciparum* being the most dangerous, in blood. The common practice is to spread blood specimen collected from the patient as a thick or thin blood smear. It is then stained with a stain, and examined with a 100X oil immersion objective. The lab technician uses visual criteria to manually identify the presence of pathogens in the blood, which is a hectic and erroneous task to be done. Thus we propose an automated technique to detect pathogens in a more efficient and accurate way.

The MMD system is analogous to human inferences. Here the system is trained with a large amount of data where the system identifies the parasitic cell in blood samples collected from the patient. The device should be kept mounted on the ocular of an optical microscope. The LCD monitor attached along with gives the result. The device is currently modeled to detect Malaria. However, it can be extended or generalized to incorporate other disease detection. This can be done with either multiple numbers of neural networks working in parallel or with a single neural network trained and powerful enough to detect multiple diseases.

System Architecture

Figure 1 represents the components utilized in the work. The components are slide, camera, Raspberry Pi 3 B, capture button and LCD 16 × 2, 7-segment display which belongs to each of the input, processing, and output units respectively.

1. **Input**
   The input to the device is a sample stained on the slide and held under a microscope, whose ocular is attached with a camera. An image of this stained sample is taken as shown in Fig. 2. The image is then sent for processing.
2. **Processing Unit**
   Raspberry Pi constitutes the processing unit. The image taken for processing undergoes different pre-processing techniques and later taken as input to the Convolutional Neural Network [8, 9, 12].
   Preprocessing of Image
   The image of the stained blood cells has to go through the pre-processing [11] stages such as contouring, segmentation, and morphological transformation.
   A bilateral filter is first applied to reduce noise in the image. It is trailed by the Canny edge detector, which converts the noise-reduced image into a black and white edge only image. This image is used to find contours. Circular contours,

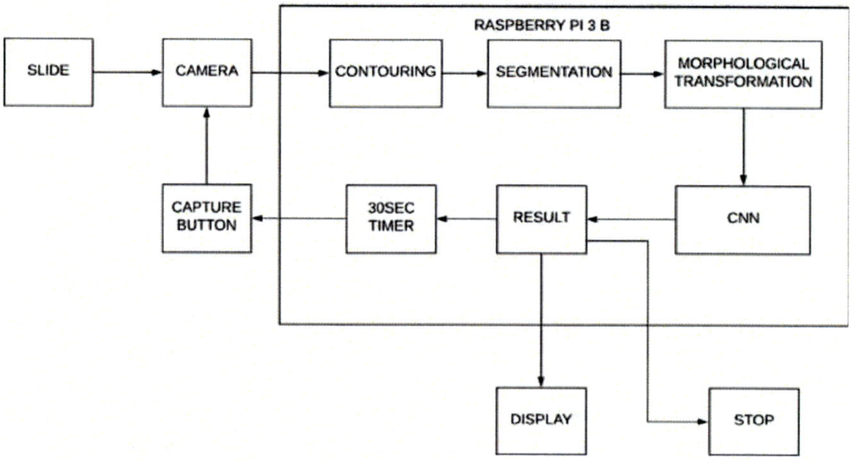

Fig. 1 System architecture

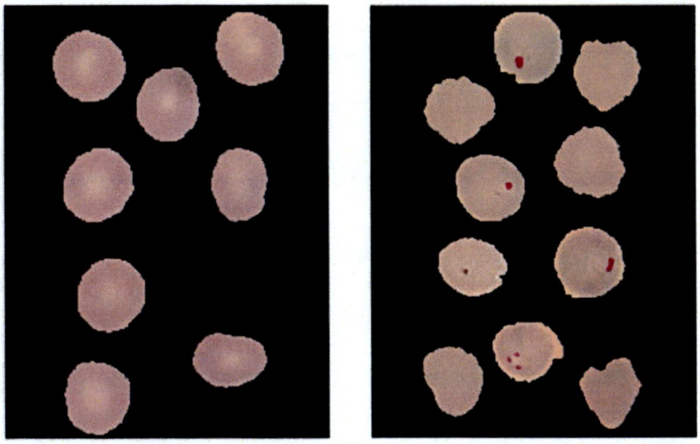

(a) Uninfected cells (b) Infected cells

Fig. 2 Sample images of stained blood cells

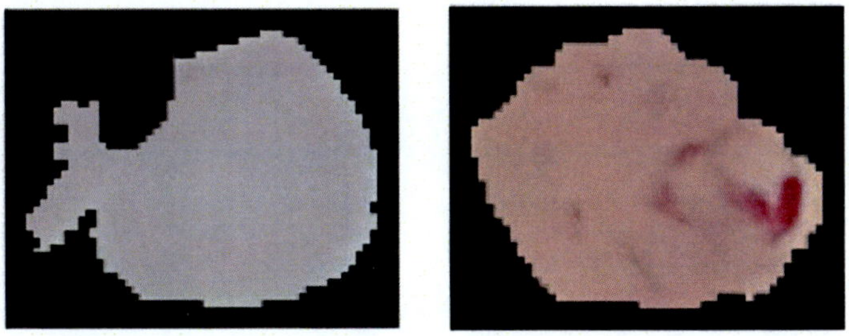

(a) Cropped uninfected cell (b) Cropped infected cell

Fig. 3 Segmented images of cells

that is what we required, is stored in a list to be used while segmentation. For each of the above-saved contours, a bounding rectangle is drawn enclosing the circular contour. Using the dimensions and location coordinates of these rectangles on the main image, each rectangular region is cropped out into a new single image. These individual images are stored in a separate folder. This gives us the image of each cell from the original image, shown in Fig. 3. Every individual image is then re-sampled to 100×100 pixel resolution, to be matched with the input requirements of our customized and pre-trained convolutional neural network. The re-sampled image is then fed to the CNN.

3. **Convolutional Neural Network**
   CNN automatically extricates the features and performs learning. The CNN greatly speeds up the prediction time and accuracy compared to a human lab technician. The model used in Fig. 4 [4] consists of an input layer followed by

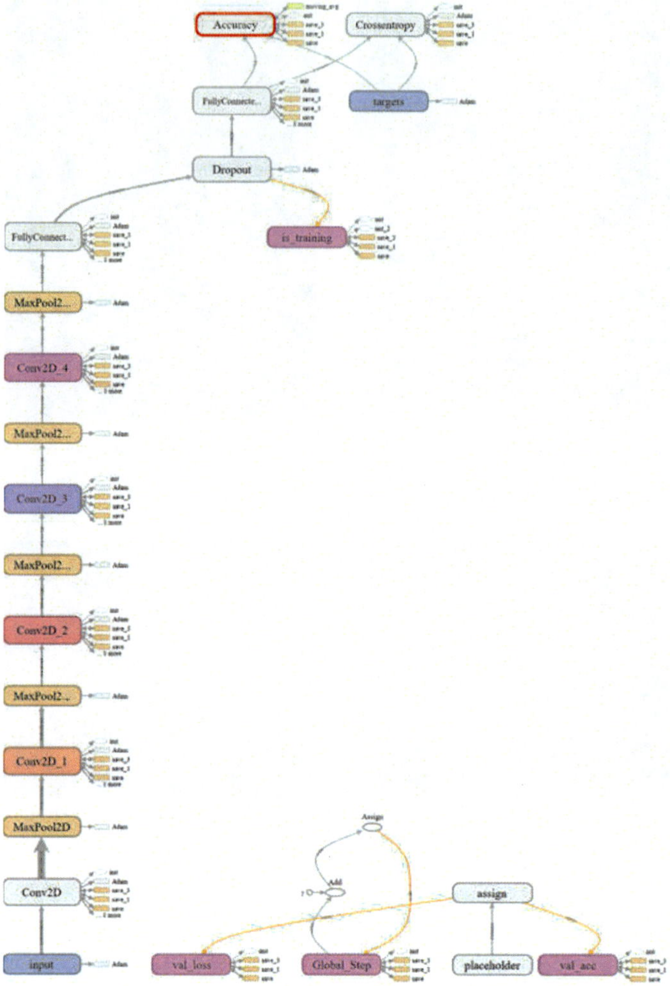

Fig. 4 Neural network model

alternate ten layers of convolution layers and maxpool layers, a fully connected layer, and a drop out layer which ends again in a fully connected layer. The activation function used is Relu. The model is then optimized using Adam optimizer which is the fastest and converges properly. The model is trained for detecting malaria for 11 epochs and is saved for further prediction of new samples with a set of CNNs working in parallel.

4. **Output Unit**
The image after processing in CNN gives the outcome which is displayed on a 16×2 LCD with HD44780 dot-matrix liquid crystal display controller/driver,

the output unit. The result, either infected or uninfected, will be appeared for 30 seconds and then the system goes idle. For proceeding with new samples, interrupt can be generated with the help of the capture button, interfaced with the Raspberry Pi. The camera and Raspberry Pi wake up to repeat the entire process once more on new test samples. On the off chance, the need is over, the system can be switched off.

5 Experimental Results and Discussion

This section represents the results obtained by applying different learning rates to the neural network model. The dataset used for testing the model is the Kaggle Malaria Cell Images Dataset, [4]. The dataset contains 27,558 cell images that are classified into parasitized and uninfected images.

The learning rates used are: 10^{-2}, 10^{-3}, 10^{-4}. The graphs are plotted using Tensorboard [4] which is the visualization tool used in machine learning experimentation provided by Tensorflow [10]. The command to access tensorboard in Windows system is:

$$Tensorboard - -logdir = .\backslash < path/to/logfile > \quad (1)$$

1. **Accuracy**
   Classification accuracy is the ratio of the number of correct predictions to the total number of input samples.

$$\text{Accuracy} = \frac{\text{Number of correct predictions}}{\text{Total number of predictions made}} \quad (2)$$

   a. **Accuracy of training data**
      The model has been trained for different learning rates. It is clear that the accuracy differs for the learning rates. The accuracy of learning rate 10^{-2} is 0.5, whereas for the learning rates 10^{-3} and 10^{-4} it is 0.97 and 0.9 respectively. Thus from Fig. 5, it can be observed that the learning rate 10^{-3} has the highest training accuracy.
   b. **Accuracy of validation data**
      The accuracy of learning rate 10^{-2} is 0.49, for 10^{-3} it is 0.94 and for the learning rate 10^{-4} it is 0.88. Thus from Fig. 6, it can be observed that the learning rate 10^{-3} has the highest validation accuracy.

2. **Loss**
   Loss is the penalty for a bad prediction. That is, the loss is a number indicating how bad the model predicts.

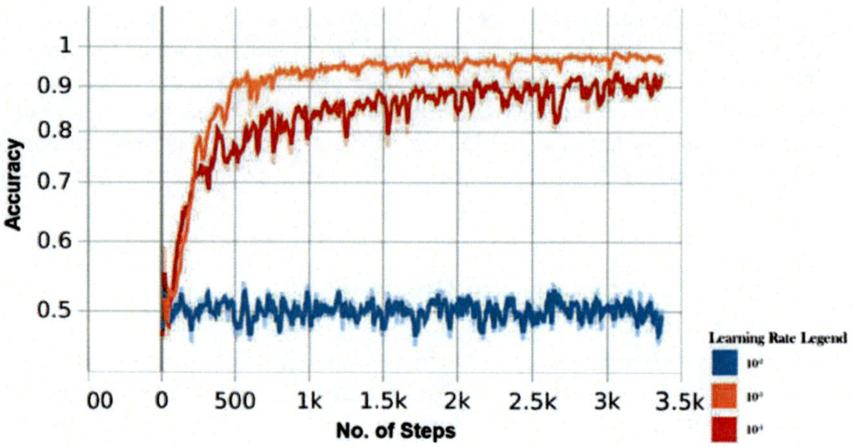

Fig. 5 Accuracy of training data

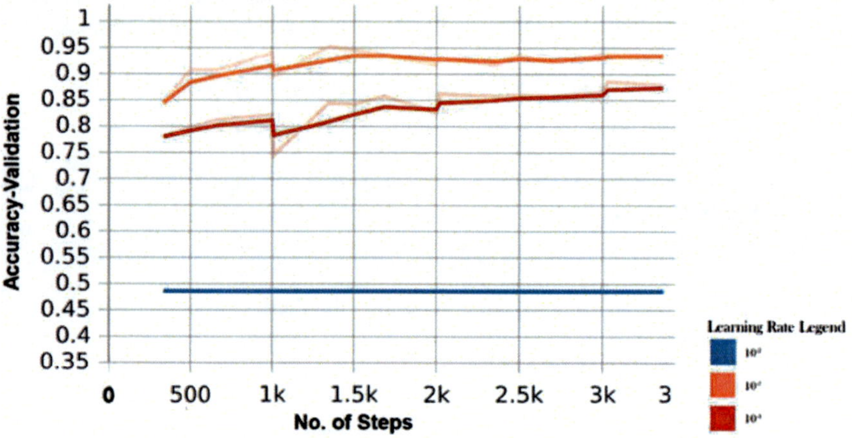

Fig. 6 Accuracy of validation data

a. **Loss of training data**
   The loss of learning rate 10^{-2} is 12, for learning rate 10^{-3} it is 0 and for learning rate 10^{-4} it is 1. Thus from Fig. 7 it is clear that learning rate 10^{-3} has the least loss.
b. **Loss of validation data**
   The loss of learning rate 10^{-2} is 12, for the learning rate 10^{-3} is 0, and for learning rate 10^{-4} is 1. Thus from Fig. 8 it can be observed that learning rate 10^{-3} has the least loss.

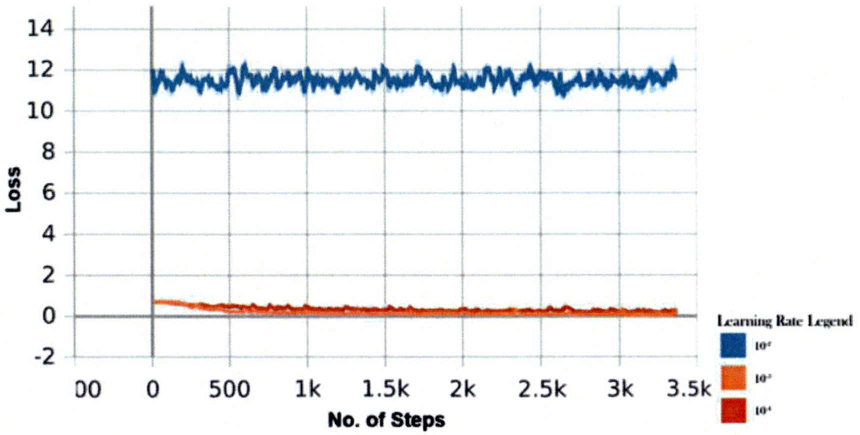

Fig. 7 Loss of training data

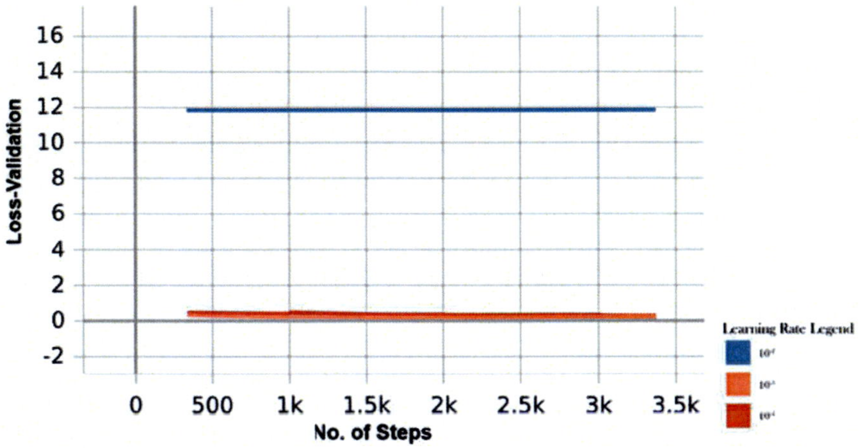

Fig. 8 Loss of validation data

3. **Adam Optimizer Loss**
   The loss of learning rate 10^{-2} is 12, for learning rate 10^{-3} is 0, and for learning rate 10^{-4} is 1. Thus from Fig. 9 it is clear that learning rate 10^{-3} has the least Adam-loss.

Many other experiments have been performed to check the accuracy of predicting microorganisms. Adedeji et al. [1] present through their work that the accuracy of Linear SVM is 85.1% accuracy, Quadratic SVM has 85.7% accuracy, Subspace KNN has 86.3%, and so on. The fit() of the Tflearn calculates the accuracy at each epoch and it reports an overall accuracy of 93%.

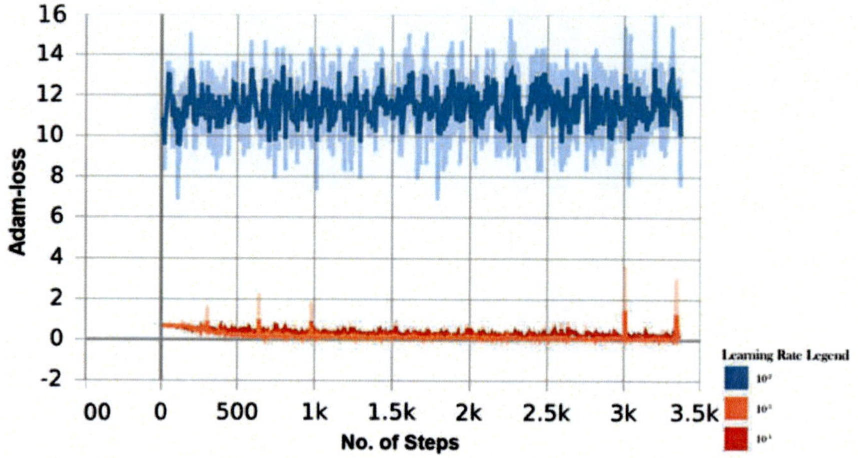

Fig. 9 Adam loss

6 Conclusion and Future Work

Microscopic Image Processing has been effectively utilized in the field of biomedical research and clinical medicine. With the advent of Artificial Intelligence, the medical diagnosis has turned out to be simple. Accuracy and precision are two parameters that are of significance in the medical field. The image-based screening methods speed up the entire procedure and are increasingly worthwhile over the conventional procedures that are inclined to errors, time-consuming, and use chemicals for the tests.

In this work, a portable device called Mobile Microorganism Detector has been developed to help the early detection of morphologically identifiable and differentiable microorganisms present in the blood. Supervised learning methodology on the images facilitate better identification of pathogens and thus presents to be a reliable and portable device with increased performance from its predecessors.

The system could be improvised to accommodate multiple disease detection by incorporating multiple CNN. Additionally, the use of this research may lead to economic solutions for the general public and can be widely accepted by anyone without having any expertise in the field of medicine. This opens up a huge territory for researchers in the field of image processing in medicine.

Acknowledgements This work was funded by the Centre For Engineering Research And Development (CERD), Govt. of Kerala, India. We would like to thank CERD for the financial support.

References

1. Olugboja A, Wang Z (2017) Malaria parasite detection using different machine learning classifier. In: 2017 international conference on machine learning and cybernetics (ICMLC). https://doi.org/10.1109/icmlc.2017.8107772
2. Xing, F., Xie, Y., Su, H., Liu, F., Yang, L.: Deep learning in microscopy image analysis: a survey. IEEE Trans. Neural Networks Learn. Syst. **29**, 4550–4568 (2018). https://doi.org/10.1109/tnnls.2017.2766168
3. Center for Disease Control and Prediction CDC - Malaria - About Malaria. [online] Available at: https://www.cdc.gov/malaria/about/. Accessed 6 Jan 2020
4. Malaria Cell Images Dataset - Cell Images for Detecting Malaria - https://www.kaggle.com/iarunava/cell-images-for-detecting-malaria. Accessed 23 Jan 2020
5. Altun AA, Taghiyev A (2017) Advanced image processing techniques and applications for biological objects. In: 2017 2nd IEEE international conference on computational intelligence and applications (ICCIA). https://doi.org/10.1109/ciapp.2017.8167235
6. John, J., Nair, M.S., Kumar, P.A., Wilscy, M.: A novel approach for detection and delineation of cell nuclei using feature similarity index measure. Biocybernetics Biomed. Eng. **36**, 76–88 (2016). https://doi.org/10.1016/j.bbe.2015.11.002
7. Tensorflow/Tensorboard. [online] Available at: https://www.tensorflow.org/tensorboard/r1/graphs. Accessed 23 Jan 2020
8. Convolutional Neural Networks For Visual Recognition. [online] Available at: http://cs231n.github.io/convolutional-networks/. Accessed 13 Dec 2019
9. Yamashita, R., Nishio, M., Do, R.K.G., Togashi, K.: Convolutional neural networks: an overview and application in radiology. Insights into Imaging **9**, 611–629 (2018). https://doi.org/10.1007/s13244-018-0639-9
10. TensorFlow. 2020. Tutorials| Tensorflow Core. [online] Available at: https://www.tensorflow.org/tutorials. Accessed 23 Jan 2020
11. Image Data Pre-Processing For Neural Networks. [online] Available at: https://becominghuman.ai/image-data-pre-processing-for-neural-networks-498289068258. Accessed 23 Jan 2020
12. Rajaraman, S., Antani, S.K., Poostchi, M., Silamut, K., Hossain, M.A., Maude, R.J., Jaeger, S., Thoma, G.R.: Pre-trained convolutional neural networks as feature extractors toward improved malaria parasite detection in thin blood smear images. PeerJ (2018). https://doi.org/10.7717/peerj.4568

Fuzzy-Based Optimization of Evaporation Process in Sugar Industries

Sebastian George and D. N. Kyatanavar

Abstract Among various unit operations in a sugar industry, the evaporation process is the most energy consuming one. The process of evaporation is highly complex and its dynamics is very much nonlinear in nature. The optimzation of evaporation process is a matter of deep interest due to its correlation with energy consumption and quality of sugar. In this paper, an optimization technique has been proposed making combined use of Taguchi technique and fuzzy logic. Modelling of a quadruple effect evaporator has been performed in MATLAB for which data collected from sugar industry has been utilized. A single objective function has been evaluated for optimization using the integrated fuzzy Taguchi approach. Optimized values of three evaporator parameters, i.e. feed temperature, flow rate of steam and flow rate of feed are determined which satisfy the objective function. Relative contribution of each of these parameters is evaluated using ANOVA technique.

Keywords ANOVA · Fuzzy Taguchi approach · Objective function · Optimization · Quadruple effect evaporator

1 Introduction

Among various agro-based industries in India, sugar industry is second from the top. In terms of cultivation of sugarcane, sugar production and sugar consumption, India's place is among the top rankers in the world. Almost 3% of India's land for agriculture is occupied by sugarcane plantation. Nearly 50 million farmers and their families in the country depend on income from cultivation of sugarcane for their livelihood [1].

Sugar production in India for 2019–20 Marketing year (October-September) has been estimated to drop down to 30.3 MMT, a decline by 8.4% compared to previous MY 2018–19. This decline is attributed to lower than expected sugar procduction

S. George (✉) · D. N. Kyatanavar
Sanjivani College of Engineering, SPPU, Kopargaon, Pune, India
e-mail: seb.jessy@rediffmail.com

© The Editor(s) (if applicable) and The Author(s), under exclusive license to Springer Nature Singapore Pte Ltd. 2021

A. P. Pandian et al. (eds.), *Proceedings of International Conference on Intelligent Computing, Information and Control Systems*, Advances in Intelligent Systems and Computing 1272, https://doi.org/10.1007/978-981-15-8443-5_44

and net reduction in the rate of national sugar recovery which was at a record high during 2018–19. For the third-consecutive turn, Uttar Pradesh will remain at the topmost position followed by Maharashtra and Karnataka as far as the production of sugar is concerned. Overall sugarcane production is also estimated to drop down by 8% to 335 MMT with sugarcane planting in 4.7 million hectares. With normal market conditions, the Indian sugar export was forcast with a rise of 3.5 MMT. The consumption of sugar in India for MY 2019–20 is expected to be 28.5 MMT [2]. Such a rise may be attributed to high demand from bulk users like sweet shops, restaurants, and food processing units. But, all these statistical figures are likely to get changed drastically in the wake of the pandemic Covid-19.

2 Sugar Manufacturing

In India, the technology for sugar manufacturing from sugarcane has been very well stablized in these years. The whole process of producing sugar can be split into a number of steps. It starts with the collection of sugarcane and weighing of the same. Finally, the white sugar is properly bagged in the jute bags. Preparation of cane, extraction of cane juice and its weighing, treatment and sulphitation of juice, evaporation and boiling of massecuite, etc., are other intermediate processes related to sugar production. Massecuite results from cane juice after multiple rounds of concentration through boiling. It is a semi-solid mixture. Through centrifugation, massecuite is separated into molasses and pure sugar. Molasses emerges as a by-product during the sugar manufacturing and it is utilized in subsidiary units like distilleries.

2.1 Sugar Plant Automation Packages

In the present era, sugar industries have been forced to reduce energy consumption, recycle materials and optimize operations because of environmental protection demands and stiff competition [1]. Plant automation packages like juice flow stabilization system, lime sulphitation pH control system, steam flow stabilization system, pan automation package, imbibition water control system, evaporator control system, etc., have been developed for improving steam fuel economy and sugar quality.

There are fluctuations in the raw juice intakes and volumes handled by juice tanks due to the variations in crushing load. It is necessary to eliminate such fluctuations to ensure a steady-state flow of juice to the juice heaters. A juice flow stabilization system is developed for the same. This system will ensure a steady-state equilibrium in the boiling house and prevent the pump from dry running.

In the juice sulphitation section, the juice is treated with milk of lime and sulphur dioxide (SO_2) to maintain its pH at 7.0. The controlled addition of milk of lime and SO_2 gas precipitates soluble and insoluble suspended non-sugars from the mixed juice. The lime sulphitation pH control system provides a very effective control of pH in the juice clarifier [6]. Majority of sugar factories in India are based on double sulphitation process for the clarification of mixed juice in which judicious addition of milk of lime and SO_2 are done depending upon the pH of mixed juice. A microprocessor-based system was developed well back in 1986 at CEERI Pilani for this [7]. Since then, considerable improvements have taken place in the state of art of such systems. Control of pH by classical means like P + I controller may be very difficult in this case due to the highly nonlinear behaviour of the process. A hybrid fuzzy logic and P + I controller will be very efficient in handling such nonlinearities [6]. The pH control will improve the quality of sugar produced by effective clarification for the liquid-phase reaction.

Specialized packages have been developed for automation of both batch and continuous pan operations. This will provide consistency in pan boiling and improve sugar grain formation. The imbibition water control system is another automation package for the optimization of tonnes of cane per day (TCD) and brix. This system will also provide substantial savings in overall steam consumption.

2.2 *Evaporation Process*

The basic principle of evaporation is to concentrate part of the solvent from the solution by employing steam as the heat source. The concentrated liquid thus obtained is the product from the evaporation process. Among the various unit operations in a sugar plant, the evaporation process is the most energy consuming one. Due to the large amount of thermal energy from the steam incurred during evaporation, the economy of sugar manufacturing is very closely related to this process. The function of evaporator in a sugar industry is to raise the concentration of sugarcane juice from a nominal value of around 20% to a reasonably high value of say, 65%. Evaporation is the process of concentrating a solution and is much different from distillation and crystallization [13].

The process of evaporation in sugar industries is highly complex and its dynamics is nonlinear in nature. Even though the evaporators have been used extensively in sugar plants, its modelling and control have been a challenging task for many decades. Traditional control techniques like the PID controllers do not yield good results due to the highly nonlinear nature of this process. Normally, evaporators used in sugar industries are of multiple effects in nature [5]. Quadruple effect evaporator, quintuple effect evaporator, etc., are very common these days. Only certain amount of water is removed in each of these effects.

Different schemes have been proposed by researchers for the control of various parameters of multiple effect evaporators. A control philosophy based on idiomatic concept was developed by Neilsen et al. [4] for a quintuple effect evaporator. It was later implemented in a Danish sugar industry. Modern control principles like Generalized Predictive Control, Internal Model Control and Linear Quadratic Gaussian were proposed by Lissane Elhaq et al. [3] for the control of a five-effect evaporator.

For controlling the brix value in an evaporator, Model Predictive Control principle was introduced by Smith et al. [8]. Pitteca et al. carried out modelling of a five-effect evaporator in MATLAB [9] and based on the model, developed a fuzzy control scheme [10]. Raghul et al. [11] carried out study on the behaviour of evaporator system used in sugar industries under an auto-tuned PID controller.

3 Modelling and Optimization of Multiple Effect Evaporator

3.1 Modelling

The multiple effect evaporator considered here has four effects as shown in Fig. 1.

Listed below are the different parameters associated with this system.

m_P—Mass flow rate of feed.
m_S—Mass flow rate of steam.

The variables m_{P1}, m_{P2}, m_{P3} and m_{P4} represent the product flow rates from different effects of the evaporator station. Similarly, m_{E1}, m_{E2}, m_{E3} and m_{E4} represent the amount of water removed and m_{C1}, m_{C2}, m_{C3} and m_{c4} denote the rate of flow of condensate. In the earlier works carried out by the authors on evaporator optimization [16], the same system had been studied and a model in Simulink had been developed based on mass and energy balance, as shown in Fig. 2.

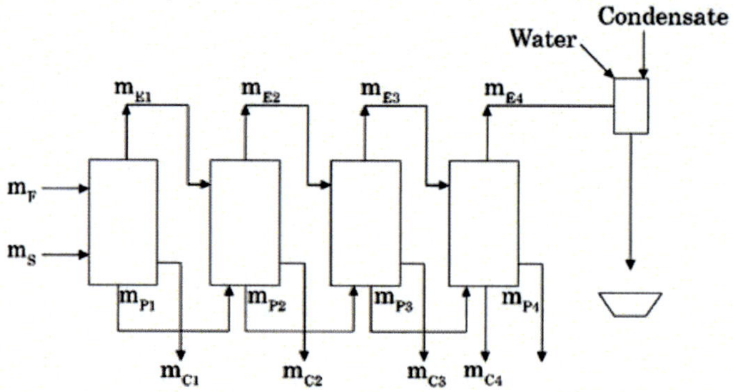

Fig. 1 Multiple effect evaporator

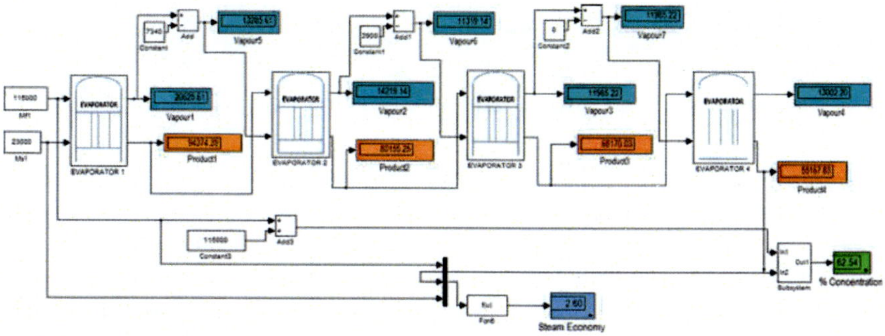

Fig. 2 Simulink model

3.2 *Taguchi Method*

In the earlier works of the authors [12], the Simulink model mentioned above was subjected to Taguchi trials based on L9 orthogonal arrays, the results of which are summarized in Table 1.

The three input parameters considered are feed temperature, feed flow rate, and steam flow rate. The performance characteristics are the steam economy and brix value.

Table 1 Taguchi trial results

Trial No.	Feed temperature (°C)	m_S (kg/hr)	m_F (kg/hr)	Steam economy	Brix (%)
1	75	22,500	100,000	1.65	51.27
2	75	23,000	115,000	1.6	44.12
3	75	23,500	130,000	1.55	39.31
4	90	22,500	115,000	2.18	52.25
5	90	23,000	130,000	2.19	46.22
6	90	23,500	100,000	2.23	67.78
7	100	22,500	130,000	2.63	51.94
8	100	23,000	100,000	2.56	78.22
9	100	23,500	115,000	2.67	64.47

4 Integration of Fuzzy Logic with Taguchi Method

4.1 Multi-response Optimization Problem

Performance characteristics of processes can be optimized by Taguchi trials through process parameter settings. It reduces system performance sensitivity to different causes of changes. Hence, the Taguchi technique is treated as a very efficient tool in Design of Experiments (DOE). But, at the same time, this method is handicapped due to its lack of effectiveness in handling optimization problems having multiple responses. For such cases, an integrated approach of fuzzy logic and Taguchi method provides a powerful solution. The concept of fuzzy logic utilizes knowledge extracted from human experts [14]. If the available information is vague and uncertain, it can be handled by fuzzy set theory. In fact, the Taguchi analyses have various performance characteristics like lower-the-better, nominal-the-best and larger-the-better which inherit certain levels of uncertainity [15]. The optimization problem of evaporation process having multiple response characteristics can effectively be solved by this integrated approach. A single objective function, namely the multi-response performance index (MRPI) is evaluated here for determining the optimum settings.

4.2 Optimization Procedure

Listed below are the various steps undertaken for the optimzation of the quadruple effect evaporator considered here.

- Using L9 orthogonal array, conduct the experiments on the Simulink model.
- Determine the signal-to-noise (S/N) ratio from the experimental results for both steam economy and brix value. It expresses deviation of response characteristics from their desired magnitudes.
- Using appropriate membership functions, fuzzify the S/N ratios.
- Define appropriate membership function for the multi-response performance index (MRPI).
- Develop fuzzy rule base.
- Through inference engine, determine the MRPI value for each trial. This will be obtained after defuzzification.
- Use MRPI values for finding the optimum settings.
- Select the optimum levels of evaporator parameters.
- Calculate the relative contribution of each of these parameters using ANOVA.

4.3 Fuzzy Implementation

In this work, the implementation of fuzzy system has been carried out in MATAB using Fuzzy logic toolbox. Tables 2 and 3 show the S/N ratios of steam economy and brix values, respectively, for different trials. For the entire course of optimization process, larger-the-better criterion has been selected for both steam economy as well as brix value. Minitab 17 software has been used for the calculation of S/N ratios. For fuzzifying the S/N ratio values, membership functions are defined as shown in Figs. 3 and 4. The membership function of MRPI is depicted in Fig. 5.

For yielding a better performance, larger S/N ratio values are required. Fuzzy rule base has been developed as per this assumption. Through fuzzy reasoning of these rules, an output has been derived which is fuzzy in nature. For fuzzy reasoning, the max–min compositional process has been followed. The developed fuzzy rule base having nine rules is pictorally represented in Fig. 6.

After firing the appropriate rule from the rule base by the inference engine, the fuzzy output obtained has to be converted into a crisp value. This is accomplished by

Table 2 S/N ratios for steam economy

S. No.	Steam economy	S/N ratio
1	1.65	4.34
2	1.6	4.08
3	1.55	3.80
4	2.18	6.76
5	2.19	6.80
6	2.23	6.96
7	2.63	8.39
8	2.56	8.16
9	2.67	8.53

Table. 3 S/N ratios for brix

S. No.	Brix (%)	S/N ratio
1	51.27	34.19
2	44.12	32.89
3	39.31	31.89
4	52.25	34.36
5	46.22	33.29
6	67.78	36.62
7	51.94	34.31
8	78.22	37.86
9	64.47	36.18

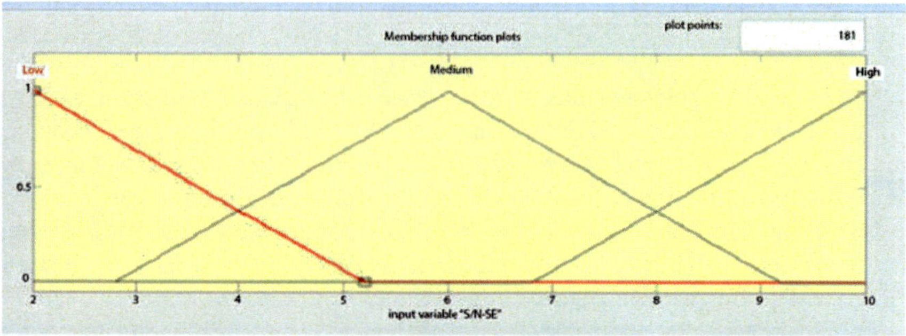

Fig. 3 Membership function for S/N ratio of steam economy (SE)

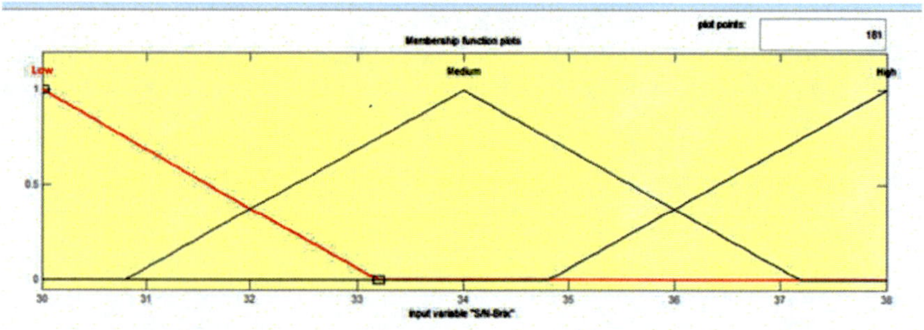

Fig. 4 Membership function for S/N ratio of brix

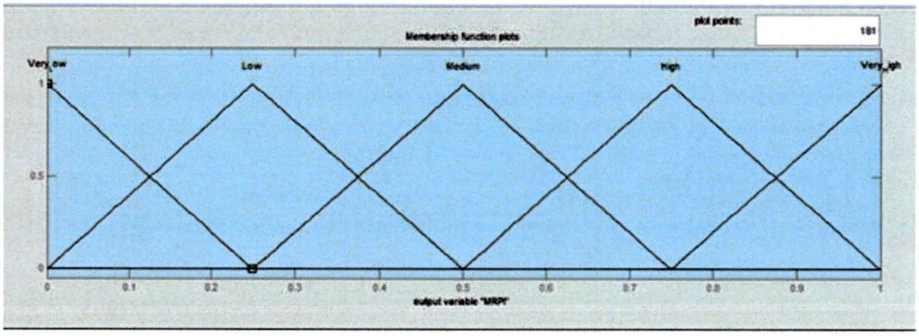

Fig. 5 Membership function of MRPI

the centre of gravity defuzzification method [14] and the crisp value thus obtained is the MRPI. Table 4 summarizes these values for different trials.

Thus, the integrated approach combining fuzzy logic and Taguchi technique has finally resulted into the conversion of a multiple response optimization problem into a single response type. The optimal combinations of evaporator parameters correspond

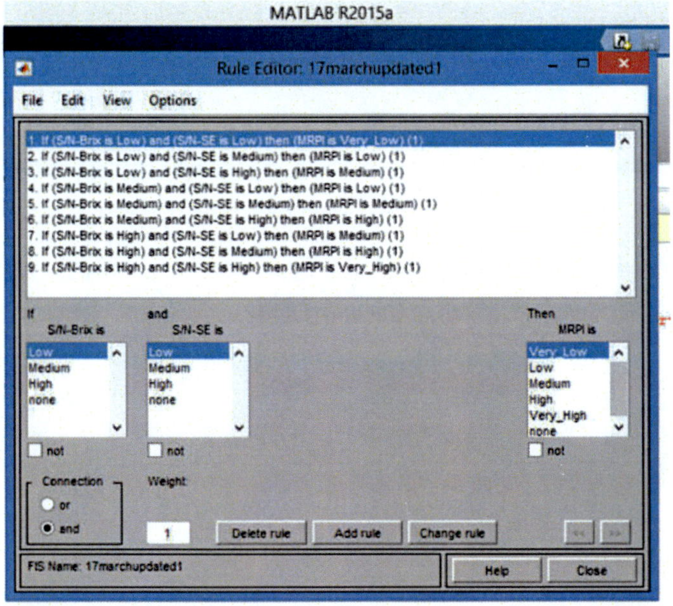

Fig. 6 Fuzzy logic toolbox operation

Table 4 MRPI values

Trial No.	MRPI value
1	0.408
2	0.38
3	0.334
4	0.5
5	0.5
6	0.687
7	0.663
8	0.787
9	0.696

to the largest value of MRPI [15]. It can be seen from Table 4 that Trial No. 8 accounts for the largest value of 0.787 for MRPI. But, in this trial run, the brix value obtained is 78.22% which is not at all recommended for good quality of sugar. Hence, Trial No. 9 having the second-largest value of MRPI is selected.

Table 5 MRPI response *t*

Input parameters	Low	Medium	High
Feed temperature (°C)	0.374	0.562333333	0.715333333
m_S (kg/h)	0.523666667	0.555666667	0.572333333
m_F (kg/h)	0.627333333	0.525333333	0.499

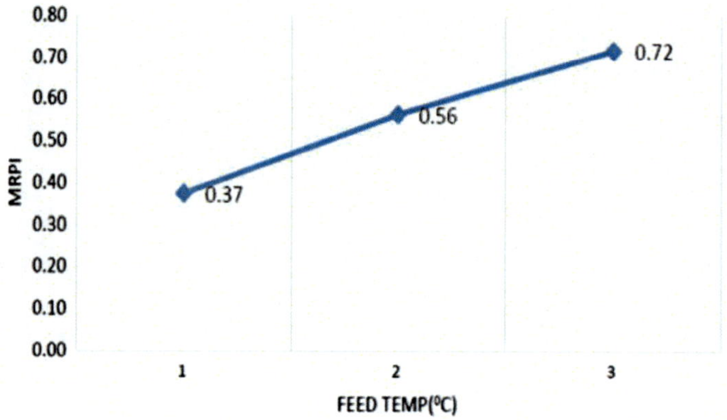

Fig. 7 MRPI versus feed temperature

4.4 *Relative Contribution and ANOVA*

Analysis of variance (ANOVA) identifies which all parameters of the evaporator significantly affect its performance [16]. This is done by separating total variability of MRPI into contribution of individual variables and error. Different levels of MRPI for all the evaporator parameters considered here are given in Table 5, the graphical representation of which are shown in Figs. 7, 8 and 9.

ANOVA results shown in Table 6 explain the relative influence of each of the evaporator parameters considered in this work. For preparation of response table, Minitab 17 software has been used.

The pie chart shown in Fig. 10 depicts the ANOVA results graphically.

5 Interpretation of Results

As stated in the previous section, Trial No. 9 has the appropriate value of MRPI which is 0.696. The brix value for this trial is 64.47% which is quite genuine as far the sugar quality is concerned. Moreover, this trial results into a reasonably fair value

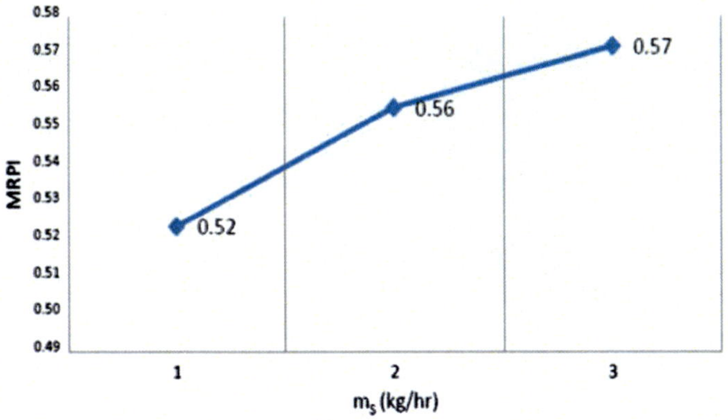

Fig. 8 MRPI versus mass flow rate of steam

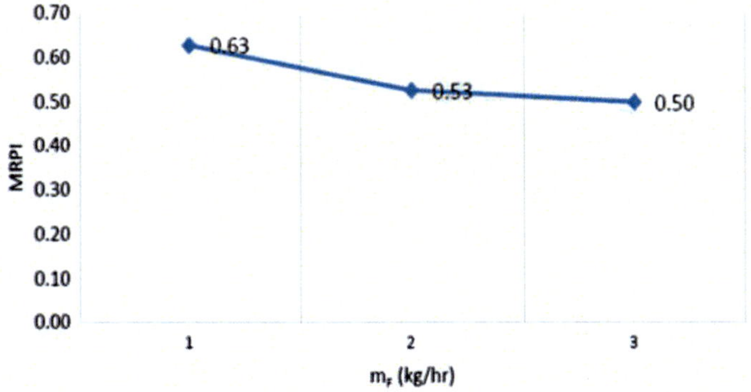

Fig. 9 MRPI versus mass flow rate of feed

Table 6 % Contribution

Input parameters	SS	DOF	MS	*F* ratio	% Contribution
Feed temperature (°C)	0.17538	2	0.087693	56.28189	83.62
m_S (kg/h)	0.00367	2	0.001835	1.177779	1.75
m_F (kg/h)	0.02756	2	0.013783	8.846252	13.14
Error	0.00311	2	0.001558		
Total	0.20974	8	0.10487		

of steam economy, i.e. 2.67. Hence, it can be concluded that the parameter settings of Trial No. 9 are appropriate for the optimum performance of the evaporator.

The relative influence of each parameter on the optimum performance of the evaporator system is clear from the ANOVA results. As seen in the pie chart, the feed temperature is the most influential parameter (87.62%) followed by feed flow

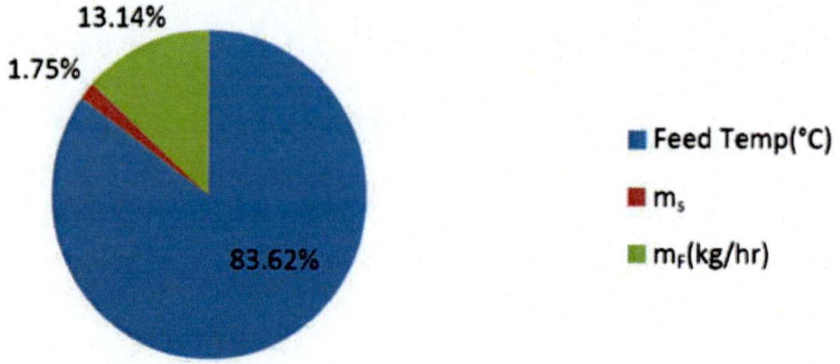

Fig. 10 Graphical representation of ANOVA results

rate (13.14%). The steam flow rate has a very negligible influence level of 1.75% only.

The last step in any optimization problem is the confirmation test. But, because of the specific nature of sugar industries, it is not possible to go for a confirmation test on the evaporator unit with the obtained optimum settings when the plant is running. As an alternative for this, the accuracy of these results has been verified from the opinion of experts working on evaporator systems installed in different sugar plants.

6 Conclusion

In this paper, optimization of a multiple effect evaporator has been attempted using Taguchi method integrated with fuzzy logic. By introducing fuzzy logic along with Taguchi technique, the multi-response optimization situation in an evaporator has been converted into a single response optimization problem [15]. ANOVA has helped in determining the relative influence of three process parameters viz. feed temperature, mass flow rate of steam and mass flow rate of feed on the evaporator performance. For easy analysis purpose, Minitab 17 software has been employed. As a continuation to this work, fuzzy integrated Grey–Taguchi method is suggested in which the Grey relational coefficients from the results of Taguchi trials are given as inputs to the fuzzy inference module. The membership value associated with each Grey relational coefficient is fuzzy in nature. Once the fuzzy rules are evaluated, the MRPI values can be obtained through defuzzification. This method will retain all the advantages of the integrated fuzzy Taguchi method and at the same time will take care of the issue related to weight assignment to the different quality characteristics in a multi-response optimization case. The technique, though quite powerful, is beyond the scope of this paper.

Acknowledgements The authors would like to acknowledge the support extended by Sanjivani (Takli) Sahakari Sakhar Karkhana Ltd., Sahajanandnagar, Kopargaon, MS, India in this work. Frequent visits were necessary to understand the operation of the evaporator station installed at the factory.

References

1. George, S., Kyatanavar, D.N.: Applications of fuzzy logic in sugar ındustries: a review. In: Int. J. Eng. Innov. Tech. **1**(6), pp. 226–231 (2012)
2. Aradhey, A.: India sugar annual 2019. In: Global Agricultural Information Network Report, USDA Foreign Agricultural Service (2019)
3. Elhaq, L.S., Giri, F., Unbehauen, H.: The development of controllers for a multiple-effect evaporator in sugar industry. https://www.cds.caltech.edu/conferences/related/ECC97/proceeds/751_1000/ECC837.PDF
4. Nielsen, K.M., Nielsen, J.F.D., Pederson, T.S.: Simulation and control of a multiple effect evaporator. https://www.control.auc.dk/prteprint/?action=abstract&abstract=4247
5. Hugot, E. In: Handbook of cane sugar engineering, 3rd edn., pp. 504–506. Elsevier Science Publishing Company Inc. (1986)
6. George, S., Kyatanavar, D.N.: Intelligent control of pH for juice clarification. In: Int. J. Electron. Electr. Eng. **7**(6), pp. 617–622 (2014)
7. Narayan, L., Rao, K.S.N., Kota, S., Acharya, G.N.: pH control system for juice clarification. In: Proceedings of Workshop on Modernization of Sugar Industry using Electronic Systems, CEERI Plilani, pp. 101–123 (1986)
8. Smith, P.D.: Cleswartz, STL Harrison: Control and Optimization of a multiple effect evaporator. In: Proc. S. Afr. Sug. Technol. Assoc. **2000**(74), pp. 274–279 (2000)
9. Pitteca, A.V., Robert, T.F., King, A., Rughooputh, H.C.: Parameter estimation of a multiple-effect evaporator by genetic algorithms. In: Proceedings of Joint 2nd International Symposium on Advanced Intelligent Systems. Yokohama, Japan (2004)
10. Pitteca, A.V., Robert, T.F., Ah King, Rughooputh, H. C.: Intelligent controller for multiple evaporator in sugar ındustry. In: Proceedings of IEEE International Conference on Industrial Technology, pp. 1177–1182 (2004)
11. Raghul, R., Shahidh, M.S.H., Luca, N.S., Singh, A.B.: Model ıdentification and fuzzy controller ımplementation for the evaporator process in sugar ındustry. In: Proceedings of International Conference on Electrical Engineering and Computer Science (ICEECS–2012), pp. 27–32 (2012)
12. George, S., Kyatanavar D.N.: Optimization of evaporation process in sugar ındustry for developing ıntelligent control strategies. In: Int. J. Mod. Trends Eng. Res. **2**(7), pp. 998–1004 (2015)
13. Shah, D.J., Bhagchandani, C.G.: Design, modeling and simulation of multiple effect evaporators. In: Int. J. Sci. Eng. Tech. **1**(3), pp. 1–5 (2012)
14. George, S., Kyatanavar, D.N.: Optimization of multiple effect evaporator using fuzzy logic ıntegrated with taguchi technique. In: International Conference on Electrical, Electronics and Optimization Techniques (ICEEOT)—2016. Chennai, India (2016). https://ieeexplore.ieee.org/document/7754917/?reload=true
15. Rajyalakshmi, G., Venkata Ramaiah, P.: Optimization of process parameters of wired electrical discharge machining using fuzzy logic integrated with Taguchi method. In: Int. J. Sci. Eng. Tech. **2**(6), pp. 600–606 (2012)
16. George, S., Kyatanavar, D.N.: Optimization of multiple effect evaporator using Taguchi method. In: International Conference on Electrical, Communication, Instrumentation and Computing (ICECEIC-2019). Chennai, India (2019)

Algorithms for Decision Making Through Customer Classification

Jesus Vargas, Nelson Alberto, and Oswaldo Arevalo

Abstract Market segmentation has distinctive characteristics, due to the different criteria that exist for this purpose. In the process of market segmentation, qualitative and quantitative variables are presented, requiring an intelligent analysis to obtain homogeneity in the information to be processed. Specifically, the segmentation criteria must be in accordance with the character that the variables take. When the variables are quantitative, it is necessary to use qualitative methods for the choice of market segments. The need to apply qualitative criteria and qualitative methods in the choice of market segments is critical for the execution of the process. For this reason, the aim of this paper is to use neutral logic for market segmentation in order to choose the target segment. Neutrosophic logic is a tool to support decision making and has a greater interpretability of linguistic terms which is useful for the analysis of qualitative information in the process of market segmentation.

Keywords Neural networks · Neutrosophic K-means · Market segmentation

1 Introduction

With the name of labor market segmentation theory (hereinafter, TSMT), a set of approaches is usually encompassed, quite diverse in terms of their origins and content,

J. Vargas (✉) · N. Alberto · O. Arevalo
Universidad de la Costa, St. 58 #66, Barranquilla, Atlántico, Colombia
e-mail: jvargas41@cuc.edu.co

N. Alberto
e-mail: nelson@unitet.ch

O. Arevalo
e-mail: oarevalo1@cuc.edu.co

N. Alberto · O. Arevalo
Universidad Tecnológica, San Pedro Sula, Honduras

© The Editor(s) (if applicable) and The Author(s), under exclusive license to Springer Nature Singapore Pte Ltd. 2021

A. P. Pandian et al. (eds.), *Proceedings of International Conference on Intelligent Computing, Information and Control Systems*, Advances in Intelligent Systems and Computing 1272, https://doi.org/10.1007/978-981-15-8443-5_45

which began to emerge in the late 1960s driven by discontent toward the neoclassical explanation of the labor market. The orthodox economy, from its perspective of equilibrium, found it difficult to explain phenomena such as the persistence of poverty, unemployment, discrimination, and above all, wage inequalities between similar individuals. In particular, for human capital theory, wage differences should reflect differences in productivity (and, ultimately, in skills); in the short term, there could be transitory inequalities or phenomena such as involuntary unemployment, but in the long term, the search for the maximization of profit and utility, and in a context of perfect information and mobility, should lead to the emptying of the market and the disappearance of inequalities [1].

However, all this collided with reality and encouraged the search for alternative explanations. Among them, various arguments emerged that had in common the conception of the labor market as a market composed of a set of different segments, with different wage formation and allocation mechanisms (and far from the mechanisms of neoclassical economics) and with obstacles to mobility between them. However, these arguments were emerging from different theoretical perspectives and showing some divergences in their content and methodology, which hinders a clear and generalizable presentation of the TSMT [2, 3].

Market segmentation begins with the recognition that the market is heterogeneous that it is divided into homogeneous groups or segments, which can be chosen as markets or targets for a company. So, the process of market segmentation requires a differentiation of needs within a market [4, 5].

It should also be noted that a segmentation study must be carried out within a framework that takes into account the objectives pursued by the company and, ultimately, its future strategy in the market. Therefore, a process of segmentation, selection of a target market, and positioning is proposed as key elements of modern strategic marketing.[6] describes the segmentation process in six stages [7]:

1) Identification of segmentation variables and market segments.
2) Development of profiles of each obtained segment.
3) Evaluate the attractiveness of each segment.
4) Select the target segment(s).
5) Identify possible concepts for positioning in the selected segments.
6) Select, develop, and create the chosen positioning concepts.

The first two stages correspond to what has traditionally been considered as market segmentation, which as mentioned above is dividing the market into groups of buyers who may require different products or commercial strategies, and then analyzing the characteristics of the segments created. The selection of one or more objective segments must be made on the basis of an economic analysis of the expected income and expenditure of each segment [8, 9].

The stage of selecting, developing, and creating the chosen positioning concepts corresponds to the search for adjustment between the real and commercial characteristics of each product and the desires of the consumers who make up the segment, as a way of achieving a favorable competitive position.

1.1 Criteria for Market Segmentation

Any segmentation model requires the selection of a dependent variable, whose behavior is to be explained, and a set of explanatory or descriptive variables for each segment. The variables used as explanatory or dependent variables correspond to the variables used to explain market behavior [10].

The variables can be classified into two groups according to market characteristics, collecting general characteristics, regardless of the product, which include demographic and socio-economic variables, personality and lifestyle characteristics, attitudes and behavior toward the media and commercial establishments in general [11].

On the other hand, the specific characteristics of the market situation include concepts such as product use, forms of purchase, attitudes toward the product and its consumption, desired benefits in the product category, and specific responses to marketing variables, such as new products, and advertising, relating to the particular product.

Some of these variables can be measured objectively, such as age and income, while other variables have to be inferred through assessments made by the consumer himself, for example, attitudes, preferences, etc. The main variables used in market segmentation studies are classified into geographic, demographic, psychographic, and behavioral variables [12, 13].

In accordance with the fact that in order to carry out a market segmentation study, the dependent variable or variable to be explained must be selected, in order to collect the information on the explanatory variables, select the segmentation technique, interpret the data or results, and propose a strategy. In this regard, [14, 15] explains that the most commonly used techniques for market segmentation are:

Crosstab method.
Belson method.
X^2 method.
Multiple regression.
Automatic interaction detection (AID)—factorial analysis.
Non-metric multidimensional analysis.

However, data mining is now frequently used as a technique that allows for the handling and classification of large amounts of data. One of the typical tasks of data mining is clustering. On such a statement, the use of a neutrosophic K-Means is proposed as a neutrosophic data mining technique for clustering that is based on the classical K-means algorithm.

Neutrosophic logic applied to the data mining process is efficient because it is robust in interpreting linguistic terms to support decision making. Neutrosophy, created by Professor Florentin Smarandache, is a new branch of philosophy which studies the origin, nature, and scope of neutrality [16].

The logic and the neutrosophic sets, on the other hand, constitute a generalization of Zadeh's logic and fuzzy sets, and especially of Atanassov's intuitionist logic, with multiple applications in the field of decision making, image segmentation, and automatic learning.

In this regard, the neutrosophic K-Means as an extension of the classic K-Means is a data mining neutrosophic technique for clustering, useful to automatically manipulate large databases in the process of market segmentation. The market segmentation process based on the neutrosophic K-means, according to the authors of this paper, is an analytical process designed to explore large amounts of data in order to detect consistent behavioral patterns or relationships between different variables to apply them to new datasets [8].

2 Materials and Methods

For the process of market segmentation, it is proposed, in the present study, the use of neutrosophic K-means, due to the diversity of the data that generates fluctuation between them, and at the same time, they are not near two clusters for labeling them. With neutrosophic K-means, an algorithm is developed to solve problems [6]. The proposed algorithm assigns each data a value of belonging within each cluster and therefore a specific data can partially belong to more than one cluster. Unlike the classical K-means algorithm that works with a hard partition, neutrosophic K-means applies a soft partition of the dataset where the data belong in some degree to all clusters; a soft partition is formally defined as *X* the dataset and *xi* represents an element belonging to *X*. According to [11], a partition $P = \{C1, C2, \ldots, Cc\}$ is said to be a soft partition of *X* if, and only if, the following conditions are met *xi*∊ *X* ∀*Cj* ∈ *P*) ≤ μ*cj*(*xi*) ≤ 1 and ∀ *xi*∊ *X* ∃*Cj* ∈ *P*) so that μ*cj*(*xi*) > 0 [9]. Where μ*cj* (*Xi*) denotes the extent to which *xi* belongs to the *Cj* cluster.

A special type of soft partition is one in which the sum of the degrees of belonging to a specific point in all the clusters is equal to 1 [5].

$$\sum_j \mu_C(x_i) = 1 \quad \forall x_i \in X \tag{1}$$

$$\mu_{C1}(x_1) = \frac{1}{\sum_{j=1}^{2} \left(\frac{\|x_1 - v_1\|^2}{\|x_1 - v_j\|^2} \right)^2} \tag{2}$$

A soft partition that meets this additional condition is called a restricted soft partition. The neutrosophic K-means algorithm produces a restricted soft partition, and to do this, the target function *J* is extended in two ways, on one hand, ∀ *xi*∊ *X* ∃*Cj* ∈ *P*) so that μ*cj*(*xi*) > 0 where the neutrosophic degrees of belonging of each data in each cluster are incorporated, on the other hand, an additional parameter *m* is introduced as an exponent weight in the belonging function, thus, the extended target function *Jm* is as shown in Eq. 2 [17].

Where P is a fuzzy partition of the X dataset formed by $C1$, $C2$, ... Ck. The parameter m is a weight that determines the degree to which partial members of a cluster affect the result [3, 7]. Like classic K-means, neutrosophic K-means also tries to find a good partition by searching for the prototypes vi that minimize the target function Jm and additionally, neutrosophic K-means should also search for the membership functions μCi to minimize Jm.

3 Results

With the application of the neutrosophic K-means, the market segmentation is made on the 11 most sold and less sold software products in the last quarter of 2018 in Guayaquil, Ecuador. To do this, a value close to 1, which indicates that the product is quite sold or is sold frequently, is wanted, depending on the case. A value close to 0 shows that the product is little sold or is not sold at all. In this regard and to perform the procedure, the dataset is separated into two groups (clusters) to see if products with special characteristics are found. For this purpose, neutrosophic K-means is used. The number of clusters is 2, the chosen m parameter is 2, the initial prototypes are defined initially in $v1 = (0.2, 0.5)$ $v2 = (0.8, 0.5)$ according to the data reported in Table 1 and the stop criterion is not taken into account because it only performs the first iteration of the algorithm.

Similarly, the values of the other membership functions are obtained, whose results are shown in Table 2.

The prototypes are then updated according to Eq. 3 [18].

Table 1 Evaluation of the 11 (software) products sold and less sold

Products	Sold	Least sold
1	0.60	0.40
2	0.90	0.12
3	0.70	0.18
4	0.12	0.45
5	0.48	0.82
6	0.30	0.85
7	0.10	0.14
8	0.76	0.12
9	0.66	0.55
10	0.09	0.71
11	0.99	0.30

Table 2 Membership value of each data item in the cluster

Products	Membership	
	Cluster1	Cluster2
1	0.3345	0.7044
2	0.2147	0.8014
3	0.2854	0.7478
4	0.9952	0.0265
5	0.5514	0.4698
6	0.8014	0.2147
7	0.8365	0.1695
8	0.2387	0.8140
9	0.1114	0.9365
10	0.9501	0.0685
11	0.1384	0.8851

$$v_1 = \frac{\sum_{k=1}^{11} (\mu_{C1}(x_k))^2 x_k}{\sum_{k=1}^{11} (\mu_{C1}(x_k))^2} \quad (3)$$

Through the use of MATLAB, it is obtained that the prototype $v1$ in the first iteration is altered little, while the coordinates of the prototype $v2$ change considerably. The 12 iterations performed and the resulting prototypes were $v1 = (0.2147, 0.6042)$ and $v2 = (0.7854, 0.2241)$. So it can be said that the second cluster brings together the products that are most sold, but not very stable on the market, while in the first cluster are the products that are not so sold, but more stable on the market.

Similarly, the values of the other membership functions are obtained, and the results are shown in Table 2.

4 Conclusions

In this paper, a theoretical description of market segmentation was made. Techniques that are often used to segment the market were mentioned and the use of neutrosophic K-means is proposed. Specifically, a neutrosophic K-means is proposed based on the classic K-means for such market segmentation.

The necessary elements are proposed as an algorithm for the use of neutrosophic K-means, then this algorithm is applied to the analysis of the most sold software products in the last quarter of 2018 in Guayaquil, Ecuador. The results obtained showed that when applying neutrosophic K-means as a neutrosophic extension of K-means in which a data can partially belong to more than one cluster. For this reason, neutrosophic K-means produces a restricted smooth partition of the dataset and is therefore useful in situations where the data has characteristics of different groups. This methodology can be applied in many fields such as: data classification, medicine, bioinformatics, economics, among others.

References

1. James, G.M., Punitha, S.C.: Tomato disease segmentation using k-means clustering. Int. J. Comput. Appl. **144**(5), 25–29 (2016)
2. Yin, S.L., Liu, J.: A K-means approach for mapreduce model and social network privacy protection. J. Inf. Hid. Multim. Sign. Process. **7**(6), 1215–1221 (2016)
3. Ou, Y., Cai, C.: Large-scale transit market segmentation with spatial-behavioural features. Transp. Res. Part C: Emerg. Tech. **90**, 97–113 (2018)
4. Knuth, M., Behe, B.K., Hall, C.R., Huddleston, P.T., Fernandez, R.T.: Sit back or dig in the role of activity level in landscape market segmentation. HortScience **54**(10), 1818–1823 (2019).
5. Boutsouki, C.: Impulse behavior in economic crisis: a data driven market segmentation. Int. J. Retail Distrib. Manag. (2019)
6. Tkaczynski, A.: Segmentation using two-step cluster analysis. In: Segmentation in Social Marketing, pp. 109–125. Springer, Singapore (2017)
7. Viloria, A. et al.: Classification of digitized documents applying neural networks. In: Bindhu, V., Chen, J., Tavares, J. (eds) International Conference on Communication, Computing and Electronics Systems. Lecture Notes in Electrical Engineering, vol 637. Springer, Singapore (2020)
8. Huerta-Muñoz, D.L., Ríos-Mercado, R.Z., Ruiz, R.: An iterated greedy heuristic for a market segmentation problem with multiple attributes. Eur. J. Oper. Res. **261**(1), 75–87 (2017)
9. Xiaowen, W., Si, S., Xu, C.: An empirical study on users' market segmentation for subject service based on k-means. Res. Libr. Sci. **9**, 15 (2017)
10. Vijay, V., Raghunath, V.P., Singh, A., Omkar, S.N.: Variance based moving K-means algorithm. In: 2017 IEEE 7th International Advance Computing Conference (IACC), pp. 841–847. IEEE (2017, January)
11. Kansal, T., Bahuguna, S., Singh, V., Choudhury, T.: Customer segmentation using K-means clustering. In: 2018 International Conference on Computational Techniques, Electronics and Mechanical Systems (CTEMS), pp. 135–139. IEEE (2018, December)
12. Syakur, M.A., Khotimah, B.K., Rochman, E.M.S., Satoto, B.D.: Integration K-means clustering method and elbow method for identification of the best customer profile cluster. In: IOP Conference Series: Materials Science and Engineering, Vol. 336, No. 1, p. 012017. IOP Publishing (2018, April)
13. Kuo, R.J., Ho, L.M., Hu, C.M.: Integration of self-organizing feature map and K-means algorithm for market segmentation. Comput. Oper. Res. **29**(11), 1475–1493 (2002)
14. Abdel-Basset, M., Mohamed, M., Smarandache, F., Chang, V.: Neutrosophic association rule mining algorithm for big data analysis. Symmetry **10**(4), 106 (2018)
15. Kamthania, D., Pawa, A., Madhavan, S.S.: Market segmentation analysis and visualization using K-mode clustering algorithm for e-commerce business. J. Comput. Inf. Tech. **26**(1), 57–68 (2018)
16. Karaşan, A., Boltürk, E., Kahraman, C.: A novel neutrosophic CODAS method: selection among wind energy plant locations. J. Intell. Fuzzy Syst. **36**(2), 1491–1504 (2019)

17. Viloria, A., Lezamab, O.B.P.: Improvements for determining the number of clusters in k-means for innovation databases in SMEs. Proc. Comput. Sci. **151**, 1201–1206 (2019)
18. Viloria, A., Varela, N., Pérez, D.M., Lezama, O.B.P.: Data processing for direct marketing through big data. In: Smys S., Tavares J., Balas V., Iliyasu A. (eds) Computational Vision and Bio-Inspired Computing. ICCVBIC 2019. Advances in Intelligent Systems and Computing, vol 1108. Springer, Cham (2020)

A Conversational AI Chatbot in Energy Informatics

Aparna Suresan, Sneha S. Mohan, M. P. Arya, V. Anjana Gangadharan, and P. V. Bindu

Abstract Energy informatics is a field of research that utilizes tools and ideas from modern information technology such as Artificial Intelligence and Machine Learning to solve energy related issues. Information systems can be used effectively to reduce energy consumption. In the present era, there is no medium for consumers to have a real-time interaction to know about their electricity consumption details. In this work, a chatbot in the field of Energy informatics is developed to pave the way for real-time interaction with the society to make them aware about their energy consumption and peak time usage. With this, the consumers can make necessary arrangements to reduce their electricity load during peak times and hence they can reduce their overall energy consumption. It also facilitates the consumers to be aware of the energy consumption level of other consumers and can compare it with their level of usage. As a result, the users could reduce their electricity consumption leading to reduction in their electricity bills. Consequently, it will result in the conservation of energy. In addition, alert mails will be sent to the registered users when their consumption exceeds a threshold level, which contain the consumption details and a graphical plot on electricity consumption during various seasons [11]. The machine learning-based dialogue management and language understanding are the main components of the chatbot system. The system is based on Recurrent Neural Networks.

A. Suresan (✉) · S. S. Mohan · M. P. Arya · V. Anjana Gangadharan · P. V. Bindu
Department of Computer Science & Engineering,
Government College of Engineering Kannur, Kerala, India
e-mail: aparna.suresan@gmail.com

S. S. Mohan
e-mail: snehasmohan23@gmail.com

M. P. Arya
e-mail: mparya1001@gmail.com

V. Anjana Gangadharan
e-mail: anjanavg29@gmail.com

P. V. Bindu
e-mail: bindupv@gcek.ac.in

© The Editor(s) (if applicable) and The Author(s), under exclusive license to Springer Nature Singapore Pte Ltd. 2021

A. P. Pandian et al. (eds.), *Proceedings of International Conference on Intelligent Computing, Information and Control Systems*, Advances in Intelligent Systems and Computing 1272, https://doi.org/10.1007/978-981-15-8443-5_46

1 Introduction

Energy informatics [1] is a field of research that utilizes Information and Communication Technology (ICT) to solve energy related issues. It focuses on improving the efficiency of energy supply systems and consumer systems by collecting and analyzing information regarding energy consumption [2]. The design of smart supply and consumer systems often utilizes tools and ideas from modern information technology such as Artificial Intelligence and Machine Learning.

The collection and analysis of information related to energy are performed to support optimization and distribution of energy and consumption networks. According to energy informatics, information systems can be used effectively to reduce energy consumption in the following way: First it requires collection of large number of consumers energy consumption details and then analyzing the collected data. There should be a real-time interaction system that enables the users to obtain the details about their energy usage trends and to make an efficient usage of energy thus contributing practical and logical ways to increase environmental sustainability. This leads to the idea of introducing a chatbot in the field of energy informatics which provides a real-time interaction to help users to be aware about their energy consumption details. It provides a platform where users can clarify their doubts regarding the usage of electricity and find the area where the usage is inefficient, then manage it and thereby reduce the electricity consumption.

Energy is extremely important to us. We need energy in our households and workplaces. Energy supply needs to be secure, stable, affordable and also sustainable. Energy informatics uses powerful tools to analyze data from different sources to find solutions to several problems related to energy. In this work, a chatbot in the field of energy informatics is developed to pave the way for real-time interaction with the society. Even though chatbots are common in many other fields such as travelling, banking, education, etc, to the best of our knowledge, there is no chatbot developed in the field of energy informatics. Bringing chatbot to this field is significant as it helps the users to know about their consumption which leads to controlling the electricity usage thereby leading to a sustainable environment. We also develop an automatic mail alert system which alerts users regarding their consumption details and a graph plot showing their level of consumption. This mail alerting system reminds the users of their increasing consumption so that they can use the chatbot for additional clarifications or tips to control their consumption.

The chatbot is developed with the open source Python libraries Rasa core and Rasa NLU using Long short-term memory (LSTM) supervised learning. The machine learning-based dialogue management and language understanding are the main components of the chatbot system. The system is based on Recurrent Neural Networks (RNN).

The rest of the chapter is organized as follows: Sect. 2 discusses the literature survey on chatbots and energy informatics. Section 3 explains the methodology adopted for developing the chatbot and Sect. 4 discusses about the hardware and software requirements for the system. The experimental results are presented in Sect. 5. Section 7 provides the conclusion and future work.

2 Literature Survey

A wide range of factors such as the environmental conditions and human behavior influence electricity consumption [3]. A retailer or network operator is the one who communicates with the end user about change in price, demand in market, and user's usage of energy, and encourages them to reduce their usage or shift their load to some other time [12]. Demand side management is a set of measures which can be effectively used to improve the energy system at the consumption side, there by leading to a efficient use of energy [4]. Hobman et al. [5] proposes that the complex and unfamiliar decision making scenario of choosing among alternative electricity pricing offers is most likely to be influenced by non-economic factors. These factors, particularly psychological and motivational influences, have the ability to decide customer choices and actions.

A chatbot has the knowledge to identify the input sentence from the user and make a decision as a response to answering the question [6]. Travel suggestions, educational assistance, medical assistance. etc. are some of the fields in which chatbots are used. Chatbots can provide better user-centric recommendations by referring to preferences ratings and act as a travel agent [7]. The drawbacks with traditional travel agents such as the lack of availability and the lack of in-depth and updated knowledge about the places can be solved. A chatbot predicts accurate and relevant answers to questions asked by the user by taking the inputs from the user. The answers to the questions are given by referring to the preferences of user, past travel history and ratings of user etc. The bot asks questions to user in case of any missing data in question until the answer fulfills the missing information.

Chatbot is used for educational assistance especially for visually impaired people [8]. The chatbot gives answers to the voice queries which are educational related. Since the chatbot works with voice queries, it is helpful for the visually impaired people as well as normal people. It uses android platform and can be launched easily by Google voice search. This chatbot was developed by gathering ideas from several existing chatbots such as chatbots that gives educational assistance to students, chatbot with text and voice output, etc. The chatbot application could give answers to any kind of questions by using information from Wikipedia and also user-defined answers. Therefore, it gives the ability to user to define their own questions and answers.

Chatbot for medical assistance [9] is developed to create a real-time interaction system with artificial intelligence. One can just enter symptoms or an ECG image to find their health problems or one can check whether the prescribed medicine is supposed to be used the way they are told to. On knowing the symptoms, the chatbot assists to deduce the problem and to verify the solution. The composition of the medicines and their prescribed uses can also be known apart from diagnosing the disease. The system also helps to take up the correct treatment.

3 Methodology

The aim of this work is to develop a conversational chatbot that facilitates a real-time platform where the users can present their queries regarding their energy consumption and obtain real-time responses. In addition, whenever the consumption of user exceeds a threshold value, a mail alert should be sent to the user showing their consumption details and a graphical representation of their usage.

The chatbot is developed with Rasa core and Rasa NLU using Long short-term memory (LSTM) supervised learning. Rasa core and Rasa NLU are open source Python libraries for creating chatbots [11]. The machine learning-based dialogue management and language understanding are the main components of the system. The system is made of a Recurrent Neural Network (RNN). The network maps raw dialogue history directly to distribution over system action. A few keywords which would be used repeatedly here are:

- Entity—Extracting useful information from user input.
- Stories—The intent and action taken by the bot will be defined by sample interactions between user and chatbot in terms of stories.
- Actions—Actions involve either asking for some more details to get all the entities or integrating with some APIs or querying the database to get/save some information which are basically the operations performed by the bot.

The detailed working of the chatbot is shown in Fig. 1. It accepts the user queries through the interaction environment. It is then fed to the normalizer that normalizes the user input to a normalized form which will be understood by the chatbot. The bot then performs spellcheck and pattern match where the user query will be analyzed for its syntax and semantics. The query is then analyzed and the intended response is retrieved from the database. This response is given back to the user. The user can either continue to ask more queries or stop. If the pattern does not match with that in the database, then an error message is displayed to the user.

Fig. 1 Flow chart

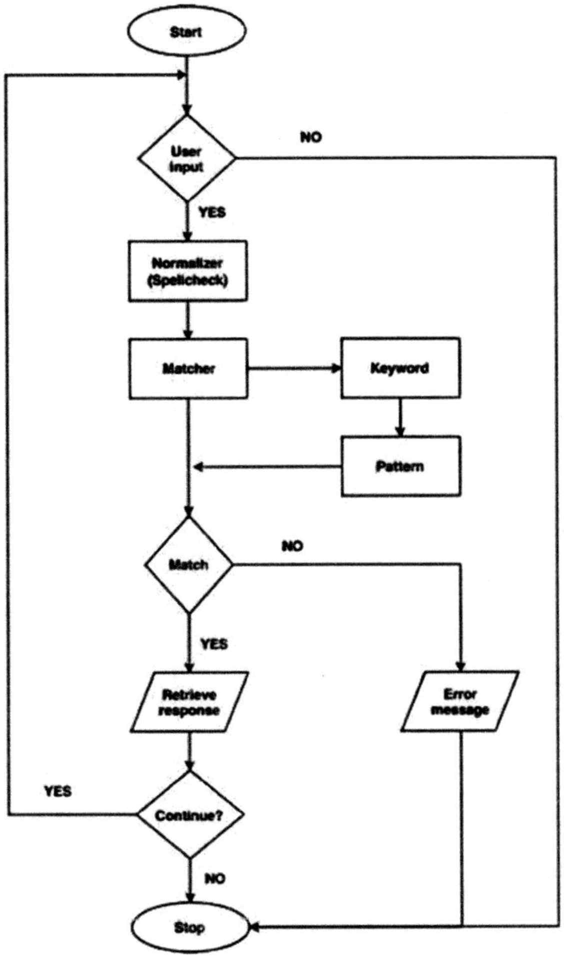

First the NLU model is trained with the inputs in a simple text format and structured data is extracted in order to teach the bot to understand the messages. The training is achieved by defining the intents and also by specifying a few ways in which users might express them. In order to make this work, we need to define some files in appropriate format. Along with the mapping of intents and entities present in each of NLU training file, they contain some training data in terms of user inputs. The bots NLU capabilities become better based on more varying examples that are provided. Stories file contains sample interactions the user and bot will have. All the intents, entities, actions, templates and some more information will be listed in the domain file.

The data set for training is stored in data.json file and stories in file stories.md. The data provided will help in training the bot. The pipeline is one through which the data will flow and intent classification and entity extraction can be done for the bot. We have used spacy sklearn pipeline and as our NLU data and pipeline are ready, the bot was trained. The model is trained and saved at models/nlu/default/chatbot. Once the model is created we keep on training the model. The training is continued until intended response is provided by the chatbot. The user interface for the chatbot is made by interfacing with slack API, an API which allows to receive and send messages in real-time.

An automatic mail alert system which keep track of user consumption is also maintained. A database containing user consumption details is maintained which keeps on incrementing. A threshold value is chosen by analysing the data and whenever the consumption exceeds the threshold value mail alert will be sent to the user. So whenever the user consumption exceed this threshold it is an indication that user has exceeded their consumption level. SMTP servers are used to send emails. We can send consumption details and graphs showing consumption during various seasons as attachment. This mail can be used to notify users regarding their consumption. The smtplib module of Python is needed for sending mails. This module is imported and then an SMTP instance is created that encapsulates an SMTP connection. It takes host address and a port number as parameters. The contact information and message templates are fetched and a mail is send separately for each of those contacts by creating a MIMEMultipart object and attaching the message body to the MIMEMultipart object as plain text. So the automatic mail alert system alerts the user with a mail which contain details regarding the users consumption as well as graphs. The graph shows the level of consumption of user during various seasons and their peak usage times. Thus the user could get to know about their consumption level.

4 System Architecture

The behavior and structure of the system is defined by the system architecture. The conversational software with machine learning is as shown in Fig. 2. The modules in the system architecture are the following:

- User: The user module adds user to the system by giving each user a unique id.
- Database: It keeps record of all the user details, user queries etc.
- Connector module: The connector module, on giving queries get connected to the conversational platform. This connector module acts as the interactive platform in our model.
- Input module: This module is used for analyzing the inputs of user. This input module act as the ears of our chatbot.
- Dialog management: This module is the brain of the system which generates the output with the help of database module. Dialog management acts as the brain of the chatbot.

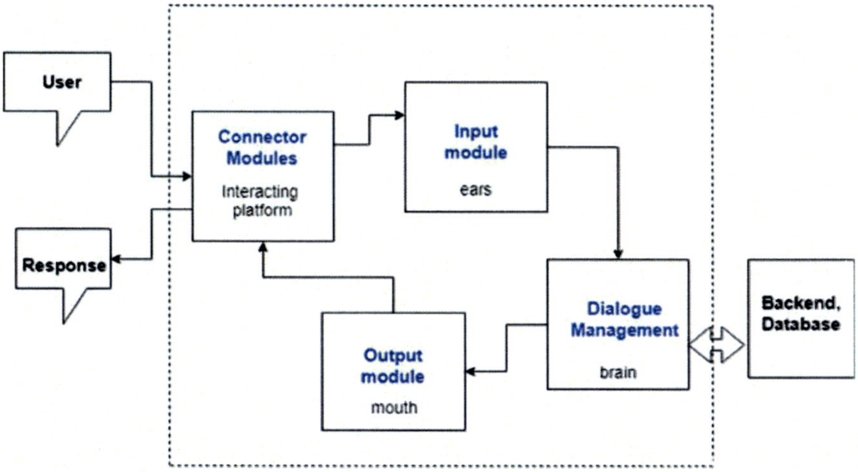

Fig. 2 RASA the OSS to build conversational software

- Output module: This module is used for providing output to users as response. This output module acts as the mouth of the chatbot.

On giving the queries, connector module connects to the conversational platform, input modules analyze inputs, and the dialogue management generates output with the help of database and gives it to the user as response. User presents the query to connector module which acts as the interacting platform. The query is then fed to the input module which acts as the ears of the chatbot. The query is then passed to dialogue management module which acts as the brain of the chatbot. The dialogue management module is connected to backend database from where the intended response is retrieved from the database. The response associated with query will be then passed to the output module which acts as the mouth of the chatbot. The response is displayed to the user through connector module.

The training procedure is as shown in Fig. 3. Sample user queries are used to train the chatbot. When the input queries are given as input, intent classification and entity extraction are performed on them. Recurrent neural networks use this intent classification and entity inputs extracted to train the model. We check the model by giving the sample actions and checking whether the chatbot identifies the entities correctly. The model gives the response once the training is complete. The query will be fed to a tokenizer where the query will get split into tokens. The tokens will be then analysed and entity will be identified and extracted. The named entity which uniquely identifies the query will be extracted and will be given as entity output. The query will be analysed for the intent and associated classification in which query belongs to will be determined. The intent along with the entity will be then fed to recurrent neural network where the training occurs. The model will be trained inorder to retrieve intended response associated with the query. Once the model is trained intended response will be given as chatbot response. Rasa NLU library package is

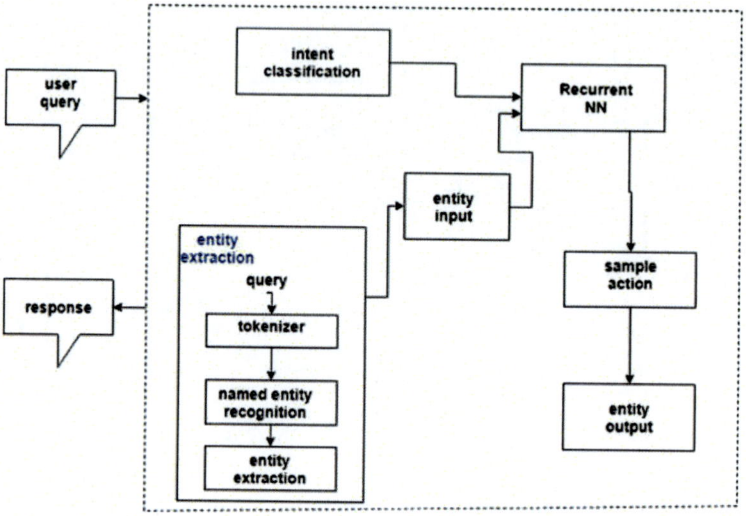

Fig. 3 Training the chatbot

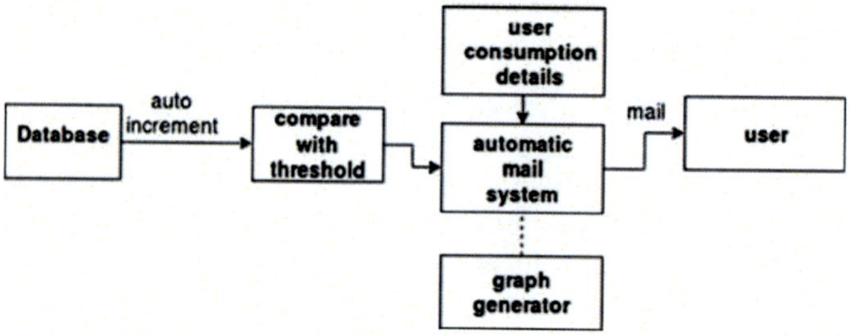

Fig. 4 Email alert system

used for entity extraction and intent classification. The input module may extract the keyword as entity which will be matched. Intent classification occurs in order to match the entity with corresponding response. Pattern matching algorithm could also be used for the matching. The following are the steps that have to be taken in order to teach our bot to understand our messages,

1. Train the NLU model by giving inputs. The inputs need to be given in a text format.
2. The structured data is extracted. This is achieved by defining intents and giving a few examples on how the users might express them.

The architecture for automatic mail alert system is shown in Fig. 4. By analyzing the auto-incrementing database, user's energy consumption details are extracted and are compared with the threshold. If the consumption exceeds the threshold value, a mail alert will be sent to the user regarding their energy consumption details.

5 System Implementation

In this section, we present the hardware and software requirements for implementing the system. The hardware requirements for this work are as follows: a computer based on Core i5 Processor, 4 GB RAM, and 32 GB Hard Disk. The software requirements for the chatbot are as follows: Ubuntu 16.04, gedit, PYTHON 3.6, Slack API, and Rasa NLU Package. The software requirements for the Automatic Mail Alerting System are as follows: Windows 10, PYTHON 2.7, Jupyter Notebook, and MySQL db.

6 Experimental Results

A comparison of typical chatbot over other applications is done through a survey. From Fig. 5, it can be observed that the chatbot is more efficient and convenient. In terms of user-friendliness and approachability, chatbot stands ahead over other applications. Chatbot provides good customer experience when compared with the other applications. Chatbot provides more detailed answers in response to the queries than given by normal applications. Even though applications stand ahead of chatbots in terms of base of communication and convenience of users, chatbots provide quick answers to both simple and complex questions and also provide 24 h service.

The dataset used for testing our work is the smart meter data from London area [10] which contains the energy consumption readings for a sample of 5,567 London

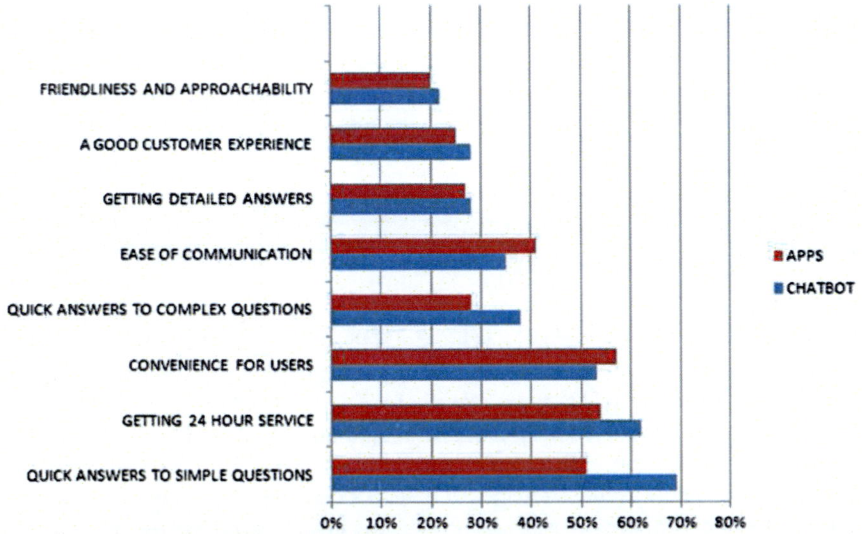

Fig. 5 Comparison of chatbot over applications

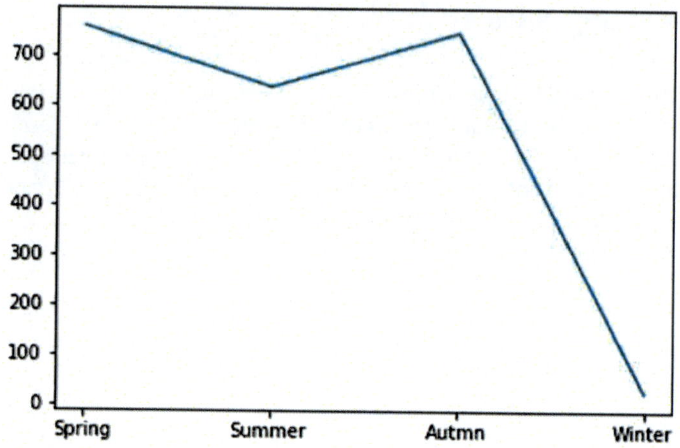

Fig. 6 Sample graph of user consumption

Households that took part in the UK Power Networks led Low Carbon London project between November 2011 and February 2014. A sample graph of energy consumption of a particular user is shown in Fig. 6. Energy consumption trend of user during various seasons is shown in the graph. Various seasons are shown in the X-axis and the energy consumed in the Y-axis. Similarly, graphs showing the energy consumption are generated for each user and are send along with the mail alert.

In a chatbot environment, user presents the query and appropriate matching response will be provided by the chatbot. A sample screenshot of our chatbot is shown in Fig. 7. In the sample shown, the conversation is started with a greeting. The user can either check their own consumption details or the details of other users by knowing their ids. Any queries of the user can be cleared by the chatbot by retrieving the correct response from the database. The conversation continues until the user says goodbye.

7 Conclusions and Future Work

In the present era of energy crisis, it is helpful if there exists a system which provides assistance to consumers regarding their energy consumption and creating awareness among them. In the digital era, people prefer for solutions that can be accessed within a touch. We have developed a chatbot and an automatic mail alerting system in the field of energy informatics. Our chatbot provides an efficient assistant for consumers to choose best energy saving strategies. It helps the consumers to analyze their energy usage, understand where the wastage occurs, and how to minimize the wastage and save energy and thereby reducing the electricity bill. Chatbot is easy to use, user friendly, real time platform for users. Along with this, an automatic mail alerting

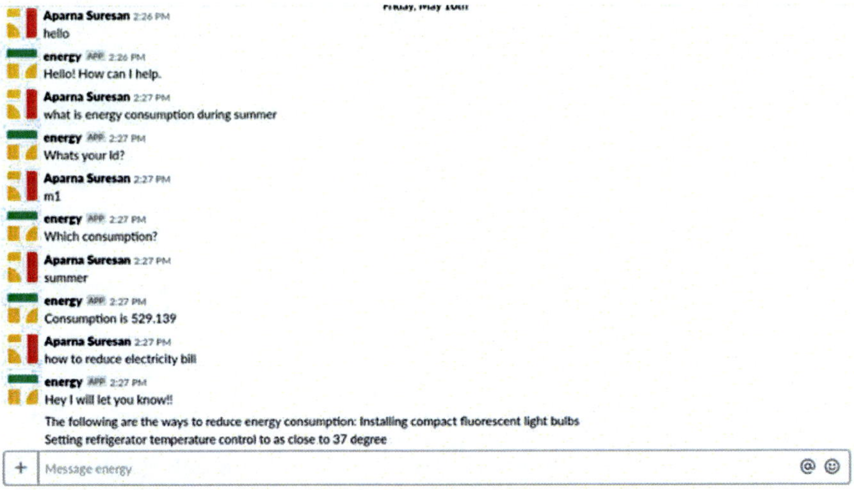

Fig. 7 Chatbot environment

system helps the user to keep track of their usage. Whenever the consumption exceeds a threshold, mail alert containing consumption units and a graphical plot showing their consumption will be sent to the user.

This work can help the society reduce energy consumption even during their busy schedule, thereby reducing their own loss and contributing their part to the society which leads to a sustainable future. The future work concentrates on focusing on latest technologies to implement various advancements. In future, chatbot conversation can be more than just text, such as voice-enabled chatbot. We can make the chatbot initiate the actions by itself by programming accordingly. The chatbot can be improved to provide alerts of the fluctuations that may happen in advance. The chatbot could have facilities for language recognition and translation.

References

1. Huang, B., Bai, X., Zhou, Z., Cui, Q., Zhu, D., Hu, R.: Energy informatics: fundamentals and standardization. ICT Express **3**(2), 76–80 (2017)
2. Watson, R.T., Boudreau, M.C., Chen, A.J.: Information systems and environmentally sustainable development: energy informatics and new directions for the IS community. In: Management Information Systems Research Center, vol. 34, pp. 23–38. University of Minnesota (2010)
3. Walker, G.R., Sernia, P.C.: Utility customer segmentation based on smart meter data: empirical study. In: IEEE Smart Grid Comm 2013 Symposium—Support for Storage, Renewable Resources and Micro-grids, vol. 19, pp. 1130–1139 (2013)
4. Palensky, P., Dietrich, D.: Demand side management: demand response, intelligent energy systems, and smart loads. In: IEEE SmartGridComm 2013 Symposium—Support for Storage, Renewable Resources and Micro-grids, pp. 381–388 (2011)

5. Hobman, E.V., Frederiks, E.R., Stenner, K., Meikle, S.: Up take and usage of cost-reflective electricity pricing: insights from psychology and behavioural economics. Renew Sustain Energy Rev 57, 455–467 (2016)
6. Setiaji, B.: Chatbot using a knowledge in database. In: Proceeding of the Royal Society A 2016 7th International Conference on Intelligent Systems, Modelling and Simulation, pp. 1721–1740 (2017)
7. Argal, A., Gupta, S., Modi, A., Pandey, P., Shim, S., Choo, C.: Intelligent travel chatbot for predictive recommendation in echo platform. In: 2018 IEEE 8th Annual Computing and Communication Workshop and Conference (CCWC), pp. 176–183. Las Vegas (2018)
8. Naveen, K.M., Linga, C.P.C., Venkatesh, P.A., Sumangali, K.: Android based educational chatbot for visually impaired people. In: IEEE International Conference on Computational Intelligence and Computing Research (ICCIC), pp. 1–4. Chennai, (2016)
9. Divya, M., Neeraj, J.C.J., Elmy, S., Shinoy, S., Anandhu, A.: A novel approach for medical assistance using trained chatbot. In: 2017 International Conference on Inventive Communication and Computational Technologies (ICICCT), pp. 243–246 (2017)
10. Smart meters in London—smart meter data from London area. https://www.kaggle.com/jeanmidev/smart-meters-in-london. Accessed on 20 Jan 2020
11. Bocklisch, T., Faulkner, J., Pawlowski, N., Nichol, A.: Rasa: open source language understanding and dialogue management. In: 31st Conference on Neural Information Processing Systems. NIPS, Long Beach, CA, USA (2017)
12. Feuerriegel, S., Strker, J., Neumann: Reducing price uncertainty through demand side management, E-business and competitive strategy. In: Thirty Third International Conference on Information Systems, Orlando (2012)

Monitoring Health of Edge Devices in Real Time

V. Meghana, B. S. Anisha, and P. Ramakanth Kumar

Abstract Monitoring all the edge devices in an enterprise environment or an edge computing environmentcan be a daunting task due to the sheer volume of edge devices deployed throughout an enterprise. As the size and complexity of computer environments grow, it is becoming more difficult to evaluate and recognize factors which restrict efficiency and scalability. Application developers and system administrators need tools that is easy to use so that they can quickly identify system bottlenecks in the edge device and configure the system that has best performance. System metrics are measurement types found within the system. Any resource that will be monitored for performance, availability, reliability, and other attributes will have one or more metrics about which data are collected at various intervals of time. CPU Usage (%), 15-min Load Average, 1-min Load Average, Available Free Memory (MB), Disk Space (MB), Available Swap Space (MB) and so on are some of the available edge device system metrics. The tool provides a way to monitor these system metrics in real time.

Keywords System metrics · Availability · Edge device · Edge computing · Real time · Monitoring · System administrators

V. Meghana (✉) · B. S. Anisha
Information Science and Engineering, RV College of Engineering (Autonomous Institution affiliated to Visvesvaraya Technological University, Belagavi), Bengaluru, India
e-mail: meghana.athreya@gmail.com

B. S. Anisha
e-mail: anisha@rvce.edu.in

P. R. Kumar
Computer Science and Engineering, RV College of Engineering (Autonomous Institution affiliated to Visvesvaraya Technological University, Belagavi), Bengaluru, India
e-mail: ramakanthkp@rvce.edu.in

© The Editor(s) (if applicable) and The Author(s), under exclusive license to Springer Nature Singapore Pte Ltd. 2021

A. P. Pandian et al. (eds.), *Proceedings of International Conference on Intelligent Computing, Information and Control Systems*, Advances in Intelligent Systems and Computing 1272, https://doi.org/10.1007/978-981-15-8443-5_47

1 Introduction

A device that provides an entry point to the enterprise environment is termed as an edge device. It is also kind of a server which accepts requests from the clients and responds to their requests. Organizations have plenty of edge devices and various applications are hosted on these devices. A single application can be hosted on more than one edge device in order to provide less response time. There are three domains which play important part in developing and maintaining these devices and applications. They are developers who develop the applications, testers who test the applications, and operations team who deploy and manage these applications and devices. An operation team consists of several system administrators who constantly monitor the devices and make sure the applications (indirectly the edge devices) are available and reliable. A device consists of an operating system and numerous system metrics which measure the system resources. In any organization, during an outage, the system administrators must login to the devices and manually check the system status by giving various commands. It requires some extra skills to analyze the attributes. It fails to give the historic performance details. The business executives like directors, senior managers need an easy way which consumes less time to analyze the system state and to monitor the state of the devices proactively to avoid any outages.

Each resource that can be observed for performance, availability, reliability, and other attributes has one or more data gathering metrics. CPU Usage (%), 15-min Load Average, 1-min Load Average, Available Free Memory (MB), Disk Space (MB), Available Swap Space (MB), etc., are some of the system metrics that are available. The model gives a way to monitor the edge devices by monitoring its metrics in real-time in order to ensure 100% availability. This eases the job of system administrators.

The main objective of the proposed design is to monitor the system metrics of all the edge devices deployed in the environment and catch the system bottlenecks quickly before the device crashes. The goal is to collect the system metrics data like CPU usage and load average. A standard threshold is fixed for each metric. If the metric reaches that threshold, then a mail will be sent to the system administrator. The system administrators need not always monitor the edge devices. The proposed model will do that job for them. They just need to act only when they get mail saying it is in the critical stage. In this way, the devices can be monitored in real time which help the system administrators in easing their work.

2 Literature Survey

Yanyan et al. [1] discuss the methods to boost the linux system performance in real time. The performance of Linux kernel is analyzed in real-time using the LmBench tool. A measurement method is implemented for measuring the results in real time. However, the model is designed to test only two versions of kernel and is assumed to be suitable for all the real-time systems. In contrast, the proposed model in the paper does not limit itself for one specific version.

Purushothaman et al. [2] discuss a way to monitor servers. Monitoring servers in an enterprise environment can be a daunting task due to the sheer volume of servers deployed throughout an enterprise. The model results in self-evolving means that identifies the most optimal threshold values for crucial/core performance parameters. However, the model does not describe what action is taken once the performance parameters reaches the threshold. On the other hand, the proposed model notifies the system administrators when the parameters reaches the threshold value.

Sumit Maheshwari et al. [3] provide an overview of the scalability and efficiency of an edge cloud network built to support latency-prone applications. The defined model can be used to decide the response time of the application, by varying the load on the computer resources. The results of the system analysis provide guidelines for selecting the best combination between the edge and core cloud services, provided a defined limit on application delay. Nevertheless, the paper concentrates on comparing cloud-core only and edge-core only systems and it concentrates more on application specific metrics like response time and not on the system metrics. On contrary, the proposed designed only concentrates on system metrics.

Yuanyuan et al. [4] analyze a way to calculate the amount of CPU utilized and the amount of memory utilized. It provides a way to test software performance of linux kernel. File system/proc is used to gather information about processes currently running and to yield a simple perspective of the kernel. Also, a way to measure time. It gives a way to calculate the CPU utilization, memory utilization, and measurement of time. The method offers a way to test the efficiency of the software programs executing on Linux which could improve the performance of the applications running on Linux. However, it is not designed for real-time monitoring of the performance of the system. Whereas, the proposed design monitors the performance in real time.

Roblee et al. [5] introduce to a new system management framework based on a modern efficient dynamic data analysis approach called Process Query Systems (PQS). PQS enables server monitoring and makes correct and swift decisions about server and service status by using advanced behavioral models. The PQS system uses a generic process to detect the software platform. It builds upon a broad range of knowledge on various levels of a system. Nonetheless, it fails to support for larger environment. The proposed system can be used for larger environment.

Yucheng et al. [6] concentrate on real-time monitoring and making it hard to understand the software performance issues more directly and conveniently through analyzing the historical data and monitoring the remote server, this presents a performance management system software for a server based on B/S mode. During the performance testing and network maintenance, the system that uses B/S mode and provides a different perspective of server to the administrators. However, the model does not describe how it notifies the administrators. On the other hand, the proposed model notifies the system administrators through e-mail.

Zeng et al. [7] introduce a simple network management protocol (SNMP) to monitor the servers. By defining MIB objects to monitor server resources, MIB resources are expanded and multi-threading technology is used for data collection and processing, which can improve the performance of collection. Nevertheless, model can be used to monitor only the network parameters and cannot be used for other system parameters, while the proposed design concentrates on the system specific parameters.

Forrest et al. [8] report a preliminary finding intended to create a self-definition for Unix processes in which normal activity is regarded as anonymous. They have shown that small succession of system calls produce a stable signature in running processes to typical behavior. However, the main goal of the paper is to avoid intrusion attack and hence, they are monitoring specific to that. Whereas, the main goal of the proposed system is to monitor the edge devices for system bottlenecks.

Bohra et al. [9] proposed a system for monitoring the computer system remotely and recovering its software state without using its processors or relying on the resources of the operating system. It may be used to detect and restore damage to the operating system state. The proposed model is not suitable for real-time monitoring, it is suitable for repairing the problem detected. Whereas, this paper aims in monitoring the device in real time.

Meira et al. [10] presents a peer-to-peer load testing approach to isolate and to scale up the load bottleneck issues related to centralized test drivers. However, the model is designed for small setup. The proposed design can be used for larger setup.

Jiang et al. [11] is an investigation into the state of research and practice of load testing. They compare current techniques used in a load test's three phases: designing a proper load, performing a load test, and analyzing the results of a load test. However, the paper is a survey and does not propose any new design.

Chandra et al. [12] provide an outline of the distributed clouds that might be better suited for applications where users, data, and computations are distributed as centralized cloud's enforce limitation in terms of cost and performance. Nevertheless, the paper just discusses how spreading the load across the distributed system is better than centralized cloud and not much on the monitoring the load on each device. The proposed design discusses on how the load can be monitored on each device.

3 Proposed Design

The performance of the server can be measured using some of the system metrics like load average, CPU, and available free memory. In order to do this, the system metric data must be collected. Collectd is a daemon process which collects the system statistics data. The data collected is pushed to influxDB. InfluxDB is a time series database. It is usually used to store data which changes with time. Each data is stored along with a timestamp. The data that is stored in influxDB can be visualized using graphs. Grafana is a visualization tool where values are plotted against the timestamp. A threshold value is fixed and configured in Grafana. Once the system metric's

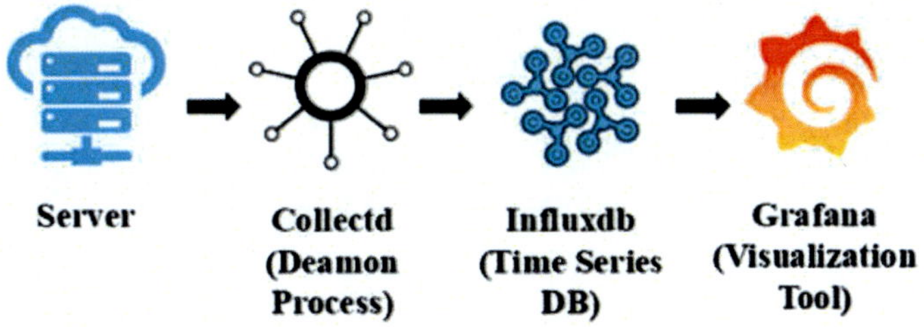

Fig. 1 System design

value reaches the threshold, a mail gets triggered and sent to the administrators. A standard threshold is usually set by the higher management by analyzing the previous performance of the systems (Fig. 1).

Figure 2 shows the complete flow of the proposed design. The details of each component are discussed below:

Collectd. Collectd is a daemon process that collects application and system statistics that are used by system administrators. The obtained data helps the administrators maintain an overview of the available resources to detect looming and existing bottlenecks. Create a daemon process, as in (1), to collect the edge device system's performance metrics like load average, CPU utilization, and available free memory from the edge device.

- systemctl start collectd.service
- systemctl enable collectd.service (1)

InfluxDB. The statistical data collected is pushed into influxDB. Influxdb is a database of time series that stores time series data. The values to be plotted must be numerical and must vary w.r.t to time. An InfluxDB database, like conventional relational databases, acts as a logical container for users, retention policies, and continuous queries. "collectd" is the database created to store the values. The measurement serves as a container for tags, fields, and the time column, and description of the data stored in the related fields is specified by the name of the measurement. Measurement names are strings, and a measurement can be treated like a table for any SQL user. Since the concentration is on load average, CPU, and free memory, respective measurements are created. Since each data is associated with timestamp, it acts as the primary key or the field key. The daemon process sends the data every second into the influxDB.

$$\#/opt/influxdb/influx$$
$$Connected\, to\, http://localhost:8086\; version\; 0.9.4.2$$
$$InfluxDB\; shell\; 0.9.4.2$$
$$> use\; collectd$$

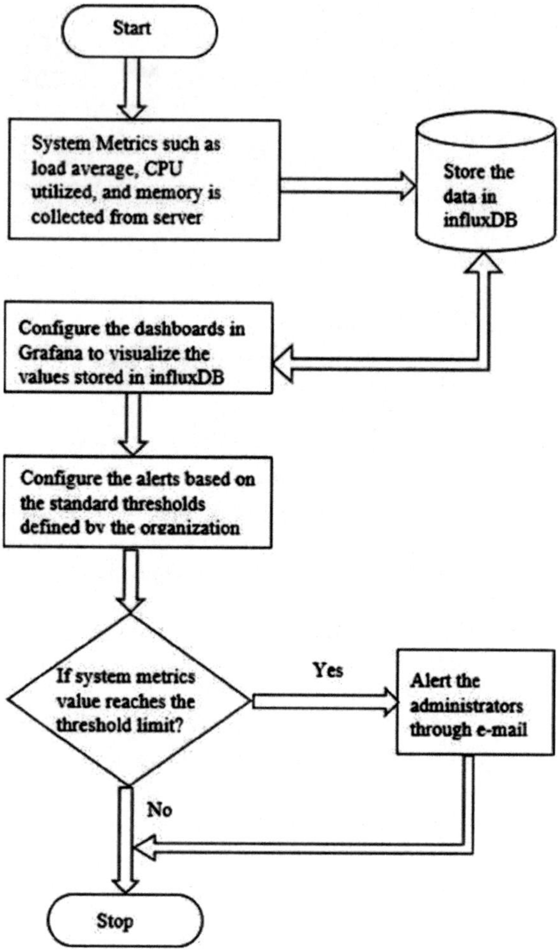

Fig. 2 Detailed flow of the proposed design

$$Using\ database\ collectd$$
$$> show\ measurements \tag{2}$$

Grafana. Grafana is a visualization tool which is used to analyze the trends in the data. It is also used to monitor and create dashboards. It can be configured to connect to influxDB in order to visualize the system metrics. Each row in the dashboard consists of various features. The graph can be configured like SQL query to display all the values that matches the condition.

Alerts. Based on the standard threshold defined by the organization's alerts are configured in Grafana. Configure SMTP server. Google provides free SMTP servers which can be used to send free e-mails. When the metrics reach the threshold limit, an e-mail is sent to respective administrators. SMTP server can be configured by

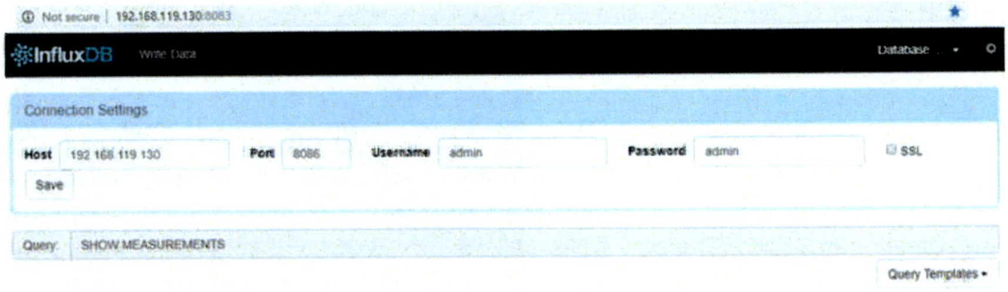

Fig. 3 Admin login page for influxDB

inserting the following lines (3) in/etc./grafana/grafana.ini file.

$$\begin{aligned}&[\text{smtp}]\\&\text{enabled} = \text{true}\\&\text{host} = \text{smtp.gmail.com:465}\\&\text{user} = \text{XYZ@gmail.com}\\&\text{password} = *********\end{aligned} \tag{3}$$

4 Results

4.1 Admin Login

The System administrators must login to influxDB and grafana once as shown in Figs. 3 and 4. Admin can login to grafana using the URL https://IP_ADDRESS:3000 and to influxdb using the URL https://IP_ADDRESS:8086.

4.2 Viewing Data in InfluxDB

The administrators can validate the data by giving SQL-like queries in influxDB. InfluxDB is known for its reliable and fast retrieval. Figure 5 shows the list of measurements in the database. Measurements are like tables for any SQL user. Query can be like (4) which retrieves the top 10 field values from measurement load_shortterm which can be viewed in Fig. 6.

$$Select * from\, load\_shortterm\, order\, by\, desc\, limit\ 10 \tag{4}$$

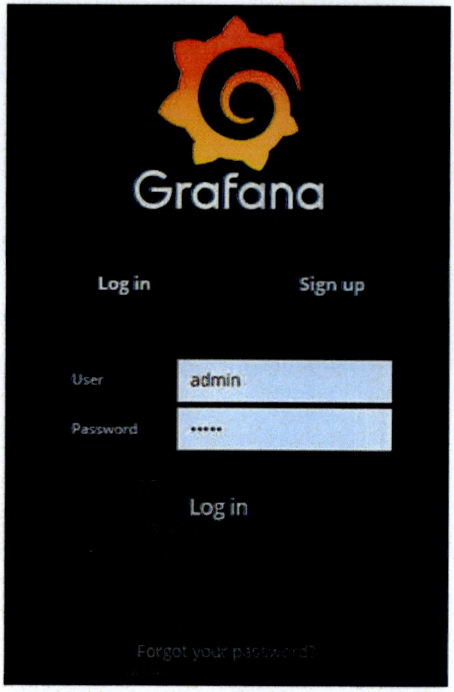

Fig. 4 Admin login page for Grafana

Figure 6 shows the values stored in load-short-term measurement. It collects the short-term load values and add it to the measurement along with the timestamp and host name.

4.3 System Metric Dashboard in Grafana

The values received from influxDB can be viewed in Grafana. The graphs are easy to understand and can be used to find trends and seasonality. The concentration of the paper is on load average, CPU, and free memory. Figure 7 shows the graph of the load average and CPU. Blue, pink, and yellow graphs depict 5, 10, and 15 min load average, respectively, on x-axis plotted against the timestamp on y-axis. The green graph depicts the CPU utilization value on x-axis versus timestamp on the y-axis. The green heart next to "Load" and "CPU" represents that the system is in good state. In this way, health of the edge device can be monitored.

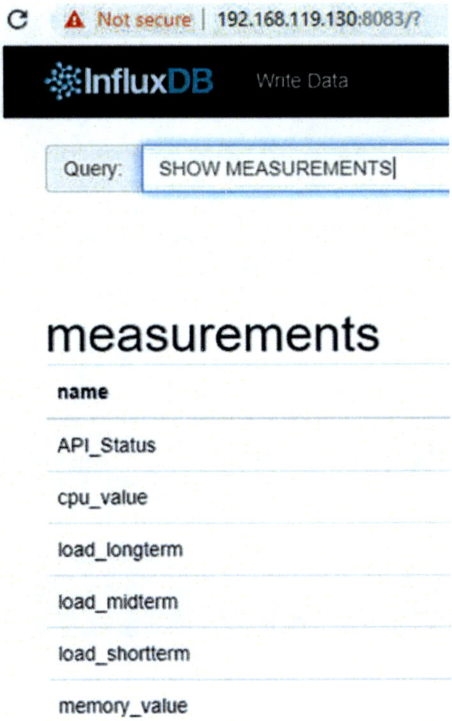

Fig. 5 Measurements in the database

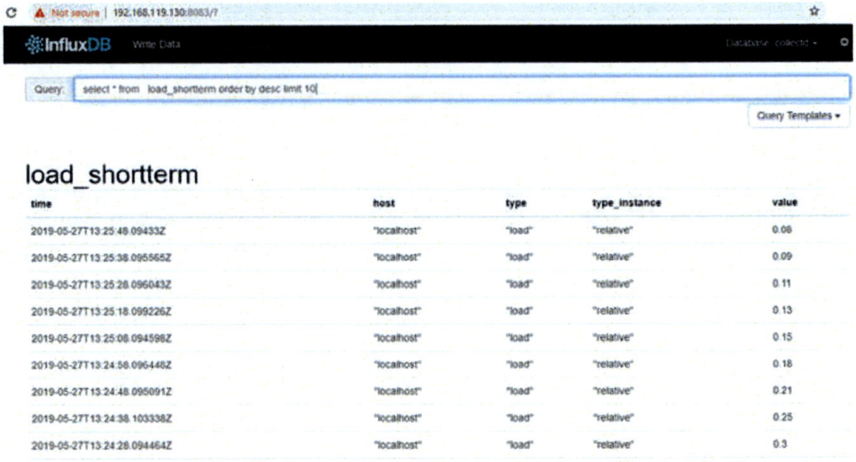

load_shortterm

time	host	type	type_instance	value
2019-05-27T13:25:48.09433Z	"localhost"	"load"	"relative"	0.08
2019-05-27T13:25:38.095565Z	"localhost"	"load"	"relative"	0.09
2019-05-27T13:25:28.096043Z	"localhost"	"load"	"relative"	0.11
2019-05-27T13:25:18.099226Z	"localhost"	"load"	"relative"	0.13
2019-05-27T13:25:08.094598Z	"localhost"	"load"	"relative"	0.15
2019-05-27T13:24:58.096448Z	"localhost"	"load"	"relative"	0.18
2019-05-27T13:24:48.095091Z	"localhost"	"load"	"relative"	0.21
2019-05-27T13:24:38.103338Z	"localhost"	"load"	"relative"	0.25
2019-05-27T13:24:28.094464Z	"localhost"	"load"	"relative"	0.3

Fig. 6 Viewing data in influxDB

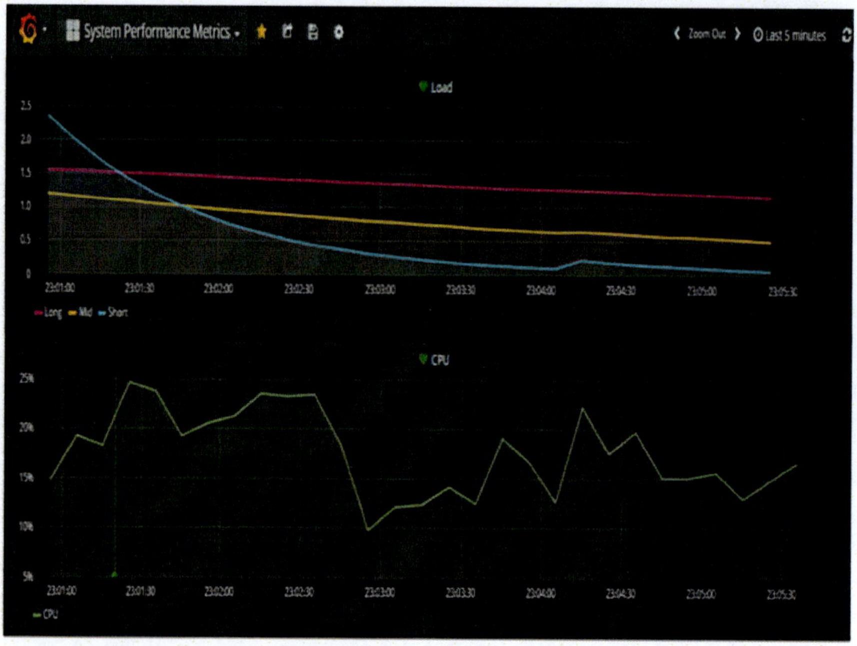

Fig. 7 System performance metrics dashboard in Grafana

4.4 E-mail Alerts

Google SMTP servers is used to send alerts to the administrators when the system metrics reaches a threshold limit as shown in Fig. 8. After 15 min, if the system is below the threshold, an OK alert is sent to administrators to notify them that the system has reached a stable state as shown in Fig. 9.

The red line in Fig. 8 is the standard threshold value configured in Grafana. Since the 5-min load average value (blue line) has exceeded the threshold limit, an alert email will be sent. The red color next to "Load" depicts the system is in bad state. Once the load decreases on the system, an OK alert will be sent to the administrator to confirm that the system is back to good state.

5 Conclusion

The monitoring module is configured to monitor a pre-determined set of edge device parameters. In this regard, all the edge devices are monitored for the same core parameters according to the same threshold limits. Load average, CPU, and memory can be monitored in real time, proactively. This model can also be implemented to monitor servers in an enterprise environment.

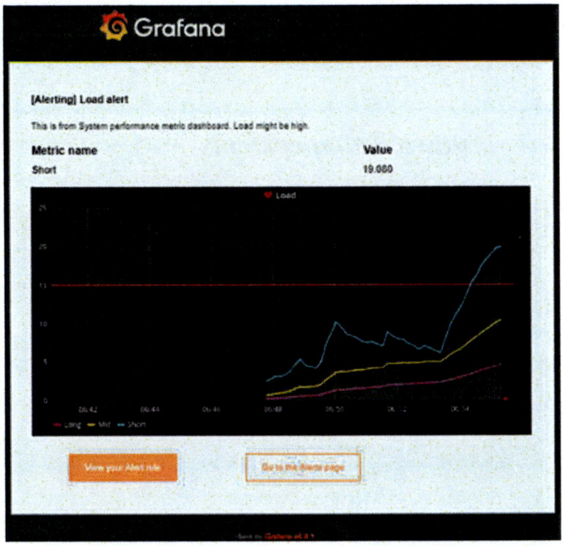

Fig. 8 E-mail alert from system performance metrics dashboard

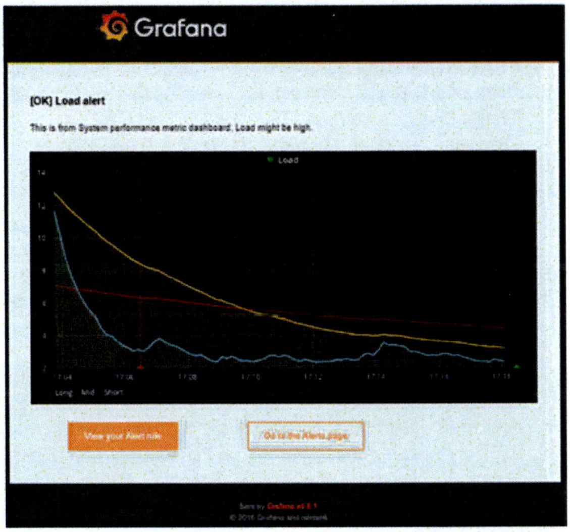

Fig. 9 OK alert when system becomes stable

5.1 Limitations

1. The services must be active and running else the data will fail to flow.
2. This has been implemented using a family of linux operating system. It can be different with other operating system, other than family of linux.

3. The server on which influxDB and grafana are hosted must be up and running always.

5.2 Future Enhancements

The tool can be enhanced by adding machine learning algorithms to predict the bottlenecks and analyze the performance of the edge device. Also, can apply algorithms to forecast the performance of the edge device and adjust the threshold based on it.

The design can also be used to check the physical state of the device. This could be done by polling the device every few seconds. If the device is not reachable for more than 10 min, then an alert could be sent to the administrators.

References

1. Yanyan, Z. et al.: Analysis of linux kernel's real-time performance. 2018 International Conference on Smart Grid and Electrical Automation, pp. 191–196 (2018)
2. Purushothaman, et al.: Optimizing the monitoring of an enterprise server environment. United States Patent (July 2018)
3. Jiang, Z.M. et al.:A Survey on Load Testing of Large-Scale Software Systems, 2015 IEEE Transactions on Software Engineering, pp. 1–32 (2015)
4. Yuanyuan, L. et al.: The method to test linux software performance. In: 2010 International Conference on Computer and Communication Technologies in Agriculture Engineering, pp. 420–423 (2010)
5. Roblee, C., Berk, V.: Implementing large-scale autonomic server monitoring using process query systems. In: IEEE Proceedings of the Second International Conference on Autonomic Computing, pp. 123–133 (September 2005)
6. Yucheng, L., Yubin, L.: A monitoring system design program based on B/S mode. IEEE International Conference on Intelligent Computation Technology and Automation, pp. 184–187 (2010)
7. Zeng, W., Wang, Y.: Design and ımplementation of server monitoring system based on SNMP. IEEE International Joint Conference on Artificial Intelligence, pp. 680–682 (2009)
8. Forrest, S., Hoffmeyr, S., Somayaji, S., Longstaff, T.: A sense of self for unix processes. In: IEEE Symposium on Security and Privacy, pp. 120–128 (1996)
9. Bohra, A., Neantiu, I., Gallard, P., Sultan, F., Iftode, L.: Remote repair of operating system state using Backdoor. IEEE Proceeding of the İnternational Conference on Autonomic computing (May 2004)
10. Meira, J.A., de Almeida, E.C., Le Traon, Y., Sunye, G.: Peer-to-peer load testing. In: IEEE Fifth International Conference on Software Testing, Verification and Validation, pp 642–646 (2012)
11. Jiang, Z.M., Hassan, A.E.: A survey on load testing of large-scale software systems. IEEE Trans. Softw. Eng. 1–32 (2015)
12. Chandra, A., Weissman, J., Heintz, B.: Decentralized edge clouds. IEEE Internet Comput. 70–73 (2013)

Statistical Evaluation of Malnutrition Status of Children in Lao Cai Province, Vietnam

Mai Van Hung, Nguyen Van Ba, and Dam Thi Kim Thu

Abstract Child development depends on many factors such as genetics and habitat, in which nutritional status has a direct and important impact on the child's growth. The anthropometric indices help to determine the weight for age and height for age and were inputted to WHO AnthroPlus software, which shows the age of children and help to assess the nutritional values and statistically evaluate the malnutrition status of children. Martin and M.F., Ashley Montagu's method was used to measure anthropometric indices and this is a cross-sectional study including mean weight, height. Population means and standard errors of the mean are indices to produce national estimates. Standard errors were estimated using SUDDAN by Taylor series linearization. Statistical evaluate anthropometric indices of the children by SPSS software. The percentage of severe underweight and underweight children accounts for a large proportion. The percentage of BMI malnutrition was large in which severe wasted and wasted rates of children under 4-year old. Although the severe wasted and wasted status of children were observed, however, the overweight and obese status of children in this research area was also found. The anthropometric parameters of preschool children in Lao Cai province followed the rules of the body growth of Vietnamese people. However, anthropometric indices of children were lower than general values of children Vietnam. Malnutrition status and severe malnutrition status of children in Lao Cai province are high compared to other provinces in Vietnam.

Keywords Weight · Height · Children · Lao Cai

M. Van Hung (✉)
Research Center for Anthropology and Mind Development, VNU University of Education, Hanoi, Vietnam
e-mail: hungmv@vnu.edu.vn

N. Van Ba
Military Medical University, Hanoi, Vietnam

D. T. K. Thu
Department of Psychology and Education, Thai Nguyen University of Education, Nguyen, Vietnam

© The Editor(s) (if applicable) and The Author(s), under exclusive license to Springer Nature Singapore Pte Ltd. 2021

A. P. Pandian et al. (eds.), *Proceedings of International Conference on Intelligent Computing, Information and Control Systems*, Advances in Intelligent Systems and Computing 1272, https://doi.org/10.1007/978-981-15-8443-5_48

1 Introduction

Nutritional status is one of the most basic factors used to assess a child's development, especially for children aged 3–6 years. According to the World Health Organization (1990), it is estimated that about 500 million children are malnourished, 150 million children under 5 are underweight, and more than 20 million children are seriously malnourished in the world [1]. In Asia and Africa, percentages of malnutrition children are the highest [2]. The standards depict normal early childhood growth under optimal environmental conditions and can be used to assess children everywhere [3].

A case control study was conducted in the maternal and child health clinics in five districts of Terengganu, Malaysia from April to August 2012. Case was a child with moderate-to-severe malnutrition with z-scores <− 2SD from the median of WHO reference [4].

A case control study was anthropometric indices of preschool children was lower than general values of Vietnam. Malnutrition percentage of children in Cao Ma Po was relatively high (underweight: 24.8%; stunting: 77.3%; wasting: 4.5%). The percentage of underweight and wasted children in this study area were lower than that of the whole Ha Giang province, however, the stunting percentage is higher [5].

Generally, in Vietnam, according to an assessment of National Institute of Nutrition, percentage of malnutrition children has been decreased in recently years. In 2007, the malnutrition percentage of children under 5-year old in Vietnam was relatively high (21.7% underweight, 33.9% stunting). In 2014, the malnutrition percentage of children under 5-year old decreased (14.5% underweight, 24.9% stunting) [6]. Lao Cai is a province in the mountain area of Northern Vietnam, economic conditions are not developed equally among localities in the province. Lao Cai has 29 ethnic groups living together stably for hundreds of years. The main causes of childhood malnutrition can be unhealthy household environment, household food insecurity, inadequate care, and lack of health care services [7].

2 Methodology

Study subjects included 3145 children (1581 boy and 1564 girl) aged 2–6 from 20 different ethnic groups in Lao Cai province.

The distribution of research samples is shown in Table 1.

Statistical analysis: Anthropometric indices including weight, height were determined. About anthropometric indices, such as date of birth, sex, and date of objectives were inputted to WHO Anthro Plus software with Standard errors were estimated using SUDDAN by Taylor series linearization which showed the age of children and the nutritional status of the children [8]. After the nutritional status for age were given, statistical evaluate anthropometric indices of the children by SPSS (Statistical Package for the Social Sciences) software [9]. It is used in market researches, health researches, education researches, etc., [10] and others.

Table 1 Number and percentage of children in age and sex

Age	boys		Girls	
	n	%	*n*	%
2	329	20.80	318	20.33
3	326	20.61	315	20.14
4	312	19.73	313	20.01
5	303	19.16	310	19.82
6	311	19.67	308	19.69
Overall	1581	99.97	1564	99.99

(*n* number of children for age and sex)

Table 2 WHO malnutrition standards [11]

Children under 5-years old			
Z-score	The growth indicators		
	Height	Weight	BMI
> 3SD	See note 1	See note 2	Obesity
> 2 SD	Normal		Overweight
> 1 SD	Normal		Possible overweight 3
0 (TB)	Normal	Normal	Normal
< −1 SD	Normal	Normal	Normal
< −2 SD	Stunted 4	Underweight	Wasted
< −3 SD	Severely stunted 4	Severely underweight	Severely wasted
Children aged 5–19			
Z-score	The growth indicators		
	Height	Weight	BMI
> 3SD	See note 1	See note 2	Severe obesity
> 2 SD	Normal		Obesity
> 1 SD	Normal		Overweight
0 (TB)	Normal	Normal	Normal
< − 1 SD	Normal	Normal	Normal
< −2 SD	Underheight4	Underweight	Wasted
< −3 SD	Severe under height	Severe underweight	Severely wasted

Statistical analysis diagram: Anthropometric indices—inputted to WHO Anthro Plus software-using the SPSS software—the data in the tables (Table 2).

Table 3 Weight for age for children

Age	boys		Girls	
	n	$X \pm SD$	*n*	$X \pm SD$
2	329	11.3 ± 1.4	318	10.1 ± 1.8
3	326	12.6 ± 1.5	315	11.5 ± 1.6
4	312	13.8 ± 1.9	313	12.5 ± 1.7
5	303	15.3 ± 1.7	310	13.8 ± 1.9
6	311	16.8 ± 1.9	308	15.9 ± 2.1

(*n* number of children for age and sex)

Table 4 Data comparison between this study and data from Vietnam Ministry of Health in 2003

Age	Mean weight of age for boys		Mean weight of age for girls	
	This study	VN Ministry of Health in 2003	This study	VN Ministry of Health in 2003
2	11.3 ± 1.4	10.56	10.1 ± 1.8	10.22
3	12.6 ± 1.5	11.55	11.5 ± 1.6	11.04
4	13.8 ± 1.9	13.34	12.5 ± 1.7	12.96
5	15.3 ± 1.7	15.03	13.8 ± 1.9	14.69
6	16.8 ± 1.9	16.27	15.9 ± 2.1	15.82

3 Results

3.1 The Reality of Weight for Age

The weight status of children from 2-to-6-years old from Lao Cai province is presented in Table 3.

As shown in Table 3, children weight increased with an increase in age in both genders. For the boy, the increase in the weight was from 11.3 ± 1.4 kg at the age of 2 to 16.8 ± 1.9 kg at the age of 6, while the increase in weight of girl was 10.1 ± 1.8 kg to 15.9 ± 2.1 kg, respectively ($P < 0.05$). The weight of boys was greater than that girls in each age ($p < 0.05$) (Table 4).

3.2 The Reality of Height for Age

The height status of children from 2-to-6-years old from Lao Cai province is presented in Table 3.

Table 5 Height for age for children

Age	Boys		Girls	
	n	$X \pm SD$	*n*	$X \pm SD$
2	329	86.12 ± 3.5	318	84.51 ± 3.6
3	326	88.22 ± 3.7	315	86.82 ± 4.2
4	312	96.76 ± 4.2	313	98.12 ± 3.6
5	303	105.35 ± 4.7	310	103.19 ± 4.4
6	311	109.6 ± 5.1	308	107.23 ± 5,2

(*n* number of children for age and sex)

The data in Table 5 show that the height of boys is always higher than girls in all age groups. This result is consistent with gender characteristics. The height for age for children at the age of 2–6 followed the rule of body growth of people (Table 6).

3.3 Malnutrition Status

A WHO Plus 2007 software is used to assess nutrition status of children in this study. As shown in Table 7.

The data in Table 7 shows that the total percentage of severe underweight children was 2.23%. The total percentage of underweight children was also quite high (9.18%). Among the different ages, at the age of 2, 3, 4, and 5, the percentage of underweight children was 1.11%, 1.46%, 1.813.6%, and 2.13%, respectively.

Malnutrition status in children's children is shown in Table 8.

The total percentage of height malnutrition of children was 69.40% in which the percentage of severe stunting children was 31.82% and that of stunting children was 37.58%. Both severe stunting and stunting percentages were high at all the age in which severe stunting rate was high at the age of 2 and 3 while stunting rate was more pronounced at the age of 4, 5, and 6.

Malnutrition status in BMI for age for children is shown in Table 9.

Table 9 shows that the percentage of BMI malnutrition was 4.15% (in which severe wasted was 3.25% and wasted was 0.9%), the percentage of BMI normal was 55.26%, the percentage of BMI possible overweight was 39.28%, the overweight was 2%, and specially the obesity was only 0.15%.

Table 6 Comparison between this study and data from Vietnam Ministry of Health in 2003 [11]

Age	Mean height of age for boys		Mean height of age for girls	
	This study	VN Ministry of Health in 2003	This study	VN Ministry of Health in 2003
2	86.12 ± 3.5	85.12	84.51 ± 3.6	83.11
3	88.22 ± 3.7	87.36	86.82 ± 4.2	83.97
4	96.76 ± 4.2	94.32	98.12 ± 3.6	93.78
5	105.35 ± 4.7	100.77	103.19 ± 4.4	100.18
6	109.6 ± 5.1	106.12	107.23 ± 5,2	105.40

4 Discussion

4.1 The Reality of Weight for Age

The result in Table 3 shows that: sex affects the weight of children from 2-to-6-years old. Because babies are not breastfed at this stage, and the increase in weight completely depends on the factors of their environment. In-depth interviews of parents' children show that there are more than 70% of the boys confirming that they eat more than girls.

In comparison with a data from VN Ministry of Heath (2003), mean children's weight in this study was higher at all of ages [4]. The reasons for this difference might be an economy development status in research area. On the other hand, from 2003 to now (2020), nearly 20 years of socio-economic conditions have changed for the better.

4.2 The Reality of Height for Age

If comparison with data of Ministry of Health (2003), the children's mean height in this study was higher at all of ages ($P < 0.05$) (Table 6). The reason for these differences was due to the difference in research areas and research time. In the present study, living standard, economic and social conditions, health care and nutrition regime of children in Lao Cai province were richer than those in the study conducted by Ministry of Health, even though that study was carried out nearly 20 years ago.

4.3 Malnutrition Status

The children under severe underweight and underweight status accounted for quite a large rate. However, the total percentage of children under 5-year old suffering from

Table 7 Malnutrition status in weight for age for children

Age	Severe underweight		Underweight		Normal	
	n	%	*n*	%	*n*	%
2	23	0.73	35	1.11	440	13.99
3	15	0.47	46	1.46	461	14.65
4	13	0.41	57	1.81	551	17.51
5	11	0.34	67	2.13	652	20.73
6	9	0.28	84	2.67	681	21.65
Overall	71	2.23	289	9.18	2785	88.53

(*n* number of children for age and sex)

both severe underweight and underweight status in the present study was higher than data from Vietnam Ministry of Health in 2003. These results indicate that although economic and social conditions in this study are better.

Mercedes de Onis et al. have been reported that though malnutrition percentage rapidly reduced in many developing countries, however, in some developing countries, this rate tended to increase [11]. The results in Table 8 imply that living standard of people in Lao Cai was quite poor.

The result in Table 9 was higher than that of the VN Ministry of Health in 2003 [11]. Although the severe wasted and wasted status of children were observed, however, the overweight and obese status of children in this research area were also found. The percentage of children under overweight condition was quite low (2.0%). So that, this situation is a dual burden of malnutrition that Lao Cai province is facing now.

Table 8 Malnutrition status in height for age for children

Age	Severe stunting		Stunting		Normal	
	n	%	*n*	%	*n*	%
2	212	6.74	301	9.57	203	6.45
3	210	6.67	298	9.47	206	6.55
4	196	6.23	205	6.51	195	6.20
5	198	6.29	190	6.04	186	5.91
6	185	5.88	188	5.97	172	5.46
Overall	1001	31.82	1182	37.58	962	30.58

(*n* number of children for age and sex)

Table 9 Malnutrition status in BMI for age for children

Age	Severe wasted		Wasted		Normal		Possible overweight		Overweight		Obesity	
	n	%	*n*	%	*n*	%	*n*	%	*n*	%	*n*	%
2	23	0.73	2	0.06	315	10.01	219	6.96	20	0.63	0	0
3	21	0.66	5	0.15	326	10.36	223	7.09	18	0.57	0	0
4	15	0.47	7	0.22	342	10.87	251	7.98	13	0.41	0	0
5	5	0.15	8	0.25	366	11.63	265	8.42	6	0.19	1	0.03
6	10	0.31	7	0.22	389	12.36	278	8.83	6	0.19	4	0.12
Overall	74	2.35	29	0.9	1738	55.26	1236	39.28	63	2.00	5	0.15

5 Conclusion

The anthropometric parameters of children in Lao Cai province followed the rules of the body growth of Vietnamese people. These indices of children were higher than general values of children Vietnam over 20 year ago. However, anthropometric indices of children were lower than general values of children Vietnam. Malnutrition status and severe malnutrition status of children in Lao Cai province are high compared to other provinces in Vietnam.

References

1. UNICEF, Situation Analysis of Woman and Children in Viet Nam, UNICEF Hanoi, pp. 108–109 (1990)
2. Mei, Z., Grummer-Strawn, L.M., Thompson, D., Dietz, W.H.: Shifts in percentiles of growth during early childhood: analysis of longitudinal data from the California child health and development study. Pediatrics **113**(6), 617–627 (2004)
3. World Health Organization—Department of Nutrition for Health and Development, WHO Child Growth Standards: Training course on child growth assessment: C. Interpreting growth indicators, Geneva (2006)
4. Wong, H.J., Moy, F.M., Nair, S.: Risk factors of malnutrition among preschool children in Terengganu, Malaysia: a case control study. BMC Public Health (2014)
5. Tam, V.V., Nhan, N.H., Tinh, H.Q., Hung, N.P.: The impacts of malnutrition status and relevant factors on preschool children in Cao Ma Po Commune, Quan Ba District, Ha Giang Province. VNU J. Sci. Natural Sci. Technol. **32**(1S), 368–375 (2016)
6. National Institute of Nutrition. Malnutrition percentage of children under 5 years old in the area of Vietnam in (2014)
7. UNICEF: Underlying causes of under nutrition: food insecurity. Food insecurity. In. Geneva
8. Tinh, H.Q., Nhan, N.H., Linh, N.T.T.: WHO software used to study some anthropometric indices. J. Milit. Med. **34**, 1–5 (2009)
9. Quintero, D. et al.: Workload optimized systems: tuning POWER7 for analytics. Abstract
10. KD nuggets Annual Software Poll: Analytics/Data mining software used? KD nuggets. May 2013
11. Ministry of Health: Biological Indicies of Vietnam in 1990s. Medical Publishing House, Hanoi (2003)

Predictive Analysis of Emotion Quotient Among Youth

Shrinivas D. Desai, Akula Revathi, S. L. Aishwarya, Aishwarya Mattur, and Aishwarya V. Udasimath

Abstract Emotion Quotient (EQ) and Intelligent Quotient (IQ) are the two most vital components for one to excel in this competitive world. These two components are key factors among youth to shape their career and their future. However, predicting young student's EQ is one of the most challenging tasks due to lack of appropriate information which directly or indirectly affects EQ. The purpose of this paper is to predict EQ among undergraduate students (age between 19 and 21 yearss) by considering all the key factors which directly or indirectly influence their EQ. Another objective of this paper is to compare and analyze the EQ of male and female students. Factors such as usage of micro-blogging sites and activeness in participating curricular activities are considered during experimentation. Participants of the study consisted of 411 (242 male, 169 female) undergraduate students selected randomly from different disciplines of engineering during even semester of the academic year 2018–2019 in KLE Tech University, Hubballi. The study also shows interesting and curious aspects of EQ among male and female students.

Keywords Emotion Quotient · Intelligence Quotient · Emotion Quotient value · Social microblogging · Financial status

---

S. D. Desai (✉) · A. Revathi · S. L. Aishwarya · A. Mattur · A. V. Udasimath
School of Computer Science & Engineering, KLE Technological University, Hubli, Karnataka 580031, India
e-mail: sd_desai@kletech.ac.in

A. Revathi
e-mail: akularevathi9@gmail.com

S. L. Aishwarya
e-mail: aishwarya4404@gmail.com

A. Mattur
e-mail: aishwaryasukhadev31@gmail.com

A. V. Udasimath
e-mail: aishwaryavudasimath@gmail.com

© The Editor(s) (if applicable) and The Author(s), under exclusive license to Springer Nature Singapore Pte Ltd. 2021

A. P. Pandian et al. (eds.), *Proceedings of International Conference on Intelligent Computing, Information and Control Systems*, Advances in Intelligent Systems and Computing 1272, https://doi.org/10.1007/978-981-15-8443-5_49

1 Introduction

Emotion Quotient (EQ) is an ability to recognize our emotions and emotions of others as well as to handle our emotions, to understand its effect and to use that knowledge to guide our thoughts and behaviors. EQ involves recognizing our emotions and emotions of others, managing emotions, being able to control our emotions effectively, developing empathy (understanding and sharing the emotions of others), motivating ourselves (self-motivation), and focusing toward our goals and achieving them. EQ is required because to succeed in one's career and to excel in this corporate world where working will be in teams (in MNC's) and emotions of others should be able to understand and react accordingly in such a way that everyone's emotions, perceptions should be respected and respond positively [1]. Hence, EQ is a leadership quality which helps in team management. The benefit of EQ includes increased self-awareness, help in managing the stress, help in self-motivation, helps us to make decisions, improves communication, and developing a better relationship with others. The objectives of the project include to build an automatic system that predicts the EQ of the students and classifies into two classes (happy, sad) based on the EQ score and to analyze the EQ scores of male and female students department wise. In today's competitive world, most of the students will be busy in attending coaching classes, tuitions, and exams and fail to understand their emotions and emotions of other, as a result, their EQ level will be less. It is proved that IQ + EQ = Success which infers that both Intelligent Quotient (IQ) and EQ are required to succeed in one's career. It is observed that during campus placement, most of the students do clear preliminary technical rounds; however, fail to clear the HR round. When investigated for this performance, it is observed that students were unable to answer well to the questions on emotional and intelligence quotient. Thus, the proposed study of analyzing influential factors of EQ among youth is the need of the hour. Youth with higher EQ level shape their career by making the proper decision [2].

2 Review of Related Work

Table 1 presents a review of the literature. Over the last five year, literature reported where authors have experimented using ANN and machine learning techniques to predict EQ and IQ among adolescents. From the observations of the review table, study is carried out using limited attributes, either they are academic performance-related or social behavior. But research gap of integrating activities on social networks with academic, cultural, and extra co-curricular activities is hardly evidenced.

Table 1 Review of literature

#	Author	Proposed methodology	Dataset	Results
1	Eduardo Fernandes et al. [3]	Gradient boosting	High school students in public schools of Brazil during 2015–2016	The most important variables for determining academic success are neighborhood, city, and age (limited attributes of academic details)
2	Samuel Peter James et al. [4]	Bagging, logistic, random forest, stacking	Total number of students are 1180 (500 girls and 680 boys)	A positive correlation between EQ and academic achievements of secondary school students (emotional based learner data has a less improvement in predictive accuracy)
3	Jayashree et al. [5]	Percentage analysis, correlation, and chi-square test	College students are 150 88 male, 62 females	Relationship between co-curricular activities and the EQ of students, which showed a positive relation (Emotional competency has a greater impact on EQ than emotional sensitivity and emotional maturity)
4	Amirah Mohammed Sahiri et al. [6]	Decision tree, ANN, Naive Bayes, KNN, SVM	Generated own dataset by surveying students	Neural network has highest prediction accuracy of 98% (focused only on academic details)
5	Joibari et al. [7]	T-chart, Pearson coefficient correlation	Number of students were 355,800, Boys—20,090 and Girls—155,710	Accuracy is 91%. It is a positive correlation (female students should be paid more attention than male students)

3 Methodology

The objective of proposed work is to "design and develop a model for predictive analysis of emotional quotient among youth."

The hypotheses defined are:

For this, the hypotheses defined were:

- Social micro-blogging Web site does not have a significant impact on EQ.
- The family background does not make a significant impact on EQ.
- Financial status does not make a significant impact on EQ.
- Friends circle does not have a significant impact on EQ.
- Gender of the students does not make a significant impact on EQ.
- Social micro-blogging Web site has a significant impact on EQ.
- Family background has a significant impact on EQ.
- Financial status has a significant impact on EQ.
- Friends circle has a significant impact on EQ.
- Significant gender differences exist for EQ.

The step by step phases involved in the study is presented below.

(1) Data collection: In this phase, a survey is carried out to collect response from almost 2000 students by following a systematic approach for data collection. These responses involve around 24 demographic questionnaires, primarily based on Technical Education Quality Improvement Program-Student Learning Assignment p(TEQIPSLA) [8–12] which is composed of several attributes related to each student, such as student characteristics, attitudes, and behavior, personal, and educational background characteristics, time spent in and out of class. Study habits (during and after class hours), participation in research and extracurricular activities, psychological traits, social networks, finance, expectations for the future work and study. Several online EQ tests' patterns are reviewed [13–16] to finalize the questions. The satisfaction level for each attribute is sort in the range of 1–10, where 1 is highly unsatisfactory and 10 being highly satisfactory. The data collected is structured in nature; however, if data available is in the unstructured pattern, then it has to be converted to proper structure then shall be employed for experimentation.
Data is collected in three phases, in the first phase, 789 students responded, in the second phase, 678 students, while in the third phase, 502 students responded. Most of the student had an inhibition to provide certain personal details and hence three phases for collecting the data. Out of all these responses, finally, 411 response data which are realistic and relevant is considered for further experimentation. All these data are from 411 adolescents (242 males, 169 females) aged 19 to 21 years (M age 20.3, SD = 1.2 years).
Dataset includes 24 (features) information, about 5 factors of EQ; addiction to social micro-blogging Web site, Family background and its influence, financial status, Academic performance, and gender bias. Features for each of these factors are listed in Table 2. A snippet of data collected from students is as shown in Fig. 1.
(2) Data preprocessing: Among data received, 55 responses were having invalid ids, and incorrect information related to social behavior. After preprocessing, the total response data is 411, and the same is considered for further experiment.
(3) Calculation of EQ: EQ is calculated based on the user score as in Eq. 1.

Table 2 List of features

S. No.	Factors affecting EQ	Feature
1	Addiction to social micro-blogging Web site	Microblogging
2		Chat application
3		Internet usage
4	Family background and its influence	Initiation
5		Sense of humor
6		Misbehavior
7	Financial status	Positive attitude
8		Seriousness
9		Self-evaluation
10		Practical knowledge
11		Stress management
12		Time management
13	Academic performance	Academic performance
14		Department activities
15		Paper presentation
16		Obedience
17		Study habit
18		Homework
19		Attendance
20		Reading habit
21	Gender bias	Mood fluctuation
22		Dependency
23		Physical fitness
24		Emotion dependency

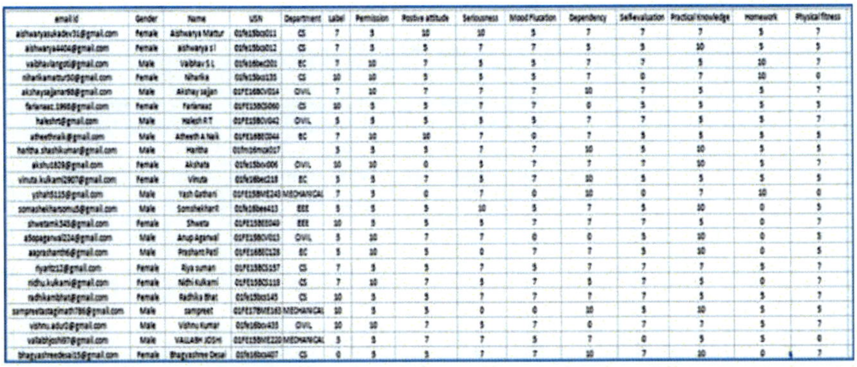

Fig 1: Snippet of the dataset

$$EQ = \sum \frac{\text{Scoreofindividualattributes}}{\text{Totalnumberofattribute} * 10} \quad (1)$$

3.1 Design

Figure 2 presents the architecture diagram of the proposed system. The process begins with structured data, which is collected through the survey (questionnaires) by adult students. In next step, preprocessing and feature extraction are carried out as explained in the previous section.

4 Results and Analysis

To verify the hypothesis claimed in the prior section, the following experiments are carried out.

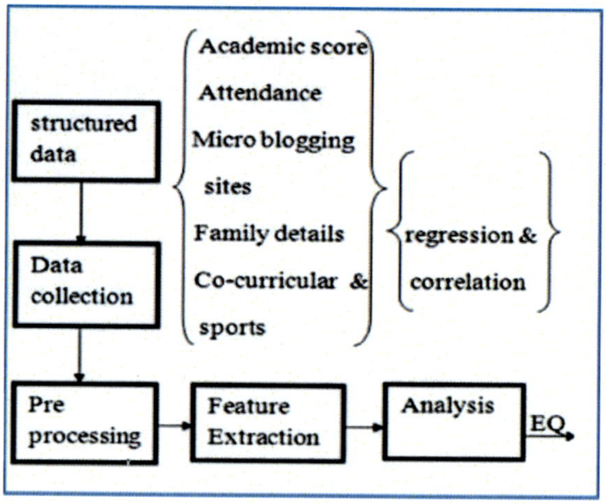

Fig 2: Architecture diagram of the proposed system

Table 3 Reliability of the factors of EQ

Factors of EQ	Cronbach Alpha
Addiction to social micro-blogging Web site	0.86
Family background and its influence	0.81
Financial status	0.82
Academic performance	0.84
Gender bias	0.80

4.1 *Reliability Assessment*

The concept of reliability has been used to measure internal consistency using Cronbach Alpha. It is test reliability indicating the extent to which individual differences in test scores are attributable to 'true' differences in the characteristics under consideration and the extent to which they are attributable to chance errors.

In this study, reliability for EQ in the form of Cronbach Alpha was found to be 0.86–0.80 for the 5 factors with an average internal consistency of 0.83. The values of Cronbach alpha for all the factors of EQ are presented in Table 3.

The factor social micro-blogging has high value, whereas gender bias showed low value. Results for factors selected showed a high index. The result indicates that the EQ possess homogeneous items of high quality, showing that the scale is a reliable instrument due to high internal consistency. Therefore, results from the scale will be highly reliable. Quinone and coworkers (2014) obtained social skills having a high reliability coefficient and obtained an overall coefficient of 0.84 indicating a high level of internal consistency.

4.2 *Assessment of Degree of Association of Factors*

To assess the relation between variables, the correlation coefficient is determined. This determines if the variables are associated and if associated, to what degree. Correlation coefficient expresses the degree of relationship between two sets of scores. To test the association of EQ and its factors, a detailed set of statistical analysis was conducted first being a confirmatory Pearson's Correlation as seen in Tables 4, 5, 6, 7, and 8.

Addiction to social micro-blogging Web site was found to be highly significant with EQ, symbolizing social network plays a vital role in emotional quotient. Applications like Twitter, Facebook, and TikTok are key players in influencing emotion component among youth. The correlation coefficient of micro-blogging site feature with a correlation of 0.367 with EQ complements to the above interpretation (Table 4). A correlation coefficient of 0.471 for the sense of humor and 0.396 for misbehavior with EQ is a clear indication of the influence of family background on EQ. Most of the students in the Indian context come from varied culture and parental

Table 4 Correlations of EQ and factors of addiction to social network

	Micro-blogging site	Chat application	Internet usage
Chat application	−0.153 0.002		
Internet usage	0.183 0.000	−0.127 0.010	
EQ	0.367 0.000	0.133 0.000	0.337 0.000

Table 5 Correlations of EQ and factors of family background and its influence

	Initiation	Sense of humour	Misbehavior
Sense of humor	0.113 0.022		
Misbehavior	0.023 0.640	0.178 0.000	
EQ	0.274 0.000	0.471 0.000	0.396 0.000

Table 6 Correlations of EQ and factors of financial status

	PA	S	SE	PK	SM	TM
Seriousness	0.102 0.040					
Self-evaluation	0.166 0.001	0.355 0.000				
Practical knowledge	0.120 0.015	−0.098 0.047	−0.011 0.817			
Stress management	0.110 0.025	0.188 0.000	0.194 0.000	0.148 0.003		
Time management	−0.069 0.163	−0.001 0.825	0.089 0.072	0.022 0.655	−0.052 0.292	
EQ	0.266 0.000	0.219 0.000	0.269 0.000	0.426 0.000	0.418 0.000	0.162 0.001

PA positive attitude, *PK* practical knowledge, *S* seriousness, *SM* stress management, *SE* self-evaluation, *TM* time management

bonding. Extreme conditions of this feature are witnessed in students as either they present a good sense of humor or misbehave in the class (Table 5).

The correlation coefficient of 0.426 for practical knowledge and 0.418 for stress management is found to be the most dominating features in the financial aspects of student and his/her family. It is observed that youth with financial constraints especially dependent on scholarship are in stress and this influences EQ of youth (Table 6).

Table 7 Correlations of EQ and factors of academic performance

	AP	DA	PP	A	O	SH	RH
Department activities	0.192 0.000						
Paper presentation	0.058 0.241	0.194 0.000					
Attendance	0.011 0.831	0.139 0.005	−0.030 0.538				
Obedience	−0.153 0.002	−0.155 0.002	−0.008 0.873	−0.154 0.002			
Study habit	0.041 0.412	−0.178 0.000	−0.057 0.247	−0.205 0.000	0.211 0.000		
Reading habit	0.001 0.978	−0.000 0.999	0.083 0.094	0.022 0.663	0.149 0.002	0.074 0.135	
EQ	0.193 0.000	0.262 0.000	0.133 0.007	0.520 0.000	0.073 0.139	0.138 0.005	0.280 0.000

AP academic performance, *A* attendance, *DA* department activities, *O* obedience, *SH* study habit, *PP* paper presentation, *RH* reading habit

Table 8 Correlations of EQ and factors of gender bias

	Mood fluctuation	Dependency	Physical fitness	Emotion dependency	Permission
Dependency	0.103 0.037				
Physical fitness	−0.066 0.183	0.050 0.313			
Emotion dependency	−0.026 0.594	0.177 0.000	-0.059 0.229		
Permission	0.057 0.247	−0.005 0.912	0.047 0.342	0.059 0.236	
EQ	0.260 0.000	0.270 0.000	0.307 0.000	0.068 0.167	0.286 0.000

Table 9 Regression analysis for factors of addiction to social networking

Variables	Coefficient	Mean	Std deviation	Sig	*T* value	*F*
Constant	0.559568			0.000	38.3743	47.2574
Micro-blogging site	0.009398	14.398	3.267	0.000	7.8838	
Chat application	0.005794	16.709	3.71	0.000	5.1544	
Internet usage	0.005790	16.006	2.637	0.000	6.9288	

$R^2 = 0.2583$ R-Sq(adj) $= 0.2529$

Table 10 Regression analysis for factors of family background

Variables	Coefficient	Mean	Std deviation	Sig	*T* value	*F*
Constant	0.465989			0.000	26.4316	80.1808
Initiation	0.012763	5.628	0.929	0.000	5.6252	
Sense of humor	0.012132	9.002	1.709	0.000	9.6773	
Misbehavior	0.009069	8.207	1.891	0.000	8.0533	

$R^2 = 0.3715$, R-Sq(adj) = 0.3668

Table 11 Regression analysis for factors of financial status

Variables	Coefficient	Mean	Std deviation	Sig	*T* value	*F*
Constant	0.435219	–	–	0.000	21.474	49.0331
Positive attitude	0.008842	34.787	4.673	0.000	4.2560	
Seriousness	0.007524	34.860	4.685	0.001	3.3958	
Self-evaluation	0.007131	35.142	4.615	0.004	2.9268	
Practical knowledge	0.008310	32.485	4.041	0.000	9.5858	
Stress management	0.006247	32.678	3.692	0.000	7.6668	
Time management	0.004917	34.376	4.587	0.000	4.4965	

$R^2 = 0.4214$, R-Sq(adj) = 0.4128

Table 12 Regression analysis for factors of academic performance

Variables	Coefficient	Mean	Std deviation	Sig	*T* value	*F*
Constant	0.368800			0.000	16.4424	55.679
Academic performance	0.003836	6.827	2.127	0.000	4.1689	
Department activities	0.006216	6.552	1.701	0.000	5.2415	
Paper presentation	0.008428	5.224	0.631	0.006	2.7385	
Attendance	0.017693	8.275	1.686	0.000	15.1923	
Obedience	0.002690	6.187	2.562	0.000	3.4372	
Study habit	0.005735	7.105	2.271	0.000	6.5086	
Reading habit	0.005758	6.586	2.060	0.001	6.1535	

$R^2 = 0.4916$, R-Sq(adj) = 0.4828%

Table 13 Regression analysis for factors of gender bias

Variables	Coefficient	Mean	Std deviation	Sig	*T* value	*F*
Constant	0.470376			0.000	13.9010	55.679
Mood fluctuation	0.013774	27.191	3.979	0.000	5.7600	
Dependency	0.005949	26.743	3.466	0.000	5.2379	
Physical fitness	0.008021	26.305	3.537	0.000	7.1760	
Emotion dependency	0.004941	28.442	4.083	0.374	0.8892	
Permission	0.005470	26.157	3.163	0.000	6.1052	

$R^2 = 0.2927$, R-Sq(adj) = 0. 2839

To academic performance, attending all classes is found to be significantly correlated with EQ with a correlation coefficient of 0.520, while physical fitness is highly correlated with EQ in gender bias factor. Youth are more conscious of their physical fitness. Fit or fat influence a lot on psychological as well as emotional quotient (Tables 7 and 8).

4.2.1 Tests of Regression

As the objective of this paper is to identify and assess the effect of factors on EQ, the method of general regression analysis has been chosen, as it helps in assessing the individual and the combined effect of independent variables (addiction to social micro-blogging Web site, Family background and its influence, Financial status, Academic performance, gender bias) on the dependent variable (EQ).

An F value for the regression analysis compares different linear model, and it can assess multiple coefficients simultaneously. Since, $p < 0.005$, they are significant. The R^2 value is less than 50% which is relatively low. It has been observed that the fields that attempt to predict human behavior such as psychology have R^2 values less than 50%. It is mainly because humans are simply harder to predict compared to any other physical processes. Hence, it is necessary to analyze the residual plots.

4.2.2 Residual Analysis

To study the best regression model for the proposed study, a residual plot is presented. If the points in a residual plot are randomly dispersed around the horizontal axis, a linear regression model is appropriate for the data; otherwise, a nonlinear model is more appropriate. Figure x presents the residual plot for all five factors affecting EQ.

From Fig. 3, it is observed that points in a residual plot are randomly dispersed around the horizontal axis for all the factors, and a linear regression model is appropriate for the data. In the figure, none of factors presented heteroscedastic data (points at widely varying distances from the line), nor data that is nonlinearly associated nor datasets with many outliers.

To summarize, in the proposed study, all the five factors considered to be most influencing factors on EQ were found to be statistically significant through Cronbach Alpha test where the alpha varied between 0.80 and 0.86.

From the regression analysis, all four major factors such as addiction to social micro-blogging Web site, family background and its influence, financial status, and academic performance found to be effective on EQ with a minimum R^2 value of 25.83 for addiction to social network and family background to 49.16 for academic performance. Emotion dependency feature with p-value 0.374 indicates that gender is not a significant factor in EQ. Residual plot justifies and complements regression analysis. To summarize all the experiment results, Table 14 is presented.

i) Addiction to Social micro-blogging website

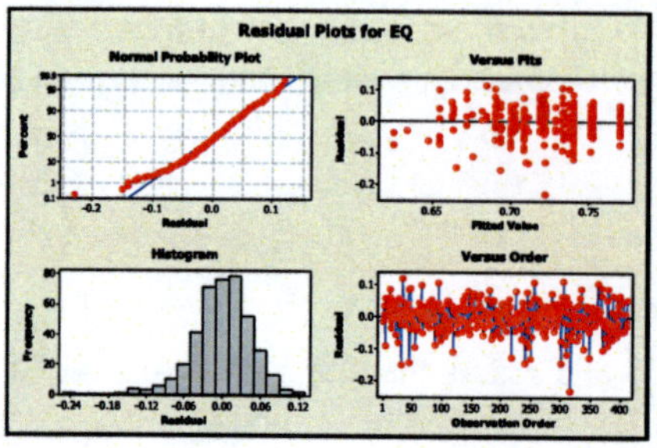

ii) Family background and its influence

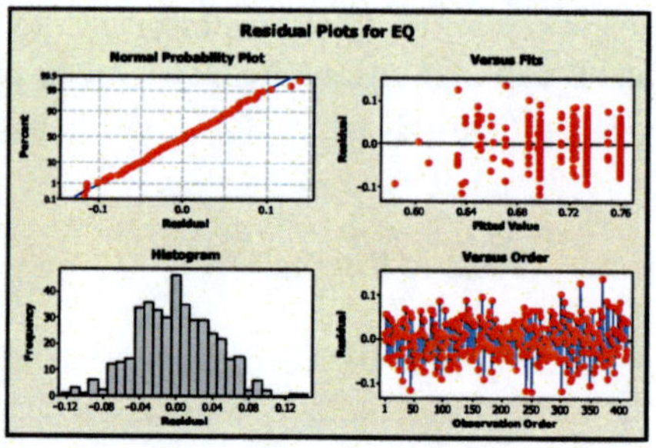

iii) Financial status

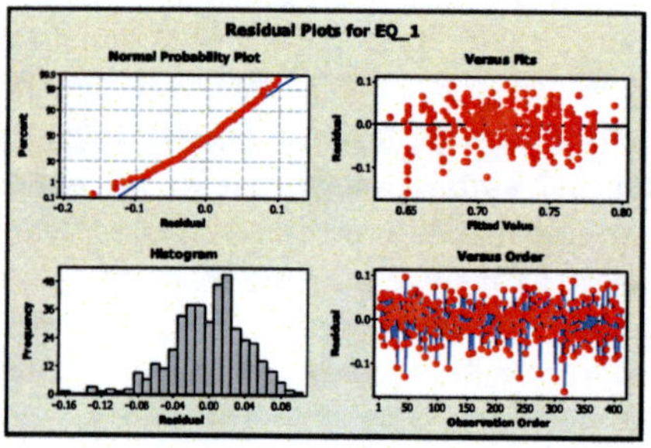

iv) Academic performance

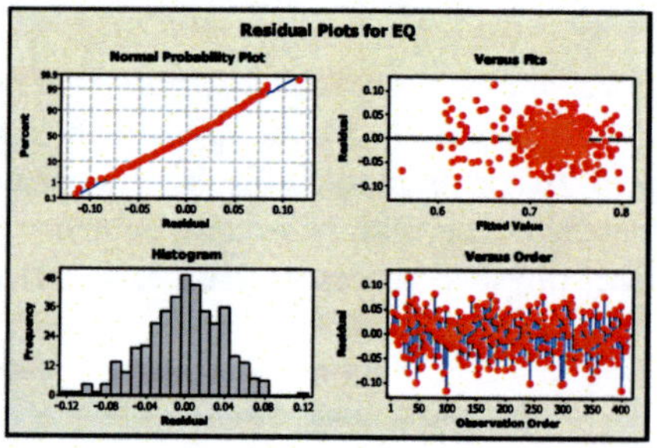

v) Gender bias

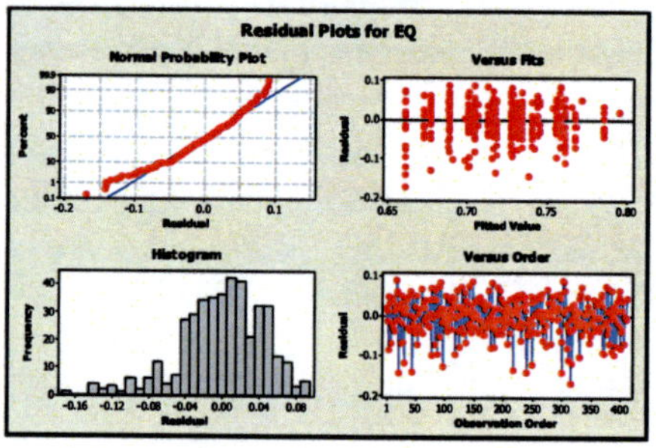

Fig. 3 Residual plots of factors affecting EQ

5 Conclusion and Future Scope

From the analysis of all experiments, it is very much evident that social networking, family background, and friends circle significantly play an important role and have an impact on EQ. Financial background of student does not significantly contribute

Table. 14: Summary of analysis

#	Hypothesis	Outcome
1	Social micro-blogging Web site has a significant impact on EQ	Significant
2	Family background has a significant impact on EQ	Significant
3	Financial status has a significant impact on EQ	Not significant
4	Friends circle has a significant impact on EQ	Significant
5	Significant gender differences exist for EQ	Not significant

to EQ. It is interesting to know the fact that EQ is not gendered biased. Considering the implications of social micro-blogging, friendship circle, and family background toward EQ among university students (youth), the current research may assist in augmenting organizational behavior in general and boost the productivity of an individual in particular which are both most premium ingredients for the growth of education system in India specifically. It is further suggested that personality development programs can periodically be initiated for personality enhancement and emotional skills of the students. It will help the university administration to achieve an enriched environment for learning, academic improvements, and developing quality personality characteristics. As a future scope, the proposed model of predicting EQ shall be integrated with student's academic monitoring system with access to university teachers/student advisors or counselors for effective counseling and progress monitoring.

References

1. Palmer, B.R., An analysis of the relationships between various models and measures of emotional intelligence Diss. Swinburne Univ. Tech. School of Biophysical Sciences and Electrical Engineering, Centre for Neuropsychology (2003)
2. Brown, C., George-Curran, R., Smith, M.L.: The role of emotional intelligence in the career commitment and decision-making process. J. Career Assess. **11**(4), 379–392 (2003)
3. Eduardo, F., et al.: Educational data mining: Predictive analysis of academic performance of public school students in the capital of Brazil. J. Bus. Res. **94**, 335–343 (2019)
4. James, S.P., Ramasubramanian, P., Angeline, M.D.: Student learning context analysis by emotion intelligence with data mining tools. Int. J. Intell. Eng. Syst. 174–183 (Dec 11, 2018)
5. Jayashree, S., Vimal Ram, S., Das Arun, M.: Emotional intelligence among college students with special reference to Kerala. Int. J. Pure Appl. Mathem. **118**(20), 4583–4589 (2018)
6. Shahiri, A.M., Husain, W.: A review on predicting student's performance using data mining techniques. Proced. Comput. Sci. **72**, 414–422 (2015)
7. Joibari, A., Mohammadtaheri, N.: The study of relation between emotional intelligence and students' academic achievement of High Schools in Tehran city. Proc. Soc. Behav. Sci. **29**, 1334–1341 (2011)
8. "TEQIP- SLA", *pec.ac.in*, https://pec.ac.in/teqip/mandatory-survey-2016
9. Carolyn, M., et al.: Emotional intelligence predicts academic performance: a meta-analysis. Psychol. Bull. **146**(2), 150 (2020)
10. Quiñones, M.M., Haddock, R.Y., Rosario-Hernández, E.: Development and validation of an emotional intelligence scale for Puerto Rican students

11. Shamira, M., Mohanty, R.P.: Factors affecting emotional intelligence: an empirical study for some school students in India. Int. J. Manag. Educ. **3**(1). 8–28 (2009)
12. Singh, T., Verma, M.K., Singh, R.: Role of emotional intelligence in academic achievement: an empirical study on engineering students. Strategic Approaches for Human Capital Management and Development in a Turbulent Economy. IGI Global, 255–263 (2014)
13. "Emotion-quotient", memordo.com: https://memorado.com/emotionalquotient.
14. Joshi, S.V., Srivastava, K., Raychaudhuri, A.: A descriptive study of emotional intelligence and academic performance of MBBS students. Proced. Soc. Behav. Sci. **69**, 2061–2067 (2012)
15. Narayana, M.S., Narasimham, B.V.: Dynamics of emotional intelligence and job performance. Int. J. Emerg. Res. Manag. Technol. **7**(1), 57–71 (2018)
16. Nath, S., Ghosh, S., Das, S.: Relation between intelligence, emotional intelligence, and academic performance among medical interns. Open J. Psych. Alli. Sci. **6**(2), 96–100 (2015)

Bots, Internet of Things and Threats to Personal Data in the Technological Era

Amelec Viloria, Nelson Alberto, and Carlos Alberto Jiménez Cabarcas

Abstract Risks to data security and integrity increase when data is managed through an explosion of mobile applications that connect users to a complex range of services, resources and data in the cloud. Privacy policies that inform about such risks and recommend safe practices under preventive criteria are the first line of defense for a company that manages a website or provides services through a mobile application. The rapid growth of opportunities linked to Big Data has occurred on what many consider a gray area for regulatory purposes, without well-established ethical standards and codes of conduct that are effective for adequately protecting consumers from the abusive use of their personal information by advertisers. Many companies lack good practice guidelines and adequate privacy policies to prevent the disclosure of confidential information. Unfortunately, illicit actors often operate on the same blocks and sets of data as legitimate users, and their annoying or dangerous behavior cannot be criminalized because of a weak or inadequate regulatory framework. This paper analyzes the emerging threats to privacy arising from the widespread use of applications and bots, in a context of growing demand for free cloud services managed through mobile devices.

Keywords Bots · Internet of Things · Technological era · Privacy

A. Viloria (✉) · C. A. J. Cabarcas
Universidad de la Costa, St. 58 #66 Barranquilla, Atlántico, Colombia
e-mail: aviloria7@cuc.edu.co

C. A. J. Cabarcas
e-mail: cjimenez40@cuc.edu.co

N. Alberto
Universidad Tecnológica, San Pedro Sula, Honduras
e-mail: nelso@unitet.edu

© The Editor(s) (if applicable) and The Author(s), under exclusive license to Springer Nature Singapore Pte Ltd. 2021

A. P. Pandian et al. (eds.), *Proceedings of International Conference on Intelligent Computing, Information and Control Systems*, Advances in Intelligent Systems and Computing 1272, https://doi.org/10.1007/978-981-15-8443-5_50

1 Introduction

Economic analyses by 2018 estimate that the global market trend in relation to services through mobile devices connected to the Internet will manage some 134 billion euros in advertising spending, almost double the estimate for 2016 and more than the total amount spent on newspapers, magazines, movies and outdoor advertising [1].

The infrastructure for this market is provided by cloud computing and data storage systems, which are accessible through certain applications installed on lightweight mobile devices with limited resources, but sufficient to interact with the systems (bots) that drive the data traffic, receive information from the physical environment and manage access to multiple services [2].

Each element integrated into this global communications infrastructure poses specific risks in terms of privacy and security (including for human users, with very diverse privacy cultures and levels of technological literacy) [3]. Among the incentives for the development of this structure is the potential for increased surveillance and control over consumers, users and citizens by state and private actors [4, 5].

The aim of this study is to identify emerging risks to the privacy of users of applications and mobile devices connected to cloud services and social networks, in a context of global services that record a high percentage of interactions between countries that lack reciprocity in the recognition of the right to privacy or that reflect very different cultures in their privacy policies.

2 App

Applications (short: apps) are programs designed to perform specific functions that can be run on desktops or on mobile devices (cross-platform). Many of them only require a browser to run (web apps), although not all require an Internet connection. The term is often used to describe any software that is not a complete program, with simpler versions intended for use on mobile devices, with small screens and touch interaction. Hybrid apps have a desktop interface and direct access to the hardware and other connected devices, but can operate offline [6].

Mobile applications are obtained from repositories associated with various platforms, from which free or paid apps can be downloaded. They are normally accessed through the devices on which they can be installed, or from websites where registered users queue them up for downloading the next time they use their mobile device. The most popular app repositories are Google Play and Appstore (Amazon), for mobile devices that work with Android. iPhone and iPad users can get them through iTunes. For MacOS computers, the place is Mac App Store, and for Windows users, the Windows Store [7, 8].

3 Bots

Used as a robot apheresis, the standard definition of the term bots (a stand-alone program or software agent, designed to mimic the behavior of human users in specific, repetitive tasks) does not provide a very precise idea of the variety of applications with which it can be associated. They are usually programmed to work in networks, especially on the Internet, and interact with other systems or users in tasks such as editing texts, moderating conversations, answering frequently asked questions—about a service, computer resource or web content search, as conversational bots do—sending e-mails or dynamizing video games, among others. The more complex bots are designed for very specific tasks, in which they compete with human users in the analysis of options and possible strategies (to win in role-playing games, for example) [9].

For complex applications and multiple services in the cloud, the services of the bots may overlap with functions of expert systems used to assist human professionals in clinical, financial, legal or insurance decisions, for example [10].

The integration of artificial intelligence and machine learning allows the design of bots capable of learning from their interactions with users and improving the quality of their responses. The key will be in the platform or database from which they can extract their answers and in the accuracy of the statistics on user needs that they can produce. But it is relatively easy to design bots to extract personal information and industry data that compromise the privacy of individuals and businesses or threaten the security of essential services.

4 Internet of Things (IoT), Datification and Big Data

The concept of the term Internet of Things (IoT), coined by Kevin Ashton, applies to a cyber-physical system where objects can be connected to the Internet through ubiquitous sensors, providing them with decisive functionalities in terms of availability, access, distribution efficiency and contextualization. It originates from the automatic identification standard (Auto-ID) for radio frequency identification (RFID) sensors. This system allows the storage and retrieval of remote data through devices called labels (stickers, cards, transponders or RFID tags). RFID technology was designed to transmit the identity of an object (something like a unique serial number) via radio waves [12].

The conceptualization suggested by Ashton was not strictly limited to the engineering aspects of the RFID standard. He was concerned about the consequences of the mismatch between the activity carried out on the Internet by human actors, with limited time, attention and precision, and the complexity of contextualizing the characteristics of the physical world from which humans extract data relevant to decision-making. This mismatch has decisive consequences in terms of economic efficiency, distribution of resources, dynamization of social life and survival strategies or adaptation to the physical environment [13].

Digitization and other technical developments that allow progress in the process of converting previously static and immovable information into dynamic and transportable resources are converging toward an Internet of Things. As a result, ideas and information will continue to be important, but better contextualized and adjusted to the characteristics of a physical world on which objects provide data of greater accuracy and scale than that achieved only by human actors.

The information that circulates between connected computers comes, for the most part, from information that has been filtered by human actors according to specific interests. According to Ashton, data collection without human assistance would allow everything to be tracked and counted, and reduce waste, losses and costs associated with multiple services. RFID technology and other sensors connected to the Internet would allow computer systems to see, hear and smell the world for themselves, in all its random glory, without the limitations of human input [2].

5 Social Bots and Risks of Distortion in the Labor Market

According to Thomas L. Friedman, an in-depth analysis of job advertisements for "traditional" jobs provides interesting clues to understanding the large-scale transformation of the labor market, in a context of increasing automation of tasks and consequent increased training requirements associated with traditional job niches. For jobs in administration and secretarial work, the requirements have evolved to the point where four-fifths of those now in the profession would not meet the requirements for work which, in essence, is the same as that which they currently perform.

The difference is that 65% of the jobs offered require a university degree, and 81% of those currently holding such jobs do not have it. The obvious distortion is to require skills that do not correspond to the demands of the job to be performed, which brings thousands of candidates closer to a worrying scenario of underemployment or unemployment. But it cannot be ignored that the supposedly required skills are now associated with the functionality of largely automated desktop applications and services or incorporated into the features of bots and certain online work platforms [4].

6 Bots and Risks of Distortion in Social Communication Flows

The dissemination of fake news leads to distorted interpretations of reality and impoverishes and intoxicates the quality of debate on matters of public interest. As many media outlets have included automatic tools managed through social bots in order to highlight and update relevant content, the risk of significant bias in corporate decision-making (e.g., by underestimating or over-generalizing certain data) increases, and new threats of social or political destabilization are generated when users and machines amplify the scope of certain hoaxes by circulating them on the main social networks [14].

Statistics on cyber-attacks often originate from a small group of countries, whose governments cover up or indirectly allow unscrupulous actors to carry out illicit actions. The target is usually strategic companies, willing to pay to keep their data safe in case of security breaches. But in the 2017 US and French election campaigns, it has been proven that intercepting messages from political opponents and controlling their dissemination can seriously distort democratic processes and promote scenarios with a high risk of social confrontation [15].

The idea of a web developed on a knowledge-based trust system is, by the moment, still a research project to reorient the indexing of web content according to endogenous criteria (veracity of the content), rather than exogenous criteria (hyperlinks to the page or website). A problem still to be solved is the filtering system in charge of feeding the knowledge base that will serve as a reference and the type of reward for sites of interest but designed for purposes other than the dissemination of facts or news.

7 Privacy Policy

The adoption of a privacy policy is not sufficient if it does not help to promote practices among users that minimize the risk of abusive behavior by those who take advantage of the gray areas of the regulatory framework. In addition, privacy policies need to be compatible with potentially conflicting objectives, such as the ease of use of certain applications, access to personal information under certain conditions, and the potential to share it with other services and connected users, in a context of trust and security. The difficulty of successfully balancing these objectives increases as working and cooperating in networks to solve complex problems and increase efficiency in tasks becomes more important [16].

For actors with malicious purposes, it is sometimes less important to access databases with personal information (age, gender, income, health, etc.) than to know the information flows associated with certain activities in real time: purchases, social interaction, file sharing, services used, data provided by sensors involved in multiple monitoring and control tasks (in the transport sector, surveillance, medical care, physical exercise, etc.). It is therefore important to specify in the privacy policies the minimum personal data required for the provision of the service, as opposed to those that would require an extension of the initial permissions to receive additional benefits or services [17].

8 Difficulties in Achieving a Reasonable Expectation of Privacy.

There is a general lack of transparency about the personal information that service companies operating on the Internet collect from their customers/users and what they do with it. This is partly because users and companies interact in countries with very different legal frameworks and privacy cultures. But the fundamental reason seems to be the willingness of users to enjoy certain services and applications without any associated cost, although it is obvious that the main incentive for their developers is the profitability linked to the strategic or market value they extract from the large-scale data and usage statistics [18].

The trend toward the use of cloud services through lightweight applications is unstoppable. As the Internet of Things (IoT) is deployed and cloud computing resources increase, a growing base of always-on devices allows the monitoring and optimization of equipment hosted in geographically dispersed facilities. In the home environment, there is an increasing demand for robots or personal assistants connected to repositories and services on the Internet, capable of processing the information needed for voice recognition, objects, navigation and real-world tasks [19].

9 Characteristics of the Best Probation Policies

There are several online resources to compare the privacy policies applied by the main companies that develop applications and provide the most demanded services on the Internet. Among those that have maintained a favorable reputation over the years are Abine, Mozilla and the DuckDuckGo search engine for their commitment to avoid tracking user activity without the need for blockers or other external tools [20].

The basic elements of a privacy policy include the purpose (what the information the user chooses to share will be used for and what purpose it will not be used for); the responsibility (who is responsible for protecting the rights and privacy of the users); the minimum number or elements of personal information required to provide the service—registration data, identity, location, etc.; categories of information that is not essential [21] for improving the functionality or convenience of the service through access to technical information and sensors of the mobile device used; information that may only be shared with express permission, if doing so would lead to specific benefits or enhanced experiences (access to photos in the gallery, contact list, microphone and camera on the device, for example) [22, 23].

Other essential aspects are the details of what information is transferred to third parties and the non-presumption of consent given by users, if it was given under stricter initial conditions. In practice, this entails keeping an up-to-date record of the conditions associated with each version of a company's privacy policy to facilitate, as far as possible, a comparison between the new requirements and its original version, at each relevant step of the authorizations that users give to update their registration details for certain services.

The current context of widespread hypervigilance, promoted by state and private actors willing to use all kinds of strategies to exploit according to their interests the smallest security breach and to access personal information from any device connected to the Internet, requires companies to specify in their privacy policies whether they always require judicial authorization to give access to the content of communications and whether they inform users about data requests received from government entities. The publication of transparency reports and compliance guidelines contributes to strengthening user confidence in those companies that do so, stand out in the fight for users' right to privacy in the courts and explicitly support legislative initiatives in this direction [24, 25].

10 Conclusion

Constitutional guarantees that contribute to effective privacy protection are not effective without robust accountability mechanisms applicable to those who abuse or misuse illegally obtained data and personal information. The current imbalance (which is hardly compatible with a reasonable expectation of privacy) is partly explained by the mismatch between technological evolution and the transformations of a regulatory framework that, in essence, responds to criteria and risk scenarios typical of the socio-technical context of the last century.

The specific risks associated with the proliferation of mobile communication devices and Internet-connected objects need to be incorporated into the social debate on the consequences of the convergence of formats in the digital medium and the automation of tasks, in a dynamic expansion of cloud services that abuse the gray areas of the regulatory framework. This infrastructure has consolidated business models, consumption patterns and leisure styles based on massive personal data traffic without due guarantees of security and privacy, which favors scenarios of abuse and vulnerability for millions of users.

References

1. Gerl, A., Meier, B.: Privacy in the future of integrated health care services–are privacy languages the key? In: In 2019 International Conference on Wireless and Mobile Computing, Networking and Communications (WiMob), pp. 312–317. IEEE (2019, October)
2. Barday, K.A., Brannon, J.B., Jones, K., Kveen, B.P., Malhotra, P., Sabourin, J.L.: U.S. Patent Application No. 16/565,265 (2020)
3. Maier, C.O., Rashad, K.G.M., Elmiligui, M.A. U.S. Patent No. 10,511,576. Washington, DC: U.S. Patent and Trademark Office (2019)

4. Bennett, C., Oduro Marfo, S.: Privacy, voter surveillance and democratic engagement: challenges for data protection authorities. In: International Conference of Data Protection and Privacy Commissioners (ICDPPC) (2019, October)
5. Casillo, M., Clarizia, F., D'Aniello, G., De Santo, M., Lombardi, M., Santaniello, D.: CHAT-bot: a cultural heritage aware teller-bot for supporting touristic experiences. Patt. Recogn. Lett. (2020)
6. Alshaikh, M., Zohdy, M., Olawoyin, R., Debnath, D., Gwarzo, Z., Alowibdi, J.: Social network analysis and mining: privacy and security on Twitter. In: 2020 10th Annual Computing and Communication Workshop and Conference (CCWC), pp. 0712–0718. IEEE (2020, January)
7. Moorthy, V., Venkataraman, R., Rao, T.R.: Security and privacy attacks during data communication in Software Defined Mobile Clouds. Comput. Commun. **153**, 515–526 (2020)
8. Patel, T., Patel, V.: Data privacy in construction industry by privacy-preserving data mining (PPDM) approach. Asian J. Civil Eng. **21**(3), 505–515 (2020)
9. Jain, R., Jain, N., Nayyar, A.: Security and privacy in social networks: data and structural anonymity. In: Handbook of Computer Networks and Cyber Security, pp. 265–293. Springer, Cham (2020)
10. Landowski, L.N., El Moujahid, K., Parunashvili, S., Blandin, W.: U.S. Patent No. 10,599,402. Washington, DC: U.S. Patent and Trademark Office (2020)
11. Meurisch, C., Bayrak, B., & Mühlhäuser, M. (2020, April). Privacy-preserving AI Services Through Data Decentralization. In Proceedings of The Web Conference 2020 (pp. 190–200).
12. Meurisch, C., Bayrak, B., Mühlhäuser, M.: Privacy-preserving AI services through data decentralization. In: Proceedings of the Web Conference 2020, pp. 190–200 (2020, April)
13. Beloussov, S., Melnikov, O., Tormasov, A., Protasov, S.: U.S. Patent Application No. 16/571,661 (2020)
14. Wrigley, S.: BOTS, Artificial Intelligence and the General Data Protection Regulation: Asking the Right Questions. Trinity CL Rev. **22**, 199 (2019)
15. Van den Hoven van Genderen, R.: Privacy and data protection in the age of pervasive technologies in AI and robotics. Eur. Data Prot. L. Rev. **3**, 338 (2017)
16. Wang, L., Near, J.P., Somani, N., Gao, P., Low, A., Dao, D., Song, D.: Data capsule: a new paradigm for automatic compliance with data privacy regulations. In: Heterogeneous Data Management, Polystores, and Analytics for Healthcare pp. 3–23. Springer, Cham (2019)
17. Das, A., Degeling, M., Wang, X., Wang, J., Sadeh, N., Satyanarayanan, M.: Assisting users in a world full of cameras: a privacy-aware infrastructure for computer vision applications. In: 2017 IEEE Conference on Computer Vision and Pattern Recognition Workshops (CVPRW), pp. 1387–1396. IEEE (2017, July)
18. Smys, S., Ranganathan, G.: Robot assisted sensing control and manufacture in automobile industry. J ISMAC **1**(03), 180–187 (2019)
19. Silva, J., Cubillos, J., Villa, J. V., Romero, L., Solano, D., Fernández, C.: Preservation of confidential information privacy and association rule hiding for data mining: a bibliometric review. In: Procedia Computer Science, Vol. 151, pp. 1219–1224. Elsevier B.V. https://doi.org/10.1016/j.procs.2019.04.175(2019)
20. Pardo, R., Le Métayer, D.: Analysis of privacy policies to enhance informed consent. In: IFIP Annual Conference on Data and Applications Security and Privacy, pp. 177–198. Springer, Cham (2019, July)
21. Yaghoub-Zadeh-Fard, M.A., Zamanirad, S., Benatallah, B., Casati, F. REST2Bot: bridging the gap between bot platforms and REST APIs. In: Companion Proceedings of the Web Conference 2020, pp. 245–248 (2020, April).
22. Gerl, A., Meier, B.: Privacy in the future of integrated health care services–are privacy languages the key? In: 2019 International Conference on Wireless and Mobile Computing, Networking and Communications (WiMob), pp. 312–317. IEEE (2019, October)
23. Papadopoulos, P., Azurmendi, I.Q., Zhang, J., Varvello, M., Nappa, A., Livshits, B.: ZKSENSE: a privacy-preserving mechanism for bot detection in mobile devices. arXiv preprint arXiv:1911.07649 (2019)

24. Mushtaq, A., Rosenberry, T., Aziz, A., Islam, A.: U.S. Patent No. 10,200,384. Washington, DC: U.S. Patent and Trademark Office (2019)
25. Silva, J., Solano, D., Fernandez, C., Romero, L., Villa, J.V.: Privacy preserving, protection of personal data, and big data: a review of the Colombia case. In: Procedia Computer Science, Vol. 151, pp. 1213–1218. Elsevier B.V. https://doi.org/https://doi.org/10.1016/j.procs.2019.04.174 (2019)

Role of Non-textual Contents and Citations in Plagiarism Detection

V. Geetha Lekshmy, R. Athira Krishnan, and S. Aparnna

Abstract Plagiarism is the act of using ideas or words of others and presenting them as one's own without due acknowledgment. Detection of plagiarism in research documents is studied extensively, and many methods have been proposed by researchers from time to time. However, due to the complexity of the problem and computational constraints, there is no single method that can be used to tackle it. This work includes a study on plagiarism detection methods based non-textual contents like images, figures, mathematical expressions, graphs, and citations. It also explores the effectiveness of citation analysis in detecting plagiarism in documents.

Keywords Non-textual plagiarism · Longest common citation sequence · Grobid · Fingerprinting

1 Introduction

In the current scenario of a digital world where information is on fingertips, academic dishonesty has become common in students and researchers. Unethical acts of stealing others intellectual work and keeping as one's own should be checked. \query{The chapter has been substantially edited for English language. Please check and approve..}In the literature, there are many survey papers that deal extensively with the classification of types and methods of plagiarism and its detection methods. Alzahrani et al. in their work[3] provide an extensive classification of plagiarism types by differentiating them as intelligent and literal plagiarism. Different types

V. Geetha Lekshmy (✉) · R. Athira Krishnan · S. Aparnna
Department of Computer Science and Applications, Amrita Vishwa Vidyapeetham, Amritapuri, India
e-mail: geethalekshmy@am.amrita.edu

R. Athira Krishnan
e-mail: athirakrishnan98rg@gmail.com

S. Aparnna
e-mail: aparnna111@gmail.com

© The Editor(s) (if applicable) and The Author(s), under exclusive license to Springer Nature Singapore Pte Ltd. 2021

A. P. Pandian et al. (eds.), *Proceedings of International Conference on Intelligent Computing, Information and Control Systems*, Advances in Intelligent Systems and Computing 1272, https://doi.org/10.1007/978-981-15-8443-5_51

of plagiarisms based on syntax, semantics, structure, and grouping of characters (N-gram) of texts are examined in detail. Vani and Gupta in [16] present an evaluation of different plagiarism detection tools using the PAN dataset in addition to the extensive classification of plagiarism techniques. They have found out that even the most commonly used commercial plagiarism detection software fails to detect translation obfuscation (modification of text by translating to another language and translating back). Foltynek et al.[8] gives a systematic review of plagiarism detection research based on papers published between 2013 and 2018. This study shows that hybrid methods including syntactic, semantic text similarity, and non-textual content analysis integrated with machine learning can give better detection systems. Non-textual feature-based detection systems are effective for documents in mathematics and physics [33]. Dr. Jacob [37] introduced a new method for text classification; it uses the capsule network to increase the expectation using routing instead of dynamic routing. Cap-Net along with multitask learning works more effectively in learning multiple of datasets in an integrated paradigm. Currently, available plagiarism detection software uses only text-based methods, which may not be effective for documents with high mathematical content. The efficiency of plagiarism detection software can be enhanced by including analysis of images, mathematical equations, charts, and citations. One of the earliest works involving examination of citations is in [19], where bibliographic coupling is introduced that finds out common references in documents. Citation-Proximity Index (CPI) value is introduced in [11] by Bela Gipp and Joran Beel for identifying documents with related content. It assigns a high value to citations appearing in the same sentence and relatively low value for those appearing in a paragraph. Vinothkanna [38] proposes a new approach by combining cryptography into the steganography for file format, i.e., cryptography incorporates steganography (CICS) to increase the robustness, capacity, and security of the information transferred. In [12], they introduced the idea of analyzing the order of citations and its application in plagiarism detection. Pertile et al. in [31] introduced a method of measuring the number of cooccurrences in citations(two documents sharing the same citation in a text fragment) and found out that documents with high co-occurrence indicate plagiarism. In [30], they present the comparison of citation and content-based detection methods to find their complementary behavior. They have used co-occurrence measures and Citing Frequency Score for finding citation similarity and WCopyFind tool for text-based similarity. Contribution of this paper:- 1. A survey of non-textual content-based plagiarism detection methods. 2. Analysis of complementing behavior of citation and text-based similarity.

This paper is organized as follows: In Sect. 2, the methodology has two parts. The first part presents a survey on different non-textual contents in scientific papers and its role in plagiarism detection. The second part of Sect. 2 explains the methods involved in the study of citation analysis and its effectiveness as an initial method for selecting similar documents. In Sect. 3, observation and findings also have two parts: The first part tabulates the results of survey and the second part explains the result of citation analysis. Finally, Sect. 4 gives conclusion and the future directions of the work.

2 Methodology

2.1 Study on Non-Textual Content Analysis for Plagiarism Detection

Plagiarism checking softwares available in the market do not fully support the analysis of images and non-textual data in the document. Non-textual feature analysis is well suited for identifying obfuscated forms of plagiarism like adopting ideas without citing original work [8]. Non-textual features included in our study are images, bar charts, line charts, flowcharts, mathematical formulae, and citations. For retrieving documents for our study, Google Scholar was queried with strings, "plagiarism detection using images", "bar chart similarity in documents", "citations in plagiarism detection", and "mathematical contents for plagiarism detection". Table 1 summarizes the papers examined.

2.1.1 Images

Feature point-based image comparison is used for image-based plagiarism detection in [18]. Among the combinations of different feature detectors and description extractors, SURF [4] proves to be effective. This method is invariant to image transformation, so the modified images can also be detected. However, charts and graphs cannot be processed with this method. Srivastava et al. in [35] propose a method that uses perceptual hashing. Perceptual hashing is a block-based feature extraction technique which producing similar multimedia content with similar hash values. Perceptual hashes of images conserve the qualitative aspects and the basic structure of the visual information content. The performance of perceptual hashing shows that accuracy for image retrieval was calculated using the using hamming distance, and it came out to be 84% which is very close to SIFT in terms of accuracy, and also, perceptual hash took lesser time than both SIFT and SURF. This method is yet to be proved for large datasets. Eisa et al. [7] use textual reference features of figure, i.e., [2] figure caption, figure reference sentence, sentences close to figure reference sentence for image plagiarism detection. This method is resilient to small changes in the figure description like using synonyms of words and changing words of figure. Petr Hurtik and Petra Hodakova in their paper [17] use the F-transform technique for storing images in the database. The images are stored as matrices representing F-transform components in the preprocessing step. The similarity between images is determined by computing distance between them. The preprocessing step is computationally intensive. However, this technique is not feasible for large images. Eisa et al. in [6] propose a method that extracts both textual and structural elements of a figure. Structural elements of the figure include basic shapes of components ,and flow among them indicates the relation between components. An integrated approach is used in [24] to cater for identifying plagiarism in different types of images including, bar charts, box plot, flowcharts, etc. A combination of perceptual hashing, text extraction, and ratio hashing is used. Ratio hashing is exclusively for bar charts. Text extraction from images includes recording positional information of characters also for comparison.

2.1.2 Flowchart

In [29], histograms containing color features extracted are used for image comparison. But the method is mentioned as applicable to only flowcharts and four block diagrams not to other kinds of images. Kuruvila et al. [20] propose a method based on image processing for detecting plagiarism in flowcharts. This method detects shapes and flow in the flowchart and represents it as a directed graph for comparison with already stored graphs in the repository. This method can detect similarity even if additional steps are added to flowcharts or changes are made to shape and orientation of its contents. This is because the basic flow represented in the graph will not change, with additional steps.

2.1.3 Bar Charts/Line Graphs

Mumtaz et al.in their paper [2] extract text data like caption, labels, and coordinate values from bar charts using OCR tool as strings. Image scanning is used for extracting numeric data (length of the bar in a bar chart). Character-based comparison finds the similarity of test data from the chart, and Euclidean distance is used to find similarity in numeric features. This method is confined to only bar charts.

2.1.4 Mathematical Formulae

Meuschke et al. in [25] propose a method for the first time that analyzes mathematical content to detect document plagiarism. They investigated on different plagiarism cases involving mathematical expression and found out that semantically, similar expressions were created by using different identifiers or function symbols, changing to equivalent forms using commutative, distributive and associative rules, splitting or merging expressions, converting expressions to the similar concept (e.g., converting integral to sum). The features selected for similarity detection are identifiers, numbers, and operators. The frequency of occurrence of features is defined for entire documents as well as different sections of the documents. A distance measure is calculated based on the absolute difference of occurrence frequencies of each feature in documents. The distance measure of variables proved to be effective in identifying plagiarized cases. An improved version of the above method is in [27]. In addition to the histogram of occurrence frequency of identifiers, a method is included considering the order of identifiers in the document. The longest common subsequence of identifiers measure is adopted, where the order of mathematical identifiers in the documents is considered. Greedy Identifier Tiles method measures the longest block of shared features in the same order. Tiles with more than five similar identifiers are considered for similarity calculation. Moritz Schubotz et al. present an elaborate study [33] on plagiarism types found in mathematics-based publications. They created a dataset of 149 documents with content reuse from zbMath. From this, they analyzed and illustrated content reuse in 11 typical documents with Hyplag [26]. A content reuse in disguise was detected in all 11 documents; however, only two were found retracted from zbMath. Hyplag is a tool that uses a combination of techniques, mathematical formula analysis, image analysis, citation analysis, and text analysis for plagiarism detection. Zanibbi et al. [36] propose a stroke-based performance evaluation metrics for handwritten mathematical expressions. The measurements are computed utilizing bipartite graphs that speak to division, classification, and spatial structure at the stroke level. These metrics compare the likeness of two elucidations of a set of online strokes, with application to pen-based arithmetic recognition.

2.1.5 Citation

Citation-based plagiarism detection method is introduced in [12] called citation order analysis where citations are extracted from the document along with its position. The similarity between documents is calculated based on the order and distance between citations. In [23], a prototype tool called Citeplag is demonstrated for citation-based plagiarism detection. The effectiveness of the method is illustrated in [10, 13, 26]. Pertile et al. [31] present a study on the co-occurrence of citations indicating the presence of plagiarism. Citations that are common in text fragments within documents refer to inter-document co-occurrence. In [30], the authors do an extensive study to evaluate the effectiveness of citation-based approach by combining it with the text-based approach. \query{The term 'Pubmed' has been changed to 'PubMed' throughout the chapter.}The datasets used for this study are from PubMed collection and papers published at the ACL conference. The percentage of common documents identified as plagiarized by both methods(citation and text-based) is 23–61% for PubMed collection and 15–46% for ACL collection. With this result, they conclude that both methods are complementary in plagiarism detection. Sidik Soleman et al. in their work [34] propose a method of plagiarism detection using citation network(set of documents that document refer to). Given suspicious and legitimate documents, this method checks for similarity between citation text fragments of documents in the citation network of legitimate document and text fragments of suspicious documents. This method uses a vector-based method for citation text fragment comparison, so the usage of paraphrases may reduce correctness. Mazov et al. [22] propose a method of querying bibliographic databases for retrieving documents that cite the same documents as the document that is checked for plagiarism. This approach is typical for the Russian scenario where the full text of the document is not freely available. The difference in citation order in documents affects the performance of this method.

2.2 Study on Complementary Behavior of Citation Analysis and Text Comparison in Document Similarity Detection

The effectiveness of citation-based plagiarism detection is studied by examining the citation overlap and text overlap in pairs of documents. The proposed system checks whether the documents having high citation-sequence overlap also have high text overlap and vice versa. This was addressed earlier in [30], and they have used co-occurrence measure and citing frequency score, using the longest common citation sequence [10] for citation-sequence comparison in documents.

The text content is compared by computing hash values and using winnowing [32] for selecting hash values. GROBID [1] is used for extracting references and citations from research papers for further processing. GeneRation Of BIbliographic Data (GROBID) is a tool for extracting structured information in XML format from scientific publications. The XML files corresponding to research papers are given as input to the proposed system. The pair of papers given as input is parsed, citation sequence is extracted, and similarity in citation sequence is determined. The system also checks for similarity between textual contents of the pair of papers. An analysis is done, to determine the number of pairs that show both citation similarity and textual similarity. Longest common citation subsequence [10] and bibliographic coupling for finding citation similarity and hashing with winnowing [32] are adopted for text finding text similarity. Figure 1 shows the overall architecture of the system. This proposed system is designed to be a part of a hybrid plagiarism detection system that combines citation-based and fingerprinting-based techniques for plagiarism detection. The system collects papers [5] from the Web based on the keywords [21, 28] of the suspicious document. A filtering of the collected documents is essential to reduce the number of candidate documents that have contents similar to the suspicious document. Only the filtered documents with citation-sequence similarity need to be checked for text similarity. For filtering purposes, bibliographic coupling(BC) and longest common citation subsequence(LCCS) are used. In BC, the references of both documents are extracted to find common references. To find the longest common citation sequence in a pair of documents, extract all the citations from documents in the order of occurrence. Remove all citations except the ones that are common on both documents. Each of the common citations in both documents is replaced by the same indicator (in this system, an alphabet is used). Then, citation sequences from both documents, where each of the common citations is represented by the same alphabet, are given as input to the LCCS algorithm. The output of the algorithm is the length of the citation subsequence that is common in both documents and that is in the same order. If the documents have an LCCS of considerable length, then there is a chance of plagiarism and both documents are subjected to the detailed text comparison. In the text comparison, the first step is preprocessing. The preprocessing [9] contains case folding, parsing, stemming, generation of n-grams for the textual content, and hashing. To select hash values, winnowing algorithm is used. The selected hash values serve as the fingerprint of the document. These fingerprints

Table 1 Number of papers studied for non-textual features

S. No.	Type of contents	Number of papers referred
1	Images/flowcharts/bar charts/line graphs	8
2	Mathematical formula	3
3	Citation	11

are used for similarity check. Jaccard similarity index (Jaccard similarity coefficient) is the metric used to measure the similarity between both documents.

2.2.1 Dataset

Since the dataset consisting of plagiarized and original documents with proper citations is not readily available, experiments are carried out with a manually created dataset consisting of papers from areas of computer science. The input to the system is a set of 25 source documents and its corresponding suspicious document. The suspicious document contains passages plagiarized from the source documents and created 25 papers that have common keywords and common references of suspicious paper.

3 Observations and Findings

3.1 Non-textual Plagiarism Detection Methods

Table 2 summarizes the observations made after the survey of non-textual methods for plagiarism detection with its current performance and future scope.

3.2 Study on Complementary Behavior of Citation-Based and Text-Based Plagiarism Detection Methods

Table 3 summarizes the observations made after study on complementary behavior of citation-based and text-based plagiarism detection methods. \query{Please check the clarity of the sentence 'The comparison of the suspicious document with 25 documents based on citation similarity and text similarity...'.}The comparison of the suspicious document with 25 documents based on citation similarity and text similarity come to the result is six documents have citation similarity and that same six documents have text similarity with input document. So, trying to say that n

Table. 2 Non-textual plagiarism methods

Non-textual features	Current work	Future work
Image	Identify exact copies, slightly altered images with changes in color, shape, orientation, size, text content, cropped images	Images created by combining parts of different images, altered images where contents are positioned differently
Bar charts/graphs	Identify exact copies, altered charts with different labels and text content	Charts that visualize the same data differently, charts formed by combining different charts or graphs
Mathematical formula	Identify exact formulae with same variables, operations, functions	Semantically similar expressions where operations or functions are transformed to equivalent ones
Citation	Identify documents, with common citations considering proximity and order	Reduce false positives, as papers having common citations may not be necessarily plagiarized

Table. 3 Complementary behavior of citation-based and text-based plagiarism

Plagiarized papers	Count
No: of documents with citation similarity	6
No: of documents with text similarity (out of 6 citation similarity documents)	6
No: of documents with citation and text similarity	6

number of documents that had citation similarity also had text similarity, i.e., it shows the complementary behavior of the two approaches.

4 Conclusion

The study on non-textual data shows that it can be combined with efficient text comparison methods for improving plagiarism detection. Based on our experiment, number of documents that had citation similarity also had text similarity. This shows the complementary behavior of two approaches: citation-based and text-based. Citation analysis can be improved by including author information and also to prevent false positives, where the document with common citation might be of the same author or co-authors. This work can also be extended to include non-textual contents like images also to improve the initial document filtering.

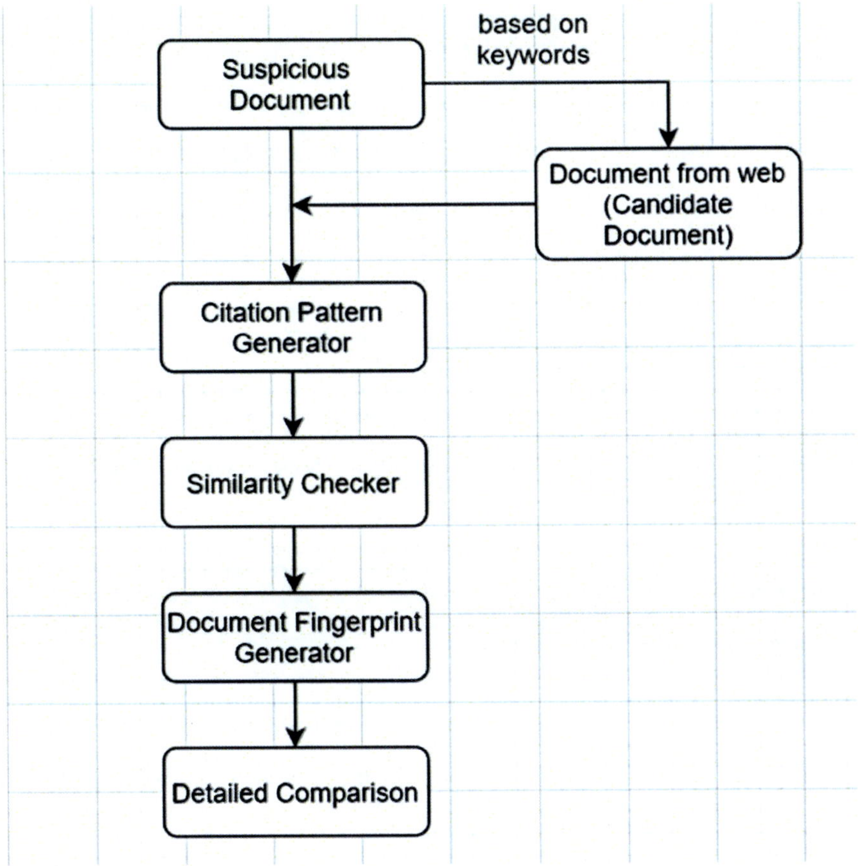

Fig. 1 System architecture

References

1. Grobid. https://github.com/kermitt2/grobid (2008–2020)
2. Al-Dabbagh, M.M., Salim, N., Rehman, A., Alkawaz, M.H., Saba, T., Al Rodhaan, M., Al-Dhelaan, A.: Intelligent bar chart plagiarism detection in documents. Sci. World J. 2014 (2014)
3. Alzahrani, S., Palade, V., Salim, N., Abraham, A.: Using structural information and citation evidence to detect significant plagiarism cases in scientific publications. J. Am. Soc. Inform. Sci. Technol. **63**(2), 286–312 (2012)
4. Bay, H., Ess, A., Tuytelaars, T., Van Gool, L.: Speeded-up robust features (surf). Comput. Vis. Image Underst. **110**(3), 346–359 (2008)
5. Adeena, K.D., Remya, R.: Extraction of relevant dataset for support vector machine training: A comparison. In: 2015 International Conference on Advances in Computing, Communications and Informatics (ICACCI), Kochi, 2015, pp. 222–227, doi: https://doi.org/10.1109/ICACCI.2015.7275613, pp. 76–85 (2015)
6. Eisa, T., Salim, N., Alzahrani, S.: Figure plagiarism detection using content based features. In: Recent Developments in Intelligent Computing, Commu nication and Devices, pp. 17–20. Springer (2017)

7. Eisa, T.A.E., Salim, N., Alzahrani, S.: Figure plagiarism detection based on textual features representation. In: 2017 6th ICT International Student Project Conference (ICT-ISPC), pp. 1–4. IEEE (2017)
8. Foltynek, T., Meuschke, N., Gipp, B.: Academic plagiarism detection: a systematic literature review. ACM Comput. Surv. (CSUR) **52**(6), 1–42 (2019)
9. Veena, G., Hareesh, H.R., J.: Relation extraction in clinical text using nlp based regular expressions. In: 2019 2nd International Conference on Intelligent Computing, Instrumentation and Control Technologies (ICICICT), Kannur,Kerala, India, pp. 1278–1282, doi: https://doi.org/10.1109/ICICICT46008.2019.8993274, pp. 76–85 (2019)
10. Gipp, B.: Citation-based plagiarism detection - detecting disguised and cross-language plagiarism using citation pattern analysis. Springer Vieweg Research (2014)
11. Gipp, B., Beel, J.: Citation proximity analysis (cpa): a new approach for identifying related work based on co-citation analysis. In: ISSI'09: 12th International Conference on Scientometrics and Informetrics, pp. 571–575 (2009)
12. Gipp, B., Beel, J.: Citation based plagiarism detection: a new approach to identify plagiarized work language independently. In: Proceedings of the 21st ACM Conference on Hypertext and Hypermedia, pp. 273–274 (2010)
13. Gipp, B., Meuschke, N., Breitinger, C.: Citation-based plagiarism detection: Practicability on a large-scale scientific corpus. J. Assoc. Inf. Sci. Tech. **65**(8), 1527–1540 (2014)
14. Gipp, B., Meuschke, N., Breitinger, C., Lipinski, M., Nurnberger, A.: Demonstration of citation pattern analysis for plagiarism detection. In: Proceedings of the 36th İnternational ACM SIGIR Conference on Research and Development in İnformation Retrieval, pp. 1119–1120 (2013)
15. Gipp, B., Meuschke, N., Breitinger, C., Pitman, J., Nurnberger, A.: Web-based ¨ demonstration of semantic similarity detection using citation pattern visualization for a cross language plagiarism case. ICEIS **2**, 677–683 (2014)
16. Gupta, D., et al.: Study on extrinsic text plagiarism detection techniques and tools. J. Eng. Sci. Tech. Rev. **9**(5) (2016)
17. Hurtik, P., Hodakova, P.: Ftip: a tool for an image plagiarism detection. In: 2015 7th International Conference of Soft Computing and Pattern Recognition (SoCPaR), pp. 42–47. IEEE (2015)
18. Iwanowski, M., Cacko, A., Sarwas, G.: Comparing images for document plagiarism detection. In: International Conference on Computer Vision and Graphics, pp. 532–543. Springer (2016)
19. Kessler, M.M.: Bibliographic coupling between scientific papers. Am. Doc. **14**(1), 10–25 (1963)
20. Kuruvila, J.S., VL, M.L., Roy, R., Baby, T., Jamal, S., Sherly, K.: Flowchart plagiarism detection system: an image processing approach. Proced. Comput. Sci. **115**, 533–540 (2017)
21. Thushara, M.G., Mangamuru, T.M.R.: A comparative study on different keyword extraction algorithms. In: 2019 3rd International Conference on Computing Methodologies and Communication (ICCMC), pp. 76–85 (2019)
22. Mazov, N., Gureev, V., Kosyakov, D.: On the development of a plagiarism detection model based on citation analysis using a bibliographic database. Sci. Tech. Inf. Process. **43**(4), 236–240 (2016)
23. Meuschke, N., Gipp, B., Breitinger, C., Berkeley, U.: Citeplag: a citation based plagiarism detection system prototype. In: Proceedings of the 5th International Plagiarism Conference. Citeseer (2012)
24. Meuschke, N., Gondek, C., Seebacher, D., Breitinger, C., Keim, D., Gipp, B.: An adaptive image-based plagiarism detection approach. In: Proceedings of the 18th ACM/IEEE on Joint Conference on Digital Libraries, pp. 131–140 (2018)
25. Meuschke, N., Schubotz, M., Hamborg, F., Skopal, T., Gipp, B.: Analyzing mathematical content to detect academic plagiarism. In: Proceedings of the 2017 ACM on Conference on Information and Knowledge Management, pp. 2211–2214 (2017)
26. Meuschke, N., Stange, V., Schubotz, M., Gipp, B.: Hyplag: a hybrid approach to academic plagiarism detection. In: The 41st International ACM SIGIR Conference on Research and Development in Information Retrieval, pp. 1321–1324 (2018)

27. Meuschke, N., Stange, V., Schubotz, M., Kramer, M., Gipp, B.: Improving academic plagiarism detection for stem documents by analyzing mathematical content and citations. In: 2019 ACM/IEEE Joint Conference on Digital Libraries (JCDL), pp. 120–129. IEEE (2019)
28. Thushara, M.G., Angali, S., Nair, M.: An analysis on different document keyword extraction methods. In: Proceedings of the 2003 ACM SIGMOD İnternational Conference on Management of Data, pp. 76–85 (2019)
29. Ovhal, P.: Detecting plagiarism in images. In: 2015 International Conference on Information Processing (ICIP), pp. 85–89. IEEE (2015)
30. Pertile, S.d.L., Moreira, V.P., Rosso, P.: Comparing and combining c ontentand c itation-based approaches for plagiarism detection. J. Assoc. Inf. Sci. Tech. **67**(10), 2511–2526 (2016)
31. Pertile, S.d.L., Rosso, P., Moreira, V.P.: Counting co-occurrences in citations to identify plagiarised text fragments. In: International Conference of the CrossLanguage Evaluation Forum for European Languages, pp. 150–154. Springer (2013)
32. Schleimer, S., Wilkerson, D.S., Aiken, A.: Winnowing: local algorithms for document fingerprinting. In: Proceedings of the 2003 ACM SIGMOD İnternational Conference on Management of Data, pp. 76–85 (2003)
33. Schubotz, M., Teschke, O., Stange, V., Meuschke, N., Gipp, B.: Forms of plagiarism in digital mathematical libraries. In: C. Kaliszyk, E. Brady, A. Kohlhase, C.S. Coen (eds.) Intelligent Computer Mathematics—12th International Conference, CICM 2019 , Prague, Czech Republic, July 8–12, 2019 , Proceedings, Lecture Notes in Computer Science, vol. 11617, pp. 258–274. Springer (2019). https://doi.org/10.1007/978-3-030-23250-418.
34. Soleman, S., Fujii, A.: Toward plagiarism detection using citation networks. In: 2017 Twelfth International Conference on Digital Information Management (ICDIM), pp. 202–208. IEEE (2017)
35. Srivastava, S., Mukherjee, P., Lall, B.: İmplag: Detecting image plagiarism using hierarchical near duplicate retrieval. In: 2015 Annual IEEE India Conference (INDICON), pp. 1–6. IEEE (2015)
36. Zanibbi, R., Pillay, A., Mouchere, H., Viard-Gaudin, C., Blostein, D.: Stroke-based performancemetrics for handwritten mathematical expressions. In: 2011 International Conference on Document Analysis and Recognition, Beijing, 2011, pp. 334–338.https://doi.org/10.1109/ICDAR.2011.75. pp. 76–85 (2011)
37. Jacob, Dr.: Performance evaluation of caps-net based multitask learning archi-tecture for text classification. J. Artif. Intell. Capsule Netw. **2**, 1–10. https://doi.org/10.36548/jaicn.2020.1.001. pp. 76–85 (2020)
38. R.V.: A secure steganography creation algorithm for multiple file formats. J. Innov. Image Process. **1**, 20–30. https://doi.org/10.36548/jiip.2019.1.003. pp. 76–85 (2019)

Leveraging Machine Learning to Augment HR Processes

Shweta Jha

Abstract Machine learning has transformed the way business decisions are being made in all sorts of organisations the world over. It has impacted all the management functions viz. marketing and sales, supply chain, finance, etc., in a big way. However, the adoption of machine learning has been a bit slow in the HR domain, which is mostly driven by subjectivity and intuition. Interestingly, machine learning has transformed HR processes quietly and gradually so much so that the top management, as well as the HR leaders, are forced to stop and take note of the emerging scenario. Machine learning has positively impacted all the functions of HR such as recruitment, selection, onboarding, learning and development, performance management, compensation, and separation management. This article critically examines how machine learning is being applied to make the HR processes all the more robust, transparent, efficient, and cost-effective. Further, the pitfalls of HR automation have also been reviewed.

Keywords Machine learning · Artificial intelligence · Automation · HR processes

1 Introduction

Machine learning, at its core, is a set of statistical methods meant to find patterns of predictability in datasets [1]. It has evolved over the years and has now taken deeper roots in almost all the spheres of human life–community, politics, businesses, healthcare, manufacturing, marketing, finance, and above all, HR. Machine learning helps us create a mathematical way to copy human learning ability over time and through a new experience [2]. While it may not take over the world anytime soon, machine learning has altered the job market today and would continue to change it on its terms in the future [2]. Changing dynamics in the job market also necessitates appropriate modifications in the HR processes in order to meet industry requirements

S. Jha (✉)
Apeejay School of Management, New Delhi, India
e-mail: shwetajha.asm@gmail.com

© The Editor(s) (if applicable) and The Author(s), under exclusive license to Springer Nature Singapore Pte Ltd. 2021

A. P. Pandian et al. (eds.), *Proceedings of International Conference on Intelligent Computing, Information and Control Systems*, Advances in Intelligent Systems and Computing 1272, https://doi.org/10.1007/978-981-15-8443-5_52

in a volatile, uncertain, complex, and ambiguous world that characterise the contemporary organisations. In the emerging scenario, machine learning can substitute the thinking skills of the employees to carry out a job [2].

Even the HR function in a large number of organisations in the world has been transformed fundamentally as a result of the intelligent use of machine learning. For example, 27% of IBM employees got new jobs or promotions thanks to machine learning application aimed at analysing employee-wise data related to the tasks performed, educational courses taken, and assessments received [3]. Organisations are already using analytics supported by machine learning for identifying the most productive employees; grouping overburdened employees and redistributing the load; categorising resource-intensive but low-value activities, routine operations that can be automated, duplication of functions, etc.; and easy disposal of inefficient workers and actions [3].

2 Methodology

This research paper is based on an integrated review of the literature. Published work on the application of machine learning in HR processes has been reviewed thoroughly for the purpose. All the relevant work published during 1990–2020 have been examined from various databases like ABI Inform (ProQuest), EBSCO Business Source Complete, Scopus, and Social Science Citation Index (Web of Science). Besides, relevant trade publications from the Google search were also included in this study. Keywords used in search of relevant literature included machine learning, artificial intelligence, HR automation, and HR processes.

3 Related Work

Machine learning is all about teaching computers to recognise patterns in the same way as human brains based on examples and experience rather than hard-coded programming rules [4]. Machine learning enables a company to reimagine end-to-end business processes with digital intelligence [5]. Also, machine learning is changing how leaders use metrics to drive business performance, customer experience, and growth [6]. It is capable of solving such business problems which require predictions rather than causal inference and are adequately self-contained or relatively insulated from outside influences [1].

Although machine learning adoption has been a bit slow in HR function in the organisations across the globe, things are changing for the better. Machine learning has a unique ability to build holistic HR system in organisations. It can objectively quantify things like efficiency, effectiveness, and employee experience by looking at hiring decisions, team climate, and personal development [7]. Table 1 provides a broad list of HR functions in which are now being supported by machine learning in a big way. Large companies have already started using machine learning in talent acquisition, talent development, compensation, and benefits as well as compliances. Still, some fence-sitters are merely waiting and watching how the machine learning scenario unfolds.

It is generally perceived that predictions made by machine learning tools are free from biases. However, the inherent fallacy about the neutrality of machine learning has been questioned by several scholars. Bias seems to be embedded in the machine learning, i.e. the system learns from data, and thus is prone to picking up the human biases that the data represents [8]. For example, many hiring algorithms may drift towards bias by default by way of institutional or historical biases, amplifying disadvantages lurking in the data points like university attendance or performance evaluation scores [9]. Further, algorithmic job boards used by employers in order to minimise their recruitment budgets usually fail to get the most suitable candidates for the role as the tool looks at not the most relevant applicant but those who might stick to the job advertisement [9]. The hiring algorithms not only discriminate against the people based on socio-economic characterisations but also inadvertently eliminate the most suitable candidates from the hiring process. However, it is possible to reduce or eliminate biases induced by gender or other types of social stereotypes [10].

4 Conceptual Model

Machine learning has emerged as the most efficient tool in augmenting HR processes encompassing the entire employee life cycle right from hiring to separation with phenomenal speed and accuracy. It facilitates agile management through perceptive use of rapid prototyping, iterative feedback, customer-focused multi-disciplinary teams, and task centred sprints [11]. It is enabling companies to expand their top-line growth and optimise while improving employee engagement and increasing customer satisfaction [5]. A step ahead, organisations are now using machine-reengineering to establish new forms of human–machine collaboration that breaks through the bottlenecks of complex digital processes [12].

Machine learning tools are deployed to attract the potential candidates to apply for the jobs in the client organisations. It has immense ability to reach out to a large pool of candidates based on demographic and other relevant characteristics at relatively lower costs as compared to traditional methods. In the automated hiring process, it is also quite easy to conduct agent-led video interviews and predict the performance of new hires. Even reference information can be generated using machine learning. Now, automated processes are there for onboarding as well. Machine learning has been successfully leveraged for augmenting learning and development. For example, Wipro in India has a valid number of training and certification programmes powered by automated learning tools for its employees. Even personal coaching supported by machine learning is available these days, and a large number of organisations are already using the same.

Machine learning can play a critical role in retention management by way of retention and attrition analytics. Besides, to keep the employees motivated, even performance management can be automated to a great extent in terms of objective setting and capacity predictions. Organisations can very well use automated engagement monitoring just as SRL Diagnostics has been doing for its employees across India. Payroll and benefits automation also helps in improving the efficiency of HR processes. In terms of exit management, organisations can benefit from automated outplacement and pension administration. Thus, it is quite possible to leverage machine learning for augmenting HR processes.

Interestingly, it is not at all necessary to have in-house machine learning capabilities which may require huge investments. There are now a good number of start-ups and other technology companies which offer all HR-related support using machine learning at an affordable cost. Hence, the costs are not the inhibiting factor. However, the top management buy-in is still an issue, especially in the wake of debates on privacy, biases, and accuracy vis-à-vis machine learning (Fig. 1).

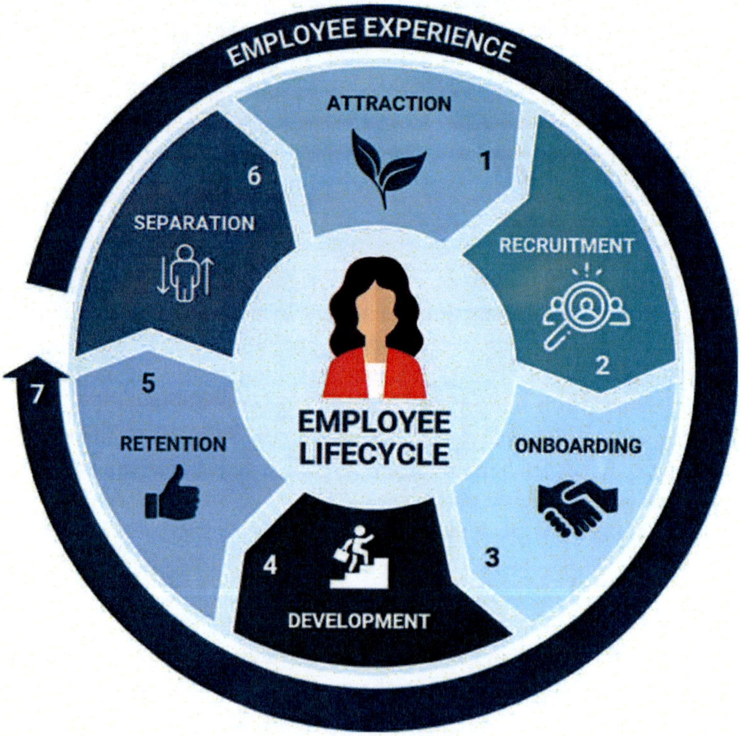

Fig. 1 Employee lifecycle augmented by machine learning. *Source* Delta partners https://www.deltapartnersgroup.com/what%E2%80%99s-next-hr-using-machine-learning-ml-talent-management

5 Results and Discussion

Machine learning is not at all infallible. It should also be noted that every value cannot always be optimised by using machine learning [8]. Moreover, machine learning is not magic, and the truth is that neither the data nor the understanding is necessary to build machines which make the decision as accurately as a human being [13] even though there has been a rapid increase in machine learning algorithms' accuracy driven by recent headway in deep learning [14]. Interestingly, the inaccuracies reported in the context of machine learning output are primarily related to imperfections in the data. Low-quality data with inherent inadequacies often distorts the reports generated by machine learning, thus leading to erroneous predictive models [15]. In one of the experiments, it was found that human beings came up with better results as compared to the machine learning algorithm even though the data was identical [16]. Nonetheless, the application of machine learning in a business context cannot be disputed merely because of certain inherent weaknesses. Researchers are already working towards overcoming such issues in future.

6 Conclusion

Machine learning indeed has wider acceptability today in a large number of organisations across the world irrespective of the questions revolving around its effectiveness, accuracy, and neutrality. Ensuring the quality of data is a significant challenge while using machine learning in organisations of all hues. Effectiveness of machine learning is highly dependent on the quality of data. Results generated by machine learning tools may be quite useless if the input data happens to be incorrect. Hence, it would be wrong to assume that machine learning is a cure for all the ills that plague an organisation. Also, there are limits to which machine learning can improve things within an organisation. Some of the problems faced by the organisation may very well require human intervention in terms of intuitive response instead of the reports generated by machine learning.

Machine learning applications in HR have compact potential to augment employee productivity and enable the HR professionals to become more knowledgeable and develop competencies to analyse, predict, diagnose, and resolve issues efficiently in order to drive high-performance work systems. Nevertheless, machine learning has certain inherent weaknesses. It is, therefore, imperative for the industry leaders to change their stance towards their employees, especially in the wake of launching a machine learning program in the organisation. At the same time, it is essential to debunk the myth that machines can never go wrong. Hence, leaders need to reorient themselves and create an organisational culture where dissent and debates are valued, and asking questions are rewarded.

References

1. Fedyk, A.: How to Tell If Machine Learning Can Solve Your Business Problem. Harvard Business Review. November 25 (2016)
2. Mellor, D.: Machine Learning for Beginners: Complete AI for Beginners with Real World Business Applications. Kindle Edition (2020)
3. Parsers: Big Data and Machine Learning in HR: 5 Examples of Successful Use of Information Technologies for Human Resource Management. https://parsers.me/big-data-and-machine-learning-in-hr-5-examples-of-successful-use-of-information-technologies-for-human-resource-management/ (2019)
4. Matyunina, J.: Best Machine Learning Application: Ideas for Mobile Apps. https://codetiburon.com/best-machine-learning-applications-mobile-apps/ (2018)
5. Wellers, D., Elliott, T., Noga, M.: 8 Ways Machine Learning is Improving Companies' Work Processes. Harvard Business Review. May 31 (2017)
6. Schrage, M., Kiron, D.: Improving Strategic Execution with Machine Learning. MIT Sloan Management Review. August (2018)
7. Faggella, D.: Machine Learning in Human Resources. https://emerj.com/ai-sector-overviews/machine-learning-in-human-resources/ (2019)
8. Weinberger, D.: How Machine Learning Pushes us to Define Fairness. Harvard Business Review. November 6 (2019)
9. Bogen, M.: All the Way Hiring Algorithms Can Introduce Bias. Harvard Business Review. May 6 (2019)
10. Feast, J.: 4 Ways to Address Gender Bias in AI. Harvard Business Review. November 20 (2019)
11. Ibarra, H., Petitti, P.: A 5-Part Process for Using Technology to Improve Your Talent Management. Harvard Business Review. July 24 (2018)
12. Wilson, J. Sachdev S., Alter, A.: How Companies are Using Machine Learning to Get Faster and More Efficient. Harvard Business Review. May 3 (2016)
13. Hodson, J.: How to Make your Company Machine Learning Ready? Harvard Business Review. November 7 (2016)
14. Schwartz, J.: Machine Learning is No Longer Just for Experts. Harvard Business Review. October 26 (2016)
15. Redman, T.C.: If Your Data is Bad, Your Machine Learning Tools are Useless. Harvard Business Review. April 2 (2018)
16. Ikudu, A., Lane, J. Staudt, J., Weinberg, B.: Occupational Classifications: A Machine Learning Approach. https://www.nber.org/papers/w24951 (2018)

A Machine Learning Approach to Analyze Marine Life Sustainability

Danish Jain, Shanay Shah, Heeket Mehta, Ayushi Lodaria, and Lakshmi Kurup

Abstract In this paper, an attempt is made to study the sustainability of marine life based on the quality of water. There are many quality parameters like 'pH', 'Temperature', 'Dissolved Oxygen (DO)', etc., which play a key role in determining the water quality index (WQI). The WQI values range from 19.3 to 99.8 which are then scaled down and classified into five categories using random forest algorithm, k-nearest neighbors (KNN), and support vector machine. By taking the WQI into consideration and can classify whether the water is competent for a fish to endure. The results obtained by the random forest algorithm provides 92.127% accuracy in classifying the water quality and 95.98% in determining the effect of WQI on marine life sustainability. The water quality is categorised to analyze its effect on aquatic life. The results were visualized on the basis of various years and states in India and were bolstered using k-fold cross-validation technique.

Keywords WQI · Random forest · Support vector machine · K-nearest neighbors · T—distributed stochastic neighbor embedding plot · K-fold cross validation · Confusion matrix · Gini index · Entropy · Information gain

*These authors contributed equally to this work.

D. Jain (✉) · S. Shah · H. Mehta · A. Lodaria · L. Kurup
Dwarkadas J. Sanghvi College of Engineering, Mumbai 400056, India
e-mail: danishjain2608@gmail.com

S. Shah
e-mail: shanayshah131@gmail.com

H. Mehta
e-mail: heeketmehta@gmail.com

A. Lodaria
e-mail: ayushilodaria@gmail.com

L. Kurup
e-mail: lakshmidkurup@gmail.com

© The Editor(s) (if applicable) and The Author(s), under exclusive license to Springer Nature Singapore Pte Ltd. 2021

A. P. Pandian et al. (eds.), *Proceedings of International Conference on Intelligent Computing, Information and Control Systems*, Advances in Intelligent Systems and Computing 1272, https://doi.org/10.1007/978-981-15-8443-5_53

1 Introduction

Water pollution causes the water quality level to degrade, thus affecting the health of marine life and subsequently of the humans who consume them. This makes it crucial to monitor the water quality and keep a check for the sustainability of marine life. The salubriousness of fish and other aquatic life depends on the water quality, measured by the water quality index (WQI) [1]. The water quality can be categorised based on WQI to determine how favorable it is for the sustainability of marine life.

Dr. Shweta Verma et al. employed artificial neural networks and radial basis function to predict the water quality based on WQI, which provides with an insight of how it is useful to analyze the water quality using sigmoid as transfer function and Levenberg Marqua as the learning rule [2]. Prabodha Kumar Meher et al. posit their research on evaluation of water quality of the Ganges River, where they take into consideration similar parameters, for analyzing water quality based on WQI values and apply Pearson correlation matrix to validate the results [3].

The paper is structured as follows: In Sect. 1, the proposed methodology has been explained where in Sect. 2, discussed the data preprocessing and cleaning steps. Further, the dataset was cleaned to bring uniformity [4]. In Sect. 3, the machine learning models like random forest classifier, k-nearest neighbor, and support vector machine have been elaborated to perform the classification of water [5]. K-fold cross validation is used to corroborate our findings and map it to marine sustainability. The comparative analysis of various machine learning techniques employed and results obtained are presented in Sect. 4 and finally concluded our paper in Sect. 5.

2 Proposed Methodology

See Fig. 1.

The above architecture depicts and describes the flow of the paper which includes cleaning of the dataset, normalizing the values of the attributes, assigning unit weights for the calculation of WQI, and comparison of various supervised classification algorithms. Subsequently, voting occurs where the algorithm with the highest accuracy is chosen and passed on for K-fold cross validation to corroborate and confirm our research.

3 Data Preprocessing

The data cleaning and the preprocessing process are done using Python programming language and its libraries. It provides with a clean dataset to perform any machine learning algorithm [6]. The dataset obtained is Indian water quality

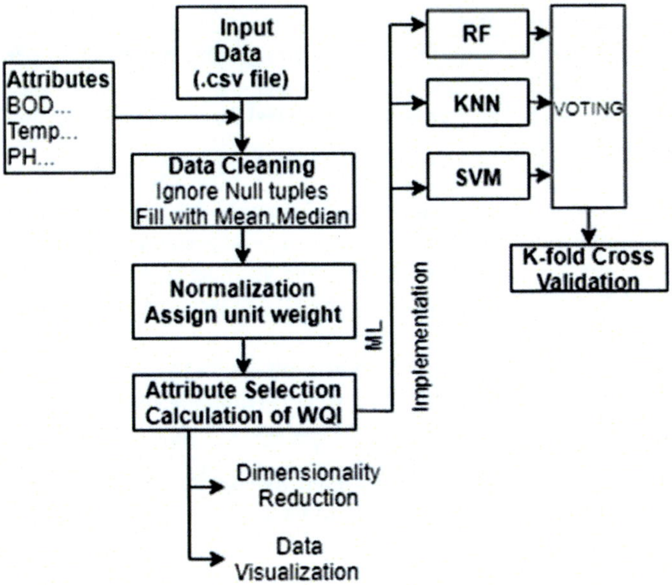

Fig. 1 System architecture

data [7] and contains pollutant measures in each column to analyze the pollution levels and temperature of water bodies across India. The dataset employed to implement the given task of analyzing the water quality contained various attributes and characteristics of numerous rivers and water bodies in multiple states of India. The dataset contains around 2000 records and 12 features, namely 'Station Code', 'Location', 'State', 'Temperature', 'D.O.(mg/l)', 'PH', 'Conductivity (μ mho/cm)', 'B.O.D.(mg/l)', 'NitrateNaNN + NitrieNaNN(mg/l)', 'Fecal-Coliform(MPN/100 ml)', 'TotalColiform(MPN/100 ml) Mean', 'Year'.

3.1 Data Cleaning

(1) *Ignore Tuples with Null Values*: At the beginning, the dataset contained three tuples in which all attributes had 'NaN' value and had to be dropped as those tuples were of no use to the analysis.
(2) *Fill in the Missing Values*: There were many tuples in which certain attributes had 'NaN' values while the other attributes had acceptable value. So, all the NaN values were first converted to an integer value zero (0) to make the dataset uniform.
(3) *Replace the Missing Values with Mean and Medians*: The values containing zero need to be replaced by a certain appropriate number relating to the attribute in order to achieve a higher accuracy. Hence, depending on the attribute, replaced

all the zero values of the dataset by their appropriate means of the attributes. In certain attributes, the mean calculated represented an outlier value to the attribute and such values lead to incorrect results and thus, could not be considered. So, those certain values of zeroes were replaced by the medians of the individual attributes [8].

(4) *Dropping Irrelevant Attributes*: Further, it was observed that the attribute station code had random four-digit numbers which were not following any serial order. These numbers had no characteristic feature in predicting the water quality of the water bodies. Hence, the station code attribute was dropped. The attribute 'Fecal Coliform' was already accounted for in the 'Total Coliform' values and thus had to be dropped from the dataset.

(5) *Normalization and Attribute Selection*: Thus, having received the cleaned dataset without anomalies, the next step is to normalize the attributes and assign unit weights to evaluate the final water quality index (Tables 1 and 2).

WQI stands for water quality index and this classification represents the water quality of the rivers and water bodies mentioned in the dataset [10].

The WQI calculations are done using the formulae given below:

Table 1 Normalization range of attributes [9]

Attributes	Excellent (1)	Good (2)	Poor (3)	Very poor (4)	Unsuitable (5)
D.O. (mg/l)	0–6	5.1–6	4.1–5	3.0–4	<3
PH	7–8.5	8.5–8.6 6.8–6.9	8.6–8.8 6.7–6.8	8.8–9.0 6.5–6.7	>9.0 <6.5
Conductivity (μ mho/cm)	0–75	75–150	150–225	225–300	>300
B.O.D. (mg/l)	0–3	3.0–6	6.0–80	6.0–80	>125
NitrateNaN N + NitrieNaNN (mg/l)	0–20	20.0–50	50.0–100	100–200	>200
Total coliform (MPN/100 ml) Mean	0–5	5.0–50	50–500	500–10,000	>10,000

Table 2 Unit weights of the attributes [9]

Parameter	Unit weight (Wi)
D.O. (mg/l)	0.281
PH	0.165
Conductivity (mho/cm)	0.009
B.O.D. (mg/l)	0.234
NitrateNaN N + NitrieNaNN (mg/l)	0.028
Total coliform (MPN/100 m) l mean	0.281

Table 3 Classification of WQI values [11]

Values of WQI	Quality of water
90–100	Excellent
70–90	Good
50–70	Poor
25–50	Very poor
0–25	Unsuitable

$$\text{WQI} = \frac{\sum w_i q_i}{\sum w_i n_i} = 1 \tag{1}$$

where q_i = quality rating of ith water quality attribute W_i = unit.
weight of ith water quality attribute; $\sum W_i = 1$.

$$q_i = 100[(V_i - V_{io})/(S_i - V_{io})] \tag{2}$$

V_i = estimated value of the ith attribute, V_{io} = ideal value of the ith attribute S_i = standard permissible value of the ith attribute. In most cases, $V_{io} = 0$ except PH where $V_{io} = 7$.

The unit weight (W_i):

$$W_i = \frac{K}{Si}, \tag{3}$$

where $K = \frac{1}{\sum \frac{1}{Si}}$.

Thus, the WQI value is calculated for all existing tuples and they are classified further into five classes, namely 'Excellent', 'Good', 'Poor', 'Very Poor', 'Unsuitable' depending on the range in which it lies given by Table 3.

Thus, the inference is arrived at 'Higher the Value of WQI, Better is the Quality of Water'. The data was further classified into numeric values, namely 1–5 in which 1 represents Excellent Quality, 2 represents Good Quality, 3 represents Poor Quality, 4 represents Very Poor Quality, whereas 5 represents Unsuitable quality. This classification in turn helps us in applying our machine learning models efficiently [12]. The next step is to extrapolate the water quality analysis to the marine life sustainability and show their correlation.

This paper aims at classifying water quality, affecting the health of marine life, into three categories, namely—(1) Desirable (2) Acceptable (3) Harmful.

The WQI values for the above-mentioned ranges can be calculated by gauging a range based on the following attribute values (Table 4).

The next step is to calculate the lower and upper WQI values for the desirable range and similarly, calculate the lower and upper WQI values of the acceptable range using the same WQI formula which was used for predicting the quality of water. On Calculation, the lowest desirable WQI value obtained is 60.90 and the lowest acceptable WQI value obtained is 48.50. These values are classified into

Table 4 Normalization ranges for health of marine life [13]

Attribute	Desirable range	Acceptable range	harmful range
Temperature (C)	20–30	15–25	<12, >35
D.O. (mg/l)	5	3–5	<3, >5
PH	6.5–9	7–9.5	<4, >11
Conductivity (μ mho/cm)	100–2000	30–50,000	<30, >50,000
B.O.D. (mg/l)	1–2	3–6	>10
NitrateNaN N + NitrieNaNN (mg/l)	0.1–4.5	0–100	>100
Total coliform (MPN/100 ml)l mean	28–10,000	0–10,000	>10,000

three classes of ranges—WQI values in the range of 60.90–100 were classified as Desirable Range for Marine Life Sustainability whereas the WQI values in the range of 48.50–60.90 were classified as Acceptable Range and all the values ranging below 48.50 were considered as a Harmful Range for Marine Life Sustainability. Thus, the Desirable Range is considered the best for Marine Life Sustainability.

4 Classification Using Machine Learning Techniques

Random Forest: Random forest algorithm is an ensemble algorithm used for classification in the field of machine learning. It is a supervised machine learning algorithm which comprises a collection of multiple decision trees. Being an ensemble algorithm, it adopts the best output, generated from the various decision trees, thus limiting, and curbing the generalization errors, as the size of the tree structure in the forest increases [14]. The decision tree uses the classification and regression tree (CART) algorithm to classify the tuples based on the target variable. This algorithm is used along with bagging, for resampling purposes, replacing the training dataset every time a novel tree grows [15].

Mathematical Representation and Equations: A tree structures is constructed, to classify the features, based on the equations and parameters given below. Gini Index (for any tuple S) can be used to generate the decision tree and is calculated as follows [16].

$$Gini(y, S) = 1 - \sum_{cj \epsilon dom(y)} \left(\frac{|\sigma_y = cj.S|}{|S|} \right), \tag{4}$$

The information gain and entropy are also useful to make a decision tree and subsequently derive its output. It can be calculated using the formulae given below [17]. For any given set *S*, *p* is the proportion of *S* belonging to class '*i*' [18].

$$\text{Entropy}(S) = \sum -p(i) \log_2 p(i) \tag{5}$$

$$\text{Gain}(S, A) = \text{Entropy}(S) - \sum \frac{|Sv|}{|S|} \text{Entropy}(S_v) \quad (6)$$

where S_v = subset of S for which attribute A has value v [19].

Implementation of Random Forest Algorithm: The random forest classifier algorithm is implemented on the dataset through Python programming language, to classify the water quality of various rivers and water bodies given, into certain ranges, based on a characteristic of water, called the water quality index (WQI). Thus, using the '*new_WQI*' attribute of our dataset, as the target variable to measure and categorise water quality, and the '*new_WQI_fishes*' to classify water quality, with respect to health of marine life, becomes our target variable as well.

(1) Water Quality Analysis Classification—The parameters like PH, conductivity, etc., have been considered as the features, to train our machine learning model against the '*new_WQI*' attribute of our dataset. Based on the rating of WQI (1, 2, 3, 4, 5), the water quality is assigned. Thus, performing the random forest algorithm on the dataset, obtained—*Number of training samples = 1391; Number of testing samples = 597*. The accuracy score, after testing the model, subsequent to the training phase, is equal to approximately 92.12%. This means that the machine learning model classified 92.12% of the testing data tuples correctly. The results of this are displayed below in the result section of this paper.
(2) Dependence of Health of Marine Life on WQI—Here, the target variable is changed, since needed to evaluate and categorize water quality index into categories into desirable water, acceptable water, and harmful water. To calculate this, the previously calculated parameter is made use '*WQI*' to determine and set the range for which the water quality may be conducive to or harmful to the health of aquatic life. These ranges have been calculated and reflected into a new column of the dataset, namely '*new_WQI_fishes*'. On training the model, obtained—*Number of training samples = 1391; Number of testing samples = 597*. The accuracy score (Accuracy) of classifying water quality into desirable, acceptable, or harmful for marine health is approximately equal to 95.97%. The results of this are displayed below in the result section of this paper.

Random Forest Versus Other Supervised Classification Algorithms: Some of the other classification algorithms, which were used to analyze the data were k-nearest neighbors (kNN) algorithm and support vector machine (SVM). The accuracy score of both was relatively lower than random forest algorithm. The comparison for this was made for classification and analysis of water quality. A comparison among the applied algorithms is tabulated below in the result section of the paper. The obtained results are presented, which convincingly demonstrates the performance of all the algorithms considered.

K-Fold Cross Validation: This technique divides the dataset into k folds and out of these k folds, $(k - 1)$ folds are used to train our model and the remaining is used for testing. This introduces our model to randomness and subsequently determines its performance, when encountered and implemented against different training and testing sets. Implementing the k-fold cross validation technique for various values of K, obtained results, which corroborate our findings by reducing and prediction/true errors [20]. Hence, making sure that our model is unbiased to data and provides cogent results. The implementation and results obtained are documented below in the results section of the paper.

5 Results and Analysis

5.1 Data Visualization

In Fig. 2, the number of WQI values termed good, increases in a nonlinear way proving that the water pollution levels in different water bodies have reduced since 2004. The purple line in the above graph indicates that the number of water bodies having unsuitable WQI is almost constant. However, between 2012 and 2013, there is a dip in the number of water bodies having excellent and good WQI while there is a slight rise in the number of water bodies having poor, very poor, and unsuitable range indicating that the water pollution levels during this time period had significantly increased in different states of India. This analysis over the years indicates how water quality deteriorates and suggests that measures should be taken to enhance the quality of water, making it suitable for marine life.

In Fig. 3, water quality indices (WQI's) of water bodies is represented in some of the major states of India suitable for aquatic life, which provides a direct visualization of the data. On comparing the WQI's of some of the states, Kerala has the highest number of water bodies with desirable WQI's for supporting marine life followed by Goa. Contrarily, Rajasthan has the maximum number of water bodies with WQI's falling in the harmful range for aquatic life. As a result, states having poor WQI's should implement stringent laws to prevent further pollution of these

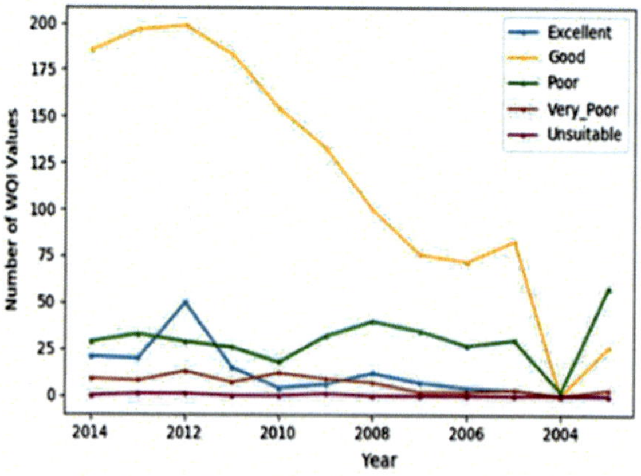

Fig. 2 Graph for WQI from 2003 to 2014

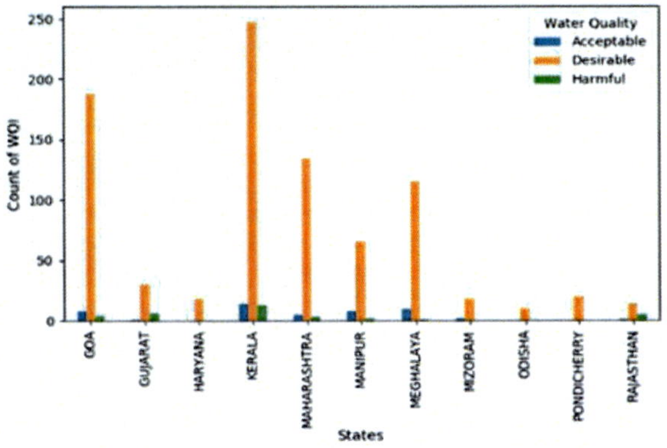

Fig. 3 New_WQI_fish count per state

water bodies, to improve its quality and make it salubrious to the health of marine life and subsequently humans, who consume them. This statewise analysis depicts how particular states should make improvements to improve water quality and make it conducive and safer for marine life sustainability.

T-Distributed Stochastic Neighbor Embedding Plot

TSNE is a novel approach used as a machine learning algorithm for data visualization. It is used for dimensionality reduction, to convert high-dimensional data into a lower (2-D/3-D) dimensional. Here, all the attributes are taken into consideration and represented it as a two-dimensional data. Figure 4 demonstrates that all the values in the same WQI range find themselves in the same or the neighboring cluster. The uniform spread of the cluster helps us infer that the accuracy of the model used is desirably high, hence corroborating to our findings.

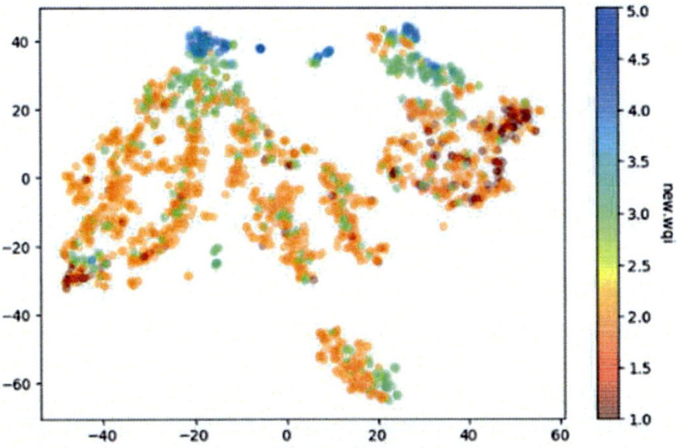

Fig. 4 T—distributed stochastic neighbor embedding plot

Fig. 5 Cross tab for water quality analysis

Predicted Outcome	Excellent	Good	Poor	Very Poor
Actual Outcome				
Excellent	44	5	0	0
Good	2	409	5	0
Poor	0	22	85	2
Unsuitable	0	0	0	1
Very Poor	0	0	10	12
ACCURACY :	92.12730318257957			

5.2 *Confusion Matrix/Cross Tab for Water Quality Assessment*

The diagonal elements of the above confusion matrix is the total number of tuples correctly classified [21, 22] is seen. The confusion matrix/cross tab can be given as follows (Fig. 5):

$$\text{Accuracy} = \frac{\text{Total number of Correctly Classified Tuples}}{\text{Total Number of Tuples}}$$
$$\text{Accuracy} = \frac{44 + 409 + 85 + 12}{597} = \frac{550}{597} = 0.92127$$
$$\text{Accuracy Score} = \text{Accuracy} * 100 = 92.127\% \quad (7)$$

Thus, given any set of attributes (feature values), the machine learning model can accurately predict and classify the quality of water into Excellent, Good, Poor, Very Poor or Unsuitable, with an accuracy of approximately 92.12%. An accuracy of 92.127% succinctly proves the efficiency of this model, for our purposes. As it can be observed from the confusion matrix, very few tuples of the testing dataset have been incorrectly classified. Thus, this satiates our requirement of analysis and determining the water quality of any water body, given merely the parameters of it. This helps us obtain a detailed and informed quality analysis of the water.

5.3 *Confusion Matrix/Cross Tab for Dependence of Health of Marine Life on WQI*

Therefore, conclude that for a given a set of values of the features considered, the water quality is classified as either desirable, acceptable, or harmful to the health of marine life with an accuracy as high as 95.97% [22]. The diagonal elements of the confusion matrix highlight the number of tuples correctly classified. Thus, this can be similarly extended to all the outcomes possible and determine the accuracy of the algorithm employed as follows (Fig. 6).

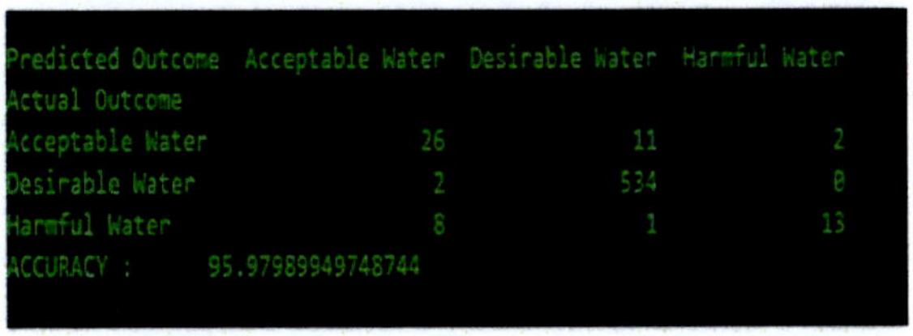

Predicted Outcome	Acceptable Water	Desirable Water	Harmful Water
Actual Outcome			
Acceptable Water	26	11	2
Desirable Water	2	534	0
Harmful Water	8	1	13

ACCURACY : 95.97989949748744

Fig. 6 Cross tab for WQI effect on marine life

Table 5 Comparison between classification algorithms

	SVM algorithm	K-nearest neighbors	Random forest
Total number of testing tuples.	597	597	597
Number of Correctly classified tuples	417	460	550
Number of falsely classified tuples	180	137	47
Accuracy score	69.85%	77.05%	92.12%

$$\begin{aligned}\text{Accuracy} &= \frac{26+534+13}{26+534+13+11+2+2+8+1} = \frac{573}{597} = 0.95979\\ \text{Accuracy Score} &= \text{Accuracy} * 100\\ \text{Accuracy Score} &= 95.98\% \end{aligned} \tag{8}$$

An accuracy score of 95.98% corroborates that the model is highly suitable and adept for determining how a particular water body might affect the health of the marine life surviving in it. The obtained accuracy states that the classification made by the random forest model is correct approximately 95.98% of the time (Table 5).

5.4 *Comparison of Various Algorithms*

Reason for Choosing Random Forest Algorithm - The above table provides a comprehensive between all three algorithms implemented. As evident from the results tabulated above, the random forest algorithm provided the best output, with the highest accuracy and hence maximum reliability. In other words, given a tuple of values corresponding to any water sample, the accuracy of predicting the quality given the set would be the highest using the random forest classifier, as compared to SVM and KNN. The presence of numerous features maximizes the performance of random forest algorithm hence, minimizing the generalization errors, while providing apt split selection between them. The voting process selects the best output of the decision tree, which computes the solution based on probability. Thus, random forest algorithm provides an analytically and statistically more cogent, efficient, and accurate result, which compels us to adhere to it, as opposed to the support vector machine and k-nearest neighbors algorithm.

5.5 *Results of K-Fold Cross Validation*

See Figs. 7 and 8.

To calculate the accuracy of the above confusion matrix—[21].

$$\begin{aligned} \text{Accuracy} &= \frac{132 + 1389 + 299 + 58}{1988} = \frac{1878}{1988} \\ \text{Accuracy} &= 0.944 \end{aligned} \tag{9}$$

To calculate the accuracy of the above confusion matrix—[21].

$$\begin{aligned} \text{Accuracy} &= \frac{132 + 1386 + 305 + 60}{1988} = \frac{1883}{1988} \\ \text{Accuracy} &= 0.947 \end{aligned} \tag{10}$$

Fig. 7 Five-fold cross validation

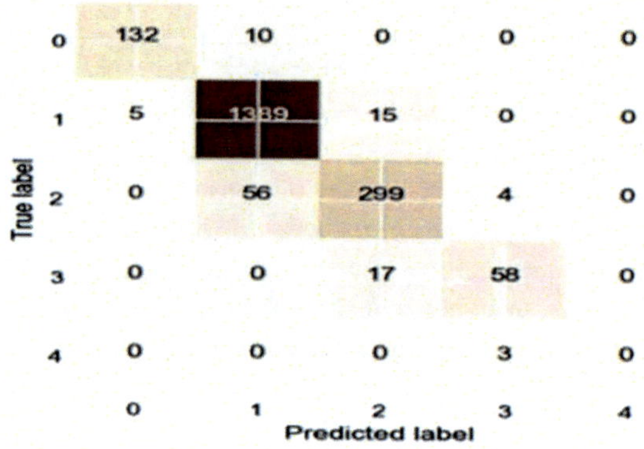

Fig. 8 Twenty-fold cross validation

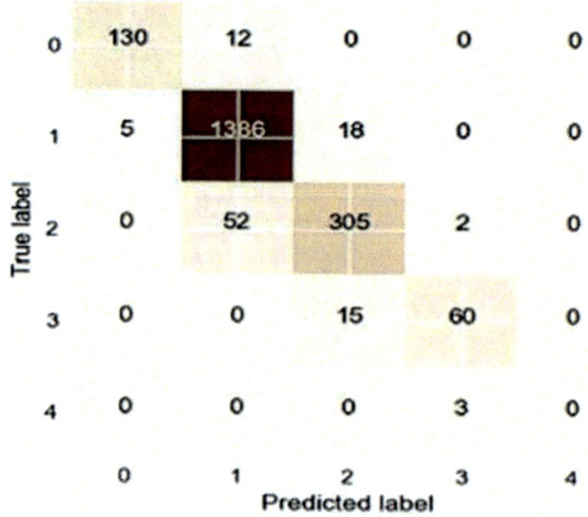

In this case, the dataset is divided into five folds and 20 fold, respectively. Thus, this type of cross validation ensures that our model can perform equivalently, given a different training and testing sample. The accuracy for 5-fold and 20-fold cross validation is 94.4% and 94.7%, respectively, which is similar to the accuracy obtained using random forest (92.12%). This indicates that our findings were cogent and negligibly swayed by any disturbance or prediction errors. This corroborates the results obtained in Eq. 7 and likewise in Eq. 8.

6 Conclusion

The purpose of doing the above classification and analysis is to gauge the water quality not only to suffice the various purposes of mankind but also analyze how the water quality affects marine life. The results obtained by our random forest-based model provides 92.127% accuracy in classifying the water quality of any given water sample, into five categories based on the above-mentioned parameters like PH, D.O., total coliform, etc. Moreover, the accuracy can be determined if the water quality is desirable, acceptable, or harmful to the salubriousness of marine life like fishes, is nearly 95.98%. Therefore, the analysis of water bodies carried out through this paper is based solely on scientific parameters and is of paramount importance in ensuring the well-being of human and aquatic life.

Future Scope: The above research methodology should be performed on the data collected for the recent years, owing to the pollution rise in recent years. To obtain a detailed analysis, a more thorough localized search and scrutiny must be conducted to accurately determine the pollution levels and its effect on marine life sustainability. Moreover, computer vision and image processing techniques can be implemented on fish tissues to gauge the freshness of the marine life. The results of our research can be extrapolated to perform the above and determine a relation of sources of pollution to marine life sustainability.

References

1. Bhardwaj, D., Verma, N.: Research paper on analysing impact of various parameters on water quality index. Int. J. Adv. Res. Comput. Sci. **8**(5), 2496–2498 (2017)
2. Prajithkumar, S., Verma, D.S., Mane, S.J.: Prediction of water quality index of Pavna river using ANN model. Int. J. Eng. Res. Tech. (IJERT) **03**(12), 121 (2014)
3. Meher, P.K., Sharma, P., Gautam, Kumar A., Mishra, K.: Evaluation of water quality of Ganges river using water quality index tool. Environ. Asia. **8**, 124–132 (2015). https://doi.org/10.14456/ea.2015.15
4. Boulder Area Sustainability Information Network—Water Quality Index [Online]. Available http://bcn.boulder.co.us/basin/watershed/wqi_info.html
5. Gajendran, C.: Water quality assessment and prediction modelling of Nambiyar River basin. Tamil Nadu, India (2011)
6. Weihs, C., Ickstadt, K.: Data science: the impact of statistics. Int. J. Data Sci. Analyt. (2017)

7. Kaggle—Indian Water Quality Data Available: https://www.kaggle.com/anbarivan/indian-water-quality-data
8. Koshley, D.K., Halder, R.: Data cleaning: an abstraction-based approach. IEEE (2015)
9. Shah, K., Joshi, G.: Evaluation of water quality index for River Sabarmati Gujarat India. Appl. Water Sci. **7**, 1349 (2019)
10. Oni, O., Fasakin, O.: The use of water quality index method to determine the potability of surface water and groundwater in the vicinity of a municipal solid waste dumpsite in Nigeria. Am. J. Eng. Res. (AJER) **5**, 96 (2016)
11. Akkaraboyina, M.K., Raju, B.S.N.: Assessment of water quality index of River Godavari at Rajahmundry. Univ J Environ Res Technol **2**, 16 (2012)
12. Bhardwaj, R.M. (2012) Scientist 'C'. Water quality monitoring. Central Pollution Control Board, India
13. Bhatnagar, A., Devi, P.: Water quality guidelines for the management of pond fish culture. Int. J. Environ. Sci.
14. Breiman, L.: Random forests (Dept. of statistics, University of California, Berkeley, CA)
15. Biau, G.: Analysis of random forests model. J. Mach. Learn. Res. **13**, 1063 (2012)
16. Rokach, L., Maimon, O.: Decision trees (2005). https://doi.org/10.1007/0-387-25465-x_9
17. Yang, Ning, Li, Tianrui, Song, Jing: Construction of decision trees based entropy and rough sets under tolerance relation. Int. J. Comput. Intell. Syst. (2007). https://doi.org/10.2991/iske.2007.258
18. Gaber, M.M., Atwal, H.S.: An entropy-based approach to enhancing random forests. Int. Dec. Tech. **7**(4), 319–327 (2013). https://doi.org/10.3233/IDT-130171
19. Liu, Y., Wang, Y., Zhang, J.: New machine learning algorithm: random forest. In: Liu, B., Ma, M., Chang, J. (eds) Information Computing and Applications. ICICA 2012. Lecture Notes in Computer Science, vol. 7473. Springer, Berlin, Heidelberg (2012)
20. Rodríguez, J., Pérez, A., Lozano, J.A.: Sensitivity analysis of k-fold cross validation in prediction error estimation. IEEE Trans. Patt. Anal. Mach. Intell. **32**, 569–575 (2010)
21. Visa, Sofia, Ramsay, Brian, Ralescu, Anca, Knaap, Esther: Confusion matrix-based feature selection. CEUR Worksh. Proc. **710**, 120–127 (2011)
22. Ting, K.M.: Confusion matrix. In: Sammut, C., Webb, G.I. (eds.) Encyclopedia of Machine Learning and Data Mining. Springer, Boston, MA (2017)

A Novel Trespassing Detection System Using Deep Networks

Harshit Gupta, Rahul Singh Yadav, Sumith M. Sree Kumar, and M. Judith Leo

Abstract Trespassing is the act when someone enters any restricted area without prior permission, and it is considered to be an act of crime. It has been a serious issue for various institutes and organizations to maintain their security and surveillance to prevent the acts of vandalism. Therefore, it is an important attribute to continuously monitor the violators. Various researches have been done so far on trespassing detection, but none of the method allows the specific object of interest to enter in a particular area. The proposed method allows us to choose the object of interest, and also it will differentiate between the types of violator (car, motorcycle, truck, pedestrian, etc.) and sends an alert text message to a designated destination with the image of the violator. Object detection is done using the YOLO algorithm which is then integrated with camera and Raspberry Pi 3 along with cloud integration. The reason for this study is to develop an efficient method that allows/restricts the object of interest in a particular area, to prevent illegal trespassing. Experimental results show that the proposed method detects the trespassers with high accuracy and speed.

Keywords Object detection · Trespassing detection · YOLO, object recognition · Vehicle detection

H. Gupta (✉) · R. S. Yadav · S. M. S. Kumar · M. J. Leo
Department of CSE, Hindustan Institute of Technology and Science, Chennai, Tamil Nadu, India
e-mail: notsoharshit@gmail.com

R. S. Yadav
e-mail: rahulsinghmp968@gmail.com

S. M. S. Kumar
e-mail: sumithkumar710@gmail.com

M. J. Leo
e-mail: judithl@hindustanuniv.ac.in.com

© The Editor(s) (if applicable) and The Author(s), under exclusive license to Springer Nature Singapore Pte Ltd. 2021

A. P. Pandian et al. (eds.), *Proceedings of International Conference on Intelligent Computing, Information and Control Systems*, Advances in Intelligent Systems and Computing 1272, https://doi.org/10.1007/978-981-15-8443-5_54

1 Introduction

Trespassing is when someone enters any confined territory without authorization, and it is viewed as an act of crime. It has been a serious issue for various institutes and organizations to maintain their security and surveillance without equipping large workforce into it. In current scenario, losses due to lack of proper surveillance result in a huge sum of money and loss of confidential organizational resources. Among various cases reported due to trespassing, lack of effective surveillance and preventive measures precedes other causes. For example, on October 2018, two Indian railway passenger trains crashed into a crowd, which nearly killed 59 people; the main cause was lack of monitoring and alerting system in the area. Moreover, a number of fatality and accidents due to trespassing are numbered more than tens of thousands every year and this figure is on a rise for past 10 years. Thus, the significance of appropriate trespassing detection and avoidance system is necessary in many parts of India, which can be for either private or government organizations. The traditional use of closed-circuit television (CCTV) camera for surveillance requires the help of human examiners to survey the video feed of the premises for 24 $\times$ 7 in order to establish security measures in possible trespassing influenced locations. This method requires very complex steps, cost, high human involvement and more time to operate. Various trespassing detection systems are available that automate the detection steps, but it is not applicable to all surveillance scenarios to monitor different types of objects.

Due to such flaws in the existing system, the proposed method is developed as an automated, flexible, versatile and hassle-free framework to tackle these trespassing activities by utilizing the state-of-the-art deep learning algorithm. The proposed method focuses on developing a versatile trespassing detection system that dynamically selects the object of interest while monitoring the secured premises. It can also be implemented in cloud as a real-time application with high processing speed and accuracy.

The paper is organized as follows: In Sect. 2, the most representative and successful approaches for trespassing detection are reviewed. Their robustness to the challenges involved and their limitations are also discussed. Section 3 details the architecture of the proposed trespassing detection system along with the clear description of the methodology involved. Section 4 discusses the results for various experimental set-up and the performance of the system with different types of selected object. Section 5 describes the conclusion of the proposed method.

2 Literature Review

This section gives a brief description about the methodology and algorithms used in the existing trespassing detection systems along with their accuracy and limitations.

In [2], Bashir et al. focus on the railroad trespassing detection system. They have implemented a railroad safety system using RCNN for trespassing detection. They successfully trained the data of rail tracks, and the system can detect the pedestrians which gives a good security mechanism for railroad crossing. However, the algorithm requires a high-quality image for trespassing detection and it has low accuracy despite using two-stage mechanisms for detection of the trespasser. The system lacks accuracy mostly in the second stage, i.e. in the object recognition phase.

Wong et al.[6] develop the trespassing detection method using a thermal camera for pedestrian detection. In this method, the algorithm converts thermal image to binary image and then the object is detected. If a trespasser (Alzheimer patient or thief) is detected, an alarm will be triggered to inform the person in charge. A drawback of this method is the thermal image is accurate only in close range. Also, the thermal camera is expensive, and this system does not work for vehicle trespassing.

In [14], Yilmaz et al. compare different deep learning approaches like RCNN, YOLO: v2/v3, faster RCNN, SSD, etc., with different datasets and find the accuracy among these algorithms. These methods are applied only to vehicle detection, and information regarding pedestrian detection is not available. Their research does not give much information about real-time implementation.

Canayaz has implemented heavy vehicle detection for the surveillance of road safety in the highway [8]. Gaussian mixture models are used for background filtration. However, in some circumstances the method also detects light vehicle but the approach cannot be applied for pedestrian detection.

Li et al. [10] have adopted the histogram of oriented gradient (HOG) method to extract vehicle-type features in images and then classified those features using the support vector machine (SVM), to achieve vehicle detection. Their system improves the accuracy of forwarding vehicle detection and has a high detection rate. However, this method is suitable only for vehicle detection and can be improved by replacing it with deep learning methods.

Maruyama et al. [13] have implemented a security system for the trespasser detection. Whenever any trespassing event occurs, the system will detect the trespasser and immediately send the notification in the email. This system detects the moving object by time-varying image processing, and it is also used for detecting different vehicles and pedestrians. However, in this system, there is no option for selecting the object of interest for surveillance.

Zhang et al. [5] have developed the method DP-SSD, a single deep neural network for vehicle detection in urban traffic surveillance. The authors have given the mean average precision (mAP) of different algorithms by implementing them with KITTI datasets and give comparisons between different algorithms based on their accuracy.

3 The Proposed Trespassing Detection System

The architecture of the proposed trespassing detection system is depicted in Fig. 1, and the system is divided into 3 layers—RPi layer, detection layer and response layer. Object detection is done on the video feed using the YOLO algorithm [7] which is

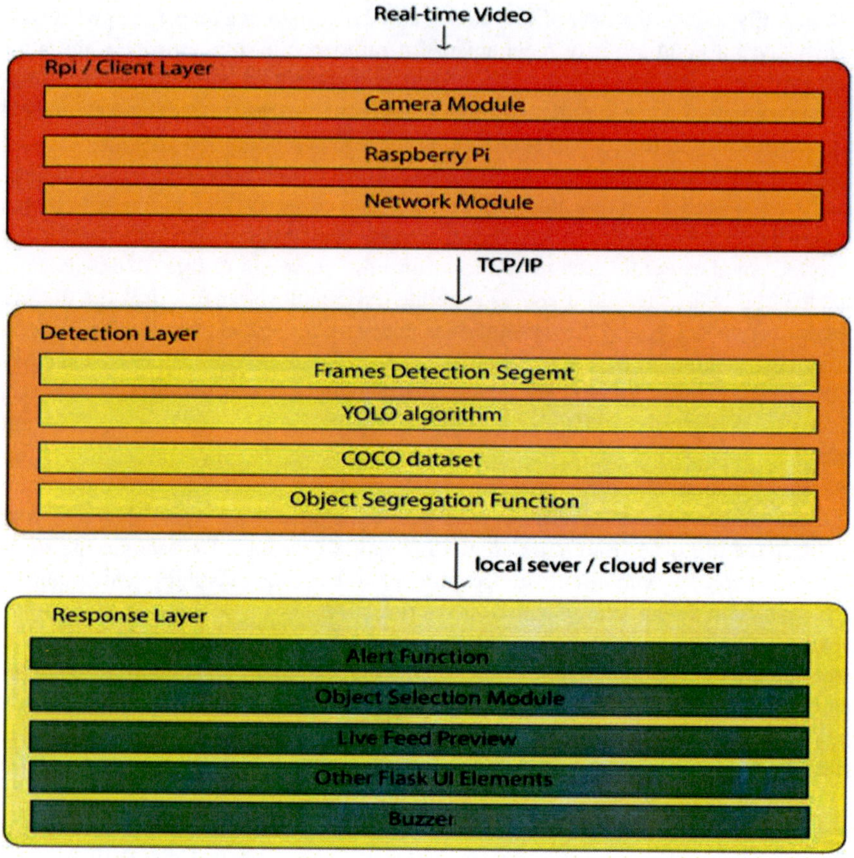

Fig. 1 Architecture of the proposed trespassing detection system

Fig. 2 Architecture of RPi layer

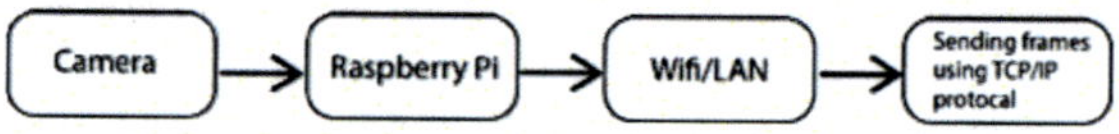

trained with the COCO dataset to detect all the objects in the frame. Finally, the object of interest for trespassing detection is selected using the proposed algorithm.

3.1 RPi Layer

RPi consists of RPi camera, Raspberry Pi and network module. The RPi camera records frames and sends it to RPi, where each frame is serialized and transmitted

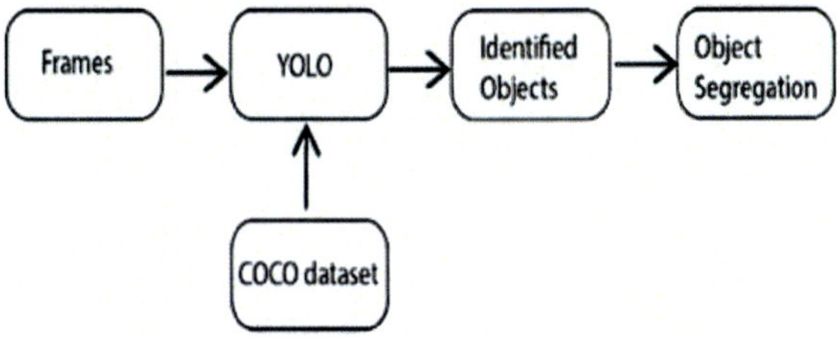

Fig. 3 Architecture of detection layer

Fig. 4 Architecture of response layer

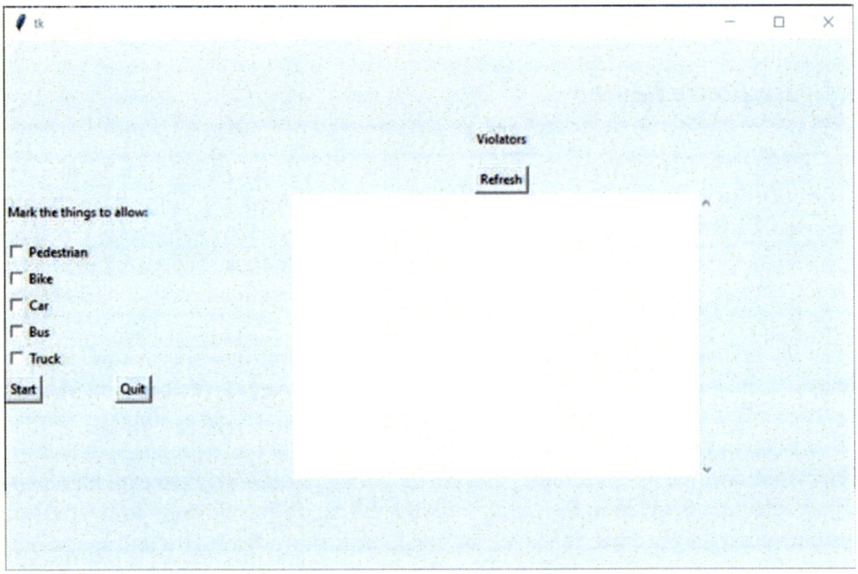

Fig. 5 Selection of object of interest

to the server using the network module. The network module can be either Wi-Fi or LAN connection. The network module specifies the mode of connectivity between the Raspberry Pi and the server (local/cloud server), which is Transmission Control Protocol/Internet Protocol (TCP/IP). By utilizing the socket client program, the frame packets can be delivered to the server IP address. Thus, RPi layer acts as a data collector and transmitter rather than processing it.

3.2 Detection Layer

The detection layer is responsible for receiving the frame packets from Rpi layer and processing it to identify the segregated objects required for the proposed system. This layer resides on either cloud server or local server. The components of detection layer are frame detection segment, YOLO algorithm trained on COCO dataset and object segregation module. This layer uses the TCP/IP to receive the frames from RPI layer. The TCP/IP is a set of networking protocols that allows two or more computers to communicate. Here, the socket concept of TCP/IP is used for client and server interaction. Utilizing the protocol, the received frames are filtered and processed with the object detection algorithm (YOLO) to detect all the objects residing in each frame. These objects will be passed on to the object segregation function, where all the objects associated with the trespassing detection system like pedestrians, heavy vehicle, light vehicle, stray animals, etc., are segregated. Objects of less importance like traffic light, cell phone, etc., will not be taken into consideration. Then, the control will be passed onto the response layer.

3.3 Response Layer

The main objective of response layer is to provide control to the user to monitor and notify the intrusion. This layer utilizes flask environment providing a hassle-free experience to the user. Flask is a lightweight Web application framework, designed to make an easy platform to scale up to complex applications. The alert function in the response layer will be activated once the object segregation function of detection layer is executed.

In the response layer, the users can customize the restricted list for trespassing detection, by selecting the object of interest (pedestrian/heavy vehicle/light vehicle, etc.) according to their needs through object selection module. For example, if entry of vehicles (heavyweight/lightweight) is restricted into the secured premises only those objects are selected for monitoring and not the pedestrians. A detailed explanation on object selection is given in Sect. 3.4. Initially, all the objects relevant to trespassing system, segregated in detection layer, are available in the restricted list and are marked as 1. After customizing the restricted list, the selected objects alone will be monitored for trespassing, and the remaining objects will be assigned 0 and removed from restricted list temporarily. This layer also provides live feed preview and other flask UI components to control various functions in the project like process termination. Finally, on detecting the object of interest in the video feed, the buzzer is triggered to raise alarm.

3.4 *Selecting Object of Interest*

This module is used to mark the "allowed" and "not allowed" items. The freedom of selection lies with the user. Given is a list of objects including pedestrians, cars, bikes, etc. The things to be allowed in the area are "checked", and unallowed objects are "unchecked". The unchecked (restricted) items are then collectively stored in a list, say *R*. On press of start button, the *ObjectDetection function* is called with the list *R* as *argument*.

3.5 *Object Detection with YOLO*

The proposed method uses You Only Look Once (YOLO) algorithm for object detection. YOLO uses convolutional neural networks (CNNs) for object detection, and it is different from other object detectors because as the name indicates it has to see each image only once; i.e. YOLO requires only one forward propagation pass through the network to make predictions. Due to this, YOLO is one of the fastest detection algorithms (although it sacrifices some accuracy) and is better at real-time object detection (up to 30 FPS). The model works by first splitting the input image into a grid of cells, where each cell is responsible for predicting a bounding box, if the centre of a bounding box falls within it. The bounding box is represented by the *x*, *y* coordinate, width, height and the confidence. These bounding boxes are weighed by the predicted probabilities.

The class prediction is also based on each cell. For example, an image may be divided into a 7×7 grid and each cell in the grid may predict 2 bounding boxes, resulting in 94 proposed bounding box predictions. The class probabilities of the bounding boxes are mapped with confidences and are then combined into a final set of bounding boxes and class labels. The image taken from the paper [7] summarizes the two outputs of the model as given in Fig. 6.

Yolo version3 [7] has been selected as the fundamental objection detection algorithm for our trespassing detection system. The YOLO algorithm is different from other method as the bounding box and class prediction are done simultaneously.

3.5.1 Bounding Box Prediction

In the bounding box prediction, input image or frame is first divided into $S \times S$ grid. The bounding box is defined in each box or cell having a confidence value. The confidence is the probability of an object that exists in each bounding box, and it is defined as

$$C = \Pr(\text{Obj})^{*}\text{IOU} \tag{1}$$

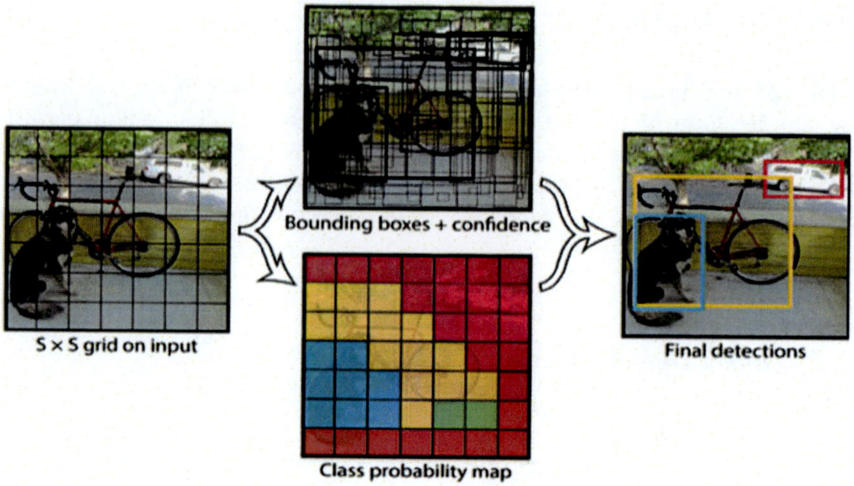

Fig. 6 Object detection using YOLO algorithm [7]

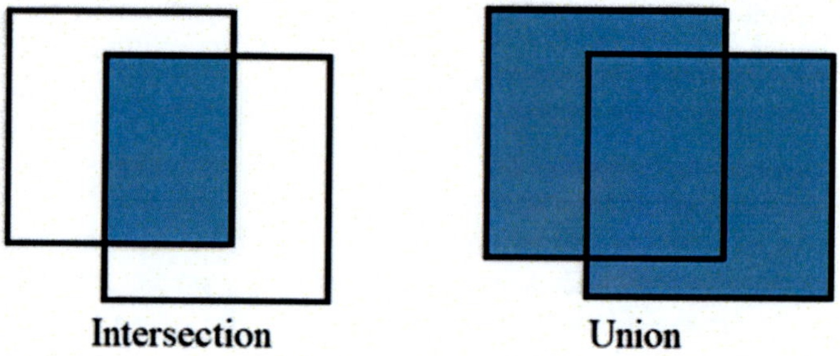

Fig. 7 Intersection over union

In this formula, C is the confidence of the object in each bounding box, Pr(obj) is the probability of the object and IOU is intersection over union. It is the fraction value which is between 0 and 1. Intersection is the overlapping area or the common region between the predicted value and the actual value, i.e. ground truth, whereas the union is the total area between the predicted and the ground truth values.

$$\mathrm{IOU} = \frac{\mathrm{AreaofIntersection}}{\mathrm{AreaofUnion}} \tag{2}$$

After the bounding box is done, simultaneously each cell will also start predicting the conditional class probability. The class probability in each cell is defined as

$$\Pr(\mathrm{Class}_i|\mathrm{Object}) * \Pr(\mathrm{Object})\mathrm{IOU}^{\mathrm{truth}}_{\mathrm{pred}}$$

$$= \Pr(\text{Class}_i) * \text{IOU}_{\text{pred}}^{\text{truth}} \tag{3}$$

YOLO uses the equation which is given below to calculate the loss, and then it optimizes the confidence.

The loss function is used for the correction of the centre of bounding box for each prediction. Each frame is being divided into $S \times S$ grids, having a bounding box for each grid. Here, b_w and b_h refer to the dimension of the bounding box; i.e. height and width and b_x and b_y are the variables that are referring to the centre of the bounding b_x grid for each prediction. The variables λ_{coord} and λ_{noobj} are used for the manipulation, i.e. to increase the emphasis of the object on the bounding box, and also it is used to decrease the emphasis on the object on the bounding box. In the above equation, C refers to the confidence in the bounding box. Lower the loss, higher will be the performance and the accuracy of the algorithm.

$$\begin{aligned} F(\text{Loss}) = {} & \lambda_{\text{coord}} \sum_{i=0}^{s^2} \sum_{j=0}^{A} 1_{ij}^{\text{obj}} \left[\left(b_{x_i} - b_{\hat{x}_i}\right)^2 + \left(b_{y_i} - b_{y_i}\right)^2 \right] \\ & + \lambda_{\text{coord}} \sum_{i=0}^{s^2} \sum_{j=0}^{A} 1_{ij}^{\text{obj}} \left[\left(\sqrt{b_{w_i}} - \sqrt{b_{\hat{w}_i}}\right)^2 + \left(\sqrt{b_{h_i}} - \sqrt{b_{\hat{h}_i}}\right)^2 \right] \\ & + \sum_{i=0}^{s^2} \sum_{j=0}^{A} 1_{ij}^{\text{obj}} \left(C_i - \hat{C}_i\right)^2 \\ & + \lambda_{\text{noobj}} \sum_{i=0}^{s^2} \sum_{j=0}^{A} 1_{ij}^{\text{noobj}} \left(C_i - \hat{C}_i\right)^2 \\ & + \sum_{i=0}^{s^2} 1_i^{\text{obj}} \sum_{c \in \text{classes}} \left(p_i(c) - \hat{p}_i(c)\right)^2 \end{aligned} \tag{4}$$

4 Results and Discussion

4.1 COCO Dataset

The COCO dataset has been used in the proposed trespassing detection system. COCO is a large-scale object detection, segmentation and captioning dataset. The dataset has been created by gathering images of complex everyday scenes containing common objects in their natural context. It includes 80 classes of objects and segmented images. Also, it is a collection of nearly 40,000 images of classes like car, truck, bike, bicycle, person and other specific animals which are labelled.

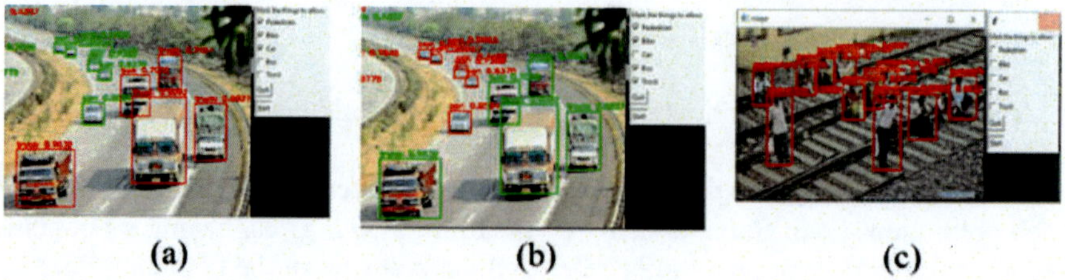

Fig. 8 Restriction of **a** heavy vehicles, **b** light vehicles and **c** pedestrians

4.2 Experimental Set-up and Results

The proposed method has been experimented by giving inputs at three different scenarios—(a) restriction of heavy vehicles, (b) restriction of light vehicles and (c) restriction of pedestrians, and the accuracy of detection is calculated.

In order to monitor or restrict the entry of heavy vehicles in any premises or in any freeway/highway, the user can select the object of interest, i.e. pedestrian, bike, car, etc., that can be allowed to enter the restricted area. The remaining heavy vehicle objects like buses and trucks will be restricted.

Figure 8a depicts an example of restriction of heavy vehicles, where trucks are marked in red. This means that trucks are not allowed (unchecked boxes) in this area. But bikes and cars are allowed (checked boxes) and hence green.

Similarly, trespassing of light vehicle and pedestrians can be detected as shown in Figs. 8b, c where the allowed objects are marked in green and restricted objects are marked in red.

4.3 Output of Notification Module

In addition to detecting the act of trespassing, all the trespassers can be identified and their image can be sent to concerned authority for taking further actions. The screenshots of trespasser information are given in Fig. 9.

The precision, recall, $F1$-score and interaction over union (IOU) are calculated for different entities like pedestrian, bike and car to find the accuracy of the system and given in Table 1.

Precision is defined as the number of true positive divided by actual result (sum of true positive and false positive).

$$\text{Precision} = \frac{(\text{TP})}{(\text{TP} + \text{FP})} \tag{5}$$

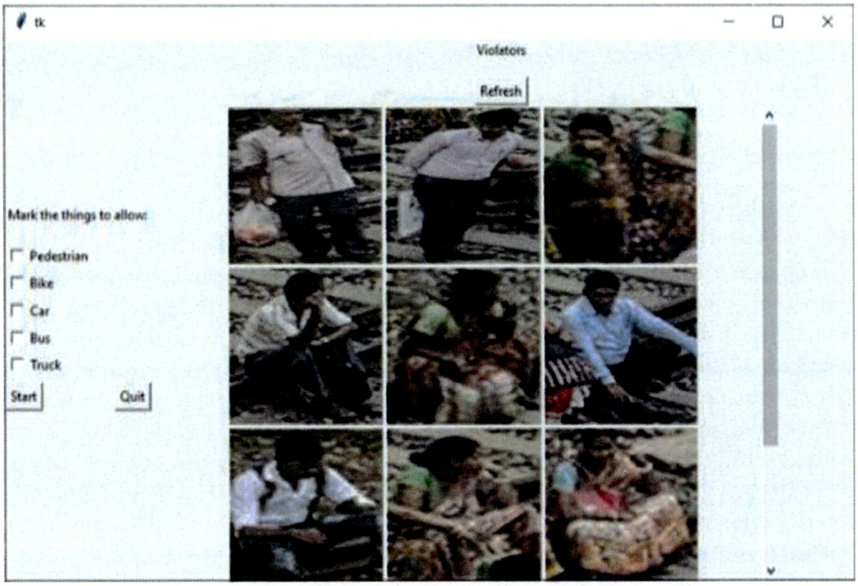

Fig. 9 Screenshot of trespassers

Table 1 Evaluation metrics for detected objects

	Precision	Recall	$F1$-score	Average IOU	Support	Loss
Empty	0.97	0.98	0.97	0.94	2443	0.3
Pedestrian	0.94	0.96	0.95	78.24	1438	0.6
Bike	0.97	0.98	0.98	0.98	80	0.3
Car	0.98	0.98	0.98	0.95	7536	0.2
Bus	0.98	0.98	0.98	0.98	4256	0.2
Truck	0.98	0.98	0.98	0.97	5546	0.2

Recall is used for measuring the success of prediction when the classes are imbalanced. It is defined as the ratio of true positive to predicted result (sum of true positive and false negative).

$$\text{recall} = \frac{(\text{TP})}{(\text{TP} + \text{FN})} \tag{6}$$

Accuracy is defined as the number of correct predictions divided by total number of predictions.

$$\text{Accuracy} = \frac{(\text{TP} + \text{TN})}{(\text{TP} + \text{TN} + \text{FP} + \text{FN})} \tag{7}$$

$F1$-score = (2 * precision * recall)/(precision + recall) (8).
Support = number of values for each class (9).
Thresh confidence = 0.1

5 Conclusion

The objective of this paper is to develop an automatic trespassing detection that allows only object of interest to enter a particular zone or area. The system imp ejects detection at a very high speed. In this paper, the limitations of the existing trespassing detection methods have discussed and highlighted the need for the versatile system that can be applied to any scenario, i.e. dynamically selecting the object of interest to enter into any premises which is more cost-effective and efficient. This method is successfully implemented, and the accuracy of restricting the selected objects is satisfactory. This work can be implemented in a wide field of applications like security service and parking system.

Reference	Features	Total data	Accuracy (%)	Error rate
Zaman et al.	Colour	6267	84.7	15.3
Bashir et al.	Railroad trespassing	12,150	87	13
Canayaz et al.	Colour	2600	75	25
Li et al.	HOG + SVM	3600	89.87	2.67
Yilmaz et al.	PCA	54,000	72.65	6.5
Our proposed model	YOLO + COCO	38,456	94.31	5.69

References

1. Shin, H.S., Lee, K.B., An application of a deep learning algorithm for automatic detection of unexpected accidents under bad CCTV monitoring conditions in tunnels. In: 2019 International Conference on Deep Learning and Machine Learning in Emerging Applications (Deep-ML)
2. Rundensteiner, E.A., Bashir, M., Ramoza, A.: A deep learning approach to trespassing detection using video surveillance data. In: 2019 IEEE International Conference on Big Data (Big Data)
3. Zaman, A., Ren, B., Xiang, L.: Automated detection of grade-crossing-trespassing near misses based on computer vision analysis of surveillance video data. www.elsevier.com/locate/safety
4. Trivedi, C., Liu, X., Zhang, Z.: Artificial ıntelligence-aided automated detection of railroad trespassing. journals.sagepub.com/home/trr
5. Zhang, F., Yang, F.: Vehicle detection in urban traffic surveillance ımages based on convolutional neural networks with feature concatenation. https://www.mdpi.com/journal/sensors
6. Loo, C.K., Lim, W.S., Wong, W.K., Chew, Z.Y.: An effective trespasser detection system using thermal camera. In: 2010 Second International Conference on Computer Research and Development
7. Farhadi, A., Redmon, J. YOLOv3: an incremental improvement. University of Washington

8. Canayaz, E., Böcekçi, V.G.: Comparison of performance of different background subtraction methods for detection of heavy vehicles. IEEE SPA 2018
9. Shang, B., Gao, C., Chengdong, W.U., Meng, T., Zhang, Y.: A data/ımage transmission device based on TCP/IP protocol. In: 2012 8th International Conference on Wireless Communications, Networking and Mobile Computing. doi: https://doi.org/10.1109/wicom.2012.6478277
10. Li, X., Guo, X.: A HOG feature and SVM based method for forward vehicle detection with single camera. In: 2013 Fifth International Conference on Intelligent Human-Machine Systems and Cybernetics
11. https://datascience.stackexchange.com/a/16813/37736
12. Cai, Y., Liu, L., Wang, X., Chen, Z., Sun, Zhang, Y.: Vehicle detection based on deep dual-vehicle deformable part models. J. Sens. 1–10 (2017). https://doi.org/10.1155/2017/5627281.
13. Maruyama, M., Taguchi, T., Hoguro, Umezaki, T.: A security system for monitoring trespass in an environment with obstacles. In: 2010 IEEE Nanotechnology Materials and device, Conference (2010). https://doi.org/10.1109/nmdc.2010.5652348
14. Yilmaz, A.A., Bostanci, E., Askerbeyli, I., Güzel, M.S.: A vehicle detection approach using deep learning methodologies

Analysis of Malicious DoS Attacks in AES-128 Decryption Module

R. Gayatri, Yendamury Gayatri, R. Karthika, and N. Mohankumar

Abstract AES is the most widely accepted standard in cryptography as it is considered to be the most secure and safest cryptographic algorithm. But in today's world even AES is susceptible to some form of malicious attacks like the Trojan horse injection. This paper proposes denial-of-service (DoS) Trojan horse injection attacks in AES-128 decryption algorithm and also discusses the effect of these DoS Trojans on the side-channel parameters, power consumption, and time delay. The proposed DoS Trojan horse injection attacks have minimal impact on the power consumed (in μW) and time delay (in ms) thus the chances of detecting these Trojans by side-channel analysis are very less.

Keywords Trojan injection · AES-128 · ChipWhisperer · Power trace · Hardware Trojans

1 Introduction

Trojan horse injection can be used in a constructive or destructive way depending on the scenario. The concept of Trojans has been adapted in the medical stream for treating tumors [1, 1].Dual twin has become the recent trends in technology. But it has

These authors contributed equally to this work.

R. Gayatri (✉) · Y. Gayatri · R. Karthika · N. Mohankumar
Department of Electronics and Communication Engineering, Amrita School of Engineering Coimbatore, Amrita Vishwa Vidyapeetham, Coimbatore, India
e-mail: gayatri896@gmail.com

Y. Gayatri
e-mail: gayatri1103@gmail.com

R. Karthika
e-mail: r_karthika@cb.amrita.edu

N. Mohankumar
e-mail: n_mohankumar@cb.amrita.edu

© The Editor(s) (if applicable) and The Author(s), under exclusive license to Springer Nature Singapore Pte Ltd. 2021
A. P. Pandian et al. (eds.), *Proceedings of International Conference on Intelligent Computing, Information and Control Systems*, Advances in Intelligent Systems and Computing 1272, https://doi.org/10.1007/978-981-15-8443-5_55

brought with it new safety and security lapses and loopholes. This has been a boon as a well as a bane. While malicious attacks of Trojans are well known, Trojans can be used constructively in a number of ways like inserting "Trojan" hardware into a system to establish a hidden communication channel to secretly transmit information or to build robust and lightweight hardware watermarks using the properties of Trojan side channels [3, 3].

Cryptographic algorithms are used with the motive to safeguard the information. Various parameters effect the performance of a cryptographic algorithm. A study has been done on the various AES models and its effect based on the key size [5]. Trojans are injected to alter the electrical characteristics by varying parameters at manufacturing process level [6]. Multi-level attacks pose a greater challenge in detection. It was demonstrated that multi-level attack has greater potency than single-level attack through analysis, simulations, and FPGA emulations [7].

Trojan circuit made up of test mode signals which leaks information related to cipher key was designed [8]. Similarly another Trojan was designed which leaks information when triggered where trigger has low trigger probability [9]. High security and lightweight (HIGHT) is a 64 bit symmetric key encryption techniques. In [10], a Trojan was designed with four register and injected in HIGHT. Also, the optimal location of the Trojan is predicted in advance through algebraic fault analysis. Microarchitectural Trojans are designed to leak information in [11]. Also, this can be used in constructive way in digital watermarking to identify the illegal manufactured processors. Dynamic partial reconfiguration (DPR) was exploited by implanting the Trojan circuit in AES encryption which recovers the secret key by mathematical analysis [12].Gate-level characterization-based HT detection using leakage power difference is proposed in [13] for Trojan attacks with less impact of side-channel parameters. Sridhar and Smys [14] propose a architecture that utilized asymmetric-key encryption to protect the system from distributed DoS attack and quantum algorithm attack.

In this paper, an effective way of Trojan horse injection attacks is processed in AES-128 decryption algorithm which has minimum impact on the side-channel parameters of power consumption and time delay.

2 Proposed Methodology

The proposed methodology comprises of two important processes, namely fault injection in AES-128 decryption module which results in DoS and extraction of the side-channel power consumption and time-delay parameters. Further statistical features are obtained from the power trace. Using these features, a comparison is performed between the Golden module and Trojan-injected module. Based on the differences calculated, the DoS Trojan attack can be easily detected by side-channel detection techniques or not. Figure 1 illustrates the proposed methodology in the form of a flowchart.

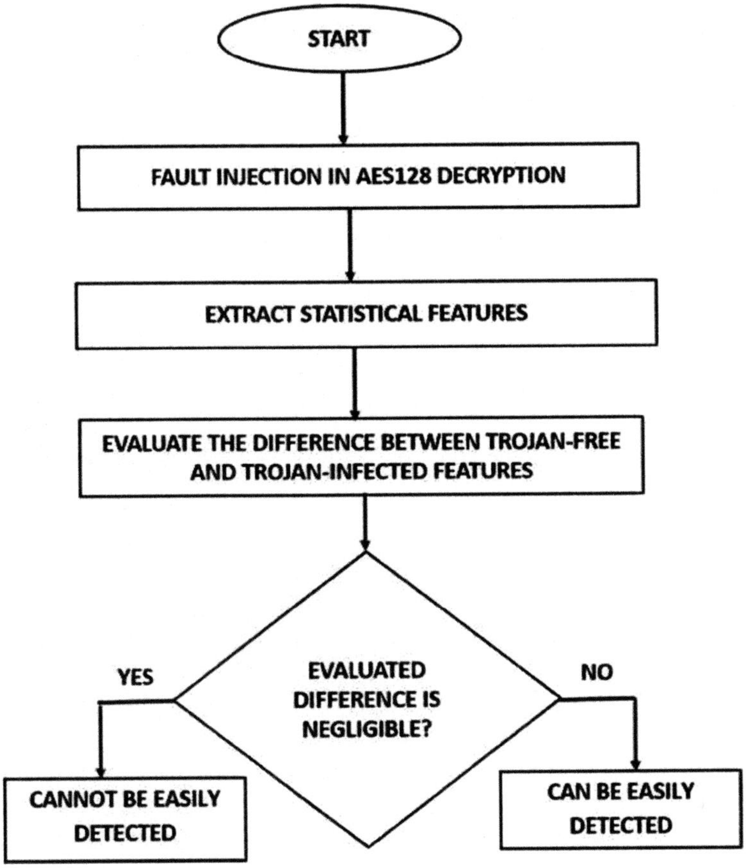

Fig. 1 Flowchart of methodology

2.1 AES-128 Decryption Process

AES-128 decryption algorithm consists of 10 rounds with the following four rounds (except the last step where Inverse MixColumns is not done). Given below is a brief explanation of the four steps involved:

AddRoundKey: This operation involves XORing of current state with secret key.

Inverse MixColumns: This operation is implemented column wise and is performed by multiplying and adding the current state values with a constant in Galois field with the help of lookup tables.

Inverse SubBytes: The bits of the current are substituted by corresponding entry in the substitution box. Nonlinear mixing up of state values results is for resistance to attacks.

Fig. 2 Steps involved in decryption process [15]

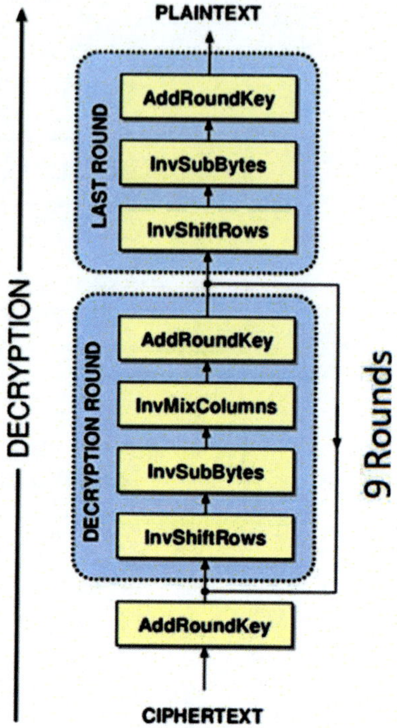

Inverse ShiftRows: The rows are cyclically shifted toward left by offsets of 0, 1, 2, and 3. Linear mixing up of state values results in diffusion of bits.

Steps involved in AES-128 decryption algorithm is depicted in Fig. 2.

2.2 Hardware Trojan Injection

The 39 denial-of-service (DoS) Trojans are inserted in the embedded c code and they are always turned on meaning that the payload does not require any trigger from the trigger circuit for it to function. Since it is coded in embedded c, it is a system-level Trojan based on abstraction. The naming of the Trojans is done in a systematic manner where in T which stands for Trojan followed by a number which stands for the step in which it is injected, then an underscore followed by a number which denotes the round in which the Trojan is inserted. This is depicted in Table 1. An example would be T3_5 which implies that the Trojan is inserted at the Inverse SubBytes step in round 5 of the AES-128 decryption algorithm.

Trojan Injection Attack 1: The injected Trojan follows always on mechanism. It is injected in each round of AddRoundKey resulting in 10 Trojan variants. The Trojan modifies the secret key by adding XOR gates. This current state matrix is

Table 1 Trojan naming

Number after Trojan	Representing attribute
1	AddRoundKey
2	Inverse MixColumns
3	Inverse SubBytes
4	Inverse ShiftRows

XORed with constant in AddRoundKey block. This attack model results in denial of service as the original state is not retained and maliciously modified.

Trojan Injection Attack 2: The injected Trojan follows always on mechanism. It is injected in each round of Inverse MixColumn resulting in 9 Trojan variants. The Trojan modifies the secret key in random fashion by mapping the current state matrix in Galois field to the output state matrix. This attack model results in denial of service as the original state is not retained and maliciously modified.

Trojan Injection Attack 3: Trojan is injected in each round of Inverse SubBytes resulting in 10 Trojan variants. The Trojan modifies the secret key by adding XOR gates. The current state matrix is XORed with a constant and mapped to substitution box. This attack model results in denial of service as the original state is not retained and maliciously modified.

Trojan Injection Attack 4: The injected Trojan follows always on mechanism. It is injected in each round of Inverse ShiftRows resulting in 10 Trojan variants. In random fashion, the current state matrix of Inverse ShiftRow is mapped to the output state matrix. Certain buffers are removed in the block which are otherwise utilized for left shift of current state matrix. This attack model results in denial of service as the original state is not retained and maliciously modified.

2.3 Effect of Trojans on Power

In order to study the effect of these Trojans on the power consumption, obtain the power profile of all the 39 Trojans injected using ChipWhisperer—Lite board using Atmel XMega microcontroller as the target board. The board is powered on with the help of PC connected via USB cable and .hex file is generated by the XMega programmer and quickly flashed into the XMega microcontroller used as the target board which runs at clock frequency of 29.538 MHz. The ChipWhisperer [16] is perfect board to perform run-time power analysis of a cryptographic algorithm as it provides clock synchronization as well as a low noise amplifier (LNA). The measurement setup is depicted in Fig. 3.

FFT is performed for both the Golden- and Trojan-infected AES-128 algorithms and depicted in Figs. 4 and 5, respectively. The FFT serves as a justification for the injection of the Trojans. Difference between power consumption of Golden- and Trojan-injected algorithm is noticeable and it clearly depicted using red marker. X-axis indicates the frequency (Hz) and Y-axis indicates the power consumption and

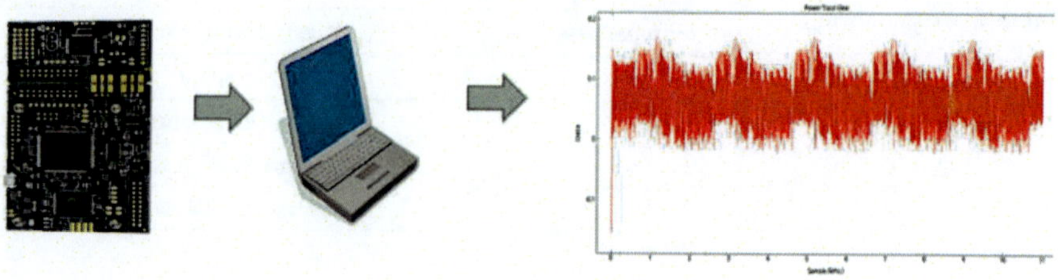

Fig. 3 Measurement setup

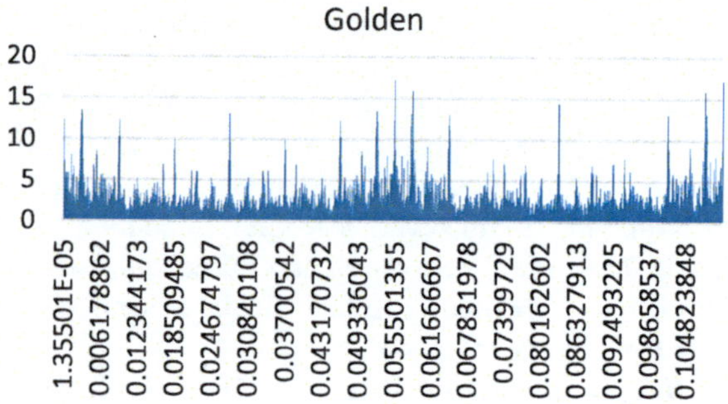

Fig. 4 Golden FFT trace

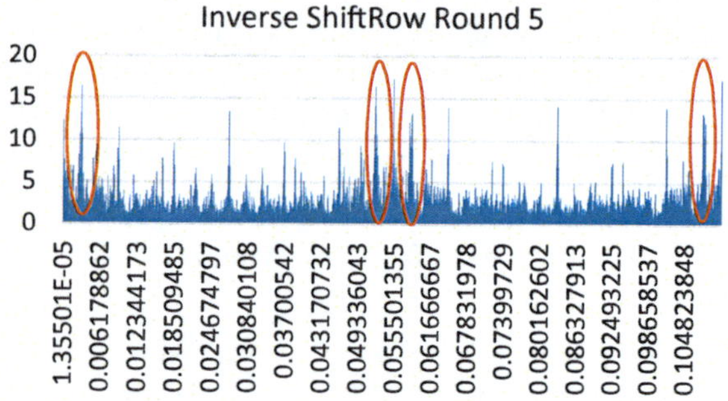

Fig. 5 Trojan-infected FFT trace

can observe that there is change power consumption. It is interesting to note that power consumption is comparatively higher when Trojan is injected. Hence, in this case, it is easier to detect the presence of Trojan in algorithm from power consumption values.

The side-channel statistical parameters for the Golden algorithm was found to be having the values listed in Table 2.

The Statistical parameters of all the Trojan injection attacks along with the Golden measurements are shown in Figs. 6, 7, 8, 9, 10 and 11, respectively.

Table 2 Golden parameters

Parameters	Values (in μW)
Average	75.05993
Median	80.07813
Range	331.0547
Standard deviation	38.38484
Variance	1.473396
Time taken for execution	251.698 (ms)

Fig. 6 Average

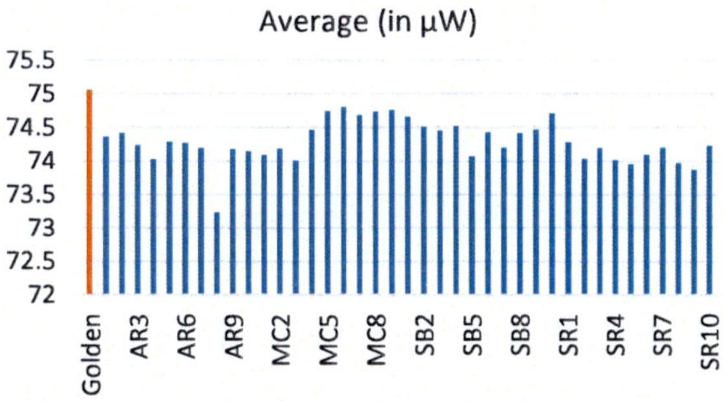

Fig. 7 Median

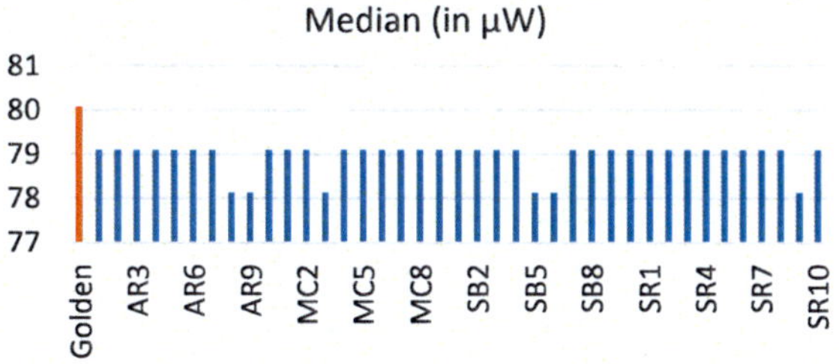

Fig. 8 Range

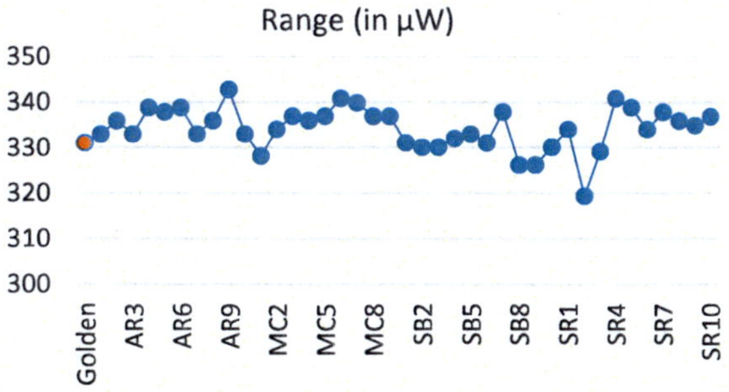

Fig. 9 Standard deviation

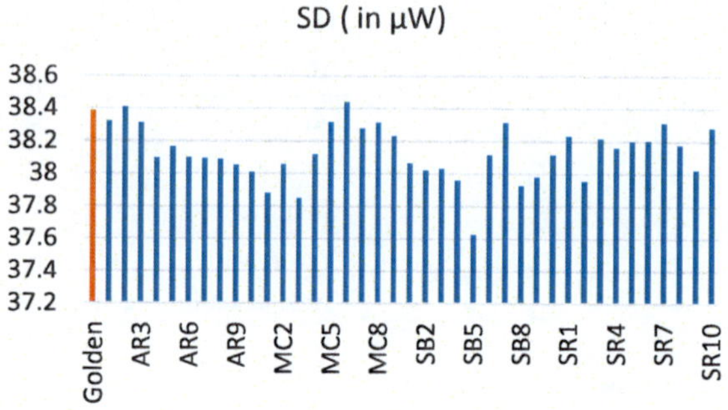

Fig. 10 Variance

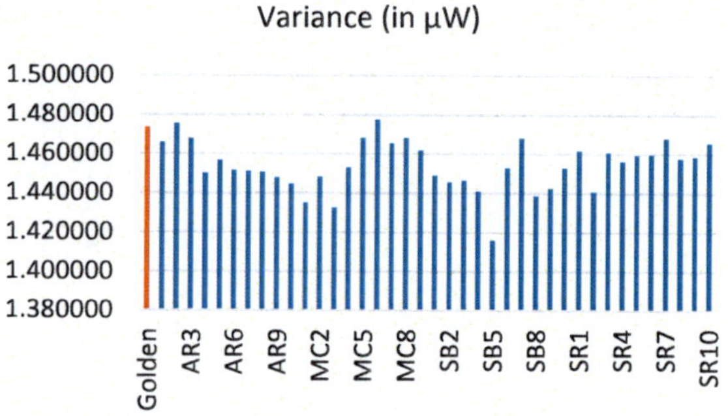

Fig. 11 Time consumed

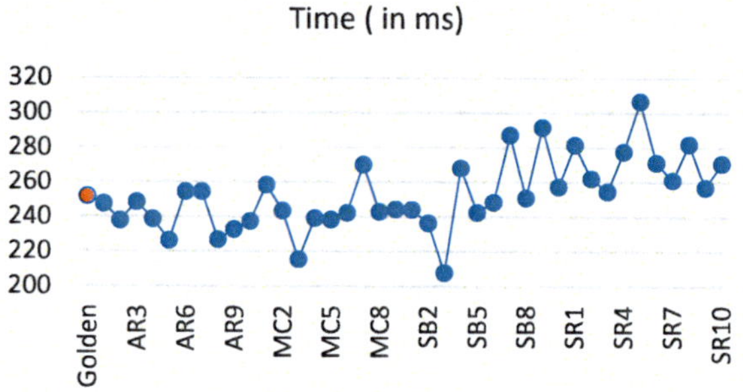

2.4 *Distance Metrics*

In this work, three distance measures are utilized to analyze the impact of Trojan in the algorithm. The dissimilarity between the input state matrix and output state matrix on occurrence of Trojan injection attack is evaluated. Given below is brief description of the distance metrics utilized.

Hamming distance: Hamming distance is a number of bit positions which are different between the input state matrix and output state matrix. It allows substitution to transform from input state matrix to output state matrix.

Levenshtein distance: Levenshtein distance is a number of bit positions which are different between the input state matrix and output state matrix. It allows insertion, deletion, and substitution to transform from input state matrix to output state matrix.

Jaro distance: Jaro distance is a number of bit positions which are different between the input state matrix and output state matrix. It allows transposition to transform from input state matrix to output state matrix.

3 Result

39 variants of DoS Trojans are injected at the system level of AES-128 decryption cryptographic algorithm creating threat models which were analyzed on the basis of the power profile and time consumption (in other words execution time) for a single trace. The difference between average, median, range, standard deviation, and variance with respect to the Golden circuit is calculated and the time delay with respect to the Golden algorithm is also obtained. These threat attacks are plotted in terms of their effect on power consumption on all the six statistical parameters. The *X*-axis indicates the round in which Trojan is injected and *Y*-axis indicates the difference in power consumption of Golden- and Trojan-injected algorithm. Refer Figs. 12, 13, 14, 15, 16 and 17 for comparative side-channel parameters analysis.

Fig. 12 Δ Average

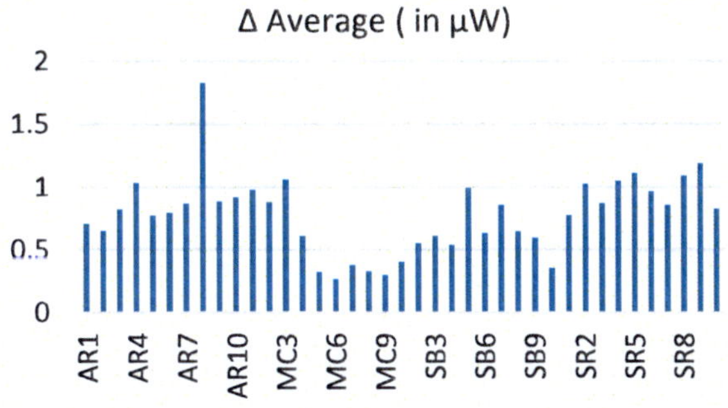

Fig. 13 Δ Median

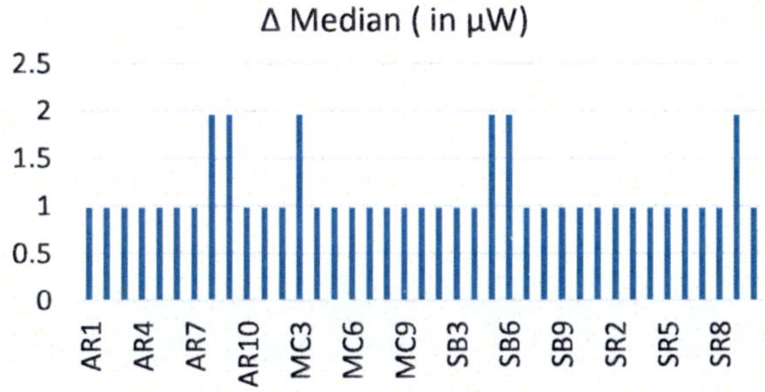

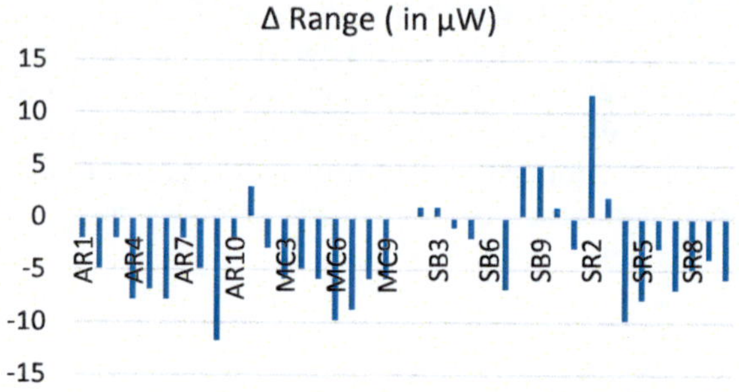

Fig. 14 Δ Range

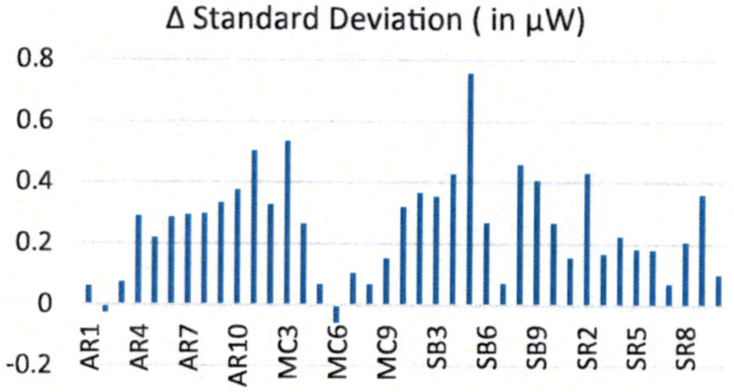

Fig. 15 Δ Standard deviation

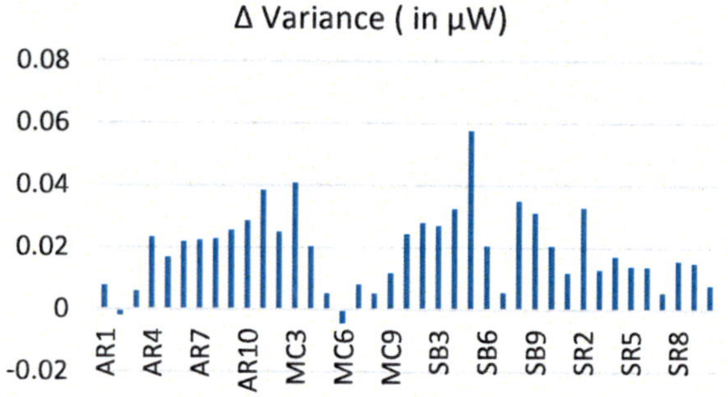

Fig. 16 Δ Variance

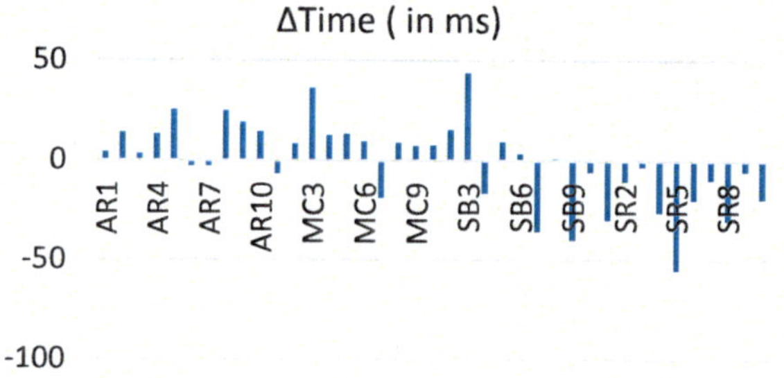

Fig. 17 Δ Time

Table 3 Result comparison

Parameters	Values (in mW)
Paper 17	
AES-128 original	677
AES-128 with HT1	668
AES-128 with HT2	634
Paper 18	
Power consumed	38.44149
Paper 19	
Original AES	20.4
AES + duplicate protection scheme	65.18
AES + linear protection scheme	30.61
AES + robust protection scheme	136.7
AES + CRC 4 Protection scheme	52.01
AES + CRC 6 protection scheme	52.97

Five statistical parameters (of power consumption) have the differences in microwatts, and the time delay is in milliseconds. The variation in power values in certain cases is due to uncorrelated state transitions. In certain cases, there is minute variation in power values due to denial of service as it denies access to service resulting of lower power consumption. The results from paper [17, 17, 17] are tabulated in Table 3 for comparison.

From Table 3, the power differences are in the range of 20–700 mW, thus inferring that the proposed DoS Trojans have higher chances of escaping existing power-based detection techniques unless very accurate power analysis with sophisticated devices having LNA with excellent SNR is carried out. Hence, all the proposed Trojans have recorded differences in power in the ranges of 0–15 microwatts which is very challenging to detect. Further even the time delay observed is in the scale of milliseconds (ms) and surprisingly the negative values in time delay means that the Trojan-infected algorithm took lesser execution time than the Golden algorithm. Thus, the detection of the proposed DoS Trojans using side-channel parameters is very low.

Further, three edit distance measures have also carried out that is Hamming distance, Levenshtein distance, and Jaro distance to demonstrate the effectiveness of our Trojans. The results are tabulated in Table 4.

The proposed Trojans partially change the decrypted output from the Golden output which is visible from the dissimilarity measures obtained. This confirms the fact that these Trojans are very efficient in denial of service (DoS).

Table 4 Distance measures

Trojans	Hamming distance	Levenshtein distance	Jaro distance
Input 1: 29c3505f571420f6402299831a02d73a *Key: 54687473206d79204b756e67204675* *Golden output: 54776f204f6e65204e696e652054776f*			
T1_2	31	27	0.505952
T1_5	32	29	0.529762
T1_8	31	27	0.505952
T2_2	31	28	0.534722
T2_5	31	29	0.550245
T2_8	30	30	0.472222
T3_2	27	26	0.589461
T3_5	32	30	0.475
T3_8	31	29	0.615741
T4_2	32	29	0.5625
T4_5	31	30	0.60636
T4_8	30	29	0.505952
Input 2: 3925841d02dc09fbdc118597196a0b32 *Key: 2b76151628aed2ababf7158809cf4f3c* *Golden output: 4d1998fe84e4c07052efa22935908470*			
T1_2	32	31	0.627976
T1_5	31	27	0.60636
T1_8	31	29	0.643849
T2_2	30	27	0.633333
T2_5	32	30	0.541667
T2_8	31	28	0.659722
T3_2	28	27	0.520833
T3_5	29	29	0.643849
T3_8	31	29	0.616667
T4_2	29	29	0.60636
T4_5	31	30	0.643849
T4_8	27	27	0.589461
Input 3: 000102030405060708090a0b0c0d0e0f *Key: 69c4e0d86a7b0430d8cdb78070b4c55a* *Golden output: 00112233445566778899aabbccddeeff*			
T1_2	32	33	0.544879
T1_5	30	29	0.63722
T1_8	30	30	0.660354
T2_2	31	28	0.63722

(continued)

Table 4 (continued)

Trojans	Hamming distance	Levenshtein distance	Jaro distance
T2_5	28	28	0.582818
T2_8	29	29	0.582818
T3_2	32	31	0.515783
T3_5	30	31	0.557449
T3_8	28	27	0.617906
T4_2	30	31	0.557449
T4_5	30	29	0.689298
T4_8	28	29	0.610059

4 Conclusion and Future Scope

Among the 39 DoS, Trojans injected can safely conclude that injecting DoS in the Inverse SubBytes step of round 1 of the AES-128 decryption is more difficult to detect as the power differences in the Golden- and Trojan-infected algorithms are extremely minute. Further, the DoS Trojans injected Trojans in the 5th round of Inverse SubBytes step, in the 2nd round of Inverse ShiftRows, and in the 8th and 9th round of AddRoundKey in the AES-128 decryption algorithm are the comparatively easier to detect due to a relatively better power differences between the Golden- and Trojan-infected algorithms. Using only time delay, the DoS Trojan inserted in the Inverse ShiftRows step of the 5th round is difficult to detect while the Trojan inserted in the 8th round of Inverse SubBytes is relatively easier to detect.

The comparative study on the effect of the various Trojan injection models was carried out only on the basis of the side-channel parameters of power profile and time delay. The future scope of this paper would be to study the effect of these DoS system-level Trojans based on other criteria like overhead. Further, a security enhanced defense model to prevent the injection of these DoS system level Trojans will also be a future scope of this work.

References

1. Nelson, D., Fisher, S., Robinson, B.: The "Trojan Horse" approach to tumor immunotherapy: targeting the tumor microenvironment. J. Immun. Res. **2014**, 789069 (2014). https://doi.org/10.1155/2014/789069
2. Choi, M., Stanton-Maxey, K.J., Plichta, J., Filgueira, C., Bardhan, R., Akin, D., Badve, S., Sturgis, J., Robinson, J., Bashir, R.: A cellular Trojan horse for delivery of therapeutic nanoparticles into tumors. Nano Lett. Nano Lett. **7**, 3759–3765 (2007)
3. Becker, G.T., Kasper, M., Moradi, A., Paar, C.: Side-channel based watermarks for integrated circuits. In: 2010 IEEE International Symposium on Hardware-Oriented Security and Trust (HOST) Anaheim, CA, pp. 30–352010). https://doi.org/10.1109/HST.2010.5513117

4. Becker, G.T., Burleson, W., Paar, C.: Side-channel watermarks for embedded software. In: 2011 IEEE 9th International New Circuits and systems conference Bordeaux, pp. 478–481 (2011)https://doi.org/10.1109/NEWCAS.2011.5981323
5. Bi Irie guy-cedric, T.: A comparative study on AES 128 bit and AES 256 Bit. Int. J. Comput. Sci. Eng. **6**, 30–33 (2018). https://doi.org/10.26438/ijsrcse/v6i4.3033.
6. Kumar, R., Jovanovic, P., Burleson, W., Polian, I.: Parametric Trojans for fault-injection attacks on cryptographic hardware. In: 2014 Workshop on Fault Diagnosis and Tolerance in Cryptography Busan, pp. 18–28 (2014)
7. Ali, S.S., Chakraborty, R.S., Mukhopadhyay, D., Bhunia, S.: Multi-level attacks: an emerging security concern for cryptographic hardware. Des. Autom. Test Eur. Grenob. **14**, 1–4 (2011)
8. Yoshimura, M., Amy, O., Toshinori, H.: A smart Trojan circuit and smart attack method in AES encryption circuits, pp 278–283 (2013). https://doi.org/10.1109/DFT.2013.6653619.
9. Wang, X., Wang, T., Li, X., Zhang, Y., Huang, C.: A smart design of hardware Trojan and differential fault analysis in AES. In: 2018 Eighth International Conference on Instrumentation and Measurement, Computer, Communication and Control (IMCCC), Harbin, China, pp. 1548–1551 (2018)
10. Chen, H., Wang, T., Zhang, F., Zhao, X., He, W., Xu, L., Ma, Y.: Stealthy hardware Trojan based algebraic fault analysis of hight block cipher. Secur. Commun. Netw. 1–15 (2017). https://doi.org/10.1155/2017/8051728.
11. Gallais, J.-F., Großschädl, J., Hanley, N., Kasper, M., Medwed, M., Regazzoni, F., Schmidt, J.-M., Tillich, S., Wójcik, M.: Hardware Trojans for inducing or amplifying side-channel leakage of cryptographic software. **6802**, 253–270 (2010). https://doi.org/10.1007/978-3-642-25283-9_17
12. Johnson, A.P., Saha, S., Chakraborty, R.S., Mukhopadhyay, D., Gören, S. Fault attack on AES via hardware Trojan insertion by dynamic partial reconfiguration of FPGA over ethernet. In: Workshop on Embedded Systems Security (WESS, part of ACM ESWEEK), New Delhi, ACM, India (2014)
13. Karunakaran, D.K., Mohankumar, N.: Malicious combinational hardware Trojan detection by gate level characterization in 90 nm technology. In: Fifth International Conference on Computing, Communications and Networking Technologies (ICCCNT), Hefei, pp. 1–7 (2014). Doi https://doi.org/10.1109/ICCCNT.2014.6963036.
14. Sridhar, S., Smys, S.: Intelligent security framework for IoT devices cryptography based end-to-end security architecture. In: 2017 International Conference on Inventive Systems and Control (ICISC), Coimbatore, pp. 1–5 (2017). Doi https://doi.org/10.1109/ICISC.2017.8068718.
15. Arrag, S., Hamdoun, A., Tragha, A., Khamlich, S.: Design and implementation a different architectures of mixcolumn in FPGA. Int. J. VLSI Des. Commun. Syst. (2012)
16. O'Flynn, C., Chen, Z.: ChipWhisperer: an open-source platform for hardware embedded security research. IACR Cryptol. EPrint Arch. **8622** (2014). https://doi.org/10.1007/978-3-319-10175-0_17.
17. Manivannan, S., Anandakumar, N.N., Devi, N.: 2019 Key retrieval from AES architecture through hardware Trojan horse. In: 6th International Symposium, SSCC 2018, Bangalore, India, September 19–22, (2018) Revised Selected Papershttps://doi.org/10.1007/978-981-13-5826-5_37
18. Ambrose, J.A., Aldon, N., Ignjatovic, A., Parameswaran, S.: Anatomy of differential power analysis for AES. In: 2008 10th International Symposium on Symbolic and Numeric Algorithms for Scientific Computing Timisoara, pp 459–466
19. Luo, P., Fei, Y., Zhang, L., Ding, A.A.: Side-channel power analysis of different protection schemes against fault attacks on AES. In: 2014 International Conference on ReConFigurable Computing and FPGAs (ReConFig14), Cancun, pp. 1–6 (2014)

Creating a 3D Model from 2D Images Using Convolution Neural Network

K. S. Vani, Rupesh Sapkota, Sparsh Shrestha, and Srujan B

Abstract In recent years, deep learning has shown outstanding performances in the field of multimedia data. Convolutional neural network is widely used for photogrammetry and 3D reconstruction. In this paper, creating a 3D model from 2D input images using convolutional neural networks is proposed. Using a set of 2D images taken from multiple viewpoints, the features of the object are extracted using CNN from which the 3D view of the object is constructed. The main reason for using CNN is its ability of feature learning. Though the use of 3D models has been enormous and convenient, developing it is still a huge challenge. The development of 3D models is made easier and more effective through the methodology used in this paper.

Keywords Convolutional neural networks · 2D to 3D · Photogrammetry · Encoder · Decoder

1 Introduction

Technology has brought people's imagination closer to reality, and the most excellent way to illustrate this is to show how 3D modeling has revolutionized the presentation world and the extraordinary innovation in the field of photogrammetry has assisted in the best possible way. It is like drawing out the multi-dimensional imagination from someone's mind and implementing it virtually to see how it works. But still to create a convenient 3D model is a huge and difficult task for many. People have to either spend a lot of time themselves using various tools to create a 3D model or to pay a huge amount of money to get it done. In this paper, a 3D model is created from 2D input images using photogrammetry with the help of machine learning techniques and convolutional neural networks.

K. S. Vani (✉) · R. Sapkota · S. Shrestha · S. B
Department of Computer Science and Engineering, Acharya Institute of Technology, Soladevanahalli, India
e-mail: vaniks@acharya.ac.in

© The Editor(s) (if applicable) and The Author(s), under exclusive license to Springer Nature Singapore Pte Ltd. 2021

A. P. Pandian et al. (eds.), *Proceedings of International Conference on Intelligent Computing, Information and Control Systems*, Advances in Intelligent Systems and Computing 1272, https://doi.org/10.1007/978-981-15-8443-5_56

Machine learning deals with the advancement of computer programs which are capable of accessing and using data to learn from them [1].

From few years, convolutional neural networks has become an important algorithm for image classification and image processing. CNN works by moving a filter across the input data and carrying out a dot product calculation between the filter and the input data. This will generate a new activation map. The activation map will become an output, or it can be the input of the next hidden layer. A hidden layer will be processed in the same way as the previous layer.

Photogrammetry falls under the broader category of geomatics, and according to the American Society for Photogrammetry and Remote Sensing, is defined as, "the art, science, and technology of obtaining reliable information about physical objects and the environment through the process of recording, measuring, and interpreting photographic images and patterns of electromagnetic radiant imagery and other phenomena".

The approach in this paper is to use CNN to build a 3D model from multiple 2D images. The generative model is divided into three different modules. The first module consists of 2D-CNN which will extract features from input 2D images, the second module consists of attention activation function which calculates attention score for each feature and aggregate function which merges all 2D features to create a single 3D deep feature, and third module consists of 3D-deconvolutional neural network which will use 3D deep feature to create a 3D model. ShapeNet dataset is used to train and test the 2D to 3D conversion model.

In 2D-CNN encoder, 2D images are given as input in the form of a matrix where each pixel is considered as a value. The input matrix is passed through different layers such as the convolution layer, normalization layer (ReLU layer), pooling layer, and fully connected layer which contains different features like filtering, pooling, and normalization which gives a set of features as output. In aggregate module, a set of features are given as input to attention activation function to calculate attention scores which are then aggregated using an aggregate function to attain a single 3D deep feature. Finally, in 3D-CNN decoder, 3D deep feature is given as input to obtain a 3D object model as output. In this module, deconvolution and unpooling are performed to decode the 3D deep feature and to obtain the 3D model (Fig. 1).

2 Literature Survey

The summary of various works on 3D model construction from 2D input images is presented below.

3D Image Reconstruction from Videos Using Patch Generated from the Tracking-Learning Detection Algorithm [2]

It was proposed by Amandeep Singh from Stanford university in 2016. The tracking-learning detection algorithm and SFM pipeline technique were used which was implemented in MATLAB. Multiple image patches P are taken which contained the interesting images or objects from a generic video stream V using the tracking-learning detection algorithm. The set of patches generated are the inputs for the SMF pipeline process and were used to produce the 3D model for the image or object of interest.

Construction and Exploration of 3D Model from 2D Image Features [3]

It was proposed by Karl Ni (MIT), Zachary Sun (MIT, Boston University), Nadya Bliss (MIT), NohaSanvely (Cornell University) in the year 2018. They used the object registration algorithm, the SFM technique, and approximate nearest neighbor algorithm. This algorithm spots the entity of interest contained by the image by recognizing if the entity belongs to this picture or not. To execute this assignment, the processing sequence depends on the 3D structure resulting from the motion which represents the feature spot in the planned model. Using the algorithm current-state-of-art, spotted objects or images are extorted, and their 2D pixels are changed to relative 3D real-world coordinates by using image data. Locations on the scene can be given by alignment distance data. This produced the result of point cloud reconstruction of Boston Skyline which has been created.

Learning-Based Depth Estimation from 2D Images [4]

It was proposed by Jose L. Herrera, Janusz Konarad, Carlos R. Del-Blanco, and Narciso Garcia from IEEE Conference on Image Processing held in 2015. They used the generalized search tree (GIST) algorithm and saliency weights which were implemented in the Kinect-NYU dataset. The deepness of the query is obtained from the dataset of colors and depth images by seeking the images from the stored data that are identical to the query. More weight is granted to the GIST descriptor for the area with peak saliency because not all the regions in the image need the same visual attraction. Thus, the depths of the most similar images are combined and filtered the output to achieve a depth approximation. The resulting output had slightly better improvement in average output over state-of-the-art method.

3D-R2N2: A Uniform Approach for Single and Multi-View 3D Object Reconstruction [5]

It was proposed by Christopher B. Choy, Danfie Xu, JunYoungGwak, Kelvin Chen, and Silvio Savarese from Stanford University in 2016. 3D-R2N2, abbreviated as 3D recurrent reconstruction neural network, long short-term memory (LSTM), and grated recurrent unit (GRU) are used in implementing this model. Using the large assortment of synthetic data, this network learns the mapping of the images of the objects to the basic 3D shapes. This given network takes one or more images of a give object instances from a capricious view point and outputs the reformation of the object in the structure of a 3D grid. This particularly does not necessitate any of the image explanation or object classes for training or testing. This framework outperforms avant-garde method and enables the renovation of the image or object in a situation where the conventional SMF/SLAM methods fail. This constructed a 3D image and model from 2D images provided which outperforms the method of Kar Et Al which was used for comparison [6].

3 Implementation

The pseudocode for the 2D to 3D conversion model is as follows:

1. Take multiple 2D images as input in matrix form.
2. Pass the input values through the convolution layer with an appropriate number of channels.
3. Select the appropriately sized filter to perform max-pooling and extract features from them.
4. Add output from step 3 and initial input.
5. Repeat actions from 1 to 4 using the output value from step 4 as input for a certain number of times.
6. The output values are then passed into fully connected layer.
7. The attention function is applied to calculate the attention score.
8. Attention scores are aggregated to get a single 3D deep feature.
9. The 3D deep feature obtained is passed through deconvolutional layer.
10. Unpool the values from deconvolutional layer to max-pool the input
11. Add step 10 and 8.
12. Repeat steps from 9 to 11 using output from 11 instead of 8 for certain number of times.
13. The model thus obtained has to be passed through the 3D SoftMax layer.
14. The output thus obtained is a 3D model.

The entire project can be divided into three separate modules namely encoder module, aggregation module, and decoder module (Fig. 2).

Encoder

First module encoder is a 2D-CNN which encodes multiple images of size [127 × 127] into sets of low dimension feature which are given input to attention module. Here, for first layer, 7 × 7 convolution layer is used, and 3 × 3 convolution layer is used for all other convolution layer in encoder. Every time convolution layer is repeated twice followed by a single pooling layer. Encoder has identity mapping connection in which after every two convolution layers output from the pooling layer and input to the convolution layer is added to obtain input for next layer. Identity mapping is not done for fourth pair. A 1 × 1 convolution is used for connection, so that number of channels after convolution matches. The flattened output of encoder is passed into fully connected layer where it is compressed to acquire 1024-dimension feature vector. The encoder algorithm used in this paper takes multiple images as input to give sets of features extracted from those images.

Aggregation Module

The aggregation module aggregates the input features that are provided by the encoder module. The prime feature of the aggregation module is the use of attention to get only the required feature out of the input provided to it.

Basic idea here in the attention is when every time the model predicts from an input, it only pays attention to the parts of the input where the most relevant information is concentrated instead of entire input image.

This module can be easily incorporated with a encode–decoder network for features extraction and processing.

Decoder

Third module decoder is a 3D-CNN where a single aggregated feature acquired from aggregated module is given as input to obtain a 3D model of given 2D images. Similar to encoder–decoder uses pairs of deconvolution layer followed by un-pooling layer. Identity mapping connection used here is same as encoder for first four times where fourth layer have three convolution layers instead of two layers. At last, obtained output is passed through a single convolution layer which provides the 3D model of object in 2D images.

4 Experimental Results

For training of the 3D model, ShapeNet dataset is used which is a gathering of 3D CAD models that are arranged according to the WorldNet hierarchy. The ShapeNet is a large-scale 3D model dataset. Only a subset of the ShapeNet dataset is used which contains 50,000 models and 13 major categories. The dataset is split into 4:1 ratio for training:testing data.

Python and libraries such as TensorFlow, NumPy, matplotlib, SciPy, and scikit learn were used to develop the 3D model. The model was trained was Nvidia GeForce GTX 950 M of 4 GB. 12 GB RAM size was used as the dataset requires a minimum of 10 GB space.

The output of this implementation is the 3D model generated from the given set of preprocessed lightweight PNG images.

Mean intersection-over-union is a common evaluation metric provided with TensorFlow. IoU is calculated for each class, and then, the average of all classes is taken which gives the mean value.

$$\text{IOU} = \text{true - positive}/(\text{true - positive} + \text{false} - \text{positive} + \text{false - negative})$$

The achieved mean IoU value for 3-view input images is 0.669 and for 24-view input images is 0.694 (Figs. 3 and 4).

5 Conclusion

A combination of the avant-garde machine learning algorithm with photogrammetry techniques has been attempted to create a 3D model from input images using neural networks. Feature learning ability, specifically from the visual or imagery data along with less preprocessing in requirement of the convolutional neural network has convinced to use it in the project. Aggregation model which uses the attention feature was the core of the proposed work which does most of the heavy lifting in the part of conversion of 2D features into 3D model. A 3D model so constructed can be used in education, 3D printing in manufacturing industries, medical science, geology and geography, astronomy, and many other fields.

Future works suggested are.

- The texture can be added to the created 3D model.
- The capability of the model can be increased to process the images with higher resolution.

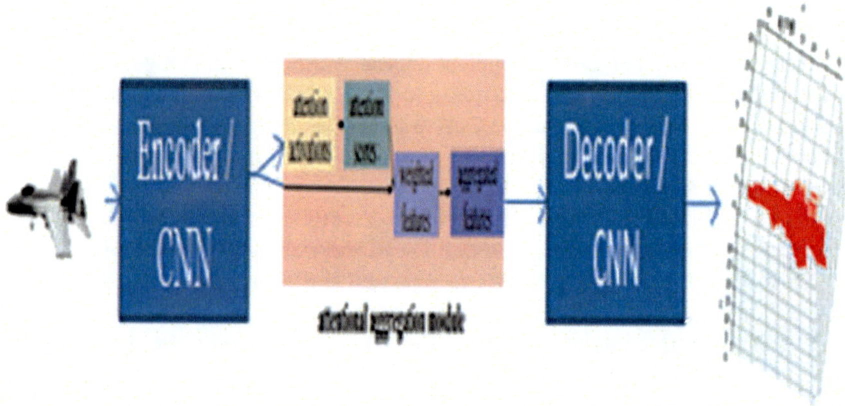

Fig. 1 General model to create the 3D model from 2D images

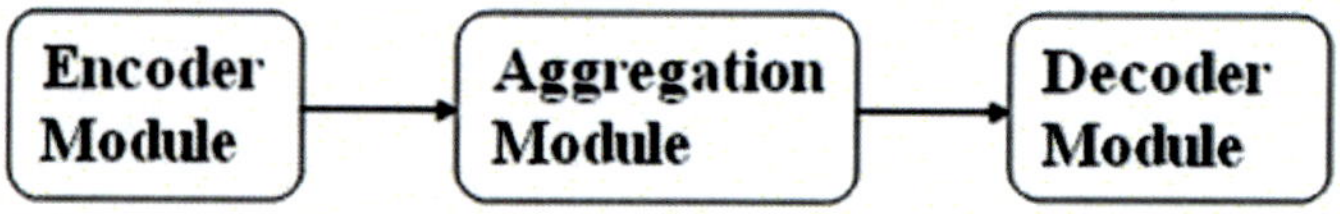

Fig. 2 Basic implementation model

Fig. 3 Input images for 3D model prediction

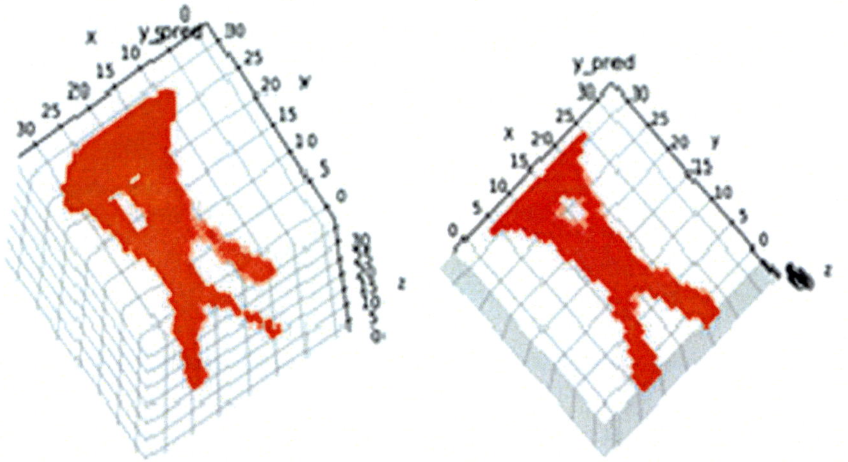

Fig. 4 Predicted 3D model from input images

References

1. Mitchell, T.M.: Machine Learning. India Edition, McGraw Hill Education (2013)
2. Singh, A.: 3D Image Reconstruction from Videos using Patch Generated from Tracking-Learning Detection Algorithm. Stanford University (2016)
3. Ni, K.: Sun, Z., Bliss, N., Sanvely, N.: Construction and Exploration of 3D Model Frim 2D Features (2018)
4. Herrera, J.L., Konarad, J., Del-Blanco, C.R., Garcia, N.: Learning based depth estimation from 2D images. In: IEEE Conference on Image Processing held in 2015
5. Choy, C.B., Xu, D., Gwak, J.Y., Chen, K., Saverese, S.: 3D-R2N2: A Uniform Approach for Single and Multi-View 3D Object Reconstruction. Stanford University (2016)
6. Hong, Y., Kim, J.: A 2D-View Depth Image and CNN based 3D Model Identification. Sangmyung University, Korea (2017)

Specular Microscopic Endothelium Image Analysis with Danielsson Morphology

Kamireddy Vijay Chandra and Bhaskar Mohan Murari

Abstract The eye is the sensitive organ in the human body. Eye consists of various sub-biological components. Every sub-organ is necessary to visualize any object. In the eye, cornea plays an important role to transfer an image to focal points on the retina. From retina, the image is transferred to the brain through the optic nerve. Optic nerve is protected by the optic disk. In the process of transferring visual data to the focal point on the retina, the cornea plays a key role, and it should be a healthy condition. The cornea may get damage when dystrophies attack the layers. Cornea internally consists of five layers. The epithelial layer is the peripheral layer, and thickness is around 50 μm, followed by the stroma layer, Bowman's layer, Descemet's layer, and endothelium layer. Stroma layer is very big among all the layers with thickness around 500 μm. And the smallest layer is Descemet's layer with thickness of 12 μm. The endothelium layer is having a thickness of around 50 μm, and it is very useful layer in the cornea. In this paper, the endothelium layer is focused on a process of transferring visual data to the optical nerve and should be healthy and watery. Fuchs dystrophy (FD), advanced Fuchs dystrophy (AFD), and irido corneal dystrophy (ICD) cause damage to the endothelium layer as well as the cornea. The ultimate result is vision loss and blind impairment. To detect these dystrophies, 'Danielsson morphology' (DM) is adopted along with this convolution filter, noise components are eliminated, and DM processing time is also minimum. Endothelium cells are estimated and compared with the existing methods. Pleomorphism, polymegathism minimum area, maximum area, average area, elongation factor, compactness factor, and circularity factor are estimated.

K. V. Chandra (✉) · B. M. Murari
Departments of Sensor and Biomedical Technology, School of Electronics Engineering, VIT, Vellore, India
e-mail: kamireddy.vijaychandra2016@vitstudent.ac.in

B. M. Murari
e-mail: bhaskarmurari@vit.ac.in

K. V. Chandra
Department of EIE, VNRVJIET, Bachupally, Hyderabad, India

© The Editor(s) (if applicable) and The Author(s), under exclusive license to Springer Nature Singapore Pte Ltd. 2021
A. P. Pandian et al. (eds.), *Proceedings of International Conference on Intelligent Computing, Information and Control Systems*, Advances in Intelligent Systems and Computing 1272, https://doi.org/10.1007/978-981-15-8443-5_57

Keywords Epithelial layer · Fuchs dystrophy · Irido corneal dystrophy · Advanced Fuchs dystrophy

1 Introduction

The cornea consists of typically 'five' layers; empirically last layer endothelium layer is the sophisticated layer among five layers; when dystrophy's like FD, AFD, ICD attacks the endothelium layer, it causes severe damage on endothelium layer and provides stains of disease [1]. The thickness of the layer is reduced when dystrophy encounters. In this paper, Danielsson morphology is used to trace out the contours of the endothelium cells and identifies the hexagonality of the cells and predicts the cell density cells/mm^2. DM also estimates the elongation of the hexagonal cell, compression of the cell, and circularity of the cell in the stream of pathology. Due to the various considerations of endothelium cell structure and its function, designing filters and algorithms to identify the stagnant of celluloid fluid in the cornea becomes stringent. The factors estimated gives the picturization of the endothelium cells and its protein concentration. DM algorithm in this paper enhances the vision toward to find accuracy level at most 100%. The complete statistical information is provided by this algorithm and magnifies the possibility of the occurrence of the dystrophies to the cornea. The methodology used in this paper for the identification of corneal dystrophies is the novel algorithm, and DM is well-established and accurate algorithm. The time consumption of the DM algorithm is very short in duration and magnifies the stability of the algorithm. DM method is useful for ophthalmologists and clinicians worked in busy clinical scheduling. DM algorithm provides the complete statistical parameters that are helpful for the clinicians which reduces time drastically. The complete statistical profile of the endothelium layer provides accurate information about the dystrophies caused to that layer. The significance of the algorithm measures concerning the statistics provided by the algorithm. The complete analysis of the problem well being executed by the proposed DM algorithm.

2 Proposed Algorithm

2.1 Methodology

Considering Fig. 1, the specular microscopic endothelium image is acquired and processed, the raw image from the specular microscope is of 32-bit in size. The color plane extraction is done to convert 32-bit size into an 8-bit image [2]. The luminance plane extraction is done to extract green plane features in images. The 8-bit size image is processed into a custom convolution filter to avoid low-frequency noise components. Convolution filter provides a noise-free image, but the image may lose quality pixels and reflects poor image illusion. To avoid blurry image illusion, the

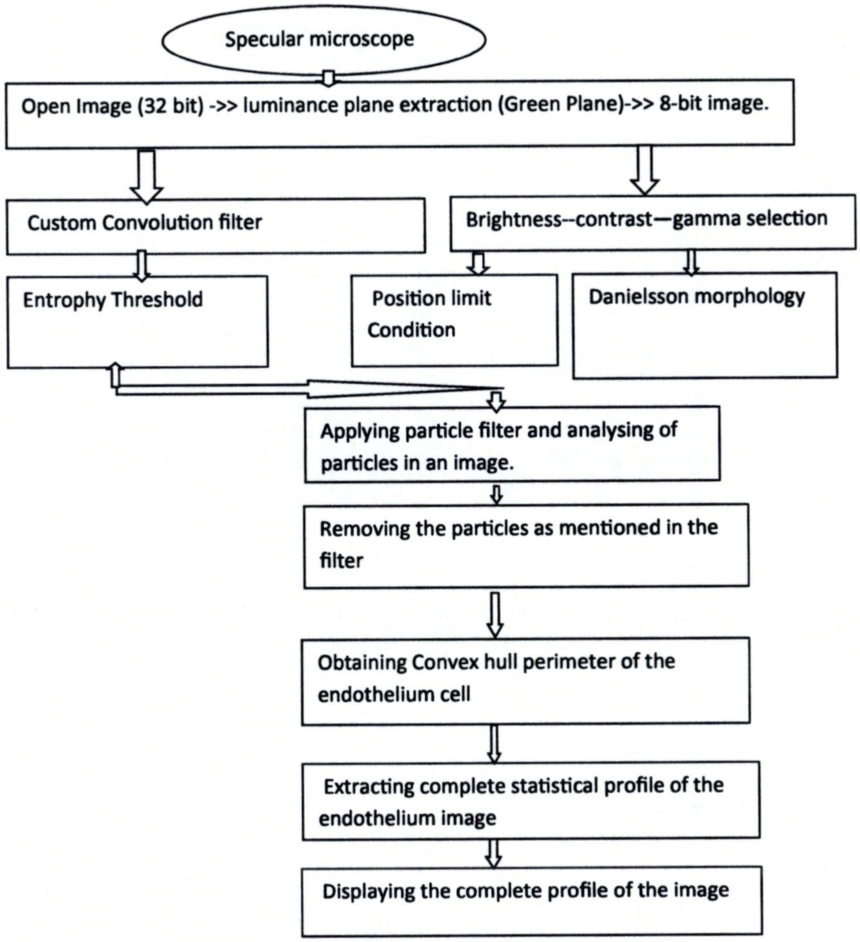

Fig. 1 Proposed algorithm for corneal endothelial cell analysis

brightness of the image is enhanced with gamma selection [3]. The image quality is enhanced and provides a high contrast picture to further process. To differentiate dark pixels in image and bight pixels in an image, the threshold barrier is applied to the image [4]. The entropy threshold is applied to the image, the value '90' is the threshold value. The pixels under '90' are clustered together to form the dark side of the image, and the threshold value above '90' are clustered together to form the bright side of the image [5]. Simultaneously, the image is centered with position limit condition and that image is applied to advanced morphology called Danielsson morphology (DM), DM enhances and triggers the cell contours, and clear hexagonality is identified. The endothelium cells are of hexagonal structure, and cell density is identified. The normal cell density in endothelium image is 2400 cells/mm^2 to 3200 cells/mm^2 and

<2400 cells/mm^2 said to be dystrophic image, >2400 cells/mm^2 are normal image. The density of cells is identified with the help of DM.

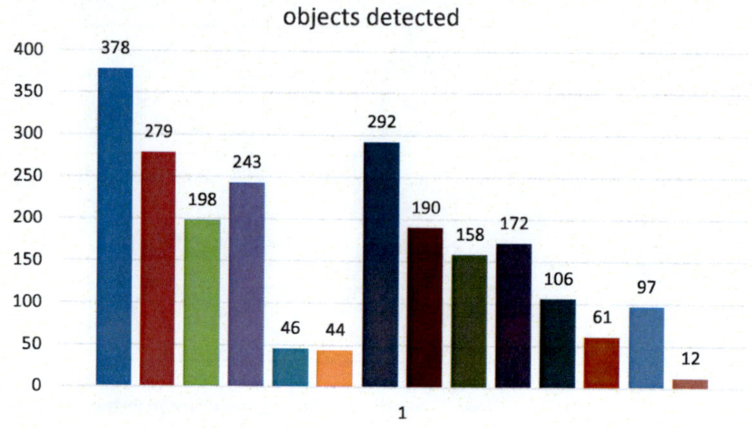

Fig. 2 Total cells detected

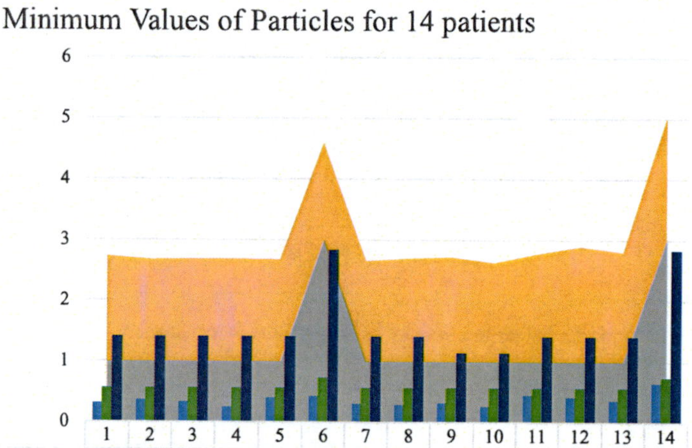

	1	2	3	4	5	6	7	8	9	10	11	12	13	14
Elongation factor	1.73	1.67	1.68	1.68	1.67	1.58	1.65	1.69	1.71	1.62	1.77	1.89	1.8	2
Particle & Holes' Area	1	1	1	1	1	3	1	1	1	1	1	1	1	3
No of Holes	0	0	0	0	0	0	0	0	0	0	0	0	0	0
%Area / Image Area	0	0.003	0.003	0.003	0.003	0.0096	0.003	0.003	0.003	0.003	0.003	0.003	0.003	0.0095
Compactness factor	0.307	0.36	0.332	0.247	0.4	0.422	0.299	0.281	0.316	0.258	0.441	0.409	0.356	0.643
Heywood circularity factor	0.564	0.564	0.564	0.564	0.564	0.739	0.564	0.564	0.564	0.564	0.564	0.564	0.564	0.737
Bounding rectngle diagonal	1.41	1.41	1.41	1.41	1.41	2.83	1.41	1.41	1.14	1.14	1.41	1.41	1.41	2.83

Fig. 3 Minimum statistical parameter of each patient

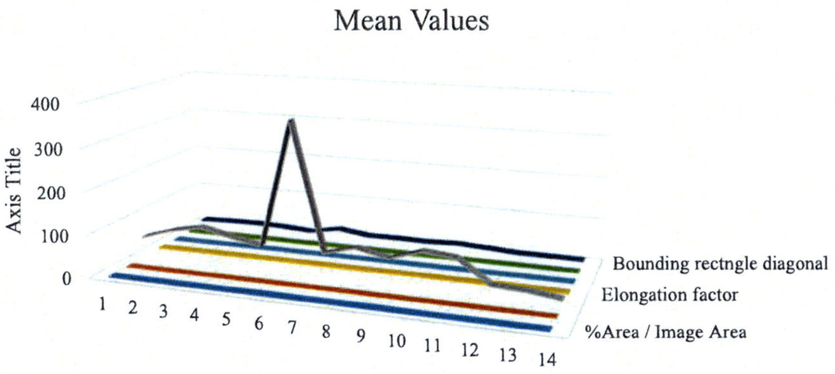

	1	2	3	4	5	6	7	8	9	10	11	12	13	14
%Area / Image Area	0.159	0.236	0.283	0.233	0.188	1.14	0.184	0.25	0.918	0.275	0.253	0.0842	0.0707	0.0413
No of Holes	0	0	0	0	0	0	0	0	0	0	0	0	0	0
Particle & Holes' Area	50.7	74.5	89	73.5	59.3	359	58	78.6	62.5	86.7	79.8	26.5	22.3	13
Elongation factor	2.34	2.19	2.35	2.37	2.3	2.19	2.35	2.28	2.45	2.47	2.31	2.32	2.43	2.32
Compactness factor	0.72	0.716	0.696	0.721	0.846	0.751	0.719	0.734	0.757	0.701	0.797	0.863	0.833	0.831
Heywood circularity factor	1.19	1.16	1.28	1.24	0.952	1.27	1.13	1.13	1.13	1.24	1.02	0.871	0.946	0.907
Bounding rectngle diagonal	11.8	14.3	15.6	13.6	8.81	22.7	12.4	13	12	15	11.1	6.28	6.26	5.29

Fig. 4 Mean statistical parameter of each patient

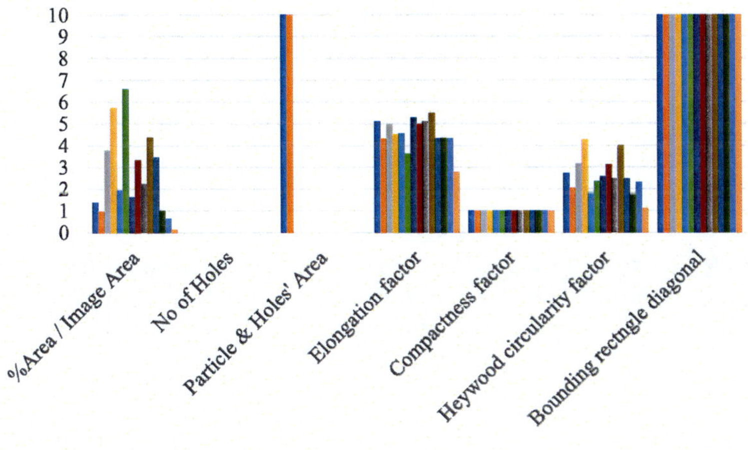

Fig.5 Maximum statistical parameter of each patient

Applying particle filter and analyzing particles in an image is an important function. To get the statistical profile of the image, the particle filter is applied and analyzed closely by all the particles in the endothelium image [6, 7]. Removing the unwanted particles as mentioned in the filter plays a predominant factor in this paper. After identifying the cell density, convex hull perimeter of the endothelium cell is also identified. The complete statistical profile of the endothelium image is extracted and displayed. The complete profile of the image is also extracted and mapped.

3 Result and Discussion

Considering Table 1, the proposed algorithm is tested on '15' images and calculated minimum area, maximum area, and the average area of each image. Along with the statistical approach, morphological factors like cell elongation, cell circularity, and cell compactness are calculated. The cell structure is completely measured, and the no.of cells per unit volume are also calculated. The no. of cells/mm^2, i.e., cell density, is also estimated. Cell density for '15' images is calculated and tabulated. Image I_1 is a normal image as it lies in the range of 2400 cells/mm^2 to 3200 cells/mm^2, remaining $I_2 - I_{14}$ images are out of normal range as they are subjected to ICD, PPCD. Due to these dystrophies, the cell density is reduced. Image morphological parameters elongation factor is more than 1, the compactness factor is =1, the circularity factor is >1 to <2.5, and the images $I_2 - I_{14}$ '13' images are in the range of >1 to $\leq$3 at maximum. Similarly, the compactness factor for the image $I_1 - I_{14}$ ranges in between >0 and <1. The proposed algorithm provides a minimum cell area value of an image, maximum cell area, and mean cell area which is also provided.

Cell densities of I_2(1758), I_3(1247), I_4(1531), I_5(277), I_6(277), I_7(1839), I_8(1197), I_9(995), I_{10}(1083), I_{11}(s667), I_{12}(384), I_{13}(611), I_{14}(490), I_{15}(780) are tabulated as given in the table.

3.1 *Particle Analysis*

A processing operational sequence for particle analysis creates an analyzing function. The produced data about a particle is organized clearly into information. Here, the area of pixels that touch with some similar states of logic or pixels that resemble whether in foreground or background is grouped.

3.2 *Morphological Functions*

Improvisation of set wise data in a regular 0 and 1's image by reframing and rejecting unnecessary information refers to morphological operations.

Eliminating noise pixels of the image that touch somewhere around borders, similar in each touching border or non-smooth and uneven edges are possible under it.

Table 1 DM method extracted statistical profile

Image. No.	Particle	%Area/Image area	Elongation factor	Compactness factor	Heywood circularity
1. High ECD	378	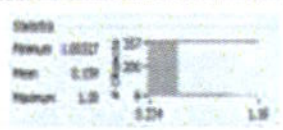			
	Min	0.00317	1.73	0.307	0.564
	Mean	0.159	2.34	0.72	1.19
	Max	1.39	5.12	1	2.06
2. Average of 01	279				
	Min	0.00317	1.67	0.36	0.564
	Mean	0.236	2.19	0.716	1.16
	Max	0.99	4.33	1	2.06
3. Average of 02	198				
	Min	0.00317	1.68	0.33	0.564
	Mean	0.283	2.35	0.696	1.28
	Max	3.79	4.99	1	3.18
4. Low 01	46	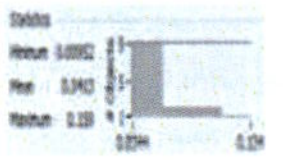			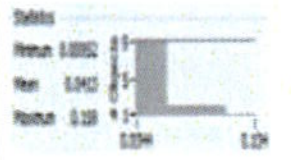
	Min	0.00317	1.67	0.40	0.564
	Mean	0.188	2.3	0.846	0.952
	Max	1.95	4.56	1	1.86
5. Low 02	44	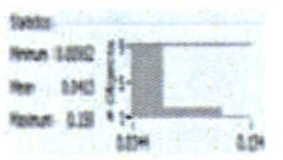	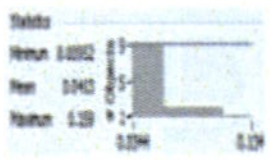	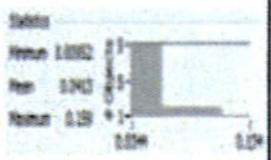	

(continued)

Table 1 (continued)

	Min	0.00317	1.58	0.42	0.739
	Mean	1.14	2.19	0.751	1.27
	Max	6.61	3.63	1	2.35
6. Surface folding	292		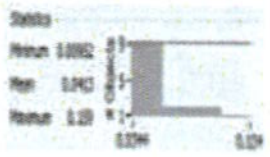	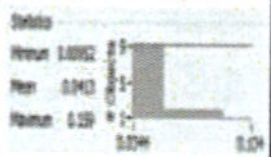	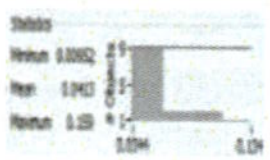
	Min	0.00952	1.65	0.29	0.564
	Mean	0.184	2.35	0.719	1.13
	Max	1.64	5.29	1	3.14
7. Mild polymegathism	190	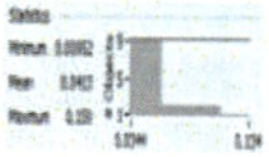	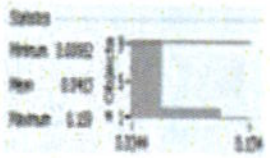	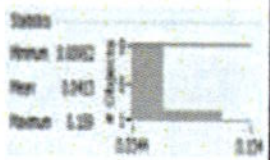	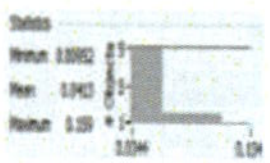
	Min	0.00317	1.69	0.28	0.564
	Mean	0.25	2.28	0.734	1.13
	Max	3.35	5	1	2.48
8. Polymegatism	158		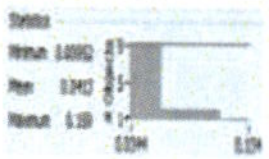	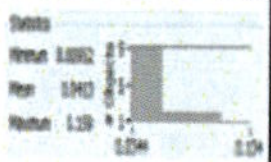	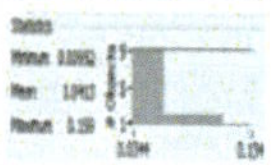
	Min	0.00317	1.71	0.32	0.564
	Mean	0.918	2.45	0.757	1.13
	Max	2.23	5.12	1	2.48
9. Poly and pleomorphism	172	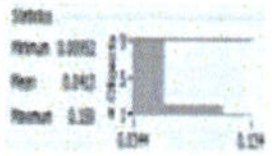		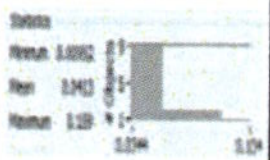	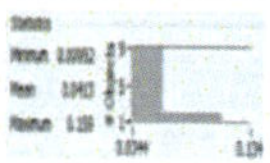
	Min	0.00317	1.62	0.26	0.564
	Mean	0.275	2.47	0.701	1.24

(continued)

Table 1 (continued)

	Max	4.38	5.5	1	4.01
10. Snail tracks	106	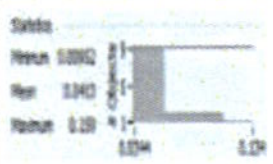			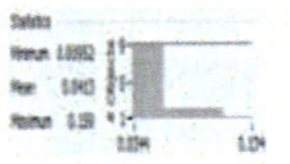
	Min	0.00317	1.77	0.44	0.564
	Mean	0.253	2.31	0.797	1.02
	Max	3.48	4.33	1	2.48
11. Corneal Guttata	61		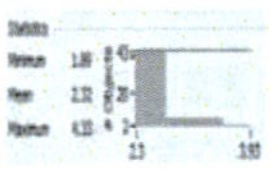		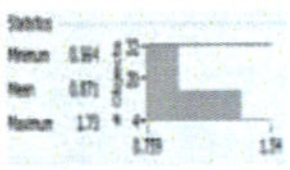
	Min	0.00317	1.89	0.41	0.564
	Mean	0.0842	2.32	0.863	0.871
	Max	1	4.33	1	1.73
12. Early dystrophy	97	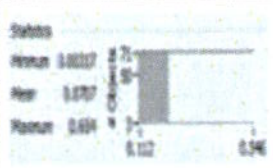	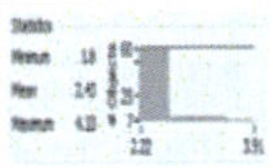	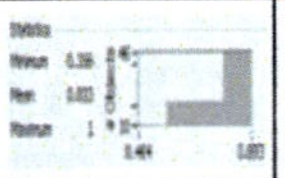	
	Min	0.00317	1.80	0.36	0.564
	Mean	0.0737	2.43	0.833	0.946
	Max	0.654	4.33	1	2.32
13. Lately diagnosed	12	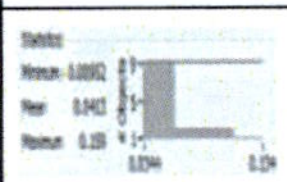	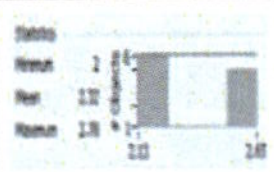	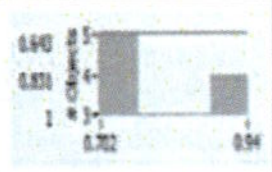	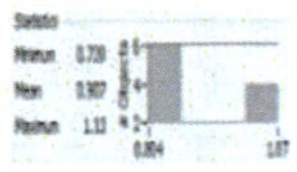
14. Fuch's dystrophy	Min	0.00952	2.00	0.64	0.737
	Mean	0.0413	2.32	0.831	0.907
	Max	0.159	2.78	1	1.13

- %Area/Image area = Percentage ratio of particle to the image of area covered.
- Particle and holes' area: If an image covers the complete area, the particle and holes' area is obtained when the particle completely covers the image.
- Elongation factor: It gives the extend to which the particle is elongated. Essentially, it is the ratio of Max Feret diameter and term called Feret given by the collective rectangle short side.
- Compactness factor: Given only between the internal of 0 and 1 inclusive of both values, i.e., [0, 1]. The term is defined using the ratio of areas obtained from the border rectangle width and bounding height of the same rectangle.
- Heywood circularity factor: The maximum value of it is 1, and it occurs when the particle is closest to the morphology of a disk.
- The factor is defined as a ratio of perimeter and circumference of the disk within the area under the same particle.
- Bounding rectangle: Initially, all particles are detected using a rectangular frame. This is fundamentally defined in the classification process. It is defined to be the diagonal of such a frame.

The bar graphs and corresponding results occur in coherence with the patient ocular conditions. The cornea affected by various dystrophies reflects the substantial decay of standard values in cell standards. The following conclusion section establishes the deviation from expected corneal endothelial cell standards.

3.3 *The Final Parameters Can Be Expressed a Graph*

The total number of particles detected in the 14 patients.

The graph for Min parameters of Morphology:

Maximum Values

Images are processed with MO1, MO2, S-PSO algorithms, and obtained maximum error as given in Table 2. The error is minimalistic when compared with existed algorithms and improves accuracy level by 2%, as well as various statistical parameters have been estimated.

Table 3 Error comparative table

S. no.	MO1	MO2	S-PSO approach	Danielsson morphology
1	13.19	22.24	12.20	11.02
2	12.28	20.93	9.55	9.5
3	10.96	18.23	7.45	6.9

4 Conclusion

Specular microscopic endothelium image analysis with Danielsson morphology is the novel algorithm to identify different corneal dystrophies like Fuchs dystrophy, advanced Fuchs dystrophy, iridocorneal dystrophy, and posterior polymorphous corneal dystrophy. The time consumption to process the specular images is below 1000 ms, and the Danielsson morphology estimates cell density cells/mm^2 of '14' different images in datasets. The convolution filter eliminates low-frequency noise components when compared with existing conventional morphological methods. Danielsson morphology is advanced morphology, and with the help of DM method various statistical methods like particle cell density, no.of particles, elongation factor, compactness factor, Heywood circularity factor are accurately estimated. These statistical parameters are very much helpful for ophthalmologists/clinicians to diagnosis the endothelium corneal layer condition accurately with a short time. DM method initiates accurate estimation of various statistical parameters which is helpful for ophthalmologists/clinicians to identify different dystrophies of endothelium corneal cells, which in turn shorten the time to identify the condition of corneal endothelium in busy clinical scheduling time.

References

1. Sharif, M.S., Qahwaji, R., Shahamatnia, E., Alzubaidi, R., Ipson, S., Brahma, A.: An efficient intelligent analysis system for confocal corneal endothelium images. Comput. Methods Programs Biomed. (2015)
2. Vincent, L.M, Masters, B.R.: Morphological image processing and network analysis of cornea endothelial cell images. **1769**, 212–226 (1992)
3. He, Z. et al.: 3D map of the human corneal endothelial cell. Sci. Rep. (2016)
4. Tsai, I.L., Hsu, C.C., Hung, K.H., Chang, C.W., Cheng, Y.H.: Applications of biomaterials in corneal wound healing, J. Chin. Med. Assoc. 78(4), 212–217 (2015)
5. Kowalewska-Zagrodnik, H., Osiagniecia, W.: Zakresie Modernizacji Kolumn Prozniowych Instalacji Destylacji Rurowo-Wiezowej W Mzrip W Plocku. Przem. Chem. **64**(5), 240–242 (1985)
6. Scarpa, F., Fiorin, D., Ruggeri, A.: In vivo three-dimensional reconstruction of the cornea from confocal microscopy images. Ann. Int. Conf. IEEE Eng. Med. Biol. Proc., 747–750 (2007)
7. Lopes, B.T., Ramos, I.C., Dawson, D.G., Belin, M.W., Ambrósio, R.: Detection of ectatic corneal diseases based on pentacam. Z. Med. Phys. **26**(2), 136–142 (2016)
8. Dyrlund, T.F. et al.: Human cornea proteome: Identification and quantitation of the proteins of the three main layers including epithelium, stroma, and endothelium. J. Proteome Res. **11**(8), 4231–4239 (2012)
9. Güell, J.L., El Husseiny, M.A., Manero, F., Gris, O., Elies, D.: Historical review and update of surgical treatment for corneal endothelial diseases. Ophthalmol. Ther. **3**(1–2), 1–15 (2014)
10. Kelliher, C. et al.: A cellular model for the investigation of Fuchs' endothelial corneal dystrophy. Exp. Eye Res. **93**(6), 880–888 (2011)
11. Zheng, T., Le, Q., Hong, J., Xu, J.: Comparison of human corneal cell density by age and corneal location: an in vivo confocal microscopy study. BMC Ophthalmol. **16**(1), 1–10 (2016)
12. Oie, Y., Nishida, K.: Corneal regenerative medicine. Regen. Ther. **5**, 40–45 (2016)

13. Thompson, A.J., Mastria, E.M., Eniola-Adefeso, O.: The margination propensity of ellipsoidal micro/nanoparticles to the endothelium in human blood flow. Biomaterials **34**(23), 5863–5871 (2013)

Sequential Workflow in Production Serverless FaaS Orchestration Platform

Urmil Bharti, Deepali Bajaj, Anita Goel, and S. C. Gupta

Abstract Today, serverless computing or Function-as-a-Service (FaaS) is an excellent way to exploit cloud computing benefits due to its simplicity, billing, and innate elasticity. FaaS providers own the responsibility of server provisioning, administration, management and patching thus making developers less burdened and more productive. Serverless applications based on complex workflows involve multiple functions which have data dependencies on each other. This brings the concept of function orchestration in picture and makes it important in implementing any serverless application. In this context Amazon Web Services (AWS Step Function), International Business Machines (IBM) Cloud Function Sequences and Azure Durable Functions are currently available in market. In this paper, different sequential composition workflows like Reflexive, Fusion, Chaining, and Client Scheduled are implemented in AWS serverless platform and compared their performance in terms of execution duration. Successively, sequential workflows in different language runtime environments in AWS Step function and IBM Cloud Function Sequences are also analyzed. Experiments are also performed to observe the cold start behavior in AWS Lambda and (IBM) Cloud functions.

U. Bharti · D. Bajaj (✉)
Department of Computer Science, Shaheed Rajguru College of Applied Sciences for Women, University of Delhi, Delhi, India
e-mail: bajajdeepali0@gmail.com

U. Bharti
e-mail: ubharti@hotmail.com

A. Goel
Department of Computer Science, Dyal Singh College, University of Delhi, Delhi, India
e-mail: goel.anita@gmail.com

S. C. Gupta
Department of Computer Science, Indian Institute of Technology, Delhi, India
e-mail: scgupta@cse.iitd.ac.in

© The Editor(s) (if applicable) and The Author(s), under exclusive license to Springer Nature Singapore Pte Ltd. 2021

A. P. Pandian et al. (eds.), *Proceedings of International Conference on Intelligent Computing, Information and Control Systems*, Advances in Intelligent Systems and Computing 1272, https://doi.org/10.1007/978-981-15-8443-5_58

Keywords AWS lambda · IBM cloud functions · Serverless trilemma · Function composition · AWS step function · IBM sequences · Function orchestration · Serverless functions · FaaS

1 Introduction

Serverless computing is a cloud computing execution model in which functions run in stateless compute containers which are event-triggered, ephemeral and completely managed by a cloud provider [1]. This is a revolutionary transition from the traditional paradigm wherein operations staff has to deploy and manage their continuously running dedicated virtual machines. With the aid of serverless technology, developers can now deploy functions that operate as event handlers, and only pay for CPU time when these functions are executing. Conceptually, scheduling of functions in reaction to events is known as reactive core [2]. Functions can be triggered by event sources including Databases, Message Queues, Web and Mobile Applications, IoT Sensors, Chatbots, Scheduled tasks (cron jobs), etc.

Serverless provides two types of services: Function-as-a-Service (FaaS) and Backend-as-a-Service (BaaS). BaaS cloud services are more relevant for Web and mobile application development [3]. It includes file storage, database storage, user account management, and social media integration. In BaaS, backend capabilities are considered as services which reduce recurring development of services. Nowadays, third-party back-end services have replaced server-side components of any cloud applications.

FaaS allows its users to merely write the code and leaves all the server provisioning and management tasks to the cloud service provider. Event-driven computing and ability to scale to thousands machines in case of concurrent invocations have encouraged cloud users to opt serverless platforms for a variety of applications like microservices, IoT, chatbot, data transformation, etc. Amazon Web Services (AWS) is the pioneer to start Function as a Service called Lambda [4] in 2014 and is significantly adopted in industry presently. Other popular cloud service providers that offer FaaS services are Google Cloud Functions [5], Azure Functions [6], and IBM Cloud function [7]. Out of these cloud service providers only IBM is based on Apache OpenWhisk [8] which is an open source distributed serverless platform.

Complex serverless applications may have multiple functions that need to communicate with each other. Further cloud function platforms are not aware of the data dependencies among the functions. This unawareness may result in sub-optimal function placement that leads to poor performance in communication [9]. For this, serverless providers are coming up with some orchestration frameworks to compose workflows that consist of a pool of cloud functions. These frameworks facilitate in building composite applications that need to pass along output of one function as input to other, for example, workflow of functions chained together as they have data dependencies among themselves [10]. Thus, independent functions which are inherently event-driven are tied together to generate complex workflows. A most popular use case is a real-world serverless application, an image editing application involving a chain of FaaS functions to re-size, grayscale, and watermark an input image. This pattern consists of a collection of independent functions that are linked together sequentially in order to process a request. One functions output becomes input to the next function. In this scenario, a colored image is input to the first FaaS and this image is converted to black and white image. This modified image is converted into a grayscale image by second FaaS and output of this is passes to the last function of the sequence to embed a watermark in the image. Popular food delivery giant, FoodPanda, uses AWS Step Functions and AWS Lambda to manage workflows of their food delivery application.

In this context, AWS Step Functions [11], Azure Durable Functions [12], and IBM Composer [13] are available. AWS Step Functions are well established and most popular for implementing complex workflows in serverless applications. IBM composer also offers all complex workflow patterns but is still in technical preview state. So, in our study, AWS Step Functions and IBM Cloud Sequences are used for comparing sequential workflows.

AWS Step Functions (ASF) is a function orchestration tool (introduced in the year 2016) that is used in designing and running a serverless workflow by coordinating together multiple AWS Lambda functions. It allows its users to implement function composition as computational graphs implemented in Amazon State Machine Language (ASL) [14]. Independent functions make decisions based on input, perform specified actions, and pass output to other functions. Programmers create their workflows in the ASL and AWS Step Function console generates a pictorial representation of the state machine to visualize whole application logic and track execution progress. Step function enables parallel execution, choice-based function (if-else statements) chaining, and sequential execution. *Parallel execution* indicates parallel branches of execution while *choice-based execution* enables its users to include branching logic to the Amazon state machine. It also allows some other useful states like *Wait* (add some delay in the execution of state machine for a specified time) or *Fail* (bring execution of a state machine to an end). ASF also supports a reflective API to track the running state or to cancel the complete workflow. Monitoring and function logging are done using AWS CloudWatch service [15].

In IBM Cloud Functions, sequence provides a mechanism to create a chain of functions (known as actions in IBM Cloud) that are invoked in a predefined order. In this chaining, output of one becomes the input to the next action. Programmers can combine existing IBM cloud actions to compose complex sequences. IBM Sequences only supports simple chaining of actions. It does not offer a reflective API to control and monitor and track actions in a workflow. Graphical view of the logs can be seen using IBM Cloud Analysis service—LogDNA [16].

Function Composition in Serverless Computing must satisfy three constraints as identified by Baldini et al. [2]. These properties determine quality of architectural design in serverless orchestration frameworks—(i) functions should be treated as black-boxes—implementation details of a function should remain abstracted from other functions (ii) their composition should follow a substitution rule with respect to synchronous invocation—composition should be used and substituted like any other function (iii) during invocations, function should not be double billed—a function waiting for response from another function should not be billed. These constraints are collectively known as Serverless Trilemma (ST) [2]. Any serverless orchestration system that satisfies all three conditions is called ST-safe [17].

Main contributions of our paper are:

- Observed cold start behavior in two most prominent Industry standards serverless platforms: AWS Lambda (proprietary) and IBM Cloud Functions (open-source)
- Compared execution duration of various sequential composition patterns like reflection, fusion, chaining, and client-based scheduling
- Evaluated performance of sequential workflows implemented in two interpreted language runtime environment (i) Python and (ii) Node.js for two leading function orchestration platforms, i.e., AWS Step function and IBM Cloud Function Sequencer.

2 Related Work

Serverless frameworks poses certain challenges like (i) preserve applications' state (ii) efficiently compose functions in order to develop complex applications (iii) design execution plan of functions (iv) monitor and debug them. These problems naturally indicate the need of a function orchestrator as more complex the serverless application, more the number of functions it involve.

Serverless orchestration is an evolving field and very few research articles are available in this area. Here, research papers related to production serverless function orchestration platforms is discussed.

Baldini et al. [2] have expressed that the event-driven mechanism of serverless is not sufficient and capable enough to implement complex workflows, i.e., compositions of independent serverless functions. They gave the concept of Serverless Trilemma that any function orchestration system should satisfy.

Lopez et al. [17] comprehend architectures of various FaaS orchestration providers, their programming and charging models, and also evaluated support for parallel execution of cloud functions. They performed a series of experiments and evaluated the runtime overhead in different serverless orchestration systems for implementing various types of workflows.

Barcelona-Pons et al. [18] demonstrated that performance of IBM Composer is better than AWS Step Functions and Azure Durable functions. They have also studied and analyzed the architecture of OpenWhisk and its orchestration features. They have also outlined its architectural issues and recommended guiding principle for orchestrating embarrassingly parallel workloads with serverless functions.

Very limited research papers are available in the domain of serverless orchestration system and have not come across any research work that implements and compares various function composition workflows. Moreover, evaluation of sequential workflows in different language runtime environments has been majorly ignored in literature. So, various experiments are performed on available serverless orchestration systems to gain useful insights on various function composition workflows.

3 Experimental Setup for Running AWS Lambda Functions and IBM Cloud Functions

The AWS Lambda Functions and IBM Cloud Functions are developed and executed to study the cold start behavior in these two serverless platforms.

Serverless platforms provide containers (sandboxed environment) to bundle function code that developer wants to execute when it is invoked. Additionally, it also consists of configuration files, classes, library, and related dependencies libraries that function code will refer on execution. Independent stateless functions are executed in these containers. When a user invokes a function for the first time, serverless platform takes a little longer in spinning the container first time [19]. This is known as Cold start problem, i.e., time taken to spin up a new container instance when there is no container available to serve a new request. For later invocations of the same function, serverless platforms keep the container active and running for a certain amount of time for future invocations of the function. If a request comes while its container is active, it is served instantaneously due to warm start. A container is said to be terminated if it has not been used for some longer duration of time. According to "Serverless Architectures with AWS Lambda" [20], a container initiation speed depends on runtime language, network configuration, and function code package size.

For our experiment, recursive and non-recursive serverless functions are run as shown in Table 1 on AWS Lambda and IBM Cloud Functions. Language runtime used in both the platforms is Python 3.7. Selection of the programs is based on increasing order of their complexity in terms of computation time. Default memory limit for a container in AWS Lambda is 128 MB and in IBM Cloud Functions is 256 MB. For the sake of uniformity and clear comparison, run our serverless functions on 256 MB memory with time-out configuration set as one second for both serverless providers. Billing interval is 100 ms in both the platforms.

All the above-mentioned functions are executed eleven times and noted their execution time. The value of first invocation's execution time is ignored while calculating the average execution time of each function. First invocation took longer time than subsequent invocations as it suffered from cold start problem. Subsequent invocations took lesser time due to warm start of containers. Cold Start Statistics in AWS Lambda and IBM Cloud Function is shown in Table 2.

The AWS Lambda does not show much variability in function execution duration for first and subsequent invocations. But in IBM Cloud Functions, first invocation takes a very long time as compared to subsequent invocations. Thus, the effect of

Table. 1 Details of recursive/non-recursive serverless functions

Program type	S. No.	Program	Detail
Nonrecursive	1	Fibonacci sequence	Function will generate first 25 Fibonacci numbers
	2	Factorial	Function will calculate 50!
	3	Matrix multiplication	Function will multiply two 4 × 4 matrix where matrix values are randomly generated and ranging from 1000 to 10,000
Recursive	4	Fibonacci sequence	Function will generate first 25 Fibonacci numbers
	5	Factorial	Function will calculate 50!
	6	Binary search	Function will search a value in an array of 100 values where values are ranging from 1 to 1000
	7	Quick sort	Function will sort an array of length 1000 where array values are ranging from 1 to 100
	8	Binary search tree	Function will created a binary search tree of 1000 nodes where values are ranging from 1 to 1000

Table. 2 Cold start statistics in AWS lambda and IBM cloud functions (total memory allocated = 256 MB, time out = 1 s)

Program type	Program	AWS lambda functions		IBM cloud functions	
		First invocation duration (ms)	Subsequent invocation duration average (ms)	First invocation duration (ms)	Subsequent invocation duration average (ms)
Nonrecursive	Fibonacci	0.28	0.35	160	1.16
	Factorial	1.51	1.40	193	1
	Matrix multiplication	0.83	0.68	150	1.33
Recursive	Fibonacci	440.05	410.59	221	72
	Factorial	9.99	3.86	215	1.28
	Binary search	0.79	0.50	200	2
	Quick sort	2.74	2.03	184	4.12
	Binary search tree	113.45	82.82	195	14

cold start is more pronounced in IBM Cloud Functions as compared to AWS Lambda. Figure 1 shows effect of cold start in AWS Lambda and IBM Cloud Functions.

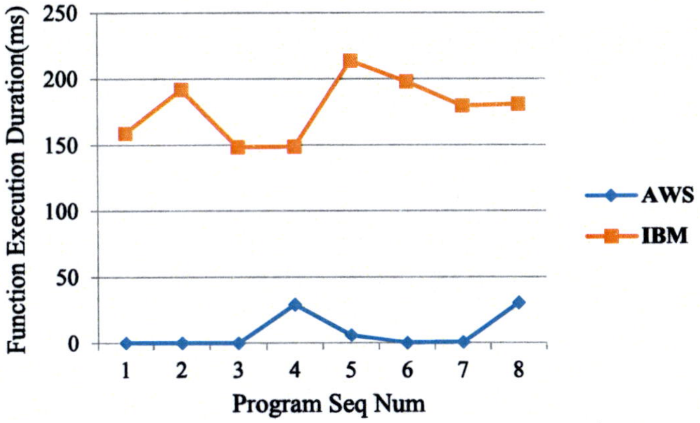

Fig. 1 Effect of cold start in AWS lambda and IBM cloud functions

4 Executing Sequential Compositions in AWS Lambda

Real world serverless applications require simple functions to be connected in some composition. Functions can be composed as parallel, sequential, or choices based. For our experiments, sequential composition of three monoglot functions named func1(), func2(), and func3() is considered. As per our experimental design, func1() is invoked first followed by func2()which in turn followed by func3(). Let func1() takes "x" as input then sequential composition is designed as

$$v = \texttt{func3(func2(func1(x)))}$$

where "v" is the final result. Sequential composition of functions can be structured in four ways based on reactive core concept [2]. One or more of the Serverless Trilemma constraints are violated by these approaches.

For experimental validation, AWS Lambda FaaS platform is opted for implementing our functions and language runtime is Python 3.7.

4.1 Composition via Reflection

In this composition structure, a serverless function synchronously invokes other functions in a sequential order. To achieve this, a main λ function called Wrapper is implemented in AWS Lambda that encapsulates other λ functions and generates a payload. Wrapper function invokes three Lambda functions λ1(), λ2() and λ3() in sequence with this payload. As all the functions are implemented as independent serverless functions so can be polyglot in implementation. Polyglot means functions

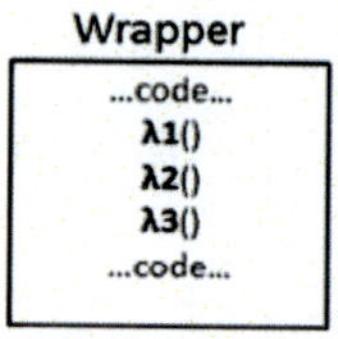


```
import boto3, json
def lambda_handler(event, context):
    invokeLam = boto3.client("lambda", region_name="us-east-2")
    payload = {"message": "Test"}
    resp = invokeLam.invoke(FunctionName="Lambda1", InvocationType = "Event",
Payload = json.dumps(payload))
    print("Lambda1 Invoked ")
    resp = invokeLam.invoke(FunctionName=" Lambda2", InvocationType =
"Event", Payload = json.dumps(payload))
    print("Lambda2 Invoked ")
    resp = invokeLam.invoke(FunctionName=" Lambda3", InvocationType =
"Event", Payload = json.dumps(payload))
    print("Lambda3 Invoked  ")
        return {
            'statusCode': 200,
            'body': json.dumps('Hello from Lambda!')
        }
```

Fig. 2 Sequential composition using reflective pattern

can be composed regardless of their language of implementation. This composition pattern suffers from double billing constraint [2] as wrapper function is also billed when all the other three functions are executing. Figure 2 shows sequential composition using Reflective Pattern.

4.2 Composition via Fusion

In this composition, a main function called Wrapper is implemented in AWS Lambda Wrapper that inlines the code of func1(), func2(), and func3(). This pattern restricts all three functions to be written in the same language as that of Wrapper function. Since code of func1(), func2(), and func3() is written inside the Wrapper, so it violates black box constraint of Serverless Trilemma [2]. Figure 3 shows sequential composition using Fusion Pattern.

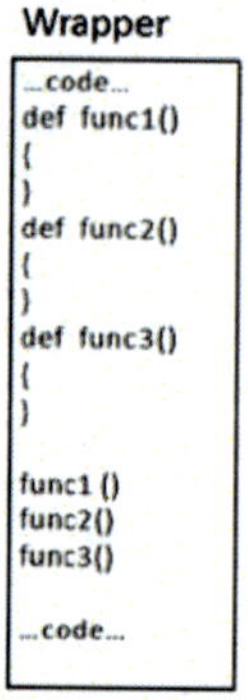


```
import json
def lambda_handler(event, context):

    def func1(param):
        print("1 Exit")
    def func2(param):
        print("2 Exit")
    def func3(param):
        print("3 Exit")

    param="Test Payload"
    func1(param)
    func2(param)
    func3(param)
       return {
         'statusCode': 200,
         'body': json.dumps('Success!')
    }
```

Fig. 3 Sequential composition using fusion pattern

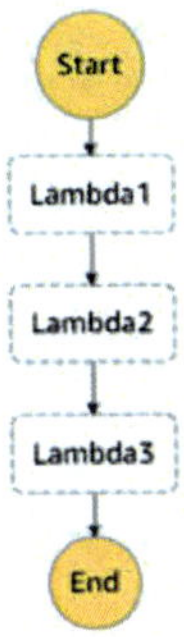


```
{
  "Comment": "My test to prove passing parameters between Lambdas",
  "StartAt": "Lambda1",
  "States": {
    "Lambda1": {
      "Type": "Task",
      "Resource": "arn:aws:lambda:us-east-2:941920168749:function:W1",
      "Next": "Lambda2"
    },
  "Lambda2": {
      "Type": "Task",
      "Resource": "arn:aws:lambda:us-east-2:941920168749:function:W2",
      "Next": "Lambda3"
    },
      "Lambda3": {
      "Type": "Task",
      "Resource": "arn:aws:lambda:us-east-2:941920168749:function:W3",
      "End": true
    }
  }
}
```

Fig. 4 Sequential composition using client in ASF

4.3 Composition via Async

The model is fire and forgets type of composition pattern. In this λ1() can asynchronously calls λ2() in such a way that λ1() does not remain active [2]. This composition pattern will violate the substitution principle as λ1() will no longer remain a composable FaaS function when invoked synchronously. Asynchronous invocation will not reflect the desired results so λ3() results are not included in final result.

4.4 Composition via Client

In this composition pattern, Wrapper function is not a serverless function and it exists outside the serverless runtime. It is considered as client scheduled structure. AWS Step Function comes under this category. Step Functions lets developers coordinate various AWS Lambda functions into serverless workflows by making finite state machines. This state machine is written in Amazon State Language (ASL). Each machine calls individual Lambda functions using their ARN (Amazon Resource Number). This type of composition violates the substitution principle as one step machine cannot be substituted in another step machine. Figure 4 shows sequential composition using Client in ASF.

4.5 Composition via Chaining

In this composition pattern, λ1() invokes λ2() which in turn invokes λ3(). In this sequential composition, one extra wrapper function is not required to encapsulate and

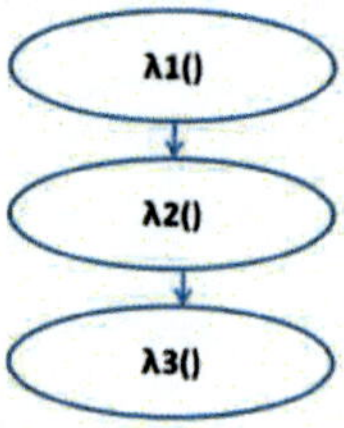


```
Lambda1
def lambda_handler(event, context):
    invokeLam = boto3.client("lambda", region_name="us-east-2")
    payload = {"message": "Test"}
    resp = invokeLam.invoke(FunctionName="Lambda2",
InvocationType = "Event", Payload = json.dumps(payload))
    print("Lambda1 Exit")
    return {
        'statusCode': 200,
        'body': json.dumps('Success 1')
    }

Lambda2
import boto3, json
def lambda_handler(event, context):
    invokeLam = boto3.client("lambda", region_name="us-east-2")
    payload = {"message": "Test"}
    resp = invokeLam.invoke(FunctionName="Lambda2",
InvocationType = "Event", Payload = json.dumps(payload))
     print("Lambda2 Exit")
    return {
        'statusCode': 200,
        'body': json.dumps('Success 2')
    }

Lambda3
import json
def lambda_handler(event, context):
    print("Lambda3 Exit")
    return {
        'statusCode': 200,
        'body': json.dumps('Success 3')
    }
```

Fig. 5 Sequential composition using chaining pattern

call other λ functions. This approach violates double billing constraint of Serverless Trilemma because λ1() is billed for the duration when λ2() and λ3() were running. This entails more execution cost which is completely undesirable. Here, λ1() is in idle state while waiting for response from λ2() and λ3(). Figure 5 shows chaining composition where λ1(), λ2() and λ3() are invoked in a sequence.

Table 3 shows observed parameters for the above-mentioned sequential composition patterns implemented in AWS serverless platform.

Our experimental results show that none of the compositions implemented in AWS platform satisfy Serverless Trilemma. Fusion has best performance in terms of execution time but with above mentioned constraints. Chaining performs worst as it's billed duration was maximum.

Table. 3 Sequential composition patterns implemented in AWS serverless platform

Composition type	Polyglot support	ST constraint violation	Duration (in ms)
Reflection	Yes	Double billing	311.03
Fusion	No	Black box	2.47
Async	N/A	N/A	N/A
External client	Yes	Substitution principle	331.83
Chaining	Yes	Double billing	1104.78

5 Experimental Setup for Comparing Sequential Composition in AWS Step Function and IBM Cloud Function Sequences

In this section, sequential compositions are compared in two serverless orchestration systems: (i) AWS Step Functions (proprietary) (ii) IBM Cloud Function Sequences (based on Apache OpenWhisk).

AWS Step Functions allows the programmer to design the state machine for any function composition. Function compositions are composed from individual and discrete Lambda functions so any updates in these functions do not require changing the state machine. This orchestration method is not ST-safe [2] as the state machine itself cannot be substituted in another state machine for more complex workflows [14].

IBM Cloud Function Sequences allows building a new sequence by adding existing actions to a sequence. In this approach, a sequence can be easily substituted in another complex workflow. Thus IBM Cloud Function Sequences satisfies all three constraints of Serverless Trilemma as discussed in introduction section and is considered ST-safe.

For our experiments, a sequential workflow (consisting of three monoglot functions) is developed in AWS Step Function and IBM Cloud Function Sequences. Sequences were developed in two language runtime: Node.js and Python. Our results show that IBM Cloud Function Sequences perform outstandingly better than ASF. Execution duration of sequences in ASF for Python was 332 ms and for Node.js 323 ms. In IBM Cloud Function Sequences, for Python, duration was 3.12 ms and for Node.js duration was 7 ms. Figure 6 shows the result of sequential workflow execution in AWS and IBM.

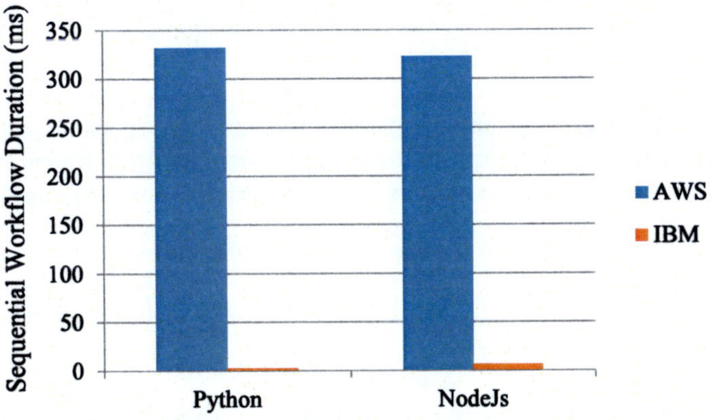

Fig. 6 Sequence in AWS Step function and IBM sequencer

6 Insights

After analysis of results of our experiments, some interesting insights and observations are presented:

- Cold start problem is more prominent in IBM cloud functions.
- AWS Lambda shows comparable result for first invocation and subsequent invocations of functions hence minimal cold start latency observed.
- IBM Cloud Function Sequences performs better than ASF sequences for both language runtime, i.e., Python and Node.js.
- AWS Step Functions is not ST-safe, i.e., all composition patterns like reflection, fusion, external client or chaining violates one or more constraints of Serverless Trilemma.
- IBM Cloud Function Sequences is ST safe, i.e., it satisfies all three constraints of Serverless Trilemma.
- As per official documentation, OpenWhisk is independent to action's language runtime (language agnostic) [21]. But in our results, execution time of sequences developed in Node.js was slightly higher than sequences developed in Python in IBM Cloud Function Sequences.

7 Conclusion

In order to run a complex application as serverless, developers need to create stateless functions for each task, and run them via a serverless orchestration system. FaaS orchestration systems are Web services that coordinate independently running functions so that they become building blocks of a larger functional pipeline. Thus, programmers may build their complex serverless applications using various function compositions. In this paper, sequential composition patterns like reflection, fusion, chaining, and client Scheduled are compared in AWS serverless platform. The effect of language runtime is also analyzed on sequential workflow implemented in AWS Step Functions and IBM Cloud Function Sequences and observed IBM Cloud Functions performs outstandingly better than ASF sequences for both language runtime (Python and Node.js). In the future, will dive into prioritization of Serverless Trilemma properties.

References

1. Baldini, I., Castro, P., Chang, K., Cheng, P., Fink, S., Ishakian, V., Suter, P.: Serverless computing: current trends and open problems. In: Research Advances in Cloud Computing, pp. 1–20 (2017). Springer, Singapore
2. Baldini, I., Cheng, P., Fink, S. J., Mitchell, N., Muthusamy, V., Rabbah, R., Tardieu, O.: The serverless trilemma: function composition for serverless computing. In: Proceedings of the 2017

ACM SIGPLAN International Symposium on New Ideas, New Paradigms, and Reflections on Programming and Software, pp. 89–103, (2017, October). ACM
3. Bharti, U., Bajaj, D., Budhiraja, P., Juyal, M., Baral, S.: Android based e-voting mobile app using google firebase as BaaS. In: International Conference on Sustainable Communication Networks and Application, pp. 231–241 (2019, July). Springer, Cham
4. https://docs.aws.amazon.com/lambda/?id=docs_gateway. Accessed on 2020–1–17
5. https://cloud.google.com/functions. Accessed on 2020–2–08
6. https://azure.microsoft.com/en-in/blog/introducing-azure-functions/. Accessed on 2020–2–26
7. https://cloud.ibm.com/docs/openwhisk?topic=cloud-functions-actions#actions_seq. Accessed on 2020–2–27
8. https://openwhisk.apache.org/documentation.html. Accessed on 2020–2–18
9. Jonas, E., Schleier-Smith, J., Sreekanti, V., Tsai, C.C., Khandelwal, A., Pu, Q., Gonzalez, J.E.: Cloud programming simplified: a berkeley view on serverless computing (2019). *arXiv preprint* arXiv:1902.03383
10. Hellerstein, J.M., Faleiro, J., Gonzalez, J.E., Schleier-Smith, J., Sreekanti, V., Tumanov, A., Wu, C.: Serverless computing: one step forward, two steps back (2018). *arXiv preprint* arXiv:1812.03651
11. https://aws.amazon.com/step-functions/. Accessed on 2020–2–24
12. https://docs.microsoft.com/en-us/azure/azure-functions/durable/durable-functions-overview?tabs=csharp. Accessed on 2020–1–23
13. https://cloud.ibm.com/docs/openwhisk?topic=cloud-functions-pkg_composer. Accessed on 2020–1–23
14. https://docs.aws.amazon.com/step-functions/latest/dg/welcome.html
15. https://aws.amazon.com/cloudwatch/
16. https://cloud.ibm.com/docs/Log-Analysis-with-LogDNA?topic=LogDNA-getting-started#getting-started
17. López, P.G., Sánchez-Artigas, M., París, G., Pons, D.B., Ollobarren, Á.R., Pinto, D.A.: Comparison of Production Serverless Function Orchestration Systems (2018). arXiv preprint arXiv:1807.11248
18. Barcelona-Pons, D., García-López, P., Ruiz, Á., Gómez-Gómez, A., París, G., Sánchez-Artigas, M. (2019, December). FaaS Orchestration of Parallel Workloads. In Proceedings of the 5th International Workshop on Serverless Computing (pp. 25–30)
19. Manner, J., Endreß, M., Heckel, T., Wirtz, G.: Cold start influencing factors in function as a service. In: 2018 IEEE/ACM International Conference on Utility and Cloud Computing Companion (UCC Companion), pp. 181–188 (2018, December). IEEE
20. https://d1.awsstatic.com/whitepapers/serverless-architectures-with-aws-lambda.pdf. Accessed on 2020–3–27
21. https://github.com/apache/openwhisk/blob/master/docs/actions.md#languages-and-runtimes. Accessed on 2020–2–28

Practical Implementation and Analysis of TLS Client Certificate Authentication

N. Rajathi and Meghna Praveen

Abstract Authentication plays a vital role in securing any system. This paper discusses a foolproof implementation of client certificate authentication in a client–server architecture consisting of a large number of clients that communicate frequently with a server in a restricted environment with less or no human intervention. The performance of the proposed TLS-based client authentication is compared with the traditional authentication mechanisms. The result shows that, despite minor increase in latency and CPU usage, it is significantly secure from multiple attacks like man-in-the-middle attack and playback attacks.

Keywords Client authentication · EST · Playback attack · Security · Sniffing attacks · SCEP · SSL · TLS

1 Introduction

SSL certificates have been widely used for authenticating servers. A vast majority of SSL certificates are the server certificates used by servers to claim its authenticity to different clients. However, SSL/TLS has more capabilities than just authenticating servers—it can also be used to authenticate the client that is trying to reach the server.

TLS has a client certificate authentication mechanism with significant advantage over the traditional form-based or token-based authentication mechanisms. When clients are dedicated software modules that automate certain tasks with less or no human interaction, then such traditional authentication mechanisms as mentioned above are not feasible. Moreover, token-based authentication involves generating a large number of unique tokens per request or per client. Such tokens should be stored

N. Rajathi (✉) · M. Praveen
Department of Information Technology, Kumaraguru College of Technology, Coimbatore 641049, India
e-mail: rajathi.n.it@kct.ac.in

M. Praveen
e-mail: meghna.16it@kct.ac.in

© The Editor(s) (if applicable) and The Author(s), under exclusive license to Springer Nature Singapore Pte Ltd. 2021

A. P. Pandian et al. (eds.), *Proceedings of International Conference on Intelligent Computing, Information and Control Systems*, Advances in Intelligent Systems and Computing 1272, https://doi.org/10.1007/978-981-15-8443-5_59

in the clients and the server they are trying to contact, thereby increasing the load on the server when the number of clients and the number of requests made by each client are very large. To overcome this, TLS client certificate authentication mechanism can be considered.

After studying the various standards of client certificate authentication, like Simple Certificate Enrollment Protocol (SCEP) and Enrollment over Secure Transport (EST), a custom implementation of client certificate authentication was done by selecting the best and suitable flows of enrollment, renewal and key exchange, and its performance was tested using Apache JMeter.

2 Literature Survey

In their white paper [1], Akamai Intelligent Platform proposes a Public Key Infrastructure (PKI) based cloud-security services. John Harris, in his whitepaper for Global Sign [2], explains certificate-based network authentication and how it compares to other leading identity and access management solutions. It also highlights the fact that certificate-based authentication requires the device to be kept secure (physically) so that the credentials cannot be removed or copied.

In [3], Trustem platform module is used to maintain the authentication token for key generation and providing certificates. It uses different encryption algorithm to make a secure communication. Here, two different keys are utilized to identify itself such as attestation identity key and binding key.

Davies [4] concludes that although password-based authentication is preferred when human interaction is involved, certificate-based authentication is better for B2B-type communications where unattended file transfers need to be performed.

Parsovs in his paper [5] discusses the issues about the certificate authentication for TLS client, and it will be analyzed from the server side based on Apache's model.

In [14], the authors proposed an authentication in the cluster-based routing to avoid the unwanted security breaches in the communication between the head of the clusters to destination.

3 General Working

Client certificate authentication happens during the SSL handshake. The client exchanges its information to the server in the form of a client certificate issued to it by the server after being signed by a trusted authority. The client exchanges its certificate information only after it authenticates the server. The process is illustrated in Fig. 1.

Step 1: Client requests a protected resource from the server.

Fig. 1 General working of client certificate authentication

Step 2: Server provides its certificate to the client. If the client has an in-built Trust Authority, it verifies the certificate chain/root of the server's certificate.
Step 3: If the server is configured for client certificate authentication, it sends a client certificate request to the client.
Step 4: If the client has an implicit or explicit Trust Authority (TA), it verifies the root/chain of the server's certificate. If the server has been successfully authenticated, the client will send its certificate to the server, along with key information or symmetrical key information.
Step 5: The server verifies the client against its TA and sends back the response finish message encrypted using the symmetric key exchanged.
Step 6: Thus, with the exchange of the client and the server certificates, mutual authentication occurs within the SSL handshake.

This type of authentication is much safer than the traditional password-based authentication. Passwords, no matter how strong can be broken using persistent brute force attacks, phishing attack and social engineering techniques.

When it comes to authentication tokens being sent from the client to the server, scalability is an issue. With the increase in the number of clients, the number of unique tokens generated per client or per request is too large. Implementing OAuth in legacy enterprise software involves too much man effort compared to implementing client certificate authentication. Hence, when it comes to effort of implementation and maintenance, client certificate authentication is preferred.

4 Protocols Available

4.1 Certificate Provisioning

Two certificate management protocols were studied, namely Simple Certificate Enrollment Protocol [6] and Enrollment over Secure Transport (RFC 7030) [7]. These protocols define ways that enable a PKI client to request or renew a certificate from the Certificate Authority. The key takeaways from studying the two protocols [8] are:

1. EST uses TLS for the secure transport of messages and certificates. SCEP runs over HTTP with pkiMessage messages that are secured in pkcsPKIEnvelope envelopes.
2. In EST, the certificate signing request (CSR) can be tied to a requestor that is already trusted and authenticated with TLS. In SCEP, the CSR is authenticated using a shared secret between the client and the CA.
3. EST supports automated certificate enrollment and re-enrollment. Such an implementation is missing in SCEP.
4. EST supports server-side key generation, whereas SCEP supports only private key being generated in the client before sending the CSR.
5. EST provides cryptographic agility. It supports elliptic curve cryptography (ECC) and secured cryptographic algorithms to come. It is computationally more efficient, in turn benefitting resource-constrained devices. SCEP does not support ECC.

4.2 Certificate Management

Certificate management protocol using Cryptographic Message Syntax (CMC), RFC 5272 [9] was defined in the year 2008, leaving certificate management over CMS, RFC 2797 [10] (also CMC) obsolete.

CMC is architecturally very similar to SCEP. However, CMC provides more options for management and has better cryptographic agility. CMC uses the Certificate Request Message Format (CRMF) or PKCS 10 for making certificate request.

The X.509 Public Key Infrastructure Certificate Management Protocol, or CMP, as defined in RFC 4210 [11] is another certificate management protocol that was introduced in between the periods RFC 5272 and RFC 2797 which were introduced. CMP provides on-line interactions between PKI components, including an exchange between a certification authority (CA) and a client system.

From the study of CMC and CMP, it is understood that EST is just another protocol build on top of CMC over a secure transport (HTTPS). While EST mainly addresses certificate provisioning, the process of enrollment and renewal is taken from CMC. The key takeaways after studying the management protocols are [8]:

1. CMC and CMP use enveloped message data and ASN.1 (Abstract Syntax Notation) [12]. This is more complicated compared to EST's message format
2. CMC does not address the renewal of CA certificates. EST combines the CMP defined renewal of CA certificates into CMC to make its own renewal of CA flow.

From the study of the certificate provisioning and enrollment protocols, it can be concluded that of all the existing standards, EST is the best choice when it comes to simplicity of development and availability of open-source resources for its implementation. The custom implementation discussed in this paper makes use of the TLS transport of messages and certificates, CSR authenticated using a shared secret between the client and the server, automated renewal of client certificate and client-side private key generation.

5 Implementation

After studying SCEP and EST, EST was found to have more benefits over SCEP. According to the EST protocol, an out-of-band distributed shared credential is used to initially authenticate to the server. An out-of-band shared credential is a shared secret (username/password, key, code, etc.) that is transferred through a stream that is independent from the main communication stream. This credential can also be used to further encrypt the contents of every request symmetrically. This credential can be a nonce value exchanged before the establishment of the SSL tunnel or a hardware-based key embedded in the client during manufacture or even embedded in the software of the client during deployment.

A nonce value is a data that is often input to cryptographic protocols and some encryption modes. Nonce values should only be used a single time with any particular cryptographic key. If at all it is being reused, the odds of reuse need to be exceptionally low. Nonce values can be generated using a cryptographically strong pseudo-random number generator. The advantage of using nonce over embedded symmetric keys is its randomness and unpredictable nature. However, embedded keys if stored safely, in a way that they cannot be reverse engineered easily, are easier to use and avoid the hassle of generating and exchanging a nonce value for every request.

5.1 Initial Enrollment

Just like/simple enroll in EST, initial enrollment involves sending a certificate signing request (CSR) to the server. This is done after the client authenticates the server using the server's SSL certificates.

The CSR request can be encrypted using a previously exchanged nonce value or an embedded symmetric key. This initial request is not authenticated using client certificate authentication, and if the server is able to decrypt it using the previously exchanged symmetric key or nonce value, then the server accepts the CSR, signs it and sends back the client certificate as response encrypted using the symmetric key or nonce value.

Once the client receives the client certificate, all subsequent requests are sent using this certificate.

5.2 Re-enrollment

Just like/simple re-enroll in EST, renewal or rekey request is an authenticated request. This request is authenticated using the previously issued certificate that is being renewed or rekeyed.

In cases where the client was down for several years and its existing certificate has expired, the abovementioned renewal or rekey method does not work. This is because, on authentication with an expired client certificate, the server responds with a 'Bad SSL Certificate' error. To renew an already expired certificate, the client has to create a new CSR and resend the initial enrollment request as explained in Sect. 5.1.

As per EST, the re-enrollment request must include the same Subject and SubjectAltName as the current certificate. Any changes to the SubjectName, ChangeSubjectName attribute should be added to the new CSR.

Similarly, if the Subject Public Key Info in the certification request is the same as the current client certificate, then the server renews the client certificate. If the public key information in the certification request is different than the current client certificate, then the server rekeys the client certificate.

5.3 Authentication and Authorization:

Nginx's ngx_http_ssl_module [13] has a variety of functionalities.

1. The 'ssl_verify_client on|off|optional|optional_no_ca;' directive enables client certificate verification.
2. The 'optional' parameter requests the client certificate and verifies if the client certificate is present.
3. The 'optional_no_ca' parameter requests the client certificate but does not require it to be signed by a trusted CA certificate. This is intended for the use in cases when a service that is external to Nginx performs the actual certificate verification.
4. The 'ssl_client_certificate' specifies a file with trusted Certificate Authority certificate to verify the incoming client certificates.
5. The results of the verification are stored in the variable '$ssl_client_verify'.
6. If the incoming client certificate is signed by the CA specified, then it sets the embedded variable '$ssl_client_verify' to SUCCESS.

7. If the verification fails, then it is set to FAILED: reason.
8. If no certificate is used, it is set to NONE.

In case of applications where authorization needs to be checked along with authentication, an upstream application server such as Glassfish, JBoss and Apache Tomcat can be used. The application server will host an application that receives the authenticated request and checks and validates the unique attribute from the client. An example of this unique attribute would be the serial number or fingerprint of the client's client certificate. If the request has been validated successfully with proper authorization, it is processed by the application server and the response is returned to the client.

6 Performance Testing

One might think all the extra steps for authentication, and authorization might cause a wide difference in the performance.

Performance testing was done using Apache JMeter. Apache JMeter is an Apache project that can be used as a load testing tool for analyzing and measuring the performance of a variety of services, with a focus on web applications. The response time of 10,000 requests with a ramp-up period of 300 s was noted. The ramp-up period specifies how long JMeter will take to 'ramp-up' to the full number of threads chosen. If 100,000 threads are used, and the ramp-up period is 300 s, then JMeter will take 300 s to get all 100,000 threads up and running. Each thread will start 0.03 (300/10,000) seconds after the previous thread was begun.

As shown in Fig. 2, the average performance variance is 10 ms between traditional authentication methods and client certificate authentication.

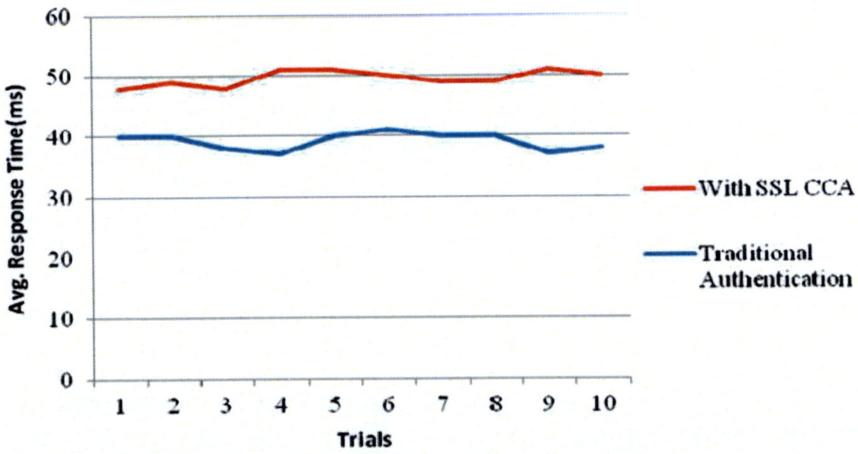

Fig. 2 Performance test results

7 Security Analysis

To establish protection against main-in-the-middle attacks, playback attacks and denial-of-service (DoS) attacks, the following have been implemented:

1. Since client certificate authentication happens during the SSL handshake, any form of interception during this process requires the attacker to be able to view the network traffic (using tools like Wireshark) by being connected to the same network (like public wifi). How safe the network one is connected to is in the hands of the user.
2. Once the SSL connection is established, man-in-the-middle attack is near to impossible, unless the malicious server/proxy's CA certificate is in the trust store of the client. This is again in the hands of the user.
3. If at all the SSL connection is established, the data being exchanged is further being encrypted by a symmetric key—a code or phrase that was exchanged using an out-of-band mechanism or nonce mechanism before the SSL communication was set up.
4. To counter playback attacks, if the encrypted body is being repeatedly sent, the server checks the decrypted content. The decrypted content usually contains the unique client attribute identifying the client. If the particular request has already been served for that client (this applies only to sensitive operations like issue, renew and update operations), then the operation is rejected. In addition, any form of DoS attacks can be prevented by limiting the number of hits from a single client to the server by Nginx itself.
5. Any fake certificate that is generated and signed by the RootCA cannot be used either. This is because the authorization as explained in Sect. 5.3 validates any unique property of the certificate against the mapping of the unique client attribute stored in a persistent or cached storage that is accessed by the application running in the application server.

8 Conclusion

After all this, the question arises—despite all the advantages, why is not it popular? The answer is the use of client certificates is just an extra step for the average Internet users. Most client–server implementations involve clients just consuming resources from the server, and authorization of the client as a device or a piece of software is not necessary, rather, as a user. This is when the use of passwords or two factor authentications comes into play. For users with less technical knowledge, storing a certificate in a trust store and certificate management is very complicated.

Client certificate authentication plays a vital role when the client itself is a device or software module that is automated to perform all the authentication, requesting and renewal process.

In another case where the client itself is a browser, the client certificate is stored in the browser. This limits the usage of one particular system. In an era where one uses several devices, the certificate becomes invalid if the device stops working or is stolen.

From the study, it can be concluded that although password-based authentication is preferred when human interaction is involved, certificate-based authentication is better and more secure for unattended file serving option done by file servers like Nginx. When an application server is behind a proxy server, these security benefits of the implementation outweigh the small acceptable amount of increase in response time.

Future enhancements for this project would be to design a protocol standard that is easier to implement and maintain especially for users with less technical knowledge. An additional mechanism to immediately invalidate the issued certificate when a device crashes or stolen can overcome the browser-based limitation.

References

1. Akami Intelligent Platform: Smart Card and Client Certificate Authentication for Web-Enabled Applications [Online]. Available https://www.akamai.com/us/en/multimedia/documents/white-paper/smart-card-and-client-certificate-authentication-for-web-applications-white-paper.pdf
2. Global Sign: Using Certificate based Authentication for Access Control [Online]. Available https://www.globalsign.com/en/resources/white-paper-certificate-based-authentication-for-access-control.pdf
3. Shiraishi, Y., Mohri, M., Kakei, S.: SSL client authentication with TPM. IEICE Trans. Inf. Syst. E99. **D**(4), 1052–1061 (2016)
4. Davies, J.: Command line fanatic. 28 Feb 2020 [Online]. Available https://commandlinefanatic.com/cgi-bin/showarticle.cgi?article=art046
5. Parsovs, A.: Practical issues with TLS client certificate authentication. Internet Society, ISBN 1-891562-35 (2014)
6. Network Working Group: Simple certificate enrolment protocol, DRAFT-GUTMANN-SCEP-15. Internet Engineering Task Force (IETF), Auckland, 4, 2020
7. Cisco Systems Inc.: Enrollment over Secure Transport. Internet Engineering Task Force (IETF) (2013)
8. Cisco Systems Inc.: PKI: Simplify Certificate Provisioning with EST. Cisco Systems Inc (2016)
9. Schaad, T.S.I.: RFC 5272—Certificate Management Over CMS (CMC).June 2008 [Online]. Available https://tools.ietf.org/html/rfc5272
10. Liu, X., Schaad, J., Weinstein, J., Myers, M.: Certificate Management Messages Over CMS.April 2000 [Online]. Available https://www.ietf.org/rfc/rfc2797.txt
11. Adams, C., Kause, T., Mononen, T.: RFC 4210—Internet X.509 Public key Infrastructure.September 2005 [Online]. Available https://tools.ietf.org/html/rfc4210
12. ITU Telecom: Introduction to ASN.1. [Online]. Available https://www.itu.int/en/ITU-T/asn1/Pages/introduction.aspx
13. Nginx: MODULE NGX_HTTP_SSL_MODULE. Nginx [Online]. Available https://nginx.org/en/docs/http/ngx_http_ssl_module.html#example
14. Smys, S., Raj, J.S.: Performance optimization of wireless adhoc networks with authentication.J. Ubiquitous Comput. Commun. Tech. **02**, 64–75

Convolutional Neural Networks in the Identification of Benign and Malignant Melanomas

Amelec Viloria, Nelson Alberto, and Isaac Kuzmar

Abstract Melanoma is one of the most lethal types of skin cancer. However, early diagnosis increases the survival rate. The way to diagnose a malignant lesion by a physician is to visually examine the suspicious skin area. Artificial intelligence and deep-learning models have been applied to help physicians who are not as experienced. Therefore, work was done on the classification of melanomas to support the possible detection of malignant melanomas. Convolutional neural networks were used.

Keywords Convolutional neural networks · Identification of benign and malignant melanomas · Cancer

1 Introduction

In the last decade worldwide, from 2008 to 2018, the annual number of melanoma cases has increased by 53% [1], partly due to increased exposure to ultraviolet (UV) rays. Prolonged exposure to inorganic arsenic, through the use and consumption of contaminated water or prepared food and food crops irrigated with arsenic-rich water, can also cause chronic poisoning, the most characteristic effects being skin lesions and skin cancer.

The Colombian health system is made up of a broad social security sector financed with public resources and a declining private sector. Its central axis is the General

A. Viloria (✉)
Universidad de la Costa, St. 58 #66, Barranquilla, Atlántico, Colombia
e-mail: aviloria7@cuc.edu.co

N. Alberto
Universidad Tecnológica, San Pedro Sula, Honduras
e-mail: nelson@unitet.edu.hn

I. Kuzmar
Universidad Bolivar, Barranquilla, Atlántico, Colombia
e-mail: isaac.kuzmar@unibolivar.edu.co

© The Editor(s) (if applicable) and The Author(s), under exclusive license
to Springer Nature Singapore Pte Ltd. 2021

A. P. Pandian et al. (eds.), *Proceedings of International Conference on Intelligent Computing, Information and Control Systems*, Advances in Intelligent Systems and Computing 1272, https://doi.org/10.1007/978-981-15-8443-5_60

System of Social Security in Health (SGSSS) [2]. Affiliation to the system is compulsory and is done through public or private health-promoting entities (EPS), which receives contributions and, through service providing institutions (IPS), offers the Mandatory Health Plan (POS) or POS-S for members of the subsidized regime (RS). The private sector is used by the upper class and a sector of the middle-income population, by caregiver coverage or in search of better access conditions offered by the SGSSS, goes to the private consultation [3].

This research focuses on the work of doctors in clinics in Colombia, where the National Health System has three levels of care, ranging from basic cases at the first level to highly complex issues such as level three [4].

This paper presents a proposal to classify melanomas to diagnose skin cancer through convolutional neural networks, for this purpose, specialized data processing is carried out.

2 The Method

For this study, images were obtained from the ISIC Dataset, where melanomas and their metadata were obtained from "https://isic-archive.com/." Two datasets were prepared: 1200 with metadata and 1200 without metadata [5].

The following three techniques were applied with the images. At the end of each technique, 80 random images were used (40 of benign melanomas and 40 of malignant melanomas), which were not present at CNN training or validation. This sample of 80 images was taken to ensure that the CNN classified images it never saw before [6].

2.1 Technique 1

For this technique, the metadata of: age, gender, and size of the melanoma within the pixel matrix were added to 1200 images, duplicating these metadata.

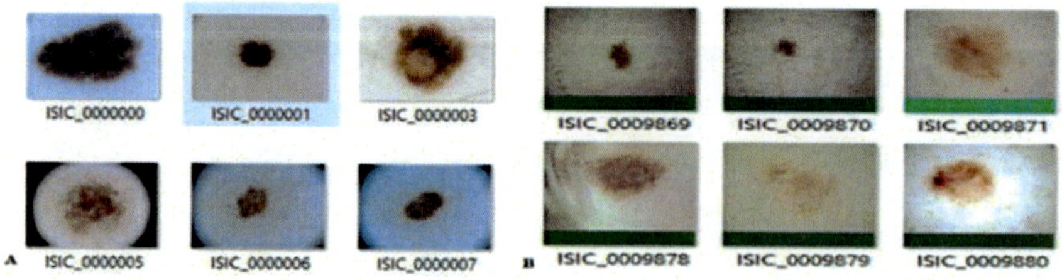

Fig. 1 **a** Image matrix without metadata. **b** Image matrix with metadata

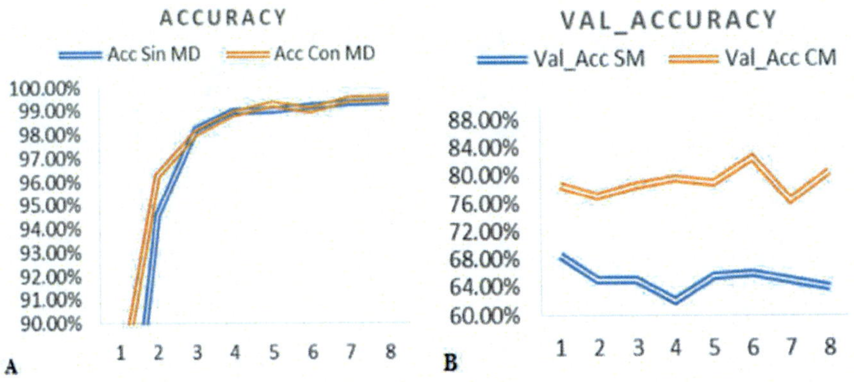

Fig. 2 Charts: **a** with metadata (with MD) and without metadata (without MD) from accuracy and **b** with metadata (CM) and no metadata (SM) Val_Accuracy

Figure 1 shows the images with and without metadata. In (a), the original images from ISIC are shown while in (b) the images with the metadata added repeatedly in the last 50 rows are shown, so, a band of colors can be seen in each image.

A CNN was executed, with the following parameters: 8 times, 350 height, 300 width, for images with metadata and 300 height, 300 width for images without metadata. Two convolution layers [7].

Once it was done in training with the CNN, the results were obtained without metadata; Acc refers to the accuracy of 99.11%, Val_Acc of 65.33% to the accuracy of the validation; on the other hand, a 0.0201 of the value of loss refers to the loss function, and a 3.002 of Val_loss to the loss function of the validation.

During the eight seasons, a better behavior can be noted, especially at Val_Accuracy. Figure 2 shows the graphs in the table showing the behavior of Accuracy and Val_Accuracy.

In the end, 40 images (20 with metadata and 20 without it, 10 malignant and 10 benign for each case) not included in the training or validation were tested and the results were 78.01% accuracy in the scenario without metadata and 78.01% in the scenario with metadata.

2.2 *Technique 2*

The skin was removed from the images, in the next technique. 800 were used for training: 400 malignant and 400 benign [8]. For the validation, 400 images were used: 200 malignant and 200 malignant. Examples of some images can be seen in Fig. 3.

The algorithm used for the detection of skin and the segmentation of melanomas. In this case instead of skin, there is a black background. The parameters for this execution were the following: Epochs 8, height 300, width 300, batch_size 32, steps

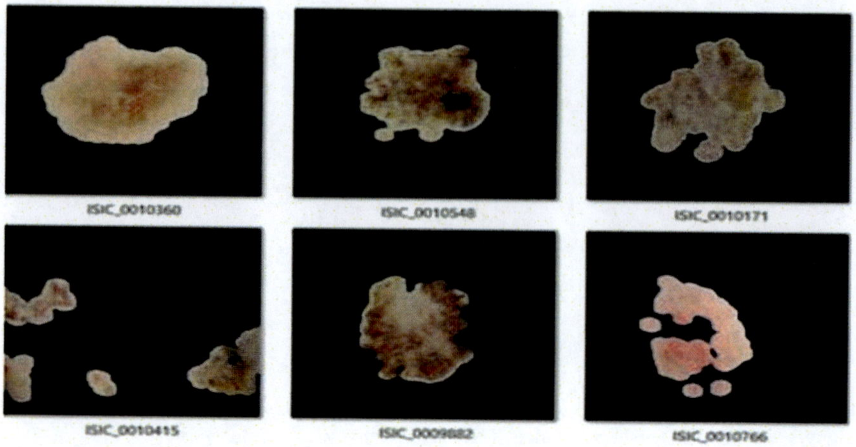

Fig. 3 Example of skinless images

1000, validation_steps 600, conv1-filters 32, conv2-filters 64, filter_size1 (3, 3), filter_size2 (2, 2), pool_size (2, 2) [9, 10].

High accuracy is observed in training, but very low during validation. By avoiding the skin, better results were expected than in technique 1. Figures 4 and 5 show the graphs showing the behavior of Accuracy and Val_Accuracy.

Accuracy's values stabilized at over 99%, while Val_Accuracy stabilized once it exceeded 81%. In Fig. 5, it can be noted that the value of loss tends to zero, but not at the time of validation. Testing was continued with 40 images (40 malignant and 40 benign) not included in the training or validation and the results were 66.98% in the Acc [11].

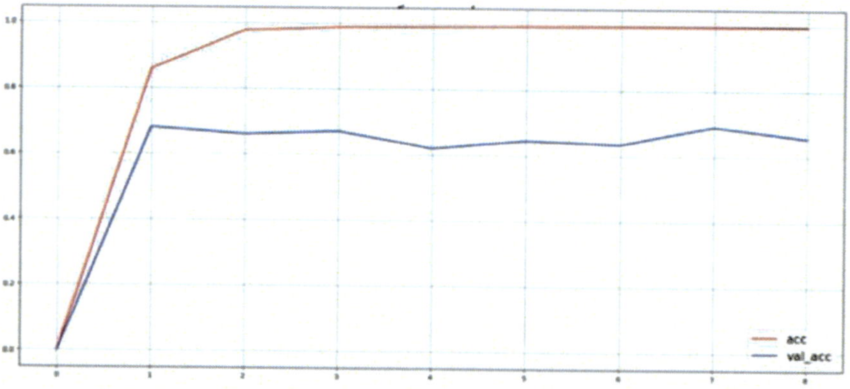

Fig. 4 Acc and val_acc chart of the 2 skinless images technique

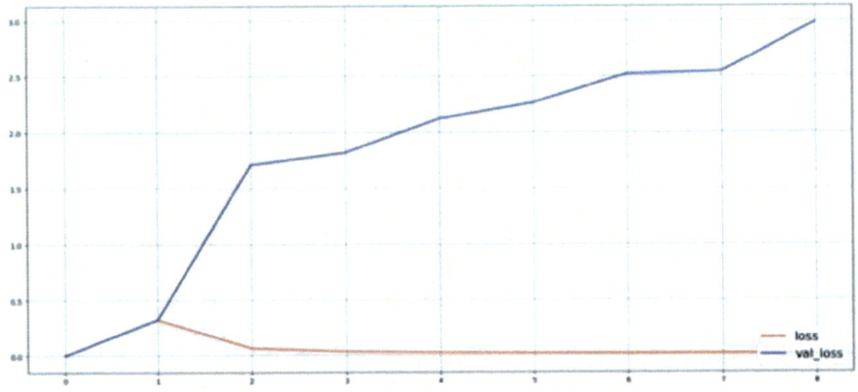

Fig. 5 Loss and val_loss graph of the 2 skinless technique

2.3 *Technique 3*

For the third technique, 1200 images were used obtaining unsatisfactory outputs, so attending to the philosophy of the CNNs that a large number of images are required. Therefore, it proceeded to double the number of images. The CNN was executed with the following parameters: 2400 images (800 malignant and 800 benign) for training. 800 validation images (400 malignant and 400 benign), the number of images was increased. Length = 300, height = 300, batch_size = 20 steps = 1000 validation_steps = 300 filtersConv1 = 32 filtersConv2 = 64. The type of images can be seen in Fig. 6 [12, 13].

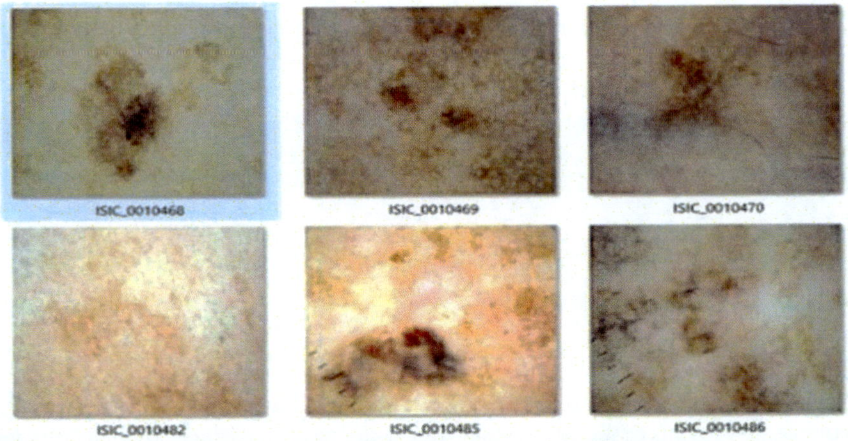

Fig. 6 Type of images used for technique 3

Table 1 Test results 3

Times	ACC	VAL_ACC	Loss	Val_loss
1	0.7231	0.8251	0.6651	0.5012
2	0.7958	0.8685	0.4754	0.3841
3	0.8214	0.8754	0.4123	0.3751
4	0.8641	0.8421	0.3451	0.4541
5	0.8985	0.8652	0.2952	0.5320
6	0.9241	0.8325	0.2221	0.6251
7	0.9485	0.8012	0.1985	0.7352
8	0.9514	0.8355	0.1412	0.7452

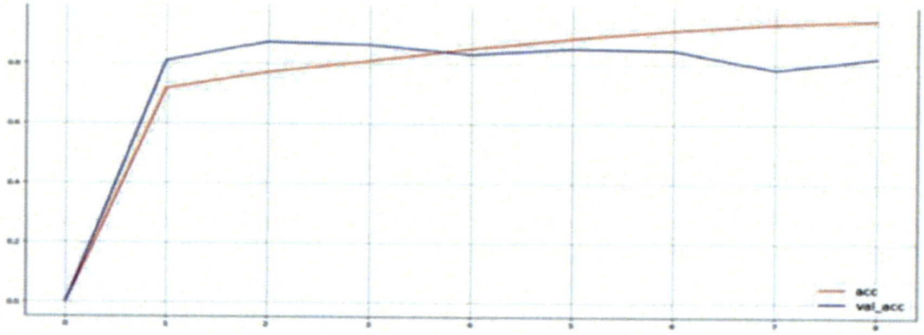

Fig. 7 Graph of acc and val_acc of technique 3

In this technique, the images were taken with any preprocessing from ISIC. The results of the test are shown in Table 1.

In this case, the highest value obtained was 95.14% and only 83.55% during validation. Figure 7 shows the behavior of Accuracy and Val_Accuracy [14].

Accuracy shows an upward trend. With more epochs, accuracy could improve; however, only 8 epochs were run to coincide with techniques 1 and 2. In this third technique, the 80 images (40 malignant and 40 benign) not included in the training and validation were tested and the results were 80% in the Acc.

3 Results and Comparison

In the case of other related studies, results were reported from Accuracy and Val_Accuracy, but no evidence was reported from other images. Table 2 shows the comparative results of the three techniques, with images never seen by CNN training or validation.

Table 2 Comparison of the three techniques

Technique	ACC
1 No metadata	76.30
1 With metadata	76.30
2 Skinless	68.40
3 Image magnification	82.10

4 Conclusions

It can be concluded that the best technique was to add more images to CNN. Furthermore, the trend in Accuracy is increasing, so that by applying more times it could improve the 82.10% obtained.

Therefore, it is concluded that the use of a CNN does not require special preprocessing in the images and that it is enough to use the greatest possible number of images as well as periods in order to find the best performance.

As future studies, it has been proposed to use more images and more epochs to improve the 80.00% of accuracy; on the other hand, pre-trained CNN will be used in order to find the best technique for its implementation at clinical level.

References

1. Kwasigroch, A., Mikołajczyk, A., Grochowski, M.: Deep neural networks approach to skin lesions classification—a comparative analysis. In: 2017 22nd International Conference on Methods and Models in Automation and Robotics (MMAR), pp. 1069–1074. IEEE (2017, August)
2. Brinker, T.J., Hekler, A., Enk, A.H., von Kalle, C.: Enhanced classifier training to improve precision of a convolutional neural network to identify images of skin lesions. PloS one **14**(6) (2019)
3. Chen, M., Chen, W., Chen, W., Cai, L., Chai, G.: Skin cancer classification with deep convolutional neural networks. J. Med. Imag. Health Inf. **10**(7), 1707–1713 (2020)
4. Petrie, T., Samatham, R., Witkowski, A.M., Esteva, A., Leachman, S.A.: melanoma early detection: big data, bigger picture. J. Investig. Dermatol. **139**(1), 25–30 (2019)
5. Pratiwi, R.A., Nurmaini, S., Rini, D.P.: Skin lesion classification based on convolutional neural networks. Comput. Eng. Appl. J. **8**(3), 203–216 (2019)
6. Zhao, S., Xie, B., Li, Y., Zhao, X.Y., Kuang, Y., Su, J., Su, J.: Smart identification of psoriasis by images using convolutional neural networks: a case study in China. J. Eur. Acad. Dermatol. Venereol. **34**(3), 518–524 (2020)
7. Guo, Y., Ashour, A.S., Si, L., Mandalaywala, D.P.: Multiple convolutional neural network for skin dermoscopic image classification. In: 2018 IEEE International Symposium on Signal Processing and Information Technology (ISSPIT), pp. 365–369. IEEE (2018, December)
8. Filali, Y., El Khoukhi, H., Sabri, M.A., Yahyaouy, A., Aarab, A.: Texture Classification of skin lesion using convolutional neural network. In: 2019 International Conference on Wireless Technologies, Embedded and Intelligent Systems (WITS), pp. 1–5. IEEE (2019, April)
9. Ali, A.A., Al-Marzouqi, H.: Melanoma detection using regular convolutional neural networks. In: 2017 International Conference on Electrical and Computing Technologies and Applications (ICECTA), pp. 1–5. IEEE (2017, November)

10. Ruiz-Castilla, J.S., Rangel-Cortes, J.J., García-Lamont, F., Trueba-Espinosa, A.: CNN and metadata for classification of benign and malignant melanomas. In: International Conference on Intelligent Computing, pp. 569–579. Springer, Cham (2019, August)
11. Raj, J.S., Vijitha Ananthi, J.: Recurrent neural networks and nonlinear prediction in support vector machines. J. Soft Comput. Parad. (JSCP) **1**(01), 33–40 (2019)
12. Viloria, A. et al.: Determination of dimensionality of the psychosocial risk assessment of internal, individual, double presence and external factors in work environments. In: Tan, Y., Shi, Y., Tang, Q. (eds) Data Mining and Big Data. DMBD 2018. Lecture Notes in Computer Science, vol 10943. Springer, Cham (2018)
13. Gowda, M.N., Rasheed, A.I.: Hardware implementation of hybrid classifier to detect cancer cells. In: 2017 14th IEEE India Council International Conference (INDICON), pp. 1–5. IEEE (2017, December)
14. Dubal, P., Bhatt, S., Joglekar, C., Patil, S.: Skin cancer detection and classification. In: 2017 6th International Conference on Electrical Engineering and Informatics (ICEEI), pp. 1–6. IEEE (2017, November)

Ant Colony Technique for Task Sequencing Problems in Industrial Processes

Noel Varela, Nelson Zelama, Ruben Hernandez, and Jeferson Rafael de Avila Villalobos

Abstract This paper presents a hybrid model of computational intelligence to solve the job shop scheduling problem, classified as full NP. It is proposed to solve it using the technique of ant colony assisted with simulated annealing. The ant colony technique was used as a global search strategy, and the simulated annealing technique was used as a local search strategy. This proposal was validated experimentally with test problems reported in the literature.

Keywords Ant colony technique · Sequencing problems · Industrial processes

1 Introduction

Job shop scheduling (JSS) problem is a well-known combinatorial optimization problem in the area of operations research and artificial intelligence. In the manufacture of a product, a series of processes are required and carried out in a determined machine [1]. This leads to different problems, among which is the sequencing of work in m machines. The aim of the sequencing is to find the order of execution of n works, which require a series of processes in m machines, so that time is optimized [2]. According to computational complexity theory, it is a complete NP problem,

N. Varela (✉) · R. Hernandez · J. R. de Avila Villalobos
Universidad de La Costa, St. 58 #66, Barranquilla, Atlántico, Colombia
e-mail: nvarela2@cuc.edu.co

R. Hernandez
e-mail: rhernandez@cuc.edu.co

J. R. de Avila Villalobos
e-mail: jdeavila8@cuc.edu.co

N. Zelama
Universidad Tecnológica, San Pedro Sula, Honduras
e-mail: nelson.li@unit.edu

© The Editor(s) (if applicable) and The Author(s), under exclusive license to Springer Nature Singapore Pte Ltd. 2021

A. P. Pandian et al. (eds.), *Proceedings of International Conference on Intelligent Computing, Information and Control Systems*, Advances in Intelligent Systems and Computing 1272, https://doi.org/10.1007/978-981-15-8443-5_61

so it cannot be solved using exact methods [3]. In this paper, the use of ant colony optimization (ACO) assisted by simulated annealing (SA) is proposed [4–6].

In recent years, according to the published literature, several hybrid algorithms have been proposed in [7] to solve this problem, but there is no history of hybrid ant colony proposals with simulated annealing.

In Sect. 2, the basic concepts of the job shop scheduling problem are presented, and Sect. 3 describes the simulated annealing algorithm; the ant colony algorithm is described in Sect. 4; in Sect. 5, the algorithm proposed in the study is presented; in Sect. 6, the tests of the proposed algorithm to problems reported in the literature are shown; in Sect. 7, the conclusions of the research are mentioned.

2 Simulated Annealing

In metallurgy, the annealing method is used to obtain stronger or more crystalline materials, in general, to determine the qualities of a material. The process consists of heating the material to a very high temperature, in that situation, the atoms get a random distribution within the structure of the material and the energy of the system is maximum [8].

Experimentally, it is found that if the temperature is lowered abruptly, or if not enough time is waited for at each stage, in the end, the structure of the material is not optimal. The simulated annealing algorithm was developed to simulate the annealing process in order to find an overall minimum of the target function. In the simulated annealing algorithm, the target function is treated as the energy function of a molten metal and an artificial temperature schedule is set to gradually cool the material, analogous to the annealing technique. This artificial temperature or set of temperatures acts as a source of randomness, which is desirable to eventually avoid a local minimum [9]. The algorithm requires an initial configuration within the solution space or set of R configurations, a cost function $C: R \rightarrow R$, a neighborhood structure^: $R \rightarrow R2$, and^(r) as a subset of^, neighbors of the configuration r, $\forall\, r \in R$ [10].

The algorithm is defined as follows: Given an initial configuration X, and a number of iterations, in each iteration Y is selected ∈^(X) and the expression $\delta = C(Y) - C(X)$ is evaluated. A state transition is performed if and only if $\delta < 0$ or using an acceptance probability is defined as $exp\text{-}\delta/T$, where T is the temperature, which suffers a decrease during the execution of the algorithm [3].

3 Ant Colony

The ant colony algorithms are metaheuristics proposed by [11], for solving combinatorial optimization problems of the full NP type [12]. ACO has its source of inspiration in the behavior of ants in the pheromone trail, a mechanism they use for communication between them.

ACO bases its operation on a colony of simple agents, artificial ants, which, similar to the biological process, will communicate indirectly through the deposit, evaporation, and monitoring of artificial pheromone.

The pheromone traces in ACO serve as information distributed among the ants to build, in a probabilistic way, solutions to the problem to be solved [13].

In order to apply the ACO algorithm, the optimization problem must be translated into a network $G = (V, L)$. V and L represent the vertices and edges of the network, respectively, and are defined with the following characteristics and notations [14]:

- A finite set of problem vertices $V = \{V_1, V_2, \ldots, V_n\}$.
- A finite set E of connections between a subset of V elements, such that $|E| \leq N_2$ (N represents the number of vertices, $N = |V|$).
- For each $ei \in E$, there is a connection fee di, which represents the cost of transiting from vertex i to vertex j, $\eta i, j = 1/\mathrm{d}i, j$ is called the heuristic distance.
- A finite number of restrictions Ω defined on the elements V and E.
- Given a set S of all possible sequences $< V_i, V_j, \ldots, V_k, \ldots >$ on the elements of V, a subset$\hat{}$ $\subseteq S$ is required, representing the feasible sequences with respect to $\subseteq$.
- One solution Ψ is a subset of$\hat{}$, $\Psi \subseteq \hat{}$.
- A cost function $\Phi\Psi(E, t)$ representing the total cost of the solution, where t represents time.

The algorithm starts by assigning a positive real value to each $ei \in E$ which will represent the pheromone trail and will evaporate from time to time. Later, artificial ants will be created along with a list associated with them denoted as *Tabuh*. They will start from an arbitrary initial position, and the initial position will be added to their Tabu list.

In step 2, the ants begin to build solutions by visiting each of the vertices V_i of the graph only once, respecting the restrictions Ω of the problem. For every ant h, the transition probability is defined as the probability that the ant will transit from a vertex V_i to a V_j in a time t shown in [15]:

$$P_{ij}^{h} = \begin{cases} \dfrac{[\tau_{i,j}(t)]^{\alpha}[\eta_{i,j}]^{\beta}}{\sum\limits_{x \in Permitidos_h} [\tau_{i,x}(t)]^{\alpha}[\eta_{i,x}]^{\beta}} & (i, j) \epsilon Tabu_k \\ 0 & \end{cases} \tag{1}$$

where $\tau i, j(t)$ is the pheromone trace on the edge of the graph (i, j) at time instant t. *Permitidosh* is the list of edges through which the ant h can transit maintaining the feasibility with respect to Ω; α and β are control parameters that determine the importance of trace with respect to the heuristic distance respectively.

Once all the ants have completed a sequence Ψ in step 3, the pheromone trace in the network will be updated using the information collected in the Tabuh list of ants using the formula:

$$\tau_{i,j}(t + n) = (1 - \rho) \cdot \tau_{i,j} + \Delta\tau_{i,j} \tag{2}$$

where $\rho \in (0, 1)$ represents a real coefficient and $1-\rho$ represents the evaporation coefficient of the pheromone on the edge (i, j) and $\Delta\tau i, j$ represents the total amount of pheromone deposited by the ants and is calculated with the formula:

$$\Delta\tau_{i,j} = \sum_{k=1}^{h} \Delta\tau_{i,j}^{k}. \quad (3)$$

4 Hybrid Model

The objective of the proposed solution is to solve the problem of job shop scheduling. To do this, a feasible sequence of operations must be found with the shortest possible total processing time, known as makespan. The following restrictions apply to the job shop scheduling problem:

- Each job is processed by a machine only once.
- Each operation of the jobs must be processed in the order in which they are given, a property known as technological constraints.
- Each machine can only process one job at a time.
- Each operation must be processed without interruption.
- Operations from the same job cannot be processed concurrently.

A job shop scheduling problem is formally denoted in the form $n/m/G/C_{\max}$ where n is the number of jobs, m the number of machines, G the technological constraints of the problem and $C_{\max}$ the target function to be minimized, in this case, the makespan.

Technological constraints G are represented by a matrix. For example, for a problem of 2 jobs and 3 machines, you would have the following matrix:

$$G = \begin{pmatrix} O_{1,3} & O_{1,2} & O_{1,1} \\ O_{2,2} & O_{2,1} & O_{2,3} \end{pmatrix} \quad (4)$$

for matrix G, the ith row represents the ith job and the elements in the row represent the operations. For example, row 1 corresponds to the first job that is processed according to the following sequence: first machine 3, then machine 2, and finally machine 1. Similarly, the sequence of job 2 established by row 2 is interpreted as machine 2, machine 1, and machine 3. In addition, there is a P matrix shown in (5) that represents the processing times of the job operations.

$$P = \begin{pmatrix} t_{1,1} & t_{1,2} & t_{1,3} \\ t_{2,1} & t_{2,2} & t_{3,3} \end{pmatrix} \quad (5)$$

for each element $t_i, j \in P$, i corresponds to the job number and j to the machine number. For applying the proposed algorithm, a proper representation of the problem is first

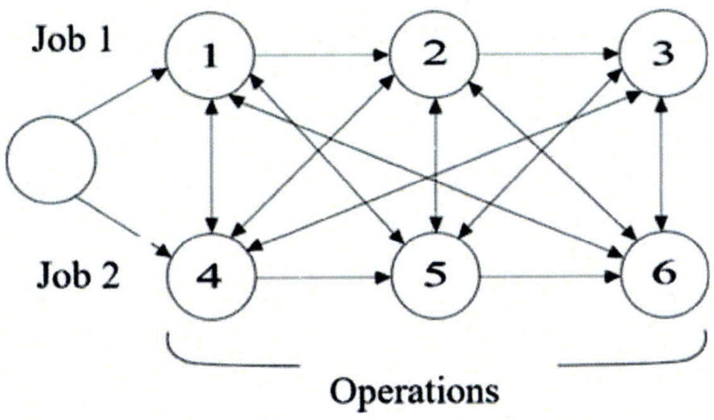

Fig. 1 Representation of the problem required for applying ACO

required. Figure 1 shows an example of the representation of the network that will be explored by the ant colony for a problem of 2 jobs and 3 machines.

For constructing the network, each node will correspond to an element in matrix G of Formula (5). For 2 nodes i, j there will be a directed arc (i, j) in the network if and only if i precedes j with respect to G, or if i and j do not belong to the same job.

The solution to the proposed job shop scheduling problem is designed to use ACO's ability to generate good solutions and take advantage of the convergence property of simulated annealing [16], to obtain the best location of the subset of the solution space (step 3) marked by the pheromone trail.

A feasible sequence of nodes traversed by the ants is used for the representation of the solutions and the initial configuration of simulated annealing. Figure 2a shows the case when a sequence of nodes is feasible, and Fig. 2b shows the case when a sequence is not feasible, that is, the set of restrictions Ω of the job shop scheduling problem is not respected.

For Fig. 2b, the sequence is correct because there is an edge from node 2 to node 5 but it is not feasible since node 5 can only be visited once node 4 is visited.

In step 6, the criteria for choosing the initial setup for simulated annealing is as follows: If the ants found a new better solution, then that sequence is used as initial configuration, otherwise one of the following criteria is chosen at random:

- The best overall sequence with probability is less than 0.75.
- Best sequence was found by ants in step 3 with probability between 0.17 and 0.55.
- Any other sequence was found by the ants in step 3 with probability greater than 0.55.

The simulated annealing algorithm applied in step 7, the proposed solution for a sequence of nodes X, a neighbor $Y = \hat{}(X)$ is obtained by choosing a position in

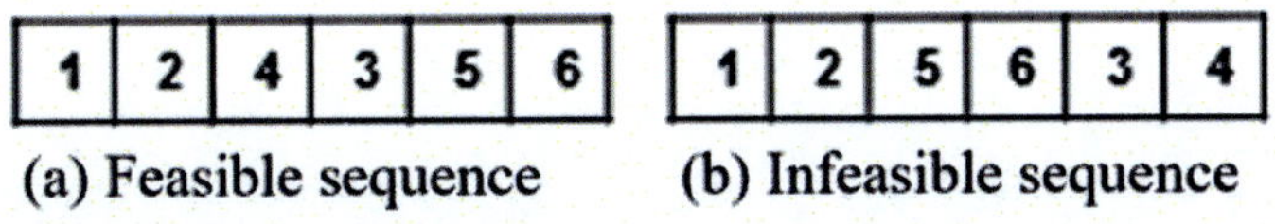

Fig. 2 Examples of sequences on the network

the sequence and exchanging it with the next one as long as the feasibility of the solution is maintained. The method used to lower the temperature is by multiplying the temperature T by a value $\propto r \in (0.1)$.

After applying simulated annealing in step 6, the best solution found by simulated annealing is updated again in step 7, and finally, in step 8, the pheromone matrix information is updated equal to step 4.

5 Experiments and Results

The proposed algorithm was validated using test problems reported in the Operation Research Library repository [2], in a machine whose characteristics are: OCTAVE programming language [8].

For each job shop scheduling test problem, the information from the published matrix is processed, since it contains sequencing and time alternately.

100 algorithm runs were performed for each test problem, with the following calibrated parameters: number of iterations 150, number of ants 500, initial pheromone $15, = 0.33, \alpha = 2, \beta = 12, Q = 1000$. In the case of the simulated annealing technique, the values of the calibrated parameters are $T = 48$, $N = 69$, $\propto r = 0.22$.

The results obtained are compared with the results reported so far as the best for each problem, although they were obtained with different techniques [12].

It can be seen that out of 35 problems, 69.7% show that the best solution has been found, and that for those problems where the best solution has not been found, the average is 98%.

6 Conclusions

With the results obtained, it is concluded that the proposed hybrid algorithm, ACO with simulated annealing as local search, converges in fifteen iterations; that is, it does not require a large number of iterations to find an optimal solution.

In 69.7% of the tested problems, the best-reported solution is reached. For the rest of the problems, the average approximation is 98%.

As future research, in simulated annealing will be tested, other cooling schedules as well as the choice of the neighbors. In the ant colony algorithm, alternative ways of updating and evaporating the pheromone will be proposed.

References

1. Deng, Q., Gong, G., Gong, X., Zhang, L., Liu, W., Ren, Q.: A bee evolutionary guiding nondominated sorting genetic algorithm II for multiobjective flexible job-shop scheduling.

Computational intelligence and neuroscience (2017)
2. Li, X., Gao, L.: An effective hybrid genetic algorithm and tabu search for flexible job shop scheduling problem. Int. J. Prod. Econ. **174**, 93–110 (2016)
3. Zhou, G., Zhou, Y., Zhao, R.: Hybrid social spider optimization algorithm with differential mutation operator for the job-shop scheduling problem. J. Industr. Manag. Optim. **13**(5) (2019)
4. Keddari, N., Mebarki, N., Shahzad, A., Sari, Z.: Solving an integration process planning and scheduling in a flexible job shop using a hybrid approach. In: IFIP International Conference on Computational Intelligence and its Applications, pp. 387–398. Springer, Cham (2018, May)
5. Li, X., Peng, Z., Du, B., Guo, J., Xu, W., Zhuang, K.: Hybrid artificial bee colony algorithm with a rescheduling strategy for solving flexible job shop scheduling problems. Comput. Ind. Eng. **113**, 10–26 (2017)
6. Nguyen, S., Zhang, M., Johnston, M., Tan, K.C.: Genetic programming for job shop scheduling. In: Evolutionary and Swarm Intelligence Algorithms, pp. 143–167. Springer, Cham (2019)
7. Wang, H., Smys, S.: Soft computing strategies for optimized route selection in wireless sensor network. J. Soft Comput. Parad. (JSCP) **2**(01), 1–12 (2020)
8. Nouri, H.E., Driss, O.B., Ghédira, K.: Solving the flexible job shop problem by hybrid metaheuristics-based multiagent model. J. Ind. Eng. Int. **14**(1), 1–14 (2018)
9. Asadzadeh, L.: A parallel artificial bee colony algorithm for the job shop scheduling problem with a dynamic migration strategy. Comput. Ind. Eng. **102**, 359–367 (2016)
10. Karimi, S., Ardalan, Z., Naderi, B., Mohammadi, M.: Scheduling flexible job-shops with transportation times: mathematical models and a hybrid imperialist competitive algorithm. Appl. Mathem. Modell. **41**, 667–682 (2017)
11. Viloria, A., Sierra, D.M., de la Hoz, L., Bohórquez, M.O., Bilbao, O.R., Pichón, A.R., Hernández-Palma, H.: NoSQL database for storing historic records in monitoring systems: selection process. In: Advances in Intelligent Systems and Computing, vol. 1039, pp. 336–344. Springer (2020). https://doi.org/10.1007/978-3-030-30465-2_38
12. Bissoli, D.D.C., Amaral, A.R.: A hybrid iterated local search metaheuristic for the flexible job shop scheduling problem. In: 2018 XLIV Latin American Computer Conference (CLEI), pp. 149–157. IEEE (2018, October)
13. Sel, Ç., Hamzadayi, A.: A simulated annealing approach based simulation-optimisation to the dynamic job-shop scheduling problem. Pamukkale Univ. J. Eng. Sci **24**(4) (2018)
14. Baykasoğlu, A., Karaslan, F.S.: Solving comprehensive dynamic job shop scheduling problem by using a GRASP-based approach. Int. J. Prod. Res. **55**(11), 3308–3325 (2017)
15. Shahgholi Zadeh, M., Katebi, Y., Doniavi, A.: A heuristic model for dynamic flexible job shop scheduling problem considering variable processing times. Int. J. Prod. Res. **57**(10), 3020–3035 (2019)
16. Viloria, A., Sierra, D.M., Duran, S.E., Rambal, E.P., Hernández-Palma, H., Ventura, J.M., Torres, L.J.J.: Optimization of flow shop scheduling through a hybrid genetic algorithm for manufacturing companies. In: Advances in Intelligent Systems and Computing, vol. 1039, pp. 20–29. Springer (2020). https://doi.org/10.1007/978-3-030-30465-2_3

Classification of Marathi Text Using Hierarchical Attention (HAN)-Based Encoder-Decoder Model

Rushali Dhumal Deshmukh and Arvind Kiwelekar

Abstract With social media explosively growing, a vast amount of data is generated, which has to be automatically analyzed. Classification assigns one of the predefined categories to given documents. Various machine learning techniques have been implemented for the classification of Marathi text documents, which is one of the popular languages spoken in India. This paper presents an attention-based hierarchical encoder-decoder model for the classification of Marathi text documents. First, important words from a sentence are computed by providing attention at the word level, using bidirectional LSTM. Then, the attention mechanism is applied at the sentence level to find important sentences. The document is classified based on important sentences from it. The experimentation is done for Marathi polarity labeled tourism dataset (MPLC) with 75 positive and 75 negative reviews and news dataset (94 positive reviews and 102 negative reviews). The accuracy of news data is 77.5%, and for MPLC dataset, accuracy is almost 100%.

Keywords Seq2seq · Encoder-decoder · Attention mechanism · Hierarchical attention mechanism (HAN) · Word-level encoder · Sentence-level encoder · Document classification

1 Introduction

Google first used seq2seq for machine translation. Seq2seq uses deep learning. It takes the current word as well as its neighborhood for translation. It uses two components encoder and decoder, hence also called encoder-decoder network. (1) Encoder: It converts input vectors to corresponding hidden vectors using deep neural network

R. D. Deshmukh (✉) · A. Kiwelekar
Department of Computer Engineering, Dr. Babasaheb Ambedkar Technological University, Lonere, Raigad (M. S.) 402103, India
e-mail: radesh19@gmail.com

A. Kiwelekar
e-mail: awk@dbatu.ac.in

© The Editor(s) (if applicable) and The Author(s), under exclusive license to Springer Nature Singapore Pte Ltd. 2021

A. P. Pandian et al. (eds.), *Proceedings of International Conference on Intelligent Computing, Information and Control Systems*, Advances in Intelligent Systems and Computing 1272, https://doi.org/10.1007/978-981-15-8443-5_62

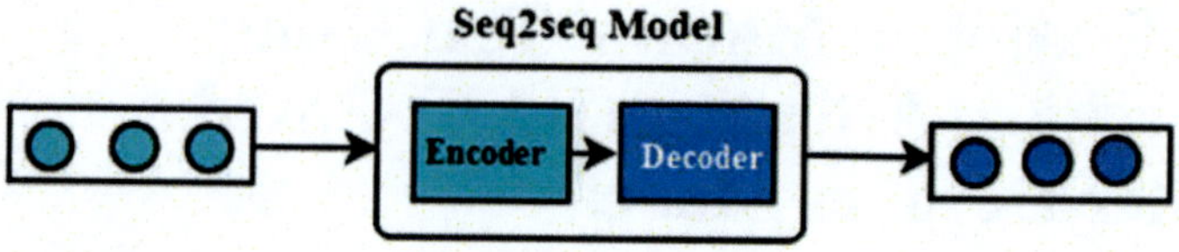

Fig. 1 Seq 2seq model

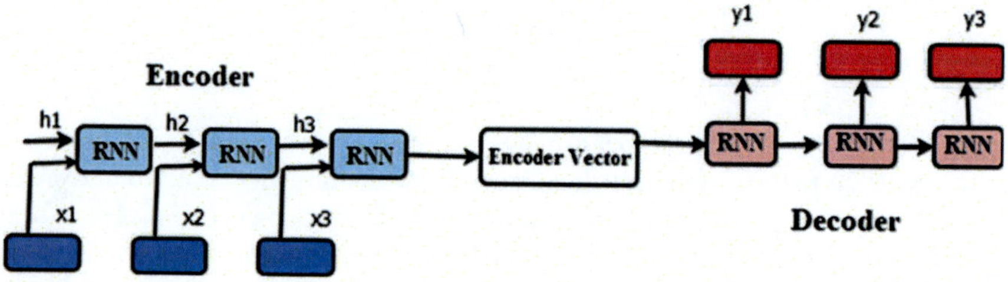

Fig. 2 Sequence-to-sequence encoder-decoder model

layers. Each vector represents the current word and the context of the word. (2) Decoder: It takes the encoder-generated hidden vectors, its hidden state, and generates the output sequence. Figure 1 shows the sequence-to-sequence model.

Figure 2 shows the encoder-decoder sequence-to-sequence model in detail. The model consists of three components: encoder, intermediate vector, and decoder.

1. **Encoder**: A stack of several recurrent elements where each element accepts a single element of the input, collects information, and propagates it forward. Here, each element LSTM or GRU cells is preferred for better performance. For example, in machine translation input is a collection of words. Each word is x_i, where i is the order of word.
   Equation 1 computes the hidden state at time step k.
   $$h_k = f\left(W^{(hh)}h_{k-1} + W^{(hx)}x_k\right) \tag{1}$$
   Here, appropriate weight $W^{(hx)}$ is applied to the input vector X_k and $W^{(hh)}$ to the previous hidden state.
2. **Encoder vector**: It is the output of the final encoder, which encapsulates the information from all the input elements. It is the initial state of the decoder part.
3. **Decoder**: It is a stack of several recurrent units. Each unit produces output y_k at time step k. Each recurrent unit accepts a previous hidden state and generates output as well as its hidden state. In the field of machine translation, the output is a sequence of target language terms. Here, y_i represents a word, where i is the order of that word.

Equation 2 computes the hidden state at time step k.

$$h_k = f(W^{(hh)} h_{k-1}) \quad (2)$$

Equation 3 computes the output y_k at time step k.

$$y_k = \text{softmax}(W_s h_k) \quad (3)$$

The output is calculated using hidden vector output with the respective weight W_s. Softmax is used to create a vector of probabilities. This model is powerful as it can map sequences of different lengths to each other.

1.1 Applications of Encoder-Decoder Models

There are many applications of encoder-decoder models. A few of the applications are as follows.

1. **Machine Translation**: One RNN is used as an encoder for input, and another RNN is used as a decoder for output.
2. **Document Summarization**: One RNN is used as an encoder to generate hidden vectors for the document, and another RNN is used to generate a summary of the document.
3. **Document Classification**: Here, RNN is used as an encoder, and Softmax is used in the decoder to generate a probability of target class.

1.2 Attention Mechanism

Decoder's input is a single vector that captures context information, which is problematic for long sequences. In the attention mechanism, at each time step, attention is given to the input sequence selectively. A weighted average of corresponding word representations could be fed to the decoder. This works better because, here, decoder is not overloaded with irrelevant information. Given enough data, the model learns attention weights. Such models work better in action than those with encoder-decoder.

An example of machine translation in Fig. 3 shows that encoder reads a sentence only once and generates an encoded vector. At each step, the decoder produces a new word using that encoder vector. However, humans do not translate like this. Humans try to produce output word by focusing only on certain words in the input. At each time step, different words may be focused.

I/P: Mi ghari jat ahe.
O/P: I am going home.

At $t1$: [1 0 0 0] to produce the first word of output, the first word from the input is focused; at $t2$: [0 0 0 1] to produce the second word of output, the fourth word from the input is focused; at $t3$: [0 0 1 0] to produce the third word of the output, the third

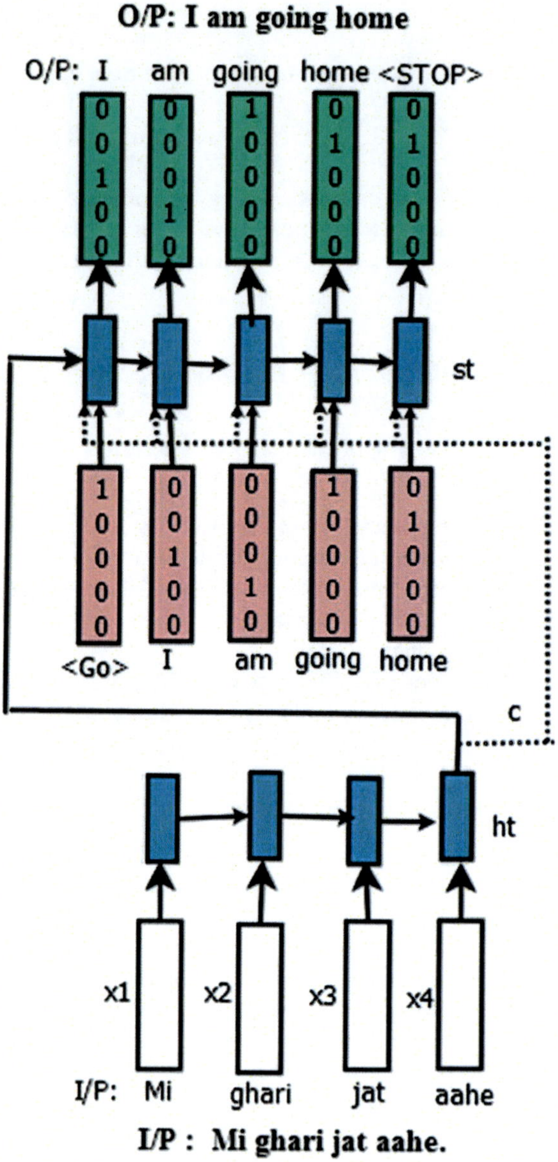

Fig. 3 Example of machine translation

word from the input is focused; at $t4$: [0 1 0 0] to produce the fourth word of output, the second word from the input is focused.

This distribution tells us how much attention to pay to each input word. Ideally, at each step, only relevant information should be fed to the decoder. In Fig. 3, encoder information is fed only once (at $s0$), or the same information can be feed at each time step.

If some oracle told which words to focus on at a given time step k, then a weighted average of the corresponding word representations could be taken and feed it to the decoder. The machine will have to learn this form of data. For this, a function is defined.

$$e_{jk} = f_{ATT}\left(s_{k-1}, c_j\right) \tag{4}$$

e_{jk} captures the importance of jth word for decoding word at time step k. Weights can normalize by using the Softmax function.

$$\alpha_{jk} = \frac{\exp\left(e_{jk}\right)}{\sum_{j=1}^{M} \exp\left(e_{jk}\right)} \tag{5}$$

α_{jk} denotes the probability of focusing on the jth word to produce the kth output word have to learn α. Let us refer decoder RNN's state as s_k at time step k and encoder RNN's state as c_j at j^{th} time step.

$$e_{jk} = V_{ATT}^{T} \tanh\left(U_{\text{att}} s_{k-1} + W_{\text{att}} c_j\right) \tag{6}$$

V_{ATT}^{T}, U_{att}, W_{att} are the additional parameters of the model. These parameters will also be learned during training, along with other parameters. Given enough data, it is possible to learn attention weights.

A two-level hierarchical RNN encoder could be think of. In every utterance, the first level RNN acts on the sequence of words and gives us a sequence of utterance representation. Now, another RNN that encodes this sequence and gives a single representation for the sequences of utterances. The decoder then can produce output based on this utterance.

Survey of attention models was done by Chaudhari et al. [1], in which they discussed attention modeling in applications like natural language generation, machine translation, question answering, generating natural text from multimedia input sequence(speech, image, and video), document classification, and sentiment analysis.

2 Literature Survey

For document classification, Yang et al. [2] proposed hierarchical network averaging, hierarchical network max pooling, hierarchical attention model. For Yelp reviews 71.0%, IMDB reviews 49.4%, Amazon reviews 63.6%, and yahoo answers 75.8% accuracy was achieved. Du et al. [3] proposed Attention LSTM Bi-RNN, AttentionGRU Bi-RNN for Chinese, and English dataset. For Xinhua News Agency, which is the Chinese dataset, a maximum F-value of 88.5 was achieved with AttentionGRU Bi-RNN, and for the English news dataset (Reuters corpus), maximum

F-value of 51.8 was obtained for Attention LSTM Bi-RNN. Gao et al. [4] experimented hierarchical attention-based network at word level and sentence level, using 1D convolutions with the exponential linear unit and tanh for feature extraction. Yang et al. [5] achieved a Normalized Discounted Cumulative Gain of 75.8 for the DBLP dataset and 78.0 for the AAN dataset using an attention mechanism. Fergadis et al. [6] implemented a bidirectional attention-based recurrent neural network for binary classification of biomedical abstracts into a relevant and non-relevant category. They achieved an $F1$-score of 0.69. Cheng et al. [7] implemented one bidirectional LSTM model at the sentence level and another bidirectional LSTM model at a document level with a convolutional layer for feature extraction. The accuracy obtained for Yelp 2015 dataset was 72.2, and for IMDB dataset, accuracy achieved was 49.8. Tang et al. [8] used forward and backward Gated Recurrent Neural Network for Sentiment Classification of Yelp and IMDB dataset. Ran Jing [9] introduced a self attention-based LSTM network for text classification documents to deal with long sentences. Wang et al. [10] combined regional CNN and LSTM to capture local information, thin sentences, and long-distance dependency across sentences and proved that their approach outperforms lexicon-based, neural network, and regression-based methods. Yan and Guo [11] implemented CNN-based attention mechanism and LSTM-based attention mechanism for feature extraction, with additional features concatenated before final output layer for Dialog Act Corpus.

3 Experimental Setup

Here, experimentation is done using a hierarchical-based encoder-decoder model for Marathi tourism and news dataset. LSTM learns faster, and it truncates gradient wherever it does not harm [12]. Bidirectional LSTM is used at the word as well as sentence level. The attention mechanism is used at both the word and sentence level.

Figure 4 shows the hierarchical attention-based model with light blue color boxes represent word-level attention. The light green color boxes represent sentence-level attention, which is then aggregated (dark blue color box) to determine the class of a document. The model for document classification is defined as follows:

Task: Document Classification

Data: $\{\mathrm{Document_i class_i}\}_{i=1}^{N}$

Word-Level (1) Encoder:

$$h_{ij} = \mathrm{RNN}\big(h_{ij-1}, w_{ij}\big) \tag{7}$$

where w_{ij} the jth word in the ith sentence is, h_{ij-1} is the state of the $j-1$ word, and h_{ij} is state of jth word.

$$u_{ij} = \tanh\big(W_w, h_{ij} + b_w\big) \tag{8}$$

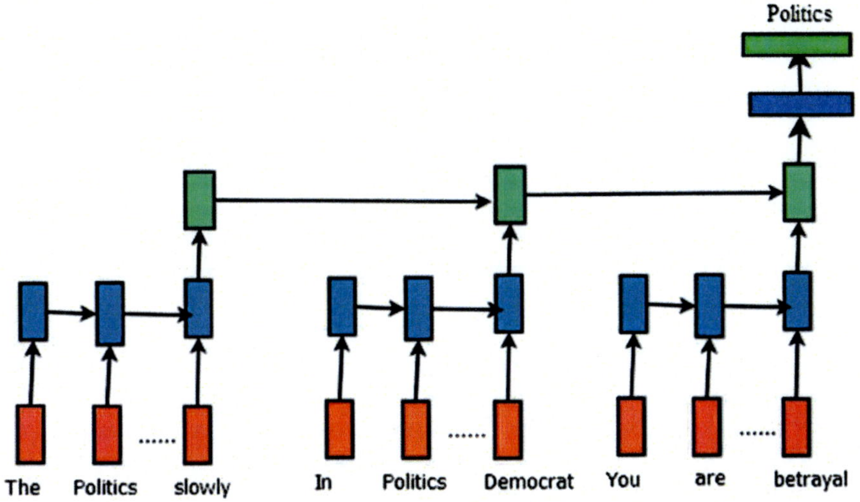

Fig. 4 Hierarchical attention-based model for document classification

where u_{ij} captures the importance of the jth word in the ith sentence. W_w is the weight matrix and b_w is the bias.

$$\alpha_{ij} = \frac{\exp\left(u_{ij}^T u_w\right)}{\sum_t \exp\left(u_{ij}^T u_w\right)} \tag{9}$$

α_{ij} is the normalized value of u_{ij}.

$$s_i = \sum_j \alpha_{ij} h_{ij} \tag{10}$$

Sentence vector s_i is computed aggregating weighted word vectors.

Sentence-Level (2) Encoder:

$$h_i = \text{RNN}(h_{i-1}, s_i) \tag{11}$$

where s_i is ith sentence, h_{i-1} is the state of $i-1$ word, and h_i is the state of ith word.

$$u_i = \tanh(W_s h_i + b_s) \tag{12}$$

where u_i captures the importance of ith sentence in the document. W_s is the weight matrix, and b_s is the bias.

$$\alpha_i = \frac{\exp(u_i^T u_s)}{\sum_t \exp(u_i^T u_s)} \tag{13}$$

α_i is the normalized value.

$$D_i = \sum_i \alpha_i h_i \tag{14}$$

Document vector D_i is computed aggregating weighted sentence vectors.

Decoder:

$$P(y|\text{document}_i) = \text{softmax}(V\ D_i + b) \tag{15}$$

where y is a class of a document$_i$, D_i is a document vector, V is weight matrix, and b is a bias.

Parameters: $W_w, V, W_s, b_w, b_s, b, u_w, u_s$

Loss function: cross-entropy

Algorithm: Gradient descent with backpropagation.

3.1 MPLC Dataset

Our system is tested for Marathi polarity labeled tourism data from CFILT, IIT Bombay. It contains 75 positive and 75 negative reviews. Sample examples for Marathi tourism data are shown in Fig. 5.

Example of negative sentiment for Marathi tourism data :

महालाच्या रस्त्यावर अतिशय घाण वास येत होता. आणि विक्रेते तर जणू लुटारू होते. आत तर इतका कुबट वास येत होता की जीव नकोसा झाला. त्यानंतर आत इतका भयाण अंधार पसरला होता की मी घाबरून गेलो. महालाचा आतला भाग जीर्ण आणि पडका आहे की आत जायची इच्छासुद्धा होत नाही. मूर्खपणाचा कहर आहे नुसता!

Example of positive sentiment for Marathi tourism data :

ह्यामुळे नव्या संधींचे मार्ग खुले होतात ज्याद्वारे पर्यटक व इच्छुक प्रवासी हे दोघेही हिमाचल प्रदेशामधील भारतातील सर्वाधिक लाडके गिरीस्थानक असलेल्या डलहौसीच्या अज्ञात मोहकतेच्या शोधासाठीची आपली भ्रमंती चालू ठेऊ शकतात. भारतातील कोणत्याही लक्षणीयरित्या प्रेक्षणीय गिरीस्थानकाच्या एखाद्या सहलीप्रमाणे डलहौसी येथे ब्रिटीश मोहकतेचे ते जुने पर्व आणि तत्संबंधित सुसंस्कृत सभ्य शिष्टाचार प्रकट होतात. ब्रिटीश युगातील डौल आणि सिमल्याच्या राजबिंडेपणा ह्यांच्या पंक्तित बसणारे डलहौसी हे निव्वळ ब्रिटीश जमान्याचा पवित्र गतकाळ अनुभवण्यासाठीही अतिशय बघण्याजोगे आहे.

Fig. 5 Example of Marathi tourism data

Here, maximum words 200 are assumed, maximum sentence length of 20, and a maximum number of sentences in each review as 6.

3.1.1 Word-Level Encoder

The first layer used is embedding layer with embedding vector of size 100 is used. The output of the embedding layer is given as input to the next bidirectional lstm layer of size 150 with l2 regularization is used. The next layer used is time distributed dense layer with size 200 is used. After a dense layer, the attention layer is used.

3.1.2 Sentence-Level Encoder

It takes input as time distributed word-level encoder's output. The next layer is the bidirectional LSTM layer of size 150 with l2 regularization. The next layer used is time distributed dense layer with size 200 is used. After that, 50% of dropout is applied to prevent the neural network from over fitting [13]. The next layer used is the attention layer. The output of the sentence-level encoder is given to decoder, which classifies given review into one of the two classes as positive or negative. For optimization, "adam" optimizer is used [14]. To reduce overfitting, "early stopping" is also used. Model parameters updating is done for a batch size of 16.

80% of the dataset is used for training, and 20% of the dataset is used as test data. From training, 20% of data is used for validation purpose. The accuracy of training data is 100%, validation data accuracy achieved is 95.83%, and test data accuracy achieved is 100%.

Figure 6 shows the attention model at the word level (1). Figure 7 shows the attention model at the sentence level (2).

Model accuracy through epochs is shown in Fig. 8. This shows that after the 20th epoch, both training and validation accuracy become 1.

Figure 9 shows that a total of 30 poems are used for testing purpose. Out of 30 tourism reviews, 17 negative reviews are classified as negative, and 13 positive reviews are classified as positive. Not a single positive review is classified as negative and vice versa. In this case, 100% accuracy is achieved. Micro-average and macro-average ROC curve area for tourism data are 1.00.

3.2 News Dataset

Here, news reviews data is collected from E-newspapers with 94 positive and 102 negative reviews. Sample examples for news reviews are shown in Fig. 10.

Here, maximum words per sentence as 200 are assumed. Word-level (1) encoder model used is the same as that of Fig. 6. The sentence-level (2) attention model of Fig. 7 is also the same for news dataset only difference is that A maximum sentence

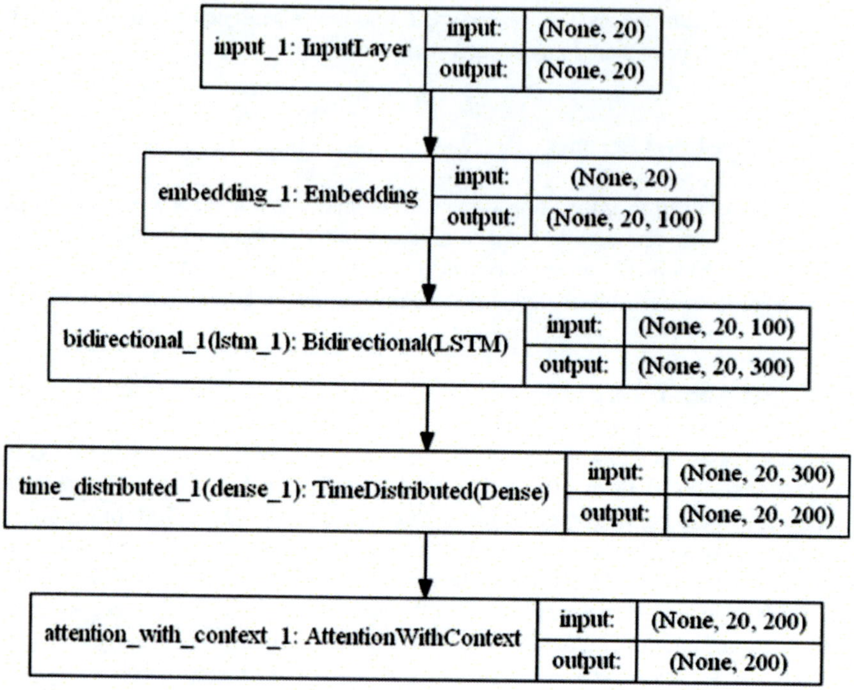

Fig. 6 Word-level (1) encoder for tourism dataset

length of 20 words and the maximum number of sentences in each review as 4. Model accuracy through epochs is shown in Fig. 11. This shows that after the 24th epoch, both training and validation accuracy are maximum.

Figure 12 shows that a total of 40 news reviews (25 negative and 15 positive reviews) are used for testing purpose. Out of 25 negative reviews, 18 negative reviews are classified as negative, and 7 are misclassified as positive. Out of 15 positive reviews, 13 are classified as positive, and 2 are misclassified as negative.

Tagwise precision, recall, $f1$-score, and support are as shown in Table 1. It shows that the average precision score, micro-averaged over all classes: 0.83.

Figure 13 shows the ROC curve, which plots the true positive rate (TPR) against the false positive rate (FPR). Classifier gives better performance for the curve which is close to the top-left corner. ROC curve shows for each class, the area is 0.79, the micro-average ROC curve area is 0.77, and the macro-average ROC curve area is 0.79.

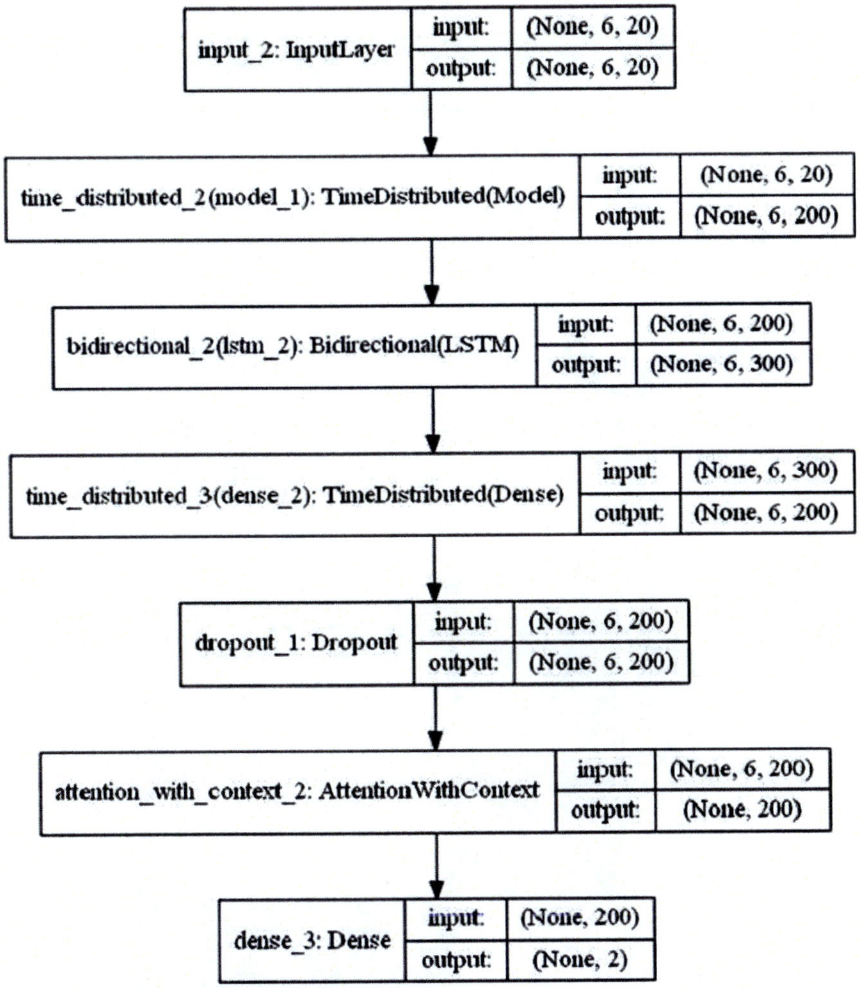

Fig. 7 Sentence-level (2) encoder for tourism dataset

4 Conclusion

This paper proposes a hierarchical attention network for the classification of Marathi documents. Our model first computes sentence vector by aggregating important word vectors. Aggregating important sentence vectors compute that document vector, which is used for classification. The model is evaluated for tourism and news reviews of the Marathi language. For tourism data, the average precision scored is 1.00, and for news review data, the average precision score is 0.83. Our approach is probably the first approach of the HAN encoder-decoder model for the classification of

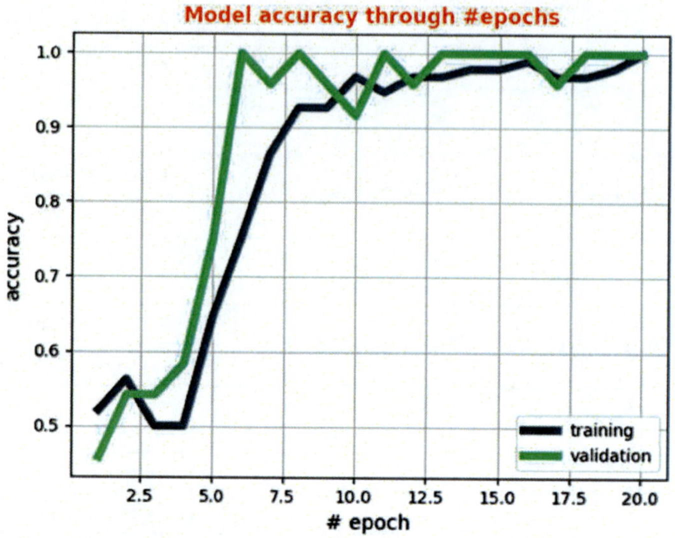

Fig. 8 Model accuracy through epochs for tourism dataset

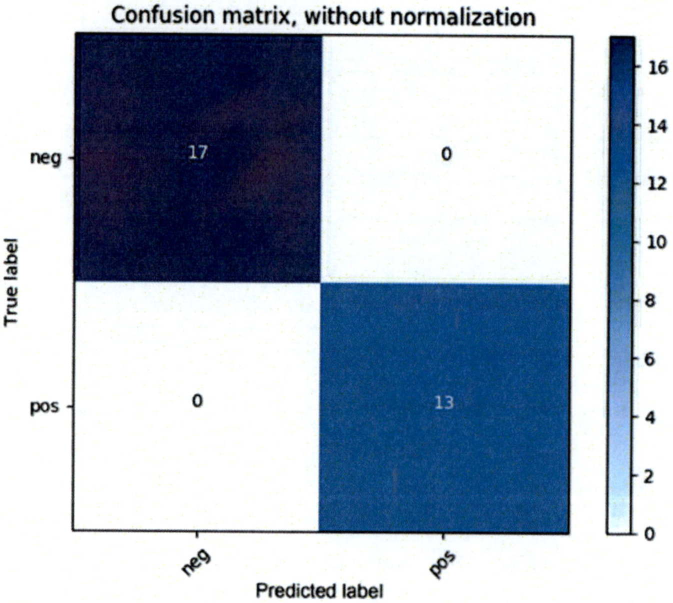

Fig. 9 Confusion matrix for tourism reviews dataset

Example of negative sentiment for News review data:
त्यांच्यामुळेच तर रेल्वे आणि लोकल अजून चालू आहेत.
आपल्या लोकांना फक्त टॅक्स आणि आरक्षण वाढवता येते.. गुणवत्ता नाही.
Example of positive sentiment for News review data:
यशवंतराव चव्हाणांनंतर महाराष्ट्राला मिळेल हे एकमेव चांगले, विचारी आणि प्रामाणिक मुख्य मंत्री आहेत. बाकी जनता जनार्धन आषाढी एकादशी च्या शुभ मुहूर्तांला २०१९ साठी योग्य मतदान करण्याचा संकल्प करेलच याची मला खात्री आहे.

Fig. 10 Example of Marathi news review data

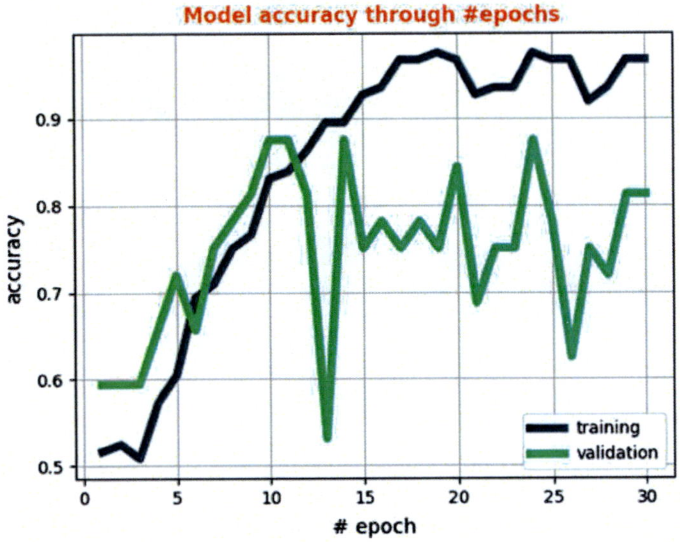

Fig. 11 Model accuracy through epochs for news dataset

Marathi text. Because of the scarcity of resources for the Marathi language, our limitation is the small size dataset used for experimentation. HAN models are complex that other machine learning models, so it takes more time for processing of data. The state of the results shows that HAN's outperform traditional machine learning, CNNs, and RNNs on large datasets, and the same is expected hold true for large datasets of Marathi. So, future work can be focused on large datasets, variations of HAN models.

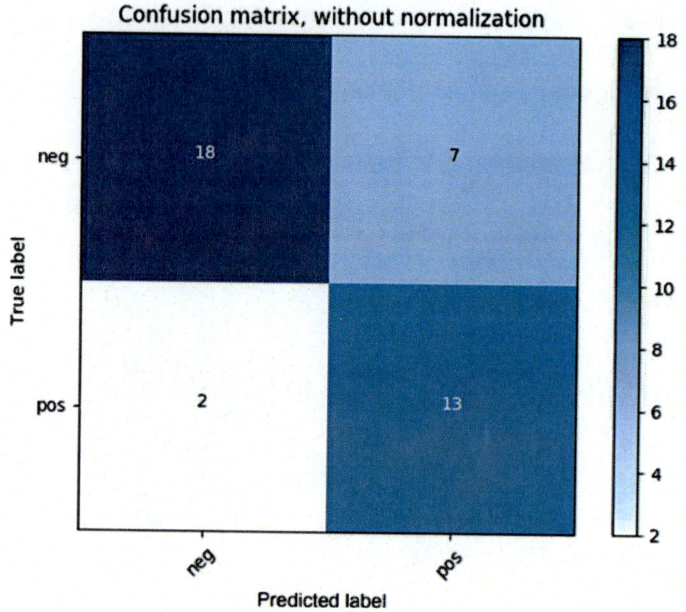

Fig. 12 Confusion matrix for news dataset

Table 1 Tagwise precision, recall, $f1$-score for news data

Category	Precision	Recall	$F1$-score	Support
Negative	0.90	0.72	0.80	25
Positive	0.65	0.87	0.74	15
Avg/total	0.81	0.78	0.78	40

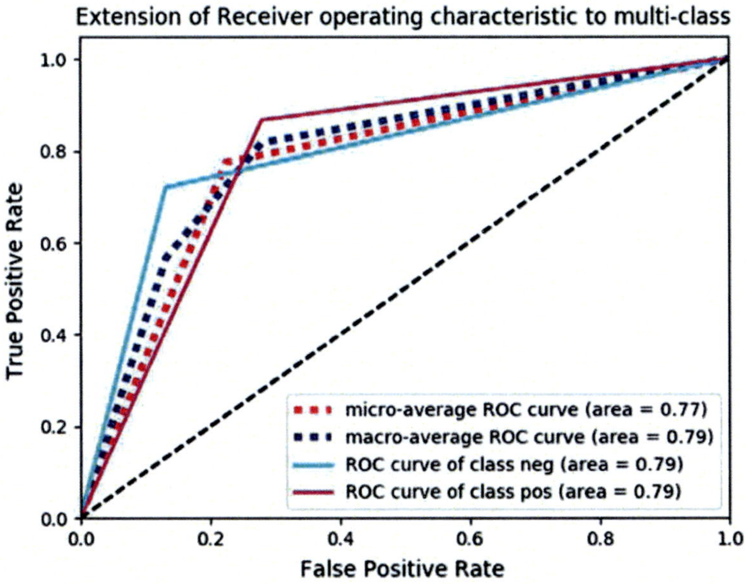

Fig. 13 ROC curve for news dataset

References

1. Chaudhari, S., Polatkan, G., Ramanath, R., Mithal, V.: An attentive survey of attention models (2019). arXiv preprint arXiv:1904.02874
2. Yang, Z., Yang, D., Dyer, C., He, X., Smola, A., Hovy, E.: Hierarchical attention networks for document classification. In: Proceedings of the 2016 conference of the North American chapter of the association for computational linguistics: human language technologies, pp. 1480–1489. (2016, June)
3. Du, C., Huang, L.: Text classification research with attention-based recurrent neural networks. Int. J. Comput. Commun. Control **13**(1), 50–61 (2018)
4. Gao, S., Ramanathan, A., Tourassi, G.: Hierarchical convolutional attention networks for text classification. In: Proceedings of the Third Workshop on Representation Learning for NLP, pp. 11–23. (2018, July)
5. Yang, L., Zhang, Z., Cai, X., Dai, T.: Attention-based personalized encoder-decoder model for local citation recommendation. Comput. Intell. Neurosci. (2019)
6. Fergadis, A., Baziotis, C., Pappas, D., Papageorgiou, H., Potamianos, A.: Hierarchical bi-directional attention-based RNNs for supporting document classification on proteinprotein interactions affected by genetic mutations. Database (2018)
7. Cheng, Y., Ye, Z., Wang, M., Zhang, Q.: Document classification based on con-volutional neural network and hierarchical attention network. Neur. Netw. World **29**(2), 83–98 (2019)
8. Tang, D., Qin, B., Liu, T.: Document modeling with gated recurrent neural network for sentiment classification. In: Proceedings of the 2015 Conference on Empirical Methods in Natural Language Processing, pp. 1422–1432. (2015, September)
9. Jing, R.: A self-attention based LSTM network for text classification. In: Journal of Physics: Conference Series, vol. 1207, No. 1, p. 012008. IOP Publishing (2019, April)
10. Wang, J., Yu, L. C., Lai, K. R., Zhang, X.: Dimensional sentiment analysis using a regional CNN-LSTM model. In: Proceedings of the 54th Annual Meeting of the Association for Computational Linguistics, vol. 2: Short Papers, pp. 225–230 (2016, August)

11. Yan, D., Guo, S.: Leveraging contextual sentences for text classification by using a neural attention model. Comput. Intell. Neurosci. (2019)
12. Hochreiter, S., Schmidhuber, J.: Long short-term memory. Neur. Comput. **9**(8), 1735–1780 (1997)
13. Srivastava, N., Hinton, G., Krizhevsky, A., Sutskever, I., Salakhutdinov, R.: Dropout: a simple way to prevent neural networks from overfitting. J. Mach. Learn. Res. **15**(1), 1929–1958 (2014)
14. Kingma, D.P., Ba, J.: Adam: amethod for stochastic optimization (2014). arXiv preprint arXiv:1412.6980

Graph-Based Hybrid Recommendation Model to Alleviate Cold-Start and Sparsity Issue

Angira Amit Patel and Jyotindra Dharwa

Abstract The main aim of this research work is to find a resolution of cold-start and sparsity issues using graph data for the development of a novel hybrid personalized recommendation model. This research work demonstrates the implementation of a graph model intended to design for recommendation model that assists in resolving the challenges of a volatile market which is very commonly found in the real world. The presented unique hybrid model here combines content-based, knowledge-based and preference-based recommendation techniques for computation tasks of recommendation. This research work puts one step forward toward the challenges of data modeling of structure, semi-structured and unstructured data. The results of the experiment show the proposed model is efficient, flexible and ready to work well in an online Web environment. More than that, the proposed model allows the integration of multiple entities together due to the flexible structure of the graph data model. This research effort works as functional guidelines for everyone who desires to implement a recommendation system using graph data as an integrated component of a real-time Web environment for their computational challenges.

Keywords Recommendation system · Hybrid recommendation model · Graph database · NoSQL · Neo4j · Property graph · Sparsity · Cold-start · Knowledge-based recommendation system

1 Introduction

The main idea behind the development of the recommendation system (RS) is to serve a shortlist of items that best match with the personalized requirements of the perspective of the customer instead of huge item catalog [1]. Recommendation

A. A. Patel (✉) · J. Dharwa
Ganpat university, Ahmedabad, Gujarat, India
e-mail: angira.it@gmail.com

J. Dharwa
e-mail: jyotindra.dharwa@ganpatuniversity.ac.in

© The Editor(s) (if applicable) and The Author(s), under exclusive license to Springer Nature Singapore Pte Ltd. 2021

A. P. Pandian et al. (eds.), *Proceedings of International Conference on Intelligent Computing, Information and Control Systems*, Advances in Intelligent Systems and Computing 1272, https://doi.org/10.1007/978-981-15-8443-5_63

systems are emerged as a tool of "decision accelerator" in the field of e-commerce [2]. The recommendation system also suggests a list of few items which are matched with the user's requirements from a wide item catalog. Recommendation systems work as "personalized information agents" [3] because it offers according to the personalized requirements of the current user.

The development of an efficient and effective recommendation system is still considered as a challenging task due to complexities of integration as well as computation complexities. The main research challenge is to develop a hybrid recommendation model which works well with sparse data and cold-start issue. This research also aims to investigate the suitability of the graph database for the development of a recommendation system which works efficiently with the online environment along with a complex data model.

2 Related Work

2.1 Various Recommendation Techniques

Many algorithms and methodologies have been invented for making recommendations. The collaborative filtering recommendation system (CFRS) is the most popular and extensively accepted technique [4] that generates a recommendation for the active user for those items which are liked by other users with similar tastes. The similarity among the taste of two users is derived based on a similar rating history of those users. So many times, it is referred to as "people-to-people correlation." It is also known as a "black-box approach" as it offers recommendations to the user without knowing much about item characteristics and not giving an insight of system logic or justifications [5].

The content-based RS (CBRS) suggests similar items, for those items that the user liked previously. The similarity among items has been calculated based on features of items or discrete characteristics of items [6], and hence, it is widely known as the "white-box approach." The underline assumption behind this approach is that the user has consistent preferences and the user has an interest in those items which are similar to items for which the user has shown interest in the past [7]. Many commercial Web sites prefer a content-based recommendation approach especially when feature-based item similarity could be possible.

Knowledge-based recommender systems (KBRSs) integrate various types of knowledge [8, 9], such as item features and/or item usability to compute recommendations and to overcome issues of data unavailability. KBRS becomes a promising mean to deal with the most critical cold-start problems [10]. The KBRS is classified as constraint-based and case-based knowledge RS. Constraint-based recommendation approach is "extended-item-search" with consideration of predefined explicit rules. These rules may be hard, soft or weighted constraints based on various kinds of knowledge [8]. The other side of the coin is it is highly domain-specific and the development of the system also depends on the availability of the expert's opinion.

Utility-based recommendation systems produce recommendations by considering the utility of each item for the active user [11]. The main challenge for the development of this kind of system is to define a utility function for each preference of the users. So, the central idea is how to learn about the user's preferences and map with the utility of each item. In majority cases, various items' features are mapped with the specific preferences of the user. Sometimes, item features like the availability of an item can also become part of the utility function but not often.

Hybrid recommendation system (HRS) normally combines two or more approaches to increase the effectiveness of the system and overcome the limitations of each other. Each of the previously discussed approaches has some shortcomings. Recent research in the field of recommendation systems has demonstrated that a hybrid approach could be more effective in some cases where it succeeds to overcome limitations of various approaches [12]. Hybrid approaches can be implemented in many different ways for real-life applications. The simplest approach is to find out outcomes of content-based and collaborative-based independently and then combine them to produce aggregated results.

3 Methodology

It is a challenging task to develop a perfect recommendation system due to various issues like integration and computation complexities. A hybrid recommendation model that integrates well with cold-start issues and sparse data is suggested in this work. To handle a complex data model in an online environment, a graph database is also recommended.

3.1 Proposed System Model

The main goal of this research work is to design, implement and evaluate a hybrid personalized recommendation system envision to overcome cold-start and sparsity issues, which is a very common phenomenon for the volatile markets. The basic idea behind the development of a hybrid personalized recommendation system is to integrate the knowledge of experts with product specifications with the use of a multilayer graph model that facilitates required computation for recommendation algorithms. The unique features of this model include advanced user–web interaction strategies, new product promotion approach, necessity-based selection facility, integrated domain knowledge and an additional advantage of expert knowledge. Another contribution of this research work is a novel and flexible similarity measurement technique proposed here. The proposed research work provides in-hand guidelines

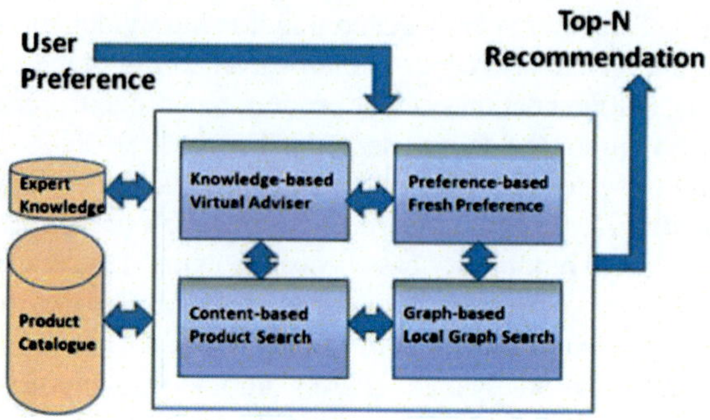

Fig. 1 System model

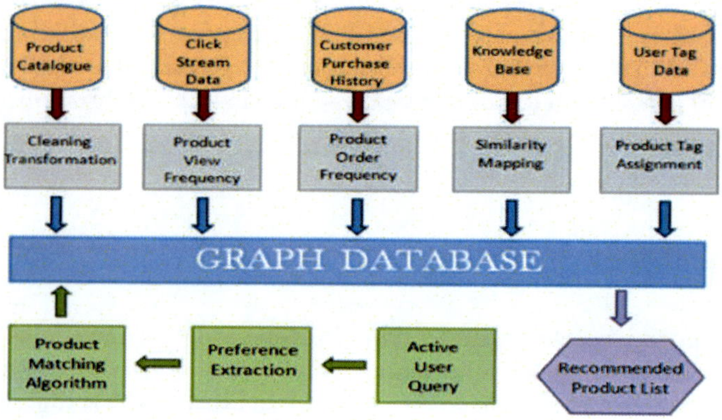

Fig. 2 System flow diagram

to the design and implementation of a personalized hybrid recommendation model for many volatile or trend-driven domains.

This research effort emphasizes the use of advanced graph database techniques, potentially useful for the management of large, complex, unstructured, time-variant and dynamic data. Figure 1 gives an idea about the system's working mechanism and its components.

3.2 A System Flow Diagram

The proposed recommendation model is very complex, and it comprises many components. For better understanding and simplicity of explanation, a system flow diagram is presented here in Fig. 2. The first most significant task is to prepare a

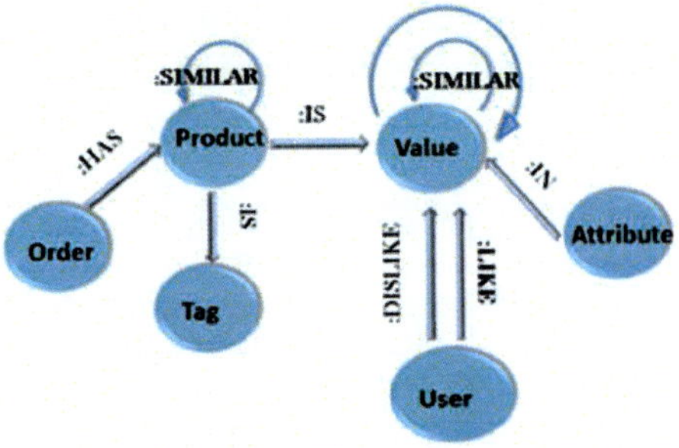

Fig. 3 A graph model

product graph. For that, the system requires to carry out mainly two processes, data preparation and data loading.

Step 1 mainly focuses on the data preparation phase. In this phase, it requires to carry out product data accumulation followed by data extraction, data cleaning and transformation process.

The graph data model supports the storage of all kinds of data, such as structured, semi-structured and even unstructured data. The product information can be available in any format. Our proposed model is flexible enough that facilitates the accumulation of any kind of product information. Here, a step-by-step process for semi-structured and unstructured data has been explained, respectively.

If the source of data is unstructured, the system has to convert unstructured data accumulated from various data sources into a semi-structured data format and then it can store into the database. For that, various data extraction techniques of natural language processing are used using the offline tool. After data completion of data extraction, two essential steps are data cleaning and transformation. The clean and transformed data helps to reduce polysemy and synonymy issues later. For that, rigorous data cleaning and transformation have to be carried out manually or using automatic tools. This ETL processes if purely domain-specific, and hence it is an external component of our system. Finally, product information becomes available in the form of keywords that describe that particular product. In this scenario, product node is created along with product detail such as unique identification number, name and price. For each new keyword, a new value node has to create and it has to connect with that previously created product node.

If the source of data is semi-structured, data is available in either a key-value format or tabular format. The system can directly load it into a graph database. In that scenario, each product is described with many key-value pairs. For example, one garment can have many key-value pairs such as size: 30, color: blue, style: low-west and fitting: skin-fit. Here, each key is nothing else but attribute and value are describing the keyword of the garment. In the above example, attributes are size, color, style and fitting whereas values are 30, blue, low-west and skin-fit. For each product, creation of a new product node and storage of product-related attributes is required to conduct as stated earlier. For each new value, which is not matching with any existing value nodes, the system stores it as a new value node and creates a connecting edge between that value node and previously created corresponding product node. In contrast to that, if that value is already present into a value node just create a new connecting edge between that value node and product node. Same way, create a new node for non-existing attributes and connect with previously created and corresponding value node.

In the case of unstructured data, one additional step is required to be carried out: the mapping between value and attribute nodes. Based on an expert's knowledge, that mapping has to be done. For the mapping process, creation of edge between various value nodes and attribute nodes is required. This step is optional, as mapping is not essential for the execution of the recommendation algorithm. Still, the mapping has one advantage; i.e., it creates readability of the graph and helps to categorize products into a treelike structure.

Step 2 mainly focuses on knowledge graph preparation with the use of an expert's knowledge. Here, the expert's knowledge about the product domain has been plotted into graphs in different ways. First, the product catalog-related knowledge is going to be fitted as a similarity of product values. For example, the expert knows that "Lycra" material has a strong similarity with "swimwear." This kind of knowledge can be represented in a graph with two edges between two "value" nodes. This step is very significant to manage polysemy and synonymy issues of the natural language. One more example is garment size can be denoted with many different values like M and XL. Looking into various size charts of many different standards, all these notations are representing to the same size.

Step 3 involves the creation of a new user node into a graph database whenever a new user visits the Web site. For each user, assign a unique identification number. More details about that user are collected, whenever user places order and it is stored into the database. For a mature system that contains details about existing customers, it can be also loaded with a graph database.

Step 4 involves loading tag-related information. New users are encouraged to tag artifacts. Even new users are encouraged to like and dislike a few sample artifacts. Store all this information into a graph database.

Step 5 involves updating a database based on current transactions related to order. Whenever a user places orders, a new node of the order has to be created with order details. This order node has to link with the corresponding product nodes. This connected edge will contain a quantity of the order.

Step 6 involves updating a database about product viewed details. The products' view history should analyze from clickstream. This process can be carried out in batch after a fixed-time interval. Based on the order frequency and product view frequency, various indexes have to be updated.

Step 7 involves calculating product similarity matching and creation of edges between two pairs of product nodes with similarity index as edge property.

In this research, the Jaccard similarity index has been used for the similarity matching of the items. The formula to find the Jaccard similarity index is as follows:

$$\text{Jaccard Similarity Index} = \frac{\text{The number of common attributes in both sets } |X \cap Y|}{\text{The number of attributes in cither set } |X \cup Y|}$$

Step 8 involves actual constraint-based product search based on the current user's preferences. An innovative algorithm has been designed to perform this constraint-based product search. For conducting search operation preferences of the current user specified in recent session has to be considered as an input of the algorithm. The system has an interactive interface specially designed for the preference accumulation.

Step 9 helps in performing scoring and ranking operation to prepare the list of top-N best matching products. A unique weighted average similarity measurement technique is proposed by this research for the ranking operation. Finally, top-N products having a maximum matching score are going to be suggested to the current user.

3.3 A Graph Model

The prime goal of this research is to develop a hybrid recommendation system using a graph database which can work in an online environment as an integrated part of the Web site. This research uses a well-known online graph database management tool called "Neo4j", which allows the creation and management of graph data using the property graph model. In this research implementation, various recommendation algorithms have been carried out with graph database management tools along with graph query language known as "Cypher."

Various types of entities and their related attributes have been stored using a multilayer graph model. Each layer contains nodes for various entities such as user, artifact, attribute, value, order and tag. Attributes of entities and context information detail are as follows:

User (U) nodes are the representative of each customer who requests for artifacts or products or services in the system. Users are encouraged to assign a tag or to give ratings for artifacts that have been viewed as well as purchased in the past. The user will connect with the product node once the user purchases that particular product.

Product (P) nodes are the representatives of objects for which recommendations are generated, commonly known as "items" or "artifacts" in recommendation systems. The ultimate goal of RS is to suggest a list of artifacts for the particular user request.

Attribute (A) nodes are comprised of the descriptive characteristics of artifacts such as size, color, weight and price. One attribute node may connect with another attribute node with an edge. This type of node structure internally creates a hierarchy of the attributes. The most significant attributes of artifacts may create a treelike structure for various features of artifacts from classification aspects. Some very important attributes of this classification tree are considered as top-level attributes and denoted as A1, A2, A3 .... AN. For searching aspects, these attributes are treated as special attributes of objects.

Value (V) nodes contain the text description of the artifact. All value nodes are always connected with any artifact nodes, as it describes any artifacts. It may be related to a particular attribute, or it may not be related to a particular attribute. Value nodes are also connected with other value nodes via edge labeled as ": SIMILAR".

Order (O) nodes contain details related to the orders such as order number, order date and order amount.

Tag (T) nodes contain details related to tags such as tag number, tag name and tag count.

3.4 Methodology

Step: 1

Input: Preference statement given by the current user.
Output: A set of keywords K consist of actual preferences of the current user.
Remove stop words and other pronouns from the preference statement. Extract each keyword to get actual requisite. Prepare a list of keywords K which user prefers in the product. Prepare a list of keywords NK which user does not prefer in the product.

Step: 2

Input: List K and NK.
Output: List K and NK.
Convert each element of both lists into lower case, and store it again in the same list.

Step: 3

Input: List K and NK.
Output: List of values V.
Search all value nodes using index which match with elements of K list. Find all distinguished and matching values of value nodes, and store into list V.

Step: 4

Input: List V.
Output: List of similar values SV.

Search all value nodes which are matching with elements of V list and having a connection with other value nodes via edge labeled as ": SIMILAR". Find all distinguished and similar values of value nodes, and store into list SV.

Step: 5

Input: List K.
Output: List of tag values T.
Search all tag nodes using an index which is matching with elements of V list. Find all distinguished and matching values of tag nodes, and store into list T.

Step: 6

Input: List K, NK, T, V, SV.
Output: List of products P.
Search among all product nodes which are connected with those value nodes having a similar value of elements of V list and SV list and T lists.
Eliminate products having value present into NK list. Eliminate product having popularity index greater than 80%. Store product details into list R for all selected products.

Step 7:

Input: List of products R.
Output: List of recommended products R.
Use algorithm as discussed earlier for finding score value for each selected product of list R. Rearrange all matching products as per the maximum scoring into list R.

4 Experiment and Evaluation

As stated above, the implementation of the system has been carried out with well-known online graph database management tool known as "Neo4j". That allows the creation and management of graph data using the property graph model. Initially, the loading of data has been done by importing from CSV file. Before importing, it is essential to create various indexes that facilitate speedup in the searching process. While importing data from CSV files, the system creates various nodes, properties of nodes and edges among nodes.

As stated earlier, this research makes use of unstructured data. The unstructured data from various Web sources has been accumulated and preprocessed. The preprocessing of data includes extraction, transformation and loading process. The preprocessed data has been stored in various CSV files. The sample CSV file of product data has been shown in Fig. 4. The dataset was prepared from various online garment trading Web sites. The figure shows a sample product data contains a product identification number, product name, associated features, tags, etc. Here, sample data is

4	Top	550	Ladies | Scallop Trim Botanical Print | Tunic Dress| XL | colourful	Fresh Fashion	9	3	2
5	Top	400	Ladies | Double sleeves overlap| Gingham Dress|XXL | plain yellow | plain yellow	Fresh Fashion	11	2	2
6	Top	500	Ladies | One Sided Exaggerated Frill | Dress|XXL | colourful	Fresh Fashion	12	4	2
7	Top	450	Ladies | Frill Trim | Surplice Wrap Crushed | Velvet Dress|XL | colourful | plain yellow	Fresh Fashion	14	3	3
8	Top	550	Ladies | Off The Shoulder | Tie-waist Ruffle | Hem Dress|XL | vintage colour	Fresh Fashion	12	4	2
9	Top	400	Ladies | Grid Print | Twiggy Dress|L | cream pink |Exclusive	Fresh Fashion	12	3	2
10	Top	600	Ladies | Mesh Insert | Bodycon Dress|M | colourful | vintage colour	Fresh Fashion	12	4	3
11	Top	500	Ladies | Navy Buttoned Keyhole Back | Flower Print | Dress|S | velvet red | plain yellow	Fresh Fashion	9	3	2
12	Top	650	Ladies | Cowl Neckline | Flare Dress|S | cream pink | grey	Fresh Fashion	10	2	3
13	Top	400	Ladies | Bardot Dip | Hem Dress|S | vintage colour	Fresh Fashion	12	3	2
14	Top	550	Ladies | Short Sleeve | Hollow Out Flippy | Dress|M | colourful | velvet red	Girly	14	4	3
15	Top	450	Ladies | Tweed Sheath Sleeve | Tunic Dress |M | grey | plain yellow|Exclusive		12	5	3
16	Top	400	Ladies | Choker One Shoulder Gigot Sleeve| Dress | M | black | grey		13	4	2
17	Top	600	Ladies | Bell Sleeve Symmetric Embroidered |Tunic Dress | M | printed | colourful		12	3	3
18	Top	650	Ladies | Crisscross | V Neck | Striped | Tee | red | S | black and white | plain yellow		11	4	4

Fig. 4 A sample data

incomplete and ambiguous. Even the expert's knowledge has been stored into a CSV file. A Cypher script has been written to load data from CSV file to graph database.

Initially, all data has to transfer into the Neo4j database, and it can be done with one-time import. Later on, continuous synchronization requires keeping the Neo4j data storage up to date with the transactional database. Synchronization can achieve by executing Cypher scripts after a fixed-time interval.

The following figures show the generated results for the sample dataset using Neo4j. Figure 5 demonstrates the working of the present model. Here, a simple scoring function has been used for measuring similarity, i.e., score component. Sorting of all products has been done based on scoring components to retrieve top-10 matching.

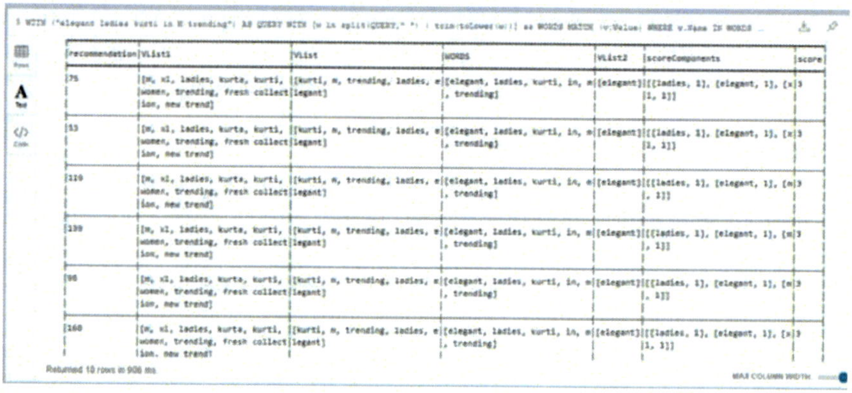

Fig. 5 A sample example of outcomes of the system

Fig. 6 An execution time measurement of the system

5 Results

A series of experiments have been carried out to test the operability of the proposed graph data model and recommendation algorithm for sample datasets. The presented model has been evaluated at each phase of the process. Very rigorous evaluation has been carried out for critical section, i.e., recommendation algorithm. The proposed recommendation algorithm had been evaluated in many ways. Table 1 shows a few test cases and derived output to measure the efficiency and effectiveness of it.

The explanation of major evaluation parameters is as follows: The number of search criteria is believed as a prime factor to affect the execution time of the algorithm. Here, all search criteria considered in extended product search are considered as an evaluation parameter. (ii) In the derived output, several matching criteria were counted. This indicates a measurement of similarity matching. Maximum and minimum matching number of nodes are noted down for each evaluation case. (iii) Here, relevant results in top-20 or top-10 recommendations have been measured by the utility in real life. For the calculation of utility, only the prime three attributes of products were considered. These prime three attributes are formally known as A1, A2 and A3. This can be derived based on the classification of products. Top-3 classifier attributes were considered as A1, A2 and A3. (iv) To measure the efficiency of the algorithm, execution time is considered as an important parameter. Here, time taken to execute (on the system with normal hardware) each time by graph engine was considered and shown.

Table 1 Outcomes of evaluation

Test case no.	No. of search nodes involved	Intersecting nodes	Relevant @ top-10	Relevant @ top-20	Execution time (ms)
1	9	3–3	10	10	3060
2	12	4–2	10	10	1052
3	08	2–1	08	08	433
4	13	3–2	10	10	576
5	16	4–2	09	09	789

The following chart shows the relation of execution time against the number of search criteria considered. It is observed that the execution time is almost in parallel to the number of search criteria.

6 Conclusion

The major findings based on the generated results based on evaluations can be listed here. (i) The result shows 100% accuracy at top-5 recommended products. Even top-10 results have almost 80% accuracy. For top-20 recommended products, the level of accuracy is reduced significantly. This helps in the promotion of novelty and is serendipitous in results. Indirectly, it helps to promote other similar products. (ii) The efficiency of the system is very satisfactory. That means the proposed model can be implemented for the real-time recommendation engine. The main reason behind this phenomenon is associative operations in the graph make it faster. The execution time of an algorithm has been significantly reduced due to the associativity of nodes. The series of experimental evaluations have been carried out to examine the execution time of algorithm against a large number of nodes and a large number of edges for the movie database. Even, the observation of that experiment has observed execution time in milliseconds. All the results presented here prove the efficiency of the results. In this research, more focus has been given to increase the efficiency of the system as it has to work in the online environment. (iii) In the absence of rating data or user preferences, it is possible to generate recommendations. This helps a new user and new product.

In other words, the presented recommendation model has been evaluated to measure efficiency and accuracy. The evaluation of the system proved that the system is accurate and efficient both.

This research work makes use of a flexible graph data model, due to which it becomes easy to represent structured, semi-structured as well as unstructured data. This flexible data model allows the storage of various kinds of entities in an integrated way. The proposed recommendation model is very advanced, unconventional and forward-looking into the future needs of the e-commerce industry. The proposed research work has a wide possibility for implementation in the real world due to the flexibility of data modeling and schema-free nature. In other words, the same model can implement for a wide range of products without modification into the schema of a data model.

References

1. Konstan, J.A., Riedl, J.: Recommender systems: from algorithms to user experience. User Model. User-Adapted Inter. **22**(1–2), 101–123. Clerk Maxwell, J.: A Treatise on Electricity and Magnetism, 3rd ed., vol. 2, pp.68–73. Clarendon, Oxford (1892)

2. Debbarma, S., Nandi, G.: Promoting E-commerce in India: main issues and challenges. Int J Comput Sci Inform Technol **5**(6), 7371–7375 (2014)
3. Sanjeevi, S.G., Grandhe, S., Shrivastav, H., Yadav, A., Mahanta, T.K., Kumar, C.N.: An adaptive clustering and ıncremental learning solution to cold start problem in recommendation systems
4. Elahi, M., Ricci, F., Rubens, N.: A survey of active learning in collaborative filtering recommender systems. Comput. Sci. Rev. **20**, 29–50 (2016)
5. Sinha, R., Swearingen, K.: The role of transparency in recommendersystems. In: CHI'02 Extended Abstracts on Human Factors in Computing Systems, pp. 830–831. ACM (2002)
6. Ricci, F., Rokach, L., Shapira, B.: Introduction to recommender systems handbook. In: Recommender Systems Handbook, pp. 1–35. Springer, New York (2011)
7. Patel, A.A., Dharwa, J.N.: An integrated hybrid recommendation model using graph database. In: International Conference on ICT in Business Industry & Government (ICTBIG), pp. 1–5. IEEE (2016)
8. Felfernig, A., Burke, R.: Constraint-based recommender systems: technologies and research issues. In : Proceedings of the 10th international conference on Electronic commerce, p. 3. ACM (2008)
9. Ghazanfar, M.A., Prugel-Bennett, A.: A scalable, accurate hybrid recommender system. In: Third International Conference on Knowledge Discovery and Data Mining (WKDD'10), pp. 94–98. IEEE (2010)
10. Iwin Thanakumar, J.S.: Survey of data mining algorithm's for intelligent computing system. J. Trends Comput. Sci. Smart Technol. (TCSST) **1**(01), 14–24 (2019)
11. Scholz, M., Dorner, V., Franz, M., Hinz, O.: Measuring consumers' willingness to pay with utility-based recommendation systems. Decision Support Syst. **72**, 60–71 (2015)
12. Bobadilla, J., Ortega, F., Hernando, A., Gutiérrez, A.: Recommender systems survey. Knowledge-Based Syst. **46**, 109–132 (2013)

An Efficient Task Scheduling Using GWO-PSO Algorithm in a Cloud Computing Environment

Avinashi Malleswaran Senthil Kumar, Parthiban Krishnamoorthy, Sivakumar Soubraylu, Jeya Krishnan Venugopal, and Kalimuthu Marimuthu

Abstract Nowadays, cloud computing becomes a major technology that provides services to users by integrating Internet technology. Scheduling of user tasks plays a vital role in a cloud environment. An efficient task scheduling method is required to manage the resources for executing user tasks in order to attain better result in QoS parameters. Task scheduling is a kind of non-deterministic polynomial (NP) hard problem, which can be solved using optimization algorithms. This paper presents a novel hybrid method by combining grey wolf optimization (GWO) and particle swarm optimization algorithm (PSO). The proposed method optimizes the result in order to maintain the makespan as a minimum. PSO improves the optimization performance of grey wolf optimization in the proposed method. The proposed algorithm performance is evaluated and analysed with GA and GWO algorithms. The proposed GWO-PSO algorithm results are better than the GWO and GA algorithms in terms of response time and makespan.

Keywords Task scheduling · Cloud computing · Makespan · Grey Wolf Optimization (GWO) · Particle Swarm Optimization (PSO)

A. M. Senthil Kumar (✉) · S. Soubraylu
Department of CSE, Koneru Lakshmaiah Education Foundation, Vaddeswaram, Guntur, Andhra Pradesh, India
e-mail: senthil1185@gmail.com

P. Krishnamoorthy
Department of CSSE, Sree Vidhya Nikethen Engineering College, Tirupathi, Andhra Pradesh, India

J. K. Venugopal
Department of CSE, Saintgits College of Engineering, Kottayam, Kerala, India

K. Marimuthu
Department of CSE, Malla Reddy Institute of Technology, Secunderabad, Telengana, India

© The Editor(s) (if applicable) and The Author(s), under exclusive license to Springer Nature Singapore Pte Ltd. 2021

A. P. Pandian et al. (eds.), *Proceedings of International Conference on Intelligent Computing, Information and Control Systems*, Advances in Intelligent Systems and Computing 1272, https://doi.org/10.1007/978-981-15-8443-5_64

1 Introduction

In a cloud computing paradigm, a group of shared computers are acting as computing resources and offer services which are demanded by the end users. Cloud computing abstracts the resources and provided as services to users from any part of the world. The business investment cost is reduced drastically when enterprises start using cloud computing-based applications [1]. Cloud computing offers three models which are infrastructure as a service (IaaS), where the entire infrastructure can provide as service based on usage basis, platform as a service (PaaS), where the platform can offer as a service to create and deploy applications knowledge about processor and memory details and software as a service (SaaS), where applications are offered as a service to the end users through the web portals [2, 3].

Scheduling the tasks is an important issue in a cloud computing environment. The goal of task scheduling algorithm is allocating the user-submitted tasks to the data centres in order to attain the optimized output parameters such as execution time, response time and makespan. [3–6]. Improper task scheduling can result in high makespan and less throughput. Hence, the efficient task scheduler is needed to allocating user tasks to the various resources by optimizing the total time. This motivated us to present a new task scheduling algorithm to perform scheduling of tasks to the various virtual machines. The core idea of proposed GWO-PSO [7, 8] algorithm is to increase the makespan of the system. To find the best match between task and resource in task scheduling is considered as non-deterministic polynomial (NP)-hard problem.

NP-hard problems can be solved using swarm intelligence algorithms efficiently. Among various optimization algorithms, GWO is interesting. GWO algorithm is simulated the hunting behaviour of the grey wolf. Due to the efficient exploring and exploiting nature, GWO gives better global optimal solutions in a faster convergence rate when compared to CSO, GA and ACO [7]. On the other hand, due to the better exploratory character of PSO, it provides optimized results when compared to ACO and GA. This motivated us to consider GWO and PSO algorithms to design a GWO-PSO-based task scheduling algorithm. In this proposed method, PSO is hybrid with GWO to provide improved results due to its higher exploratory nature of finding optimal solutions [9–15].

This proposed paper contributes the following,

(1) This work proposes a hybrid task scheduling based on grey wolf optimization-particle swarm optimization algorithm (GWO-PSO) which assigns the tasks to the suitable data centres in order to keep the makespan minimum.
(2) This work proposes a PSO to improve the grey wolf optimization results. High exploratory nature of PSO algorithm helps GWO algorithm to attain better results.

The proposed algorithm is executed and analysed with standard GWO and PSO algorithms. The remaining content is planned as follows: Sect. 2 describes literature works, and Sect. 3 details the problem statement. Section 4 presents the proposed framework, and Sect. 5 shows simulation results. Section 5 describes the conclusion of the work.

2 Related Work

Gobalakrishnan et al. [16] presented task scheduling method using genetic-grey wolf optimization algorithm (GGWO) in a cloud environment. This model focuses and optimizes energy, time and load parameters. In this algorithm, genetic algorithm is combined with the GWO to optimize the results. Results show that the GGWO method performs better than GA and GWO algorithms.

In [17], Ebadifard et al. presented PSO-based scheduling method to optimize the resource utilization and makespan. Modified PSO algorithm is used to schedule independent and non-preemptive tasks to the suitable resources. The proposed technique outperforms the results when compared to PSO. Gupta et al. [18] presented a GWO for scheduling in the cloud environment. The proposed algorithm focuses request and schedule time. The proposed algorithm results outperform GA results.

In [19], Natesan et al. presented a new GWO to improve system performance. Mean GWO method minimizes the makespan and energy consumption parameters. The proposed algorithm provides better results than PSO and GWO algorithm. Natesan et al. [20] presented an effective multi-objective task scheduling method using grey wolf optimization algorithm. This algorithm focuses to manage data centre load by optimal utilizing of cloud resources. Results show that the multi-objective grey wolf optimization algorithm outperforms the conventional algorithms in terms of energy consumption, resource utilization and makespan.

The hybrid task scheduling algorithm proposed in the current study gives better results than GWO and PSO algorithm. Hybrid form of algorithms can be used to overcome the failures of single algorithm [21–25]. This study focuses on hybrid meta-heuristic-based task allocation in a cloud computing environment.

3 Problem Formulation

The aim of the task scheduling is to minimize the makespan. The makespan is the time taken by all tasks to execute on the recourses. Makespan is calculated by adding execution time and response time. Execution time is the difference between finishing time and starting time of the tasks. Response time is total of submission and waiting time.

$$\mathrm{ET} = \sum_{c=1}^{M} (ft_c - st_c) \tag{1}$$

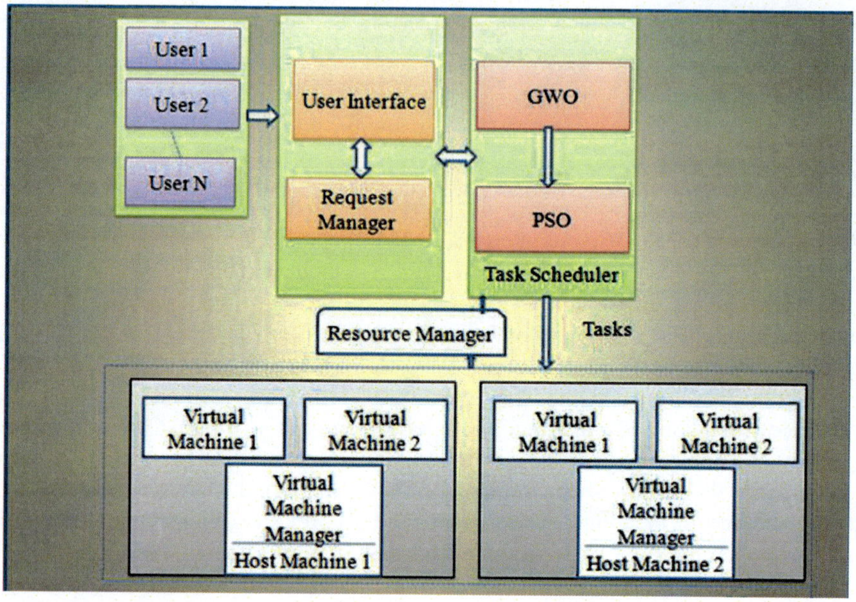

Fig. 1 Architecture diagram

$$\mathrm{RT} = \sum_{c=1}^{M} \sum_{d=1}^{K} (subt_{cd} + wt_{cd}) \tag{2}$$

$$\mathrm{MS} = \sum_{c=1}^{M} \sum_{d=1}^{N} (\mathrm{RT}_{cd} + \mathrm{Et}_{cd}) \tag{3}$$

$$\mathrm{MakeSpan} = \min\{\mathrm{RT}_{j,}, \mathrm{ET}_{j} | j = 1, 2, \ldots M\} \tag{4}$$

Subject to

$$c, d \leq 1$$
$$c \in M$$
$$d \in K$$

where RT is a response time, ET is a execution time, and MS is a makespan.

4 The Proposed Task Scheduling Model

The proposed architecture is shown in Fig. 1. The user submits a task to the user interface module where tasks managed. Request manager forwards task to the scheduler. In task scheduler, tasks are allocated to the suitable virtual machines using GWO-PSO algorithm. Resource manager handles the resources effectively.

GWO-PSO algorithm is given below.

```
Input:
The GWO-PSO algorithm parameters
Input Si, Execution time, Response time, make span
Output:
Efficient scheduling
Initialization:
Initialize the tasks Ti, virtual machines Vi,
Start:
        Initial population pi,
for all pi, do
        Evaluate the population against fitness function using Equation (4)
end for
        Initialize the iteration number to 1
Repeat
        choose the bestagent α
        choose the next best agent β
        choose the third best agent γ
  until maximum loop
    for each search
                Update current agent position
    end for
Update a, X, Y
Update α , β and γ
Apply PSO update procedure to GWO solutions
        Initialize particle's position vectors, velocity vectors
for each particle
        Calculate fitness function using
        Calculate pbest and gbest using
end for
If current value > pbest and gbest
        Update pbest and gbest
End if
Update position and velocity using Equation (13)
return pα
stop
```

Algorithm 1. GWO-PSO algorithm

The proposed GWO-PSO algorithm is shown in algorithm, and it is used to achieve optimized task scheduling using GWO and PSO. In this hybrid approach, PSO is used to improve the optimization efficiency. This hybrid algorithm (GWO-PSO) focuses to optimize the makespan of the scheduling in a cloud environment.

4.1 Basics of GWO

The highest success hunting probability wolf is considered as fittest wolf alpha α to hunt the prey. Based on the success probability, a second highest wolf is beta β wolf, and third and fourth highest wolves are considered as delta δ and omega ω. During hunting the prey, beta, delta and omega wolves help alpha wolf to hunt a prey. In each iteration, other wolves adjust their hunting positions towards the prey with respect to the alpha wolf. GWO algorithm includes four major phases such as initialization, tracking, harassing and hunting the prey [3, 7].

4.2 *Solution Encoding*

Initially, tasks are assigned to the virtual machines randomly. For example, let assume task list $T = \{T_1, T_2, T_3, T_4, T_5\}$ and virtual machines list $R = \{VM1, VM_2, VM_3, VM_4, VM_5\}$. All the tasks in the task list need to assign in the given virtual machines. The dimensions of the virtual machines are denoted by $D = (VM1, VM_2, VM_3, VM_4, VM_5)$.

4.3 *Fitness Function*

This step evaluates the fitness values of all individuals in the solution. The aim of introducing fitness function (3.4) is to optimize the makespan. Equation (3.4) is used to evaluate the fitness values of each virtual machine in the resource list. The maximum fitness value of the virtual machine is considered as best solution for the given tasks.

4.4 *Calculating the Fitness*

Fitness is calculated using following equations,

$$G = |\vec{Y} \cdot E(k+1) - |E(k)|| \tag{5}$$

$$E(k+1) = |E(k) - \vec{x} \cdot G| \tag{6}$$

$E(K+1)$ is position of wolf, and $E(k)$ is position of prey. Co-efficient vectors are $\vec{x}$ and $\vec{y}$.

Vectors are given below,

$$\vec{x} = 2\vec{a}r1 + \vec{a} \quad \text{and} \quad \vec{y} = 2r2 \tag{7}$$

r_1 and r_2 are random vectors, and $\vec{a}$ is a linear vector.

Here, *E(t)* is considered as function of total makespan,

$$E(t) = \min\{\mathrm{RT}_j, \mathrm{ET}_j | j = 1, 2, \ldots M\} \tag{8}$$

where RT_j is a response time, and ET_j is an execution time.

4.5 *Evaluating the Best Solutions*

Best solutions are evaluated based on fitness values. The top three solutions are identified and named as α, β and γ. α, β and γ are the three levels of best solutions. Based on α, β and γ, the hunting is guided to remaining candidates.

4.6 *Harassing and Hunting the Prey*

The fitness of best solutions $\vec{S}\alpha$, $\vec{S}\beta$, $\vec{S}\delta$ with respect to fitness function $E(t)$ is evaluated using Eq. 4.6.

$$\begin{aligned}
\vec{S}\alpha &= \left|\vec{Y}1 \cdot E_\alpha(t) - E(t)\right| \\
\vec{S}\beta &= \left|\vec{r}_2 \cdot E_\beta(t) - E(t)\right| \\
\vec{S}\delta &= \left|\vec{Y}3 \cdot E_\delta(t) - E(t)\right|
\end{aligned} \tag{9}$$

The top three solutions are represented as E_1, E_2 and E_3 and evaluated using Eq. (10)

$$\begin{aligned}
E_1 &= |E_\alpha - \vec{X}_1 \cdot (\vec{S}\alpha)| \\
E_2 &= |E_\beta - \vec{X}_2 \cdot (\vec{S}\beta)| \\
E_3 &= |E_\delta - \vec{X}_3 \cdot (\vec{S}\delta)|
\end{aligned} \tag{10}$$

Based on the best three solutions obtained, average of the three best search agents is calculated. Other agents update their new positions based on the average value. The process is repeated until the required fitness level is reached or no change in the fitness results.

4.7 *Updating Solution Using PSO Update Procedure*

Particle swarm optimization algorithm's update procedure Eq. (12) is called to update the solutions obtained from the GWO algorithm. PSO particles and vector positions are initialized. The fitness value is evaluated for all the particles by applying fitness function. The pbest, gbest and velocity vector values are evaluated at each iteration. If current values are greater than the existing population's values, pbest, gbest and velocity vector values are updated.

Velocity is defined as below

$$\vec{V}e_i(t-1) = \omega * \vec{V}e_i(t) + C1r1\left(\vec{l}_i - \vec{Z}_i(t)\right) + C2r2\left(\vec{g}_i - \vec{Z}_i(t)\right) \quad (12)$$

where inertia weight is ω, cognitive parameters are $C1$, $C2$ and social learning parameter, and random values $r1$, $r2$, $\vec{l}_i$ are local best and $\vec{g}_i$ are global best.

Finally, optimal solution is obtained from PSO. In proposed algorithm, PSO improves the GWO algorithm's optimization performance.

5 Experimental Results

Simulation is done on CloudSim, and cloudlets are considered as tasks. Execution time, response time and makespan required by the cloudlets (tasks) on the virtual machines are recorded when changing the parameters like cloudlet, bandwidth, RAM and MIPS. The time unit is in milliseconds. Table 1 shows the simulation parameters such as cloudlet and VM parameters.

From Fig. 2, it could be observed that the response time of GWO-PSO is lower than GWO and PSO, and the time taken by GWO is lesser than PSO. When the number of tasks is 500, the response time of the proposed GWO-PSO is lower by 607 ms and 2300 ms compared to GWO and PSO algorithms.

Figure 3 compares the execution time of GWO-PSO, GWO and PSO algorithms. The execution time represents the time taken by the tasks to be completed. The execution time unit is in milliseconds. When the number of tasks is 500, the execution time of the proposed GWO-PSO is lowered by 1607 and 4300 ms compared to GWO and PSO.

Figure 4 presents the comparison of makespan between the proposed GWO-PSO algorithm and GWO and PSO algorithms. The makespan is the summation of response time and execution time. When tasks are 50, makespan of GWO-PSO is lesser by 1007 ms and 1600 ms than that with GWO and PSO.

Table 1 Parameters

Type	Parameters details	
Data centre	Resource	0–5
	Ram	2048 MB
	Bandwidth	3000 B/S
Cloudlets	Count	100–500
	Input	400
	Output	800

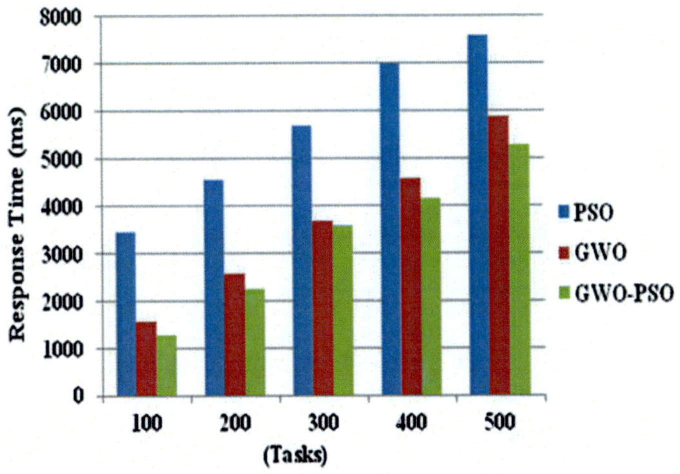

Fig. 2 Response time

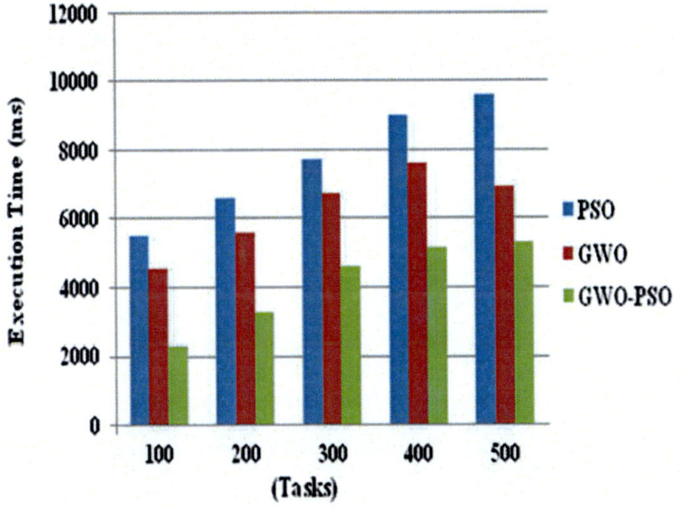

Fig. 3 Execution time

6 Conclusion

The proposed work has addressed a scheduling issue and sought its solution by applying a novel GWO-PSO algorithm. The simulation results prove that task

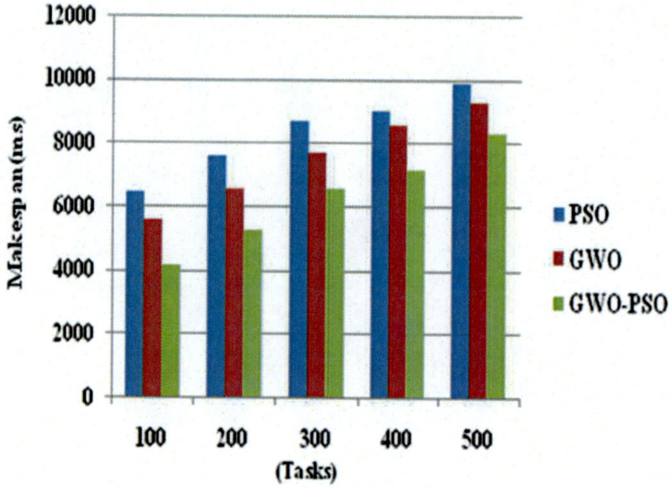

Fig. 4 Makespan

scheduling proposed GWO-PSO algorithm is better than the GWO and PSO algorithms in all cases. In GWO-PSO algorithm, due to exploratory nature of PSO algorithm it improves the GWO algorithm results. The proposed algorithm minimizes the makespan by optimizing the response time and waiting time in the system. In future, resource management issues can be focused in a real-world cloud environment.

References

1. Senthil Kumar, A.M., Venkatesan, M.: Task scheduling in a cloud computing environment using HGPSO algorithm. In: Cluster Computing (2018)
2. Avram, M.G.: Advantages and challenges of adopting cloud computing from an enterprise perspective. Procedia Technol. **12**, 529–534 (2014)
3. Shi, H.L.: Research of Job Scheduling on Cloud Computing. Nanjing University of Science and Technology, Nanjing (2012)
4. Senthil Kumar, A.M., Venkatesan, M.: Multi-objective task scheduling using hybrid genetic-ant colony optimization algorithm in cloud environment. Wireless Personal Commun. (2019)
5. Senthil Kumar, A.M., Venkatesan, M., Rajivkannan, A.: ERAM2-energy based resource allocation with minimum reckon and maximum reckon. J. Adv. Chem. **12**(23), 5485–5493 (2016)
6. Senthil Kumar, A.M., Venkatesan, M.: A novel based resource allocation method on cloud computing environment using hybrid differential evolution algorithm. J. Comput. Theor. Nanosci. **14**(11), 5322–5326 (2017)
7. Mirjalili, S., Mirjalili, S.M., Lewis, A.: Grey Wolf Optimizer. Adv. Eng. Softw. **69**, 46–61 (2014)
8. Kennedy, J., Eberhart, R.: Particle swarm optimization. In: Proceedings of IEEE International Conference on Neural Networks, Perth, Australia (1995)

9. Wang, M., Zeng, W., IEEE: A comparison of four popular heuristics for task scheduling problem in computational grid. In: The 6th International Conference on Wireless Communications Networking and Mobile Computing (2010)
10. Hongbo Liua, Ajith Abraham, and A.E. Hassaniend, *Scheduling jobs on computational grids using a fuzzy particle swarm optimization algorithm.* Future Generation Computer Systems, 2010. 26: p. 1336–1343
11. Wang, J., Li, F., Zhang, L.Q.: Apply PSO into cloud storage task scheduling with QoS preference awareness. J. Commun. **35**(3), 231–238 (2014)
12. Yang, Z., et al.: Optimized task scheduling and resource allocation in cloud computing using PSO based fitness function. Inf. Technol. J. **12**(23), 7090–7095 (2013)
13. Chitra, S., et al.: Local minima jump PSO for workflow scheduling in cloud computing environments. In: Lecture Notes in Electrical Engineering, pp. 1225–1234 (2014)
14. Zhan, S., Huo, H.: Improved PSO-based task scheduling algorithm in cloud computing. J. Inform. Comput. Sci. **9**(13), 3821–3829 (2012)
15. Verma, A., Kaushal, S.: Bi-criteria priority based particle swarm optimization workflow scheduling algorithm for cloud. In: Recent Advances in Engineering and Computational Sciences (RAECS 2014) (2014)
16. Gobalakrishnan, N., Arun, C.: A new multi-objective optimal programming model for task scheduling using genetic Gray Wolf Optimization in cloud computing. Comput. J. **61**(10), 1523–1536 (2018)
17. Ebadifard, F., Babamir, S.M.: A PSO-Based Task Scheduling Algorithm Improved Using a Load Balancing Technique for the Cloud Computing Environment, pp. 1–13 (2017)
18. Gupta, P., Ghrera, S.P., Goyal, M.: QoS aware Grey Wolf Optimization for task allocation in cloud infrastructure. In: Somani, A., Srivastava, S., Mundra, A., Rawat, S. (eds.) Proceedings of First International Conference on Smart System, Innovations and Computing. Smart Innovation, Systems and Technologies, vol. 79 (2018)
19. Gobalakrishnan, N., Arun, C.: Task scheduling in heterogeneous cloud environment using mean grey wolf optimization algorithm. ICT Express **5**(2), 110–114 (2019)
20. Natesha, B.V., Kumar Sharma, N., Domanal, S., Reddy Guddeti, R.M.: GWOTS: Grey Wolf Optimization based task scheduling at the green cloud data center. In: 2018 14th International Conference on Semantics, Knowledge and Grids (SKG), Guangzhou, China, pp. 181–187 (2018)
21. Senthil Kumar, A.M., Venkatesan, M., Rajivkannan, A.: A novel approach for multiple object resource allocation using hybrid algorithm. Middle-East J. Sci. Res. **23**(10), 2586–2591 (2015)
22. Senthil Kumar, A.M., Venkatesan, M.: An efficient multiple object resource allocation using hybrid GA-ACO algorithm. Austral. J. Basic Appl. Sci. **9**(31), 53–59 (2015)
23. Zukhri, Z., Paputungan, I.V.: A hybrid optimization algorithm based on genetic algorithm and ant colony optimization. Int. J. Artif. Intell. Appl. (IJAIA) **4**(5) (2013)
24. Tao, F., Feng, Y., Zhang, L., Liao, T.W.: CLPS-GA: a case library and Pareto solution-based hybrid genetic algorithm for energy-aware cloud service scheduling. Appl. Soft Comput. **19**, 264–279 (2014)
25. Wang, J., Li, F., Zhang, L.: QoS preferenceawareness task scheduling based on PSO and AHP methods. Int. J. Control Autom. **7**(4), 137–152 (2014)

Spike Correlations and Synchrony Affect the Information Encoding of Neurons

Manjusha Nair, Richard Laji, and Reshma Mohan

Abstract Coordination in neural activity is thought to be a key element in propagation of information. Coordinated spiking in neurons is an effective signaling mechanism and helps in self-organization of the network over time. Using computational techniques, this study investigates the parameters affecting the coordinated activity of a population of granular layer neurons of the cerebellum. This study investigates the spike train coordination using two basic measures: spike correlations and spike synchrony. The values were quantified from modeled granule neurons of the cerebellum under various excitatory and inhibitory synaptic balances and under induced plasticity states. Increased excitation from mossy fibers (MFs) increased the output correlation of granule neurons, while increased inhibition from Golgi neuron reduced the output correlations. Synchronous firing of granular layer population under in-vivo like inputs was increased with the increase in excitation and inhibition. Synchrony was also estimated under different induced plasticity states of the modeled neurons. The firing was more synchronous under the long-term potentiation (LTP) state compared to control state and was more asynchronous under long-term depression (LTD) state compared to control. Since plasticity signifies the learning activity in the neurons and is affected by temporal coordination in spiking, these results have greater relevance in understanding the nature of neural code.

Keywords Computational neuroscience · Spike train analysis · Correlation · Synchrony · Granule neuron

1 Introduction

A nerve cell (neuron) in the brain carries electrical impulses and fire signals to communicate with each other through the tiny gaps between them called synapses

M. Nair (✉) · R. Laji · R. Mohan
Department of Computer Science and Applications, Amrita Vishwa Vidyapeetham, Amritapuri, India
e-mail: manjushanairm@gmail.com

© The Editor(s) (if applicable) and The Author(s), under exclusive license to Springer Nature Singapore Pte Ltd. 2021

A. P. Pandian et al. (eds.), *Proceedings of International Conference on Intelligent Computing, Information and Control Systems*, Advances in Intelligent Systems and Computing 1272, https://doi.org/10.1007/978-981-15-8443-5_65

[1]. Depending on the nature of the neuron receptor present in the postsynaptic side, the postsynaptic membrane either becomes more electrically positive (excitatory) or becomes more electrically negative (inhibitory) and produces postsynaptic potentials (PSP) called excitatory postsynaptic potential (EPSP) or inhibitory postsynaptic potential (IPSP). PSPs thus formed are nonlinearly integrated across time (signals from the same presynaptic neuron, one after the other) and space (signals from various presynaptic neurons, received at the same time) in the axon hillock and result in producing a signal called action potential.

Even though there are minor variations in the amplitude or shape of different such signals, these variations are assumed to be irrelevant. Neurons are assumed to use stereotyped and pulsed variants of such action potentials or spikes to communicate over long distances. A sequence of spikes from a neuron is represented as a spike train, consisting of a sequence of their spike times. Each spike typically lasts for 1–3 ms approximately and is separated from other spikes by a few milliseconds [2]. These spike trains can be modeled as random processes since the same neuron produces a wide range of responses for the multiple invocation of identical stimuli. Neurophysiologists use multi-electrode arrays to record simultaneous activity of several neurons within a particular brain region. Simultaneously recorded data from multiple neurons represent multi-dimensional point process time series.

Various coding schemes are used to study the transformation of information from one processing stage to another in a neural circuit. Rate coding assumes that the firing frequency of the neuron carries information. Since the shape of the action potential is not assumed to be significant in carrying information, the time of arrival of the action potential is used in many studies to measure the information content of the signal [3]. Hence, time code considers the temporal patterns as information carriers. Inter-spike interval-based and first spike latency-based coding schemes also exist [4]. In frequency coding, neurons act as an analog-to-frequency converter while in latency code [5], neurons act as an analog-to-delay converter [6]. All these above-mentioned coding schemes study the neuron from a single-neuron perspective and fail to represent the emergent properties of the joint activities of a group of neurons. Population-level coding schemes are used to estimate the relationship between groups of neurons. For example, correlation coding studies suggest that correlations or synchrony between neurons could convey information in neuronal population level which is not contained in other coding schemes.

Correlation coding schemes are in practice from the early 1960s [7], and several studies demonstrated the functional correlates of this measure with a multitude of parameters such as shared inputs, shared presynaptic partners, distance between neurons and receptive fields, intrinsic properties such as synaptic time constants [8]. It has been suggested that the neural responses are influenced not only by the firing rate of the inputs that it receives but also by the input correlations [9]. Spike train distance-based estimates also exist to measure the synchrony in neural firing [10].

This study investigates the correlation and synchrony between specific types of tiny neurons in the cerebellum, called granule neurons. The granular layer of the cerebellum has more than 50% of the total number of the neurons in the brain. Granule neurons are present in the input layer of the cerebellum and receive inputs from various brain regions through long axons called mossy fibers. These mossy fiber inputs are excitatory in nature, while the input connection that the granule neuron receives from another type of neurons called Golgi neurons is inhibitory in nature. Each granule neuron, on an average, receives one to four excitatory and inhibitory connections. Using computational models, this study investigates the coincident activity of multiple granule neurons under different synaptic conditions. The functional significance of these very tiny and densest neurons in the brain has greater relevance in computational modeling studies.

2 Methods

2.1 Computational Modeling of Neurons

In computational modeling studies, a neuron is abstracted as an RC circuit describing the electrical properties of the neuron [11]. The RC circuit equivalent model is used to mathematically model the current flow in the circuit. In simulation studies, this mathematical framework is converted into a computational model. The biophysics of a single neuron is often abstracted as resistance–capacitance (RC) circuit with low-pass filtering properties, wherein the lipid bilayer of the neuronal membrane is modeled as capacitance (C), the voltage-gated ion channels in the membrane as electrical conductance (g) and the electrical gradient across the membrane as reversal potentials. Ohm's and Kirchhoff's laws are used to calculate the total current flowing and the voltage generated across the membrane of the neuron. The lipid bilayer of the membrane acts as a capacitor by accumulating charges over its surface. The intracellular and extracellular solutions of the membrane act as the conducting plates separated by the non-conducting membrane. Total current flowing across the membrane is the sum of the capacitive and ionic currents:

$$I_{\text{Total}} = I_{\text{c}} + I_{\text{ion}} \tag{1}$$

When the capacitor is further charged with the input current I_{ext} which is applied into the cell, the current leaks through the membrane. Open membrane pores or channels in the membrane give rise to more resistance components in parallel. Conductance of the channel increases with the increase in number of open ion channels in the membrane. Ionic current flow through a neuron with sodium, potassium and leak channels can thus be modeled as,

$$I_{E\pi} = C\frac{\mathrm{d}V}{\mathrm{d}t} + G_{\text{Na}}(V - E_{\text{Na}}) + G_{\text{K}}(V - E_{\text{K}}) + G_{\text{L}}(V - E_{\text{L}}) \tag{2}$$

2.2 *Multi-compartmental Modeling of Granule Neurons of the Cerebellum*

In multi-compartmental neuron modeling, each neuron is abstracted as a cylinder which is divided into different RC compartments. Each compartment is modeled with voltage-gated ion channels and synaptic channels. The equations model close correspondence with the biophysical properties of neurons, ion channels and synapses. They also consider the propagation of membrane voltage change over axons and dendrites. Cable theory is used to model the variation of voltage across the axon. Every compartment is assumed to be iso-potential; i.e., the voltage within each compartment is assumed to be constant and is modeled as an RC circuit. Multi-compartmental model of granule neuron used in this study contains 52 compartments [12]: N_A—no. of axonal compartments, N_D—no. of dendritic compartments, N_S—no. of somatic compartments, N_H—no. of hillock compartments (Fig. 1a). The membrane voltage V_m is estimated separately for each of the compartments separately.

$$\frac{dV}{dt} = \frac{1}{\tau_m}\left(\frac{\sum_i g_i(V - V_i) + \sum_{syn} g_{syn}(V - V_{syn}) + \sum_{br} g_{br}(V - V_{br})}{g_{tot}}\right) \tag{3}$$

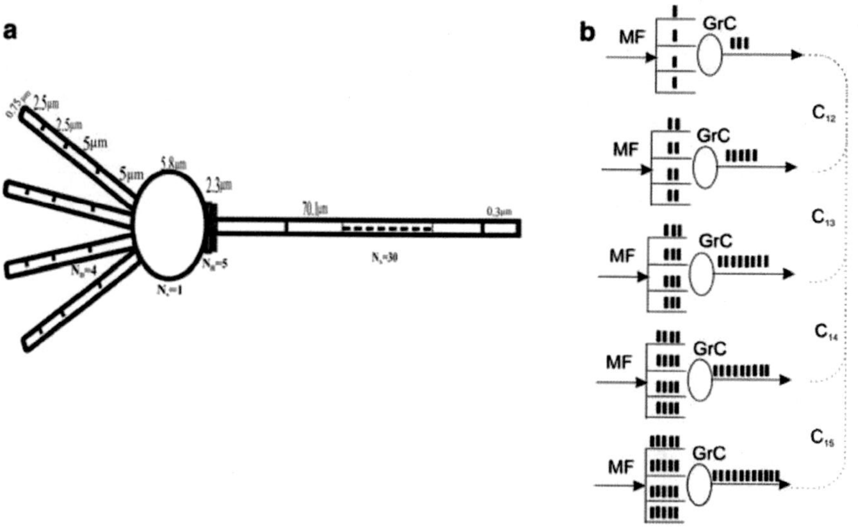

Fig. 1 Computational modeling of neuron. **a** The granule neuron model used in this study included 52 compartments: 1 for the soma, 4 each for the four dendrites, 5 hillock compartments and 30 compartments for the axon. **b** Input of varying spike rate was applied to the granule neuron model (GrC) through mossy fibers (MFs). Cross-correlation coefficients between paired neurons (C_{12}, C_{13}, C_{14}, C_{15}) were estimated as shown

where the conductance of ion channel i is modeled as g, syn (synaptic dynamics), br representing the neighboring attached branch and g_{tot} (total) is the total conductance. The time constant is represented as $\tau_m = R_m C_m R_m$ is the membrane resistance, and C_m is the membrane capacitance. The multi-compartmental GrC model used included stochastic synapses where each synapse was constituted by few releasing sites comprising presynaptic dynamics and release mechanisms. The granule cell model includes 1–4 excitatory and 1–4 inhibitory connections. Excitatory postsynaptic mechanisms were shown as AMPA and NMDA receptor components as seen in granule neurons [13]. AMPA receptor dynamics was modeled using a three-state scheme that was activated by a direct component of glutamate, modeled as pulse, and an indirect component representing the spillover of glutamate from nearby synapses in the glomerulus. Similarly, the NMDA receptors were modeled as five-state kinetic scheme, activated by the same direct and indirect glutamate components, and magnesium block was modeled by a Boltzmann equation [12]. The $GoC-GrC$ inhibitory synapses were modeled as in [13]. The granule neuron model was provided with 1–5 input spikes through each of the four MFs (Fig. 1b), and the simulations were performed in the presence and absence of inhibitory inputs to the neurons.

Plasticity states were also introduced in the neuron model. LTP and LTD were introduced by modifying the intrinsic excitability of the cell and also by modifying the release probability of the synapses [14].

2.3 Modeling the Neural Responses

Individual responses to the same stimulus are highly variable in the neurons. The process of action potential generation is a stochastic process. Poisson point processes on real line are used to model these spiking responses of neurons which are random spiking events in time. Point processes describe random events whose occurrence is timed. A Poisson point process is defined as a collection of random elements located on some mathematical space which can be used as mathematical models of some phenomena. The inter-spike interval (ISI) of spike trains follows exponential distribution (Poisson distribution).

$$P(x) = \lambda e^{-\lambda t} \tag{4}$$

where λ is the inter-spike interval. Refractory periods modify this Poisson process to form gamma distributions. Simultaneously recorded spike trains from multiple neurons are considered in our study as stochastic–dynamic multi-dimensional point process time series whose properties probabilistically change over time. Standard signal processing techniques used to analyze continuous valued data are not suitable to analyze such discrete datasets. The following sections describe the different techniques used in this study for spike train data analysis (Fig. 2).

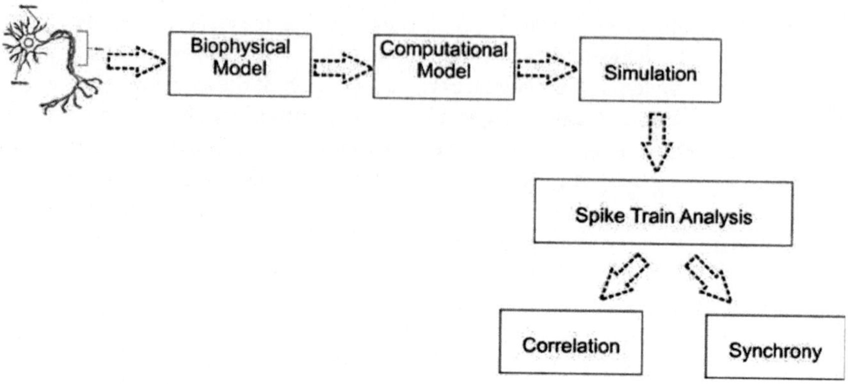

Fig. 2 Analysis of spike trains. The neuron is abstracted into a biophysical model and is converted into computational model. The spike train responses of neural simulations are then used to quantify the coordination between neurons using the estimates of correlation and synchrony

2.4 Measures of Spike Coordination

2.4.1 Cross-correlation

Paired association between neurons (say a reference neuron and a target neuron) is tested traditionally using cross-correlation measures. It shows the dependence of firing events of two neurons. The degree of correlation between neurons is quantified generally using various correlation measures. This study uses standard measures of correlation.

Given two time series,

$$X = \{X_1, X_2, X_3 \ldots, X_n\} \quad \text{and} \quad Y = \{Y_1, Y_2, Y_3, \ldots Y_n\}$$

The correlation coefficient r measures the linear "similarity" between them.

$$r = \frac{\sum_{i=1}^{n} \left((X_i - \bar{X})(Y_i - \bar{Y}) \right)}{\sqrt{\sum_{i=1}^{n} \left((X_i - \bar{X} \right)^2} \sqrt{\sum_{i=1}^{n} \left((Y_i - \bar{Y} \right)^2}} \tag{5}$$

Cross-correlation measures the degree of similarity between spike trains as a function of its time shift. The values range between 0 and 1 where 1 represents increased correlation strength.

2.4.2 Spike Train Synchrony

Since the standard measures of correlations work on a rate-based framework, neural coordination is also estimated by measuring relative timing in spiking events of two neurons.

This study used a spike pattern distance measure called Victor–Purpura distance to estimate the synchrony in the neural responses [10, 15].

Consider p and q as two spike trains with spike times $p_1, p_2..p_n$ and$q_1 q_2..q_{n'}$. A measure of spike train synchrony maps the pair to a positive real number. A high value corresponds to two closely related spike trains and is expressed as $s(p, q)$, and a low value corresponds to very similar spike trains. In the case of dissimilarity measures expressed as the distance between spike trains$d(p, q)$, a measure of dissimilarity is a metric, so that $d(p, q) = 0$ implies $p = q$.

The Victor–Purpura Distance Measure

It is a measure of the amount of change required to transform one spike train into the other [15]. It uses a cost-based metric, measured while changing one spike train into another using three basic operations: spike insertion, spike deletion and spike shift. Spikes can be added or deleted at a cost of one for each operation, and spikes can be moved at a cost of $q|\delta t|$ for a cost per time unit q and a temporal distance δt.

Population measures were calculated in two ways in this study: those which measure the similarity (or dissimilarity) between a pair of responses and those that measure the overall similarity of a set of responses and are calculated by averaging single-neuron measures.

3 Results and Discussion

3.1 Neural Correlation Affected by Synaptic Excitation and Inhibition

In order to estimate the effect of excitation and inhibition on coordination between neurons, two modeled neurons with differing synaptic balances were chosen and the paired correlation coefficient was estimated. In order to change the effect of input firing rate on correlation, MF inputs with 1 to 5 spikes through the four synapses were applied (Fig. 1b). Cross-correlograms were generated for paired spike trains.

It has been observed that cross-correlation was reduced as the neurons more differed in their firing frequency. The number of mossy fiber inputs was also changed from 2 to 4 and found that the correlation increased with the increase in mossy fiber inputs to the neuron (Fig. 3a). The granule neuron responses were also recorded in the presence of Golgi neuron inhibitory inputs and found that the correlation was high in the absence of inhibition and reduced with the introduction of more and more inhibition into the neurons (Fig. 4a). These correlation estimates suggested that the

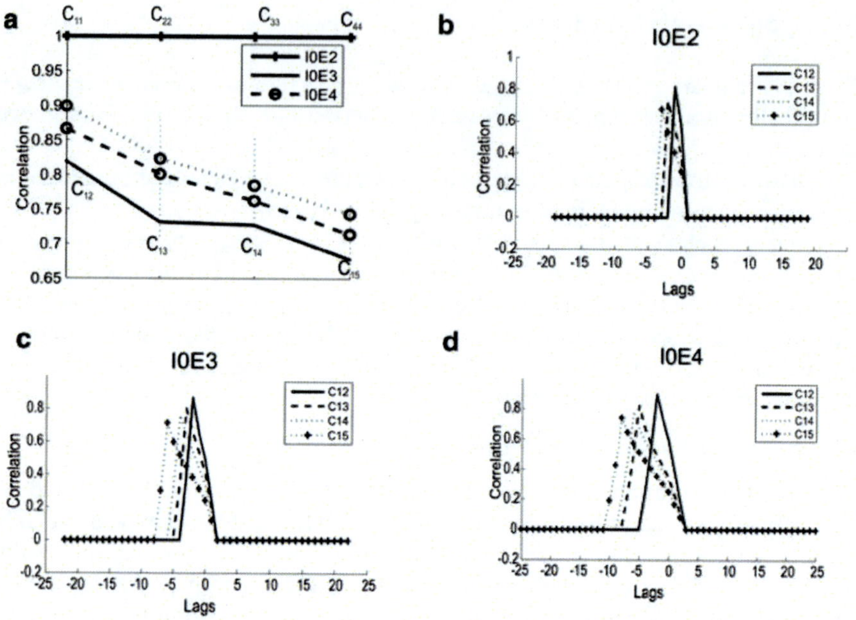

Fig. 3 MF excitation and granule neuron response correlation. **a** Cross-correlation reduced with the increase in firing rate and increased with the increase in mossy fiber inputs. **b–d** Cross-correlograms plotted for three different synaptic conditions: I0E2—inhibition 0 excitation 2, I0E3—inhibition 0 excitation 3, I0E4—inhibition 0 excitation 4

paired association between neurons is increased with more excitation and reduced with more inhibition into the neurons.

Correlation for time lags of ±25 ms, centered on the first spike train, was calculated, and the peaks in the correlogram indicated the highest value of correlation between the spike trains. The correlation measures are the maximum values from the cross-correlograms of the spikes. Peaks in the cross-correlograms represent the simultaneous firing events of target and reference neurons. The width of the correlogram increased with increase in excitation, and the correlation peaks were more shifted toward the left (Fig. 3b–d). The width of correlogram was reduced with increase in inhibition, and the correlation peaks were more shifted toward the center (Fig. 4b–d).

3.2 *Neural Synchrony Affected by Synaptic Excitation and Inhibition*

The synchronous firing of granule neurons under different synaptic conditions was estimated using a spike distance measure. MF input spikes of 1–5 were applied

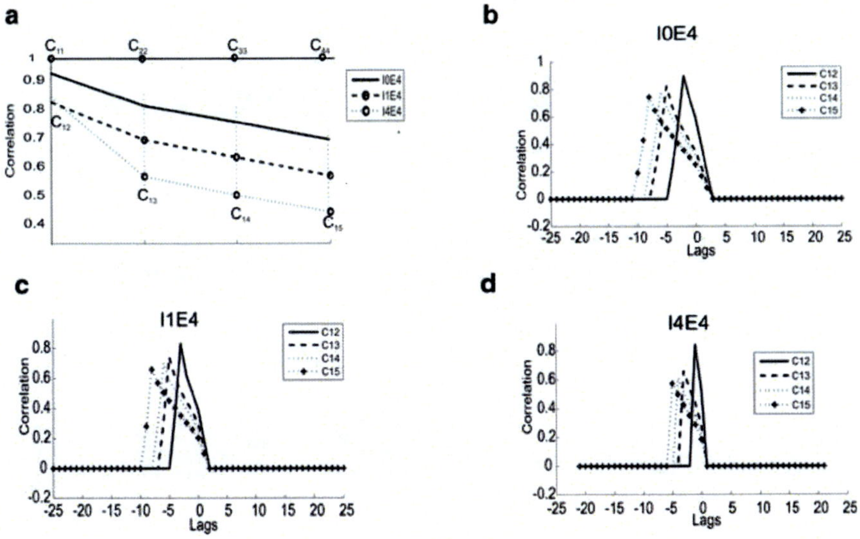

Fig. 4 Golgi neuron inhibition and granule neuron response correlation. **a** Cross-correlation reduced with the increase in firing rate and with the increase in inhibitory inputs. **b–d** Cross-correlograms plotted for three different synaptic conditions: I0E4—inhibition 0 excitation 4, I1E4—inhibition 1 excitation 4, I4E4—inhibition 4 excitation 4

to a granule neuron population, and the multi-neuronal responses were recorded. Average spike distance between all pairs of responses (D_{Mean}) in the response set was estimated (Fig. 5a). This measure was observed to be decreased with the increase in MF excitatory inputs to the neuron, suggesting the increased synchronous firing with increased excitation.

D_{Mean} was also estimated with and without inhibition into the neurons in the population (Fig. 5b). Reduced distance measure with the increase in inhibition suggested the increased synchrony between firing responses.

3.3 Firing Synchrony During Induced Plasticity States

Simulations were performed under induced plasticity states of the neurons [16]. Long-term potentiation (LTP) was introduced by increasing the intrinsic excitability of the neuron and by increasing the release probability of MF synapses above 0.41. Long-term depression (LTD) was induced by reducing the intrinsic excitability of the neuron and by decreasing the release probability of the MF synapses below 0.41. D_{Mean} was estimated for LTP, control and LTD conditions of the neuron, and it was observed that D_{Mean} was decreased during LTP from control while D_{Mean} was increased during LTD from control (Fig. 5c). These results suggested that the neurons

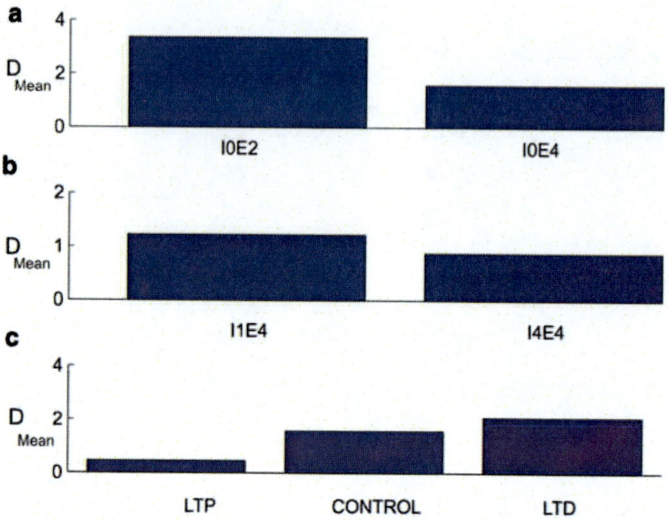

Fig. 5 Average spike distance as a measure of synchrony. **a**, **b** Average spike distance increased with increase in excitation (**a**) and inhibition (**b**). **c** Average spike distance was decreased in LTP compared to control and increased in LTD compared to control

fired more synchronously in their LTP state compared to the control state and the firing was more asynchronous during LTD state compared to control state.

4 Conclusion

The observation from the study suggested how the inhibitory and excitatory synaptic balance modulates the postsynaptic responses of a neuron. Also, this work proposed that coordinated activity is controlled in the neuronal level using such balances. Enhanced firing synchrony during plasticity conditions also reflects its influence on functional connectivity between neurons.

Due to the high dimensionality of the multi-neuronal data, population coding studies are less advanced compared to studies on information coding in single neurons [17]. Even though cross-correlation represents relationships that exist between the spiking activities of two neurons, this method assumes that the spike trains are stationary. The time varying adaptation of the spike trains in the time course of the stimulus is not addressed in the method, and it is assumed that the statistics of the spike trains are constant over time.

Spike distance-based measures are able to capture the temporal structure and synchrony in the responses and are a powerful tool in mutual information-based studies [18]. Spike correlations and spike synchrony may function as an additional information transfer mechanism in neurons, and further work is required in this direction to study the effect of such coordinated activities in the information encoding properties of the neurons.

Acknowledgements This work derives direction and ideas from the Chancellor of Amrita University, Sri Mata Amritanandamayi Devi. The authors of this work also acknowledge the support received from Computational Neuroscience Lab, School of Biotechnology, Amrita Vishwa Vidyapeetham.

References

1. Purves, D., Augustine, G.J., Fitzpatrick, D., Hall, W., LaMantia, A.-S., White, L.: Neurosciences. De Boeck Supérieur (2019)
2. Rieke, F., Warland, D.: Spikes: Exploring the Neural Code. MIT Press, Cambridge (1999)
3. Tateno, T., Robinson, H.P.C.: Rate coding and spike-time variability in cortical neurons with two types of threshold dynamics. J. Neurophysiol. **95**(4), 2650–2663 (2006)
4. Mita, A., Mushiake, H., Shima, K., Matsuzaka, Y., Tanji, J.: Interval time coding by neurons in the presupplementary and supplementary motor areas. Nat. Neurosci. **12**(4), 502 (2009)
5. Heil, P.: First-spike latency of auditory neurons revisited. Curr. Opin. Neurobiol. **14**(4), 461–467 (2004)
6. Woolley, S.M.N., Casseday, J.H.: Response properties of single neurons in the zebra finch auditory midbrain: response patterns, frequency coding, intensity coding, and spike latencies. J. Neurophysiol. **91**(1), 136–151 (2004)
7. Gerstein, G.L., Clark, W.A.: Simultaneous studies of firing patterns in several neurons. Science **143**(3612), 1325–1327 (1964)
8. Dettner, A., Münzberg, S., Tchumatchenko, T.: Temporal pairwise spike correlations fully capture single-neuron information. Nat. Commun. **7**, 13805 (2016)
9. Budnik, V., Salinas, P.C.: Wnt signaling during synaptic development and plasticity. Curr. Opin. Neurobiol. **21**(1), 151–159 (2011)
10. Kreuz, T.: Measures of spike train synchrony. Scholarpedia **6**(10), 11934 (2011)
11. Dabrowski, K.M., Castaño, D.J., Tartar, J.L.: Basic neuron model electrical equivalent circuit: an undergraduate laboratory exercise. J. Undergrad. Neurosci. Educ. **12**(1), A49 (2013)
12. Diwakar, S., Magistretti, J., Goldfarb, M., Naldi, G., D'Angelo, E.: Axonal Na+ channels ensure fast spike activation and back-propagation in cerebellar granule cells. J. Neurophysiol. **101**(2), 519–532 (2009)
13. Nieus, T., Sola, E., Mapelli, J., Saftenku, E., Rossi, P., D'angelo, E.: LTP regulates burst initiation and frequency at mossy fiber-granule cell synapses of rat cerebellum: experimental observations and theoretical predictions. J. Neurophysiol. **95**(2), 686–699 (2006)
14. Roggeri, L., Rivieccio, B., Rossi, P., D'Angelo, E.: Tactile stimulation evokes long-term synaptic plasticity in the granular layer of cerebellum. J. Neurosci. **28**(25), 6354–6359 (2008)
15. Houghton, C.: Population measures of spike train synchrony. Scholarpedia **8**(10), 30635 (2013)
16. Nair, M., Melethadathil, N., Nair, B., Diwakar, S.: Information processing via post-synaptic EPSP-spike complex and model-based predictions of induced changes during plasticity in cerebellar granular neuron. In: Proceedings of the 1st Amrita ACM-W Celebration on Women in Computing in India, p. 5 (2010)
17. Diwakar, S., Medini, C., Nair, M., Parasuram, H., Vijayan, A., Nair, B.: Computational neuroscience of timing, plasticity and function in cerebellum microcircuits. In: Computational Neurology and Psychiatry. Springer, Berlin, pp. 343–371 (2017)
18. Nair, M., Madhu, P., Mohan, V., Rajendran, A.G., Nair, B., Diwakar, S.: GPGPU implementation of information theoretic algorithms for the analysis of granular layer neurons. In: 2015 International Conference on Computing and Network Communications (CoCoNet), pp. 18–26 (2015)

A Novel Design of Dual Band Twin Tag H-Shaped Antenna for the Satellite Applications

Aylapogu Pramod Kumar, D. Venkatachari, J. Siddarthavarma, and Donga Madhu

Abstract In recent trends antennas are playing crucial role in the broadcasting system such as mobile phones, radar, TV and satellite communications. Because of low cost and easily fabrication, the tagged slot in the suggested antenna improvises the gain with less interference due to truncated ground plane. This article proposed a dual band twin tagged H-shaped antenna with the truncated ground plane. The primary function of this article is to develop high gain and bandwidth in order to support the multi band operation and to increase the speed. The monopole action of the antenna skeleton will boost the gain; the notches on the antenna promote the multiband action. The proposed design is to conduct in UWB interval from 2 to 12 GHz with multi band process. The antenna is encouraging the orbicular emission at more advanced the frequencies and can be suitable for the forthcoming 5G mobile utilizations. Moreover, the recommended structure offers fewer conflict at the receiver, because of standalone actions at various numerous frequencies. The antenna is developed with Rogers Substrate with a dielectric consistent of 4.4, fall tangent of

A. Pramod Kumar (✉)
Electronics and Communication Engineering, Vardhaman College of Engineering, Hybderabad, India
e-mail: pramodvce@gmail.com

D. Venkatachari · J. Siddarthavarma
Electronics and Communication Engineering, Lendi Institute of Engineering and Technology, Vizianagaram, India
e-mail: Venkatachari409@gmail.com

J. Siddarthavarma
e-mail: siddartha_varma@yahoo.co.in

Electronics and Communication Engineering, MVGR College of Engineering, Vizianagaram, India

D. Madhu
Electronics and Communicsation Engineering, VIGNAN'S Institute of Information Technology, Visakhapatnam, India
e-mail: madhudonga@gmail.com

© The Editor(s) (if applicable) and The Author(s), under exclusive license to Springer Nature Singapore Pte Ltd. 2021

A. P. Pandian et al. (eds.), *Proceedings of International Conference on Intelligent Computing, Information and Control Systems*, Advances in Intelligent Systems and Computing 1272, https://doi.org/10.1007/978-981-15-8443-5_66

0.0009 and possess a density of 1.6 mm. The skeleton is found to be minor in dimensions, comfortably build at low cost and can be unified into any compact wireless device. The skeleton can also be used for X-Band function for distinct applications of traffic regulation, climate anticipate and automobile speed disclosure systems. The simulation results are verified with the aid of Ansys HFSS software. This design obtained the return losses are −40.60 dB and −31.14 dB at respective frequencies 3.9 and 8.6 GHz, and the gain is 2 dB.

Keywords 5G mobile communications · Twin tagged H- shaped · High gain · High directivity · Band width · Dual band · Satellite applications

1 Introduction

Big challenging issues are running in front of every academician and industrial researcher, because of the rising technology, the communication systems lead to rapid growth in the development of wireless devices. Various types of electronic gadgets are used for our communication. These types of gadgets are developed with antennas; particularly the Ultra-wide band antennas are playing a key part in the WLAN and WiMAX applications [1]. Band notch UBW consists of U-shaped and C- shaped slots [2–4] to minimize the potential interference and to get the stable gain, better impedance matching by colliding a pair of T-shaped stubs. The bandwidth level and radiation emission can be enhanced [5]. The monopole action of the antenna skeleton will enlarge the gain; it covers the frequency from 2.45 to 11 GHz [6] especially by adding proper notches on the radiating patch like inverted T-shaped stub. Wider impedance bandwidth can be achieved and circular polarization fields with the aid of a truncated ground plane at higher frequencies [7–9]. The modified ground plane structure will enhance the radio band at the same time minimize the interference with less cross polarization and excellent radiation pattern [10]. In extension to stepped rectangular tag added in the ground plane at 45° to upgrade the port isolation [11–12] and the antenna performance characteristics such as band rejections and risen bandwidth obtained by the notches [13] (L-type slot) introduced in the emission patch. The evaluation of each and every parameter can be reduce easily at the same time the size of the antenna, cost and fabrication process will be reduced for better gain and efficiency [14–16] however UWB technology has observable factor including high data transmission rate, low complexity and low power transmission levels. Usually these types of antennas are used to minimize the multipath fading and enhance the channel capacity [17] by applying tuning procedure to obtain the dual frequency band for the applications WLAN [18–19] and WiMAX. Particularly higher gains are used at long distance communication. In this regard various types of applications such as wide band, space applications, X-band and radar applications [20–21]. The proposed structure also is expected to lower the interference levels at the receiver because of the independent operation of frequencies because of the usage of twin tagged H-Shaped stub in the design. The parametric analysis of the antenna structure is performed to improve the operational characteristics of the proposed structure.

The recommended antenna design is simulated with the aid of High Frequency Structure Simulator (HFSS). The obtained simulation issues are observed and verified. The following Sect. 2 describes the antenna design parametric and Sect. 3 explains the simulation result and measurement analysis. Section 4 discusses the conclusion of the proposed structure.

2 Design of Dual Band H-Shaped Antenna

The given Fig. 1 represents the schematic diagram of the proposed dual band twin tag H-shaped antenna structure. Basically the antenna structure is fabricated with dielectric constant of 4.4, loss tangent 0.0009 and having a thickness 0.8 mm. To obtain the best characteristics impedance 50 Ω, the dimensions of the transmission line is selected as 2 mm. The suggested antenna consists of twin tagged slots with H-shaped. To achieve the consistent input impedances, radiation pattern and the bandwidth, the modified truncated ground is more helpful. The strip cut symptoms of the antenna is obtained by etching the H-fashioned slot on the elliptical tag. Strip cut symptoms can be restrained by proper tuning of H-shaped slot criterion. The structured parameters are chosen SW = 32 mm, SL = 26 mm, PL = 13 mm, PW = 11.5 mm, Fl = 14.5 mm Wf = 2 mm, HL = 7 mm, Tg = 8 mm, Tg1 = 9.7 mm, Tg2 = 4.4 mm. The edge cuts were introduced in the design to resonate the structure with dual band operation. The stubs in the slot will improve the gain of the antenna. The wavelength of the slot can be calculated as.

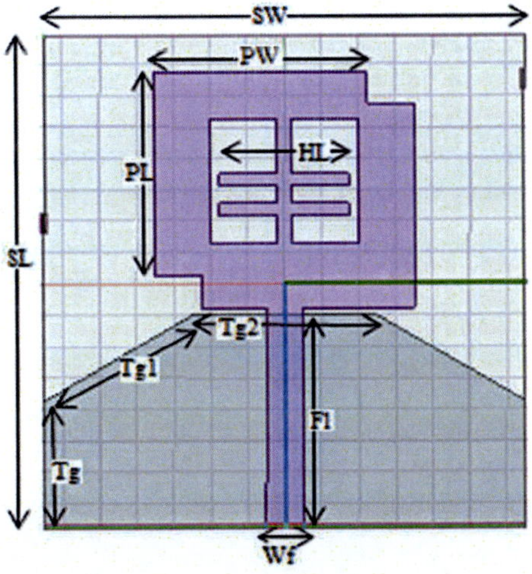

Fig. 1 Twin tagged H-shaped proposed antenna

$$\lambda_g = \frac{\lambda_0}{\sqrt{\varepsilon_{\text{eff}}}} \quad (1)$$

where

λ_g and λ_0 guided wavelengths.
ε_{eff} effective substrate dielectric relative constant.

The patch length is derived from the following equation

$$L = L_{\text{eff}} - 2\Delta L \quad (2)$$

where

$$L_{\text{eff}} = \frac{c}{2f_o\sqrt{\varepsilon_{\text{reff}}}} \quad (3)$$

L_{eff} the effective length of the patch antenna.
ΔL fringing field effect

$$\Delta L = 0.412 * h\left(\frac{\varepsilon_{\text{reff}} + 0.3}{\varepsilon_{\text{reff}} - 0.253}\right)\left(\frac{\frac{w}{h} + 0.26}{\frac{w}{h} + 0.8}\right) \quad (4)$$

where Width of the strip

$$W = \frac{c}{2f_o\sqrt{\frac{\varepsilon_r+1}{2}}} \quad (5)$$

where

C velocity of the light (3×10^8 m/s)
f_o Operating Frequency
ε_r Dielectric constant (4.4).

The above Eq. (4) is stated that the unwanted length of the fringing field effect.

The approximated dimensions of the ground and substrate values are mentioned in below equations respectively.

$$L_g = L + 6h \quad (6)$$

$$W_g = W + 6h \quad (7)$$

Here

h substrate height
L Length of the patch

Table 1 Proposed antenna geometry dimensions

Parameter	Value (mm)	Parameter	Value (mm)
SL	26	Tg	8
SW	32	Tg1	9.7
PL	13	Tg2	4.4
PW	11.5	Fl	14.5
HL	7	Wf	2

W Width of the patch.

The antenna structure design parameters are shown in Table 1.

3 Simulation Results and Measurement Analysis

In this section, the proposed antenna describes the brief explanation about simulation results and its measurement analysis. The suggested antenna operating radio bands are with dual frequency 3.9 and 8.6 GHz. Here the measurement of simulation results and its analysis are studied.

3.1 S-Parameter Loss

It is one of the crucial parameters in the antenna design theory. Generally power loss is occurred by conflict of impedance in the communication lines. It is in decibels (dB). The suggested antenna has a reappear loss of −40.60 dB and −31.14 dB at corresponding operating frequencies 3.9 GHz and 8.6 GHz respectively. The schematic waveform is shown in the given Fig. 2.

3.2 Voltage Standing Wave Ratio (VSWR)

The term Voltage standing wave ratio is defined as the ratio of conflict between the generated signal and re bounce signal due to the conflict of impedance in the communication lines. The value of VSWR should be in the range of 1–2. If it is closer to 1 that shows the best results. In this article the VSWR value is 1.05 at 8.6 GHz frequency. The graphical illustration is displayed in Fig. 3.

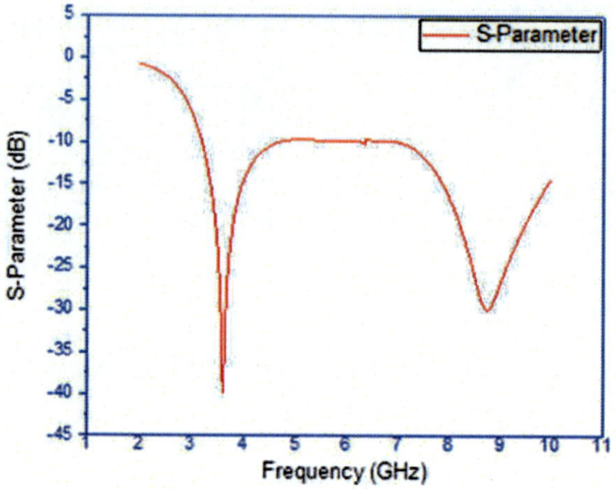

Fig. 2 Characteristics of S-parameter

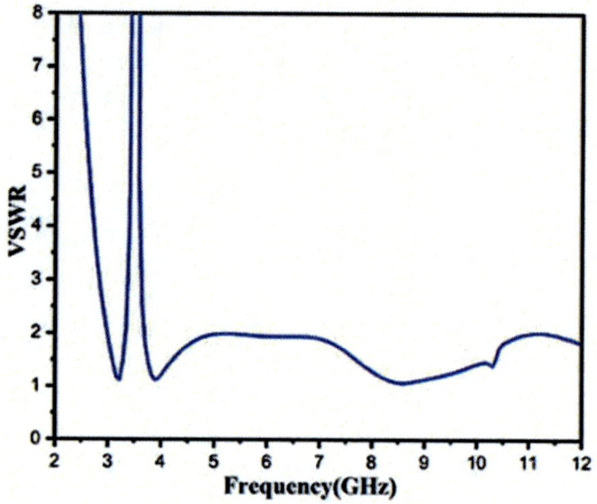

Fig. 3 VSWR graphical representation

3.3 Gain

The gain decides how much of power is delivered in the required direction. The power emitted by the proposed antenna in different paths is measured at different planes, The E-plane exists at 0 deg and H-plane occurred at 90 deg with the best results. The radiation characteristics for the given frequencies are shown in the Fig. 4a at 3.9 GHz and Fig. 4b at 8.6 GHz for both Phi = 0 and 90 deg.

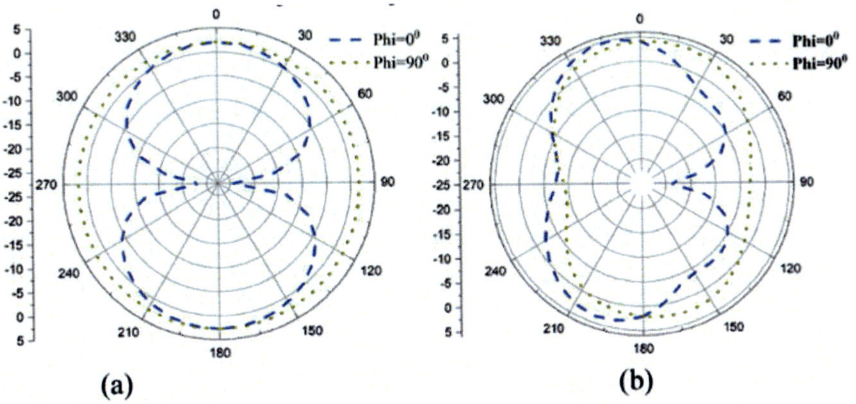

Fig. 4 **a, b** Radiation pattern characteristics for both Phi = 0 and 90 deg at 3.9 and 8.6 GHz

3.4 Directivity (D)

The directivity is one of the fundamental parameters. It is clearly stated that how much energy concentrates in a required direction. In this regard the proposed antenna offers the circular diffusion at greater frequency. The same thing can be detected in the return loss waveform. The Axial ratio at 8.6 GHz is shown in the Fig. 5.The suggested structure gain versus frequency characteristics plot is represented in the given Fig. 6.

Fig. 5 Axial ratio at 8.6 GHz

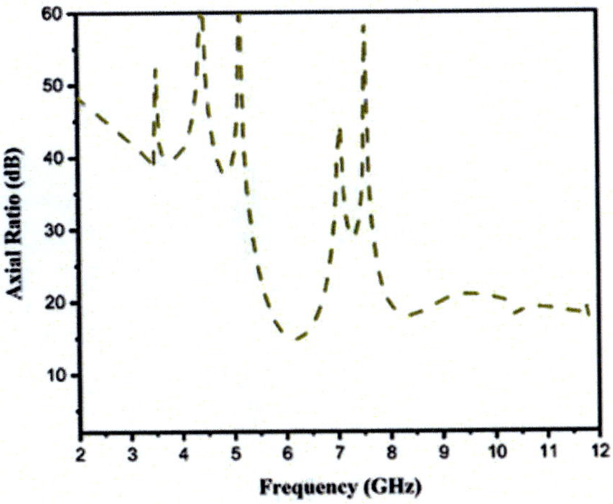

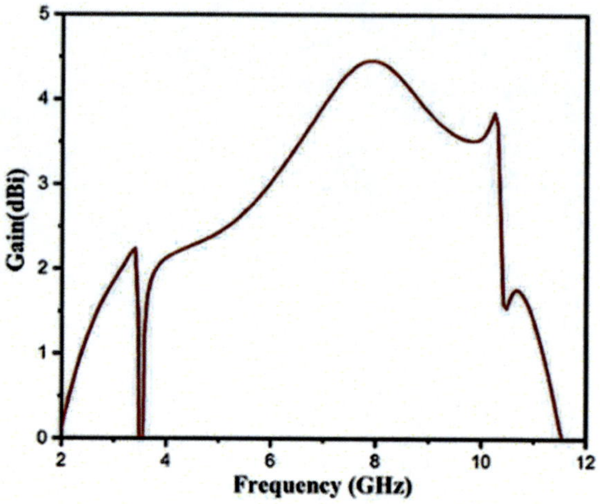

Fig. 6 Gain versus frequency plot

3.5 3-D Radiation Pattern

The 3-Dimensional graphical view of the transmission arrangement of the suggested antenna is illustrated in Fig. 7.The comparison among different parameter in the presented antenna work with the existing antenna models are shown in Table 2.

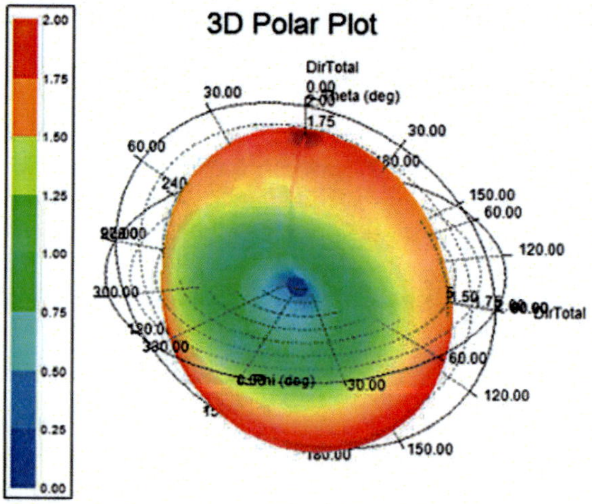

Fig. 7 3-D radiation polar plot

Table 2 Performance evaluation between different designs

Parameter	Proposed work	Ref. [1]	Ref. [2]	Ref. [3]	Ref. [19]
Return loss(S_{11})	−34.60	−28.4	−18.87	−30.51	−30.73
VSWR	1.05	1.5	3.8	1.6	1.063
Gain in dB	2	2	1.8	2	1.5
Resonant frequency (GHz)	8.86	5.5	4.4	8.4	3.69

4 Conclusion

This article proposes a dual band twin tagged H-shaped antenna with the truncated ground plane. The notches on the antenna promote the multiband action. The suggested design is to operate, duration from 2 to 12 GHz with the dual band action. The propound antenna is encouraging the orbicular emission at more advanced frequencies and can be suitable for the forthcoming 5G mobile utilizations. Moreover, the recommended structure offers less conflict at the receiver. So, this type of antenna structure is more applicable for far distance communications. The proposed design is simulated by using the Ansys HFSS software and the results are analyzed to study the performance of the antenna structure and presented. The return losses are −40.60 dB and −31.14 dB at respective frequencies 3.9 and 8.6 GHz, VSWR is 1.05 and the gain is 2 dB. With the help of this work further the designs can be implemented at higher frequencies for the millimeter wave and drone type radar applications.

References

1. Jiang, W., Che, W.: A novel UWB antenna with dual notched bands for WiMAX and WLAN applications. IEEE Antennas Wirel. Propag. Lett. **11** (2012)
2. Nagre, S.S., Shirsta, A.S.: Design of Compact CPW-fed printed monopole UWB antenna for band notched applications. In: International Conference on Automatic Control and Dynamic Optimization Techniques (ICACDOT), Pune, pp. 394–397 (2016)
3. Azarmanesh, M., Soltani, S., Lotfi, P.: Design of an ultra-wideband monopole antenna with WiMAX, C and wireless local area network band Notches. IET Microwaves Antenna Propag. **5**(6), 728–733 (2011)
4. Mei, Z., Fu, G., Gong, J.-G., Li, Q., Wang, J.: Printed monopole UWB antenna with dual-band notched characteristics. In: Proceedings of IEEE international conference on ultra-wideband (ICUWB 2010), pp. 1–3 (2010)
5. Abdullahvand, M., Dadashzadeh, G., Mostafa, D.: Compact dual-band notched printed monopole antenna for UWB application. IEEE Antennas Wirel. Propag. Lett. **9**, 1148–1151 (2010)
6. Jingfang Wu, Jiusheng Li. "Compact ultra-wideband antenna with 3.5/5.5GHz dual band-notched Characteristics" IEEE International Conference 2013, pp 446–450
7. Ojaroudi, N., Ojaroudi, M.: Novel design of dual band-notched monopole antenna with bandwidth enhancement for UWB applications. IEEE Antennas Wirel. Propag. Lett. **12**, 698–701 (2013)

8. Zhou, Y.-D., Jiao, Y.-C., Weng, Z.B., Ni, T.: A novel single-fed wide dual-band circularly polarized dielectric resonator antenna. IEEE Antennas Wirel. Propag. Lett. **10**, 1–4 (2015)
9. Vendik, I.B., Rusakov, A., Kanjansit, K., Hong, J., Filonov, D.: Ultra-wideband (UWB) planar antenna with single-, dual-, and tripple band-notched characteristics based on electric ring resonator. IEEE Antennas Wirel. Propag. Lett. **1**, 1–4 (2017)
10. Li, M., Wang, R., Yasir, J.M., Jiang, L.: A miniaturized dual-band dual-polarized band-notched slot antenna array with high isolation for base station applications. IEEE Trans. Antennas Propag. **1**, 1–9 (2019)
11. Huang, H., Liu, Y., Zhang, S., Gong, S.: Uni planar ultra-wideband polarization diversity antenna with dual band-notched characteristics. IEEE Antennas Wirel. Propag. Lett. **13**, 1745–1748 (2014)
12. Feng, B., Li, L., Cheng, J.-C., Sim, C.-Y.-D.: A dual-band dual-polarized stacked microstrip antenna with high-isolation and band-notch charactersitcs for 5G microcell communications. IEEE Trans. Antennas Propag. **1**, 1–11 (2019)
13. Jyothi Singh, K.M., Mishra, R.: A circular micro strip patch antenna with dual band notch characteristics for UWB applications. In: International Conference on Power Energy, Environment and Intelligent Control (PEEIC),Greater Noida, UP, pp. 153–156, April 2018
14. Shrikanth Reddy, G, Kamma, A., Mishra, S.K., Mukherjee, J.: Compact bluetooth/UWB dual-band planar antenna with quadruple band-notch characteristics. IEEE Antennas Wirel. Propag. Lett. **10**, 1–4 (2014)
15. Li, W.T., Hei, Y.Q., Feng, W., Shi, X.W.: Planar antenna for 3G/bluetooth/WiMAX and UWB applications with dual band-notched characteristics. IEEE Antennas Wirel. Propag. Lett. **11**, 61–64 (2012
16. Soliman, M.S., Al-Dwairi, M.O., Hendi, A.Y.: A compact ultra-wideband patch antenna with dual band-notch performance for WiMAX/WLAN Services. In: IEEE Jordan International Joint Conference on Electrical Engineering and Information Technology (JEEIT), pp 831–834 (2019)
17. Zhang, Y., Zhang, Y.H., Fan, Y., Li, D.: Band-notched filtering crossed dipole antenna without extra circuit. In: IEEE International Conference, AP-S-2019, pp 1135–1136 (2019)
18. Li, B., Hong, J., Wang, B.: Switched band-notched UWB/dual-band WLAN slot antenna with inverted S-shaped slots. IEEE Antennas Wirel. Propag. Lett. **11**, 572–575 (2012)
19. Boutejdar, A., Salamin, M.A., Bennani, S.D., El Hani, S.: Design of compact monopole antenna using double UDMS resonators for WLAN, LTE and WiMAX applications. Telkomnika **15**(4), 1693–1700 (2017)
20. Iqbal, A., Sani, F.R., Ullah, Z., Khattak, M.I., Khattak, M.A., Saqib, N., Masood, H.: Comparative study of micro strip patch antenna for X-band using micro strip line feed and coaxial feed. In: IEEE International Conference Publications (2018)
21. Maulana, Y.Y., Wahyu, Y., Oktafiani, F., Saputra, Y.P., Setiawan, A.: Rectangular patch antenna array for radar applications. Telkomnika **14**(4), 1345–1350 (2016)

Analysis of Sleep Apnea Considering Biosignals from Peripheral Capillary Oxygen Saturation Level and Electrocardiogram Data

Senthilnathan Ramasubbu, R. Vijaya Lakshmi, S. Lakshmi Priya, R. Prakash, A. Balaji Ganesh, A. Lakshmi Sangeetha, and Senthil Kumar Thangavel

Abstract Sleep apnea is a disorder, where the obstructed upper airway affects breathing during sleep time. This ailment causes disruptions in sleep, leading \sleep deprivation. Age has a great impact on sleep deprivation due to apnea. Furtherance, it has been observed that as age progresses, severity of sleep apnea increases. This apnea minute's increases, as age progresses, found in literature, is proved with the analysis performed in this paper. Apnea detection has been done using combined data of ECG and SpO_2 for the users at different ages, and tested with various machine learning algorithms like K-nearest neighbor, Logistic regression, Random forests, and Ada-boost. From the results, it is observed that old age people are affected from sleep apnea more than 100% than youngsters. Also, the analysis has also proved that the sample time and combination of SpO_2 and ECG influence the accuracy.

Keywords Sleep apnea · Old age · ECG · SpO_2 · Machine learning

1 Introduction

Sleep apnea is a disruptive serious sleep disorder, in which breathing repeatedly, stops and starts. The 10–20 s pauses in breathing may occur more than 30 times per hour. In its most common form, Obstructive Sleep Apnea (OSA), a partial or

S. Ramasubbu · S. K. Thangavel
Department of Computer Science and Engineering, Amrita School of Engineering, Amrita Vishwa Vidyapeetham, Coimbatore, India
e-mail: r_senthilnathan@cb.students.amrita.edu

S. K. Thangavel
e-mail: t_senthilkumar@cb.amrita.edu

R. V. Lakshmi · S. L. Priya · R. Prakash · A. B. Ganesh (✉) · A. L. Sangeetha
Electronic System Design Laboratory, Department of TIFAC-CORE, Department of Electronics & Instrumentation, Velammal Engineering College, Chennai, India
e-mail: abganesh@velammal.edu.in

© The Editor(s) (if applicable) and The Author(s), under exclusive license to Springer Nature Singapore Pte Ltd. 2021

A. P. Pandian et al. (eds.), *Proceedings of International Conference on Intelligent Computing, Information and Control Systems*, Advances in Intelligent Systems and Computing 1272, https://doi.org/10.1007/978-981-15-8443-5_67

complete collapse of the upper airway occurs, ensued by relaxation of the muscles controlling the soft palate and tongue. Denoted.

There are 3 types of apnea which affect individuals: Central Sleep Apnea (CSA), Obstructive Sleep Apnea (OSA) and Mixed Sleep Apnea. During CSA, respiratory drive is absent or inhibited, as noted, with OSA collapse of the upper airway is responsible for disrupted respiration. MSA is manifested by effects of both OSA and CSA.

Polysomnography (PSG), a type of sleep study, is a test used to diagnose sleep disorders. It records brain waves, the oxygen level in the blood, heart rate and breathing, as well as movements of the eye and leg, during the study. PSG tests are performed at a sleep disorders unit within a hospital, or at a sleep center. PSG tests entail the discomfort of being attached to numerous wires, and sleeping in an unfamiliar environment.

A device consisting of a Personal Activity Assisting & Reminding (PAAR) band and a bio cradle deployed at home, to detect apnea, is discussed in [1]. Peripheral oxygen saturation (SpO_2) is an estimate of the oxygen saturation level in humans, measured with a pulse oximeter device. SpO_2 measurements and apnea diagnosis can be conducted in a home environment. Detection of sleep apnea has been simplified by application of machine learning algorithms—Linear regression and AdaBoost in [2].

The amount of oxygen measured in the human blood specified as peripheral capillary oxygen saturation and it is the abbreviation for SpO_2. The detailed explanation of this could be the percentile of hemoglobin in human blood cells with oxygen to the sum of all the hemoglobin blood cells present in the complete human body. SpO_2 signal is the signal generated by any Oximeter, a device to monitor the SpO_2 level in blood. SpO_2 signal is commonly used for detection apnea [3]. Random Forest classifiers have been used for apnea detection SpO_2 signal. Electrocardiogram (ECG) is a routine test that is used to look at the electrical activity of the heartbeat. Normal ECG Signal has list of characteristics including Heart Rate (number of times a person's heart beats per minute), PR Interval, QRS Interval, QT Interval and ST Segment. The period of time from the onset of the P wave to the beginning of the QRS complex is termed the PR interval. This interval represents the time between the onset of atrial depolarization and the onset of ventricular depolarization. In ECG, the P wave represents initiating pressure in atrial, which pushes the blood to ventricles. It is the depolarization of human heart. The electrical signal generated in heart passes to the ventricles, and creates ventricular depolarization. This captured as QRS complex wave in ECG. This wave starts as the ventricular starts to contract. It is not necessary for every QRS wave to have each Q, R along with S wave. It may miss one of them. After the P wave, the following downward is Q wave. After the Q wave, the upward is R wave. Next to the R wave, the following downward wave is S wave. After the S wave, T wave captured in ECG. At times, there could be an additional wave, known as U wave comes next to the T wave. This corresponds to the contraction of both the ventricles by ventricular muscles, initiating the depolarization in human heart. The normal duration (interval) of the QRS complex is QRS Interval. The time taken for depolarization, repolarization and along with ventricular activity represented as QT interval. The QT interval calculated from the ECG graph, by measuring starting of QRS complex wave to the point where T wave ends. The ventricular depolarization, repolarization time duration represented as ST segment. ECG signals also used for apnea detection. The R wave is providing the electrical signal propagation status in walls of ventricular system of human heart. The maximum amplitude in the R wave known as "R peak amplitude," or just "R peak". The respiratory-induced amplitude modulation built from height of (amplitude) time based series of R peaks along with timings of the R peaks from ECG signal as in [4].

Overnight (nocturnal) home oximetry uses a small device to monitor and record oxygen levels in blood. Nocturnal pulse oximetry has also been used for sleep apnea detection [5]. Quantitative characteristics of the blood oxygen saturation level have been extracted and used for detection. The classification algorithm is trained, and performance was evaluated using leave-one-patient-out cross-validation. The leave-one-patient-out cross-validation used to evaluate the performance for the trained classification algorithm. The Machine Learning (ML) algorithm process the selected features from the classification algorithms and determines the quantification of Apnea Hypopnea Index (AHI), a measurable indication of apnea severity as documented in [6].

The features of wavelet derived from ECG signal, using both discrete and continuous wavelet transforms [7]. In [8], the efficacy of AdaBoost algorithm in the detection of apnea was discussed. Airflow signals are taken as inputs and feature extraction and selection is done. Spectral (frequency based features) and non-spectral features are extracted from the input signal. The waves can be selected based on the windowing method. The RR Intervals are varying, and window level for Windowing method are determined based on this. For the effective monitoring of heart conditions in patients requires ECG timing intervals as in [9].

The different kind of classifiers process the statistical features reviewed and extracted as in [10]. Random Forest algorithm applied for the Features extraction, the complete data separated in to two data sets, one is for training, and the other data set is for testing. The sleep apnea detection efficiency improved with; both SpO_2 and ECG data as used in [11]. Classifiers were combined to improve accuracy.

Geriatric patients tend to be more susceptible to medical complications, which get aggravated by mental agonies and depressions. These deleterious symptoms affect sleep quality as age progresses, and enjoyment of a restful, sound sleep becomes an intractable challenge. Nowadays many user friendly apps have come up to detect apnea. This paper aims to detect apnea and infer that apnea severity progresses with age and old patients are more likely to experience sleep apnea.

2 Materials and Methods

The graphs below show the variations in SpO_2 (Fig. 1) and ECG (Fig. 2) for middle age and old age patients.

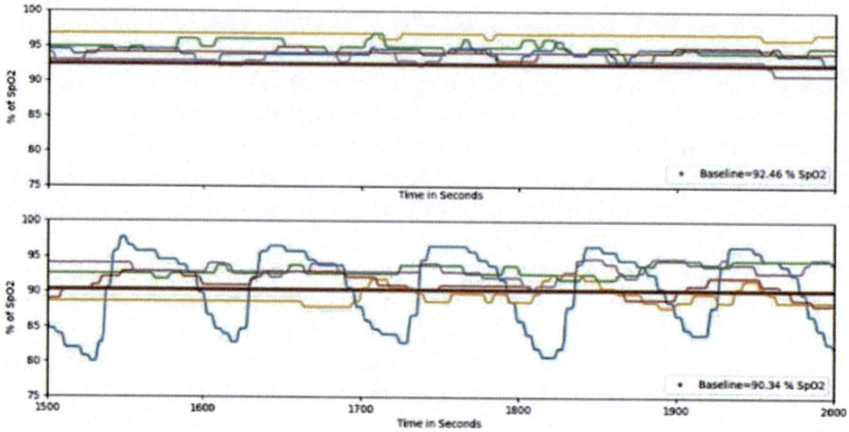

Fig. 1 SpO_2 signal variations (middle vs. old)

Fig. 2 ECG signal variations (middle vs. old)

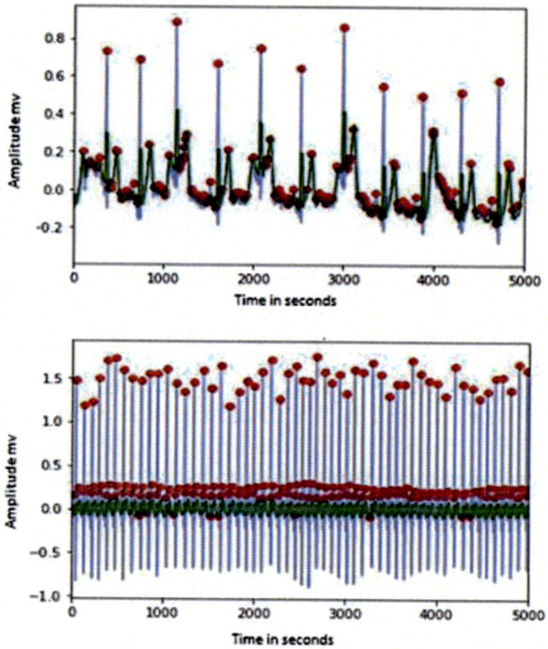

The sleep apnea detection process involves a series of steps, as represented in Fig. 3

Fig. 3 Block diagram of the process to detect sleep apnea

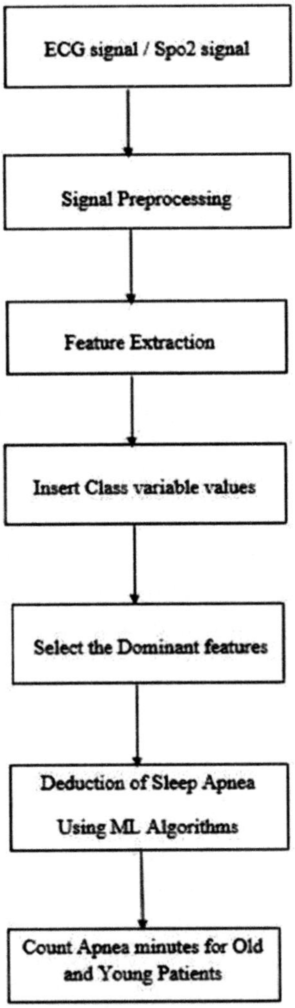

2.1 Data Used

The overnight SpO_2 and ECG signal data of 9 middle age (30–50) years and 9 old age (50–70) years, patients are taken from the Physiobank UCD sleep apnea database. Normally, SpO_2 data is above 60% even when apnea is present. Hence, values below 60% are removed. ECG and SpO_2 data are divided into periods of 1-min samples.

2.2 Feature Extraction

Every signal analysis involves an important step called as feature extraction from medical data as discussed in [12]. The signal is approximated into a set of features which is given to machine learning algorithms. The features related to ECG were derived from the R-R intervals. The time measured between two consecutive R waves from the QRS wave defined as RR interval in ECG signal. It calculates based on heartbeat annotations.

The features extracted from ECG signal were:

(1) Square root of the average of the sum of the squares of differences between adjacent RR intervals. The time difference across consecutive RR interval vales squared and then added. The average calculated for that value. Square root of that value considered as RMSSD.
(2) The average value of RR interval is subtracted from a period of RR interval values denoted as (MAD) Mean absolute deviation values.
(3) *Mean* of the time measured between two consecutive R waves in ECG
(4) The middle of the sorted consecutive RR intervals is calculated and denoted as *Median* of RR interval.
(5) The difference of 25th percentile and 75th percentile from RR Interval distribution denoted as (IQR) Inter-quartile range.
(6) Standard deviation applied to the time variances measured between two consecutive R Waves intervals (RR Intervals), also noted as SDSD

The features extracted from SpO_2 signal are:

(1) *Moments of the signal*

- Mean—mn
- Variance—var
- Skew—sk
- Kurtosis—kur
- Minimum—min

(2) *Power Spectrum Density related features*—Using Welch method

- Moments of PSD—Mean-mnpxx
- Variance—varpxx
- Skew—skpxx
- Kurtosis—kurpxx

(3) *Wavelet based features*—Using HAAR wavelet-level 9 Coefficients. A Type of Wavelet Formed from a Series of Resized Square-Shaped Functions is Haar Wavelet. It is Used Here to Capture the Rapidly Changing SpO_2 Values

- Moments-Mean—mnwav
- Variance—varwav
- Skew—skwav

- Kurtosis—kurwav
- Maximum amplitude—maxwav
- Energy—enwav
- Entropy—entropywav

(4) *ODI3*—Count of number of seconds the SpO_2 value was below 3% of the baseline of the signal-counter

2.3 Feature Selection

It is not necessary that all features contribute to the classification in a good way. Hence, it is required to identify the set of dominant features that help in the classification. Values are assigned for the class variables and given to feature selection algorithms in [13].

Here 4 algorithms are used, and the common dominant features are selected:

(1) *Using trees*—This technique implements a meta estimator designed and determined by various decision trees that could fit on variety of samples, subset of samples from many datasets. This technique also controls the over-fitting and improves the accuracy of prediction. It gives the importance of each feature for the given input accurately.
(2) *Recursive Feature Elimination (RFE) method*—This algorithm builds a model with the remaining attributes after eliminating many attributes by recursive removal method. During the process, it ranks, prioritizes the features with accuracy metric applied on the importance of the features.
(3) *Embedded technique*—This technique of feature selection uses Lasso regularization. If the feature is not, lasso penalizes it's coefficient and make it 0. Non-zero features are selected.
(4) *Using backward elimination*—In this approach, performance of the model is checked and the worst performing features are iteratively removed, one by one. The performance metric used to evaluate feature performance is p value.

2.4 Classification

The set of dominant features of ECG and SpO_2 are combined and given to the classifiers. Actual features are also given and the performance is compared. The feature set was split into training (70%) and test set (30%). It is ensured that both the sets contained apnea and non-apnea minutes data so that the classifier performs well. This is given to 4 classifiers and the performances were compared:

(1) *K-nearest neighbor classifier*—Here, k-closest neighbors (using distance formula) are picked and analyzed, where most of these neighbors are classified in. The new item is classified there.

(2) *Random forests*—generates decision trees by choosing random samples from the dataset, and then for each tree the prediction is calculated. After this based on higher number of votes, it chooses the best solution. Also provides a significant indicator of importance of the selected features.
(3) *Logistic Regression* (LR)—The probability of definite dependent variables predicted with LR. The logistic regression, calculates probability of Y, where $(Y = 1)$ as a function of X with the dependent variable having data values as 1 for "yes" and 0 for "No", a binary variable.
(4) *AdaBoost (AB) classifier*—AB begins applying a classifier on the actual data set. After that, it starts to fit additional variations of the classifier on the same data set. All those classifier instance weights with lower fitting rate are adjusted in a way that the next cycle classifiers tests the fitting for difficult use cases.

3 Results and Discussion

First, only SpO_2 signal features or ECG features set are given for detecting apnea. It is observed that as the number of datasets increases the ECG signal and SpO_2 give better predictions, which are shown in Table 1. It is presumed that combination of ECG and SpO_2 will give better results than individual biosignals. It is subsequently proved that more than 90% better results would be possible when two biosignals are given as input rather than single into a learning algorithm. However, sampling time period improves significantly over all performance of algorithm. When the sample of time period 360 min considered the accuracy is improved in all cases, individual signal, with or without feature selection and also combination of two bio signals. The results are shown in Table 2. Classification is also done by giving the complete feature set considered and the selected dominant feature set.

It is also seen that each algorithm gives a different efficiency for the same input. Using this, the best algorithm for this application can be selected. It is seen that KNN and random forest algorithms give better accuracy.

The apnea moments, detected by the classifiers, for each patient, were counted and found that old people (50–70) years had more apnea minutes i.e., more than double the amount of time as compared to middle age patients (30 to 50) years. It is inferred that old people are more susceptible to sleep apnea as compared to the young. It is also observed that middle aged patients recovered faster from apnea phase as compared to older patients. Older patients took more than double the amount of time to recover from an apnea episode.

4 Conclusion

From the above implementation it is clear that ECG and SpO_2 data immensely help in detection of apnea. It is inferred that if the dataset has maximum minutes ECG

Table 1 Classification results for 120 min

Age	Signal used	Without feature selection				With feature selection				Apneic minutes (out of 120 min)	Recovery time taken (in min)
		KNN	RF	LR	AB	KNN	RF	LR	AB		
Young age	ECG	66.6	66.6	83.3	33.3	83.3	66.6	83.3	50	8	10
	SpO_2	88.8	83.3	82.8	61.1	88.8	84.5	83.3	66.6		
	ECG + SpO_2	93.1	90.0	93.46	90.45	94.4	92.7	94.47	94.52		
Old age	ECG	49.01	48.05	49.01	50	50.01	50	50.11	66.6	17	19
	SpO_2	89.3	87.3	88.5	88.02	89.66	88.02	89.01	89.4		
	ECG + SpO_2	91.02	91.02	89.52	89.02	91.7	92.26	90.07	90.5		

Table 2 Classification results for 360 min

Age	Signal used	Without feature selection				With feature selection				Apneic minutes. (out of 360 min)	Recovery time taken (in min)
		KNN	RF	LR	AB	KNN	RF	LR	AB		
Young age	ECG	94.76	95.01	95.51	94.26	97.01	95.76	96.75	95.01	16	10
	SpO_2	95.76	95.01	95.76	94.26	95.6	95.01	95.5	95.01		
	ECG + SpO_2	95.26	96.76	96	94.51	96.01	96.25	96.75	96.50		
Old age	ECG	91.58	90.34	91.58	89.6	90.59	90.8	92.32	91.83	35	19
	SpO_2	90.27	89.52	90.27	89.27	94.01	93.51	92.76	94.76		
	ECG + SpO_2	90.27	91.52	89.27	88.77	94.76	94.76	93.76	95.76		

signal or SpO_2 signal can give better results. But, if the dataset is minimum minutes, then combination gives good results in predicting sleep apnea. It can also be inferred that as the age increases apnea severity also increases.

In order to increase the detection efficiency of sleep apnea, body position and airflow data can also be added. Future enhancements will be developed using cloud environment for deploying effective monitoring analysis and detection of apnea, which will enable quick and easy way for Doctors to get connected with patients and use their data for real time as discussed in [14].

Acknowledgements The authors respectfully acknowledge Department of Science and Technology (DST), India for sanctioning the project "Wireless body sensor network supported Sleep Disorder Analysis system", DST No. SEED/TIDE/2018/20/G, dated 08th May 2018 to Velammal Engineering College, Chennai. The study utilizes the resources from the sanctioned project.

References

1. Jeong, Y.J., Kang, S.J.: Wearable sleep care kit: analysis and prevention of sleep apnea symptoms in real time. IEEE Access, May 2019
2. Gutiérrez-Tobal, G.C., Daniel, Á., Crespo, A., del Campo, F., Hornero, R.: Evaluation of machine-learning approaches to estimate sleep apnea severity from at-home oximetry recordings. IEEE J. Biomed. Health Inform. **23**(2) (2019)
3. Deviaene, M., Testelmans, D., Buyse, B., Borzée, P., Van Huffel, S., Varon, C.: Automatic screening of sleep apnea patients based on the SpO_2 signal. IEEE J. Biomed. Health Inform. **23**(2), 607–617 (2019)
4. Jarchi, D., Sanei, S., Prochazka, A.: Detection of sleep apnea/hypopnea events using synchrosqueezed wavelet transform. In: ICASSP 2019—2019 IEEE international conference on acoustics, speech and signal processing (ICASSP)
5. Jung, D.W., Hwang, S.H., Cho, J.G., Choi, B.H., Baek, H.j., Lee, Y.J., Jeong, D., Park, K.S.: Real time automatic apneic event detection using nocturnal pulse oximetry. IEEE Trans. Biomed. Eng. **65**(3), 706–712 (2018)
6. Iwin Thanakumar, J.S.: Survey of data mining algorithm's for intelligent computing system. J. Trends Comput. Sci. Smart Technol. **1**(01), 14–24 (2019)
7. Gualsaquí Miranda, M.V., Vizcaíno Espinosa, I.P., Flores Calero, M.J.: ECG signal features extraction. In: IEEE Ecuador Technical Chapters Meeting (ETCM) (2016)
8. Gutierrez-Tobal, G.C., Alvarez, D., del Campo, F., Hornero, R.: Utility of adaboost to detect sleep apnea-hypopnea syndrome from single-channel airflow. IEEE Trans. Biomed. Eng. **63**(3), 636–646 (2016)
9. Umer, M., Bhatti, B.A., Tariq, M.H., Zia-ul-Hassan, M., Yaqub Khan, M., Zaidi, T.: Electrocardiogram feature extraction and pattern recognition using a novel windowing algorithm. Adv. Biosci. Biotechnol. **5**, 886–894
10. Jezzini, A., Ayache, M., Elkhansa, L., al abidin Ibrahim, Z.: ECG classification for sleep apnea detection. In: 2015 International Conference on Advances in Biomedical Engineering (ICABME) (2015)
11. Xie, B., Minn, H.: Real-time sleep apnea detection by classifier combination. IEEE Trans. Inform. Technol. Biomed. **16**(3), 469–477 (2012)
12. Dr. Senthil Kumar, T., Suresh, A., Pai, K. Kiron, and Chinnaswamy, P.: Survey on predictive medical data analysis. J. Eng. Res. Technol. **3**, 2283–2286 (2014)

13. Sankar, A., Bharathi, P.D., Midhun, M., Vijay, K., Dr. Senthil Kumar, T.: A conjectural study on machine learning algorithms. Adv. Intell. Syst. Comput. **397**, 105–116 (2016)
14. Dr. Senthil Kumar, T., Pandey, S.: Customization of recommendation system using collaborative filtering algorithm on cloud using mahout. Adv. Intell. Syst. Comput. **321** (2015)

Convergence of a Finite-Time Zhang Neural Network for Moore–Penrose Matrix Inversion

G. Sowmya and P. Thangavel

Abstract Moore-Penrose or pseudo inverse is examined for a time varying matrix. The Finite time model for Moore-Penrose left and right inverse is proposed and investigated. Infinite time for convergence in the original Zhang network is said to converge in finite time when the proposed Finite time Moore-Penrose inverse is employed. Nonlinear activation functions are tested to quicken the speed of the Finite time Zhang Neural Network. Theoretical and experimental results show that Finite time Zhang Neural network converges much faster than Zhang neural network. In addition the hyperbolic sine activation function applied to the proposed system converges faster than the linear activation function.

Keywords Finite time Zhang Neural Network (FTZNN) · Activation functions · pseudo inverse · Moore-Penrose

1 Introduction

Moore-Penrose matrix inversion has marked its importance for solving problems in extreme learning machine [1], Image Processing [2], Inverse kinematics [3], Markov Chain [4], and so on. Moore-Penrose inverse takes $O(n^3)$ arithmetic operations for computation. Therefore certain parallel algorithms for computation where introduced known as the Gradient Neural Network [5] and Zhang Neural Network [6]. Zhang Neural Network proves its superiority over Gradient Neural Network [7]. Different Zhang models for solving generalized inverse was formulate [8, 9]. To effectively solve matrix inversion in a stipulated time [10] was proposed. Further the dynamics was extended for Sylvester Equation [11]. Therefore this research work utilizes Finite

G. Sowmya (✉) · P. Thangavel
University of Madras, Guindy Campus, Chennai 600025, India
e-mail: sowmya.sureshkumar@gmail.com

P. Thangavel
e-mail: pthang.cs@unom.ac.in

© The Editor(s) (if applicable) and The Author(s), under exclusive license to Springer Nature Singapore Pte Ltd. 2021
A. P. Pandian et al. (eds.), *Proceedings of International Conference on Intelligent Computing, Information and Control Systems*, Advances in Intelligent Systems and Computing 1272, https://doi.org/10.1007/978-981-15-8443-5_68

time Zhang Neural Network to solve Moore-Penrose inverse. The focus of the paper is as follows,

- Maximum time required to calculate Moore-Penrose inverse using ZNN was found.
- Comparative analysis between Original ZNN and the proposed system for Moore-Penrose matrix inverse was depicted.
- To accelerate the speed of convergence for Finite time Zhang Neural Network, nonlinear activation functions where applied.

2 Preliminaries on Moore-Penrose Matrix Inversion

A Moore-Penrose inverse is used to solve an over or under determined systems. Let $M_p(t_m) \in R^{m\times n}$ then, there exists a Moore-Penrose inverse $M_p(t_m)^\dagger \in R^{n\times m}$ if it satisfies all of the Moore-Penrose conditions.

1. $M_p(t_m)M_p^\dagger(t_m)M_p(t_m) = M_p(t_m)$,
2. $M_p^\dagger(t_m)M_p(t_m)M_p^\dagger(t_m) = M_p^\dagger(t_m)$,
3. $(M_p(t_m)M_p^\dagger(t_m))^T = M_p(t_m)M_p^\dagger(t_m)$,
4. $(M_p^\dagger(t_m)M_p(t_m))^T = M_p^\dagger(t_m)M_p(t_m)$,

where, T denote the transpose of the matrix. If only condition 1 satisfies then it is termed as generalized inverse. When both condition 1 and 2 satisfied then it is called generalized reflexive inverse. Generalized inverse are not generally unique. Conditions 3 and 4 guarantee unique solution. Therefore a Moore-Penrose inversion has one solution and Frobenius norm is minimum than other inverse.

If $M_p(t_m) \in R^{m\times n}$ with no equal values of m and n, then $rank(M_p(t_m)) = min(m, n)$ implies $M_p^T(t_m)M_p(t_m) \in R^{n\times n}$ or $M_p(t_m)M_p^T(t_m) \in R^{n\times n}$ is non singular and thus $(M_p^T(t_m)M_p(t_m))^{-1}$ or $(M_p(t_m)M_p^T(t_m))^{-1}$ exists. Moore-Penrose inverse $M_p^\dagger(t_m)$ is given as,

$$M_p^\dagger(t_m) = \begin{cases} M_p^T(t_m)(M_p(t_m)M_p^T(t_m))^{-1}, & if \quad m < n \\ (M_p^T(t_m)M_p(t_m))^{-1}M_p^T(t_m), & if \quad m > n \end{cases} \tag{1}$$

Thus $M_p^\dagger(t_m)$ fulfils the conditions for full rank matrix $M_p(t_m)$.

$$M_p^\dagger(t_m)M_p(t_m)M_p^T(t_m) = M_p^T(t_m), \quad if \quad m < n \tag{2}$$

$$M_p^T(t_m)M_p(t_m)M_p^\dagger(t_m) = M_p^T(t_m), \quad if \quad m > n \tag{3}$$

Equations 2 and 3 are used in the formulation of Zhang's right and left Moore-Penrose inverse. Further, the system of FTZNN is described.

3 FTZNN Description

This section deals with the formulation of FTZNN for Moore-Penrose inverse for both right and left inverse. The proposed method differs from the traditional Zhang network while choosing the derivative of the error.

$$\dot{Err}(t_m) = -\gamma \Phi(\alpha_1 Err(t_m) + \alpha_2 Err^{b/a}(t_m)) \tag{4}$$

where γ, α_1 and α_2 are positive values. a and b are positive odd integers with $a > b$. Φ is the activation function applied to the network.

3.1 FTZNN to Solve Right Pseudo Inverse

The error from Eq. 2 is given as

$$Err(t_m) = M_p^{\dagger}(t_m)M_p(t_m)M_p^T(t_m) - M_p^T(t_m) \in R^{n\times m} \tag{5}$$

Replacing the Moore-Penrose inverse $M_p^{\dagger}(t_m)$ with the calculated inverse $X(t_m) \in R^{n\times m}$, the error function can be rewritten as $Err(t_m) = X(t_m)M_p(t_m)M_p^T(t_m) - M_p^T(t_m)$. According to the traditional Zhang neural network the derivative of error is defined as $\dot{Err}(t_m) = -\gamma\Phi(Err(t_m))$ where $\gamma > 0$ and Φ is the activation function applied to the neural network and $\Phi{:}R^{n\times m} \to R^{n\times m}$. Traditional Zhang neural network for time varying Moore-Penrose right inversion is given as,

$$\begin{aligned}\dot{X}(t_m)M_p(t_m)M_p^T(t_m) &= -\gamma\Phi(X(t_m)M_p(t_m)M_p^T(t_m) - M_p^T(t_m)) \\ &\quad - X(t_m)(\dot{M}_p(t_m)M_p^T(t_m) + M_p(t_m)\dot{M}_p{}^T(t_m)) + \dot{M}_p{}^T(t_m)\end{aligned} \tag{6}$$

The FTZNN for right Moore-Penrose inverse is formulate by substituting $Err(t_m)$ in Eq. 4 and is given as,

$$\begin{aligned}\dot{X}(t_m)M_p(t_m)M_p^T(t_m) &= -\gamma\Phi\alpha_1(X(t_m)M_p(t_m)M_p^T(t_m) - M_p^T(t_m)) \\ &\quad - \gamma\Phi\alpha_2(X(t_m)M_p(t_m)M_p^T(t_m) - M_p^T(t_m))^{b/a} \\ &\quad - X(t_m)(\dot{M}_p(t_m)M_p^T(t_m) + M_p(t_m)\dot{M}_p{}^T(t_m)) + \dot{M}_p{}^T(t_m)\end{aligned} \tag{7}$$

As defined in the earlier section design parameters $\gamma > 0$, $\alpha_1 > 0$, $\alpha_2 > 0$, $a > 0$, $b > 0$ and Φ is the activation function.

3.2 FTZNN to Solve Left Pseudo Inverse

Likewise the error is formulated with 3 for the left Moore-Penrose inverse,

$$Err(t_m) = M_p^T(t_m)M_p(t_m)X(t_m) - M_p^T(t_m) \tag{8}$$

Using the definition formulated by Zhang where $\dot{Err}(t_m) = -\gamma\Phi(Err(t_m))$, the time varying Moore-Penrose left inverse is given as

$$\begin{aligned} M_p^T(t_m)M_p(t_m)\dot{X}(t_m) = &-\gamma\Phi(M_p^T(t_m)M_p(t_m)X(t_m) - M_p^T(t_m)) \\ &- (\dot{M_p}^T(t_m)M_p(t_m) + M_p^T(t_m)\dot{M}_p(t_m))X(t_m) + \dot{M_p}^T(t_m) \end{aligned} \tag{9}$$

To obtain FTZNN for Moore-Penrose Inverse, the proposed work uses the definition for derivative of error given in Eqs. 4 and 8.

$$\begin{aligned} M_p^T(t_m)M_p(t_m)\dot{X}(t_m) = &-\gamma\Phi\alpha_1(M_p^T(t_m)M_p(t_m)X(t_m) - M_p^T(t_m)) \\ &- \gamma\Phi\alpha_2(M_p^T(t_m)M_p(t_m)X(t_m) - M_p^T(t_m))^{b/a} \\ &- (\dot{M_p}^T(t_m)M_p(t_m) + M_p^T(t_m)\dot{M}_p(t_m))X(t_m) + \dot{M_p}^T(t_m) \end{aligned} \tag{10}$$

4 FTZNN Model with Activation Functions for Accelerating Convergence

For faster convergence of the FTZNN the value of Φ is applied. To test the efficiency of the system, our discussion is limited to one of the three activation functions such as a linear function $\Phi(U_i)_{li}$, Power Sigmoidal Function $\Phi(U_i)_{ps}$ and an Hyperbolic Sine Function $\Phi(U_i)_{hs}$ where as many activation function are found in the literature.

4.1 Linear Activation Function

It is given as follows,

$$\Phi(U_i)_{li} = U_i, \tag{11}$$

On the application of a linear activation function to the finite time left and right Moore Penrose Zhang Network Equations 7 and 10 can be rewritten as follows,

FTZNN Right Moore Penrose Inverse with linear activation

$$\begin{aligned}\dot{X}(t_m)M_p(t_m)M_p^T(t_m) = &-\epsilon_1(X(t_m)M_p(t_m)M_p^T(t_m) - M_p^T(t_m))\\ &- \epsilon_2(X(t_m)M_p(t_m)M_p^T(t_m) - M_p^T(t_m))^{b/a}\\ &- X(t_m)(\dot{M}_p(t_m)M_p^T(t_m) + M_p(t_m)\dot{M}_p{}^T(t_m)) + \dot{M}_p{}^T(t_m)\end{aligned} \tag{12}$$

FTZNN Left Moore Penrose Inverse with linear activation

$$\begin{aligned}M_p^T(t_m)M_p(t_m)\dot{X}(t_m) = &-\epsilon_1(M_p^T(t_m)M_p(t_m)X(t_m) - M_p^T(t_m))\\ &- \epsilon_2(M_p^T(t_m)M_p(t_m)X(t_m) - M_p^T(t_m))^{b/a}\\ &- (\dot{M}_p{}^T(t_m)M_p(t_m) + M_p^T(t_m)\dot{M}_p(t_m))X(t_m) + \dot{M}_p{}^T(t_m)\end{aligned} \tag{13}$$

where $\epsilon_1 = \gamma\alpha_1$, $\epsilon_2 = \gamma\alpha_2$ and ϵ_1, ϵ_2 are positive integers.

4.2 *Power Sigmoidal Activation Function*

Power Sigmoidal activation function is described as following,

$$\Phi(U_i)_{ps} = \begin{cases} \frac{1+e^{(-\xi)}}{1-e^{(-\xi)}} \cdot \frac{1-e^{(-U_i\xi)}}{1+e^{(-U_i\xi)}}, & if \ \ |U_i| < 1 \\ U_i^p, & if \ \ |U_i| \geq 1 \end{cases}$$

where p is an odd integer and $\xi > 0$.

4.3 *Hyperbolic Sine Activation Function*

Hyperbolic Sine activation function is given as,

$$\Phi(U_i)_{hs} = \frac{e^{(U_i m)}}{2} - \frac{e^{(-U_i m)}}{2}, \tag{14}$$

where m is an odd integer.

5 Theoretical Results

Theorem 1 *[12] When a linear activation function is applied to Zhang's dynamics $\dot{Err}(t_m) = -\gamma\Phi(Err(t_m))$, starting from any randomly generated initial state $X(0)$, state matrix $X(t_m)$ of the original right Moore-Penrose ZNN model 6 and left Moore-Penrose ZNN model 9 converges to the theoretical time-varying solution of 5 and 8 exponentially with time.*

Theorem 2 *A system with $\dot{Err}(t_m)$ designed as 4 with a linear activation function, starting from any initial random error $Err(0)$, the error $Err(t_m)$ converges to zero in finite time t_{ft}.*

$$t_{ft} = \frac{a}{\alpha_1(a-b)} \ln \frac{\alpha_1 el^{+}(0)^{(a-b)/a} + \alpha_2}{\alpha_2} \tag{15}$$

where $el^{+}(0)$ is the largest element in the initial error matrix $Err(0)$

Theorem 3 *Given time-varying matrix $M_p(t_m)$, state matrix $X(t_m)$ in right FTZNN model 7 and left FTZNN model 10 with linearly activation function, starting from any randomly generated initial state $X(0)$ converges to the theoretical time-varying solution of Moore-Penrose equation 5 and 8 in finite time t_c*

$$t_c \le \max\{\frac{a}{\alpha_1(a-b)} \ln \frac{\alpha_1 el^{+}(0)^{(a-b)/a} + \alpha_2}{\alpha_2}, \frac{a}{\alpha_1(a-b)} \ln \frac{\alpha_1 es^{-}(0)^{(a-b)/a} + \alpha_2}{\alpha_2}\} \tag{16}$$

where $el^{+}(0)$ is the largest element and $es^{-}(0)$ is the smallest element in the initial error matrix $Err(0)$

6 Computer Stimulations and Comparison

Further to examined the correctness of FTZNN for Moore Penrose inverse, it is examined with three examples as given below. The Stimulation of the model is done using MATLAB and the graph shows the speed of convergence as time goes to ∞.

Since Eq. 7 is a matrix form differential equation, we convert it into vector using Kronecker product and vectorization,

$$\begin{aligned}
&((M_p(t_m)M_p^T(t_m))^T \otimes I)\mathbf{vec}\,\dot{X}(t_m) = \\
&-\gamma\Phi\alpha_1(((M_p(t_m)M_p^T(t_m))^T \otimes I)\mathbf{vec}\,X(t_m) - M_p^T(t_m)) \\
&-\gamma\Phi\alpha_2(((M_p(t_m)M_p^T(t_m))^T \otimes I)\mathbf{vec}\,X(t_m) - M_p^T(t_m))^{b/a} \\
&-((\dot{M}_p(t_m)M_p^T(t_m))^T \otimes I) + (M_p(t_m)\dot{M}_p^T(t_m))^T \otimes I))\mathbf{vec}\,X(t_m) + \mathbf{vec}\,\dot{M}_p^T(t_m)
\end{aligned} \tag{17}$$

Left Moore-Penrose equation 10 takes the following the vector differential equation,

$$
\begin{aligned}
&(I \otimes (M_p^T(t_m)M_p(t_m))^T)\mathbf{vec}\,\dot{X}(t_m) = \\
&\quad -\gamma\Phi\alpha_1((I \otimes (M_p^T(t_m)M_p(t_m))^T)\mathbf{vec}\,X(t_m) - M_p^T(t_m)) \\
&\quad\quad -\gamma\Phi\alpha_2((I \otimes (M_p^T(t_m)M_p(t_m))^T)\mathbf{vec}\,X(t_m) - M_p^T(t_m))^{b/a} \\
&\quad -((I \otimes \dot{M_p}^T(t_m)M_p(t_m))^T) + I \otimes (M_p^T(t_m)\dot{M}_p(t_m))^T))\mathbf{vec}\,X(t_m) + \mathbf{vec}\,\dot{M_p}^T(t_m)
\end{aligned}
\tag{18}
$$

Assuming the value of design parameters are set as $a = 2, b = 1, \alpha_1 = \alpha_2 = 10$ and $\gamma = 1$. For power sigmoidal activation function $p = 3$ and $\xi = 3$. The value of $m = 3$ for Hyperbolic Sine activation function.

Example 1 Let $M_p(t_m)$ be a the given matrix along with its right Moore Penrose theoretical inverse

$$
M_p(t_m) = \begin{pmatrix} \sin(2t_m) & \cos(2t_m) & -\sin(2t_m) \\ -\cos(2t_m) & \sin(2t_m) & \cos(2t_m) \end{pmatrix} \tag{19}
$$

$$
M_p^{\dagger}(t_m) = \begin{pmatrix} 0.5\sin(2t_m) & -0.5\cos(2t_m) \\ \cos(2t_m) & \sin(2t_m) \\ -0.5\sin(2t_m) & 0.5\cos(2t_m) \end{pmatrix} \tag{20}
$$

Figures 1 and 2 shows the computational value plotted against theoretical inverse ZNN and the Finite time Zhang model for Example 1. The computed inverse is generated with 8 initial states $X(0) \in R^{3\times 2}$ jotted with different colour in Figure.

Example 2 The following example is taken into consideration with $M_p(t_m)$ and $M_p^{\dagger}(t_m)$ to be theoretical inverse

$$
M_p(t_m) = \begin{pmatrix} \sin(2t_m) & \cos(2t_m) \\ -\cos(2t_m) & \sin(2t_m) \\ \sin(2t_m) & \cos(2t_m) \end{pmatrix} \tag{21}
$$

$$
M_p^{\dagger}(t_m) = \begin{pmatrix} 0.5\sin(2t_m) & -\cos(2t_m) & 0.5\sin(2t_m) \\ 0.5\cos(2t_m) & \sin(2t_m) & 0.5\cos(2t_m) \end{pmatrix} \tag{22}
$$

The transient behaviour of the calculated inverse and theoretical inverse is jotted using the Original Zhang Network for Example 2 is shown in Fig. 3 with eight initial valves for $X(0) \in R^{3\times 2}$ and Finite time Zhang Network is shown in Fig. 4. On the comparative analysis of both the figures, it is clearly concluded that Fig. 4 has faster convergence than that of Fig. 3.

Example 3 Let $M_p(t_m)$ and its theoretical right Moore Penrose $M_p^{\dagger}$ given as.

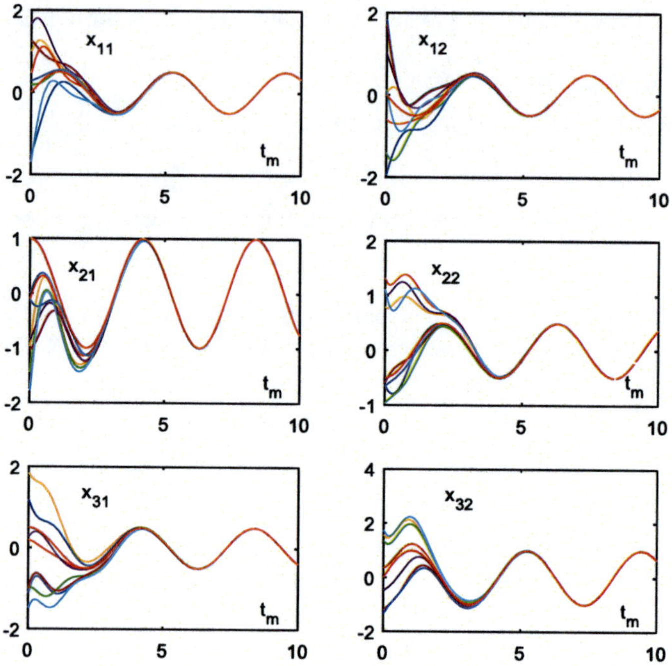

Fig. 1 Matrix elements using Original ZNN 6 with linear activation function for Example 1

$$M_p(t_m) = \begin{pmatrix} \sin(t_m) & \cos(t_m) & -\sin(t_m) \\ -\cos(t_m) & \sin(t_m) & \cos(t_m) \end{pmatrix} \tag{23}$$

$$M_p^{\dagger}(t_m) = \begin{pmatrix} 0.5\sin(t_m) & -0.5\cos(t_m) \\ \cos(t_m) & \sin(t_m) \\ -0.5\sin(t_m) & 0.5\cos(t_m) \end{pmatrix} \tag{24}$$

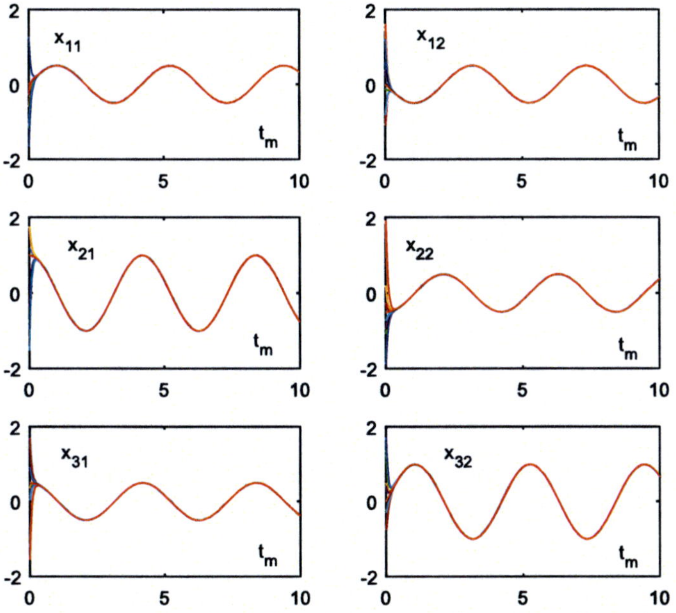

Fig. 2 Matrix elements using Finite time ZNN 7 with linear activation function for Example 1

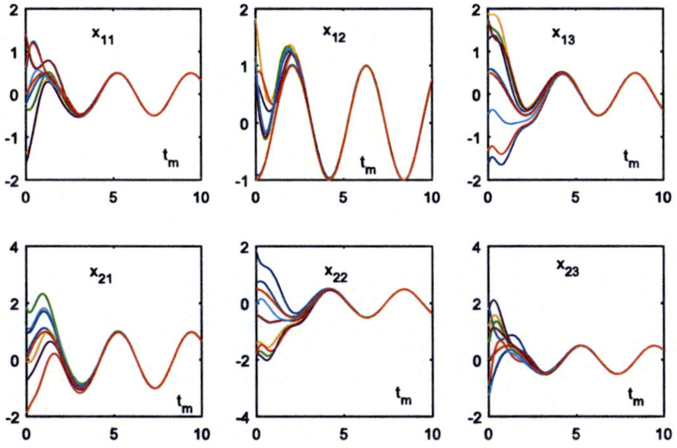

Fig. 3 Matrix elements using Original ZNN 9 with linear activation function for Example 2

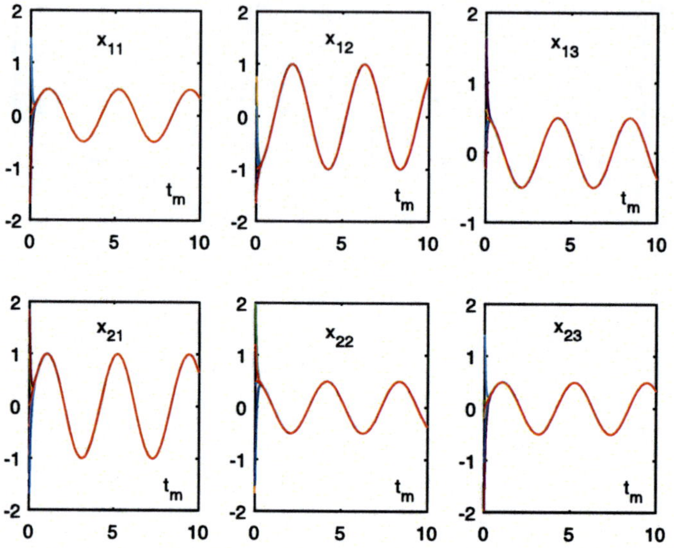

Fig. 4 Matrix elements using Finite time ZNN 7 with linear activation function for Example 2

The Frobenius norm (i.e., $\|X(t_m) - M_p^{\dagger}(t_m)\|_F$) for Examples 1, 2 and 3 is plotted in the Fig. 5. Figure 5a, c and e shows the comparison between the Original Zhang network and the Finite time Zhang network. Figure 5b, d and f describes the performance of nonlinear activation functions upon Finite time Zhang network.

7 Conclusions

This paper has proposed a finite time Moore-Penrose inverse that is different from the Zhang's model of describing the derivative of the error. The Right and Left Moore-Penrose FTZNN are formulated with suitable examples. Further to accelerate the speed of convergence upon FTZNN certain non-linear activation functions where applied to the network. It is found from the theoretical and experimental results that FTZNN globally asymptotically converges in finite time than the Zhang neural network. Computer stimulation depict that Power Sigmoidal and Hyperbolic Sine activation function converges faster than a linear activation function for finite time Zhang Moore-Penrose inversion.

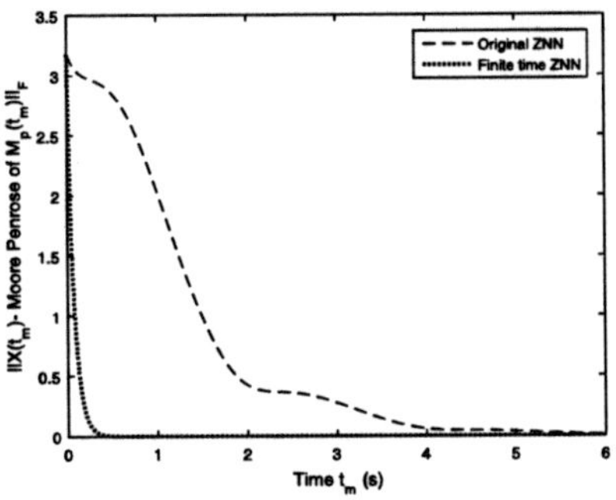

(a) Comparison between Original ZNN 6 and Finite time ZNN 7 for Example 1.

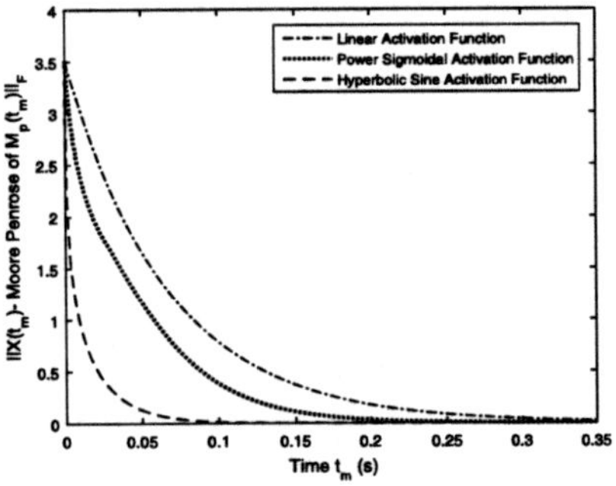

(b) Activation functions applied to Finite time ZNN 7 for Example 1.

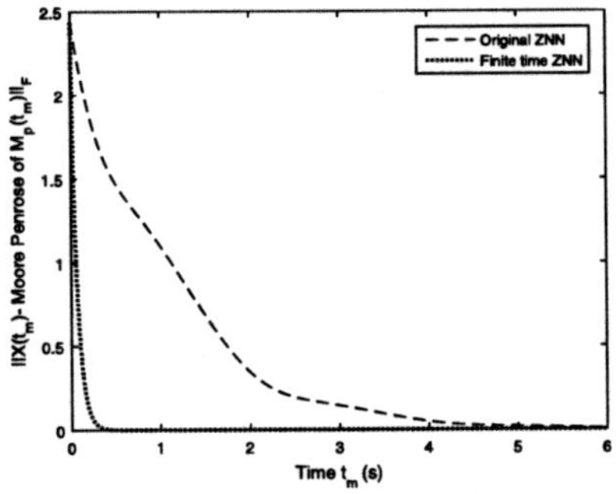

(c) Comparison between Original ZNN 9 and Finite time ZNN 10 for Example 2.

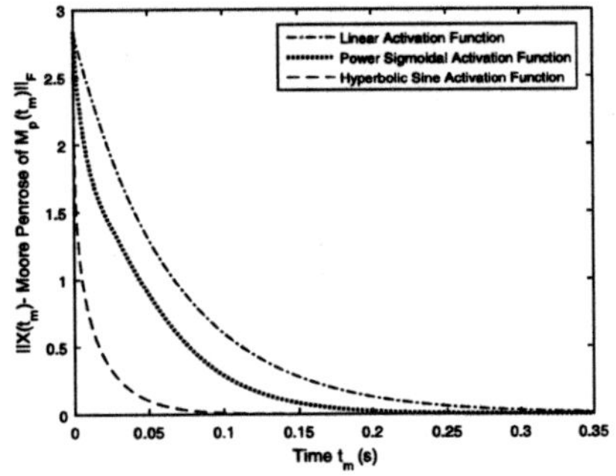

(d) Activation functions applied to Finite time ZNN 10 for Example 2.

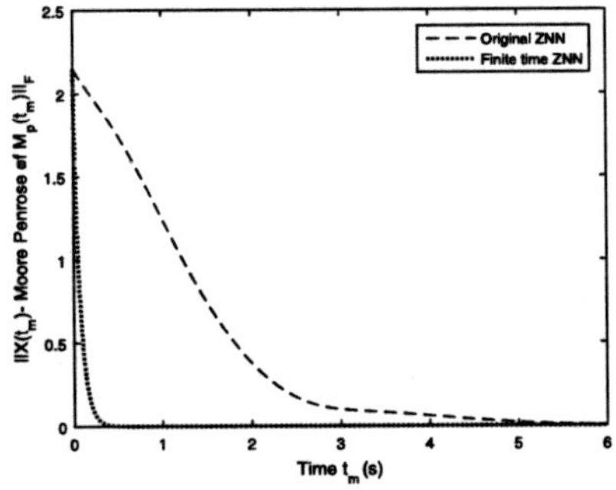

(e) Comparison between Original ZNN 6 and Finite time ZNN 7 for Example 3.

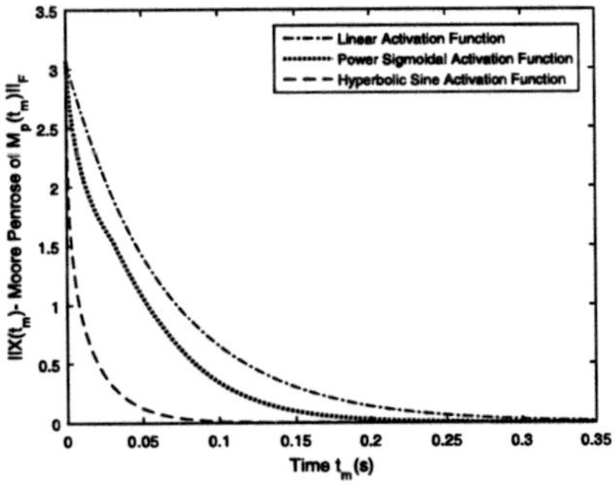

(f) Activation functions applied to Finite time ZNN 7 for Example 3.

Fig. 5 Convergence computation in terms Frobenius norm of $||X(t_m) - M_p^{\dagger}(t_m)||_F$

References

1. Sharina Huang, M.C., Zhao, G.: Tensor extreme learning design via generalized Moore–Penrose inverse and triangular type-2 fuzzy sets. Neural Comput. Appl. **31**, 5641–5651 (2019)
2. Sheng, X.: Computation of weighted Moore–Penrose inverse through Gauss-Jordan elimination on bordered matrices. Appl. Math. Comput. **323**, 64–74 (2018)
3. Sturges, R.H.: Analog matrix inversion (robot kinematics). IEEE J. Robotics Autom. **4**(2), 157–162 (1988)
4. Hunter, J.J.: Generalized inverses of Markovian kernels in terms of properties of the Markov chain. Linear Algebra Appl. **447**, 38–55 (2014) (special Issue - Memorial Issue for Michael Neumann and Uriel Rothblum)
5. Lv, X., Xiao, L., Tan, Z., Yang, Z., Yuan, J.: Improved gradient neural networks for solving Moore–Penrose inverse of full-rank matrix. Neural Process. Lett. **50** (2019)
6. Zhang, Y., Yang, Y., Tan, N., Cai, B.: Zhang neural network solving for time-varying full-rank matrix Moore–Penrose inverse. Computing **92**, 97–121 (2011)
7. Zhang, Y., Tan, N., Cai, B., Chen, Z.: Matlab simulink modeling of Zhang neural network solving for time-varying pseudoinverse in comparison with gradient neural network. In: 2008 Second International Symposium on Intelligent Information Technology Application, vol. 1, Dec 2008, pp. 39–43
8. Yunong Zhang, H.T., Xie, Y.: Time-varying Moore–Penrose inverse solving shows different Zhang functions leading to different ZNN models. In: Wang, J., Yen, G.G., Polycarpou, M.M. (eds.) Advances in Neural Networks– ISNN 2012. ISNN 2012. Lecture Notes in Computer Science, vol. 7367 (2012)
9. Liao, B., Zhang, Y.: Different complex ZFS leading to different complex ZNN models for time-varying complex generalized inverse matrices. IEEE Trans. Neural Networks Learn. Syst. **25**(9), 1621–1631 (2014)
10. Xiao, L.: A new design formula exploited for accelerating Zhang neural network and its application to time-varying matrix inversion. Theor. Comput. Sci. **647**, 50–58 (2016)
11. Xiao, L.: A finite-time recurrent neural network for solving online time-varying Sylvester matrix equation based on a new evolution formula. Nonlinear Dyn. **90**, 1581–1591 (2017)
12. Guo, D., Yi, C., Zhang, Y.: Zhang neural network versus gradient-based neural network for time-varying linear matrix equation solving. Neurocomputing **74**(17), 3708–3712 (2011)

Estimation of Differential Code Bias and Local Ionospheric Mapping Using GPS Observations

Yogesh Lingwal, Fateyh Bahadur Singh, and B. N. Ramakrishna

Abstract Dual frequency GPS receiver provides measurements at two frequencies, one at $L1$: 1575.42 MHz another at $L2$: 1227.60 MHz, over worldwide. Measurements of any GPS receiver are affected by instrument biases of satellite and receiver. Differential Code Bias (DCB) is the instrumental biases between two GNSS code observations at same or different frequencies. Knowledge of Precise Differential code bias for satellite and receiver end is essential for code based positioning, remote sensing of ionosphere, and other applications. This paper presents the estimation of DCB for GPS satellites and receiver along with the Local TEC estimation using spherical harmonic function. Measurements of single and multiple stations were taken for processing. DCB results obtained after the processing were compared with the results published by IGS Analysis center for validation, and found an agreement of ~±1 ns between them while processing with single station and ~ ±0.5 ns while processing with multiple receivers at different locations.

Keywords Differential code bias (DCB) · Vertical total electron content (VTEC) · Global positioning system (GPS) · International GNSS service (IGS) · Spherical harmonic function (SH) · Timing group delay (TGD)

Y. Lingwal (✉)
Scientist/Engineer, NSA, ISTRAC/ISRO, Jodhpur, Rajasthan, India
e-mail: yogesh_lingwal@istrac.gov.in; gbpec.yogesh@gmail.com

F. B. Singh
DGM, IGSA/NSA, ISTRAC/ISRO, Bengaluru, India
e-mail: fbsingh@istrac.gov.in

B. N. Ramakrishna
Deputy Director, Navigation System Area, ISTRAC/ISRO, Bengaluru, India
e-mail: ramki@istrac.gov.in

© The Editor(s) (if applicable) and The Author(s), under exclusive license to Springer Nature Singapore Pte Ltd. 2021

A. P. Pandian et al. (eds.), *Proceedings of International Conference on Intelligent Computing, Information and Control Systems*, Advances in Intelligent Systems and Computing 1272, https://doi.org/10.1007/978-981-15-8443-5_69

1 Introduction

Global Positing system is a space based navigation system, currently having 32 satellites, orbiting the earth at an inclination of 55° and at the altitude of about ~20,000 km. Each GPS satellite transmits the navigation signal at $L1$ (1575.42 MHz denoted as f_1) and $L2$ (1227.60 MHz denoted as f_2). Each GPS receiver collects the measurements using these navigation signals depending upon its code tracking capability [1]. Most of the receivers provide code measurements $C1, P1, P2$ (known as pseudoranges from C/A code, $P1$ code and $P2$ code) and phase measurements ($L1$ and $L2$). All these measurements have bias because of the hardware of satellite and receiver. Estimation of hardware bias for Satellite and receiver is difficult in absolute sense using the observation, but in relative sense it is possible to measure the bias. Bias between the two code at same frequency or different frequency, can be estimated on relative basis, and called as Differential code Bias. So Differential Code Bias (DCB) may be explained as the systematic bias between two codes at same frequency or different frequency. Bias between two codes ($P1$-$C1$) at same frequency is known as intra Frequency bias and at different frequency ($P1$-$P2$, $C1$-$P2$) is known as inter frequency bias.

The group Delay of Satellite hardware, known as Timing Group Delay (TGD) is measured during the launch of the satellite, but it does not remain constant over a period. Different studies show a change of ~±0.2 ns in satellite hardware bias over a month [2, 3]. So, Satellite DCB can be considered constant for a month, however the present work has processed the data by assuming the DCB of satellite and receiver constant for a day. Group Delay of satellite is also broadcasted in navigation message in the term of Timing Group Delay (TGD) computed or estimated by JPL, follows almost same pattern as the DCB of satellite (comparison done in results section). However Broadcasted TGD value is referred as absolute group delay between two codes at same or different frequency whereas DCB is defined in relative basis [4].

Receiver DCB depends on the bias due to the Antenna, Cable, and Receiver Electronics. Antenna has significant contribution in Receiver DCB as it is mounted in open sky. The DCB of Antenna is generally provided by manufactures at particular temperature (usually room temperature). Cable is generally treated as non-dispersive medium, however some of studies show minimum impact on overall DCB because of cable of different length [5]. DCB of Receiver may be measured by simulator [6]. All possible GPS signal may be generated by simulator without Ionospheric and tropospheric delay, with satellite signal delay set to zero, to estimate the precise receiver DCB in absolute sense. But when the proposed work is assuming the DCB of receiver, constant over a day, it is not cost effective to measure the DCB of receiver (of any regional network) using simulator. It makes important the estimation of Receiver DCB using the collected observation of receiver.

DCB is not only essential for knowing the behavior or any change in the hardware of satellite and receiver but also required for the extracting precise TEC content at ionosphere. Ionosphere ranges from 90 to 1000 km from the surface of earth. Ionosphere is a dispersive medium where wave of two different frequency travels with different speed. So the signal of different frequency suffers different delay because of ionosphere. Ionospheric Delay is generally proportional to total electron content and inversely proportional to the square of frequency. DCB of Satellite and receiver typically ranges from −10 ns to 10 ns, ignoring DCB may cause an error of around ±18 TECU at $L1$ band while extracting TEC using the GNSS observations.

On June 1998, International GNSS services (IGS) had established Ionospheric associated Analysis Centers (IAAC) in order to provide the DCB of GNSS satellites along with the DCB of IGS station receiver. These centers (CODE, ESA. JPL, UPC, CAS) have developed different techniques in order to provide Global Ionospheric Map (GIM) in a same format called IONosphere Map EXchange (IONEX) on regular basis [3, 7]. These centers are providing GIM with a 2 h of temporal resolution and 2.5 × 5° of spatial resolution.

Here paper presents the estimation of DCB of GPS satellites and receiver using Spherical Harmonic function of order 4–15. Processing is implemented using MATLAB tool. Here the Vertical TEC is represented by Spherical Harmonic function. Section 2 explains the estimation theory adopted for processing and Steps of Processing are discussed in Sect. 3, and Results and Discussion are summarized in Sect. 4. Conclusion and Future scope is explained in Sect. 5.

2 Estimation Theory

As earlier explained the GPS receiver provides observations consist of carrier phase and pseudoranges. The pseudoranges and carrier phase observations may be expanded as follows:

$$P_1 = P_o + \frac{I}{f_1^2} + T + c(\partial t_r - \partial t^s) + D_1^r + D_1^s + \varepsilon_p^1 \tag{1}$$

$$P_2 = P_o + \frac{I}{f_2^2} + T + c(\partial t_r - \partial t^s) + D_2^r + D_2^s + \varepsilon_p^2 \tag{2}$$

$$L_1 = P_o - \frac{I}{f_1^2} + T + c(\partial t_r - \partial t^s) + \lambda_1(N_1 + \alpha_1^r + \alpha_1^s) + \varepsilon_\emptyset^1 \tag{3}$$

$$L_2 = P_o - \frac{I}{f_2^2} + T + c(\partial t_r - \partial t^s) + \lambda_2(N_2 + \alpha_2^r + \alpha_2^s) + \varepsilon_\emptyset^2 \tag{4}$$

Here P: GPS pseudorange, L: GPS carrier phase measurements. P_o represents geometric range between satellite and receiver, I: Ionospheric delay, T: tropospheric delay, ∂t_r: receiver clock bias, ∂t^s: satellite clock bias, D^r receiver hardware delay, D^s: satellite delay, N: carrier ambiguity, α^r and α^s: receiver and satellite phase delay, ϵ: includes error due to receiver noise and multipath. Here 1 and 2 stands for frequencies.

Both pseudorange and carrier phase measurements have error added with geometric range because of troposphere, ionosphere, satellite clock and receiver clock. As ionosphere is dispersive media, so signal of different frequency travels with different speed. Ionosphere offers delay in Code measurements whereas it advances the carrier as indicated in Eqs. 1–4.

Now taking the Geometry free combination of pseudorange (Eqs. 1 and 2) and carrier phase measurement (Eqs. 3 and 4) will remove all error because of troposphere, satellite clock, and receiver clock, it may be represented as follows:

$$P_4 = P_1 - P_2 = I\left(1/f_1^2 - 1/f_2^2\right) + \text{DCBSat} + \text{DCBRx} \tag{5}$$

$$\begin{aligned} L_4 = L_1 - L_2 = -I\left(1/f_1^2 - 1/f_2^2\right) + (\lambda_1 N_1 - \lambda_2 N_2) \\ + \left(\lambda_1 \alpha_1^r - \lambda_2 \alpha_2^r\right) + \left(\lambda_1 \alpha_1^s - \lambda_2 \alpha_2^s\right) \end{aligned} \tag{6}$$

where $\text{DCBSat} = D_1^s - D_2^s$ represents DCB of Satellite and $\text{DCBRx} = D_1^r - D_2^r$ is DCB of receiver. In equation number (6) the second term is the difference of ambiguity of carrier phase at $L1$ and $L2$. This second term will not change as long as receiver maintains lock with signal (in f_1 and f_2 frequency). As pseudorange difference $P4$ is noisier (unambiguous but less precise) in comparison to carrier phase difference (ambiguous but more precise), so pseudorange is smoothed with carrier phase measurement using Hatch filter. Before smoothing, cycle slip is detected and measurements are repaired [8]. Melbourne-Wubeena (M-W) combination is represented as follows, used for the detection of cycle slip [9]:

$$L6 = \frac{1}{f_1 - f_2}(f_1 L_1 - f_2 L_2) - \frac{1}{f_1 + f_2}(f_1 P_1 + f_2 P_2) \tag{7}$$

M-W combination uses both dual frequency carrier phase and code measurements [8, 10]. M-W has wide lane combination of carrier phase which offers advantage of wider wavelength. Measurements with wider wavelength provide wider spacing between the ambiguities, in case of cycle slip, so detected easily. On other hand the noise is reduced by narrow lane combination of code measurements [11]. Here if data gap is identified in code/carrier measurements, it will be treated as cycle slip. Otherwise the current value of L6 combination is compared with the mean of previous epochs, if it found greater than the multiplication of standard deviation and sensitivity factor ($k = 4$), then cycle slip is declared. Sensitivity factor should be small to lead higher fault detection probability [10]. Combination is free from ionospheric change, so avoid false slip detection as of sudden change in Ionosphere. However in case of more receiver noise or multipath noise, it may result the false detection.

After cycle slip detection, the smoothing of pseudorange may be done easily using carrier phase data. Hatch filter is used for this purpose as follows [11]:

$$P_{4\mathrm{Prd}}(t) = P_{4\mathrm{sm}}(t-1) + (L_4(t) - L_4(t-1)) \quad \text{If } k > 1 \tag{8}$$

$$P_{4\mathrm{sm}}(t) = \frac{1}{n}P_4(t) + \left(1 - \frac{1}{n}\right)P_{4\mathrm{Prd}}(t) \tag{9}$$

Here $P_{4\mathrm{Sm}}$ is representing the smoothed pseudorange. Here '*n*' stands for epoch number. Ionospheric Delay may be defined by first order component of Taylor series; higher order components contain less than 1% of ionospheric delay, so avoided here:

$$\mathrm{Ion} = \frac{40.3}{f^2}\mathrm{STEC} \tag{10}$$

Here STEC stands for slant total electron content in the line of sight of satellite and receiver, so replacing same on Eq. (5), the smoothed pseudorange difference may be represented as:

$$P_{4\mathrm{sm}} = 40.3\left(1/f_1^2 - 1/f_2^2\right)\mathrm{STEC} + \mathrm{DCBSat} + \mathrm{DCBRx} \tag{11}$$

There are different models or methods for translating the STEC into VTEC (vertical total electron Content), Ionosphere can be assumed as a thin layer shell at the height of $H = 450$ km [12], and so the VTEC can be written as:

$$\mathrm{VTEC} = \mathrm{MF}(z)^*\mathrm{STEC} \tag{12}$$

where MF: multiplying factor, and can be defined as:

$$\mathrm{MF} = \cos\left(\sin^{-1}\left(\frac{R}{R+H}\sin(\alpha z)\right)\right) \tag{13}$$

Here R: radius of earth (~6371 km) and H: height of Ionospheric layer. Now the precision of TEC estimate will depend upon what model or function used for representing TEC. Here a spherical harmonic function is used for that purpose [1, 13]. Ionosphere can be mapped globally and regionally using spherical harmonic function. SH function basically depends on Ionospheric pierce point (IPP), point where the line of sight between satellite and receiver intersect the thin shell of ionosphere of height $H = 450$ km [3, 14].

$$\mathrm{VTEC} = \sum_{n=0}^{N}\sum_{m=0}^{n} P_{nm}(\sin(\beta))[\alpha_{nm}\cos(ms) + b_{nm}\sin(ms)] \tag{14}$$

IPPs are defined in solar geomagnetic reference frame, β and s are the latitude and longitude of IPPs at given epoch, P_{nm} represents normalized Legendre polynomial function with the degree n and order m. N is the order of SH function ranges from 4 to 15. α_{nm} and b_{nm} are the coefficient of SH function. Now combining Eqs. (11), (12), (13), and (14) a normal equation may be written as:

$$\sum_{n=0}^{N}\sum_{m=0}^{n} P_{nm}(\sin(\beta))(\alpha_{nm}\cos\ ms + b_{nm}\sin\ ms)$$
$$= \cos\left(\sin^{-1}\left(\frac{R}{R+H}\sin(z)\right)\right)$$
$$\left[-\frac{f_1^2 f_2^2}{40.3(f_1^2 - f_2^2)}(P_{4\text{sm}} - \text{DCBSat} - \text{DCBRx})\right] \quad (15)$$

In Eq. (15), α_{nm} and b_{nm} represents the Ionospheric coefficients, and need to be estimated along with DCBSat and DCBRx. Order and degree (N) of function defines the number of these Ionospheric coefficients. Normal Equation (Eq. 15) can be written in matrix form, and the unknown parameters were estimated using Least Square method. Here in order to separate DCBSat and DCBRx a zero mean condition is used on satellites DCB, it is assumed that the mean of satellite DCB is equal to zero, in this way the computed DCB of Satellites and receiver gives a relative value rather than absolute. Further it is known that there is lot of variation in Ionosphere over a day so the SH coefficient, (IONO coefficients) are estimated for every two hours, However the DCB of satellite and Receiver were assumed constant throughout a day. In this way taking 4th order SH function, there will be 300 Ionospheric coefficients (25 coefficients for each 2 h of a day) to be estimated along with the DCB of 32 satellites and n_r numbers of receiver in a network. While processing the Data of single station or small regional network, it is better to take small order of SH function (min 4) but for Global network a higher order SH function (max 15) is recommended for precise DCB estimation and TEC mapping. Matrix Form of normal equation is represented as:

$$
\begin{bmatrix}
A & A & \dots & A & P_{00} & P_{10} & P_{11}\cos(s(i)) & \dots & P_{nm}\sin(ms(i)) \\
A & \dots & A & 0 & P_{00} & P_{10} & P_{nm}\cos(ms(i+1)) & \dots & P_{nm}\sin(ms(i+1)) \\
\dots & \dots & \dots & \dots & \dots & \dots & \dots & \dots & \dots \\
\dots & \dots & \dots & \dots & \dots & \dots & P_{nm}\cos(ms(i+m)) & \dots & P_{nm}\sin(ms(i+m)) \\
A & 0 & A & \dots & P_{00} & P_{10} & P_{nm}\cos(ms(i+n)) & \dots & P_{nm}\sin(ms(i+n)) \\
0 & 1 & \dots & 1 & 0 & \dots & 0 & \dots & 0
\end{bmatrix}
\begin{bmatrix}
\mathrm{DCBRx} \\ \mathrm{DCBSat}(1) \\ \dots \\ \dots \\ \mathrm{DCBSat}(32) \\ a_{00} \\ a_{10} \\ \dots \\ \alpha_{nm} \\ b_{nm}
\end{bmatrix}
=
\begin{bmatrix}
AP_4(i) \\ AP_4(i+1) \\ AP_4(i+2) \\ \dots \\ \dots \\ AP_4(i+k) \\ \dots \\ \dots \\ \vdots \\ \dots \\ \dots \\ \dots \\ AP_4(i+n) \\ 0
\end{bmatrix}
\tag{16}
$$

3 Step of Processing

IGS observation Data was processed for 6 different days (DOY 1st, 15th, 30th, 46th, 59th and 69th of 2018) using MATLAB. Observations of IGS Stations KAT1, TOW2, SYDN, MOBS, CEDU, NNOR YAR2, and HOB2 were used for processing individually and together for DCB Estimation. Steps of processing are as follows in Fig. 1.

The observation data from different IGS station is available on ftp://cddis.gsfc.nasa.gov/gnss/data/daily. Sp3 files, contains the XYZ of satellites for every 15 min and satellite clock bias, computed by different analysis center, is available on ftp://cddis.gsfc.nasa.gov/gnss/products/. XYZ from sp3 file is further processed using Lagrange interpolation method for obtaining XYZ of satellite for every 30 s (sampling period of observation data). IPP of each epoch was computed using precise XYZ of satellites and receiver, and fitted in normal equation with the data obtained from observation file. After processing Normal Equation with LS estimate, obtained DCB of satellite and receiver were compared with the DCB published by IGS

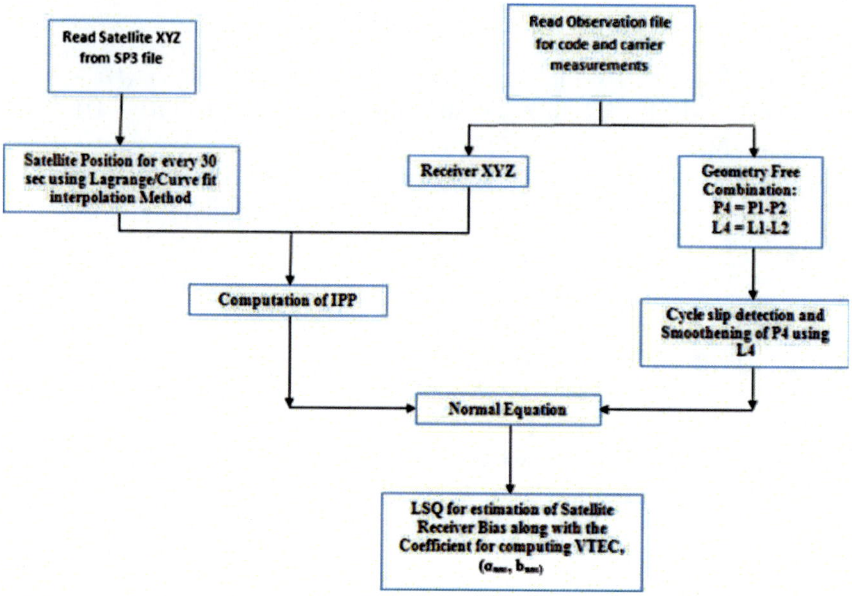

Fig. 1 Block diagram

in IONEX file for that day. IONEX file for given day may be downloaded from ftp://cddis.gsfc.nasa.gov/gnss/products/ionex/.

4 Results and Discussion

4.1 Results by Processing Single Station Data

Data from each IGS station mentioned above processed individually. Maximum of ~±1 ns of difference was seen in the estimated DCB of satellite and receiver with respect to IGS published DCB. For a given six days, variation was observed high on the DCB of both satellite and receiver, same is represented in Figs. 2 and 3.

4.2 Results Using the Multiple Station Data

Based on the DCB results from single station processing, observations from all station were processed together to obtain the more precise values of DCB for satellite and receiver. IGS Stations KAT1, TOW2, SYDN, MOBS, CEDU, NNOR YAR2, and HOB2 were used together for processing. DCB was estimated for DOY 1st, 15th,

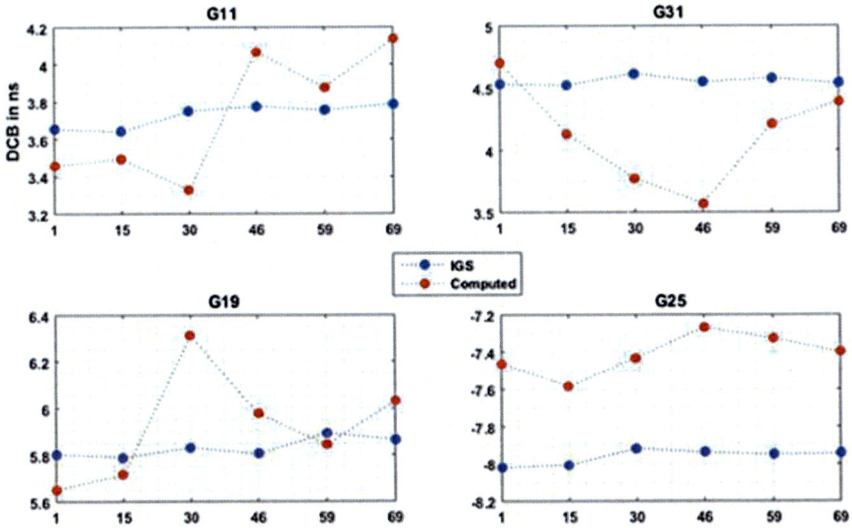

Fig. 2 Comparison of satellite DCB processed using single station data wrt IGS

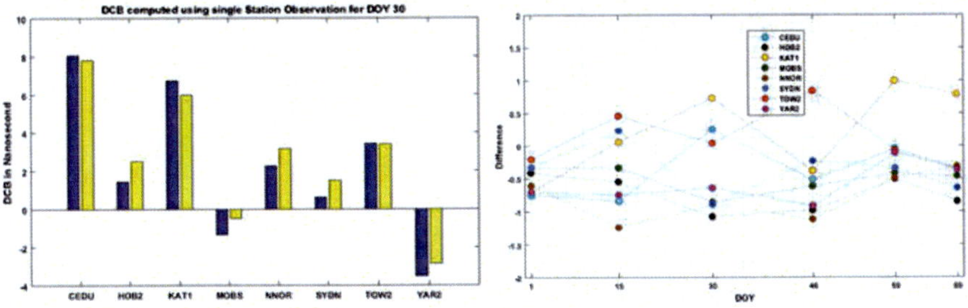

Fig. 3 Difference of receiver DCB computed by data of single station wrt IGS published DCB (left for single day in bar form, right for 6 days)

30th, 46th, 59th and 69th of 2018. Using multi-station the Difference of Satellite and receiver was seen ~±0.5 ns with respect to DCB published in IONEX file (as observed in Figure:4) for most of days. Standard deviation was also seen less than 0.25 ns for all satellite DCBs and receiver DCBs (observed in Figs. 4, 5, 6 and 7). From Figures it is clear that the variation in Satellite DCB is less than the receiver one.

While comparing with the results of Single station based DCB, the Multi-station Processing provide more precise estimate of DCB, and it is confirmed for both satellite and receiver. Similar results were obtained for VTEC estimation, explained in next part of this section. So for obtaining the better and precise estimate of DCB of any receiver along with precise TEC mapping, it is recommended to process the data with other nearby receivers in the network. However the precision depends on the

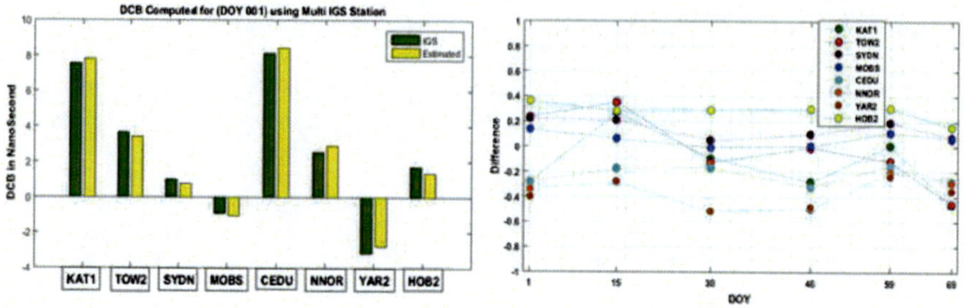

Fig. 4 Difference of receiver DCB computed by Multi Ststion processing wrt IGS published DCB (left for single day in bar form, right for 6 days)

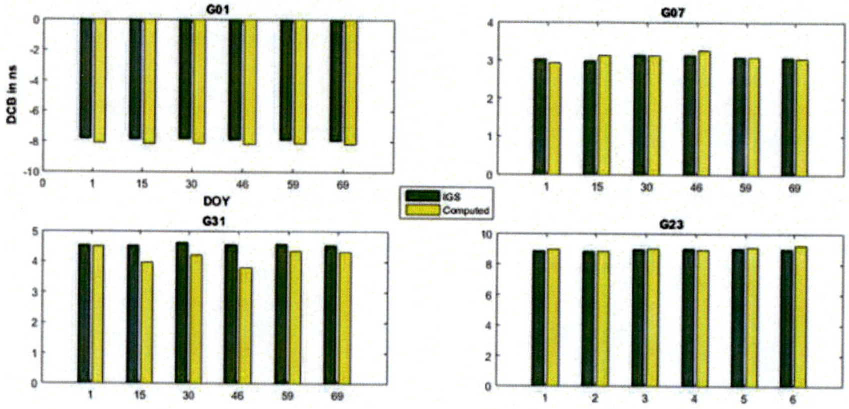

Fig. 5 Comparision (using bar plot) of satellite DCB using multi-station data and IGS published results

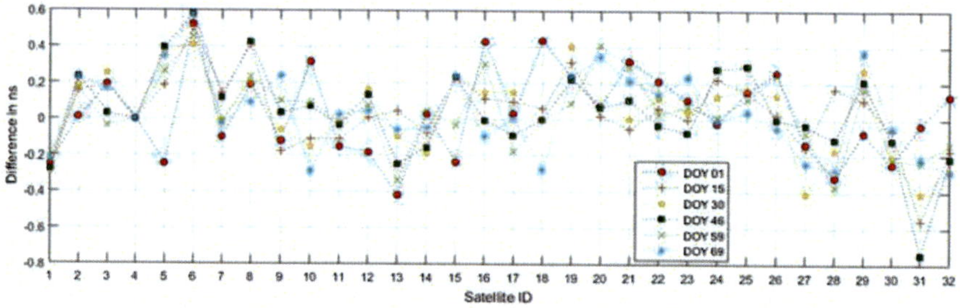

Fig. 6 Difference of satellite (estimated) DCB wrt IGS for all 6 DOY (G04 DCB is set to zero)

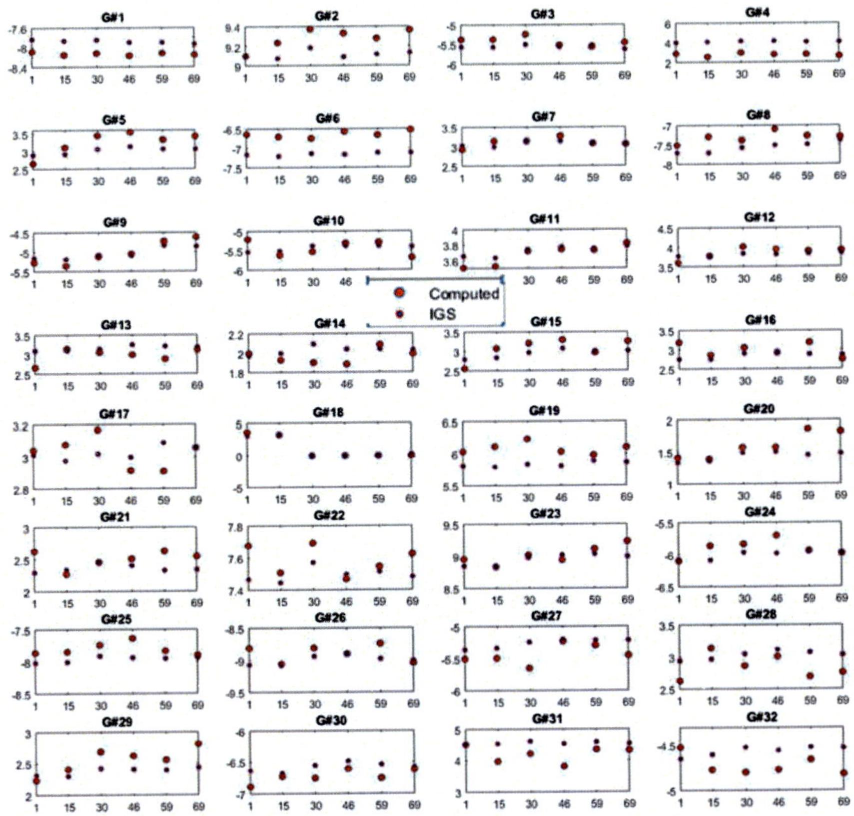

Fig. 7 Comparision plot satellite DCB computed using multi IGS Station wrt IGS published results

location of the station and data quality from that station. Data quality of the station may be degraded by Interference and multipath effect, and subsequently reduces the accuracy of DCB.

Figure 7, is showing the Computed DCB plot of all 32 satellite along with the IGS published value for six days (DOY 1st, 15th, 30th, 46th, 59th and 69th of 2018), almost all the satellite have a difference ~±0.5 ns with respect to IGS results, except the G04 and G18 satellite. G04 satellite is set unusable (However observation data of G04 satellite is available from few stations). G18 was not available for DOY 30, 46 and 59, a jump was observed in the DCB of Satellite G18 on DOY 69 wrt DOY 15. It is actually because of the SVN54 satellite, which was transmitting signal on PRN 18, was set unusable on January 23, 2018, and decommissioned from active constellation on March 5, 2018. SVN54/PRN18 was replaced with SVN34/PRN18 on March 8, 2018 so a jump was observed in DCB value of Satellite (Source: https://www.navcen.uscg.gov) [15]. G06 and G31 were observed in low elevation from most of the station for most of time. Figure 6 indicate the estimated satellite DCB difference

Table 1 Difference of receiver DCB estimated by multiple sations data wrt IGS for DOY 15

Station	Difference in ns
KAT1	0.2685438
TOW2	0.35230996
SYDN	0.21048431
MOBS	0.06305598
CEDU	−0.1743715
NNOR	−0.2865553
YAR2	−0.27649
HOB2	0.28163128

Table 2 Difference of satellite DCB estimated by multiple sations data wrt IGS for DOY 15

Satellite	Difference in ns	Satellite	Difference in ns
G01	0.28168818	G17	−0.0986665
G02	−0.156812	G18	−0.060068
G03	−0.188246	G19	−0.3172412
G04	1.46917808 (unusable satellite)	G20	−0.0188409
G05	−0.1877846	G21	0.05320586
G06	−0.5042369	G22	−0.0615578
G07	−0.1455412	G23	0.00116016
G08	−0.4077485	G24	−0.2229592
Go9	0.18029341	G25	−0.1555809
G10	0.11077183	G26	−0.0247796
G11	0.11217007	G27	0.15506378
G12	−0.0085269	G28	−0.1704074
G13	−0.0446434	G29	−0.1055954
G14	0.07299587	G30	0.06081657
G15	−0.2403858	G11	0.5561648
G16	−0.1151435	G32	0.17325759

with respect to IGS for all six days, most of the satellites are under the boundary of ~±0.5, difference is also indicated in Table 2 for DOY 15. Difference of computed DCB of receiver, used in a processing, wrt is IGS for DOY 15 is in Table 1.

4.3 VTEC Results

As discussed above in the Sect. 2, Ionosphere is represented by Spherical harmonic (SH) function and the Ionospheric coefficient representing SH function (SH coefficients) were estimated for every 2 h along with the Satellites and receiver DCB using Multi-Station observations. Estimated Ionospheric coefficients were used to extract the total electron content in the area covered by these receivers. The receivers are located at the Latitude of range from −14° to −42°, and the longitude of 115°–151°. VTEC is computed, with a temporal resolution of 2 h, for different location of Ionospheric layer at the height of 450 km and compared with the VTEC for same location in the GIM map of IONEX file, found a difference of ~1TECU between them. Figure 8 represent the TEC comparison plots for four different IPPs, these plots shows good agreements between computed and IGS published results.

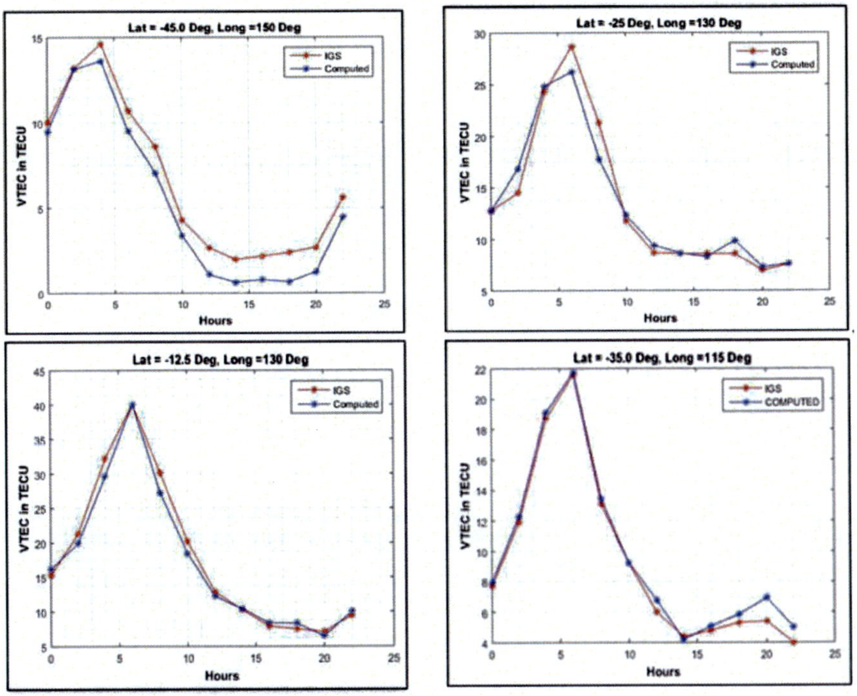

Fig. 8 Comparision of estimated total electron contents

5 TGD and DCB Comparison

Each GPS Satellite Broadcast the TGD known as timing group delay for single frequency user in real time for satellite clock correction as broadcasted clock correction contains linear combination of satellite biases at f_1 and f_2 frequency [4, 9]. TGD is computed on regular basis using the observations of calibrated reference stations. As reference stations are calibrated on routine basis so absolute satellite bias can be computed and represented in the form of TGD, whereas in case of DCB, the satellite and receiver DCB are separated by assuming the mean DCB of satellite in constellation is equal to zero, So DCB is computed on relative basis and reflect the differential hardware delay between two different code observations, on the same or two different frequencies [13]. On the basis of pseudo range equations and Iono free pseudorange equation the relationship between TGD and DCB may be explored as follows [4]:

$$\text{TGD} = \left(\frac{1}{1-\gamma}(D_1^s - D_2^s)\right) = \frac{1}{1-\gamma}\text{DCB} \tag{17}$$

where

$$\gamma = f_1^2/f_2^2$$

Figure 9, is representing the relationship of TGD and DCB. A constant bias is observed for all the satellite while taking the difference of the broadcasted TGD with TGD computed from DCB (using Eq. 17). Constant Difference is because of relative nature of DCB. Equation 17 is showing that TGD is not a mean satellite group delay differential but actually mean satellite group delay differential multiplied by factor $1/(1-\gamma)$.

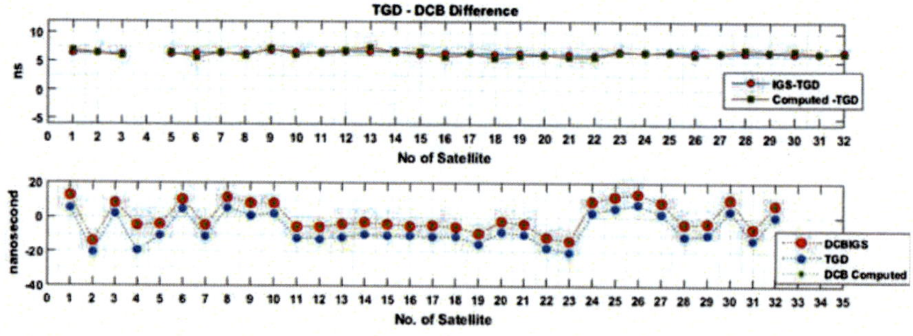

Fig. 9 TGD versus DCB comparision

6 Summary and Conclusion

As per the results shown above, DCB value of receiver and satellite have a good agreement with the IGS provided DCB of satellite and receiver. It was seen that estimation of DCB for any receiver along with TEC content may be computed more precisely using Multi-Station method rather than single station. Using a grid of receiver a DCB can be estimated with a difference of ~±0.5 ns with respect to the IGS data, where as a close agreement of ~±1 ns was observed with single station processing. Similarly the VTEC mapping may be done in global or regional fashion using the method with the accuracy of 1 TECU. Further GPS Ionospheric/DCB estimation accuracy is depend on the mathematical model describing the VTEC. Here a processing is done using 4th order spherical harmonic function, however IGS is taking the weighted sum of different Ionospheric Associated Analysis Center. Using a SH function of higher order or degree can fit the VTEC better however higher order is essential for Global Network. Estimation further depends on the density of GPS station and mapping function. More GPS stations for processing shall improve the accuracy for estimation. The GPS data processing model can also affect the accuracy of DCB/VTEC. The Carrier smoothening and taking the data with higher elevation (reducing the multipath noise here, cut off is taken greater than 15°) may be beneficial for enhancing the accuracy of estimation. The Accuracy of receiver DCB is location dependent. Estimation accuracy could be affected in equatorial anomaly region, because of more complicated TEC distribution in that region. In short a bigger and distributed network of GPS/GNSS station offers more observation, better elevation cut off and subsequently improves the accuracy of estimation.

Acknowledgements This work was carried out at ISTRAC/ISRO. Authors would like to be grateful to Director ISTRAC, Associate Director ISTRAC and entire IGSA/NSA team for their kind support, motivation, and encouragement. Authors would like to be thankful to IGS community for providing quality Data and precise products for GPS satellite.

References

1. Schaer, S., Steigenberger, P.: Determination and use of GPS differential code bias values. In: IGS workshop, Darmstadt, Germany, May 8–11
2. Li, Z., Yuan, Y., Li, H., Ou, J., Huo, X.: Two-step method for the determination of the differential code biases of BDS satellites. J. Geodesy. **86**, 1059–1076 (2012); Chen, W.-K.: Linear networks and systems (book style), pp. 123–135. Wadsworth, Belmont, CA (1993)
3. Zhang, Q., Zhan, Q.: Global Ionosphere mapping and differential code bias estimation during low and high solar activity periods with GIMAS software. Remote. Sens. **10**, 705 (2018). https://doi.org/10.3390/rs10050705
4. Gun, F., Zhang, X., Wang, J.: Timing group delay and differential code bias corrections for BeiDou positioning. J Geod **89**, 427–445 (2015)
5. Sanz, J., Juan, J.M.: GNSS measurements and their combinations. gAGE/UPC
6. www.aibu.unibe.ch

7. Ammar, M., Aquino, M., Vadakke Veettiiil, S., Andreotti, M.: Estimation and analysis of multi-GNSS differential code biases using a hardware signal simulator. GPS Sol. (2018)
8. https://gssc.esa.int/navipedia/
9. Sardonand, E., Zarraoa, N.: Estimation of total electron content using GPS data: how stable are the differential satellite and receiver instrumental biases? Radio Sci. **32**(5), 1899–1910 (1997)
10. www.igs.org IGS. IONEX: The IONosphere Map Exchange format version 1. ESA/ESOC, Darmstadt, Germany, 25 Feb 1998
11. Silva, P.: Cycle slip detection and correction for precise point positioning
12. Montenbruck, O., Hauschild, A.: Differential code bias estimation using multi-GNSS observatios and global ionosphere maps. ION-ITM-2014, 27–29 Jan 2014, San Diego, USA
13. Zhang, D.H., Zhang, W., Li, Q., Shi, L.Q., Hao, Y.Q., Xiao, Z.: Accuracy analysis of the GPS instrumental bias estimated from observations in middle and low latitudes. Ann. Geophys. **28**(8), 1571–1580 (2010)
14. Schaer, S.: Mapping and predicting the earth's ionosphere using the global positioning system. In: Ph.D dissertation, Astronom. Institute of University Bern, Berne, Switzerland (1999)
15. Navigation Center: https://www.navcen.uscg.gov, GPS Almanacs

Design and Development of a Formation Control Scheme for Multi-robot Environment

Harika Pudugosula and M. Rajesh

Abstract Coordination and control are the main central essential requirements for the successful accomplishment of a task in a multi-robot system. Some of the coordinated missions are localization, exploration etc., and the control mission includes positioning, formation control between the robots. The formation control problem consists of designing control inputs for the robots and the localization maintains a predefined geometric shape in space even in the presence of external disturbances. The positioning is designed in such a way that the robots correct their position looking at the position of their neighbors. One of the techniques used is the leader–follower approach. This paper guarantees that the robot avoids the static and dynamic obstacles using onboard sensors and form a geometrical shape based on the requirement of the task.

Keywords Autonomous system · Formation control · Leader–follower algorithm · Webots simulator

1 Introduction

Nowadays, for the attainment of complex tasks, deploying a single robot system will consume more time which will lead to delay in the sharing of data. To deal with this, multi-robot system has gained a lot of attention. The system in which a team of robots, that can interact, coordinate, and communicate with each other and fragments a complex task into subtasks and instructs these subtasks to the individual robots can be termed as multi-robot system. Such an environment is a multi-robot environment. The coordination and control of every robot in such an environment

H. Pudugosula · M. Rajesh (✉)
Department of Computer Science and Engineering, Amrita School of Engineering, Bengaluru, Amrita Vishwa Vidyapeetham, Bengaluru, India
e-mail: m_rajesh@blr.amrita.edu

H. Pudugosula
e-mail: harikapudugosula06@gmail.com

© The Editor(s) (if applicable) and The Author(s), under exclusive license to Springer Nature Singapore Pte Ltd. 2021

A. P. Pandian et al. (eds.), *Proceedings of International Conference on Intelligent Computing, Information and Control Systems*, Advances in Intelligent Systems and Computing 1272, https://doi.org/10.1007/978-981-15-8443-5_70

can be done within the time premises. So the multi-robot system has become feasible in the implementation of real-time environment.

In the past two decades, for solving complex problems in several applications, i.e., enclosing and tracking a target, exploring an area, data sharing, sensing grids, etc., multi-robot system has been taken the place of a single robot system. The formation of the pattern by a team of a robot is the main central to all the before-mentioned tasks. In real-time implementation of multi-robot system, robots will run into some external disturbances, i.e., sudden blocks, irregular surfaces, external forces like sun heat, wind, etc., which decay the performance of the entire system. So to master it, the controller has to be designed in such a way that it can overwhelm external disturbances and able to radiant point to the desired geometric shape. In this paper, to achieve coordination and control of multi-robot system centralized coordination has been considered where the leader robot commands the fellow robots to execute the task, respectively. To develop and design a multi-robot system, i.e., theoretically and practically, quite a few assumptions have been considered. The experimental test bench has been established for control and monitoring.

2 Related Work

In the branch of robotics, numerous researchers have been working to increase robotic technology in our day-to-day life. Nowadays, in many of the rescue operations like data mining and military operations, humans have been replaced by robots. Not only in those areas but also in industries and factories, robots have been used to reduce the human effort. Some of the applications where robots have been taken the place of humans were in exploring, medical applications, transportation, etc. To implement all these applications in real time or simulation, a robot has to follow a path which is nothing but a 'pattern formation.' Pattern formation is an elementary activity to master all these applications. The design and implementation for pattern formation can be done in either of the approaches ranging from a decentralized algorithm which was termed also as 'leader-less' approach, centralized coordination, i.e., 'leader–follower' approach which was based to a virtual leader. Each one of the approaches has its own merits and demerits. In case of a decentralized algorithm, the movement of an individual robot has been calculated by themselves with the help of sensors, equipped on them. But in a centralized algorithm, the movement of a robot and distribution of a task were decided by the fictitious leader. The tasks are fragmented into smaller tasks and carried out by a team of the robot in the presence of the leader; simultaneously the motion plan of an individual robot has been accomplished by the leader to fellow robots. The vital role of the leader was to communicate/command individual robot within the time premises to neat up the task successfully.

Jisha [1] used the rigid graph theory for describing the multi-robot pattern formation in which it uses graph theory concepts and rigidity concepts. In rigid graph theory, the multi-vehicle systems are modeled by a graph in which agents are represented as vertices and as edges, inter-agent distance is considered. To track the trajectory and to acquire information about control inputs, the back-stepping control technique has been used. Gautam [2] proposed an experimental setup for positioning multi-robot system in an indoor environment using a low-intensity camera. The author has considered two distributed algorithms, i.e., round-robin and dining philosopher, compared each among them. The first algorithm has been proven with assumptions that are not feasible in the real world, the second algorithm outperforms the first one. The main drawback of the system is while communicating, the system has been affected with external noises which led the system in loss of data and there is a possibility in the failure of the entire system.

A multi-robot system has been proposed [3] to avoid an obstacle in an unknown environment. The author has used Pioneer3 mobile robot platform to simulate the multi-robot system. By using the leader–follower algorithm [4], various pattern formations have been done, and by using artificial potential field algorithm, robots avoid an obstacle. The main drawback of this model was a leader robot does not consider the movement of the following robot during the calculation of its path. If in case the leader moves too fast or a follower robot avoided by an obstacle, the system will be failed. Bhavana et al. [5] proposed a coordination and control scheme for multiple mobile robots to do a task by avoiding obstacles [6]. Leader–follower approach which was considered a centralized technique has been used to communicate and provide information between a team of robots. In this work, the leader robot was equipped with ultrasonic sensors to avoid and to detect an obstacle. The communication between robots has been done through Bluetooth technology which uses universal asynchronous receiver and transmitter (UART). Here, follower robot has equipped with an infrared sensor so that they can sense the leader robot. The communication between robots has been done by Bluetooth technology which is of short. If in case robots are far away from each other, the communication becomes difficult at times it may lead to failure in the entire system. Madhevan et al. [7] proposed tracking algorithm for a multi-robot system by using the leader–follower algorithm (LFA). Improved control techniques were introduced to enhance applications of wheeled mobile robots [8]. The tracking algorithm has been implemented in a known environment where the trajectory path was assigned in prior.

Thus, the implementation of this system in a known environment was a major drawback. Unlike the work done in [3], a follower robot has also been considered to calculate any point to track the destination. The problem in a short range of communication that has occurred in [5] has been avoided by using nRF24L01 which has high range of communication. The model has been implemented in an unknown environment.

The organization of the paper is as follows, Sect. 3 deals with an implementation of the system where the model of the entire system and the simulation results of obstacle avoidance and hardware output of the model in the presence of obstacles. Section 4 deals with the test cases and explains how effectively the model performs. Finally, the paper is concluding in Sect. 5.

3 Implementation

The implementation phase of this project is divided into two categories; they are software implementation where the process of designing and development of a formation control scheme for multi-robot environment that avoids the static and dynamic obstacles using onboard sensors and to form a geometrical shape based on the requirement of the task has to be done. The next phase is to implement the above scheme in the real-time environment where hardware implementation plays an important role. The detailed description of the two phases has been discussed in the following sections.

The architecture of the framework for the entire implementation of this project is shown below in Fig. 1.

The leader robot module has user interface where a user provides commands through the code and classification of task where the tasks get classified, object identification block where the recognition of an object takes place, based on these two, decisions can be achieved by using decision on shape formation, obstacle avoidance used to avoid obstacles in an environment. The communication can be achieved by using wireless technology. In the follow robot module, there are two blocks which are used to avoid obstacles and another is to follow commands from the leader robot module.

In the classification of task block, the task has been predefined. The task includes lifting an object, pushing an object, pulling an object. In the object recognition block,

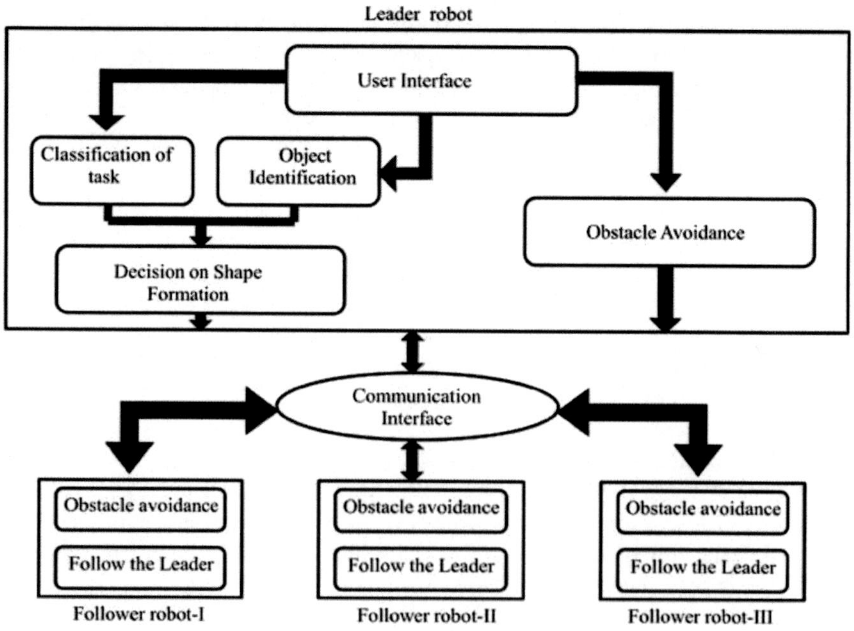

Fig. 1 Architecture of framework

the object that has to distinguish will be provided earlier to the leader robot module. The objects to differentiate are ball, box, and a string. Depending on the output of object recognition block and classification of task block, the decision will be decided by the decision of shape formation block.

An obstacle avoidance block is used to avoid an obstacle. The entire information will be communicated to follow robot modules with the help of a communication interface and this is achieved by using nRF24L01. The follow robot modules will only communicate with the leader robot module; there will be no communication between one follow robot module and another follow robot module. Once follow robot modules receive commands from the leader robot module through the communication interface, they follow the commands properly. If anyone of the robots gets fail, physically the robot has to be replaced. There will be no other algorithm internally takes place to replacement of failed robot.

3.1 *Software Implementation*

The software implementation of this work is done by using Webots [9, 10], a robotic simulator, which has been widely used for educational purposes. The e-puck robot has been used to implement the model. The obstacle has been avoided by both the leader robot and follower robots.

Figure 2 illustrates the entire project setup which is done in Webots simulator. It explains that the robots are moving independently in a rectangular arena of size (1X1). Figure 3 describes that the robot avoids the wall and turns toward the obstacle-free zone and their movement can be easily visible in their respective cameras. Figure 4

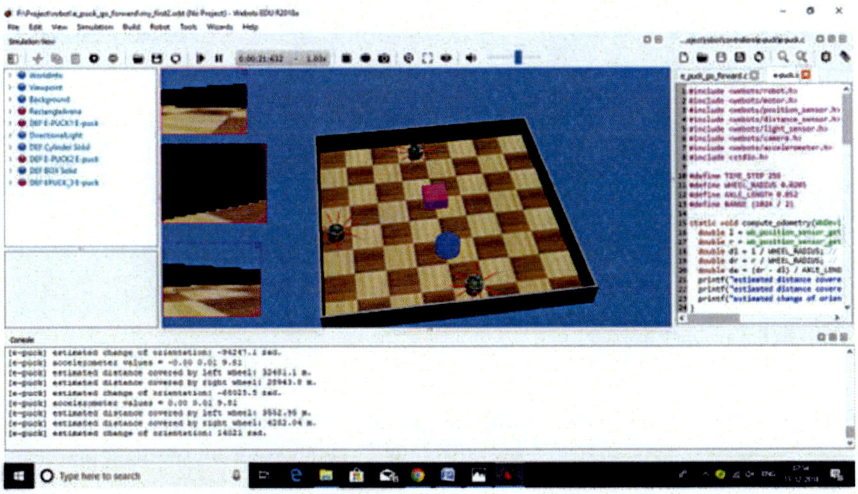

Fig. 2 Simulator window of the Webots simulator

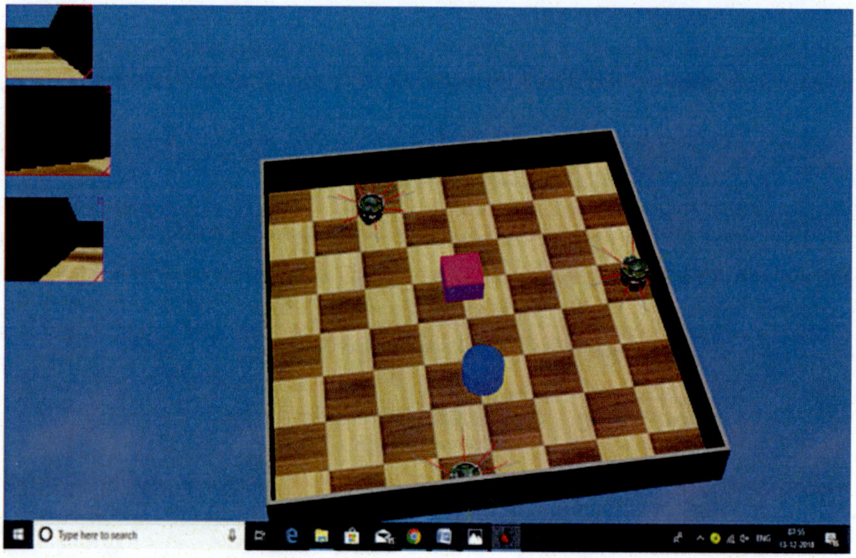

Fig. 3 Robots are avoiding the wall

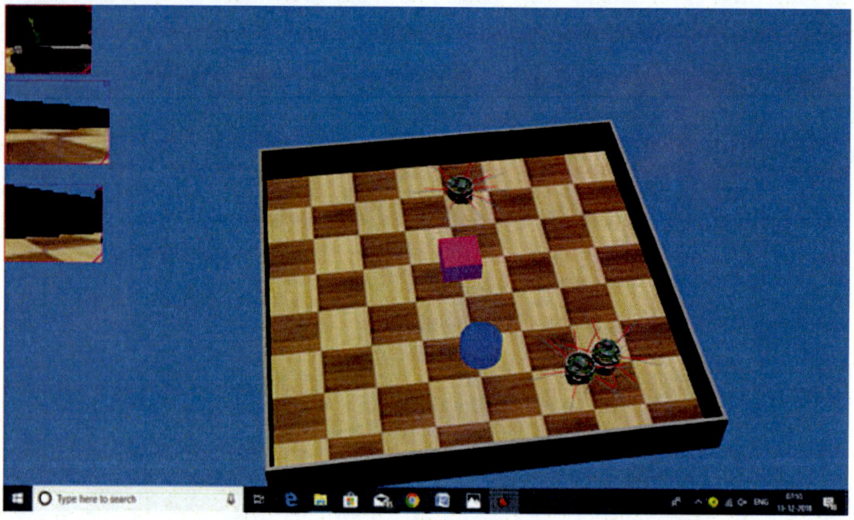

Fig. 4 Robot avoiding another robot

shows how the robot will distinguish one robot to another robot and how it avoids the robot, respectively. Figure 5 explains how a robot avoids an obstacle and takes path away from an obstacle.

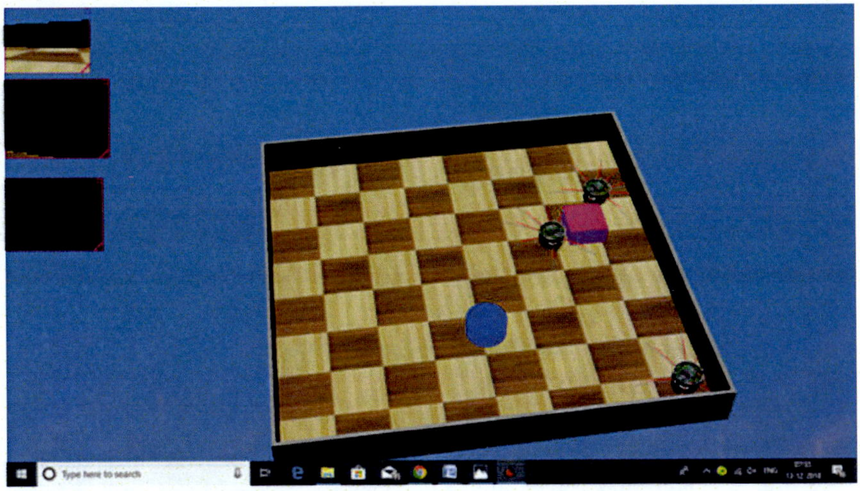

Fig. 5 Robot avoiding an obstacle

3.2 Hardware Implementation

The robots are equipped with Arduino microcontroller [11] in the hardware implementation which is based on Arduino Uno. It has six analog pins, 14 digital pins, a power jack, and a reset button. The communication between robots is done by using nRF24L01 wireless communication technology. The nRF24L01 which is used in implementation is shown in Fig. 6.

The entire system is embedded in the four-wheeled robot chaos. DC motors are used to drive the wheels to which the requirement of extra 9 V rechargeable battery has equipped. The module setup consists of Arduino Uno, robot chaos, DC motors, nRF24L01, ultrasonic sensors, and a 9 V rechargeable battery.

The serial communication has achieved among robots by using the nRF24L01 module. This wireless technology uses a half-duplex mode of communication. The

Fig. 6 nRF24L01 communication module

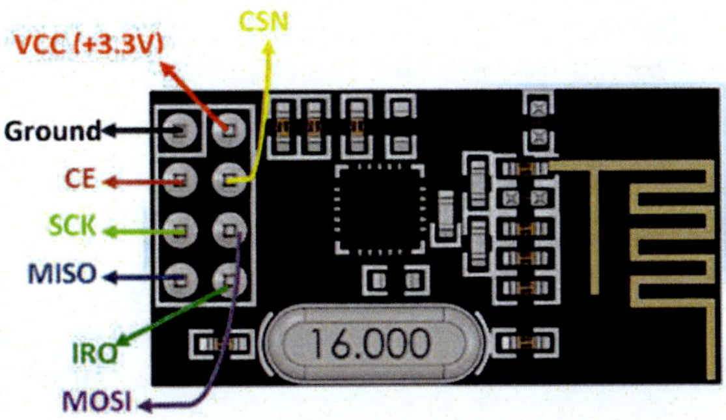

Fig. 7 Diversion from an obstacle and the follow robots follow the leader robot

transmission of data between the leader and follower robot happens and vice versa will not happen.

Figure 7 explains how the leader robot avoids an obstacle and take diversion path away from it. The follow robots follow the commands from the robot and follow along with the leader robot.

4 Test Cases

The entire has been implemented and tested by using Webots robotic simulator. A rectangular arena of size (1X1) has been created. To implement the robotic environment, an e-puck robot has been used, and an external disturbance has also been created. Now a particular factory kind of environment has been designed and developed by using Webots simulator. The leader–follower approach has been successfully tested and implemented both in the hardware and in the software. The simulation is done using Webots simulator. The hardware implementation is achieved primarily in Arduino IDE. Table 1 describes the test cases that have been implemented in the system.

Test case 1: The robot has to avoid the obstacles.

This test case has been successfully tested and implemented both in the hardware and in the software.

Test case 2: The robot has to distinguish between robots and obstacles.

This test case has been successfully tested and implemented in software but in hardware, this will be implemented in further.

Table. 1 Test cases

Test cases	Software	Hardware
Avoidance of obstacles	✓	✓
Distinguish between robot and obstacles	✓	✗
Leader–follower algorithm	✗	✓
Communication	✗	✓

Test case 3: To implement leader–follower algorithm.

The leader–follower algorithm has been implemented in hardware but not in the software part.

Test case 4: To establish communication between robots.

Effective communication has been done in case of hardware but not in the software part.

5 Conclusion and Future Scope

The design and development of a formation control scheme for multi-robot system have been implemented. By using the leader–follower algorithm, the entire project has been implemented. The communication was established between the robots and successfully implemented in both the software part and hardware part.

The development of a multi-robot has been done. The formation of various patterns can be done in future. Several functionalities like lifting an object, pushing an object, pulling an object can also be done.

Reference

1. Revathi, B.R., Jisha, V.R.: Formation control of mobile robots in the presence of disturbance uncertainties. In: International Conference on Emerging Technological Trends (2016)
2. Gautam, A.: Positioning multiple mobile robots for geometrical pattern formation. In: International Conference on Contemporary Computing (2014)
3. Zhang, T., Li, X. et al.: Formation and obstacle avoidance in the unknown environment of multi-robot system. In: International Conference on Robotics and Biomimetics (2009)
4. Dai, Y., Lee, S.G.: Formation control of mobile robots with obstacle avoidance based on GOACM using on board sensors. Int. Control Autom. Syst. (2014)
5. Bhavana, T, Nithya, M.: Leader-follower coordination of multiple robots with obstacle avoidance. Smart Tech. Con. (2017)
6. Nippun Kumaar, A.A., Amrita, L.P.: Bluetooth RSSI based collision avoidance in multirobot environment. In: International Conference on Advances in Computing, Communications and Informatics (2016)
7. Madhevan, B., Sree Kumar, M.: Tracking algorithm using leader follower approach for multi robots. In: International Conference on Design and Manufacturing (2013)
8. Ge, X., Han, Q.-L.: Distributed formation control of networked multi-agent systems using a dynamic event-triggered communication mechanism. IEEE Trans. Ind. Electron. (2017)
9. Sharma, H., Rajesh, M.,: Design and Simulation of WSN for Zigbee based communication in multi-robot system. In: International Conference on Energy, Communication, Data Analytics and Soft Computing (2017)
10. Magyar, B., ForhCcz, Z., Korondi, P.: Developing an efficient mobile robot control algorithm in the webots simulation environment. In: IEEE International Conference on International Technology (2004)
11. Rajesh, M, Joseph, R., Sudarshan, T.S.B.: Fully distributed and decentralized map building for multi-robot exploration. In: International Conference on Embedded Systems (2014)
12. Cyberobotics webpage. https://www.cyberbotics.com

High Throughput FIR Filter Architecture Using Retiming and Fine-Grain Pipelining

K. P. Heena

Abstract A high throughput FIR filter architecture having low critical path delay, area-delay product, and power-delay product for various DSP applications is proposed in the paper. Architecture uses square root carry select (CSLA)-based adders and radix-4 booth multiplier to perform high-speed addition and multiplication operation, respectively. Here, retiming and fine-grain pipelining are the two techniques used to improve the throughput. To reduce the critical path length, retiming technique is applied to the direct-form FIR filter structure without increase in latency. In this paper, fine-grain pipelining scheme is applied to the more time-consuming path to improve the throughput. Pipelining of final adder section in retimed tree-based filter structure had reduced the average power dissipation and latency. The design and implementation are done using Cadence software with generic process design kit (gpdk) of 45 nm technology.

Keywords FIR filter · Carry select adder · Retiming · Fine-grain pipelining · Cadence

1 Introduction

In signal processing, a filter is a device that involves the suppression of some unwanted aspect of the signal. Among analog and digital filters, the most precise and the filter that has better signal-to-noise ratio (SNR) is the digital filter. FIR and IIR are the two digital filters that are widely used. FIR filters have more advantages when compared with IIR filter such as it has finite duration unit sample response, they are stable, requires no feedback, finite memory requirement, whereas IIR filter has infinite duration unit sample response, infinite memory requirement. FIR filter structure consists of an adder, multiplier, and a delay unit. Designing of FIR filter

K. P. Heena (✉)
Electronics and Communication Engineering Department, LBS College of Engineering, Kasaragod, India
e-mail: kpheenakp96@gmail.com

© The Editor(s) (if applicable) and The Author(s), under exclusive license to Springer Nature Singapore Pte Ltd. 2021

A. P. Pandian et al. (eds.), *Proceedings of International Conference on Intelligent Computing, Information and Control Systems*, Advances in Intelligent Systems and Computing 1272, https://doi.org/10.1007/978-981-15-8443-5_71

having high throughput and low power is an important point of discussion in DSP applications. In order to improve the performances of FIR filter, various techniques and modifications have been developed and incorporated by many researchers. Techniques like pipelining and parallel processing can be used to improve the throughput by reducing the critical data path. The idea behind pipelining is that new data can be processed before the prior data has finished. Advantage is that it reduces the longest critical data path and increases the clock speed (or sampling speed). Parallel processing involves duplicating of functional blocks to operate different tasks simultaneously and to produce multiple outputs at the same time. As it involves L-fold, it requires replication of architecture, thus increasing the area requirement.

The three main factors that a VLSI technology requires are less power, small area, and high speed. By applying delay transformation (retiming) to digital signal block, the computation time of digital circuits can be reduced. Computation time is greatly affected by the critical path delay. Retiming is a technique which can be used to improve circuits critical path by better balancing of delay elements between different pipelining stages, to improve its performance. It does not change the latency of the system. Introducing delays at the input followed by retiming is same as that of pipelining. Generally, retiming is done to meet the clock period minimization (minimizing the clock period of the circuit) and register minimization that is required to implement the circuit.

Design used for implementing adaptive filters makes use of large number of multipliers and more cycle periods. In paper [1], a design that uses a set of multipliers and weight increment block for generating the filter outputs is proposed. These computations are then divided and mapped to a folded structure which had reduced the multiplier resources. Weight increment term makes use of error value for determining the weight vector required for the next iteration. Author of [2] developed a structure to make area-delay-power efficient to achieve lower adaptation delay and also used partial product generator and pipelining technique which had improved throughput. In paper [3], a pipelined LMS adaptive filter is developed. Generally, for a filter length L, the filter output is delayed by L-samples. Here, throughput is independent of the filter taps. Author of [4] used both direct and transpose form filter structures and are combined to reduce the latency, but increased the area. In [5], author had used Wallace tree multiplier for multiplication in FIR filter, as it is an efficient high-speed multiplier and an improved tree-based multiplier architecture. It uses carry save addition logic to reduce the overall latency. It had improved the efficiency of the filter by reducing the delay by an increase in area.

Many of the applications require high-speed processing of data. Unfolding is a technique that can be used to improve the throughput of the system by replicating the functional blocks, without increasing the number of delay elements. This concept was used by the author in paper [6]. The author had applied the unfolding technique to the adder section and Vedic multiplier was used for multiplication. As it involves duplicating of blocks, it increased the area requirement. A study on different FIR filter structures was made in [7]. Implemented filter using these structures and concluded that direct-form structure has high operating frequency and for an optimized area filter, fine-grain structures can be used.

Power dissipation is another factor that needs to be considered while designing the filter. In combinational and sequential circuits, the major cause for power dissipation is the switching activity. Author of [8] used retiming technique to reduce power dissipation in sequential circuits and selected a set of nodes in the pipelined circuit based on the glitches present at the output having flipflops. This had reduced power dissipation to a greater extent. In [9], DSP filter block is retimed by using the technique like cutset and by reducing the clock period and improved performance of the filter by using high-speed adders in the circuits. By using retiming technique, the number of registers used is minimized along the clock period. Author of [10] had used sequential optimization retiming technique to achieve reduced clock period and used less number of registers and by retiming the sequential circuits the critical path was reduced.

In [11], a scheme for low power implementation of high throughput FIR filter is used. It makes use of a multiplier architecture, which employed coefficient block to reduce the power consumption and also incorporated sign magnitude data representation which further reduced the power consumption with a slight increase in area. In [12], fine-grain pipelining technique is applied to both multiplier and adder, and it used two-dimensional pipelining technique and achieved improvement in power awareness and reduced the overall latency. As adders lie in the critical path and in the non-truncated FIR filters, they have the greater word size, hence, delay of adders greatly affects the speed of the design and achieved improvement in power awareness and reduced latency with increase in area. In [13], author has done pipelining to the adder. Throughput of the adder improved with an increase in area requirement.

Pramod and Shahana [14] have brought the concept of retiming in filter section in order to improve the throughput. Retiming is the repositioning of delay elements in the circuit without changing the input–output characteristics. They proposed two designs, one with retiming and three-level pipelining and the other with retiming and fine-grain pipelining. Filter taps are grouped and by changing the position of delay elements achieved improvement in critical path delay. In [15], two versions of pipelined adders using ripple carry adder (RCA) are implemented. Ripple carry adders involve rippling of carry which leads to an increased delay. As the bit width increases, the delay incurred will be more. By generating carries independently, the speed of operation can be increased, this concept was used in [16], but increased the area. Bahadori [17] had made improvements to carry skip adder by using increment adder block and used gates having reduced power for skip logic operation. The delay performance in adder structure is improved, but had increased hardware complexity.

Pramod and Shahana [18] had proposed two versions of adder structures and used concatenation and incrementation schemes in order to achieve improved delay and energy efficiency. By improving the performances of multiplier, throughput can be further improved. In [19], approximate 4–2 compressor and approximate tree structure were used which had increased the speed and area requirement to implement approximate booth multiplier. Multiplier proposed by the author [20] has shorter critical path delay and more power efficient and requires less area. As odd multiples of the multiplicands are to be generated, they are slow. In [21], modifications are made to the first stage of the conventional radix-4 booth multiplier to decrease the power-delay product. Booth multipliers of radix-4 are more efficient due to the generation of $n/2$ partial products, and the number of hardware implementations required is reduced.

Pramod and Shahana [22] had focused mainly on retiming technique and modifications in adder and multipliers. Two adder structures were proposed, and modified square root carry select adder (CSLA) was used for addition. Radix-4 booth multiplier is used for multiplication operation. Square root adder is incorporated in multiplier to improve the delay performance. Two filter structures, filter 1 and filter 2, were proposed. Filter 1 was designed using RCA and radix-4 booth multiplier and filter 2 with square root CSLA and improved multiplier (multiplier with square root CSLA-based adder). Improvement in delay was achieved by small increase in area.

This paper provides a technique to improve the throughput of the retimed-based filter structure. To reduce the critical path and to improve throughput, retiming and fine-grain pipelining techniques are used. Fine-grain pipelining technique is applied to the final adder section.

The paper is systematized as: Filter design is discussed in Sect. 2, performance comparison in Sect. 3, and conclusion in Sect. 4.

2 Filter Design

Latency is the time required by the input samples to produce its corresponding output, and throughput of a system is the output produced per unit time. Several techniques are used to improve the critical data path length. To implement high throughput FIR filter, two-pipelined structures can be considered: pipelined direct-form FIR and transposed form structure. An N-tap filter consists of N multiplier and $(N-1)$ adders. Let the delay incurred by each multiplier and adder be t_{m} and t_{a}, respectively. Then the maximum sampling frequency is given by:

$$f_{\text{sampling}} = \frac{1}{t_{\mathrm{m}} + (N-1)t_{\mathrm{a}}} \tag{1}$$

In direct-form FIR filter, critical path length increases as the filter length increases, thus throughput decreases (number of adders in the critical data path increases). By pipelining the direct-form FIR, the critical path can be reduced to $t_{\mathrm{m}} + t_{\mathrm{a}}$. Generally, for a pipelined design, the critical path is the longest path between input and pipeline register or two-pipeline register or pipeline register and output or input and output. In transposed form filter, input is broadcasted to all multiplier at the same time, which is in contrast to direct form (input reaches the multiplier at different clock cycles). It requires less number of registers. Here, critical path length is reduced to the delay incurred by a single adder and a multiplier unit and is given by, $t_{\mathrm{m}} + t_{\mathrm{a}}$. Direct-form filter structure has a latency from input to output but it is not in the case of transposed filter structure. Considering the number of pipeline registers and latency, transposed filter structure is preferred.

This paper provides a technique to improve the throughput of the retimed-based filter structure [22]. In [22], retiming is applied to the direct-form FIR filter structure. Retimed FIR filter structure is designed using square root carry select adder [22] and ripple carry adder (RCA) for addition and radix-4 booth multiplier [21] for multiplication. Multiplier with square root carry select adder and RCA for partial product additions is used. In order to have reduced critical path delay, square root carry select adder and multiplier (with square root carry select adder) are placed along long delay paths. Hence, the critical path delay of the retimed filter is reduced by:

$$t_r = t_{m1} + t_{af} \tag{2}$$

where t_{m1} represents the computation time taken by the first multiplier and t_{af} is the computation time taken by the final adder that is the critical path is reduced to a single multiplier and an adder unit. In order to further reduce the delay and to improve the throughput, fine-grain pipelining technique is used.

Fine-grain pipelining is a method of dividing a component into two subsections and delay units are inserted in such a way that the critical data path and execution time are reduced. As adder has greater word size and lies in the critical path, the delay incurred by the adder greatly affects the speed of the design. Hence, the technique is applied to the final adder (a_f). Figure 1 shows the general block diagram of fine-grained final adder. Each stage [stage0, stage1, stage $(n - 1)$] represents the n-bit adder consisting of k-bit carry propagate–generate block, NAND network carry block, block for selecting the carry depending upon the input carry, and block for generating the final output and intermediate carry generation block. Fine-graining of adder is done by delaying long critical path from input to output. Here, first section of the adder input starting from the most significant bit (MSB) is delayed and then the remaining output section of the adder in least significant bit (LSB). Hence, critical path delay of the fine-grained filter is reduced to less than the delay incurred by the final adder (a_f):

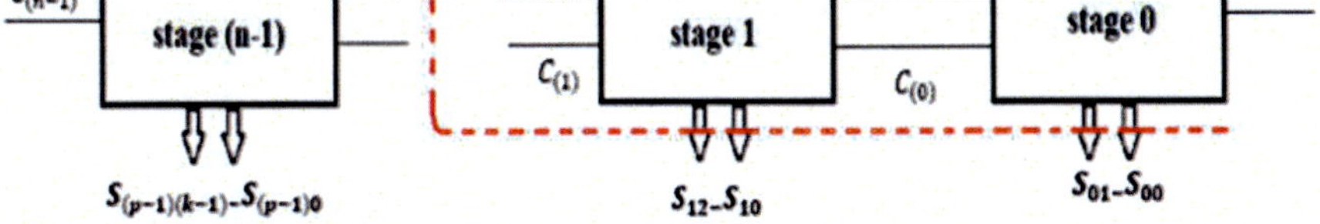

Fig. 1 General block diagram of fine-grained final adder

$$t_{\mathrm{rf}} < t_{\mathrm{af}} \tag{3}$$

3 Performance Comparison

Fine-grain pipelining technique is applied to 16, 32, and 64 taps retimed filter structures. Table1 shows performance comparison among various filters for all the 3 taps.

The filter structures are numbered as:

1. Proposed filter represents the filter in which the final adder is fine-grained.
2. Filter 2 (HTR-FIR2) represents high throughput retimed filter 2 [22].
3. Filter 3 (HTR-FIR1) represents high throughput retimed filter 2 [22] using RCA.
4. Filter 4 (HTAF) represents high throughput adaptive filter [14].
5. Filter 5 (FIR–DWN) represents filter designed using Wallace multiplier [5].
6. Filter 6 (FIR-UFT) represents filter designed using unfolding technique [6].

Filter is synthesized using Cadence software RTL compiler v11.10. Compared critical path delay (CPD), area-delay product (ADP), and power-delay product (PDP) of various filters and its results are shown in Table 1. In table, total power represents the sum of static power and dynamic power. Static power and dynamic power are also called as leakage power and switching power, respectively. As the result of leakage current, CMOS devices have very low static power consumption. Switching is dissipated when charging and discharging internal and net capacitances.

From the above comparison table, it is clear that the proposed filter had showed better results in terms of delay, power, PDP, and ADP when compared with other filter structures. Modifying the HTR-FIR2 filter showed improvement in performances when compared with its original structure. CPD is reduced by 18.19%, 24.14%, 28.60%, PDP by 32.55%, 24.76%, 26.94%, and ADP by 17.55%, 18.78%, 19.35%, for 16, 32, and 64 taps, respectively. Figure 2 shows performance comparison of various filters in terms of CPD, PDP, and ADP. In comparison with filter 6, CPD of proposed filter is reduced by 46.33%, 57.25%, and 66.13%, PDP by 51.67%, 75.46%, 86.94%, and ADP by 48.95%, 66.11%, and 77.434% for 16, 32, and 64 taps, respectively. For $N = 64$, power is reduced by 61.22% and 9.71%, 42.56% for $N = 16$ and 32, respectively. When filter 6 is compared with filter 5, it achieved improvement in CPD, PDP, and ADP with a slight increase in area by 6.87%, 14.37%, and 15.91% for filter taps 16, 32, and 64, respectively. When proposed filter is compared with filter 5, CPD is reduced by 71.29%, 82.71%, 90.30%, power by 40.48%, 39.44%, 62.23%, PDP by 83.12%, 91.80%, 96.33%, and ADP by 70.98%, 83.99%, 92.27% for 16, 32, and 64 taps, respectively. When compared with filter 4, improvement in CPD, PDP, and ADP with a small increase in power is achieved, CPD by 37.32%, 42.72%, 55.26%, PDP by 28.65%, 28.64%, 61.04%, ADP by 26.21%, 39.28%, 49.27% for 16, 32, and 64 taps, respectively, are reduced, and power by 12.34% and 21.76% for 16 and 32 taps, respectively, is increased. CPD of the proposed filter is reduced by

Table 1 Various filter structure results in terms of CPD, area, power, PDP, and ADP

Filter length (N)	Filter design	CPD (ns)	Area (μm^2)	Total power (mW)	PDP (fJ)	ADP (μm^2 × ns)
16	Proposed filter	0.821	14,366	0.316	259	11,794
	HTR-FIR2 [22]	1.05	13,674	0.366	384	14,305
	HTR-FIR1 [22]	1.22	13,313	0.33	403	16,242
	HTAF [14]	1.31	12,202	0.277	363	15,985
	FIR-DWN [5]	2.89	14,064	0.531	1535	40,645
	FIR-UFT [6]	1.53	15,102	0.350	536	23,106
32	Proposed filter	0.842	28,508	0.657	553	24,003
	HTR-FIR2 [22]	1.11	26,626	0.662	735	29,555
	HTR-FIR1 [22]	1.28	26,181	0.62	796	3352
	HTAF [14]	1.47	24,661	0.514	775	39,534
	FIR-DWN [5]	4.87	30,786	1.085	6745	149,928
	FIR-UFT [6]	1.97	35,953	1.144	2254	70,827
64	Proposed filter	0.859	57,016	1.140	979	48,976
	HTR-FIR2 [22]	1.15	52,806	1.165	1340	60,727
	HTR-FIR1 [22]	1.32	52,235	1.129	1490	68,950
	HTAF [14]	1.92	50,283	1.309	2513	96,543
	FIR-DWN [5]	8.86	71,566	3.019	26,748	634,075
	FIR-UFT [6]	2.55	85,112	2.94	7497	217,036

32.70%, 34.21%, 34.92%, PDP by 35.73%, 30.52%, 34.29% when compared with filter 3 for 16, 32, and 64 taps, respectively, with a small increase in power by 5.63% ($N = 32$) and 0.96% ($N = 64$). Filter 2 achieved improvement in CPD by an increase in area of 10.76%, 7.38%, and 4.77% with respect to filter 4 for 16, 32, and 64 taps, respectively.

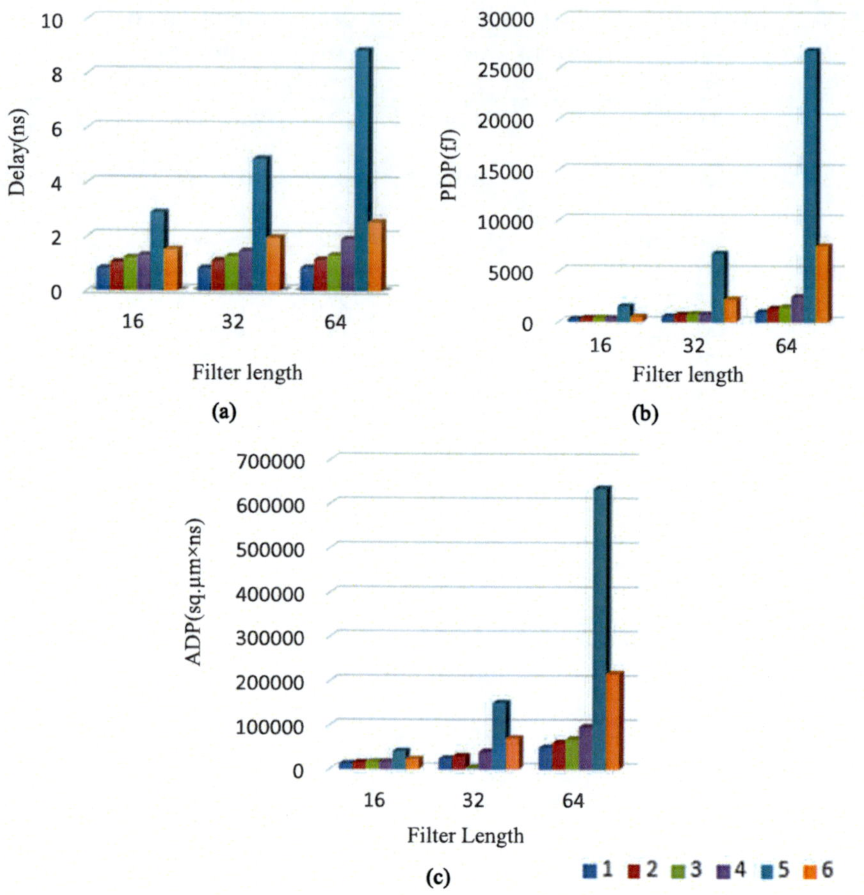

Fig. 2 Performance comparison of various filters **a** delay versus filter length, **b** PDP versus filter length, and **c** ADP versus filter length

4 Conclusion

Latency and throughput are the two challenges considered while designing the filter structure. A filter structure having low power and high throughput, applicable for DSP applications using retiming and fine-grain pipelining techniques, is proposed. Retimed-based filter structure is implemented using square root carry select adder and radix-4 booth multiplier. Fine-grain pipelining technique is applied to the final adder in the critical path to reduce the path length and hence achieved improvement in throughput. Pipelining the adder had improved the power consumption and reduced the overall latency of the designed FIR filter structure. Proposed filter's CPD is decreased by 18.19%, 24.14%, 28.60%, PDP by 32.55%, 24.76%, 26.94%, and ADP by 17.55%, 18.78%, and 19.35% for 16, 32, and 64 taps, respectively, with respect to

HTR-FIR2. Designing the filter with retiming and pipelining technique had showed better results in terms of CPD, PDP, and ADP when compared with all other filter structures. Future scope is that the proposed filter structure can be used for various DSP applications. Filter can be designed for eliminating unwanted noises from the signal such as ECG signal denoising.

References

1. Safarian, C., Ogunfunmi, T., Kozacky, W.J., Mohanty, B.: FPGA implementation of LMS-based FIR adaptive filter for real time digital signal processing applications. In: 2015 IEEE International Conference on Digital Signal Processing (DSP) (2015)
2. Meher, P.K., Park, S.Y.: Area-delay-power efficient fixed-point LMS adaptive filter with low adaptation delay. IEEE Trans. Very Large Scale Integr. (VLSI) Syst. **22**(2), 362–371 (2014)
3. Zhu, Q., Douglas, S.C., Smith, K.F. (n.d.): A pipelined architecture for LMS adaptive FIR filters without adaptation delay. In: 1997 IEEE International Conference on Acoustics, Speech, and Signal Processing
4. Lee, H.R., Jen, C.W., Liu, C.M.: New hardware-efficient architecture for programmable FIR filter. IEEE Trans. Circuits Sys. **43**(9), 637–644 (1996)
5. Swati, C., Himanshu, J.: FIR filter designing using wallace multiplier. Int. J. Eng. Tech. Res. **3**(6), 276–278 (2015)
6. Thakral, S., Goswami, D., Sharma, R.S.: Design and implementation of a high speed digital FIR filter using unfolding. In: Proceedings of IEEE 7th International Conference on Bikaner, India, Nov. 2016
7. Bar, J., Mukherjee, M.: A Comparative study on direct form-1, broadcast and fine grain structure of FIR digital filter. Int. J. Electron. Commun. Soft Comput. Sci. Eng. **2**(5). ISSN: 2277-9477
8. Monteiro, J., Devadas, S., Ghosh, A. (n.d.): Retiming sequential circuits for low power. In: Proceedings of 1993 International Conference on Computer Aided Design (ICCAD)
9. Yagain, D., Vijaya Krishna, A., Chennapnoor, S.: Design optimization platform for synthesizable high speed digital filters using retiming technique. In: 2012 10th IEEE International Conference on Semiconductor Electronics (ICSE) (2012)
10. Mehra, H., Bhat, M.S.: High level optimization methodology for high performance DSP systems using retiming techniques. In: 2018 IEEE Distributed Computing, VLSI, Electrical Circuits and Robotics (DISCOVER)
11. Arslan, T., Erdogan, A.T. (n.d.): Low power implementation of high throughput FIR filters. In: 2002 IEEE International Symposium on Circuits and Systems. Proceedings (Cat. No.02CH37353)
12. Di, J., Yuan, J.S., Demara, R.: High throughput power-aware FIR filter design based on fine-grain pipelining multipliers and adders. In: IEEE Computer Society Annual Symposium on VLSI (ISVLSI'03) 0-7695-1904-0/03 (2003)
13. Mathias, F., Martin, K., Chip-Hong, et al.: Efficient structural adder pipelining in transposed form FIR filters. In: Proceedings of IEEE International Conference on Digital Signal Processing (DSP), Singapore, July 2015, pp. 21–24
14. Pramod, P., Shahana, T.K.: High throughput adaptive filter architecture using modified transpose form FIR filters. J. Adv. Res. Dyn. Control Syst. **10**(15), 68–82 (2018)
15. Singh, M., Nowick, S.M. (n.d.): Fine-grain pipelined asynchronous adders for high-speed DSP applications. In: Proceedings IEEE Computer Society Workshop on VLSI 2000. System Design for a System-on-Chip Era
16. Bedrij, O.J.: Carry-select adder. IRE Trans. Electron.Comp. **EC-11**(3), 340–344 (1962)
17. Bahadori, M., Kamal, M. Afzali- Kusha, A.: High-speed and energy-efficient carry skip adder operating under awide range of supply voltage levels. IEEE Trans. Very Large Scale Integr. Syst. **24**(2), 421–433 (2016)

18. Pramod, P., Shahana, T.K.: Delay and energy efficient modular hybrid adder for signal processor architectures. IETE J. Res. (2019). Print ISSN: 0377-2063 Online ISSN: 0974-780X. https://doi.org/10.1080/03772063.2019.1627917
19. Qian, L.,Wang, C., Liu,W., et al.: Design and evaluate-on of an approximate wallace-booth multiplier. In: IEEE International Symposium of Circuits and Systems(ISCAS), Montreal, QC,Canada, May 2016, pp. 1974–1977
20. Jiang, H., Han, J., Qiao, F., et al.: 'Approximate radix-8 booth multipliers for low-power and high-performance operation' . Trans. Comput. **65**(8), 2638–2644 (2016)
21. Xue, H., Patel, R., Boppana, N.V.V.K., Ren, S.: Low-power-delay-power-product radix-4 8*8 booth multiplierinCMOS. Electron. Lett. 344–46 (2018)
22. Pramod P., Shahana T.K.: High throughput FIR filter architectures using retiming and modified CSLA based adders. IET Circ. Dev. Syst. (2019)

Line Stability Index-Based Voltage Stability Assessment Using CSO Incorporating Thyristor Controlled Series Capacitor

Poonam Upadhyay and S. Ravikumar

Abstract Voltage stability is the main concern for the transfer of power to the longer distances. Generally, instability of the voltage arises due to the sudden increase of the load side which involves sag in reactive power. Also, different contingencies occur due to the instability that makes the vulnerable bus which prone to the overall collapse of the voltage. In this paper, the optimal location for placing thyristor controlled series capacitor (TCSC) in series with the line using line voltage stability factor as the value ranges nearer or equal to one was found and also depicts vulnerable bus of a particular system performing under N-1 and N-2 contingencies. Also, the size of the compensation device was found using cat swarm optimization technique under contingency cases for IEEE 14 bus system.

Keywords CSO · LVSF · Optimal sizing · Optimal location · Compensation

1 Introduction

Due to the enormous increase in the power usage at the load side, which is observed as the complex power system network demands continuous power supply for various needs to the customers. This rise in power demand involves maintaining the power system in a secure state which is observed under the security constraints like equality and inequality constraints [1]. Power system stability involves steady, transient and dynamic security of stability where steady-state stability involves a major role that comes under line outages and transformer outages so on. Whenever line outages occurs, due to the voltage instability having lagging loads makes the system to a voltage collapse state.

P. Upadhyay (✉) · S. Ravikumar
Department of EEE, VNR Vignana Jyothi Institute of Engineering and Technology, Hyderabad, India
e-mail: upadhyay_p@vnrvjiet.in

S. Ravikumar
e-mail: suramravikumar94@gmail.com

© The Editor(s) (if applicable) and The Author(s), under exclusive license to Springer Nature Singapore Pte Ltd. 2021

A. P. Pandian et al. (eds.), *Proceedings of International Conference on Intelligent Computing, Information and Control Systems*, Advances in Intelligent Systems and Computing 1272, https://doi.org/10.1007/978-981-15-8443-5_72

Voltage instability leads to the damage of the entire network for not having enough reactive power supply for the increase in the loads. So the concept of providing reactive power can be achieved by the facts devices [2]. These devices play a major role in providing compensation by without disturbing the operating condition of the line series compensation like TCSC, GCSC is used for providing compensation connected in series with the line and also can reroute the power by using the devices like IPFC and UPFC's [3].

In this paper, load flow was carried out for the IEEE 14 bus system using Newton–Raphson method [4]. Using line voltage stability factor under different contingencies optimal location for the placement of TCSC is found to reduce the size of the compensating device. Cat swarm optimization technique is used where the cats were traveled in search space of different velocities that involve two stages like seeking mode and the tracing mode for finding the best solution to know the exact size of the device of specific compensation. The entire analysis is done by MATLAB for which an algorithm is developed.

2 Methods Applied

2.1 Line Voltage Stability Factor

Generally, many stability factors decide the vulnerability lines due to the voltage instability of the system. Voltage instability is caused due to the sudden increase in the loads which makes the insufficient supply of the reactive power to the line. Finding the vulnerable line is hard without knowing the proper indices value. So here line voltage stability factor places a major role in finding the critical line which is going to collapse decided based on the index value ranges zero (no load) to one (voltage stability) [5].

In this paper, the line voltage stability factor (LVSF) is used for the optimal placement of series compensation TCSC [6]. Here, under N-1 and N-2 contingency cases, the value of LVSF ranges between zero(no load) to one (voltage stability) is ranked accordingly. The greater the value for particular line indicates the optimal location for the placement of TCSC where compensation is provided so that the LVSF decreases which depict the line in a stability mode maintaining under the limits. Here, when applying CSO as an optimization technique based on the highest value of LVSF, the CSO formulation should be applied providing required reactive power compensation for the reduction of LVSF value thereby indicates the size of the device based on the reactive power provided (Fig. 1).

The derived method is shown [7]

$$L_{QP} = 4\left(X/V_i^2\right)\left(\left(X/V_i^2\right) + P_i^2 + Q_i\right)$$

L_{QP} is the stability index of the line

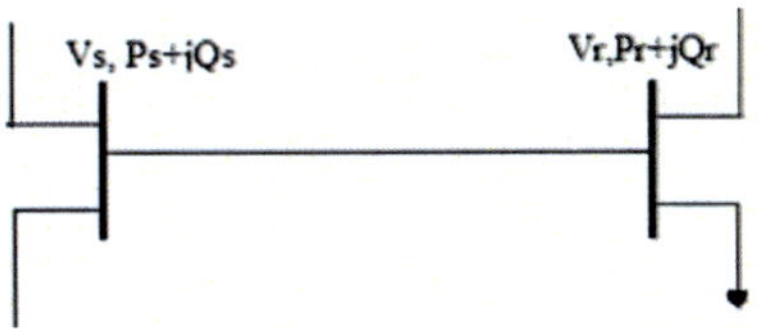

Fig. 1 Single line diagram of the transmission system

where x-line reactance

p_i real power at receiving end
Q_i Reactive power at receiving end
V_i voltage at the sending end.

2.2 *Cat Swarm Optimization*

Many optimization techniques were evolved from the real life so as CSO from the cats point of behavior. Generally, many optimization techniques are available to find the optimal size are of GA, ACO and PSO [8]. But CSO results depicted is one of the best optimization techniques. It is used for knowing the optimal size of the compensation device which is to be placed in series with the line. Cat swarm optimization technique is used to find the optimal size of the TCSC device which ensures the cost reduction of the compensating device. Without CSO, the compensating device costs more because of not knowing the exact reactive power injection to the line. CSO helps by choosing the exact reactive power injection by doing fewer iterations in the search space when in comparison with other optimization techniques. So in this paper, CSO is used for the optimal sizing of the TCSC device placed at the required location of the line based on the LVSF factor value.

Algorithm for CSO
It is known that the actions of the cat are like searching for the prey to the gaining of the prey. So this involves two stages like seeking mode and the tracing mode [3].

Seeking mode
Here, in this stage, cats will take rest which is waiting to jump around by looking at different directions in the search space as a candidate. It has a particular velocity at particular dimensions only involves four factors like SRD, i.e., it seems range for a particular dimension, SPC, i.e., it considers the proper position, CDC, i.e., it changes the no. of dimensions. The above factors play an important role for the cat to be stayed on or to move next position. Some of the steps are as follows:

Step 1 choose all cats which are required
Step 2 here a constant velocity is observed for every cat
Step 3 fitness value to be known for all cats
Step 4 choose required cats for applying in the seeking mode

Tracing mode

Actually when the cat comes into this mode says that it does not has particular velocity rather has its very own dimensions of velocities

Step 1 change the values as of required to get the best solution for each dimension
Step 2 see all the velocities are within the range if not make it to have in the limits
Step 3 make sure to update the status of the cat till to get the best fitness ultimately the location of the cat.

2.3 *Process Flow Chart for CSO*

The explanation for the flow chart was given above in the algorithm of both seeking and tracing mode of operations (Fig. 2).

Fig. 2 Flow chart for CSO

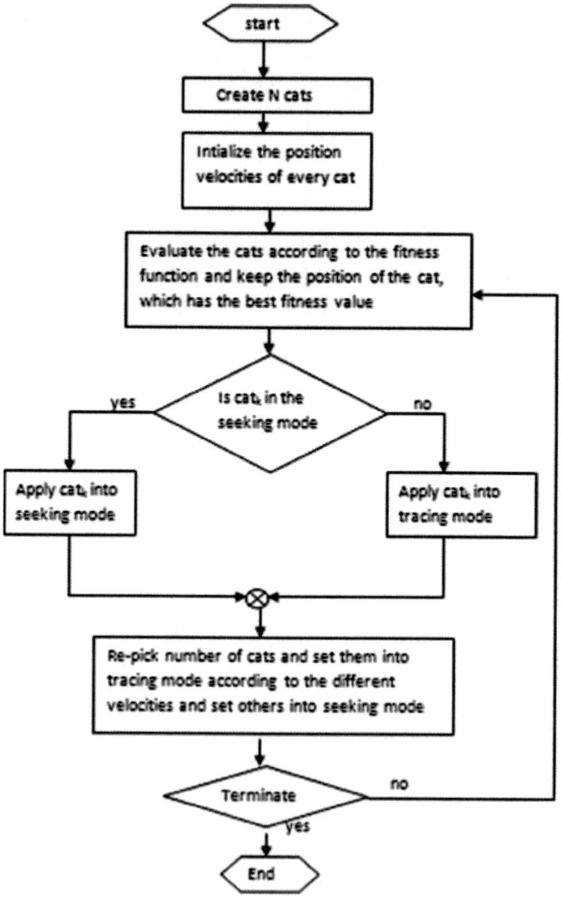

2.4 CSO Formulation

The cats were introduced into the search space based on the line voltage stability factor. So here the cat swarm optimization has two modes in which the cats behave which is of the seeking mode and the tracing mode. In seeking mode, the optimization function is formulated based on the reactive power injection as to get the desired solution. The reactive power values will be updated to the previous updated value to get the best solution needed which depends on the fitness value provided to approach for the desired solution. In the seeking mode, the velocities of the cat which is the LVSF values will be in the specified limit as depicted. The device provides 90% of the compensation chosen from the device modeling to which the values get updated in the seeking mode as the best solution to which it compares with other iteration best solution value to get the global best solution to which it is updated for every iteration with the positive or the negative fitness value depending upon the increase or decrease in the voltage till get the best optimization solution

- Here vulnerable buses are nothing but cats, i.e., opted for the required reactive power injection
- Make sure LVSF as a constraint required to be maintained under the limit with minimization
- Put the fitness value as fixed which is taken care of considering LVSF values in the limit as maintaining specified limits given by the equation of objective function as:

$$Q_i^{(n)} = Q_{inj}^{(n-1)} - \text{fit}(n-1) * \text{Rand}(1-10) * n$$

Here, n is the number of lines which is to be considered as critical for choosing the placement of compensation. If the velocities out of the limit, the cats will jump into the tracing mode where the optimization function is formulated based on the reactive power updating with the increased fitness value. Here, in the tracing mode, the velocity of the cats can get controlled based on the various fitness values regardless of the limits provided which again goes to the seeking mode if the LVSF value was within the limits. If the LVSF values range out of the limits moving in its velocities, then the following equation of objective function as:

$$Q_i^{(n)} = Q_{inj}^{(n-1)} + \text{Rand}(1-10) * \text{fit}(n-1)$$

Here, the Q indicates the reactive power which is to be injected according to the fitness logic rely on the voltage levels on the bus as shown below. It is like a fuzzy kind of logic rely on voltage level which best-fitted value desires to the solution at speeder pace.

	Voltage	Fitness	
EL	Extremely low	Highly positive	HP
VL	Very low	Positive medium	PM
L	Low	Positive low	PL
M	Medium	Zero	Z
H	High	Negative low	NL
VH	Very high	Negative medium	NM
EH	Extremely high	Highly negative	HN

And the values confirmed for each case as follows (Fig. 3).

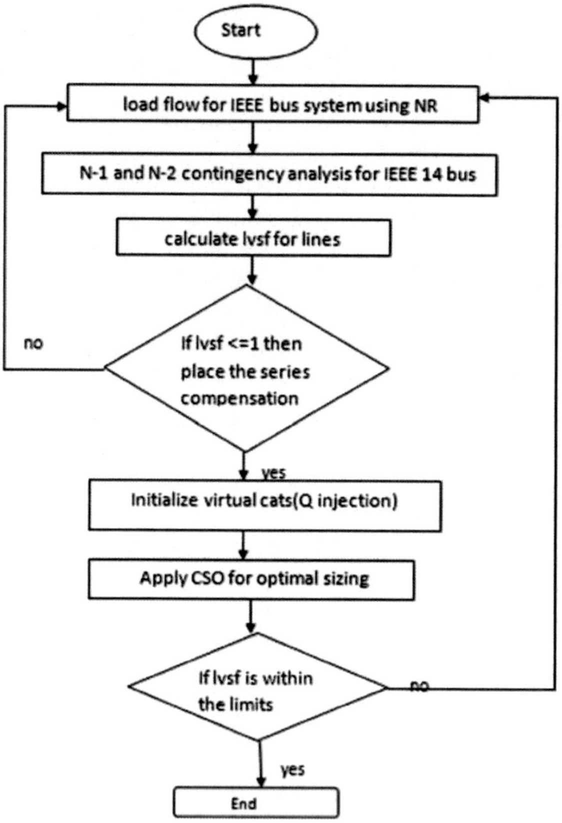

Fig. 3 Planned methodology

Voltage range (P.U)	Fitness	Percentage variation
Less than 0.8 P.U	1.5–2	150–200% increase
0.8–0.85	1.2–1.5	120–150% increase
0.85–0.95	1–1.2	Up to 120% increase
0.95–1.05	1	No change
1.05–1.15	0.8–1	20% decrease
1.15–1.2	0.6–0.8	20–40% decrease
Above 1.2	0.4–0.6	40–60% decrease

The planned methodology is a process to maintain the reactive power within the limits using a compensating device for maintaining the voltage stability. So, by considering the IEEE bus system, load flow analysis should be done using Newton–Raphson method [9] for finding the power flows of the system, then create an outage of the line to know the critical line in the system by calculating the LVSF for the lines. If LVSF value is less than or equal to one indicates the critical line and the bus associated with the line is a critical bus which is going to collapse. Place the TCSC compensating device for the particular line. If the LVSF values are within the limits, then do for other creating outages. The critical LVSF value line should be placed with the series compensation with the help of CSO technique; the optimal sizing of the device can be found by checking the required injection of the reactive power. At certain iterations, the optimal sizing value can be known which is as the eighth iteration by making the LVSF value in the limits with the required reactive power injection to gain voltage stability of the line. If LVSF value is not within the limits, do the load flow for the system.

2.5 Algorithm for Planned Methodology

- Step 1 Read the line data and the bus data of the given system
- Step 2 Do the load flow using the NR method
- Step 3 Do the N-1 and N-2 contingencies for IEEE test bus system
- Step 4 calculate LVSF for all the transmission lines
- Step 5 If any of the line LVSF values are within the limits then go to step 2
- Step 6 If any of the line LVSF values is leaa than or equal to one is considered as the critical line to be placed the series compensation device
- Step 7 Initialize the virtual cats (Q injection)
- Step 8 Find the optimal size by applying the CSO formulation technique
- Step 9 If LVSF of the particular line is not within the limits then go to step 2
- Step 10 If LVSF of the particular line is within the limits at particular iteration is considered as the optimal size of the device

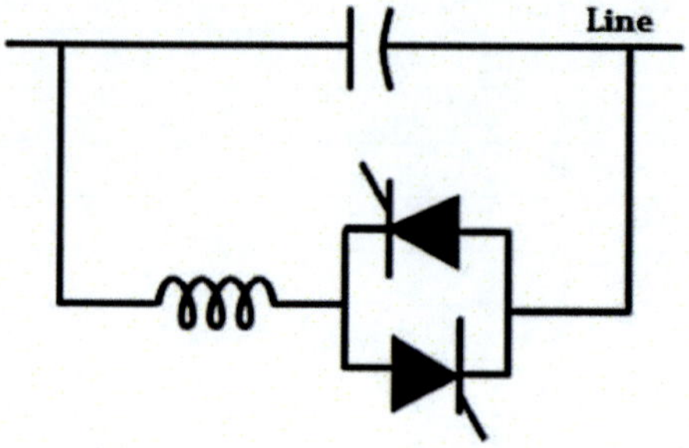

Fig. 4 Schematic diagram of TCSC

3 Thyristor Controlled Series Capacitor

It is one of the series compensating device known as thyristor controlled series capacitor provides the compensation which is used for to be placed in series with the line [10]. The modeling of the TCSC is given as [11], which adjusted the value to be injected for the enhancement of the voltages.

It operates in inductive mode and the capacitive mode as follows [12] (Fig. 4).

Inductive mode
TCR of $XL(\alpha)$ at firing angle is shown with the variation as from 0 to 90 is of

$$X_L(\alpha) = \frac{X_L(\pi)}{\pi - 2\alpha - \sin 2\alpha}$$
$$X_C = 1/2\pi f c$$

Capacitive mode
When reactor connected in series of a capacitor, then the reactance can be varied as shown above.

4 Results

Here, considering IEEE 14 bus test system for different contingency cases is considered to apply CSO for enhancement of voltages by taking LVSF values of individual lines (Fig. 5).

In this IEEE test bus system under base case conditions, the values are depicted at Tables 1 and 2, when considering N-1 contingency case 1–5 which involves critical line as 2–4 where the bus present to the line prone to the collapsing of voltage due to the voltage instability based on the severity index of the line 0.370 and can be reduced by placing series compensation as an optimal location TCSC, i.e., 90% of compensation is provided based on the modeling of the TCSC by considering the

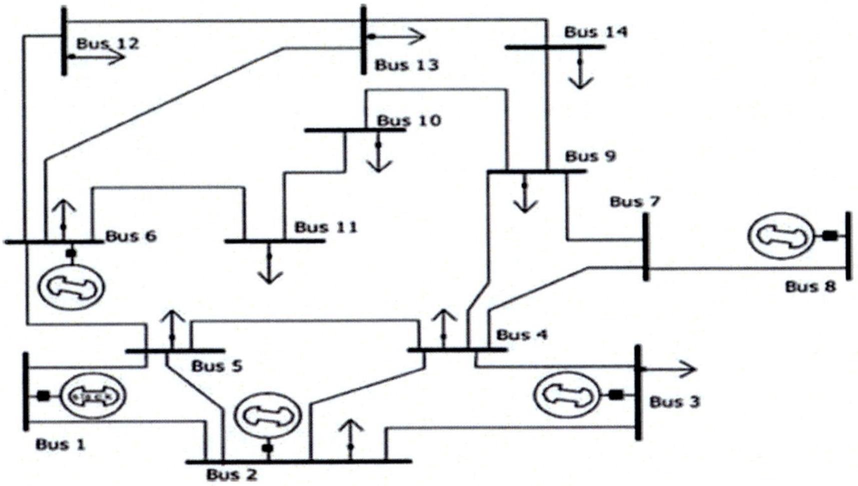

Fig. 5 IEEE 14 bus system

Table 1 Variation of *x*tcsc (α) by varying α

S. No.	Delay angle (α)	*X*tcsc (α)
1	0	0.055
2	30	0.1090
3	45	0.1857
4	60	0.6228
5	65	Resonance
6	75	−0.4605
7	80	−0.2915
8	85	−0.2133
9	90	−0.1687

reactance of the line thereby effects the reactive power which is stated as the optimal size of the device and is achieved by CSO technique, thereby decreasing the index value as 0.020.

when considering N-2 contingency case 1–5 and 2–4 which involves critical line as 2–5 where the bus present to the line prone to the collapsing of voltage due to the voltage instability based on the severity index of the line 0.579 has reduced to 0.061 by placing series compensation with an optimal size using CSO technique.

Table 3 indicates the voltages before applying the CSO technique for the base case and after applying to it. At 14 buses, the voltages were low as depicted above but after applying for the TCSC compensation with the optimal size, the voltages were enhanced within the limits of + or −0.05 above 1 at all the buses.

Table 4 indicates the LVSF values before and after applying the CSO technique at all the lines of the base case. As seen, for all the lines, the LVSF values were decreased

Table 2 Line data of iEEE 14 bus system

S. No.	Lines	Resistance	Reactance	Shunt admittance
1	1–2	0.01938	0.05917	0.02640
2	1–5	0.05403	0.22304	0.02190
3	2–3	0.04699	0.19797	0.01870
4	2–4	0.05811	0.17632	0.02460
5	2–5	0.05695	0.17388	0.01700
6	3–4	0.06701	0.17103	0.01730
7	4–5	0.01335	0.04211	0.00640
8	4–7	0	0.20912	0
9	4–9	0	0.55618	0
10	5–6	0	0.25202	0
11	6–11	0.09498	0.1989	0
12	6–12	0.12291	0.25581	0
13	6–13	0.06615	0.13027	0
14	7–8	0	0.17615	0
15	7–9	0	0.11001	0
16	9–10	0.03181	0.0845	0
17	9–14	0.12711	0.27038	0
18	10–11	0.08205	0.19207	0
19	12–13	0.22092	0.19988	0
20	13–14	0.17093	0.34802	0

Table 3 Voltage performance of iEEE 14 bus system at base case

Buses	Voltages before applying CSO	Voltages after applying CSO
1	0.98	0.98
2	0.99	0.99
3	1.0	1.0
4	0.97	1.0
5	0.98	1.03
6	0.96	1.05
7	0.95	1.03
8	0.95	1.02
9	0.93	1.04
10	0.93	1.04
11	0.94	1.01
12	0.94	1.01
13	0.93	1.01
14	0.95	1.05

Table 4 LVSF values for base case

S. No.	Lines	LVSF values before applying CSO	LVSF values after applying CSO
1	1–2	0.231	0.001
2	1–5	0.299	0.058
3	2–3	0.343	0.095
4	2–4	0.251	0.003
5	2–5	0.085	0.022
6	3–4	0.107	0.001
7	4–5	0.699	0.000
8	4–7	0.115	0.005
9	4–9	0.170	0.051
10	5–6	0.213	0.031
11	6–11	0.048	0.001
12	6–12	0.065	0.001
13	6–13	0.077	0.000
14	7–8	0.000	0.000
15	7–9	0.056	0.000
16	9–10	0.013	0.001
17	9–14	0.037	0.001
18	10–11	0.081	0.020
19	12–13	0.008	0.000
20	13–14	0.056	0.020

after compensation which means the critical lines came to the normal state and the highest value is of line 4–5 almost reduced to 0 indicating the non-vulnerable line at the eighth iteration and the remaining line values were also reduced bringing to the limits at final iteration.

Table 5 indicates the voltages before applying the CSO technique for the contingency case of line 1–5 and after applying to it. At 14 buses, the voltages were low as depicted above but after applying for the TCSC compensation with the optimal size, the voltages were enhanced within the limits of + or −0.05 above 1 at all the buses.

Table 6 indicates the LVSF values before and after applying the CSO technique at all the lines of the contingency case of line 1–5. As seen, for all the lines, the LVSF values were decreased after compensation which means the critical lines came to the normal state and the highest value is of line 2–4 almost reduced to 0.020 indicating the non-vulnerable line at the eighth iteration and the remaining line values were also reduced bringing to the limits at final iteration.

Table 7 indicates the voltages before applying the CSO technique for the contingency case of lines 1–5 and 2–4 and after applying to it. At 14 buses, the voltages were low as depicted above but after applying for the TCSC compensation with the

Table 5 Voltage performance for contingency 1–5 case

Buses	Voltages before applying CSO	Voltages after applying CSO
1	0.98	0.98
2	0.99	0.99
3	1.0	1.0
4	0.93	1.03
5	0.93	1.0
6	0.94	1.06
7	0.94	1.03
8	0.93	1.04
9	0.94	1.02
10	0.97	1.02
11	0.88	1.05
12	0.88	1.05
13	0.87	1.04
14	0.85	1.02

Table 6 LVSF values for contingency 1–5 case

S. No.	Lines	LVSF values before applying CSO	LVSF values after applying CSO
1	1–2	0.351	0.001
2	2–3	0.361	0.190
3	2–4	0.370	0.020
4	2–5	0.341	0.015
5	3–4	0.041	0.004
6	4–5	0.029	0.000
7	4–7	0.126	0.026
8	4–9	0.193	0.123
9	5–6	0.214	0.176
10	6–11	0.042	0.002
11	6–12	0.067	0.000
12	6–13	0.079	0.001
13	7–8	0.000	0.000
14	7–9	0.069	0.000
15	9–10	0.015	0.000
16	9–14	0.096	0.000
17	10–11	0.009	0.003
18	12–13	0.001	0.002
19	13–14	0.001	0.013

Table 7 Voltage performance for contingency 1–5 and 2–4

Buses	Voltages before applying CSO	Voltages after applying CSO
1	0.98	0.98
2	0.99	0.99
3	1.01	1.04
4	0.80	0.99
5	0.84	0.96
6	0.85	0.95
7	0.80	1.01
8	0.80	0.99
9	0.77	0.98
10	0.89	1.05
11	0.91	1.06
12	0.90	1.06
13	0.90	1.06
14	0.87	1.04

optimal size, the voltages were enhanced within the limits of + or −0.05 above 1 at all the buses.

Table 8 indicates the LVSF values before and after applying the CSO technique at all the lines of the contingency case of lines 1–5 and 2–4. As seen for all the lines, the LVSF values were decreased after compensation which means the critical lines came to the normal state and the highest value is of line 2–5 almost reduced to 0.061 indicating the non-vulnerable line at the eighth iteration and the remaining line values were also reduced bringing to the limits at final iteration.

The below-showing plots are the bar chart performance of the results of the above table. The values which got are represented with the bar graphs for easy understanding as follows:

Base Case
See Figs. 6 and 7.

Contingency 1–5 Condition
See Figs. 8 and 9.

Contingency 1–5 and 2–4 Condition
See Figs. 10 and 11.

Table 8 LVSF values for contingency 1–5 and 2–4 case

S. No.	Lines	LVSF values before applying CSO	LVSF values after applying CSO
1	1–2	0.364	0.004
2	2–3	0.517	0.160
3	2–5	0.579	0.061
4	3–4	0.123	0.072
5	4–5	0.093	0.028
6	4–7	0.135	0.000
7	4–9	0.206	0.200
8	5–6	0.254	0.180
9	6–11	0.058	0.044
10	6–12	0.078	0.001
11	6–13	0.094	0.001
12	7–8	0.002	0.001
13	7–9	0.075	0.000
14	9–10	0.015	0.000
15	9–14	0.092	0.001
16	10–11	0.029	0.002
17	12–13	0.001	0.001
18	13–14	0.059	0.000

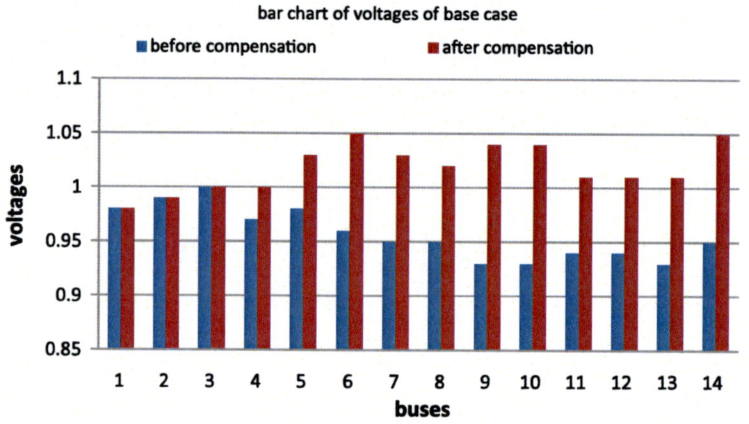

Fig. 6 Voltage performance bar chart for buses

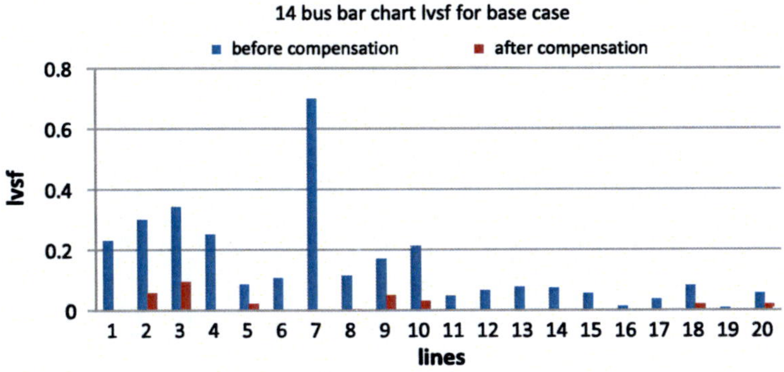

Fig. 7 LVSF values bar chart for lines

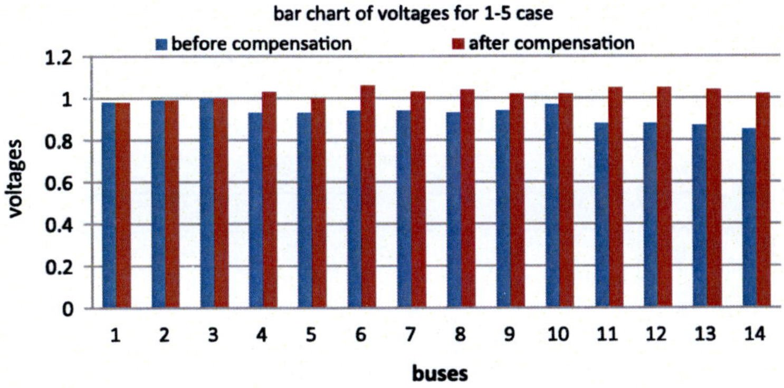

Fig. 8 Voltage performance bar chart for buses

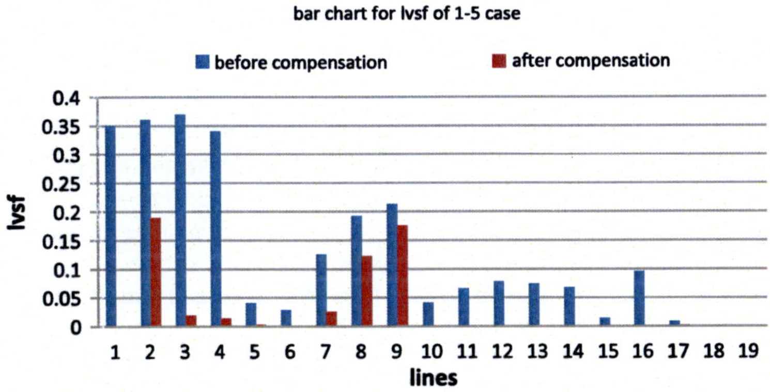

Fig. 9 LVSF values bar chart for lines

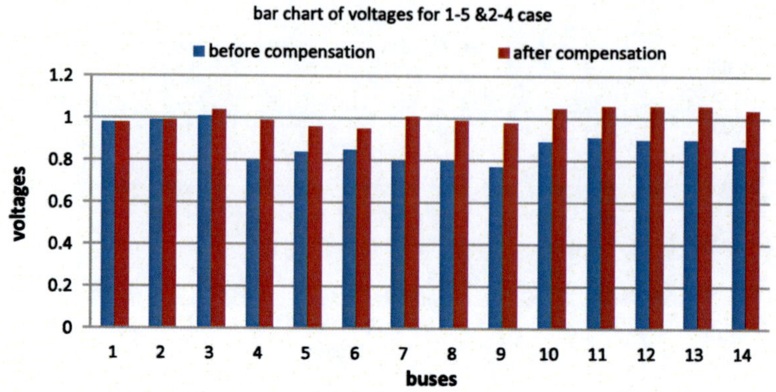

Fig. 10 Voltage performance bar chart for buses

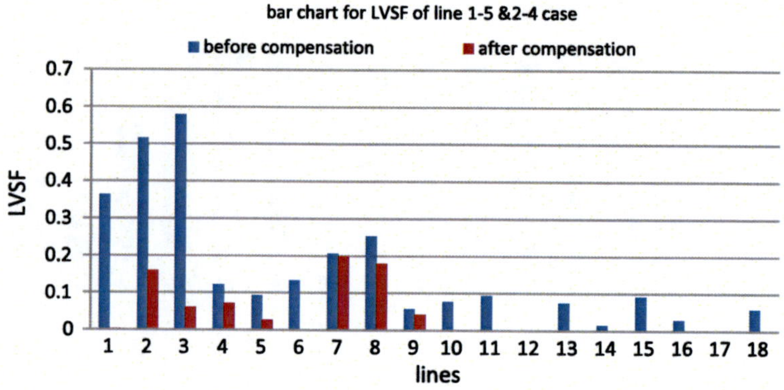

Fig. 11 LVSF values bar chart for lines

5 Conclusion

The CSO is one of the best optimization techniques when compared to other optimization techniques that can be observed finding solution under fewer iterations with faster speed. Voltage enhancement is done for 14 bus system under different contingency conditions by placing the series compensation with the help of LVSF as an optimal location wherever there is a vulnerable bus. To find an optimal size, CSO is used as seeking mode and the tracing mode under different iterations using LVSF whereby increasing reactive power leads to voltage stability. Earlier optimization techniques can be easily replaced by CSO because of its advantages and future work can be done by using different facts devices where rerouting of the power can be easily possible.

References

1. Taylor, C.W.: Power System Voltage Stability. McGraw-Hill, New york, p. 273 (1994)
2. Hingorani, N.G., Gyugi, L.: Understanding Facts: concepts and technology of flexible ac transmission systems. IEEE Press, Newyork (2000)
3. Naveen Kumar, G., Surya Kalavathi, M.: Cat swarm optimization for the multiple placement of UPFC's in voltage stability enhancement under contingency. Electr. Power Energy Syst. 97–104 (2014)
4. Sekhar, P., Mohanty, S.: Power system contingency ranking using Newton Raphson load flow method. IEEE India Conference (2013)
5. Yang, C.-F., Lai, G.G., Lee, C.-H., Su, C.-T., Chang, G.W.: Optimal setting of reactive compensation devices with an improved voltage stability index for voltage stability enhancement. Int. J. Electr. Power Energy Syst. **37**, pp. 50–57 (2012)
6. Rajyalakshmi, P.: A comparision of transmission line voltage stability indices. In: 2nd IEEE Conference Electrical Electronics Information, Communicationand Bioinformatics, pp. 44–48 (2016)
7. Mohamed, A., Jasmon, G.B., Yusoff, S.: A static voltage collapse indicator using line stability factors. J. Ind. Technol. **145**, N5 (1998)
8. CSO and PSO to Solve Optimal Location Capacity for High Tension Customers. IEEE, PEDS (2009)
9. Naik, P.: Power System Contingency Ranking using Newton Raphson Load Flow Method and its Prediction using Soft Computing Techniques. Thesis of Master of Technology in Industrial Electronics, National Institute of Technology, Rourkela, India (2014)
10. Sureshbabu Dharam, V.: TCSC incorporated voltage stabilty assessment under contingency condition. Int. J. Grid Distrib. Comput. **10**(7), 27–40 (2017)
11. Ghosh, A., Ledwich, G.: Modeling and control of thyristor controlled series compensators. Proc. Inst. Elect. Eng. Gen. Transm. Distrib. **142**, 297–304 (1995)
12. Meikandasivam, S., Nema, R.K., Jain, S.K.: Selection of TCSC parameters: capacitor and inductor. In: 2010 India International Conference on Power Electronics, pp. 1–5 (2011)

Portable Multifunction Tester Design to Check the Continuity of Wires and to Measure the Electrical Parameters

A. Kunaraj, J. Joy Mathavan, and K. G. D. R. Jayasekara

Abstract The objective of this paper is to design a single device for checking the continuity of single wire and multiwired cables with ease, measurement of current and voltage in an electrical circuit and providing light to the experimentation area by a single device. The device is named as portable multifunction tester. The status of the errors found by this device appears on the LCD display for visual observation. Buzzers are also provided to indicate the errors by making identical sounds. This is a non-destructive testing mechanism because no need to unmount or break any circuit or structures to troubleshoot and rectify the error.

Keywords Continuity checking of wires · VGA cable · RJ45 cable · Portable multifunction tester

1 Introduction

Electrical and electronic equipment used in various applications is prone to faults and defects. There are different testing methods and equipments available to troubleshoot and fix the problems in electrical and electronic-related applications. But they all are individual testers for faults. The design of research work presented here is the combination of four troubleshooting related operations. They are

1. Continuity testing in a single wire.
2. Continuity testing in multiple wires.
3. Check the continuity of current carrying conductor without breaking slab.
4. Voltage and amphere testing.

A. Kunaraj (✉) · J. Joy Mathavan · K. G. D. R. Jayasekara
Department of Engineering Technology, Faculty of Technology, University of Jaffna, Jaffna, Sri Lanka
e-mail: kunaraj12@gmail.com

J. Joy Mathavan
e-mail: joymathavan1991@gmail.com

© The Editor(s) (if applicable) and The Author(s), under exclusive license to Springer Nature Singapore Pte Ltd. 2021

A. P. Pandian et al. (eds.), *Proceedings of International Conference on Intelligent Computing, Information and Control Systems*, Advances in Intelligent Systems and Computing 1272, https://doi.org/10.1007/978-981-15-8443-5_73

5. Providing light to the testing space.

Usually, the defects associated with electrical and electronic appliances are minor and are in the form of broken wires. Single and multiwire continuity tester are subunits in this device, which is used to check the errors in VGA and network cables. There are incidents where the fault in one individual cable in a network cable or VGA cable causes mulfunctioning of the entire cable. In such situation, this device can check the number of cables and the error associated with the exact cable. There are individual LED indicators in main controller module which can indicate which wire is broken in between. In industrial point of view, time is very important and equivalent to money. If the fault detected quickly, it can be fixed quickly and the process can be started immediately once the fault is fixed. Since most of the current equipment consists of bunch of wires, the fault finding has become pretty tough and difficult to diagnose. The small, inexpensive portable multifunction tester can be used to detect such defects. Uninstalling the entire machinery inorder to find the error in a wire is prevented by this equipment. The probes of the continuity tester can be connected to both the ends of the wire and can be tested for its continuity. If the wire is continuous without any break in between, the circuit makes a sound and that is the indication. In addition to that, when the electrical components are fixed on the printed circuit board or on the breadboard, there is a possibility that the components get attached to the printed circuit board or breadboard due to the mistakes committed while assembling the circuit. The continuity tester helps to debug the circuit with ease under such situation. The tester consists of an indicator in series with a source of electrical power—normally a battery, terminating in two test-leads. If a complete circuit is established between the test-leads, the indicator will be activated. The indicator is a light and a buzzer, which would be the output of continuity test. Usually, audible buzzers or beepers are built into some models of multimeter, and the continuity testing is normally annexed with the ohmmeter setting. Continuity tester is divided into two circuits, namely transmitter circuit and receiver circuit. There are two terminal probes on the transmitter circuit for the continuity testing purpose while the receiver circuit has LCD for visual indication and buzzer for sound indication of the result.

Wirless conductor checking is used to identify the errors in conductor wires without damaging the slab area. Electricity carrying wires are installed through wall or slab in a house wiring. In case of any irregularity, the fault finding in the conductor is very difficult without breaking the wall or conductor. But with the help of wireless conductor checking function available in multifunction tester, the fault finding is easy. When the antenna of the tester is kept near the conductor wire, the LED in the tester blings and buzzer makes sound if there is no fault.

The multifunction tester can check the voltage and give the output ranging between 5 and 40 V and the relavant current measurement will also be displayed. The device can also provide light to the area under inspection in case of poor lighting condition.

The organization of this paper is as follows: Sect. 1 depicts the introduction, Sect. 2 depicts the literature review, Sect. 3 depicts the subsystems, namely single wire continuity testing, mutiwire continuity testing, wireless conductor testing, current and amphere testing and providing light to the testing area along with circuit diagrams and working principle. The results are presented and discussed in Sect. 4 followed by conclusion and future scopes in Sect. 5.

2 Literature Review

John E.Fletcher and john O'Reilly [1] explained about short circuit and ground fault analysis. Also they explained about troubleshooting the underground cables based on voltage source converter method (VSC). This proposed method was based on simulation results with the range of fault resistances, distances and operational conditions considered. Kedia et al. [2] explained about underground cable fault detection and distance measurement using AT mega 328 microcontroller. This design used for the three-phase supply and it can individually measure the resistance of the cables and find the broken cable. The resistance and distances were measured based on current sensing data values. According to current sensing data, ATmega328 processor processes the sensor data, and the output will be supplied according to program. Kesim [3] explained about automated continuity testing using a continuity tester. Military computers have been widely used by ASELSANT has flexible cable, which has number of wires (nearly 100–160). Since this cable is used to connect processor unit with computer, large number of cables are going through this electrical component. This design was designed to check all the wires at the same time and quickly. Roy et al. [4] designed an electronics circuit to measure the polarity and check for the continuity. Nandhini et al. [5] used Arduino technology to find the fault and the exact location of the fault in the underground cable. Padmanaban et al. [6] designed a circuit using Arduino technology and Ohm's law to identify the fault and abnormality occurs in the underground cables. Choi et al. [7] used microcontroller techniques to find the fault and the distance of the fault from any end and LCD to display the finding. Chandra et al. [8] also found short-circuit fault in underground cables using advanced microcontroller techniques. Tarlochan et al. [9] used two different type of algoritham to identify fault in the underground cable. One algoritham is used to identify the fault induced-transient and the other algoritham is used to identify the single-line-to-ground fault. Sampathraja et al. [10] used PIC16F877A controller to identifiy the distance of fault of the cable from the base station in km. Pal et al. [11] found the fault in the underground cable based on the ohms law principle. Apisit et al. [12], in their research, find the fault and identify the phase using a decision algorithm.

It can be noticed that most of these researches were targeted on designing a device for checking the continuity of an underground cable. Some worked on single wire continuity testing and some worked on multiwire continuity testing. The difference between the past researches and the current research presented here is the past researches concentrated on finding the solution to a particular problem while the current research intended to address more than one problem by a single device. In other words, the device is designed to answer the problems which have high probability to occur while testing a circuit.

3 Methodology

Technicians currently facing time delay in testing a circuit for any irregularities. This is due to the usage of various equipment for fault finding in the industry, which the technicians are not comfortable with. This paper aims to minimize this problem by providing a single device to check and solve four different problems. The main controller module design consists of testing of four functions which are frequently used in industry. In case if one needs additional measurement or testing, there are several sensors and ports available, which can be connected with the main controller modular. Therefore, the design is not limited to address the above mentioned issues only.

The specifications of ATmega328 microprocessor [13] are shown in Table 1. This microcontroller IC is very small, cheap and have high clocking speed. It has more number of input ports and output ports for external connections as shown in Fig. 1.

The circuit of portable multifunction tester is designed using this particular IC, which is the main controller unit. Different sensor modules are attached here with the main controller unit for controlling and switching. Display module also connected with ATmega328 microprocessor IC to display the processed information as visual based output. The block diagram of multifunction tester is shown in Fig. 2.

3.1 Power Supply

AC to DC converter inside the power supply module is used to convert AC voltage to DC voltage. Power supply module is a separate module which can be connected and disconnected with main module, since the power supply module is not required once the multitester is charged. As shown in Fig. 3., the input voltage is 230 V/50 Hz (120VAC–240VAC), and output voltage is 13.5VAC 1000 mA.

The power supply is used to charge the battery. Since 13.5VAC is not suitable for this circuit, it is converted to DC voltage using center tap transformer, diode bridge and capacitor as shown in Fig. 4. The average input voltage of the circuit is 230v/50 Hz and output voltage is 12 V. Figure 5 shows the graph of signal after passing through diode and Fig. 6 shows the signal after passing through capacitor.

Table 1 Specifications of Atmega328 microcontroler IC

Parameters	Range
Microcontroller	Atmega328
Operating voltage	5 V
Digital I/O pins	14 (of which 6 provide PWM output)
Clock speed	16 MHz
DC current per I/O pin	40 mA
DC current for 3.3 V pin	50 mA

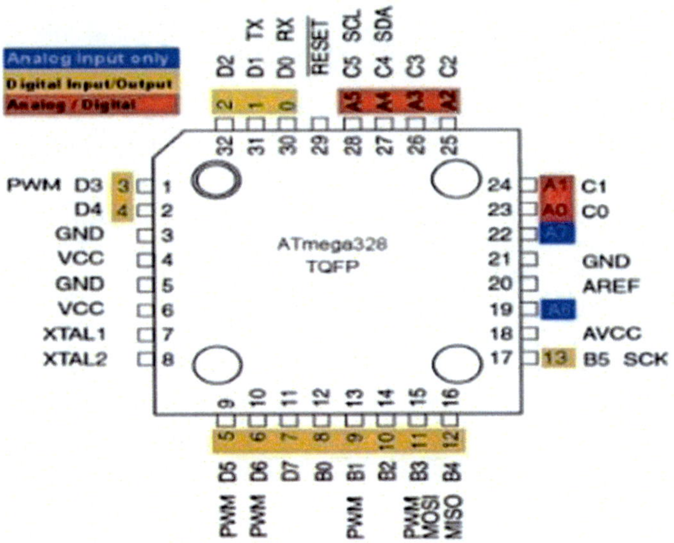

Fig. 1 ATmega328 microprocessor

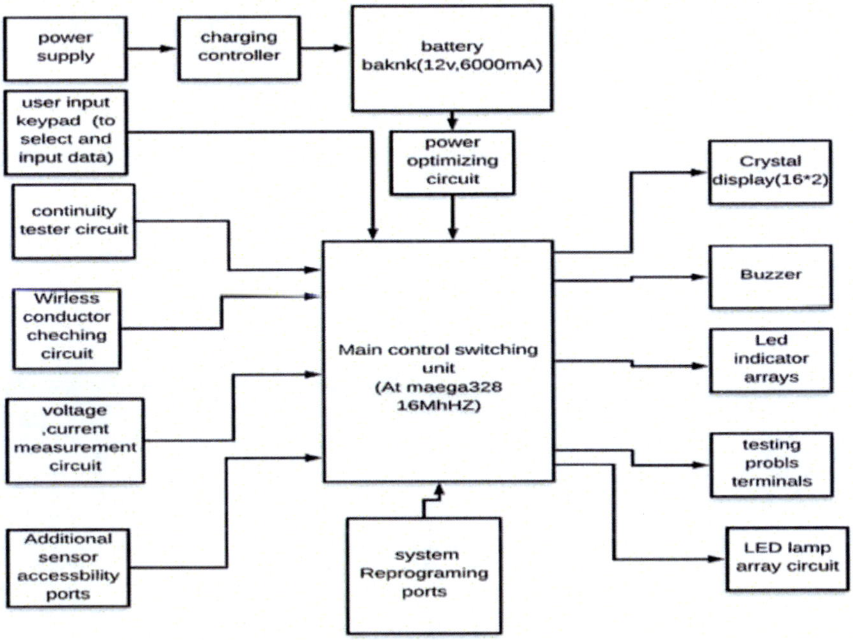

Fig. 2 The block diagram of multifunction tester

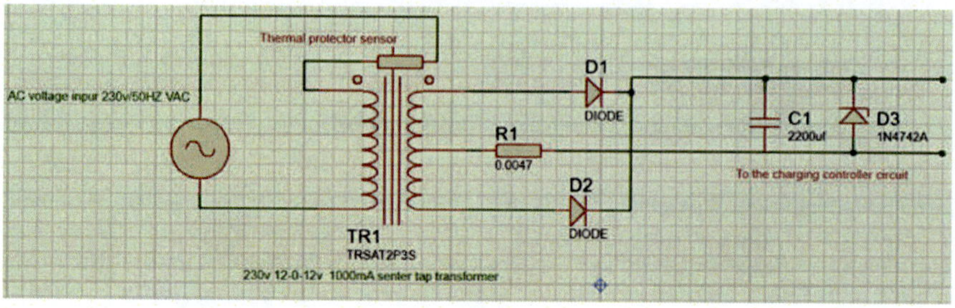

Fig. 3 Power supply circuit consists of AC–DC converter

Fig. 4 Signal generated by AC power source

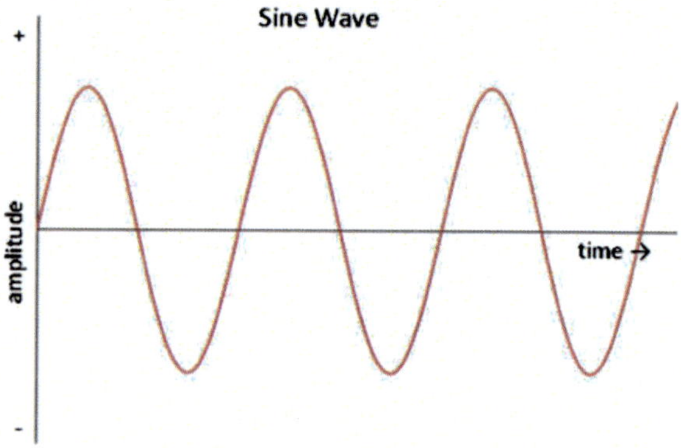

Fig. 5 Signal after passing through diode

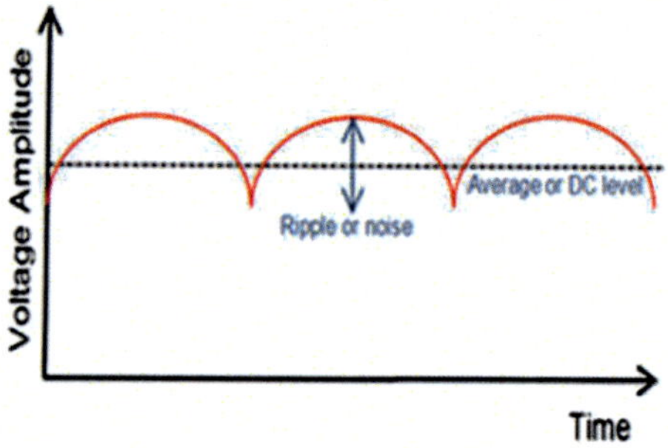

Fig. 6 Signal after passing through capacitor

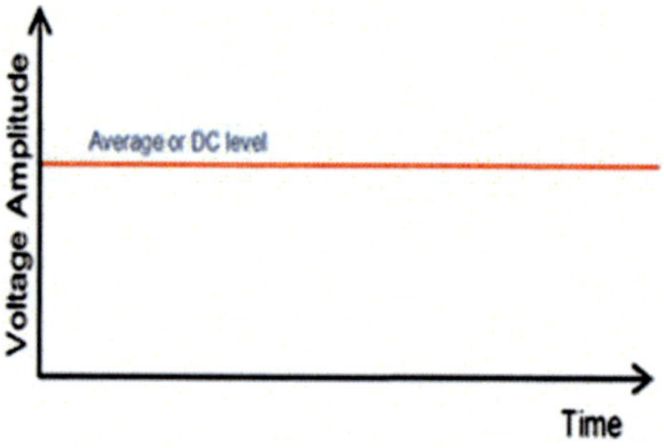

Due to less initial cost while purchasing and less maintainance cost while operating, center tap transformer is used. Thermal protection sensor is used to cutoff the supply voltage, if the temperature exceeds 40 °C, because there are chances that center tap transformer may be over heated and the copper windings of it get damaged. Primary and secondary coil of center tap transformer is protected by a fusible resistor connected in series with the negative terminal. In case of any short circuit, the fusible resistor breaks the circuit and protects other electrical components. 2200 μf capacitor is used to get smooth DC output voltage and 12 V sensor diode is used to regulate the output voltage.

3.2 Charging Controller

12 V supply voltage is supplied from power supply to battery through charging controller. The charging controller circuit is used to cut off the connection between power supply and battery once the battery is fully charged. Charging controller protects the battery from over charging. BC547 NPN transistor is used in charging controller circuit to sense the voltage level when charging the battery. The emitter terminal is connected with the negative terminal of the battery. The base–emitter (BE) junction of the transistor is forward biased and current flow through the collector–emitter (CE) junction, if the voltage level increases above 12 V while charging the battery. At this point, the relay is on and the battery connection to positive terminal will be cut off. Purpose of D4 diode is to avoid the back EMF of relay and protect the transistor. The green LED indicates fully charged state and red LED indicates the charging state of the charging controller. D7 diode is connected to avoid the discharging of battery once it is fully charged and disconnected from supply. The charging controller circuit is shown in Fig. 7.

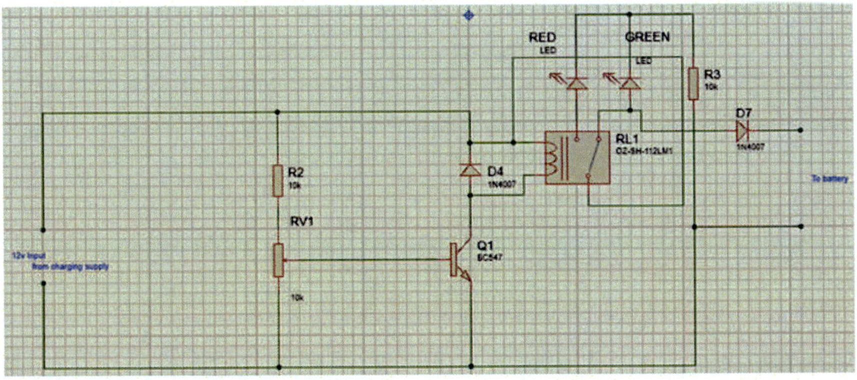

Fig. 7 Auto-cutoff charging controller circuit

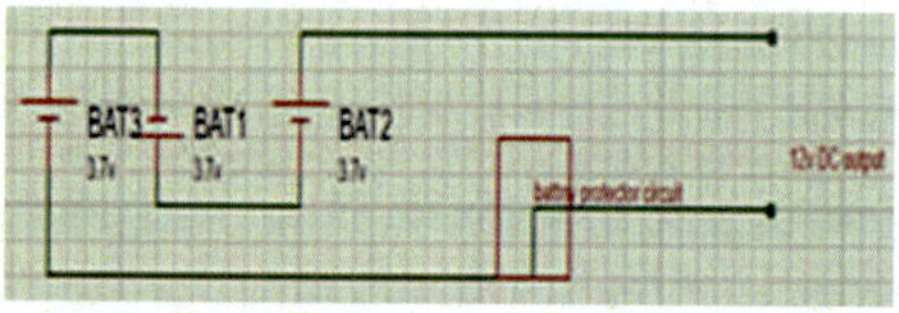

Fig. 8 Circuit diagram of battery connection

Battery bank is the power source to store the power and energize the multitester module. Lithium ion battery is used here and one lithium ion battery is of 3.7 V. Since 12 V power is needed as the source, three 3.7 V batteries were connected in series to get the necessary power as shown in Fig. 8.

3.3 User Input Key Pad Array

The key pad is created using three push buttons: an up button, a down button and one select button. 5 V supply is supplied to each button terminal and the last terminal is the signal output terminal. The signal input terminal is connected with 10 kΩ resistor in series. The signal pin is grounded and it is connected with analog pin of the ATmega328 microcontroller. The user input key pad is shown in Fig. 9 The buzzer and LED circuit are used to indicate the output as sound and visual output as shown in Fig. 10.

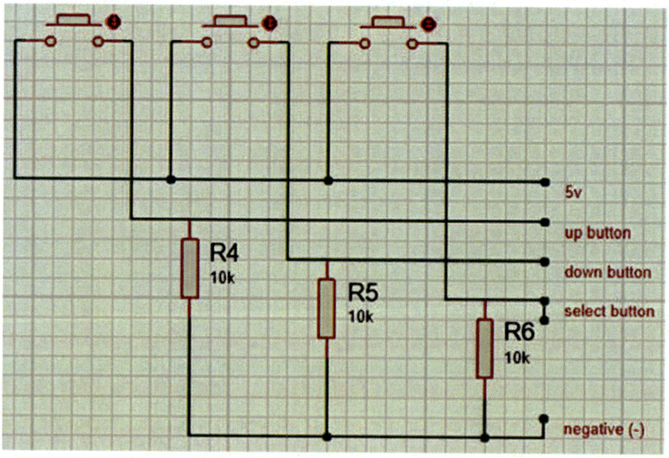

Fig. 9 Keypad array

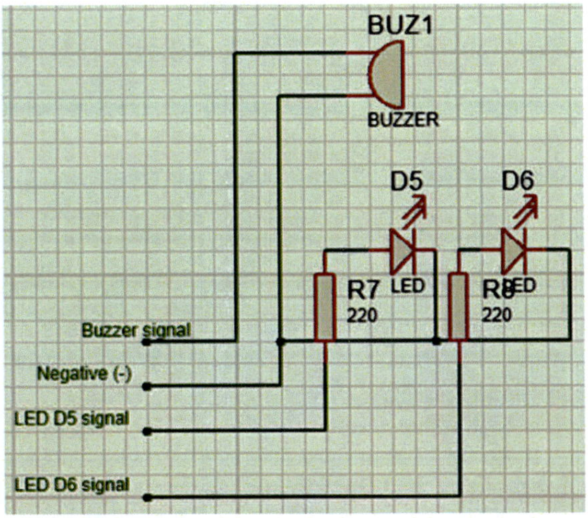

Fig. 10 Buzzer and LED indicators

3.4 Single Wire Continuity Tester

Continuity testing is devided into two major parts: single wire testing and multiwire testing. BC547 NPN transistor switches the single wire tester and voltage supply to the particular circuit. The opperating voltage is 5 V. The output terminal of the IC connected with the positive terminal of the buzzer and the negative terminal of the buzzer is connected with test probe. The output of the IC is connected with one end of the LED through the resistor R2 and the other end of the LED is connected with the tset probe. If the test probe is continueous, the current flows through the circuit without any interuption. Under this condition, buzzer makes sound and LED lit up and this method is used to check and verify the continuity of a single wire. LM7805 regulator IC was used to regulate the voltage. LM7805 IC is one of the best ICs in electrical industry. Maximum output current of LM7805 is 1A and if the IC is overheated due to high current flow through it, the heat sink dissipates the heat and transfers the air very quickly. Any other IC would be burnt out if it is overheated. The circuit diagram of single wire continuity tester is shown in Fig. 11.

3.5 Multiple Wire Continuity Tester

This circuit is designed to check the continuity of multiple cables (VGA or RJ45). There are number of current conducting wires inside VGA and RJ45 cable. Checking each wire one by one is difficult and time-consuming task. The portable multifunction tester circuit is deigned to avoid such delays in trouble shooting. Two type of ICs were used to design the circuit. CD4017 IC is used to count the number of wires and send the output voltage signal. NE555 IC generates chocking pulse, if it encounters

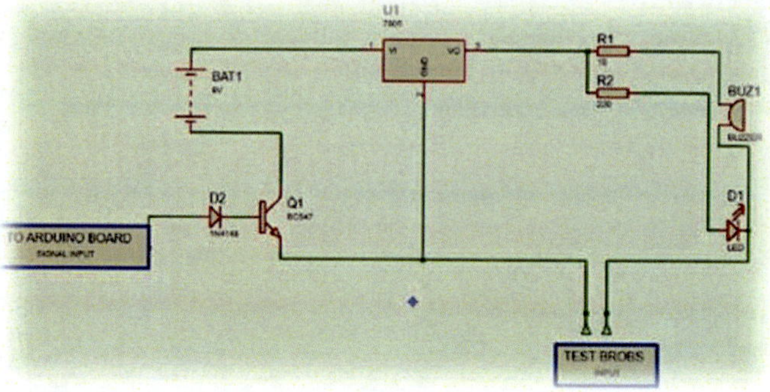

Fig. 11 Single wire tester

any problems. The chocking pulse generated by NE555 IC is given by pin number 3 to clock pin of CD 4017 IC. According to the received pulse signal, CD 4017 IC sends voltage signal to pin 1–12. The circuit diagram of multiple wire tester is shown in Fig. 12.

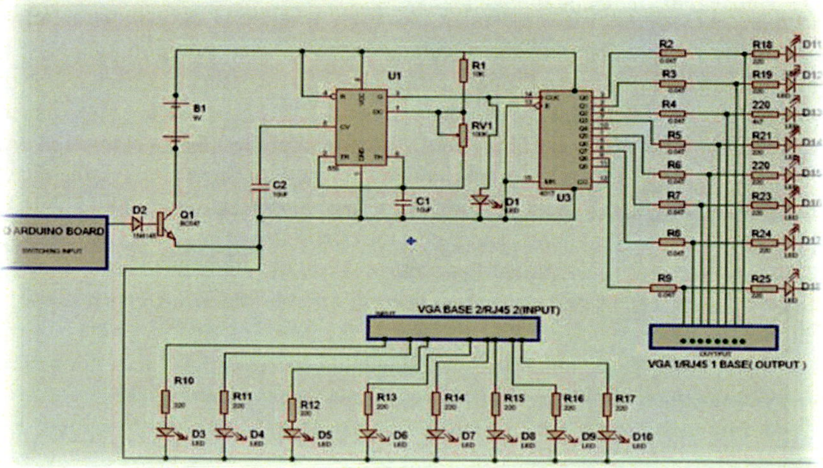

Fig. 12 Multiple wire continuity tester

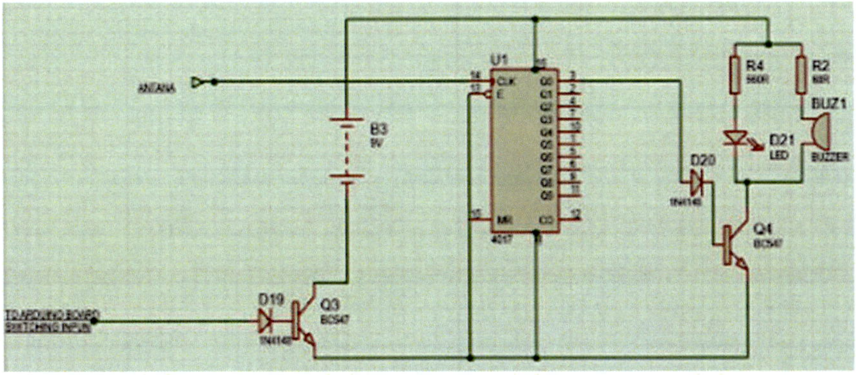

Fig. 13 Circuit diagram for wireless conductor testing

3.6 *Wireless Conductor Checking Circuit*

The wireless conductor checking circuit is shown in figure. When 9 V power is supplied to the transistor Q3, it switches on the wireless testing circuit. The 4017 IC senses the current carrying conductor through the antenna. The antenna is used to magnify the signal strength. When the conductor kept near to the antenna, a small current is induced in the antenna and it flow to CD4017 IC. This current is detected by the IC and the output signal sends to the transistor Q4 through the diode. When the output signal is high, the Q4 transistor would be biased and it supplies the voltage to the buzzer and LED and they make the indication. The circuit diagram of wireless conductor testing circuit is shown in Fig. 13.

3.7 *Voltage and Current Measurement*

Various supply voltages are required to work in the industries. There are situations where up to 12 V supply may not be available for most of the test equipment. But this tester can supply up to 40 V with the help of the inverter (5–40 V). The circuit diagram of voltage and amphere tester is shown in Fig. 14. Supply voltage and ampheres are displayed on the liquid crystal display. A 5 V relay is used for switching operation because there would not be any voltage drop through it while using it for switching operation. If a transistor is used for this purpose, there will be some voltage drop across the transistor. A step down transformer also not advisable in this circuit, since it is too big to this situation.

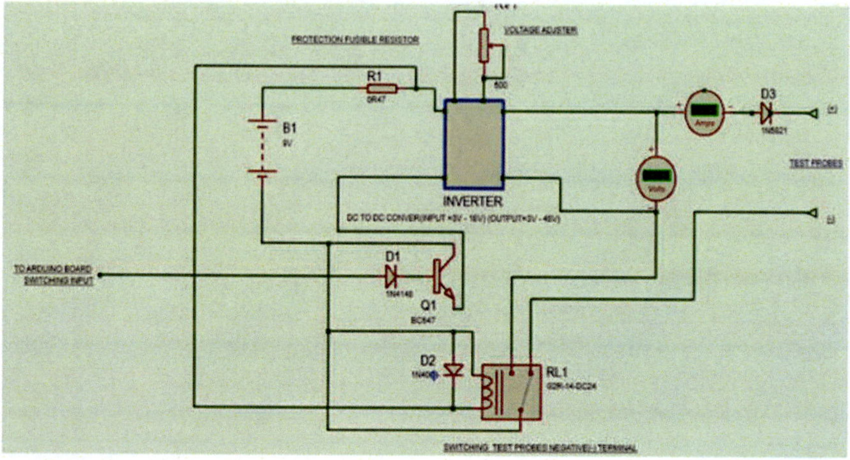

Fig. 14 Circuit tester

3.8 LED Array Circuit

Light emiting diodes are used at troubleshooting time to light up the testing area in case of poor lighting condition. The circuit shown in Fig. 15 switches based on BC547 transistor. The supply voltage is given to the circuit and LEDs lit up. There are two LED arrays and seven LEDs in each array. All the LEDs are connected in parallel to the supply voltage. R2 and R3 resistors are connected in series to control the current for LED array.

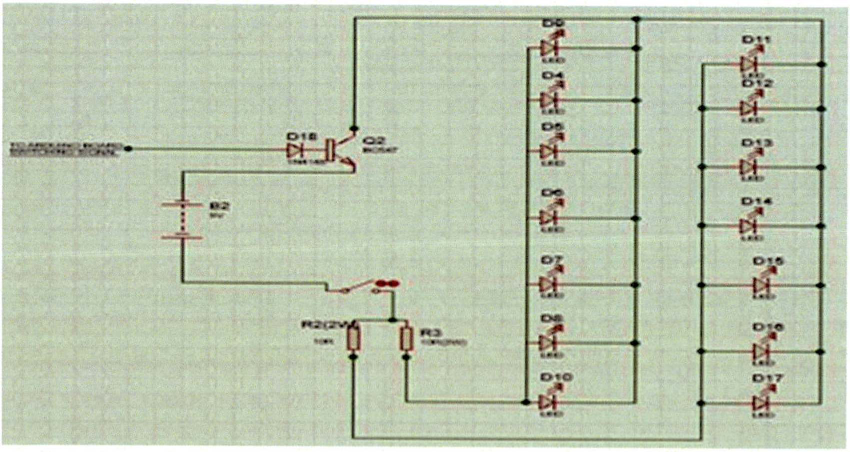

Fig. 15 LED array circuit

In the similar way, more client sensors can be connected with main module through the extra ports provided. Based on the requirement, additional sensors can be connected with the main module and read the information from display fixed on the main module.

4 Results and Discussion

Troubleshooting and fault finding are very important parts in electrical and electronics technology. Proper fault finding device is necessary to finish the troubleshooting process on time and with less labor effort. Eventhough different equipments are used to find and rectify faults in the industry, it would be beneficial if it is possible to compact more number of testing devices in one small portable device. In other words, time and other resources will also be managed in very effective manner. The multifunction tester is designed based on these ideas. There is one main master module. Frequently used functions are already in build with master module. Additional secondary modules can be connected with the master module based on the requirement.

Broken wires passing through slabs or walls could be identified without demolishing the slab or wall. Instead of dismantling the whole system in a machine just to find a broken wire, that particular wire can be identified with ease, if this method is used. The replacement of broken wire can also be achieved by using the lighting provider attached in this device, in case if the broken wire is identified in some highly complexed circuit. Exact broken wire can be identified in the cables like VGA cable and RJ 45 cable which has more than one wire. Since individual LEDs are allocated for each wire, the fault in the exact wire can be identified easily. Through this finding, the broken wire can be changed instead of disposing the whole wire. Voltage and amphere measurement submodule is used to check the voltage and amphere of circuits. In a testing environment, a voltage ranged between 3 and 40 V can be supplied by this device. So the experimentation could be continued without plugging the machinary into an external circuit to obtain the neceaary voltage. The experimentation of each module is shown in Figs. 16, 17, 18, 19, 20 and 21. But it is not limited to only these submodules. Based on user requirements, suitable secondary sensor modules can be connected with the main module.

5 Conclusion and Future Scope

This single device called multifunction tester can be used to

1. Check the continuity of a single wire.
2. Check the continuity of multiple wires in VGA cable, RJ 45 cable, etc.

Fig. 16 Continuity testing of single wire

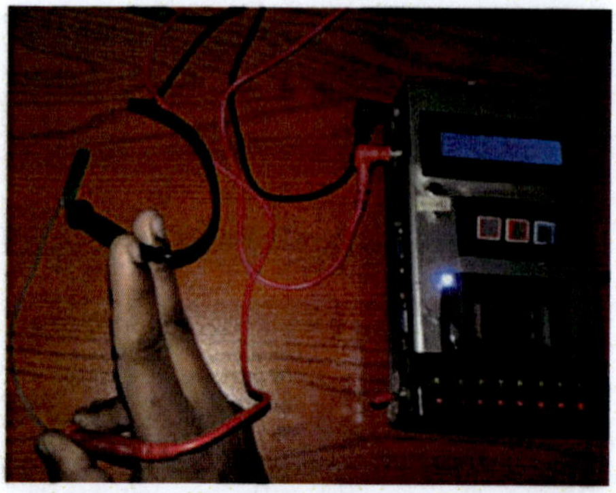

Fig. 17 Continuity testing of multiple wires

Fig. 18 Wireless conductor testing

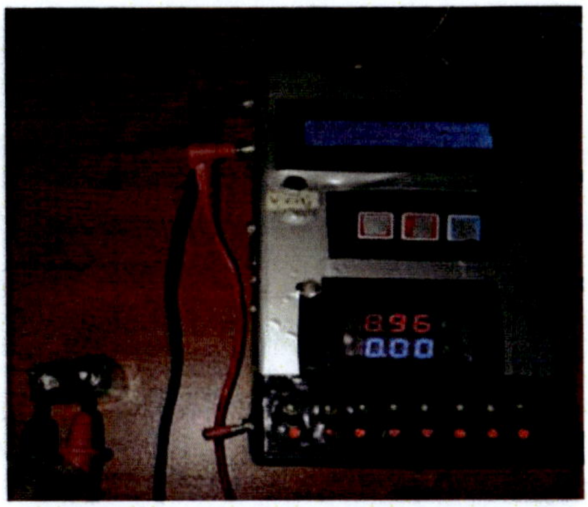

Fig. 19 Voltage testing

Fig. 20 Amphere testing

Fig. 21 Portable multifunction tester

3. Check the continuity of current carrying conductor without breaking the slab, without connecting any wire when it is in a circuit.
4. Measure and display voltage and amphere of equipment under test.
5. Provide light to the testing space inside a machine where external light finds difficulty to enter in.

It is highly beneficial to people like electrical engineers, electrotechnicians and underground cable operators. Instead of using several devices to troubleshoot various issues in a circuit, a single multifunction tester which can analyze different parameters can be used. The basic unit of the master circuit of the multifunction tester is processing, functioning and controlling based on microcontroller ATmega328. There can be some limitations with input and output pins, frequency, accuracy, speed and storage capacity. To improve the speed, storage, processing delay and multiprocessing, can go for high speed and high accuracy PLC technology. Programmable logical controller unit (PLC) is considered as more efficient than the microcontroller. But the cost should also be taken care if the PLC is used.

References

1. Yang, J., Fletcher, J.E., O'Reilly, J.: Short-circuit and ground fault analysis and location in VAC – based DC network cable. IEEE Trans. Ind. Electron. **59**(10), 3827–3837.
2. Kedia, T., Sahare, V., Bauri, K.K., Sahu, R.K., Kumar, S., La, A.: Underground cable fault distance detector using Atmega328 Microcontroller. IJAREEIE **7**(10) (2017)
3. Kesim, H.: Automated continuity testing of flexible backplanes using a cable tester, test systems and process design Department Defense Systems Technologies Sector, ASELSAN Inc Ankara, TURKEY
4. Roy, P., Karmakar, S., Basak, P., Das, S., Khatua, S.: Designing of a device for checking the polarity and continuity of any electrical and electronics circuit. IRJET **3**(7), 170–172
5. Nandhini, T., Shalini, J., SaiSangeeetha, T., Gnanaprakasam, D.: Underground cable fault detection using Arduino. IJESC **7** (2017)
6. Padmanaban, K., SanjanaSharon, G., Sudharini, N., Vishnu Varthini, K.: Detection of Underground cable fault using Arduino. IRJET **4** (2017)
7. Choi, M.S., Lee, D.S., Yang, X.: A line to ground fault location algorithm for underground cable system. KIEE Trans. Power Eng. 267–273 (2005)
8. Chandra, D., Singh, P., Rajguru, P.P., Namekar, S.A.: Detection of short circuit fault in underground cable using microcontroller. IJIREEICE **5**(6) (2017)
9. Sidhu, T.S., Xu, Z.: Detection of ıncipient faults in distribution underground cables. IEEE Trans Power Del. **25**(3) (2010)
10. Sampathraja, N., Ashok Kumar, L., Kirubalakshmi, V., Muthumaniyarasi, C., Vishnu Murthy, K.: IoT based underground cable fault detector. IJMET **8**(8) (2017)
11. Pal, S., Verma, S., Yadav, S.S., Kumar, D., Kaur, A.: Development of a prototype of a GSM based underground cable fault detector. Int. J. Res. Educ. Sci. Methods IJARESM **5**(4), 4 (2017). ISSN: 2455-6211
12. Apisit, C., Ngaopitakkul, A.: Identification of fault types for underground cable using discrete wavelet transform. In: Proceeding of the International Multi Conference of Engineers and Computer Scientist 2010, vol. 11. IMECS 2010, Hong Kong, Mar 17–19, 2010
13. Atmel-8271-8-bit-AVR-Microcontroller-ATmega48A-48PA-88A-88PA-168A-168PA-328–328P_datasheet_Complete, Microchip

Kinematic Performance Analysis of 3-DOF 3RRR Planar Parallel Manipulator

Himam Saheb Shaik and G. Satish Babu

Abstract In this article, the 3RRR planar parallel manipulator (PPM) is considered for modeling and performance analysis. As the revolute joints have better performance compared to prismatic joints because of frictional losses, in this paper, an attempt is made to calculate all possible performance indices. The 3RRR PPM is modeled, and performance indices like condition number, singularity, transmissibility, manipulability, stiffness and velocity indices are calculated. The three revolute joints (*R*) are considered, and the *R* indicates the actuated joint. Performance indices measurement is useful to the manipulator design, analysis, understanding and application of robotic manipulators to specific fields. Various different performance indices have been used to study the behavior of robotic manipulators from the inception of robots. This work is useful to have a clear understanding of 3RRR planar parallel manipulators.

Keywords 3RRR planar parallel manipulator · Performance indices

1 Introduction

Parallel and serial manipulators have been receiving a lot of consideration from researchers over the recent decades, because of their easiness in using, accurate results and specifically the parallel robots have clear attention by the researchers due to their advantages over the serial manipulators, like the high accuracy, high load carrying capacity with less deflection in other words good manipulator mass

H. S. Shaik (✉)
Department of Mechanical Engineering, Faculty of Science and Technology, The ICFAI Foundation for Higher Education, Hyderabad 501203, India
e-mail: himam.mech@gmail.com

G. Satish Babu
Department of Mechanical Engineering, GVP College of Engineering (Autonomous), Visakhapatnam 530048, India
e-mail: satishbabu.jntuhm@gmail.com

© The Editor(s) (if applicable) and The Author(s), under exclusive license to Springer Nature Singapore Pte Ltd. 2021

A. P. Pandian et al. (eds.), *Proceedings of International Conference on Intelligent Computing, Information and Control Systems*, Advances in Intelligent Systems and Computing 1272, https://doi.org/10.1007/978-981-15-8443-5_75

ratio and better robustness, as well as they have taken parallel robotic manipulators with less than and equal to six degrees of freedom as most of the applications need less than the six degrees of freedom. Six degrees of freedom include three rotations and three translations, rotations in all three planes and translation in three directions. The degrees of freedom can be calculated by Grublers criteria [1]. As well as constrained and over-constrained parallel manipulators clearly explained. Over-constrained parallel robotic manipulators include the Gosselin's 3RRR manipulator [2].

Beginning in the 1990's, planar parallel manipulators (PPMs) have begun either to be placed being used in different fields, for example, precise accurate positioning machine parts and some of the potential applications, machine tools, battery swapping applications, food processing, space applications, telescope positioning, and many medical applications starting from da vinci robot to cardio pulmonary resuscitation, even though this many applications and advantages the parallel manipulators are taking less usage because of the hardware used in the parallel robot [3].

2 Planar 3RRR Parallel Manipulator

3 Geometry of the 3RRR Manipulator

From the above figure, the origin point of the fixed ground coordinate point is taken at point P. From Fig. 1, the X-axis variables are along the way of PQ, and the Y-axis is orthogonal to PQ. By considering both the moving and fixed platform, ABC and

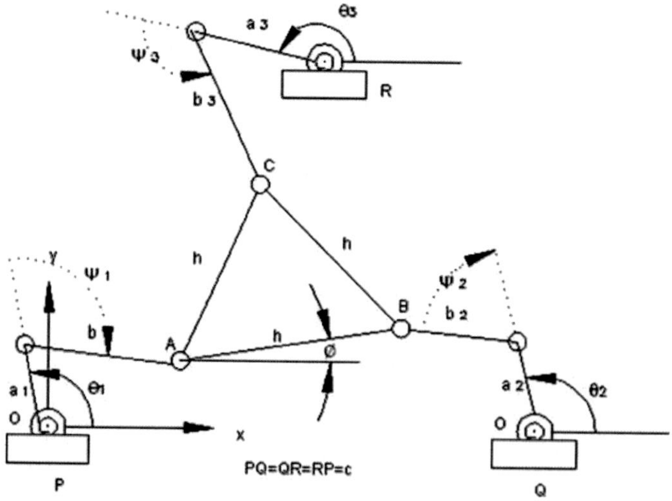

Fig. 1 Kinematic model of planar 3RRR parallel manipulator

PQR respectively are equilateral triangles that indicate $AB = BC = AC = h$ and $PQ = QR = RP = c$. The moving platform of manipulator is connected with three links to the manipulator fixed platform, total six binary joints are used to connect fixed and moving platform on the base of the fixed platform, the motors are fixed for actuation purpose, and the manipulator is getting input instructions and same executed on the moving platform, where the end effector is located.

Form the loop closure equations, the following equations can be written

$$x_B = x_A + h\cos\emptyset, \tag{1}$$

$$y_B = y_A + h\sin\emptyset \tag{2}$$

and

$$x_C = x_A + h\cos\left(\emptyset + \frac{\Pi}{3}\right), \tag{3}$$

$$y_C = y_A + h\sin\left(\emptyset + \frac{\Pi}{3}\right) \tag{4}$$

With the help of the figure geometry, the loop closure equation can be written easily as shown in Eq. (5)

$$\overline{\mathrm{OA}} = \overline{\mathrm{OP}} + \overline{\mathrm{PD}} + \overline{\mathrm{DA}} \tag{5}$$

Rearranging the vector-loop closure path of the coordinate frame gives the following equations

$$x_A = a_1\cos\theta_1 + b_1\cos(\theta_1 + \varphi_1), \tag{6}$$

$$y_A = a_1\sin\theta_1 + b_1\sin(\theta_1 + \varphi_1) \tag{7}$$

As φ_1 is joint angle which is passive, to be eliminated from the above equations

$$x_A^2 + y_A^2 - 2x_A a_1\cos\theta_1 - 2y_A a_1\sin\theta_1 + a_1^2 - b_1^2 = 0 \tag{8}$$

4 Inverse Kinematic Analysis of 3RRR Manipulator

In inverse kinematics, x_A, y_A and $\emptyset$ are given, and the corresponding joint angles θ_1, θ_2 and θ_3 to be calculated. These are taken on a limb-to-limb basis. For limb 1, Eq. (9) is arranged in the following form (Fig. 2):

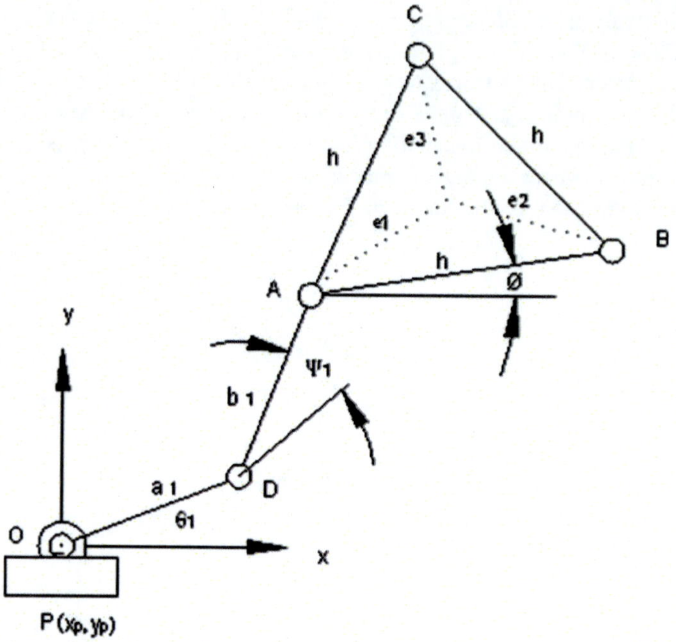

Fig. 2 One single loop of the parallel manipulator

$$K_1 \sin\theta_1 + K_2 \cos\theta_2 + K_3 = 0 \tag{9}$$

where

$$K_1 = -2y_A a_1,$$
$$K_2 = -2x_A a_1,$$
$$K_3 = x_A^2 + y_A^2 + a_1^2 - b_1^2.$$

$$(K_3 - K_2)t_1^2 + K_1 t_1 + (K_3 + K_2) = 0. \tag{10}$$

Solving Eq. (10) for t_1 yields

$$\theta_1 = 2\tan^{-1}\left(\frac{\left(-K_1 \pm \sqrt{K_1^2 + K_2^2 - K_3^2}\right)}{(K_3 - K_2)}\right) \tag{11}$$

Subsequently, comparing to every moving platform location, there are generally two different solutions of θ_1 and accordingly two configurations of limb 1. If Eq. (11)

has double root, the two links PD and DA are in a completely stretched-out (folded-back configuration) also called as singular (position) configuration. When Eq. (11) has no real root, the specified moving platform location is out of reach. Once θ_1 is solved, φ_1 can be found from Eq. (11).

Following the same procedure, the other two limb configurations can be easily solved and also can conclude that there are eight robotic manipulator positions related to a given end-effector location.

5 Jacobian of a 3RRR Planar Parallel Manipulator

The moving platform of the robotic manipulator is having pivoted ends *A*, *B* and *C*, similarly the fixed platform of the robotic manipulator which is defined by the fixed pivoted ends *P*, *Q* and *R*. Point *G* is a point in a moving platform where generally the end effector placed, and a_i and b_i denote the vectors of the first and second moving links of the *i*th limb, while θ_i and φ_i denote the orientation angles of a_i and b_i with reference to the *X*-axis.

The loop closure equation is to be written for each limb for position equation. For example, the first limb has below the following:

$$\overline{\mathrm{PG}} + \overline{\mathrm{GA}} = \overline{\mathrm{PD}} + \overline{\mathrm{DA}} \tag{12}$$

Upon differentiating Eqs. 6 and 7

$$\dot{x}_A = -a_1 \sin(\theta_1)\,(\dot{\theta}_1) - b_1 \sin(\theta_1 + \varphi_1)\big(\dot{\theta}_1 + \dot{\varphi}_1\big) \tag{13}$$

$$\dot{y}_A = a_1 \cos(\theta_1)\,(\dot{\theta}_1) + b_1 \cos(\theta_1 + \varphi_1)\big(\dot{\theta}_1 + \dot{\varphi}_1\big) \tag{14}$$

$$\cos(\theta_1 + \varphi_1).\dot{x}_A = -a_1 \sin(\theta_1)\,(\dot{\theta}_1).\cos(\theta_1 + \varphi_1) - b_1 \sin(\theta_1 + \varphi_1)\big(\dot{\theta}_1 + \dot{\varphi}_1\big) \\ \cos(\theta_1 + \varphi_1) \tag{15}$$

$$\cos(\theta_1 + \varphi_1).\dot{y}_A = a_1 \cos(\theta_1)\,(\dot{\theta})_1.\cos(\theta_1 + \varphi_1) + b_1 \cos(\theta_1 + \varphi_1)\big(\dot{\theta}_1 + \dot{\varphi}_1\big) \\ .\cos(\theta_1 + \varphi_1) \tag{16}$$

As this is a planar kinematic mechanism, the velocity of all link point is in the positive *Z*-direction.

$$\boldsymbol{v}_g + \dot{\emptyset}(\boldsymbol{k} \times \boldsymbol{e}_i) = \dot{\boldsymbol{\theta}}_i(\boldsymbol{k} \times \boldsymbol{a}_i) + \big(\dot{\boldsymbol{\theta}}_i + \dot{\boldsymbol{\varphi}}_i\big)(\boldsymbol{k} \times \boldsymbol{b}_i), \quad \mathrm{i} = 1, 2, 3 \tag{17}$$

where $\boldsymbol{v}_g$ is velocity of the point G and $\boldsymbol{k}$ is a unit vector which is positioning towards the positive Z-axis direction. Since $\dot{\boldsymbol{\varphi}}_i$ is a passive variable, this needs to be removed from Eq. (17). To achieve this output, dot-multiply both sides of Eq. (17) by $\boldsymbol{b}_i$, and this leads to

$$\boldsymbol{b}_i.\boldsymbol{v}_g + \dot{\emptyset}\boldsymbol{k}.(\boldsymbol{e} \times \boldsymbol{b}_i) = \dot{\boldsymbol{\theta}}_i\boldsymbol{k}.(\boldsymbol{a}_i \times \boldsymbol{b}_i) \tag{18}$$

Writing Eq. (18) for three links, once for each link $i = 1, 2,$ and 3, becoming three scalar equations, which is to be arranged in matrix form

$$\boldsymbol{J}_x\dot{\boldsymbol{x}} = \boldsymbol{J}_q\dot{\boldsymbol{q}} \tag{19}$$

where

$$\boldsymbol{J}_x = \begin{bmatrix} b_{1x}\ b_{1.y}\ K_{1x}.b_{1y} - K_{1y}.b_{1x} \\ b_{2x}\ b_{2.y}\ K_{2x}.b_{2y} - K_{2y}.b_{2x} \\ b_{3x}\ b_{3.y}\ K_{3x}.b_{3y} - K_{3y}.b_{3x} \end{bmatrix}$$

$$\boldsymbol{J}_q = \begin{bmatrix} a_{1x}.b_{1y} - a_{1y}.b_{1x} & 0 & 0 \\ 0 & a_{2x}.b_{2y} - a_{2y}.b_{2x} & 0 \\ 0 & 0 & a_{3x}.b_{3y} - a_{3y}.b_{3x} \end{bmatrix}$$

And where

$$\dot{\boldsymbol{x}} = \left[v_{gx}, v_{gy}, \dot{\emptyset}\right]^{\mathrm{T}} \text{and}\ \dot{\boldsymbol{q}} = \left[\dot{\theta}_1, \dot{\theta}_2, \dot{\theta}_3\right]^{\mathrm{T}}$$

The Jacobian of the 3RRR manipulator is given by

$$\boldsymbol{J} = \boldsymbol{J}_q^{-1}\boldsymbol{J}_x$$

6 Jacobian Matrix

Jacobian matrix in the robotics explains the connection between joint speeds/velocities (A) and end-effector velocities (B) of a robotic manipulator.

The robotic arm is moving with certain speed to analyze the robot, to calculate the performance of robot, and the proper relation between joint velocity and end-effector velocity is compulsory, and this Jacobean matrix format is as shown below and useful to calculate the robot performance as well as to design robotic manipulator

$$B = J \times A$$

•

7 Workspace of Manipulators

The research works in the literature are concentrated on the enlargement of the manipulator workspace [4–8], and still, it continues as a emerging research topic as workspace a major role in the robotic manipulator selection [9–14]. Many numerical solutions and analytical methods are used to calculate the different workspaces, namely dexterous workspace, workspace index, reachable workspace and constant orientation workspace of manipulators from past two decades, but, as of today, a satisfactory and mathematically cost-effective generalized workspace solution need to be achieved to reduce maintenance cost of the robot [15]. In some of the research works, the workspace of a robotic manipulator is referred as work envelope [9]. To design a parallel manipulator for any specific application it is compulsory to find its boundaries, workspace, holes and contours of manipulability and dexterity [10] based on the degrees of freedom of the parallel robotic manipulator, the relative arrangement of the robotic joints in the manipulator frame, and number of links, specifications of links, based on these all the robot workspace is may be flat, cylindrical, triangular or any intricate shape let take and example a Three degrees of freedom revolute (3R) planar parallel manipulator [16]. The workspace is looking like a triangle as all the three links can rotate with certain degrees of freedom [17] to compare manipulators; workspace is also one of the important parameters [9]; many researchers studied the performance indices of manipulators [18] to optimize manipulator for better output. Workspace of 3$\underline{R}$RR planar parallel manipulator can be calculated by drawing three circles, and the radius of circles is the maximum link length of the robotic arm. The common area of the three circles indicates the dexterous workspace where robotic manipulator can be used more effectively as this space consumes less energy (Fig. 3).

8 Condition Number

Utilization of the condition number means the amplification of the manipulator error on the pose of the end effector. The small number of condition number gives the best results, and for this reason, small number is desirable. The condition number is calculated by singular value decomposition theorem (SVD) of the Jacobian matrix. SVD refers to expanding the Jacobian matrix into three components as shown below

$$J = U\Sigma V^{\mathrm{T}}$$

The Jacobian is calculated by $\boldsymbol{J} = \boldsymbol{J}_q^{-1}\boldsymbol{J}_x$. The condition number of a Jacobian matrix is expressed as the ratio of the maximum singular value of the singular value

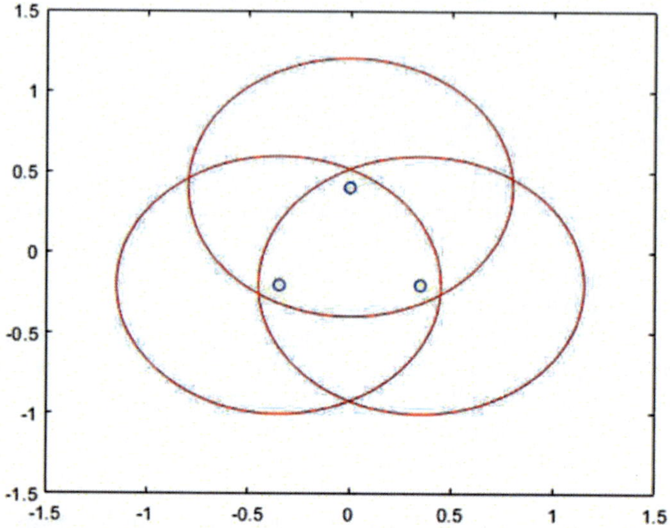

Fig. 3 The MATLAB solution of workspace of 3RRR planar parallel manipulator

matrix, $\sigma_{\max}$, to the smallest singular value of the same matrix, $\sigma_{\min}$, and from this, we can write, $k = \sigma_{\max}/\sigma_{\min}$ which explains the change in the end effector for change of the actuators. Condition number is a big number. Using the reciprocal condition number helps to be presentable in the graph, and this reciprocal value is called dexterity indices which ranges from 0 to 1. Attempting to move toward a condition number near to one helps the manipulator end effectors to have exactness near the precision of the actuators and to keep up a good stiffness which are required characteristics of the robotic manipulator (Fig. 4).

9 Manipulability Indices

The manipulability indices (MI) [19] shows that the manipulator how much dexterous to do any task in the workspace. In [20], it is clearly estimated as the product result of ellipsoid's axes or to be specific is the product of the singular value of the Jacobian matrix, which is mathematically represented as

$$[\mu] = \sqrt{\det\left(J J^{\mathrm{T}}\right)} = \sqrt{\lambda_1, \lambda_2, \lambda_3, \ldots \lambda_n} = \sigma_1, \sigma_2, \sigma_3 \ldots \sigma_m$$

where $\lambda_1, \lambda_2, \ldots, \lambda_m$ are the eigenvalues of JJT and the $\sigma_1, \sigma_2, \ldots, \sigma_m$ are the singular values of J.

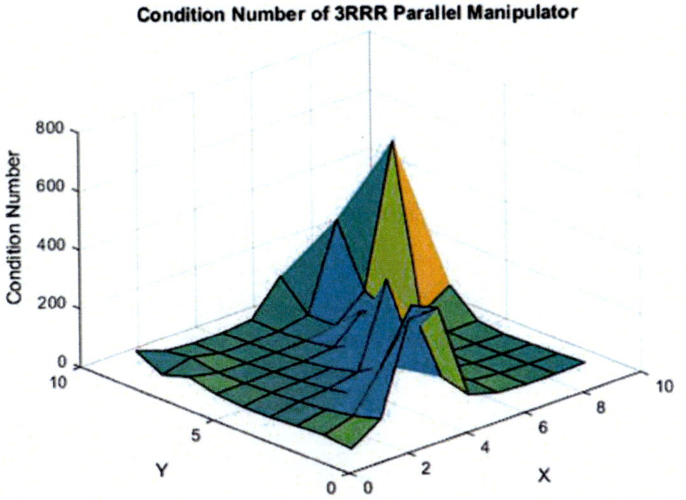

Fig. 4 The condition number of 3RRR planar parallel manipulator

The manipulability value measures an estimation of containing actuated motors, output links the closeness of the exactness in the velocity between the joints and the end effectors (Fig. 5).

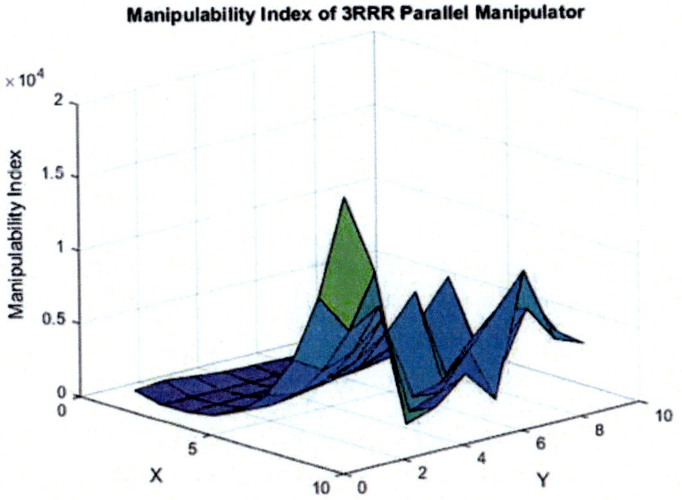

Fig. 5 The manipulability distribution of the planar parallel manipulator

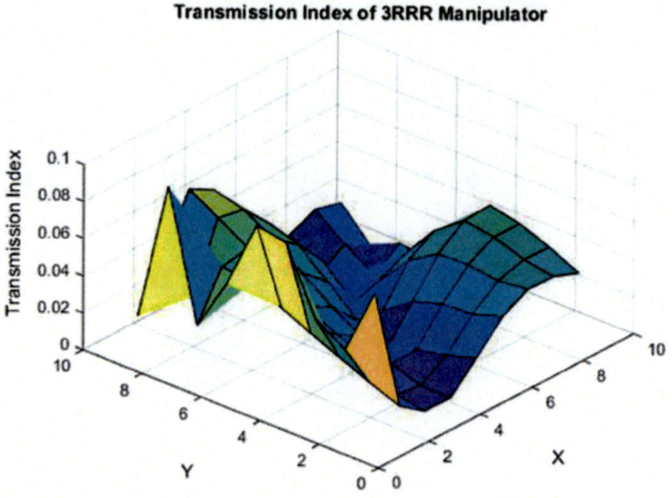

Fig. 6 The transmission indices of planar parallel manipulator

10 Transmission Indices

The important and main application of the planar parallel manipulator is to transmit force and movement from input links to its output links. In general ordinarily, the input links are containing actuated motors, and output links are moving platforms which carries end effector. The motion and force transmissibility indices were calculated, and the detailed analysis of the 3RRR planar parallel manipulator is presented, which can be further used as base data to determine the singularity indices of the robotic manipulator. The value of transmission index is ranging from 0 to 0.1, and the higher the value indicated that the better transmission possible with the manipulator, vice versa (Fig. 6).

11 Stiffness Indices

As the stiffness is affecting the repeatability and accuracy of the robotic manipulator, the stiffness is expressed as the capacity of the robotic manipulator to bear the deformation under load. If the structure has high value of stiffness, it means that the flexibility of robot is less, flexibility is the complete inverse of stiffness, and stiffness of the manipulator is depending on the figure geometry, robot configuration and robot topology. The robotic manipulator stiffness is effecting the position accuracy and dynamic behavior of the manipulator that is the reason the robotic manipulator stiffness analysis is considered to be a major performance index. As reported by Wu,

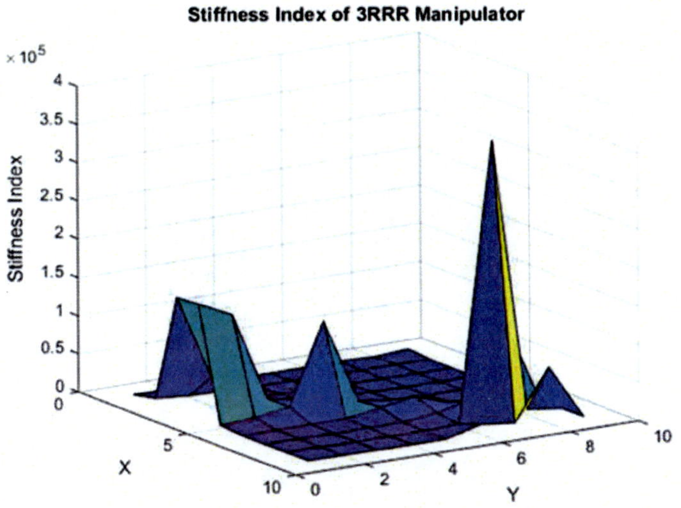

Fig. 7 The stiffness indices of 3RRR planar parallel manipulator

X, [21], the parallel manipulator stiffness matrix can be written as the eigenvalues of $\left(K^{-I}\right)^{\mathrm{T}}.K^{-I}$.

where $K = J^{\mathrm{T}}.P.J$

$P_1 = P_2 = P_3 = 1$ (Fig. 7).

12 Velocity Indices

For general parallel manipulator, the mathematical relation between velocity of moving platform $\dot{x}$ and the velocity of the active joints $\dot{q}$ is expressed as

$$\dot{q} = J\dot{x} \tag{31}$$

As the velocity performance also has an important key role in the performance analysis of planar parallel manipulators, from the above equation, the relation between velocity of moving platform and the velocity of actuated joints can be expressed as below equation

$$\dot{x} = J^{-I}\dot{q}$$

Xiaoyong et al. [18] presented the velocity indices as the eigenvalues of the matrix $\gamma = \left(J^{-I}\right)^{\mathrm{T}} * \left(J^{-I}\right)$ and the output velocity at extreme values are expressed as

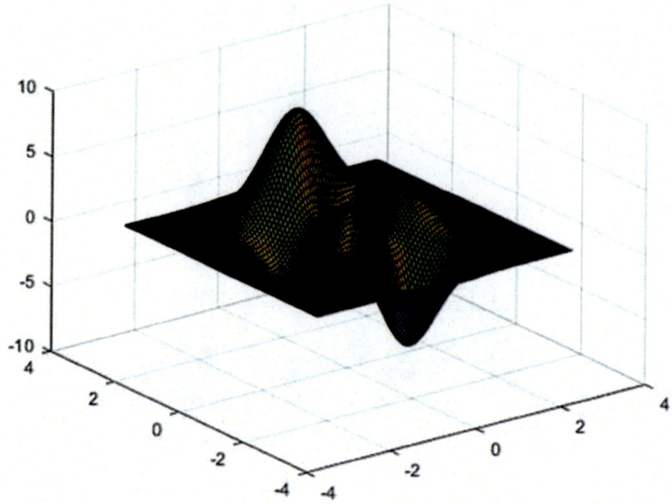

Fig. 8 The velocity distribution of 3RRR planar parallel manipulator

$$\text{Velocity}_{\text{min}} = \sqrt{\min\left(:\gamma:\right)}$$

$$\text{Velocity}_{\text{max}} = \sqrt{\max\left(:\gamma:\right)}$$

For 3RRR planar parallel manipulator, from the above Jacobian matrix, eigenvalues are calculated in the workspace and maximum and minimum velocities are plotted as shown in Fig. 8

13 Singularity Indices

Singularity indices of planar parallel robotic manipulator have been explored by different researchers because this is a major significant issue [22]. At singularity, the robotic manipulator loses its control as it will not give response to the input variables that condition is called singularity, and this condition generally occurs in the following cases that are (1) the robotic manipulator attains several (more than one) unwanted degrees of freedom (DOF) that leads to loses or decrease of the stiffness and less load capacity in the way of the unwanted additional degrees of freedom, and (2) the robotic manipulator loses required degrees of freedoms and stable at dead point as it has no input arguments [23]. In practice, the possible singular configurations are to be removed for the best output of the robotic manipulator.

In the conclusion, the singularity indices are the result of Jacobian matrix rank deficiency; as shown in the figures, the singularity occurs in robot, whenever the robot gains some extra, uncontrollable DOF or lost some of required DOF. At the same time, the force transmission capacity of the planar parallel manipulator is too poor at singular configurations of the manipulator as mentioned in [24]. Merlet [25] reported that all possible singular configurations are identified with the help of line geometric functions. In recent years, different approaches to calculate singularity points are explained by Kim and Park [26], Ma and Angeles [27], Davidson and Hunt [28] and Sai [29]. The best method to calculate the singularity [30] is determinant of $[J]=0$ (Figs. 8, 9, 10, 11 and 12).

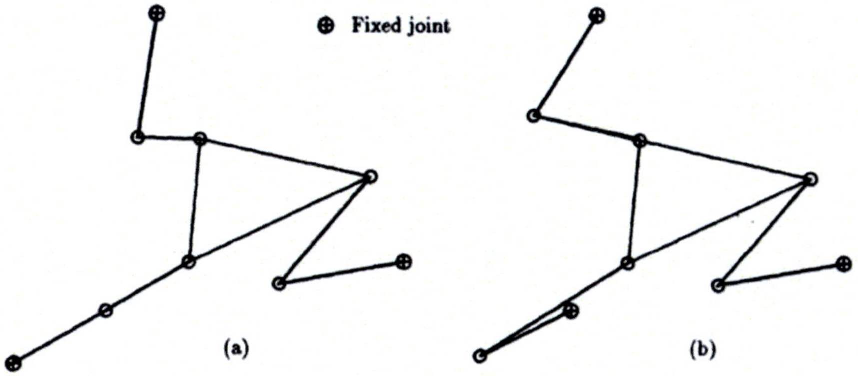

Fig. 9 The I type of singularity for the 3RRR planar parallel manipulator **a** the one leg fully extended singularity and **b** the one leg fully folded singularity

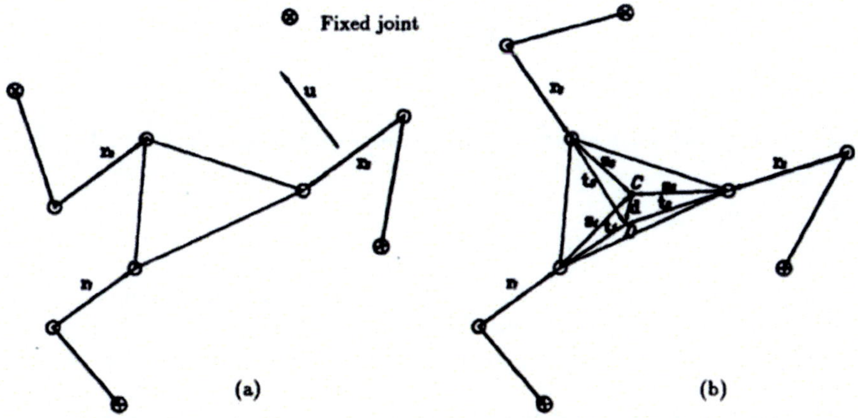

Fig. 10 The II type of singularity for the 3RRR planar parallel manipulator **a** that three vectors are parallel and **b** three vectors are intersected other

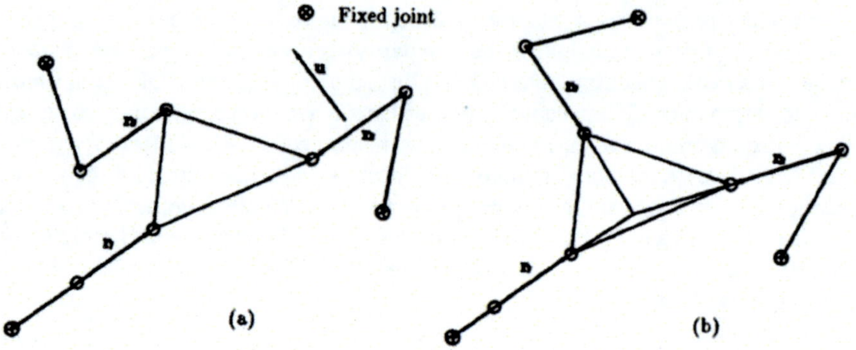

Fig. 11 The III type of singularity for the 3RRR planar parallel manipulator **a** that the three vectors are parallel singularity and **b** three vectors are coinciding points

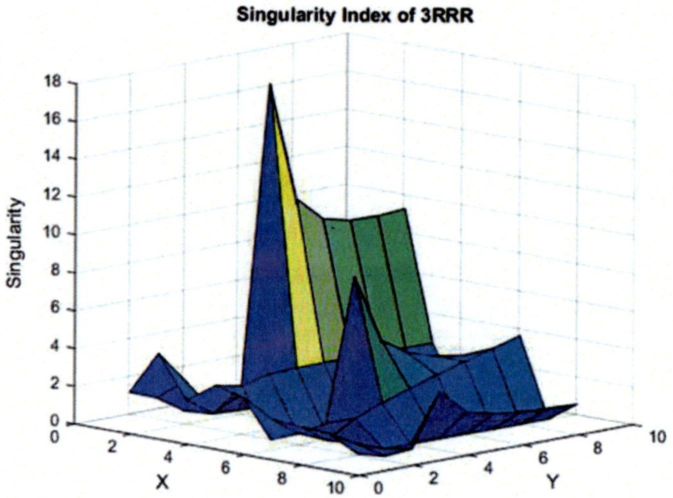

Fig. 12 The singularity distribution of the 3RRR planar parallel manipulator

14 Conclusion

In this present article, the 3RRR planar parallel manipulator is modeled and analyzed. This work is useful to the researchers for selecting the best position and orientation of the manipulator for positioning the end effector; in this work, each limb contains three joints which revolute in nature and one joint is actuated with motor; the actuated joint is connected to the ground and represented as *R*; after proper modeling, Jacobian matrix is formulated, and performance indices are calculated. The detailed analysis of these indices is presented in the paper for better understanding of planar parallel manipulator.

References

1. Gosselin, C.: These de Doctorat Mcgill University. Ph.D. dissertation, McGill University, Montreal, Canada (1988)
2. Di Gregorio, R.: Statics and singularity loci of the 3-upu wrist. IEEE Trans. Robot. **20**(4), 630–635 (2001)
3. Merlet, J.P., Daney, D.: Appropriate design of parallel manipulators. In: Wang, L., Xi, J. (eds.) Smart Devices and Machines for Advanced Manufacturing. Springer, London (2008)
4. Kumar, A., Waldron, K.J.: The workspace of mechanical manipulator. J. Mech. Des. Trans. ASME **103**, 665–672 (1981)
5. Gupta, K.C., Roth, B.: Design considerations for manipulator workspace. ASME J. Mech. Des. **104**, 704–711 (1982)
6. Yang, D.C., Lai, Z.C.: On dexterity of robotic manipulators—service angle. J. Mech. Transm. Autom. Des. **107**(2), 262–270 (1985)
7. Kumar, A., Patel, M.S.: Mapping the manipulator workspace using interactive computer graphics. Int. J. Robot. Res. **5**(2), 122–130 (1986). https://doi.org/10.1177/027836498600500213
8. Alameldin, T., Sobh, T.: On the evaluation of reachable workspace for redundant manipulators. In: Paper Presented at the Proceedings of the 3rd International Conference on Industrial and Engineering Applications of Artificial Intelligence and Expert Systems, vol. 2 Charleston
9. Kucuk, S., Bingul, Z.: Robot workspace optimization based on a novel local and global performance indices. In: Proceedings of the IEEE International Symposium on Industrial Electronics, ISIE, pp. 1593–1598 (2005)
10. Chang-Hwan, C., Hee-Seong, P., Seong-Hyun, K., Hyo-Jik, L., Jong-Kwang, L., Ji-Sub, Y., Byung-Seok, P.: A manipulator workspace generation algorithm using a singular value decomposition. Int. Conf. Control Autom. Syst. ICCAS **2008**, 163–168 (2008)
11. Jin, Y., Chen, I.M., Yang, G.: Workspace evaluation of manipulators through finite-partition of SE(3). Robot. Comput. Integr. Manuf. **27**(4), 850–859 (2011). https://doi.org/10.1016/j.rcim.2011.01.006
12. Wang, Z., Ji, S., Li, Y., Wan, Y.: A unified algorithm to determine the reachable and dexterous workspace of parallel manipulators. Robot. Comput. Integr. Manuf. **26**(5), 454–460 (2010). https://doi.org/10.1016/j.rcim.2010.02.001
13. Yang, J., Abdel-Malek, K., Zhang, Y.: On the workspace boundary determination of serial manipulators with nonunilateral constraints. Robot. Comput. Integr. Manuf. **24**(1), 60–76 (2008). https://doi.org/10.1016/j.rcim.2006.06.005
14. Tsai, K.L.: Determining the workspace boundary of 6-DOF parallel manipulators. Robotica **24**(5), 605–611 (2006). https://doi.org/10.1017/s0263574706002682
15. Alameldin, T., Badler, N., Sobh, T., Mihali, R.: A computational approach for constructing the reachable workspaces for redundant manipulators. Comput. J. **2**(1), 48–52 (2003)
16. Abdel-Malek, K., Yeh, H.J., Othman, S.: Interior and exterior boundaries to the workspace of mechanical manipulators. Robot. Comput. Integr. Manuf. **16**(5), 365–376 (2000). https://doi.org/10.1016/S0736-5845(00)00011-9
17. Abdel-Malek, K., Adkins, F., Yeh, H.J., Haug, E.: On the determination of boundaries to manipulator workspaces. Robot. Comput. Integr. Manuf. **13**(1), 63–72 (1997)
18. Xiaoyong, Wu., Yan, R., Tan, R.: Performance analysis and comparison of three planar parallel manipulator MeTrApp. MMS **79**, 270–279 (2020). https://doi.org/10.1007/978-981-15-0142-5_27
19. Yoshikawa, T.: Manipulability of robotic mechanisms. Int. J. Robot. Res. **4**(2), 3–9 (1985)
20. Zhang, P., Yao, Z., Du, Z.: Global performance index system for kinematic optimization of robotic mechanism. Contributed by the Mechanisms and Robotics Committee of ASME for Publication in the Journal of Mechanical Design
21. Wu, X.: Optimal design and singularity analysis of a spatial parallel manipulator. Symmetry **11**, 551 (2019)

22. Bonev, I.A., Zlatanov, D., Gosselin, C.M.: Singularity analysis of 3-DOF planar parallel mechanisms via screw theory. ASME J. Mech. Des. **125**(3), 573–581 (2003)
23. Chen, X., Xie, F., Liu, X.-J.: Singularity Analysis of the Planar 3-RRR Parallel Manipulator Considering the Motion/Force Transmissibility, pp. 250–260 (2012). https://doi.org/10.1007/978-3-642-33509-9_24
24. Mohammadi Daniali, H.R.: Instantaneous center of rotation and singularities of planar parallel manipulators. Int. J. Mech. Eng. Educ. **33**(3), 251–259 (2005)
25. Merlet, J.P.: Singularity configurations of parallel robots and Grassmann geometry. Int. J. Robot. Res. **8**(5), 45–56 (1989)
26. Park, D.C., Kim, J.W.: Singularity analysis of closed kinematic Chains. J. Mech. Des. **121**(1), 32–38 (1999)
27. Ma, O., Angeles, J.: Architecture singularities of parallel manipulators. Int. J. Robot. Autom. **7**(1), 23–29 (1992)
28. Davidson, J.K., Hunt, K.H.: Robots and Screw Theory: Applications of Kinematics And Statics To Robotics. Oxford University Press, London (2004)
29. Tsai, L.W.: Robot Analysis: The Mechanics of Serial and Parallel Manipulators. WileyInter-science Publication, New York (1999)
30. Patel, K., Nayakpara, V., Patel, Y., Patel, Y.: Workspace and Singularity Analysis of 3-RRR Planar Parallel Manipulator (2013)

Constant Q Cepstral Coefficients and Long Short-Term Memory Model-Based Automatic Speaker Verification System

Aakshi Mittal and Mohit Dua

Abstract Automatic speaker verification is an advancement in authentication property of a security system like biometric at organizations, IVR system at call centers, an unlocking system of smartphones, etc. Significant components of these sensitive systems are speech feature extraction unit as frontend and a classification model for backend. This paper provides the development details of an ASV system that is trained by ASVspoof 2019 dataset. Knowledge of speech signal processing is essential to design the frontend of any speech-based system. Process of constant Q cepstral coefficient (CQCC) feature extraction is discussed with the spectrographic views of outcomes of signal processing operations. A long short-term memory (LSTM) model with the time distributed wrapper layers is used for the backend of the proposed system. Proposed system is achieving 0.42% EER for logical access (LA) set and 0.51% for physical access (PA) set of the ASVspoof 2019 dataset.

Keywords Speech signal processing · CQCC features · LSTM · ASVspoof 2019 dataset

1 Introduction

Speech-driven smart devices are being popular these days due to the availability of low-cost voice sensors and ability of speech making interaction easiest with living beings as well as with the machines. Similarly, research community and industry are also attracted to use speech signals of individuals to build the authentication part of the security mechanisms. Automatic speaker verification (ASV) systems are used to verify the genuineness of the speech of the claimer in authentication systems.

A. Mittal (✉) · M. Dua
Computer Engineering Department, National Institute of Technology Kurukshetra, Kurukshetra, Haryana, India
e-mail: aakshi1317@gmail.com

M. Dua
e-mail: er.mohitdua@gmail.com

© The Editor(s) (if applicable) and The Author(s), under exclusive license to Springer Nature Singapore Pte Ltd. 2021

A. P. Pandian et al. (eds.), *Proceedings of International Conference on Intelligent Computing, Information and Control Systems*, Advances in Intelligent Systems and Computing 1272, https://doi.org/10.1007/978-981-15-8443-5_76

There are two types of ASV systems, namely text-dependent ASV system and text-independent ASV system. In the case of former systems, claimer needs to speak the preregistered text password correctly before the system, but in the case of the later systems, claimer can freely speak the speech content before the system as the system verifies the originality of the speech of the claimer. This paper discusses the development of text-independent ASV system.

ASV systems are comprised of frontend and backend. Frontend of the system involves various speech signal processing tasks and extracts comparable information from the input speech signal. Speech signal processing is applicable in frontend design of each voice-based system like speech recognition, speaker recognition and verification, speech synthesis, voice conversion, speech coding, speech enhancement, etc. [1, 2]. Availability of various readymade libraries in Python, MATLAB, etc. makes it easier to examine the effects of various transformation operations on speech signal [3, 4]. Aspects of speech signal can reflect the human auditory system, shape of human vocal tract, etc. that can be extracted by a series of operations on the signal called feature extraction techniques. Feature extraction techniques working in cepstrum domain use fast Fourier transformation (FFT), discrete cosine transformation (DCT), logarithm operators, filters of different shapes of various domains, etc. [5–7]. Mel-frequency cepstral coefficients (MFCCs), inverse mel-frequency cepstral coefficients (IMFCCs), linear frequency cepstral coefficients (LFCCs), constant Q cepsral coefficients (CQCCs), etc. [8, 9]. are the speech feature extraction techniques that can model the human auditory and shape of human vocal tract very well. Currently, CQCC features are being most promising and suitable for ASV systems as their design was aimed to strengthen ASV systems [10, 11]. This work discusses the series of operation applied to the speech signal in CQCC feature extraction technique. And these features are being used to design the frontend of the proposed system. Features extracted on the frontend are applied to the classification model of the backend that accepts or rejects the claimer after verifying the genuineness of his/her input speech signal. Backend is comprised of a machine learning model. Hidden Markov model (HMM) and Gaussian mixture model (GMM) are used in various speech-related systems since the practical work in the area of speech [2, 5, 12], and from about a last decade, deep learning is stepping in this field [14]. In the field of ASV systems also, GMM has been used over the years [13], but with the enhancements in deep learning techniques, ASV community is attracted to deep learning models [15]. Deep learning-based models like convolutional neural network (CNN), recurrent neural network (RNN), long short-term memory network (LSTM), etc. are being more suitable and efficient to the ASV systems [16]. In the proposed system, LSTM model with time distributed layers is used.

ASV systems have brought a lot of convenience in human life even from their early stages. But these systems are highly transparent to risks. Speech synthesis (SS), voice conversion (VC), replay, twins and mimicry are the spoofing attacks that are degrading the quality and performance of these systems. SS attacks are known as text-to-speech (TTS) attacks also as they are generated by written statement converting into the legitimate user's voice by various synthesis techniques, and VC attacks are generated by converting the voice of a fraudster into the voice of a legal user. These attacks are injected into the ASV system via the channel and so-called logical access (LA) attacks [17]. Replay attacks are generated by recording the legal user's voice via any recording device. These are the type of the most common attacks to the ASV system and so are the most researched attacks. In the case of mimicry attacks, a skilled person or professional generates a similar voice to a user by modifying his/her voice features, and twins attacks can also be categorized into the mimicry attacks as twins have similar vocal tract. These attacks are applied to the system via microphone and therefore categorized into physical access (PA) attacks [18]. Due to these LA and PA spoofing attacks, a fraudster can achieve unauthorized access to a legitimate user's data.

The proposed ASV system is being trained with the ASVspoof 2019 dataset that is partitioned into LA and PA sets. After, training system is evaluated by equal error rate (EER).

2 ASVspoof 2019 Dataset

ASVspoof 2019 dataset is designed to cover the both LA and PA attack types. LA set of this dataset is having SS and VC attacked utterances, and PA set is having only replay attacked utterances. This dataset is enriched with the five known and five unknown attack types. All the training, development and testing sets of the dataset are recorded by different groups of male and female speakers. Most of the utterances of the dataset are 2–4 s in length and recorded in the English language.

3 Proposed ASV System Architecture

The proposed ASV system architecture is using CQCC features at the frontend, and a LSTM model with time distributed layers is using at the backend. And this system is trained with the ASVspoof 2019 dataset. Firstly, audios from the dataset are applied to the frontend of the system where the 13 CQCC features from each frame of the signal are extracted by various signal processing transformations and operations. All extracted features along with their corresponding labels (Bonafide/spoofed) are applied to the LSTM model of the backend. Backend model classifies the input audio into their corresponding categories. This whole arrangement is evaluated by the EER. Detailed discussion on the process of frontend and backend provided below.

3.1 Frontend of ASV System

Front of the proposed ASV system uses the CQCC feature extraction technique. CQCC feature extraction process that is discussed below is applied to whole speech data. This process is used to extract 13 CQCC features from every audio frame.

3.2 Speech Signal Processing Steps for CQCC Feature Extraction Process

CQCC feature extraction process is the series of different signal processing steps that includes conversion of the time-varying speech signal into the frequency domain, resampling to change the geometrically spaced bins into the linearly spaced bins, etc. Figure 1 shows the spectrogram of a speech signal taken from ASVspoof 2019 dataset. This is the initial signal on which whole CQCC feature extraction technique is applied, and spectrographic views of the signal after each operation are provided in the below discussion.

Calculating Constant Q Transform (CQT)
Speech signal's time domain can be converted into the frequency domain so that it can become handy to work with it. Firstly, constant Q transform (CQT) is applied to the initial speech signal. CQT can convert time domain to frequency domain while maintaining the constant Q factor across the signal [19, 20]. Figure 2 shows the spectrogram of the signal after applying CQT.

Squaring and Log Operation on Spectrum
After applying, the CQT power of the spectrum is calculated that is followed by the logarithm operation of the spectrum. Before finding, the logarithm absolute of squared values are taken [11]. Figure 3 shows the spectrogram for the absolute of the squared spectrum, and Fig. 4 shows the spectrum after applying the logarithm.

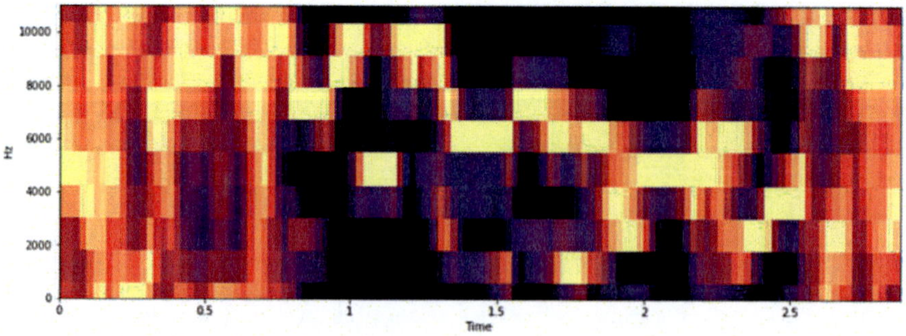

Fig. 1 Spectrogram of initial speech signal

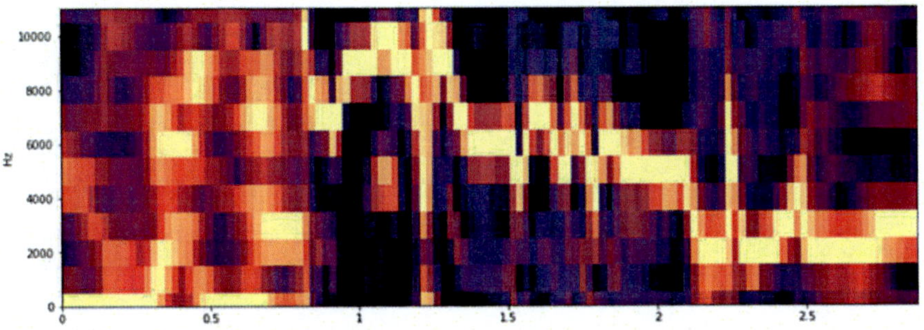

Fig. 2 Spectrogram of the signal after application of CQT

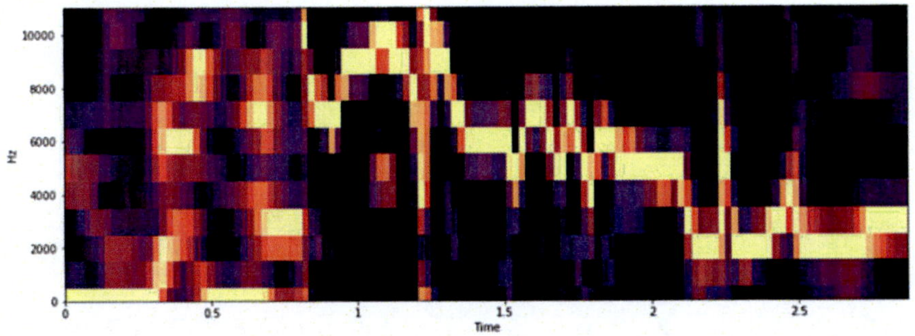

Fig. 3 Spectrographic view of the absolute squared spectrum applied after CQT

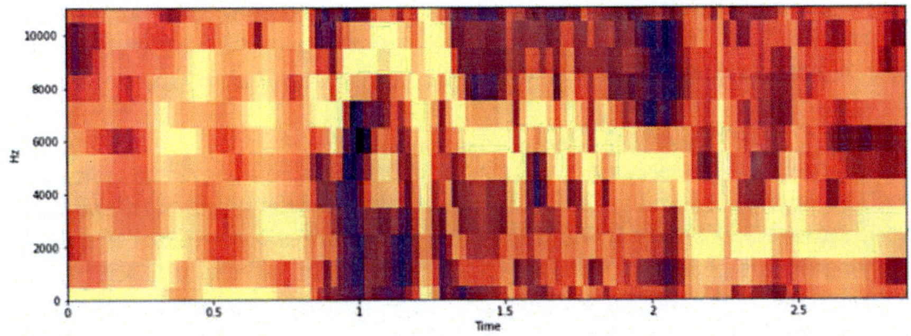

Fig. 4 Spectrogram of the signal after applying log to the absolute squared spectrum

Resampling

Resampling is an important step in CQCC feature extraction technique because CQT offers the geometrically spaced frequency bins that are converted into linearly spaced bins by applying resampling [21]. In the applied process, 16 kHz resampling

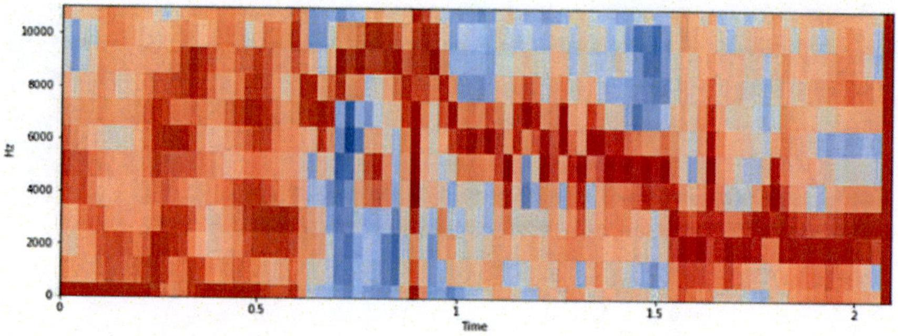

Fig. 5 Spectrographic view of the previous signal after resampling at 16 kHz

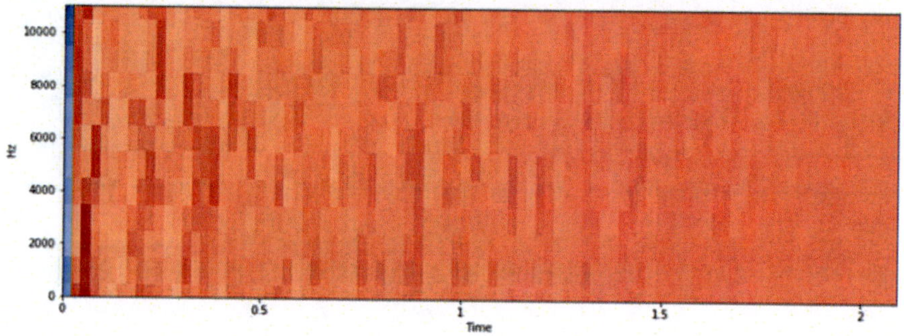

Fig. 6 Spectrogram of the signal after applying DCT on resampled spectrum

frequency is used. Figure 5 shows the spectrographic view of the previous signal after resampling at 16 kHz frequency.

Discrete Cosine Transform (DCT)

The resampling to the speech signal is applied and then discrete cosine transformation (DCT) is applied. Basically, DCT is applied to compress the speech signal so that it can fit into the less number of bits [22]. Figure 6 shows the spectrogram of signal after applying DCT, which are the desired CQCC features. In the case of CQCC feature extraction technique, step of applying DCT ends the process of feature extraction. This whole process provides the 13 CQCC features.

3.3 *Backend Model: Long Short-Term Memory (LSTM) Model with Time Distributed Wrapper*

Long short-term memory (LSTM) model can be categorized as a special type of RNN model. It can hold the information in the logic gates for a long period [23]. Proposed

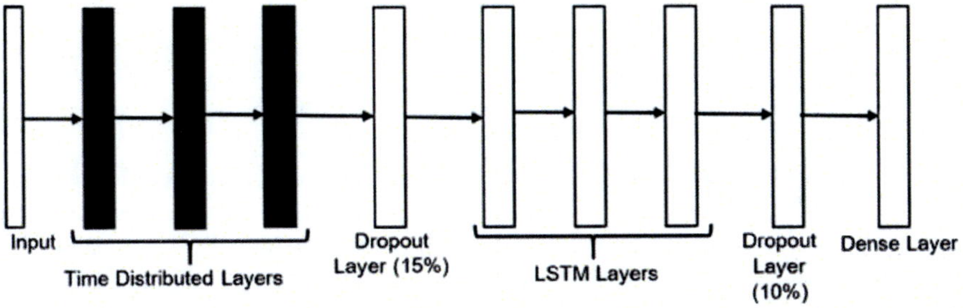

Fig. 7 Architecture of backend long short-term memory (LSTM) model with time distributed wrappers

long short-term memory (LSTM) network (Fig. 7) is comprised of three time-distributed dense layers, each having ReLU activation function. Time distributed wrapped layers are especially suitable for time-varying data frames like audio, video, etc. Proposed model has 32, 16 and 10 units in time distributed dense layers in this order. After that, 15% dropout is applied to disuse the effect of some randomly selected neurons. Addition of dropout layer prevents the model from overfitting. In the proposed model, this operation is followed by three LSTM layers each having 100, 200 and 300 units in this order. Again these layers are followed by 10% dropout, and the result of dropout is passed to a dense layer having sigmoid activation function in it.

4 Equal Error Rate (EER)

EER is used to evaluate the proposed ASV arrangement. EER is one of the popular evaluation metrics for speech-based biometrics. It checks the false rejection rate (FRR) and false acceptance rate (FAR) at some predefined threshold [24]. The point where the FRR and FAR become equal is considered as the EER for the system. FRR is calculated by dividing the count of all bonafide utterances having score less than the threshold value by the total number of bonafide utterances (Eq. 1), whereas the FAR is found out by dividing the number of all spoofed utterances having score more than or equal to the threshold by total number of spoofed utterances (Eq. 2).

$$\text{FRR} = \frac{\text{number of bonafide utterances with sore} < \Delta}{\text{number of bonafide utterances}} \tag{1}$$

$$\text{FAR} = \frac{\text{number of spoofed utterances with score} \geq \Delta}{\text{number of spoofed utterances}} \tag{2}$$

Here, Δ is denoting threshold.

5 Experimental Setup

At the frontend, 13 CQCC coefficients are extracted by the signal transformation and operation involved in CQCC feature extraction techniques. Dynamic speech features can be calculated by taking first order and second order derivative of static coefficients [25] that are computed after finding the static CQCC features. All the signal processing tasks, and finally, used features are extracted by the Octave on Linux operating system. All the spectrograms, shown in this paper, are plotted with the help of different Python libraries. Backend LSTM model with time distributed wrappers is developed in Python. Backend is trained by the ASVspoof 2019 speech dataset. For the LA and PA sets of used dataset, backend is trained up to 50 epochs with the batch size of 512. Early stopping criteria is used with the patience of 2. Adam optimizer is used for the optimizing purpose, and binary cross-entropy loss is used as the loss function.

6 Results

The proposed system is evaluated by the EER that is calculated with the help of scores provided to the utterances by LSTM model and a balanced threshold value. EER for the development data is 0.29% and 0.42% for evaluation data LA set of the ASVspoof 2019 dataset. And for the development and evaluation data of PA set are 0.38% and 0.51%, respectively. Cai et al. are training deep neural network (DNN) and ResNet with only PA set [6]. Kumar et al. are training a time delay shallow neural network (TDSS) with LA and PA. Baseline model provided by the ASVspoof 2019 challenge is a GMM that is trained with LA and PA sets of ASVspoof 2019 dataset. Table 1 shows the comparative study of these models with the proposed model.

Table 1 % EER for different architectures taken from the literature for development and training data of LA and PA sets of ASVspoof 2019 dataset, best performances are shown in bold

System	LA data		PA data	
	Development	Evaluation	Development	Evaluation
DNN, ResNet [6]	–	–	0.24	0.66
TDSNN [16]	0.4	5.7	7.5	6.4
GMM [26]	–	0.43	–	9.87
Proposed system	0.29	0.42	0.38	0.51

7 Conclusion

ASV systems are successfully being used in the real world. ASV community is working to reduce their transparency from spoofing attacks in parallel. Study of different speech signal processing operations can empower the frontend of speech-based systems. Proposed LSTM is overtaking the classical machine learning as well as other deep learning models. It is achieving 0.42 and 0.51% EER for LA and PA sets of the ASVspoof 2019 dataset. For the future, mimicry and twins attacks should also be taken care.

References

1. Dua, M., Aggarwal, R.K., Biswas, M.: Performance evaluation of Hindi speech recognition system using optimized filterbanks. Int. J. Eng. Sci. Technol. **1**(3), 389–398 (2018)
2. Dua, M., Aggarwal, R., Kadyan, V., Dua, S.: Punjabi speech to text system for connected words, pp. 206–209 (2012)
3. Glover, J.C., Lazzarini, V., Timoney, J.: Python for audio signal processing (2011)
4. Cheuk, K.W., Anderson, H., Agres, K., Herremans, D.: nnAudio: an on-the-fly GPU audio to spectrogram conversion toolbox using 1D convolution neural networks. www.arXiv:1912.12055 (2019)
5. Chakroborty, S., Saha, G.: Improved text-independent speaker identification using fused MFCC & IMFCC feature sets based on Gaussian filter. Int. J. Signal Process. **5**(1), 11–19 (2009)
6. Cai, W., Wu, H., Cai, D., Li, M.: The dku replay detection system for the asvspoof 2019 challenge: on data augmentation, feature representation, classification, and fusion. www.arXiv:1907.02663 (2019)
7. Balamurali, B.T., Lin, K.E., Lui, S., Chen, J.M., Herremans, D.: Toward robust audio spoofing detection: a detailed comparison of traditional and learned features. IEEE Access **7**, 84229–84241 (2019)
8. Sahidullah, M., Kinnunen, T., Hanilçi, C.: A comparison of features for synthetic speech detection (2015)
9. Dua, M., Aggarwal, R.K., Biswas, M.: Discriminative training using heterogeneous feature vector for Hindi automatic speech recognition system. Int. Conf. Comput. Appl. ICCA: 158–162 (2017)
10. Jelil, S., Das, R.K., Prasanna, S.M., Sinha, R.: Spoof detection using source, instantaneous frequency and cepstral features. In: Interspeech, pp. 22–26 (2017)
11. Todisco, M., Delgado, H., Evans, N.: Constant Q cepstral coefficients: a spoofing countermeasure for automatic speaker verification. Comput. Speech Lang. **45**, 516–535 (2017)
12. Dua, M., Aggarwal, R.K., Biswas, M.: Discriminative training using noise robust integrated features and refined HMM modeling. J. Intell. Syst. **29**(1), 327–344 (2018)
13. Stefanus, I., Sarwono, R.J., Mandasari, M.I.: GMM based automatic speaker verification system development for Forensics in Bahasa Indonesia. In: 5th International Conference on Instrumentation, Control, and Automation (ICA), pp. 56–61 (2017)
14. Dua, M., Aggarwal, R.K., Biswas, M.: Discriminatively trained continuous Hindi speech recognition system using interpolated recurrent neural network language modeling. Neural Comput. Appl. **31**(10), 6747–6755 (2019)
15. Shim, H.J., Jung, J.W., Heo, H.S., Yoon, S.H., Yu, H.J.: Replay spoofing detection system for automatic speaker verification using multi-task learning of noise classes. In: Conference on Technologies and Applications of Artificial Intelligence (TAAI), pp. 172–176 (2018)

16. Kumar, M.G., Kumar, S.R., Saranya, M.S., Bharathi, B., Murthy, H.A.: Spoof detection using time-delay shallow neural network and feature switching. In: IEEE Automatic Speech Recognition and Understanding Workshop (ASRU), pp. 1011–1017 (2019)
17. Wu, Z., Evans, N., Kinnunen, T., Yamagishi, J., Alegre, F., Li, H.: Spoofing and countermeasures for speaker verification: a survey. Speech Commun. **66**, 130–153 (2015)
18. Chettri, B., Stoller, D., Morfi, V., Ramírez, M.A.M., Benetos, E., Sturm, B.L.: Ensemble models for spoofing detection in automatic speaker verification. www.arXiv:1904.04589 (2019)
19. Brown, J.C., Puckette, M.S.: An efficient algorithm for the calculation of a constant Q transform. J. Acoust. Soc. Am. **92**(5), 2698–2701 (1992)
20. Brown, J.C.: Calculation of a constant Q spectral transform. J. Acoust. Soc. Am. **89**(1), 425–434 (1991)
21. Todisco, M., Delgado, H., Evans, N.W.: Articulation rate filtering of CQCC features for automatic speaker verification. In: Interspeech, pp. 3628–3632 (2016)
22. Kaur, S., Er, G.K.: Enhancement of speech recognition algorithm using DCT and inverse wave transformation. J. Eng. Res. Appl. **3**(6) (2013)
23. Mobiny, A., Najarian, M.: Text-independent speaker verification using long short-term memory networks. www.arXiv:1805.00604 (2018)
24. Huang, L., Pun, C.M.: Audio replay spoof attack detection using segment-based hybrid feature and DenseNet-LSTM network. In: ICASSP 2019–2019 IEEE International Conference on Acoustics, Speech and Signal Processing (ICASSP), pp. 2567–2571. IEEE (2019)
25. Sun, Y., Zhang, X.Y., Ma, J.H., Song, C.X., Lv, H.F.: Nonlinear dynamic feature extraction based on phase space reconstruction for the classification of speech and emotion. Math. Probl. Eng. (2020)
26. ASVspoof 2019: Automatic Speaker Verification Spoofing and Countermeasures Challenge Evaluation Plan. https://www.asvspoof.org/ (2019)

Hardware-in-the-Loop Simulation of Induction Heating System for Melting Applications Using Xilinx System Generator

Darshana N. Sankhe, Rajendra R. Sawant, and Y. Srinivasa Rao

Abstract Hardware-in-the-loop (HIL) simulation is increasingly used in development and testing of industrial systems and many more applications. This paper presents HIL simulation of induction heating (IH) system used for metal melting applications, wherein a resonant inverter with IH load and control logic, both, are modeled using high-level Xilinx System Generator (XSG). A phase shift control technique with continuous resonant frequency tracking mechanism is deployed for power regulation, which improves the efficiency of the IH system. Performance analysis of the complete system is carried out using HIL simulation, and results are validated for varying power conditions. Hardware resource utilization, timing, and power requirements are analyzed by real-time implementation on Zynq XC7Z020-1clg484 FPGA board. This complete system modeling with the HIL simulation approach allows through testing of the system with different control algorithms before the commencement of physical systems, which further reduces risk of device damage, controller development time, initial cost and supports rapid prototyping of control strategy on high-speed FPGA platforms.

Keywords Hardware-in-the-loop · Digital controller · FPGA · Induction heating · Xilinx System Generator

D. N. Sankhe (✉) · R. R. Sawant · Y. S. Rao
Sardar Patel Institute of Technology, Mumbai, India
e-mail: darshanasankhe78@gmail.com

R. R. Sawant
e-mail: rajendra.sawant@spit.ac.in

Y. S. Rao
e-mail: ysrao@spit.ac.in

© The Editor(s) (if applicable) and The Author(s), under exclusive license to Springer Nature Singapore Pte Ltd. 2021

A. P. Pandian et al. (eds.), *Proceedings of International Conference on Intelligent Computing, Information and Control Systems*, Advances in Intelligent Systems and Computing 1272, https://doi.org/10.1007/978-981-15-8443-5_77

1 Introduction

An exemplary growth in microelectronics and power electronics devices has accelerated the use of digital controllers in industrial applications over their counterpart analog controllers. Conventionally, microcontrollers, microprocessors, and digital signal processors are commonly used digital controllers in industrial applications. These controllers have limited function and low computational speed, thus exhibits high latency for complex control strategies [1–3]. Today, the field-programmable gate array (FPGA) is in great demand due to its fast response time, parallel processing capabilities, reconfigurable, and reprogrammable characteristics. Hence, they are used in applications like resonant inverters, photovoltaic systems, delta inverters, and many more applications [4–9]. These FPGAs are programmed using hardware description language (HDL), and it is a known fact that developing HDL code for real-time control application is a challenging task and requires expertise in HDL. Further with increase in control algorithm's complexity, it becomes time-consuming. In order to reduce development time, a model-based design approach, Xilinx System Generator (XSG), is used. This XSG-based design allows HDL code generation that can be ported on FPGA board. In all, XSG supports HIL simulation [3, 10, 11].

Nowadays, HIL simulation is used in analysis and testing of power systems over experimental verifications, due to its distinct advantages listed as follows:

- Allows real-time testing of control strategy on a software system/plant model before applying it on a physical system. This reduces risk of plant damage.
- Facilitates development of novel control algorithms without building physical systems. This reduces system development cost and accelerates the development of controller design pace/period during initial design and development phase.
- Real-time testing supports rapid design iterations to get optimized design, which in turn helps in rapid prototype of proposed system hardware setup.
- FPGA's parallel processing capability not only provides the simulation in cycle-accurate manner but also accelerates the software simulation.
- HIL simulation results are more close to physical/experimental system results.

Various HIL platforms are available, such as *eTap, Opal-RT, Typhoon HIL, and dSPACE*. These platforms are commercial with restricted/limited functionality in the demo/lower versions. In this paper, Xilinx HIL simulation platform is used, wherein power system model and entire control logic are developed using MATLAB/Simulink-XSG block sets. Further Xilinx's *Vivado design suite* is used for real-time testing and implementation on FPGA platform [12–15].

Today, IH technique has become leading technology for industrial applications as well as in domestic cooking due to its improved efficiency, safety, cleanness, and smooth power control, which outperform other heating techniques [16]. IH systems are commonly used in shrink-fitting, brazing, annealing, soldering, pre and post-heating, melting, and hardening applications. Basic components of IH system are high-speed inverters and a resonant load, wherein resonating components (L and C) are connected in series or parallel configuration, termed as series or parallel resonant inverters. Commonly used inverter topologies are half-bridge inverter, full-bridge inverter, multi-inverter topologies, and single switch inverter, wherein full-bridge topology is preferred for medium-to-high power applications. A voltage source, full-bridge series resonant inverter (SRI) is preferred for IH applications due to its finer inverter utilization, simple design, and high efficiency.

In IH systems, an essential requirement is accurate and smooth power control. To achieve this, various control strategies are discussed in the literature such as pulse density modulation, phase shift control (PSC), square wave control, asymmetrical pulse width modulation, and asymmetrical voltage cancellation control. [17–19]. Further, in IH systems, maximum power is transferred to the workpiece when resonant inverters are operated at resonating frequency, f_0. This operation at f_0 increases switching stresses on active devices and can damage IGBT switches. Thus to achieve better efficiency and at the same time to ensure safety of IGBTs, these inverters are operated at switching frequency f_S slightly higher than f_0. This operating condition ($f_S > f_0$) is termed as zero voltage switching (ZVS)/soft switching. In this paper, a PSC technique is used for power control along with continuous tracking of resonant frequency without phase-locked loop, to achieve soft switching [20–22].

Considering above discussion/facts, this paper focuses on HIL simulation of IH system for melting applications, whose performance is validated through XSG-based design, in real time on FPGA platform. Section 2 covers power circuit and time domain analysis of IH system to investigate the dependences of different input–output variables. XSG-based IH system model and control logic design are explained in Sect. 3. In Sect. 4, comparative analysis is carried out between software simulation and HIL simulation, and the performance of controller is validated for step change in reference power. Also the power, timing, and resources required for implementation are summarized. Section 5 concludes the paper.

2 Analysis of Series Resonant Inverter-Based IH System

The power stage of voltage source SRI used for IH applications is shown in Fig. 1. Here, mains supply is rectified, filtered, and connected to inverter stage. A full-bridge

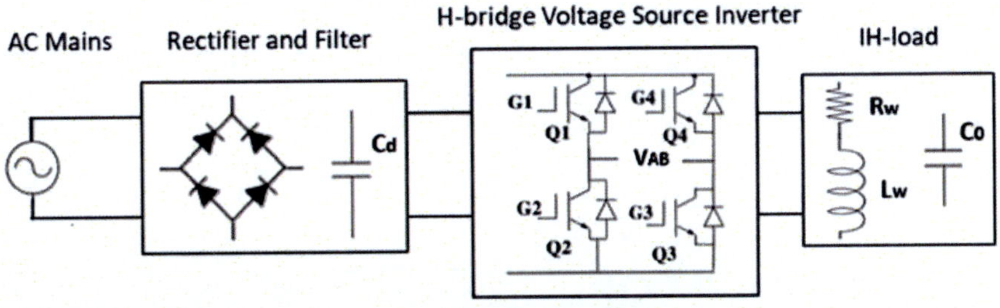

Fig. 1 Power stage of series resonant inverter for IH applications

(H-bridge) SRI with high switching frequency is implemented using IGBT switches ($Q_1 - Q_4$). These switches, Q_1 to Q_4, are operated with a 50% duty cycle, and the switches in each leg (Q_1, Q_2 or Q_3, Q_4) are switched 180° out of phase. Inverter output voltage, V_{AB}, is connected through an isolation transformer to the IH load. R_W and L_W are the equivalent resistance and inductance of the workpiece (to be heated/melted) plus heating coil. C_0 is an external capacitor connected in series with R_W to form a RLC series network [23]. The resonant frequency of this network is given as

$$f_0 = \frac{1}{2\pi\sqrt{L_W C_0}} \tag{1}$$

These resonant inverters are always operated above the resonance frequency to achieve ZVS. Operating inverter high above resonance results in large circulating current, which decreases efficiency and demands for large VA rating of power components. Thus, operate inverters slightly above resonance, which ensures ZVS, reduces switching losses and device (IGBT) stresses, with improved efficiency. When inverter is operated above f_0, the inverter output voltage (a quasi-square wave, V_{AB}) leads the load current (i_0) as shown in Fig. 2. This output V_{AB} appears across RLC network has duty ratio d given as

$$d = \frac{2 \times t_d}{T} \tag{2}$$

Here t_d and T are inverters on time and switching period, respectively. The inverter output power is regulated by modifying the duty ratio (d), i.e., by controlling the phase shift between switches Q_1 and Q_3. For constant switching frequency (f_S) operation, ZVS will be lost for decrease in duty ratio d. Now to retain the ZVS condition, f_S should be increased with decrease in d[21]. In order to analyze these dependencies of f_0, f_S, d, and output power, the IH system is modeled as shown in Fig. 3. The inverter output V_{AB} is connected across RLC network, wherein $R, L, and\, C$ are

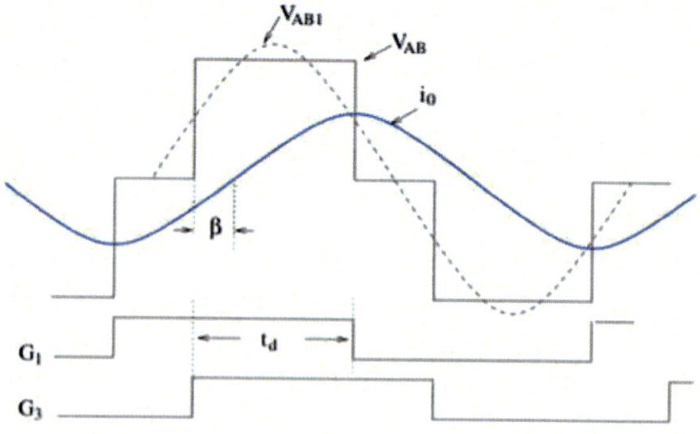

Fig. 2 Inverter output voltage, current, and gate control pulses

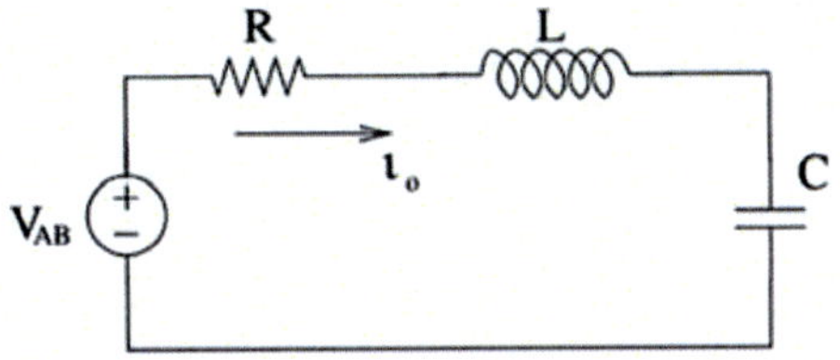

Fig. 3 Equivalent circuit representing power stage of IH system

equivalent resistor, inductor, and capacitor reflected at the primary of $n : 1$ turns ratio transformer. They are given as follows: $R = n^2 R_W$, $L = n^2 L_W$, and $C = C_0/n^2$.

By applying KVL to above circuit,

$$V_{AB}(\mathrm{t}) = i_0(t)R + L\frac{\mathrm{d}i_0(t)}{\mathrm{d}t} + \frac{1}{C}\int i_0(t)\mathrm{d}t \quad (3)$$

Now, considering the fundamental frequency component of quasi-square wave, V_{AB} is represented in terms of d as

$$V_{AB}(t) = \frac{4V_{\mathrm{dc}}}{\pi} Sin\left(\frac{\pi}{2}d\right) Sin(\omega t) \quad (4)$$

Here, V_{dc} is DC input voltage of inverter. By solving Eqs. (3) and (4), an expression for current $i_0(t)$ is obtained; using this average power (P_{avg}) is evaluated as

$$P_{\mathrm{avg}} = \frac{1}{T}\int_0^T i_0^2(t) * R\mathrm{d}t \quad (5)$$

$$P_{\mathrm{avg}} = \frac{R\omega^2}{2L^2} * \frac{1}{x^2 + y^2} * \left[\frac{4V_{\mathrm{dc}}}{\pi}\sin\left(\frac{\pi}{2}d\right)\right]^2 \quad (6)$$

where

$$x = a^2 + b^2 - \omega^2; \quad y = 2 * a * \omega; \quad a = R/2L \text{ and } b = \omega_0\sqrt{1 - \frac{1}{4Q^2}}$$

In this expression, $\omega_0 = 2\pi f_0$ and Q is the heating coil quality factor, $Q = \omega_0 L/R$. For IH systems, during the heating cycle Q varies in the range 5–20. An effect of variation of Q on peak current, as function of duty ratio d is shown in Fig. 4. Peak current increases with increase in duty ratio and quality factor. Similarly, the output power also increases with increase in d and Q. For specific duty ratio, if Q changes from 5 to 20, the power increases drastically. Here, controller has to reduce the duty ratio for maintaining the required output power. Table 1 gives details of the circuit parameters, used in time domain analysis.

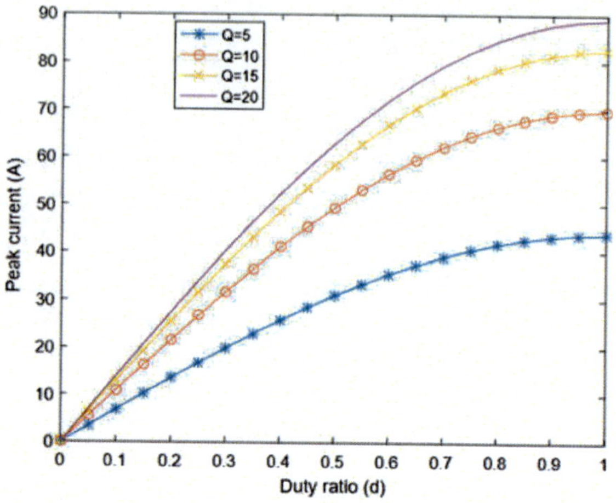

Fig. 4 Plot of peak current versus duty ratio for different values of Q

Table 1 IH system parameters

Description	Parameter	Value
DC supply voltage	V_{dc}	600 V
Load resistor	R	8.33 Ω
Coil inductance	L	675 μH
Resonant capacitor	C	0.11 μF
Resonant frequency	f_0	18.5 kHz

The voltage gain of IH system is evaluated by considering active power across resistor R. By referring to Fig. 3, if V_R is the output voltage across R, then voltage gain of equivalent circuit is given as

$$G(S) = \frac{V_R(S)}{V_{AB}(S)} = \frac{R}{R + SL + 1/SC} \tag{7}$$

Substituting $S = j\omega_S$ in above equation and rearranging the terms, the magnitude of voltage gain is obtained as

$$|G(j\omega_S)| = 1/\sqrt{1 + Q^2\left(\frac{\omega_S}{\omega_0} - \frac{\omega_0}{\omega_S}\right)^2} \tag{8}$$

Here, ω_S *and* ω_0 are switching and resonant frequencies ($\omega_S > \omega_0$), and Q is the quality factor of a coil. An effect of variation of switching frequency on voltage gain for different values of Q is shown in Fig. 5. Here f_n is the normalized frequency, $f_n = f_s/f_0$. It is observed that maximum gain can be achieved by operating the IH system at resonant frequency (i.e., $f_n = 1$), but to maintain ZVS condition it is always operated slightly above the resonant frequency with small reduction in gain.

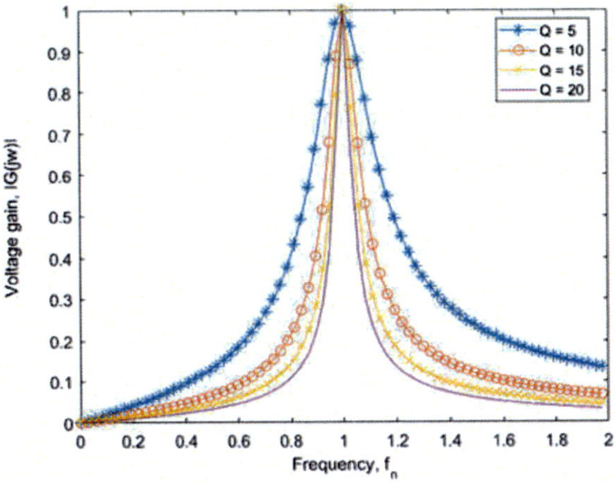

Fig. 5 Plot showing voltage gain as function of normalized frequency, f_n

An increase in a quality factor increases selectivity of RLC network with further reduction in the gain.

3 XSG-Based IH System and Control Logic Design

In order to develop XSG-based model of IH system, the discrete-time equations of the equivalent circuit (Fig. 3.) have to be developed. Consider V_R, V_L, and V_C as corresponding outputs across the equivalent R, L, and C, respectively, in Fig. 3. At any time instance t, time domain expressions can be written as follows:

$$V_{AB}(t) = V_R(t) + V_L(t) + V_C(t) \tag{9}$$

$$V_R(t) = i_O(t) * R; \quad V_L(t) = L\frac{\mathrm{d}i_O(t)}{\mathrm{d}t} \text{ and } i_O(t) = C\frac{\mathrm{d}V_C}{\mathrm{d}t} \tag{10}$$

By solving these expressions (9), (10), discrete-time equations are obtained as follows:

$$Io(n+1) = Io(n) + \left(\frac{\mathrm{d}t}{L}\right) * V_L(n) \tag{11}$$

$$V_R(n+1) = I_O(n) * R \tag{12}$$

$$V_C(n+1) = V_C(n) + \left(\frac{\mathrm{d}t}{C}\right) * I_O(n) \tag{13}$$

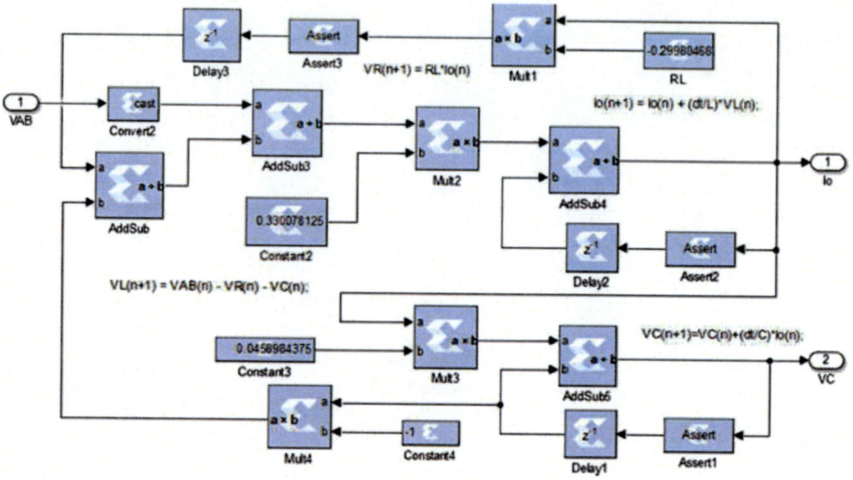

Fig. 6 IH system power circuit model using XSG block sets

$$V_L(n+1) = V_{AB}(n) - V_R(n) - V_C(n) \tag{14}$$

Using expressions (11)–(14), IH system model is developed using XSG block sets as shown in Fig. 6. In this design, various arithmetic operators and constant blocks are used to build the required expressions. *Cast* block converts received input samples into a required arithmetic data type. *Assert* block allows user define sample rate required for proper functioning of delay and register blocks. Outputs from this model are used by control logic to generate gate control pulses ($G_1 - G_4$), as shown in Fig. 7. The control block has to perform these functions: (i) regulate output power, (ii) achieve ZVS at all load conditions by continuously tracking the resonant frequency and operating inverter slightly above the resonance. Load current, i_0, is passed through frequency tracking network. It has a RC network with single pole, comparator, and attenuator sub-systems.

Using Euler's integration discretization method, the discrete-time equation for output square wave v_{sq} is written as follows:

$$v_{sq}(n+1) = K * \left(v_{sq}(n) + \frac{T_s}{\tau}[i_0(n+1) - v_{sq}(n)] \right) \tag{15}$$

Here, $n, T_s, \tau,$ and K represent discrete-time step, sampling time, RC time constant, and attenuation factor, respectively.

This signal is passed through ramp generator to get v_{ramp} signal as shown in Fig. 7. Further, instantaneous output power is passed through RC network, and average output power P_{avg} is evaluated. This P_{avg} is compared with P_{ref} to get an error output signal e. This error signal is processed by PI controller to obtain v_{PI} output. An expression of PI block is obtained using *bilinear transformation* discretisation

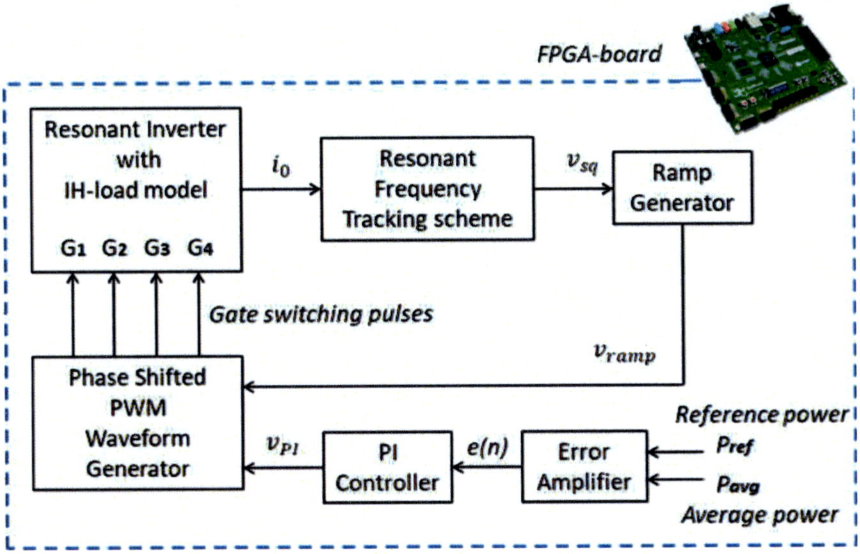

Fig. 7 Controller block diagram

method and given as

$$v_{PI}(n+1) = K_P * e(n+1) + K_I * \left(v_{PI}(n) + \frac{T_s}{2}\{e(n) + e(n+1)\} \right) \quad (16)$$

Here, $e(n)$ is an error input, and K_P, K_I are gains of the PI controller. The discrete-time expressions for an error amplifier and ramp generator blocks are derived in similar manner, and the entire control system is implemented using XSG block sets. Figure 8 shows XSG-based implementation for ramp generator block. The PI block output and ramp signal are used for generating gate control pulses as shown in Fig. 9. The square wave output is used as switching pulse G_1. Ramp generator's output is continuously compared with PI controller's output, to generate switching pulse G_3.

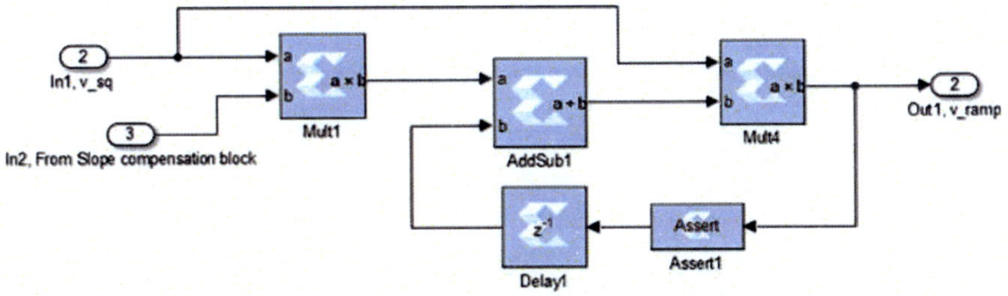

Fig. 8 Xilinx block sets for ramp generation

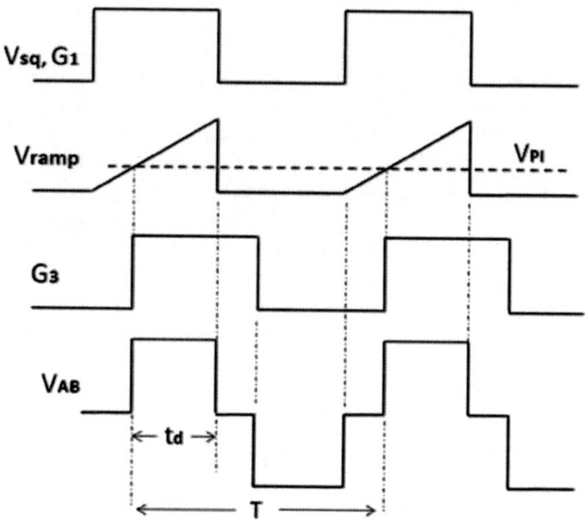

Fig. 9 Generation of switching pulses G1 and G3

Phase delay between G_1 and G_3 decides the pulse width (t_d) of inverter output voltage V_{AB}. By varying the PI controller's output and hence modifying PWM duty ratio, the controller continuously tracks reference power. Pulses G_2 and G_4 are complements of G_1 and G_3, respectively.

Here, various logical operators and a network of D-flip-flop are used. State diagram of implemented DFF network is shown in Fig. 10, where Q_1 *and* Q_2 are states of the flip-flop, triggered at both positive and negative levels, respectively. The phase shift between respective gate pulses is varied to maintain the required reference power level.

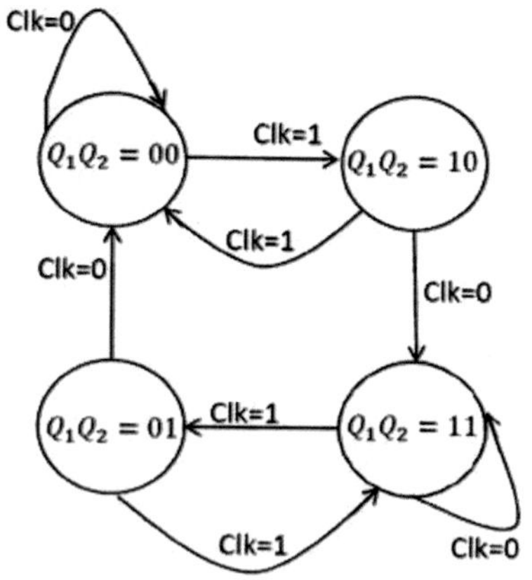

Fig. 10 State diagram of D-flip-flop

4 Simulation Results

To validate the IH system load model, first a complete system MATLAB/Simulink model is developed, and performance is verified using simulations. Further, the entire system is developed using XSG blocks and again performance is verified. The electrical system parameters considered for simulation are given in Table 1. HIL simulation facility of XSG allows validation of any system design on a high-speed FPGA platform. The IH system model and control logic developed here are validated on Zynq XC7Z020-1clg484 Evaluation and Development Board. Figure 11 shows setup of HIL simulation, consisting of XSG model (host PC) connected to FPGA board through JTAG interface. XSG allows automatic generation of HDL code, through Xilinx *Vivado Design Suit 2017.4* software, which is ported on FPGA board using JTAG cable. Resource utilization is summarized in Table 2, while path delay is observed as 0.51 ns and power dissipation is 0.31 W. The power dissipation increases post place-and-route implementation. FPGA clock frequency was 100 MHz.

In order to verify the performance of XSG-based IH system model and controller design, the reference input power is changed from 12.5 to 18.5 kW as shown in Fig. 12. Initially, reference power was set as $P_{ref} = 12.5$ kW, here controller gradually tracks the reference power and attains the set level in 12.5 ms. Further at $T = 26.5$ ms, again the reference power is modified to 18.5 kW and observed that from initial 12.5 kW level to desired 18.5 kW level is attained by the controller in 5 ms. The inverter output voltage and current waveforms for both the power levels were observed and found that duty ratio gets modified by the controller, in accordance with the set reference power levels.

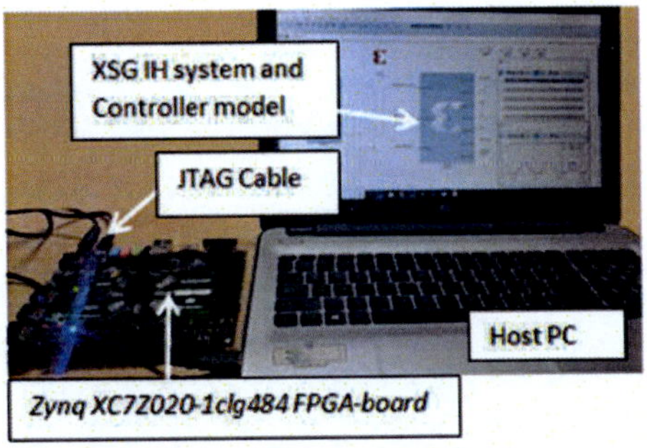

Fig. 11 HIL simulation setup

Table 2 FPGA resource utilization

Resources	Frequency tracking loop	PI controller	IH model	Other blocks	Utilization
DSPs	06	16	20	16	58 (26.4%)
LUTs	242	293	342	1399	2276 (4.3%)
Registers	48	72	95	153	368 (0.35%)

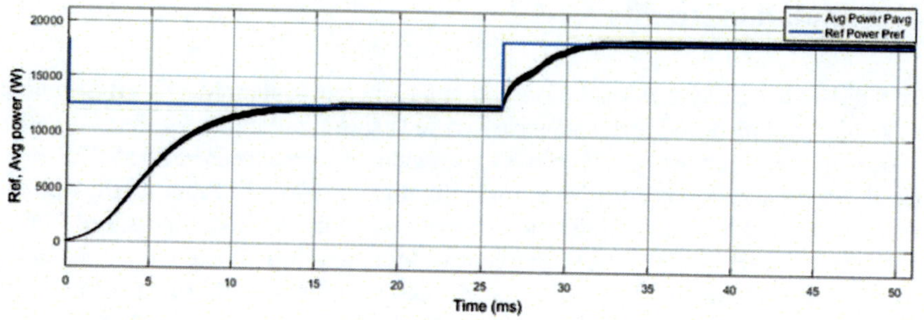

Fig. 12 Step transient response of controller with IH load model

To verify real-time HIL system performance under presence of FPGA hardware, the HIL simulation results are compared with software simulation. Figure 13 represents inverter average power, linearly increasing to track the reference power level. Here HIL and software simulation results are similar hence tracking error is almost zero. Figure 14 depicts load voltage and current waveforms for reference power set to 15 kW. Both software simulation and HIL simulation results are perfectly matching, ZVS is maintained, and switching frequency is around 20 kHz.

In all HIL, results perfectly match with the software simulation results and can be verified at the output of each sub-system. This validates the effectiveness of XSG-based system modeling and its HIL validation on FPGA platform.

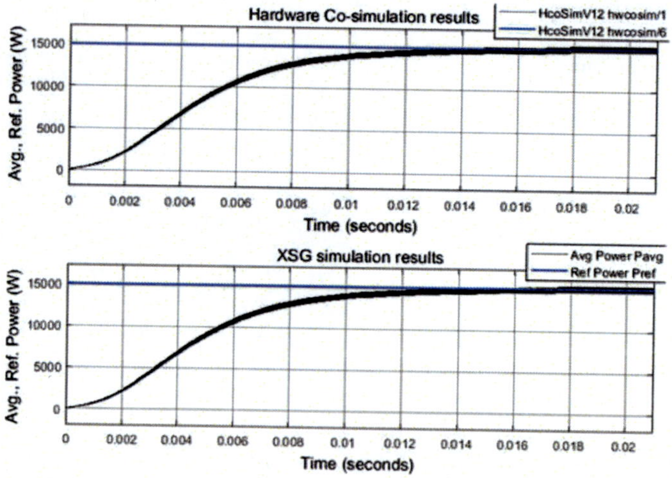

Fig. 13 Reference power tracking using HIL simulation and software simulation

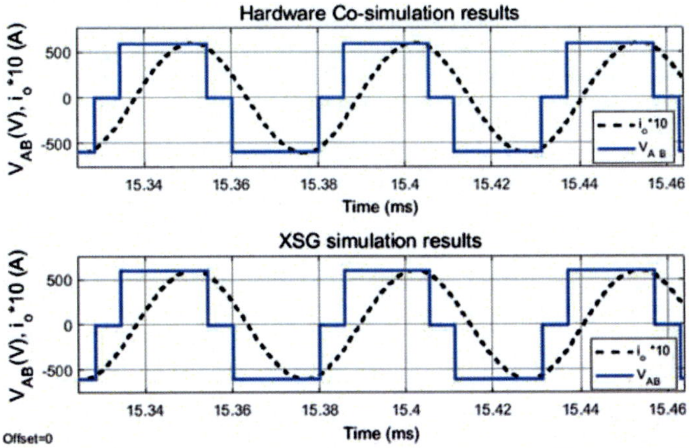

Fig. 14 Load voltage and current waveforms using HIL simulation and software simulation

5 Conclusion

HIL simulation methodology implemented in this paper is found to be an effective method for design and testing of any electrical system along with control logic, on high-speed FPGA platform. Xilinx XSG block set supports automatic generation of HDL code, which easies and accelerates the FPGA-based controller design. The IH load model's time domain analysis showed dependencies of output power with duty ratio and is verified with HIL simulation result. Further, the controller is efficiently tracking the resonant frequency in real time and maintains ZVS in entire power range; this ensures high efficiency. This controller uses minimal hardware resources, thus can be implemented on lower-version FPGA boards. Above methodology can be easily used for testing of other similar systems, while discrete-time controller design can be readily used in resonant inverter-based applications such as induction cooking, battery charging system for electrical vehicles, and SMPS.

References

1. Sepulveda, C.A., Munoz, J., Espinoza, J.R., Figueroa, M., Baier, C.R.: FPGA v/s DSP performance comparison for a VSC-based STATCOM control application. IEEE Trans. Ind. Informat. **9**(3), 1351–1360 (2013)
2. Zumel, P., Fernandez, C., Sanz, M., Lazaro, A., Barrado, A.: Step-by-step design of an FPGA-based digital compensator for DC/DC converters oriented to an introductory course. IEEE Trans. Educ. **54**(4), 599–609 (2011)
3. Selvamuthukumaran, R., Gupta, R.: Rapid prototyping of power electronics converters for photovoltaic system application using Xilinx system generator. IET Power Electron. **7**(9), 2269–2278 (2014)

4. Nagarajan, B., Sathi, R.R.: FPGA-based automatic frequency-controlled resonant inverter for induction heating system. J. Inst. Eng. India Ser. B. **100**(2), 131–141 (2019)
5. Morales-Caporal, M., Rangel-, J., Peregrina-Barreto, H., Morales-Caporal, R.: FPGA-in-the-loop simulation of a grid connected photovoltaic system by using a predictive control. Electr. Eng. **100**(3), 1327–1337 (2018)
6. Rajkumar, M.V., Manoharan, P.S.: FPGA based multilevel cascaded inverters with SVPWM algorithm for photovoltaic system. Sol. Energy **87**, 229–245 (2013)
7. Alouane, A., Rhouma, A.B., Hamouda, M., Khedher, A.: Efficient FPGA-based real-time implementation of an SVPWM algorithm for a delta inverter. IET Power Electron. **11**(9), 1611–1619 (2018)
8. Liu, C., Guo, X., Ma, R., Li, Z., Gechter, F., Gao, F.: A system level FPGA-based hardware-in-the-loop test of high-speed train. IEEE Trans. Transp. Electr. **4**(4), 912–921 (2018)
9. Ingole, D., Kvasnica, M.: FPGA implementation of explicit model predictive control for closed loop control of depth of anesthesia. IFAC-Pap. Online. **48**(23), 483–488 (2015)
10. Pinto, S.J., Panda, G., Peesapati, R.: An implementation of hybrid control strategy for distributed generation system interface using xilinx system generator. IEEE Trans. Ind. Informat. **13**(5), 2735–2745 (2017)
11. Sankhe D.N., Sawant R.R., Srinivas Rao Y.: Implementation of harmonic oscillator using xilinx system generator. In: Abraham A., Cherukuri A., Melin P., Gandhi N. (eds.) Intelligent Systems Design and Applications. ISDA 2018. Advances in Intelligent Systems and Computing, vol. 941. Springer, Cham (2020). https://doi.org/10.1007/978-3-030-16660-1_49
12. Singh, V.K., Tripathi, R.N., Hanamoto, T.: HIL co-simulation of finite set-model predictive control using FPGA for a three-phase VSI system. Energies **11**(4), 909 (2018)
13. Matar, M., Paradis, D., Iravani, R.: Real-time simulation of modular multilevel converters for controller hardware-in-the-loop testing. IET Power Electron. **9**(1), 42–50 (2016)
14. Fernandez-Alvarez, A., Portela-Garcia, M., Garcia-Valderas, M., Lopez, J., Sanz, M.: HW/SW co-simulation system for enhancing hardware-in-the-loop of power converter digital controllers. IEEE J. Emerg. Sel. Top. Power Electron. **5**(4), 1779–1786 (2017)
15. Chen, S., Chen, Y., Zhang, S., Zheng, N.: A novel integrated simulation and testing platform for self-driving cars with hardware in the loop. IEEE Trans. Intell. Veh. **4**(3), 425–436 (2019)
16. Lucia, O., Maussion, P., Dede, E.J., Burdio, J.M.: Induction heating technology and its applications: past developments, current technology, and future challenges. IEEE Trans. Ind. Electron. **61**(5), 2509–2520 (2014)
17. Kumar, A., et al.: A survey on high-frequency inverter and their power control techniques for induction heating applications. J. Power Technol. **97**(3), 201–213 (2017)
18. Siwakoti, Y.P., et al.: Impedance-source networks for electric power conversion part II: review of control and modulation techniques. IEEE Trans. Power Electron. **30**(4), 1887–1906 (2015)
19. Meziane, B., Zeroug, H.: Comprehensive power control performance investigations of resonant inverter for induction metal surface hardening. IEEE Trans. Ind. Electron. **63**(10), 6086–6096 (2016)
20. Roy, M., Sengupta, M.: Design, real-time modelling, simulation and digital implementation of phase-locked loop-based auto synchronising current-sourced converter for an induction heating prototype. Sadhana **42**(6), 879–888 (2017)
21. Grajales, L., et al.: Design of a 10 kW, 500 kHz phase-shift controlled series-resonant inverter for induction heating. In: Conference on Record of IEEE Industry Applications Conference on Twenty-Eighth IAS Annual Meeting, vol. 2, pp. 843–849 (1993)
22. Sawant, R.R., Rao, Y.S.: A discrete-time controller for phase shift controlled load-resonant inverter without PLL. In: Proceedings of IEEE International Conference on Power Electron. Drives and Energy Systems (PEDES), Mumbai, pp. 1–4 (2014)
23. Dominguez, A., Barragan, L.A., Artigas, J.I., Otin, A., Urriza, I., Navarro, D.: Reduced-order models of series resonant inverters in induction heating applications. IEEE Trans. Power Electron. **32**(3), 2300–2311 (2017)

A Grid-Connected Distributed System for PV System

Md. Fahim Ansari and Anis Afzal

Abstract Renewable energy like solar energy, wind energy and hydropower energy may be used to connect with grid-connected distributed system. The excess energy available by renewable energy sources may be supplied to grid and to get some positive cash flow. In this paper, application of power electronics has been explained and discussed and also providing some simulation results to be used with grid-connected distribution system. To achieve a reliable working of grid-connected distribution system, active and reactive power are suggested as compensation for decrease of grid voltage due to the reactive power injected to the grid. Absorption of reactive power has been accompanied with increase in active power injection. Also, increase in active power absorption has been accompanied reactive power injection. Rated active power has been supplied to the grid voltage. The active power injection has been decreased till minimum optimized value of active power being achieved.

Keywords Active and reactive power · Grid voltage · Power electronics · Multilevel inverter

1 Introductions

In present days, active and reactive power switching between grid and distributed generation has to be managed due to massive integration of various renewable energy sources [1, 2]. Distributed renewable energy sources provide a cheaper option than high-cost grid power system. Main sources of renewable energy are wind energy,

Md. Fahim Ansari (✉)
Department of Electrical Engineering, Graphic Era Deemed to be University, Dehradun, Uttrakhand, India
e-mail: engineer.19781234@gmail.com

A. Afzal
Department of Electrical Power and Control Engineering, Adama Science and Technology University, Adama, Ethiopia
e-mail: anisafzal.786@gmail.com

© The Editor(s) (if applicable) and The Author(s), under exclusive license to Springer Nature Singapore Pte Ltd. 2021

A. P. Pandian et al. (eds.), *Proceedings of International Conference on Intelligent Computing, Information and Control Systems*, Advances in Intelligent Systems and Computing 1272, https://doi.org/10.1007/978-981-15-8443-5_78

solar photovoltaic energy, municipal solid waste, biomass, biofuels, wave energy and tidal energy. All these energy sources are available in abundance and are also sustainable. The main challenge is to harness energy out of these sources. Integration of these renewable energy sources is crucial and enormous task. Consumption of coal as energy generation source in the world is decreasing day by day, whereas consumption of natural gas and non-hydro renewable energy are going up. In the coming days, efforts must be there to minimize consumption of natural gas and maximize the use of non-hydro renewable energy sources. However, all predictions depend upon local government's policies and priorities. Figure 1 depicts the use renewable energy sources and other sources of energy in India.

Solar power in India is a fast developing industry; its installed capacity has been increased to 20 GW in February 2018 [3]. Solar power capacity has been increased up to eight times from 2650 MW from May 2014 to 20 GW till February 2018 [4, 5]. Major solar power plant in India is depicted in Table 1.

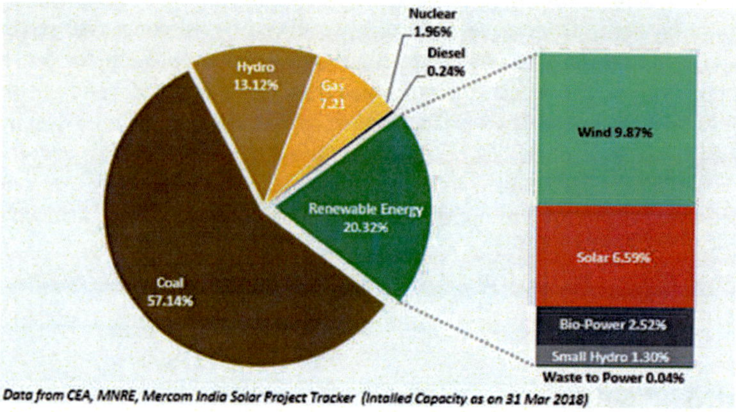

Fig. 1 Indian scenario of energy production by different means

Table 1 Major solar power plant in India

Place/State	Capacity (in MW)	Land occupied in acres
Kurnool ultra-mega solar park, Andhra Pradesh	1000	5932.32
Kamuthi solar power project, Tamil Nadu	648	2500
Bhadla Solar Park, Rajasthan	2255	10,000
Charanka village solar park, Gujarat	600	5384
Sakri Photovoltaic Solar Energy Project	125	400

2 Methodology

Distributed generation of energy from various resources such as solar photovoltaic panels, wind, hydropower, biomass, municipal solid waste, fuel cell, etc., are generated at a place used at the same place without sending it to the centralized place [1, 6]. Due to the use of place of generation, it has got a lot of advantages such as transmission and distribution lasses are reduced, and due to local positioning of power renewable waste can be utilized at the same place, local positioning also permits the utilization of single-phase or three-phase power [7–9]. In spite of so many advantages, it has also got many severe disadvantages such as dispatching of energy is very complicated, charges of energy calculation is also very difficult, system protection is also difficult because of two-way exchange of power and safety distracts due to multiple sources of generation of power [10, 11].

Characteristics of different types of renewable energy system (RES)-based distributed generations (DGs) have been discussed below [12];

The effect of solar insolation and temperature is given in Figs. 2 and 3 [13]

Two major types of renewable energy source (RES)-based DGs and DGs with bidirectional power capability [14] RES-based DGs are shown in Fig. 4.

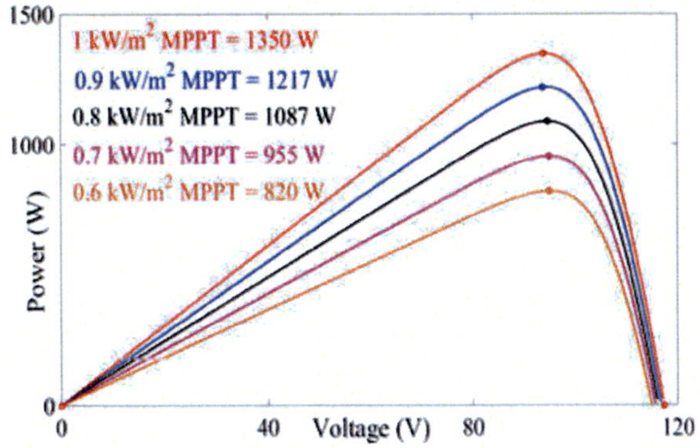

Fig. 2 Reduction in solar irradiation leads to reduction in panel's current and power

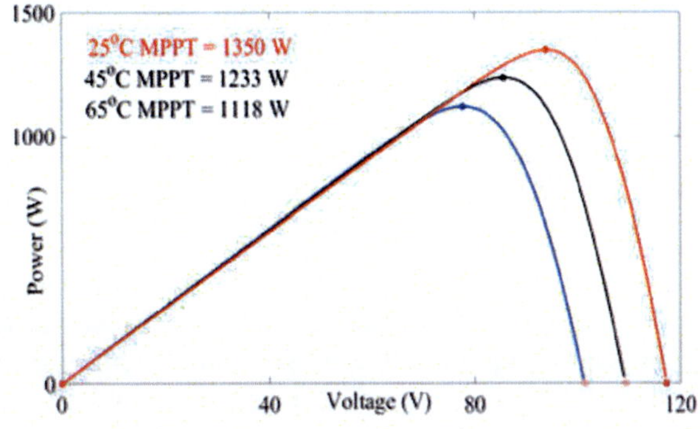

Fig. 3 Variation of temperature: Increase in temperature leads to reduction in panel's voltage and power

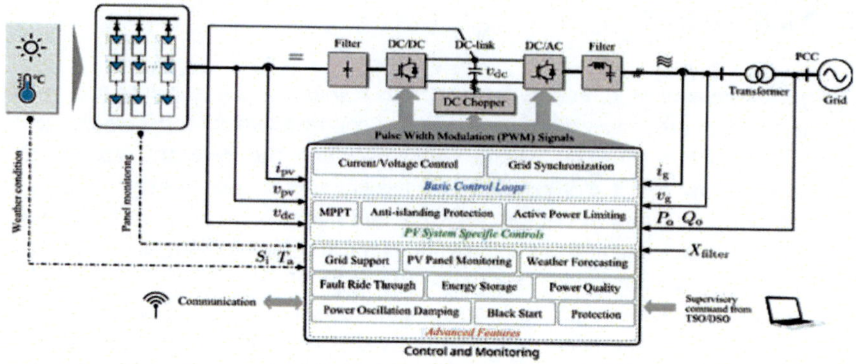

Fig. 4 Power electronics-based distributed renewable system

3 Results

Voltage magnitude frequency and phase difference have been managed for grid synchronization, and the value of grid voltage here considered is $V\text{grid} = 230\sqrt{}2\sin(2\pi f t)$ and filter resistance = 0.1 Ω and filter inductance = 5 mH. It has been seen from Fig. 4 that the synchronization with grid is good promising to actual voltages and frequency.

Voltage magnitude, frequency and phase difference have been managed for grid synchronization using the mathematical relation given in Eq. (1). The filter used here consists of resistance 0.1 Ω and inductance 0.005 H, frequency $f = 50$ Hz

$$V\text{grid} = 230\sqrt{}2\text{Sin}(2\pi f t) \tag{1}$$

The inverter voltage with different peak voltage and phase differences are given in Eqs. 2 and 3. The plot of these equations is shown in Figs. 6 and 7.

$$Vinv = 200\sqrt{}2\sin(2\pi f t) \tag{2}$$

$$Vinv = 200\sqrt{}2\sin(2\pi f t - \pi/6) \tag{3}$$

4 Conclusion

Existing global energy scenarios reflecting importance of renewable energy source-based distributed generators are discussed. Characteristics of different distributed generators and relevant power electronic circuitry are presented. Modern standards and guidelines for integration and operation of DGs are discussed. Concepts related

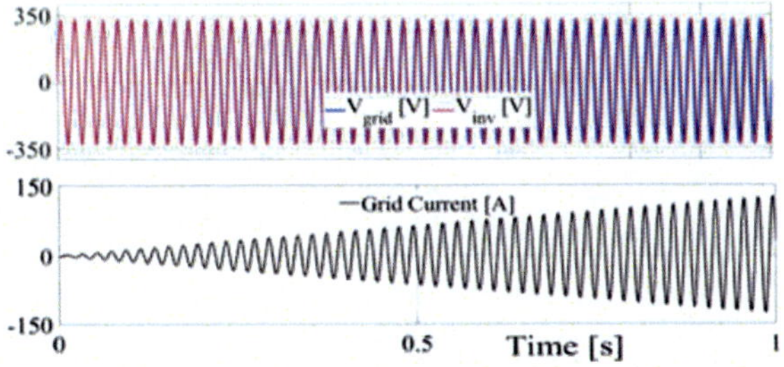

Fig. 5 Filter voltage and frequency response for grid synchronization

to grid synchronization of three—phase system are presented. Performance of DC grid-connected solar PV-based DC-DC converter is discussed. It could be seen in Figs. 5 and 6 that the filter voltage and frequency are quite close to the desired voltage and frequency of the grid. Figure 7 also depicts the desired phase angle of grid voltage and inverter voltage.

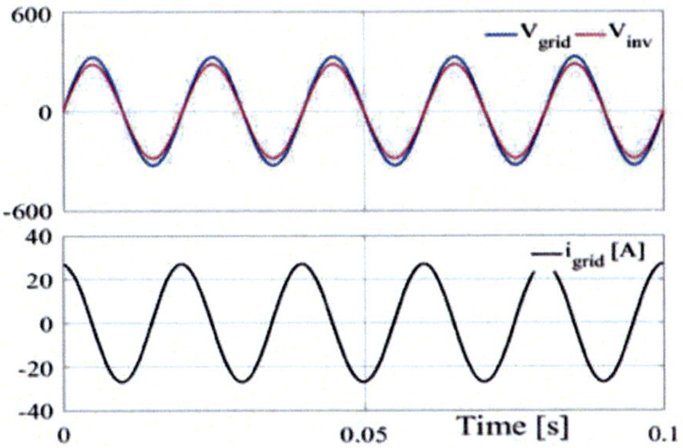

Fig. 6 Difference in voltage magnitude of grid voltage and inverter voltage

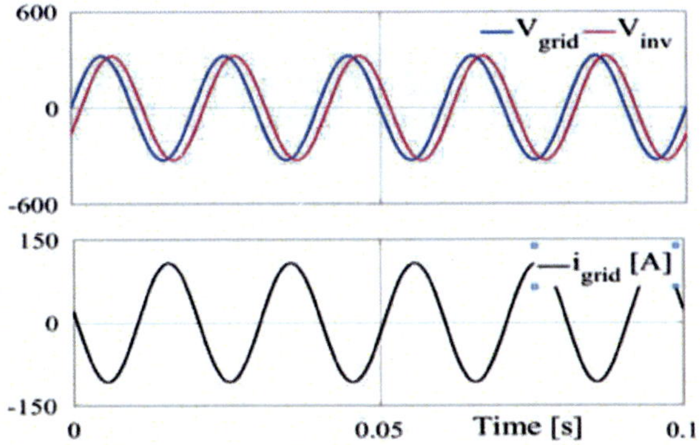

Fig. 7 Difference in phase angle of grid voltage and inverter voltage

References

1. Haque, A., Zaheeruddin: Research on Solar Photovoltaic (PV) energy conversion system: An overview. In: Third International Conference on Computational Intelligence and Information Technology (CIIT 2013), Mumbai, 2013, pp. 605–611
2. Wu, X., et al.: A two-layer distributed cooperative control method for Islanded networked microgrid systems. IEEE Trans. Smart Grid **11**(2), 942–957 (2020)
3. Yuen, C., Oudalov, A., Timbus, A.: The provision of frequency control reserves from multiple microgrids. IEEE Trans. Ind. Electron. **58**(1), 73–183 (2011)
4. Farajollahi, M., Shahsavari, A., Mohsenian-Rad, H.: Topology identification in distribution systems using line current sensors: an MILP approach. IEEE Trans. Smart Grid **11**(2), 1159–1170 (2020)
5. Wang, X., Blaabjerg, F.: Harmonic stability in power electronic-based power systems: concept, modeling, and analysis. IEEE Trans. Smart Grid **10**(3), 2858–2870 (2019)
6. Olivares, D.E., et al.: Trends in microgrid control. IEEE Trans. Smart Grid **5**(4), 1905–1919 (2014)
7. Kordestani, M., Mirzaee, A., Safavi, A.A., Saif, M.: Maximum power point tracker (MPPT) for photovoltaic power systems-a systematic literature review. In: 2018 European Control Conference (ECC), Limassol, 2018, pp. 40–45
8. Salam, Z., Ramli, M.Z.: Distributed maximum power point tracker for additional energy harvesting during partial shading of PV system. In: 2015 First Workshop on Smart Grid and Renewable Energy (SGRE), Doha, 2015, pp. 1–5
9. Ishaque, K., Salam, Z.: Dynamic efficiency of direct control based maximum power point trackers. In: 2014 5th International Conference on Intelligent Systems, Modelling and Simulation, Langkawi, 2014, pp. 429–434
10. Cherchali, N.O., Morsli, A., Boucherit, M.S., Barazane, L., Tlemçani, A.: Comparison of two maximum power point trackers for photovoltaic systems using robust controllers. In: 3rd International Symposium on Environmental Friendly Energies and Applications (EFEA), St. Ouen, 2014, pp. 1–5
11. Zanotti, J.W., Martins, D.C.: Input characteristic impedance technique of power converters circuits applied to the maximum power point tracker of photovoltaic panels. In: 2014 IEEE Asia Pacific Conference on Circuits and Systems (APCCAS), Ishigaki, 2014, pp. 292–295
12. Murtaza, A.F., Sher, H.A., Chiaberge, M., Boero, D., De Giuseppe, M., Addoweesh, K.E.: Optimization of the Perturb and Observe Maximum Power Point Tracker for a Distributed Photovoltaic System, pp. 77–82. Lahore, INMIC (2013)
13. Cristaldi, L., Faifer, M., Laurano, C., Leone, G., Toscani, S.: Model based maximum power point tracker stability over time. In: 2017 6th International Conference on Clean Electrical Power (ICCEP), Santa Margherita Ligure, 2017, pp. 591–595
14. Ramakumar, R., Allison, H.J., Hughes, W.L.: Solar energy conversion and storage systems for the future. IEEE Trans. Power Apparat. Syst. **94**(6), 1926–1934 (1975)

Real-Time Implementation of Multi-model Reference-Based Fuzzy Adaptive PI Controller for a Liquid-Level Process

A. Ganesh Ram and S. Meyyappan

Abstract The process industries require accurate, efficient, reliable, and flexible control systems to achieve a better performance. So, the use of advanced control techniques becomes the need of the hour by considering the extreme complexity of the processes. The new trends in control technology are adaptive control and intelligent control. These control schemes can be visualized as strategic solutions that intend to mimic the significant characteristics that reflect the thought processes of humans when it comes to decision making. The first aspect is to select the nonlinear spherical tank-level process, and the second approach is to develop a mathematical model of the selected nonlinear process and the real-time test-rig models at different operating regions. The open-loop transfer function models are held from the real-time observational data by using iterative predictive error minimization (PEM). Finally, the research concern with the feasibility of design, implementation, and testing of multi-model reference-based fuzzy adaptive PI (MMR-FAPI) controller for a spherical tank-level process.

Keywords MMR-FAPI · PEM · VMAT-01

1 Introduction

The most commonly used controllers in process industries are proportional integral derivative (PID) controllers. Even though they are easy to tune and implement, they provide satisfactory performance only if the mathematical model of the system is linear and accurate, which in turn makes the implementation of traditional PID control strategy for time varying and complex processes. The controller is proposed in this work namely MMR-FAPI controller, and it is the combination of fuzzy and conventional PI controller with reference models [1–4]. The reference models are

A. Ganesh Ram (✉) · S. Meyyappan
Department of Instrumentation Engineering, MIT Campus, Anna University, Chennai, Tamil Nadu 600044, India
e-mail: agram72@gmail.com

© The Editor(s) (if applicable) and The Author(s), under exclusive license to Springer Nature Singapore Pte Ltd. 2021

A. P. Pandian et al. (eds.), *Proceedings of International Conference on Intelligent Computing, Information and Control Systems*, Advances in Intelligent Systems and Computing 1272, https://doi.org/10.1007/978-981-15-8443-5_79

considered to be second-order state-space models of the system at different operating points. The online controller parameters are estimated subject to the error and tracking error where the tracking error is the difference between $y(t)$ and $y_m(t)$. The real-time implementation and control scheme is done using MATLAB software using VMAT-01 Data Acquisition Module.

2 MMR-FAPI Controller

In some control tasks, the measurement systems may have parameter uncertainty. Unless such parameter uncertainty is gradually reduced online by an appropriate adaptation or estimation mechanism, it may cause inaccuracy or instability of the control systems. The fuzzy adaptive control methods [5, 6] have become popular and are being used widely in many practical and industrial situations. The structure of MMR-FAPI controller is shown in Fig. 1.

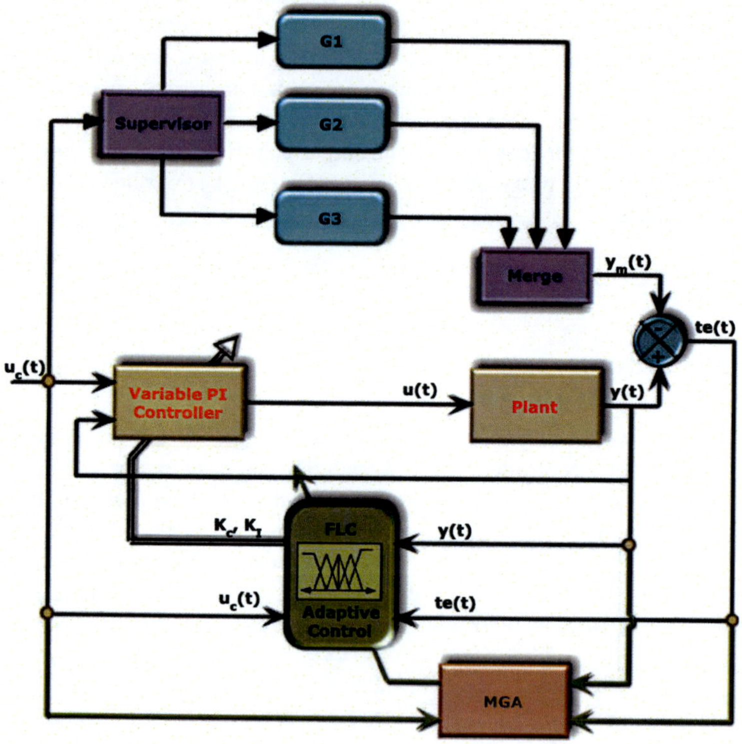

Fig. 1 Structure of MMR-FAPI controller

The proposed MMR-FAPI controller uses a model reference adaptive control structure, but the adaptive mechanism used here is the adaptive fuzzy logic controller. Here, the fuzzy controller plays the role of continuous scheduler to update the PI controller parameters, such that at each operating point there exists a controller that achieves the model following objective. The fuzzy controller uses the error $[e(t) = \mathrm{sp} - \mathrm{pv}]$ and tracking error $[te(t) = y(t) - y_m(t)]$ as its inputs and the outputs are updating the PI controller parameters K_c and K_I. The online estimated PI controller parameters are updated with the help fuzzy controller, and the fuzzy controller is tuned off-line by a modified genetic algorithm (MGA).

A. *MGA—Optimization of Fuzzy Logic Control*
The two-input and two-output (TITO) [7] fuzzy inference system's [8] all membership functions are trapezoidal membership functions with five lingustic variables such as NB, NS, ZE, PS, and PB.
B. *MGA Parameters Initialization*
Initial value of the MGA parameters like chromosome size (fuzzy rules, MFs, input/output gain), crossover rate (single point, two point, and multipoint), and mutation rate, etc., are given by

MGA Generation Index (gen) = 1;
Population size (N) = 20;
Chromosome size (L) = 65;
Base = 2;
Crossover (pc) = 0.7; (two point).
Mutation (pm) = 0.04;
Maximum generation (max.gen) = 100;
MGA-chromosome at generation—1.

Rule-base	*te*-MF	*e*-MF	K_c-MF	K_I-MF	*te*-scale	*e*-scale	K_c-scale	K_I-scale
100110011	1010001	0010010	1000011	1101010	0110010	0011001	1100101	1000110

C. *Termination Criteria*
The termination of MGA algorithm can take place when number of iterations is over. The best parent fit can be achieved by the minimum objective function (i.e., cost function) from every generation within the population size. The best solution of input/output scaling gain, membership functions, and fuzzy rules are determined from the best pattern fit with minimum cost function. The inputs and output scaling gains and its corresponding best fitness for every generation are as shown in Fig. 2.

After the termination criteria, the optimum fuzzy logic controller is constructed by the minimum cost function from the parent fitness.

The control mechanism consists of the fuzzy adaptive controller together with the PI [9] controller constituently the closed-loop system with adjustable parameters. The basic system comprises of the reference models, process plant, feedback controller, and adaptive fuzzy controller. The adaptive regulation process of

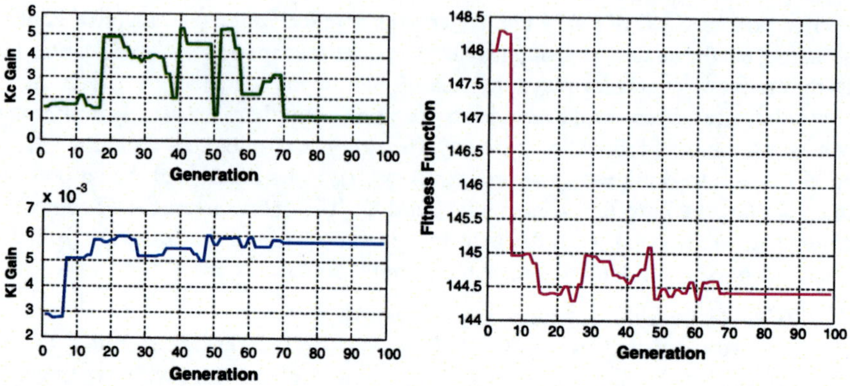

Fig. 2 Outputs scaling gains and its fitness Function

the controller parameters, when the input value $u_c(t)$ is set into the adaptive fuzzy controller and the supervisor, the supervisor is used to select the desired model based on the command signal. The difference between the reference model output $y_m(t)$ and plant output $y(t)$ produces the tracking error $te(t)$. When $te(t)$ and $e(t)$ are applied to the fuzzy adaptive regulation loop, the FLC produces the appropriated updating parameters of PI controller, and the PI controller produces the proper dynamic signal to make the $y(t)$ get approaching to $y_m(t)$.

D. Design of Controller Algorithm
Considering the nonlinear system

$$y(t+1) = g(t) + h(t) * u_c(t) \tag{1}$$

$$\begin{aligned} g(t) = g[&y(t), t(t-1), \ldots y(t-n_1+1) \\ &u(t), u(t-1), \ldots u(t-n_2+1)] \end{aligned} \tag{2}$$

$$\begin{aligned} h(t) = h[&y(t), t(t-1), \ldots y(t-m_1+1) \\ &u(t), u(t-1), \ldots u(t-m_2+1)] \end{aligned} \tag{3}$$

where $k \geq n_1, n_2, m_1, m_2$; $u(t) \in R$ and $y(t) \in R$.

For the system as illustrated in (1), when $g(t)$ and $h(t)$ are known, applying the control law

$$u(t) = \frac{1}{h(t)}[-g(t) + y_m(t+1)pe(t)] \tag{4}$$

where $p > 0$,

The output tracking error of the system is

$$te(t+1) = y(t+1) - y_m(t+1) \tag{5}$$

$$te(t+1) = [g(t) + h(t)u_c(t)] - y_m(t+1) \tag{6}$$

$$te(t+1) = \left[g(t) + h(t)\frac{1}{h(t)}(-g(t) + y_m(t+1) + pe(t))\right] - y_m(t+1) \tag{7}$$

$$te(t+1) = y_m(t+1) - y_m(t+1) + pe(t) \tag{8}$$

$$te(t+1) = pe(t) \tag{9}$$

Then, it is well known that if $|p| < 1$, the output system $y_m(t+1)$ is able to track asymptotically.

As $g(t)$ and $h(t)$ are unknown continuous functions, when they are substituted using the fuzzy inference system $\overline{g}(t)$ and $\overline{h}(t)\overline{h}\ (t)$, respectively, the control law becomes

$$u(t) = \frac{1}{h(t)}[-\overline{g}(t) + y_m(t+1) + pe(t)] \tag{10}$$

$$y(t+1) = g(t) + h(t)\frac{1}{h(t)}[-\overline{g}(t) + y_m(t+1) + pte(t)] \tag{11}$$

According to Eq. (10)

$$u(t) = \frac{1}{\overline{h}(t)}[-\overline{g}(t) + y_m(t+1)] + \frac{1}{\overline{h}(t)}pte(t) \tag{12}$$

$$u(t) = u'(t) + \frac{p}{\overline{h}(t)}te(t) \tag{13}$$

where

$$u'(t) = \frac{1}{h(t)}[-\overline{g}(t) + y_m(t+1)] \tag{14}$$

Further, the system error is

$$te(t+1) = y(t+1) - y_m(t+1) \tag{15}$$

$$te(t+1) = g(t) + h(t)\frac{1}{\hbar(f)}[-\overline{g}(t) + y_m(t+1) + pte(t)] - y_m(t+1) \tag{16}$$

$$te(t+1) = g(t) + h(t)u_c'(t) + h(t)\frac{p}{h(t)}te(t) - y_m(t+1) \tag{17}$$

$$te(t+1) = [g(t) - \overline{g}(t)] + [h(t) - \overline{h}(t)]u_c'(t) + \quad h(t)\frac{p}{h(t)}te(t) \tag{18}$$

Let $\Delta g(t) = [g(t) - \overline{g}(t)]$, $\Delta h(t) = [h(t) - h(t)]$,
Then

$$te(t+1) = \Delta g(t) + \Delta h(t)u_c'(t) + p\frac{h(t)}{h(t)}te(t) \tag{19}$$

The fuzzy adaptive system [10 ,11] at each sampling instant n reads the tracking error signal $te(t+1)$ which is defined as $te(t+1) = y(t+1) - y_m(t+1)$ and error signal which is defined as $e(t+1) = u_c(t) - y(t)$, to find the new PI parameters ΔK_c, ΔK_I using fuzzy rule base system.

E. *Fuzzy Reasoning and Defuzzification*
The TITO fuzzy inference system [12 ,13] is possible to get $25 \times 2 \times 2$ fuzzy rules that may be described by the following statements.

If $te = A_i$ and if $e = B_j$, then $K_c = C_{ij}$, and $K_I = D_{ij}$

where
A_i, B_j are the input fuzzy sets and C_{ij}, D_{ij} are the output fuzzy sets.
So, the MF of U_R is

$$u_R(x_1, x_2, y_1, y_1) = u_{A_1}(x_1) \cap u_{B_1}(x_2) \cap u_{C_1}(y_1) \cap u_{D_1}(y_2) \tag{20}$$

MF of u_{K_c} and u_{K_I} are given by

$$u_{K_c} = u_R(x_1, x_2, y_1, y_1) \cap u_{B_j}(x_2) \cap u_{C_1}(y_1) \tag{21}$$

$$u_{K_I} = u_R(x_1, x_2, y_1, y_1) \cap u_{B_1}(x_2) \cap u_{D_1}(y_2) \tag{22}$$

Consequently, the FLC output K_c and K_I are

$$K_c = \frac{\sum_{j=-5}^{j=5} j.u_{K_C}(j)}{\sum_{j=-5}^{j=5} u_{K_C}(j)} \tag{23}$$

$$K_I = \frac{\sum_{j=-5}^{j=5} j.u_{K_I}(j)}{\sum_{j=-5}^{j=5} u_{K_I}(j)} \tag{24}$$

The mathematical model of the spherical tank-level process is represented by

$$\frac{dh}{dt} = \frac{F_{in}}{\pi(2rh - h^2)} - \frac{b\alpha\sqrt{2gh}}{\pi(2rh - h^2)} \tag{25}$$

Table 1 Reference models for spherical tank process

	State-space models			
	A	*B*	*C*	*D*
M1	$\begin{bmatrix} -0.02521 & -0.005016 \\ 0.007813 & 0 \end{bmatrix}$	$\begin{bmatrix} 0.0625 \\ 0 \end{bmatrix}$	$\begin{bmatrix} -0.02663 & 0.08028 \end{bmatrix}$	[0]
M2	$\begin{bmatrix} -0.05098 & -0.00625 \\ 0.007813 & 0 \end{bmatrix}$	$\begin{bmatrix} 0.0625 \\ 0 \end{bmatrix}$	$\begin{bmatrix} -0.01563 & 0.1 \end{bmatrix}$	[0]
M3	$\begin{bmatrix} -0.0674 & -0.0066 \\ 0.00781 & 0 \end{bmatrix}$	$\begin{bmatrix} 0.0625 \\ 0 \end{bmatrix}$	$\begin{bmatrix} -0.01239 & 0.1057 \end{bmatrix}$	[0]

Let F_{in} be the inlet flow rate to the tank (m^3/min), b is the outlet valve constant, α specifies the outlet valve stem position, h indicates the height of the liquid level at any time $t(m)$, and r is the radius of the spherical vessels at a particular level of height $h(m)$.

The second-order state-space reference models at different operating levels model1 is rage between 20 and 39%, model 2 is 40–59%, and model 3 is the 60–75% of level, and they are tabulated in Table 1.

3 Real-Time Implementation

The MMR-FAPI control strategy is implemented and validated through VMAT-01 data acquisition module with a sampling time of 0.1 using MATLAB software.

To study the real-time servo response of spherical tank process with MMR-FAPI controller for a set point change of 30% from 0% level applied at 0th s, it is shown in Fig. 3a, and alongside the corresponding controller output is presented in Fig. 3b.

Figure 4a shows the real-time experimental servo response of spherical tank-level process in presence of MMR-FAPI controller for a set point change of 50% from 0% level applied at 0th s. The corresponding controller output is shown in Fig. 4b.

Figure 5a shows the real-time experimental servo response of spherical tank-level process in the presence of MMR-FAPI controller for a set point change of 60% from 0% level applied at 0th s. The corresponding controller output is shown in Fig. 5b.

Similarly the real-time experimental servo response of spherical tank-level process with MMR-FAPI controller for a set point change of 70% from 0% level applied at 0th sec is shown in Fig. 6a, and the corresponding controller output is shown in Fig. 6b.

The real-time experimental servo response of spherical tank process in presence of MMR-FAPI controller for a set point change of 20% from steady-state operating point of 10% is applied at 500th s. After the level reaches the new set point, another

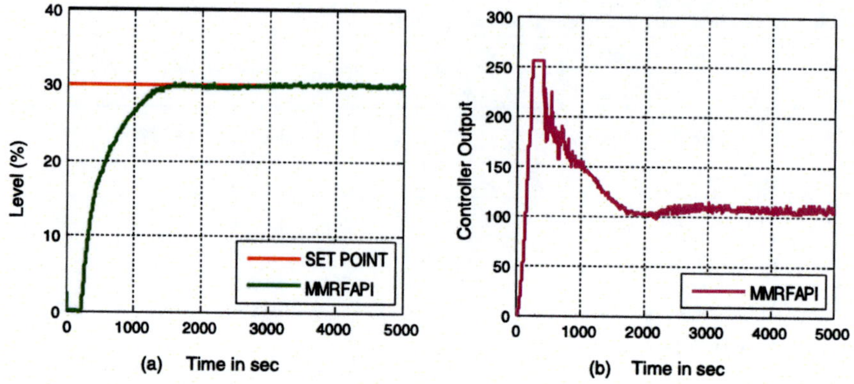

Fig. 3 Real-time response for spherical tank process at 30% level **a** process output, **b** controller output

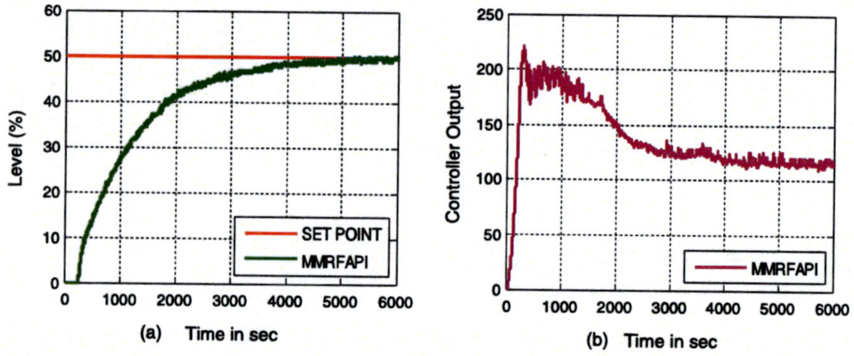

Fig. 4 Real-time response for spherical tank process at 50% level **a** process output, **b** controller output

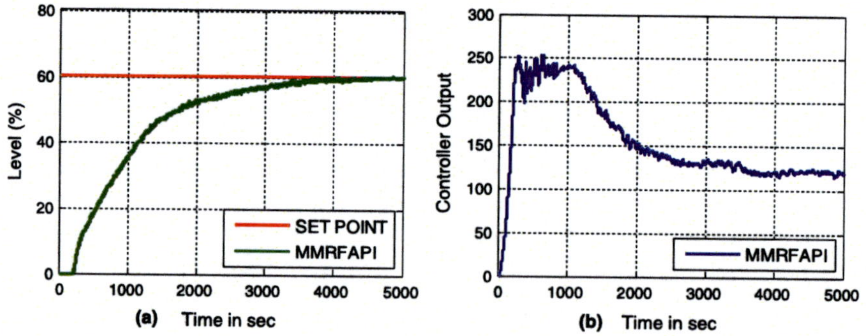

Fig. 5 Real-time response for spherical tank process at 60% level **a** process output, **b** controller output

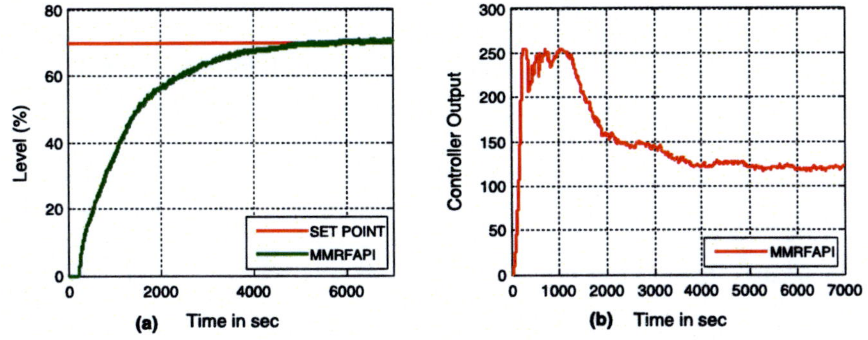

Fig. 6 Real-time response for spherical tank process at 70% level **a** process output, **b** controller output

step change of 20% (from 30 to 50%) has been introduced to the process at 5001th s as shown in Fig. 7a, and the corresponding controller output is presented in Fig. 7b.

The real-time experimental servo response of spherical tank process in presence of MMR-FAPI controller for a set point change of 40% from steady-state operating point of 10% is applied at 500th s. After the level reaches the new set point, a load disturbance of 200 LPH inflow has been introduced to the process at 5226th s as

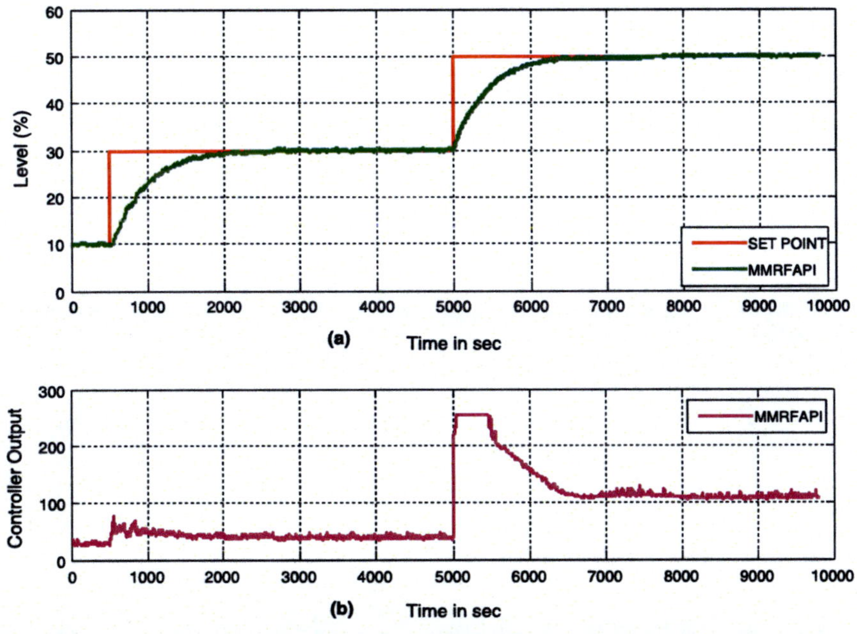

Fig. 7 Real-time servo response for spherical tank process **a** process output, **b** controller output

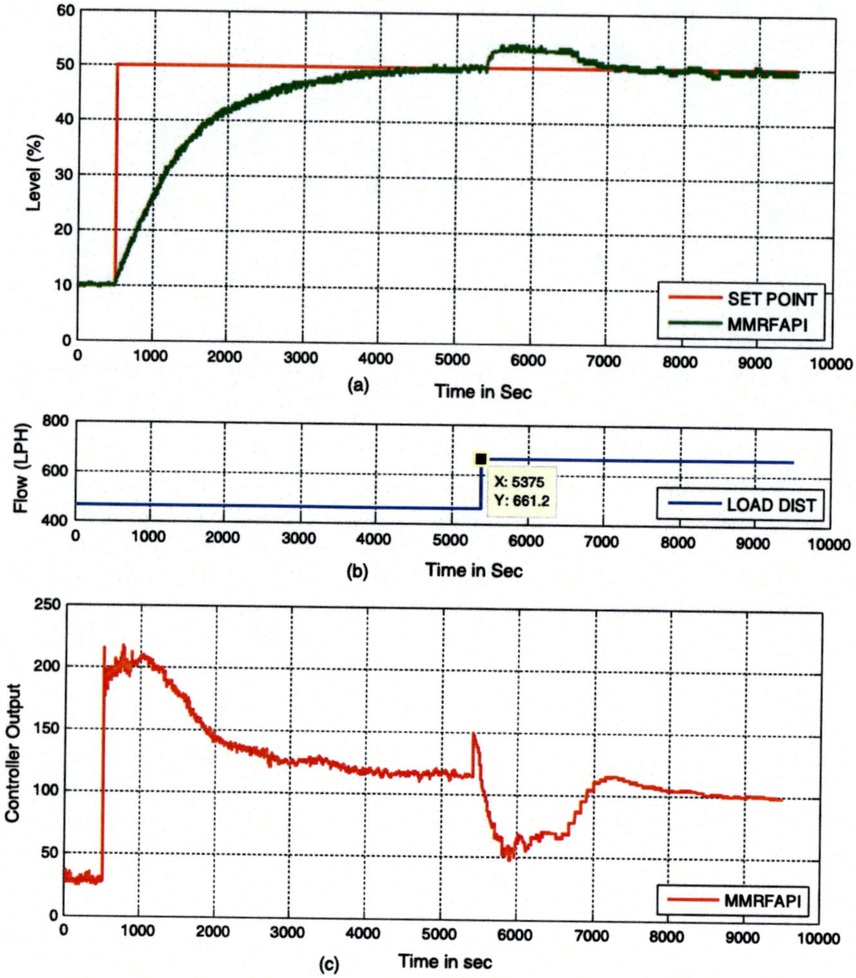

Fig. 8 Real-time response for spherical tank process **a** regulatory response, **b** load disturbance, **c** controller output

shown in Fig. 8b, the process output is shown in Fig. 8a, and the corresponding controller output is presented in Fig. 8c.

4 Conclusion

In this work, a MMR-FAPI controller is designed and implemented using MATLAB Simulink, and the real-time experimental test-rig is interfaced with the help of a

VMAT_01 data accusation module. The performance of the MMR-FAPI controller for various operating regions of the spherical-tank liquid-level process is found to be highly satisfactory. It gives no overshoot and exhibits slow settling time because of adaptive in nature. Moreover, when the load disturbances are applied to the process, the proposed MMR-FAPI controller takes fast action and tracks the set point immediately.

References

1. Harikrishnaa, K., Satheesh Kumar, J., Shaik, M.: Desigh and development of model based controller for a spherical tank. Int. J. Curr. Eng. Technol. **2**(1), 374–376 (2012)
2. Harikrishnaa, K., Satheesh Kumar, J., Shaik, M.: Real time implementation of model based controller for a sperical tank. Int. J. Curr. Eng. Technol. **3**(2), 587–590 (2013)
3. Angeline, D., Vivetha, K., Gandhimathi, K., Praveena, T.: Model based controller design for conical tank system. Int. J. Comput. Appl. **85**(12), 12–15 (2014)
4. Joseph, A., Samson Isaac, J.: Real time implementation of model reference adaptive controller for a conical tank. Int. Conf. Mech. Ind. Eng. **2**(1), 57–62 (2013)
5. Ganesh Ram, A., Abraham Lincoln, S.: Real time implementation of fuzzy based adaptive PI controller for a spherical tank system, Int. J. Simul. Syst. Sci. Technol. **14**(6), 1–8 (2013)
6. Abdullah, J.H., Gizi, A., Mustafa, M.W.: Improve Fuzzy-PSO PID controller by adjusting transfer function parameters. Int. J. Sci. Res. Publ. **2**(11), 1–5 (2012)
7. Karamimohammadi, A.: A variable structure model reference adaptive control for MIMO systems. Proc. Int. MultiConf. Eng. Comput. Sci. **II**(5), 19–21 (2008)
8. Adrian-Vasile, D., Stefan, G.H.: Design of an adaptive fuzzy-PD controller. Sci. Int. Conf. Interdis. Eng. 15–16 (2007)
9. Dong. H., Li, W., Zhao, Y., Lin, K.: Design and simulation a fuzzy-adaptive PI controller based on MRAS. Int. Conf. Nat. Comput. ICNC IEEE **5**, 2321–2324 (2010)
10. Ehsani, M.: Adaptive control of servo motor by MRAC method. In: IEEE International Conference on Vehicle Power and Propulsion Conference, PP. 78–83 (2007)
11. Qin, G., Ma, Y., Zhang, X., Zhang, M.: Design of fuzzy adaptive PID temperature controller based on FPGA. TELKOMNIKA 11(10), 6008–6016 (2013)
12. Kamalasadan, S., Ghandakly, A.A.: Multiple fuzzy reference model adaptive controller design for pitch-rate tracking. IEEE Trans. Instrum. Meas. **56**(5), 1797–1808 (2007)
13. Li, Z.: Model reference adaptive controller design based on fuzzy inference system. J. Inf. Comput. Sci. 1683–1693 (2011)

Intelligent Learning Control Strategies for Speed Control of DC Motor

M. Vijayakarthick, N. Vinoth, S. Sathishbabu, and S. Ramesh

Abstract Recently, the intelligent controllers like learning controller and repetitive controller has gained significant research attention in the control process in which the input commands are periodic signals with fixed period need to be tracked. This paper has made an attempt to implement the control strategies for incorporating speed control on DC motor. A second order DC motor system model is analyzed through the specifications of the motor. The design parameters of the proposed algorithm such as learning filter L(s) and robustness filter Q(s) are identified as ZPETC algorithm and plot. Finally, the DC motor process with intelligent controllers is executed for the periodic reference inputs. Conventional controller is also analyzed by performing comparative studies. The efficiency of intelligent controllers is estimated by analyzing its tracking error. An evidence of robustness is also tested.

Keywords DC motor · ILC · MRC · PI

1 Introduction

DC motors are the electrical drives that achieves high popularity due to their high competence, superior dynamic performance, and low maintenance. If the disturbances are periodic in DC motor system, the conventional PID controller does not promise to deliver a satisfactory performance, which is assumed with known process parameters and operating factors. To overcome this problem, repetitive controller

---

M. Vijayakarthick (✉) · N. Vinoth
Department of Instrumentation Engineering, MIT Campus, Anna University, Chennai, Tamil Nadu 600044, India
e-mail: vijayakarthick@yahoo.co.in

S. Sathishbabu
Department of Electronics and Communication Engineering, TPGIT, Vellore, 632002, India

S. Ramesh
Department of Electronics and Instrumentation Engineering, Annamalai University, Annamalainagar, 608002, India

© The Editor(s) (if applicable) and The Author(s), under exclusive license to Springer Nature Singapore Pte Ltd. 2021

A. P. Pandian et al. (eds.), *Proceedings of International Conference on Intelligent Computing, Information and Control Systems*, Advances in Intelligent Systems and Computing 1272, https://doi.org/10.1007/978-981-15-8443-5_80

and iterative learning control (ILC) method is projected [1–3] for the control of periodic process that needs to carry out the same work over and over again. The ILC is considered based on the previous control history and a learning method depending on the knowledge of the number of the periodic tasks required to improve the control performance. The ILCS can be used in many industrial processes such as manufacturing, robotics, and chemical processing, where the large manufacturing of an assembly line involves repetition. Arimoto [4, 5] proved the robustness of P-type learning controls based on the passivity robot dynamics analysis and observed the properties of the P and D-type learning algorithms for linear time-invariant systems in the occurrence of noise.

In view of the robustness improvement of the DC motor control system, an Iterative Learning Control (ILC) is suggested here. The proposed structure consists of feedback controller and feed-forward controller.

The feedback controller make sure the closed loop stability and feed-forward gives the enhanced tracking ability over a specific path with minimum tracking error, fast convergence in the minimum trials than the simple ILC and conventional controllers.

In the same way, Repetitive Control is an IMP based control approach in which the infinite dimensional model $\frac{1}{1-e^{-Ls}}$ provides rise to infinite poles on the imaginary axis. The fundamental design of RC is the cancelation view on the Internal Model Principle (IMP) [6, 7]. It is initially developed for continuous SISO LTI systems in, for high precision tracking of a periodic reference with known periods. The repetitive control is begun to get more awareness and applications, and become an extraordinary topic in control theory. By the use of frequency domain techniques, the theories and applications of RC in LTI systems have been developed splendidly. On the other hand, RC for nonlinear process has established limited research attempt. Taking these into account, the applications of RC for electrical systems are provided in this report.

The major aids of the work in this paper are especially real time attempt of the ILCS in a DC motor and analysis of the tracking achievements. The process description of DC Motor system is summarized in Sect. 2. The design and construction of ILC and RC Strategies are given in Sects. 3 and 4; Results are investigated in Sect. 5. Finally, conclusion of the proposed work is presented in Sect. 6.

2 Mathematical Model of DC Motor and Controller Settings

The speed control of DC motor and its equivalent circuit are given in Fig. 1a, b.
where

- R_a armature resistance of in Ω;
- L_a armature inductance in H;
- i_a current through armature winding (A);
- V_a voltage across armature coil (V);
- e_b back emf;

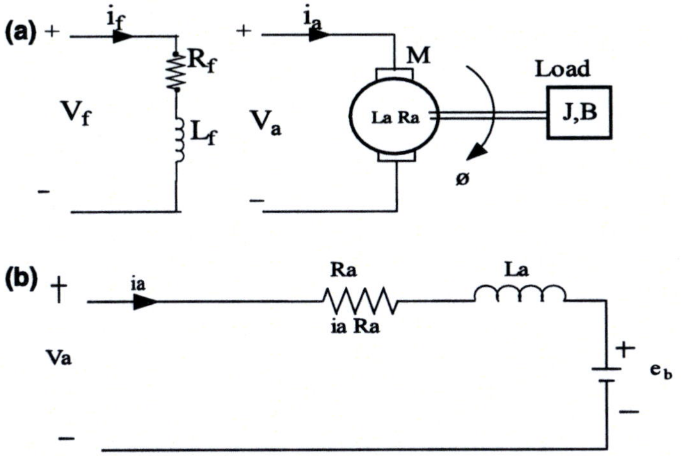

Fig. 1 **a** Schematic representation of DC motor, **b** equivalent circuit of armature

T motor Torque in kg cm;
ω rotational displacement in rad/s;
J inertia moment in kg m^2/rad;
K_b back emf constant;
K_t torque constant;
B frictional coefficient in NM rad/s.

By applying Kirchhoff's Law to the Equivalent circuit of armature as shown in Fig. 2b

$$i_a + L_a \frac{di_a}{dt} + e_b = V_a \tag{1}$$

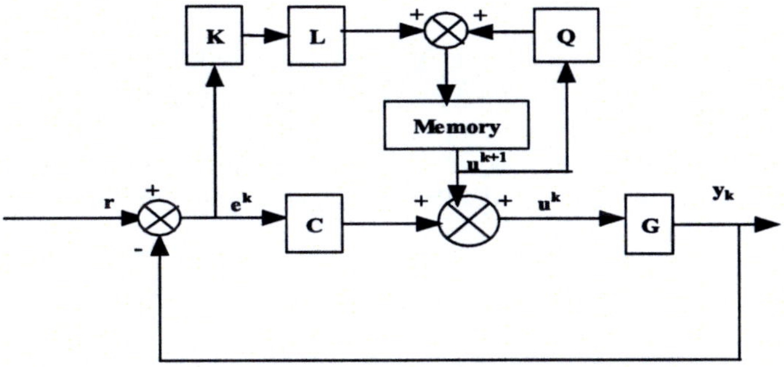

Fig. 2 Structure of ILC

Table 1 DC motor parameters specifications

J	0.03 kg m^2
b	0.019 Nms
K	$K = 0.1331$ Nm/A
R	$R = 6\ \Omega$
L	$L = 4.5$ mH

The DC motor torque is related to the flux and current in armature. As flux is invariable, the torque is related to i_a.

$$\text{(i.e.)}\quad T = K_t i_a \tag{2}$$

The equation governed the mech. system of DC motor is shown in Fig. 2a is given by

$$J\frac{d^2\theta}{dt^2} + B\frac{d\theta}{dt} = T \tag{3}$$

Since the back emf is related to shaft speed, it may be defined as

$$e_b = K_b\frac{d\theta}{dt} \tag{4}$$

From Eqs. (1) to (4), the DC motor transfer function is derived and as follows:

$$G_P(s) = \frac{\omega(s)}{V(s)} = \frac{K}{\left[(R + Ls)(Js + B) + K^2\right]} \tag{5}$$

The specifications of DC motor considered in this study and are given in Table 1. By taking Eq. 5 and values given in Table 2, the 2nd order transfer function for the DC motor is identified as

$$G(s) = \frac{1.01}{0.001025s^2 + 1.367s + 1}$$

Based on the DC motor model, Conventional P mode controller parameter ($K_c = 1.59$) is calculated by considering relay technique and Z-N [8] closed loop tuning rule (ZNTR).

3 Iterative Learning Controller (ILC)

Figure 2 shows the ILC structure. It involves the design of Learning filter (L) and Robustness filter (Q). The learning filter is vital role in error minimization and tracking ability. Also the stability boundary is greatly minimum in the simple ILC structure.

3.1 Design of the L

In Fig. 2, according to small-gain type of argument $r_k = 0$ and all initial conditions are zero.

$$e_k = \frac{-G}{1 + \mathrm{GC}} u_k$$

By applying learning up-date rule, $U_{k+1} = Q(u_k + L_{ek})$ and eliminate u.

$$e_{k+1} = Q\left(1 + \frac{-G}{1 + \mathrm{GC}} L\right) e_k$$

For convergence condition:

$$\text{if} \quad \left| Q\left(1 - \frac{G}{1 + \mathrm{GC}} L\right) \right| < 1$$

where $L = \frac{1 + \mathrm{GC}}{G}$ and process-sensitivity $T = \frac{G}{1 + \mathrm{GC}}$

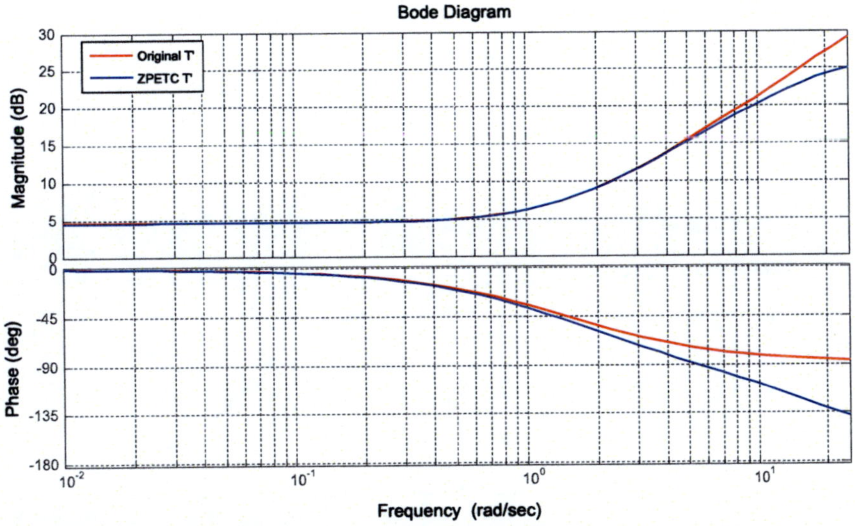

Fig. 3 Frequency plot of inverse complementary sensitivity

Figure 3 shows the bode plot of Inverse Complementary sensitivity (Original vs. ZPETC [9]).

3.2 *Q Filter*

A first order low pass filter is taken here. i.e. $Q(s) = \frac{\omega_c}{s+\omega_c}$. From the frequency plot (refer Fig. 4), the cut-off frequency (ω_c) is found to be 0.606.

4 Repetitive Controller (RC)

Repetitive controller is suitable for tracking a periodic input trajectories and eradicating a periodic disturbance signal. And it is based on the IMP proposed by Wonham and Francis.

The RC contains the factor $\frac{e^{-Ls}}{1-e^{-Ls}}$ which has poles at $jk\frac{2\pi}{L}$, $k = 0, \pm 1, \ldots, \pm\infty$, the controller can track any periodic signal and reject any disturbance of period L. Based on this concept, RCS is built with a model of $\frac{e^{-Ls}}{1-e^{-Ls}}$.

The Repetitive controller structure is shown in Fig. 5. The sensitivity function is of great importance in this study because this function gives the relation between the disturbances d and the tracking error *e*.

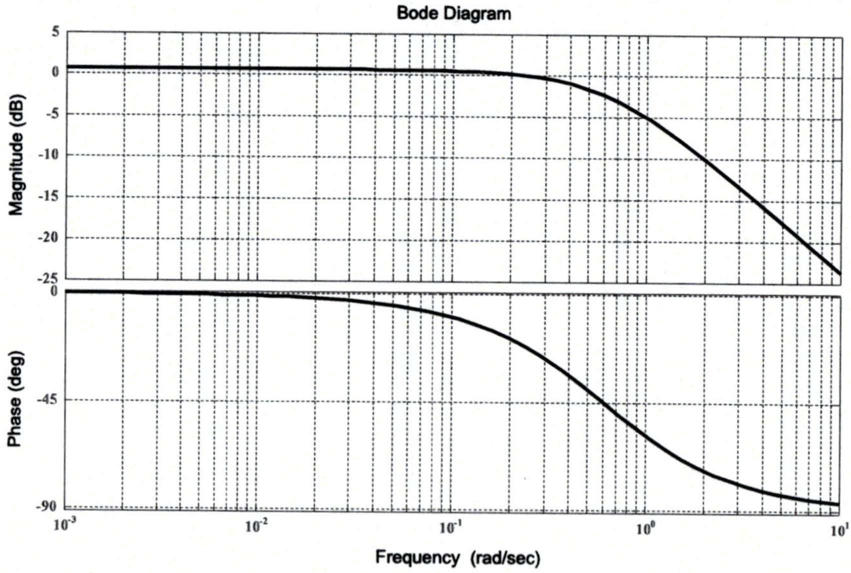

Fig. 4 Bode plot of the DC motor system

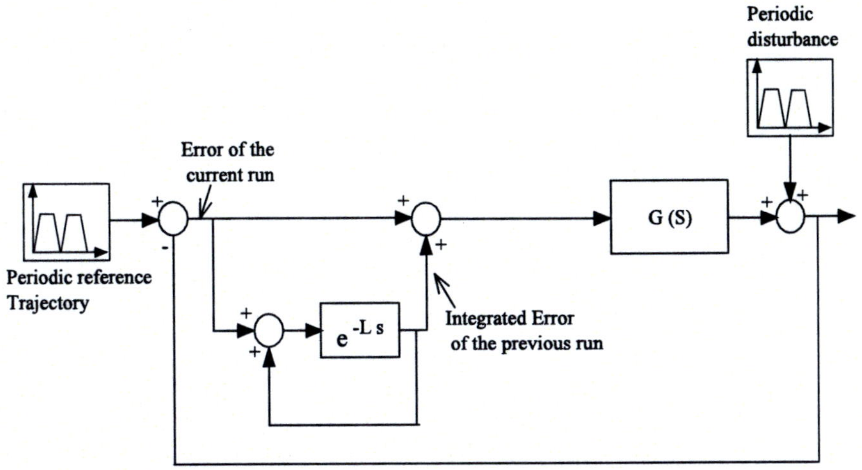

Fig. 5 Repetitive controller

$$S(j\omega) = \frac{e}{d} = \frac{-1}{1 + \text{PC}(1 + M)} = \frac{-1}{1 + \text{PC}\left(1 + \frac{\text{e}^{-j\omega Td}}{1-\text{e}^{-j\omega Td}}\right)} \tag{6}$$

If there is no memory loop's is zero and the standard sensitivity function appears again. Looking at the sensitivity, when M goes to infinity (at $\omega = k2\pi/\text{Td}$), S goes to zero. In this way, the disturbances at those frequencies will be suppressed (almost) completely. The frequencies of the main disturbances are at $\omega = k2\pi/\text{Td}$ and thus reduced dramatically.

4.1 Modified Repetitive Control (MRC)

Due to uncertainty in the process, noise will play massive role in the frequency response and cannot promise the stability condition. To meet out this, a $Q(s)$ is incorporated to the presented RC structure and to make sure the system permanence. This structure is named as Modified Repetitive Control (MRC) as shown in Fig. 6.

5 Simulation Results and Discussion

A periodical sinusoidal signal with known period (Tp = 62) and Peak to Peak (V_{pp} = 5) is produced by using a Periodic Signal Generator as shown in Fig. 7.

The design parameters of ILC are summarized by using ZPETC and frequency technique as

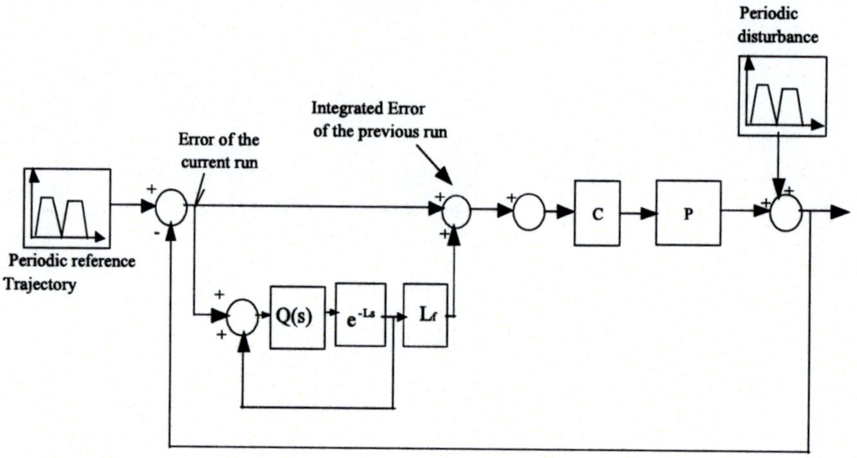

Fig. 6 Modified repetitive controller

Fig. 7 Repetitive signal generator

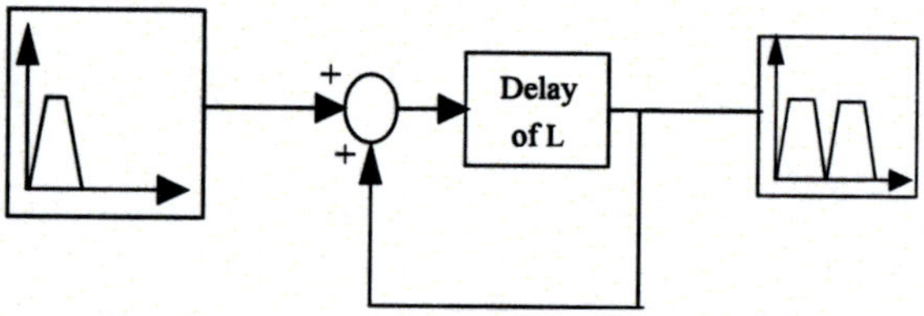

$$L(s) = \frac{-30.67s + 45.54}{s + 20} \quad Q(s) = \frac{0.606}{s + 0.606}.$$

The DC motor operating at 40% speed with ILC is interrupted by the generated sine wave signal. The tracking response is recorded in Fig. 8. Similarly, a sine wave tracking disturbance is given to DC motor which runs at the same operating point with MRC and conventional PI control respectively. The simulation results are plotted in Figs. 9 and 10. From the tracking responses as shown in Figs. 8, 9 and 10 the tracking errors are computed with respect to time and given in Fig. 11. It is noted that MRC fetches minimum tracking error than that of the other controllers.

To investigate the robustness of intelligent controller, an experiments with different characteristics of input signal Tp = 62 & V_{pp} = 6 and Tp = 45 & V_{pp} = 5 are conducted. The responses are recorded from Figs. 12, 13, 14, 15, 16 and 17 and the tracking errors are plotted from Figs. 18 and 19. The Tp and V_{pp} distinctions in control loop have obviously confirmed the merits of the intelligent controller that is modified repetitive Control Structure.

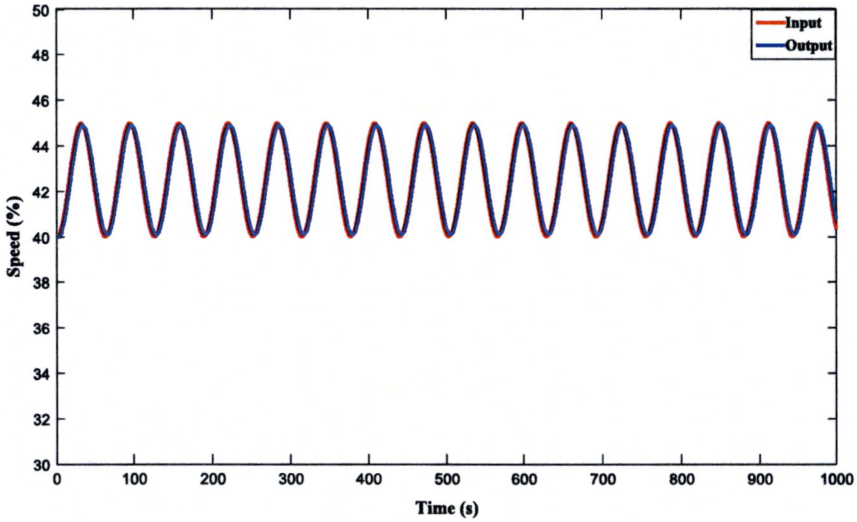

Fig. 8 Tracking response DC motor with ILC [Tp = 62, $V_{pp} = 5$]

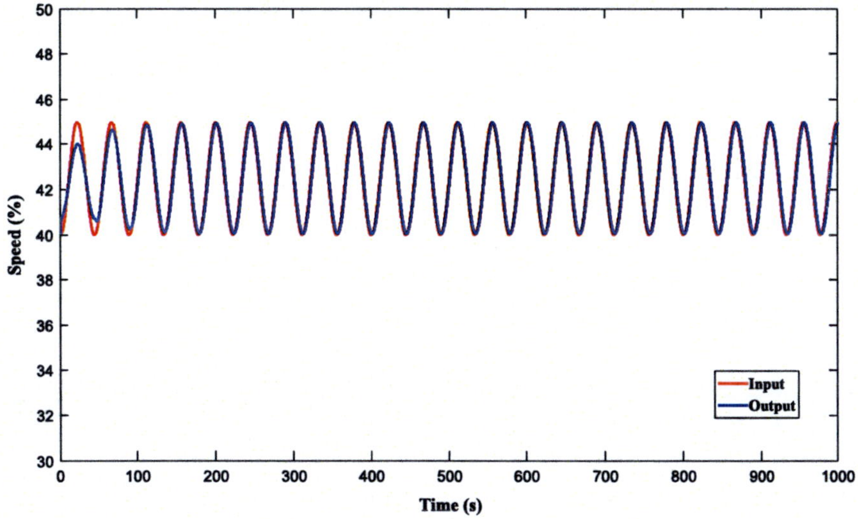

Fig. 9 Tracking response DC motor with MRC [Tp = 62, $V_{pp} = 5$]

6 Conclusion

This paper has presented the novel design and development of Iterative Learning Controller (ILC), and Modified Repetitive Controller (MRC) for achieving the

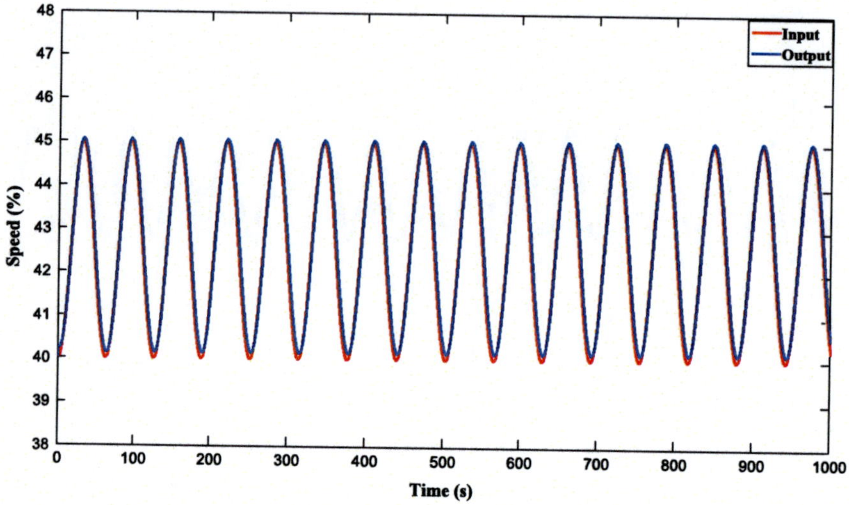

Fig. 10 Tracking response DC motor with Conventional PI [Tp = 62, $V_{pp} = 5$]

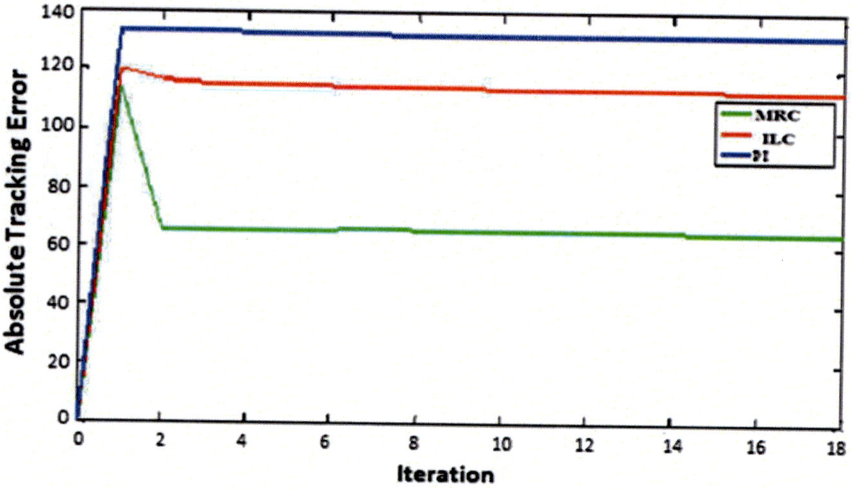

Fig. 11 Tracking error responses of all three control strategies [Tp = 62, $V_{pp} = 5$]‘

tracking improvement in the speed control of DC motor system. These controllers are always exceptional to improve the system performance significantly, particularly in the case of applications that involve repeated motion trajectories. This controller can guarantee the speed tracking with a resistance to periodic disturbance at any desired field point of operation. Comparative studies have shown the best control performance than that of conventional controller.

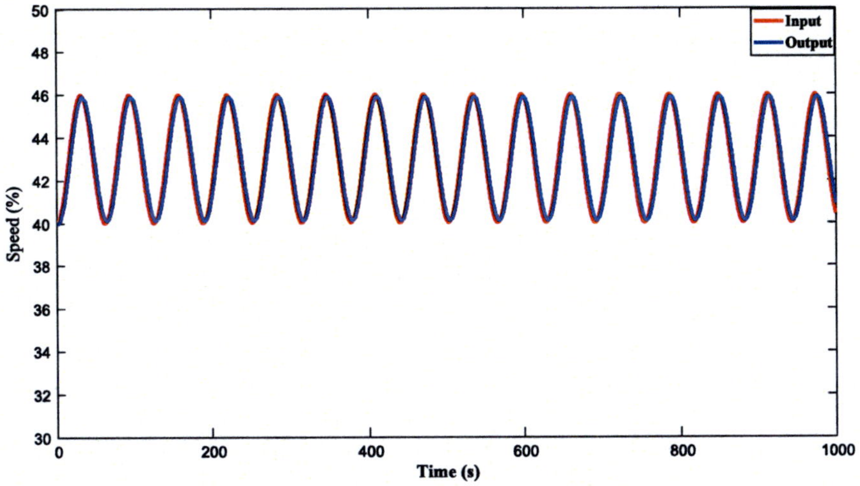

Fig. 12 Tracking response DC motor with ILC [Tp = 62, $V_{pp} = 6$]

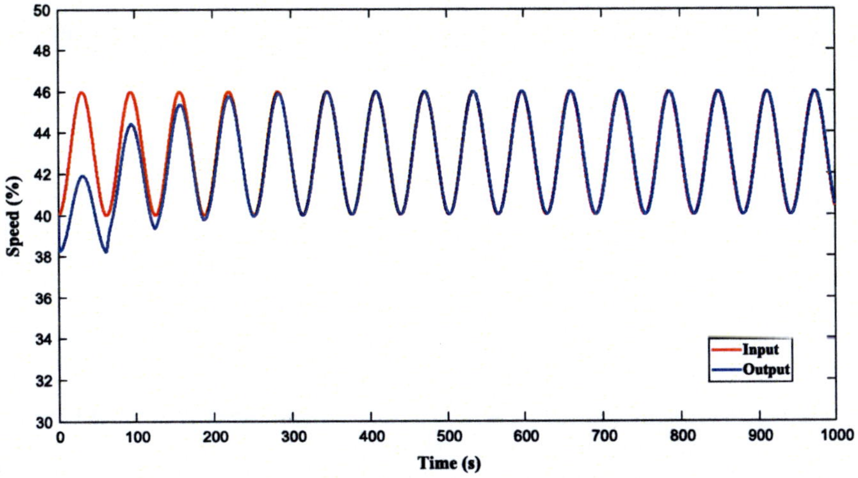

Fig. 13 Tracking response DC motor with MRC [Tp = 62, $V_{pp} = 6$]

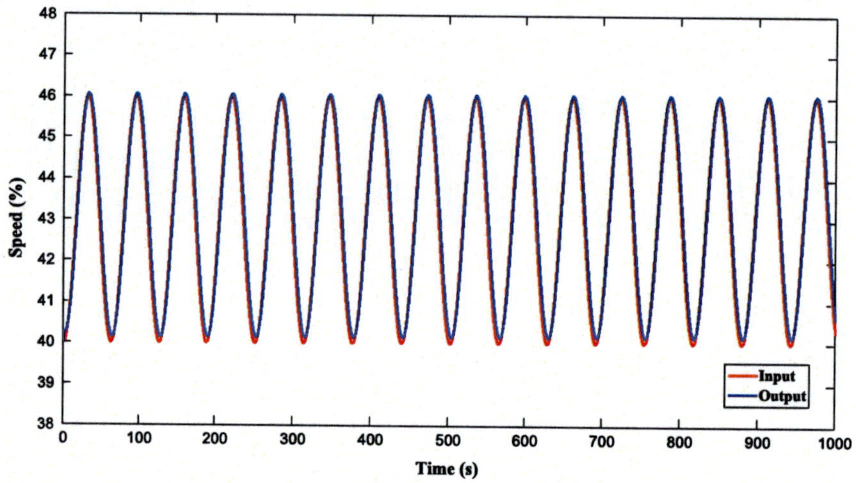

Fig. 14 Tracking response DC motor with conventional PI [Tp = 62, $V_{pp} = 6$]

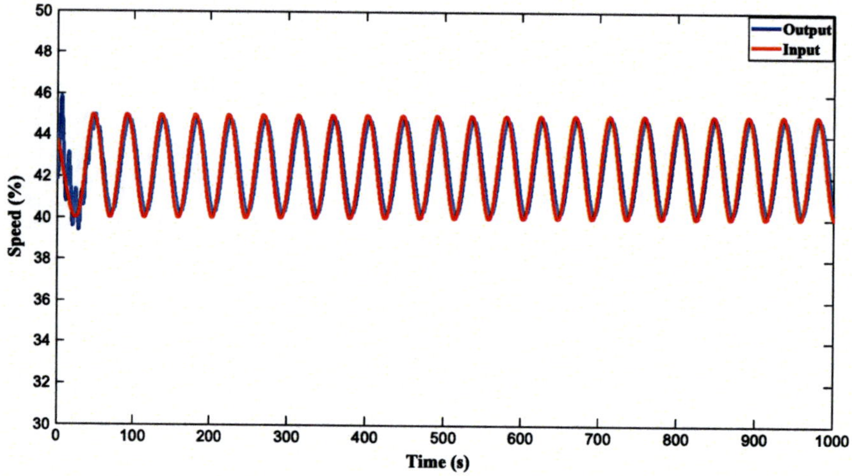

Fig. 15 Tracking response DC motor with ILC [Tp = 45, $V_{pp} = 5$]

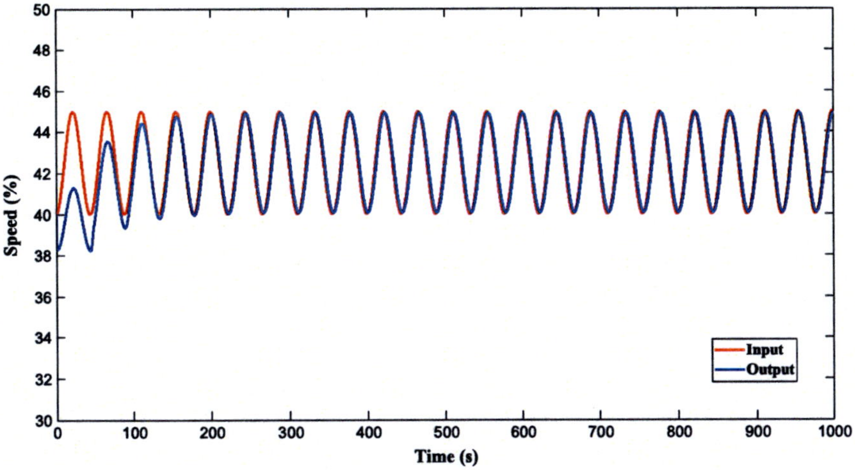

Fig. 16 Tracking response DC motor with MRC [Tp = 45, $V_{pp} = 5$]

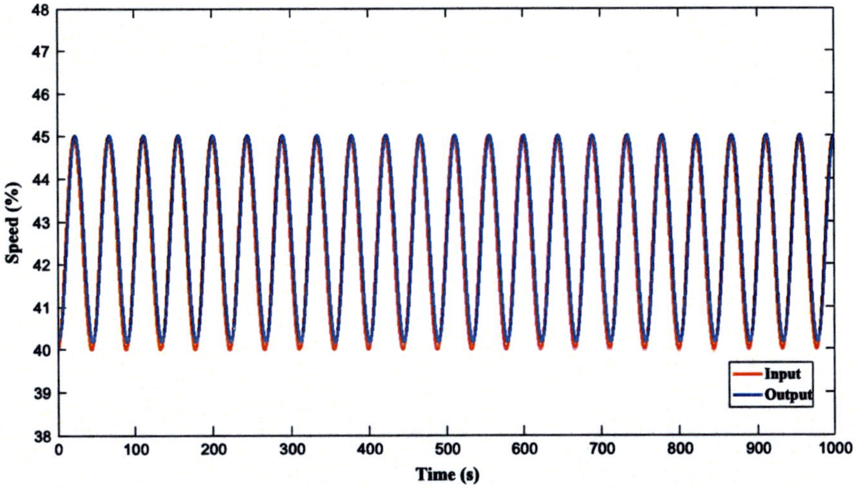

Fig. 17 Tracking response DC motor with conventional PI [Tp = 45, $V_{pp} = 5$]

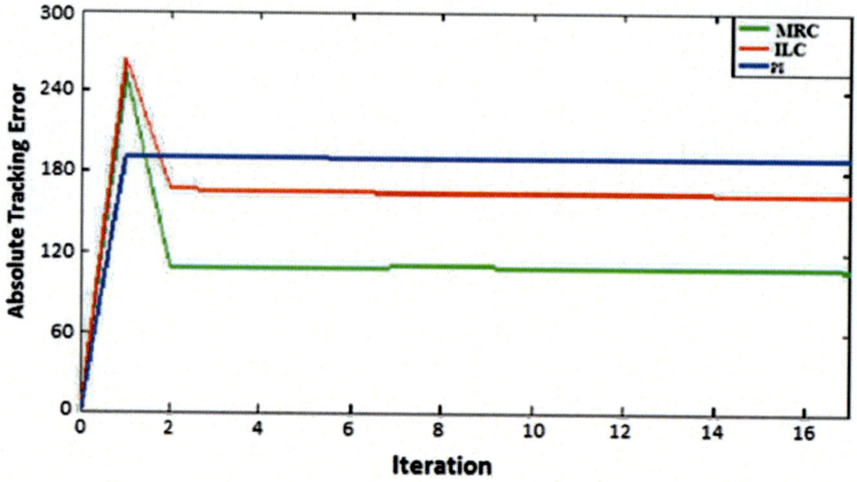

Fig. 18 Tracking error responses of all three control strategies [Tp = 62, $V_{pp} = 6$]

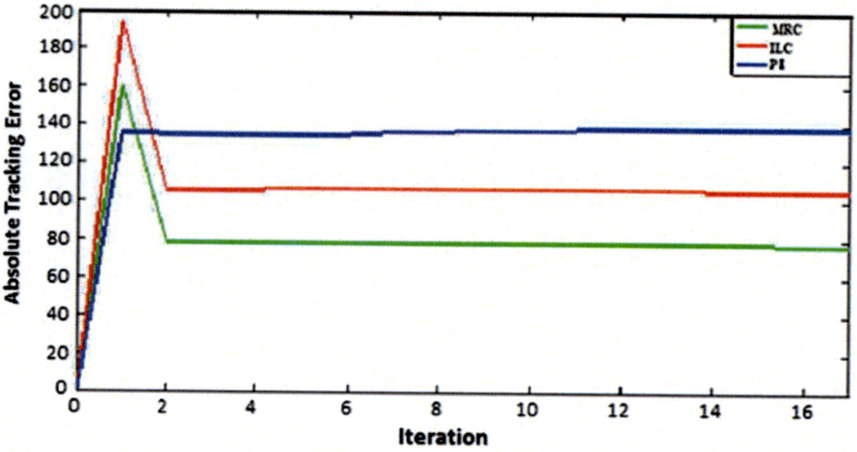

Fig. 19 Tracking error responses of all three control strategies [Tp = 45, $V_{pp} = 5$]

References

1. Arimoto,S., Kawamura,S., Miyazaki, F.: Learning control theory for dynamical systems. In: Proceedings of the 24th IEEE Conference on Decision and Control (1985)
2. Kim, D.I., Kim, S.: An iterative learning control method with application for CNC machine tools. IEEE Trans. Ind. Appl. **32**(1), 66–72 (1996)
3. De Roover, D., Bosgra, O.H.: Synthesis of robust multivariable iterative learning controllers with application to a wafer stage motion system. Int. J. Contr. **73**(10), 968–979 (2000)
4. Arimoto, S.: Learning control theory for robotic motion. Int. J Adaptive Contr. Sig. Process **4**, 544–564 (1990)

5. Arimoto, S., Naniwa, T., Suzuki, H.: Robustness of P-type learning control with a forgetting factor for robotic motions. In: Proceedings of the 29th Conference Decision Control, Honolulu, Hawaii, pp. 2640–2645 (1990)
6. Francis, B.A., Wonham, W.M.: The internal model principle of control theory. Automatica **12**(5), 457–465 (1976)
7. Hara, S., Yamamoto, Y., Omata, T., Nakano, M.: Repetitive control system: a new type servo system for periodic exogenous signals. IEEE Trans. Autom. Control **33**(7), 659–668 (1988)
8. Ziegler, J.B., Nichols, N.B.: Optimum settings for automatic controllers. ASME Trans. **64**, 759–768 (1942)
9. Tomizuka, M.: Zero-phase error tracking algorithm for digital control. J. Dynam. Syst. Measure. Control Trans. ASME **109**(1), 65–68 (1987)

Fabrication of Variable Speed Lamina Cutting Machine

Gajavalli Venkata Pavan Gopi Nikhil and Nadendla Srinivasababu

Abstract Cutting plays a key role in manufacturing. The present work focus on development of lamina cutting machine has variable speed which is achieved by connecting the electronic speed control unit, where it consists of different electronic parts. Basic jigsaw J-819 machine was purchased from local market, and DC motor is used for operating the machine by interfacing both the pulleys of it and machine with help of V belt. Speed of the DC motor is regulated by pulse width modulation (PWM) method which is generated by the micro controller ATmega328p exist in Arduino nano is coded as per the requirement. IR sensor is used to the detect the speed of machine pulley and gives it as a feedback to the microcontroller in arduino and it regulate the pulses and sent to motor controller which helps the motor to rotate and maintain stability at the given speed through keyboard. Further, the machine is operated at different input speeds and all input and output speeds of machine pulley and motor pulley were measured and reported. Further, mechanical test specimens are machined from the developed composite lamina as per ASTM standards at different speeds on fabricated variable speed cutting machine to evaluate the ability of the developed cutting machine.

Keywords Variable speed DC motor · Lamina cutting machine · Arduino-based control unit · PWM method · Sensing · Monitoring · Controlling and machaning at different speeds

1 Introduction

A machine which is used for removal of unwanted material from workpiece with the help of cutting tool is known as cutting machine. And over a century ago there was no cutting machine available for cutting any kind of material and the first machine was

G. V. P. G. Nikhil (✉) · N. Srinivasababu
Department of Mechanical Engineering, K L E F Deemed to be University, Green Fields, Vaddeswaram, India
e-mail: nikhilnani3112@gmail.com

© The Editor(s) (if applicable) and The Author(s), under exclusive license to Springer Nature Singapore Pte Ltd. 2021

A. P. Pandian et al. (eds.), *Proceedings of International Conference on Intelligent Computing, Information and Control Systems*, Advances in Intelligent Systems and Computing 1272, https://doi.org/10.1007/978-981-15-8443-5_81

developed by George Eastman in 1888 further the development in cutting machines has been goes on till today. These machines are used for performing the cutting operations with high cutting efficiency and cutting quality and reducing the labor intensity of operators. Operational speed of the machine, geometry and dimensions of the cutting tool, feed and depth of cut during cutting plays a key role in obtaining the surface finish and geometrical accuracy of the final product. So in this work an attempt was made to fabricate / develop a new platform for machining of plastic / composite lamina, where the composites are currently engaged in a broad range of industries for various applications. So, a modification for an existing jigsaw machine is done with electronic-based control system which varies the speed of the machine. In order to understand the electronic speed control, various parts of the setup, i.e., key pad, PCB, Arduino, LCD, battery, motor controller, SMPS, DC motor, were studied. Then, the machine was fabricated and tested by cutting mechanical test specimens from developed composite lamina as per ASTM standards at different speeds. Further input and output speeds were measured during experimentation.

Properties of the composites such as mechanical properties (tensile strength, impact strength, flexural strength, etc.), dimensional accuracy and surface finish are the major parameters needed to be tested before using the these composites in the real time application to get rid of from the unexpected damages. As operational speed of the cutting machine is one of the major parameters responsible for the dimensional accuracy and surface finish, this motivates me to fabricate a variable speed composite lamina cutting machine and examine the responsibility of the parameter speed in dimensional accuracy and surface finish by machining the mechanical test specimens from the composite lamina as per ASTM standards. This fabricated variable speed cutting machine helps the small-scale industries and laboratories of scientific and educational institutions due to tremendous advantages such as it maintains good efficiency in cutting, fast and suitable to cut any shape of the specimen/product, low investment and easy to operate, easy to maintain accuracy even for small components, suitable for small scale cutting, simple in construction and easy to maintain and repair, the production speed is good and rough work can be easily done by hands.

2 Survey of Literature

2.1 Arduino

Hardware and software prototyping was complex before the development of hardware platform, i.e., Arduino. Arduino was introduced in 2005 at IVREA Institute, Italy. At present, Arduino boards are available with large variety of boards come various processors, sizes and features of connectivity. Hence, Candelas et al. [1] developed different experiments like temperature control for 3D printer hot end, Cartesian robot automation, programming of humanoid robot and follower robot.

By considering the views of students, they concluded that the experiments were running properly.

2.2 Torque and Speed Control of DC Motor Using PWM

Using P89V51RD2 microcontroller, DC motor speed was controlled with the support of PWM. The motor is fed by IC L239D and is driven by PWM signal generated from microcontroller [2]. The speed control of DC motor was done with PWM method using infrared remote control based on ATmega16 microcontroller [3]. A new approach of DC motor PWM speed control was done by using AT89S51single chip microcomputer [4].

In order to achieve forward, reverse, speed up, slow down, stop, etc., operations were obtained by DC motor and were governed by PWM. Sun et al. have used Proteus software for design/simulation of DC motor [5].

Robotics and other industries widely use the gripping mechanisms. Similar concept is used in conventional gripper mechanism which consists of DC motor and actuator. When the gripper reaches the end of travel/last position, the problems of over current in a motor or actuator occur. So, algorithms to control the torque of DC motor via motor driver were developed. This is achieved by the variations in pulse width modulation (PWM).

Therefore, How et al. [6] have developed and tested different algorithms like on–off controller algorithm (bang–bang controller), proportional algorithm (P controller), proportional and derivative algorithm (PD controller), proportional and integral algorithm (PI controller) and proportional, integral and derivative algorithm (PID controller) and based on the responses of the algorithms from the graphical view. They have concluded that PID controller is the optimal method to control torque of the DC motor.

Applications of the DC motors are huge in various industries such as electrical vehicles, robotic manipulators, home appliances, washers, driers and compressors. An important application of DC drive is in wind energy. Pulse width modulation (PWM) is a very accurate and reliable method to control the speed of DC motor. With 18F25K20 digital signal processor the speed of DC motor was controlled. Circuits used in the experiment were designed and simulated using Proteus program. From the experiments of Dubey et al. [7], it was concluded that the applicability of different motor sizes and the capability of controlling motor speed with precision.

As the applications of the speed-controlled DC motor are in wide range in industries. AT89S52 microcontroller was used by Priyanka et al. [8] to generate PWM. The motor was driven by L239D IC and is made up of 2 H-bridges in order to sense the speed of Dc motor 555 IC was used along with opto coupler.

A mathematical model was created for DC motor, and with the help of frequency shaping method, PWM speed control of it was done [9].

Tripathi et al. [10] have made a review on different controllers like (1) proportional and integral controller, (2) PID controller, (3) Ziegler–Nichols, (4) neural network controller, (5) fuzzy logic controller, (6) linear quadratic regulator (LQR) technique which are widely used in different areas like process control, manufacturing, automation, aerospace, etc., and concluded that PID controller for the speed control of DC motor is the best method to optimize the best result.

2.3 Types of DC Motor, SMPS, Keypad and LCD Display

Shrivastava et al. [11] have reviewed various types of DC motors like permanent magnet, DC series, DC shunt, compound motor and the necessity of starter for regulating the speed.

A universal switch mode power supply was developed using fly-back topology and operated by either AC or DC input power [12].

Using two full bridge diode converter circuits and one full bridge MOSFET converter circuit, a high efficiency AC–DC switch mode power supply was developed by Kwon et al. [13].

Security through electronic based is of major interest to the public to satisfy various day-to-day needs. So, a digital code lock was designed and implemented by Oommen et al. [14].

The hardware components like a keypad and the digital display were used in the present work for input speed entering and output speed display, respectively.

2.4 Jigsaw Machine

In wood working industries jigsaw machines are mostly used for cutting. But their use is limited due to imprecision cutting resulted from blade deflection. So, Manap [15] developed a double-guided roller for minimizing the problem.

In order to cut accurate, precise, intricate contours, a computer assisted jigsaw machine was developed [16].

3 Experimental Work

The present section describes about various electronic parts used in the lamina cutting machine and their specifications and functions. Further the steps involved in fabricating the machine are also described along with planned tests.

Table 1 Specifications of jigsaw machine

S. no.	Parameter	Description
1	Throat depth	24 inches
2	Cutting capacity	1.25 inch
3	Table size	14 inches × 12 inches
4	Engine speed	525/625 rpm
5	Power required	1/1440 hp/rpm

3.1 Jigsaw Machine

Wood working Indian made jigsaw J-819 model is purchased from Sri Aditya Engineering Company, Vijayawada, Andhra Pradesh. The specifications of the machine are given in Table 1.

3.2 Electronic Parts for Jigsaw Machine and Their Modification

Different parts considered for modifying the machine are given in Table 2 along with their respective specifications and functions.

Figure 1 shows the schemetic representation of the arrangement of different electronic parts in the machine, and it also represents the line diagram / work flow diagram of the fabricated variable speed lamina cutting machine.

3.3 Working of Lamina Cutting Machine

Initially the DC motor is modified into a variable speed motor with the help of the motor driver unit / motor speed control unit and it is assembled on the motor rail of machine with the help of bolt and nut assembly. Then, the motor rail is adjusted on the tie rod of the machine in such a way that both the pulleys of the motor and machine are connected with the help of V belt to transmit the rotary motion of motor to machine.

When the power supply is given to the machine initially the micro controller receives the power, acts as a brain of the machine and instructs all the electronic components according to the instruction set / code dumped into the microcontroller; further, the micro controller instructs the motor to start rotating at the desired speed which is calculated internally by the micro controller through the given code; this calculation is performed with reference to the speed of machine pulley given through keypad, it transfers the motion to the machine pulley, the machine pulley starts rotating, and then engine body converts the rotary motion of the machine pulley to

Table 2 Electronic parts of lamina cutting machine

S. no.	Name of the electronic component/part, model, manufacturer	Specifications	Functions
1	DC motor, MY1016, Bombay Electronics, Mumbai, Maharashtra	Voltage: 24 V DC Rated speed: 2650 rpm Rated current: 13.7 Amp Output: 250 W	Converts DC electrical energy to mechanical energy
2	Motor controller, R5B4AB, Bhosari MIDC, Pune, Maharashtra	Voltage: 24 V Rated power: 500 W	Controls the speed of DC motor up to 250 W, 24 V with help of microcontroller ATmega328p on platform of Arduino Nano
3	SMPS, Invento Sales Corporation, Delhi	Input: AC 220 V, 50/60 hz Output: DC24 V, 20 Amp	Converts AC power supply to DC power supply
4	Arduino Nano, Sri Sai Srinivasa Electronics, Guntur, Andhra Pradesh	Operating voltage: 5 V Input voltage (recommended): 7–12 V Input voltage (limits): 6–20 V DC: 40 mA	Interlinks with every electronic module in the circuit, communicates the given program by the instructor to each module totally and acts like a heart of the circuit
5	Digital display, Sri Sai Srinivasa Electronics, Guntur, Andhra Pradesh	Operating voltage: 5 V	To display the speed of the machine pulley
6	IR Sensor, Sri Sai Srinivasa Electronics, Guntur, Andhra Pradesh	Operating voltage: 4.5–5.5 V Average current consumption: 33 mA Output type: analog voltage Update period: 38 ± 10 ms	To detect the no. of revolutions of the shaft of the machine
7	Keypad, Sri Sai Srinivasa electronics, Guntur, A.P	Operating voltage: 5 V	To give the speed input

the reciprocating motion of the cutting tool with the help of the piston and connecting rod assembly in the engine body, where the cutting tool is assembled to the ram set in between the upper tool holder and lower tool holder. And this reciprocating motion of the machine helps to cut the lamina.

As a pulse width modulation method is a best method for generating an analog signal from the digital source, here the PWM method is used to control the speed of DC motor while the speed of the machine pulley is given in the digital form through keypad and the microcontroller ATmega328p present in arduino Nano will generate the analog signal, i.e., no. of pulses according to the instruction set/code given to it; i.e., it calculates the required no. of pulses to be generated according to the given input speed of machine pulley; further, the desired/calculated no. of pulses is instructed to the motor controller where it is used for functioning of motor at the

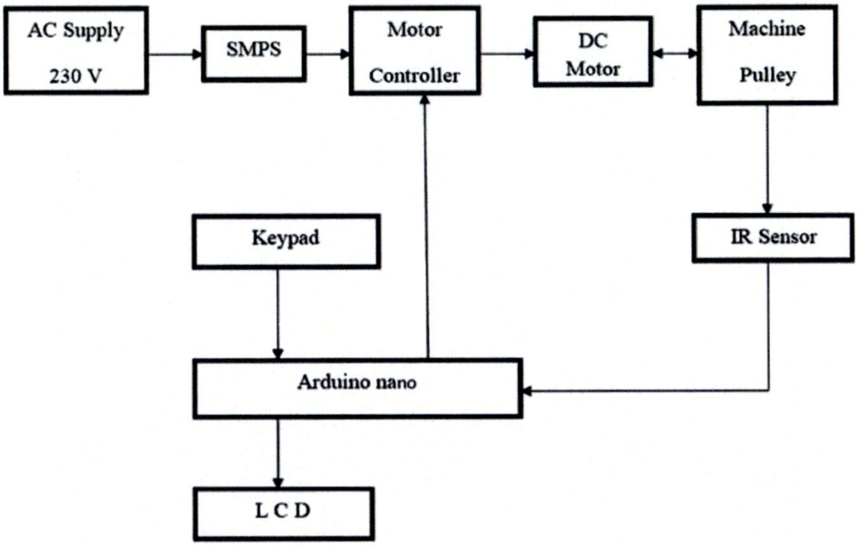

Fig. 1 Block diagram of different electronic parts in cutting machine

given speed and maintains the stability in the machine pulley speed, frequency of the PWM is adjusted according to the given input, i.e., machine pulley speed, to detect the speed of the machine pulley the IR sensor is mounted near to the machine pulley as its size is larger, and the time taken for one complete rotation is high so the sensor measures the speed accurately when compared to the IR sensor mounting at motor pulley. As the motor and the machine pulleys are interfaced with V belt, the speed of the motor and machine is dependent on each other.

4 Experimental Work Results and Discussion

Using various electronic components, a cutting machine was manufactured and is shown in Fig. 2. And an assembly of various electronic components/electronic speed control unit of DC motor is shown in Fig. 3. When the cutting speed is entered to Arduino Nano through keypad, then output speed of the motor is measured with the help of digital tachometer. And a theoretical speed of the motor (N_2) is calculated with reference to the output speed of machine pulley (N_1) by the well-known formula $N_1/N_2 = D_2/D_1$ whereas both the diameters of machine (D_2) and motor (D_1) pulleys are known. Correlation between output speeds of the motor with the input displayed speed is given in Table 3. Further to evaluate the accuracy of the developed speed control unit, the graphs are plotted for displayed output speed of machine pulley for their respective given input speed through keypad as shown in Fig. 4. And a variation between the theoretical and actual information is also shown in the graph plotted for

Fig. 2 Electronic-based lamina cutting machine

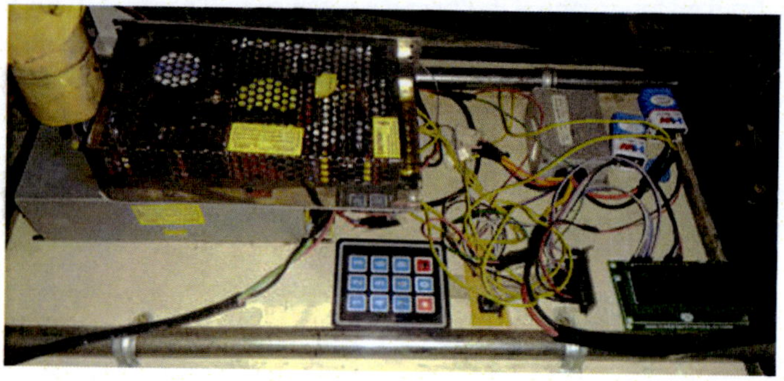

Fig. 3 Electronic part assembly (keypad, PCB, Arduino, LCD, battery, motor controller, SMPS)

the theoretical and actual speed of the motor pulley with reference to the displayed speed of machine pulley as shown in Fig. 5.

Further the machin ability is tested by cutting polymer composite lamina at different speeds with reference to ASTM standards of ASTM D638 and ASTM D790 which are shown in Fig. 6. The machined specimens are measured through an electronic vernier caliper for the dimensions, the recorded dimensions are compared to the standard dimensions as per ASTM, and the specimens machined are also checked optically for the surface finish by surface roughness testers, i.e., 3D optical microscopes.

Table 3 Speed measurement, calculation in electronic lamina cutting machine

S. no.	Input speed of machine pulley through keypad (rpm)	Output speed of machine pulley through digital display (rpm), N_1	Theoretical speed of motor pulley with respect to output speed of machine pulley (rpm), N_2	Actual speed of motor pulley measured through digital tachometer (rpm)
1	20	36	105	115
2	45	60	175	190
3	59	72	210	230
4	81	96	280	300
5	113	127	371	386
6	124	132	385	395
7	148	156	362	377
8	162	158	461	477
9	177	180	526	537
10	198	192	561	577
11	213	216	631	646
12	234	240	701	721
13	261	264	770	785
14	278	276	806	826
15	310	312	911	926
16	334	336	981	991
17	351	348	1016	1036
18	367	360	1051	1071
19	376	384	1121	1130
20	401	408	1191	1210
21	420	420	1226	1240
22	446	446	1302	1317
23	473	475	1387	1407
24	481	486	1419	1439
25	490	494	1442	1462

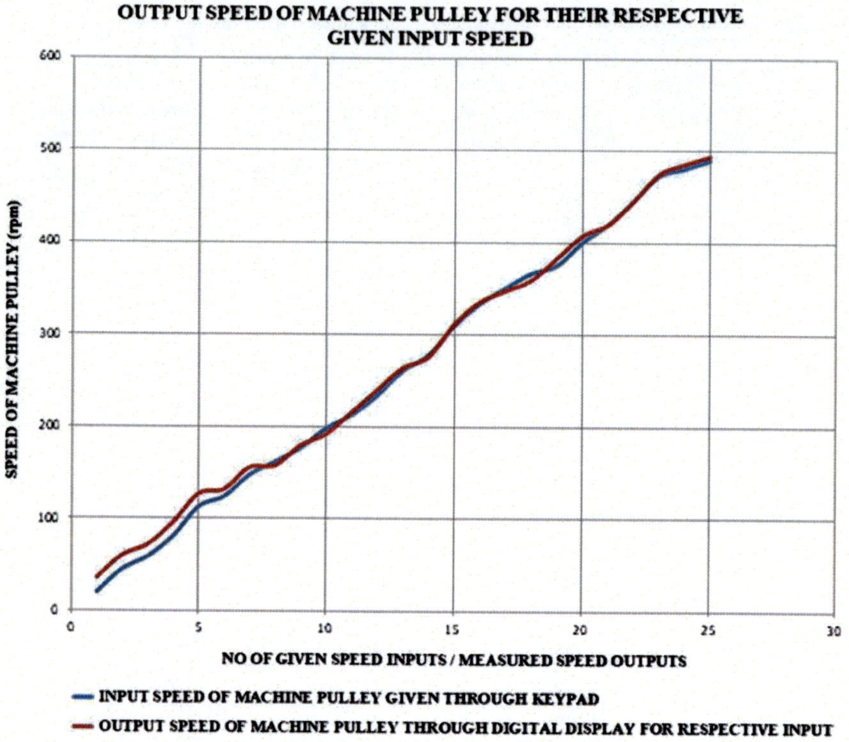

Fig. 4 Comparison of output speed displayed for respective input speed given by keypad of machine pulley

5 Conclusions and Scope for Further Work

A variable speed lamina cutting machine is manufactured successfully. Further the machine is operated at different input speeds to evaluate the accuracy of the developed speed control unit, all input and output speeds of machine pulley and actual and theoretical speed of motor pulley with reference to the output speed of machine pulley were measured and reported, and the graph plotted for the input and output speeds of the machine pulley shows the accuracy of the developed speed control unit; this evaluation from graph demonstrates it is well capable, feasible and reliable as there is a negligible variation between both the input and output speeds of machine pulley. And the analysis graph plotted for theoretical and actual speed of motor pulley with reference to the output speed of machine pulley shows that there is a negligible variation between both speeds of motor pulley, so it is proven that the theoretical information derived from the proper analysis is approximately equivalent to the actual information. The machin ability is tested by cutting polymer lamina with reference to ASTM standards of ASTM D638 and ASTM D790. But further analysis, modifications are needed for the current developed cutting machine in order to obtain the

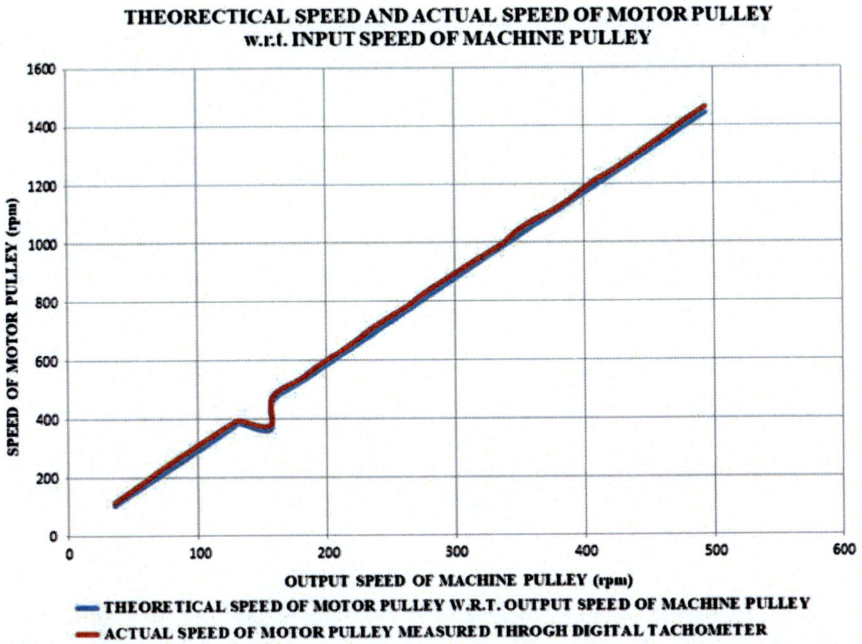

Fig. 5 Variation between the calculated theoretical speed and actual speed of motor pulley w.r.t. output speed of machine pulley

specimens with more accurate dimensions and surface finish by avoiding the inconveniences exists in the current developed cutting machine such as the deflection of lamina results in inaccurate specimens and products during machining, requirement of a special assistance to operate them as it may harm people on the site and also a necessity of a skilled worker to perform an accurate machining of the lamina.

Fig. 6 Composite mechanical test specimens machined as per ASTM standards on fabricated variable speed lamina cutting machine at different speeds

Acknowledgements Words are often less to reveal one's deep regards. An understanding of the work life can never be outcome of a single person without one's support. Hence, the first author takes this opportunity to express our profound sense of gratitude and respect to all those who helped us throughout the duration of this work along with the second author given guidance in Design/Fabrication Project in B.Tech program. First author would like to extend his heartfelt thanks to beloved parents whose blessings and encouragement were always there as a source of strength and inspection.

References

1. Candeals, F.A., Garcia, G.J., Puente, S., Pomares, J., Jara, C.A., Perez, J., Mira, D., Torres, F.: Experiences on using arduino for laboratory experiments of automatic control and robotics. In: IFAC Papers Online, pp. 105–110 (2015)

2. Ravi Kumar, K.S., Jaideep, Rohit, Vikas: Microprocessor based closed loop speed control of DC motor using PWM. In: International Conference on Control, Instrumentation, Communication and Computational Technologies, pp. 255–257 (2015)
3. Raka, A., Huda, S., Arta Wijaya, I.W.: Speed control for DC motor with pulse width modulation (PWM) method using infrared remote control based on ATmega 16 microcontroller. In: International Conference on Smart Green Technology in Electrical and Information Systems, pp. 108–112 (2014)
4. Liu, Z.,Jiang, L.: PWM speed control system of DC motor based on AT89S51. In: International Conference on Electrical & Mechanical Engineering and Information Technology, pp. 1301–1303 (2011)
5. Sun, J., Sun, Q.: Design and Simulation of PWM DC motor speed regulator based on Proteus. In: International Conference on Fluid Power and Mechatronics, pp. 1210–1213 (2015)
6. Dickson Neoh Tze How, Mohd Zafri Baharuddin, Syed Sulaiman Kaja Mohideen, Khairul Salleh Mohammed Sahari and Adzly Anuar.: Modular motor driver with torque control for gripping mechanism. In: International Symposium on Robotics and Intelligent Sensors, pp. 1476–1482. 2012.
7. Dubey, S., Naidu, H., Ghutke, P.: Speed control of DC motor by using PWM. Int. J. Res. Sci. Eng. 115–120 (2017)
8. Priyanka, K., Mariyammal, A.: DC motopr speed control using PWM. Int. J. Innov. Sci. Res. Technol. 584–587 (2018)
9. Mikova, L., Virgala, I., Kelemen, M.: Speed cotrol of DC motor. Am. J. Mech. Eng. 380–384 (2016)
10. Tripathi, N., Singh, R., Yadav, R.: Analysis of speed control of DC motor—a review study. Int. Res. J. Eng. Technol. 1616-1621 (2015)
11. Shrivastava, K., Pawar, M.D.: A review on types of DC motors and the necessity of starter for its speed regulation. Int. J. Adv. Res. Comput. Commun. Eng. 61–63 (2016)
12. Jagdish, Jayapal, R.: Design and development of universal switch mode power supply. Int. J. Adv. Res. Electr. Electron. Instrum. Eng. 6296–6302 (2017)
13. Kwon, S.-H., Yoo, D.-H., Jeong, G.-Y.: High-efficiency AC–DC switch-mode power supply using full bridge converter circuits. Int. J. Control Autom. 189–200 (2014)
14. Oommen, A.P., Rahul, A.P., Pranav, V., Ponni, S., Nadeshan, R.: Design and implementation of a digital code lock. Int. J. Adv. Res. Electr. Electron. Instrument. Eng. 7604–7607 (2014)
15. Manap, H.: Design and analysis of a double guided roller for a jigsaw. ARPN J. Eng. Appl. Sci. 7221–7224 (2016)
16. Dahake, P.H., Patil, V.V.: Development of computer assisted jig saw machine. Int. J. Recent Adv. Eng. Technol. 25–27 (2013)

Electrocardiogram QRS Complex Detection Based on Quantization-Level Population Analysis

Hamdi M. Mohamed and Akula Rajani

Abstract Electrocardiography is the principle of recording the heart muscle's action potentials. It is a primary method in detecting the electrical abnormality of the heart. The graph generated by an electrocardiograph shows the temporal characteristics of the voltage across the heart and different body locations. The graph is composed of many parts designated as waves (P, Q, R and S waves), segments (QT segment) and intervals (PR and RR intervals). Detecting these components is important to extract the features of the electrocardiogram and consequently determine whether the electrocardiogram is normal or abnormal. The first component to be detected in the electrocardiogram is the R wave because of its distinct morphology of sharp deflections. This paper proposes a method to localize the R waves by setting a static threshold value that its magnitude depends on the electrocardiogram's statistical properties. The threshold value is unique to every rhythm, and it is determined by calculating the population of each quantization level and subsequently finding the most repeated population figure and the quantization level associated with it. This level stands as the threshold to detect the R waves.

Keywords ECG · Frequency · Histogram · Magnitude · Mode · Occurrence · Quantization · Level · QRS · Threshold · Wave

1 Introduction

Electrocardiography exists since a century; it is a primary test to assess the heart. The electrocardiogram (ECG) is the signal acquired by recording the voltages across the heart which are sensed noninvasively via the chest surface, and statistics reveal

H. M. Mohamed (✉) · A. Rajani
Electronics and Communications Engineering, Jawaharlal Nehru Technological University, Hyderabad, India
e-mail: hamdi.med1991@gmail.com

A. Rajani
e-mail: rajani.akula@jntuh.ac.in

© The Editor(s) (if applicable) and The Author(s), under exclusive license to Springer Nature Singapore Pte Ltd. 2021
A. P. Pandian et al. (eds.), *Proceedings of International Conference on Intelligent Computing, Information and Control Systems*, Advances in Intelligent Systems and Computing 1272, https://doi.org/10.1007/978-981-15-8443-5_82

that around 40 million electrocardiography tests were performed in the USA in 2015 [1]. Biomedical signal processing specialists process such cardiac signals to quantify various components that constitute the signals. Many methods aiming to detect the components of the ECG were developed since the incorporation of computer sciences to medicine. The traditional methods to detect the R waves are based on setting a threshold value and classifying peaks higher than this magnitude only as R waves. Such methods fail in many cases as they rely on a static threshold that is determined a priori at the time the ECG varies from subject to subject in terms of peak amplitudes and wave morphology. Later, dynamic (time-dependent) thresholding methods were introduced using adaptive processing concepts but the limitations of such methods arise from the increased power and time required to analyze the signal due to the iterative nature of adaptive algorithms.

A method is proposed that deploys an improved static thresholding technique in which the threshold is predetermined based on the statistical properties of the respective ECG; in other words, the threshold is a function of the histogram of the ECG being processed. This approach—more or less—combines adaptive thresholding and static thresholding to avail their advantages. An adaptive (varying) threshold changes continuously as a function of signal's past values; on the other hand, a static threshold is predefined and constant throughout the rhythm. This paper details a modified technique presented earlier in [2]. The threshold in this method is static but is determined analytically from the statistical properties of the ECG being processed.

The morphology of the ECG shown in Fig. 1 illustrates that the signal is vertically composed of two parts, the lower part which contains low-amplitude, low-frequency waves (P and T waves) as well as low-amplitude, high-frequency waves (Q and S waves), and the higher part which contains the high-amplitude multiple deflection R waves. The morphology of the R waves causes the population of samples in the quantization levels to be constant in some regions (the region of quantization levels above the highest amplitude T wave and below the lowest amplitude R wave). This morphological characteristic in the ECG represents the physiological phenomenon that the ventricles of the heart contract more forcefully than the atria. In other words, the quantization levels of equal population occur during the R wave elevation progression as it is of high amplitude and sharp rise morphology compared to other ECG components. The R wave morphology is like a unidirectional triangular wave (positive and negative ramps). The total quantization levels that quantify P, Q and S waves are less than those which quantify T and R waves, and the total quantization levels that quantify the T waves are less than those which quantify the R waves. The proposed method detects the quantization levels which belong to the R waves in every ECG rhythm by counting the sample population of all quantization levels and subsequently picking the most repeated population figure which should belong to a couple or more of quantization levels above the highest T wave and below the shortest R wave. This region of quantization levels in the quantization axis is inhabited by only the R wave samples; hence, it is of the same sample population count throughout the rhythm. The algorithm calculates the mode of sample populations of the quantization levels, finds the corresponding quantization levels of the mode

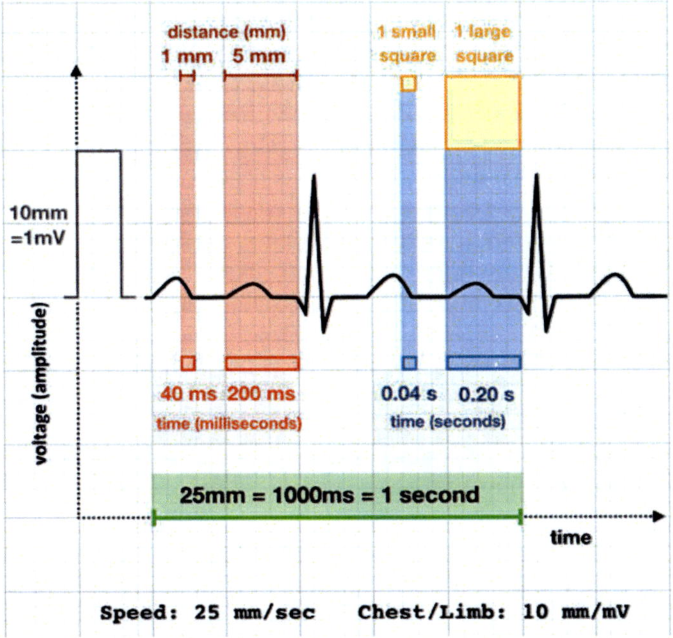

Fig. 1 Temporal components of ECG

value and assigns the resulting minimum and maximum levels as two thresholds for detecting the R waves. The two thresholds may be distinct or repeated.

Section 2 presents the preprocesses to condition the signal prior to algorithm application, Sect. 3 explains the algorithm in detail, Sect. 4 provides the results of applying the proposed method on 50 ECG records obtained from Physionet arrhythmia databasc [3], Mendeley database [4] and a personal ECG collection, Sect. 5 discusses the results and flaws of the algorithm, and Sect. 6 provides conclusion remarks.

2 Preprocessing

Preprocessing is the first stage in signal analysis applications. It includes elimination of unwanted components, and it is attained by using filters or by—alternatively—applying mathematical operations which lead to the same effects. The mathematical operations are opted to apply that lead to the desired effects. The algorithm chops off the first and last 5% of samples because the samples at the extremities may contain spurious noise due to the discontinuity effect generated by the ECG recording system. Then, the algorithm clamps the signal down/up to make the signal's minimum practical amplitude value at the first quantization level. Clamping down/up is done by adding or subtracting the minimum value of the signal, respectively, to avoid negative sample values as well as including uninhabited quantization levels in the processing (due to the DC component), respectively. During the aforementioned conditioning processes, the signal amplitude values are rounded to the nearest integer to have non-rational quantities in the quantization-level step sizes, and then the signal is scaled down by dividing it by 100 for preparation to subsequent squaring operation.

The R waves are high-amplitude, high-frequency components and need to be emphasized—relatively—compared to other components prior to the application of the algorithm. Squaring is equivalent to high-pass filtering; however, squaring cannot be applied because it prolongs the execution time of the algorithm as the algorithm depends on the analysis of quantization levels, and such levels would increase if all the signal sample values are squared. In order to enhance the R waves, selective squaring is applied, i.e., selecting a set of points which would contain all the R waves certainly—in addition to some other points—and squaring only the amplitudes of those points. Our intuition for detecting this set of points is to find the highest amplitude of 100 points in the signal, and all the R waves belong to this set at a high probability as the maximum recordable heart rate is 300 beats per minute which is equivalent to observing 50 R waves in the 10-s rhythm strip. Some non-R wave points may be among the selected points, but there exists a difference in the ratios of amplification of both the types of points as squaring is a nonlinear operation. The higher amplitude points, i.e., R waves, would be amplified greater than other points. Figure 2 illustrates the preprocessing procedure of clamping and subsequently amplifying the highest amplitude of 100 points in an ECG (figure 100 is selected based on trial and error approach, the theoretical limit is 50 points as the maximum heart rate is 300 beats per minute, the number to guarantee is doubled to a higher success rate due to the flat apex problem which is discussed in the next paragraph).

There is another variant of the algorithm which took the highest amplitude of 100 peaks instead of the highest amplitude of 100 points in the selective squaring set. The disadvantage of the second variant is that it misses some R waves in the ECG as some R waves are not considered technically as peaks as per the definition of a peak (the point higher than the preceding and succeeding points in amplitude), and this disadvantage is due to the problem that sometimes the R wave apex is flat on the infinitesimal (sample-wise) scale. On the other hand, the disadvantage of the first variant is that the aforementioned flat apex samples may be of enough quantity to block the detection of any later distinct R wave apex as the number of selected points in the set is limited to 100 points only. However, the probability of missing a distinct R wave apex after exceeding the 100 point limits due to flat apexes (quantitative problem) is less than the probability of missing an R wave due to not getting qualified as a peak as per the strict mathematical definition of a peak (qualitative problem). The former problem can be solved by varying the number of points/length of selective squaring set based on trial and error approach, but the solution of the latter problem may require complicated techniques. Figure 3 highlights the selected 100 points in an ECG which are amplified as per the direct square law. The selected points may contain some T waves as in Fig. 3 or may contain R waves only as in Fig. 2. If some T waves belong to the selected points, the difference between their amplitudes and

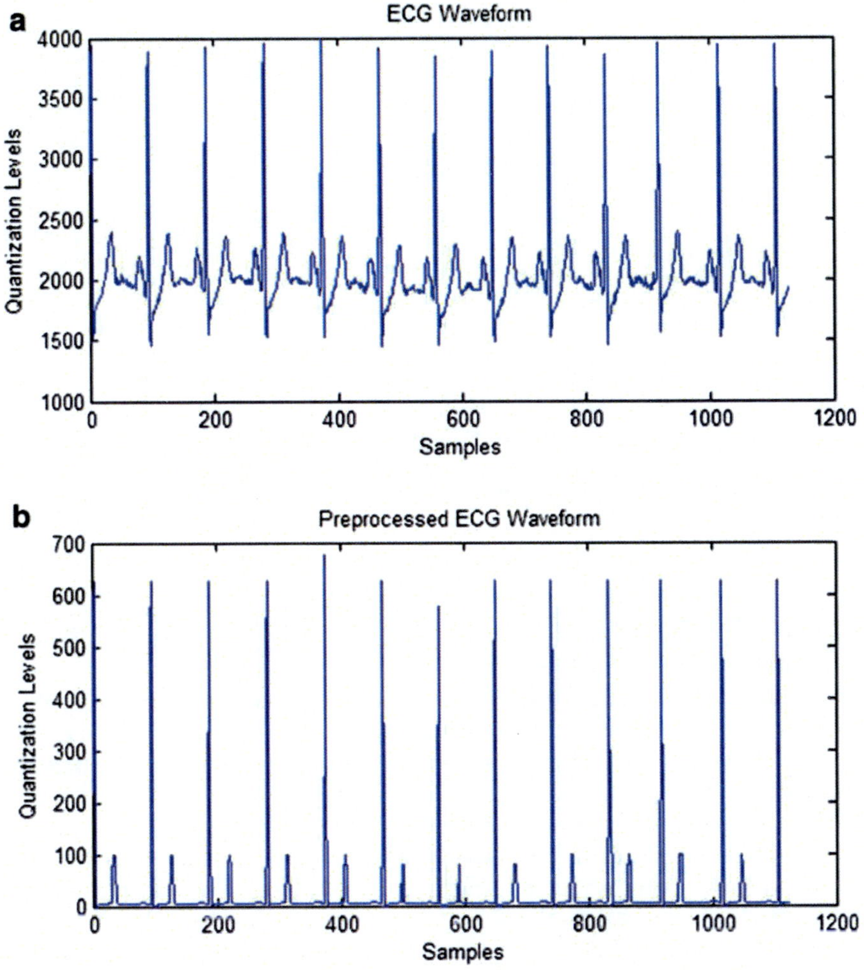

Fig. 2 **a** Original ECG and **b** preprocessed ECG where the DC component is removed and the R waves are amplified more than the other points after the selective squaring process (selective squaring follows scaling the signal down by a factor of 100)

the amplitudes of the R waves before the selective squaring plays a key role—as the square law proportionality implies—in creating a considerable difference in height between the two components after the selective squaring.

The effects of flat apexes are discussed in blocking some R waves in subsequent sections. The three T waves in the selected points in Fig. 3 are neither comparable to the R waves after selective squaring process nor higher than the threshold which is highlighted in green.

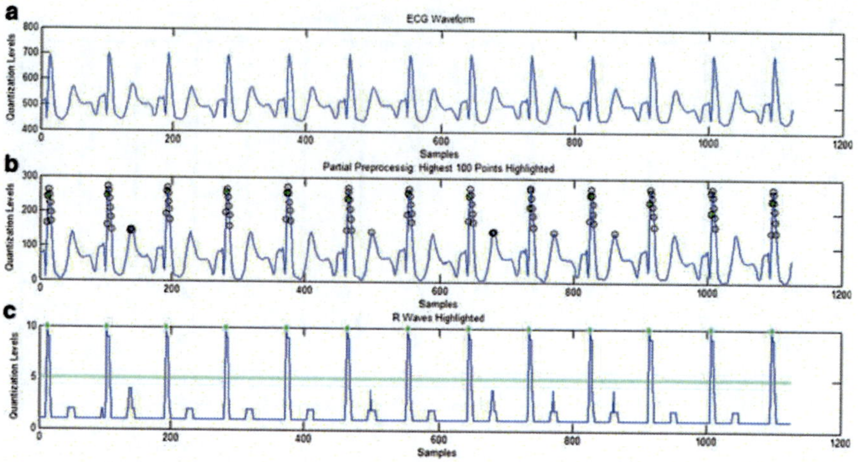

Fig. 3 **a** Original ECG and **b** ECG with the DC voltage removed and highest amplitude of 100 points marked and **c** ECG after compression (division by 100) and selective squaring (squaring the highest amplitude of 100 points)

3 Algorithm

Algorithms for R wave detection in ECG processing depend on the intuition which emerges from the techniques, the technique is determined by the approach, and the approach is dictated by the aim of the signal processing practitioner as illustrated by the model in Fig. 4 (terminology elaboration is in Sect. 6).

The aim may be to find the R wave from its morphology, or it may be to find the R wave by its height. The morphology approach is generally more complicated than the height approach because the morphology approach enables the practitioner to differentiate between equal height R and T waves (even between shorter height R wave and a hyper-T wave); on the other hand, the height approach focuses principally on the general difference in the amplitudes between the R waves and other waves. The

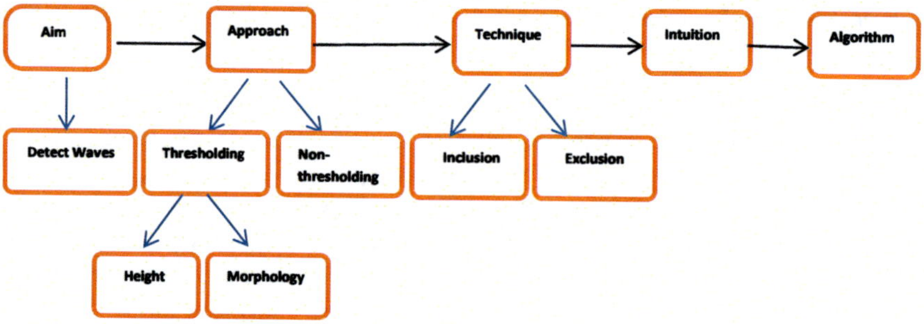

Fig. 4 Two-dimensional model of ECG processing

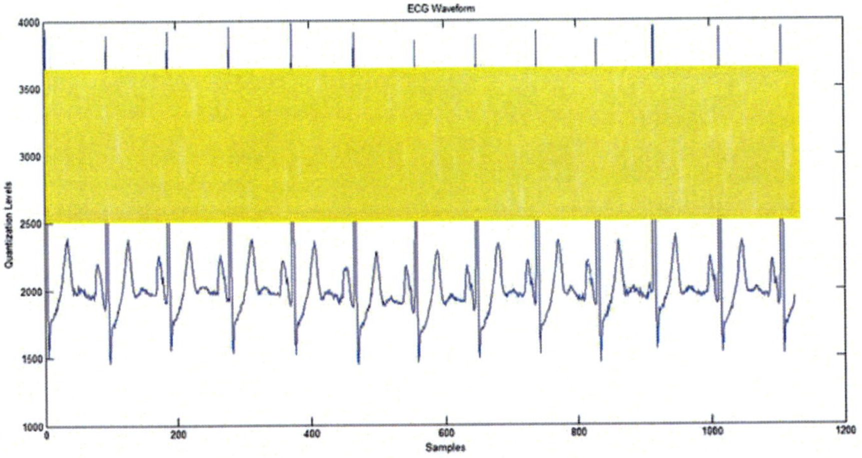

Fig. 5 ECG waveform with the quantization levels theoretically containing the same number of samples (population) shaded in yellow

proposed method is of a height-based approach, and its technique of thresholding is exclusion. Exclusive thresholds exclude invalid or pseudo-R waves from a primarily selected candidate point. The intuition of the method in finding a reliable threshold (threshold which differentiates between the R wave and other waves) is to utilize the information regarding the quantization-level sample populations by analyzing the histogram of the ECG as illustrated in Fig. 5. Finally, the algorithm is the practical demonstration of the intuition in MATLAB platform. The algorithm calculates the frequency of occurrence of every quantization level in the signal throughout the rhythm (histogram) and represents the outcome visually. The population graph plots the quantization levels (hereinafter sometimes called population levels) as the independent variable versus the frequency of occurrence as the dependent variable as shown in Fig. 6, and subsequently the algorithm generates a cumulative histogram by calculating how many total samples are above every population level (instead of how many samples are at every population level which was done in the histogram shown in Fig. 6) as shown in Fig. 7. The highest value of the cumulative population curve corresponds to the lowest quantization level (the level of S waves). The accumulative population decreases as go higher through the quantization levels till the lowest population level—which corresponds to the highest quantization level—is reached (the level of R waves). The accumulative population graph makes the population inversion level (the quantization level where the highest curve slope takes place) clear which takes place in the region of the isoelectric line. The population inversion level is the level of maximum derivative value in the accumulative population curve. Figure 8 illustrates an interesting observation that the population inversion level in the accumulative graph corresponds to the highest population level in the sample distribution graph in Fig. 6 that implies that once the quantization-level pointer crosses the baseline region toward the positive direction (the direction of the R wave apexes) the

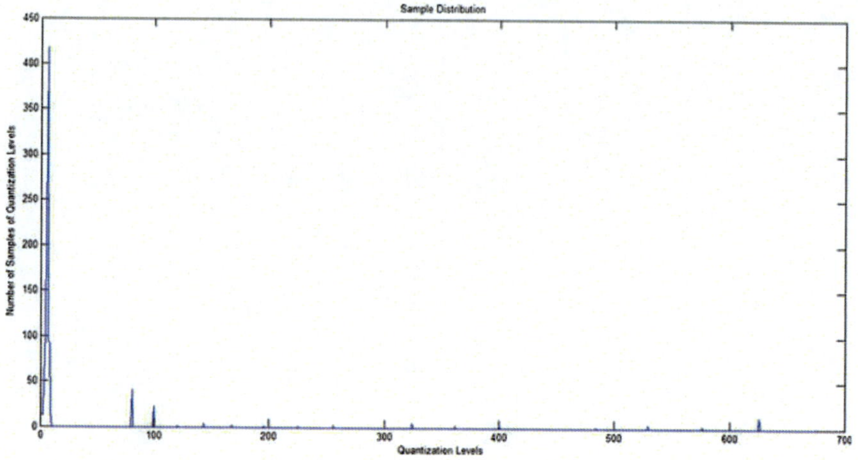

Fig. 6 Sample distribution of the signal. The horizontal and vertical axes represent quantization levels and populations, respectively

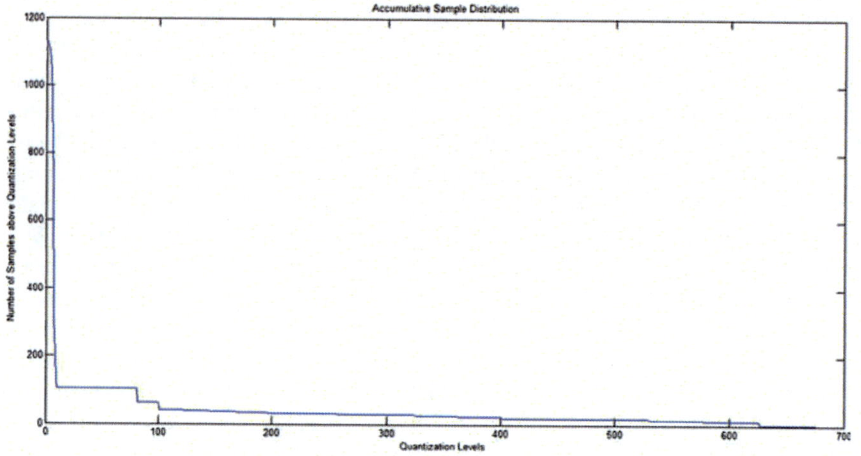

Fig. 7 Accumulative sample distribution. The vertical axis represents the accumulative population of all the succeeding levels, i.e., higher quantization levels

population inversion takes place. Hence, rely on this observation in the population distribution curve to determine the population inversion point in the accumulative population curve. In fact, the two curves are mathematically related to each other as the accumulative curve in Fig. 7 is the integral of the distribution curve in Fig. 6 and the distribution curve in Fig. 6 is the derivative of the accumulative curve in Fig. 7; therefore, either one of them can be obtained from the other.

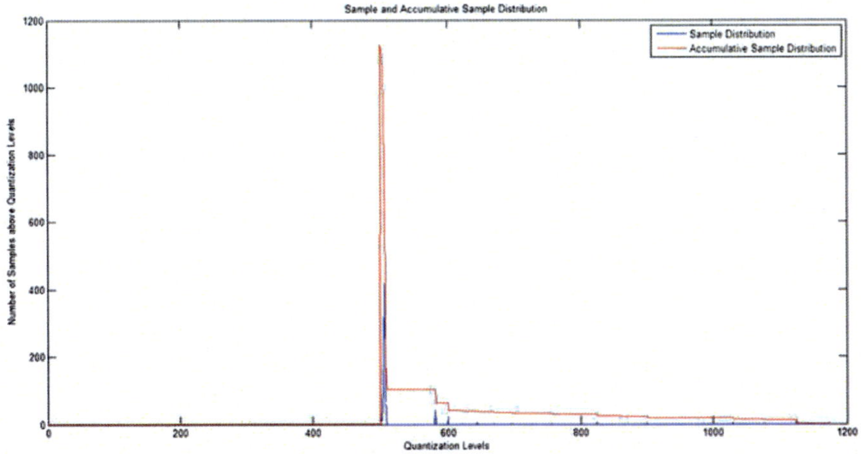

Fig. 8 Sample distribution and accumulative sample distribution curves merged in one plot (the curves are delayed to be centered in the plot for clarity of observation)

The algorithm focuses on the segment of the accumulative population curve that is after the population inversion level. This segment is generally decreasing in amplitude with different slopes. There is a flat part (constant slope) in the higher quantization-level region in the graph which implies that the signal morphology and population are uniform throughout the rhythm in this amplitude region, and this observation corresponds to the occurrence of the parts of the signal above the highest amplitude T wave and below the lowest amplitude R wave. The intuition of the method is to detect this value consistency in the accumulative population curve and subsequently locate its minimum and maximum boundaries in terms of respective quantization (amplitude) levels.

For detecting the range of quantization levels that corresponds to the zero-derivative part of the accumulative population curve (even in case of multiple existence), two procedures are used; the first procedure is based on the mathematical mode operation which finds the highest frequency of occurrence value in the accumulative population curve; if multiple values exist, the mode operator qualifies the lowest value of them (e.g., if a sequence 'A' is composed of the following values [47, 47, 3, 3, 2, 9], the mode operation picks '3' instead of '47' because it is the lowest value one). This procedure does not rely on a complex determination process for picking one of the equally frequent values. The second procedure resolves this flaw by adopting a different process in determining the value to be picked among the equal frequency repeated values, and it is based on counting the frequency of occurrences and generating a vector that holds those frequencies (e.g., if 'A' = [2, 2, 2, 2, 1, 1], it selects the earliest occurring value which is '47' whereas the first procedure detects the latest occurring value '3' because it is lower in magnitude). Subsequently, the magnitudes of the corresponding quantization levels are determined for both

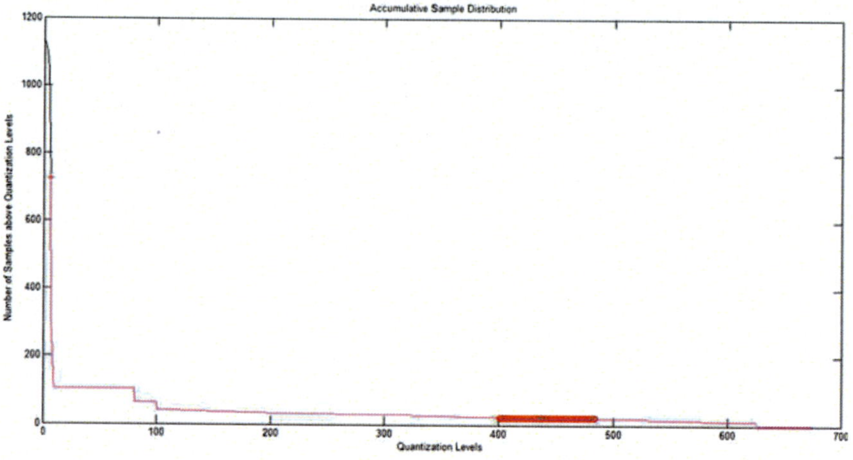

Fig. 9 Post-population inversion segment in the cumulative curve with the longest flat portion marked in red. The asterisk separates the red and black sections of the curve and highlights the point and level of population inversion. In case of multiple occurrence of equal flat portions, the red and green (not visible for being masked by the red color) marks highlight the outcomes of the first and second procedures (discussed in Sect. 3), respectively

approaches. In the first procedure, the maximum magnitude quantization level—among the obtained levels—is adopted, and in the second procedure the minimum magnitude quantization is set to be the threshold. The process is illustrated in Fig. 9.

In case of multiple occurrence of equal length flat portions in the accumulative curve, the lowest magnitude quantization level is taken in the first approach and the highest magnitude quantization level is taken in the second approach.

In summary, the proposed method generates two thresholds based on the two approaches and if there is no consistency in the cumulative curve at all; the algorithm assigns the levels of the population inversion and the highest amplitude signal peak in the ECG as the thresholds for the first and second approaches, respectively.

4 Results

Fifty digital ECGs are processed, 27 rhythms from Physionet arrhythmia database, 18 rhythms from Mendeley database and 5 rhythms from a personal ECG collection. The algorithm generates two thresholds where the success rate (ratio of number of detected R waves to number of actual R waves) of the lower threshold is 96% in the first group of ECGs, 89% in the second group and 93% in the third group. The success rate (ratio of waves detected as R waves to actual R waves present in the ECG) of the higher threshold is 92% in the first group, 374% in the second group and 115% in the third group. The lower threshold is more successful than the higher one. The

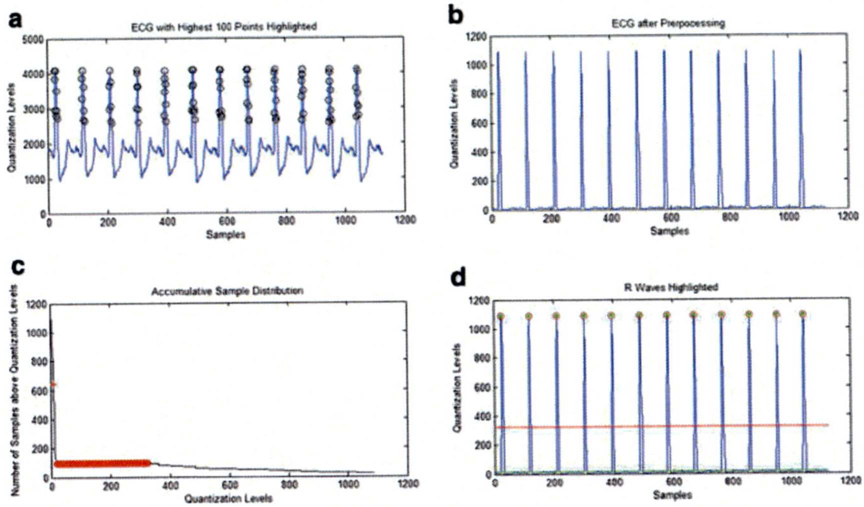

Fig. 10 Post-population inversion segment in the cumulative curve with the longest flat portion marked in red. The asterisk highlights the point and level of population inversion. In case of multiple occurrence of equal flat portions, the red and green (not visible for being masked by the red color) marks highlight the outcomes of the first and second approaches (discussed in Sect. 3), respectively

detection rates calculated are with respect to the preprocessed ECG waveforms and not the original ECG waveforms. The rates may vary if the reference is the original ECG waveforms as some R waves might get dropped or some pseudo-R waves might get considered due to the preprocessing stage effects. The ninth column in the tables holds the number of R waves in the original ECGs. The table shows the number of detected waves with no information about whether the location is correct or not, observed in few cases that a real R wave gets replaced with a pseudo-one without affecting the success ration. The first 4 columns in the tables are reserved for 4 types of thresholds to give the reader an impression of how the determined thresholds are close or far from the optimum one. The average processing time of the algorithm is 4 s for the first group, 5.2 s for the second group and 6 s for the third group.

Pan and Tompkins' algorithms are applied [5] implementation code [6] to the same records and find that it detects all the R waves in all the records except in rhythm C36m and rhythm ecg32 where it drops one R wave in each one, and the difference is also in rhythm 207 m (4) where the preprocessed signal according to Pan and Tompkins' algorithm is observed to contain 35 R waves instead of 32 R waves contained in the original rhythm and among which 23 are detected. The proposed method detects all the R waves in rhythm C36m when the length of the selected points set for selective squaring stage is equal to 50 points (Tables 1, 2 and 3).

Table 1 Results of applying the proposed method on 25 ECGs from Physionet arrhythmia database

ECG rhythm group 1	Minimum threshold (visual)	Maximum threshold (visual)	Determined threshold 1	Determined threshold 2	Max amplitude	R wave detection rate 1	R wave detection rate 2	R waves in original ECG	Time (s)
C00m	100	500	401	484	677	13/13	13/13	13	3
C01m	50	1000	19	324	1090	12/12	12/12	12	5
C03m	300	600	15	196	842	18/14	15/14	14	4
C07m	200	700	530	625	730	13/13	13/13	13	4
C09m	400	800	20	361	1157	15/14	15/14	11	5
C11m	50	350	362	400	400	6/15	6/15	15	2
C12m	50	900	14	169	1025	11/11	11/11	11	4
C14m	4	8	5	8	9	13/10	24/10	13	3
C15m	150	350	15	196	401	24/23	24/23	12	3
C18m	25	150	257	288	289	1/12	1/12	12	3
C21m	25	500	19	361	577	18/18	1/18	18	3
C22m	150	600	325	441	842	12/12	12/12	12	4
C29m	50	350	257	324	485	10/10	10/10	10	3
C32m	300	1000	15	225	1090	12/10	10/10	10	5
C35m	400	900	13	144	1297	28/15	26/15	15	4
C36m	600	1200	22	441	1297	18/10	17/10	10	7
C37m	12	175	197	224	225	1/13	1/13	3	13
C39m	50	250	13	144	401	14/14	14/14	14	3
C40m	50	1100	29	784	1157	15/15	15/15	15	7
C41m	30	180	145	195	196	10/10	14/10	10	3

(continued)

Table 1 (continued)

ECG rhythm group 1	Minimum threshold (visual)	Maximum threshold (visual)	Determined threshold 1	Determined threshold 2	Max amplitude	R wave detection rate 1	R wave detection rate 2	R waves in original ECG	Time (s)
C45m	400	1500	626	1024	1682	9/9	9/9	9	10
C47m	50	500	170	256	786	10/10	10/10	10	4
C51m	100	400	530	575	576	3/14	3/14	14	3
C53m	100	700	677	784	901	16/16	14/16	16	3
C55m	50	1000	17	289	1157	11/11	1/11	11	5
C58m	50	800	28	729	901	7/8	18/8	8	6
C60m	150	600	530	625	730	10/10	6/10	10	4
Totals and success rate	–	–	–	–	–	330/342 96%	314/342 92%	321	–

Table 2 Results of applying the proposed method on 5 ECGs from a personal digital collection

ECG rhythm group 2	Minimum threshold (visual)	Maximum threshold (visual)	Determined threshold 1	Determined threshold 2	Max amplitude	R wave detection rate 1	R wave detection rate 2	R waves in original ECG	Time (s)
Ecg6	1.5	3.5	2	3	4	8/11	78/11	11	5
ecg8	1.5	3.5	2	3	4	13/13	2/13	14	5
ecg32	1.5	3.5	2	3	4	3/3	2/3	23	5
ecg38	3	8	4	8	9	9/9	2/9	11	5
ecg84	1.5	3.5	2	3	4	9/11	92/11	11	6
Total	–	–	–	–	–	42/47 89%	176/47 374%	70	–

Table 3 Results of applying the proposed method on 18 ECGs from Mendeley database

ECG rhythm group 3	Minimum threshold (visually)	Maximum threshold (visually)	Determined threshold 1	Determined threshold 2	Max amplitude	R wave detection rate 1	R wave detection rate 2	R waves in original ECG	Time (s)
100m(0)	4	14	10	15	16	11/11	1/11	11	6
100m(2)	4	14	10	15	16	9/12	1/12	12	6
105m(0)	5	15	5	16	26	13/13	1/13	13	6
106m(0)	5	25	6	25	26	13/12	7/12	8	6
107m(19)	10	80	10	81	101	12/12	7/12	10	6
109m(10)	5	20	6	24	25	13/13	38/13	13	6
114m(0)	10	40	8	49	101	8/13	13/13	13	6
114m(1)	10	40	7	36	82	8/9	19/9	9	6
118m(6)	5	35	6	25	37	12/12	11/12	12	6
118m(7)	5	35	6	25	50	12/12	11/12	12	6
119m(0)	10	60	8	49	82	9/9	13/9	9	6
201(0)	2	8	5	8	9	12/12	26/12	12	6
202(0)	2	14	10	15	16	19/19	40/19	19	6
207m(4)	5	15	5	15	16	14/14	1/14	32	6
207m(9)	4	14	5	15	16	8/8	37/8	9	6
209m(0)	5	25	5	16	37	22/22	21/22	22	6
209m(1)	5	25	5	16	37	16/16	15/16	16	6
230m(0)	5	20	37	48	49	3/9	1/9	9	6
Total	–	–	–	–	–	214/228 93%	263/228 115%	241	–

5 Challenges and Discussion

The method has 3 versions; it can be applied to the original signal directly, i.e., without the preprocessing stage, a specially decimated signal, i.e., the signal composed of the peak points as per the mathematical definition of peaks, and it can be applied to a preprocessed signal as in Sect. 2. The reader can walk through various cases of inaccurate results along with their associated problems and how to resolve them in the following paragraphs.

Rhythm C01m is a classical successful case of the method, and it is processed to have all the R waves detected. The longest population consistency part is in the region of lower quantization levels, and the algorithm generates two thresholds: one at each end of the flat part in the cumulative curve as shown in Fig. 11.

Rhythm C03m is processed to have 6 points which are not R waves detected as R waves. The points get amplified and then considered by the thresholds generated after the cumulative histogram analysis of the preprocessed signal. The thresholds are at the lower region because the fluctuations of the amplitudes in the higher region of the signal disturb the consistency of quantization-level populations in the higher region as shown in Fig. 12. If the length of the selected points in the selective squatting set is made 50, none of the pseudo-peaks are included in the R waves set except the T wave succeeding the ectopic (irregular) beat.

Rhythm C09m is processed to have many pseudo-R waves due to the baseline wandering effect which makes many T and P wave points included in the selected squaring set as shown in Fig. 13.

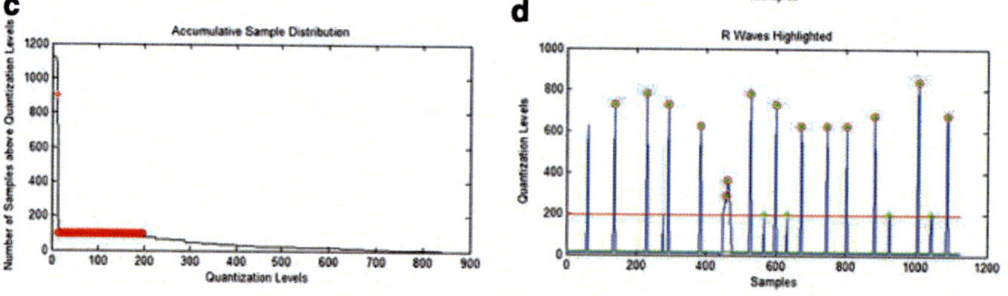

Fig. 11 Top **a** ECG with the set of selected 100 points highlighted, **b** preprocessed ECG by selective squaring, bottom **c** cumulative population curve with the longest consistency portion highlighted, **d** preprocessed ECG with R waves detected based on the two thresholds. Thresholds are delineated in green and red

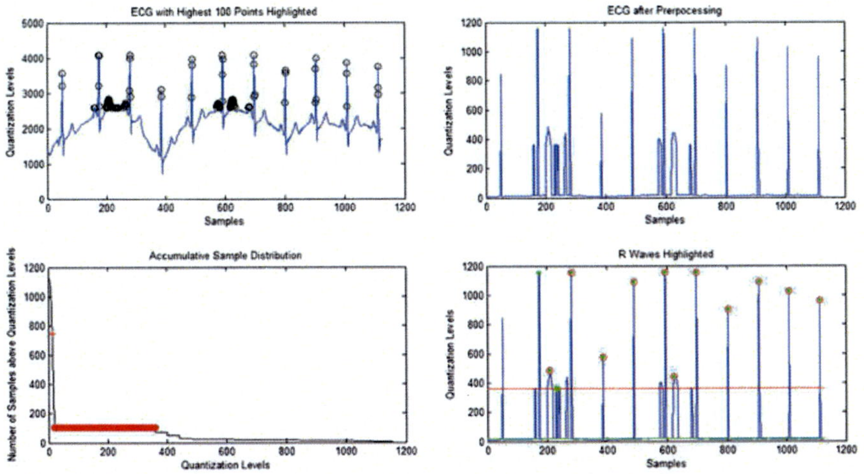

Fig. 12 Some pseudo-R waves are detected when the length of selective squaring set is 100 and eliminated when the length is 50

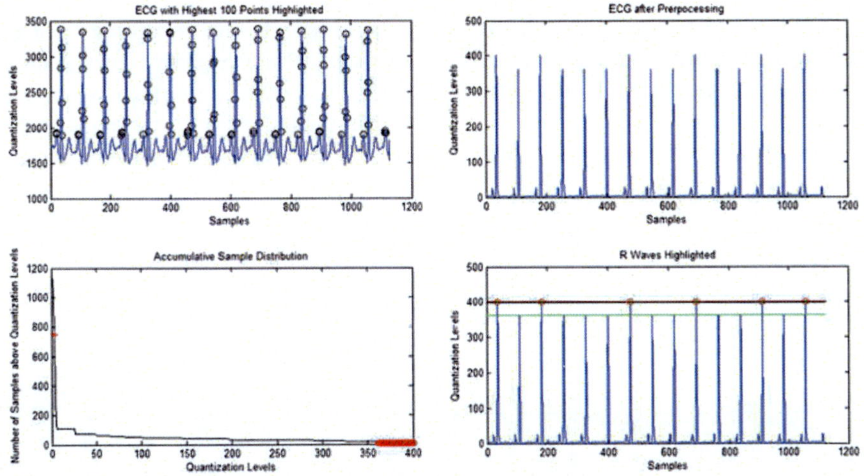

Fig. 13 Some pseudo-R waves are detected when the length of the selective squaring set is 100 as well as when the length is 50

Rhythm C11m is processed to have all non-R wave peaks which are selected in the 100 points of selective squaring set attenuated considerably after the selective squaring. However, both thresholds are high enough to drop most of the R waves as the population consistency takes place late at the end of the histogram as shown in Fig. 14.

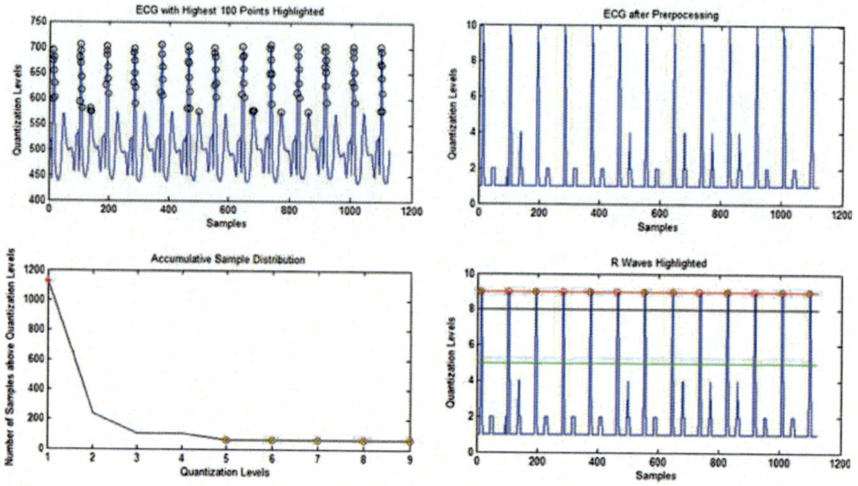

Fig. 14 Population consistency takes place at the end of the cumulative curve which makes both thresholds inappropriately high

Rhythm C14m is processed to have the higher threshold high for all the R waves and the lower one suitable for them to be detected. There is an exceptional third and intermediate threshold (shown in black in Fig. 15) derived from the higher one in some exceptional cases. When the higher threshold—which corresponds to the higher boundary level of the population consistency region—is too high to detect any

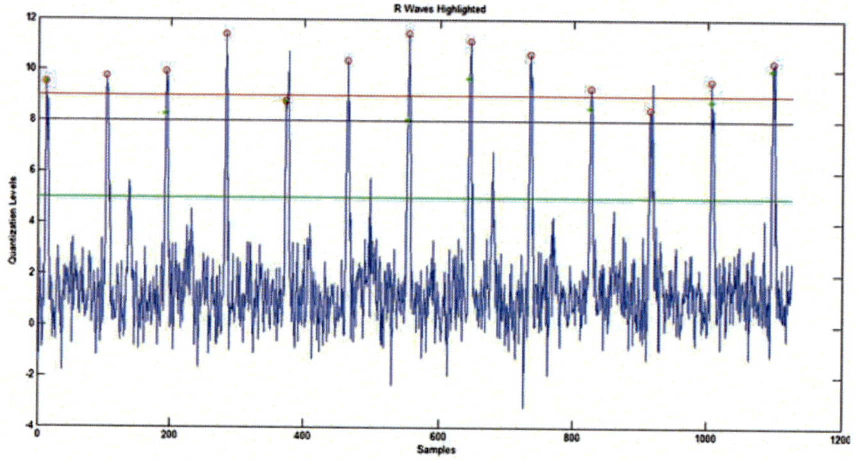

Fig. 15 The higher threshold (red) is high enough to detect none of the R waves which interrupts the algorithm flow to carry out subsequent stages. The threshold then gets lowered automatically by one quantization level. Some R waves may still be not detected due to the flat apex problem

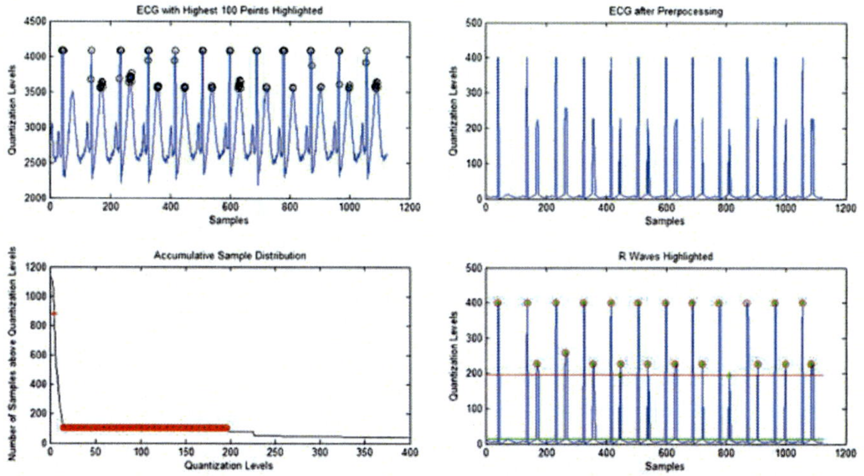

Fig. 16 ECG after contamination with random noise to create variations in the flat apexes of the R waves

of the R waves due to the absence of consistency (such absence might be caused by infinitesimal variations in the signal amplitudes) in the R wave region, the threshold is lowered by one quantization level automatically which is sufficient to contain at least one of the missed R waves. This allows the program to continue execution and not halt. Sometimes, the algorithm still is unable to detect any R wave, and the R waves are missed due to being flat in their apexes. In this case, the algorithms infect the signal by random noise to introduce variations throughout the signal including the apexes so that peaks—as per the mathematical definition—are formed in apexes. This resolution succeeds but may result in the creation of pseudo-peaks in the vicinity of R wave apexes. This scenario is observed in other cases like 109m(10), 201m(0) and 202m(0) and is illustrated in Figs. 15 and 16.

Rhythm C15m is processed to have the majority of the T waves included in the 100 points of selective squaring set and also qualified by the thresholds as the T waves are comparable to the R waves in height after preprocessing. The difference of amplitudes between the T and R waves that the square law proportionality is dependent on is not sufficient—in this case—to suppress the T waves relative to the R waves. Figure 17 illustrates this problem. Do not know why the sample population in the cumulative curve is calculated to be stable in the region marked in red in Fig. 17 though there is no consistency observed visually, that may be due to the limitation of the programming platform capacity in number representations and memory allocations.

Rhythm C58m is processed to have the T waves detected as the R waves because the morphology of the ECG is abnormal. This case makes the R waves small and the S waves deep, and consequently the T waves are considered as R waves. The thresholds are correct (as per the preprocessed signal) but when the locations of the

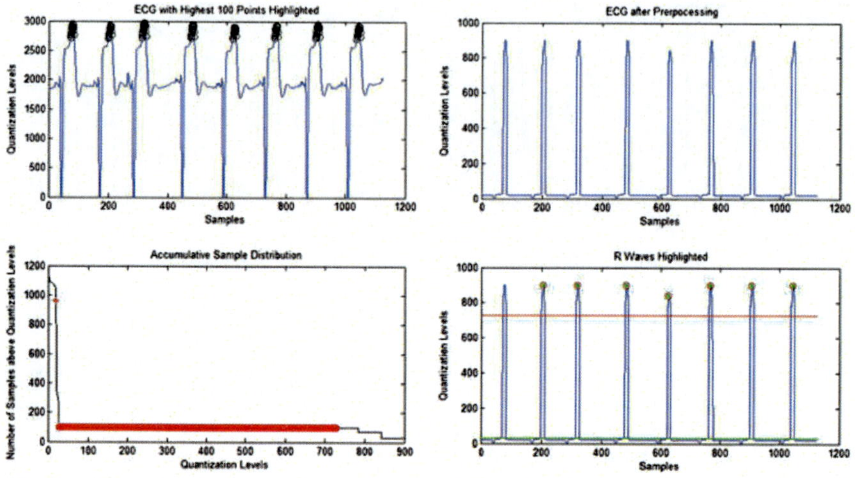

Fig. 17 The T waves are detected along with the R waves

detected waves are projected in the original ECG, the detected waves are noted to be the T waves. Figure 18 elaborates this issue.

Rhythm C207m(4) is processed to have 14 R waves out of 32 waves detected. This is because the heart rate of the ECG is high (32 R waves). The predetermined length of 50 in the selective squaring set is not sufficient to include all the R waves as there are pseudo-peaks in the vicinity of every early R wave peak, those pseudo-peaks are

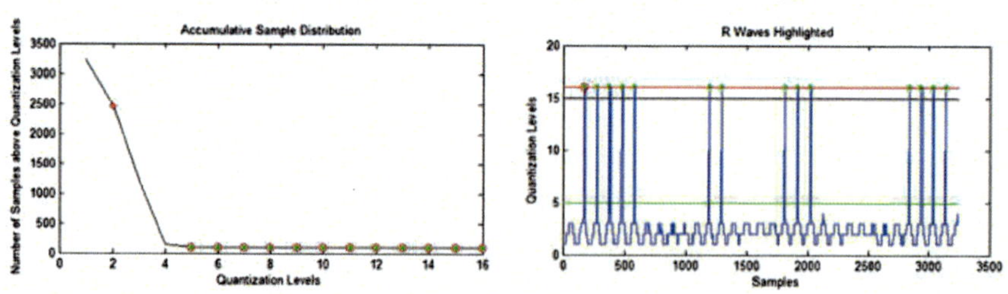

Fig. 18 The hyper-T waves are detected by the algorithm as the R waves

included as well in the selective squaring set and this makes the probability of later R wave being blocked higher. Figure 19 visualizes this issue.

6 Conclusion

In this paper, the graphs of the ECGs in the time domain show the labels of 'Quantization Levels' and 'Samples,' respectively, in the amplitude and time axes purposefully, that is because ECGs are used from many sources and may not be able to know precisely about the specifications of the respective recording systems in terms of sampling rates and quantization depths. The proposed method depends on the characteristic morphology of the ECG and its property of waveform periodicity. It makes use of the statistical properties observed in the waveform and deduced from the histogram. There are many quantization levels that read the same sample population throughout the rhythm; therefore, detecting those quantization levels gives a hint about the suitable threshold of detecting the R waves. Figure 4 illustrates the ECG processing model.

The method succeeds in rhythms where there is a considerable difference between the amplitudes of the highest T wave and the lowest R wave, and vice versa it is less successful in cases of short R waves or high T waves; this results in shrinking the number of quantization levels between the R and T waves which disturbs the length of the population consistency region. If the signal is noisy, there is less consistency in the sample populations of the region above the T waves and below the R waves; hence, the determined threshold may be inaccurate. The algorithm processes successfully the rhythms where the R waves are upward oriented, and this orientation may be in lead II or any other lead depending on the pathological situation. The algorithm fails to deal with high heart rate cases because the former R waves block the later ones as the length of the selective squaring set is limited to 100. The algorithm sometimes contaminates the signal by noise—if necessary—to resolve the problem of flat apex R wave peaks, but this resolution may bring additional problems as it increases the pseudo-peaks in the vicinity of the R waves. Improve the proposed method to deal with the above challenges as well as the flaws discussed in Sect. 4.

All roads lead to Rome, and the proposed method is not necessarily better than—in terms of results—other methods of the same category or different categories as per the model in Fig. 4; rather, it is a new perspective tending to diversify the algorithms aiming to detect the QRS complexes; the method is developed on an intention to approach the same problem from a different engineering perspective. The scientific community is proposed the 2-D model in Fig. 4 which simplifies the categorization of R wave detection methods in practice. Ambiguous terminology is used distinctively to precisely design the model. The definition of the terms is as follows:

Aim: Ultimate goal.
Approach: Idea to fulfill the aim.
Technique: Logic to achieve the approach.

Intuition: Concept to implement the technique.
Algorithm: Practical demonstration of the intuition.

The aim of ECG processing is more often to detect the R wave in the QRS complex as it is in this work, but it might be finding other characteristic features. The approach to fulfill the above aim is mostly by either amplitude analysis or morphology analysis where the goal for both the procedures is to differentiate between R and T waves and adopted a height-based approach in this paper. The technique to achieve this approach depends on the signal processing practitioner, and it might be inclusion or exclusion technique. Exclusive thresholds exclude points from candidate R wave points obtained a priori based on some intuition, whereas inclusive thresholds include points from the ECG into confirmed R wave points based on some intuition. The intuition is the concept to implement the technique, and it depends on the mathematical strength of the programmer; our intuition is to find out the zero-derivative part of the cumulative histogram and the tool, using is the histogram through which carried out the similarity analysis of quantization-level populations. Finally, the algorithm is the practical demonstration of the intuition.

This method is compared with Pan and Tompkins' method in brief statements in Sect. 5, and the detailed comparison of the method with other existing methods in terms of results, challenges, complexity of calculations and processing time is beyond the scope of this paper.

References

1. National Ambulatory Medical Care Survey: 2015 State and National Summary Tables
2. Akula, R., Mohamed, H.: Automation algorithm to detect and quantify electrocardiogram waves and intervals. Procedia Comput. Sci. **151**, 941–946 (2019)
3. Goldberger, A.L., Amaral, L.A.N., Glass, L., Hausdorff, J.M., Ivanov, P.Ch., Mark, R.G., Mietus, J.E., Moody, G.B., Peng, C.K., Stanley, H.E.: PhysioBank, PhysioToolkit, and PhysioNet: components of a new research resource for complex physiologic signals. Circulation **101**(23), e215–e220 (2000)
4. Plawiak, P.: ECG signals (1000 fragments). Mendeley Data, v3 (2017)
5. Pan, J., Tompkins, W.J.: A real-time QRS detection algorithm. IEEE Trans. Biomed. Eng. **BME-32**(3) (1985)
6. Hooma, S.: MATLAB implementation of Pan Tompkins ECG QRS detector. https://doi.org/10.13140/RG.2.2.14202.59841

Author Index

A

Adarsh, S., 195
Aeri, Manisha, 485
Afzal, Anis, 919
Aishwarya, S. L., 577
Alberto, Nelson, 367, 421, 535, 591, 705
Anandakumar, H., 359
Ananya, R. K., 175
Anisha, B. S., 555
Anitha, S., 441
Ansari, Iffat Rehman, 67
Aparnna, S., 601
Arevalo, Oswaldo, 535
Arun Jayakar, S., 125
Arya, M. P., 543
Athira Krishnan, R., 601
Avila Villalobos de, Jeferson Rafael, 713

B

Bajaj, Deepali, 681
Balakrishnan, Athul, 107
Bansal, Nayan, 93
Ba Van, Nguyen, 567
Belapurkar, Gayatri, 107
Bharathi, D., 147
Bharti, Urmil, 681
Bhatia, Meenu, 297
Bhatia, Parul, 161
Bhengura, Rajpreet Singh, 107
Bindu, P. V., 509, 543
Bommannan, Sathish Raja, 39
Borda, Jorge, 57
B, Srujan, 661

C

Cabarcas, Carlos Alberto Jiménez, 591
Chaithanya, D. J., 441
Chakraborty, Sanjay, 461
Chandra, Kamireddy Vijay, 669
Charris, Milvio, 1
Choyyan, Vinayak Prakashan, 509

D

Das, Sukhen, 461
Dehalwar, Vasudev, 9
Desai, Shrinivas D., 577
Deshmukh, Rushali Dhumal, 721
Dharwa, Jyotindra, 737
Dhondiyal, Shiv Ashish, 485
Dhuru, Saloni, 283
Dua, Mohit, 895
Dushyanth Reddy, B., 359

F

Fahim Ansari Md., 919

G

Ganesh, A. Balaji, 785
Ganesh Ram, A., 925
Gangadharan, V. Anjana, 543
Gautam, Rohit, 93
Gautam, Tarika, 321
Gayatri, R., 647
Gayatri, Yendamury, 647
Geetha Lekshmy, V., 601
Geetha, P., 147

© The Editor(s) (if applicable) and The Author(s), under exclusive license
to Springer Nature Singapore Pte Ltd. 2021

A. P. Pandian et al. (eds.), *Proceedings of International Conference on Intelligent Computing, Information and Control Systems*, Advances in Intelligent Systems and Computing 1272, https://doi.org/10.1007/978-981-15-8443-5

George, Sebastian, 521
Goel, Anita, 681
Gulati, Paras, 485
Gupta, Harshit, 633
Gupta, Priya, 161
Gupta, Rahul, 207
Gupta, S. C., 681
Gupta, Shruti, 23

H
Heena, K. P., 835
Hegde, Pragathi, 175
Hegde, Vinayak, 263
Hernandez, Ruben, 713
Hung Van, Mai, 567
Hurtado, Juan, 255
Husain, Maaz, 497

J
Jahnavi, Ambati, 359
Jain, Danish, 619
Jangale, Smita, 107
Jani Anbarasi, L., 335
Januzaj, Ylber, 83
Jayalakshmi, V., 397
Jayasekara, K. G. D. R., 863
Jetti, Venkatateja, 225
Jha, Ankit, 9
Jha, Shweta, 613
Jimenez, Jose Eduardo, 313
Joy Mathavan, J., 863

K
Kadam, Sayali, 23
Kapoor, Cheshtha, 207
Karsh, Ram Kumar, 225
Karthika, R., 647
Karthi, R., 147
Kilambi, Bhargav Ram, 449
Kiwelekar, Arvind, 721
Kokare, Manesh, 375
Kolhe, Mohan Lal, 9
Komatwar, Rupali, 375
Kommineni, Madhuri, 359
Krishnamoorthy, Parthiban, 751
Kumar, P. Ramakanth, 555
Kumar, Rajat, 461
Kumar, Rohan, 461
Kumar, Sumith M. Sree, 633
Kumar, Vishal, 461
Kumari, Pinki, 461
Kunaraj, A., 863
Kurup, Lakshmi, 619
Kuzmar, Isaac, 705
Kyatanavar, D. N., 521

L
Laji, Richard, 763
Lakshmikantha, B. R., 429
Lakshmi, R. Vijaya, 785
Leo, M. Judith, 633
Lezama, Omar Bonerge Pineda, 1, 57, 117, 187, 255, 313
Lingwal, Yogesh, 809
Lodaria, Ayushi, 619
Luma, Artan, 83

M
Madhu, Donga, 775
Mamatha, H. R., 345
Marimuthu, Kalimuthu, 751
Mattur, Aishwarya, 577
Meghana, V., 555
Mehta, Heeket, 619
Meyyappan, S., 925
Mittal, Aakshi, 895
Mohamed, Hamdi M., 967
Mohana Roopa, Y., 389
Mohankumar, N., 647
Mohan, Reshma, 763
Mohan, Sneha S., 543
Moharil, Ambarish, 239
Motwani, Dilip, 297
Murari, Bhaskar Mohan, 669

N
Naaz, Sameena, 67
Nair, Manjusha, 763
Narendra, Modigari, 335
Nayak, Vasath, 175
Nikhil, Gajavalli Venkata Pavan Gopi, 953
Niranjan, B. S., 263

O
Ogura, Koki, 9
Otalora, Jorge, 477

P
Panuganti, Ajay, 497
Paradkar, Abhishek, 23

Parankusham, Anantha Rohan, 449
Patel, Angira Amit, 737
Payares, Karen, 117
Pinillos-Patiño, Yisel, 421
Prajoth SenthilKumar, A. L., 335
Prakash, R., 785
Pramod Kumar, Aylapogu, 775
Praveen, Meghna, 695
Priya, S. Lakshmi, 785
Pudugosula, Harika, 825
Pushpa, B., 275

R
Rajani, Akula, 967
Rajathi, N., 695
Raj, Benson Edwin, 335
Rajesh, M., 825
Rajesh, T., 125
Ramakrishna, B. N., 809
Ramasubbu, Senthilnathan, 785
Ramesh, S., 937
Rana, Deepak Singh, 485
Rao, Ashwini, 23
Rao, Y. Srinivasa, 905
Raufi, Bujar, 83
Ravikumar, S., 845
Revathi, Akula, 577
Roshith, K. K., 509

S
Sadhana, B., 175
Sana, Ehtesham, 67
Sangeetha, A. Lakshmi, 785
Sankhe, Darshana N., 905
Sapkota, Rupesh, 661
Sathishbabu, S., 937
Satish Babu, G., 879
Sawant, Rajendra R., 905
Selimi, Besnik, 83
Senthil Kumar, Avinashi Malleswaran, 751
Shah, Shanay, 619
Shaikh, Ashfaq, 497
Shaik, Himam Saheb, 879
Shaji, Airin Elizabath, 509
Shenoy, Chitra, 175
Shenoy, P. Vanditha, 175
Shetty, Neha, 175
Shrestha, Sparsh, 661
Shrivas, Avinash, 283
Siddarthavarma, J., 775
Singh, Alka, 93
Singh, Fateyh Bahadur, 809
Singh, Mukhtiar, 321
Singh, Prateek, 497
Singh, Shreya, 239
Singh, Sumeshwar, 485
Soubraylu, Sivakumar, 751
Sowmya, G., 797
Sree Yeshvathi, G., 195
Sri HarshaVardhan Goud, P., 389
Srinivasababu, Nadendla, 953
Sri Ritvik, R., 389
Sri Vidya, Boyanapalli, 39
Sundararajan, T. V. P., 125
Suresan, Aparna, 543

T
Tadepalli, Satya Kiranmai, 449
Tamayo, Diana Garcia, 187
Tamilselvan, G. M., 125
Thangavel, P., 797
Thangavel, Senthil Kumar, 785
Thapa, Surendrabikram, 93
Tharol, Sreya, 509
Thu, Dam Thi Kim, 567
Thuwal, Harish, 411
Tiwari, Rishabh, 93
Trishala, G., 345
Turriago, John Rhenals, 367

U
Udasimath, Aishwarya V., 577
Uma Hema Sri, Mylavarapu, 39
Upadhyay, Poonam, 845

V
Vani, K. S., 661
Varela, Noel, 1, 57, 477, 713
Vargas, Jesus, 117, 187, 313, 535
Vasantha, Bhavani, 359
Vashishtha, Utkarsh, 411
Venkatachalam, S., 195
Venkatachari, D., 775
Venkat Hruday, Kaza, 195
Venkat, Karthik, 321
Venugopal, Jeya Krishnan, 751
Vidhya, S., 39
Vijayakarthick, M., 937
Vijayalakshmi, K., 397
Vijay Sunder, R., 195
Viloria, Amelec, 255, 367, 421, 591, 705
Vineeth, Chennuru, 39

Vinoth, N., 937
Vuyyuru, Srija, 389

W
Wajhal, Gaurav, 9

Y
Yadav, Jayesh, 207
Yadav, Mohit, 321
Yadav, Rahul Singh, 633
Yashaswini Gowda, N., 429

Z
Zelama, Nelson, 477, 713